Lecture Notes in Civil Engineering

Volume 225

Lecture Notes in Civil Engineering (LNCE) publishes the latest developments in Civil Engineering - quickly, informally and in top quality. Though original research reported in proceedings and post-proceedings represents the core of LNCE, edited volumes of exceptionally high quality and interest may also be considered for publication. Volumes published in LNCE embrace all aspects and subfields of, as well as new challenges in, Civil Engineering. Topics in the series include:

- Construction and Structural Mechanics
- Building Materials
- Concrete, Steel and Timber Structures
- Geotechnical Engineering
- Earthquake Engineering
- Coastal Engineering
- Ocean and Offshore Engineering; Ships and Floating Structures
- Hydraulics, Hydrology and Water Resources Engineering
- Environmental Engineering and Sustainability
- Structural Health and Monitoring
- Surveying and Geographical Information Systems
- Indoor Environments
- Transportation and Traffic
- Risk Analysis
- Safety and Security

To submit a proposal or request further information, please contact the appropriate Springer Editor:

- Pierpaolo Riva at pierpaolo.riva@springer.com (Europe and Americas);
- Swati Meherishi at swati.meherishi@springer.com (Asia - except China, and Australia, New Zealand);
- Wayne Hu at wayne.hu@springer.com (China).

All books in the series now indexed by Scopus and EI Compendex database!

More information about this series at https://link.springer.com/bookseries/15087

Stefanus Adi Kristiawan ·
Buntara S. Gan · Mohamed Shahin ·
Akanshu Sharma
Editors

Proceedings of the 5th International Conference on Rehabilitation and Maintenance in Civil Engineering

ICRMCE 2021, July 8–9,
Surakarta, Indonesia

Set 1

Springer

Editors
Stefanus Adi Kristiawan
Department of Civil Engineering
Universitas Sebelas Maret
Surakarta, Indonesia

Buntara S. Gan
College of Engineering
Nihon University
Tokyo, Japan

Mohamed Shahin
Department of Civil Engineering
Curtin University
Perth, WA, Australia

Akanshu Sharma
Institute of Construction Materials
University of Stuttgart
Stuttgart, Germany

ISSN 2366-2557 ISSN 2366-2565 (electronic)
Lecture Notes in Civil Engineering
ISBN 978-981-16-9350-2 ISBN 978-981-16-9348-9 (eBook)
https://doi.org/10.1007/978-981-16-9348-9

This Springer imprint is published by the registered company Springer Nature Singapore Pte Ltd.
The registered company address is: 152 Beach Road, #21-01/04 Gateway East, Singapore 189721, Singapore

Foreword

The International Conference on Rehabilitation and Maintenance in Civil Engineering (ICRMCE) is a triennial conference that aims to provide a forum for researchers, academicians (professors, lecturers, and students), government agencies, consultants, and contractors to exchange experiences, technological advancement, and innovations in the world of civil engineering, specifically in the fields of rehabilitation and maintenance. The previous four ICRMCE conferences took place successfully in 2009, 2012, 2015, and 2018. Hundreds of researchers worldwide attended these events to present their scientific papers in various areas of civil engineering such as material engineering, structural engineering, geotechnical engineering, transportation engineering, and construction management.

This year's conference was organized by Sebelas Maret University in collaboration with Mataram University. The conference was initially scheduled offline in Mataram, Indonesia. However, due to the escalating coronavirus (COVID-19) outbreak and the need for social distancing, we decided to hold the conference online. Some reputable universities and institutions are participating in the current ICRMCE as partners. Among them are Nihon University, University of Stuttgart, National Taiwan University, TU Delft, Hiroshima University, Diponegoro University, Muhammadiyah University of Yogyakarta, Jenderal Soedirman University, University of Jember, UPN Veteran East Java, the National Center for Research on Earthquake Engineering (NCREE) Taiwan, Himpunan Ahli Konstruksi Indonesia (HAKI), and Himpunan Ahli Teknik Tanah Indonesia (HATTI).

The ICRMCE 2021 was successfully held on July 8–9. Presenters who joined this conference came from Japan, Singapore, Malaysia, China, Vietnam, Taiwan, England, the Netherlands, Kuwait, and Indonesia. Furthermore, several outstanding keynote speakers gave a presentation of the state-of-the-art findings in the field of civil engineering. Our esteemed speakers are Prof. Shyh-Jiann Hwang (National Taiwan University), Prof. Buntara Sthenly Gan (Nihon University), Dr. Edgar Bohner (VTT Technical Research Centre of Finland), and Prof. Mohamed Shahin (Curtin University).

In the process of organizing this conference, we received invaluable motivation, advice, and support from several individuals and institutions. I intend to express my gratitude and appreciation to all of them. First, my most profound appreciation goes to all organizing committee members who worked day and night preparing this conference. Special thanks to the conference and media partners for their generous support. We also express our gratitude to Prof. S.A. Kristiawan (Sebelas Maret University), Dr. Ing. Akanshu Sharma (University of Stuttgart), Prof. Mohamed Shahin (Curtin University), and Prof. Buntara Sthenly Gan (Nihon University) for their willingness to serve as the editors of the 5th ICRMCE proceedings.

Halwan Alfisa Saifullah
The 5th ICRMCE Chairman

Preface

Civil engineering infrastructures are the backbone for the continuous development of civilization. Managing these infrastructures is essential in keeping the quality of services they provide to the community. A decline in the performance of key infrastructure will have an impact on the quality of these services, which in turn can cause social and economic problems. A variety of factors affects the performance of infrastructure. In each case, the declining performance of infrastructure requires an appropriate and adaptive response to offer effective solutions. Protection, maintenance, repair, and retrofitting are part of the various solutions that can be implemented. All of these solutions are assisted by technological developments related to repair materials, methodologies, systems, management, and operational efficiency, as well as economic and social considerations.

Infrastructure performance is also inevitably affected by exposure to hazards originating from natural and environmental conditions such as earthquakes, landslides, and floods, among others. Therefore, hazard mitigation is also an interesting topic of discussion. In addition, risk reduction and safety are among the most important issues of infrastructure management. Finally, various perspectives on sustainability in civil engineering are also covered in this conference.

This book is a collection of papers presented at the 5th International Conference on Rehabilitation and Maintenance in Civil Engineering (ICRMCE) 2021 that deals with the issues stated above. The papers are grouped into sequential themes representing the structure of this book:

- Part I: Factors affecting performance of buildings and infrastructures
- Part II: Assessment, protection, maintenance, repair, and retrofitting of buildings and infrastructures
- Part III: Maintenance management of buildings and infrastructures
- Part IV: Hazard mitigation
- Part V: Risk reduction and safety management
- Part VI: Sustainability aspects in transportation engineering
- Part VII: Sustainability aspects in construction projects

- Part VIII: Sustainability aspects in water resources management
- Part IX: Construction materials for sustainable infrastructures

Postgraduate students, researchers, and practitioners who would like to update their knowledge on the topics above will find this book very useful.

Surakarta, Indonesia

Stefanus Adi Kristiawan
Chief Editor

Contents

Factors Affecting Performance of Buildings and Infrastructures

A Review on Application of Machine Learning in Building Performance Prediction

R. W. Triadji, M. A. Berawi, and M. Sari

Abstract Designers usually use Building Performance Simulation (BPS) to support decision making in facing design requirements and expected building performance. However, the fact is that BPS still experiences several limitations, such as BPS requires high computation time in assessing various design options. Machine learning is considered capable of solving the problem that the existing BPS has. Research on this problem has been conducted to provide solutions and prove the reliability of machine learning in predicting building performance. Therefore, this paper aims to discuss the research and overview of how machine learning has been used in predicting building performance. The results show that, performance prediction using machine learning has been developed on energy and environmental performance. Also, machine learning can significantly reduce the prediction time without reducing its accuracy.

Keywords Machine learning · Building performance · Energy performance · Environmental performance

1 Introduction

The importance of digital technology and automation systems in playing a role in industry 4.0 makes every device now equipped with machine learning [1]. As the core of artificial intelligence [2], machine learning is a process where machines can learn various things by themselves based on their experiences [3]. Machines learn the relationship between input and output data received as information and turn it

R. W. Triadji (✉) · M. A. Berawi · M. Sari
Civil Engineering Department, Faculty of Engineering, Universitas Indonesia, Depok, Indonesia
e-mail: rizka.wulan@ui.ac.id

M. A. Berawi
e-mail: maberawi@eng.ui.ac.id

M. Sari
e-mail: mustikasari01@ui.ac.id

S. A. Kristiawan et al. (eds.), *Proceedings of the 5th International Conference on Rehabilitation and Maintenance in Civil Engineering*, Lecture Notes in Civil Engineering 225,
https://doi.org/10.1007/978-981-16-9348-9_1

into an experience to offer solutions that can adapt according to given situations [4, 5]. Machine learning has been developed in the construction industry to solve complicated and difficult problems [6]. It is shown by the potential of machine learning in improving building performance [3].

Building performance is considered since the design stage; this step is taken as an effort for the building to perform its functions optimally when it is operating [7]. Designers usually use Building Performance Simulation (BPS) to support decision making in facing design requirements and expected building performance [8]. However, BPS is still experiencing several deficiencies in its application; first, BPS requiring a high computation time to assess various design options [8, 9]. Moreover, some factors have not been considered in implementing the simulation, thus making the predictions offered by BPS less accurate [7]. Machine learning is considered to have the potential and capability to answer these deficiencies [8]. Its ability to predict various designs in a second and incorporate factors not covered by existing BPS will improve the prediction results and replaced the existing Building performance Simulation [7, 9]. Below, we made the conceptual diagram to give a glance of insight (see Fig. 1). Building performance no longer predicts by the existing BPS; it replaced by machine learning and its algorithm to predict those predictions such as energy performance and environmental performance.

Therefore, this paper aims to examine studies related to machine learning in building performance and an overview of how machine learning has been used to predict building performance. The review expected can provide more extensive reference and insight in involving machine learning in the construction industry to enhance building performance prediction.

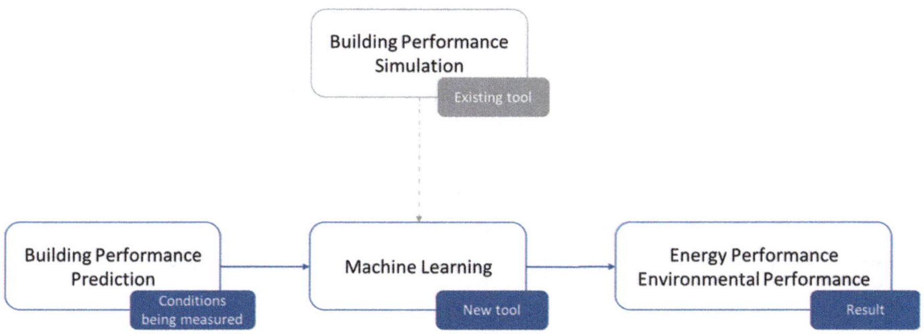

Fig. 1 Conceptual diagram of machine learning and existing BPS

2 Methods

In this study, we collected literature through Scopus as the central database for academic publications. The keywords used are "machine learning", "building performance", "energy performance", "energy consumption", "environment" "indoor environment", "thermal comfort", "visual comfort", the selected documents are limited to journal articles and conference papers that have been published in the last ten years (2011–2021). The search structure example for Scopus is as follows:

TITLE-ABS-KEY ("machine learning" AND "building performance") AND (LIMIT-TO (DOCTYPE, "ar") OR LIMIT TO (DOCTYPE, "cp")) AND (EXCLUDE (PUBYEAR, 2010))

The authors conducted the first screening by reviewing abstracts and matching predetermined keywords so irrelevant documents could be separated from the search results. 37 documents are deemed relevant to the topics we discussed. Then, the second screening will be carried out by reading the entire document. 17 papers were not selected to be discussed, after the second screening. Thus, 20 papers will be discussed and grouped based on two primary performances that become the main concern in a building: energy performance and environmental performance [10, 11]. The algorithm of machine learning in each paper will also be discussed in this review.

3 Result and Discussion

3.1 Energy Performance

Several building energy performance tools such as TrnSys, EnergyPlus, ESP-r, DOE-2 and Modellica have been widely used, but these tools still require high computation time for various designs with multiple inputs; this also makes simulation tools difficult in delivering precise result [9, 12, 13]. In this section, the discussion will be grouped into two main parts, which are thermal and electricity; due to the current study that we discussed in this section focuses on thermal performance and follows by electricity performance.

Thermal Energy

In previous studies, machine learning has been developed to predict the thermal (cooling and heating energy) performance of buildings using an Artificial Neural Network (ANN) [9, 12, 14–16]. Comesaña et al. [12] developed a machine learning model based on outdoor and indoor temperatures and radiation levels. Meanwhile, Geyer and Singaravel's [14] model based on building components such as walls, windows and floors. Geyer and Singaravel [14] stated that the computation time to make predictions using machine learning is reduced drastically. In line with this, Ascione et al. [16] ensure a computation time savings of 98% can be achieved using machine learning, this has been proven in their research. They develop a model for predicting energy performance and the energy retrofitting scenario for each building element. Similarly, Seyedzadeh et al. [17] also made a model to predict energy performance and support decision-making about energy retrofit scenarios, the algorithm used is gradient boosted regression trees.

Singh et al. [9] make energy predictions by creating four schematic methods for making predictions. The first three using EnergyPlus, and the last method is using machine learning. The result is that machine learning can achieve the objectives of reducing computation time than EnergyPlus. Westermann et al. [18] using the Convolutional Neural Network (CNN) algorithm to develop a machine learning model to displace a building energy simulation tool. It can predict thermal energy based on input from various design and climate variations. The model is representative enough to be used in various climates of building locations and does not affect its accuracy. On the other hand, Ciulla and D'Amico [19] developed a model without the intention of replacing the existing simulation model. The model was developed with the Multiple Linear Regression algorithm, built as simple as possible, so even non-expert users can use it.

Chakraborty and Elzarka [20] developed their model using three different algorithms (XGBoost, ANN and Degree-day-based OLS regression); the result is XGBoost the most accurate algorithm in predicting thermal performance. In contrast to other studies, Attanasio et al. [21] developed a model only to predict heating energy by comparing four machine learning algorithms. Meanwhile, Yu et al. [13] combine two algorithms, genetic algorithm (GA) and back-propagation (BP), to optimize energy performance predictions and thermal comfort based on the building design and design envelope in residential building. Similarly, Robinson et al. [22] developed a model for commercial building using CBECS (Commercial Building Energy Consumption Survey) data released by the US Energy Information Administration (EIA). It is intended that the proposed model can be used in any city in the US.

Electricity Energy

Zeng et al. [23] developed machine learning to predict electrical energy consumption in a building using the Gaussian Process algorithm. The model developed is based on three types of buildings: offices, shopping centres, and hotels with various design configurations. Later, the model can be used in several kinds of buildings. Moreover, Pangaribuan et al. [24] also build a model to predict electrical energy using a support vector regression algorithm for residential homes.

3.2 Environmental Performance

Environmental performance is also one of the criteria that need to be considered in the design to balance each building's performance and achieve optimal sustainability goals [25, 26]. In this case, the environmental performance consists of an indoor or outdoor environment. The indoor performance focuses on the comfort of occupants in the building [13]. Meanwhile, outdoor performance pays more attention to the impact caused by buildings on the outdoor environment, such as CO_2 gas emissions [11]. Mazuroski et al. [27] developed a recurrent neural network-based model to predict indoor temperature based on density, specific heat and thermal conductivity of each building material that will be used. The same idea was done by Kamel et al. [28]. The difference between these two studies is the data used as input to predict indoor temperature, i.e. humidity, radiations, airspeed, door open/close status, and motion from sensors installed in the room. In their research, the model developed by Yu et al. [13] with GA-BP algorithm is not only to predict energy performance but also to predict thermal comfort.

On the other hand, Symonds et al. [29] used an artificial neural network to predict indoor environmental quality. The model developed can make predictions in overheating metric, PM2.5 ratio, relative humidity and heating energy use with a reasonable degree of accuracy. In his study, Chen [30] said that difference results between the building performance analysis predictions and reality could not be avoided. Machine learning was developed to improve the accuracy of green BIM by using an artificial neural network algorithm and daylight luminance level as an outcome of the prediction. The model can also predict climate analysis, thermal comfort, energy calculation, and other BPA dimensions. Chatzikonstantinou and Sariyildiz [31] focused their study on office space's visual comfort; a model developed can predict Daylight Glare Probability and Daylight Autonomy. Algorithms feed-forward networks (FFNs) were chosen as the algorithm with the highest level of accuracy compared to support vector machines (SVMs) and random forests (RFs).

While some of the studies focused on indoor performance, Feng et al. [11] developed a model to discover the environmental impact from the building design. They were using GWP/unit area as an indicator of environmental performance. The input of the model is the early design parameter, and fuzzy c means algorithm approach was used to perform the model. The model expected can help designers in testing several design scenarios and the resulting performance.

4 Conclusion

Machine learning has been used to predict buildings performance in terms of energy and the environment. From our review, energy performance is a performance that is widely used in developing machine learning models. Also, artificial neural networks are found as an algorithm that is widely used in model development. The potential of machine learning in predicting performance in a second and accurately can help designers make decisions of a wide variety of designs. It reveals that machine learning can be a promising and powerful tool, mainly when used at the design stage. However, while machine learning has promising potential, it still has its limitations. First, to make good predictive, machine learning requires big data to train the algorithm [3]. Moreover, some studies apply machine learning to only one building type, and this makes the model unable to be used for other building types. It is expected that with these reviews, the construction industry can take advantage of machine learning capabilities in predicting building performance.

Acknowledgements The author is grateful to Universitas Indonesia for giving support in this research.

References

1. Muhuri PK, Shukla AK, Abraham A (2019) Industry 4.0: a bibliometric analysis and detailed overview. Eng Appl Artif Intell 78:218–235. https://doi.org/10.1016/j.engappai.2018.11.007
2. Jordan MI, Mitchell TM (2015) Machine learning: trends, perspectives, and prospects. Science (80-)349:255–260. https://doi.org/10.1126/science.aaa8415
3. Hong T, Wang Z, Luo X, Zhang W (2020) State-of-the-art on research and applications of machine learning in the building life cycle. Energy Build 212.https://doi.org/10.1016/j.enbuild.2020.109831
4. Dimiduk DM, Holm EA, Niezgoda SR (2018) Perspectives on the impact of machine learning, deep learning, and artificial intelligence on materials, processes, and structures engineering. Integr Mater Manuf Innov 7:157–172. https://doi.org/10.1007/s40192-018-0117-8
5. Yücel M, Nigdeli SM, Bekdaş G (2021) Artificial intelligence and machine learning with reflection for structural engineering: a review. In: Studies in systems, decision and control. Springer Science and Business Media Deutschland GmbH, Department of Civil Engineering, Istanbul University-Cerrahpaşa, Avcılar, Istanbul, 34320, Turkey, pp 23–72
6. Darko A, Chan APC, Adabre MA et al (2020) Artificial intelligence in the AEC industry: scientometric analysis and visualization of research activities. Autom Constr 112:103081. https://doi.org/10.1016/j.autcon.2020.103081
7. Chokwitthaya C, Zhu Y, Mukhopadhyay S, Collier E (2020) Augmenting building performance predictions during design using generative adversarial networks and immersive virtual environments. Autom Constr 119:103350. https://doi.org/10.1016/j.autcon.2020.103350
8. Singaravel S, Suykens J, Geyer P (2018) Deep-learning neural-network architectures and methods: using component-based models in building-design energy prediction. Adv Eng Inform 38:81–90. https://doi.org/10.1016/j.aei.2018.06.004
9. Singh MM, Singaravel S, Klein R, Geyer P (2020) Quick energy prediction and comparison of options at the early design stage. Adv Eng Inform 46:101185. https://doi.org/10.1016/j.aei.2020.101185
10. Kim J, Hong T, Hastak M, Jeong K (2020) Intelligent planning unit for the artificial intelligent based built environment focusing on human-building interaction. J Asian Archit Build Eng 00:1–18. https://doi.org/10.1080/13467581.2020.1803076
11. Feng K, Lu W, Wang Y (2019) Assessing environmental performance in early building design stage: an integrated parametric design and machine learning method. Sustain Cities Soc 50:101596. https://doi.org/10.1016/j.scs.2019.101596
12. Comesaña MM, Febrero-Garrido L, Troncoso-Pastoriza F, Martínez-Torres J (2020) Prediction of building's thermal performance using LSTM and MLP neural networks. Appl Sci 10:1–16. https://doi.org/10.3390/app10217439
13. Yu W, Li B, Jia H et al (2015) Application of multi-objective genetic algorithm to optimize energy efficiency and thermal comfort in building design. Energ Build 88:135–143. https://doi.org/10.1016/j.enbuild.2014.11.063
14. Geyer P, Singaravel S (2018) Component-based machine learning for performance prediction in building design. Appl Energ 228:1439–1453. https://doi.org/10.1016/j.apenergy.2018.07.011
15. Hwang JK, Yun GY, Lee S et al (2020) Using deep learning approaches with variable selection process to predict the energy performance of a heating and cooling system. Renew Energ 149:1227–1245. https://doi.org/10.1016/j.renene.2019.10.113

16. Ascione F, Bianco N, De Stasio C et al (2017) Artificial neural networks to predict energy performance and retrofit scenarios for any member of a building category: a novel approach. Energy 118:999–1017. https://doi.org/10.1016/j.energy.2016.10.126
17. Seyedzadeh S, Pour Rahimian F, Oliver S et al (2020) Machine learning modelling for predicting non-domestic buildings energy performance: a model to support deep energy retrofit decision-making. Appl Energ 279:115908. https://doi.org/10.1016/j.apenergy.2020.115908
18. Westermann P, Welzel M, Evins R (2020) Using a deep temporal convolutional network as a building energy surrogate model that spans multiple climate zones. Appl Energy 278:115563. https://doi.org/10.1016/j.apenergy.2020.115563
19. Ciulla G, D'Amico A (2019) Building energy performance forecasting: a multiple linear regression approach. Appl Energ 253:113500. https://doi.org/10.1016/j.apenergy.2019.113500
20. Chakraborty D, Elzarka H (2019) Advanced machine learning techniques for building performance simulation: a comparative analysis. J Build Perform Simul 12:193–207. https://doi.org/10.1080/19401493.2018.1498538
21. Attanasio A, Piscitelli MS, Chiusano S et al (2019) Towards an automated, fast and interpretable estimation model of heating energy demand: a data-driven approach exploiting building energy certificates. Energies 12.https://doi.org/10.3390/en12071273
22. Robinson C, Dilkina B, Hubbs J et al (2017) Machine learning approaches for estimating commercial building energy consumption. Appl Energ 208:889–904. https://doi.org/10.1016/j.apenergy.2017.09.060
23. Zeng A, Ho H, Yu Y (2020) Prediction of building electricity usage using Gaussian Process Regression. J Build Eng 28:101054. https://doi.org/10.1016/j.jobe.2019.101054
24. Pangaribuan AB, Octa A, Pradnyana IWW, Afrizal S (2019) Predictive analytic for estimating electric consumption of smart grid platform in residential single-family building using support vector regression approach. J Phys Conf Ser 1196.https://doi.org/10.1088/1742-6596/1196/1/012022
25. Geng Y, Ji W, Wang Z et al (2019) A review of operating performance in green buildings: energy use, indoor environmental quality and occupant satisfaction. Energy Build 183:500–514. https://doi.org/10.1016/j.enbuild.2018.11.017
26. Jalaei F, Jalaei F, Mohammadi S (2020) An integrated BIM-LEED application to automate sustainable design assessment framework at the conceptual stage of building projects. Sustain Cities Soc 53.https://doi.org/10.1016/j.scs.2019.101979
27. Mazuroski W, Berger J, Oliveira RCLF, Mendes N (2018) An artificial intelligence-based method to efficiently bring CFD to building simulation. J Build Perform Simul 11:588–603. https://doi.org/10.1080/19401493.2017.1414880
28. Kamel E, Javan-Khoshkholgh A, Abumahfouz N et al (2020) A case study of using multi-functional sensors to predict the indoor air temperature in classrooms. In: ASHRAE Transactions, pp 3–11
29. Symonds P, Taylor J, Chalabi Z et al (2016) Development of an England-wide indoor overheating and air pollution model using artificial neural networks. J Build Perform Simul 9:606–619. https://doi.org/10.1080/19401493.2016.1166265
30. Chen SY (2019) Enhancing validity of green building information modeling with artificial-neural-network-supervised learning—taking construction of adaptive building envelope based on daylight simulation as an example. Sens Mater 31:1831–1845. https://doi.org/10.18494/SAM.2019.2147
31. Chatzikonstantinou I, Sariyildiz S (2016) Approximation of simulation-derived visual comfort indicators in office spaces: a comparative study in machine learning. Archit Sci Rev 59:307–322. https://doi.org/10.1080/00038628.2015.1072705

The Effect of P-Delta and P-Delta Plus Large Displacements Modelling on Lateral and Axial Displacement

Jonie Tanijaya and Robby S. Kwandou

Abstract The structural analysis programs have become increasingly sophisticated along with the advancement of science. However, engineers need to be more careful and understand the use of the options in the program. One program that can incorporate non-linear effects is SAP2000. In the SAP2000 program, non-linear modeling options are classified into two options, namely non-linear and non-linear plus large displacements. The SAP2000 program is a finite element-based program. The finite element method is a method that solves problems by dividing a large element into several small element segments or commonly referred to as meshing. The smaller the meshing segment used, the more accurate the output will be. However, its significance remains to be studied further. Therefore, this study was conducted to determine the significance of differences in modeling options (linear, non-linear, non-linear plus large displacements), as well as the effect of segment division in the analysis. The analysis results show that the linear model cannot capture the effect of the lateral displacement that changes due to the incremental of the compression force. The compression force has significant effect to moment value especially for the higher compression force. Therefore, the P-Delta effect should be analyzed carefully, especially for element with high compression force since the linear model could not capture this effect. The meshing by divide segment does not provide significance difference for both P-Delta and P-Delta plus large displacement model in this case.

Keywords P-Delta · P-Delta plus large displacements · Lateral displacement

J. Tanijaya (✉)
Civil Engineering Department, Universitas Kristen Indonesia Paulus Makassar, Makassar, Indonesia

R. S. Kwandou
Postgraduate Alumni of Hasanuddin University, Makassar, Indonesia

S. A. Kristiawan et al. (eds.), *Proceedings of the 5th International Conference on Rehabilitation and Maintenance in Civil Engineering*, Lecture Notes in Civil Engineering 225,
https://doi.org/10.1007/978-981-16-9348-9_2

1 Introduction

The structural analysis programs have become increasingly sophisticated along with the advancement of science. However, engineers need to be more careful and understand the use of the options in the program. The assumptions of modelling need to be considered carefully. The effect of lateral load on horizontal displacement should be checked in analysis. The elastic-linear analysis condition can produce an underestimated result, especially in the case of structural elements that experience large deformation, such as thin glass panel structures and mono-poles that are exposed to the wind load. A structural analysis that can take into account the large deformations effect will certainly provide a more accurate result. When using linear elastic analysis, the effect of changes in geometry due to large deflection cannot be taken into account in the analysis [1]. One program that can incorporate non-linear effects is SAP2000. In the SAP2000 program, non-linear modeling options are classified into two options, namely non-linear and non-linear plus large displacements. The SAP2000 program is a finite element-based program. The finite element method is a method that solves problems by dividing a large element into several small element segments or commonly referred to as meshing. The smaller the meshing segment used, the more accurate the output will be. Different output values would be provided when including the meshing effect. However, its significance remains to be studied further. Therefore, this study was conducted to determine the significance of differences in modeling options (linear, non-linear, non-linear plus large displacements), as well as the effect of segment division in the analysis.

2 Literature Review

The P-Delta effect occurs due to gravity load (P) that produces the increasing of horizontal displacement (Δ) [2]. The eccentricity of the gravitational load (P) against the column axis causes an increase in horizontal displacement when the structure is subjected to lateral loads. Thus, the P-Delta effect will cause an increase in the moment value as well as the structural drift [3]. Indonesia is surrounded by active tectonic faults so that the lateral load due to earthquakes needs to be considered [4]. The lateral force due to earthquake loads which is influenced by the P-Delta effect becomes critical, especially when the P-Delta effect needs to be considered in the analysis. Based on the research results of Isitono and Ramadhan, the increasing moment value is about 10% on the building structure model [2]. The average percentage increase in moment value that occurs due to P-Delta is between 19 and 27% based on the results of Suhana and Pello's research on a 15-story building [5].

P-Delta effect as one type of geometric nonlinearity, typically produce relatively small additional displacement due to large external force. If deformations become

Fig. 1 P-Delta on column

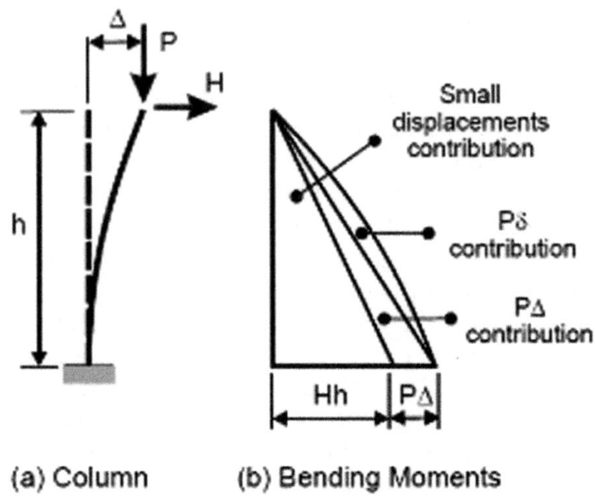

sufficiently large as to break from linear compatibility relationships, then Large-Displacement analyses may become necessary [6]. Typically, there are two P-Delta effect, namely P-Δ or P-big delta and P-δ or P-small delta (see Fig. 1). P-δ is associated with local deformation between end nodes while P-Δ is associated with the member ends displacements. The difference between P-Delta and P-Delta plus large displacements is that the P-Delta option only capture P-Δ effect while P-Delta plus large displacements captures both P-Δ and P-δ.

The geometric stiffness matrix is used in SAP2000 for nonlinear analysis [7]. The use of the geometric stiffness matrix is a general approach to include secondary effects in the static and dynamic analysis of all types of structural systems. Geometric stiffness in this case is also refer to geometrical nonlinearity. Secondary order effect is not presented in linear analysis results [8]. Hence, the nonlinearity option should be activated to account this effect in analysis. In civil structural engineering, this geometric stiffness is also commonly referred to P-Delta Analysis. The additional moment resulting from the P-Delta effect is called the secondary moment or second-order moment (see Fig. 2).

The elastic force deformation relationship for a prismatic bending element without shearing deformations is given by the following equation:

$$F_E = k_E v$$
$$\begin{bmatrix} F_i \\ M_i \\ F_j \\ M_j \end{bmatrix} = \frac{EI}{L^3}\begin{bmatrix} 12 & 6L & -12 & 6L \\ 6L & 4L^2 & -6L & -2L^2 \\ -12 & -6L & 12 & -6L \\ -6L & -2L^2 & -6L & 4L^2 \end{bmatrix}\begin{bmatrix} v_i \\ \Phi_i \\ v_j \\ \Phi_j \end{bmatrix} \tag{1}$$

The force–displacement relationship of bending element is given by the following equation:

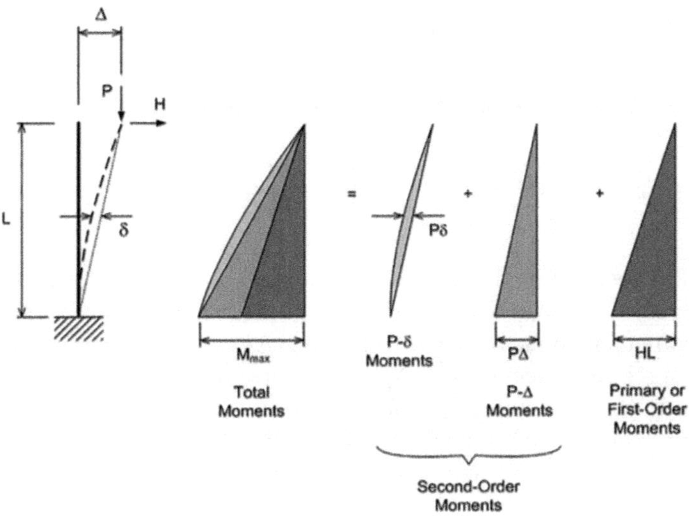

Fig. 2 Secondary moment

$$F_G = k_G v$$

$$\begin{bmatrix} F_i \\ M_i \\ F_j \\ M_j \end{bmatrix} = \frac{T}{30L} \begin{bmatrix} 36 & 3L & -36 \; 3L \\ 3L & 4L^2 & -3L \; -L^2 \\ -36 & -3L & 36 \; -3L \\ 3L & -L^2 & -3L \; 4L^2 \end{bmatrix} \begin{bmatrix} v_i \\ \Phi_i \\ v_j \\ \Phi_j \end{bmatrix} \tag{2}$$

Therefore, the total forces acting on the bending element will be the summary of linear and nonlinear force:

$$\mathrm{F_T = F_E + F_G = [k_E + k_G]v = k_T v} \tag{3}$$

P-Delta effect is correlated with compression axial force. For concrete structures as stated in [9, 10], the compression force should be taken less than 0.75 critical buckling load (0.75Pcr). The theoretical value of Pcr is given by the formula:

$$P_{cr} = \frac{\pi^2 EI}{(kL)^2} \tag{4}$$

$$\text{For k} = 2,\ P_{cr} = \frac{\pi^2 EI}{4L^2} \tag{5}$$

Equation (5) is used for cantilevered element where the effective length factor k = 2 [11]. In calculating the critical buckling load, effective stiffness $(\mathrm{EI})_{\mathrm{eff}}$ is used for reasonable approximation due to cracking and nonlinearity effect of concrete. Effective stiffness is presented in [10], which is given by the formula:

$$(EI)_{eff} = \frac{E_c I}{1 + \beta_{ns}} \tag{6}$$

The inertia I in Eq. (5) for column element is expressed as:

$$0.35 I_g \leq I = \left(0.80 + 25\frac{A_{st}}{A_g}\right)\left(1 - \frac{M_u}{P_u h} - 0.5\frac{P_u}{P_o}\right) I_g \leq 0.875 I_g \tag{7}$$

where, E_c = concrete modulus (MPa), I_g = moment of inertia of gross concrete section (mm^4), L = span length (mm), β_{ns} = maximum factored axial sustained (dead) load/maximum factored axial total load, A_{st} = total area of non-prestressed longitudinal reinforcement (mm^2), A_g = gross area of concrete section (mm^2), Mu = factored moment (N mm), P_u = factored axial load (N), h = depth of member (mm), P_o = nominal axial strength at zero eccentricity (N).

3 Methodology

The analysis model is a single cantilevered column with a height of 5 m with a free-fixed condition (see Fig. 3). A simple model is used in this case in order to speed up the analysis process and also to review different options that would take a large amount of time process if a more complex model is used. The model is analyzed using SAP2000 program. There are three analysis options used, which is linear, P-Delta and P-Delta plus large displacements. The compression force (P) and horizontal force (F) are applied to the column for all models. In this case, the horizontal force F = 1 ton is applied as the same final horizontal load while the final compression force is varied for each model. SNI 2847:2019 [9] article 18.6 states that the beam is an element with a compression force that does not exceed 0.10Agf'c while the it exceeds 0.10Agf'c for column element. For beam element, the compression force can be neglected due to its small value (less than 0.10Agf'c). Therefore, the minimum compression force

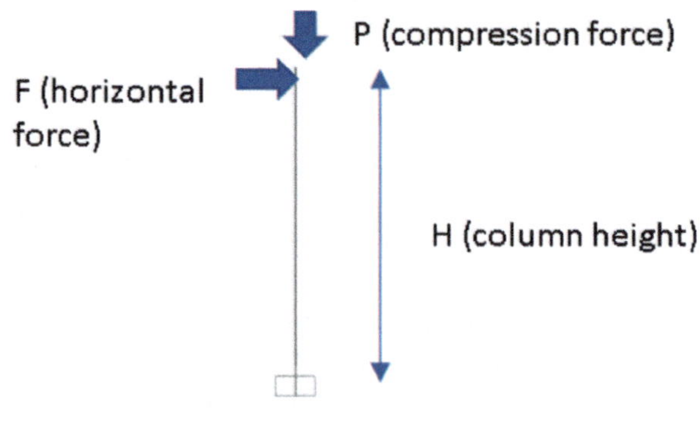

Fig. 3 Analysis model

Table 1 Parameter model

Item	Units	Value
Concrete compression strength f'_c	MPa	25.00
Concrete modulus E_c	MPa	23,500.00
Column dimension	mm	400 × 400
Concrete gross area A_g	mm^2	160,000
Column reinforcement A_{st}	mm^2	3402.34 (12D19)
Ratio A_{st}/A_g	%	2.13
Column length L	mm	5000.00
Gross stiffness = $(EI)_g$	$N.mm^2$	5.0133E+13
Effective stiffness calculated $(EI)_{eff} = 0.84(EI)_g$	$N.mm^2$	4.2178E+13
Hand calculation for critical axial buckling load P_{cr} using $(EI)_{eff}$	N	4,162,804.39
	ton	424.49

in this case is taken as 0.10Agf'c = 40 ton to match the column definition in SNI 2847:2019.

The parameter used in model can be seen in Table 1. The critical buckling load P_{cr} needs to be calculated by trial_error method to match the limit $P_u = 0.75P_{cr}$ since the effective stiffness value is varied due to the change in compression force and moment value. The model is then analyzed using buckling analysis in SAP2000 to validate the critical axial buckling load P_{cr} between hand calculation with model analysis. It can be seen that the critical buckling load produce almost the same value with the difference is smaller than 1% (see Table 2). Hence, it has no significance difference between hand calculation with model analysis. For P_{cr} = 422.50 ton, 0.75 P_{cr} is 316.87 ton. Therefore, the P applied in the model should be taken less than 316.87 ton. The compression forces and horizontal force applied in model can be seen in Table 3. The effective stiffness for P1, P2, P3 and P4 is calculated using Eq. (6). P2, P3 and P4 has the same effective stiffness since it is limited by 0.875 I_g as given in Eq. (7).

Table 2 Buckling analysis result

Stiffness	Buckling factor calculate output	$P_{initial}$ (ton)	$P_{cr\ model}$ = buckling factor × P (ton)	P_{cr} using hand calculation (ton)	$P_{cr\ model}/P_{cr}$
Using $(EI)_{eff}$	422.50	1.00	422.50	424.49	0.9953

Table 3 Force and stiffness applied in model

Notation	Compression force P (ton)	Horizontal force F (ton)	Correlation with $A_g f'_c$	Correlation with P_{cr}	Effective stiffness $(EI)_{eff}$
P1	40.00	1.00	$0.10A_g f'_c$	$0.09P_{cr}$	$0.852(EI)_g$
P2	120.00	1.00	$0.30A_g f'_c$	$0.28P_{cr}$	$0.875(EI)_g$
P3	200.00	1.00	$0.50A_g f'_c$	$0.47P_{cr}$	$0.875(EI)_g$
P4	280.00	1.00	$0.70A_g f'_c$	$0.66P_{cr}$	$0.875(EI)_g$

4 Result and Discussion

Nonlinear result (P-Delta or P-Delta plus large displacement) refers to as NL while the linear result refers to as LIN (see Figs. 4 and 5). The lateral displacement value for linear case has the same value between P2, P3 and P4 due to its similar effective stiffness value. It indicates that the lateral displacement in linear analysis is not affected by different compression forces. Figures 4 and 5 also show that the curve line is formed as the increasing of compression force due to the effect of geometric nonlinearity while the straight line is formed for linear case. It can also be seen that the same moment value is obtained for all models in linear analysis (see Table 4). This indicates that the moment value is only provided by the horizontal force acting on the top column in linear case, while the compression force does not give contribution to it.

The analysis results show that the linear model cannot capture the effect of the lateral displacement that changes due to the incremental of the compression force.

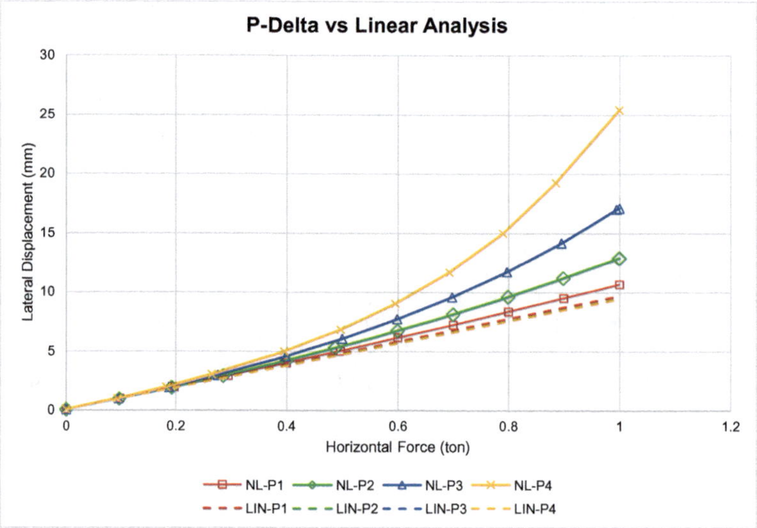

Fig. 4 Lateral displacement for P-Delta model

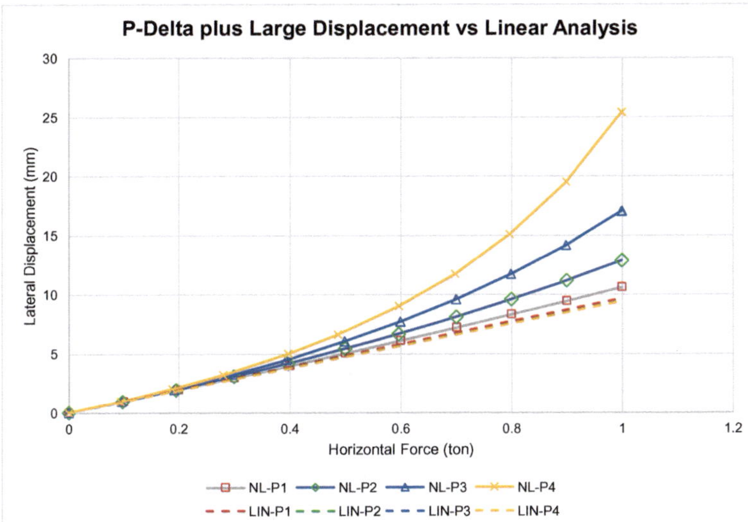

Fig. 5 Lateral displacement for P-Delta plus large displacement model

The increasing of lateral displacement due to larger compression force can only be seen when activating geometric nonlinearity option for both compression and lateral force. Table 4 also shows that the second-order moment increases as the increasing of compression force. It can also be seen that the moment for nonlinear case is about 1.1–2.4 times the moment value for linear case. This concludes that the compression force has significant effect to moment value especially for the higher compression force. In this case, the moment value of P-Delta plus large displacement is larger than the moment value of P-Delta. However, the difference is smaller than 3%. The larger moment of P-Delta plus large displacement comparing to P-Delta indicates that the large displacement effect takes account to the increasing of moment value.

Table 5 shows that there is no significant difference between lateral displacement of P-Delta plus large displacement with P-Delta model (smaller than 0.10%). However, the difference of lateral displacement is quite significance when comparing between nonlinear with linear model. The difference ratio of lateral displacement when comparing P-Delta plus large displacement to linear case is about 1.10–2.71. Therefore, the P-Delta effect should be analyzed carefully, especially for element with high compression force since the linear model could not capture this effect. Table 6 shows that axial displacement between linear and nonlinear model (P-Delta and P-Delta plus large displacement) has no significant difference (smaller than 3%). This indicates that P-Delta effect only affect the lateral displacement but has no significant effect for the axial displacement case.

Figures 6 and 7 shows the axial displacement value for the same incremental horizontal force with variation in compression force. It can be seen that the graph for nonlinear model for both P-Delta and P-Delta plus large displacement has almost coincides with the gradient line for linear case. However, it is quite different for the

Table 4 Second-order moment for P-Delta dan P-Delta plus large displacement model

Notation	$M_{first\text{-}order}$ (t m)	P-Delta			P-Delta plus large displacement		
		$M_{nonlinear}$ (t m)	$M_{second\text{-}order}$ (t m)	$M_{nonlinear}/M_{first\ order}$	$M_{nonlinear}$ (t m)	$M_{second\text{-}order}$ (t m)	$M_{nonlinear}/M_{first\ order}$
P1	5.000	5.433	0.433	1.0866	5.433	0.433	1.0866
P2	5.000	6.544	1.544	1.3088	6.550	1.550	1.3100
P3	5.000	8.376	3.376	1.6752	8.408	3.408	1.6816
P4	5.000	11.859	6.859	2.3718	12.115	7.115	2.4230

Table 5 Lateral displacement difference between P-Delta and P-Delta plus large displacement

Notation	Lateral displacement (mm)			Difference ratio	
	Linear	P-Delta	P-Delta plus large displacement	P-Delta plus large disp/P-Delta	P-Delta plus large disp/linear
P1	9.604	10.604	10.598	0.9994	1.1035
P2	9.352	12.836	12.825	0.9992	1.3714
P3	9.352	17.039	17.023	0.9991	1.8202
P4	9.352	25.329	25.354	1.0010	2.7110

Table 6 Axial displacement difference between P-Delta and P-Delta plus large displacement

Notation	Axial displacement (mm)			Difference ratio	
	Linear	P-Delta	P-Delta plus large displacement	P-Delta plus large disp/P-Delta	P-Delta plus large disp/linear
P1	0.534	0.534	0.545	1.0209	1.0209
P2	1.577	1.579	1.594	1.0095	1.0107
P3	2.621	2.635	2.653	1.0071	1.0125
P4	3.664	3.664	3.732	1.0185	1.0185

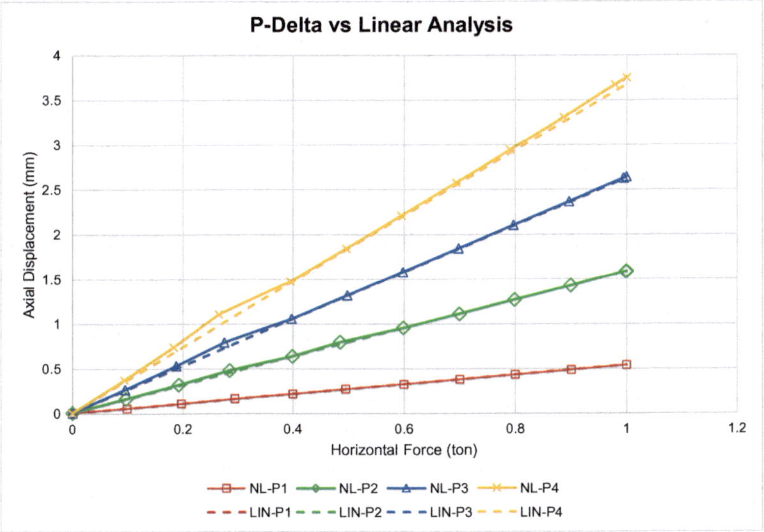

Fig. 6 Axial displacement for P-Delta model

lateral displacement graph (see again for Figs. 4 and 5). This confirms that P-Delta effect has significant effect to lateral displacement but not significant for the axial

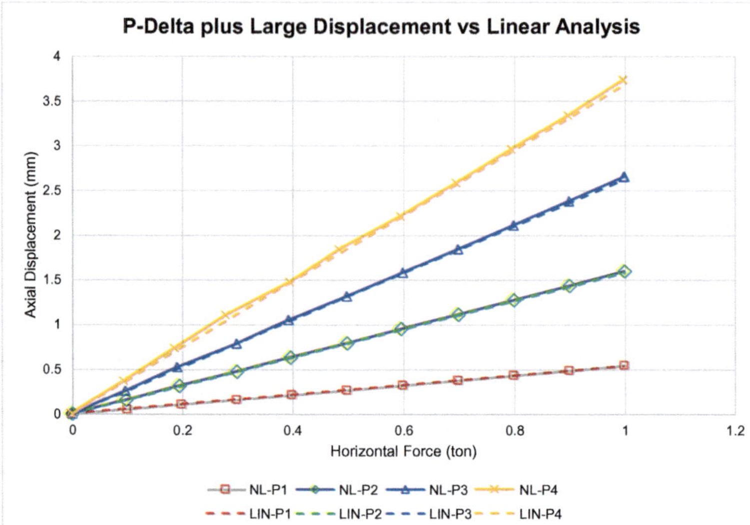

Fig. 7 Axial displacement for P-Delta plus large displacement model

displacement when comparing between nonlinear and linear model as we can see for each load step has no significant difference.

As stated before, the smaller meshing used in finite element program could produce different output, but it needs to be studied further in this case. To observe the meshing effect, the single line model is then divided into 5 segments with a length per segment of 1 m for the both P-Delta and P-Delta plus large displacement model. Figures 8 and 9 shows that the meshing by divide segment does not provide significance difference for both P-Delta and P-Delta plus large displacement model in this case. The marking point (data for meshing by divide frame) almost coincides with the line (data for no meshing frame) for each load step. The difference ratio is smaller than 1% for top lateral displacement (see Table 7). Hence, it can be concluded that the meshing option has no significant effect comparing to no mesh option in this case.

Figures 10 and 11 produce a different curve due to the difference compression force versus horizontal force applied in each load step. As stated before, the same final horizontal force 1 ton is applied while the final compression force is different for P1, P2, P3 and P4. The graph lines show that the larger the value of the compression force, the more visible the curve line due to the nonlinear effect. When the small compression force is applied (for P1 case), the line formed tends to be almost straight (linear). Then, the curve line takes forms as the increasing of compression force. It can also be seen that for linear model, the lines formed tend to coincide with each other even when applied with different compression force. This concludes that the lateral displacement is not affected by the compression force in linear model analysis.

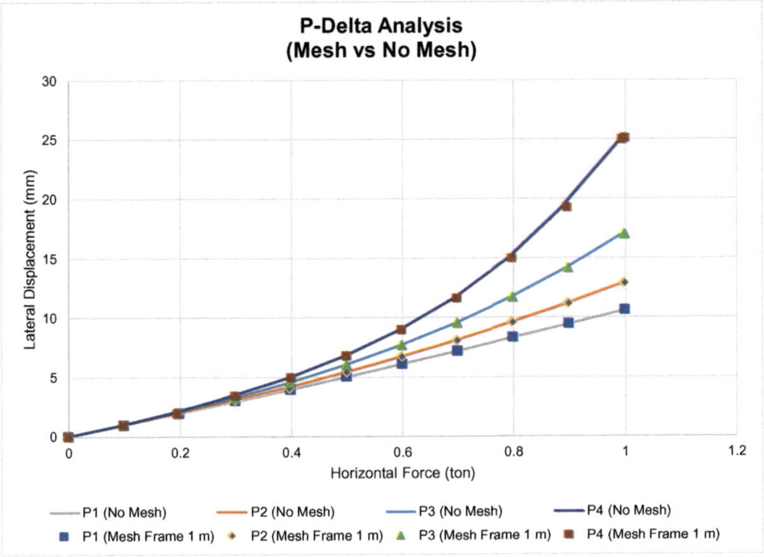

Fig. 8 Meshing effect for P-Delta model

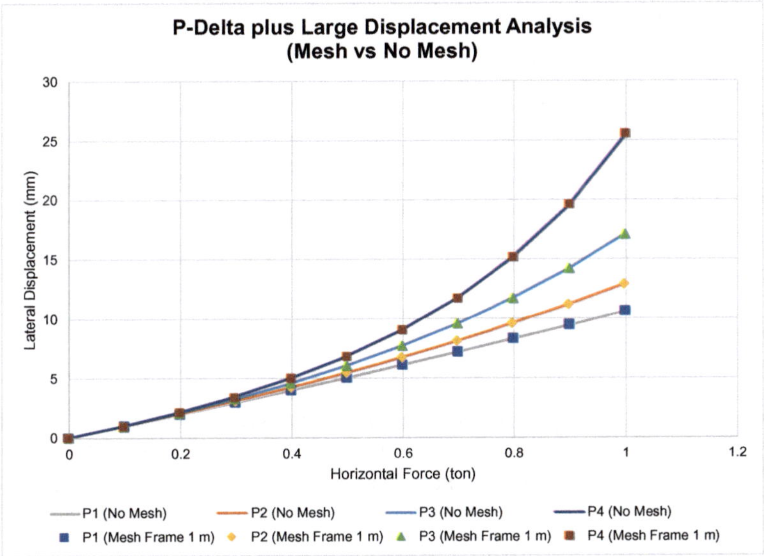

Fig. 9 Meshing effect for P-Delta plus large displacement model

Table 7 Meshing effect for P-Delta and P-Delta plus large displacement

	Lateral displacement for P-Delta model (mm)			Lateral displacement for P-Delta plus large displacement (mm)		
Notation	No mesh (mm)	Mesh by divide segment (mm)	No mesh/mesh	No mesh (mm)	Mesh by divide segment (mm)	No mesh/mesh
P1	10.604	10.595	1.0009	10.598	10.595	1.0003
P2	12.836	12.820	1.0012	12.825	12.827	0.9999
P3	17.039	16.991	1.0028	17.023	17.046	0.9986
P4	25.329	25.072	1.0103	25.354	25.525	0.9933

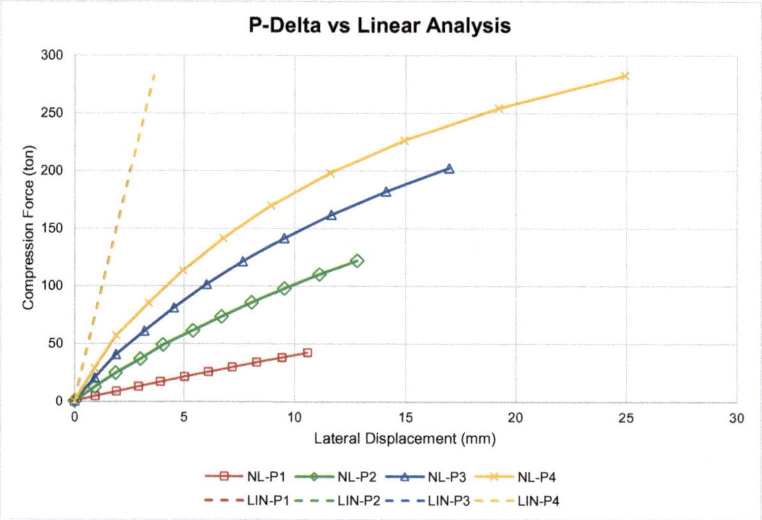

Fig. 10 P-Delta model (compression force versus lateral displacement)

The meshing model by divide frame for 1 m is used to observe the P-Delta plus large displacement effect for local bending moment at each joint (see Tables 8 and 9). It can be seen that the bending moment at each local joint increase as the increasing of compression force. Hence, the P-Delta plus large displacement effect tends to have more effect for the larger compression force. However, the difference is not significant in this case. The largest difference is about 2.65% for the largest compression force applied (P4), while it has the same value for the smallest compression force applied (P1). It can also be seen that each local joint has different moment ratio, this indicates that the P-Delta plus large displacement effect is taking account for local deformation.

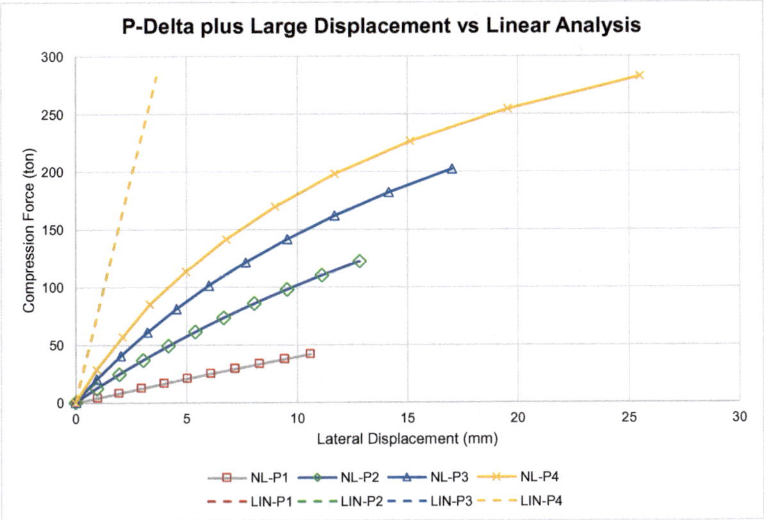

Fig. 11 P-Delta plus large displacement model (compression force versus lateral displacement)

Table 8 Local moment at joint for P1 and P2

		P1			P2		
Distance from support	Linear M (t m)	P-Delta M (t m) (a)	P-Delta plus large disp M (t m) (b)	Ratio a/b	P-Delta M (t m) (a)	P-Delta plus large disp M (t m) (b)	Ratio a/b
0 m (base)	5.000	5.431	5.431	1.0000	6.541	6.548	1.0011
1 m	4.000	4.406	4.406	1.0000	5.457	5.463	1.0011
2 m	3.000	3.340	3.340	1.0000	4.225	4.230	1.0012
3 m	2.000	2.243	2.243	1.0000	2.880	2.884	1.0014
4 m	1.000	1.126	1.126	1.0000	1.459	1.461	1.0014
5 m (top)	0.000	0.000	0.000	1.0000	0.000	0.000	1.0000

5 Conclusion

1. The lateral displacement in linear analysis is not affected by different compression forces. It can also be seen that the same moment value is obtained for all models in linear analysis. This indicates that the moment value is only provided by the horizontal force acting on the top column in linear case, while the compression force does not give contribution to it.

Table 9 Local moment at joint for P3 and P4

		P3			P4		
Distance from support	Linear M (t m)	P-Delta M (t m) (a)	P-Delta plus large disp M (t m) (b)	Ratio a/b	P-Delta M (t m) (a)	P-Delta plus large disp M (t m) (b)	Ratio a/b
0 m (base)	5.000	8.380	8.419	1.0047	11.903	12.166	1.0221
1 m	4.000	7.199	7.236	1.0051	10.543	10.793	1.0237
2 m	3.000	5.697	5.729	1.0056	8.531	8.744	1.0250
3 m	2.000	3.943	3.966	1.0058	5.995	6.149	1.0257
4 m	1.000	2.016	2.028	1.0060	3.091	3.173	1.0265
5 m (top)	0.000	0.000	0.000	1.0000	0.000	0.000	1.0000

2. The analysis results show that the linear model cannot capture the effect of the lateral displacement that changes due to the incremental of the compression force. The increasing of lateral displacement due to larger compression force can only be seen when activating geometric nonlinearity option for both compression and lateral force.
3. The moment for nonlinear case is about 1.1–2.4 times the moment value for linear case. This concludes that the compression force has significant effect to moment value especially for the higher compression force.
4. The moment value of P-Delta plus large displacement is larger than the moment value of P-Delta. However, the difference is smaller than 3%. The larger moment of P-Delta plus large displacement comparing to P-Delta indicates that the large displacement effect takes account to the increasing of moment value.
5. There is no significant difference between lateral displacement of P-Delta plus large displacement with P-Delta model (smaller than 0.10%). However, the difference of lateral displacement is quite significance when comparing between nonlinear with linear model. The difference ratio of lateral displacement when comparing P-Delta plus large displacement to linear case is about 1.10–2.71. Therefore, the P-Delta effect should be analyzed carefully, especially for element with high compression force since the linear model could not capture this effect.
6. The axial displacement between linear and nonlinear model (P-Delta and P-Delta plus large displacement) has no significant difference (smaller than 3%). This indicates that P-Delta effect only affect the lateral displacement but has no significant effect for the axial displacement case.
7. The meshing by divide segment does not provide significance difference for both P-Delta and P-Delta plus large displacement model in this case. The difference ratio is smaller than 1%.
8. The bending moment at each local joint increase as the increasing of compression force. Hence, the P-Delta plus large displacement effect tends to have more

effect for the larger compression force. However, the difference is not significant in this case. The largest difference is about 2.65% for the largest compression force applied (P4), while it has the same value for the smallest compression force applied (P1). It can also be seen that each local joint has different moment ratio, this indicates that the P-Delta plus large displacement effect is taking account for local deformation.

References

1. Dewobroto W, Chendrawan W (2011) Effect of modeling and analysis on behavior of glass structures against wind pressure. J Nat Seminar-1 BMPTTSSI—KoNTekS 5 USU, Medan
2. Istiono H, Ramadhan AY (2020) Analysis of the P-Delta effect due to the height difference in earthquake resistant buildings (Case Study: Non-Highrise Building). J REKAYASA SIPIL 14(3). ISSN 1978-5658
3. Avinash T, Pandian G (2017) Investigation of the effects of P-Delta on tubular tall buildings. J IJCIET 8:487–495
4. Adityawarman G (2014) Planning of evacuation buildings in earthquake and tsunami prone areas. J Kajian Teknologi 10(2)
5. Suhana N, Pello O (2015) P-Delta effect analysis on fifteen stories building due to the increasing of helipad load. J Rekayasa Infrastruktur 1(2):44–105. ISSN 2460-335X
6. CSI technical knowledge base. https://wiki.csiamerica.com/display/kb/P-Delta+effect. Last accessed January 2021
7. Wilson EL (2002) Three-dimensional static and dynamic analysis of structures, 3rd edn. Computers and Structures Inc., California
8. SCIA support. https://www.scia.net/en/support/faq/calculation-non-linear-dynamics/second-order-analysis-questions-part-2. Last accessed May 2021
9. National Standardization Agency (2019) SNI 2847:2019, Structural Concrete Requirements for Buildings with Commentary. BSN, Jakarta (2019).
10. Computers & Structures Inc (2019) Concrete frame design manual ACI 318-14 SAP2000. Computers and Structures, Inc., California
11. Computers & Structures Inc (2017) CSI analysis reference manual for SAP2000. ETABS, SAFE and CSIBridge, California

Effect of Regulatory Change in Earthquake Load Analysis on Structures with Irregular Shapes

Hendramawat Aski Safarizki and Dini Ayu Saputri

Abstract The effect of the earthquake must be reviewed in the design of the building structure. The development of ASCE 7-10 and ASCE 7-16 regulations was followed by the development of earthquake-resistant building regulations in Indonesia, namely SNI 1726-2019. The developments of Indonesia SNI 1726-2019 was accompanied by the publication of the Indonesian Spectra Design 2021, which replaced the Indonesian Spectra Design 2011. The development of this regulation has raised concerns that the existing buildings designed by referring the previous regulation would behave differently when they were analyzed with the latest SNI 1726-2019 regulations. In this study, a response spectrum analysis was carried out based on the Indonesian Spectra Design 2021 for buildings that were designed according to the Indonesian Spectra Design 2011. The building under study has an L shape with a difference in length on the side of the X direction and the Y direction. The shape of a building falls into a category of high irregularity. The results of this analysis indicate that there is a change in the level of safety in the structure. It is necessary to conduct a study on the planned building before the implementation of the SNI 1726-2019 regulations which are accompanied by Indonesian Spectra Design 2021.

Keywords Earthquake · Regulation · Response spectra

1 Introduction

Indonesia is an archipelago whose geographic position is on the most active earthquake route in the world because it is surrounded by the Pacific Ring of Fire (Ring of Fire) which is above three collisions of continental plates, namely the Indo-Australian, Eurasian and Pacific plates. This geographical condition makes Indonesia a region prone to natural disasters such as volcanic eruptions, earthquakes, and tsunamis.

H. A. Safarizki (✉) · D. A. Saputri
Universitas Veteran Bangun Nusantara, Sukoharjo, Indonesia

S. A. Kristiawan et al. (eds.), *Proceedings of the 5th International Conference on Rehabilitation and Maintenance in Civil Engineering*, Lecture Notes in Civil Engineering 225,
https://doi.org/10.1007/978-981-16-9348-9_3

The basic shape of the structure in a building generally has to contribute to resisting the lateral forces caused by earthquakes. The magnitude of the earthquake load is greatly influenced by the condition of the building structure such as the building plan and the mass of the building itself [1]. Recent research states that the higher the horizontal irregularity, the higher the shear force and max drift [2–4]. Based on research on a building with an L plan, there will be a critical area located at the back end of the building [5].

The effect of the earthquake must be reviewed in the planning of the building structure. The structure as a whole must remain standing even though it is on the verge of collapse in the event of an earthquake. In general, structural analysis of earthquakes is divided into 2 types, namely equivalent static analysis, namely in the form of horizontal force (Px, Py) applied to the floor of the structure, and dynamic analysis (time history and response spectrum) in the form of waves based on previous earthquake data. Which is applied to the base structure, and analyzed under non-linear conditions. Spectrum response analysis has been widely applied in earthquake-resistant structure analysis [6–12]. The dynamic analysis aims to determine the distribution of the level shear forces due to ground motion by earthquakes and can be done by analyzing the response spectrum [13].

The development of ASCE 7-10 and ASCE 7-16 regulations was followed by the development of earthquake-resistant building regulations in Indonesia, namely SNI 1726-2019. On developments in Indonesia SNI 1726-2019 was followed by the publication of the Indonesian Spectra Design in 2021 which replaced the Indonesian Spectra Design in 2011. The development of this regulation was followed by concerns that buildings designed before this regulation were implemented would behave differently when they were tested with the latest SNI 1726-2019 regulations.

In this study, a response spectrum analysis was carried out based on the Indonesia Spectra Design 2021 against the planned building using the Indonesian Spectra Design 2011. It is hoped that the results of this study can illustrate changes in the behavior of earthquake-resistant building structures with the implementation of SNI 1726-2019 followed by the 2021 Spectra Design.

2 Methods

2.1 *Structure Plan*

The structure reviewed in this research is an L-shaped 12-story building in Surakarta on moderate soil conditions which is planned as a high-rise building. The structure consists of 12 floors with a total height of floors 1 to 12 is $\pm$38.10 m.

The loading of the building can be determined by adjusting the use of the building, that is, the function of the room in each different building is determined based on the determination of the load following SNI 03-1727-2013 Minimum Load for

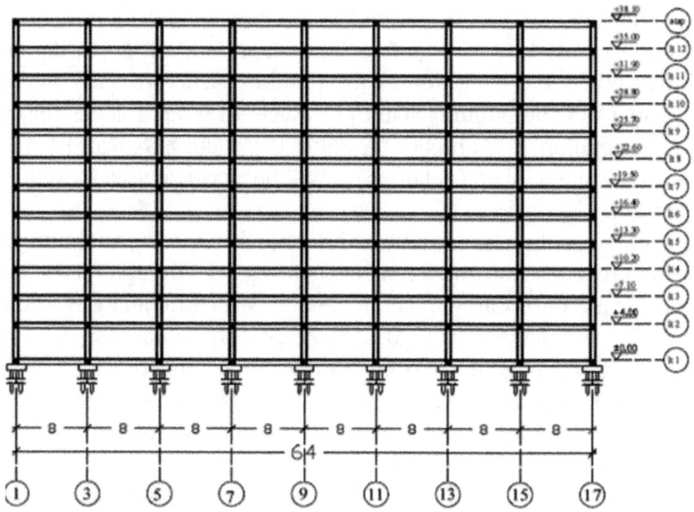

Fig. 1 Typical floor plan 1–12

Designing Buildings and Other Structures and SNI 03-2847-2013 Concrete Requirements Structural for Buildings. Effect of an earthquake and earthquake-resistant planning for building structures following the Guidelines for Earthquake Resistance Planning based on SNI 03-1726-2012 with Indonesian Spectra 2011 and SNI 03-1726-2019 with Indonesian Spectra Design 2021. The plan of the building being reviewed appears in Fig. 1.

The building under study has an L shape with a difference in length on the side of the X direction and the Y direction. The shape of a building like this has a high irregularity. Irregularity considerably affects the seismic and the seismic response of the structures [3].

2.2 *Structural Modeling*

Structural modeling using SAP 2000 software is used to facilitate analysis of the building structure under study. In the model created, material specifications and other data are also applied to calculate the load on the building. The planned building structure is an L-shaped 12-story apartment building with a building height of ±38.10 m (Fig. 1).

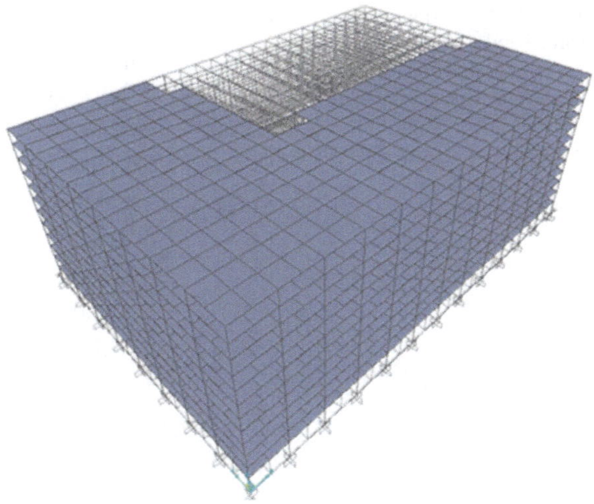

Fig. 2. 3D modeling of building structures

2.3 *Load on Structure*

The planning of loading is guided by SNI 03-1727-2013 concerning the Planning of Loads for Houses and Buildings. The load calculation includes dead load, live load, and wall load. Based on SNI 03-1726-2012, the building risk category is included in group II with the earthquake priority factor (Ie) of 1.0. Based on the acceleration response parameter in the short period (S) of 0.602 with the class II risk category being the seismic design category D (KDS D). The seismic design category is based on the acceleration response parameter in the 1 s (S) period of 0.570 with the class II risk category being the seismic design category D (KDS D). So that the seismic design category based on SDS, SD1, and risk category II is included in KDS D.

Based on SNI 03-1726-2019, the building risk category is included in group II with the earthquake priority factor (Ie) of 1.0. Based on the acceleration response parameter in the short period (S) of 0.672 with the class II risk category being the seismic design category D (KDS D). The seismic design category based on the acceleration response parameter in the 1 s period (S) is 0.633 with the class II risk category being the seismic design category D (KDS D). So that the seismic design category based on SDS, SD1, and risk category II is included in KDS D. Table 1 describes the differences in spectra design from the Indonesian Spectra Design in 2011 and 2021. If Table 1 is then made in the form of a response spectrum curve, it will look like in Fig. 3. In this study did not consider the torque magnification factor.

Site Coefficients and Risk-Targeted Maximum Considered Earthquake (MCER) Spectral Response Acceleration Parameters for short periods (SMS) and at 1 s (SM1), adjusted for Site Class effects, shall be determined by Eqs. 11.4.1 and 11.4.2 [14], respectively.

$$\mathrm{SMS} = \mathrm{FaSS} \tag{1}$$

Table 1 The difference in value between Indonesian Spectra Design 2011 and Indonesian Spectra Design 2021

Variable	Value	
	Indonesian Spectra Design 2011	Indonesian Spectra Design 2021
PGA (g)	0.3560	0.3503
SS (g)	0.7360	0.8009
S1 (g)	0.3090	0.3883
FA	1.2270	1.2593
FV	2.7630	2.4468
SMS (g)	0.9040	1.0086
SM1 (g)	0.8540	0.9501

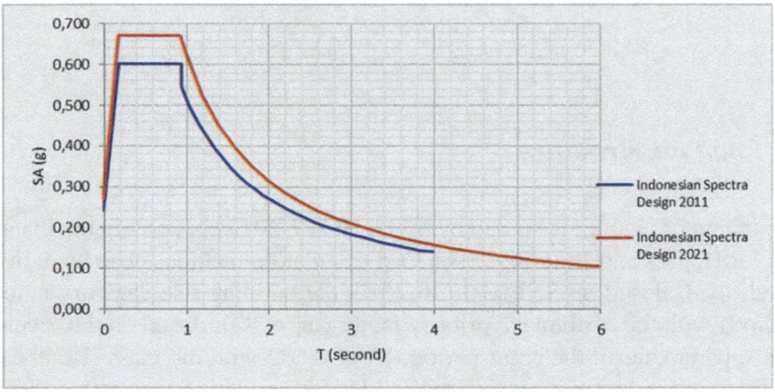

Fig. 3 Indonesian Spectra Design 2011 and 2021 response spectrum curve

$$SM1 = FvS1 \tag{2}$$

where:

SS = the mapped MCER spectral response acceleration parameter at short periods as determined by Section 11.4.1 ASCE 7-10, and

S1 = the mapped MCER spectral response acceleration parameter at a period of 1 s as determined by Section 11.4.1 ASCE 7-10

3 Results

3.1 Based on Regulations SNI 03-1726-2012

The deviation control between design floors (Δ) shall be calculated as the difference in deflection at the center of mass at the top and bottom levels under review. Based on SNI 03-1726-2012, the allowable deviation between floors for risk category II (other structures) for this structure is 0.020 hsx. This type of structure is not a structure that uses shear walls.

$$\begin{aligned}\Delta_a &= 0.020h_x \\ &= 0.020 \times 3100 \\ &= 62 \text{ mm} \rightarrow 0.062 \text{ m}\end{aligned}$$

From Table 9 on SNI 03-1726-2012, the Cd factor in the earthquake resistance system, the designed structure is a special moment-bearing reinforced concrete frame system, the value of Cd = 5.5. Calculating the deviation that occurs, take floor 12 joint number 6233 and floor 11 joint number 5936 viewed in the X-direction. Points 6233 and 5936 are the same point on different floors. Number joint 6233 δ2 = 0.030.153 m, number joint 5936 δ1 = 0.029.492 m. The complete calculation results for all floors are in Table 2. The control of the maximum deviation between floors for the X direction based on calculating using SNI 03-1726-2012 is 0.00847 m < 0.062 m (the allowable deviation limit between floors) and the drift ratio on the top floor (0.12%) which is calculated based on the deviation shows that the drift ratio that occurs is still smaller than the drift. The maximum required in SNI 1726-2019 is 2% so that it still meets the requirements.

Table 2 The deviation between the floors of the design level (Δ) in the X direction based on calculating using SNI 03-1726-2012

Floor	Height (m)	No. joint	Deflection/δ (m)	Deviation/Δ (m)	Drift ratio (%)
12	3.1	6233	0.030.153	0.003.635	0.12
11	3.1	5936	0.029.492	0.005.797	0.19
10	3.1	5639	0.028.438	0.008.201	0.26
09	3.1	5342	0.026.947	0.010.516	0.34
08	3.1	5045	0.025.035	0.012.678	0.41
07	3.1	4748	0.022.730	0.014.647	0.47
06	3.1	4451	0.020.067	0.016.423	0.53
05	3.1	4154	0.017.081	0.017.969	0.58
04	3.1	3857	0.013.814	0.019.239	0.62
03	3.1	3560	0.010.316	0.020.026	0.65
02	3.1	3263	0.006.675	0.036.713	1.18
01	4.0	33	0	0	0

Table 3 The deviation between the floors of the design level (Δ) in the X direction based on calculation using SNI-03-1726-2019

Floor	Height (m)	No. joint	Deflection/δ (m)	Deviation/Δ (m)	Drift ratio (%)
12	3.1	6233	0.062.562	0.0.084.700	0.27
11	3.1	5936	0.061.022	0.0.126.775	0.41
10	3.1	5639	0.058717	0.0.175.835	0.57
09	3.1	5342	0.055520	0.0.224.180	0.72
08	3.1	5045	0.051444	0.0.269.500	0.87
07	3.1	4748	0.046544	0.0.311.080	1.00
06	3.1	4451	0.040888	0.0.348.260	1.12
05	3.1	4154	0.034556	0.0.380.325	1.23
04	3.1	3857	0.027641	0.0.404.690	1.31
03	3.1	3560	0.020283	0.0.414.700	1.34
02	3.1	3263	0.012743	0.0.700.865	2.26
01	4.0	33	0	0	0

3.2 *Based on Regulations SNI 03-1726-2019*

From Table 12 (page 50) on SNI 03-1726-2019, the Cd factor for earthquake resistance systems, the designed structure is a special moment-bearing reinforced concrete frame system, the value of Cd = 5.5. Calculating the deviation that occurs as shown in Table 3, take floor 12 joint number 6233 and floor 11 joint number 5936 viewed in the X-direction. Points 6233 and 5936 are the same point on different floors. The control of the maximum deviation between floors for the X direction is 0.00847 m < 0.062 m (the allowable deviation limit between floors) and the drift ratio on the 2nd floor (2.26%) which is calculated based on the deviation shows that the drift ratio that occurs is greater than the maximum drift which is required in SNI 1726-2019 which is 2% so it does not meet the requirements.

4 Discussion and Conclusions

If Tables 2 and 3 are made in the form of a graph of the relationship between the drift ratio analysis results with SNI 1726-2012 and SNI 1726-2019, it will appear that there is an increase in the deviation of each floor (Fig. 4). This increase in deviation causes the drift ratio of each floor to increase above 90% if the analysis results with SNI 1726-2019 are compared with the planning using SNI 1726-2012.

The results of the calculation of the critical area as reviewed by SNI 1726-2012 are located at the back end of the building with maximum control of the deviation between floors for the X direction of 0.036713 m < 0.062 m (allowable deviation between floors) and has a drift ratio of 0.12% < 2% (maximum drift required in SNI 1726-2012) indicates that the building structure has met the required limits.

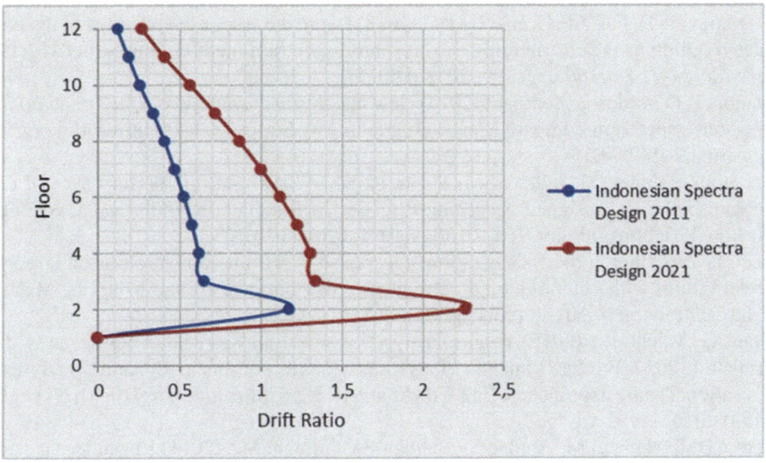

Fig. 4 Comparison of drift ratio values

The results of the calculation of the critical area as reviewed by SNI 1726-2019 are located on the 2nd floor with a maximum control of the deviation between floors for the X direction of 0.00847 m < 0.062 m (allowable deviation between floors) and has a drift ratio of 2.26% > 2% (maximum drift required in SNI 1726-2019). This indicates that the building structure does not meet the limits set out in SNI 1726-2019.

The results of this analysis indicate that there is a change in the level of safety in the structure. It is necessary to conduct a study on the planned building before the implementation of the SNI 1726-2019 regulations which are accompanied by Indonesian Spectra Design 2021.

References

1. Batu ML, Dapas SO, Wallah SE (2016) Efisiensi Penggunaan Dinding Geser untuk Mereduksi Efek Torsi pada Bangunan yang Tidak Beraturan. Sipil Statik
2. Prijasambada, Hafifah V (2018) Analisa Gaya Diafragma, Kord Dan Kolektor Pada Bangunan Gedung Sesuai Dengan Sni 1726:2012. IKRAITH-TEKNOLOGI. 2:41–49
3. Naveen SE, Abraham NM, Kumari ASD (2019) Analysis of irregular structures under earthquake loads. Procedia Struct Integr 14:806–819. https://doi.org/10.1016/j.prostr.2019.07.059
4. Ahmed MMM, Abdel Raheem SE, Ahmed MM, Abdel Shafy AGA (2016) Irregularity effects on the seismic performance of L-shaped multi-story buildings. JES J Eng Sci 44:513–536. https://doi.org/10.21608/jesaun.2016.111440
5. Saputri DA, Safarizki HA (2021) Perencanaan Struktur Gedung Apartemen 12 Lantai Berbentuk L di Surakarta. In: Simposium Nasional Teknologi Infrastruktur Abad ke-21, Yogyakarta, pp 201–206

6. De Domenico D, Falsone G, Ricciardi G (2018) Improved response-spectrum analysis of base-isolated buildings: a substructure-based response spectrum method. Eng Struct 162:198–212. https://doi.org/10.1016/j.engstruct.2018.02.037
7. Cacciola P, D'Amico L, Zentner I (2014) New insights in the analysis of the structural response to response-spectrum-compatible accelerograms. Eng Struct 78:3–16. https://doi.org/10.1016/j.engstruct.2014.07.015
8. Mostafijur Rahman M, Jadhav SM, Shahrooz BM (2018) Seismic performance of reinforce concrete buildings designed according to codes in Bangladesh, India and U.S. Eng Struct 160:111–120. https://doi.org/10.1016/j.engstruct.2018.01.010
9. Cacciola P, Zentner I (2012) Generation of response-spectrum-compatible artificial earthquake accelerograms with random joint timefrequency distributions. Probabilistic Eng Mech 28:52–58. https://doi.org/10.1016/j.probengmech.2011.08.004
10. Milani G, Valente M (2015) Failure analysis of seven masonry churches severely damaged during the 2012 Emilia-Romagna (Italy) earthquake: non-linear dynamic analyses versus conventional static approaches. Eng Fail Anal 54:13–56. https://doi.org/10.1016/j.engfailanal.2015.03.016
11. Islam ABMS, Jameel M, Ahmad SI, Salman FA, Jummat MZ (2011) Engendering earthquake response spectra for Dhaka region usable in dynamic analysis of structures. Sci Res Essays 6:3519–3530. https://doi.org/10.5897/sre11.218
12. Hassaballa AE (2013) Seismic analysis of a reinforced concrete building by response spectrum method. IOSR J Eng 03:01–09. https://doi.org/10.9790/3021-03930109
13. Mulyo A, Utami D, Dinamik A (2012) Metode Respon Spectrum Ditinjau Pada Drift Dan Displacement Menggunakan Software ETABS (Performance Analysis of Building Structural with Spectrum Response Method Reviewed on Management and Control Using ETABS Software). J Infras 4:65–71
14. ASCE (2016) Minimum Design Loads for Buildings and Other Structures (ASCE 7-10)

Seismic Performance of Tall and Slender Minaret Structure with Hexagonal RC Wall Section by Means Fragility Curve Development

Erik Wahyu Pradana, Senot Sangadji, and Angga Destya Navara Noor

Abstract Currently, there are 278,629 mosques registered in the information system of the Indonesian Ministry of Religion. This number is believed to keep growing. One of them is a mosque that was being built in Surakarta, namely Taman Sriwedari Mosque. This building is equipped with 114 m height minaret and is claimed to be the tallest in Indonesia. Considering that Indonesia is located in an earthquake-prone region, it is important to assess the seismic performance of this structure. This research aims to evaluate the seismic performance of the minaret using the fragility function. Fragility function is a function that correlates the damage level with an intensity measure. The structural model of the minaret was developed using Seismostruct. Furthermore, the structural response is obtained from the structural model analyzed using static nonlinear pushover. Based on this analysis result, a structural capacity curve is obtained which is then used in constructing the fragility curve. The fragility curve shows that the damage probability of minaret structure on spectral acceleration, Sa = 0.098 g based on the material strain limits criteria for damage states DS1, DS2, DS3, and DS4 is 57%, 29%, 19%, and 12% respectively. On the other hand, based on the maximum base shear criteria for the damage states DS1, DS2, and DS3 is 50%, 20%, and 12% respectively. Perceiving the fragility curve for the structure under review, building owners can predict the probability of structural damage due to other earthquake intensity scenarios and carry out evaluations to determine retrofitting strategies rationally.

Keywords Seismic performance · Minaret structure · Fragility curve

E. W. Pradana (✉) · S. Sangadji · A. D. N. Noor
SMARTQuake Research Group, Civil Engineering Department, Faculty of Engineering, Universitas Sebelas Maret, Surakarta, Indonesia
e-mail: erikwpradana@staff.uns.ac.id

S. Sangadji
e-mail: s.sangadji@ft.uns.ac.id

S. A. Kristiawan et al. (eds.), *Proceedings of the 5th International Conference on Rehabilitation and Maintenance in Civil Engineering*, Lecture Notes in Civil Engineering 225,
https://doi.org/10.1007/978-981-16-9348-9_4

1 Introduction

An earthquake with a magnitude of 9.1 that occurred in Northern Sumatra, Indonesia in 2004 was the third-largest earthquake in the world, recorded since 1900 [1]. In the list, 7th and 11th positions are earthquakes that also occurred in Indonesia, namely the Northern Sumatra earthquake with a magnitude of 8.6 and the Banda Sea earthquake with a magnitude of 8.5. A number of major earthquakes that happened in Indonesia followed by large tsunamis, such as the 2004 Aceh; 2005 Nias; 2006 Yogyakarta, West Java, and West Sumatra; and 2007 Bengkulu and West Sumatra earthquake, cause loss of life and structural damage [2]. These facts indicate that Indonesia is a country with a high risk of seismicity.

Most earthquakes in Indonesia are caused by tectonic activity. The relative movement of tectonic plates causes the collision of the Indo-Australian Plate, the Eurasian Plate, and the Pacific Plate. This tectonic activity led to the formation of a volcano series along with the islands of Sumatra, Java-Bali-Nusa Tenggara, northern Sulawesi-Maluku, to Papua which became known as the ring of fire. In the forearc where the two tectonic plates meet, many active faults often cause earthquakes.

Currently, there are 278,629 mosques registered in the information system of the Ministry of Religion in Indonesia [3]. This number is believed to keep growing. One of them is a mosque that was being built in Surakarta, namely the Taman Sriwedari Mosque. This building is equipped with five minarets. One of the minarets is 114 m in height and is claimed to be the tallest in Indonesia. Considering that Indonesia is located in an earthquake-prone region, it is important to assess the seismic performance of the minaret. This research aims to evaluate the seismic performance of the minaret by developing its structural fragility curve. The fragility function elucidates the probability of exceedance for multiple damage states as a function of ground-motion intensity [4].

The structural model of the minaret was developed using Seismostruct [5]. Furthermore, the structural response is obtained from the structural model analyzed using a nonlinear static procedure (pushover analysis). Based on this analysis, a structural capacity curve is obtained which is then used to construct the fragility curve. The fragility curve obtained shows the probability of structural damage (exceed the predetermined damage states) related to the intensity of ground motions during its service life. The fragility curve shows that the damage probability of minaret structure on spectral acceleration, Sa = 0.098 g based on the material strain limits criteria for damage states DS1, DS2, DS3, and DS4 is 57%, 29%, 19%, and 12% respectively. On the other hand, based on the maximum base shear criteria for the damage states DS1, DS2, and DS3 is 50%, 20%, and 12% respectively. Perceiving the fragility curve for the structure under review, building owners can predict the probability of structural damage due to other earthquake intensity scenarios and carry out evaluations to determine retrofitting strategies rationally.

2 Research Method

In this research, the structural model of the minaret was developed using Seismostruct (see Fig. 1a). Seismostruct is a finite element-based structural analysis software that has the capability for predicting the large displacement behavior of 3D frames under static or dynamic loadings by considering both geometric nonlinearities and material inelasticity [6]. The load–displacement relationship (capacity curve) of the modeled structure is obtained from static nonlinear pushover analysis. Furthermore, the capacity curve obtained is used to determine the fragility curve of the structure under study.

The minaret structure has been built in 2019 using the slip-form construction method. The minaret structure consists of two layers (inner and outer) of reinforced concrete shear walls with a hexagon geometry. The height of the minaret is 114 m. The material properties of the concrete are as follows: compressive strength $\left(f_c'\right)$ is 30 MPa and the modulus of elasticity is taken as $4700\sqrt{f_c'}$. While the material properties of the reinforcing steel are as follows: yield strength $\left(f_y\right)$ is 390 MPa and the modulus of elasticity is 200,000 MPa. Detailed dimensions and reinforcement of the minaret structure are presented in Fig. 1b–g.

The live load design of this structure is 4.79 kN/m^2 in accordance with SNI 1727:2013 [7]. Meanwhile, the earthquake load is determined based on SNI 1726:2012 [8] with the Peak Ground Acceleration value (Sa) = 0.602 g. This structure is located in Surakarta with SD (medium soil) site class, the risk-targeted maximum considered earthquake (MCER) spectral response acceleration at short period (S_s) = 0.754 g, and at a 1 s period (S_1) = 0.315 g. Based on the structural model, the fundamental period of the structure is 3.782 s. Furthermore, this value corresponds to the spectral acceleration value, (Sa) = 0.098 g, in the Response Spectrum Design for the Surakarta region based on SNI 1726:2012. The obtained spectral acceleration value is then used to evaluate the seismic performance of the minaret using the fragility function.

3 Result and Discussion

3.1 Capacity Curve and Capacity Spectrum

A capacity curve is a curve that describes the relationship between base-shear and displacement at the top of the observed structure. In this research, the capacity curve of the minaret structure is obtained through static nonlinear pushover analysis (see Fig. 2). Furthermore, this curve is converted into a capacity spectrum in ADRS (Acceleration Displacement Response Spectrum) format which is used to determine the median spectral displacement (See Fig. 3). The capacity spectrum describes the

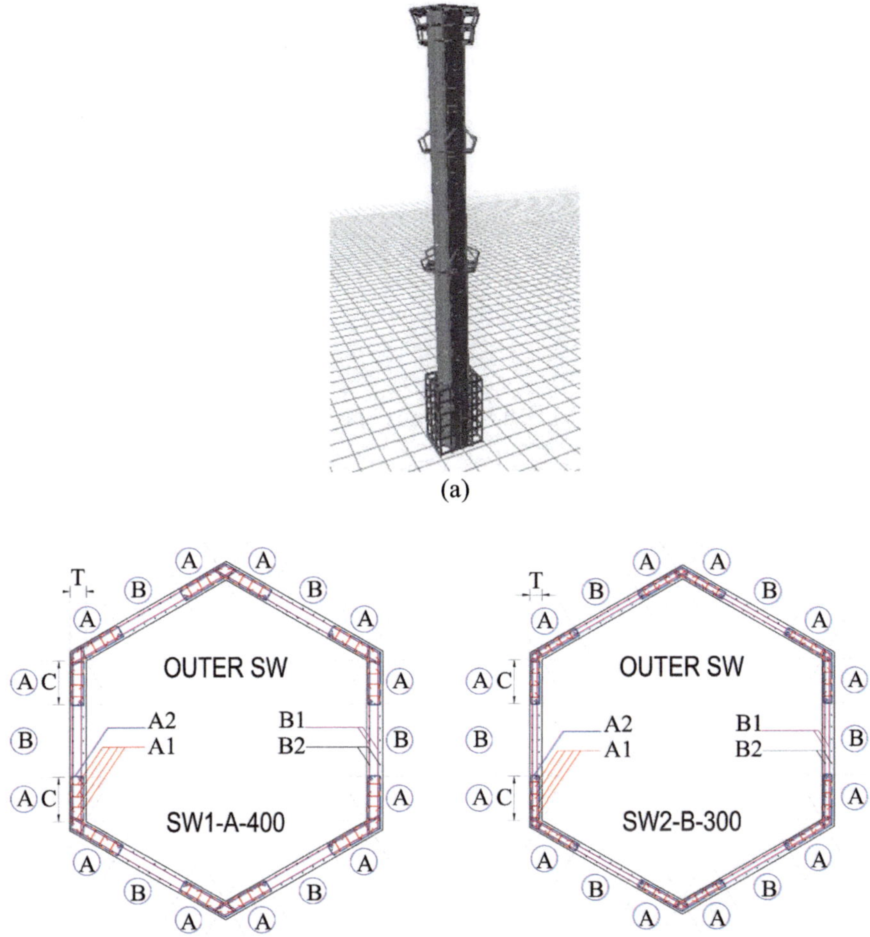

(b) SW1-A-400 (-4.50 to +45.60)

A1	=	D19-100 Vertical (Crossties)
A2	=	2D19-100 Vertical (Hoops)
B1	=	D25-300 Horizontal
B2	=	D19-300 Vertical
C	=	12D25 (L=1100 mm)
T	=	400 mm

(c) SW2-B-300 (+46.80 to +74.40)

A1	=	D19-100 Vertical (Crossties)
A2	=	2D19-100 Vertical (Hoops)
B1	=	D22-300 Horizontal
B2	=	D19-300 Vertical
C	=	12D22 (L=1100 mm)
T	=	300 mm

Fig. 1 **a** Structural model in seismostruct, **b** reinforcement details of SW1-A-400, **c** reinforcement details of SW2-B-300, **d** reinforcement details of SW2-C-300, **e** reinforcement details of SW3-A-250, **f** reinforcement details of SW3-B-250, and **g** reinforcement details of SW3-C-250

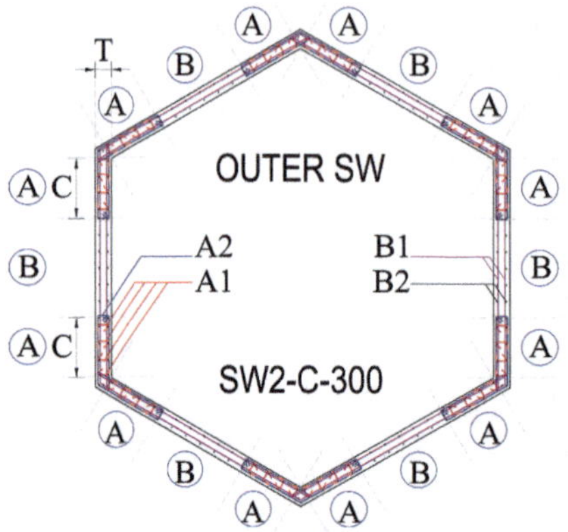

(d) SW2-C-300 (+75.60 to +104.40)

A1 = D19-100 Vertical (Crossties)
A2 = 2D19-100 Vertical (Hoops)
B1 = D19-300 Horizontal
B2 = D19-300 Vertical
C = 12D19 (L=1100 mm)
T = 300 mm

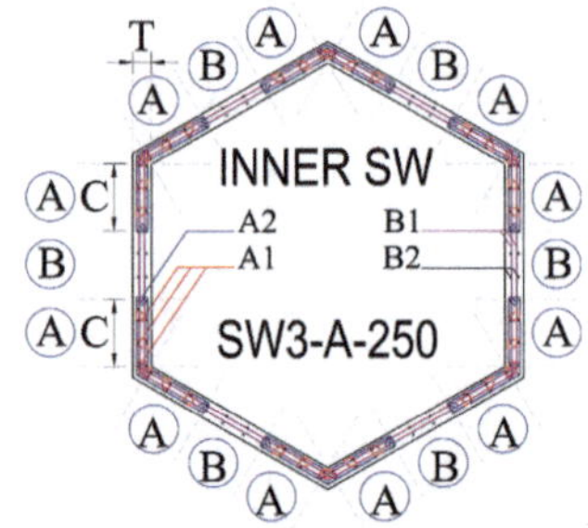

(e) SW3-A-250 (-4.50 to +45.60)

A1 = D16-100 Vertical (Crossties)
A2 = 2D16-100 Vertical (Hoops)
B1 = D25-300 Horizontal
B2 = D16-300 Vertical
C = 8D25 (L=900 mm)
T = 250 mm

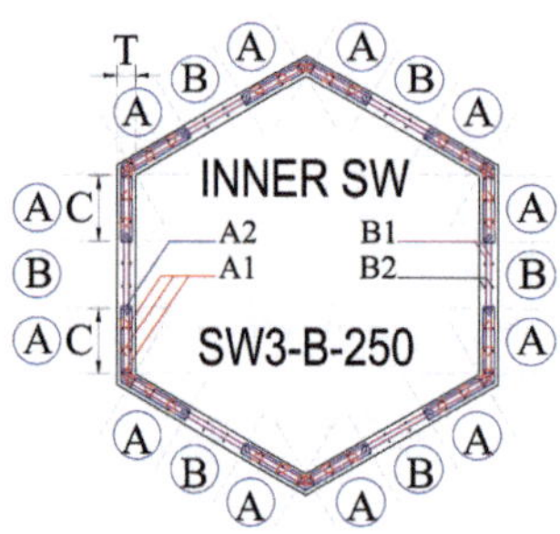

(f) SW3-B-250 (+46.80 to +74.40)

A1 = D16-100 Vertical (Crossties)
A2 = 2D16-100 Vertical (Hoops)
B1 = D22-300 Horizontal
B2 = D16-300 Vertical
C = 8D22 (L=900 mm)
T = 250 mm

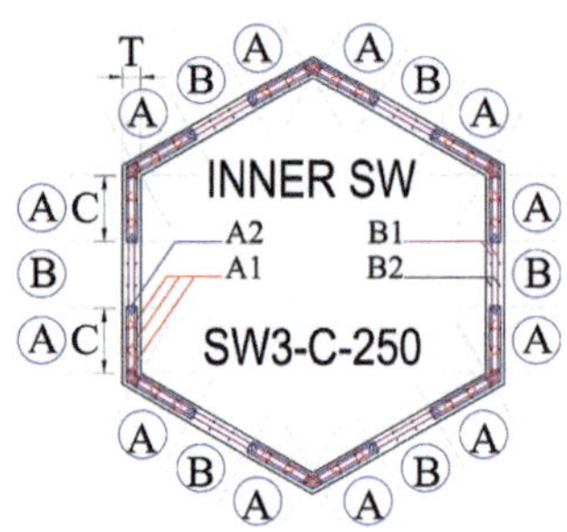

(g) SW3-C-250 (+75.60 to +104.40)

A1 = D16-100 Vertical (Crossties)
A2 = 2D16-100 Vertical (Hoops)
B1 = D19-300 Horizontal
B2 = D16-300 Vertical
C = 8D19 (L=900 mm)
T = 250 mm

Fig. 1 (continued)

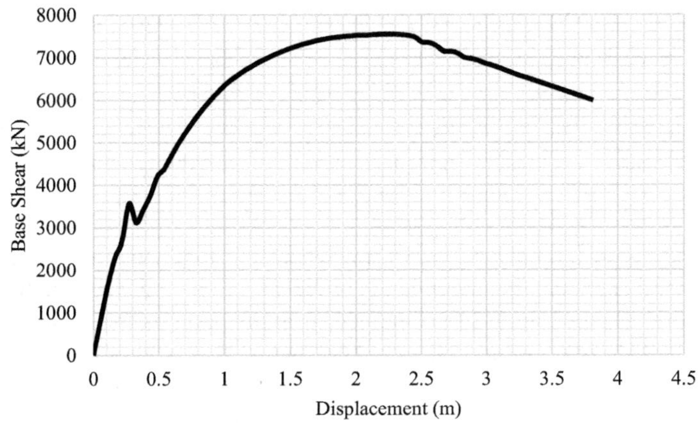

Fig. 2 Capacity curve of minaret structure

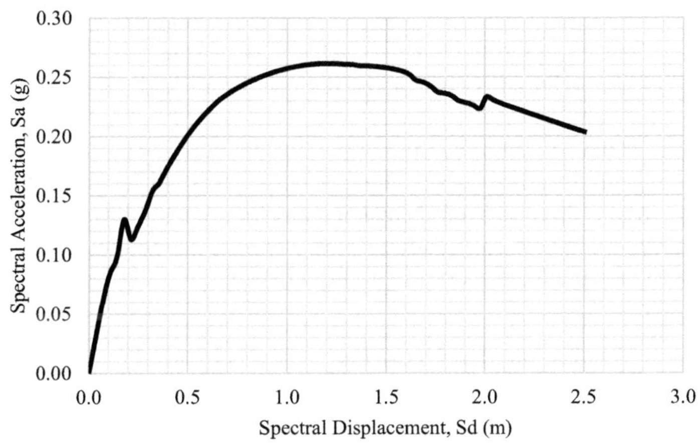

Fig. 3 Capacity spectrum of minaret structure

relationship between spectral displacement (Sd) and spectral acceleration (Sa) that occurs in a structure when resisting a certain ground acceleration.

3.2 Damage States and Uncertainty Parameters

In this research, the damage states are determined based on material strain limits and maximum base shear criteria. The material strain limits are determined by the strain limits of the concrete and reinforcing steel materials. In general, these damage states are classified into DS1, DS2, DS3, and DS4. DS1, DS2, and DS3 were obtained

Table 1 Damage states based on material strain limits and maximum base shear criteria

	Based on material strain limits				Based on maximum base shear		
	DS1	DS2	DS3	DS4	DS1	DS2	DS3
$\overline{S_a}$, ds	Sa	Sa	Sa	Sa	Sa	Sa	Sa
	0.0840	0.1538	0.2013	0.2610	0.0974	0.1925	0.2586

from the points in the capacity spectrum where the first yielding occurred in the reinforcement, cracks in the unconfined concrete section, and cracks in the confined concrete core. Meanwhile, DS4 is obtained from the point on the capacity spectrum where the first chord rotation occurs on the structure that is automatically calculated by the Seismostruct [6].

Damage states based on maximum base shear are determined based on research conducted by Silva et al. [9]. These damage states are classified into DS1, DS2, and DS3. DS1 is obtained from the condition when the base shear reaches 75% of the maximum base shear. DS2 is obtained from the condition when the base shear reaches its maximum condition. Furthermore, DS3 is obtained from the condition when the base shear drops 20% from its maximum condition. Table 1 provides details of the damage states based on the material strain limits and maximum base shear criteria.

The standard deviation of uncertainty parameters needs to be taken into account in constructing the fragility curve. This standard deviation includes the spectrum demand uncertainty (β_d), structural capacity spectrum uncertainty (β_c), and damage states value uncertainty ($\beta_{M(ds)}$). Furthermore, the total uncertainty based on Hazus-MH MR5 can be calculated according to Eq. (1) [10]:

$$\beta_{ds} = \sqrt{[(CONV[\beta_c, \beta_d])]^2 + \left[\left(\beta_{M(ds)}\right)\right]^2} \tag{1}$$

where:

β_c = standard deviation for structural capacity spectrum uncertainty
β_d = standard deviation for spectrum demand uncertainty
$\beta_{M(ds)}$ = standard deviation for damage states value uncertainty.

The standard deviation for structural capacity spectrum uncertainty can be calculated using Eq. (2):

$$\beta_c = \sqrt{\ln\left(\frac{s^2}{m^2} + 1\right)} \tag{2}$$

where:

m = average spectral structure acceleration capacity
s = standard deviation of spectral structure acceleration capacity.

Table 2 Uncertainty response parameter

	Damage States	$\overline{S_a}, ds$ (g)	Parameter			
			β_C	β_D	$\beta_{M(ds)}$	β_{ds}
Based on material strain level	DS1	0.084	0.551	0.45	0.4	0.816
	DS2	0.154	0.551	0.45	0.4	0.816
	DS3	0.201	0.551	0.45	0.4	0.816
	DS4	0.261	0.551	0.45	0.4	0.816
Based on max. base shear	DS1	0.097	0.551	0.45	0.4	0.816
	DS2	0.193	0.551	0.45	0.4	0.816
	DS3	0.259	0.551	0.45	0.4	0.816

Furthermore, the calculation results of uncertainty parameters for each damage state of the minaret structure are presented in Table 2.

3.3 Fragility Curve

The fragility function elucidates the probability of exceedance for multiple damage states as a function of ground-motion intensity [4]. In this research, the fragility curve was developed using the fragility function according to Hazus-MH MR5 [10]. This function is presented in Eq. (3):

$$P[\mathrm{ds}|S_a] = \Phi\left[\frac{1}{\beta_{ds}}\ln\left(\frac{S_a}{\overline{S_a}, \mathrm{ds}}\right)\right] \tag{3}$$

The fragility curve of the minaret structure under study is presented in Fig. 4. The fragility curve depicts the probability of structural damage (beyond the predetermined damage states) related to the intensity of ground shaking over its service life for the specified damage states. In this research, the damage states are determined based on the material strain limits and maximum base shear criteria. Based on the material strain limits criteria, damage states are classified into DS1, DS2, DS3, and DS4. Meanwhile, based on the maximum base shear criteria, the damage states are classified into DS1, DS2, and DS3.

The fragility curve may be understood as a building fingerprint, which helps to assess the vulnerability of a structure during its service life. For example, in this research, the minaret structure with a natural period of 3.782 s has a spectral acceleration, Sa = 0.098 g according to the spectrum response design [8]. Based on the fragility curve, the damage probability of the minaret structure for DS1, DS2, DS3, and DS4 is 57%, 29%, 19%, and 12% respectively, based on the material strain limits criteria. Meanwhile, based on the maximum base shear criteria, the damage probability of the structure for DS1, DS2, and DS3 is 50%, 20%, and 12% respectively.

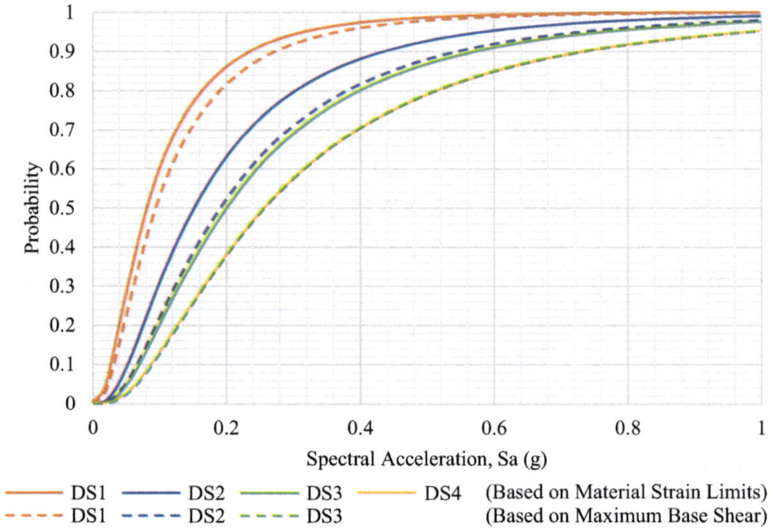

Fig. 4 Fragility curve of minaret structure

By comparing the two damage states criteria used, it can be concluded that the damage for the material strain limits criteria provides a more sensitive prediction with four damage levels compared to the maximum base shear criteria which only provides three damage levels. In general, the maximum base shear criteria presents a lower probability of damage prediction than the material strain limits criteria. The damage probability difference of the minaret structure for spectral acceleration, Sa = 0.098 g for the damage level DS1, DS2, DS3 of both damage states is 7%, 9%, and 7% respectively. Perceiving the fragility curve for the structure under review, building owners can predict the probability of structural damage due to other earthquake intensity scenarios and carry out evaluations to determine retrofitting strategies rationally.

4 Conclusion

In this research, a fragility curve has been developed for the minaret structure under study. The fragility curve is developed using two damage states criteria, based on material strain limits and maximum base shear. The natural period for the minaret structure is 3.782 s. By plotting the natural period of the structure in the spectrum response design, the acceleration spectrum is obtained, Sa = 0.098 g. Based on the developed fragility curve, it is concluded that the damage probability of the minaret structure on spectral acceleration, Sa = 0.098 g based on the material strain limits criteria for damage states DS1, DS2, DS3, and DS4 is 57%, 29%, 19%, and 12% respectively. On the other hand, based on the maximum base shear criteria for the damage states DS1, DS2, and DS3 is 50%, 20%, and 12% respectively. Perceiving the fragility curve for the structure under review, building owners can predict the probability of structural damage due to other earthquake intensity scenarios and carry out evaluations to determine retrofitting strategies rationally.

Acknowledgements This research was financially supported by Universitas Sebelas Maret, Indonesia through Hibah Penelitian Unggulan, 2020.

References

1. Villaverde R (2009) Fundamental concepts of earthquake engineering, 1st edn. CRC Press, United States of America
2. Pribadi K, Kusumasturi D, Rildova (2008) Learning from recent indonesian earthquakes: an overview to improve structural performance. In: 14th world conference on earthquake engineering, Beijing, China
3. Mosque Information System (The Indonesian Ministry of Religion) Homepage. https://simas.kemenag.go.id/. Last accessed 2021/6/16
4. Elnasai AS, Sarno LDi (2015) Fundamentals of Earthquake Engineering (From Source to Fragility). 2nd ed. Wiley, United Kingdom
5. Seismosoft Homepage. https://seismosoft.com/product/seismostruct/. Last accessed 2021/3/18
6. Seismosoft (2020) Seismostruct User Manual. Seismosoft Ltd., Italy
7. Badan Standardisasi Nasional (BSN) (2013) 1727:2013 Beban Minimum untuk Perancangan Bangunan Gedung dan Struktur Lain. BSN, Indonesia
8. Badan Standardisasi Nasional (BSN) (2012) SNI 1726:2012 Tata Cara Perencanaan Ketahanan Gempa untuk Struktur Bangunan Gedung dan Non Gedung. BSN, Indonesia
9. Silva V, Varum H, Crowley R, Sousa R, Pinho R (2012) Evaluation of analytical methodologies to derive vulnerability functions. In: 15th world conference on earthquake engineering, Lisboa, Portugal
10. Federal Emergency Management Agency (FEMA) (2010) HAZUS-MH MR5 Technical and User's Manual. FEMA, United States of America

Role of Diagonal Bars in Reinforced Concrete Deep Beams Tested Under Static Load

Erwin Lim and Rahmat Ramli

Abstract Four deep beam specimens were tested to investigate the role of diagonal bars. This study shows that the presence of diagonal bars increases the shear strength and ductility of deep beam specimens. In addition, test results also show that specimens loaded through stub, instead of bearing plate, generally gives higher shear strength due to steeper strut inclination angle. The strength of a deep beam can be predicted using ACI 318 strut-and-tie model with conservatism.

Keywords Deep beam · Diagonal bar · Strut-and-tie

1 Introduction

The use of diagonal bars has been widely adopted in the seismic design of a coupling beam [1]. The seismic behavior of a coupling beam is greatly enhanced due to the presence of diagonal bar, especially for that with span-to-depth ratio less than 2.0 where shear behavior is dominant [2–6]. The main reason for the successful application of diagonal bars being that they increase the beam's shear strength and therefore allowing the full development of its ductile flexural behavior.

The structural element susceptible to shear failure is not merely limited to that under seismic load. Some elements under static gravity load might also experience similar issue. However, the use of diagonal bars in those elements is not common and limited to some tests of corbel [7]. Therefore, in this paper, the authors examine the use of diagonal bars in deep beam specimens under static loading. As many as four deep beam specimens were tested in the laboratory and the test results are discussed and evaluated using the strut-and-tie model.

E. Lim (✉) · R. Ramli
Institut Teknologi Bandung, Ganesha 10, Bandung, Indonesia
e-mail: erwinlim@si.itb.ac.id

E. Lim
Lab. Rekayasa Struktur, Pusat Rekayasa Industri ITB, Ganesha 10, Bandung, Indonesia

S. A. Kristiawan et al. (eds.), *Proceedings of the 5th International Conference on Rehabilitation and Maintenance in Civil Engineering*, Lecture Notes in Civil Engineering 225,
https://doi.org/10.1007/978-981-16-9348-9_5

2 Experimental Study

2.1 Design of Test Specimens

As many as four deep beam specimens were designed and tested under static loading test. Two of them, labeled with "C" are specimens tested through column stub representing real column in the building, while the other two specimens, labeled with "B" are specimens tested through bearing plate. In each group (both "C" and "B"), one specimen was designed with the presence of diagonal bars, while the other one is none.

Specimens without diagonal bars (B-15–0 and C-15–0) were deliberately designed to fail in shear so that the beneficial effect of a diagonal bar can be more distinct. The main flexural reinforcement was 3D19 and 3D16 (flexural reinforcement ratio ρ equals 1.73%). Meanwhile, the compressive longitudinal reinforcement ratio, ρ' equals 0.47% and the vertical (ρ_v) and horizontal (ρ_h) reinforcement ratios were kept to be 0.37% and 0.12%, respectively as shown in Fig. 1a and b.

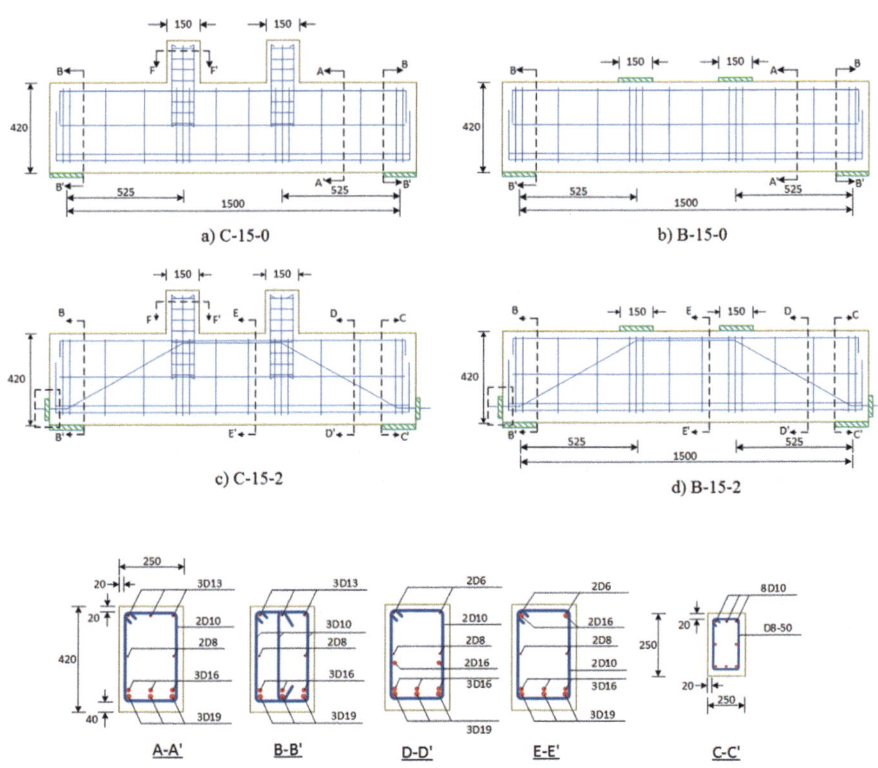

Fig. 1 Specimen's geometry and reinforcement layout

Table 1 Parameters of deep beam specimens

No	Spec. ID	f'_c (MPa)	a/d	d (mm)	ρ (%)	ρ' (%)	ρ_v (%)	ρ_h (%)	ρ_d (%)
1	C-15–0	21.19	1.56	336	1.73	0.47	0.37	0.12	–
2	B-15–0	20.21	1.56	336	1.73	0.47	0.37	0.12	–
3	C-15–2	19.63	1.56	336	1.73	0.47	0.37	0.12	0.48
4	B-15–2	24.84	1.56	336	1.73	0.47	0.37	0.12	0.48

The hypothesis of this study is that the presence of diagonal bars could alter the failure mode of a shear dominated deep beam becoming a flexure-dominated beam. So, for specimens with diagonal bars (C-15–2 and B-15–2), the flexural reinforcement ratio was maintained similar to those without diagonal bars. The only difference was simply the presence of diagonal bars (2D16) as shown in Fig. 1c and d. All specimen parameters are summarized in Table 1.

2.2 *Test Setup and Results*

Before testing of a deep beam specimen, tensile test of reinforcement bars were carried out. For each bar diameter, as many as two coupon tests were tested and the average yield strength is reported in Table 2.

Several strain gages were attached to the flexural, vertical, horizontal, and diagonal bars of each specimen as indicated in Fig. 2a as monitoring device. After that, the vertical load was applied slowly using displacement controlled actuator as shown in Fig. 2b. Several LVDTs were also mounted as presented in Fig. 2b.

The load–deflection curve of each specimen is given in Fig. 3a. The plotted load is the force measurement from actuator with respect to displacement at mid span. Specimen C-15–0 showed elastic behavior when the load was still below 590 kN with displacement of 6.19 mm. The first crack on this specimen occurred at the load of 210.28 kN and formed flexure crack pattern with yielding of longitudinal reinforcement as well. When the load reached the maximum load, a large crack occured and formed a diagonal crack and then the load started to drop suddenly. The maximum load achieved was 748.2 kN with corresponding displacement of

Table 2 Yield strength of rebars

No.	Diameter (mm)	f_y (MPa)
1	6	338.28
2	8	387.07
3	10	564.15
4	13	412.04
5	16	370.81
6	19	386.36

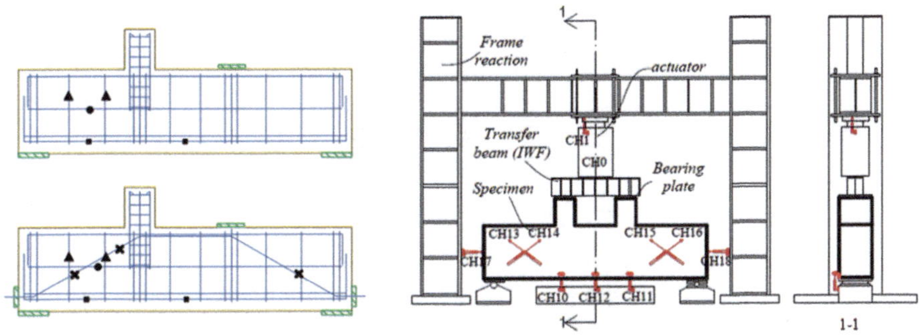

a) strain gage locations b) LVDT locations and test setup

Fig. 2 Instrumentation layout and test setup

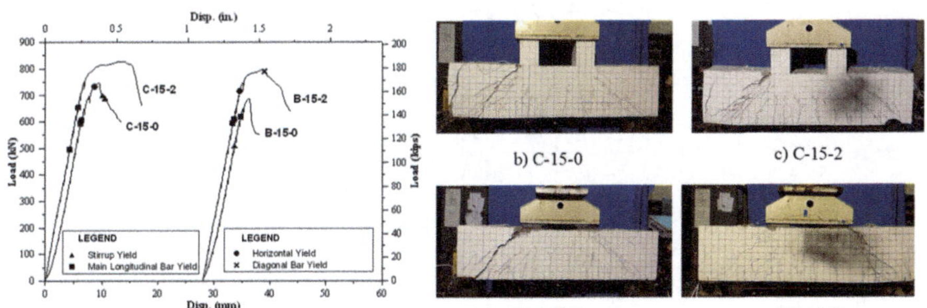

a) load deflection curves b) C-15-0 c) C-15-2 d) B-15-0 e) B-15-2

Fig. 3 Load displacement curve testing of deep beams

9.41 mm. The dominant crack pattern that occurs was shear crack. The failure mode of this specimen was shear with diagonal splitting after yielding of longitudinal reinforcement as shown in Fig. 3b. Due to the presence of 2D16 diagonal bars, specimen C-15–2 attained the maximum load at 827.4 kN and the corresponding displacement was 13.91 mm. The failure mode of this specimen was flexural failure with slight indication of concrete crushing at the vicinity of column stub as indicated in Fig. 3c.

Specimens loaded through bearing plate in general show similar behavior, but with lower maximum attained load than specimens loaded through stub. The maximum load of specimen B-15–0 and B-15–2 was 687.9 kN and 796.8 kN, respectively. Again, specimens with diagonal bars possessed higher strength than that without diagonal bars. The failure mode of B-15–0 is immediate shear with diagonal splitting after flexural yielding of longitudinal reinforcement (Fig. 3d), while that of B-15–2 was flexural failure (Fig. 3e). Summary of test results is given in columns (b)–(d) of Table 3.

Comparing specimens with and without diagonal bars, Fig. 3a clearly shows that the presence of diagonal bar increased the maximum strength and ductility of a deep

Table 3 Strength results

Spec. ID	Test values			Strut-and-tie model						
	V_{test} (MPa)	Δ_{test} (mm)	FM_{test}	θ (deg)	V_{ns} (kN)	V_{nn} (kN)	V_{nt} (kN)	V_{STM} (kN)	V_{test}:2/V_{STM}	$FM_{predicted}$
(a)	(b)	(c)	(d)	(e)	(f)	(g)	(h)	(i)	(j)	(k)
C-15–0	748.2	9.41	FS* with diag. splitting	31.3	186.9	249.2	341.4	186.9	2.00	Upper strut crushing
B-15–0	687.9	8.18	FS* with diag. splitting	27.3	222.2	296.3	289.4	222.2	1.55	Upper strut crushing
C-15–2	827.4	13.91	Flexure	30.9	259.8	323.9	335.2	259.8	1.59	Upper strut crushing
B-15–2	796.8	10.99	Flexure	28.3	317.1	400.2	302.4	302.4	1.32	Tie yielding
							Average		1.62	
							COV		0.18	

*FS indicates: immediate shear after flexural yielding of longitudinal reinforcement

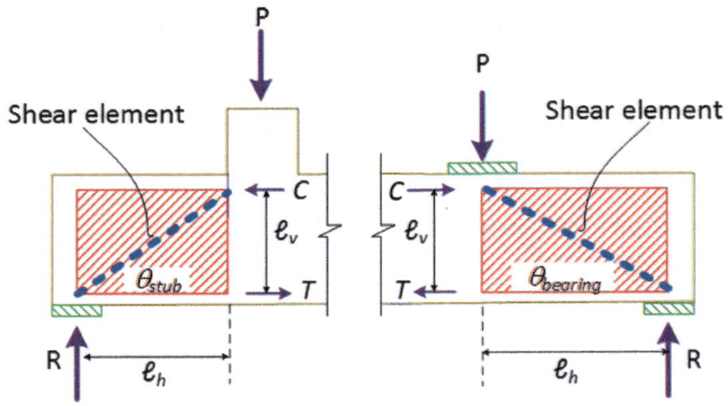

Fig. 4 Illustration of shear element for specimens with column stub and bearing plate

beam specimen, despite some early yielding of main longitudinal reinforcement in all four specimens. The failure mode also changed from flexural shear failure to ductile flexural failure. The major crack of specimen with column stub (θ_{stub}) was observed to occur mainly at the vicinity of the stub. Meanwhile, major inclination crack of specimens with bearing plate ($\theta_{bearing}$) was observed to occur under the point load as seen in Fig. 3b–e and illustrated in Fig. 4. Therefore, the crack inclination angle of specimens loaded through column stub in general was steeper than those loaded through bearing plate. This could be the reason why the maximum load of specimens with column stub was higher than those with bearing plate.

3 Analytical Study

In this study, the strength of a deep beam is evaluated analytically using the strut-and-tie provision in ACI 318-14 [1]. The strength of a deep beam can be taken as the minimum among the strength of concrete strut (V_{ns}), nodal zone (V_{nn}), or yielding of tie (V_{nt}) as shown in Eq. (1):

$$V_{STM} = \min\{V_{ns}; V_{nn}; V_{nt}\} \tag{1}$$

Before determining the strength of respective components, it is important to firstly determine the size of shear element through the macro model [8]. The size of shear element also represents the inclination angle of diagonal strut, θ as shown in Eq. (2):

$$\theta = \tan^{-1}\left(\frac{\ell_v}{\ell_h}\right) \tag{2}$$

where ℓ_v is the lever arm determined from the compressive force to the tensile force, and ℓ_h is the horizontal boundary of a shear element. For deep beam loaded under column stub, the ℓ_h should be bounded by the geometry discontinuity of the concrete stub, while for beams loaded under bearing plate, the ℓ_h is bounded by the force discontinuity as illustrated in the following Fig. 4. Similar determination of shear element size was also proposed by Lu and Chu [9]. The calculated strut angle for all four specimens can be seen in column (e) of Table 3.

Strength of a Strut

The strength of a diagonal strut, F_{ns} can be determined from the contribution of concrete strut and diagonal bar (if presence) as shown in the Eq. (3):

$$F_{ns} = 0.85 f_c' \beta_s A_{cs} + A_d f_{yd} \tag{3}$$

where β_s is the ACI 318 strut efficiency factor, A_{cs} is area of concrete strut, A_d is area of diagonal bar, and f_{yd} is yield strength of a diagonal bar.

Due to the difference of discontinuity as illustrated in Fig. 4, the area of concrete strut (A_{cs}) at upper part for specimens with column stub and bearing plate can be calculated as shown in Eqs. (4a) and (4b):

$$\text{Column stub}: A_{cs} = (w_s \cos\theta)b \tag{4a}$$

$$\text{Bearing plate}: A_{cs} = (w_s \cos\theta + a_p/2 \sin\theta)b \tag{4b}$$

Since the lower part of the specimen uses bearing plate for all specimens, therefore the area of concrete strut is determined using Eq. (5):

$$A_{cs} = (w_t \cos\theta + l_b \sin\theta)b \tag{5}$$

The width of strut at upper part, w_s and width of tie at lower part, w_t can be determined from horizontal force equilbirum as suggested by Tjhin and Kuchma [10]. Meanwhile, a_p and ℓ_b equals to 150 mm is the depth of bearing plate and column stub, respectively. These parameters are illustrated in Fig. 5.

Finally, the shear strength of a deep beam due to crushing of a diagonal strut is presented in Eq. (6):

$$V_{ns} = F_{ns} \sin\theta \tag{6}$$

Strength of a Nodal Zone

The strength of a nodal zone, F_{nn} can be determined from the contribution of concrete strut and diagonal bar (if presence) as shown in the Eq. (7):

$$F_{nn} = 0.85 f_c' \beta_n A_{cn} + A_d f_{yd} \tag{7}$$

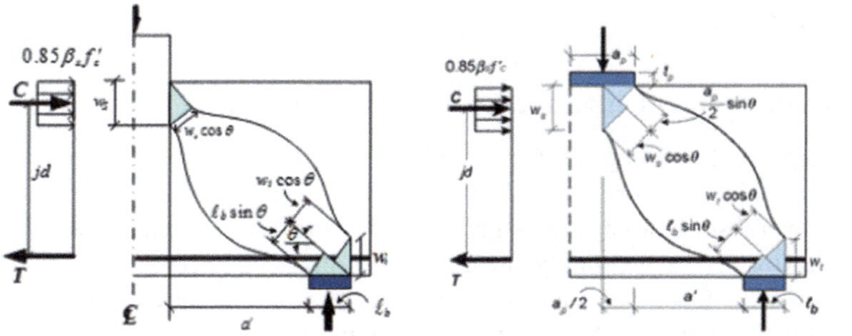

a) Strut area for beam with stub

b) strut area for beam with bearing plate

Fig. 5 Illustration of strut area

where β_n is the ACI 318 nodal efficiency factor, A_{cn} is area of nodal zone similar to area of concrete strut.

Therefore, the strength of a deep beam due to crushing of nodal zone is shown in Eq. (8):

$$V_{nn} = F_{nn} \sin\theta \tag{8}$$

Normally, since the lower node governs because it is considered as a CCT node.

Strength of a Yielding Tie

The strength of a yielding tie can be determined using Eq. (9):

$$F_{nt} = A_{st} f_y \tag{9}$$

where A_{st} is the area of longitudinal flexural reinforcement, f_y is the yield strength of main longitudinal beam reinforcement.

The shear strength of a deep beam due to yielding of longitudinal bar can be written as:

$$V_{nt} = F_{nt} \tan\theta \tag{10}$$

Finally, Table 3 summarizes the calculated strengths for all four specimens along with the strength ratio and predicted failure mode in columns (j) and (k), respectively. The average strength ratio is 1.62 with COV of 0.18, which gives conservative strength estimation. This conservatism is expected because the ACI 318 STM might not properly capture the overall behavior which leaves possibility of further refinement of this model.

4 Conclusion

Based on the results of this study, it can be concluded that:

1. The presence of diagonal bars could alter the failure mode of deep beam specimens which were originally designed to fail in shear. The diagonal bars increase the overal shear strength and therefore allowing development of beam's flexural behavior. Correspondingly, the ductility of the deep beam is also enhanced.
2. Specimens loaded through stub generally showed higher capacity because the inclination angle of the strut is steeper than specimens loaded through bearing plate
3. ACI 318 strut-and-tie model gives conservative strength estimation of the tested four specimens.

Acknowledgements This research was carried out using the grant from Program Penelitian, Pengabdian kepada Masyarakat, dan Inovasi (P3MI) ITB of which the authors are grateful.

References

1. American Concrete Institute (2016) Building code requirements for structural concrete (ACI 318–14) and commentary (ACI 318M-14). Farmington Hills, pp 623
2. Paulay B, Binney JR (1974) Diagonally reinforced coupling beams of shear walls. ACI J 579–598
3. Naish D, Fry A, Klemencic R, Wallace J (2013) Reinforced concrete coupling beams—Part I: testing. ACI Struct J 110:1057–1066
4. Lim E, Hwang SJ, Wang TW, Chang YH (2016) An Investigation on the seismic behavior of deep reinforced concrete coupling beams. ACI Struct J 113:217–226
5. Park WS, Kang THK, Kim S, Yun HD (2020) Seismic performance of moderately short concrete coupling beams with various reinforcements. ACI Struct J 117:141–154
6. Ameen S, Lequesne RD, Lepage A (2020) Diagonally reinforced concrete coupling beams with grade 120 (830) high-strength steel bars. ACI Struct J 117:199–210
7. Foster SJ, Powell RE, Selim HS (1996) Performance of high-strength concrete corbels. ACI Struct J 93:555–563
8. Lim E, Hwang SJ (2016) Modeling of the strut-and-tie parameters of deep beams for shear strength prediction. Eng Struct 108:104–113
9. Lu WY, Chu CH (2019) Tests of high-strength concrete deep beams. Mag Concr Res 71:184–194
10. Tjhin TN, Kuchma DA (2002) Example 1b: alternative design for non-slender beam (deep beam). ACI Special Publication (SP-208). Ed. Karl-Heinz Reineck, pp 242

Analysis of Reinforced Concrete Capacity for Irregular Cross-Sections Using Numerical Methods

Nuroji

Abstract Although reinforced concrete member usually forms the rectangular shape, for architectural reasons or an optimization purpose the section may be formed on the nonrectangular or irregular shape. Analysis of the sectional capacity in the case of non-square or irregular sections is very complex and takes time. This paper offers a cross-sectional analysis with a numerical approach that is implemented in a computer program. Several cross-sectional forms from the results of previous studies were adopted for validation purposes. The selection of the cross-section considers the representation of the cross-section shape and the configuration of the reinforcement. In addition to the cross-sectional shape, the method of analysis is also a consideration for selection. From the results of the analysis using numerical methods and comparison of the analysis of previous researchers, it shows a fairly good level of accuracy with an average deviation of 2.35%. The largest deviation was in Section 7 and 8 with a deviation of 5.03 and 8.62%, respectively. This deviation is more due to the analysis method. Analysis using finite element method for sections, 9 and 10 show slightly higher than the numerical method due to neglecting tensile strength of concrete and strain hardening of steel on the numerical method. On the other side, the analysis with the cross-section conversion approach shows lower results. This software may be used to solve the nonrectangular or irregular sections.

Keywords Irregular section · Numerical · Flexural

1 Introduction

In contrast to steel materials that have the same strength between compressive and tensile conditions. Generally, the tensile strength of concrete does not exceed 10% of the compressive strength [1, 2]. The low tensile strength of the concrete makes this material often combined with reinforcing steel to withstand the tension so that it forms a reinforced concrete composite structure. In general, the structural elements

Nuroji (✉)
Diponegoro University, Semarang, Indonesia
e-mail: nuroji@lecturer.undip.ac.id

S. A. Kristiawan et al. (eds.), *Proceedings of the 5th International Conference on Rehabilitation and Maintenance in Civil Engineering*, Lecture Notes in Civil Engineering 225,
https://doi.org/10.1007/978-981-16-9348-9_6

of reinforced concrete are in the form of a square section. However, due to architectural rea-sons or structural optimization considerations, reinforced concrete sections can be non-square, such as circle, octagonal, hexagonal, trapezoidal, and triangular. Moreover, a square section that is subjected to the biaxial moment can also behave as a non-square section due to the rotation of the section in receiving the biaxial resultant moment.

In a bending element, the internal forces acting in a section will take balance, so that the compressive forces will be the same as the tensile forces. The compressive force in the cross-section is generated from compressed concrete and compressive reinforcement, while the tensile force is generated from tensile reinforcement. The compressive force of the concrete is the resultant stress of the concrete which forms a stress block. By assuming the plane section remains plane before and after bending, the strain distribution from the neutral axis to the extreme compression fiber can be considered a linear function of the neutral axis distance. Therefore, the stress distribution will also be identical to the shape of the stress–strain relationship curve. A stress block in a reinforced concrete section is a volume formed by the compression area of the concrete and its stress. In a square section where the width of the compression area is constant, the concrete compressive force of the stress block can be simplified to be an equivalent stress block in a square shape [3]. However, for non-square sections, the simplification of square stress blocks to calculate the compressive force of concrete may be wrong.

The calculation of non-square sections will be more appropriate if using the numerical approach than the stress block simplification. This paper discusses the moment capacity analysis on irregular sections of reinforced concrete beam by using a numerical approach implemented in a computer program to overcome design problems, especially in the case of non-square beam sections or biaxial bending cases.

1.1 Material Model

1.1.1 Concrete Compressive Stress–Strain

The compressive strength of concrete is generally determined from the compressive test of a concrete cylinder with 150 mm diameter and 300 mm length in the longitudinal direction at the age of 28 days. Numerous approaches are used to determine the shape of the concrete compressive stress–strain relationship curve before maximum stress in a second-order parabolic [4]. Hognestad proposed the compressive stress–strain curve of concrete as a parabolic function up to the maximum stress and subsequently decreasing linearly until it reaches the ultimate strain as shown in Fig. 1a [5]. The concrete compressive Stress–strain curve is described in two regions as shown in Eqs. 1 and 2.

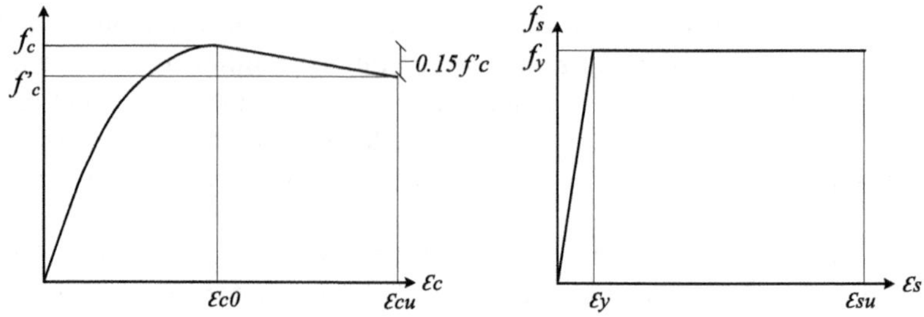

Fig. 1 Stress–strain relationship. **a** Concrete in compression. **b** Steel

$$f_c = f_c'\left[\frac{2\varepsilon_c}{\varepsilon_0} - \left(\frac{\varepsilon_c}{\varepsilon_0}\right)^2\right], 0 \geq \varepsilon_c \geq \varepsilon_{c0} \tag{1}$$

$$f_c = \left(\frac{0.15\varepsilon_c + 0.85\varepsilon_{co} - \varepsilon_{cu}}{\varepsilon_{co} - \varepsilon_{cu}}\right) f_c', \varepsilon_{c0} \geq \varepsilon_c \geq \varepsilon_{cu} \tag{2}$$

Where, $\varepsilon_0 = \frac{2f_c'}{E_c}$ and $\varepsilon_{cu} = 0.0038$.

1.1.2 Steel

In general, the behavior of the steel is determined from the stress–strain relationship of the uniaxial tensile test results in the laboratory. The stress–strain relationship curves of steel for the compressive conditions are considered to be the same and identical to the tensile [4]. Chen determined the stress–strain relationship of steel by dividing three parts, the linear elastic region starting from the point of origin until the steel reaches the yield strain ε_y, horizontal plateau starting from yield-strain ε_y up to 8–15 times its elastic range ε_y, and strain hardening where the steel stress increases and reaches a maximum at the ultimate strain then gradually decreases until the failure strain [6]. In this study, the behavior of the stress–strain relationship of reinforcing steel is modeled as a bilinear function with neglecting the strain hardening effect as shown in Fig. 1b.

1.1.3 Section Model and Reinforcement Configuration

The irregular sections are modeled by multilinear from nodal points coordinate forms closed polygon. The coordinates of each polygon boundary point are denoted ($\mathbf{X_{(i)}}$, $\mathbf{Y_{(i)}}$), where $\mathbf{i} = 1, 2, 3, \ldots$ n, and **n** is the number of polygon coordinates. The configuration and location of the reinforcement in the section are also determined by coordinates ($\mathbf{X_{r(i)}}$, $\mathbf{Y_{r(i)}}$), where $\mathbf{i} = 1, 2, 3, \ldots$ nr, and **nr** is the number of reinforcements in the section. From the cross-sectional coordinates, it can be determined Y_{max} which is the outermost compressive point that undergoes the maximum compressive

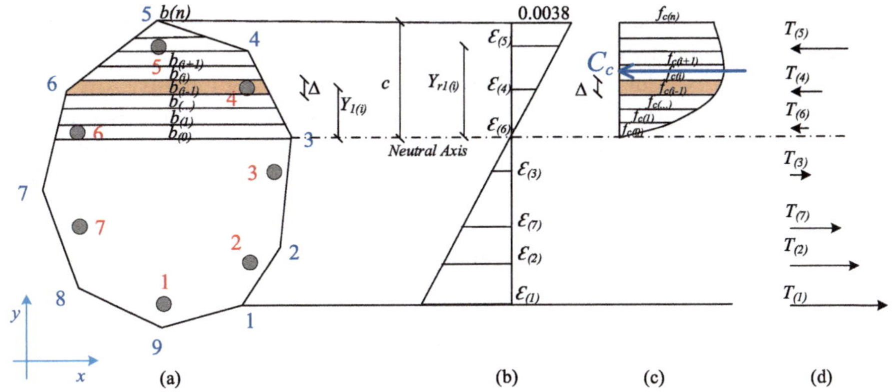

Fig. 2 **a** Section. **b** Strain diagram. **c** Concrete stress block. **d** Forces of reinforcements

strain. Meanwhile, from the coordinates of the location of the reinforcement can be determined the outermost tensile reinforcement. Irregular cross-sectional shape and reinforcement configuration as shown in Fig. 2a.

2 Cross Section Analysis

In the analysis of flexural beam with a rectangular section, the compressive force of concrete, and the tensile force of the reinforcement can be obtained from the equilibrium of forces, where the compressive force $\boldsymbol{C_c}$ is defined based on a simplified equivalent stress block of concrete. While the irregular cross-section, determination of compressive force $\boldsymbol{C_c}$ becomes very complex because it has to be solved through the integration of the stress–strain relationship function on the compression concrete area. Therefore, it will be more accurate in case irregular cross-section analysis undertakes numerically. The analysis of reinforced concrete sections in this study is based on several basic assumptions on the following flexural sections [7].

1. The first presumption is plane sections remain plane before and after bending as Bernoulli's principle. This assumption implies that the strain in the section is proportional to the distance to the neutral axis. The result of this assumption is that the distribution of compressive stress in the compressive region is identical to the stress-strain relationship curve.
2. The bond between concrete and steel reinforcement is assumed as full-bonded, and there is no slip between both concrete and steel. It means that the strain of reinforcement is the same as concrete strain at the same level.
3. The tensile stress of the concrete is not more than10% of its compressive strength, thus the force of the tensile concrete below the neutral axis is also small. The lever arm of the concrete tension force to the neutral axis is small. So that, the contribution of concrete tension force to the bending capacity of

the cross-section is considered very small. For this reason, the tensile stress of concrete can be neglected.

4. Concrete is considered only to be able to withstand compressive stress until the ultimate strain. ACI sec. 10.2.3 defined the ultimate compressive strain of the concrete to 0.003 for design [8]. However, in this study, the ultimate strain of compression concrete was determined according to Hognestad's model, i.e., 0.0038.

Resultant of internal forces in a section can be derived by Eq. 3.

$$P_t = C_c + \sum_{i=1}^{nr} T_{(i)} \quad (3)$$

Where:

C_c: Concrete compression force.

$\sum_{i=1}^{nr} T_{(i)}$: Sum of the reinforcement forces, where $\boldsymbol{nr}$ is number of reinforcement.

The resultant of internal forces $\boldsymbol{P_t}$ in Eq. 3 are obtained by determining the value of the outer tensile reinforcement as shown in Fig. 3. The process of determining $\boldsymbol{P_t}$ is carried out with the following procedure.

1. Define the strain of the outer tensile reinforcement $\boldsymbol{\varepsilon_s}$.
2. Assuming the strain distribution is linear, the depth of the neutral axis $\boldsymbol{c}$ can be defined. (see Fig. 2b).
3. The compressed concrete area is divided into small slices extending in the **X** direction in a number of $\boldsymbol{ns}$ slices. The thickness of each slice is $\boldsymbol{\Delta = c/ns}$.
4. Calculate the coordinates of the intersection points between the edge of slice and the polygon to calculate the width of the slice. $b_{(i)}$, $\boldsymbol{i} = 0, 1, 2, \ldots, \boldsymbol{ns}$, and $\boldsymbol{ns}$ is number of slices.

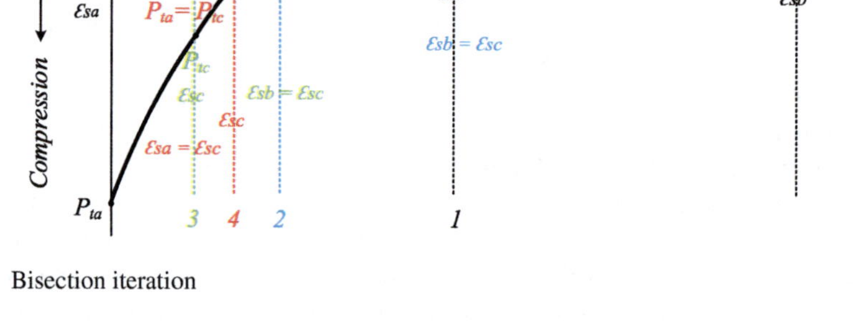

Fig. 3 Bisection iteration

5. Average slice width can be derived using Eq. 4.

$$b_{1(i)} = \frac{b_{(i-1)} + b_{(i)}}{2} \tag{4}$$

6. The center of gravity of each slice can be assumed exist at midpoint of slice thickness, thus the arm of these points to neutral axis is approximated by Eq. 5.

$$Y_{(i)} = (i - 0.5)\Delta \tag{5}$$

7. The concrete compressive strain on each slice edge is proportional to the distance of the edge to neutral axis, and the strain can be defined by Eq. 6.

$$\varepsilon_{c(i)} = \frac{i}{n} \times 0.0038 \tag{6}$$

Where: $\boldsymbol{i} = 0, 1, 2, \ldots, \boldsymbol{ns}$, $\boldsymbol{ns}$ is number of slices

8. Furthermore, the compressive stress of the concrete on each slice edge can be determined based on the stress-strain relationship according to Eqs. 1 and 2.
9. The average stress of slice can be calculated with Eq. 7.

$$f_{c1(i)} = \frac{f_{c(i-1)} + f_{c(i)}}{2} \tag{7}$$

10. Compression force for each slice can be defined by Eq. 8.

$$C_{c1(i)} = \left(b_{1(i)} \times \Delta\right) f_{c1(i)} \tag{8}$$

11. The resultant of concrete compressive force $\boldsymbol{C_c}$ is the sum of all the compressive forces of the on each slice as shown in Eq. 9.

$$C_c = \sum_{i=1}^{n} C_{c1(i)} \tag{9}$$

12. The space of the reinforcement to the neutral axis is determined based on Eq. 10. The negative sign indicates the location of the reinforcement is below the neutral axis.

$$Y_{r1(i)} = Y_{r(i)} + c - Y_{max} \tag{10}$$

13. The reinforcement strain proportional with respect to the distance of the reinforcement to the neutral axis, the reinforcing strain can be calculated by Eq. 11. In this case the negative sign indicates the tensile strain.

$$\varepsilon_{s(i)} = \frac{0.0038}{c} Y_{r1(i)} \tag{11}$$

14. Base on Eq. 11, then reinforcement stresses in each reinforcing bar is obtained using expression of Eq. 12.

$$f_{s(i)} = \varepsilon_{s(i)} \times E_s, \text{ where } E_s = 2 \times 10^5 \text{Mpa} \tag{12}$$

if $\varepsilon_{s(i)} > \varepsilon_y$, then $f_{s(i)} = f_y$.
if $\varepsilon_{s(i)} < -\varepsilon_y$, then $f_{s(i)} = -f_y$.

15. The forces on each reinforcement are the product of the reinforcement area and the stress as shown in Eq. 13.

$$T_{(i)} = A_{s(i)} \times f_{s(i)} \tag{13}$$

16. From the compression force C_c and the reinforcement forces $T_{(i)}$ can be calculated the resultant internal forces of section using Eq. 3.

2.1 Internal Forces Equilibrium

To achieve a balance of internal forces of the cross section can be obtained iteratively by using the bisection method. The concrete strain at top fiber is set 0.0038, and the outer reinforcing steel strain is entered twice, i.e., $\boldsymbol{\varepsilon}_{sa} = 0$ and $\boldsymbol{\varepsilon}_{sb} = 0.04$ as the initial strain. Iteration process of bisection method as shown in Fig. 3.

1. The first step is to enter $\boldsymbol{\varepsilon}_{sa} = 0$. By using the calculation procedure to find the resultant of internal forces for $\boldsymbol{\varepsilon}_s = \boldsymbol{\varepsilon}_{sa}$ will be found $\boldsymbol{P}_{ta}$ that produce the compressive force, (see Fig. 4c).

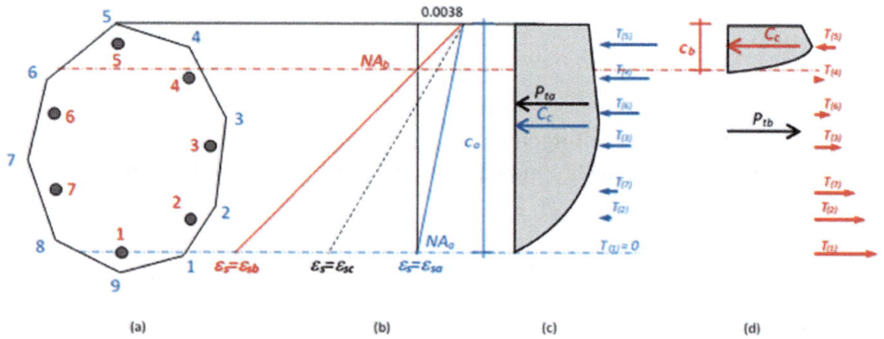

Fig. 4 **a** Section. **b** Strain iteration. **c** Internal forces at $\varepsilon_s = \varepsilon_{sa}$. **d** Internal forces at $\varepsilon_s = \varepsilon_{sb}$

2. By using the same procedure enter $\boldsymbol{\varepsilon}_{sb} = 0.04$ as $\boldsymbol{\varepsilon}_s$ to get $\boldsymbol{P}_{tb}$ that produce the force in tension, (see Fig. 4d). Where $\boldsymbol{\varepsilon}_{sa}$ and $\boldsymbol{\varepsilon}_{sb}$ are the initial strain given to outer reinforcement which has not produce the balance condition ($\boldsymbol{P}_t = 0$).
3. The Next Step Using $\boldsymbol{\varepsilon}_{sc}$ as Mid-Point Between $\boldsymbol{\varepsilon}_{sa}$ and $\boldsymbol{\varepsilon}_{sb}$, $\varepsilon_{sc} = \left(\frac{\varepsilon_{sa}+\varepsilon_{sb}}{2}\right)$. Then Calculate the Resultant of Internal Forces for $\boldsymbol{\varepsilon}_s = \boldsymbol{\varepsilon}_{sc}$, i.e., $\boldsymbol{P}_{tc}$.
   If $P_{tc} \times P_{ta} < 0$, then $\boldsymbol{\varepsilon}_{sb} = \boldsymbol{\varepsilon}_{sc}$.
   If $P_{tc} \times P_{tb} < 0$, then $\boldsymbol{\varepsilon}_{sa} = \boldsymbol{\varepsilon}_{sc}$.
4. If $\left|\frac{(\varepsilon_c - \varepsilon_a)}{\varepsilon_c}\right| \leq Tolerance$ or $\left|\frac{(\varepsilon_c - \varepsilon_b)}{\varepsilon_c}\right| \leq Tolerance$, then go to step 5. If No, back to step 3.
5. The $\boldsymbol{\varepsilon}_s = \boldsymbol{\varepsilon}_{sc}$, in this iteration the equilibrium occurs.

The balanced condition occurs when the strain of outer tension reinforcement ε_s produces internal forces resultant P_t which very small or less than the tolerance value defined earlier. Then calculate the concrete compressive moment from the number of slice forces multiplied by its lever arm to the neutral axis as shown in Eq. 14, and the moment of reinforcement which is the result of an addition of moment of each reinforcement to the neutral axis as shown in Eq. 15.

$$M_c = \sum_{i=1}^{ns} C_{c1(i)} \times Y_{1(i)} \tag{14}$$

$$M_r = \sum_{i=1}^{nr} T_{(i)} \times Y_{r1(i)} \tag{15}$$

The flexural capacity can be determined by using Eq. 16

$$M = M_c + M_r \tag{16}$$

Numerical solution to determinate the reinforced concrete capacity presented in Fig. 5.

3 Program Validation

Before the program is used, a validation process is required to ensure whether the program is correct or still requires improvement. In this study, the validation process was undertaken by comparing the results of program running and analysis from previous studies. The sections used for validation represent the geometric shape of the square and irregular sections. Reinforcement ratio and reinforcement configuration are also considered in section selection. The selected sections in this study as shown in Fig. 6 (Table 1).

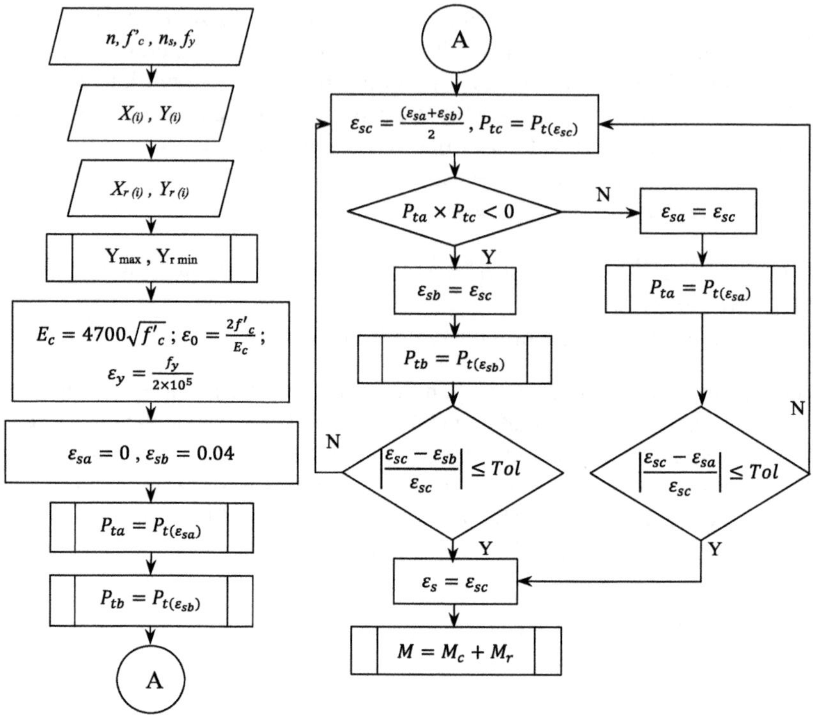

Fig. 5 Flow chart of irregular reinforced concrete beam section analysis

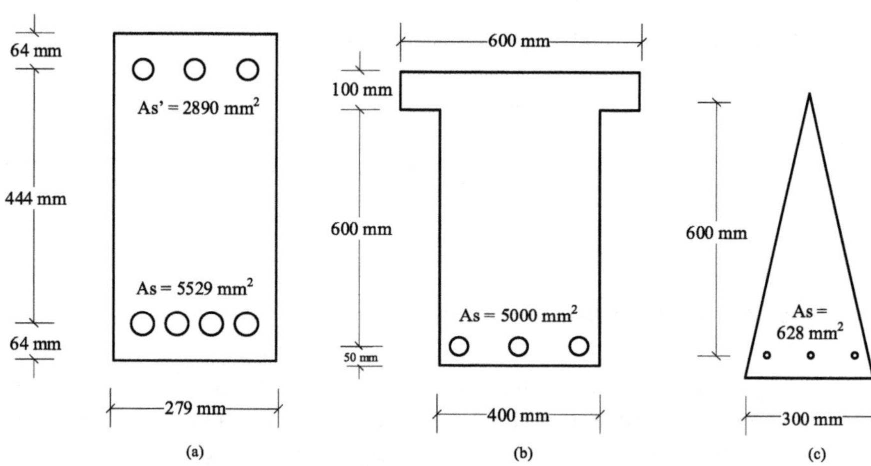

Fig. 6 Selected cross section of beam

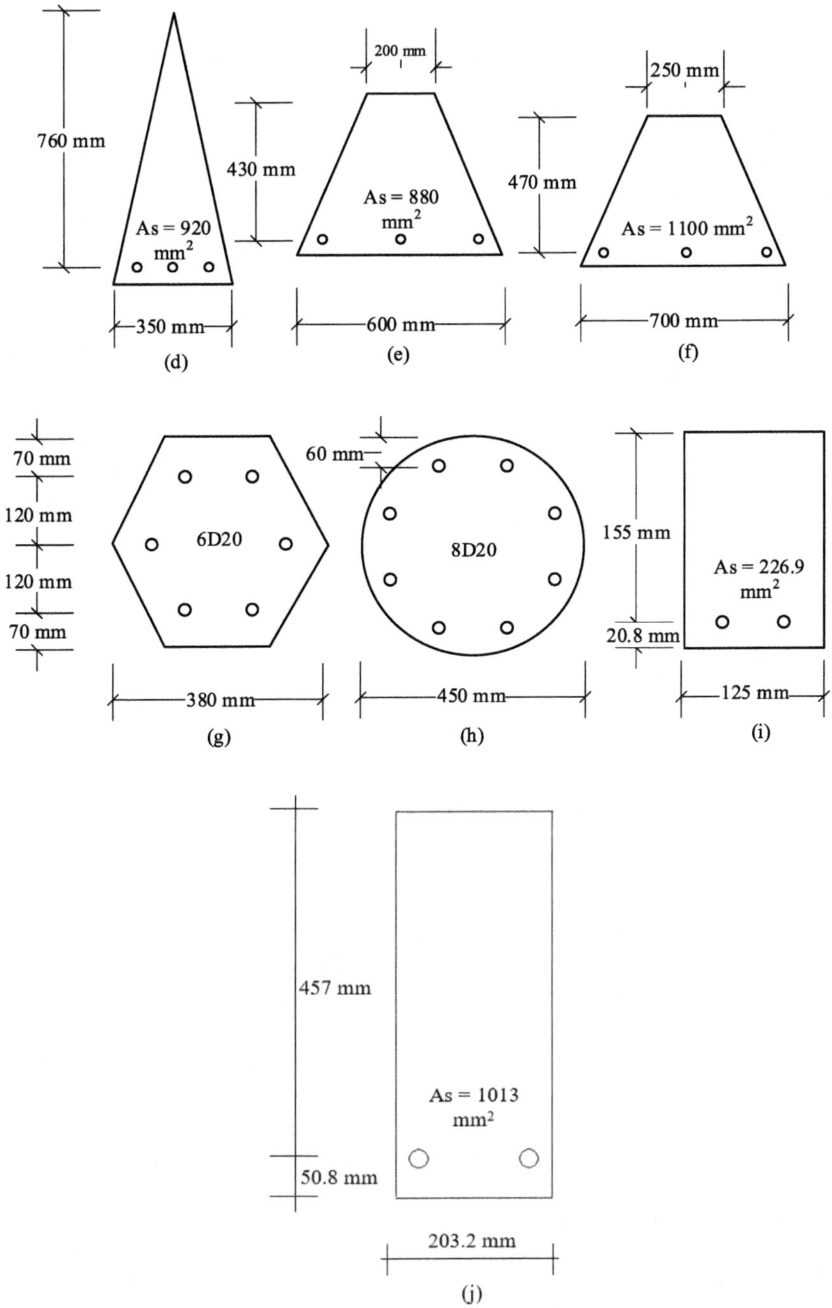

Fig. 6 (continued)

Table 1 The results of analysis

No	Section	f'_c (MPa)	f_y (MPa)	ϕ μ (kN.m)		Deviation (%)	Analysis method
				Analysis	Reference		
1	Figure 6a	20.7	276	603.31	595.00	1.40	Approximation analysis [4]
2	Figure 6b	30	400	1141.14	1139.00	0.19	Flexural equation [9]
3	Figure 6c	30	420	108.56	109.00	0.41	Flexural equation [9]
4	Figure 6d	30	420	202.22	203.00	0.38	Flexural equation [9]
5	Figure 6e	30	420	131.99	132.00	0.00	Flexural equation [9]
6	Figure 6f	30	420	181.45	180.00	0.81	Flexural equation [9]
7	Figure 6g	30	400	93.38	88.90	5.03	Equivalent square shape [9]
8	Figure 6h	30	400	152.72	140.60	8.62	Equivalent square shape [9]
9	Figure 6i	41.8	400	11.88	12.10	1.84	Finite element analysis [10]
10	Figure 6j	33.23	309.5	122.36	128.50	4.78	Finite element analysis [10]

4 Result and Discussion

Based on the numerical analysis results and comparing with the previous studies, it is apparent that the cross-sectional analysis using the numerical method implemented in a computer program is very accurate with an average deviation of 2.35%.

The largest deviation in this study is the comparison between cross-sections 7 and 8 analysis in this study and reference with deviation 5.03 and 8.62%, respectively. These deviations are more due to the analysis method distinction, where hexagonal and circular sections are simplified with square sections. Cross-sections analyzed with FEM i.e., 9 and 10 indicate greater capacity than this study. This is because the numerical analysis of cross-section ignores the concrete tensile stress as stated in one of the analysis assumptions. The cross-sections 7 and 8 analyzed with the section area approximation, where the hexagonal and circular sections are approached by converting the square section to simplify the analysis. The other cross-sections analyzed with flexural section analysis show a very small deviation i.e., 0.36%.

A similar numerical simulation was also developed by Dundar and Sahin that assume the irregular cross-sections as a closed polygon [11]. Even though the simulation model can analyze arbitrary reinforced concrete sections subjected to axial load and biaxial bending. But, the model still uses an equivalent stress block to compute concrete compression force that may be inappropriate for the non-square section. On the other hand, Oscar Fitrah Nur undertook a numerical analysis of rectangular reinforced concrete beam with divide compression area into small slices to generate the stress block which appropriates with actual stress–strain [12].

The simulation in this study considers the cross-section as a closed polygon and the stresses in compression concrete as a function of strain which linear to the distance of the neutral axis. So, this study indicates that the numerical simulations implemented in a computer program show progress in research.

5 Conclusion

The numerical method by using a multi-slice approach with bisection iteration to calculate the moment capacity of the reinforced concrete section shows a good result with an average deviation of 2.35% compared with other methods. Implementation of this method as a computer program is very precise to analyze reinforced concrete beams for the irregular section including circular, hexagonal, T shapes, etc.

References

1. O'Neil EF, Neeley BD, Cargile JD (1999) Tensile properties of very-high-strength concrete for penetration-resistant structures. Shock Vib 6:237–245, ISSN 1070-9622
2. Arιoglu N, Girgin ZC Arιoglu E (2006) Evaluation of ratio between splitting tensile strength and compressive strength for concretes up to 120 MPa and its application in strength criterion. ACI Material Journal/January-February 2006
3. Whitney CS (1937) Design of reinforced concrete members under flexure or combined flexure and direct compression. J ACI 33(3):483–498
4. Park R, Paulay T (1975) Reinforced concrete structures. Jhon Wiley & Sons, Canada
5. Hognestad E (1951) A study of combined bending and axial load in reinforced concrete. University of Illinois Engineering Experiment Station Bulletin Series No. 399 University of Illinois, Urbana
6. Chen WF (1982) Plasticity in reinforced concrete. McGraw-Hill, New York
7. MacGregor JG (1997) Reinfoced concrete mechanics and design. Inc. USA, Prentice Hall International
8. ACI 318–05: building code requirements for structural concrete (ACI 318–05) and commentary (ACI 318R-05), 2005
9. Al-Ansari MS, Afzal MS (2019) Simplified irregular beam analysis and design. Civ Eng J 5(7):1577–1589
10. Nuroji, Besar MS, Imran I (2010) Rotated discrete crack model for reinforced concrete structures. In: 35th conference on our world in concrete & structures. 25–27 August 2010, Singapore
11. Dundar C, Sahin B (1993) Arbitrarily shaped reinforced concrete members subject to biaxial bending and axial load. Comput Struct 49(4):643–662
12. Nur OF (2009) Analisa Pengaruh Penambahan Tulangan Tekan Terhadap Daktilitas Kurvatur Balok Beton Bertulang. Jurnal Rekayasa Sipil 5(1)

Assessment of Retaining Wall Design in Harris Skyline Tower's Basement, Surabaya

Siti Nurlita Fitri and Ahmad Soimun

Abstract The project of Harris skyline tower in Surabaya was designed as a hotel and office tower concept. The total area of this project is 38.821 m^2 with 29 floors and 2nd levels of basement. The substructure design is necessary to 7.5 m depth of basement construction. Furthermore, Surabaya has a high level of groundwater table which affects the risk of the project with the depth excavation. This condition encourages the retaining wall construction design to prevent the structure from collapse and failure in overall stability. The cost for basement projects included in the cost of the superstructure, thus it became essential to reduce this cost by investigating pile varieties and settling on the most cost-effective material type. This study aims to know the suitable design of the retaining wall in Harris Skyline tower. The material structure was considered in optimum result (safety factory and deformation) and best cost value. Substructure calculation of this research was conducted in 3 alternatives; sheet pile, steel pipe pile, and bored pile with the high level of groundwater table design. The deformation analysis was carried out in Plaxis simulation modeling. The result of this paper is a 21.5 m length calculation of a retaining wall with a 14 m embedded structure. The final displacement is 3,7 cm as SF = 2.66. The appropriate alternative design is a bored pile with an analysis of material price Rp 11.369.860.000.

Keywords Retaining wall · Sheetpile · Bored pile · Plaxis · Displacement

1 Indtroduction

The development of Surabaya as the capital city of east java rises rapidly with the establishment of high-rise buildings in all of the area. Harris Skyline Tower is a prestigious project from PT. Bumi Sejahtera Tower combines the concept of hotel

S. N. Fitri (✉)
Civil Engineering, Universitas Sebelas Maret, Surakarta, Indonesia
e-mail: sitinurlitafitri@staff.uns.ac.id

A. Soimun
Polytechnic of Land Transportation, Bali, Indonesia
e-mail: soimun@poltradabali.ac.id

S. A. Kristiawan et al. (eds.), *Proceedings of the 5th International Conference on Rehabilitation and Maintenance in Civil Engineering*, Lecture Notes in Civil Engineering 225,
https://doi.org/10.1007/978-981-16-9348-9_7

and office tower. The Harris Skyline Tower project location is on Jl. Maj. Gen. Sungkono No. 178, Dukuh Pakis, Surabaya City, East Java.

The construction of office buildings, apartments, and public areas require a large area of parking to accommodate visitor vehicles. This is a problem for developers to make parking zones; therefore a substructure construction (basement) is the best alternative to provide a parking area. Basement construction has its own requirements criteria and stages of construction implementation. The design of retaining walls for the material of basements under the construction method used has a considerable affected. The structure method for the basement influences the implementation schedule for the project because the first process of building is substructure construction.

Harris skyline tower was designed with 2 levels of basement and 29 floors. The basement construction is 7.5 m depth of levels. Moreover, Surabaya has a high level of the groundwater table. Hence, the project with deep excavation has a high-risk condition. Design of the basement construction is necessary to prevent the failure and sliding of the soil behind the substructure [1, 2].

There are many alternatives basement construction designs; sheet pile, bored pile, steel pipe pile, and many more. Many studies in Indonesia success to design the basement construction with the selected material [3–6]. The other reinforcement to prevent the sliding slope is geotextile and terracing [7, 8]. However, the last improvement does not suitable for this case. The decision of retaining walls to support deep excavations for basement construction is affected by a combination of variables, including soil types, allowable movement, construction methods, and material and equipment availability [9].

This research aims to determine the best alternatives of retaining wall design to Harris Skyline towers basement based on the material prices. According to Lateef and Burhan [10], The material price of pile foundations to basement project is included in the cost of the superstructure, and it became necessary to decrease this cost by studying pile types and then deciding on the best pile type in terms of cost. The output of this study is expected as guidance to a similar construction basement project.

2 Methodology

This study has simulated the performance design of 3 alternatives basement materials. The embedded pile has been carried out in manual calculation. Next, the Safety factor value in order to installation the basement pile is designed in Plaxis calculation. Moreover, The maximum deformation value has been obtained as Plaxis output. Furthermore, to conclude the best alternative structure material was look in the lowest material cost.

Table 1. N-SPT test for Harris skyline tower

Depth (m)	Soil	Average SPT
0–6	Brown clay	3
6–10	Brown silt	6
10–25	Silt sand	40
25–35	Loamy sand	50

Table 2. Soil data calculation for Harris skyline tower

Depth (m)	Soil	Average SPT	Cu (t/m^2)	Gs	γb	γsat	e	n	Φ	K	E (t/m^2)
0–6	Brown clay	3	1.5	2.62	1.66	1.704	1.29	0.57	3	2×10^{-10}	400
6–10	Brown loam	6	2.6	2.69	1.641	1.67	1.49	0.62	6	2×10^{-10}	900
10–25	Clayed sand	40	–	2.68	1.987	2.00	0.78	0.44	38	2×10^{-9}	2000
25–35	Loamy sand	50	–	2.68	2.057	2.25	0.72	0.3	41	2×10^{-9}	4000

2.1 *Soil Data*

The SPT investigation is described in Table 1. According to Table 1, The SPT test was conducted in Ahmad soimun and Arif [11]. The dominant soil in clay with different consistency (0–10 m) and Sand until 35 m depth. After the result of SPT test analysis, the next step is to calculate the other parameters (Specific gravity, unit weight, and shear stress based on Bowles [12]. The detail of soil data is shown in Table 2.

2.2 *Lateral Earth Pressure*

The lateral earth pressure is analyzed according to Rankine and Coulomb Theory [13]. The calculation considered two conditions of the groundwater table; normal and maximum level. The model was shown in Fig. 1.

Figure 1 describes the model of lateral earth pressure calculation. The first picture (a) illustrates the maximum level, this condition realizes the maximum range of the embedded pile requirement, while the second (b) considers the lowest result. The soil data were analyzed in 3 types. Soil 1 was model in 6 m depth, Soil 2 in 4 m, and 15 m for the third soil. The external load 1 t/m2 represents a load of heavy equipment and the material. Do is the calculation result of sheet pile embedded depth.

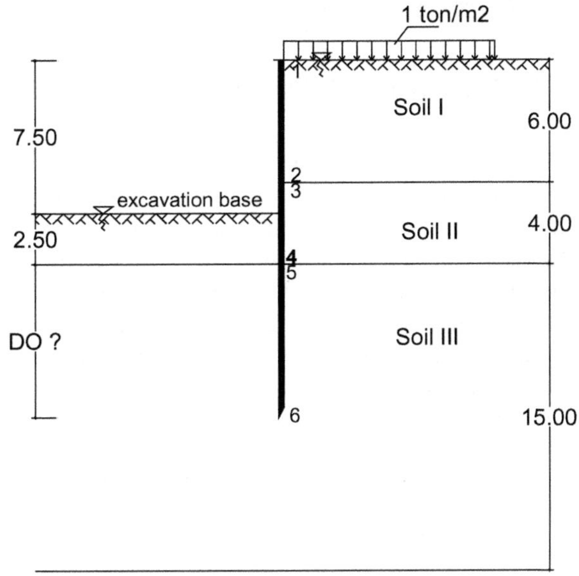

(a)

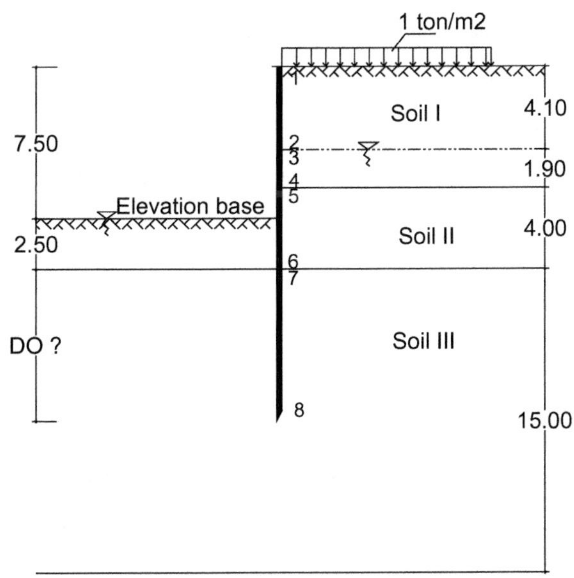

(b)

Fig. 1. Model of lateral earth pressure calculation. **a** Normal groundwater table. **b** Maximum ground water table

3 Result and Discussion

3.1 Lateral Earth Pressure Analysis

The analysis of lateral earth pressure was conducted by manual calculation. The process result is shown in Fig. 2 in a normal level of groundwater table and Fig. 3 in maximum level. The figure depicts the diagram of active and passive force from Tables 3, 4, 5 and 6. The active force is located on the right side of the graph (behind the sheet pile), and the passive force is in the front.

After the calculation of active and passive force, the next step is to determine the moment from force and distance based on the diagram. The simulated was carried out to how long the embedded depth of substructure requirement (do). Based on the calculation in Tables 3 and 4, the analysis describes as follows:

$$\sum \text{MPassive} = 28.84 + 27.37\text{do} + 4.795\text{do}^2 + 0.863\text{do}^3$$

$$\sum \text{Mfinal} = \sum \text{Mactive} - \sum \text{Mpassive}$$

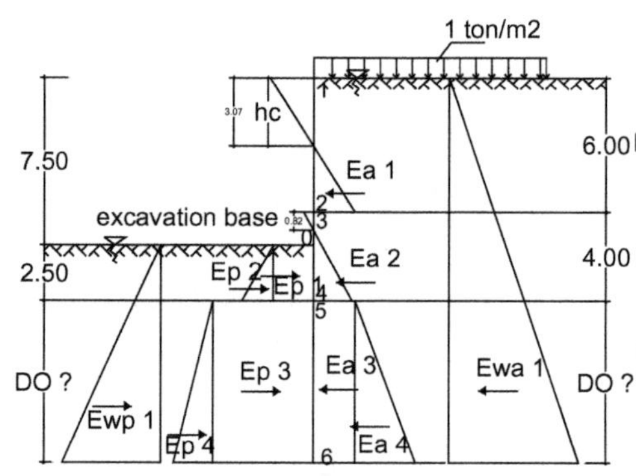

Fig. 2. Lateral earth pressure diagram in maximum level of ground water table

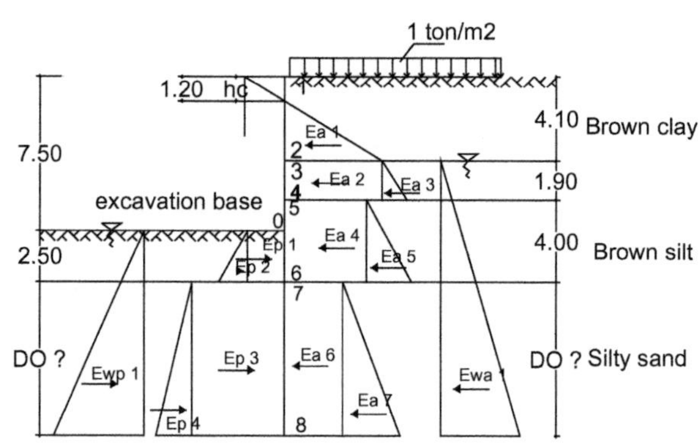

Fig. 3. Lateral earth pressure diagram in normal level of ground water Table

Table 3. Active force calculation in maximum level of ground water table

σ_v' (t/m^2)	σ' H (t/m^2)	H (m)	Active force (ton)	Distance (m)	Moment (tm)
1.00	−1.95	3.07			
5.22	1.86	2.93	2.721229595	4.98+do	13.54+2.72do
5.22	−0.45	0.82			
7.92	1.74	3.18	2.769944132	1.06+do	2.94+2.77do
7.92	1.88	do	1.88xDo	0.5 do	0.94 do^2
7.92+ (1x do)	1.901+0.24do	do	0.001Do+0.06Do^2	0.33do	0.00033 do^2+0.0198 do^3
	0.5 x (10+do)^{2}x1x1	10+do	50+10Do+0.5Do^2	3.33+0.33do	166.5+49.80do+4.97 do^2+0165 do^2

Table 4. Passive force calculation in maximum level of ground water table

σ_v' (t/m^2)	σ' H (t/m^2)	H (m)	Passive Force (ton)	Distance (m)	Moment (tm)
00	5.78	2.5	14.43796269	1.25 + do	18.05 + 14.44do
1.69	7.86	2.5	9.820811	0.833 + do	8.18 + 9.82do
1.69	7.09	Do	7.09do	0.5do	3.545 do^2
1.69 + do	7.09 + 4.2do	Do	2.1do^2	0.33do	0.693 do^3
	0.5 x (2.5 + do)^2 × 1 × 1		3.13 + 2.5do + do^2	0.833 + 0.33do	2.61 + 3.11do + 1.25 do^2 + 0.17do^3

$$= 154.14 + 27.92\text{do} + 1.12\text{do}^2 - 0.678\text{do}^3$$
$$\text{or} = -0.678\text{do}^3 + 1.12\text{do}^2 + 27.92\text{do} + 154.14$$

Hence,

do = 9.012 m
and total the depth of embedded substructure is 9.012 + 2.5 = 11.512 m
for safety reason, the result must times with Safety factor
Thus,
11.512 m × 1.2 = 13.81 m = 14 m

The final result of substructure is 14 m + 7.5 m(open cut) = 21.5 m
Next, the maximum momen is calculated as follow :

$$\begin{aligned}\sum \text{Mfinal} &= -0.678\text{do}^3 + 1.12\text{do}^2 + 27.92\text{do} + 154.14\\ &= -0.678(4.29)^3 + 1.12(4.29)^2 + 27.92(4.29) + 154.14\\ &= 240.905\ \text{Ton.m} = 2409.05\ \text{Kn.m}\end{aligned}$$

Table 5. Active force calculation in normal level of ground water table

σ_v' (t/m^2)	σ' H (t/m^2)	H (m)	Active force (ton)	Distance (m)	Moment (tm)
7.81	4.18	2.9	6.07	6.87+do	41.65+6.07do
7.81	4.18	1.9	7.948511606	4.95+do	39.35+7.95do
9.14	5.39	1.9	1.144326245	4.63+do	5.302+1.14do
9.14	2.73	4	10.9261824	2+do	21.85+10.93do
11.84	4.92	4	4.377928192	1.33+do	5.84+4.38do
11.84	2.82	do	2.82do	0.5do	1.41do^2
11.84+do4.97+0.42do	do	1.24do+0.105do^2		0.33do	0.409do^2+0.035do^3
5.9+do	5.9+do	5.9+do	17.41+5.9do+0.5Do^2	0.33do	34.3+17.4do+2.37do^2+0.17do^3

Table 6. Passive force calculation in normal level of ground water table

σ_v' (t/m^2)	σ' H (t/m^2)	H (m)		Passive force (ton)	Distance (m)	Moment (tm)
0.00	5.78		2.5	14.43796269	1.25+do	18.05+14.44do
1.69	7.86	2.5		9.820811367	0.83+do	8.18+9.82do
1.69	7.09	do		7.09do	0.5do	3.545do^2
1.69+do	7.09+4.2do	do		2.1do^2	0.33do	0.693do^3
2.5+do	2.5+do	2.5+do		3.13+2.5do+o.5do^2	0.833+0.33do	2.61+3.11do+1.25do^2+0.17do^3

The same procedure was applied in the calculation in the normal level of the groundwater table. The result analysis of the embedded depth and the maximum moment is shown as follows:

$$\sum \text{MPassive} = 28.84 + 27.37\text{do} + 4.795\text{do}^2 + 0.863\text{do}^3$$

$$\begin{aligned}\sum \text{Mfinal} &= \sum \text{Mactive} - \sum \text{Mpassive} \\ &= 119.45 + 20.47\text{do} - 0.606\text{do}^2 - 0.658\text{do}^3\end{aligned}$$

Hence do = 7.185
Consired the SF do = 9.685
Final do = 7.5 + 12 = 19.5 m.

The maximum moment :

$$\begin{aligned}\sum \text{Msum} &= -0.658\text{Do}^3 - 0.606\text{Do}^2 + 20.47\text{do} + 119.45 \\ &= -0.658(2.92)^3 - 0.606(2.92)^2 + 20.47(2.92) + 119.45 \\ &= 157.647\ \text{Ton.m} = 1576.47\ \text{Kn.m}\end{aligned}$$

3.2 Retaining Wall's Structure Calculation

The basement substructure analysis was conducted in 3 types of material. The first is steel pipe pile with diameter = 0.9 m, thick = 0.015 m, the area of section = 417 cm^2, inertia moment = 409000 cm^4, and section modulus = 9080 cm^3. The profile determines by the calculation as follows :

$$\sigma = 310000\text{kN}$$

$$Zo = \frac{\text{Mdesign}}{\sigma a} \times 1^6 = \frac{2409.05}{310000} \times 10^6 = 7771.12\text{cm}^3$$

Hence, the applied profile with w = 9080 cm^3 > 7771.12 cm^3

The second calculation is bored pile with fy = 400 Mpa, fc = 30 Mpa, diameter = 1 m, and result of bored pile analysis :

$$\begin{aligned}\rho_s &= 0,45\left(\frac{Ag}{Ach} - 1\right)\frac{fç}{f_{yt}} \\ &= 0,45\left(\frac{785398.16}{664761.01} - 1\right)\frac{30}{250} = 0.00979 \\ s &= \frac{a_s\pi(D_c - d_b)}{(\pi/4)D_c^2\rho_s} = \frac{132.73\pi(920 - 13)}{(\pi/4)920^2x0.00979} = 58.05\ \text{mm}\end{aligned} \quad (1)$$

Thus the reinforcement is D13-60mm
The last is sheet pile Z-BOX Piles type CAZ Z-Box

σt = 265000 knE = 200000 Mpa
B = 1400 mm I = 475810 cm^4
H = 922 mm S = 408 cm^3
W = 10285 cm^3
$Zo = \frac{\text{Mdesign}}{\sigma a} \times 10^6 \frac{2409.05}{265000} \times 10^6 = 9090.745\,\text{cm}^3$

Control Rowe reduction : w = 13140 cm^3 > 9090.745 cm^3

The result of the basement substructure design will be the basic calculation to analysis the material price.

3.3 *Plaxis Modeling Analysis*

The Finite element method analysis was conducted to know the deformation of the basement model. The result is shown in Figs 4, 5 and 6.

Based on Fig. 4, the total displacement of steel pipe pile design is 4.1 cm. Figure 5 describes the output of total displacement of the bored pile, the value is 5.05cm, and sheet pile analysis is 4.12 cm of displacement in Fig. 6. Meanwhile, the result of Safety factor value is 1.66, 1,56 and 1.6 respectively. Based on Bowles [12], all of the SF result value indicated a safety condition.

The highest deformation is design of bored pile, and the others have a similar value. while, in SF assessment, the lowest is also bored pile structure. This result appropriate with [16], which the SF is corresponding with the deformation analysis.

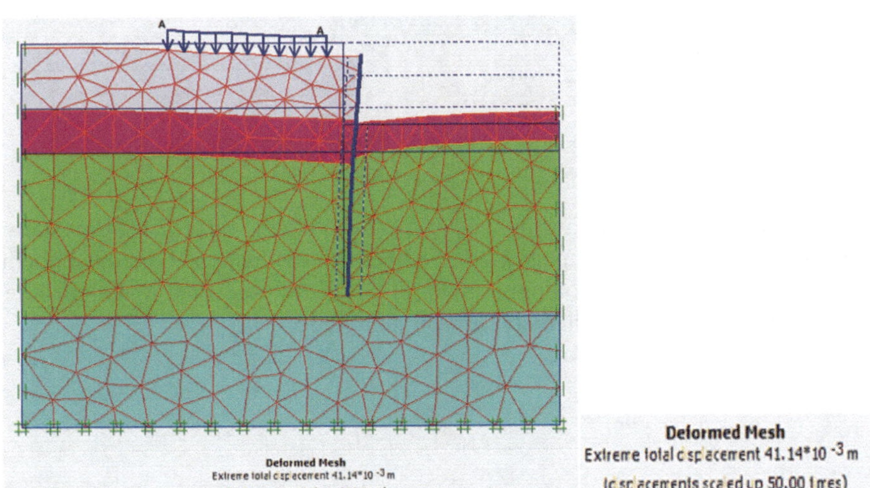

Fig. 4. Total displacement steel pipe pile

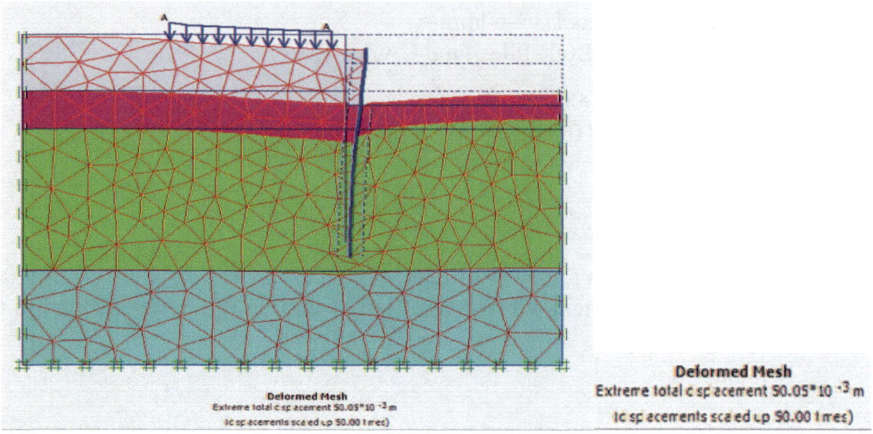

Fig. 5. Total displacement bored pile

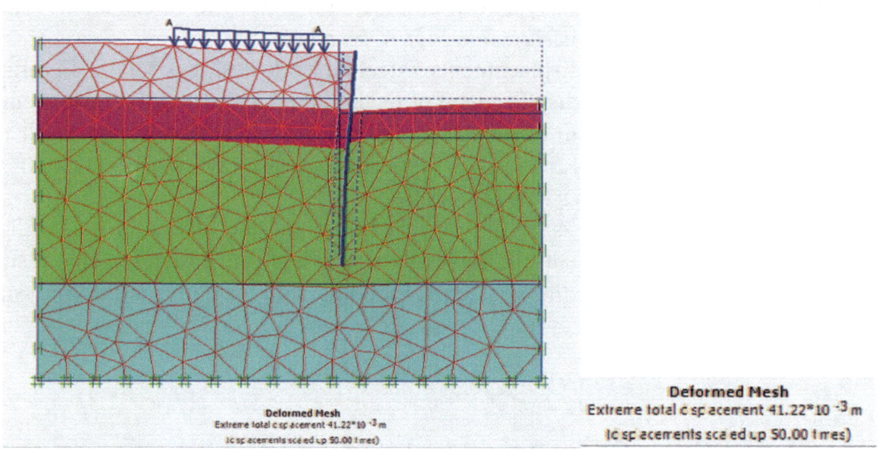

Fig. 6. Total displacement sheet pile

The analysis use Mohr-Coulomb model, according to Mahmood et al. [17] has shown in Table 1. The structure is represented as a linear elastic material, and the model is simulated using the Mohr-Coulomb model. It has been discovered that embedding the pile reduces the maximum displacement amplitude.

Table 7. Material price result

Type	Price
Steel pipe pile	Rp 31.234.158.000,-
Bored pile D = 1.0 m	Rp 11.369.860.000,-
Sheet pile Z-box	Rp 17.747.111.000,-

3.4 Material Price Analysis

The material volume of the basement structure design was analyzed with the material requirement. The calculation is considered to cover all the area of the basement site. The result of the final material price has shown in Table 7.

Table 7 depicts a material price with 3 type categories. The bored pile material price is the lowest material cost, and the steel pipe pile is the highest. The steel pipe pile price is expensive in the market construction

4 Conclusions

The result shows different parameters in 3 alternative retaining wall structures. The Steel Pipe pile has the lowest deformation and highest Safety factor calculation. However, the material cost is very expensive. The bored pile is the lowest material cost with Safety Factor value 1.56 and 5.05 cm deformation model. In the case of the Harris skyline basement project, the considerable structure that applicable to construct is a retaining wall with a bored pile structure.

References

1. Ng WM, Khor EL, Tiong LK, Lee J (1998) Simulation modeling and management of large basement construction project. J Comput Civ Eng 12(2):101–110. https://doi.org/10.1061/(asce)0887-3801(1998)12:2(101)
2. Wen YW, Liu SY, Hu ML, Zhang GZ (2013) Deformation control techniques for existing buildings during construction process of basement. Yantu Gongcheng Xuebao/Chinese J Geotech Eng 35(10)
3. Ray N, Setia L, Wibowo B, Cahyono MSD (2018) Perencanaan Dinding Penahan Tanah Pada Basement Grand Dharmahusada Lagoon Surabaya, pp 1–6
4. Widyarini EL (2016) Perencanaan Basement 5 Lantai Dan Pondasi Gedung Spazio Tower Di Surabaya Barat. Tugas Akhir Jur. Tek. Sipil ITS
5. Kurniawan DA (2017) Perencanaan Dinding Penahan Tanah Pada Basement Midtown Point And Ibis Styles Hotel Jakarta. Tugas Akhir Jur. Tek. Sipil Its
6. Livando R, Kawanda A (2020) Perancangan Dinding Penahan Tanah Pada Basement Dengan Dukungan Strut-Beam. JMTS J. Mitra Tek. Sipil 3(3). https://doi.org/10.24912/jmts.v3i3.8748
7. Surjandari NS et al (2021) Slope stability analysis in various Terraces model (case study : Sendangmulyo, Tirtomulyo District, Wonogiri Regency). J Phys Conf Ser 1858:2021. https://doi.org/10.1088/1742-6596/1858/1/012005

8. Nurlita Fitri S (2021) Perkuatan Oprit Jembatan Kali Jubang Jalan Tol Pejangan-Brebes Timur menggunakan Geotekstil. BENTANG J Teor dan Terap Bid Rekayasa Sipil 9(1):37–46
9. Tan Y, Chow C (2008) Design of retaining wall and support systems for deep basement construction–A Malaysian experience. Deep Excav Retaining Walls
10. Lateef HH, Burhan AM (2019) Time-cost-quality trade-off model for optimal pile type selection using discrete particle swarm optimization algorithm. Civ Eng J 5(11):2461–2471. https://doi.org/10.28991/cej-2019-03091424
11. Ahmad Soimun YL, Arif M (2016) Perencanaan dinding penahan tanah untuk pembuatan basemnet Harris Skyline Tower Surabaya. Tugas akhir Inst. Teknol. Sepuluh Nop
12. Bowles JE (1996) Foundation analysis and design 5th edn. McGraw-Hill handbooks
13. Das BM (2006) Principles-of-geotechnical-engineering. Thomson, pp 1–10
14. Kanoun F, Haffoudhi S, Bouassida M (2017) Design and Follow up of 20 m Depth Nailed Wall. In: ICSMGE 2017—19th international conference on soil mechanics and geotechnical engineering, vol 2017
15. Jasmine Nisha J, Muttharam M (2017) Deep excavation supported by diaphragm wall: a case study. Indian Geotech J 47(3):373–383. https://doi.org/10.1007/s40098-017-0230-1
16. Srivastava A, Kumar P, Sivakumar GL (2012) Stability analysis of 18 M deep excavation using micro piles. In: Proceeding Indian geotechnology conference
17. Mahmood M, Al-Wakel S, Abdulwahhab I (2018) Three-dimensional analysis for piled raft machine foundation embedded in sand. In: MATEC Web of conferences, vol 162. https://doi.org/10.1051/matecconf/201816201023

Additional Vertical Movement of the Single Pile Foundation with Combined Loads

Sumiyati Gunawan, Niken Silmi Surjandari, Bambang Setiawan, and Yusep Muslih Purwana

Abstract Pile-foundation support the upper structure to the subgrade, it generally supports vertical loads, occasionally when the horizontal loads are found to be more dominant, it becomes an important consideration in the design. Pile foundations with combined loads are generally analyzed separately, however on the site, the two loads work simultaneously. In-Indonesia, particularly, loading tests do not performed once. Therefore, it does not account for the additional vertical displacement and lateral deflection. This research will investigate the analysis and experimental laboratory, effect of vertical displacement on single pile foundation, as the effect of combined loads, and get a relationship that can be used as a reference for analyzing additional vertical displacement due to combined loads on single pile foundations. The process was in three stages with the first being the the preliminary analysis through 2D Finite Element Method (2D-FEM) Plaxis2D8.6 while the second was an experimental test in the laboratory, and third, being the analysis through 3DFEM Plaxis3D1.1 for validation with soil data and loading test on several projects in the field. The results obtained show that after adding vertical load (Pu) about 5 $\times$ lateral load (Hu), that on (Pu/Hu > 5), the vertical displacement stays, (as the result of the loading test) until it collapses, when (Pu/Hu $\leq$ 5), the lateral load will cause a significant increase in vertical movement due to the combination of the load.

Keywords Vertical-horizontal load · Single pile · Vertical displacement

S. Gunawan (✉) · N. S. Surjandari · B. Setiawan · Y. M. Purwana
Department of Civil Engineering, Sebelas Maret University, Jawa Tengah, Indonesia
e-mail: sumiyati.gunawan@uajy.ac.id

N. S. Surjandari
e-mail: nikensilmisurjandari@staff.uns.ac.id

B. Setiawan
e-mail: bambangsetiawan@staff.uns.ac.id

Y. M. Purwana
e-mail: ymuslih@ft.uns.ac.id

S. A. Kristiawan et al. (eds.), *Proceedings of the 5th International Conference on Rehabilitation and Maintenance in Civil Engineering*, Lecture Notes in Civil Engineering 225,
https://doi.org/10.1007/978-981-16-9348-9_8

1 Introduction

The vertical loads of a structure are usually carried by pile-foundation but horizontal loads are dominant in some other structures. This means it is very important to calculate the horizontal loads in pile foundations [1]. Combined loads are seldomly analyzed in simultaneity. This involves calculating the axial load first to determine the axial carrying capacity and vertical displacement followed by determining the lateral load to evaluate the lateral bearing capacity and deflection. Meanwhile, the two loads work simultaneously on the site [2]. In Indonesia, these loading tests are not usually conducted simultaneously and this means the additional lateral deflection due to the combined loads is not considered in construction as required by the ASTM D3966-07 [3].

Previous studies reported the reduction of lateral deflection in pile foundation under combined loads due to axial loads [4] while lateral loading was discovered not to be causing any vertical movement but has the ability to increase the movement in combined loads [5]. A study also showed lateral bearing was reducing as the embedded part of the pile decreased while lateral deflections reduced with the increase in vertical load on the pile head [6]. A three-dimensional finite element analysis was also conducted to determine the influence of combined axial and lateral loads on homogeneous clay and sandy soils. The results showed a significant increase in the effect of axial load on the lateral bearing capacity in sandy soil and a slight decrease in clay soil but a substantial influence of axial load was recorded for sandy soils, even for piles with 30D in length, and a less significance impact was found with clay soils for piles above 15D in length [7].

Another test on poorly graded sand with variations in the pile and loading also showed an increase in the lateral bearing capacity as the vertical load was increased [8]. A numerical study analysis of pile-soil interactions which were subjected to axial and lateral loads simultaneously using LPILE, a finite element (FE) model with Abaqus/Cae and SAP 2000, showed an increase in axial load caused a reduction in the induced bending moment and lateral displacement and this subsequently increased its capacity to withstand lateral forces [9]. It was also discovered in another research that the influence of axial loads on the lateral bearing capacity pile increased significantly in sandy soils but less significant in loamy soils [10].

Vertical loads were also reported to have less effect on lateral resistance in sandy soil but the progression was observed to be increasing as the soil density increased [11, 12]. Moreover, the influence of axial load on lateral bearing capacity significantly increased in sandy soils and slightly in clay soils while the square-shaped pile was found to have the ability to withstand 1.3 times more load than the round pile [13]. A numerical analysis also showed the effects of combined loads are beneficial but the interactions were very complex and, depending on the load conditions, there is a possibility of a contrary effect on system rigidity and max load [14]. A limited experimental research discovered the application of a static axial load has a minimal effect on the lateral behavior of micropiles fixed in rigid clay soil [15]. Furthermore, some studies also provided evidence that lateral loads were decreased in combined loads even though deflection was reduced by axial loads due to their presence.

Although previous studies reported that the axial loads reduce deflection due to lateral loads, lateral loads also increase vertical movement in the load combination, this has been proven by the preliminary 2D Finite Element Method test (2D-FEM). This research will investigate the analysis and experimental laboratory, effect of vertical displacement on single pile foundation, as the effect of combined loads, and get a relationship that can be used as a reference for analyzing additional vertical displacement due to combined loads on pile foundations.

The process was divided into three stages, the first being the preliminary analysis conducted using the 2D Finite Element Method (2D-FEM) *Plaxis2D8.6* approach, the second was the Laboratory Model Experimentation test. The model used was a (1,5 × 1,5 × 1,2) m3 box filled with silty-sand soil, pile foundation steels at 0,5 and 0,6 m lengths and 0,015 and 0,002 m diameters, load variations. The third with being the analysis conducted using the **3D** Finite Element Method (**3D**-FEM) *Plaxis3D1.1* approach based on the results of loading tests in several projects in the field.

2 Metodology

2.1 Stage I, Analysis with 2D Finite Element Method (2D-FEM)

Keywords: The preliminary analysis with 2D Finite Element Method (2D-FEM), modeling of a single pile foundation using the axisymetry menu on 2D Finite Element Method (2DFEM) comprised of variation on loading direction (axial, lateral, combination of axial and lateral), variation on length and diameter of the pile, variations on soil type (soft, medium, dense, homogeneous and layered), variations on groundwater table condition, (submerged and not submerged). Variation of soil parameter and loading direction in this research which was conducted using 2D-FEM presented in Tables 1 and 2.

2.2 Stage II, Laboratory Model Experimental Test

In this study, The model a test box with a size of (1,5 × 1,5 × 1,2) m3, it filled with silt sand soil, as a description of conditions in the field (samples taken in Karang, Kalitirto, Berbah, Sleman, Special Region of Yogyakarta), the single pile foundation using steel with length of 0,5 and 0,6 m, diameter of 0,015 and 0,02 m. Given loads, axial, horizontal, and combined loads. The laboratory model can be seen in Fig. 1.

The variation of the length and diameter of the pile, and combined load is described in Table 3,

Table 1 Variation of soil parameter on 2D-FEM

Name	T3		T5	T6				Pile
	Layer 1	Layer 2	Layer 1	Layer 1	Layer 2	Layer 3	Layer 4	
Material model	MC	MC	MC	MC	MC	MC	MC	LE
Condition	D	D	D	D	D	D	D	Non Porous
γk kN/m^3	16	17	16,5	11,86	12,22	12,86	13,57	24
γsat kN/m^3	20	21	20	17,50	18,03	17,67	17,83	
E kN/m^2	1,2.10^5	1,2.10^5	8.10^4	1,6.10^6	1,6.10^6	1,6.10^6	1,6.10^6	25,33
Poisson Ratio, μ	0,3	0,3	0,3	0,3	0,3	0,3	0,3	0,3
c kN/m^2	1	1	1	2,05	2,30	0,05	0,02	
φ(°)	30	33	31	30	30	35	40	
Interface, R_{inter}	1	0,7	1	0,4	0,35	0,7	0,7	

Note D = Drained, NP = Nonporous, MC = Mohr Coulomb, LE = Linear Elastic

Table 2 Variation of load on 2D-FEM

Type of soil	Pile dimensions		Load Kn, until it collapses		
	D (m)	L (m)	Lateral (5–600) KN	Axial (5–12.000) kN	Combined (kN)
T5 (soft soil) 1 layer	0,65	17	√	√	√
		30	√	√	√
T3 (medium) 2 layer	0,65	22	√	√	√
		26	√	√	√
		30	√	√	√
T6 (dense) 4 layer	0,6	17	√	√	√

2.3 Stage III, Analysis with 3D Finite Element Method (3D-FEM)

The analysis using the 3D Finite Element Method (3D-FEM) comprised of modeling a single pile foundation soil data from several projects in the field, namely: Citarum Bridge, South Jakarta Cikampek II; Dompak Bridge, Tanjung Pinang, Riau; PLTU Batang, Central Java, The variables used including axial, lateral, and combined loads, of when it collapses, are presented in Tables 4 and 5.

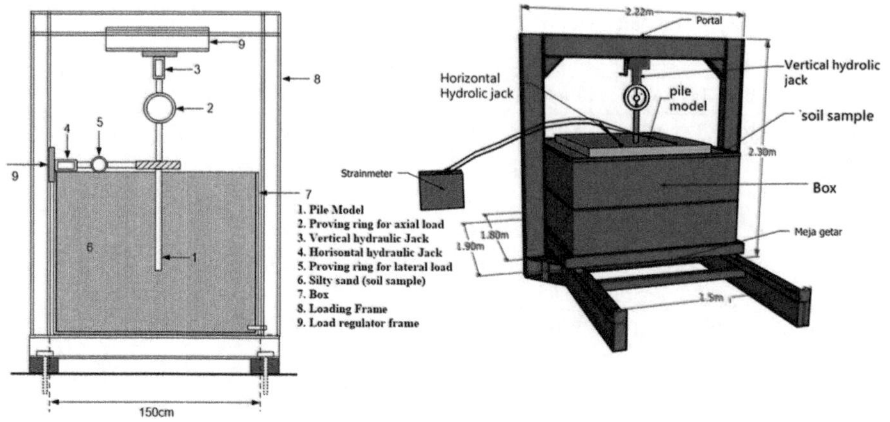

Fig. 1 Laboratory test model

Table 3 Variation of load on laboratory test model

Type of soil	Pile dimensions		Load, until it collapses		
	D m	L m	Lateral gr	Axial gr	combined loads gr
Silty sand	0,015	0,5	√	√	√
Silty sand	0,02	0,6	√	√	√

Table 4 Variation of soil parameter on 3D-FEM

Project	Depth (m)	Type of soil	Gs	w	γb	γk	γsat	Soil parameters	
				(%)	(kN/m^3)	(kN/m^3)	(kN/m^3)	φ (°)	C (kN/m^2)
PLTU Batang	1 to 8	Silty sand	2,6	40,0	16,6	11,86	17,50	25,0	2,05
	8 to 12	Silty sand	2,6	35,0	16,5	12,22	18,03	25,0	2,30
	12 to 20	Sand	2,7	40,0	18,0	12,86	17,67	30,0	0,05
	20 to 30	Sand gravel	2,7	40,0	19,0	13,57	17,83	35,0	0,02
Dompak Bridge	1 to 6	Silty sand	2,5	20,0	15,5	12,92	19,62	15,0	2,20
	6 to 12	Sandy silt	2,5	20,0	15,8	13,17	19,62	20,0	2,10
	12 to 22	Sand	2,7	20,0	16,6	13,83	20,39	25,0	0,05

(continued)

Table 4 (continued)

Project	Depth (m)	Type of soil	Gs	w	γb	γk	γsat	Soil parameters	
				(%)	(kN/m^3)	(kN/m^3)	(kN/m^3)	φ (°)	C (kN/m^2)
	22 to 34	Sand gravel	2,7	20,0	19,5	16,25	20,64	35,0	0,02
Citaru Bridge	0 to 1	Top soil	2,5	20,0	15,5	12,92	19,62	15,0	2,20
	1 to 4	silty clay	2,5	20,0	15,8	13,17	19,62	17,0	3,00
	4 to 14	Sand	2,7	20,0	19,7	16,42	20,64	35,0	0,02
	14 to 60	silty sand	2,6	20,0	17,6	14,67	20,14	35,0	0,02

Table 5 Variation of the load on 3D-FEM

Project	Pile dimensions		Load (kN) (until collapsed)		
	D (m)	L (m)	Lateral	Axial	Combined
Citarum bridge	1,2	20,0	√	√	√
Dompak bridge	1,0	18,0	√	√	√
PLTU batang	0,6	17,0	√	√	√

3 Result and Discussion

3.1 Result

The results of the preliminary analysis with 2D-FEM, with axial loading, lateral loading, combined loads (axial and lateral), variation of length and diameter of the pile, soft soil type, medium and dense, homogeneous soil and layered, submerged and not submerged in water, illustrates movement at the head of the pile due to axial and lateral loading are presented in Figs. 2 and 3.

Experimental tests of laboratory scale models, with silty sand as a description of conditions in the field, while the single pile foundation uses steel with lengths of 0,5 and 0,6 m, diameters of 0,015 and 0,02 m, movement at the head of the pile with variations in axial loads and lateral loads in Fig. 4 and vertical movement because of the combination of loads in Fig. 5.

Analysis with 3D-FEM, with axial loading, lateral loading, and combined loads (axial and lateral), based on soil data from several projects in the field, movement at the head of the single pile due to axial and lateral loading as in Figs. 6 and 7.

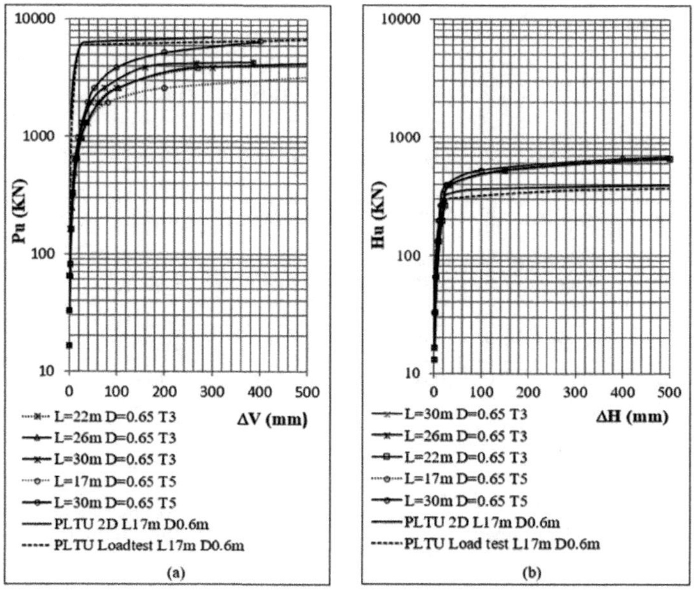

Fig. 2 Movement at the head of the pile (2D-FEM). **a** Displacement vertical without lateral load. **b** Deflection horizontal without axial load

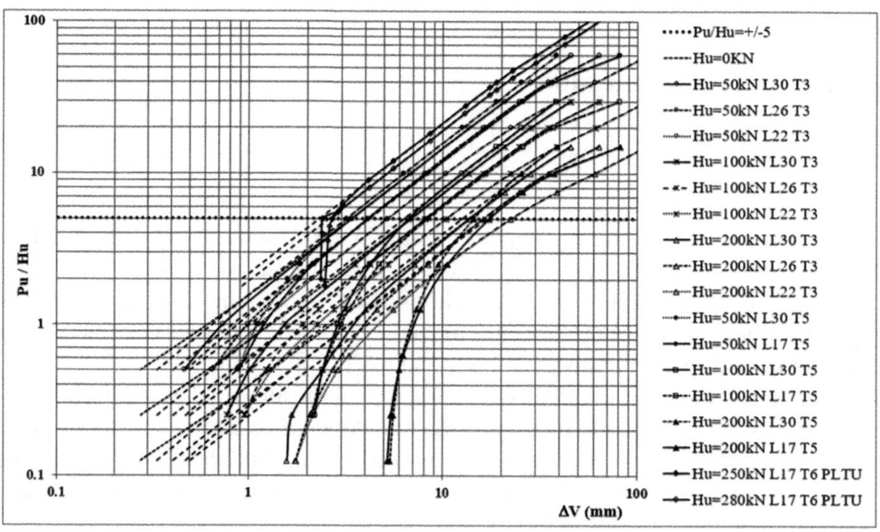

Fig. 3 Vertical displacement With Combined Loads (2D-FEM)

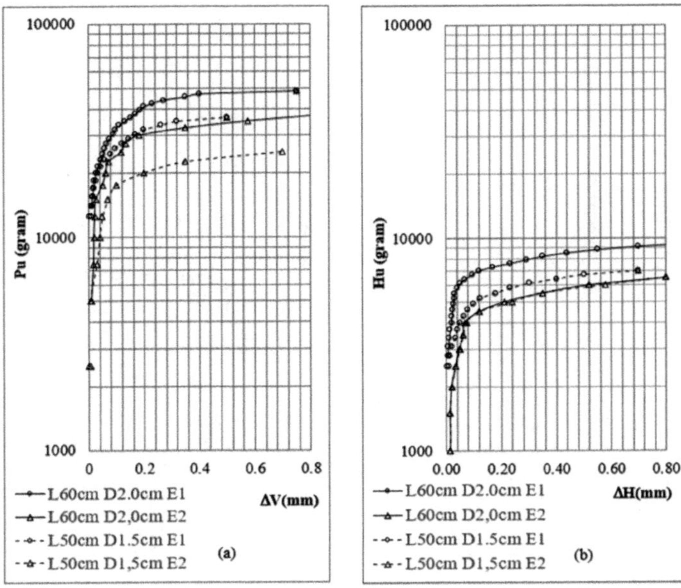

Fig. 4 Movement at the head of the pile (laboratory). **a** Displacement vertical without lateral load. **b** Deflection horizontal without axial load

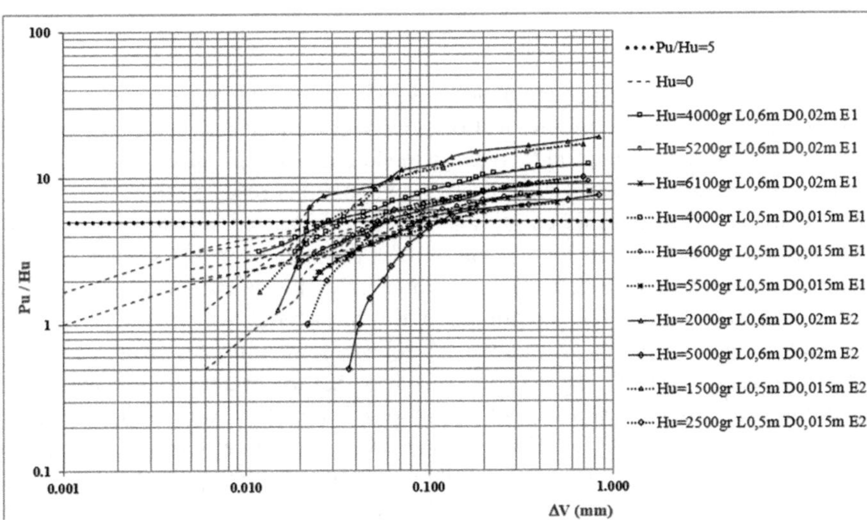

Fig. 5 Vertical displacement with combined loads silty sand (laboratory)

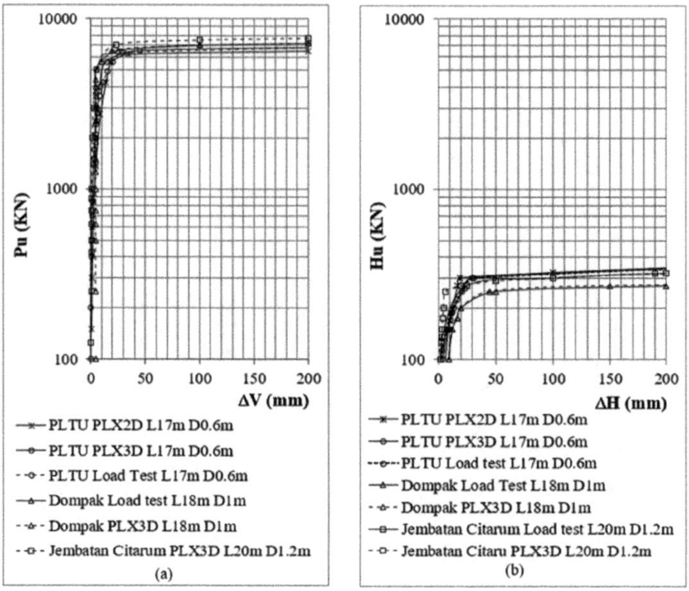

Fig. 6 Movement at the head of the pile (3D-FEM). **a** Displacement vertical without lateral load. **b** Deflection horizontal without axial load

3.2 Discussion

Previous studies reported the la teral deflections reduced with the increase in vertical load on the pile head [6], another test on poorly graded sand with variations in the pile and loading also showed an increase in the lateral bearing capacity as the vertical load was increased [8]. It was also discovered in another research that the influence of axial loads on the lateral bearing capacity pile increased significantly in sandy soils but less significant in loamy soils [10]. Although previous studies reported that the axial loads reduce deflection due to lateral loads, lateral loads also increase vertical movement in the load combination, this has been proven by the preliminary 2D Finite Element Method test above. Moreover, something new was found with the preliminary FEM 2D analysis and this was that at certain comparisons of Pu (axial load) and Hu (lateral load), whereas the addition of the vertical load averages 5.0 × the lateral load, at (Pu/Hu > 5.0) the vertical decrease tends to remain ((as the result of the loading test) until it collapses, when (Pu/Hu $\leq$ 5), the lateral load will cause a significant increase in vertical movement due to the combination of the load. Following up on the preliminary 2D-FEM analysis test above, the research was continued with a experimental test in the laboratory and validation by modeling the 3DFEM using based on loading test data on several projects in the field. It shows things that confirm the same thing at certain comparisons of Pu (axial load) and Hu (lateral load), whereas the addition of the vertical load averages 5.0 × the lateral load, at (Pu/Hu > 5.0) the vertical decrease tends to remain ((as the result of the loading

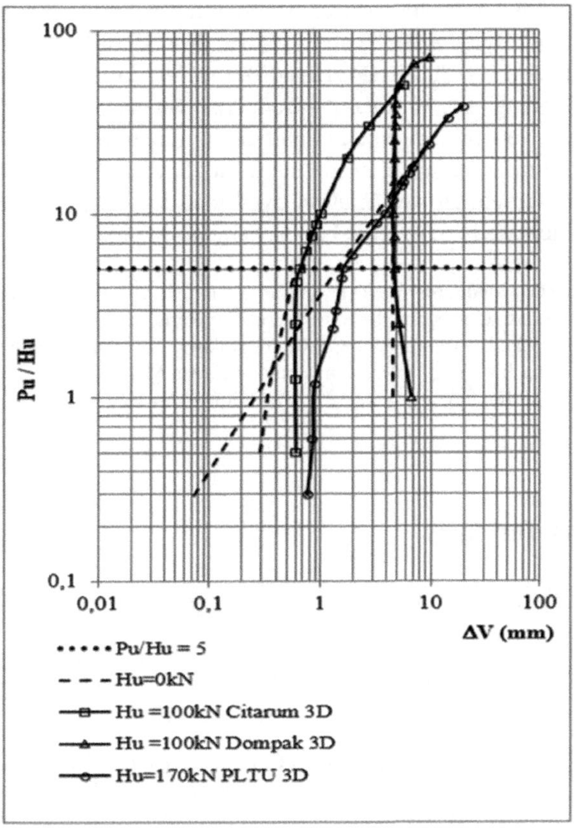

Fig. 7 Vertical displacement with combined loads (3D-FEM)

test) until it collapses, when (Pu/Hu $\leq$ 5). This, therefore, needs to be considered in calculating and planning the pile foundations' bearing capacity.

Full-scale models also need to be verified through further studies to obtain satisfactory results.

4 Conclusion

This research focuses on investigate the analysis and experimental laboratory, effect of vertical displacement on single pile foundation, as the effect of combined loads, and get a relationship that can be used as a reference for analyzing additional vertical displacement on single pile foundations. Following the preliminary 2D-FEM analysis test, the analysis was continued laboratory experiment and further validated by 3D-FEM model using loading test data of several projects in the field. The test results for experiment in the laboratory and vali-dation by modelling the 3D-FEM not much different from 2D-FEM, that when the vertical movement was increased significantly at (Pu/Hu > 5.0), the lateral load has no effect on vertical movement. Hence the behaviour, needs to be considered in designing the pile foundations bearing capacity. The results obtained showed that after adding vertical load (Pu) of about 5 × lateral load (Hu), that is (Pu/Hu > 5), the vertical displacement remain constant (like the loading test) until it collapsed. Also when (Pu/Hu ≤ 5), the lateral load will increase the vertical movement significantly due to the load combination, with axial load is 5 times greater than the lateral load, there is no increase in vertical displacement observed. However, at an axial load less than 5 times the lateral load, there is an increase in the vertical displacement that needs to be taken into account.

Acknowledgements The authors appreciate the Soil Mechanics Laboratory, Faculty of Engineering, Universitas Sebelas Maret Surakarta, and Soil Mechanics Laboratory, Faculty of Engineering, Universitas Atma Jaya, Yogyakarta, for supporting the research as well as the Universitas Atma Jaya Yogyakarta, Universitas Sebelas Maret Surakarta, and The Slamet Riyadi Foundation for all funding allowances for education and research.

References

1. Brown DA, Morrison C, Reese LC (1988) Lateral load behavior of pile group in sand. J Geotech Eng 114(11):1261–1276
2. Mandolini GR, Viggiani C (2005) Pile foundations: experimental investigations, analysis and design. In Proceedings of the international conference on soil mechanics and geotechnical engineering, vol 16, no 1, p 177
3. Standard A, "D3966 (2007) Standard test methods for deep foundations under lateral load. ASTM International, West Conshohocken, pa, 2007. https://doi.org/10.1520/d3966-07
4. Ibrahim SF, Al-Soud MS, Al-Asadi FI (2018) Performance of a single pile under combined axial and lateral loads in layered sandy soil. J Eng Sustain Dev 22(1):121–136
5. Anagnostopoulos C, Georgiadis M (1993) Interaction of axial and lateral pile responses. J Geotech Eng 119(4):793–798
6. Zhu M-X, Zhang Y, Gong W-M, Wang L, Dai G-L (2017) Generalized solutions for axially and laterally loaded piles in multilayered soil deposits with transfer matrix method. Int J Geomech 17(4):04016104
7. Karthigeyan S, Ramakrishna V, Rajagopal K (2007) Numerical investigation of the effect of vertical load on the lateral response of piles. J Geotech Geoenviron Eng 133(5):512–521
8. Nugroho SA et al Pengaruh beban vertikal terhadap daya dukung lateral pondasi tiang. Ph.D. dissertation, Riau University
9. Khodair Y, Abdel-Mohit A (2014) Numerical analysis of pile-soil interaction under axial and lateral loads. Int J Concr Struct Mater 8(3):239–249
10. Rajagopal K, Karthigeyan S (2008) Influence of combined vertical and lateral loading on the lateral response of piles. In: International association for computer methods and advances in geomechanics (IACMAG), pp 3272–3282
11. Hazzar L, Hussien MN, Karray M (2016) Investigation of the influence of vertical loads on the lateral response of pile foundations in sands and clays 2
12. Hazzard (2017) Influence of vertical loads on lateral response of pile foundations in sands and clays. J Rock Mech Geotech Eng 9(2):291–304
13. Abbas JM, Chik Z, Taha MR (2018) Modelling and assessment of a single pile subjected to lateral load. Studia Geotechnica et Mechanica 40(1):65–78
14. Achmus M, Thieken K (2010) On the behavior of piles in non-cohesive soil under combined horizontal and vertical loading. Acta Geotech 5(3):199–210
15. Kershaw KA, Luna R (2014) Full-scale field testing of micropiles in stiff clay subjected to combined axial and lateral loads. J Geotech Geoenviron Eng 140(1):255–261

The Influence of the Number and Height Steps of Terraced Model on Slope Stability Analysis

Niken Silmi Surjandari, Siti Nurlita Fitri, and Fenty Madani

Abstract Wonogiri is an area with a high volume of private, public, and logistics vehicles because it is located in the border connecting East Java, Central Java, and the Special Region of Yogyakarta. This area also has high susceptibility to landslide due to its topography. These landslides occur due to the weight of the slope and excess external traffic loads. Therefore, this research aims to determine the influence of the number and height steps of the terraced model on slope stability analysis in Wonogiri. Data were collected from 3 types of traffic loads, namely Own Weight (BS) + pavement + truck's load on two paths, BS + pavement + truck's load on one path, and BS + pavement. The road class was designed with an assumed average daily traffic (ADT) of 3000 pcu/day (local road) with a width of 3 m (Direktorat Jenderal Bina Marga in Tata Cara Perencanaan Geometrik Jalan. Antar Kota, No.038/T/BM/1997. Badan Penerbit Pekerjaan Umum, Jakarta, 1997 [1]). Furthermore, the slope stability analysis was determined using the limit equilibrium method. The result showed that the addition of steps increases the value of the safety factor (SF) significantly. The greater the steps of the terracing model, the higher the SF value of slope stability. The height variation with the same number of steps proves that it affects the safety factor value. Therefore, the more step height designs, the higher the steps cloud the ability to resist the traffic load.

Keywords Terracing · Steps · Safety factor · Slope

N. S. Surjandari (✉) · S. N. Fitri · F. Madani
Universitas Sebelas Maret, Surakarta, Indonesia
e-mail: nikensilmisurjandari@staff.uns.ac.id

S. N. Fitri
e-mail: sitinurlitafitri@staff.uns.ac.id

S. A. Kristiawan et al. (eds.), *Proceedings of the 5th International Conference on Rehabilitation and Maintenance in Civil Engineering*, Lecture Notes in Civil Engineering 225,
https://doi.org/10.1007/978-981-16-9348-9_9

1 Introduction

Landslide is one of the natural disasters that often occur in Indonesia due to internal as properties of a slope and external factors found outside, such as seismicity, rainfall, vegetation, rock/soil morphology, soil moisture level, seepage, and geological activity [2].

Therefore, this research determined the effect of landslides in various terracing models in Sendangmulyo Village, Tirtomoyo District, Wonogiri Regency and a continuation of the research carried out by Lugut Tri Pramudo et al. [3]. Numerous methods of slope reinforcement have been carried out in Indonesia, such as soil nailing, gabions, vegetation, geotextiles, grouting, and terracing. This study was developed using traffic loads with the most critical safety factor in previous studies used as a reference to improve the new model. Angles 30° and 60° were used to determine the slope before and after the rain. Finally, the model was simulated by two-dimensional slope modeling analysis.

2 Literature Review

Listyawan et al. analyzed the slope stability of the Gondang Dam Spillway in Karanganyar Regency using the Fellenius Method. The slope was divided into 25 slices with a landslide angle φ of 42°. The results of the analysis obtained a safety factor (SF) of 0.742, which indicated that the slope is unsafe hence reinforcement is needed. After terracing was made for slope improvement, the safety factor (SF) value was raised to 1.609 and was safe from landslides ($SF \geq 1.5$) [4].

Haryadi and Razali analyzed slope stability using terracing with Fellenius Method analysis in Geostudio Slope Program. The type 1 variation had 3 steps with the effect of pore water pressure, at a slope of 19° to obtain $SF = 1.608 > 1.5$ (safe slope). Type 2 with a slope angle of 29° obtained $SF = 1.601 > 1.5$ (safe slope), while type 3 with a slope angle of 29° obtained $SF = 1.539 > 1.5$ (safe slope). Slope terracing with 3 steps and without the effect of pore water pressure on types 1, 2, and 3 with slope angles of 19°, 29°and 29° obtained SF values of 2.833 > 1.5, 2.499 > 1.5, and 2.697 > 1.5 [5].

Terracing is soil and water management systems mechanically built to decrease the slope length by digging and filling the dirt. Terraces are useful to increase water penetration into the soil, decrease the amount of surface runoff, and reduce the possibility of water erosion on the soil [6].

According to [7], the highway is classified based on its function, class, and status. The classification of roads based on their function is divided into the following:

1. Arterial roads, such as public roads that serve major transportation with the characteristics of long-distance trips, high average speed, and the number of efficiently limited entrances.
2. Collector roads, such as public roads that serve collectors or dividers with the characteristics of moderate travel, moderate average speed, and a limited number of roads.
3. Local Roads, such as public roads that serve local transportation with the characteristics of short trips, low average speed, and an unlimited number of entrances. This road network is usually used for regional activities and to connect different groups.

The type of pavement model used is asphalt concrete, with a homogeneous mixture of aggregate and asphalt as a binder at a certain temperature, and is used to accept high traffic loads. Construction of Hot Mix Asphalt Pavements [8] stated that asphalt concrete is a mixture consisting of hard asphalt as a binder and coarse, fine, and filler aggregates by mixing and compaction under hot conditions at certain temperatures. Table 1 is the Guidelines for Planning for Flexible Pavement Thickness, which consists of a minimum thickness of the asphalt concrete surface layer and the aggregate base layer.

The materials used for asphalt concrete pavement need to comply with established regulations, such as a List of the weight of contents (γ). Load weight for dead load (kN/m^3) based on [1] for asphalt concrete is 22.0 kN/m^3 and for concrete is 22.0–25.0 kN/m^3.

Table 1 Minimum thickness of asphalt concrete surface and subbase course

Traffic (ESAL)	ACWC		Surface course		AC-Base		Base course	
	inci	cm	inci	cm	inci	cm	inci	cm
< 50,000*	1.0*	2.5	2	5	2	5	4	10
50,001–150,000	2.0	5.0	–	–	–	–	4	10
150,001–500,000	2.5	6.25	–	–	–	–	4	10
500,001–2,000,000	3.0	7.5	–	–	–	–	6	15
1,000,001–7,000,000	3.5	8.75	–	–	–	–	6	15
> 7,000,000	4.0	10.0	–	–	–	–	6	15

* or surface maintenance condition

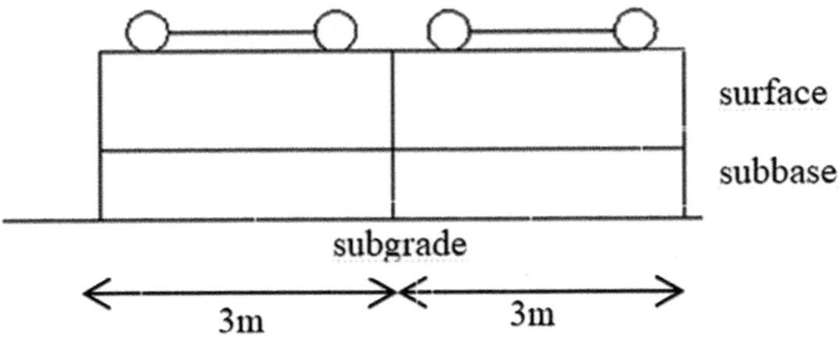

Fig. 1 Pavement model structure

3 Method

3.1 Data

The soil parameter data used [9, 10] was conducted in Sendangmulyo Village, Tirtomoyo District, Wonogiri Regency with its analysis consisting of γb, γsaturated, φ, and c. The road class model has an ADT assumption of 3000 pcu/day with a width of 3 m [1]. The road structure model is shown in Fig. 1.

3.2 Pavement Model

The pavement in this study is asphalt concrete (γbeton = 24 kN/m^3), while the lower foundation is mashed concrete with a bottom thickness foundation of 10 cm.

3.3 Vehicle

The design is assumed when two vehicles are parallel with the double-axle MST (2 axles) of 8 tons used to ensure the load for each wheel is 40 kN [1].

3.4 Traffic Load Model

This study simulated 3 types of traffic loads that refer to the calculations above.

A. Type 1
   The type 1 load used the road load plus the load of 2 trucks on both roads as shown in Fig. 2.

$$\begin{aligned}\text{Load type 1(2 truck + pavement)} &= q_{pavement} + 4(p\prime)\\ &= 2.95 + 4(172.7513)\\ &= 693.9552\ \text{kN/m}^2\end{aligned}$$

B. Type 2
Load type 2 is the same model as the previous type, with an additional truck-load in one path (Fig. 3).

$$\begin{aligned}\text{Load Type 2(1 truck + pavement)} &= q_{pavement} + 2(p\prime)\\ &= 2.95 + 2(172.7513)\\ &= 348.4526\ \text{kN/m}^2\end{aligned}$$

C. Type 3
Type 3 load uses road loads only as shown in Fig. 4.

$$\begin{aligned}\text{Type 3(Pavement)} &= q_{pavement}\\ &= 2.95\ \text{kN/m}^2\end{aligned}$$

Fig. 2 Type 1 traffic model

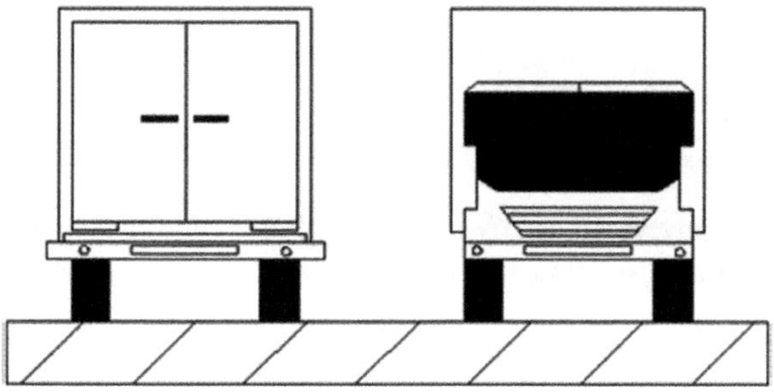

Fig. 3 Type 2 traffic model

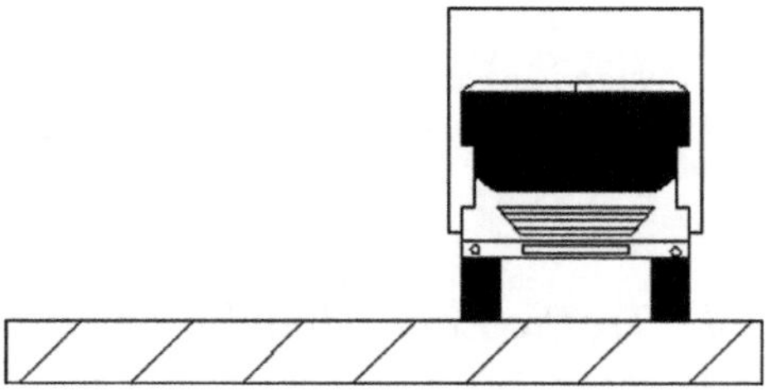

Fig. 4 Type 3 traffic model

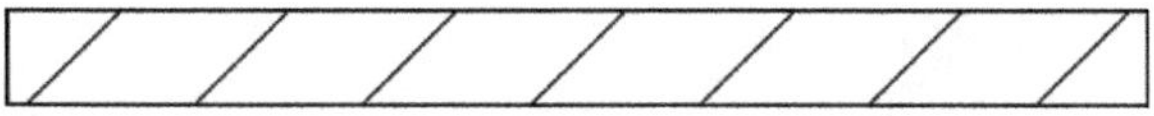

4 Result and Discussion

The results of the original 30° slope angle modeling output and the Type 1 traffic load (BS + pavement + truckload on 2 roads) before rainfall conditions are shown in Fig. 5.

Figure 6 shows that at 30° the slope terraces with the addition of Type 1 traffic loads before the conditions.

4.1 *SF Relationship with Existing Slopes, Load Type, and Season*

The slope stability analysis was carried out in 2 types of modeling conditions. The first was before and after rainfall, while the second was before and after terracing at 30 and 60 slope degrees. The types of loading due to traffic loads are as follows: own weight, type 1 load (BS + pavement load + truckload on two roads), type 2 (BS + pavement load + truck on one section, and type 3 (BS + pavement load) as shown in Fig. 7.

Based on the analysis in Fig. 7, there is an increase in traffic loads during the rainy season with a decrease in SF when the slope becomes steeper. The analysis carried out in rainy conditions shows that all safe numbers are below the safe value required by Bowles (SF = 1.0). Therefore, the vulnerability of landslides in the study area increases in accordance with the amount of rainfall.

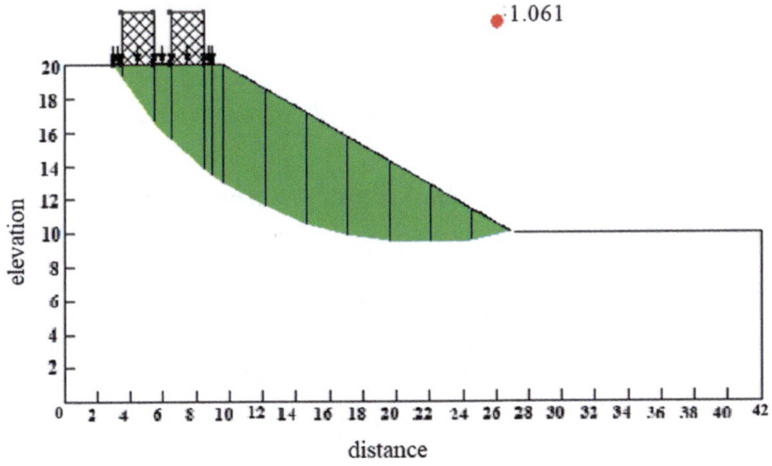

Fig. 5 Analysis of existing slope of 30° + traffic load Type 1 before rain

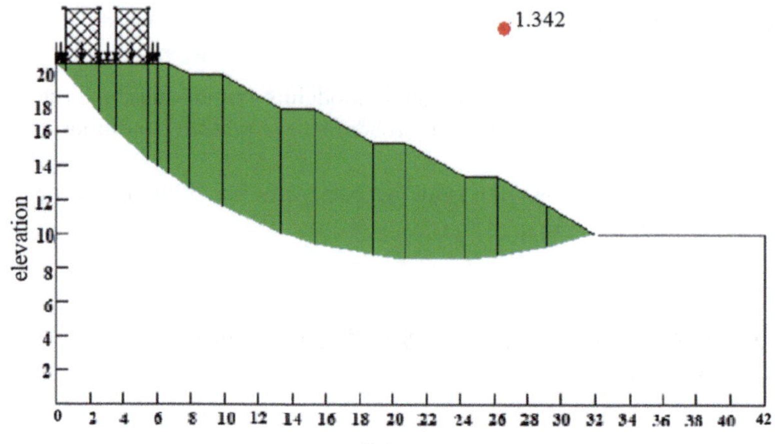

Fig. 6 Analysis of 30° slope with Type 1 traffic load terracing before rain

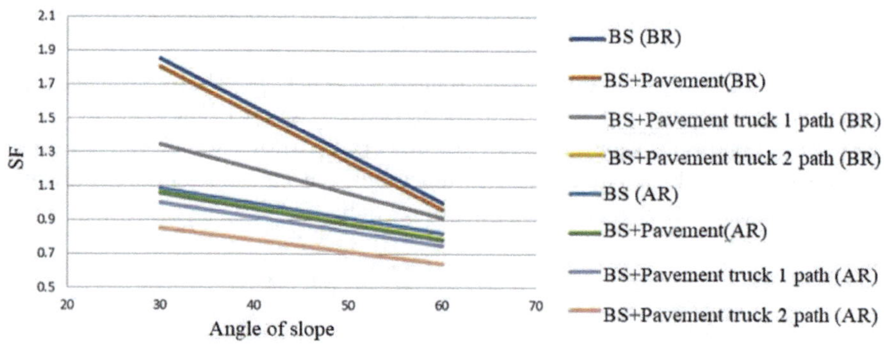

Fig. 7 Relationship between SF and slope under several loads and season conditions

4.2 SF Relationship Between SF and Number of Step with Variation Terraces

In this research, the terracing used 10 variations with 2, 3, 4, and 5 numbers of steps and varying trap heights. The relationship between the safety factor (SF) value and the number of steps in many variations are shown in Figs. 8 and 9.

Furthermore, the addition of steps affects the value of the safety factor (SF), as indicated by the trend line which always increases in each graph. The greater the number of steps in the slope, the increase in the soil masses on the slopes, therefore the vulnerability of landslides is also low. In rainy conditions, the number of steps affects the speed of rainwater flow through the slopes. The more steps in the terracing model, the slower the speed of rainwater flow, therefore it is not easily eroded by rainwater. Terracing is proven to increase the slope safety factor with traffic loads as evidenced by the increase in the Safety Factor with each increase in the number of steps.

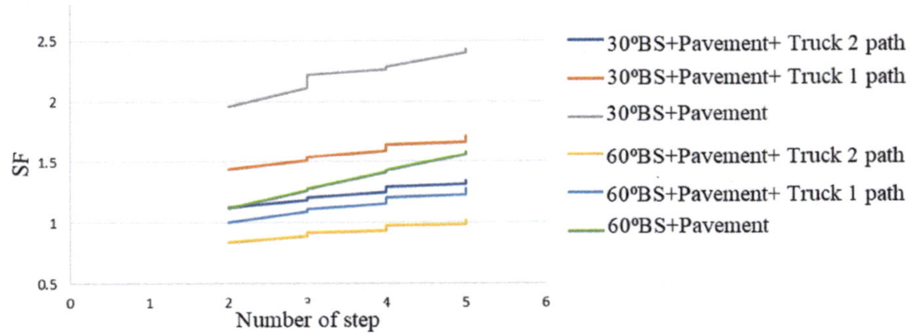

Fig. 8 The relationship between SF and the number of steps in several load conditions (before rainfall)

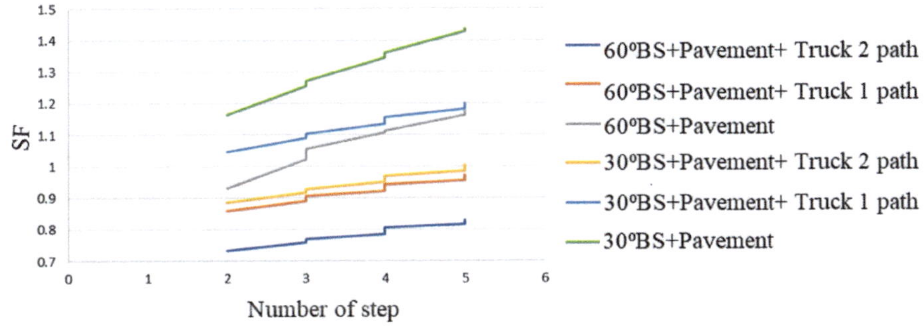

Fig. 9 The relationship between SF and the number of steps in several load conditions (after rainfall)

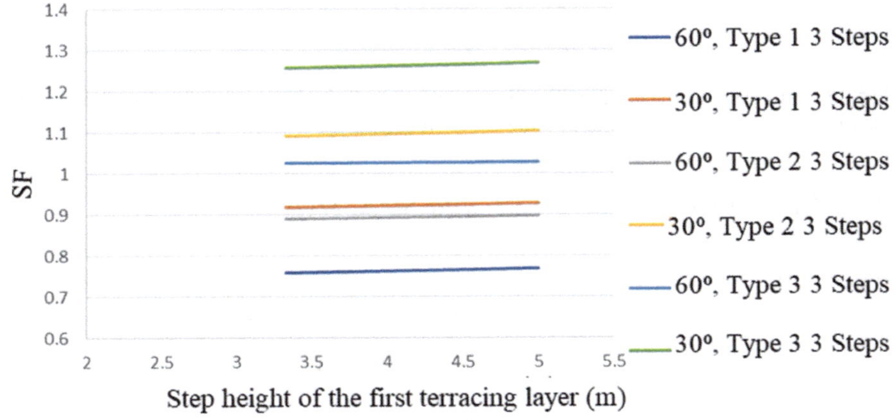

Fig. 10 The relationship between SF and the step height of the first terracing layer in the 3 steps, before the rain for slope angles of 30° and 60° (before rainfall)

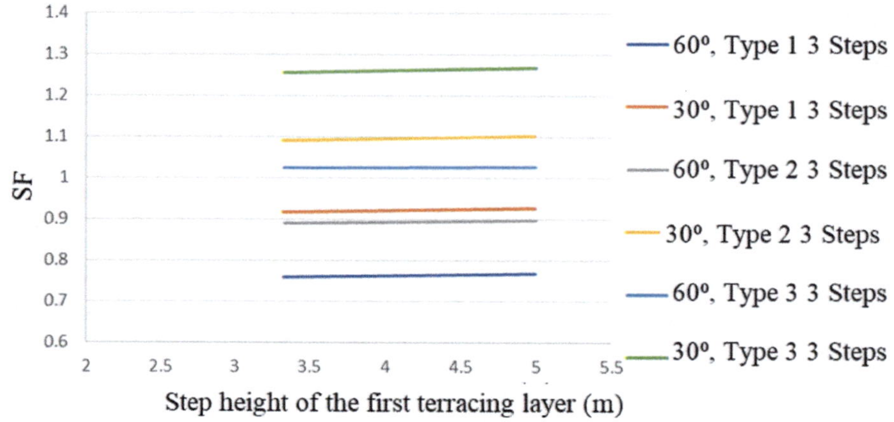

Fig. 11 The relationship between SF and the step height of the first terracing layer in the 3 steps, before the rain for slope angles of 30° and 60° (after rainfall)

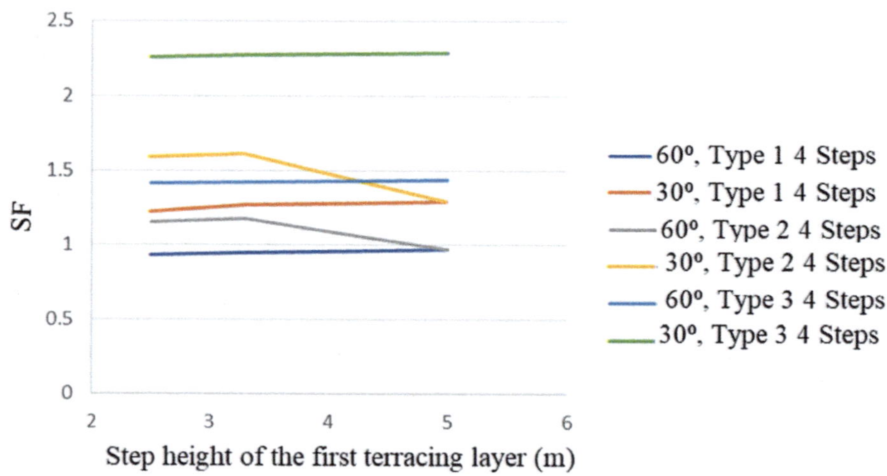

Fig. 12 The relationship between SF and the step height of the first terracing layer in the 4 steps, before the rain for slope angles of 30° and 60° (before rainfall)

4.3 Relationship Between SF and Steps Height

Figures 10, 11, 12, 13, 14 and 15 compare the heights of the first layer of terraces in slope angles 30 and 60 under several load types and seasons.

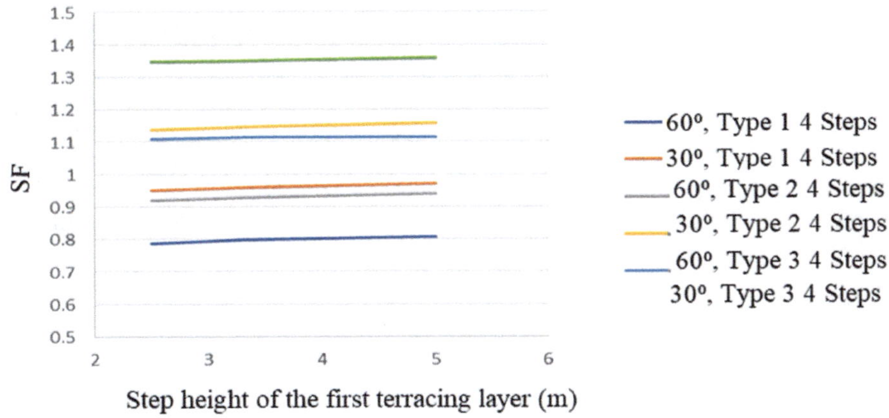

Fig. 13 The relationship between SF and the step height of the first terracing layer in the 4 steps, before the rain for slope angles of 30° and 60° (after rainfall)

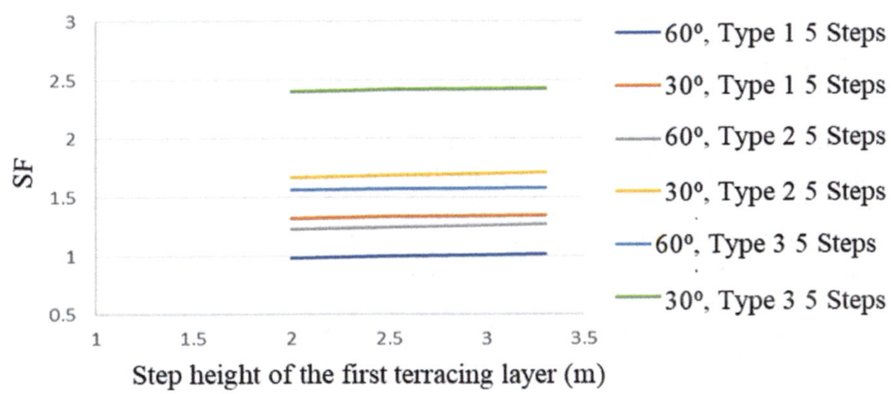

Fig. 14 The relationship between SF and the step height of the first terracing layer in the 5 steps, before the rain for slope angles 30° and 60° (before rainfall)

Figures 10, 11, 12, 13, 14 and 15, shows that the same variation of the step height significantly affects the safety factor (SF) value and the number of steps. The height of the first layer of the step is higher than the next trap with the ability to withstand vertical loads in this case traffic loads.

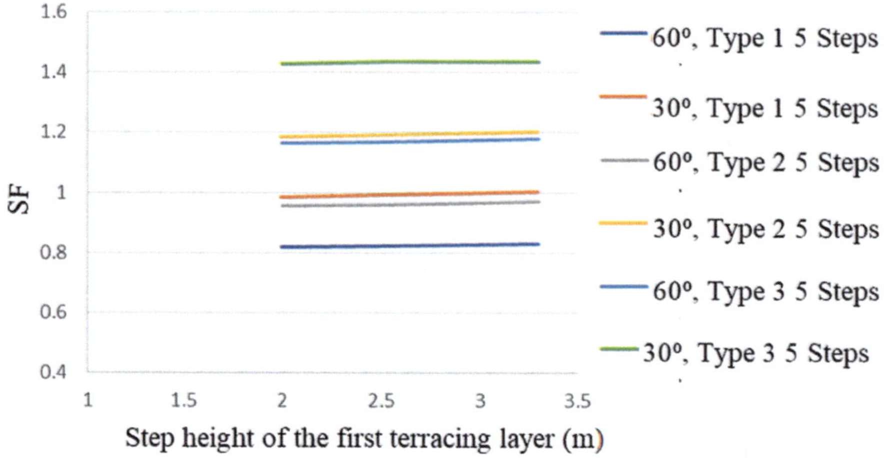

Fig. 15 The relationship between SF and the step height of the first terracing layer in the 5 steps, before the rain for slope angles of 30° and 60° (after rainfall)

5 Conclusion

The addition of steps significantly affects the value of the safety factor of slopes with a significant number unable to reached the Bowles's Safety Number (SF = 1.0). The greater the steps, the higher the slope safety factor value. Furthermore, the variation in height in the same number of steps proves that the height affects the safety factor value with the ability to withstand vertical loads in traffic loads.

Acknowledgements The authors are grateful to LPPM UNS for funding this research through Research Group grants.

References

1. Direktorat Jenderal Bina Marga (1997) Tata Cara Perencanaan Geometrik Jalan. Antar Kota, No.038/T/BM/1997. Badan Penerbit Pekerjaan Umum, Jakarta
2. dan Kesumadhama A (1991) Konstruksi Jalan di daerah Pegunungan tropis, Makalah Ikatan Ahli Geologi Indonesia, PIT ke-20, Desember 1991, pp 471–481
3. Lugut Tri Pramudo H, Djarwanti N, Surjandari NS (2016) Analisis stabilitas lereng dengan terasering di desa sendangmulyo, Tirtomoyo, Wonogiri, pp 470–475
4. Listyawan AB, Wiqoyah Q, Rahmawan RA (2019) 155–272, pp 155–162
5. Haryadi D, Mawardi M, Razali MR (2019) Analisis Lereng Terasering Dalam Upaya Penanggulangan Longsor Metode Fellenius Dengan Program Geostudio Slope. Inersia, Jurnal Teknik Sipil 10(2):53–60
6. Sukartaatmadja S (2004) Perencanaan dan Pelaksanaan TeknisBangunan Pencegah Erosi. Institut Pertanian Bogor (IPB), Bogor
7. Indonesian Law no. 22 of 2009 about traffic and road

8. Construction of Hot Mix Asphalt Pavements (1970) Manual ... The Asphalt Institute: thickness design—full depth asphalt pavement
9. Hawin (2014) Pengaruh Hujan 2 Harian Terhadap Stabilitas Lereng Di Das Tirtomoyo Wonogiri (Studi Kasus: Desa Sendang Mulyo, Tirtomoyo, Wonogiri). Universitas Sebelas Maret, Surakarta
10. Surjandari NS et al (2021) Slope stability analysis in various Terraces model (case study: Sendangmulyo, Tirtomulyo District, Wonogiri Regency). J Phys Conf Ser 1858:2021. https://doi.org/10.1088/1742-6596/1858/1/012005

Validation of TRMM Rainfall Data on Slope Stability in Karanganyar, Indonesia

David Raja Simare Mare, Rr Rintis Hadiani, and Raden Harya Dananjaya

Abstract The use of TRMM rain data is one source of solving the lack of rain data sources in Indonesia. TRMM satellite data has not been widely used, so the validation of the data needs attention. TRMM GSMap_NRT is one type of satellite data that can be used as a source of rain data. However, research on the validation of TRMM GSMap_NRT rain data has not been widely discussed. This study presents the validation of the GSMap_NRT TRMM satellite data against the rain gauge's regional rain data in the Karanganyar area, Central Java. The data validation was also carried out on soil stability at the site, considering that the research location is landslide-prone. This validation compares the safety factor (SF) value from calculating the two types of rain data (TRMM GSMap_NRT and rain gauge). Rain indicators for regional rain are Jatipuro, Tawangmangu, and Karangpandan. Rainfall processing used the Thiessen Polygon method. Infiltration processing used the SCS-CN method. The data validation used the r correlation equation approach and the calibration coefficient was obtained. The correlation coefficient and error values are 0.76 and 251.63 mm/month (20.97 mm/day). The TRMM rainfall data were then analyzed to find the value of slope stability by analyzing the annual maximum two-day rainfall. The results of this study are a map of the SF for 7 years (2014–2020). The conclusion of this study is the SF map for seven years does not show a significant difference. However, further research in this regard is highly recommended.

Keywords TRMM validation · TRMM GSMap_NRT · Rainfall · Landslide

D. R. S. Mare (✉) · R. R. Hadiani · R. H. Dananjaya
Department of Civil Engineering, Sebelas Maret University, Ir. Sutami 36A Street, Surakarta 57126, Indonesia

S. A. Kristiawan et al. (eds.), *Proceedings of the 5th International Conference on Rehabilitation and Maintenance in Civil Engineering*, Lecture Notes in Civil Engineering 225,
https://doi.org/10.1007/978-981-16-9348-9_10

1 Introduction

1.1 Preliminary

Disaster Information Data of Indonesia The National Disaster Management Agency (BNPB) of the Republic of Indonesia (2018) noted that 151 landslides occurred in Central Java Province. One of the mountainous areas that often occurs in landslides is the area around Mount Lawu, precisely in Karanganyar district, an area located in the arch of volcanic mountains [1].

Landslides can be caused by rainfall. Water falling down the slope will become a burden to the soil, affecting its stability. The need for rainfall data is essential in this case. The limited availability of rain gauges is a significant problem in Indonesia. Therefore, other data sources are needed to be processed in particular case studies; one of the data sources is the Tropical Rainfall Measuring Mission (TRMM) data. TRMM is a joint project between NASA, America, and the Japan Aerospace Exploration Agency. The TRMM theory was developed to find a rain plan in a watershed with a rain register station [2].

However, satellite data is considered less representative of the actual conditions in the field. Therefore, validation of rain data needs to be done. The point of view carried out in this study is to validate the TRMM data to be used for safety factor analysis.

2 Research Significant

2.1 TRMM and Rainfall Gauge Data

Research conducted by Syaifullah [3] shows that GSMap_NRT has a better correlation value to data on rainfall at breeding stations compared to data from 3B42RT. Therefore, this study will use TRMM JAXA GSMap_NRT rain data [3]. The rain gauges used came from three stations around the study area: Karangpandan, Tawangmangu, Jatipuro. Rainfall data was taken from 2014 to 2020 in the wet months (November–April).

2.2 Homogeneity

Test Data homogeneity test was carried out to ensure that the data used were classified as homogeneous. According to Buishand (1982), homogeneity is based on the number of partially adjusted or mean deviations [4].

Table 1 Critical value of *Q_RAPS* and *R_RAPS*

n	$Q/\sqrt{n}$			$R/\sqrt{n}$		
	CI = 90%	CI = 95%	CI = 99%	CI = 90%	CI = 95%	CI = 99%
10	1.05	1.14	1.29	1.21	1.28	1.38
20	1.10	1.22	1.42	1.34	1.43	1.60
30	1.12	1.24	1.46	1.40	1.50	1.70
40	1.13	1.24	1.50	1.42	1.53	1.74
50	1.14	1.27	1.52	1.44	1.55	1.78
100	1.17	1.29	1.55	1.50	1.62	1.86
∞	1.22	1.36	1.63	1.62	1.75	2.00

Source Harto [5]

Rescaled Adjusted Partial Sums (RAPS) is a method that determines a relatively homogeneous method by comparing the calculated value with the value of criticism by the table of values criticism. QRAPS and RRAPS data are classified as homogeneous if the calculated value is smaller than the critical value according to the Confidence Interval (CI). The critical value of QRAPS and RRAPS is provided in Table 1.

2.3 The Coefficient of Rain Regional Thiessen

The rainfall intensity is the depth of rainwater per unit time [6]. This study uses the Thiessen Polygon method to calculate rainfall intensity. The Thiessen method is also known as the weighted mean method. The area of influence is formed by drawing the axes perpendicular to the two nearest measuring posts. It is assumed that the variation of rainfall between stations is linear and that any post is considered to represent the nearest area [6].

2.4 Calculation of the Depth of Saturated Soil

The use of the SCS-CN method is considered possible to analyze the infiltration because this method also considers soil conditions in determining the Curve Number (CN). The basic equation for calculating the depth of adequate rain is formulated in the following equation [7]:

$$Pe = \frac{(P - 0.2S)^2}{P + 0.8S} \tag{1}$$

The maximum potential retention value has the following equation:

$$S = \frac{25{,}400}{CN} - 254 \tag{2}$$

CN or Curve Number is a function of watershed characteristics such as soil type, land use, cover crops, and soil management methods [8].

The CN value varies from 0 to 100. The higher the CN value, the more impermeable the land surface. An increase in the CN value indicates a decrease in the land's ability to store precipitation water. The impact of this is a reduced infiltration volume, thereby increasing the volume of surface runoff or increasing peak discharge [9].

Research conducted by Amalia [10] concluded that the CN value would affect the flood discharge. In his research in the Progo watershed, the CN value change in 2006 was 83.5–86.2 in 2020, making the measured hydrograph unit value 16.3%. The most dominant CN value change is caused by land-use assuming the soil type will be relatively constant [10].

2.5 *The Validity of TRMM Rainfall Data*

The analysis was conducted by comparing the annual maximum two-days. The analysis was performed by calculating the correlation coefficient value and root mean square error (rmse). Research conducted by Raharjo [2] revealed a close relationship between the use of TRMM data and local rain stations [2]. This research found that this value has a pattern that follows the actual rainfall even though the value tends to be below the estimate. The correlation coefficient (R) value is expressed by a coefficient that states a linear relationship between two variables. The equation is as follows:

$$R = \frac{n\sum xy - \sum x \sum y}{\sqrt{n\sum x^2 - \left(\sum x\right)^2}\sqrt{n\sum y^2 - \left(\sum y\right)^2}} \tag{3}$$

If the value (R) is approaching 1, the relationship between the two and the closer to the actual value. The error that occurs is defined as the difference between the two variables, the estimated value, and the observed value. Root mean square error (rmse) shows the degree of bias in the estimation model's estimation. The rmse equation can be seen as follows:

$$rmse = \sqrt{\frac{\sum_{i=1}^{n}(yi - y')^2}{n}} \tag{4}$$

If the rmse value between the two is getting smaller, the difference between the two is getting smaller, and then the data is getting more accurate.

2.6 *Slope Stability with the Method* **Infinite Slope**

Alternatively, soil stability can be shown by the value of the safety factor (SF). In Hardiyatmo Soil Mechanics (2007), Terzaghi divides slope landslides' causes consisting of internal and external influences. What is meant by external influence here is any factor that causes an increase in shear forces without a change in the soil's shear strength. The safety factor (SF) is defined as the value of the ratio between the holding force and the driving force [11].

$$SF = \frac{\tau}{\tau d} \tag{5}$$

Generally, the SF value is taken to be greater than or equal to 1.2–1.5. Table 2 is the value of the safety factor according to Bowles [12].

This study uses the infinite slope method. This method assumes the variable value of soil b has the same size in each part and the forces acting on each side. Figure 1 is shown to illustrate the model [13].

Based on the model above, the equation can be modified to be:

$$SF = \frac{c}{\gamma sat\ Zw\ Sin\alpha\ Cos\alpha} + \frac{\gamma' tan\varphi}{\gamma sat\ tan\alpha} \tag{6}$$

Table 2 Classification of the value of the safety factor

Value of the	Intensity or occurrence of landslides
SF < 1.07	Landslides are common/frequent (unstable slopes)
1.07 < SF < 1.25	Landslides have occurred (critical slopes)
SF > 1.25	Landslides rarely occur (slopes are relatively stable)

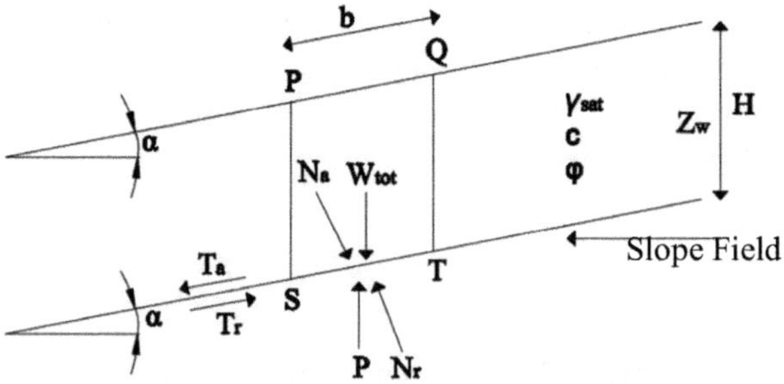

Fig. 1 Infinite slope method model

3 Research Methods

The intended research location is in the Karanganyar area. The taken soil data consists of four points locations spread over three districts: Matesih, Karangpandan, and Tawangmangu. The central point of the location, which is the reference for TRMM rain data collection, is at coordinates (– 7.657557, 111.081424). Soil sample data was taken at four points around the center point, and the samples were processed at the Soil Mechanics Laboratory, Sebelas Maret University, Surakarta.

Secondary data goes through a data similarity process first to determine its homogeneity. Rainfall data TRMM and rainfall gauge experienced homogeneity test with the RAPS method with a CI value of 95%. Regional rainfall analysis uses the Thiessen Polygon method with the Quantum GIS application's help to obtain regional coefficients. Rainfall data of TRMM is not analyzed for regional rainfall. Infiltration and saturated soil depth values for safety factor analysis using the SCS-CN method consider land use in the study area. The validation test is validating rainfall data and use that valid data to analyze the safety factor.

4 Analysis and Discussion

4.1 Soil Data Analysis

Results show that the soil is categorized as sandy soil. Based on the undisturbed sample analysis, the soil has an average cohesion value and an average shear angle of 0.073 kN/m^2 and 35.84°.

4.2 Rainfall Data

Analysis Homogeneity analysis shows that the data for rainfall gauge and TRMM rain are homogeneous. Thiessen Polygon method is used to get the area coefficient. The polygon coefficient has a value of 0.65; 0.25; 0.10 in Karangpandan Station, Tawangmangu Station, and Jatipuro Station.

Jatipuro Station is considered to have a too-small coefficient so that the coefficient of Jatipuro Station is combined with the nearest station, namely Tawangmangu Station. The latest coefficient data obtained are 0.65 and 0.35 at Karangpandan Station and Tawangmangu Station. Two consecutive daily rainfall analyses were carried out with the latest coefficient, and two daily rainfall data were obtained. The maximum annual two-day rainfall values are obtained as shown in Table 3.

Table 3 Annual maximum two-day rainfall value

Year	Rainfall gauge (mm)	TRMM rainfall (mm)
2014	68.85	93.64
2015	73.90	81.30
2016	90.35	72.80
2017	101.67	159.63
2018	96.46	52.10
2019	132.65	125.64
2020	128.65	52.30

4.3 Rainfall Station Data

The next analysis is to process the maximum consecutive two-day TRMM rain data based on the annual maximum two-day rainfall data. The recapitulation of station rainfall data and TRMM two-day maximum annual rainfall show in Table 3.

4.4 Rainfall Validation and Calibration

Validation analysis uses maximum monthly data from 2014 to 2020. The daily rainfall data had a value that tends to vary between satellite and actual data. Analysis of the two-days rainfall data shows the correlation coefficient 0.22 and error value 46.47 mm/day. The Analysis of total monthly rainfall data shows better value than two-days rainfall data. The correlation coefficient and error values are 0.76 and 251.63 mm/month (20.97 mm/day).

By eliminating data that is far apart from other data, data is obtained with a tendency to approach the linear line formed. The equation you get is $y = 1.0811x + 31.205$. With the equation above, the calibration is carried out on each annual maximum rainfall data so that the maximum annual two-day rainfall data is obtained as follows (Fig. 2; Table 4).

4.5 Safety Factor Analysis Based on Calibrated TRMM Rainfall Data

Slope stability analysis using a newly developed web-based application called UNSlide. The required parameter values are the cohesion value, internal shear angle, slope, saturated soil depth, and the value of the safety factor. Saturated soil depth depends on the curve number of the field. The curve number value if the field is 74.23.

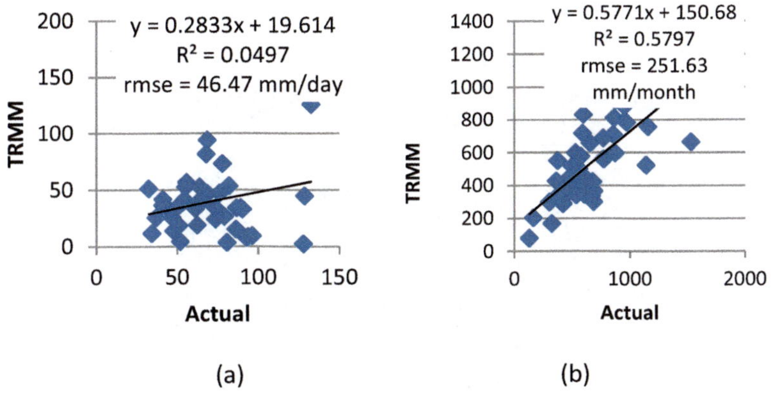

Fig. 2 **a** Maximum two-days rainfall data. **b** Monthly rainfall data

Table 4 Annual maximum two-day rainfall value after calibration

Year	Before calibration (mm)	After calibration (mm)
2014	93.64	132.44
2015	81.30	119.10
2016	72.80	109.91
2017	159.63	203.78
2018	52.10	87.53
2019	125.64	167.03
2020	52.30	87.75

The results show that the difference in the value of rain that occurs each year does not give much difference to slope stability. However, the greater the rain that occurs each year and occurs in a long period, the greater the potential for landslides to occur. Figures 3 and 4 show the 2017 and 2018 SF map, which are the map with the peak and lowest rainfall values.

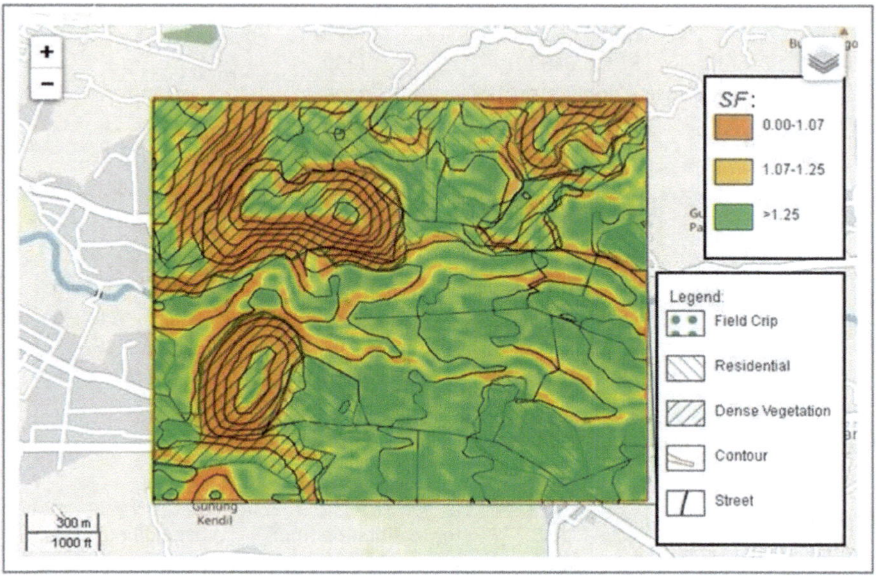

Fig. 3 Safety factor map in 2017

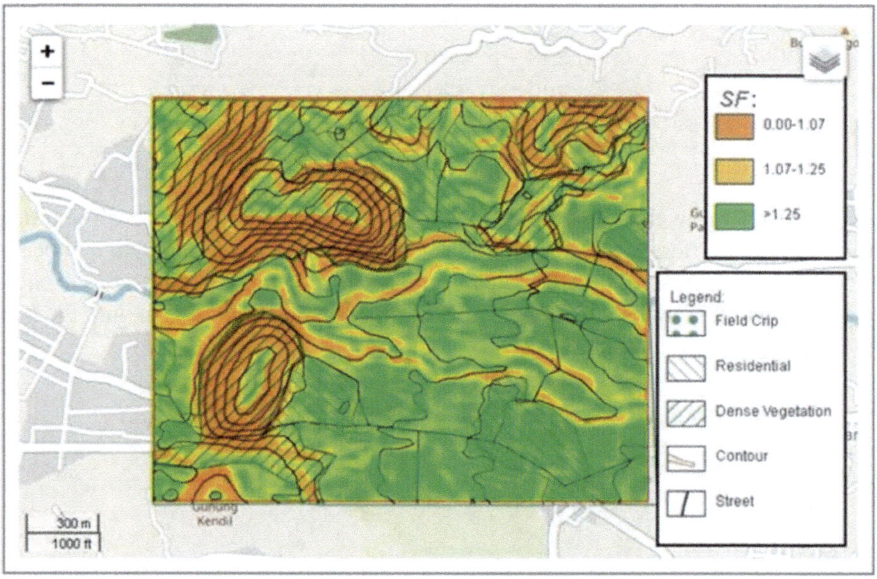

Fig. 4 Safety factor map in 2018

5 Conclusion

TRMM rain data can be used to overcome the problem of limited sources of rain data. If the study area does not have reliable rain data, TRMM data can be used. Validation based on the total monthly rainfall data has a high enough correlation coefficient and a minimal error than the validation based on two-days rainfall data. The author suggests validating with more comprehensive rainfall data and various methods for its utilization. It is expected that more comprehensive rainfall data will provide more reliable accuracy and validation. The use of various methods to validate the TRMM rain data is expected to provide a new approach to lead to the most reliable results in the future.

References

1. Karnawati D (2011) Landslide hazard and community-based risk reduction effort in Karanganyar and the surrounding area, Central Java, Indonesia. J Mt Sci 8(2):149–153
2. Raharjo SR, Suprapto M, Ikhsan C (2018) Rain design analysis using TRMM (tropical rainfall measuring mission) method (case study: Cisadane watershed). In: *AIP Conference proceedings*, vol 2014. American Institute of Physics Inc.
3. Syaifullah MD (2014) Validation of TRMM data with actual rainfall data in three watersheds in Indonesia. J Meteorol Geophys 15(2):109–118
4. Buishand TA (1982) Some method for testing the homogeneity of rainfall records. J Hydrol 58(1–2):11–27
5. Harto S (1993) Hydrological analysis. Gramedia Pustaka Utama, Jakarta
6. Suripin (2004) Sustainable urban drainage systems. Indonesia
7. Chow (1988) Applied hydrology. McGraw Hill
8. Triatmodjo B (2010) Applied hydrology. Indonesia
9. Dasanto dan Risyanto BD (2006) Evaluation of landuse change impact on run-off volume case study: Ciliwung Hulu watershed, West Java. J Agromet Indonesia (2):1–13
10. Amalia M (2011) Analysis of increasing curve number value on flood debit in Progo watershed. Infoteknik 12(2)
11. Hardiyatmo HC (2007) Soil mechanics 2, 4th edn. Indonesia
12. Bowles JE (1989) Physical properties of soil geology (soil mechanics). Erlangga, Jakarta
13. Dananjaya RH, Djarwanti N, Gunawan BJ (2021) Development of web-based mapping program for spatially distributed slope stability analysis. In: IOP Conference series: materials science and engineering, vol 1144. IOP Publishing

Rubberized Asphalt Pilot Road Trial in Kuwait

H. Al-Baghli, Z. S. Awadh, and S. E. Zoorob

Abstract Results of an investigation carried out at the Kuwait Institute for Scientific Research (KISR) are presented showing the advantages of incorporating Pellet-Pave (a rubber-bitumen additive in pelletized form) as partial substitute to conventional bitumen in rubberized hot mix asphalt (rubberized-HMA). The Pelletpave was composed of 18% tire derived crumb rubber blended with 60/70pen grade bitumen and processed by USA technology provider Phoenix Industries LLC into easy to handle granular/pellet form. Laboratory mix designs were followed up with full scale asphalt plant trials and the construction of a pilot trial (mill and replace an old wearing course) to test out the technology. The trial comprised 3 different test sections to compare the performance of rubberized-HMA with polymer modified Marshall asphalt and polymer modified SuperPave asphalt mix formulations. The test sections were laid on a low-medium trafficked road and one year following the trial, the performance of the rubberized-HMA appears to be on par with Polymer Modified HMAs.

Keywords Hot mix asphalt · Rubberized asphalt · PelletPave · Pelletized

1 Background and Justification

Conventional rubberized hot mix asphalt (rubberized-HMA) is a process of blending waste tire derived crumb rubber (minimum 15% by mass of bitumen) into hot bitumen

H. Al-Baghli (✉) · Z. S. Awadh · S. E. Zoorob
Kuwait Institute for Scientific Research, P.O. Box 24885, 13109 Safat, Kuwait
e-mail: hbaghli@kisr.edu.kw

Z. S. Awadh
e-mail: Zawadh@kisr.edu.kw

S. E. Zoorob
e-mail: szoorob@kisr.edu.kw

S. A. Kristiawan et al. (eds.), *Proceedings of the 5th International Conference on Rehabilitation and Maintenance in Civil Engineering*, Lecture Notes in Civil Engineering 225,
https://doi.org/10.1007/978-981-16-9348-9_11

(wet process) to produce a highly resilient rubber modified binder (rubberized-bitumen) for use in road paving. Rubberized-HMAs have an excellent record in relation to rutting, fatigue, reflective and thermal cracking and environmental durability [1].

In 2019, the State of Kuwait introduced new and improved specifications (QCS-2014) for asphalt materials that encompass performance specifications, and encourage the adoption of performance graded polymer modified binders (PMBs) and SuperPave mix design methodology [2].

In this investigation it was decided to formulate rubberized-HMA using locally available 60/70pen grade Kuwaiti bitumen and crumb rubber sourced from discarded tyres. The objective was to utilize local materials and introduce a viable competitor to high performance polymer modified HMAs.

Crumb rubber was hence blended with 60/70pen grade bitumen to produce a rubberized-bitumen blend. The optimum blend, based on a target PG grade of 76 °C, maximum elastic recovery and minimum non-recoverable creep performance (using Dynamic Shear Rheometer testing) was obtained at 18% rubber content by mass of bitumen. This blend (along with a small quantity of hydrated lime stiffener) was subsequently converted into a pelletized form (PelletPave) via a patented process by USA technology provider Phoenix Industries LLC [3].

One major advantage of PelletPave is that it can be introduced directly into the mixer at the asphalt plant during the HMA production process as a partial substitute to the bitumen content, furthermore, unlike polymer and wet mix rubber modified bitumens, PelletPave is designed to be stored and handled as granular matter [4, 5].

2 PelletPave Rubberized Asphalt Mix Design Methodology

In this investigation, several Gabbro aggregate trial blends were formulated by KISR team and investigated at a local asphalt producer's laboratory which has an ISO/IEC Standard 17025:2017 (General requirements for the competence of testing and calibration laboratories) valid certificate. The source of all the Gabbro aggregates (coarse aggregate and crushed sand) used in all the trials was the United Arab Emirates (Ahgar Al-Fujairah). An example of key properties of the 20 mm Gabbro coarse aggregates are shown in Table 1. Properties of all coarse and fine aggregates satisfy local QCS-2014 highway materials specifications.

The design gradations were all in compliance with Type III wearing course HMA and with nominal maximum aggregate size of 19 mm. For each of the trial asphalt mix blends, the following key properties were analyzed: Aggregate Effective (G_{se}) and Bulk (G_{sb}) specific gravities, Volume of absorbed binder (V_{ba}), and Voids in mineral aggregate (VMA). The blend that satisfied all asphalt mix volumetric requirements had the following particle size distribution (% passing): 25 mm (100%), 19 mm (99.7%), 12.5 mm (86%), 9.5 mm (77.5%), 4.75 mm (53.1%), 2.36 mm (35%), 1.18 mm (23.7%), 0.6 mm (15.4%), 0.3 mm (11.3%), 0.15 mm (8%), 0.075 mm

Table 1 Physical properties 20 mm crushed Gabbro test results

Test method	Results
Relative density; (oven dry), (SSD)	2.913, 2.926
Apparent relative density	2.952
Water absorption (%)	0.5
% of Flat and elongated particles (ASTM D4791-10 method B)	0
Aggregate crushing value (%) (BS 812-1990)	12
Los Angeles abrasion (%) (ASTM C535-16)	11
Soundness for coarse aggregate (%) (ASTM C88-18)	1
Particles with the specified no. of fracture faces (ASTM D5821-13)	2 or more100%
Flakiness index (%), (BS 818-105:1) and (BS 818-105:2)	8, 22

(6.5%). The filler was composed entirely of Gabbro powder from asphalt plant dust collectors (Baghouse fines).

In the following stage, PelletPave plus 60/70pen grade bitumen was introduced to the optimum aggregate blend to produce rubberized-HMAs. The objective at this stage was to determine the optimum combination of PelletPave and 60/70pen bitumen to satisfy key volumetric and compaction properties. A gyratory compactor was used to compact the HMAs which facilitated monitoring the reduction in volume of each asphalt specimen as the number of compaction gyrations were increased [6].

The laboratory compaction effort selected for this part of the investigation was suitable for HMAs expected to carry "Medium to Heavy Traffic (i.e., 3 to < 30 million standard axles (ESALs)". The Initial, Design and Maximum gyratory compaction effort was pre-set at $N_{ini} = 8$, $N_{des} = 100$ and $N_{max} = 160$ gyrations respectively. The mixing temperature was specified at 180 °C whilst the target compaction temperature range was 157–158 °C.

In compliance with the SuperPave based local HMA specifications (QCS-2014), the required gyratory compacted mix density (%G_{mm}) values and their equivalent (% voids) were as follows:

- $N_{ini} \leq 89\%$ (i.e., 11% voids at initial compaction), $N_{des} = 96\%$ (i.e., 4% voids for design), and $N_{max} \leq 98\%$ (i.e., 2% voids at refusal density).
- At N_{des}; Voids in Mineral Aggregate (VMA) $\geq$ 13%, and Voids Filled with Bitumen (VFB) = 65–75%.

Figure 1 shows gyratory mix densification curves of 3 rubberized-HMAs, all containing 3% PelletPave by mass of total mix. Each curve represents the average results obtained from 2 gyratory compacted specimens. The 60/70pen bitumen content was gradually increased from 1.75 to 2.5% in order to assess impact on mix volumetrics.

Close observation during laboratory mixing showed that for all 3 mixes, the Pellet-Pave did not disperse adequately and uniformly throughout the asphalt mix, even after

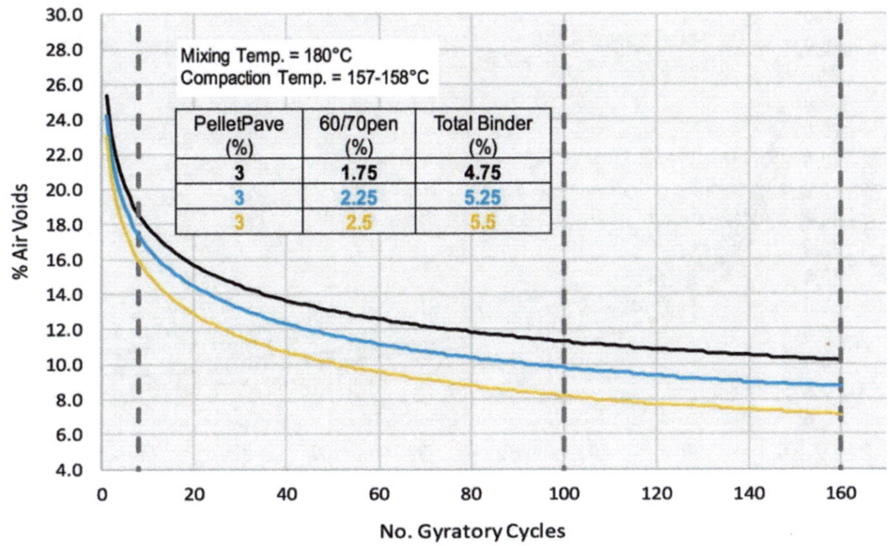

Fig. 1 Gyratory compaction profiles of 3% PelletPave content mixes at various 60/70pen contents. Each densification line shows the average result from 2 gyratory specimens

several minutes of continuous high temperature mixing. This was a cause for concern, as the probability of having a non-uniform mix on the road was unacceptable.

More importantly, none of the 3% PelletPave blends shown in Fig. 1 had acceptable compaction profiles (voids too high) and they all failed to satisfy the QCS-2014 volumetric requirements at N_{des} and N_{max}. It was therefore decided to produce alternative blends at a reduced PelletPave content. Figure 2 shows the gyratory compaction profiles of 2.5% PelletPave mixes at varying 60/70pen content ranging from 3.25 to 4.85% by mass of total mix.

Based on the data shown in Fig. 2, the rubberized asphalt mix with 2.5% PelletPave plus 4.55% 60/70pen bitumen was found to satisfy voids requirements at N_{des} and N_{max}. This mix was reproduced and was shown to result in volumetric properties that were highly repeatable.

Figure 3 presents a comparison of the average density values at N_{ini}, N_{des} and N_{max} for all rubberized asphalt mix trials as a percentage of the maximum theoretical density values (%G_{mm}) for each formulation. As a reminder, the QCS-2014 requirement are as follows; $N_{ini \leq}$ 89%, $N_{des =}$ 96%, $N_{max} \leq 98\%$, and mix (2.5% PelletPave + 4.55% 60/70pen) was found to possess optimum volumetric properties.

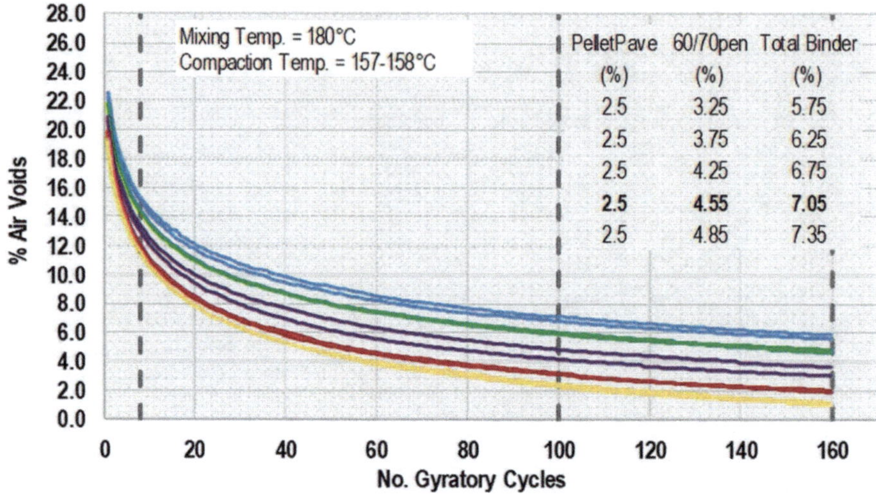

Fig. 2 Gyratory compaction profiles of 2.5% PelletPave content mixes at various 60/70pen contents. Each densification line shows the average result from 2 gyratory specimens

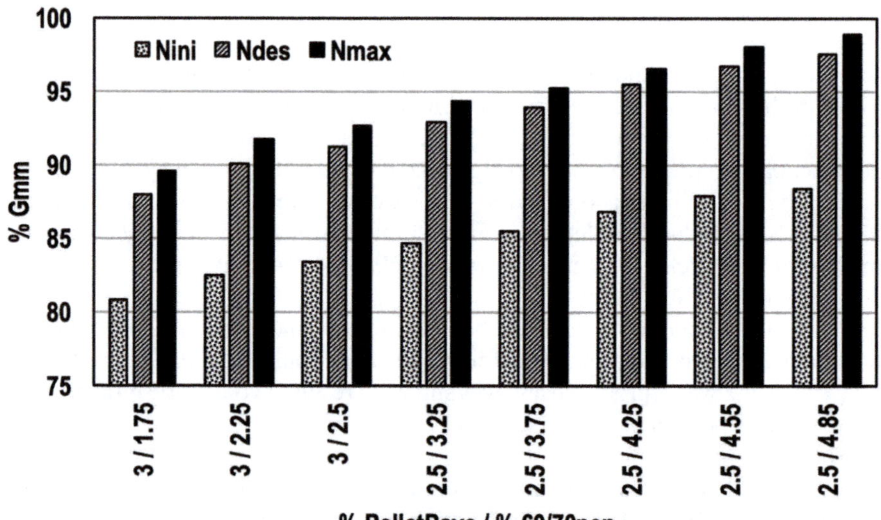

Fig. 3 %Gmm versus rubberized asphalt mix composition

3 Discussion on the Performance of Rubberized Asphalt Mix

It must be noted that the total binder content (i.e., 2.5% PelletPave + 4.55% 60/70pen bitumen) for the optimized rubberized asphalt mix was equal to 7.05%. To ensure adequate volumetric properties (i.e., acceptable voids, VMA, VFB), typical non-rubber modified continuously graded mixes have optimum binder contents in the range of 5–6%, thus the rubberized-HMA designed in this investigation can be classified as a binder rich mix. Binder rich mixes are more durable to environmental exposure (i.e., better resistance to water stripping and oxidation) due to the thicker binder layer coating the aggregates, on the other hand, a thicker binder layer has the potential to reduce the mix resistance to externally applied shear stresses (i.e., reduced rutting resistance).

In order to assess the mix performance to combined water resistance and rutting, Immersion wheel tracking tests at 60 °C (20,000 cycles) in compliance with (AASHTO T324) were carried out on the optimized rubberized-HMA described above. Figure 4 shows the result on the (2.5% PelletPave + 4.55% 60/70pen) as shown by the black line. The pass/fail criteria in QCS-2014 is a maximum of 12 mm rut depth at 20,000 cycles and the results (see black line) indicate that unfortunately the mix is susceptible to high temperature rutting (though visual inspection of the tracked specimens did not reveal any binder stripping which was a very positive indicator).

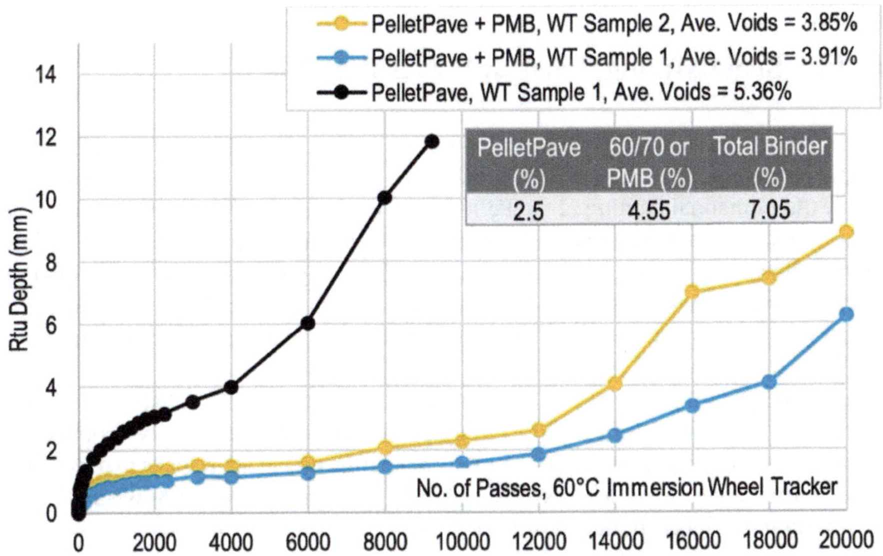

Fig. 4 Immersion wheel tracking test results on optimized 2.5% PelletPave mix plus 4.55% 60/70pen bitumen (black line) in comparison to 2 runs of 2.5% PelletPave plus 4.55% PMB mix (yellow and blue lines)

Thus far it was possible to conclude that the optimized rubberized asphalt mix (2.5% PelletPave + 4.55% 60/70pen) would not be ideal for medium to heavy volume trafficked roads (> 3 million standard axles). On the other hand, for low intensity trafficked roads which the mix was primarily designed for, the mix is expected to perform to a very high standard with no stripping or fretting or raveling which are primary causes for concern with local asphalt mixes utilizing Gabbro aggregates.

An additional variation on the optimized mix (2.5% PelletPave + 4.55% 60/70pen) was investigated by replacing the 60/70pen bitumen component by the same mass of polymer modified bitumen (PMB). The objective was to assess the capability of PelletPave mixes to be designed for heavily trafficked roads. The polymer used was Elvaloy and the performance grade of the PMB was PG 76H-10. The mix designation thus became (2.5% PelletPave + 4.55% PMB).

Figure 4 shows wheel tracking results of the (2.5% PelletPave + 4.55% PMB) mix (colored lines). Compared to the (2.5% PelletPave + 4.55% 60/70pen) (black line), a significant improvement in rutting performance was observed. The PMB modified rubberized-HRA readily satisfied QCS-2014 SuperPave rutting and stripping requirements.

The resistance of the optimized rubberized-HMA was also tested using Tensile Strength Ratio (TSR) test (AASHTO T283). Six nearly identical gyratory specimens were produced, 3 were tested in a dry state whilst the other 3 were conditioned in water at 60 °C for 24 h and all specimens tested for tensile strength at 25 °C. For SuperPave mix designs, the minimum TSR must be 80% which was readily satisfied by the (2.5% PelletPave content + 4.55% 60/70pen bitumen) rubberized-HMA formulation.

4 Full Scale Mixing Trials at the Asphalt Plant

Preparation work for a full-scale asphalt plant production run were initiated with careful batching of PelletPave (20 kg batches) into polyethylene quick melt bags to enable direct manual feeding of the PelletPave into the mixer during production. The plant operators were faced with the problem of PelletPave granules adhering into clumps (during material stacking and storage under high ambient temperatures). A team of workers spent an entire week manually sieving out the clumps and retaining the correct sized granules in order to weigh the material into the required 20 kg batches.

Eventually, a mixing trial was carried out at the contractor's Asphalt Plant. In the first step, 2 batches of dry graded aggregates (no added bitumen) were produced to ensure that the asphalt plant drying drum were heating up the aggregates to the correct target temperature range (approx. 190 °C). The aggregate gradation was identical to the optimum designed in the laboratory designed as described earlier.

In the second step, 2.5% PelletPave was manually introduced into the mixer in 20 kg pre-batched bags and mixed with the graded hot aggregates in the temperature range between 180 and 190 °C. Seven trial batches were produced with different mixing times, namely 40, 50, 60 and 70 s. Each completed (PelletPave + hot aggregate) batch was dumped on the ground for inspection. The temperature of each trial batch was measured and close visual inspection carried on each batch out to identify the presence of PelletPave semi-molten dark clumps. Clumping is an indicator that the particular combination of mixing time and/or temperature was insufficient to evenly and uniformly disperse the PelletPave throughout the mix. This process was repeated until everyone were satisfied with the outcome of the "semi-dry" PelletPave-aggregate mix.

The final optimum Plant mixing temperature was achieved when batches were maintained at 187–190 °C. The hot graded aggregate and 2.5% PelletPave were initially dry mixed for 75 s, this stage was followed by introducing 4.55% 60/70pen bitumen followed by 30 s wet mixing, resulting in a laying temperature of around 177 °C. This protocol ensured a well dispersed PelletPave fully coated mix.

A trial strip (25 m × 3.5 m) of rubberized-HMA was next laid and compacted adjacent to the asphalt plant. No problems were encountered laying the material and no issues with binder adherence to the steel compactors (which were lightly lubricated with vegetable oil). Visual inspection of cores taken from the trial show that the rubberized asphalt layer to be very well compacted with no visible mix segregation or large voids.

5 Pilot Road Trial

In the following phase, a full-scale pilot road trial was conducted in order to assess the performance of PelletPave rubberized-HMA during full scale production, transportation, paving and compaction operations. The secondary objective was to compare the long-term performance of the PelletPave mix with other high-performance polymer modified HMAs. An old internal road carrying medium–low volume traffic was selected for this trial. The aged wearing course on a stretch 450 m in length was milled to a depth of 5 cm, the milled sections was then divided into 3 sections and these were subsequently re-laid with 3 new HMA wearing course formulations as follows:

Section 1 (100 m in length): Rubberized-HMA (2.5% PelletPave + 4.55% 60/70pen),
Section 2 (140 m in length): SuperPave HMA (PG 76H-10 PMB binder),
Section 3 (210 m in length): Conventional Marshall HMA (PG 77H-10 PMB binder).

Details of the final gradations and binder contents of the Rubberized-HMA, Marshall and SuperPave mixes as laid during the KISR Pilot Road trial is shown in Fig. 5.

The 60/70pen bitumen used in the rubberized-HMA mixes had the following properties: Density at 15 °C (1.0508 kg/L), Cleveland open cup flash point (318 °C), Loss on heating (0.05% by mass), Ductility at 25 °C (> 150 cm), Penetration at

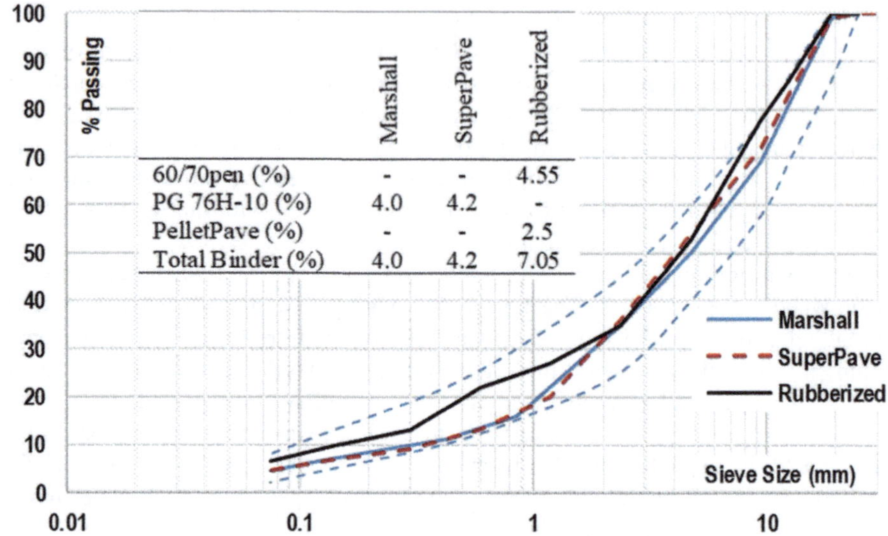

Fig. 5 Gradations of Marshall, SuperPave and rubberized asphalt mixes laid during the KISR pilot trial. The dashed lines are the Kuwait QCS-2014 upper and lower limit specifications

25 °C (65 dmm), Softening point (53 °C), Solubility in trichloroethylene (99.85% by mass).

Rheologically the PMB produced was equivalent to a PG grade 76H-10 and primary test results obtained from the unaged PMB, which were all within QCS-2014 limits are as follows: Density at 15 °C (1.037 kg/L), Cleveland open cup flash point (317 °C), Viscosity at 135 °C (2.17 Pa s), G*/sinδ @ 76 °C and 10 rad/s (1.98 kPa), Separation test (ASTM D7173-20) difference in G* @ 76 °C between top and bottom of oven conditioned binder specimen (2.8%).

For the SuperPave and Modified Marshall mixes, the same PMB binder grade was utilized for both mix types. The polymer type was Elvaloy 5160 (Dupont Elvaloy RET) at a concentration of 1.7%. At the asphalt plant, blending of the polymer with bitumen was carried out using a Marini PMB High Shear Mill unit using 2 passes through the mill at 5000 rpm followed by 10 min in an agitation tank. From the agitation tank each batch of PMB was next fed into the storage tanks and kept under additional circulation for a minimum of 35 min. The entire blending/agitation/storage process was maintained at 185 °C. For each working day, the PMB was kept at high temperature in storage tanks (with constant circulation) for up to 6 h.

Tables 2 and 3 show the key volumetric and mechanical properties of the SuperPave and Modified Marshall wearing course mixes, both designed at optimum binder contents using the same Gabbro aggregates and PMB. It is interesting to note that the Superpave mix designed at 4% voids is 0.2% richer in binder content compared to the Marshall mix at 6.8% voids, despite both mixes having very similar aggregate gradations (see Fig. 5). Compared to the SuperPave mix, the VMA result was higher

Table 2 SuperPave mix design summary, Type III wearing course

Test	Design value	QCS-2014
$N_{initial}$, N_{design}, N_{max}	8, 100, 160	Design ESALs 10 to < 30 msa
Design binder content (%)	4.2	–
SG of combined aggregate (G_{sb})	2.847	–
Effective SG of combined aggregate (G_{se})	2.861	–
Effective binder content (P_{be})	4.0	Report
Max. SG of mixture (G_{mm})	2.664	–
% G_{mm} @ N_{ini}	87.7	≤ 89.0
% G_{mm} @ N_{des}	96.3	96.0
VMA @ N_{des} (%)	13.6	≥ 13.0
VFB @ N_{des} (%)	73	65–75
Test	Design value	QCS-2014
% G_{mm} @ N_{max}	96.8	≤ 98.0
Dust proportion based on effective b.c. (%)	1.1	0.6–1.2
Indirect tensile strength dry set (IDT) (kPa)	828	Report
Tensile strength ratio (TSR) (%)	87	≥ 80
Immersion wheel tracking rut depth (mm)	2.91	12.5

by 1.8% for the Marshall mix, whilst the VFB was lower by 17%. Hence, the SuperPave mix design skeleton was slightly denser, whilst the interstitial spaces within the aggregate skeleton were significantly more saturated with binder, approaching the upper limit of specifications.

6 Results of Core and Loose Mix Testing

For each mix type, hot loose mix asphalt samples were collected from behind the paver for specific gravity determinations and following the compaction process, cores were extracted from the 3 trial sections and the bulk density was measured and air voids calculated. The range of measured compacted voids for the various mix types were as follows: PMB Marshall Mix (4.4–5.9% voids), PMB SuperPave Mix

Table 3 Marshall PMB Type III asphalt mix design, key mix properties

Item	o.b.c.	QCS-2014 limits
Design bitumen content (% by mass of total mix)	4.0	3.4–4.4
Effective binder content (P_{be}) (%)	3.5	–
Air voids (V_a) (%)	6.8	5.0–8.0
Voids in mineral aggregate (VMA) (%)	15.4	14.0 min
Voids filled with bitumen (VFB) (%)	56	50–75
Marshall stability (kN)	19.1	11.5 min
Marshall flow (mm)	2.71	2–4
Air voids after 400 BPF @ o.b.c. (%)	4.5	4.0 min
Retained stability (%)	87	75 min
Tensile Strength ratio (TSR) at 25 °C, at o.b.c.	76	Reported
(Filler/Binder) ratio	1.12	0.75–1.35
Dust proportion (filler/effective binder content)	1.26	–
Theoretical maximum specific gravity (G_{mm})	2.693	–
Compacted mixture bulk specific gravity (G_{mb})	2.511	–
Immersion wheel tracking rut depth (mm)	5.48	12.5

(6.1–7.3% voids), PelletPave Rubberized Asphalt (7.3–7.6% voids). All results were within acceptable specification values for field cores.

Loose hot mix samples obtained from the various trial sections were laboratory compacted. For the PMB Marshall mix compacted at (75 blows Marshall hammer), the stability obtained was 21.7 kN. For the PelletPave rubberized asphalt mix (75 blows), the stability was 13.5 kN. The rubberized mix was also compacted to refusal (400 blows Marshall hammer) to assess the effect of densification on the mix mechanical performance, and the stability measured was 22.2 kN. The stability of the rubberized asphalt mix increased significantly with increased compaction effort and hence the mix is expected to react positively to extra heavy compaction during laying operations.

The 3 trial sections have been visually monitored for over 1 year and partly as a result of the low-medium traffic levels, no visible signs of distress were observed. Falling weight deflectometer (FWD) testing was also carried out after 1 year of trafficking and no significant difference in deflection values were observed between the 3 sections. Thus far, the performance of the 3 trial sections have been indistinguishable.

7 Conclusions

An investigation was carried out to evaluate the advantages of incorporating PelletPave (a pelletized rubber-bitumen additive) as partial substitute to conventional bitumen in the formulation of rubberized hot mix asphalt (rubberized-HMA). A formulation composed of 2.5% PelletPave plus 4.55% of 60/70pen bitumen satisfied all volumetric requirements according to Superpave based specifications. The binder rich mix was shown to have excellent resistance to water stripping as evidenced by both immersion wheel tracking and Tensile Strength Ratio tests. On the other hand, rutting resistance was compromised making the mix more suitable for up to medium traffic levels.

Test results on loose mix samples and cores obtained from a road trial showed that volumetric properties were all within specifications. Marshall stability of the rubberized-HMA was sensitive to compaction level and hence the mix is expected to react positively to heavy compaction during laying operations.

Following one year of visual monitoring, no visible signs of distress were observed as a result of the low-medium traffic levels on this pilot road and falling weight deflectometer testing revealed comparable deflection values between the rubberized-HMA section and two other polymer modified HMA sections laid as part of the same trial.

References

1. Rubber Asphalt Foundation (RAF). http://www.ra-foundation.org/about-raf/. Accessed 29 Mar 2021
2. Qatar Construction Specifications (QCS-2014). Section 06: roadworks, Part 05; Asphalt works. Fully adopted by Ministry of Public Works, State of Kuwait
3. Amirkhanian S, Sockwell K (2012) Development of polymerized asphalt rubber pelleted binder for HMA mixtures. In: Asphalt rubber conference 2012, Munich. http://www.ra-foundation.org/library/
4. Zoorob SE, Al-Bahar SK, Al-Otaibi SF (2018) Design & optimization of a rubber-bitumen blend in preparation for a rubberized-asphalt road trial in the State of Kuwait. In: 4th International conference on rehabilitation & maintenance of civil engineering, Solo, Indonesia, 11–12 July 2018. MATEC Web of conferences, vol 195
5. Zoorob S, Al-Bahar S (2018) Preliminary rheological characterization of tyre derived crumb rubber blended with Kuwaiti bitumen. In: 4th International conference on civil, offshore & environmental engineering (ICCOEE-2018), 13–15 Aug 2018, Malaysia. MATEC Web of conference, vol 203
6. Al-Baghli H (2020) Characterization and design of a crumb rubber modified asphalt mix formulation. World Acad Sci Eng Technol Int J Civ Environ Eng 14(2)

Application of the Updated PSHA on the Stability Analysis of the Meninting Diversion-Spillway Tunnel in Lombok Island—Indonesia

Didi S. Agustawijaya, Ria R. Marlaningtyas, Suryawan Murtiadi, Mudji Wahyudi, Muhajirah, Hartana, and Ausa R. Agustawijaya

Abstract The Meninting diversion-spillway tunnel is a part of the Meninting dam project located in West Lombok District. The construction commenced in 2017; then, a series of severe Lombok earthquakes halted the construction of the tunnel for a while in 2018, where some landslides occurred around the site. This possibly caused rock masses around the tunnel to be sheared off during the earthquake events. Updated probability seismic hazard analysis shows increased seismic parameters, including peak ground, short spectrum and long spectrum accelerations, which then resulted in increased stresses around the tunnel. Thus, the stability of the tunnel decreased in terms of its factors of safety. Although, the overall factor of safety was about 2.5, the tunnel was still stable in current conditions; the tunnel certainly requires more stability improvements to face possible similar severe earthquakes in future.

Keywords Lombok earthquake · Meninting tunnel · Updated PSHA · Stress around tunnel · Factor of safety · Stability improvement

---

D. S. Agustawijaya (✉) · R. R. Marlaningtyas · S. Murtiadi · M. Wahyudi · Muhajirah · Hartana
Post-Graduate Study Program, Department of Civil Engineering, University of Mataram, Mataram, Indonesia
e-mail: didiagustawijaya@unram.ac.id

S. Murtiadi
e-mail: s.murtiadi@unram.ac.id

Muhajirah
e-mail: muhajirah@unram.ac.id

Hartana
e-mail: hartana@unram.ac.id

D. S. Agustawijaya
Geo-Engineering Research Group, Department of Civil Engineering, University of Mataram, Mataram, Indonesia

A. R. Agustawijaya
Department of Geomatics Engineering, ITS, Surabaya, Indonesia

S. A. Kristiawan et al. (eds.), *Proceedings of the 5th International Conference on Rehabilitation and Maintenance in Civil Engineering*, Lecture Notes in Civil Engineering 225,
https://doi.org/10.1007/978-981-16-9348-9_12

1 Introduction

The Mininting dam project is located in Dasan Gria, West Lombok District, which is only at 50 km distance from the epicenter of the 5th August 2018 earthquake. During a series of shallow earthquakes from July to September 2018, the seismic events have ruined thousands of houses and buildings in Lombok Island [1–3]. These current events occurred at depths of less than 30 km, which were triggered by the Flores Fault located at the north part of Lombok Island [4]. The intensity of the earthquake event was from MMI III to IX, and was MMI VII at the project site [5] (Fig. 1), which caused some collapsed area around the site [2].

Agustawijaya et al. [6] have updated probabilistic seismic hazard analysis (PSHA) for Nusa Tenggara Region based on earthquake data of 1922–2018, which resulted in higher peak ground acceleration, spectral accelerations SS and S1, compared to those of the SNI 1726:2019 [7]; whereas, the Meninting tunnel was possibly designed based on the older standards for calculating its design earthquake forces [8]. Therefore, this paper recalculated the seismic design of the Meninting diversion-spillway tunnel according to the current PSHA.

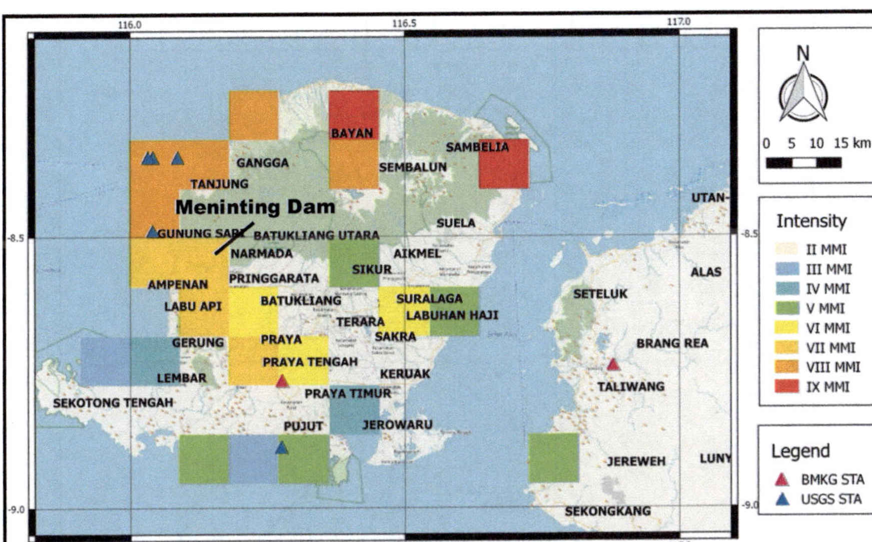

Fig. 1 The intensity of the earthquake event in 5th August 2018 around Lombok Island, the location of Meninting dam included into MMI VII [5]

2 Tunnel Design

The Meninting dam located at Dasan Gria, where the three rock formations: Kalipalung, Kalibabak, and Lekopiko covered the area [9]. These sedimentary rock materials contained volcanic breccias, lava and poor compacted volcanic rock materials of agglomerates and colluviums. Most rock materials were from erupted rock materials of the Rinjani volcano [10].

The Meninting dam was designed to have two tunnels: intake, and diversion-spillway. The diversion-spillway tunnel was designed to be of two types: Type 1 diversion tunnel, and Type 2 spillway tunnel. The type 1 was connected to the type 2 with a connecting shaft to become a diversion-spillway tunnel.

As be seen Fig. 2, the particular shoe-shaped type 2 was analyzed. The tunnel had a dimension of 9.40 × 9.40 × 252.5 m. The elevation of the tunnel base would be at + 147.8 m above the MSL, while the top surface of the tunnel would be at + 211.9 m above the MSL. The tunnel was planned to have fully support systems, including wiremesh with a 0.15 m thickness of shotcrete, steel H-beams installed with a spacing

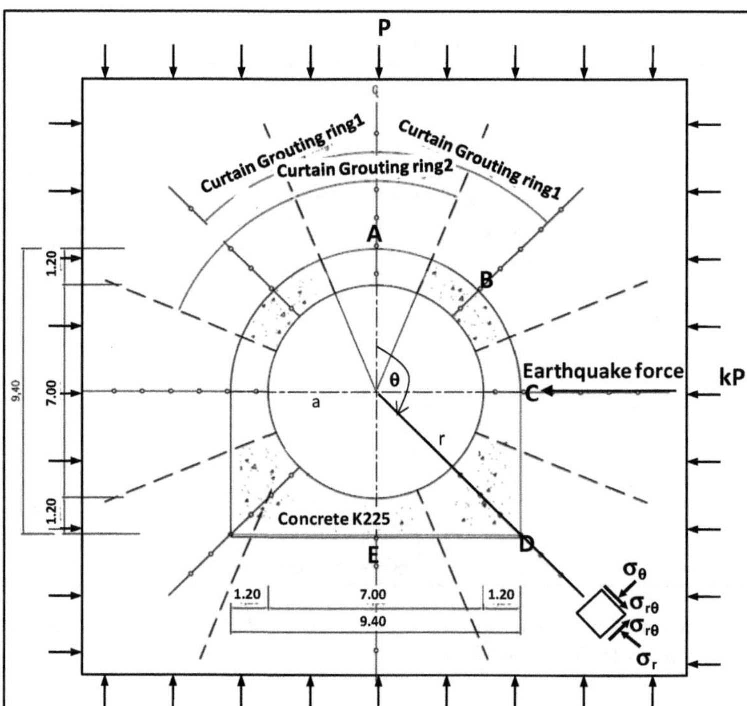

Fig. 2 Typical type 2 of diversion tunnel with stresses working around tunnel: Crown A, Crown B, Wall C, Base D and Base E are parts of tunnel being analyzed. Earthquake forces are horizontally working perpendicular to the tunnel axis, where a is radius and r is a point distance of estimations [8]

of 1 m, a concrete lining of 0.6 m thickness, consolidation and curtain grouts. Then, far field stresses, P on vertical and kP on horizontal directions, were estimated on each part of the tunnel: A, B, C, D, E; and additional stresses of earthquake were estimated working on perpendicular to tunnel axis.

3 Method

The construction of the Meninting tunnel followed the New Austrian Tunneling Method (NATM) [11]. The support systems of the tunnels utilized the geomechanics classification by Bieniawski [12], and suggested geological data by the ISRM [13].

For the current stability analyses in this paper, the estimations of rock mass strength followed the non-linear Hoek–Brown (HB) [14, 15]:

$$\sigma_1 = \sigma_3 + \sigma_{ci}\left(m_b\frac{\sigma_3}{\sigma_{ci}} + s\right)^a \tag{1}$$

According to Eq. (1), the strength of rock masses will depend on parameters m_b, s and a, which can be obtained from the rock mass rating (RMR) proposed by Bieniawski [12].

For a seismic area, additional uncertainties will be from earthquake forces. When earthquake forces work on rock, the rock is under dynamic shear conditions, which may cause a stress drop up to 10% from the static conditions [16].

Then, the estimation of stresses around the tunnel followed the Kirsch solutions [14]. Within these estimations of Eq. (2), the far field stresses are P in the vertical direction, and kP in the horizontal direction. Then, stresses at a point (r, θ) around the tunnel will be calculated as radial (σ_r), tangential (σ_θ) and shear stresses ($\sigma_{r\theta}$), where θ is the angle between the vertical and r line from the axis point, as in Fig. 2:

$$\begin{aligned}
\sigma_r &= 0.5P\left[(1+k)\left(1-\frac{a^2}{r^2}\right)+(1-k)\left(1+\frac{3a^4}{r^4}-\frac{4a^2}{r^2}\right)\cos 2\theta\right]\\
\sigma_r &= 0.5P\left[(1+k)\left(1+\frac{a^2}{r^2}\right)-(1-k)\left(1+\frac{3a^4}{r^4}\right)\cos 2\theta\right]\\
\sigma_{r\theta} &= 0.5P\left[-(1-k)\left(1-\frac{3a^4}{r^4}+\frac{2a^2}{r^2}\right)\sin 2\theta\right]
\end{aligned} \tag{2}$$

A specific earthquake design standard for tunnels has not been established yet by the BSNI; but, two national standards area available for embankment dam (Pd T-14-2004-A) [17], and for building and non-building [7]. These standards might be adapted for the estimations of earthquake forces for the Meninting diversion-spillway tunnel. Equation (3) is adapted from the Pd T-14-2004-A to estimate earthquake forces from peak ground acceleration on the surface and weight of rocks around the tunnel. Then, the equation uses the earthquake zone coefficient Z, which is depending on the location on the map of Indonesia earthquake zone [17].

$$F = a_1 \times \frac{a_d}{g} \times W \tag{3}$$

F = horizontal earthquake force
a_1 = correction factor for construction type
a_d = peak ground acceleration on the surface
. = $Z \times a_c \times V$
Z = earthquake zone coefficient
a_c = peak ground acceleration
V = correction factor for rock/soil type
g = gravity
W = weight.

However, the SNI 1726:2019 [7] has already applied the seismic parameters of peak ground and spectral accelerations to calculate shear forces. The consideration in utilizing the standards [7] to the tunnel is that a shoe-shaped underground structure (Fig. 2) is constructed into rocks; then, seismic forces are transmitted to the structure through surrounding rocks.

Subsequently, Agustawijaya et al. [6] updated seismic parameters for Lombok and surrounding islands based on the earthquake data of up to 2018. The updated PSHA for the event period of 1922–2018 determined parameters for peak ground acceleration of bedrock (PGA), spectral acceleration at $T = 0.2$ s (S_s), and spectral acceleration at $T = 1.0$ s (S_1) for the applied exceedance probability of 2% in 50 years [6]. Results shows that the PGA values of bedrock were 0.6–0.7 g; the S_s values were in the range of 1.0–1.2 g, and the S_1 values were 0.25–0.4 g for Lombok and surrounding islands. These values were therefore applied for the current stability analysis of the Meninting tunnel.

4 Results

4.1 Rock Mass Strength

The Meninting diversion-spillway tunnel was mainly excavated into volcanic breccias; although on the top of the tunnel, rocks contained lava, agglomerates and colluviums. The RMR should be different for each rock type; and of course, other rock material parameters should also be different, such as cohesion, friction, and unit weight. In general, these rocks had a RMR of 40, which was included into poor rock mass quality, thus the stand up time for the tunnel was one week without any support [14].

Table 1 Results of the rock mass strength of the Mininting tunnel type 2

Parameter	Crown A	Crown B	Wall C	Base D	Base E
Rock	V Breccia	V Breccia	V Breccia	V Breccia	V Breccia
Unit weight, γ (MN/m^3)	0.021	0.021	0.021	0.021	0.021
Depth of axis, H (m)	26.6	26.6	26.6	26.6	26.6
σ_{ci} (MPa)	5.9	5.9	5.9	5.9	5.9
σ_3 (MPa)	0.06	0.05	0.06	0.07	0.07
m_i	19	19	19	19	19
RMR	40	40	40	40	40
σ_1 (1) (MPa)	0.36	0.32	0.36	0.41	0.39

Using this RMR value, m_i of 19 and confining pressures of 0.06 MPa, the rock mass strength of Eq. (1) was within the range of 0.32–0.41 MPa depending on the rock position on the tunnel (Table 1). Such low σ_1 values were very much influenced by low confining pressures of 0.06 MPa. Also, a RMR of 40 did not really increase the rock mass strength, as it should be [18].

4.2 *Stresses Around the Tunnel*

Each part of the tunnel had a different stress concentration working on each part. On the boundary when the radius a equals the distance of the point estimated r, the radial stress σ_r and the shear stress $\sigma_{r\theta}$ were zero, so the only tangential stress σ_θ had a non-zero value. When $r = a + 1$ m, stress values on each part of the tunnel can be seen in Table 2.

The tangential stresses worked significantly on the Crown B, Wall C and Base D; shear stress concentrations should be on the Crown B and Base D; while, the Crown A should be tensioned, and some heave might occur on the Base E.

Table 2 Stresses working on each part of the tunnel for $r = a + 1$ m

Stress (MPa)	Crown A	Crown B	Wall C	Base D	Base E
σ_r	0.01	0.08	0.18	0.07	0.01
σ_θ	−0.09	0.22	0.63	0.29	−0.09
$\sigma_{r\theta}$	0.00	−0.20	0.00	0.20	0.00

Table 3 Stresses after earthquake working on each part of the tunnel for r = a + 1 m

Stress (MPa)	Crown A	Crown B	Wall C	Base D	Base E
Estimations from Eq. (3)					
σ_r	0.013	0.10	0.20	0.08	0.01
σ_θ	−0.09	0.29	0.69	0.31	−0.09
$\sigma_{r\theta}$	0.00	−0.26	0.00	0.21	0.00
Estimations from Eq. (3) with updated PGA [6]					
σ_r	0.013	0.12	0.25	0.09	0.01
σ_θ	−0.09	0.34	0.87	0.37	−0.09
$\sigma_{r\theta}$	0.00	−0.31	0.00	0.25	0.00
Estimations from [6, 7]					
σ_r	0.013	0.14	0.29	0.11	0.01
σ_θ	−0.09	0.39	1.01	0.42	−0.09
$\sigma_{r\theta}$	0.00	−0.35	0.00	0.29	0.00

4.3 Earthquake Stresses

When earthquake forces working on the tunnel were considered to be horizontally perpendicular to the tunnel axis, the influence of the earthquake forces (3) on the tunnel depended on the position on the tunnel. The earthquake stresses on the peak Crown A and Base E were minimum; otherwise, on the Wall C were maximum of 0.04 MPa. Thus, these earthquake forces resulted in the increase of stresses working around the tunnel (Table 3).

However, when the tunnel has been fully constructed as an underground structure, the earthquake force estimate procedures of the SNI 1726:2019 [7] may still be relevant to investigate the influence of earthquake forces on the tunnel. Then, updated PSHA [6] provided higher peak ground, short and long acceleration values than that of [7], which were also applied to estimate earthquake stresses working around the tunnel. Results show higher estimations than that of Eq. (3), (Table 3).

From all estimations, the earthquake stresses certainly increased radial and tangential stresses working on the Crown B, Wall C and Base D. Particularly, the Wall C was under higher radial and tangential stresses, such tangential stresses were up to 1.01 MPa.

4.4 Factor of Safety

The stability of the tunnel was described in terms of a factor of safety (FoS) of each part of the tunnel. The FoS of the tunnel after excavation and under the influence of earthquakes was calculated in terms of a ratio between rock mass strength and forces working around the tunnel subject to Kirsch formulations [14], for each part

Table 4 Factor of safety on each part of the tunnel

Factor of safety	Crown B	Wall C	Base D
Estimations from [17]			
σ_1/σ_r	3.04	1.78	5.10
σ_1/σ_θ	1.10	0.52	1.30
Estimations from [6, 17]			
σ_1/σ_r	2.57	1.42	4.31
σ_1/σ_θ	0.93	0.41	1.09

of the tunnel: Crown B, Wall C and Base D, where stresses under the influence of earthquakes were concentrated.

When earthquake forces were considered to work on the horizontal line perpendicular to the tunnel axis, the most deformed part of the tunnel was the Wall C. The wall C had only an FoS of 0.52 against the σ_θ stress. When using the updated data of PGA [6], the FoS value for the Wall C was higher for about 27%, from that of [17], which was 0.41 (Table 4).

According to Hoek and Brown [14], considering the RMR of 40, rocks around the tunnel with such an FoS might have a stand up time of up to 1 week prior to the installation of tunnel supports. However, this estimated stand up time would be too long; and it would shorten, probably down to be 25 min when earthquake forces were added. Thus, supports should be immediately installed as primary and secondary supports [11, 19].

After the completion of the construction, the tunnel had an increased FoS to become over 2.0; although, the Wall C had still a lower value than other parts of the tunnel (Table 5). The FoS value for the Wall C estimated from the updated PGA was slightly lower than that from the older PGA value [7], which reduced for about 7%.

Overall FoS values may still be relevant to the tunnel, since all tunnel supports were fully installed, and the underground structure had fairly stability under earthquakes. The overall FoS of the structure was about 2.5.

Table 5 Factor of safety for fully constructed Meninting tunnel

Factor of safety	Crown B	Wall C	Base D
Estimations from [6, 7]			
σ_1/σ_r	3.65	1.93	6.01
σ_1/σ_θ	1.32	0.56	1.52
Estimations from [7]			
σ_1/σ_r	3.84	2.06	6.32
σ_1/σ_θ	1.39	0.60	1.60

5 Discussion

The stability of the Meninting tunnel was influenced by earthquake forces. The FoS of the tunnel reduced significantly, particularly on the tangential directions of 0° on the horizontal line, 45° up and down the horizontal line, when earthquakes forces worked on horizontal direction. The HB strength criterion seems to have less sensitive to the influence of earthquakes forces, rather depends on rock properties. One suggested stability improvement of many suggestions [16] would be the application of grouting and rock bolting [20]. This could increase the shear strength; also reduce deformation of rock masses [21, 22].

The Meninting tunnel construction might be sufficient to stand earthquake stresses, but since it was excavated into a hill of various volcanic rock materials, the stability of rock masses would be a problem [23, 24]. On the top of the tunnel, loose bouldery agglomerates and colluviums had low strength, particularly with a high coefficient of permeability; they could be easily to lose their shear strength under earthquake forces. As the events in 2018, the surrounding area of the Meninting dam collapsed, where many landslides occurred during the events [2]. Estimations show that the upper tunnel cross section could be more vulnerable than the lower part; but, it might due to the PGA < 0.3 g [25]. Therefore, the stability of the tunnel would not only depend on the support systems installed to the tunnel; but, it also could depend on the stability of residual rock mass strength around the tunnel. The residual shear strength of the rocks could drop 41% after the earthquake events in 2018 [3], this should be problems for the tunnel stability in future [26].

6 Conclusions

The stability of Meninting diversion-spillway tunnel has been recalculated using the updated data of peak ground acceleration, short and long spectral accelerations of Lombok Island. The recalculations resulted in the reductions of the factor of safety of the tunnel for all calculation procedures of earthquake forces for embankment dam and building and non-building designs. The FoS reduction could be down to 35%; although, overall stability of the tunnel should be fine regarding the overall FoS of 2.5. For future stability, the tunnel certainly requires some improvements since the current stability should be depending on the residual rock strength against possible similar severe earthquakes.

Acknowledgements Authors acknowledge BWS Nusa Tenggara I (NTB) for data supports and access to the Meninting dam site.

References

1. Pramono S (2018) Perencanaan Rekonstruksi Wilayah Lombok Berbasis Mitigasi Gempabumi. Badan Meteorologi, Klimatologi dan Geofisika, Jakarta (2018) (in Indonesian). https://cdn.bmkg.go.id/web/Artikel_20181002115022_u1dpuf_-Perencanaan-Rekontruksi-Wilayah-Lombok-Berbasis-Mitigasi-Gempabumi.pdf#viewer.action=download. Last accessed 22 Oct 2018
2. Irsyam M, Hanifa NR, Djarwadi D (2018) Kajian Rangkaian Gempa Lombok, Provinsi Nusa Tenggara Barat. Pusat Studi Gempa, Pusat Litbang Perumahan dan Pemukiman, Kementerian Pekerjaan Umum dan Perumahan Rakyat, Jakarta (2018) (in Indonesian)
3. Agustawijaya DS, Sulistyowati T, Layli BA, Agustawijaya AR (2019) The ground deformation of the Luk Barat Bridge after Lombok earthquakes 2018. IOP Conf Ser Earth Environ Sci 389(012051). https://doi.org/10.1088/1755-1315/389/1/012051
4. Pranantyo IR, Cummins PR (2019) Multi-data-type source estimation for the 1992 Flores earthquake and tsunami. Pure Appl Geophys. https://doi.org/10.1007/s00024-018-2078-4[lastaccessed23/1/2020]
5. Agustawijaya DS, Taruna RM, Agustawijaya AR (2022) Earthquake forces on the stability of the Meninting diversion tunnel design based on current seismic conditions of Lombok Island. Indonesian J Geosci 9(1):105–117
6. Agustawijaya DS, Taruna RM, Agustawijaya AR (2020) An update to seismic hazard levels and PSHA for Lombok and surrounding islands after earthquakes in 2018. Bull N Z Soc Earthq Eng 53(4):215–226
7. Badan Standardisasi Nasional (BSNI) (2019) SNI 1726:2019: Tata Cara Perencanaan Ketahanan Gempa Untuk Struktur Bangunan Gedung dan Non Gedung (in Indonesian)
8. Balai Wilayah Sungai (BWS) (2017) Sertifikasi Disain Bendungan Meninting di Kabupaten Lombok Barat. Laporan Utama, Satuan Kerja Balai Wilayah Sungai Nusa Tenggara I, Mataram (in Indonesian)
9. Andi Mangga S, Atmawinata S, Hermanto B, Setyogroho B (1994) Geological map of the Lombok sheet, West Nusa Tenggara, scale 1:250.000. Geological Research and Development Centre, Bandung
10. Lavigne F, Degeai JP, Komorowski JC, Guillet S, Robert V, Lahitte P, Wassmer P (2013) Source of the great AD 1257 mystery eruption unveiled, Samalas volcano, Rinjani volcanic complex, Indonesia. Proc Natl Acad Sci 110(42):16742–16747
11. Harjomuljadi S (2010) Terowongan dengan NATM. Mediatama Saptakarya, Jakarta, 90 pp (in Indonesian)
12. Bieniawski ZT (1989) Engineering rock mass classifications. Wiley, New York, p 272
13. International Society for Rock Mechanics (ISRM) (1981) Rock characterization, testing and monitoring. In: Brown ET (ed) ISRM suggested methods. Pergamon Press, Oxford
14. Hoek E, Brown ET (1994) Underground excavations in rock. Chapman & Hall, London
15. Hoek E, Carranza-Torres C, Corkum B (2002) Hoek-Brown failure criterion. In: Proceedings of NARMS-TAC conference, 2002 edn, Toronto, pp 267–273
16. Brady BHG, Brown ET (1993) Rock mechanics for underground mining, 2nd edn. Chapman and Hall, London
17. Department of Public Work (2004) Analisis stabilitas bendungan tipe urugan akibat beban gempa, Pedoman Konstruksi dan Bangunan Pd T-14-2004-A (in Indonesian)
18. Priest SD (1993) Discontinuity analysis for rock engineering. Chapman & Hall, London
19. De Farias MM, Moraes AH Jr, De Assis AP (2004) Displacement control in tunnels excavated by the NATM: 3-D numerical simulations. Tunn Undergr Space Technol 19:283–293
20. Aldiamar F, Kartikasari S, Desyanti (2015) Metode perencanaan penggalian dan sistem perkuatan terowongan jalan pada media campuran tanah-batuan. Pedoman Bahan Konstruksi Bangunan Dan Rekayasa Sipil, Kementerian Pekerjaan Umum, Jakarta (in Indonesian)
21. Zhang B, Li S, Xia K, Yang X, Zhang D, Wang S, Zhu J (2016) Reinforcement of rock mass with cross-flaws using rock bolt. Tunn Undergr Space Technol 51:346–353

22. Agustawijaya DS, Sulistiyono H, Wahyudi M, Yasa IW, Ashari KF, Agustawijaya AR (2020) Grouting performance of the Pandan Duri dam shacked by Lombok earthquake 2018. In: MATEC Web of conference, vol 331(02004). https://doi.org/10.1051/matecconf/202033102004
23. Agustawijaya DS (2019) Practical applications of strength criteria in civil engineering design for shallow tunnel and underground structures in weak rocks. Int J Technol 5(2):93–105
24. Agustawijaya DS (2018) Influence of rock properties in estimating rock strength for shallow underground structures in weak rocks. Indonesian J Geosci 5(2):93–105
25. Wen K, Shimada H, Sasaoka T, Zhang Z (2018) Numerical study of plastic response of urban underground rock tunnel subjected to earthquake. Int J Geo-Eng 8(28). https://doi.org/10.1186/s40703-017-0066-7. Last accessed 2/11/2018
26. Stiros SC, Kontogianni VA (2009) Coulomb stress changes: from earthquake to underground excavation failures. Int J Rock Mech Min Sci 46:182–187

Risk and Stability Evaluation of Klego Dam, Boyolali, Indonesia

Suharyanto, Kresno Wikan Sadono, Rizqi Iqbal Maulana, P. Arie Bagus, and Dyah Ari Wulandari

Abstract Klego Dam was constructed in the 1990s and has been operated since 1993. Its main function is to supply Irrigation water of 13.53 km^2. It is located right at the boundary between Bengawan Solo Watershed and Jratunseluna Watershed. Previously, the dam was managed by BBWS Bengawan Solo. In the 2010s, the dam's management is relinquished to Pemali Juana River Basin Authority. This management's change contributes to some discontinuity in the dam's operation and maintenance, which contributes further to the loss of some important historical data and information of the dam. This unfavorable situation for proper operation and maintenance of the dam is further aggravated by the publication of Indonesian's new ground speed (earthquake) map for seismic design. It is, therefore, urgent to evaluate the stability and safety of Klego Dam with the new earthquake map even under limited available data and information. The paper presents the evaluation on the status of the dam's safety considering the seepage and slope stability. It uses Geostudio software to evaluate stability of the dam under various conditions. It also performed seepage analysis uses limited Piezometers' records. The analysis shows that the safety of Klego Dam is barely sufficient. However the dam needs some rehabilitation on its embankments, dam's crest, and its instrumentations as well as the proper implementation of Maintenance and Operation procedures. The availability on the SOP and its implementation will reduce the risk of dam's failure.

Keywords Klego Dam · Stability analysis · Seepage analysis

Suharyanto (✉) · K. W. Sadono · R. I. Maulana · P. A. Bagus · D. A. Wulandari
Civil Engineering Department, Engineering Faculty, Diponegoro University, Semarang, Indonesia
e-mail: suharyanto@lecturer.undip.ac.id

K. W. Sadono
e-mail: kresnowikansadono@lecturer.undip.ac.id

D. A. Wulandari
e-mail: dyah@lecturer.undip.ac.id

S. A. Kristiawan et al. (eds.), *Proceedings of the 5th International Conference on Rehabilitation and Maintenance in Civil Engineering*, Lecture Notes in Civil Engineering 225,
https://doi.org/10.1007/978-981-16-9348-9_13

1 Introduction

Klego Dam was built in the 1990s and operated since 1993. Its main function is to supply irrigation water of 13.53 km^2. Its location is right at the boundary between Bengawan Solo Watershed and Jratunseluna Watershed. Previously, the dam was managed by BBWS Bengawan Solo, while its water distribution is operated by Public Works Office in Boyolali Regency. In the 2010s, the dam's management is relinquished to BBWS Pemali Juana, however the Public Works office in Boyolali still responsible for water allocation and distribution. This management change contributes to some discontinuity in its operation and maintenance, which contributes further to the loss of some important historical data and information of the dam.

This unfavorable situation for proper operation and maintenance of the dam is further aggravated with the publication of new ground speed graph for seismic design (Indonesian's earthquake map, 2010). It is, therefore, urgent to evaluate the stability and safety of Klego Dam using the new earthquake map even under limited data and information.

2 Literature Review

The inherent risk of a dam's failure, has stipulated that a dam be evaluated its state of safety regardless of its limited data/information. Under this limited data/information, the safety state cannot be analyzed using more rigorous statistical methods, rather a more simplified one. It uses a method of total risk factor, TRF [1]. The method is simple and does not require adequacy of historical data or complete probability of hazards. It uses the risk factors accounting for risk contributing by the dam (Capacity, Height, and Age), the Downstream hazard (population and property at risk), and the predicted Damage factor (the dam type and the site-specific seismic hazard), respectively [1]. In Indonesia, it adopts similar principle and modified into: Guideline for stability analysis of earth and rock fill dams under earthquake condition [2] which is referred to PD-T-14-2004 A.

$$\mathrm{TRF} = \mathrm{CRF} + \mathrm{HRF} + \mathrm{ERF} + \mathrm{DRF}$$

where:

TRF: Total risk factor,
CRF: Influence of Dam's Capacity to risk factor,
HRF: Influence of Dam's Height to risk factor,
ERF: Influence of Potential Evacuated Persons to risk factor,
DRF: Influence of Downstream damages to risk factor.

2.1 Risk Classification

In the analysis of total risk, it considers four risk factors in determining risk level of a dam in Indonesia. They are factors related to dam's capacity, dam's height, potential evacuated persons, and downstream damages. Table 1 defines each factor scores [2].

2.2 Risk Level

Based on the TRF, the dam's risk level can be classified as low risk (class I), moderate risk (class II), high risk (class III), and extreme risk (class IV) and defined as follow (Table 2).

This risk level classification is used to determine the scale of design earthquake such as shown in the Table 3. The earthquake design uses MDE (Maximum Design Earthquake) condition.

Table 1 Risk factor and its classification

Risk factor	Classification (in the brackets is weight)			
	Extreme	High	Moderate	Low
Dam Capacity (10^6 m^3), CRF	>100 (6)	100–1.25 (4)	1.00–0.125 (2)	<0.125 (0)
Dam Height (m), HRF	>45 (6)	45–30 (4)	30–15 (2)	<15 (0)
Evacuation Need (person), ERF	>1000 (12)	1000–100 (8)	100–1 (4)	0 (0)
D/S damage DRF	VH (12)	H (10)	M (4)	Nul (0)

Table 2 Risk level for a dam safety [2]

TRF	Risk level
0–6	I (low)
7–18	II (moderate)
19–30	III (high)
31–36	IV (extreme)

Table 3 Earthquake design level [2]

<table>
<tr><th rowspan="2">Risk class</th><th colspan="2">No damage allowed</th><th colspan="2">Allowable damage condition</th></tr>
<tr><th>T (years)</th><th>Analysis method</th><th>T (years)</th><th>Analysis method</th></tr>
<tr><td>IV (Ext.)</td><td>100–200</td><td rowspan="4">Earthquake Coeff.</td><td>10.000 (MDE)</td><td rowspan="4">Earthquake Coef. Or Dynamic</td></tr>
<tr><td>III (High)</td><td rowspan="3">50–100</td><td>5.000 (MDE)</td></tr>
<tr><td>II (Mod)</td><td>3.000 (MDE)</td></tr>
<tr><td>I (Low)</td><td>1.0 (DE)</td></tr>
</table>

Table 4 Design ground acceleration [2]

Return period, T (years)	a_c (gal)
10	90
20	120
50	160
100	190
200	220
500	250
1000	280
5000	330
10,000	330

2.3 Design Ground Acceleration (ac)

The ground acceleration (ac) is determined based on the return period of the earthquake used in the analysis. The ground acceleration is presented as follow (Table 4):

2.4 Loads of Earthquake

The design load of earthquake is usually represented as horizontal forces acting on the structures, which is called coefficient approach.

Coefficient Approach

The horizontal force due to an earthquake's dynamic is represented using the following expressions:

$$\mathrm{ad} = \mathrm{Z} * \mathrm{ac} * \mathrm{v}$$
$$\mathrm{Kh} = \mathrm{ad}/\mathrm{g}$$

where:

Z = Earthquake Zone Coefficient.
ac = Basic ground acceleration (gal).
ad = corrected ground acceleration (gal).
v = correction factor due to type of soil/rock at site.
Kh = horizontal earthquake coefficient (gal).
G = gravitational acceleration (cm/s^2) = 981 cm/s^2.

In this approach, it assumes that the structures' components, on the base or either at higher position, will subject to equal earthquake forces horizontally. This assumption is not quite realistic. Another approach is to apply segmental coefficient.

Modified Coefficient for Earthquake

These modified coefficients are endorsed in [2] as well as in Seismic design guideline for earth and rock fill dams [3]. The design earthquake coefficient Kh = a/g. The design earthquake coefficient at each height can be calculated as follow:

$$\mathrm{Ko} = \alpha 2 \times \mathrm{Kh}$$

where:

Ko: design earthquake coefficient at the basis of the structure (a dam).
α2: correction based on the type of structures (for earth-fill dam α2 = 0.5).
Kh: basic earthquake coefficient (depend on the return period of the analysis).

The design earthquake coefficient at each depth of the dam is calculated for Y = 0.25H; 0.50H; 0.75H; (where Y is the depth measured from the crest of the dam and H is the dam's height) using formula as follow.

$$\text{For } 0 < \mathrm{Y/H} \leq 0.4:$$
$$\mathrm{K} = \mathrm{Ko} \times (2.5 - 1.85 \times (\mathrm{Y/H}))$$
$$\text{For } 0.4 < \mathrm{Y/H} \leq 1.0:$$
$$\mathrm{K} = \mathrm{Ko} \times (2.0 - 0.60 \times (\mathrm{Y/H}))$$

3 Stability Analysis

Stability analysis of dams has long been conducted using mathematical model of finite element method subject to earthquake dynamic forces [4]. Since then, the consideration of earthquake forces has also been practiced [5, 6]. The conditions of analysis are: at (1) the end of construction, (2) the steady seepage, and (3) rapid drawdown. Early example is by [7] who analyzed the during-construction and early impounding stage.

In this study, it needs to collect secondary and primary data. The secondary data available is very limited and mostly on water distribution. The data on cross section of the dam, its zonation, and their properties are not available. Therefore, it needs to perform field measurement, dam boring and samplings.

The dam boring and sampling was conducted at six locations in the dam, i.e., 3 at upstream slope and 3 at downstream slope such as shown in Fig. 1 and Table 5. The results of laboratory analysis on undisturbed samples taken at several depths are

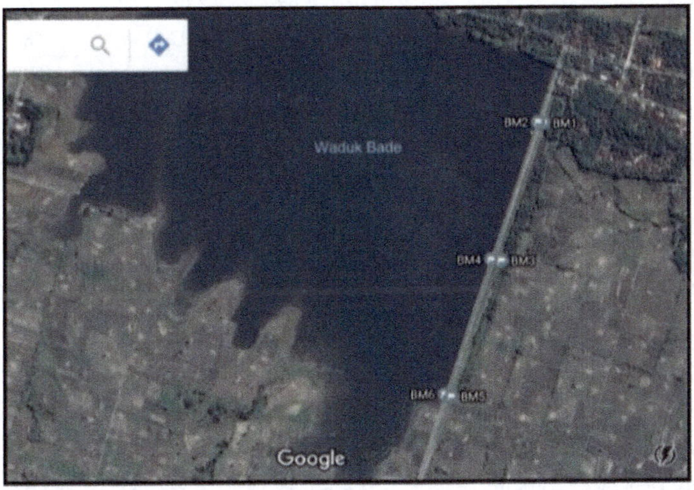

Fig. 1 Location of Bor machine along the Dam

Table 5 Location for soil boring

Bor No.	Location	Coordinat	
		X	Y
BM1	D/S slope, north	467,477	9,186,402
BM3	D/S slope, mid	467,385	9,186,105
BM5	D/S slope, south	467,276	9,185,813
BM2	U/S slope, north	467,464	9,186,404
BM4	U/S slope, mid	467,364	9,186,108
BM6	U/S slope, south	467,262	9,185,818

shown in Tables 6 and 7. From Table 8, the embankment material of Klego Dam is hard clay, sandy clay, and clay with fine sand. Therefore, the embankment of the dam is uniform (non zonation) (Fig. 2).

From topographic survey and detail measurement combined with the result of embankment boring, a typical cross section of Klego Dam is shown in Fig. 3 which is the highest cross section. It is shown that there is origin hard rock formation at the downstream side which was kept and thus adding more stability.

During the survey, at Klego Dam, there are 17 standpipe piezometers installed, two of which are not functioning. The piezometer readings are performed once a

Table 6 Embankment material at downstream slope

No.	Parameter	Unit	BM.1	BM.3	BM.5
1	Wet density	gr/cm^3	1.71–1.73	1.62–1.63	1.60–1.72
2	Dry density	gr/cm^3	1.43–1.48	1.36–1.39	1.35–1.44
3	Wc	%	17.30–19.57	17.58–19.28	19.13–19.28
4	SG		2.67–2.69	2.68–2.69	2.68–2.69
6	E		0.82–0.87	0.93–0.97	0.86–0.99
7	Finer # 200	%	87.34–94.44	89.23–92.28	89.83–90.53
8	Cc		0.80–0.85	0.28–0.32	0.22–0.37
9	Cv	$\times 10^{-2}$ cm^2/s	Stiff	0.11–0.63	0.11–0.24
10	Mv	$\times 10^{-2}$ cm^2/kg	Stiff	6.31–9.17	7.92–9.06
11	Cohesion c	kg/cm^2	Stiff	0.09–0.26	0.14–0.147
12	Phi	°	>35	19.42–19.28	19.69–20.03
13	Permeability	$\times 10^{-6}$ cm/s	8.49–3.75	8.44–3.69	7.21–6.20

Table 7 Embankment material at upstream slope

No.	Parameter	Unit	BM.2	BM.4	BM.6
1	Wet Density	gr/cm^3	1.66–1.74	1.66–1.74	1.61–1.72
2	Dry Density	gr/cm^3	1.40–1.48	1.39–1.47	1.35–1.44
3	Wc	%	17.30–18.57	18.14–18.85	18.99–19.86
4	SG		2.67–2.69	2.68–2.69	2.68–2.68
6	E		0.80–0.91	0.82–0.92	0,86–0,99
7	Finer # 200	%	89.34–89.64	87.44–93.04	89.63–91.52
8	Cc		0.21–0.33	0.22–0.315	0.23–0.38
9	Cv	$\times 10^{-2}$ cm^2/s	1.14–1.30	0.013–0.099	0.016–0.08
10	Mv	$\times 10^{-2}$ cm^2/kg	9.06–9.08	8.92–9.99	1.08–9.06
11	Cohesion c	kg/cm^2	0.09–0.31	0.05–0.15	0.305–0.311
12	Phi	°	19.30–>35	18.57–>35	19.44–19.69
13	Permeability	$\times 10^{-6}$ cm/s	6.48–3.66	8.77–7.23	6.94–6.33

Table 8 Embankment material at Klego Dam

Bor location	Depth of Boring (m)	Depth from surface (m)	Description
BM1 X: 467,477 Y: 9,186,402	10.00 m (Water table at −4.0)	0.00–2.20	Clay with fine sand, brown, hard
		2.20–4.60	Clay hard sand, coral, brown and hard
		4.60–6.50	Sandy clay, coral, greyish brown, hard to very hard
		6.50–10.00	Sandy clay, coral, grey, very hard
BM2 X: 467,464 Y: 9,186,404	10.00 m (Water table at −4.0)	0.00–4.20	Clay with fine sand, reddish brown, rather hard
		4.20–5.40	Clay with fine sand, reddish, brown hard
		5.40–6.00	Clay with fine sand, fine, reddish brown, hard
		6.00–7.20	Sandy clay, grey, hard
		7.20–8.20	Hard clay, sand, brown, very hard
		8.20–10.00	Sandy clay, coral, grey, very hard

Fig. 2 Boring at Klego Dam

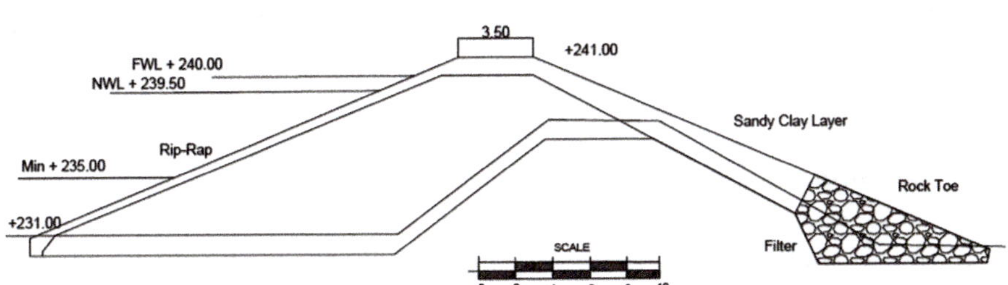

Fig. 3 Typical cross section adopted of the Klego Dam

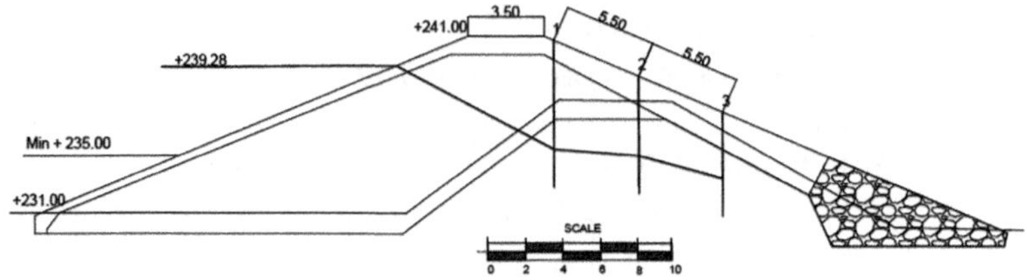

Fig. 4 Observed phreatic line based on piezometric data

Table 9 Values for factor of risk (RTD Klego 2016)

Normal Storage	2.33×10^6 m^3
Height of the Dam	10 m
People requires evacuation	9802 persons
Potential damages or losses	113 Billion Rupiah (Very High)

month. Observation on the piezometer no 1, 2, and 3 at the highest recorded water level (+239.28) is shown in Fig. 4. It can be seen that the water pressure falls at the existing hard rock formation downstream. It indicates that the adopted typical cross section of Klego Dam is represented.

3.1 Risk Evaluation

Based on Table 1 and data for Klego Dam, the risk level for the dam is:

CRF = 4 (capacity 2.33×10^6m^3)
HRF = 0 (Height is 10 m)
ERF = 12 (potential evacuee 9000 persons)
DRF = 12 (potential loss Rp. 113 billion)

$$\text{TRF} = \text{CRF} + \text{HRF} + \text{ERF} + \text{DRF} = 28$$

Based on the total risk factor (TRF) = 28, the risk level of Klego Dam is considered High (19–30) (Table 9).

3.2 Earthquake Load

Based on the High risk level (III), the earthquake analysis is performed with: (1) design ground acceleration of 100 year return period using OBE for no damages analysis (a_c = 190 gal) and (2) design ground acceleration of 5.000 year return period using MDE for allowed damages with no failures (a_c = 330 gal).

OBE Design Ground Acceleration

$a_{d} = Z * a_c * v = 0.7 * 190 * 1 = 133$ gal
$K_{h} = a_d/g = 133/981 = 0{,}136$ g
$K_{0} = \alpha_2 \times K_h = 0.5 \times 0.136 = 0.068$ g
$y/H = 0.25 : K = K0 \times (2.5–1.85\ y/H) = 0.13855$ g
$y/H = 0.50 : K = K0 \times (2.0–0.60\ y/H) = 0.1156$ g
$y/H = 0.75 : K = K0 \times (2.0–0.60\ y/H) = 0.1054$ g
$y/H = 1.00 : K = K0 \times (2.0–0.60\ y/H) = 0.0952$ g

MDE Design Ground Acceleration

$a_{d} = Z * a_c * v = 0.7 * 330 * 1 = 231$ gal
$K_{h} = a_d/g = 133/981 = 0.235$ g
$K_{0} = \alpha_2 \times K_h = 0.5 \times 0.235 = 0.1175$ g
$y/H = 0.25 : K = K0 \times (2.5–1.85\ y/H) = 0.2395$ g
$y/H = 0.50 : K = K0 \times (2.0–0.60\ y/H) = 0.1998$ g
$y/H = 0.75 : K = K0 \times (2.0–0.60\ y/H) = 0.1821$ g
$y/H = 1.00 : K = K0 \times (2.0–0.60\ y/H) = 0.1645$ g

3.3 Stability Analysis Using Geostudio Software

The stability analysis uses Geostudio software. The input for the model is material of embankment and idealized as in Fig. 5. The embankment is homogenous provided with toe drain along its downstream end. The embankment lies over a stiff clay bed (layer 1). The properties of the material in the model is shown in Table 10. The analysis is performed for normal condition, OBE earthquake loads, and MDE earthquake loads.

Factor of Safety during normal condition

See Table 11.

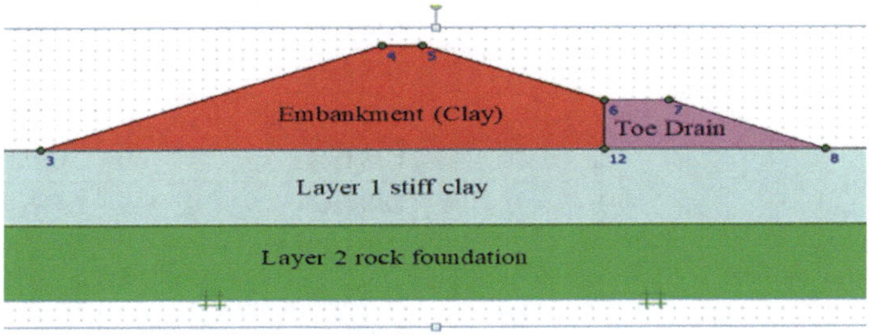

Fig. 5 Embankment material at Klego Dam

Table 10 Properties of embankment material

Zona	γ_d (kN/m^3)	γ_{sat} (kN/m^3)	C′ (kPa)	Φ (°)	E (kPa)
Embankment	13.95	18.72	30	25	15
Toe drain	17	20	0	45	30
Stiff clay	14.315	18.96	40	27	30
Fondation	16	18	40	27	30

Table 11 Factor of safety under normal condition

No	Condition	FS
		Geostudio 2007
1	*Steady state*	
A	U/S Face	2.287
B	D/S Face	3.229
2	*Rapid draw down*	
A	U/S Face	2.858
B	D/S Face	1.949

Factor of Safety During Earthquake (OBE)

See Table 12.

Factor of Safety Under Earthquake (MDE)

See Table 13.

Downstream face

See Figs. 6, 7, 8 and 9.

Table 12 Factor of safety during earthquake (OBE)

No	**Y/H; K**	**FS**
1	0.25; 0.139 g	
a	U/S face	3.274
b	D/S face	3.329
2	0.5; 0.116 g	
a	U/S face	2.659
b	D/S face	2.516
3	0.75; 0.105 g	
a	U/S face	2.255
b	D/S face	2.202
4	1.0; 0.095 g	
a	U/S face	2.212
b	D/S face	2.232

Table 13 Factor of safety during earthquake (MDE)

No	Y/H; K	SF Geo Slope 2012
1	0.25; 0.240 g	
A	U/S face	2.399
B	D/S face	2.647
2	0.5; 0.20 g	
A	U/S face	2.012
B	D/S face	2.046
No	Y/H; K	SF Geo Slope 2012
3	0.75; 0.182 g	
A	U/S face	1.781
B	D/S face	1.895
4	1.0; 0.165 g	
A	U/S face	1.777
B	D/S face	1.899

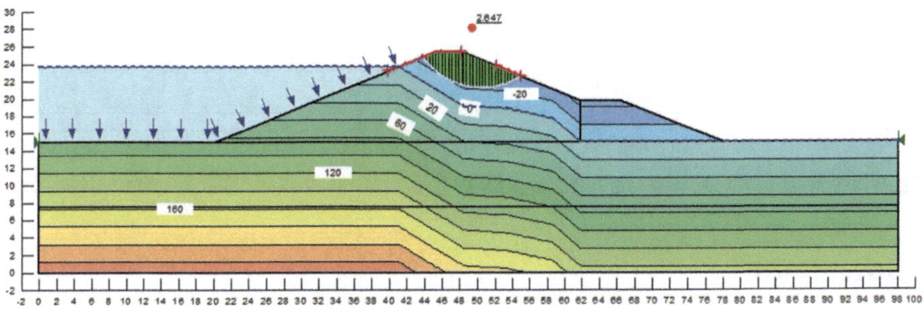

Fig. 6 FS at y/H = 0.25; K = 0.240 g (SF = 2.647)

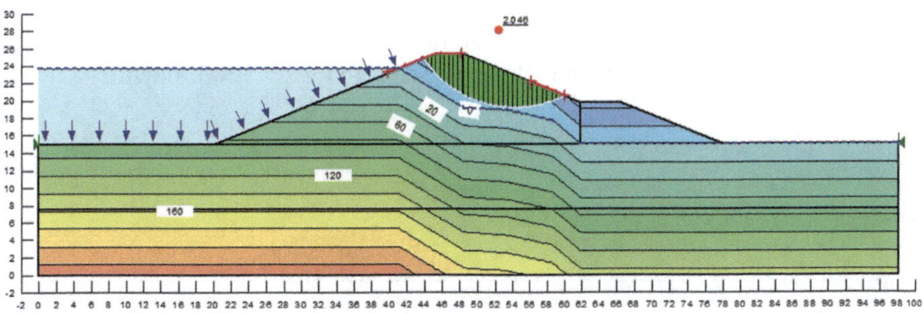

Fig. 7 FS at y/H = 0.5; K = 0.20 g (SF = 2.046)

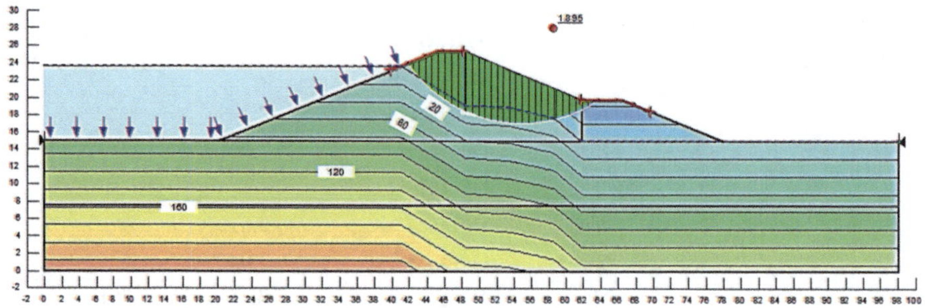

Fig. 8 FS at y/H = 0.75; K = 0.182 g (SF = 1.895)

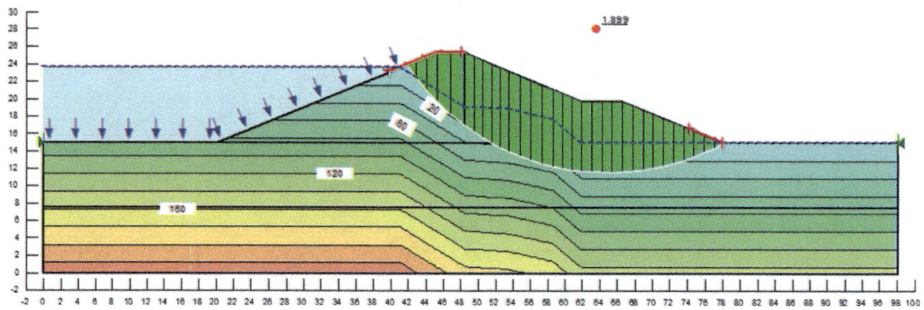

Fig. 9 FS at y/H = 1; K = 0.165 g (SF = 1.899)

Upstream face

See Figs. 10, 11, 12 and 13.

Seepage Rate

The result of the model gives the seepage rate is 171.87 × 10^{-9} m^3/sec/m (=14.84 L/day/m), which is still below the allowable seepage rate = 100 L/day/m.

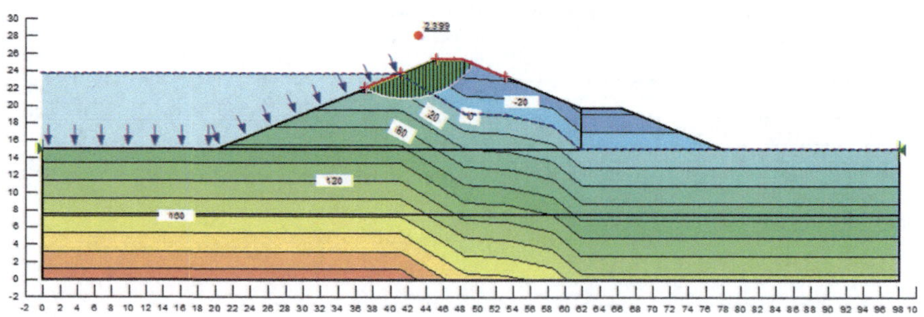

Fig. 10 At y/H = 0.25; K = 0.240 g (SF = 2.399)

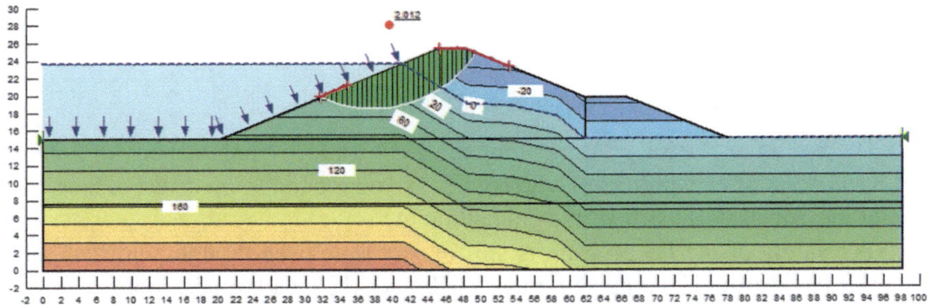

Fig. 11 At y/H = 0.5; K = 0.20 g (SF = 2.012)

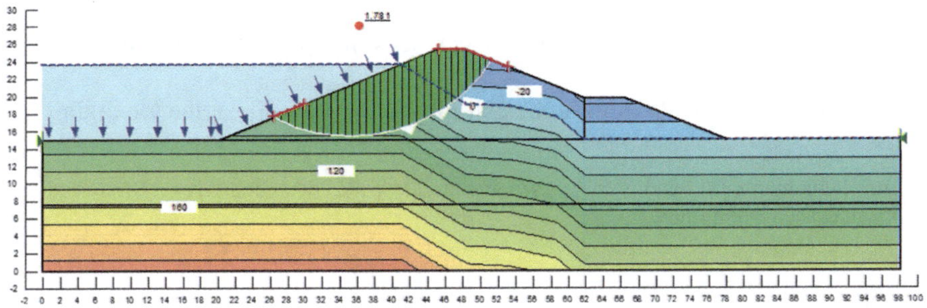

Fig. 12 FS at y/H = 0.75; K = 0.182 g (SF = 1.781)

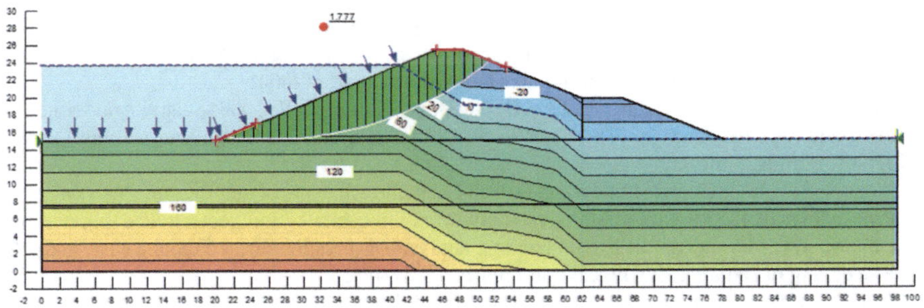

Fig. 13 FS at y/H = 1; K = 0.165 g (SF = 1.777)

Phreatic Line

The model also shows that the outlet of the phreatic line lies at the toe drain, which is rock fill formation. This shows that the seepage exit from the dam through its designated rock toe.

4 Conclusion

From the analysis it can be concluded as follows:

1. The boring on the Klego Dam along the crest confirms that the body of the dam is uniform of clay protected with rip rap of stones and the existence of hard rock at the downstream side of the embankment,
2. The safety of the Klego Dam is still in safe state even though analyzed using the new earthquake map and under OBE and MDE earthquake loads,
3. The lowest factors of safety obtained are: 1.949 at downstream face under rapid drawdown; 2.212 at the upstream face under OBE loadings; and 1.777 at upstream face under MDE loadings, which are surpassed the minimum allowable FS of 1.5.
4. The seepage is 14.8 L/day/m' which is still lower than the allowable of seepage 199 L/day/m'.
5. The phreatic line also shows the exit falls in the safe side at the toe drain.

Acknowledgements The authors express their gratitude to the Pemali Juana River Basin Authority for the data used in the study.

References

1. Chen Y, Lin P (2018) The total risk analysis of large dams under flood hazards. Water 10(2)
2. Ministry of Housing and Regional Infrastructures No. 360/2004 on: Stability Analysis of Embankment type Dam under earthquake Condition. Jakarta (2004)
3. Hernández UM, Nava-Tristán OE, Liu XL, Sandoval EM, Hach F (2008) Seismic guidelines for earth and rock fill dams. In: The 14th World Conference on Earthquake Engineering October 12–17, Beijing, China
4. Maaty AS, Rothenburg L, Lelievre B (1985) Dynamic Analysis of stability of Earth dams. In: Proceedings of ISSMG: international society for soil mechanics and geotechnical engineering
5. Chhatre MV, Muralidhar B (2010) Dynamic analysis of matatila Earthed dam: a case study. In: Proceeding of Indian geotechnical conference, 16–18 Dec
6. Paul DK (2004) Assessment of damage and rehabilitation of small and medium Earth dams. In: 13th world conference on earthquake engineering, Vancouver, BC, Canada, 1–6 Aug 2004, Paper No. 2160
7. Rashidi M, Mohsen Haeri S (2017) Evaluation of behaviors of earth and rockfill dams during construction and initial impounding using instrumentation data and numerical modeling. J Rock Mech Geotech Eng 9:709–725

Impact of Climate Change on Dam Safety

Heri Sulistiyono, Ery Setiawan, and Humairo Saidah

Abstract Dams are built to overcome surface water problems. These problems include excess water in the rainy seasons and scarcity of water in the dry seasons. As the occurrence of climate change has been recognized by people around the world, engineers have to be able to consider its impact on the safety of dams. This study proposes a procedure for studying the impact of climate change on the safety of gated spillway dams. The procedure consists of a statistical consistency test, selection of global climate change variables, modelling of local climate change, calculating flood discharge due to climate change, and evaluation of dam safety. This study was conducted at the Batujai Dam, Central Lombok Regency. Global climate change data from 1998 to 2100 taken from the Intergovernmental Panel on Climate Change (IPCC) and maximum daily rainfall data from 1998 to 2020 from the Pengadang Station were used to demonstrate the application of the proposed procedure. The results show that the proposed procedure can be used to evaluate the impact of climate change on dam safety. It is also known that climate change has a significant effect on the tendency of increasing flood discharge entering the Batujai Reservoir. The four spillway gates: P_1, P_2, P_3, and P_4 have to be opened 1.0, 1.2, 1.2, and 1.0 m, respectively to anticipate the 1000-year return period of a flood event due to climate change, at a rate of 1833.54 m^3/s.

Keywords Climate change · Maximum daily rainfall · Gated spillway dams · Safety of dam

H. Sulistiyono (✉) · E. Setiawan · H. Saidah
Department of Civil Engineering, University of Mataram, Mataram, Indonesia
e-mail: h.sulistiyono@unram.ac.id

E. Setiawan
e-mail: ery.setiawan@unram.ac.id

H. Saidah
e-mail: h.saidah@unram.ac.id

S. A. Kristiawan et al. (eds.), *Proceedings of the 5th International Conference on Rehabilitation and Maintenance in Civil Engineering*, Lecture Notes in Civil Engineering 225,
https://doi.org/10.1007/978-981-16-9348-9_14

1 Introduction

Dams are built for many benefits. However, there are potential dangers if floods overtop the dam embankments. The dam will break and a flash flood will occur downstream. In the Dam Design guidelines, embankment heights should be designed based on at least a 1000-year return period of flooding. Since the consensus on climate change in Kyoto in 1992, engineers must evaluate the impact of climate change on dam safety.

Global climate change predictions have been simulated by agencies around the world under the coordination of the IPCC. The resolution of the global simulation is still very rough, however, and cannot be used directly for local designs. The results of global climate change simulation have to be downscaled following the local historical data to obtain local scaled climate change data that can be used in local designs and evaluations [1].

This study proposes the procedure for dam safety evaluation based on the impact of climate change, to develop a local rainstorm model based on climate change, to calculate the 1000-year return period of flooding due to climate change, and to obtain the opening height of the spillway gates required to secure the dam.

2 Literature Review

The problem of the increase in water demand caused by the increase in population in the Tasikmalaya Regency can be solved through the construction of a water reservoir, namely the Manonjaya Dam [2]. The Manonjaya Dam is an earth-fill type dam with a vertical core on the Citanduy River with a catchment area of 590,422 km^2. The purpose of the Manonjaya Dam is to meet the needs for domestic water and irrigation in the Tasikmalaya Regency area.

The problem of flooding in the rainy season and the lack of clean water during the dry season in Ponorogo Regency, East Java, can be overcome through the construction of the Bendo Dam [3]. A 1000-year return period of a flood of 676.37 m^3/s was used to design the dam height. The design life of the dam is 50 years.

Climate change refers to permanent changes in the statistical characteristics of climate factors such as solar radiation, temperature, wind, humidity, rainfall, air pressure, and so on, over a long period such as tens of years to millions of years [4–6]. Currently, the world is experiencing global climate change [7]. The impacts of climate change in Indonesia include an increase in the average air temperature from January 1971 to December 2006 of 0.5 °C; an increase in the maximum air temperature of 0.7 °C; a decrease in the minimum air temperature of about 1.2 °C; change in the maximum rainy season from January–March to October–December; and the occurrence of drought in some areas. In addition, many small islands such as Lombok, Sumbawa, Flores, Sumba, Timor, Solor, and Alor will be more vulnerable to climate change than the large islands [8].

3 Methodology

The study procedure consists of the following five strategic steps: Preparation, Standard Statistical Screenings, Climate Change Modelling, Hydrological Frequency Analysis, and Hydraulic Analysis to check for overtopping. The procedure is shown in Fig. 1.

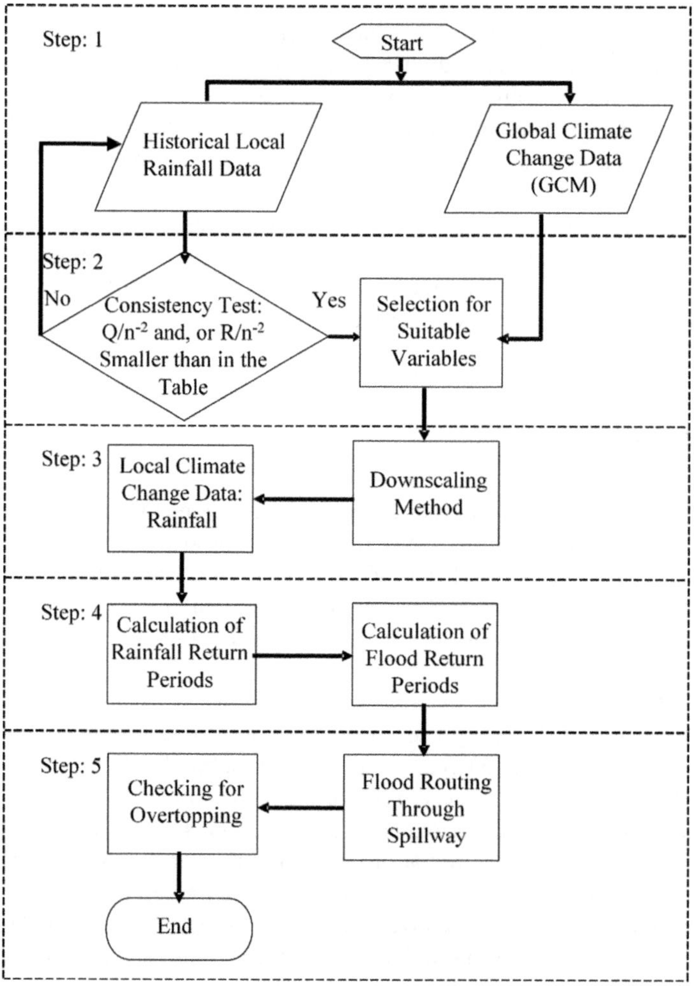

Fig. 1 The procedure of the study

3.1 Step: 1 Preparation

In this step, both historical local rainfall data and global climate change data are collected. The range of historical local rainfall data (HLRD) has to be within the range of the global climate change data (GCCD). The period of HLRD has to be at least 20 years to ensure statistical validity in the calibration and verification processes. HLRD is collected from all rainfall stations around the catchment area of the reservoir.

The GCCD is the result of global climate models (GCM). In the Intergovernmental Panel on Climate Change (IPCC), GCCD of various GCMs from various countries are stored. One has to register before downloading the data. The GCCD cannot be applied directly for local designs because the resolution is very coarse. The GCM variable needs to be downscaled to get better-scaled variables that can be used for local designs. Downscaling models can be grouped into the three following approaches: dynamic models, statistical models, and change factor methods [1, 9]. Regression models are proposed in this study procedure for the climate change downscaling model.

3.2 Step: 2 Standard Statistical Screening

Statistical screening tests are used to ensure all the data are ready. The term 'ready' refers to data that are consistently collected in the expected condition. The Rescaled Adjusted Partial Sums (RAPS) method [10] is applied in this procedure for the standard statistical screening test. As RAPS has been frequently explained in many hydrological references, this method is not described further in this paper.

3.3 Step: 3 Climate Change Modelling

This study applies the three most suitable GCM variables as independent variables to simulate local climate variables. The selection of suitable GCM variables is made using common statistical analysis: correlation coefficients (CC) [11].

In this study, the best multiple linear regression is selected and used to downscale the GCM variable into local rainfall variables.

3.4 Step: 4 Hydrological Frequency Analysis

In this study, the frequency analysis is used to obtain the estimated return period of rainstorms due to climate change. Then, the flood return period is calculated based on the return period of rainstorms. The Nakayasu Synthetic Unit Hydrograph is used to simulate the flood [10]. The Unit Hydrograph as shown in Fig. 2 is calculated using Eq. (1) to Eq. (11):

$$Q_P = \frac{AR_0}{3.6(0.3t_P + T_{0.3})} \tag{1}$$

where: Q_p = peak flow (m^3/s), A = catchment area (km^2), R_o = unit rainfall (mm), t_p = time lag from the initial rain to the peak flow (hours), $T_{0.3}$ = the time it takes for the discharge to decrease, from peak discharge to 30% of peak discharge (hours).

In Fig. 2, T_p and $T_{0.3}$ are calculated using Eqs. (2) and (3):

$$T_p = tg + 0.8\,tr \tag{2}$$

$$T_{0.3} = \alpha \cdot tg \tag{3}$$

where tg is calculated based on Eq. (4) or Eq. (5), and tr is calculated using Eq. (6):

$$tg = 0.21L^{0.7} \text{ for } L < 15\,km \tag{4}$$

$$tg = 0.40 + 0.058\,L \text{ for } L > 15\,km \tag{5}$$

$$tr = (0.5 \sim 1.0)tg \tag{6}$$

where tr = the effective duration of rain.

The rising curve for the range of $0 \leq t \leq T_p$

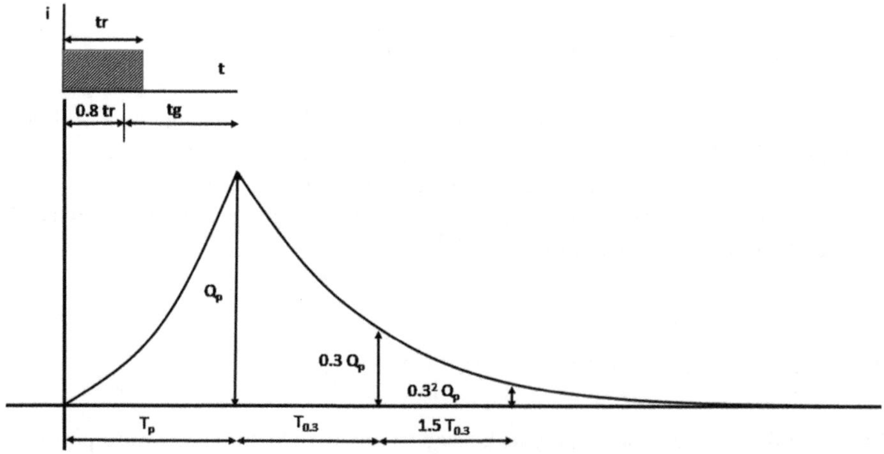

Fig. 2 The illustration of Nakayasu synthetic unit hydrograph

$$Q_a = Q_P \left[\frac{t}{T_P} \right]^{2.4} \tag{7}$$

The declining curves for the range of
(1) $Q_d > 0.3\ Q_P$ or $T_P \le t < T_{0.3}$

$$Q_d = Q_P 0.3 \left[\frac{t - T_P}{T_{0.3}} \right] \tag{8}$$

(2) $0.3\ Q_P > Q_d > 0.32\ Q_P$ or $T_{0.3} \le t < 1.5\ T_P$:

$$Q_d = Q_P 0.3 \left[\frac{(t - T_P) + 0.5 T_{0.3}}{1.5 T_{0.3}} \right] \tag{9}$$

(3) $0.3^2\ Q_P > Q_d$ or $t \ge 1.5\ T_{0.3}$

$$Q_d = Q_P 0.3 \left[\frac{(t - T_P) + 1.5 T_{0.3}}{2 T_{0.3}} \right] \tag{10}$$

The relationship between the shape of the drainage area and $T_{0.3}$ can be stated as a constant of α:

$$\alpha = \frac{0.47 (AL)^{0.25}}{tg} \tag{11}$$

where: Q_a = runoff before reaching peak discharge (m^3/s), Q_d = runoff after reaching peak discharge (m^3/s), t = time (h), L = length of river channel (km), tg = concentration time (h), and α = constant.

3.5 Step 5: Hydraulic Analysis for Overtopping

The flood routing through the spillway aims to simulate the capacity of the spillway concerning changes in the inflows of the dam. Equations (12) and (13) were used in the simulation. The simulation must consider the upstream and downstream water conditions.

$$Q_s = \mu \cdot a_s \cdot b_s \cdot \sqrt{2gh_1} \tag{12}$$

where: Q_s = discharge (m^3/s), μ = discharge coefficient (given: 0.8), a_s = height of the opening (m), b_s = width of opening (m), g = gravitational acceleration (9.8 m/s^2), and h_1 = hydraulic energy at the opening (m).

$$I + \frac{S_1}{\Delta t} - \frac{O_1}{2} = \frac{S_2}{\Delta t} + \frac{O_2}{2} \quad (13)$$

where: I = Inflow (m^3/s), S_1 = Storage at time 1, S_2 = Storage at time 2, Δt = routing interval (h), O_1 = Outflow at time 1 (m^3/s), O_2 = Outflow at time 2 (m^3/s).

4 Case Study

To demonstrate the procedure, this study was applied in the Batujai Dam in Penujak Village, Central Lombok Regency, West Nusa Tenggara Province, Indonesia, as shown in Fig. 3.

4.1 Step: 1 Preparation

In this study, 23 years of maximum daily rainfall data from 1998 to 2020 from Pengadang Rain Stations were used as local historical data and Global Climate Change data from 1998 to 2100 were downloaded from the IPCC website. The data are: X1 = mean 2 m wind speed (m/s), X2 = evaporation (mm/day), X3 = precipitation (mm/day), X4 = screen (2 m) temperature (°C), X5 = screen spec. humidity (kg/kg), X6 = sea level pressure (hPa), X7 = surface temperature (°C), X8 = solar flux at surface (W/m^2), X9 = surface pressure (hPa).

Fig. 3 Study location

4.2 Step: 2 Standard Statistical Screening

From the calculations of the RAPS test of the annual rainfall data from the Pengadang Station, it is found that these are consistent.

The next step is to select the three most suitable GCM variables, which are air temperature (X4), specific air humidity (X5), and air temperature at the Earth's surface (X7). The three GCM variables are then considered as independent variables in the multiple linear regression model for modelling local rainfall due to climate change.

4.3 Step: 3 Climate Change Modelling

The multiple linear regression model is expressed as $Y = 0.453 + 0.44\ X4 + 0.253\ X5 - 0.81\ X7$. The proportion of the variance for a dependent variable that can be explained by the variables is quite high, being approximately 80.7%.

The local rainfall data due to climate change are generated using the regression model. The results are shown in Fig. 4.

Figure 4 shows that the predicted mean local maximum daily rainfall in the area of study has an increasing trend. The next step is to look for the 1000-year return period of rainfall and the 1000-year return period of flooding.

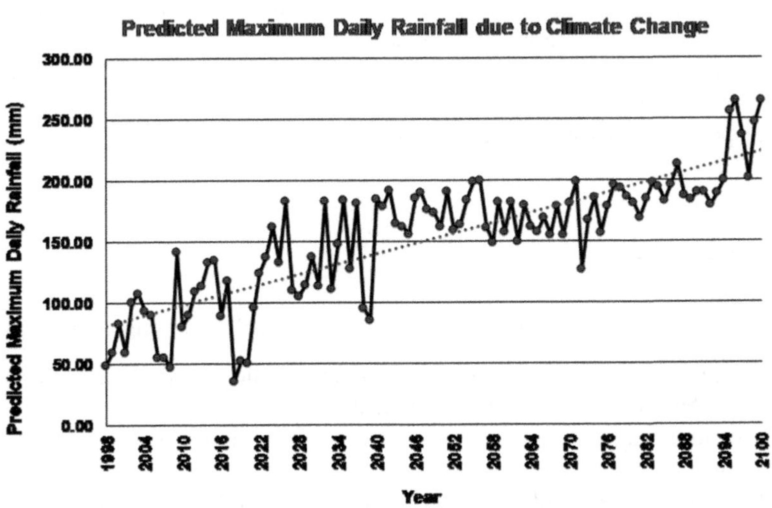

Fig. 4 Predicted local maximum daily rainfall due to climate change

Table 1 Results of flood routing simulation

Simulation: Routing output for Batujai Reservoir			
Q-inflow	1833.5	m^3/s	
Q-outflow	875.24	m^3/s	
Reduction—Q	52.26	%	4 × @ 11 m
Opening (m)	Gate #2	Gate #3	Gate #4
Gate #1	1.20	1.20	1.00

4.4 Step: 4 Hydrological Frequency Analysis

It is found that the statistical distribution of the data is a 2-parameter exponential. The amount of the 1000-year return period of a rainstorm is 797.2 mm. This amount is used to simulate the inflow of the Batujai Dam using the Nakayasu Hydrograph. From the simulation, it is obtained that the peak of the 1000-year return period of flooding that enters the Batujai reservoir is 1833.538 m^3/s.

4.5 Step 5: Hydraulic Analysis for Overtopping

In this step, flood routing simulation through a gated spillway is conducted to obtain whether the 1000-year return period of flooding overtops the dam or not, and how to arrange the gates of the spillway to avoid overtopping. The results of the simulation using Eqs. (1)–(13) are presented in Table 1.

Table 1 shows that when the amount of the 1000-year period of flooding of 1833.54 m^3/s comes into the reservoir, the four gates have to be opened by 1.0, 1.2, 1.2, and 1.0 m respectively to avoid overtopping of the dam.

5 Conclusion

Based on the results of the analysis, the following conclusions can be drawn:

(1) The proposed procedure can be reasonably applied for the evaluation of dam safety based on the impact of climate change;
(2) The study successfully developed a local rainstorm model affected by climate change;
(3) The amount of the 1000-year return period of flooding due to climate change is 1833.54 m^3/s; and
(4) The opening heights of the four spillway gates required to secure the dam are 1.0, 1.2, 1.2, and 1.0 m respectively.

References

1. Wilby RL, Charles SP, Zorita E, Timbal B, Whetton P, Mearns LO (2004) Guidelines for use of climate scenarios developed from statistical downscaling methods. Supporting Material of the IPCC. http://www.ipcc-data.org/guidelines/dgm_no2v1_09_2004.pdf
2. Shofa K, Hidayat M, Wahyuni S, dan Darsono S (2013) Perencanaan Bendungan Manonjaya Kabupaten Tasikmalaya, Jawa Barat. Jurnal Karya Teknik Sipil 2(3):189–204
3. Busiri A, Fajar R, Sangkawati S, dan Budieny H (2016) Perencanaan Bendungan Bendo Ponorogo. Jurnal Karya Teknik Sipil 5(2):159–169
4. IPCC (2001) Climate Change 2001. The scientific basis; impacts, adaptation, and vulnerability; mitigation; and model evaluation. The Third Assessment Report (TAR) of the Intergovernmental Panel on Climate Change (IPCC). Cambridge University Press, Cambridge, UK and New York, USA
5. IPCC (2007) The synthesis report and the physical science basis—summary for policymakers. IPCC Plenary XXVII, Valencia, Spain
6. Pryor SC (2009) Understanding climate change: climate variability, predictability, and change in the midwestern United States. Indiana University Press, Bloomington
7. Gleick PH (1997) Water planning and management under climate change. J Am Water Works Assoc
8. Hilman M, Hatta GM, Hilman D, Gunawan A, Suryani Y, Irwani S, Setiawan G, Von Luepke H, Hertz T, Panjiwibowo C, Pasaribu K, Suroso SA, Hadi TW, Sofian I, Latief H, Abdurahman O, Halil, Julianto H, Riawan E, Muhammad A, Anggraini L (2010) A synthesis report: risk and adaptation assessment to climate change in Lombok Island, West Nusa Tenggara Province. Ministry of Environment of the Republic of Indonesia
9. Sulistiyono H, Lye LM (2011) Predicted regional hydrologic and climatic variables under climate change scenarios using statistical downscaling techniques for future water resource studies in Lombok, Indonesia. In: The 64th Canadian Water Resources Association (CWRA), Canada
10. Harto S (1993) Analisa Hidrologi. Jakarta, Gramedia Pustaka Utama
11. Prak JL (2011) An empirical study on the reliability of correlation coefficients. Br J Psychol 21(4). https://doi.org/10.1111/j.2044-8295.1931.tb00602.x

Assessment, Protection, Maintenance, Repair, and Retrofitting of Buildings and Infrastructures

Investigating Materials for Refurbishment Strategies of Heritage Buildings: A Case Study of *Soesman Kantoor*, Semarang

Ferry Hermawan, Didi Wibowo Tjokro Winoto, Ismiyati, Bambang Purwanggono, and Robby Soetanto

Abstract The management of heritage buildings presents complex challenges for local authorities regarding repair methods, the selection of materials and post-repair impacts. Restoring the function of heritage buildings is often still constrained by commercialisation and the problem of maintaining sociocultural values. This raises an important question, what approach strategy is most suitable for maintaining the characteristics of heritage building in the long term? Here, we present a case study with a materials laboratory test strategy for a heritage building in Semarang. The parties involved include local authorities, architects, heritage building experts and academics. Three technical assessments are conducted, namely, materials analysis using X-ray fluorescence spectroscopy and brick compression and paint adhesion tests. This study provides an overview of the strategic approach from the aspects of measuring, interpreting and providing the best solutions for the material elements of heritage buildings in the future. The results promote improving heritage buildings on the basis of stakeholder mapping and building materials improvement strategies. This research contributes to local authorities in arranging heritage buildings with technical considerations in line with the characteristics of the building and development programs that consider local knowledge.

Keywords Strategy · Refurbishment · Heritage

F. Hermawan (✉) · D. W. T. Winoto · Ismiyati · B. Purwanggono
Diponegoro University, Semarang 50275, Indonesia
e-mail: ferry.hermawan@live.undip.ac.id

B. Purwanggono
e-mail: bambangpurwanggono@lecturer.undip.ac.id

R. Soetanto
Loughborough University, Loughborough, LE 113TU, UK
e-mail: R.Soetanto@lboro.ac.uk

S. A. Kristiawan et al. (eds.), *Proceedings of the 5th International Conference on Rehabilitation and Maintenance in Civil Engineering*, Lecture Notes in Civil Engineering 225,
https://doi.org/10.1007/978-981-16-9348-9_15

1 Introduction

1.1 Background

The Old Town area of Semarang (*Kota Lama* Semarang) has high historical value with many buildings from the Dutch colonial era. Some of these heritage buildings are listed as cultural heritage buildings according to Law No. 11 of 2020 on Cultural Heritage [1]. *Soesman Kantoor* is one of these buildings and is located in *Kota Lama* Semarang. According to law, buildings that have existed for at least five decades can be classified as cultural heritage buildings [1]. In addition to this criterion, these buildings are expected to have essential historical, scientific or cultural value.

In the modern era, the attentiveness of cultural heritage buildings is often neglected. The European Charter of Architecture Heritage 1975 Article 6 states that architecture's cultural heritage is being unsettled in Indonesia and around the world [2]. In mid-January and April 2011, two cultural heritage buildings collapsed in Kepodang, *Kota Lama* Semarang. This was caused by the buildings' lifetime and natural factors, such as flooding, seawater intrusion and extreme weather [3]. As reviewed through financial, information and technological aspects, each city's evolution will always be the same from time to time. Cultural heritage buildings have unique features, the significant elements of a culture that reflect an area's identity. Cultural values grow and live within society and have their own uniqueness that makes each stand apart. If these values are maintained, each region will have an identity that differentiates it from others, thus representing its unique charm [4].

The refurbishment and improvement of building function are essential for the aim that cultural heritage can function as a new attraction for tourism and bring an area back to life. The repair of cultural heritage buildings is different from modern buildings. Professionals must handle cultural heritage buildings to avoid irreversible damage. The repair of existing structures must be carried out attentively and by not changing or spoiling any characteristics or details, although changes might be needed [5].

Research on the correct method for cultural heritage building repairs is essential because incorrect action could cause even more damage. The cultural values might be lost if damage to a single component of a cultural heritage building occurs. This research leads to the important question of what is the best method for repairing such buildings?

1.2 Research Method

The refurbishment of cultural heritage buildings in Semarang must be carried out by paying particular attention to the materials and procedures. Laboratory tests can be used to identify the material specification of existing building components. Laboratory test results are used as a benchmark to obtain a better specification of building

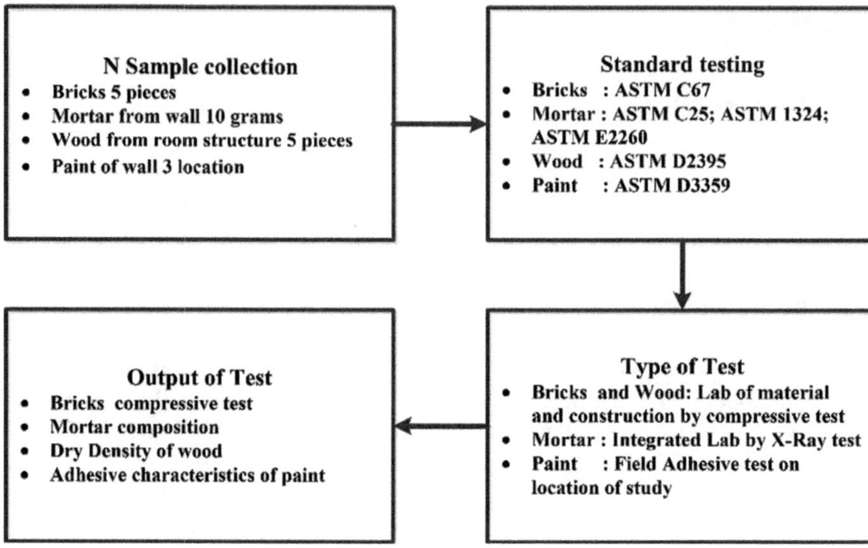

Fig. 1 Laboratory and material test activities

repair or at least at the same grade as the required criteria. The analysis referred to various standards. There are four tested cultural heritage building components, i.e., bricks, mortar, wood and paint.

The source of quantitative data came from the results of the laboratory tests. The tests were conducted on existing building material to obtain information regarding cultural heritage building characteristics. The tests were conducted on materials collected from an existing building, *Soesman Kantoor*.

In the case study building, there are four primary testing materials. i.e., bricks, mortar, wood and paint. The sample collection and testing were conducted by referring to the American Society for Testing and Materials (ASTM). The flow of materials testing activity from materials collection to laboratory tests and the output can be seen in Fig. 1.

2 Results and Discussion

2.1 Materials Test of Soesman Kantoor's *Building*

a. Brick Compressive Strength Test Results
   The compressive strength of brick samples taken from *Soesman Kantoor* was tested. The compressive strength test aims to acquire the required minimum specifications of brick compressive strength to replace the new bricks during the reparation. The selected bricks should have a greater or equal compressive

strength to avoid even greater damage to the existing building. The compressive strength test used automatic cement concrete compression machines with a 3000 kN capacity. The compressive strength test results were obtained from the load-receive brick cross-section area divided by the maximum load from the compression machine. The compressive strength test resulted in the compressive strength of each brick [6].

Older bricks are generally larger than modern ones. The dimensions of the bricks in *Soesman Kantoor* are 25 cm × 12 cm × 5 cm for length, width and height, respectively. Bricks with the same dimensions can preserve the building's aesthetics. In addition to dimensions, colour and texture are also important factors to consider [7]. The bricks in *Soesman Kantoor* are of a rectangular prism shape, with a reddish-orange colour and a smooth surface.

b. Mortar Composition Test Results

Mortar was tested to reveal the composition of the aggregate. The test used a petrographic method to find the chemical content inside the components of the mortar [8, 9]. Petrographic mortar with microscopic and stereoscopic irradiation principles that involved X-ray and electron microscopy was used. The petrographic test could be tested with wavelength dispersive X-ray fluorescence (WDXRF) spectroscopy.

WDXRF spectroscopy is an analytical technique used to detect chemical elements in solid and liquid materials. X-ray fluorescence (XRF) is a "secondary" characteristic X-ray emission (or fluorescence) of materials excited through high-energy X- or gamma rays. WDXRF is known for its high accuracy, precision, and reliability.

The mortar test sample was taken from the indoor wall of *Soesman Kantoor*. The minimum mass of samples tested was 10 g. The mortar composition test was conducted with a petrographic examination utilising WDXRF spectroscopy to analyse the metal contents in solid and liquid materials, both metal elemental and metal oxide. Generally, lime mortar consists of several primary chemical compositions, including CaO, MgO, SiO_2, Fe_2O_3, Al_2O_3 and CO_2 [10].

c. Wood Specific Gravity Test Results

The testing of wood samples was completed using a wood specific gravity test. This test aims to discover the minimum wood specific gravity that is to be used in the building component repair of the existing cultural heritage building. Wood specific gravity is generally proportional to wood strength or its mechanical properties [11]. The greater the wood specific gravity, the stronger the wood. Wood specific gravity compares the oven dry wood weight at 105 °C and the weight of water with a volume equal to that of the wood. The specifications of new materials must be better or equal to avoid further damage.

After being weighed, the wood sample was dried in a laboratory bench oven and with digital control to eliminate moisture from the wood samples. The final wood sample gravity was obtained from the weighing after being dried in an oven under 103 ± 2 °C for 24 hours The results of wood specific gravity were obtained from the wood sample dry weight divided by moisture level and volume. The mean specific gravity from the five wood samples was 0.571 gr/cm^3.

d. Paint Adhesive Power Test Results
The paint adhesive power test was directly conducted on the *Soesman Kantoor* exterior wall utilising a cross cut tape test method. The test was completed on three walls in different locations of *Soesman Kantoor*. The part of the wall to be tested must be smooth and clean. It was then cut longitudinally and circularly, each with six lines of a 2 cm length and a 2 mm gap between each line. We observed the test area surface on the cutting line area. Sticky tape with a 1-inch width was then applied until the whole area was closed and the tape was flattened with fingers or additional tape. The tape was taken off with a near 180° direction, with constant speed and at once. The condition of the paint surface was observed [12]. In location 1 (old wall), the test result showed that the paint's grade was zero because more than 65% of the paint surface was peeled. For the side wall (location 2), the test result showed a grade 2 paint indicated from the peeled paint at the side of the cutting line and some parts of the square area. The peeled paint area was ~15–35%. For the third location (front wall), the test result showed a grade 5 paint, as seen from the zero-peeled paint on the test area where the cutting edge also remained intact.

2.2 *Refurbishment Strategy Formulation for* Soesman Kantoor

Brick Repair

The repair was completed by first determining the correct replacement materials. The criterion for alternative brick materials was materials with a minimum compressive strength of 100.63 kg/cm^2. Brick wall repair that included an area of more than 0.19–0.27 m^2 or more than 20% of an existing wall, or exceeding 2.4–3.0 m length, as a required consideration to meet the physical strength criteria of the brick. It needs to meet the physical strength criteria to level the building movements due to the load distribution and expansion and shrinkage of the existing and new bricks. The installation of new bricks with different specifications might cause cracks and fracturing of the bricks.

Another aspect to consider is the appearance of the bricks, which includes their dimensions, colour and texture. The old brick dimensions are different from the size of the contemporary brick. It is suggested to ask the brick manufacturer for the specific brick size to make it more economical. The most crucial must-match dimension is for the brick parts to be exposed [7]. For the other unexposed parts, a variation in size is allowed as long as it does not obstruct the repair. The dimensions of bricks known as the government's model are $26 \times 12.4 \times 52$ cm.

Contemporary bricks with the same colour and texture, but larger than the existing bricks, could be used by cutting it with a wet saw to the desired size. This is an economical way for the small needs of bricks. Using materials other than bricks is not recommended due to the building's visual aspect. Repair work on new brick components must be marked to identify the components that have been repaired.

Mortar Repair

Mortar repair on the cultural heritage building wall was completed by stripping the old mortar layer, a process known as repointing [11]. Repointing increases weather resistance, reduces water seepage, maintains structural strength and appearance, and extends the building's lifetime. Field inspection was essential to point the damaged mortar. If the previous mortar repair was not matched to the original mortar, it needed to be stripped. Original mortar identification is crucial to obtain a suitable new mortar. The colour and texture of the original mortar had to be investigated visually, while the mortar composition was observed in the laboratory.

After getting the correct mortar composition, a test must be carried out to compare to the original mortar, in terms of colour, texture and the finishing of the surface. This assessment was carried out once the mortar hardened. The new mortar trial assessment must represent the conditions and surface material to be peeled and have proper lighting and the same environmental condition. The mortar repair must use compositions similar to the existing conditions so it can sufficiently stick. The composition was acquired from the laboratory test results with WDXRF to find the ratio of the correct mortar constituent components. The most dominant component of mortar test results in the laboratory were calcium oxide (CaO), which represents lime material with ~85.2%, ferric oxide (Fe_2O_3)/clay with ~32.4%, silica (SiO_2)/sand with ~16.4% and aluminium oxide (Al_2O_3)/clay with ~3.96%. The composition of the mortar mixture according to the existing test results were lime:clay:sand with a ratio of 8:3:1. The laboratory test results show the composition of the old building mortar, but it is still in the percentage of mass per molecule that significantly involves oxygen and causes the mass to exceed 100%. The mass per element is acquired through the calculation that involves atom mass and molecular weight.

The simple lime mortar process was completed by burning the lime in a furnace to form quicklime. The lime was then mixed with water to form slaked lime, which forms a hydrated lime putty or lime powder. The powder was then mixed with sand and water to form mortar. The type of lime mortar that is known as non-hydraulic has a prolonged setting time through the reaction with carbon dioxide in the air. The setting speed can be increased by using limestone, which is burned in a furnace, to form a hydraulic lime that breaks down if it comes into contact with water. The application of mortar to the gap of the bricks must be compact and should not leave any cavities. The application must consist of double layers, where each is compacted beforehand.

Wood Repair

The wood component repair was carried out by determining the material with a specific gravity that was better or at least equivalent to the existing wood component. The wood specific gravity that must be met is 0.571 gr/cm^3 or equivalent to wood strength grade level II, according to SNI 7973:2013 [12].

Structural reinforcement in cultural heritage buildings was carried out by replacing wood materials with the same quality. There are 17 various types of wood in Indonesia with various specific gravities that can be used as replacements in heritage buildings, including timber (*Lagerstroemia speciosa*), teak tree (*Tectona grandis*), rubber (*Hevea brasiliensis*) and cinnamon (*Cinnamomum purrectum*). Knowledge of the various types of wood would reduce the dependency on a specific type of wood, so that other types of wood that have not been used could increase. Wood material repair for construction elements could use alternative wood types, but it is better to use wood with a better or at least equivalent grade.

In addition to structural strengthening, damages caused by termites must also be replaced by replacing wood materials of the same quality. To avoid termites, maintenance is essential by spraying with anti-termite liquid and traditional cleaning using cloves, tobacco and banana fronds with the ratio of 1:1:1.

Differently coloured wood components could be repaired by means of camouflage, i.e., change the shape and materials to make it look like the original. Maintenance that could be carried out includes monitoring climatological conditions with an extreme thermometer [13].

Paint Repair

Repainting cultural heritage buildings requires special paint that allows water vapor to move freely through walls. Painting with a mixture of cement and regular sand or cement plaster will only cause cracks and will not bond well with the old cement-sand mixture that sticks to the wall. Plaster made of regular sand cement and cement plaster would be too stiff and hard, thereby potentially causing cracks on the wall.

Paints must be breathable, but it could not be found easily in paint products in the modern era because modern buildings tend to seek waterproof paints to prevent rain from getting into buildings. Coloured mortar could be acquired by adding pigments into the mortar mixture. This method can only be achieved if the desired colour can be achieved using aggregates that can provide that colour. The breathable paint was made from a lime-based mixture. Lime could also provide a white colour if it is the desired one. The paint composition used referred to the BOW analysis. The mixture for a 10 m^2 paint was one part *Tras Muria*, one part limestone and three parts sand.

The pigment added into the mortar must be smooth to be mixed evenly, but it should not cause a reaction that can damage the mortar. This kind of pigment could be achieved through pigments with a metal-oxide nature, i.e., lime (CaO). The pigment addition must be carried out in moderation to maintain the composition and mortar performance. The suggested pigment addition is a maximum of 10% of the total mixture.

The advantage of using lime is its flexibility and its resistance to shrinkage cracking from wall movements. The cracking that occurs due to the structural movement can occasionally "self-heal" with water circulation. The coloured mortar also does not need to worry about adhesion because it is homogeneous with the mortar. The disadvantage of lime is that it requires a long time to harden. It could turn soft and crumble if continuously exposed to water. The summary of refurbishment strategies as presented in Table 1.

Table 1 Summary of refurbishment methods

Component and characteristics	Visual	Repair method
Bricks Dimensions 25 cm × 12 cm × 5 cm Colour Dark orange Compressive Strength	Long rectangular prism brick, flat and no-cracked surface, and angled corners. Dark orange bricks or with as close as possible colour to the existing bricks if it is exposed	Repair using bricks with a minimum compressive strength of 105.03 kg/cm^2. The new brick dimension must be made as close as possible to the old one, i.e., 25 cm × 12 cm × 5 cm The colour and texture of the bricks are adjusted to the existing one to maintain the aesthetic value, i.e., reddish orange with a flat surface Alternative material is not recommended because it might damage the appearance of the new brick wall
Mortar Colour Composition	The colour is dominantly dark white due to the mixture of lime and clay paste	Making mortar with a composition significantly similar to the existing mortar, i.e., the mixture of lime:clay:sand with the ratio of 8:3:1. Application
Wood Dimension Colour Specific gravity	The wood dimension depends on the repaired components The colour of the wood is adjusted to the original condition	The structural wood repair must meet the conditions of minimum wood specific gravity of 0.571 gr/cm^3. The type of wood may vary as long as the specific gravity is sufficient. Joinery could be done if the desired wood dimension is unavailable, by paying attention on the technical guidelines of *SNI* 7973, 2013 Non-structural component repair
Paint Composition Waterproof paint is not allowed Grade 5 is the expected adhesive power, in which no paint is peeled during the tape test	White paint or adjusted to the colour of the original paint	Repair using pigment-coloured mortar. The pigment used could come from one of the aggregates with the desired colour. Lime aggregate works as an adhesive and pro-vide white colour to the paint. The adhesive power in paint is not a problem because the paint itself is the mortar's homogeneous part

3 Conclusions

It is strategic to conduct a comprehensive investigation before repairing cultural heritage buildings to acquire thorough information regarding the buildings. The laboratory test on *Soesman Kantoor* building components is a way to obtain the characteristics of each material. The testing was carried out on the elements with the most damaged materials, i.e., bricks, mortar, wood and paint.

The differences in materials and manufacturing methods of two time periods show the technology development that becomes more efficient each day. However, the material authenticity and work methods must be maintained as the original; thus, modern materials and manufacturing methods cannot be carelessly applied.

There is a wide range of cultural heritage building components, so it could not be generalised. The repair must be done per case because each issue needs a different solution and could not be equated to other building components. The repair guidelines of a cultural heritage building with *Soesman Kantoor* as the case study focuses on brick, mortar, wood and paint components. All components could be repaired by following the minimum criteria for replacement materials. If a material cannot be repaired as well as the original, it can be replaced as long as it is not invasive and reversible.

Acknowledgements Thankful to research group of Heritage from Laboratory of Construction Management, Department of Civil Engineering, Diponegoro University and support funding from Directorate General Empowering Research and Development Ministry of Research and Technology, National Research Council and Innovation-Budgetary Year 2020, Scheme of Master Thesis Research Contract Number 225-52/UN7.6.1/PP/2020.

References

1. Regulation No.11 regarding Cultural Heritage. Jakarta. President Republic of Indonesia (2010)
2. ICOMOS (1975) European charter of the architecture heritage 1975 Art. 6
3. Utami P (2013) Bangunan Tua di Kota Lama Semarang Roboh. Kompas, 13 Januari
4. UNESCO (2016) The HUL guidebook managing heritage in dynamic and constantly changing urban environments
5. Rasmussen TV (2014) Model for refurbishment of heritage buildings. ICVEST 2014 building for a changing world, 2
6. ASTM C62-17 (2017) Standard specification for building brick (Solid masonry units made from clay or shale). ASTM International, West Conshohocken, PA
7. The Brick Industry Association (2011) Brick brief: brick replacement, Virginia, USA (gobrick.com). Accessed Oct 2020
8. ASTM C141/C141M-14 (2014) Standard specification for hydrated hydraulic lime for structural purposes. ASTM International, West Conshohocken, PA
9. ASTM D2395-07ae1 (2007) Standard test methods for specific gravity of wood and wood based materials. ASTM International, West Conshohocken, PA
10. ASTM D3359-09e2 (2009) Standard test methods for measuring adhesion by tape test. ASTM International, West Conshohocken, PA

11. ASTM E2260-03(2012)e1 (2012) Standard guide for repointing (tuckpointing) historic masonry. ASTM International, West Conshohocken, PA
12. SNI 7973:2013 (2013) Spesifikasi Desain untuk Konstruksi Kayu, BSN, Jakarta
13. Ardian M (2018) Manual Pemeliharaan Bangunan Menara, Masjid dan Makam Sunan Kudus, Kudus. (docplayer.info). Accessed Aug 2020

Self-monitoring and Localization of Crack of Concrete Beam with Fibers and Carbon Black Subjected to Bending

Genjin Liu and Yining Ding

Abstract In order to endow concrete an ability of sensing cracks, conductive materials (carbon fiber, steel fiber and carbon black) are added into concrete. The effect of the conductive materials on the post crack performance of concrete subjected to bending is studied. In order to localize the crack, multiple sections of a concrete beam under bending are monitored. The influence of the conductive materials on the self-sensing property of the crack opening displacement (COD) of fiber reinforced concrete was investigated. The results illustrated that there is a linear relationship between COD and FCR. The self-localization of the flexural cracks will be realized based on the fractional change in resistance of concrete.

Keywords Self-localization of crack · Self-sensing of concrete crack · Fiber · Fractional change in resistance

1 Introduction

Rehabilitation and maintenance in civil engineering is not less important than planning and designing. The excessive load, natural hazard, lack of construction procedures and material degradation are causes of infrastructure defects. It is well-known, reinforced concrete works with cracks in the service stage. Concrete structure may lose its serviceability if excessive cracking happens.

Rehabilitation and maintenance help to extend the service life of RC. Structure repairing and strengthening belong to this topic. Localizing the crack and sensing the crack widening of concrete are prerequisite of the rehabilitation of RC. Some literatures have studied the strain sensing ability of short carbon fiber (CF) cementitious materials; However, the monitoring ability range is limited in the elastic period before cracking because the short carbon fibers are incapable to bridge the macro

G. Liu · Y. Ding (✉)
State Key Laboratory of Coastal and Offshore Engineering, Dalian University of Technology, Dalian 116024, China
e-mail: ynding@dlut.edu.cn

S. A. Kristiawan et al. (eds.), *Proceedings of the 5th International Conference on Rehabilitation and Maintenance in Civil Engineering*, Lecture Notes in Civil Engineering 225,
https://doi.org/10.1007/978-981-16-9348-9_16

crack [1–3]. In addition, carbon fibers are brittle material, which hardly improve the toughness of cementitious matrix obviously. One way to enhance the crack-bridge ability and toughness of cementitious materials is to add macro steel fibers (SF) into the matrix [4–6].

The self-sensing ability of cementitious materials with coarse aggregate are less focused. The experimental investigations on concrete reinforced with carbon fiber (CF), carbon black (CB) and macro steel fiber (SF) for crack sensing. The combination of two distinct fibers (SF, CF) as admixtures in concrete are effective for enhancing the electric conductivity due to the formation of a continuous conducting path way [3]. In addition, the carbon black with its high conductivity, low cost and fine filler effect can be used as ideal admixture for conductive concrete before cracking [7]. In the composite, CB exhibits the short-range contacts, whereas CF and SF exhibit a long-range conductive bridging effect [8].

This paper mainly focuses on the post-crack self-sensing behavior of deflection softening and deflection hardening concrete beams added with CF, CB and SF. The results provide compelling evidence that a strong relationship exists between resistivity and the load bearing capacity change and crack propagation. Furthermore, how to localize of concrete crack is still an open problem, and the localization of crack is a prerequisite for accurately self-sensing in crack widening of concrete. The purpose of this article also develops a method for sensing the crack location and development of concrete under bending.

2 Experimental Program

2.1 Materials and Mixture Design

The base mix design of the concrete is listed in Table 1. The CB, CF and SF content is between 0–1, 0–2 and 40–60 kg/m^3. The diameter of CB is smaller than 90 nm. The diameter of CF is 12 μm and length is 6 mm. The diameter of macro SF is 0.55 mm and the length is 35 mm. Methylcellulose (0.4%wt) and defoamer (0.19%vol) are added for CF reinforced specimens. The content of the conductive materials for the concrete samples are listed in Table 2.

Table 1 Mix design (kg/m^3)

Cement CEM I 42.5	Fly ash	Water	Fine aggregate (0–5 mm)	Coarse aggregate (5–10 mm)	Superplasticizer
390	155	272.5	848	822	5.5

Table 2 Comparison of content of conductive admixtures

Specimen	SF (kg/m^3)	CB (kg/m^3)	CF (kg/m^3)
SF40	40	–	–
SF60	60	–	–
SF40B1	40	1	–
SF60B1	60	1	–
SF40CF2B1	40	1	2
SF60CF2B1	60	1	2

2.2 Test Set-up

The concrete mixtures with conductive materials are produced in a laboratory mixer. Specimens designed for bending test are beams with the size of 100 mm × 100 mm × 400 mm. Four-probe method is adopted for electric resistance measurement, outer electrical contacts A and D (130 mm apart) are designed for passing current while the inner contacts B and C are prepared for measuring of voltage. Voltmeter V_1 and V_2 are designed to measure the resistance of the pure bending part of the beam and fixed resistor R_f, respectively.

The Relationships between FCR (Fractional change in resistance) and COD (crack opening displacement) are discovered and established by the experimental investigation. Details of the beams and electrical measuring system are shown in Fig. 1a.

2.3 Testing Methods

The test set up are shown in Fig. 1b. Based on ASTM C1609, the experiment tests a simply supported beam under third-point loading [9]. The deformation rate of mid-span is 0.2 mm/min. LVDTs are used to measure the deflection of mid-span. An extensometer is attached at the mid-span of the beam to measure the COD during the test.

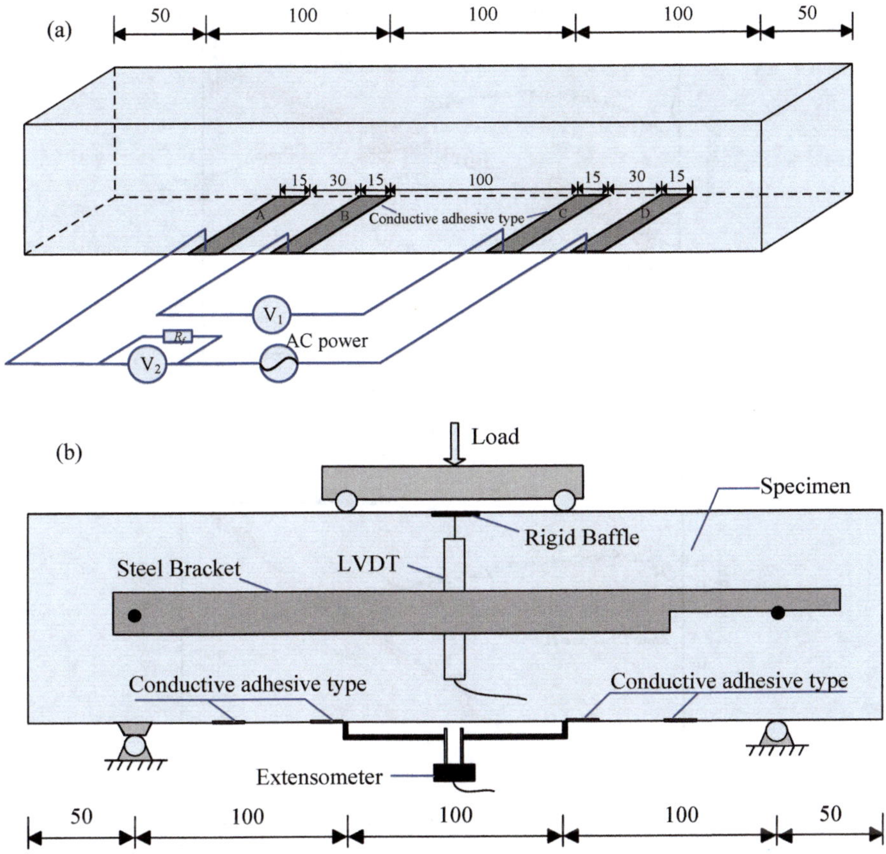

Fig. 1 Sample configuration for measuring the resistivity of the specimen, **a** probes; **b** schematic of the experimental set-up

3 Results and Discussion

3.1 Crack Sensing

Figure 2a–f display the relationship between FCR and COD on the tension side of the beams. From Fig. 2 it can be observed that the relationships between FCR and COD for beams with 40 kg/m^3 SF (series 40 specimens) are different to that of beams with 60 kg/m^3 SF (series 60 specimens).

For series 40 specimens, a monotone linearity increasing relationship between FCR and COD can be found. In order to analysis the effect on the self-sensing performance of different specimens, the gauge factor (GF), defined as the fractional change in resistance per unit crack opening displacement, is introduced to quantify the slop of the FCR–COD relationships. The GF values can be acquired from the

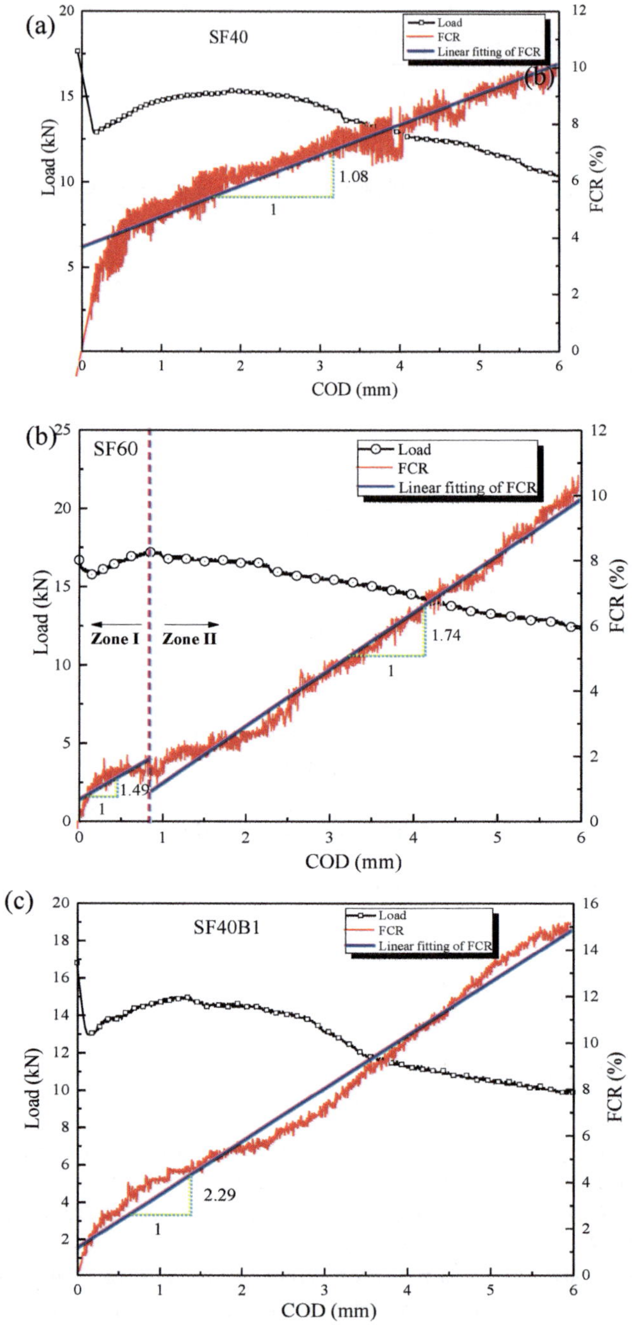

Fig. 2 Relationship between load and COD, FCR and COD of **a** SF40, **b** SF60, **c** SF40B1, **d** SF60B1, **e** SF40CF2B1, **f** SF60CF2B1 [10]

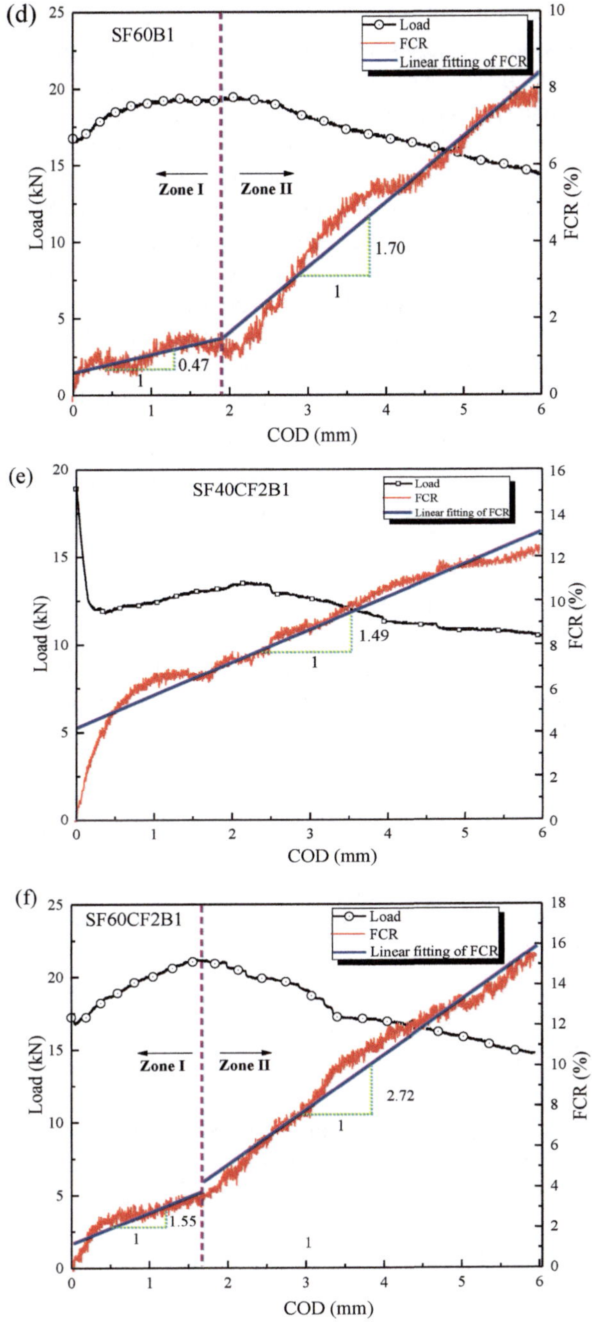

Fig. 2 (continued)

slope of the regression line fitted to the data from FCR–COD curve for the beams (Fig. 2a, c, e). However, for series 60 specimens, bi-linearity increasing relationships between FCR and COD are observed instead (see in Fig. 2b, d, f). The slope of FCR and COD curve changes when the load reaches to the peak load. GF_1 is the slope of the regression line fitted to those data from FCR–COD curve in zone I, corresponding to multiple cracks initial and propagation period before the peak load. And GF_2 is the slope of the regression line fitted to those data from FCR–COD curve in zone II, in this region crack localization happened after post peak load.

The GF index are exhibited in Fig. 2, the GF varies from 0.47 to 2.72 for the specimens. And it can be concluded that, GF_2 is larger than GF_1 for series 60 specimens. Using specimen SF60B1 as a representative specimen, GF_2(1.70) is 3.6 times GF_1. It means that, crack sensing ability of the specimen is better in crack localization period than that in multiple cracks initial and propagation period.

- Compared with beam SF40, the GF of SF40B1 increased by 112%. And GF of SF60B1 is 68.4% less than that of SF60 in zone I; however, the GF values are almost the same with each other in zone II for the two specimen. It can be concluded that 1 kg/m^3 of CB will enhance the GF of concrete beam reinforced with 40 kg/m^3 steel fibers, but for beams with 60 kg/m^3 steel fibers if added with 1 kg/m^3 of CB the GF will reduce in zone I.
- Compared with SF40B1, the GF of SF40B1CF2 decreased by 34.9%. Whereas, GF of SF60CF2B1 is 229.8 and 60% higher than that of SF60B1 in zone I and zone II, respectively. It indicates that no obvious effect on the sensitivity with the addition of 2 kg/m^3 CF.
- SF40B1 doped with 40 kg/m^3 SF and 1 kg/m^3 CB possess the highest GF when compared with the other beams of series 40 specimens.
- SF60CF2B1 doped with 60 kg/m^3 SF, 2 kg/m^3 CF, and 1 kg/m^3 CB possess the highest GF when compared with the other beams of series 60 specimens. In addition, specimen with triphasic conductive materials (SF60CF2B1) provides more stable GF in both zone during the test than specimen with diphasic conductive materials (SF 60B1).

3.2 Crack Location

In order to realize the crack region location, concrete beam with 40 kg/m^3 steel fiber and 1.5 kg/m^3 carbon black is produced. Monitoring zone of the specimen is divided into two parts, namely section AB (in red colour) and section BC (in blue colour), see in Fig. 3.

Figure 4 exhibits the load-time-FCR curves of the SF40B1.5 under flexural load. From Fig. 4, it can be seen that the load-time curve of the specimen increases up to the peak load corresponding to the cracking point. Followed that, a great drop down of loading (30% of the peak load) is observed during crack happening.

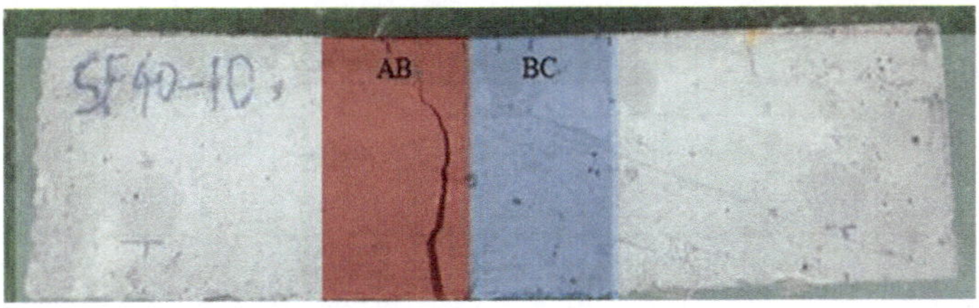

Fig. 3 Cracking pattern of SF40B1.5

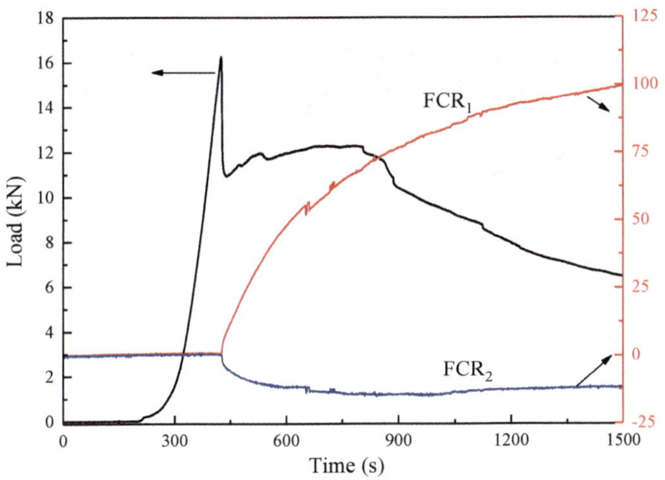

Fig. 4 Load-time-FCR curve of specimen SF40B1.5

According to Fig. 4, FCR of each section of the specimen almost equals to zero before crack. In addition, FCR-Time curve of section AB (red curve) increase rapidly after crack (at 426th seconds), however, the FCR-Time curve of section BC (blue curve) decreased slightly (up to 10%) after crack.

The re-distribution of the potential of the specimen leads to the difference of FCR curves for the cracked section (section AB) and the un-cracked section (section BC). Part of the conductive paths can be cut off by the crack appeared in section AB, meanwhile, the electrical current can only flow through the un-cracked concrete area and the way bridged by the steel fibers.

A large beam with size of 120 mm × 200 mm × 1400 mm is also made attached with different distance of inner electrodes. The distance of the inner electrodes are 200 mm (A_1–A_2), 400 mm (B_1–B_2), 500 mm (C_1–C_2), 600 mm (D_1–D_2) and 800 mm (E_1–E_2). Figure 5 illustrates the schematic diagram of electrodes. Figure 6 displays the electrodes of the smart concrete beam with 20 kg/m^3 and 1 kg/m^3 carbon black. The FCR-time-load curves of the smart concrete beam under bending is shown in Fig. 7. From Fig. 7, it can be seen that the value of FCR for all the sections (besides section A1–A2) increase rapidly after the crack appears. However, section A1–A2

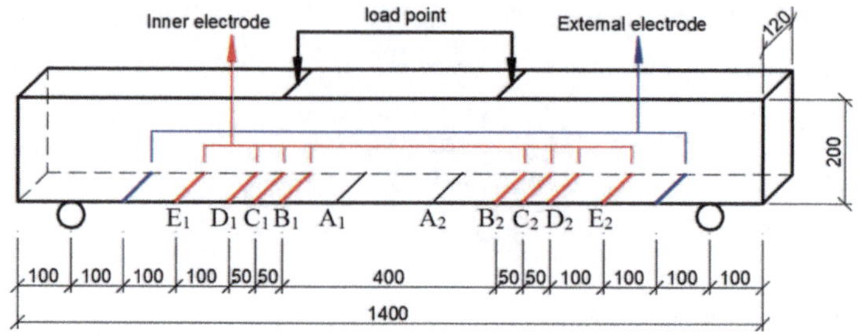

Fig. 5 Schematic diagram of electrodes

Fig. 6 Electrodes of the smart concrete beam

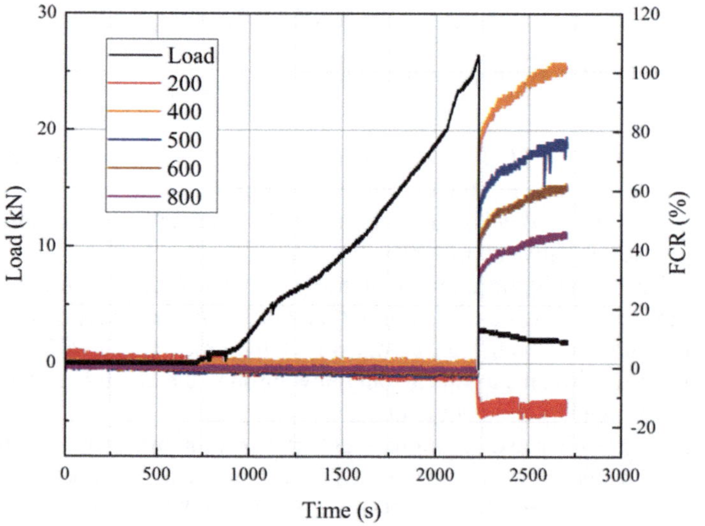

Fig. 7 FCR-time-load curves of the smart concrete beam

shows a slightly reduction of FCR. Thus, it can be judged that the crack is outside the section A1–A2 and inside the section B1–B2. From the experiment, a crack is obeserved in section A2–B2. That implies that the crack location method is effective to a certain extent.

4 Conclusion

(1) Self-sensing of concrete crack can be achieved by adding conductive materials like SF/CB/CF into concrete. As the load bearing capacity declines rapidly at the first cracking point, the FCR of all beams shows a clear increasing trend.
(2) A linearly increasing of FCR–COD correlation can be found for series 40 specimens. And, a bi-linearly increasing of FCR–COD are observed, and GF_2 in zone II is larger than GF_1 in zone I.
(3) The self-localization of the flexural cracks can be realized based on the fractional change in resistance of concrete.

References

1. Ding Y, Chen Z, Han Z et al (2013) Nano-carbon black and carbon fiber as conductive materials for the diagnosing of the damage of concrete beam. Constr Build Mater 43(Jun):233–241
2. Ding Y, Huang Y, Zhang Y et al (2015) Self-monitoring of freeze–thaw damage using triphasic electric conductive concrete. Constr Build Mater 101:440–446
3. Ding Y, Han Z, Zhang Y et al (2016) Concrete with triphasic conductive materials for self-monitoring of cracking development subjected to flexure. Compos Struct 138:184–191
4. Banthia N, Nandakumar N (2003) Crack growth resistance of hybrid fiber reinforced cement composites. Cement Concr Compos 25(1):3–9
5. Banthia N, Sappakittipakorn M (2007) Toughness enhancement in steel fiber reinforced concrete through fiber hybridization. Cem Concr Res 37(9):1366–1372
6. Buratti N, Mazzotti C, Savoia M (2011) Post-cracking behaviour of steel and macro-synthetic fibre-reinforced concretes. Constr Build Mater 25(5):2713–2722
7. Wen S, Chung D (2007) Partial replacement of carbon fiber by carbon black in multifunctional cement–matrix composites. Carbon 45(3):505–513
8. Wu S, Mo L, Shui Z et al (2005) Investigation of the conductivity of asphalt concrete containing conductive fillers. Carbon 43(7):1358–1363
9. ASTM C 1609 (2012) Standard test method for flexural performance of fiber-reinforced concrete (Using beam with third-point loading)
10. Ding Y, Liu G, Hussain A et al (2019) Effect of steel fiber and carbon black on the self-sensing ability of concrete cracks under bending. Constr Build Mater 207:630–639

Seismic Performance Analysis of Multi-story Buildings with Addition of Bracing Based on SNI 1726: 2019 (Case Study: Airlangga University Parking Building)

Krisnamurti, Willy Kriswardhana, and Achmad Wahyu Ramadiyan

Abstract There are various alternatives to improve the performance of a structure in resisting lateral forces. One of them is the addition of types X, V, Z, and inverted V braces or stiffeners to the elements of the portal structure. The research aims to study the effect of X and V's bracing types on the behavior of special moment-resisting frame systems of building using the response spectrum analysis method. For analysis purposes, 3 models of the frame were used, namely without bracing, with type X braced frame, and type V braced frame. The three models can be used to fulfill the requirement of service limit performance, ultimate limit performance, and control of the P-delta effect. The biggest drift between floors occurs in modeling without bracing. Significant changes occur in modeling with bracing type X and type V with a drift reduction of 51.5 and 63.8%. The result shows that the use of bracing can increase the stiffness and strength of the structure. The optimal bracing performance in this study is bracing with type X.

Keywords Moment-resisting frame systems · Bracing · Response spectrum analysis

1 Introduction

Indonesia has a high potential for earthquakes, one of which is in East Java. There are two sources of earthquakes in the province of East Java, namely a plate collision in the south of East Java (crack of the Indo-Australian plate) and an active fault in Java (Baribis Kendeng Fold Thrust Zone) [1], both sources threaten most areas in East Java and potentially a tsunami. The active fault passes through several cities, one of which is Surabaya [2]. The large earthquake potential has a direct impact on the buildings in the city, especially high-rise buildings. The little resistance to

Krisnamurti (✉) · W. Kriswardhana · A. W. Ramadiyan
Department of Civil Engineering, University of Jember, Jalan Kalimantan 37 Kampus Bumi Tegalboto, Jember 68121, Indonesia
e-mail: murti_krisna.teknik@unej.ac.id

S. A. Kristiawan et al. (eds.), *Proceedings of the 5th International Conference on Rehabilitation and Maintenance in Civil Engineering*, Lecture Notes in Civil Engineering 225,
https://doi.org/10.1007/978-981-16-9348-9_17

earthquake loads can cause the building to sway and even collapse and endanger human safety.

The design and analysis of earthquake-resistant building structures in Indonesia are guided by the latest regulation, namely SNI 1726-2019 concerning "Earthquake Resistance Planning System for Buildings and Non-Buildings." SNI 1726-2019 contains several methods for analyzing earthquake resistance in a building, namely response spectrum analysis methods. This analysis method used to determine the behavior and seismic forces acting on the structure [3] dan to calculate the displacement and drift of the building [4]. Several structural systems can be used as earthquake resistance, as shear wall, braced frame, and moment-resisting frame system. The use of shear walls and bracing in the same building have the same performance level [5]. It stated that adding bracing with the right location can significantly improve the structure's performance [6]. In general, bracing is placed diagonal using various configurations in the structure's portal. The stiffness increased efficiently using bracing because a diagonally mounted brace can only withstand axial forces while receiving horizontal shear forces. The shape of concentric bracing can be the Z (diagonal), X, V, and inverted V (Λ) which can be seen in Fig. 1 [7].

In the comparison of bracing with the types X, V, Z, and inverted V, it is known that the optimum structural performance was obtained from bracing type X with the smallest displacement value compared to other types of bracing [8]. Meanwhile, the type V bracing placed on both building axes (x and y axes) will increase the performance of the structure and there is a decrease in displacement compared to structures without bracing [9].

Airlangga University Parking Building is a 10-story building with a ± 36,340 m^2 and a height of 30,8 m. The dimension of building is 112,8 × 32,3 m. The type of soil in the location is medium soil. The building is planned to contain 414 motorbikes on the first floor and 124 cars on the 2nd to 10th floors. With ten floors and large external loads, the building that can accommodate up to 1000 cars in its planning does not use shear walls and is guided by the SNI 1726:2012 regulations. This problem is the basis for this research to review the suitability of planning against regulations of SNI 1726-2019 and to add bracing as a substitute for shear walls. Based on the explanation above, the aim of this research is to analyze the seismic performance of the structure using the response spectrum analysis method in the parking building of Airlangga University using the braced frame variation.

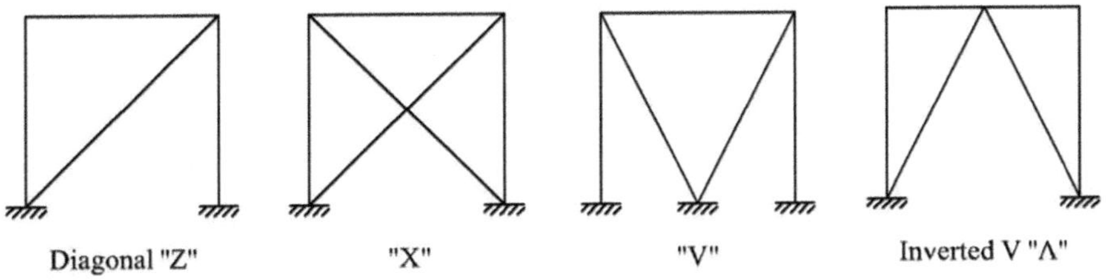

Fig. 1 Various forms of concentric braces [7]

2 Research Methodology

The structural data using a 10-story parking building, which is for 3D modeling. The structural data consist of floor plans, beams, columns, and plates along with the quality of concrete and reinforcing steel, as well as cross-sections and longitudinal sections to determine the height of each floor. The structural model used in this study is a 10-story parking building, which consists of 3 models. Model 1 is an existing building structure. Models 2 and 3 are a modification of model 1. Bracing is used as an element to resist lateral forces (earthquake forces). Model 2 is a model with type X bracing, while model 3 is a model with type V bracing. In practice, the bracing elements in this structure must be prepared as precast elements to avoid difficulties in the casting process (Fig. 2).

The materials quality are:

- the beam and slab concrete (*f*'c) = 25 MPa.
- the column and bracing concrete (*f*'c) = 25 MPa.
- the reinforcing steel (fy) = 240 MPa (BJTP 24).

$$\begin{aligned} \text{the reinforcing steel } (fy) &= 240 \text{ MPa(BJTP 24)}. \\ &= 400 \text{ MPa(BJTD 40)}. \end{aligned}$$

The 3D model of the building structure is in the Fig. 3, and Fig. 4. Dead load (DL), additional dead load (SDL), live load (LL), earthquake load for X direction (EX), and earthquake load for Y direction (EY) use as action load. Life and death loads refer to PPIUG 1987 (regarding building loading) and SNI 1727: 2013. The earthquake load refers to SNI 1726: 2019, and The Surabaya earthquake area is using with moderate ground conditions. Earthquake loading calculations using response spectrum analysis methods.

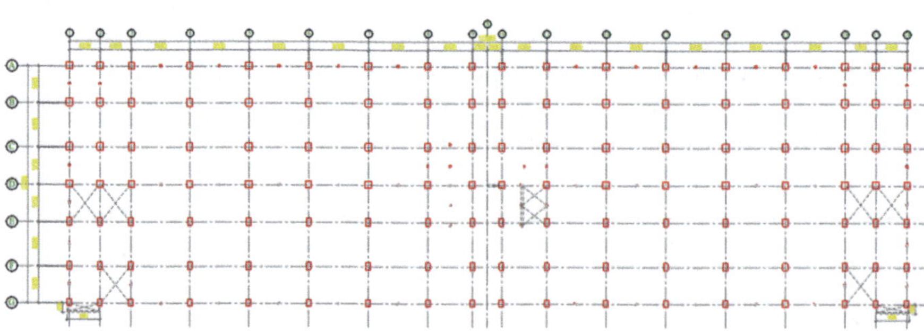

Fig. 2 Airlangga University parking building plan campus B

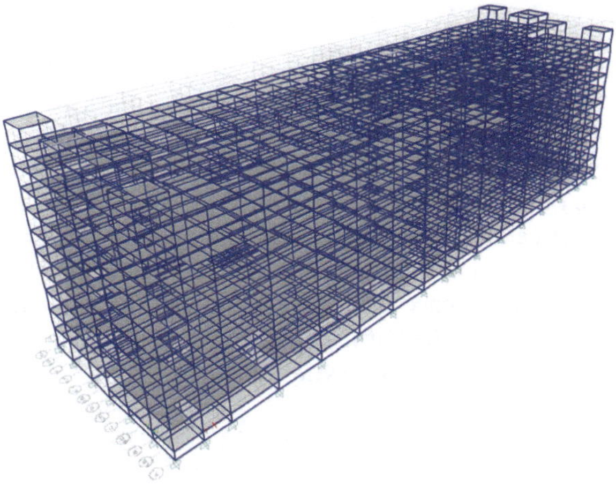

Fig. 3 3D building modeling without bracing

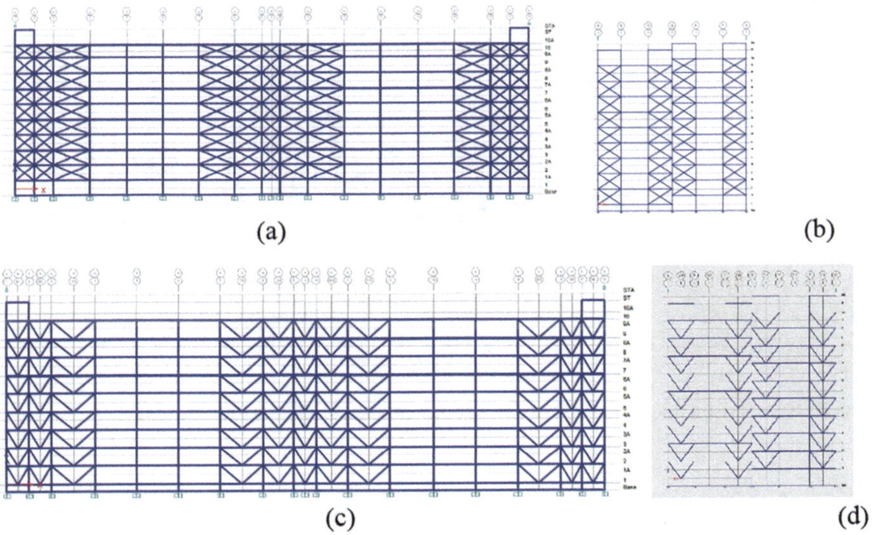

Fig. 4 **a** Front View of Type X Brace Placement in Building. **b** Side View of Type X Braces in the Building. **c** Front View of Type V Brace Placement in Building. **d** Side View of Type V Braces in the Building

3 Result and Discussion

Story drift is a determinant of a building's service limit performance. To determine the amount of story drift, a structural analysis calculation tool is used. The allowable limit for the change in the deviation between levels must not exceed the following equation:

$$\Delta_S < 0,03/R \times h \quad (1)$$

where Δ_S = storey drift displacement (mm), R = modification coefficient, and h = storey height. From Eq. (1) it is obtained the service limit performance value of 12.75 mm. Figure 5 and Fig. 6 shows the graph Δ_S value of each model and the direction of the earthquake, Figs. 5 and 6 show that the three models that have been made have met the service limit performance controls. The highest Δ_S value is experienced by building models without bracing while building models with bracing type X experience the lowest Δ_S value. The ultimate limit performance is regulated in SNI 1726: 2019 article 7.12.1 [10], with the deviation limit value between floors determined by the following equation:

$$\text{Storey drift limit } (\Delta_{limit}) = 0,015h_{sx} \quad (2)$$

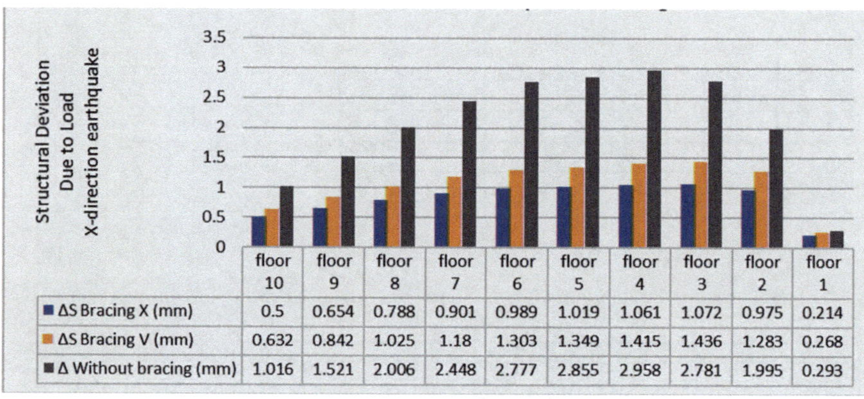

	floor 10	floor 9	floor 8	floor 7	floor 6	floor 5	floor 4	floor 3	floor 2	floor 1
ΔS Bracing X (mm)	0.5	0.654	0.788	0.901	0.989	1.019	1.061	1.072	0.975	0.214
ΔS Bracing V (mm)	0.632	0.842	1.025	1.18	1.303	1.349	1.415	1.436	1.283	0.268
Δ Without bracing (mm)	1.016	1.521	2.006	2.448	2.777	2.855	2.958	2.781	1.995	0.293

Fig. 5 Comparison diagram of service limit performance on X-direction earthquake

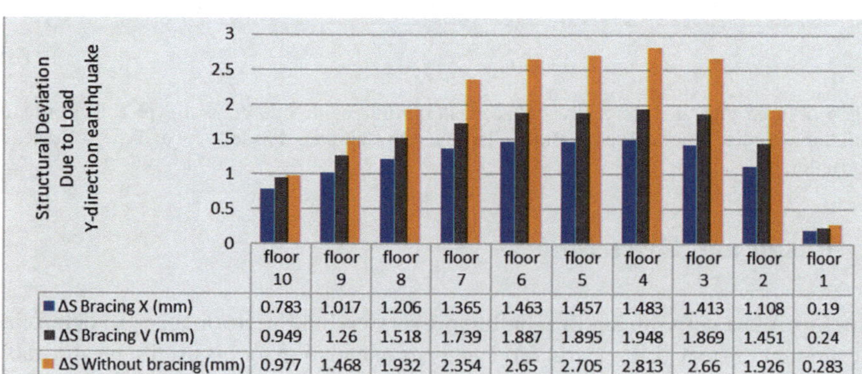

	floor 10	floor 9	floor 8	floor 7	floor 6	floor 5	floor 4	floor 3	floor 2	floor 1
ΔS Bracing X (mm)	0.783	1.017	1.206	1.365	1.463	1.457	1.483	1.413	1.108	0.19
ΔS Bracing V (mm)	0.949	1.26	1.518	1.739	1.887	1.895	1.948	1.869	1.451	0.24
ΔS Without bracing (mm)	0.977	1.468	1.932	2.354	2.65	2.705	2.813	2.66	1.926	0.283

Fig. 6 Comparison diagram of service limit performance on Y-direction earthquake

$$\text{Storey drift maximum } (\Delta_x) = \Delta_S \times \text{ Cd} \quad (3)$$

$$\Delta_x < \Delta_{limit} \quad (4)$$

where h_{sx} = height of the level below the x-level; Δ_S = storey drift displacement (mm); Cd = Deflection Factor. From Eq. (2) it is obtained the Ultimate Limit Performance value of 55 mm. Figure 8 and Fig. 9 show the graph Δ_x value of each model and the direction of the earthquake.

Figure 7 and Fig. 8 shows that the three models have met the Ultimate Limit Performance controls. The highest Δx value is experienced by building models without bracing while with bracing type X experience the lowest Δx value. Control of P-delta effects performance refers to the SNI 1726: 2019 clause 7.8.7, to determines

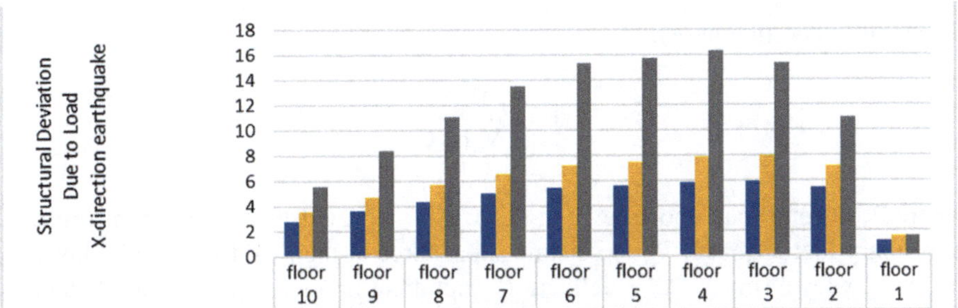

	floor 10	floor 9	floor 8	floor 7	floor 6	floor 5	floor 4	floor 3	floor 2	floor 1
Δx Bracing X (mm)	2.75	3.597	4.334	4.9555	5.4395	5.6045	5.8355	5.896	5.3625	1.177
Δx Bracing V (mm)	3.476	4.631	5.6375	6.49	7.1665	7.4195	7.7825	7.898	7.0565	1.474
Δx Without bracing (mm)	5.588	8.3655	11.033	13.464	15.2735	15.7025	16.269	15.2955	10.9725	1.6115

Fig. 7 Comparison diagram of ultimate limit performance on X-direction earthquake

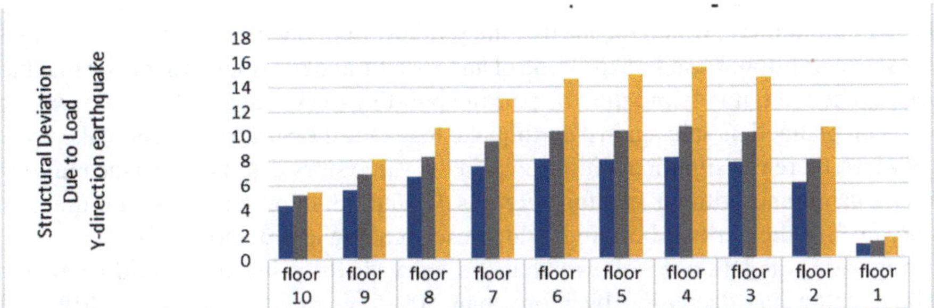

	floor 10	floor 9	floor 8	floor 7	floor 6	floor 5	floor 4	floor 3	floor 2	floor 1
Δy Bracing X (mm)	4.3065	5.5935	6.633	7.5075	8.0465	8.0135	8.1565	7.7715	6.094	1.045
Δy Bracing V (mm)	5.2195	6.93	8.349	9.5645	10.3785	10.4225	10.714	10.2795	7.9805	1.32
Δy Without bracing (mm)	5.3735	8.074	10.626	12.947	14.575	14.8775	15.4715	14.63	10.593	1.5565

Fig. 8 Comparison diagram of ultimate limit performance on Y-direction earthquake

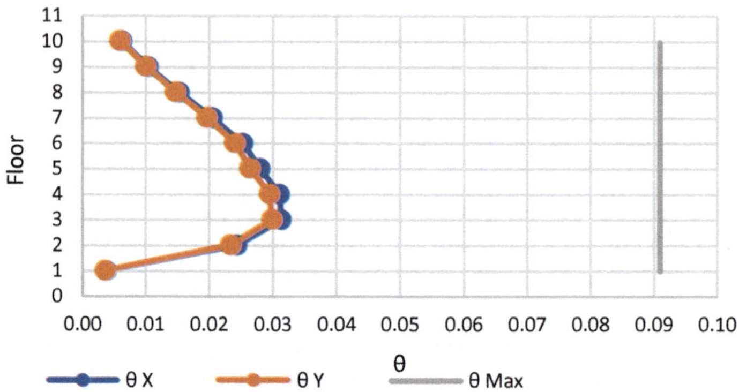

Fig. 9 Control diagram of P-Delta effect of buildings without bracing in the Surabaya zone earthquake with x and y directions load

the stability coefficient (θ):

$$\theta = \frac{P_x \Delta I_e}{V_x h_{zx} C_d} \tag{5}$$

where P_x = total load above level x, (kN); Δ = storey drift displacement (mm); I_e = primary factor quake; V_x = seismic shear force at the x and x-1 levels (kN); h_{zx} = height x level (mm); Cd = deflection magnification factor. The value of stability coefficient (θ) should not exceed θ_{max}. The value of θ_{max} determined by:

$$\theta_{max} = \frac{0.5}{\beta C_d} \tag{6}$$

From Eq. (6), the θ_{max} value is 0.09091.

Figures 9, 10, and 11 shown that the three models have θx and θy values below the maximum theta, which means that the p-delta effect that occurs in the building is in a safe condition. Based on the control and structural performance analysis that has been carried out, it is found that the parking building of Airlangga University, which has been planned, is safe against earthquake loads for the Surabaya earthquake zone. However, there is a significant difference in the value story drift between the building model using and without bracing. Figures 12 and 13 show a comparison diagram story drift of the parking building, which uses bracing and without bracing.

Figure 12 and Fig. 13 shown that the story drifts value of a building model with bracing has decreased by more than 50% compared to the story drift of a building model without bracing. This shows that bracing has a major effect on the performance of the building structure in resisting earthquake forces. Previous studies also mentioned the same statement. The performance of bracing structures is better compared to structures without bracing [6]. The picture above shows the structure's optimal performance value is a structure with the addition of X-type bracing when the

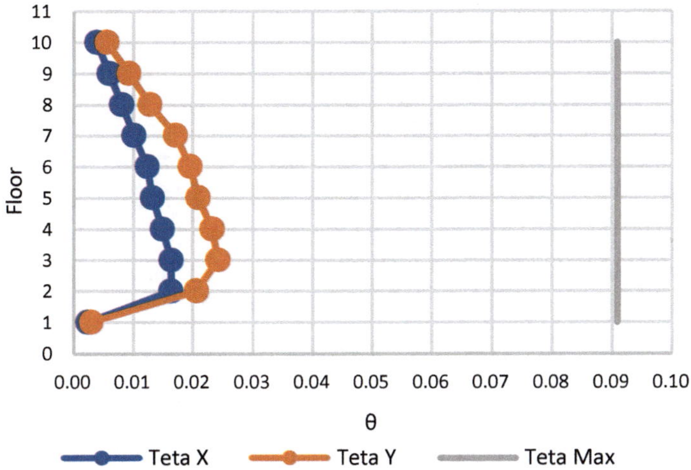

Fig. 10 Control diagram of P-Delta effect of buildings with bracing type X in the Surabaya zone earthquake with x and y directions load

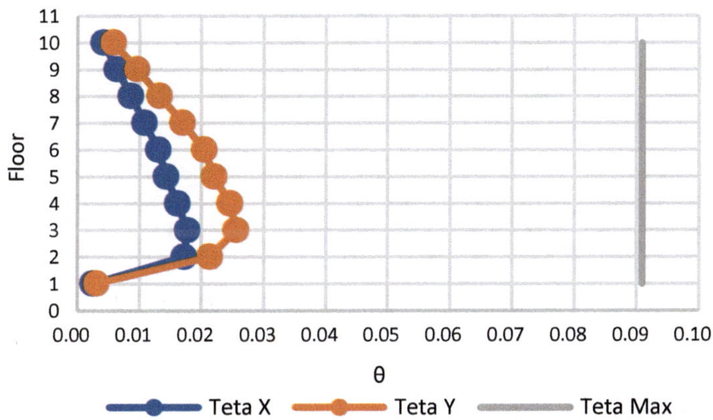

Fig. 11 Control diagram of P-Delta effect of buildings with bracing type V in the Surabaya zone earthquake with x and y directions load

value of "deviation between floors" is used as a reference. Previous studies obtained the same results. The structural model with type X bracing has the best performance compared to other types of bracing models [8].

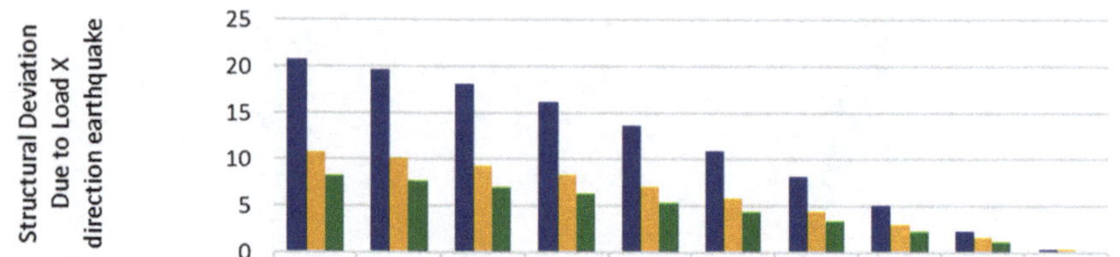

	floor 10	floor 9	floor 8	floor 7	floor 6	floor 5	floor 4	floor 3	floor 2	floor 1
Without bracing (mm)	20.65	19.634	18.113	16.107	13.659	10.882	8.027	5.069	2.288	0.293
Bracing type V (mm)	10.733	10.101	9.259	8.234	7.054	5.751	4.402	2.987	1.551	0.268
Bracing type X (mm)	8.173	7.673	7.019	6.231	5.33	4.341	3.322	2.261	1.189	0.214

Fig. 12 Comparison diagram of the total deviation of buildings with and without braces for earthquake loads in the X direction of the Surabaya zone

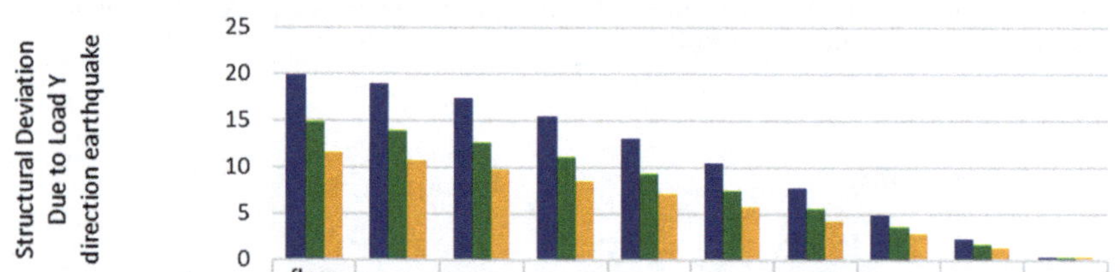

	floor 10	floor 9	floor 8	floor 7	floor 6	floor 5	floor 4	floor 3	floor 2	floor 1
Without bracing (mm)	19.768	18.791	17.323	15.391	13.037	10.387	7.682	4.869	2.209	0.283
Bracing type V (mm)	14.756	13.807	12.547	11.029	9.29	7.403	5.508	3.56	1.691	0.24
Bracing type X (mm)	11.485	10.702	9.685	8.479	7.114	5.651	4.194	2.711	1.298	0.19

Fig. 13 Comparison diagram of the total deviation of buildings with and without braces for earthquake loads in the Y direction of the Surabaya zone

4 Conclusion

(1) The story drift of the Airlangga University parking building has decreased significantly after being given bracing. The value story drift of the building model without bracing is 16,269 mm, but after bracing is used, it has decreased to 10,714 mm for the V type brace and 8,156 mm for the X type brace; (2) Based on the performance control, the building model structure with bracing and without bracing has met the requirements for the structure's performance limits. The optimal type of bracing in the Airlangga University parking building is type X bracing. The increase in building performance of type X bracing reached 68.6% for service limit performance and 73.6% for ultimate limit performance, while type V bracing increased 62.11% for service limit performance and 67.3% for ultimate limit performance.

References

1. Tim PSGN (2017) Peta Sumber Dan Bahaya Gempa Indonesia Tahun 2017. PuslitbangPP, Balitbang Kemen PUPR, Bandung 40393
2. Kuncoro H, Kartini GAJ, Meilano I, Susilo S (2019) Identification of fault mechanism in Eastern part of Java Island using Continuous GNSS data from 2010 to 2016. In: Seminar Nasional Geomatika, vol 3. Bandung, pp 805–812
3. Rendra R, Kurniawandy A, Djauhari Z (2015) Kinerja Struktur Akibat Beban Gempa Dengan Metode Respon Spektrum dan Time History (Studi Kasus: Hotel SKA Pekanbaru). JOM FTEKNIK 2(2):1–15
4. Afrida I, Wahyuningtyas WT, Krisnamurti (2020) Analisis Ketahanan Gedung Apartemen Surabaya dengan Menggunakan Metode Respon Spektrum, Berkala Sainstek 8(4):132–139
5. Hutahaean SG, Aswandy (2016) Kajian Pemakaian Shear Wall dan Bracing pada Gedung Bertingkat, Reka Racana Jurnal Online Institut Teknologi Nasional 2(4):100–111
6. Haryono S, Arumningsih D, Purnamawanti D (2015) Penggunaan Struktur Bresing Konsentrik Tipe X Untuk Perbaikan Kinerja Struktur Gedung Bertingkat Terhadap Beban Lateral Akibat Gempa, Jurnal Teknik Sipil dan Arsitektur 16(20)
7. Smith BS, Coull A (1991) Tall building structures: analysis and design. Willey, Inc., New York
8. Aryandi D, Herbudiman B (2017) Pengaruh Bentuk Bracing terhadap Kinerja Seismik Struktur Beton Bertulang. Reka Racana Jurnal Online Institut Teknologi Nasional 3(1):1–11
9. Repadi JA, Sunaryati J, Thamrin R (2016) Analisis Kinerja Struktur Beton Bertulang Dengan Variasi Penempatan Bracing Inverted V. Jurnal Rekayasa Sipil 12(2):32–39
10. SNI 1726:2019 (2019) Tata cara perencanaan ketahanan gempa untuk struktur bangunan gedung dan non gedung. Jakarta

Various Strut—Macro Modelings for Infilled Frame Analysis

Isyana Ratna Hapsari, Marwahyudi, Edy Purwanto, Senot Sangadji, and Stefanus Adi Kristiawan

Abstract Equivalent diagonal strut is a simple macro model used to represent the behavior of masonry walls as an infilled frame structure when exposed to lateral loads. The model has been shown capable to produce the response of an infilled frame structure that is comparable to the experiment results. Several important parameters need to be considered when initiating diagonal strut modeling. The selection of strut type related to the number of struts, i.e. single, double, or triple struts will greatly affect the behavior of infill walls. In addition, the residual area after crack (Am2) is a very important parameter to be defined properly when modeling the strut. This study shows that the double strut (DS) and triple strut (TS) give response of infilled frame structure close to testing, with an error is around 9–35% when the response is assessed based on peak load, displacement at peak load, and initial stiffness of the structure. For DS model, further improvement can be made by assigning the Am2 to around 55 and 78%.

Keywords Macro modeling of masonry · Diagonal strut · Number of struts · Residual area of strut

I. R. Hapsari (✉) · Marwahyudi
Civil Engineering Department, Faculty of Engineering, Universitas Sebelas Maret, Surakarta, Indonesia
e-mail: isyanahapsari@student.uns.ac.id

Marwahyudi
e-mail: yudhie_dsg@student.uns.ac.id

E. Purwanto · S. A. Kristiawan
SMARTCrete Research Group, Department Faculty of Engineering, Universitas Sebelas Maret, Civil Engineering, Surakarta, Indonesia
e-mail: edypurwanto@student.uns.ac.id

S. A. Kristiawan
e-mail: s.a.kristiawan@ft.uns.ac.id

S. Sangadji
SMARTQuake Research Group, Civil Engineering Department, Faculty of Engineering, Universitas Sebelas Maret, Surakarta, Indonesia
e-mail: s.sangadji@ft.uns.ac.id

S. A. Kristiawan et al. (eds.), *Proceedings of the 5th International Conference on Rehabilitation and Maintenance in Civil Engineering*, Lecture Notes in Civil Engineering 225,
https://doi.org/10.1007/978-981-16-9348-9_18

1 Introduction

Research on masonry infill frame structures has been carried out for a long time. This is not surprising since masonry infill frame is a common structure for residential buildings in many countries. Previous studies have shown that masonry infill walls contribute to the increase of stiffness and strength of structures [1]. Therefore, the contribution of infill walls should be considered in the modeling to describe the structure's response accurately when exposed to earthquake loads. However, the actual behavior of infill is very complex. Non-standard characteristics of masonry in each region due to workmanship contributes an additional complexity to the behavior of this infill wall. Furthermore, the bonding between bricks and mortar also affects the behavior of the masonry structure.

According to Crisafulli [2] there is a grouping of parameters that can affect the response of the masonry infilled frame:

1. Masonry unit: dimension, mechanical properties, type of brick, type of brick surface character
2. Mortar joints: mechanical properties, lime content, the thickness of the species, if any reinforcement is used
3. The structural frame: mechanical properties of the concrete or steel used, dimension, the amount of longitudinal and transverse reinforcement used.
4. General i.e. type of excitation loading, the number of floors, the number of bays, the presence of openings or not, the position of the openings, and the construction technique.

In previous studies, several macro models have been introduced to describe the infill panels. One of such models is the equivalent diagonal strut. The type of diagonal strut its self has transformed; starting from what was introduced by Crisafulli et al. [3] in the form of a single strut. The single strut has the advantage of the simplicity of mathematical formulas but it is still capable to provide sufficient estimation of the stiffness value. However, the single strut is unable to predict the shape of the deflection that occurs in the column. According to Kumar [4], a single strut gives an overestimate result of the displacement of the column. In addition, this model produces a bending moment and shear forces that are very low or unrealistic when compared to the strut model quantification through FEM Software. Meanwhile, the double strut model is more capable of illustrating the similarity of the model crack pattern to the test object. El-Dakhakhni [5] and Tanganelli et al. [6] stated that triple struts produce the most rational bending moment values if compared to other strut models. In addition, the principle that the walls do not immediately fail simultaneously can capture more accurately when it is estimated by using triple strut modeling.

The purpose of this paper is to identify parameters of the equivalent diagonal compressive strut, which will determine the suitability of the model to represent the behavior of the infilled frame structure. Three types of diagonal compressive strut models will be studied i.e. single strut (SS), double strut (DS), and triple strut (TS). The test results of the infilled frame by Kakaletsis et al. [7] will be used as a validation for the numerical simulation. Numerical modeling of the infilled frame with the diagonal compressive strut approach is carried out using the SeismoStruct 2020 [8]. Struts are built-in SeismoStruct by defining 32 parameters covering the geometric, mechanical, and empirical parameters. Modeling in this software considers the behavior of the fiber-based cross-section about the uniaxial stress-strain relationship. The results of this study can be used as a reference for simple analysis of masonry infilled frame yet keeping the accuracy of the numerical model in predicting the behavior of the actual structure.

2 Description of Model Structure

2.1 *Geometry and Materials*

Several data to build a numerical model in this study will use the existing two-dimensional portal model researched by Kakaletsis [7]. There are two types of portals in Kakaletsis's study i.e the open frame reinforced concrete (OF) model and the masonry infilled reinforced concrete frame (IF) model. The OF modeling and validation are needed to ensure that OF behavior is as expected, so that in the subsequent modeling i.e infilled frame (IF), we can quantify the contribution of infill walls to the behavior of the structure. The constitutive model of concrete follows that of Mander [9] for both confined and unconfined concrete sections. As for the steel material, the model of Menegotto-Pinto [10] is selected. The boundary conditions of a model are defined as fixed support. These structural models can be described as shown in Fig. 1. The geometric parameters and mechanical properties of the model structures can be seen in Tables 1 and 2. Especially for geometric dimensions using a model with a size of 1/3 of the prototype.

Columns use the longitudinal reinforcement 6D100 and stirrup reinforcement D80-200. Whereas for beams use the longitudinal reinforcement 6D80 and stirrup

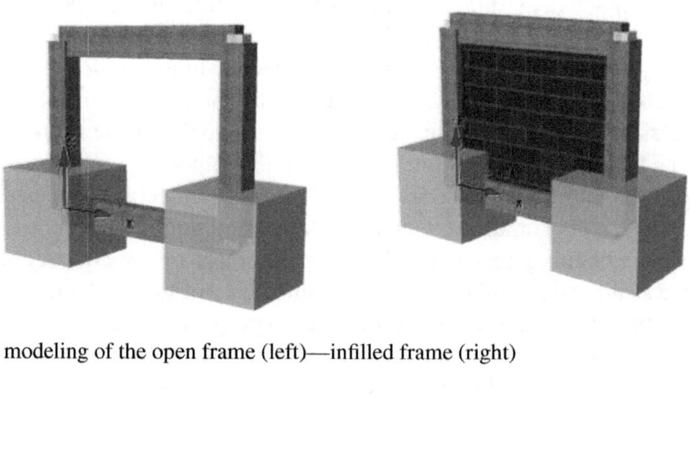

Fig. 1 3D modeling of the open frame (left)—infilled frame (right)

Table 1 The geometry of structure elements

Model parameters	Dimension (mm)
Column section	150 × 150
Column height	1150
Beam section	100 × 200
Beam length	1350
Wall thickness	60

Table 2 Mechanical properties of the model

Mechanical properties	Values
Compressive strength of concrete (Mpa)	28.51
Specific weight of concrete (kg/m^3)	2200
Modulus of elasticity—concrete (Mpa)	23692
Diagonal compressive strength of masonry (Mpa)	5.11

reinforcement D80-200. Based on additional information through numerical validation by Sukrawa [11] using strut and shell modeling, he showed that the modulus of elasticity of the masonry (Em) obtained from the experiment i.e. 670.3 MPa is too small when compared with the test results of other studies. For this reason, Sukrawa used the value Em = 2810 MPa, referring to the equation presented by FEMA 356 [12], i.e. Em = 550 fcm. This modulus of elasticity will be applied to the current study.

2.2 *The Simplified Infilled Frames—Macro Model*

There are 2 (two) types of concepts for modeling the masonry infilled frame. The first is a micro model where the infill structure is divided into many elements in detail using the discretization method to take into account the local effects when subjected to earthquake loads. This complex micro model requires large computerization efforts and time from the modeling process to the analysis stage. The second type of infilled frame modeling is the macro model, a simplification of the masonry model based on a physical understanding of the infill behavior in order to approximate the actual behavior of the infilled frame when subjected to loads. The main objectives of making this macro model are to reduce the complexity and the need for very large computation.

Crisafulli's research related to infill walls proved that when the infilled frame is subjected to increased lateral loads, some parts of the walls or connection between masonry and frames suffered cracks. If the lateral load continues to increase then part of the wall will separate (gap) from the frame. After the existence of the gap, the stress at the corner of the tensile area decreases while those near the corner of the compressive area increase significantly. The stress distribution that occurs after separation produces a diagonal stress trajectory. Based on these results, Crissafuli proposed a compressive diagonal strut to simplify the model of the masonry infill panel. Crisafulli and Crowley et al. [13] stated that diagonal compressive single strut can represent the behavior of infilled frame. However, further, development has been proposed to improve the prediction of the behavior by introducing double strut and triple strut [3, 5].

In this current study, 3 (three) types of diagonal struts for masonry infill panels modeling will be discussed. The principle differences underlying the three types of diagonal strut models can be seen in Fig. 2 [3]. Which is the *hz* value is the vertical distance between the nodes on the strut, and Am1 is the strut area 1 or initial strut. The model parameter of the infilled frame based on the equivalent diagonal compressive strut in SeismoStruct 2020 uses 3 main criteria. The first criterion is the geometric parameter related to the dimensions of the infilled frame to be modeled, according to Fig. 3 below (a) illustrate the real wall model, (b) masonry modeling built by the geometric parameters, and (c) strut approach for macro modeling as the infilled frame in SeismoStruct 2020. The second criterion is mechanical parameters, which are parameters related to the material that makes up the infilled frame. And the third criterion is the empirical parameters, which are the parameters needed to represent the behavior of infilled frame due to cyclic loading.

The general parameters related to the 3 (three) criteria mentioned above are presented in Table 3. The parameters used for modeling the diagonal strut refers to the values and formulas of the previous research on infilled frame [2, 14–16].

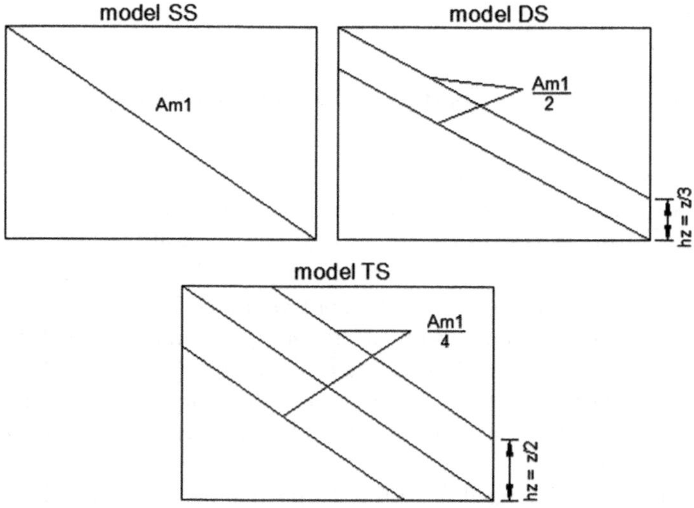

Fig. 2 Macro modeling of masonry infilled using various strut types [3]

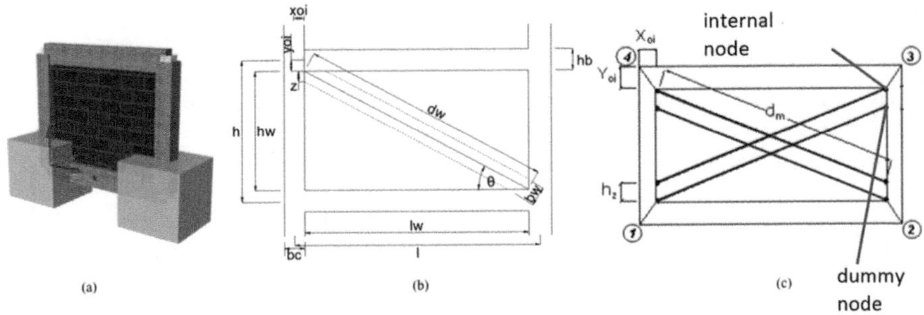

Fig. 3 **a** Masonry visual. **b** Geometrical parameters of masonry infilled for strut modeling. **c** Macro modeling of masonry infilled in SeimoStruct 2020

Table 3 Infilled parameters, equations, and references

Infilled panel model parameters	Unit	Used value	Equation	Ref
1. Mechanical properties				
a. Tensile Strength, *ft*	MPa	0		[14]
b. Bond shear strength, τ_0	MPa	0.08		[7]
c. Coefficient of friction, μ		0.77		[7]
d. Maximum shear stress, τ_{max}	MPa	1		[14]
e. Strain at max stress, (ε_m)	MPa	0.0012		[14]
f. Ultimate strain, (ε_u)	MPa	0.024	$\varepsilon_u = 20 \times \varepsilon_m$ (1)	[14]
g. Closing strain, (ε_{cl})	MPa	0.003		[14]
h. Specific Weight, *W*	N/mm^3	1.7E-005		
2. Empirical Properties				
a. Starting Unloading Stiffness Factor, (γ_{un})		1.7		[14]
b. Strain Reloading Factor, (α_{re})		0.2		[14]
c. Strain Inflection Factor, (α_{rh})		0.7		[14]
d. Complete Unloading Strain Factor, (β_a)		2		[14]
e. Stress Inflection Factor, (β_{ch})		0.9		[14]
f. Zero Stress Stiffness Factor, (γ_{plu})		1		[14]
g. Reloading Stiffness Factor, (γ_{plr})		1.1		[14]
h. Plastic Unloading Stiffness Factor, e_{x1}		3		[14]
i. Repeated Cycle Strain Factor, e_{x2}		1		[14]
j.Reduction Shear Factor, (α_s)	MPa	1.43		[2]
k. Out-of-plane Failure Drift	%	1		[15]

(continued)

Table 3 (continued)

Infilled panel model parameters	Unit	Used value	Equation	Ref
1. Proportion of stiffness assigned to shear, γ_s	%	70		[14]
3. Geometrical Properties				
a. Dimensionless relative stiffness, λ_h			$\lambda_h = h\sqrt[4]{\frac{E_m t_w \sin(2\theta)}{4E_c I_c h_w}}$ ....(2)	[2]
b. Strut Area Reduction Strain, (ε_1)		0.0006		[16]
c. Residual Strut Area Strain, (ε_2)		0.001		[16]
d. Horizontal Offsets, x_{oi}	%	Varies	$xoi = \frac{0.5 \times column\ width}{width\ of\ infill\ panel}$ (3)	[14]
e. Vertical Offsets, y_{oi}	%	Varies	$yoi = \frac{0.5 \times beam\ depth}{height\ of\ infill\ panel}$ (4)	[14]

3 Numerical Simulation

3.1 *Open Frame Model for Validation*

The cyclic loads employed in this Kakaletsis study were replaced by lateral loads which were increased incrementally per 10 kN until the maximum load was reached. Before conducting a parametric study on the modeling of infilled wall structures using a diagonal compressive strut, we validated the open frame structure modeling first. The open frame modeling and analysis process is assisted by the SeismoStruct 2020. The numerical modeling is validated by the results of Kakaletsis [7]. The resulting capacity curve (see Fig. 4) can provide an overview of the fitness of the model. It can be concluded that the result of the open frame modeling in SeismoStruct is comparable to the experiment result.

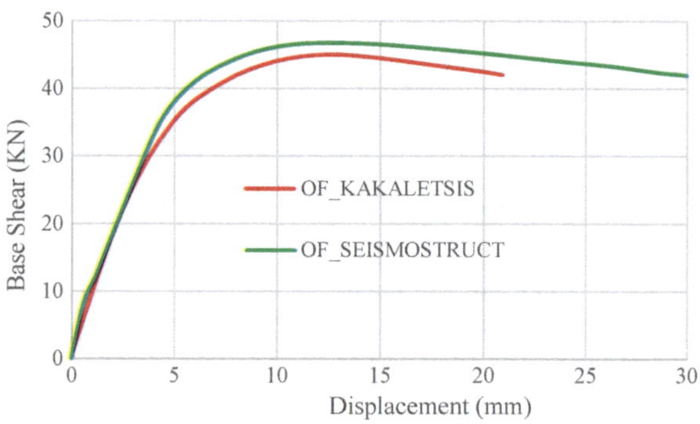

Fig. 4 Capacity curve open frame between experiment and numerical modeling

3.2 Numerical Model with Various Types of Struts

In this chapter, the numerical study is carried out to compare the results of the infill panel testing conducted by Kakaletsis with several types of infill panel models. The following stages of modeling are conducted in this study: the first is modeling the masonry infilled frame using 3 types of diagonal struts, the second is assigning variations in the elastic modulus of masonry (Em), and the third is defining various residual strut area (Am2). From these stages of numerical modeling, it can show the most suitable diagonal compressive strut model in simulating experimental test results.

In the first stage of modeling, the fundamental parameter that distinguishes the three strut types is strut area 1 (Am1). The formula for getting the value of strut area 1 (Am1) in single strut (SS) model is:

$$Am_{1single} = bw_1.t_w \tag{5}$$

Where t_w is the masonry thickness and bw_1 is the width dimension of the diagonal strut, as recommended by FEMA 356 as follows:

$$b_{w1} = 0,175(\lambda_h)^{-0,4}d_w \tag{6}$$

The definition of d_w is the length of the diagonal strut being modeled. For dimensionless relative stiffness, λ_h is a parameter that expresses the relative stiffness of the infill panel to the frame structure. This parameter is sufficient to describe the elasticity behavior of the infill panel. Meanwhile, to get the area of the strut area 1 (Am1) in the double strut (DS) and triple strut (TS) model are as follows:

$$A_{m1double} = \frac{bw1.tw}{2} \tag{7}$$

$$A_{m1triple} = \frac{bw1.tw}{4} \tag{8}$$

The definition of the *hz* value is the vertical distance between the nodes on the strut in double strut (DS) modeling with the following formula:

$$h_z = {}^1/_3.z \tag{9}$$

Meanwhile, the *hz* value in triple strut (TS) modeling uses the following formula:

$$h_z = {}^1/_2.z \tag{10}$$

where z is the actual contact length value determined using the formula (11). Meanwhile, the actual contact length (z) in single strut modeling is defined as 0.

$$z = \frac{\pi}{2\lambda} \tag{11}$$

The capacity of curves obtained from numerical modeling of masonry infilled frame (IF) structure using various strut models is shown in Fig. 5. This curve is constructed by first setting the displacement control value = 30 mm, and iteration is carried out in 50 steps. From Table 4 and Fig. 5, it can be concluded that double strut (IF_DS) and triple strut (IF_TS) give the closest to the structural response of the experimental results. Variances of response values in IF_DS and IF_TS are relatively smaller than the single strut model (IF_SS). This can be seen through the quantification of model errors based on the peak load (IF_DS = 11%, IF_TS = 16%) and initial stiffness (IF_DS = 35%, IF_TS = 9%).

In the second study, the mechanical properties of the double strut i.e the elastic modulus of masonry (Em) will be adjusted by the results of the Kakaletsis (670.3 MPa). The double strut model with this Em value is assigned as IF_DS_EM670. The resulting capacity curve (see Fig. 6) shows that the Em value contributes to the added strength of the overall structure. The IF_DS_ EM670 model can withstand a maximum lateral load of 95 kN, an increase of 8% from the initial

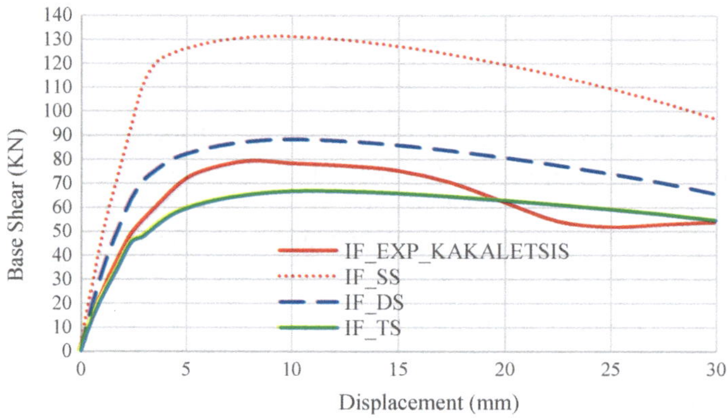

Fig. 5 Capacity curve infilled frame between experiment and single, double, triple strut

Table 4 Parameters of capacity curves determined by various struts models

Parameters quantification	Models				Model Error		
	IF_EXP_KAKALETSIS	IF_SS	IF_DS	IF_TS	IF_SS (%)	IF_DS (%)	IF_TS (%)
Peak load (kN)	79	130.96	88.08	66.70	66	11	16
Displacement at peak load (mm)	8	9	10.2	10.8	13	28	35
Initial stiffness (kN/mm)	19.38	39.92	26.22	21.12	106	35	9

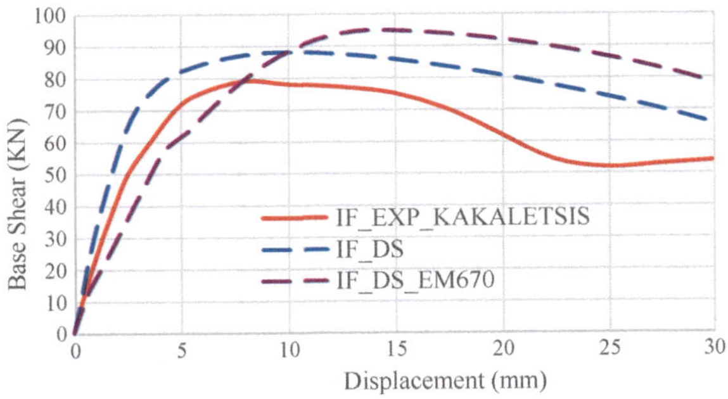

Fig. 6 Capacity curve infilled frame between experiment, double strut (DS) and double strut with Em adjustment (DS_EM670)

DS modeling which uses an Em value of 2810 MPa. The smaller Em produces a greater value of the strut width (bw_1), because the value of the strut width referring to formula (6). Hence, this strut width (bw_1) value will lead to the wider area of strut 1 (Am1), according to the calculation in formula (7). This is the reason why the smaller Em will influence the strength of infilled frame structure.

Table 5 shows the parameters of capacity curves determined using a double strut model with various Em values. IF_DS_EM670 produces a larger error value, where the peak load and displacement at peak load indicate error values of 20% and 80%, respectively, in comparison to the experimental result. Thus, it is reasonable to use Em estimated from FEMA 356 (2810 MPa) for constructing the diagonal strut model.

The geometric parameter is one of the influencing parameters to the behavior of infilled frame against the lateral loads according to Tanganelli et al. [6]. Thus, in this third stage of the study, the contribution of the residual area of the strut (Am2) to the infilled frame behavior will be explored. Crisafulli [2] and Smyrou [14] stated that the area of the residual strut (Am2) occurs due to cyclic or increased lateral loads, causing the contact length between the frame and the infill panel to decrease. This will automatically affect the area after the panel has cracked. Despite the lack of information about the practical value to estimate this parameter, Crisafulli and Smyrou used the experimental results report from Decanini [17] which suggests that the strut equivalent width is reduced by 20–50% due to the cracking of the infill panel.

Table 5 Parameters of capacity curves determined from double strut model with various Em

Parameters quantification	Models			Model error	
	IF_EXP_KAKALETSIS	IF_DS	IF_DS_ EM670	IF_DS (%)	IF_DS_ EM670 (%)
Peak load (kN)	79	88.08	94.71	11	20
Displacement at peak load (mm)	8	10.20	14.4	28	80
Initial stiffness (kN/mm)	19.38	26.22	19.6	35	1

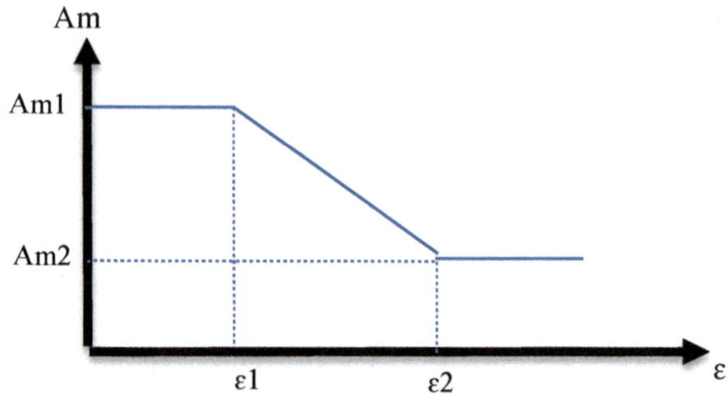

Fig. 7 Variation of the strut area related to the alteration of axial strain value

According to Meslem [16], when the elastic limit of the infill panel is exceeded due to the cracks the assumption used between the area of strut 1 and 2 has linear variation as a function of axial strain (see Fig. 7). And after doing the parametric study, Meslem chooses the residual area of strut (Am2) = 55%.

Meanwhile, according to Al-Chaar [18], a reduction factor used to quantify the strut area is 0.7 and 0.4 for moderate and heavy damage infill panel, respectively. Kumar [4] argues that Am2 has varied values in the range of 60 and 80%, and based on his parametric study, Am2 = 70% was able to give results that were almost alike experimental. Wang et. al. [19] and Totoev [20] give Am2 value about 40–50% for their proposed 2 infill wall models.

In this current study, we estimated the residual value of the strut area (Am2) based on the stiffness degradation data of the test results from Kakaletsis. The resulting stiffness degradation is shown in Figure 8. This graph can show the value of stiffness in elastic conditions of 19.38 kN/mm corresponds to the displacement of 2.58 mm where the first crack occurs. The comparison of the stiffness at the first crack with the initial stiffness is equal to 78%. Meanwhile, the residual strength of the infill panel occurs at a displacement of 22.5 mm (see Fig. 9) corresponding to a stiffness of 2.40 kN/mm (see Fig. 8). At this residual strength, the stiffness ratio remains 10%. For detailed information regarding the behavior of infill walls in an elastic state, see Fig. 9. Through several literature studies and consideration of the value of

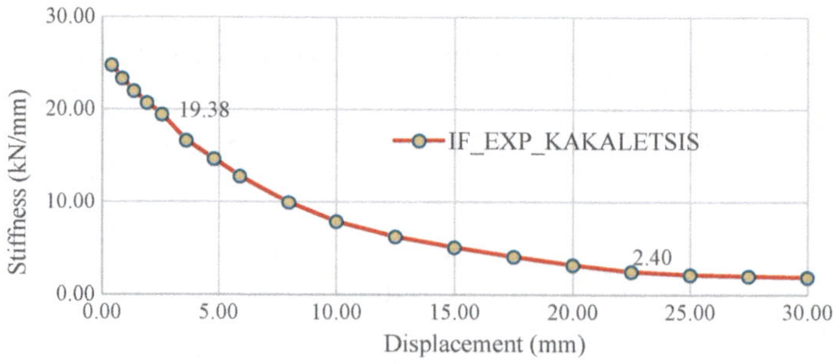

Fig. 8 Stiffness degradation of Kakaletsis's infilled frame experiment

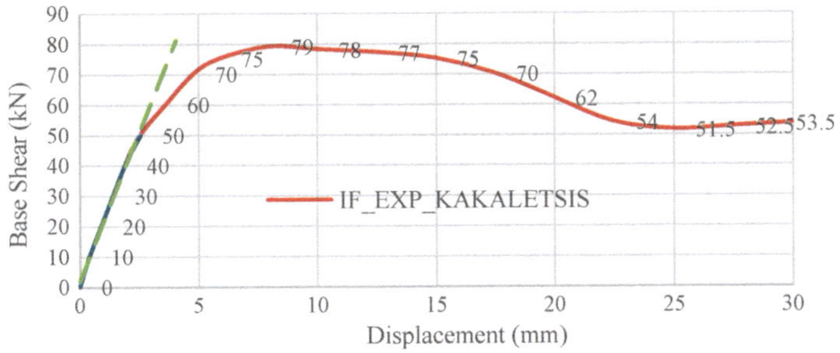

Fig. 9 Plastic—elastic condition for identifying first crack and residual strength condition of infilled frame

stiffness degradation, we propose the value of residual strut area (Am2) in the range of 44–78%. This range of Am2 will be used to modify the IF_DS (see Table 6).

A comparison of the resulting capacity curves between the double strut model with various Am2 values can be seen in Fig. 10. Because at this modeling phase only the Am2 variable is adjusted, the character and initial stiffness values have a similarity. From this reason, we do not need to show this comparison of values. The graph shows IF_DS_AM2_55% gives the most similarity values of peak load achievement. And IF_DS_AM2_78% shows the closest values of stiffness at peak load (see Table 7).

Table 6 Type of double strut model with proposed residual strut area (Am2)

Infilled frame model	Am2 (%)	Reason for proposing
IF_DS	70	Initial setting
IF_DS_AM2_78%	78	First crack condition
IF_DS_ AM2_55%	55	Estimation
IF_DS_ AM2_44%	44	Mean values between first crack—residual condition

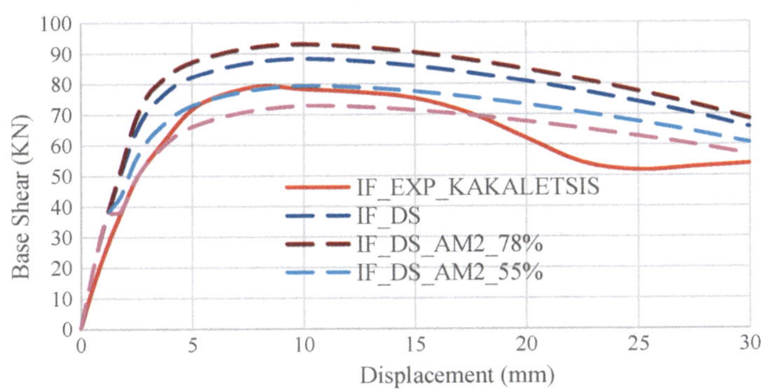

Fig. 10 Capacity curve infilled frame between experiment, double strut (DS) and double strut with Am2 variation (DS_AM2_78%, DS_ AM2_55% and DS_ AM2_44%)

Table 7 Parameters of capacity curve determined from double strut model with various Am2

Parameters quantification	Models					Model error			
	IF_EXP_KAKALETSIS	IF_DS	IF_DS_AM2_78%	IF_DS_AM2_55%	IF_DS_AM2_44%	IF_DS (%)	IF_DS_AM2_78% (%)	IF_DS_AM2_55% (%)	IF_DS_AM2_44% (%)
Peak Load (kN)	79	88.08	92.86	79.14	72.60	11	18	0	8
Displacement at Peak Load (mm)	8	10.20	9.60	10.20	10.80	28	20	28	35
Stiffness at Peak Load (kN/mm)	9.88	8.64	9.67	7.76	6.72	13	2	21	32

Based on the numerical study, it is shown that the smaller the Am2 value will greatly affect the decrease in the strength and stiffness of the infilled frame. From this analysis, we can conclude that the determination of the residual strut area after cracked (Am2) is very important in initiating infill panel modeling as a diagonal compressive strut. Based on experiments conducted by Kakaletsis, it is stated that nonlinear behavior is initiated by infill cracking. These cracks start as a 45° diagonal line and then join with several horizontal shear cracks. The failure mode of the numerical simulation shows that the masonry loses its contribution when the masonry stiffness decreases about 40% remaining from the initial condition.

4 Summary and Conclusions

Three types of diagonal struts i.e. single strut (SS), double strut (DS), and triple strut (TS) have been used in this study to represent the influence of masonry infilled panel on the behavior of frame structure. It is shown that the double strut (DS) and triple strut (TS) models have the capability to estimate the structural response of masonry infilled frame structures. The corresponding model errors for DS and TS are 11 and 16% when assessed on the basis of peak load, and 35 and 9% based on initial stiffness.

The modulus elasticity masonry (Em) influences the overall behavior of the structure. A smaller value of Em increases the strut width (*bw1*), thereby increasing the strength of the structure as a whole. IF_DS_EM670 with Em of 670.3 MPa can withstand a maximum lateral load of 95 kN, which represents an increase of 8% compared to IF_DS model with Em of 2810 MPa. However, the IF_DS_EM670 has bigger (more deviant) model errors i.e. 20% and 80% when assessed based on its peak load and displacement at peak load, respectively.

The Am2 has been determined in this study based on the stiffness degradation. The proposed Am2 is in the range of 44–78%, representing a range of conditions from initial stiffness to the stiffness at residual strength. IF_ DS_ AM2_55% model with Am2 equal to 55% provides the best estimate of the infilled frame behavior with model errors 0% when viewed from the peak load achievement. Whereas IF_ DS_ AM2_78% model with Am2 = 78% can show the value of stiffness at peak load with model error in 2%.

References

1. Stavridis A, Koutromanos I, Shing PS (2012) Shake-table tests of a three-story reinforced concrete frame with masonry infill walls. Earthq Eng Struct Dyn
2. Crisafulli FJ (1997) Seismic behaviour of reinforced concrete structures with masonry infills. In: Thesis, University of Canterbury
3. Crisafulli FJ, Carr AJ (2007) Proposed macro-model for the analysis of infilled frame structures. Bull New Zeal Soc Earthq Eng 40(2)

4. Kumar M, Khalid F, Ahmad N (2018) Macro-modeling of weak masonry infilled reinforced concrete frame for seismic analysis. NED Univ J Res
5. El-Dakhakhni WW, Elgaaly M, Hamid AA (2003) Three-strut model for concrete masonry-infilled steel frames. J Struct Eng 129(2):177–185. https://doi.org/10.1061/(asce)0733-9445(2003)129:2(177)
6. Tanganelli M, Rotunno T, Viti S (2017) On the modelling of infilled RC frames through strut models. Cogent Eng 4(1). https://doi.org/10.1080/23311916.2017.1371578
7. Kakaletsis DJ, Karayannis CG (2009) Experimental investigation of infilled reinforced concrete frames with openings. ACI Struct J (106-S14):132–141
8. SeismoSoft (2020) Seismostruct-a computer program for static and dynamic nonlinear analysis of framed structures, [Online]. Available: http//www.seismosoft.com
9. Mander JB, Priestley MJN, Park R (1988) Theoretical stress-strain model for confined concrete. ASCE J Struct Eng 114(8):1804–1826
10. Menegotto M, Pinto PE (1973) Method of analysis for cyclically loaded R.C. plane frames including changes in geometry and non-elastic behaviour of elements under combined normal force and bending
11. Sukrawa M (2015) Earthquake response of RC infilled frame with wall openings in low-rise hotel buildings. Procedia Eng 125:933–939. https://doi.org/10.1016/j.proeng.2015.11.118
12. FEMA356 (2000) Prestandard and commentary for the seismic rehabilitation of buildings. vol 36, no 17
13. Crowley H, Pinho R (2006) Simplified equations for estimating the period of vibration of existing building. In: First European conference on earthquake engineering and seismology, no. 1122
14. Smyrou E (2006) Implementation and experimental verification of models for nonlinear analysis of masonry infilled R.C. frames. In: Dissertation, University of Pavia, Italy, pp 1–108
15. Da Porto F, Guidi G, Dalla Benetta M, Verlato N (2013) Combined in-plane/out-of-plane experimental behaviour of reinforced and strengthened infill masonry walls. In: 12th Canadian masonry symposium, pp 1–11
16. Meslem A, D'Ayala D (2012) Toward worldwide guidelines for the development of analytical vulnerability functions and fragility curves at regional level. In: Proceedings 15th WCEE
17. Decanini LD, Fantin GE (1986) Modelos simplificados de la manposteria incluida en portieos, Caracteristicas de rigidez y resistencia lateral en estado limite (in Spanish). Jornadas Argentinas Ing. Estructural 2:817–836
18. Al-chaar G (2002) Evaluating strength and stiffness of unreinforced masonry infill structures. ERDC/CERL TR-02-1
19. Wang Z, Totoev YZ, Page A, Sher W, Lin K (2015) Numerical simulation of earthquake response of multi-storey steel frame with SIM infill panels. Adv Struct Eng Mech
20. Totoev YZ, Williamson D, Wang Z (2014) Vibrations of multi-storey RC frame with SIM panels; numerical simulation. In: Procedings 9th international mason conference

Strengthening of Non-engineered Building Beam-Column Joint to Increase Seismic Performance with Variation of Steel Plate Width

Edy Purwanto, P. Amarta Adri, S. A. Kristiawan, Senot Sangadji, and S. Halwan Alfisa

Abstract Most of the residential buildings were built without following the requirements specified in the design Code of earthquake-resistant building. Such non-engineering buildings (NEB) are susceptible to seismic hazard. Beam-column joint failure is a typical example of structural failure in NEB induced by seismic load. Strengthening of the beam-column joint by mean of external steel plate is expected to increase the seismic performance of NEB. This study aims to find the most effective plate width as a strengthening material for the beam-column joint. ATENA software will be used for the investigation. The structural model is portal structure without strengthening as a control, namely the PU-200 model and with an external steel plate, namely PUP-200 with various width of plates. From the results of the analysis, it is found that strengthening joint with a plate width of 50 mm has better performance than plate widths of 100 mm and 150 mm. The increase performance is quantified in term of increase of load by 14.60%, reduction of displacement by 18.43% and reduction of crack width by 23.65%.

Keywords Strengthening · Steel plate · Non-engineered building

E. Purwanto (✉) · P. A. Adri · S. A. Kristiawan · S. Sangadji · S. H. Alfisa
Civil Engineering Department, Universitas Sebelas Maret, Surakarta, Indonesia
e-mail: edypurwanto@student.uns.ac.id

P. A. Adri
e-mail: amartaadri@student.uns.ac.id

S. A. Kristiawan
e-mail: s.a.kristiawan@ft.uns.ac.id

S. Sangadji
e-mail: s.sangadji@ft.uns.ac.id

S. H. Alfisa
e-mail: halwan@ft.uns.ac.id

S. A. Kristiawan et al. (eds.), *Proceedings of the 5th International Conference on Rehabilitation and Maintenance in Civil Engineering*, Lecture Notes in Civil Engineering 225,
https://doi.org/10.1007/978-981-16-9348-9_19

1 Introduction

Indonesia is located on top of three major tectonic plates in the world, namely the Eurasian, Australian and Pacific plates which produce frequent earthquakes due to their relative movement. Several types of hazard may be associated with the earthquake, however, ground shaking with high intensity may damage structures like building and bridges.

Observation from the several past earthquake event exhibit dire situation where most of non-engineered residential buildings were severely damaged and collapse, even in the moderate strength of earthquake. This non engineered buildings (NEB), typically built by and for the middle to lower economic class, were built spontaneously and informally without any input or supervision from a qualified engineer. Often built with poor workmanship whose rely only on past experience, these typical non-engineered buildings have not met engineering criteria for earthquake resistant structure [1]. Quite large number of this typical structures demonstrate poor performance during earthquake shaking resulted in people having no place to live and many other secondary losses; such as business interruption.

Complete collapse of structure often occurred as the structure have unable to withstand additional lateral seismic load, exhibit large enough horizontal deformation resulting in the detachment of elements in the NEB due to, for example, failure of beam-column joints, as well as the removal of roof frames and building walls.

Generally, failure of beam-column joint may take one of the following modes: joint shear failure without beam reinforcement yielding (J), joint shear failure with beam reinforcement yielding (BJ), beam flexural failure (BF), column flexural failure (CF), beam reinforcement pull-out failure (P), and beam reinforcement 90 hook anchorage failure (A) [2].

Several factors involved in the beam-column joint failure including properties of concrete and reinforcement, dimension of beam and column, ratio of longitudinal reinforcement, anchorage of reinforcement, the presence of joint shear reinforcement, spacing of shear reinforcement, etc. [2–6].

The poor performance of beam-column joint in NEB to resist seismic load can be related to the lacks implementation of engineering principle specifically in the design and construction of this part. Observation in the field shows these insufficient engineering principles implementation include the use of low concrete strength, the use of plain reinforcement instead of deformed bar, no shear reinforcement within the joint, not enough anchorage development due to insufficient reinforcement detailing, considerable spacing between shear reinforcement in the column and beam, etc.

It is noted that the degradation of joint due to seismic load will reduce the rigidity of the joint. Consequently, the load transfer between elements connected by the joint becomes less effective. Hence, the capacity of structure to resist lateral load decreases [7, 8].

Improving the seismic performance of existing NEB becomes urgent to reduce the vulnerability of such building in the earthquake-prone area. Truong et al. investigated various retrofitting technique for enhancing beam-column joint performance.

They showed head re-bar anchoring, CFRP wrapping, haunch, and steel jacketing could partially enhance the strength and deformation capacity of the beam–column joints. Several investigators [9–11] used carbon-fiber-reinforced-plastic (CFRP) for reinforcing beam-column joints to increase the ability to accept seismic loads so that it has better load capacity, deformation capacity, and ductility. Meanwhile Thirumalini [12] used glass-fiber reinforced-plastics (GFRP) and Rahmi [13] used steel plates and props to strengthen the beam-column joint in order to increase the seismic performance of building.

This study proposes the use of steel plates to externally strengthen the beam-column joints of NEB. The L-shape steel plates were glued and bolted in the external side of the beam column joint. The main consideration of using this strengthening material and method is that the installation is easy and the cost is affordable to the owner of the buildings.

The efficacy of this material and method for such strengthening will be numerically examined using ATENA 3D v.5 (Advanced Tool for Engineering Nonlinear Analysis) software. The parameters investigated to quantify performance include load capacity, crack pattern, ductility and energy dissipation. There are 4 variations of the test object model, namely PU-200 as the control model and PUP-200 as the reinforcement test model with plate width variations of 5, 10 and 15 cm.

2 Test Object Model

The test object model to be studied is a simple building construction that represents non-engineered buildings widely used by the community, especially those that are prone to earthquake loads. The quality of the concrete, steel and reinforcements is in accordance with the conditions in the field. Furthermore, a nonlinear analysis was performed using ATENA software. The concrete used is f'c 17 MPa, the yield strength of reinforcing steel fy 400 MPa with the reinforcement lay out as in Figs. 1 and 2.

There are two models of test objects, namely PU-200 as a control, where the test object model has column dimensions 150 × 150 mm with the number of main

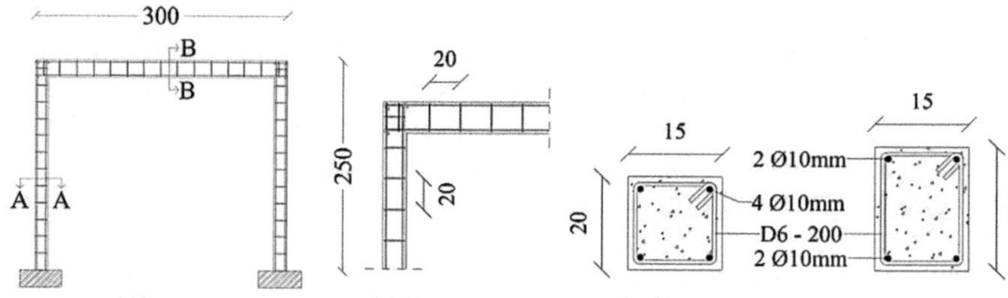

Fig. 1 Test object model PU-200 (control)

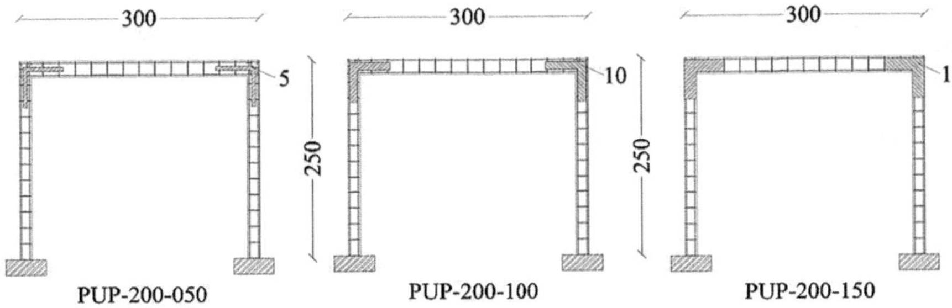

Fig. 2 Test object model PUP-200 with variations in steel plate width

reinforcement 4-D10 with stirrup D6-200 mm, while the beam has dimensions of 150 × 200 mm with the number of bending reinforcement 4-D10 with stirrup D6-200 mm.

Meanwhile, the second model is retrofitted with L-shape steel plates. This model, denoted PUP-200 has the same specifications as PU-200 but with the addition of reinforcing steel plates on the beam-column joint with variations in plate widths of 50, 100 and 150 mm. PU-200 and PUP-200 modeling can be seen in Figs. 1 and 2. In numerical testing with ATENA 3D v5, the concrete material is modeled as 3D Nonlinear Cementitious 2, reinforcing steel as bilateral reinforcement, and reinforcing steel plates as 3D Bilinear Steel Von Mises. The global element size of the macro elements is 2.5 cm. The supports at both ends of the specimen column are restrained in the direction of the X, Y and Z axes. Each macro elements has a perfect connection. The beam-column connection is made as a rigid connection because the reinforcement of the beam and column is fully joined into the joint.

3 Test Result and Analysis

3.1 Load Displacement Relationship

ATENA software was used to numerically analyze the model with the Newton Raphson method was used as solution parameters. Load capacity, displacement, stress, and crack width were obtained from the analysis as shown in Figs. 3, 4 and 5.

In this study, the forces inputted into the load-cases have been determined to be 5 × 10–4 MN or 50 kg to the –X direction at the top of the beam column linkages, while the point of monitoring for the displacement of the model specimen is at the other end. The number of load steps for analysis in this study is determined to be a maximum of 50 steps and will stop when the frame model is not strong enough to withstand lateral loads or collapses. The relationship between lateral loads and displacement is shown in Fig. 6.

Figure 3 shows that the PU-200 specimen has the largest cracks in the joint area that this will trigger a total failure while the PUP-200 specimen with reinforcement

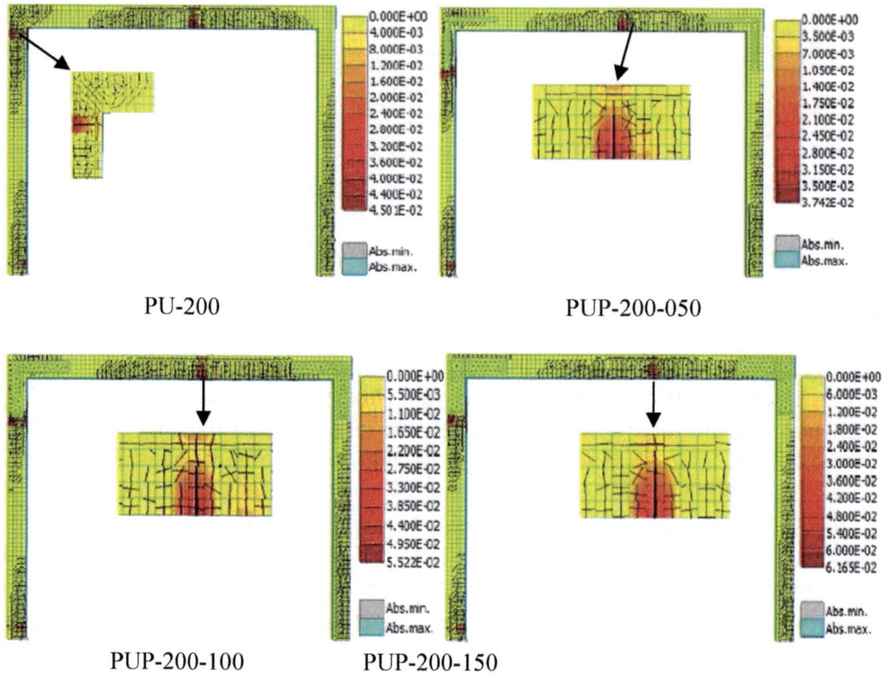

Fig. 3 Crack pattern in the test object model PU-200, PUP-200-050, PUP-200-100, PUP-200-150

plates with a width of 50, 100 and 150 mm has the largest cracks in the area of beam, hence that the collapse pattern will be partial in the beam section only.

Figure 4 shows that the PU-200 steel reinforcement will yield in the column around the joint, while the PUP-200 test object with reinforcement plates with a width of 50, 100 and 150 mm, yielding of reinforcement will occur first on the beam. This can be inferred that the retrofit model demonstrate a better performance as the brittle failure pattern will be change in more acceptable ductile failure pattern.

Figure 5 shows that the retrofitted model PUP-200 with variations in plate widths of 50, 100 and 150 mm shows reduction in concrete stress compared to non-retrofitted model PU-200. This is in agreement with the failure pattern in Fig. 3 and stress reinforcement in Fig. 4.

The main parameters that can be deduced from Fig. 6 are presented in Table 1.

Based on Table 1, the strengthening of the non-engineered building model with a plate width of 50 mm (PUP-200-050) will increase by 14.60%, while for plate widths of 100 mm (PUP-200-100) and 150 mm (PUP-200-150), it increases by 9.91% and 8.05%, respectively. Hence that the reinforcement of a L-shape steel plate with a width of 5 cm provides the greatest value of increase. 50 mm wide L-shape steel plate is also the most effective retrofitting material at reducing the displacement of the test object model, which is 18.43%. Meanwhile the reinforcement of 100 mm and 150 mm wide steel plates has a greater displacement than the model, respectively

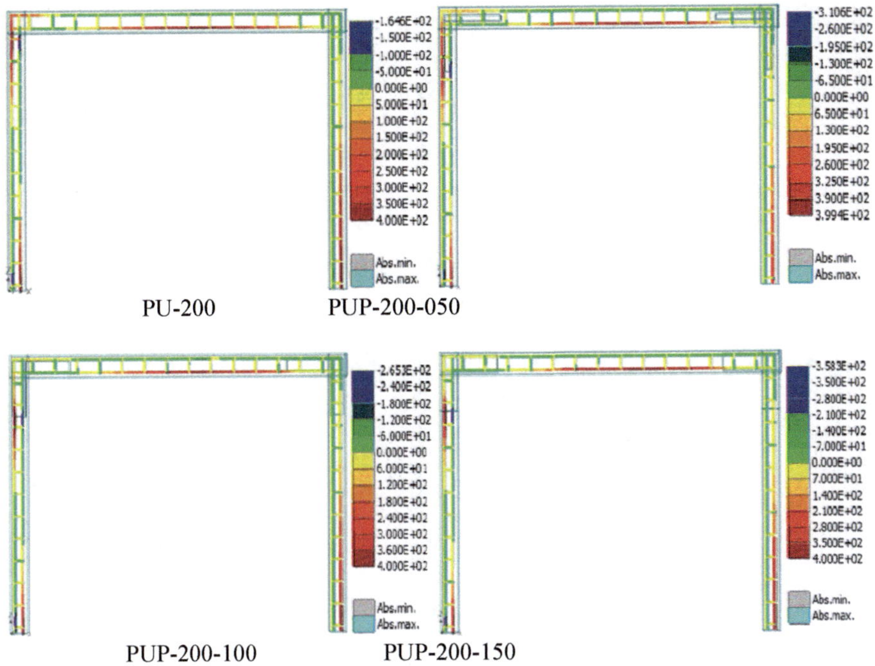

Fig. 4 Reinforcement stresses in the test object model PU-200, PUP-200-050, PUP-200-100, PUP-200-150

14.44% and 15.90%. 50 mm wide L-shape steel plate is also the most effective retrofitting material at reducing the crack width, which was 23.65%, while for the reinforcement of 100 mm and 150 mm wide steel plates had a crack width that was greater than the control specimen model, respectively equal to 28.08% and 17.49%. Therefore, 50 mm wide L-shape steel plate provides the best performance for strengthening the beam column joint.

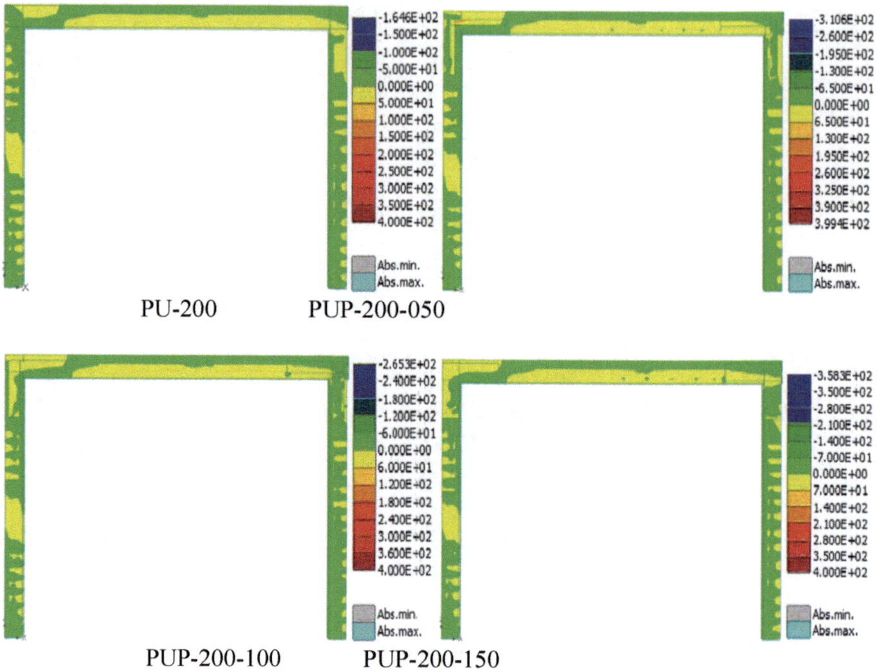

Fig. 5 Concrete stress on the test object model PU-200, PUP-200-050, PUP-200-100, PUP-200-150

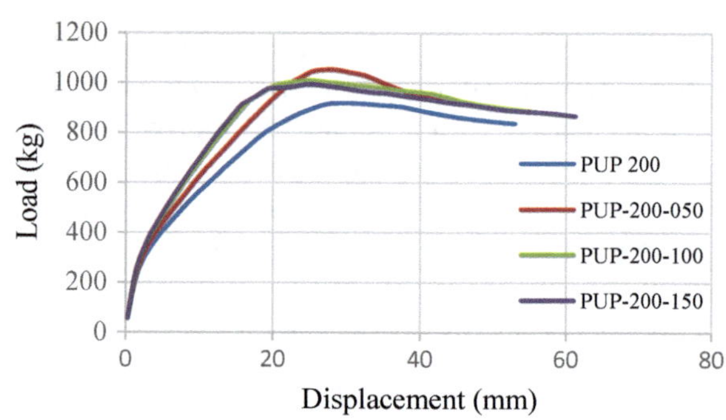

Fig. 6 Graph of the load–displacement relationship on the test object model

Table 1 Recapitulation of test result

Test object model	Load		Displacement		Crack	
	Maximum (Kg)	Increase %	Maximum (mm)	Increase %	Width (mm)	Increase %
PU-200	915.30	–	52.96	–	4.06	–
PUP-200-050	1049.00	14.61	43.20	−18.43	3.10	−23.65
PUP-200-100	1006.00	9.91	60.61	14.44	5.20	28.08
PUP-200-150	989.00	8.05	61.38	15.90	4.77	17.49

Table 2 The ductility value of test object model

Test object model	Displacement (mm)		Ductility	Increase %
	Yield	Maximum		
PU-200	14.61	52.96	3.63	–
PUP-200-050	15.14	43.20	2.85	−21.30
PUP-200-100	11.88	60.61	5.10	40.73
PUP-200-150	11.23	61.38	5.47	50.18

3.2 Ductility

Ductility of the frame can be calculated from Fig. 3 by obtaining the displacement value at ultimate condition (Δu) and yield condition (Δy). Where Δy is determined from the displacement at 75% of the peak lateral load. From these data, the ductility (μΔ) of the test object model can be calculated using the formula:

$$\mu_\Delta = \frac{\Delta_u}{\Delta_y} \tag{1}$$

where μΔ is ductility, Δu is displacement at maximum condition and Δy is displacement at yield condition.

Table 2 shows that the increase in the ductility value of the steel plate reinforcement model with a width of 50 mm, 100 mm and 150 mm respectively is −21.30%, 40.73% and 50.18%, so that the largest ductility value will be achieved in the test specimen model with reinforcing a steel plate width of 150 mm.

3.3 Energy Dissipation

Based on the graph in Fig. 3, the calculation of the energy dissipation in each model of the test object is carried out by calculating the area under the graph. The results of the energy dissipation calculation are presented in Table 3.

Table 3 The energy dissipation value of the test object model

Test object model	Energy dissipation	Increase
	(kg.mm)	%
PU-200	39.647,05	–
PUP-200-050	34.954,45	−11,84
PUP-200-100	51.268,69	29,31
PUP-200-150	51.583,00	30,11

From Table 3 above shows that the energy dissipation in the steel plate reinforcement model with a width of 50 mm, 100 mm and 150 mm is −11.84%, 29.31% and 30.11% respectively. Hence that the greatest energy dissipation will be achieved in the test object model with the reinforcement of a steel plate width of 150 mm.

Based on the analysis, it can be recapitulated that retrofitting the beam column joint of the test object with 50 mm L-shaped steel plate demonstrates the most effective seismic performance while also cost effective as it used the least retrofitting material. This strengthening technique will dissipate more earthquake energy by performing more ductile controlled failure pattern and reduce lateral displacement within certain allowable displacement requirements.

4 Conclusion

From the test results of the PU-200 and PUP 200 test object models with plate variations, it can be concluded that the increase in the load on the strengthening of the beam column joint with plate width variations of 50 mm, 100 mm and 150 mm is respectively 14.60%, 9.91%. and 8.05% so that the strengthening of 50 mm wide steel plate has the greatest value of increase. For a review of the displacement, the strengthening of 50 mm wide steel plates will effectively reduce the displacement by 18.43% while for the widths of 100 mm and 150 mm still has a displacement value that is greater than the test object model before strengthening, respectively 14.44% and 15.90%. If viewed from the reduction in crack width, the strengthening of a steel plate with a width of 50 mm will also be able to reduce the crack width by 23.65%, while for the crack width in the test specimen model with reinforcement of a steel plate with a width of 100 mm and 150 mm, it is greater than with the test object model before strengthening, each increased by 28.08% and 17.49%. When viewed from the ductility and energy dissipation values in the steel plate reinforcement model with a width of 150 mm, it has a higher value than the strengthening for steel plates with a width of 100 mm and 50 mm.

In the event of earthquake, the ability to carry additional lateral seismic load and perform the smallest possible displacement is important factors of structural performance. It will prevent the collapse of the building. NEB does not have these characteristics as it has not been designed properly. It has been demonstrated numerically that strengthening the beam column joint with 50 mm wide L-shaped steel plate provide the most effective seismic performance by reducing lateral displacement and increase ductility. Applying this simple strengthening technique will help community to more resilient toward earthquake hazard by making these typical vulnerable non-engineered buildings earthquake resistant.

References

1. Boen T (2001) Earthquake resistant design of non-engineered buildings. In: Indonesia1 Earthquake Resistant Design of Non-Engineered Buildings In Indonesia 1
2. Park S, Mosalam KM (2012) Parameters for shear strength prediction of exterior beam-column joints without transverse reinforcement. Eng Struct 36:198–209. https://doi.org/10.1016/j.engstruct.2011.11.017
3. Alaee P, Li B (2017) High-strength concrete exterior beam-column joints with high-yield strength steel reinforcements. Eng Struct 145:305–321. https://doi.org/10.1016/j.engstruct.2017.05.024
4. Faleschini F, Hofer L, Zanini MA, dalla Benetta M, Pellegrino C (2017) Experimental behavior of beam-column joints made with EAF concrete under cyclic loading. Eng Struct 139:81–95. https://doi.org/10.1016/j.engstruct.2017.02.038
5. Ricci P, De Risi MT, Verderame GM, Manfredi G (2016) Experimental tests of unreinforced exterior beam-column joints with plain bars. Eng Struct 118:178–194. https://doi.org/10.1016/j.engstruct.2016.03.033
6. Barbhuiya S, Choudhury AM (2015) A study on the size effect of RC beam-column connections under cyclic loading. Eng Struct 95:1–7. https://doi.org/10.1016/j.engstruct.2015.03.052
7. Unal M, Burak B (2013) Development and analytical verification of an inelastic reinforced concrete joint model. Eng Struct 52:284–294. https://doi.org/10.1016/j.engstruct.2013.02.032
8. Shafaei J, Zareian MS, Hosseini A, Marefat MS (2014) Effects of joint flexibility on lateral response of reinforced concrete frames. Eng Struct 81:412–431. https://doi.org/10.1016/j.engstruct.2014.09.046
9. Deng Y et al (2021) Experimental and analytical investigation on flexural behaviour of RC beams strengthened with NSM CFRP prestressed concrete prisms. Compos Struct 257:113385. https://doi.org/10.1016/j.compstruct.2020.113385
10. Hadi MNS, Tran TM (2014) Retrofitting nonseismically detailed exterior beam–column joints using concrete covers together with CFRP jacket. Constr Build Mater 63:161–173. https://doi.org/10.1016/j.conbuildmat.2014.04.019
11. Esmaeeli E et al (2015) Retrofitting of interior RC beam–column joints using CFRP strengthened SHCC: cast-in-place solution. Compos Struct 122:456–467. https://doi.org/10.1016/j.compstruct.2014.12.012
12. Thirumalini S, Shanmugapriya, Atchaya (2021) Experimental and numerical investigation of short GFRP-reinforced steel tubular short columns. Mater Today Proc. https://doi.org/10.1016/j.matpr.2020.11.103
13. Rahmi Y, Saputra A, Siswosukarto S (2017) Numerical modelling of interior RC beam-column joints for non-engineered buildings strengthened using steel plates. In: MATEC Web Conference, vol 138. https://doi.org/10.1051/matecconf/201713802007

A Proposed Method of FRP Anchorage for FRP Confined Over-Reinforced Concrete Beam

Nuroji, Ay Lie Han, Sri Tudjono, Lena Tri Lestari, and Tiara Murtisari

Abstract Fiber Reinforced Polymer (FRP) is a material for strengthening, retrofitting or confining concrete elements. Research on the FRP performance demonstrated that debonding was one of the major failure modes, reducing the ultimate capacity of the composite action between the FRP and concrete. Debonding between FRP and concrete is mostly characterized in the interfacial transition zone (ITZ). Increasing the bond performance in the ITZ is conducted by applying an anchorage system to prevent premature bond loss. The aim of this paper is to propose three concepts of FRP anchorage systems and evaluate their usefulness and theoretical effectiveness. For evaluation, an over-reinforced flexural member is externally strengthened using carbon fiber wraps, the compression area of this member is confined using a u-shape configuration. The three types of FRP anchorages introduced in this study are the spike, insertion, and stitch anchor. It is expected that the anchors will increase the bond performance between the concrete and the CFRP, preventing premature failure and increasing the concrete strength and deformation behavior due to the confinement effect. At further stages, full size elements will be tested to prove this hypothesis, and to evaluate which type of the three anchors is the most effective.

Keywords Fiber-reinforced polymer · Anchorage · Confinement

Nuroji (✉) · A. L. Han · S. Tudjono · L. T. Lestari · T. Murtisari
Universitas Diponegoro, Semarang 50275, Indonesia
e-mail: nuroji@lecturer.undip.ac.id

A. L. Han
e-mail: hanaylie@lecturer.undip.ac.id

S. Tudjono
e-mail: sritudjono@lecturer.undip.ac.id

L. T. Lestari
e-mail: lenatrilestari@students.undip.ac.id

T. Murtisari
e-mail: tiaramurtisari@students.undip.ac.id

S. A. Kristiawan et al. (eds.), *Proceedings of the 5th International Conference on Rehabilitation and Maintenance in Civil Engineering*, Lecture Notes in Civil Engineering 225,
https://doi.org/10.1007/978-981-16-9348-9_20

1 Introduction

Improving and strengthening of existing structural elements in the field are often needed as a result of additional building loads in the form of live, dead, or earthquake loads. The use of Fiber Reinforced Polymers (FRP) for the external reinforcement of structural elements is one method that is being researched extensively, since the FRP has a very high tensile strength exceeding the tensile strength of commonly reinforcing steel [1]. Further, the composite action between the FRP and concrete provides an excellent compatibility behavior in transferring both stress and strain between the two materials. FRP can be used to increase the shear [2–6], the bending and the concrete strength due to the confinement effect [7–15] of a concrete element, leading in an overall enhancement of the member. One of the problems in related prior studies was that the beam failed before the FRP reached its maximum strength due to debonding [16]. Previous studies have found that FRP sheets are debonding by average of 50% of their tensile capacity. Therefore, additional anchorage systems are required to improve the bond behavior in the interfacial transition zone (ITZ) [17–22]. Anchorage system for externally bounded FRP aims to: (I) prevent or delay interfacial crack opening; (II) increase the total available interfacial shear stress transfer; and (III) to provide a stress mechanism where no bond length is available beyond the critical section [18, 23]. The effectiveness of CFRP wraps can be improved by applying multiple layers of wraps [24], but this method will not increase the bond capacity between the wraps and the concrete.

The effective strain of the FRP reinforcement must be limited by the debonding strain, εfd, to avoid cracking due to debonding failure (Eq. 1) [25].

$$\varepsilon\, fd = 0.41 f c' n E f t \leq 0.9 \varepsilon\, f u \tag{1}$$

where f_c', n, Ef, tf, and εfu are concrete compression strength, number of FRP layers, elastic modulus tensile of FRP, thickness of FRP for each layer, and ultimate strain of FRP, respectively. Several studies have shown that anchorage FRP is viable method to delay or prevent FRP debonding to concrete [19, 23, 26, 27].

In the research conducted by [3, 16, 28], experimental testing was carried out on over-reinforced beams that were reinforced more with external FRP reinforcement, but debonding occurred in the stress area due to the beam bending response followed by the shear strain between the concrete and FRP. The members were utilized with u-wraps in the compression zone of the member. U-wraps are commonly used for shear reinforcement of flexural elements [3, 29–32].

2 Concept and Method

2.1 Specimen and Anchorage Details

Figure 1 shows a CFRP wrap-confined beam without anchorage. Three types of FRP anchorage used in this study are shown in Fig. 2.

Over-reinforced beams are used to obtain a wide compressive area of the beam. CFRP Wrap is installed in the stressed area of the beam and a U-shape is used to consider that the FRP cannot be wrapped completely due to the presence of a slab at the existing structure.

In this study, three types of FRP anchor are introduced, they are spike, insertion, and stitch. Anchors will improve the strength of the far end fibers of the FRP, and directly prevent the shear-bond strains to increase [2, 33–37]. The spike anchor is made of CFRP string which is partially hooked into the concrete cover and the rest is fan-shaped spread over the surface of CFRP wrap (Fig. 2a). This anchor model uses the model proposed by researchers and proven highly effective [19, 20, 38, 39] with slight adjustments to the beam dimensions used in this study. Another method of anchoring is an insertion (Fig. 2b), the CFRP wrap edge is inserted into the concrete cover opening and filled with epoxy. The opening groove is curved to avoid the CFRP wrap being torn at the outside corner of the opening due to a stress concentration. In the stitch anchor, CFRP string [4, 40] is used which is formed as shown in Fig. 2c.

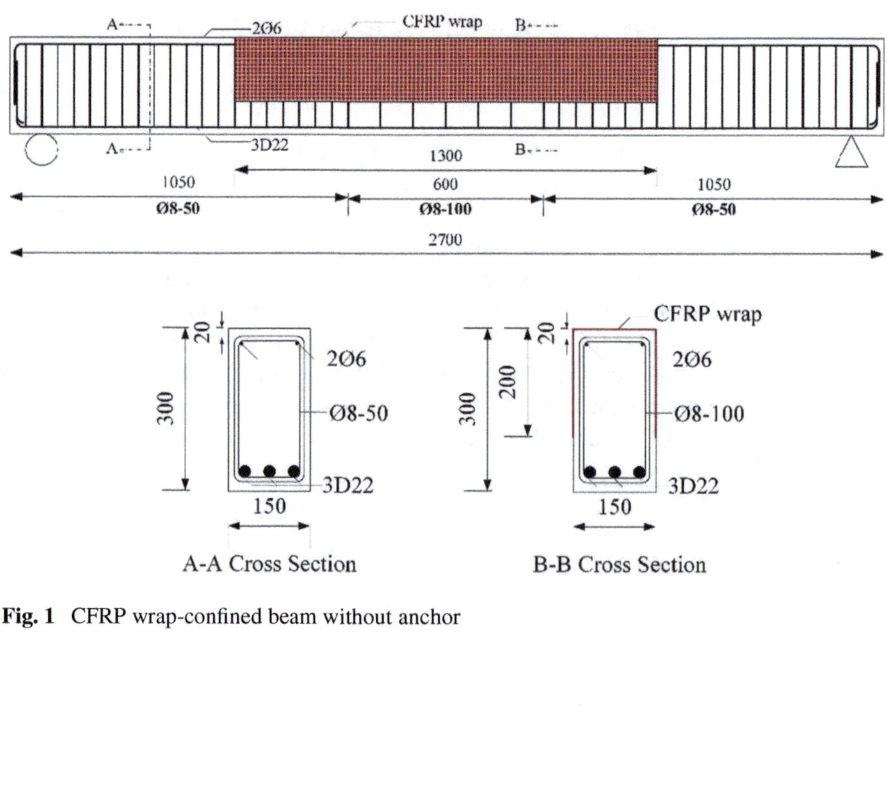

Fig. 1 CFRP wrap-confined beam without anchor

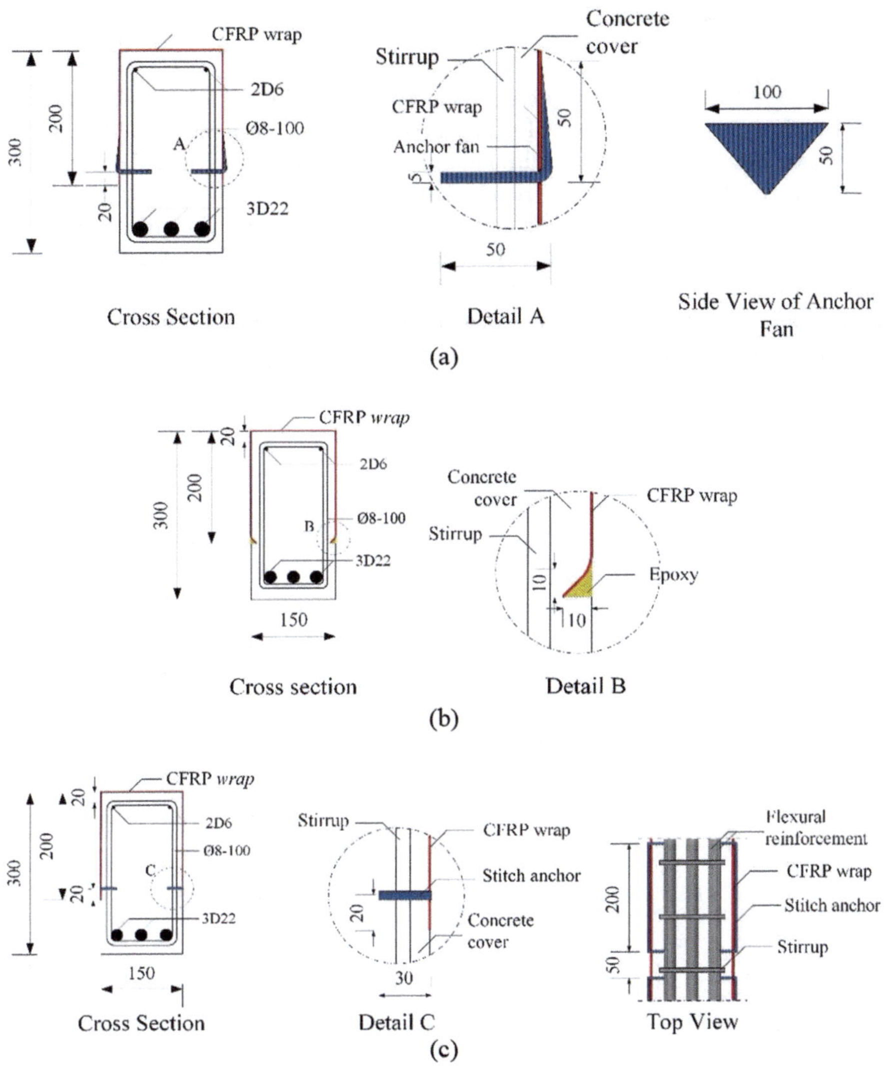

Fig. 2 Variety of FRP anchorage **a** Spike **b** Insertion **c** Stitch

The concrete is perforated using a drill then the anchor stitch is inserted and glued using epoxy.

2.2 Experimental Test Set-Up

The test will be carried out when the concrete reaches 28 days with a 2-point-load system (Fig. 3). The two-point loading system is used so that at middle span of the beam, pure bending occurs without any shear intervention. The FRP is installed 200 mm over the load point to avoid excessive stresses on the ends of the CFRP wrap. The compression strain gauge is installed horizontally and transversally for both concrete and FRP to assess the confinement. Strain gauge for steel is only installed on the tensile reinforcement due to the impossibility to be attached on small diameter compression reinforcement. In order to obtain beam deflection and curvature, one LVDT is installed at the middle of the beam span and two LVDT with a yoke for each other are installed next to it.

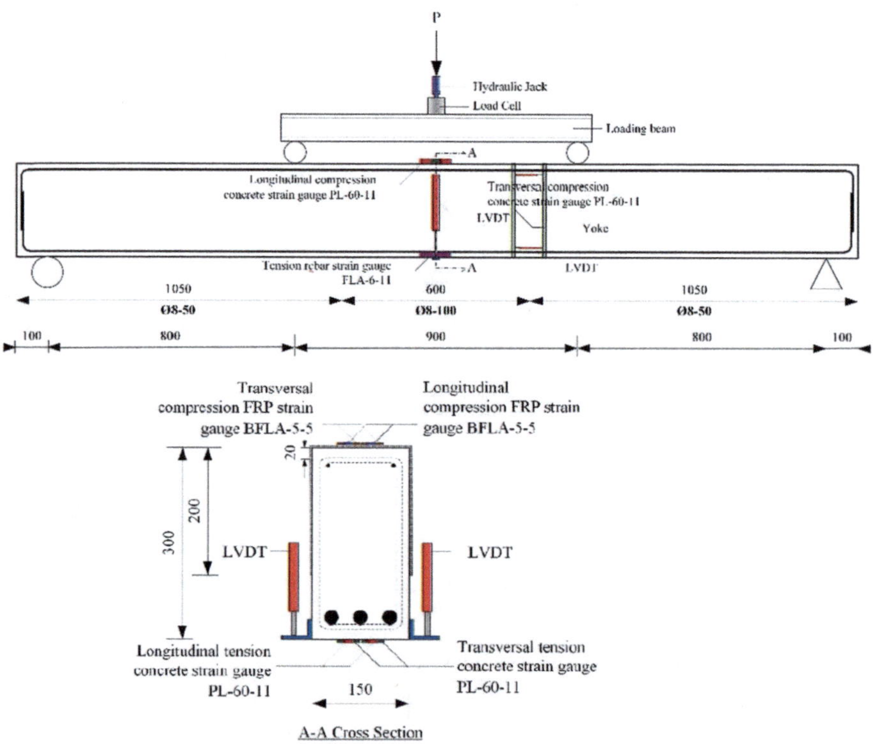

Fig. 3 Setup of testing

3 Discussion

The three types of FRP anchors are expected to be able to overcome debonding by providing confinement to the compressed concrete region of the beam. In over-reinforced beams, failure will occur due to the maximum compressive strain of the concrete before the reinforcing steel yields. As a result of FRP anchor, U-shaped FRP can hold the compressive stress after maximum compressive strain of concrete is reached so that the ultimate load increases until the tensile reinforcement yields. FRP anchorage has a role to anticipate U-shaped FRP debonding phenomenon which is caused by shear stress along the interface of U-shaped FRP and concrete. For spike and stitch anchor, the anchor space/distance is important to make sure the anchors work fine and do not lead stress concentration around the anchor hook. By comparing the three types of FRP anchors, it can be seen which types of anchoring are more effective to use in beam reinforcement using FRP.

References

1. Rasheed HA (2014) Strengthening design of reinforced concrete with FRP
2. Bae S-W, Belarbi A (2013) Behavior of various anchorage systems used for shear strengthening of concrete structures with externally bonded FRP sheets. J Bridg Eng 18(9):837–847. https://doi.org/10.1061/(asce)be.1943-5592.0000420
3. Tee HH, Al-Sanjery K, Chiang JCL (2018) Behaviour of over-reinforced concrete beams with double helix and double square confinements related to ultimate bending and shear strength. J Phys Sci 29. https://doi.org/10.21315/jps2018.29.s2.7
4. Eslami A, Moghavem A, Shayegh HR, Ronagh HR (2020) Effect of FRP stitching anchors on ductile performance of shear-deficient RC beams retrofitted using FRP U-wraps. Structures 23:407–414. https://doi.org/10.1016/j.istruc.2019.11.007
5. Lavorato D, Bergami AV, Fiorentino G, Fiore A, Santini S, Nuti C (2018) Experimental tests on existing RC beams strengthened in flexure and retrofitted for shear by C-FRP in presence of negative moments. Int J Adv Struct Eng 10(3):211–232. https://doi.org/10.1007/s40091-018-0193-1
6. Mahrenholtz P, Cho JY, Park JM, Eligehausen R (2018) Characterization of shear strength of FRP anchors. In: MATEC web conference. vol 199. pp 4–8. https://doi.org/10.1051/matecconf/201819909008
7. Chaallal O, Hassan M, Shahawy M (2003) Confinement model for axially loaded short rectangular columns strengthened with fiber-reinforced polymer wrapping. ACI Struct J 100(2):215–221. https://doi.org/10.14359/12485
8. Fam A, Qie FS, Rizkalla S (2004) Concrete-filled steel tubes subjected to axial compression and lateral cyclic loads. J Struct Eng 130(4):631–640. https://doi.org/10.1061/(asce)0733-9445(2004)130:4(631)
9. Madas P, Elnashai AS (1992) A new passive confinement model for the analysis of concrete structures subjected to cyclic and transient dynamic loading. Earthq Eng Struct Dyn 21(5):409–431. https://doi.org/10.1002/eqe.4290210503
10. Priastiwi YA, Imran I, Nuroji Hidayat A (2014) Behavior of ductile beam with addition confinement in compression zone. Proc Eng 95:132–138, Elsevier. https://doi.org/10.1016/j.proeng.2014.12.172

11. Priastiwi YA, Imran (2015) Nuroji: the effect of different shapes of confinement in compression zone on beam's ductility subjected to monotonic loading. Proc Eng 125:918–924, Elsevier https://doi.org/10.1016/j.proeng.2015.11.098
12. Tee HH, Al-Sanjery K, Chiang JCL (2018) Behaviour of over-reinforced concrete beams with double helix and double square confinements related to ultimate bending and shear strength. J Phys Sci 29:77–98. https://doi.org/10.21315/jps2018.29.s2.7
13. Nguyen DH, Hong WK (2019) Part I: the analytical model predicting post-yield behavior of concrete-encased steel beams considering various confinement effects by transverse reinforcements and steels. Materials (Basel) 12(14). https://doi.org/10.3390/ma12142302
14. Aylie H, Antonius Okiyarta AW (2015) Experimental study of steel-fiber reinforced concrete beams with confinement. Proc Eng 125, Elsevier. https://doi.org/10.1016/j.proeng.2015.11.158
15. Nematzadeh M, Fazli S (2020) The effect of active and passive confining pressure on compressive behavior of STCC and CFST. Adv Concr Constr 9(2):161–171. https://doi.org/10.12989/acc.2020.9.2.161
16. Suprapto H, Tudjono S, Susilorini RMR (2020) Study of the role of CFRP shear on increased bending capacity of reinforced concrete beams. In: Engineering, information and agricultural technology in the global digital revolution, pp 134–138
17. Aljaafreh T, Beneberu E, Yazdani N (2018) Anchorage effect on flexural fiber reinforced polymer (FRP) laminate strengthening of lightweight concrete beams. J Eng Archit 6(1):14–25. https://doi.org/10.15640/jea.v5n2a2
18. Jumaah R, Kalfat R, Al-Mahaidi R, Abdouka K (2017) Anchorage systems used in FRP strengthening of concrete members. In: High tech concrete: where technology and engineering meet—proceedings of the 2017 fib symposium, June, pp 877–886. https://doi.org/10.1007/978-3-319-59471-2_102
19. Villanueva Llauradó P, Ibell T, Fernández Gómez J, González Ramos FJ (2017) Pull-out and shear-strength models for FRP spike anchors. Compos Part B Eng 116:239–252. https://doi.org/10.1016/j.compositesb.2017.02.029
20. del Rey Castillo E, Kanitkar R, Smith ST, Griffith MC, Ingham JM (2019) Design approach for FRP spike anchors in FRP-strengthened RC structures. Compos Struct 214:23–33. https://doi.org/10.1016/j.compstruct.2019.01.100
21. del Rey Castillo E, Griffith M, Ingham J (2018) Straight FRP anchors exhibiting fiber rupture failure mode. Compos Struct 207:612–624. https://doi.org/10.1016/j.compstruct.2018.09.073
22. Li W, Liu W, Yang X, Xing F (2020) Experimental study on FRP-to-concrete bonded joints with FRP sheet anchor system. Adv Mater Sci Eng. https://doi.org/10.1155/2020/2514313
23. Grelle SV, Sneed LH (2013) Review of anchorage systems for externally bonded FRP laminates. Int J Concr Struct Mater 7(1):17–33. https://doi.org/10.1007/s40069-013-0029-0
24. Khoso AR, Akhund MA, Meghwar SL, Hussain F (2018) Effect of multiple layers of carbon fiber reinforced polymer on flexural strength of reinforced concrete. 1–11. https://doi.org/10.20944/preprints201807.0368.v1
25. American Concrete Institute and ACI Committee 440: Guide for the design and construction of externally bonded FRP systems for strengthening concrete structures (2017). American Concrete Institute
26. Orton SL, Jirsa JO, Bayrak O (2008) Design considerations of carbon fiber anchors. J Compos Constr 12(6):608–616. https://doi.org/10.1061/(asce)1090-0268(2008)12:6(608)
27. del Rey Castillo E, Dizhur D, Griffith M, Ingham J (2019) Strengthening RC structures using FRP spike anchors in combination with EBR systems. Compos Struct 209:668–685. https://doi.org/10.1016/j.compstruct.2018.10.093
28. Mohamed HA (2018) Improvement in the ductility of over-reinforced NSC and HSC beams by confining the compression zone. Structures 16:129–136. https://doi.org/10.1016/j.istruc.2018.09.005
29. Lee J, Lopez MM (2016) Characterization of FRP Uwrap anchors for externally bonded FRP-reinforced concrete elements: an experimental study. J Compos Constr 20(4). https://doi.org/10.1061/(asce)cc.1943-5614.0000642

30. Kim YJ, Bhiri M (2020) Grid U-wrap anchorage for reinforced concrete beams strengthened with carbon fiber-reinforced polymer sheets. ACI Struct J 117(1):3–16. https://doi.org/10.14359/51716772
31. Huang L, Zhang C, Yan L, Kasal B (2018) Flexural behavior of U-shape FRP profile-RC composite beams with inner GFRP tube confinement at concrete compression zone. Compos Struct 184:674–687. https://doi.org/10.1016/j.compstruct.2017.10.029
32. Ali AM, Tarkhan MA (2015) Experimental investigation of confining the compression zone in over-reinforced beams. Int J Eng Sci Res Technol 4(11):611–617
33. Cortez Flores IA, Fernández Gómez J, Villanueva Llauradó P (2019) Influence of multiple anchor arrangement in the behaviour of FRP-to-concrete anchored joints. Compos Struct 230:111528. https://doi.org/10.1016/j.compstruct.2019.111528
34. Singh A, del Rey Castillo E, Ingham J (2019) FRP-to-FRP bond characterization and force-based bond length model. Compos Struct 210:724–734. https://doi.org/10.1016/j.compstruct.2018.12.005
35. Del Rey Enrique C, Michael G, Jason I (2016) Fibre rupture tensile capacity of FRP straight anchors
36. Aules W, Saeed YM, Rad FN (2020) A novel anchorage system for strengthening slender RC columns with externally bonded CFRP composite sheets. Constr Build Mater 245:118423.https://doi.org/10.1016/j.conbuildmat.2020.118423
37. Zhang HW, Smith ST, Kim SJ (2012) Optimisation of carbon and glass FRP anchor design. Constr Build Mater 32:1–12. https://doi.org/10.1016/j.conbuildmat.2010.11.100
38. del Rey Castillo E, Dizhur D, Griffith M, Ingham J (2019) Experimental testing and design model for bent FRP anchors exhibiting fiber rupture failure mode. Compos Struct 210:618–627. https://doi.org/10.1016/j.compstruct.2018.11.091
39. Sun W, Liu S, Zhang C (2020) An effective improvement for enhancing the strength and feasibility of FRP spike anchors. Compos Struct 247:112449. https://doi.org/10.1016/j.compstruct.2020.112449
40. Saeed YM, Aules WA, Rad FN, Raad AM (2020) Tensile behavior of FRP anchors made from CFRP ropes epoxy-bonded to uncracked concrete for flexural strengthening of RC columns. Case Stud Constr Mater 13:e00435. https://doi.org/10.1016/j.cscm.2020.e00435

Experimental Investigation on the Shear Behavior of Patched RC Beams Without Web Reinforcements: Efficacy of Patching Position with Respect to the Shear Span

Adji P. Abrian, Stefanus A. Kristiawan, Halwan A. Saifullah, P. Muhammad Rafi, R. Muhammad Hafizh, Andreas M. Simanjuntak, and D. Abel Bismo

Abstract Spalling on concrete covers is a type of degradation that commonly appeared in reinforced concrete structures. This form of degradation could occur at the web of shear-span due to, for example, corrosion of reinforcements within the shear span. The degradation will eventually affect the shear strength of that particular section, resulting in un-conservative shear capacity. Patch repair may be an option for restoring the strength of the damaged RC beam. The shear behavior of patched RC beams without web reinforcements is investigated in this study, with a particular focus on determining the effectiveness of patching position with respect to the shear span. Unsaturated Polyester Resin (UPR)-mortar was used in this investigation as a patch repair material. UPR-mortar was applied to patch RC beam web damage at 0.25 and 0.5 of the shear span (measured from the loading point). The RC patched beams accompanied by some normal beams were loaded until failure in the laboratory. The results indicate that patched RC beams exhibit a higher shear capacity than the original undamaged RC beam. Furthermore, UPR-mortar is more effective when applied at 0.25 of the shear span rather than 0.5 of the shear span. This is demonstrated by the evidence that after an initial diagonal crack was formed, the RC beam with patching at 0.25 of the shear span still carries a substantial load before the diagonal crack propagates, causing the beam to fail in shear. Whereas in RC beam with patch repair position at 0.5 of the shear span, the RC beam immediately fails in shear once the significant diagonal crack was formed. However, in NC beams the shear failure occurred earlier than in patched RC beams.

A. P. Abrian · S. A. Kristiawan · H. A. Saifullah (✉) · P. M. Rafi · R. M. Hafizh · A. M. Simanjuntak · D. A. Bismo
Department of Civil Engineering, Faculty of Engineering, Universitas Sebelas Maret, Surakarta, Indonesia
e-mail: halwan@ft.uns.ac.id

S. A. Kristiawan
e-mail: s.a.kristiawan@ft.uns.ac.id

A. M. Simanjuntak
e-mail: andre230410@gmail.com

S. A. Kristiawan et al. (eds.), *Proceedings of the 5th International Conference on Rehabilitation and Maintenance in Civil Engineering*, Lecture Notes in Civil Engineering 225,
https://doi.org/10.1007/978-981-16-9348-9_21

Keywords Crack · Patch repair · RC beam · Shear · UPR-mortar

1 Introduction

A reduction in the shear capacity of RC beams can be expected if there is damage to that particular beams. One type of damage often found in RC beam elements is spalling of concrete covers. Spalling is a type of damage that occurs in the concrete cover which can be triggered by, for example, corrosion of reinforcements. The damage raises a concern about the structural component's safety, especially when it occurs within the shear span of the beam. If spalling is not repaired, it will decrease the shear strength of the respected RC beams, which eventually resulting in un-conservative shear capacity. Shear failure is more dangerous than flexural failure because it typically occurs unexpectedly [1–4]. According to the literature review, the flexural strengthening performance of the RC beam has been fully resolved, but its mechanism is not as complicated as shear [5–7]. For this reason, any attempts to repair a deficient shear capacity of RC beam are encouraged.

Diagonal cracks must occur in the beam before failure in all cases of shear failure mode. Several factors that influence the shear failure of RC beam without web reinforcements are concrete properties, longitudinal reinforcement properties, longitudinal reinforcement ratio, shear span-to-depth ratio, and beam size [8]. For RC slender beams with a ratio between shear span a to effective depth d is between 2.5 and 6.5, it has been determined that three components contribute to the shear transfer mechanism, i.e., un-cracked concrete in the flexural compression zone, V_{cy}; the vertical component of the aggregate interlock, V_{ay}; and dowel action, V_d [9, 10]. An attempt has been carried out by Kristiawan et al. [11] to recover the shear capacity of the damaged RC beam using UPR-mortar. The results show that after an initial diagonal crack was formed, the patched RC beam can still withstand a greater load before the beam fails in shear. The shear force capacity is increased by 14% compared to the normal RC beam. Interestingly, the diagonal cracks did not penetrate the UPR mortar. As a result, the UPR mortar remains un-cracked when the repaired RC beam fails in shear [11]. This could be the reason why by using UPR-mortar, the shear capacity of the patched RC beam can be increased.

The position of patching can affect the efficacy of the repair. This is because the effective contribution of UPR-mortar to interfering with the propagation of diagonal cracks is location-dependent. The current study examines the influence of UPR-mortar repaired positions on the shear performance of RC beam. In this inquiry, two repaired places are chosen, which are 0.25 and 0.50 of the shear span measured from the loading point.

2 Experimental Program

2.1 Detail Specimen

Three different types of RC beam specimens with a shear span to effective depth ratio of 3.0 were prepared and tested. The first beam specimen, NC (see Fig. 1a), was used as a control beam, which was not patched with UPR-Mortar. While the remaining two beams were patched RC beams, PR-1 (see Fig. 1b) and PR-II (see Fig. 1c), with the repair position at 0.25 and 0.50 of the shear span, respectively. The beams were made of ordinary concrete with a compressive strength of 22.5 MPa and tensile strength of 2.5 MPa at the loading test's age (63 days). On the other

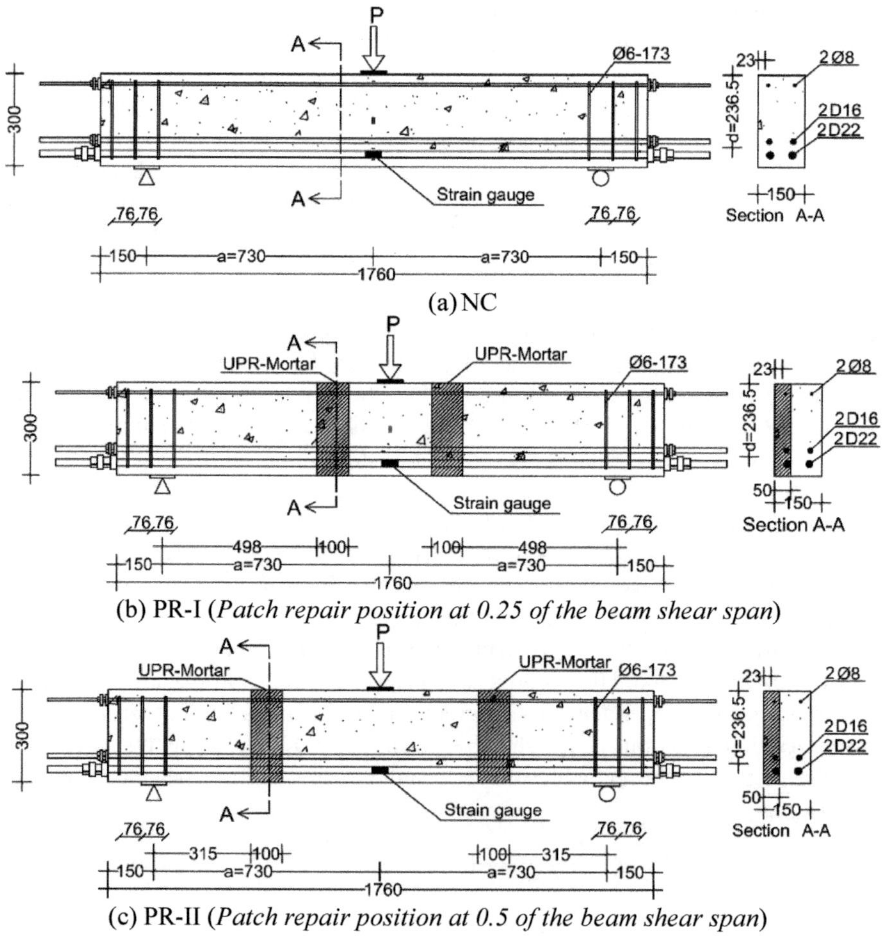

(a) NC

(b) PR-I (*Patch repair position at 0.25 of the beam shear span*)

(c) PR-II (*Patch repair position at 0.5 of the beam shear span*)

Fig. 1 Details of reinforcement and cross sections (units in mm)

hand, UPR-mortar used had a compressive strength of 67 MPa and tensile strength of 5.09 Mpa (3 days). A relatively high tensile reinforcement (with a yield strength of 399 MPa) ratio was used to achieve a bending capacity of at least 40% greater than the shear capacity ensuring the shear failure mode. The detail of RC beams can be seen in Fig. 1.

2.2 Testing of the RC Beams

The loading test was carried out at the Structural Engineering Laboratory, Universitas Sebelas Maret. Figure 2 shows the setup of beam testing. A hydraulic jack with a capacity of 500 kN was installed on the loading frame for imposing load. The load cell, which was located in direct contact with the hydraulic jack, was used to measure the applied load. During the loading test, the RC beam deflection and crack development were monitored. A linear vertical displacement transducer (LVDT) was positioned at midspan to measure the maximum deflection of RC beam. Demountable mechanical strain gauges (DEMEC) with a gauge length of 200 mm were also attached to the RC beam's surface to track the diagonal crack development. Additionally, strain gauges were installed on the tension reinforcement at a location that

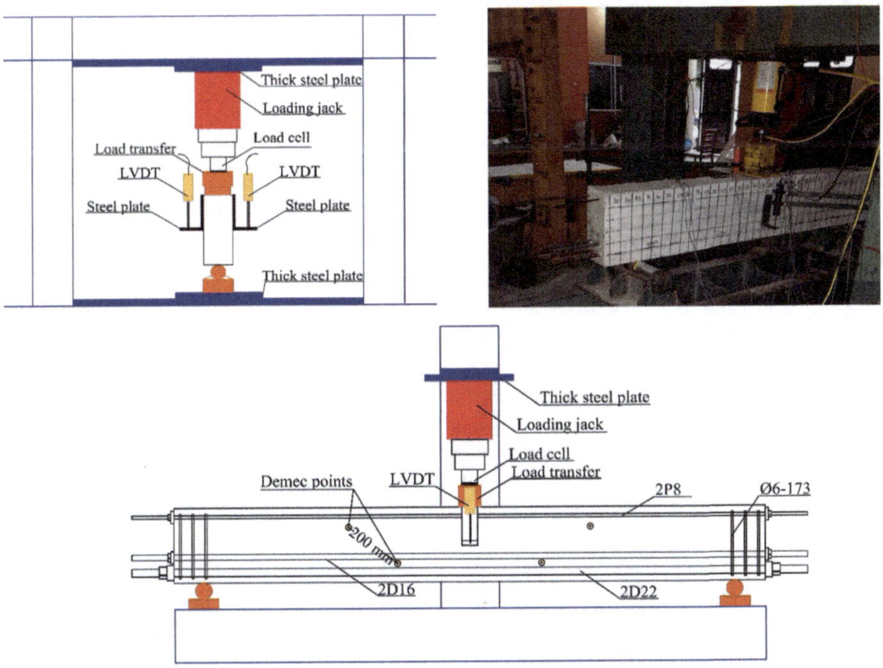

Fig. 2 The setup of beam testing

coincides with the beam's midspan. The development of crack was marked at each load increment of 5 kN.

3 Behavior of Test Specimens

3.1 Failure Mode

For all RC beams, the maximum longitudinal reinforcement strains at the ultimate load (923 μ at the maximum recorded in the NC beam) did not exceed the reinforcement bar's yield strain of approximately 1995 μ. This indicated that shear failure modes occurred without longitudinal reinforcement yielding. The failure mode of all beam specimens is typical of diagonal tension (see Figs. 3, 4 and 5). After the initial diagonal cracking formation, which indicated the beginning of shear cracks,

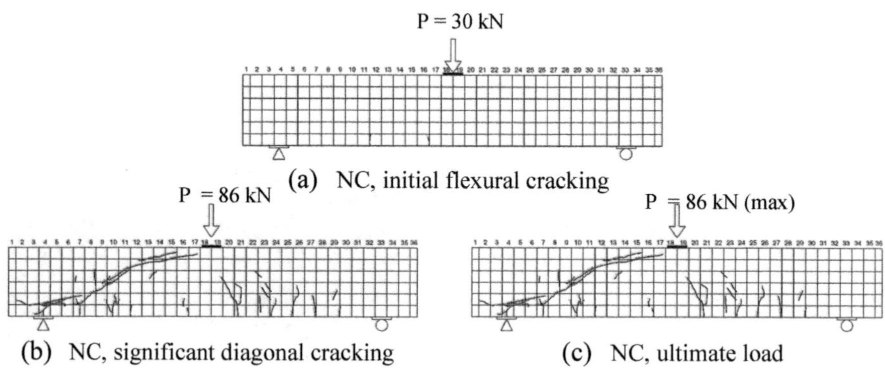

Fig. 3 Crack patterns of the NC beam

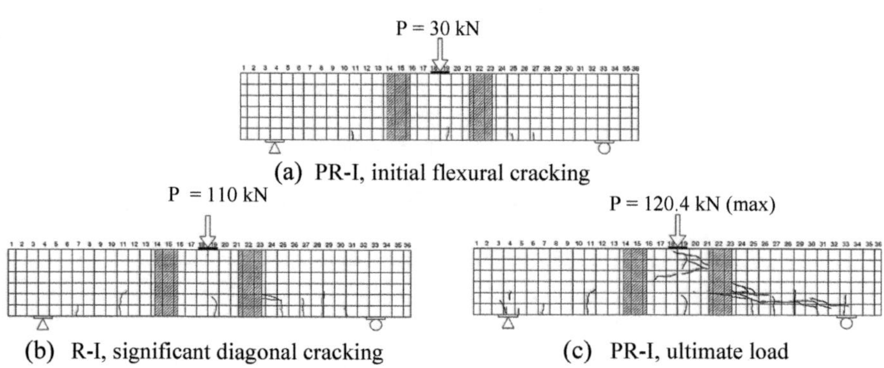

Fig. 4 Crack patterns of the PR-I beam

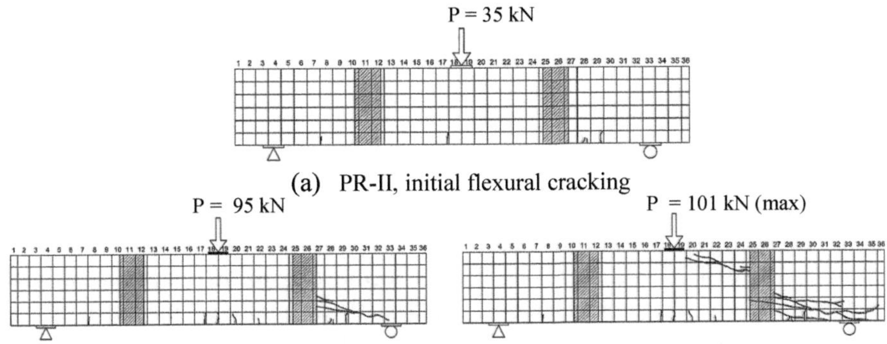

(a) PR-II, initial flexural cracking

(b) PR-II, significant diagonal cracking

(c) PR-II, ultimate load

Fig. 5 Crack patterns of the PR-II beam

the patched RC beams resisted more shear force accompanied by further development of shear cracks. The diagonal crack penetrating the compression zone near the loading points causes shear failure.

3.2 *Crack Pattern*

Figures 3, 4 and 5 shows the crack patterns exhibited in the beam specimens. Generally, initial flexural cracks were located around the mid-span where the flexural stress at maximum. Flexural cracks propagated toward the loading point as the applied load increased. When the load was sufficient to induce diagonal tension stress greater than the concrete's tensile strength, diagonal cracks will appear. When the applied load was increased until a certain level, more flexural cracks were formed, and the previous flexural cracks extend in length and width. Once the significant diagonal cracks had formed, a further increase in load would cause the beams would fail immediately.

In the NC beam case, a continuous propagation of the diagonal crack is observed from the lower surface of the concrete toward the loading point. Distinct phenomena were observed in the PR-I dan PR-II cases. The presence of UPR-mortar could resist the diagonal crack propagation. Obviously, this is due to the high tensile strength of the UPR-mortar. The resistance against the propagation of diagonal crack by the patching material causing an improvement in the shear resistance of patched RC beams. The patching location also affects the extent of the propagation of the crack pattern. The crack pattern on the PR-I beam with patching location at 0.25 of the shear span gives a more significant effect in resisting the diagonal crack than the patching location at 0.5 of the shear span. This may be due to the contribution of UPR-mortar to preserve the compression zone at a high load level when this repair material close to the loading point (the PR-I beam). The intensity of cracks at this compression zone of PR-I prior to shear failure indicates the above argument (see Fig. 4).

3.3 Load–Deflection Response

Figure 6 shows the relationship between the load and midspan deflection. The applied load-midspan deflection curves could only be recorded up to the peak load under the load control system's implementation. The beam's flexural cracking capacity was exceeded at a load of approximately 30–35 kN, resulting in the stiffness reduction of the RC beam. The amount of stiffness reduction appears to be different for the three types of beam, with the NC beam having the most significant decrease. It indicates the influence of patching repair becomes more pronounced after the first flexural cracking occurrence.

The parameters describing the characteristic of load–deflection curves are summarized in Table 1. The ultimate load for the NC, PR-I, and PR-II beams were 86, 120.4 and 101 kN, respectively. Although the corresponding deflections had almost the same value, the patched beams exhibited a higher ultimate load (40 and 17% for

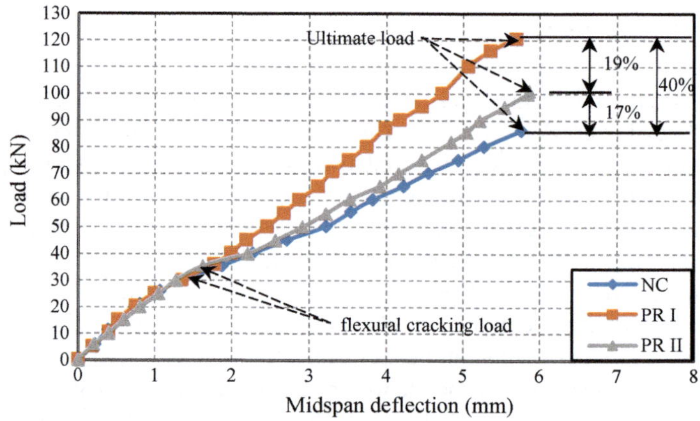

Fig. 6 Load and mispan deflection of RC beams

Table 1 Flexural cracking load, ultimate load, and deflection at the ultimate load of RC beams

Specimen	Flexural cracking load (kN)	Ultimate load (kN)	Deflection at the ultimate load (mm)
NC	30	86	5.75
PR-I	30	120.4	5.69
PR-II	35	101	5.84

PR-I and PR-II, respectively) than the normal beam. The efficacy of patching position with respect to the shear span is also apparent. The PR-I beam with a patching position at 0.25 of the shear span shows the largest shear capacity. The effectiveness of UPR-mortar to reduce the intensity of cracks at similar load levels (see Figs. 3, 4 and 5) could be the reason for such behavior. This phenomenon has also been identified in the flexural behavior of patched RC beams [12].

3.4 Diagonal Crack

The development of diagonal cracks was measured by the DEMEC gauge every load increment of 5 kN. Figure 7 shows the relationship between the load and diagonal crack width. In the present study, a significant diagonal cracking load was identified by a sudden increase in the diagonal crack's width. All RC beams indicated almost the same significant inclined crack width. However, the patched beams exceeded the NC beam in terms of significant diagonal cracking load, ultimate load, and maximum diagonal crack width, as shown in Table 2. Additionally, the patching position affects the resistance of RC beams to the applied loads. This is demonstrated by the more

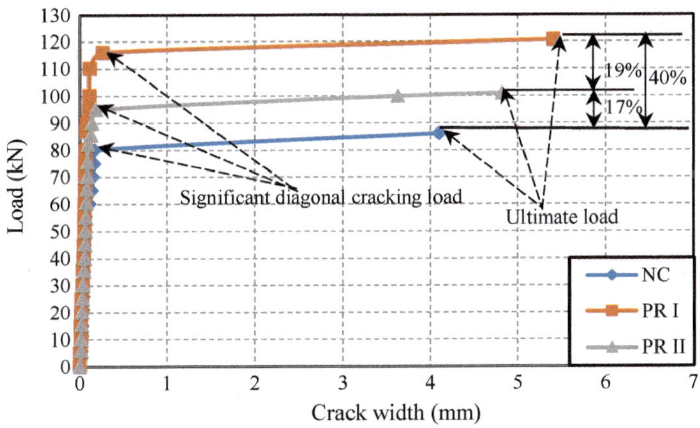

Fig. 7 Beam specimens diagonal crack width

Table 2 Significant diagonal cracking load, ultimate load, and maximum diagonal crack width of RC beams

Specimen	Significant diagonal cracking load (kN)	Ultimate load (kN)	Maximum diagonal crack width (mm)
NC	80	86	4.1
PR-I	115.9	120.4	5.4
PR-II	95	101	4.8

excellent shear resistance and the higher corresponding maximum diagonal crack width in the PR-I case compared to the PR-II.

The presence of patching can influence the shear transfer mechanism. The high tensile strength of the patch repair material effectively prevents the diagonal crack from propagating. Consequently, the beam holds a sufficient compression zone to provide a considerable V_{cy}. Furthermore, in the diagonally cracked beam, the un-cracked UPR-mortar provides slip resistance, promoting better aggregates interlocks on the surface of diagonal crack and improving V_{ay}.

4 Conclusion

The failure mode of all beam specimens is typical of diagonal tension. UPR-mortar can be applied to repair the damaged RC beam and increase its shear capacity. The un-cracked feature of UPR-mortar affects the shear transfer mechanism by arresting diagonal crack propagation and providing slip resistance. The position of UPR-mortar with respect to the shear span influences the efficacy of the repair material to recover the shear resistance. The closer the UPR-mortar to the loading point will be more effective in preserving the compression zone, thus increasing the shear strength.

Acknowledgements This research has been funded by Sebelas Maret University through PNBP-UNS Grant No. 452/UN27.21/PN/2020. The specimens were built and tested at the Structural Laboratory of the Department of Civil Engineering, Sebelas Maret University, with the help of the laboratory technicians and undergraduate students (Atep Nuryana, Alfian Muhammad, Fendy Prayugo, Inggrid Monalita, Ummi Sarah, and Elisabeth Nathasya).

References

1. Joint ACI-ASCE Committee 445 (2009) Recent approaches to shear design of structural concrete : ACI manual of concrete practice (ACI 445R-99)
2. Jayaprakash J, Abdul Samad AA, Anvar Abbasovich A, Abang Ali AA (2008) Shear capacity of precracked and non-precracked reinforced concrete shear beams with externally bonded bi-directional CFRP strips. Constr Build Mater 22(6):1148–1165
3. Ali N, Samad AAA, Mohamad N, Jayaprakash J (2015) Shear behavior of pre-cracked continuous beam repaired using externally bonded CFRP strips. Proc Eng 53:129–144
4. Khalifa A, Nanni A (2000) Improving shear capacity of existing RC T- section beams using CFRP composites. Cement Concr Compos J 22(2):165–174
5. Saadatmanesh H, Ehsani RM (1991) RC beams strengthened with GFRP plates I: experimental study. Struct Eng J 117(11):3417–3433
6. GangaRao HVS, Vijay PV (1998) Bending behavior of concrete beams wrapped with carbon fabrics. Struct Eng J 124(1):3–10
7. El-Mihilmy Mahmoud T, Joseph WT (2000) Analysis of reinforced concrete beams strengthened with FRP laminates. Struct Eng J 126(6):684–691
8. Thamrin R, Tanjung J, Aryanti R, Nur OF, Devinus A (2016) Shear strength of reinforced concrete T-beams without stirrups. J Eng Sci Technol 11(4):548–562

9. Hassoun MN, Al-Manaseer A (2015) In: Structural concrete: theory and desain. 6th edn.
10. Wright JK, MacGregor JG (2011) In: Reinforced concrete: mechanics and design. 6th edn. Prentice Hall
11. Kristiawan SA, Agus S, Senot S, Hapsara BW (2017) Shear failure of patched reinforced concrete beam without web reinforcements. Key Eng Mater 737 KEM:441–447
12. Kristiawan SA, Supriyadi A, Sangadji S, Santosa D (2017) Cracking behavior and its effect on the deflection of patched-reinforced concrete beam under flexural loading. In: MATEC web of conferences. vol 138

Secondary AE Analysis of Pre-corroded Concrete Beam

Ahmad Zaki and Zainah Ibrahim

Abstract Acoustic emission technique of non-destructive testing (NDT) method is the more successful method of corrosion assessment of reinforced concrete (RC) structures for the evaluation of pre-corroded concrete beam specimens under load testing. The tests were used for small scales of pre-corroded concrete beams subject to cyclic loading before the first visible cracks for secondary AE source. The secondary AE source caused by the cyclic load at the maximum load level is half of the estimated first visible cracks (i.e., 4 kN). To study the impact of significant damage due to corrosion before the first visible cracking occurs, the AE parameters included the cumulative signal strength (CSS) and *I*b-value are proposed. The results showed that the high CSS ratio (more than 69%) is related to the damage growth of the beam specimens. The *I*b-value tends to decrease from loading cycle 2 to cycle 3, indicates that the occurrence of damage.

Keywords AE · NDT · Corrosion · Assessment · Secondary AE

1 Introduction

Steel corrosion of reinforced concrete (RC) structures is a global problem that causes the deterioration of RC structures. Corrosion damage typically needs sufficient repairs and maintenance. Moreover, the Non-destructive testing (NDT) approach is practically a necessity in this study to rehabilitate the corroded RC structures efficiently and cost-effectively. Over the past few years, the acoustic emission (AE)

A. Zaki (✉)
Department of Civil Engineering, Faculty of Engineering, Universitas Muhammadiyah Yogyakarta, Bantul 55183, Daerah Istimewa Yogyakarta, Indonesia
e-mail: ahmad.zaki@umy.ac.id

Z. Ibrahim
Department of Civil Engineering, Faculty of Engineering, University of Malaya, 50603 Kuala Lumpur, Malaysia
e-mail: zainah@um.edu.my

S. A. Kristiawan et al. (eds.), *Proceedings of the 5th International Conference on Rehabilitation and Maintenance in Civil Engineering*, Lecture Notes in Civil Engineering 225,
https://doi.org/10.1007/978-981-16-9348-9_22

technique more useful techniques for structural health monitoring and damage assessment of the corroded RC structures [1–7]. The AE technique can be operated without disturbing any progressions associated with the RC structures [8–11]. The conversion of transient elastic waves (ultrasonic frequency range) from a confined source within a material into electrical signals via connected piezoelectric sensors is referred to as AE technique [12–15]. The sources of AE are deformation processes such as crack growth, corrosion, and other material degradation. Localised energy release gives rise to elastic waves that are detected by sensors placed on the concrete surface [16, 17].

The AE technique successfully analyses the energy signals produced by the concrete cracking induced by the corrosion of steel reinforcement by the AE sensors mounted on the concrete surface. The AE technique has been widely used for corrosion, especially the monitoring of corroded RC structure. AE parameters, including AE hits, energy, amplitude, signal strength, etc. were successfully used to describe the corrosion processes for reinforcement of RC structures [5, 6, 18–24]. The AE technique enables damage localization of steel corrosion and for detecting the micro-cracking and the macro-cracking [21, 25–33]. However, there is a limited study of the impact of the significant damage due to steel corrosion before visible cracking occurs. In this study, the AE technique is the proposed technique to monitor the pre-corroded concrete specimens subjected to cyclic loading before the first visible cracking occurs (at early stage). Early warning could be given to allow for repair work before the concrete structure is seriously damaged. The CSS and *I*b-value of AE data have been carried out to characterize the corrosion damage of the pre-corroded specimens.

2 Materials and Methods

2.1 Fabrication of Concrete Specimens

The deformed steel reinforcement of sizes 12 mm in compliance to BS 4449 [34] was used as tension reinforcement. The pre-corroded specimens were prepared using corroded steel reinforcement before concrete casting The steel reinforcements were corroded before concrete casting. The steel reinforcements were immersed in a 5% sodium chloride (NaCl) solution in a plastic container. A direct current (DC) power supply was used as the electrolysis. As the anode, the copper wire connected the steel reinforcement to the positive terminal—meanwhile, the steel bar connected to the negative terminal as the cathode. Therefore, after the corrosion process, to remove the corrosion product (rust) of the steel reinforcements were cleaned in accordance to ASTM G1-03 [35]. The corroded steel reinforcement weighed to assess the mass loss. Three pre-corroded specimens with a dimension of 100 × 100 × 500 mm^3 were fabricated. The details of the pre-corroded specimen are shown in Fig. 1.

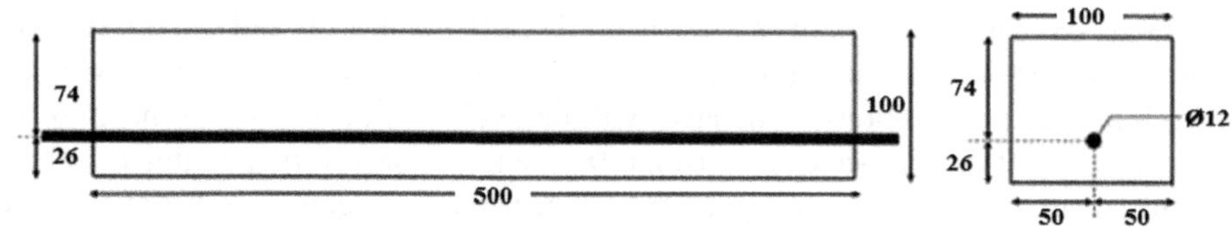

Fig. 1 The details of the beam specimen (in mm)

The beam specimens with 28 MPa of compressive strength were designed in accordance to BS 8110–1 [36]. The reinforced concrete properties are 2192.91 kg/m^3 of dry density, 392,95 kg/m^3 of cement content, 0,5429 of water to cement ratio, 67 mm of slump value, and 200.000 MPa of modulus elasticity of reinforcement. The specimens with steel reinforcements of 0% (S0), 9.06% (S9), and 27.56% (S28) of corrosion levels were prepared for pre-corroded beam specimens. The corrosion level (%) of the beam specimen could be measured as the mass loss of the steel reinforcement before and after accelerating corrosion [37, 38]. The bottom cover was 26 mm from the concrete surface. Ordinary Portland cement (OPC), sand from the river, and 10 mm of crushed type granite were used to prepare the specimens.

2.2 Data Acquisition

The specimens were subjected to three-point load testing, as shown in Fig. 2. The flexural load testing using 100 kN of Universal Testing Machine (INSTRON Satec Series). Each specimen was supported at both ends with distance between each support to the loading point was 200 mm. The data acquisitions of the specimens were conducted using the AE technique throughout the cyclic load testing. A PCI-2 AE System of MITRAS Group, Inc. was connected to the six AE sensors. The sensors were mounted using wax to the specimen surface as the coupling agent. After testing the specimens with the AE technique, the AE data will be analyzed for AE parameters analysis.

Fig. 2 Data acquisitions using INSTRON Satec series of 100 kN and AE technique

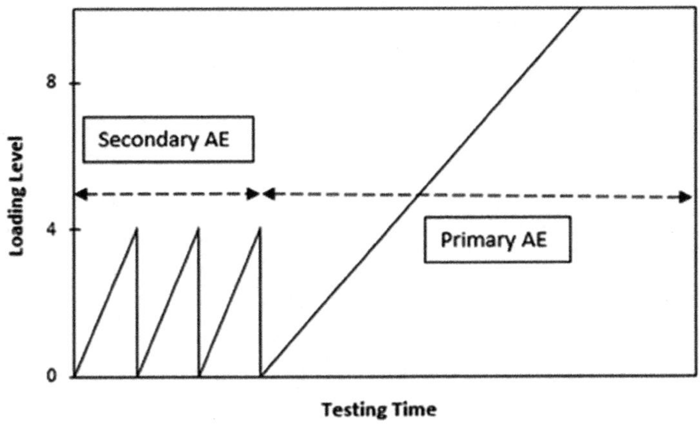

Fig. 3 Cyclic load testing for secondary AE

The source types of AE monitoring of the specimens can be classified as primary AE and secondary AE [39]. The primary AE is related to crack growth and fracture mechanic parameters, while in secondary AE comprises all other sources, including the micro-fracture. The secondary AE sources induced by the cyclic load testing with a maximum load level is half of the estimated first visible cracks, that is 4 kN, as shown in Fig. 3. The loading cycle load is intended to study the effect of significant damage before the first visible cracking occurs. Furthermore, the number of cycles was chosen based on many references, such as Paul and Adam [40].

3 Results and Discussion

Table 1 presents a summary of the secondary AE parameters of the pre-corroded beam specimens along with the three different cycles (i.e., Cycles 1, 2, and 3) of cyclic load testing. The loading rate of secondary AE is 0.1 mm/min. The secondary AE parameters are duration, amplitude, average frequency (AF), hits, and counts.

3.1 Cumulative Signal Strength (CSS)

Figure 4 shows the curve of signal strength and load with time for all the three beam specimens under three cycles. Figure 5 shows the curve of CSS with time for all the three beam specimens. In general, the CSS rate decreases from cycle 1 to cycle 3. It means that the AE activities decrease due to the limitation of AE activities during uncracked beam as the beam specimen was subjected up to 4 kN load, at which it is 50% of the theoretical first visible flexural cracking load at middle span of the beam

Table 1 Secondary AE parameters

Beam specimens	Cycle	Duration (× 1000 μs)	Amplitude	AF (kHz)	Hits	% Of total hits	Counts
S0	1	23.572	29.58	14.08	8992	52.87	2,986,932
	2	24.848	23.23	7.26	5148	30.27	986,366
	3	24.023	22.61	6.77	2868	16.86	532,277
S9	1	24.033	34.28	19.25	14,667	61.59	8,640,423
	2	16.408	23.27	5.55	5595	23.50	531,306
	3	14.821	23.70	5.91	3551	14.91	311,610
S28	1	38.696	29.49	22.50	4837	35.94	6,012,840
	2	33.668	23.54	11.23	4182	31.07	1,879,292
	3	30.228	22.71	9.09	4440	32.99	1,440,196

Note Duration: Time between signal start and signal end, Amplitude: Largest voltage peak in the waveforms, Average Frequency (AF): Nature of source event, Hits: Detection of an AE signal, Counts: Number of times AE signal crosses threshold

specimen. Therefore, to evaluate the damage of beam specimens, the CSS ratio is proposed.

A study done by Paul and Adam [33] proposed an evaluation criterion based on the cumulative signal strength (CSS) ratio of RC beams that were strengthened with carbon fiber-reinforced polymer (CFRP). It is defined as the peak CSS at the end of the reload hold period divided by the peak CSS at the end of the initial load hold period. The study suggested that a peak CSS ratio between 30 and 50% was an appropriate value to select as an indication of significant damage. Table 2 shows the CSS ratio for the specimens. The suggested ratio of CSS ratio based on the study done by Paul and Adam [33] is not suitable in this case due to the CSS ratios obtained for the specimens that are not absolutely within the range of 30–50% for significant damage due to steel corrosion. It can be attributed to the sensitivity of individual sensors, attachment methods of sensors, concrete material, and steel corrosion of pre-corroded specimens of this study that are different from the study by Paul and Adam [33].

However, from the results obtained, by comparing with the S0 beam specimen, it can be deduced that S9 and S28 beam specimens with higher corrosion levels experienced very significant damage. Due to the specimen having a much higher CSS ratio compared to the S0 beam specimen, especially for the CSS ratio obtained using cycle 2 and 3 values, which are more than 60%. This result is supported by the findings from the mechanical loading test, where these specimens failed at the lowest peak load for S28 beam specimens. On the other hand, the result for S9 beam specimens, their first CSS ratios were the lowest with less than 18%, but their second CSS ratio was very significant, with a percentage of more than 69%. The high value of CSS ratio is related to the damage growth of the specimens [33]. The lowest CSS ratio might imply the steel corrosion of the specimen does not affect the damage growth of the beam specimens at the first CSS ratio.

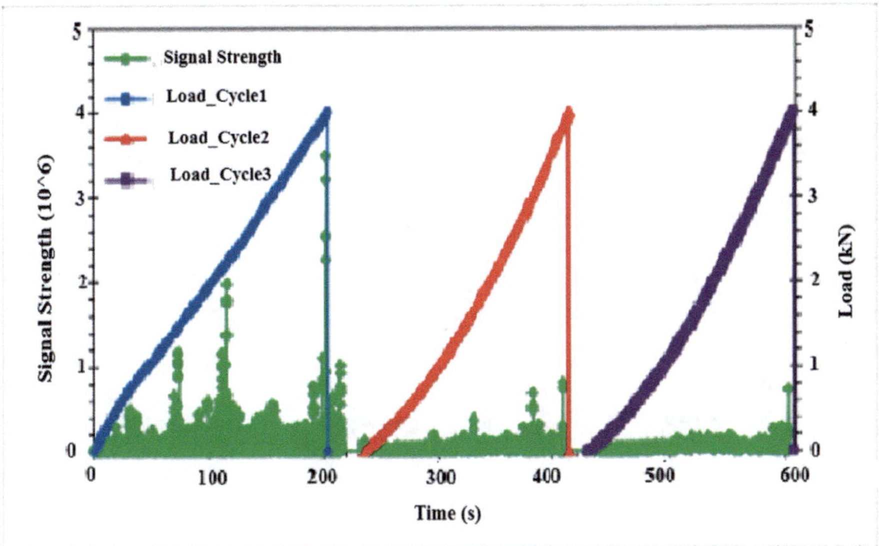

(a) S0 beam specimen

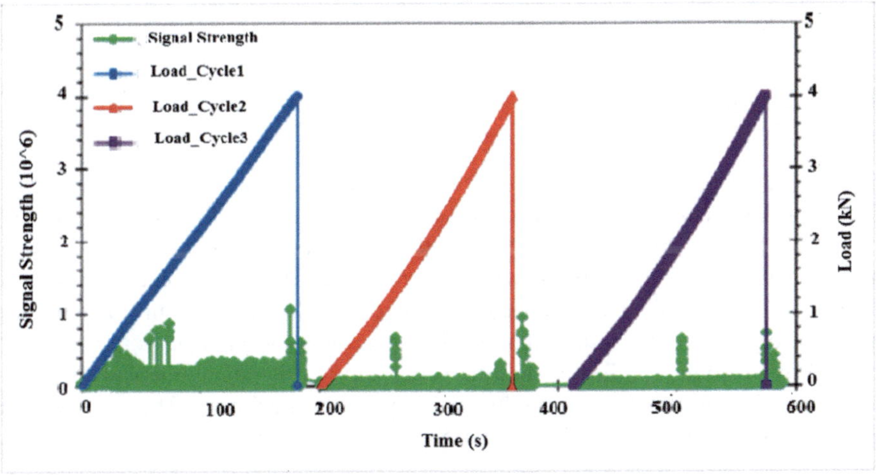

(b) S9 beam specimen

Fig. 4 Signal strength (pVs) and load versus time (s)

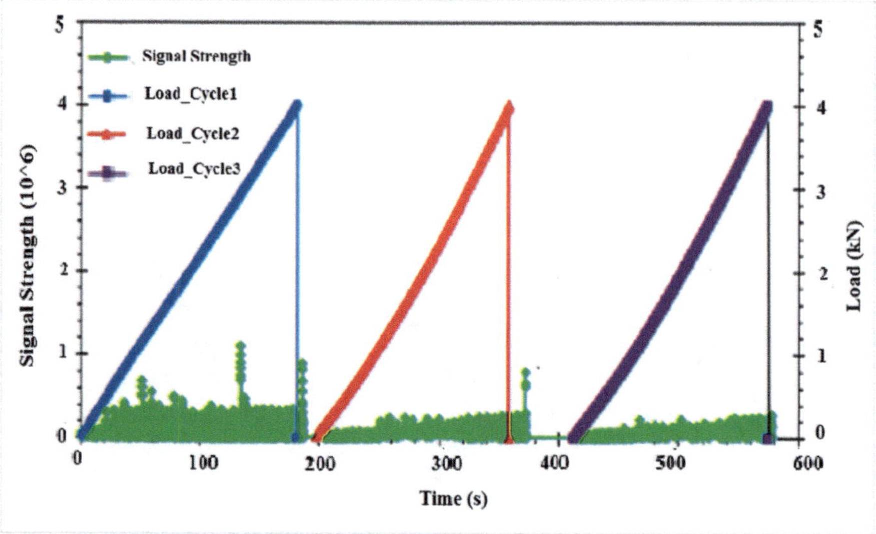

(c) S28 beam specimen

Fig. 4 (continued)

3.2 *I*b-Value

Figure 6 shows the *I*b-value for the specimens, i.e., S0, S9, and S28 beam specimens. The beam specimens experienced steel corrosion has a trend different from S0 beam specimen. For the S0 specimen, the trend established that the *I*b-value tends to increase from cycle 1 to cycle 3 of loading. However, for S9 and S28 beam specimens, the *I*b-value tends to decrease from cycle 2 to cycle 3 of loading in terms of increasing in corrosion level. This result signifies the occurrence of damage from cycle 2 to cycle 3 of loading. It indicates that the occurrence of damage is generated due to steel corrosion (loss in the cross-section) of the beam specimens. The cross-section of the S28 beam specimen is lower than S9 beam specimen.

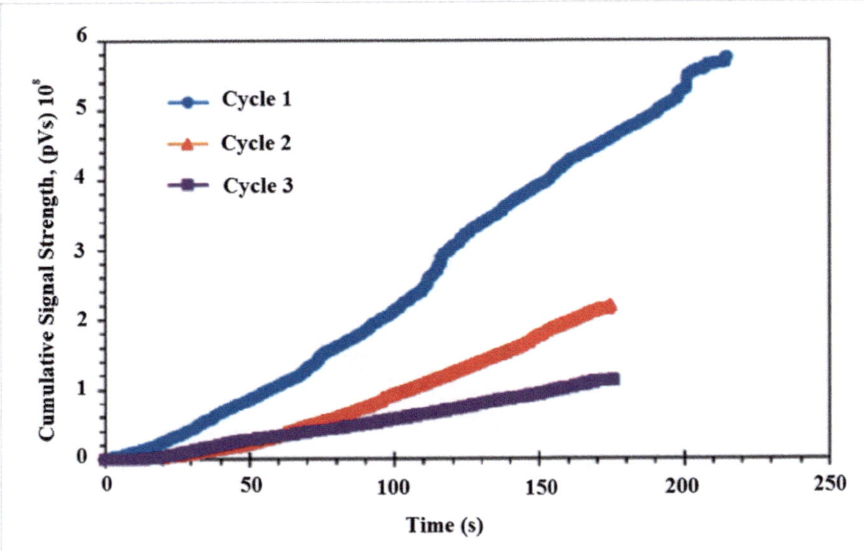

(a) S9 beam specimen

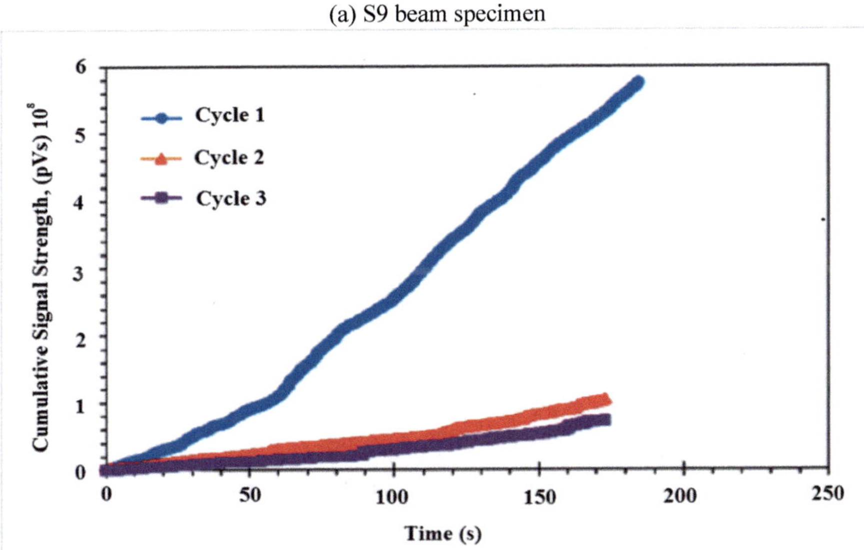

(b) S28 beam specimen

Fig. 5 Cumulative signal strength (pVs) against time (s) of the beam specimens

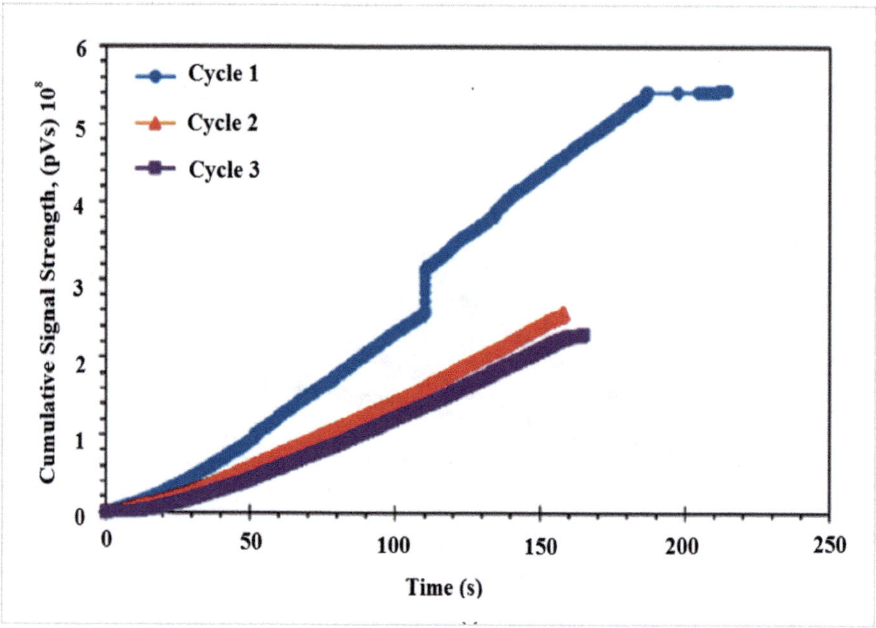

(c) S28 beam specimen

Fig. 5 (continued)

Table 2 CSS ratio

Beam specimens	Cycle	CSS	CSS ratio 1 (%)	CSS ratio 2 (%)
S0	Cycle 1	5.75×10^8		
	Cycle 2	38.09		
	Cycle 3	1.13×10^8		
S9	Cycle 1			69.96
	Cycle 2	1.04×10^8		
	Cycle 3	7.25×10^7		
S28	Cycle 1	5.43×10^8	47.33	89.49
	Cycle 2	2.57×10^8		
	Cycle 3	2.30×10^8		

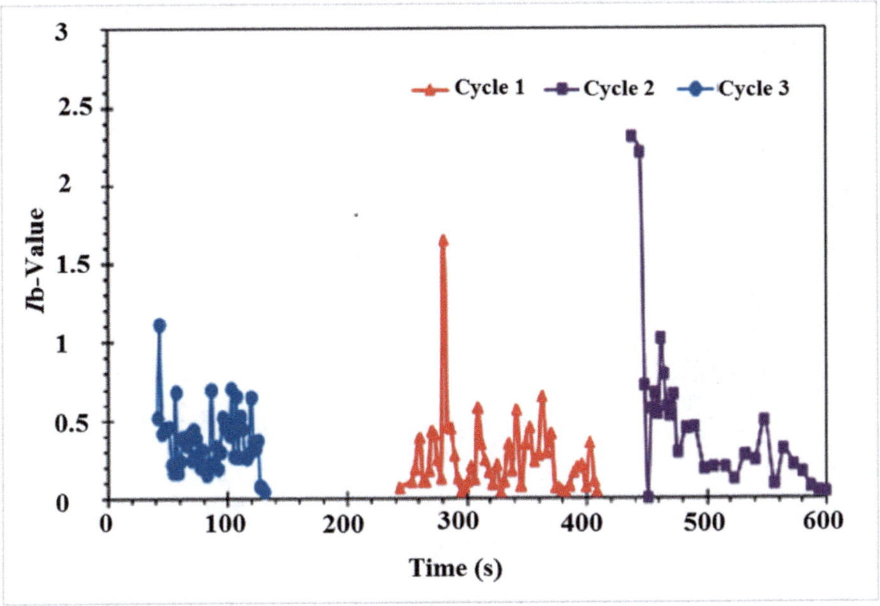

(a) S9 beam specimen

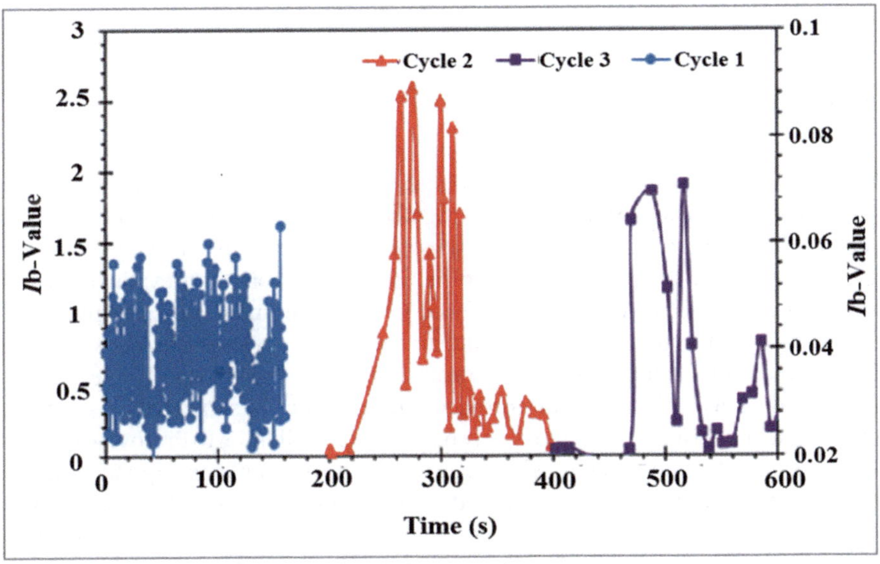

(b) S28 beam specimen

Fig. 6 Ib-value against time graph of the beam specimens

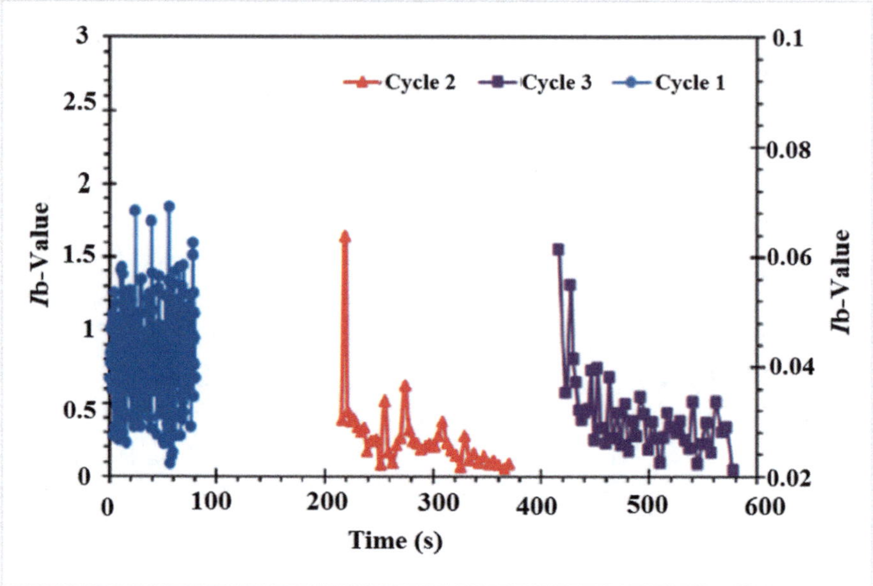

(c) S28 beam specimen

Fig. 6 (continued)

4 Conclusion

This experiment deliberates the mechanical behaviour of the RC beam specimens with pre-corroded steel reinforcement, together with the AE monitoring results for secondary AE signals analysis. The results obtained in this experiment showed the great potential of the AE technique to be used in the corrosion evaluation of the beam specimens. The AE parameters used in this experiment provide some useful information and trend that could be illustrated on the corrosion level experienced by the corrosion beam specimens, as discussed in this experiment. However, these findings do not apply to a real corroded structure that is subjected to water and chloride attack. In addition, there is only one factor contributing to the loss of concrete strength, i.e., flexural strength, shear strength, load-bearing capacity etc., which is the decrement in the cross-section of the steel reinforcement of the pre-corrosion beam specimens. Thereafter, based on the limitation gained in the experiment, the next experiment was utilized for post-corrosion beam specimens.

References

1. Malhotra VM, Carino NJ (1991) Handbook on nondestructive testing of concrete. CRC Press, Florida

2. Ohtsu M, Yuyama S (2001) Recommended practice for in-situ monitoring of concrete structures by acoustic emission. J Acoust Emiss 184–190
3. Carpinteri A, Lacidogna G, Niccolini G (2011) Damage analysis of reinforced concrete buildings by the acoustic emission technique. Struct Control Health Monit 660–673
4. Calabrese L, Campanella G, Proverbio E (2013) Identification of corrosion mechanisms by univariate and multivariate statistical analysis during long term acoustic emission monitoring on a pre-stressed concrete beam. Corros Sci 161–171
5. Zaki A, Tan JY, Chai HK, Aggelis DG (2015) Assessment of corrosion damage using acoustic emission technique under load testing. In: The 7th Asia Pacific young researchers and graduates symposium, University of Malaya, Kuala Lumpur
6. Zaki A, Chai HK, Behnia A, Aggelis DG, Tan JY, Ibrahim Z (2017) Monitoring fracture of steel corroded reinforced concrete members under flexure by acoustic emission technique. Constr Build Mater 609–618
7. Zaki A, Chai HK, Aggelis DG, Alver N (2015) Non-destructive evaluation for corrosion monitoring in concrete: a review and capability of acoustic emission technique. Sensors 19069–19101
8. Ohtsu M (1987) Mathematical theory of acustic emission and moment tensor solution. J Soc Mater Sci 1025–1031
9. Colombo S, Main IG, Forde MC (2003) Assessing damage of reinforced concrete beam using "b-value" analysis of acoustic emission signals. J Mater Civ Eng 280–286
10. Grosse CU, Reinhardt HW, Finck F (2003) Signal-based acoustic emission techniques in civil engineering. J Mater Civ Eng 274–279
11. Ohno K, Ohtsu M (2010) Crack classification in concrete based on acoustic emission. Constr Build Mater 2339–2346
12. Beattie AG (1983) Acoustic emission, principles and instrumentation. J Acoust Emission 34
13. Shiotani T (2006) Evaluation of long-term stability for rock slope by means of acoustic emission technique. NDT & E International 217–228
14. Alvarez MG, Lapitz P, Ruzzante J (2012) Analysis of acoustic emission signals generated from SCC propagation. Corrosion Sci 5–9
15. Lapitz P, Ruzzante J, Alvarez MG (2007) AE response of A-brass during stress corrosion crack propagation. Corros Sci 3812–3825
16. Wells D (1970) An acoustic apparatus to record emissions from concrete under strain. Nuclear Eng Design 9
17. Luo X, Haya H, Inaba T, Shiotani T, YN.: Experimental study on evaluation of breakage in foundations using train-induced acoustic emission. In: Structural engineering world congress, T9–1
18. Yoon DJ, Weiss WJ, Shah SP (2000) Assessing damage in corroded reinforced concrete using acoustic emission. J Eng Mech 273–283
19. Ohtsu M, Tomoda Y (2007) Acoustic emission techniques for rebar corrosion in reinforced concrete. In: Advances in construction materials, Springer, Switzerland, pp 615–621
20. Kawasaki Y, Tomoda Y, Ohtsu M (2010) AE monitoring of corrosion process in cyclic wet–dry test. Constr Build Mater 2353–2357
21. Ohtsu M, Tomoda Y (2007) Corrosion process in reinforced concrete Identified by acoustic emission. Mater Trans 1184–1189
22. Ohtsu M, Tomoda Y (2008) Phenomenological model of corrosion process in reinforced concrete identified by acoustic emission. ACI Mater J 194–199
23. Kawasaki Y, Wasada S, Okamoto T, Izuno K (2014) Evaluation for RC specimen damaged from rebar corrosion by acoustic emission technique. Constr Build Mater 157–164
24. Zaki A, Ibrahim Z (2021) Corrosion assessment of pre-corrosion concrete specimens using acoustic emission technique. J Eng Technol Sci 210211
25. Leelalerkiet V, Shimizu T, Tomoda Y, Ohtsu M (2005) Estimation of corrosion in reinforced concrete by electrochemical techniques and acoustic emission. J Adv Concr Tech 137–147
26. Aggelis DG, Soulioti DV, Sapouridis N, Barkoula NM, Paipetis AS, Matikas TE (2011) Acoustic emission characterization of the fracture process in fibre reinforced concrete. Constr Build Mater 4126–4131

27. Behnia A, Chai HK, Yorikawa M, Momoki S, Terazawa M, Shiotani T (2014) Integrated non-destructive assessment of concrete structures under flexure by acoustic emission and travel time tomography. Constr Build Mater 202–215
28. Kawasaki Y, Wakuda T, Kobarai T, Ohtsu M (2013) Corrosion mechanisms in reinforced concrete by acoustic emission. Constr Build Mater 1240–1247
29. Behnia A, Chai HK, Shiotani T (2014) Advanced structural health monitoring of concrete structures with the aid of acoustic emission. Constr Build Mater 282–302
30. ElBatanouny MK, Ziehl PH, Larosche A, Mangual J, Matta F, Nanni A (2014) Acoustic emission monitoring for assessment of prestressed concrete beams. Constr Build Mater 46–53
31. Behnia A, Chai HK, GhasemiGol M, Sepehrinezhad A, Mousa AA (2019) Advanced damage detection technique by integration of unsupervised clustering into acoustic emission. Eng Fract Mech, 212–227
32. Guzman C, Torres D, Hucailuk C, Filipussi D (2013) Analysis of the acoustic emission in a reinforced concrete beam using a four points bending test. In: International congress of science and technology of metallurgy and materials. Elsevier Procedia, Amsterdam, pp 148–154
33. Aggelis DG, Tsangouri E, Van Hemelrijck D (2015) Influence of propagation distance on cracking and debonding acoustic emissions in externally reinforced concrete beams. Meccanica 1167–1175
34. BS (2005) BS 4449: Specification Steel for the reinforcement of concrete. Weldable reinforcing steel. Bar, coil and decoiled product. British Standards Institution, London
35. ASTM (2003) ASTM G1–03: Standard practice for preparing, cleaning, and evaluating corrosion test specimens, ASTM International, West Conshohocken, PA
36. BS (1997) BS 8110–1: Structural use of concrete. Code of practice for design and construction, British Standards Institution, London
37. Apostolopoulos CA, Michalopoulos D (2007) Impact of corrosion on mass loss, fatigue and hardness of BSt500s steel. J Mater Eng Perform 63–67
38. Majhi S, Mukherjee A, George NV, Uy B (2019) Corrosion detection in steel bar: a time-frequency approach. NDT E Int 102150
39. Holford KM (2000) Acoustic emission–basic principles and future directions. Strain 51–54
40. Paul Z, Adam R (2006) Evaluation of strengthened RC beams: cyclic load test and acoustic emission methods. ACI Struct J 832–841

Cable Force Prediction Technique Using Subspace and Effective Vibration Length Method

Muhammad Ibnu Syamsi, Hao-Lin Wang, and Chung-Yue Wang

Abstract Inevitably, cable structure is often used as a structural part of either a middle or long-span bridge, so its existing tension force is essential to be evaluated and monitored periodically. This research aims to demonstrate the use of covariance type of Stochastic Subspace Identification (SSI-COV) to identify the dynamical system properties of bridge's cable and estimate the cable tension. The concept of effective vibration length proposed by previous researchers is implemented in this research to overcome the unknown rotational stiffness of the cable's supporting system. During the measurement, multiple-synchronized vibration measuring devices are deployed on a cable to get the corresponding eigenvector and its effective length. The cable force of an existing bridge is identified using two different approaches, the least-square approximation and the two-frequency method. It is found that the deviation between those two mentioned approaches is very small. The overall process evidence that this non-destructive testing procedure is practicable to estimate the internal force of a cable in real structures.

Keywords Cable force · Subspace method · Effective vibration length

1 Introduction

Assessing the existing condition of infrastructure is a need to maintain its performance and operational condition. Non-destructive testing (NDT) is the most widely used technique to be used and developed among some measurement methods. As a part of NDT, the vibration-based approach has been successfully applied in many fields to detect the dynamic tension force of a stayed cable of structure [1–4]. Consequently, a signal processing technique is required to extract the dynamical properties

M. I. Syamsi (✉)
Universitas Muhammadiyah Yogyakarta, Bantul 55183, Indonesia
e-mail: syamsibnu@umy.ac.id

M. I. Syamsi · H.-L. Wang · C.-Y. Wang
National Central University, Taoyuan 32001, Taiwan
e-mail: cywang@cc.ncu.edu.tw

S. A. Kristiawan et al. (eds.), *Proceedings of the 5th International Conference on Rehabilitation and Maintenance in Civil Engineering*, Lecture Notes in Civil Engineering 225,
https://doi.org/10.1007/978-981-16-9348-9_23

of a system. Operational Modal Analysis (OMA) is the most efficient method to reveal the system properties because of its advantage, which does not require any knowledge of exiting input. Compared to the traditional vibration test, this technique is more feasible for a large structure. One of the OMA methods developed in the time-domain approach is Stochastic Subspace Identification (SSI).

SSI method starts its popularity since a paper published by Van Overschee and De Moor [5] along with the book [6]. A well-reviewed paper reported by Brincker and Andersen [7] helps engineers better understand the difficulties of the theory behind this technique. Besides, SSI has been proved to be the most powerful output-only modal analysis technique in many applications, including civil engineering [8, 9]. Many researchers have also performed this technique in a bridge's structure. Magalhaes, Cunha [10] proposed an automated subspace identification by combining this method with machine learning and applied it in a long span arch bridge. Further application of the previous researcher's automated subspace technique is demonstrated by Ubertini, Gentile [11]. Moreover, the comparison with robust OMA in the frequency domain, Frequency Domain Decomposition (FDD), shows a promising result.

Meanwhile, many researchers have developed the vibration-based approach of a cable to assess the tension. The most traditional approach uses the string theory, where the relationship is expressed in terms of natural frequency and its corresponding mode order. However, the application of this theory is very limited and mostly unsuitable to be implemented in the real cable structure. Thus, the modern cable formula is developed from the concept of axially-loaded Euler–Bernoulli beam theory [12]. Some researchers reported the application of the formula for long-term cable force monitoring [13, 14]. Since the cable's end support system in the real case lies between hinged and clamped, then the application of the formula became challenging. The concept of effective vibration length is then proposed to overcome the uncertainty end-restrain of cable and demonstrated in the real stayed cable of bridges [2, 15–17]. Besides, an alternative determination for cable force even with the absence of flexural rigidity using a combination of two frequencies is introduced by Cheng [3].

It is known that vibration-based is one of the non-destructive testing (NDT) methods requiring signal processing knowledge. Studies in taking advantage of the vibration-based approach with the robust SSI technique to measure and monitor the bridge cables have also been done by previous researchers [1, 18]. However, the covariance-driven type of SSI has not been widely applied in determining effective vibrational length. Besides, two different approaches in predicting cable forces are presented, which is the least-square approximation and the two-frequency methods in which effective cable lengths are adopted in those two. Therefore, this research aims to demonstrate the mentioned methods in a real arch-cable railway bridge. Besides, a challenge of using the subspace method for determining the effective length is also discussed.

2 Subspace Method

The SSI method is started by considering the state-space form of the second-order differential equation of a system. It is assumed that the input process and measurement process follows the gaussian distribution to represent the noisy signal in the real condition. Therefore, the discrete-time stochastic state-space model of a system is expressed as

$$\boldsymbol{x}(k+1) = \boldsymbol{A}\boldsymbol{x}(k) + \boldsymbol{w}(k) \tag{1}$$

$$\boldsymbol{y}(k) = \boldsymbol{C}\boldsymbol{x}(k) + \boldsymbol{v}(k) \tag{2}$$

where $\boldsymbol{A} \in \mathbb{R}^{n\times n}$ describe the discrete-time system matrix, $\boldsymbol{x}(k) \in \mathbb{R}^n$ is the state vector in time step k, $\boldsymbol{C} \in \mathbb{R}^{l\times n}$ stands for the allocation matrix output, $\boldsymbol{y} \in \mathbb{R}^l$ is output measurement, while both $\boldsymbol{w} \in \mathbb{R}^n$ and $\boldsymbol{v} \in \mathbb{R}^l$ are the noise input and measurement noise, which both are white noise, zero-mean.

There are two types of SSI: SSI-Data, which directly works on the raw time series data and SSI-COV requires preprocessing the raw response data into the covariance matrices [19]. The post-processing data in SSI-COV let this type of SSI runs faster than the SSI-DATA [10]. Once the covariance about the time lag i are obtained, then rearrange them into a Toeplitz matrix, $\boldsymbol{T}_{1|i}$, as shown in Eq. (3)

$$\boldsymbol{T}_{1|i} = \begin{bmatrix} \boldsymbol{R}_i & \boldsymbol{R}_{i-1} & \dots & \boldsymbol{R}_1 \\ \boldsymbol{R}_{i+1} & \boldsymbol{R}_i & \dots & \boldsymbol{R}_2 \\ \vdots & \vdots & \ddots & \vdots \\ \boldsymbol{R}_{2i-1} & \boldsymbol{R}_{2i-2} & \dots & \boldsymbol{R}_i \end{bmatrix} \tag{3}$$

Factorize this huge matrix by taking the SVD and gain the system matrix by involving the Moore–Penrose pseudo-inverse as explained in Van Overschee and De Moor [6]. Once the system matrix, $\boldsymbol{A}$, is identified, the eigenvalue decomposition is performed to attain the dynamical system properties through the following well known complex system poles formula.

$$\boldsymbol{A} = \Psi\,\Lambda\,\Psi^{-1} \tag{4}$$

$$\mu_j = e^{\lambda_j} \tag{5}$$

$$\lambda_j = -\zeta_j\omega_j + \mathrm{i}\omega_j\sqrt{1-\zeta_j^2} \tag{6}$$

where $\boldsymbol{\Lambda}$ is a diagonal matrix containing eigenvalues μ_j, $\boldsymbol{\Psi}$ stands for the eigenvector, λ_j and μ_j are the continuous and discrete system poles, respectively, ω_j is the circular natural frequency, and ζ_j represents the system damping ratio. The mode shape, **Φ**, can be obtained can be through Eq. (7).

$$\Phi = \boldsymbol{C}\,\Psi \tag{7}$$

A stabilization diagram is performed to distinguish between the physical system properties among the noisy signal by varying the order of system poles under specific stable criteria reported in Peeters, Van der Auweraer [20].

3 Straight Cable Theory and Effective Vibration Length

Some successful vibration-based technique to attain the force of cable has reported [1, 2, 14, 18, 21]. The analytical formula of cable tension force derived from a simply supported axially-loaded Euler–Bernoulli beam is shown in Eq. (8). This equation is one of the most often equation to be used for estimating the cable force [4].

$$T = \frac{4\overline{m}L^2 f_n^2}{n^2} - \frac{n^2\pi^2 EI}{L^2} \tag{8}$$

where T stand for the force, L is the cable length, $\overline{m}$ is the linear mass, EI is the bending stiffness, and f_n is the frequency of mode n. Without removing the cable's bending stiffness effect, Cheng [3] propose a direct tension force formula shown in Eq. (9). In this equation, flexural rigidity is implicitly replaced through the combination of two-mode frequencies, mode p and q thus, the cable force estimation can be performed with the absence of bending stiffness information. The equivalent two-mode frequency is then denoted as f_c.

$$T = 4\overline{m}L^2\left(\frac{q^4 f_p^2 - p^4 f_q^2}{p^2 q^2 \left(q^2 - p^2\right)}\right) = 4\overline{m}L^2 f_c^2 \tag{9}$$

However, in a real case, cable's support system commonly involves special support, which neither belongs to hinged nor fixed. Hence, the application of the formula leads to some error due to the complexity of rotational stiffness. Chen [15], Wu [17] and Chen [2] proposed the concept of effective vibration length (L_n^{eff}) to be applied in any cable end-restraint condition. The corresponding vibration length is obtained through the mode shape of a simply-supported case in which expressed in the form of the sine wave function. Therefore, by obtaining the cable's effective vibration length and replacing the actual length, the tension force can be identified through the same relationship.

The cable mode shape can be categorized into two groups according to their movement about the mid-axis, symmetrical mode shapes ($n = 1, 3, 5, \ldots$) which are belong to the cosine function and unsymmetrical ones ($n = 2, 4, 6, \ldots$), which belong to the sinusoidal form. And the mode shape vector for r measured location can be formed as shown in Eq. (10) where the $n = n_1, n_2, \ldots, n_m$ are the m major modes [15, 17].

$$\phi_n = \begin{Bmatrix} \phi_{1n} \\ \phi_{2n} \\ \vdots \\ \phi_{rn} \end{Bmatrix} = a_n \begin{Bmatrix} \text{cosin}\frac{n\pi x_1}{L} \\ \text{cosin}\frac{n\pi x_2}{L} \\ \vdots \\ \text{cosin}\frac{n\pi x_r}{L} \end{Bmatrix}, n = n_1, n_2, \ldots, n_m \tag{10}$$

where

$$\text{cosin}(\cdot) = \begin{cases} \cos(\cdot) \text{ if } n = \text{odd} \\ \sin(\cdot) \text{ if } n = \text{even} \end{cases} \tag{11}$$

4 Implementation

A steel-arch railway bridge located in Taiwan is taken as the object in this research to illustrate the theories as mentioned in the previous section, shown in Fig. 1a. A total of 16 cables on each side support the bridge deck. According to the basic information of the cable, it is recorded that the cable diameter is 9.5 cm consisting of 19 strands in

Fig. 1 **a** Bridge A **b** velocity-meter deployment **c** measurement signal

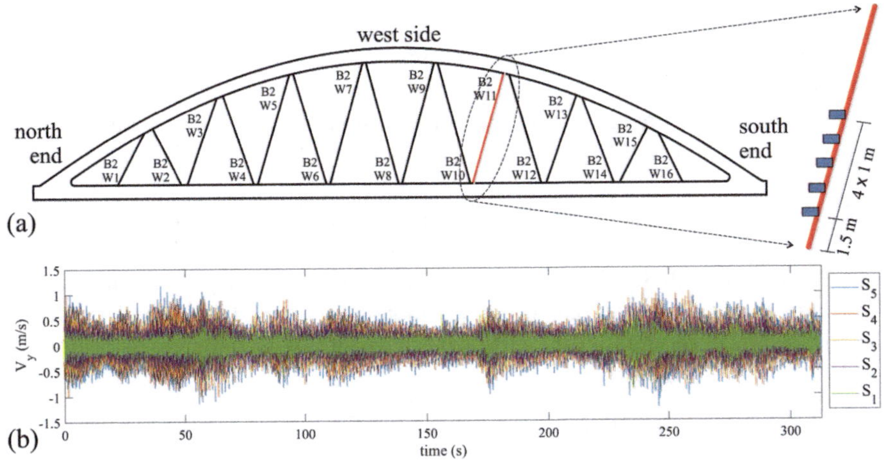

Fig. 2 Sensor deployment configuration

which the area is 140 mm^2 per each and covered by HDPE material. By considering all section and material properties, it is estimated that the linear mass of the cable is 23.26 kg/m. This mass information is required for detecting the cable force through the vibration-based formula.

Five vibration transducers are used to measure the vibration of a cable deployed on a different position along the longitudinal axis, as shown in Fig. 1b. The first sensor position is 1.5 m from the bridge deck inline to the cable's local axis, while the gap between each of them are set to be equal distance, which is 1 m. The first sensor location is different from others to accommodate the uncertainty boundary condition. The sensors placement configuration can be observed in Fig. 2a. Each sensor must be adjusted such that they have the same alignment about the horizontal axis.

Through five synchronized vibration sensors, the vibration signals of a cable are recorded in the same period. The sampling frequency is set to 1 kHz, and the data capturing duration is about 5–10 min under ambient vibration. Onsite random signals from ambient vibration excitation in a certain short period are shown in Fig. 1c. By taking an example of the cable around the middle span, named B2W11, the vibration responses of the inspection are shown in Fig. 2b.

The vibration responses obtained through the five sensors are used to perform the modal analysis using SSI. Since the preprocessing of the data required in the covariance-driven SSI type, the correlation between signals is computed and arranged in the Toeplitz matrix. In this process, the vibration data is scaled down become 100 Hz, and the time lag is set to be 40 s. This scaling down step is performed to save the memory during the computation without losing the important information. Once the Toeplitz matrix is established, SVD is taken to factorize the system and obtained the system matrix. The classical modal analysis of the system poles is then performed, and the stabilization diagram can be established, as shown in Fig. 3.

The stabilization diagram is constructed under the stable criteria $\Delta f < 1\%$, $\Delta\zeta < 5\%$, and $\Delta \text{MAC} < 2\%$ [20]. Any poles satisfying all three parameters are denoted as stable poles, while the poles that are not stable in all criteria are named the new pole. There are also obtain poles that just stable in frequency and damping ratio, frequency and MAC, or even just the frequency. The physical dynamic system poles can generally be figured out by observing the stable poles' vertical line.

As shown in Fig. 3, under the 50 Hz frequency range, there are many stable lines detected, which indicated by a green-circle mark. It is shown in Fig. 4, 15 modes are extracted through SSI, but not all the detected modes belong to the cable modes. There are many possibilities that the modes obtained are mixed with other elements of the bridge. Since the vibration-based approach requires the cable frequency, thus it is

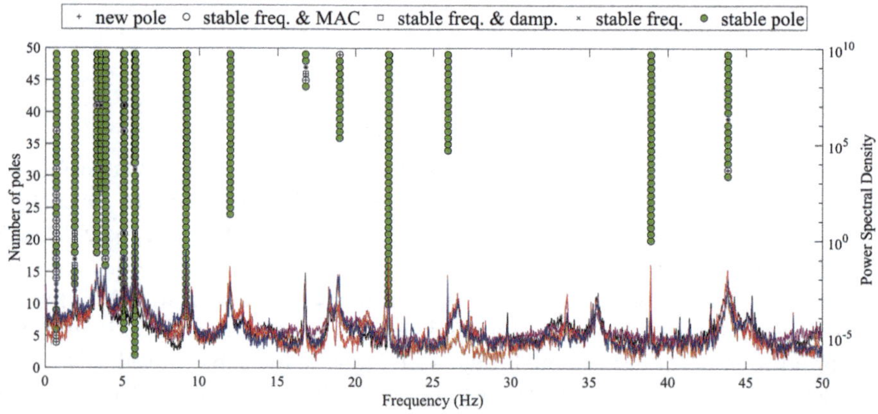

Fig. 3 Stabilization diagram

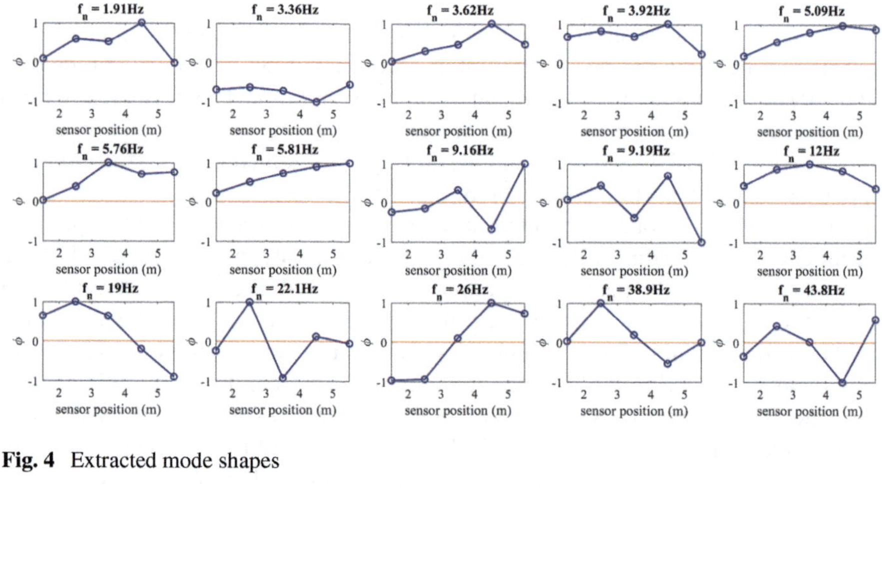

Fig. 4 Extracted mode shapes

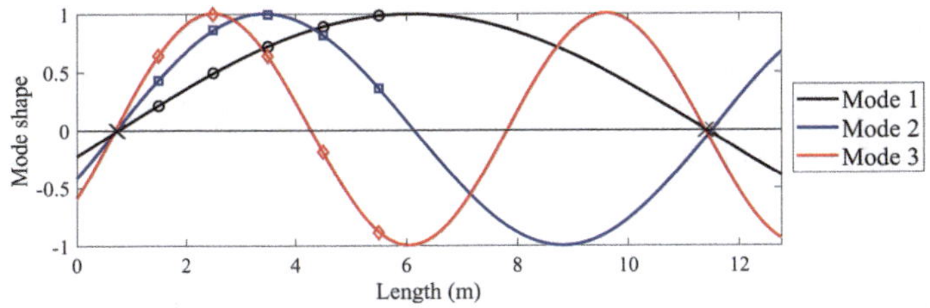

Fig. 5 Effective vibration of the first three cable mode

needed to select the right frequency which belongs to the cable. The selection of cable frequency can be executed by matching detected modes to the cable modes' natural behaviour. The fitting curve is then taken to the extracted mode shape following the trigonometric function shown in Eq. (10) within the range of real cable length. Cable effective lengths are identified by observing the zero-crossing distance between the maximum and minimum.

As shown in Fig. 5, three cable frequencies are detected along with the effective vibration modes. It is seen that there is a gap between the end-zero-crossing to the corresponding end-boundary support, so the effective length is shorter than the real length ($L = 12.77$ m). This indicates that the supporting system is not hinged, i.e. the rotational stiffness exists at both ends. The effective lengths of these three modes are obtained by calculating the distance of their corresponding end-zero-crossing. Through these processes, it is found that the frequency of mode 1, mode 2, and mode 3 are 5.81, 11.98, and 18.89 Hz, respectively; and their corresponding effective vibration length are 10.64, 10.77, 10.65 m. With the obtained mode frequencies (f_n) and their corresponding effective lengths (L_n^{eff}), the remaining unknown remains the tension force (T) and flexural rigidity (EI). Three selected modes are then substituted into Eq. (8), resulting in three linear functions with T and EI as the unknowns. The optimal values of these unknowns can be solved through the least squared approximation [2]. Finally, the tension force (T) obtained is 351.093 kN and the bending stiffness (EI) is 89.18 kNm2.

Besides using least-squares approximation, tension force can also be attained through the two-frequency approach introduced by Cheng [3]. But since it is found that the cable effective length shorter than its actual length, so it is not a hinged cable system. As a result, the application of Eq. (9) cannot be directly adopted. It requires computing not only the equivalent frequency of two modes but also its corresponding equivalent vibration length. After computing these two equivalencies, the force of cable B2W11 obtained using the two-mode frequency method is 351.457 kN.

Overall in the final step, two different approaches to find out the tension forces are used. Even though the force values produced through these two methods have a gap of about 0.1% to each other, but this small deviation is acceptable.

Meanwhile, the minimum breaking strength for one strand is 260.7 kN. If the maximum allowable force is assumed to be 85% of the breaking strength, the allowable cable force is 4210.3 kN. Consequently, it is computed that the measured cable force is only 8.3% of the allowable one; in other words, the cable is safe.

5 Conclusion

This research presents a cable force estimation approach applied on a real steel arch bridge through a vibration-based approach. Five sensors are mount at a different location along the cable axis to get the vibration responses. Random signals are obtained under the ambient environment, and SSI-COV is used to extract the cable dynamical properties. Three cable natural frequencies (f_n) are obtained along with their effective vibration modes, where f_1, f_2, and f_3 are 5.81, 11.98, and 18.89 Hz, respectively, and their corresponding effective vibration lengths are 10.64, 10.77, 10.65 m. Two different approaches to reveal the force are utilized, the first uses the least-squares approximation, and the second is the two-frequency approach. The force obtained using the first method is 351.093 kN, while the second one is 351.457 kN, so the gap between these two results is 0.1%. Compared to the allowable force, the measured cable force indicates that it is still under a safe level. Besides, it is also attained from the first method that the bending stiffness of cable B2W11 is 89.18 kNm2.

Overall, three main concerns can be pointed out from this study. First of all, the use of SSI helps in extracting the mode frequency of the system under operational conditions, and cable modes selection is required as the additional post-procedure since the extracted modes are mixed with other bridge elements. Secondly, the application of effective cable length can successfully reveal the cable force through two different approaches with a slight deviation. Lastly, the cable internal force disclosure is performed under uncertain end-restraint condition, indicating that this procedure is a proper method to estimate an existing stayed cable internal force on a real structure.

References

1. Haji Agha Mohammad Zarbaf SE et al. (2017) Stay force estimation in cable-stayed bridges using stochastic subspace identification methods. J Bridge Eng 22(9):04017055
2. Chen C-C et al (2016) Tension determination of stay cable or external tendon with complicated constraints using multiple vibration measurements. Measurement 86:182–195
3. Yu C-P, Cheng C-C, Chiang C-H (2011) Alternative determination of cable forces using flexural theory of axially loaded member. In: Non-destructive characterization for composite materials, aerospace engineering, civil infrastructure, and homeland security. International Society for Optics and Photonics, pp 79832R
4. Kim BH, Park T (2007) Estimation of cable tension force using the frequency-based system identification method. J Sound Vibr 304:660–676

5. Van Overschee P, De Moor B (1996) Continuous-time frequency domain subspace system identification. Signal Process 52(2):179–194
6. Van Overschee P, De Moor B (1996) Subspace identification for linear systems: theory—Implementation—applications. Kluwer Academic Publishers, London
7. Brincker R, Andersen P (2006) Understanding stochastic subspace identification. In: Proceedings of the 24th IMAC, St. Louis 126
8. Döhler M, Andersen P, Mevel L (2011) Data merging for multi-setup operational modal analysis with data-driven SSI. In: Conference proceedings of the society for experimental mechanics series. Springer, New York, NY, pp 443–452
9. Giraldo DF et al (2009) Modal identification through ambient vibration: comparative study. J Eng Mech 135(8):759–770
10. Magalhaes F, Cunha A, Caetano E (2009) Online automatic identification of the modal parameters of a long span arch bridge. Mech Syst Signal Process 23(2):316–329
11. Ubertini F, Gentile C, Materazzi AL (2013) Automated modal identification in operational conditions and its application to bridges. Eng Struct 46:264–278
12. Humar J (1990) Dynamics of structures. Prentice-Hall, Englewood Cliffs, New Jersey
13. Sim S-H et al (2013) A wireless smart sensor network for automated monitoring of cable tension. Smart Mater Struct 23(2):025006
14. Kim S-W et al (2013) Vision-based monitoring system for evaluating cable tensile forces on a cable-stayed bridge. Struct Health Monit 12(5–6):440–456
15. Chen C-C et al (2013) Determination of stay cable force based on effective vibration length accurately estimated from multiple measurements. Smart Struct Syst 11(4):411–433
16. Wu W et al (2012) Determination of stay cable force based on multiple vibration measurements to consider the effects of unsymmetrical boundary constraints. In: 6th European workshop on structural health monitoring. NDT. net Bad Breisig, Dresden, Germany, pp 1049–1056
17. Wu W et al (2010) Determination of stay cable force based on multiple vibration measurements to consider the effects of uncertain boundary constraints. In: Proceedings of 5th European workshop on structural health monitoring. Naples, Italy, pp 282–287
18. Geier R, De Roeck G, Petz J (2005) Cable force determination for the Danube channel bridge in Vienna. Struct Eng Int 15(3):181–185
19. Peeters B (2000) System identification and damage detection in civil engeneering
20. Peeters B et al (2007) Operational modal analysis for estimating the dynamic properties of a stadium structure during a football game. Shock Vib 14(4):283–303
21. Zui H, Shinke T, Namita Y (1996) Practical formulas for estimation of cable tension by vibration method. J Struct Eng 122(6):651–656

Implementation of Microbially Induce Calcite Precipitation (MICP) by Bacillus Subtilis and Adding Sand in Repairing Shear Strength Parameters of Peat

Firman Syarif and Dian M. Setiawan

Abstract These days, the utilization of microbially induced calcite precipitation (MICP) has gotten mainstream as a ground improvement strategy for sandy soil. In any case, this strategy was not all that much applied to peat soil. This research focused on bacterial calcium carbonate precipitation from Bacillus Subtilis and its impact on shear strength parameter of peat soil. An uncommon injection framework was ready for actuating bacterial solution for the examples. The bacterial arrangement provided to the examples by gravity for 7 days in explicit molds intended for this work. From this exploration. The most noteworthy compressive strength esteem is found in the expansion of 15% cementation arrangement of 0.42 kg/cm^2 of the first soil test + 5% sand, which has a free compressive strength worth of 0.30 kg/cm^2.

Keywords Bacillus subtilis · Organic soil · Peat soil · Shear strength

1 Introduction

In the previous 18,000 years there was an arrangement of peat soil in southeast Asia. While Indonesia's peat conformed to 5000–8000 years prior. This peat soil is framed in territories that have for quite some time been lowered in water. Indonesia has 20 million hectares of peatland which is number four after Canada, Russia, and America in the broadest class of peat land on the planet. The circulation of peatlands in Indonesia is in four huge islands, in particular Sumatra 35%, Kalimantan 32%, Sulawesi 3%, and Papua 30%. On the island of Sumatra, peatlands are typically found in the marshes along the east coast with a space of 7.2 million hectares. On

F. Syarif
Universitas Islam Riau, Pekanbaru 28284, Riau, Indonesia
e-mail: firmansyarif@eng.uir.ac.id

D. M. Setiawan (✉)
Universitas Muhammadiyah Yogyakarta, Bantul 55183, Yogyakarta, Indonesia
e-mail: diansetiawanm@ft.umy.ac.id

Texas A&M University, College Station, TX 77843, USA

S. A. Kristiawan et al. (eds.), *Proceedings of the 5th International Conference on Rehabilitation and Maintenance in Civil Engineering*, Lecture Notes in Civil Engineering 225,
https://doi.org/10.1007/978-981-16-9348-9_24

Sumatra Island, the Province with the most extensive peatland is Riau Province with a space of ±4.04 million Ha or 56.1% of the absolute peatland territory in Sumatra [1–5].

Peat soil is a problem that often occurs in the construction world. Peat soils have a low bearing capacity and can cause large land settlement during loading, Peat soil is very soft soil (very soft soil) with a very low bearing capacity and is easily compressible if the load is working on it [6]. There is a fundamental problem that arises due to the loading on the peat soil layer, namely the higher the organic content, the lower the bearing capacity and shear strength and the greater the compression [7]. Peat soil compression is very different from clay soil, where the compression that occurs on peat soil is a long process of compression. The amount of compression due to the consolidation of a soil layer is very dependent on the size of the load given and the thickness of the soil layer that is loaded [8]. So that in this study, the addition of bacillus subtilis bacteria to peat soil using bio-grouting technique method to determine the effect of the addition of these bacteria on the compressive strength of peat soil.

This repair technique uses the biogrouting method with Bacillus subtilis bacteria. Biogrouting is a soil stabilization technique involving microorganisms that are induced by calcium carbonate ($CaCO_3$) precipitation. Precipitation of calcium carbonate acts as a crystallizer between cells that stimulates the cementation process between the soil grains. Biogrouting also means injection of adhesive in soil or rock that passes water with the aim of closing pores in the soil, biogrouting also functions as a penetration or penetration and compacting or compaction in the soil [9].

Bacillus are gram-positive rod-shaped bacteria with the optimal temperature for growth between 25–35 °C. Although bacillus was considered strictly aerobic, it was discovered later that they could live anaerobically under the prescribed conditions. Bacillus subtilis has a relatively different physiology from other non-pathogenic bacteria, which is relatively easy to genetically manipulate and easy to breed so that it can be developed on an industrial scale [10].

Aside water, cementitious materials are the most eaten up substances on Earth. Along these lines, they should be concentrated, to show adequate execution, worried about the presentation of a couple of works trying to improve the materials' mechanical properties. Besides, in the point of view of formative difficulties on improving their characteristics and durability, increases of minute creatures, including the class Bacillus, are being read for break filling, consequently, growing compressive characteristics, through calcium carbonate ($CaCO_3$) precipitation. Consequently, leading this interaction is moreover called biocementation, subsequently, it should be seen that organisms exist for billions of years, and countless their biotechnological applications are not yet settled. Furthermore, Albeit· various experts are enthusiastic about the experimentation of bioprecipitation of $CaCO_3$, albeit the whole strategy isn't exceptionally clear and portrayed. In the meantime, microorganisms are fit for hurrying calcium carbonate creation through a couple of frameworks, the physiology and innate contemplations required are puzzling and difficult to grasp. A couple of microorganisms with the ability to make minerals were used in fixing limestone tourist spots and furthermore in filling pores and parts in concrete and

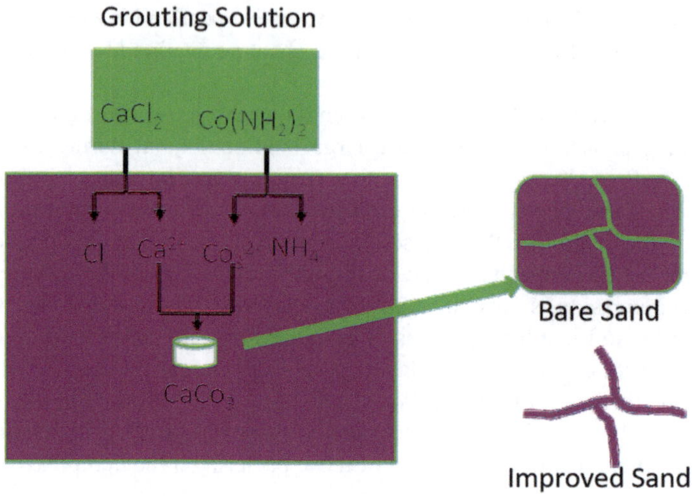

Fig. 1 Process of calcite precipitation process on MICP technique [14]

different cementitious materials, fit for improving mechanical properties and sturdiness. Among the microorganisms used in the investigation on biocementation, the living being B. subtilis sticks out, in view of their capacity to convey calcite, a translucent sort of $CaCO_3$. This cycle is sped up within the sight of a medium with a calcium source [11].

An enzyme-reagent mixed solution from bacteria (i.e., purified urease and $CaCl_2$-urea), created the hastened calcite after the substance response, is infused into the dirt. The hastened calcite may give spans between the grains of sand, confining their development, and thus, improving the solidness and the strength of the dirt [12–14]. A schematic of the entire interaction recorded above, and grouting instrument expected are represented in Fig. 1.

2 Material and Method

2.1 Material

Peat Soil and Sand. The soil sample (see Fig. 2) that the researchers will use in this research is peat soil whose condition has been disturbed, where the soil is taken at a depth of +50 cm from the topsoil surface using a hoe, then the soil is transported to the laboratory to be dredged by drying it in the open sun. Then the soil is sifted to pass sieve no 4. The sand used by researchers for this research is white sand that has been sieved until it passes using sieve number 40.

Fig. 2 Peat soil and sand

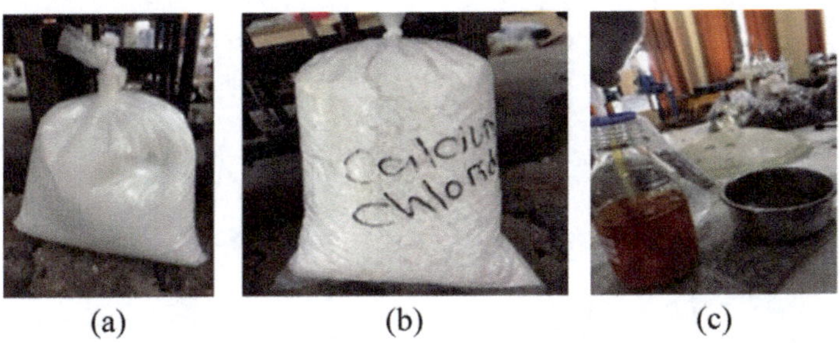

Fig. 3 Reagents include Bacillus Subtilis (**a**), CO (NH_2) urea (**b**), and $CaCL_2$, and the microorganism (**c**)

Reagent. Reagent used in this research are CO (NH_2) urea, $CaCL_2$, and the microorganism, Bacillus Subtilis as a grouting material seen in Fig. 3.

2.2 *Method*

Preliminary Testing. The initial research process was carried out by a preliminary test which was a test to obtain the physical properties of the soil. This test aims to check the soil conditions to match the ground conditions in the field. Water content testing (ASTM D 2216-98) was carried out to obtain the percentage of water weight against the dry soil weight. Examination of specific weight (ASTM D 854-02) test was carried out to obtain the specific weight of the soil which is the ratio of soil weight to water weight. Soil mechanical properties test was carried out with the Proctor test (ASTM D 698) to obtain the optimum moisture content and maximum density, then the data was used as a comparison to the soil density in the test.

Main Testing. This test is carried out after the preparation of the cementation solution and the preliminary testing has been completed. This testing process begins with mixing the cementation solution in which bacillus subtillis is already present in the peat soil that has been added with sand. The cementation solution and water are adjusted to the optimum amount of water content, then mixed with peat soil that has been mixed with sand to test its Free Compressive Strength which is the main test of this study. The amount of sand added to the compressive strength sample is 5% of the weight of the soil. The amount of cementation solution mixed in the sample of this specimen has 6 variations, namely sample 1 = 0%, sample 2 = 5%, sample 3 = 10%, sample 4 = 15%, sample 5 = 20%, sample 6 = 25%. The percentage of the cementation solution is the optimum moisture content of the test for the free compressive strength of the peat soil seen in Figs. 4 and 5.

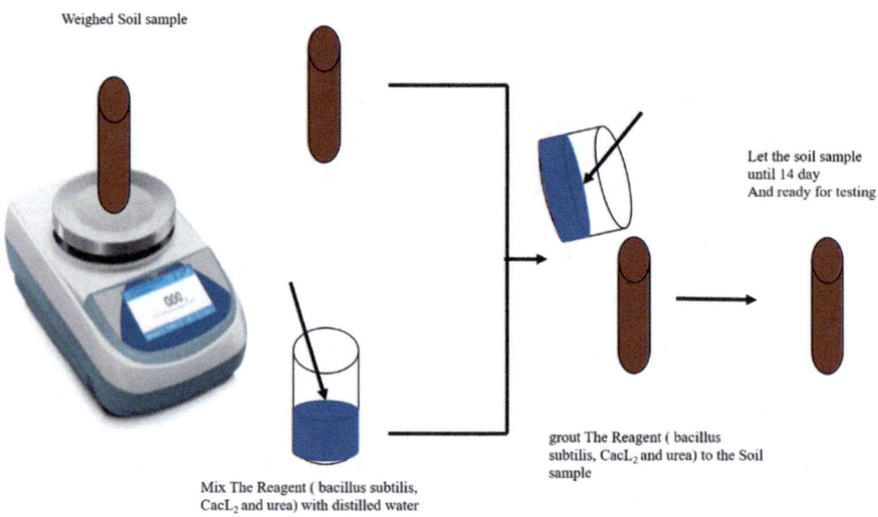

(a)

(b)

Fig. 4 Reagent manufacturing process (**a**) and equipment (**b**)

The UCS is perhaps the most ordinarily utilized stone designing boundaries for its mass arrangement, or strength assurance. Also, the mean worth and its changeability are frequently expected to dependably address material property, despite the fact that it once in a while addresses the flawless structures. Besides, the test essentially records the breakdown of burden during uniaxial stacking of a tube-shaped example, and many recommended this appraisal as a list, as opposed to just an interesting designing boundary. Subsequently, it is an intermediary for rock strength, subject to the stacking rate, example size, and other various components. Moreover, it is not equivalent to the Hoek–Brown strength basis boundary σ_{ci} [11].

3 Result and Discussion

Compaction testing is carried out to obtain the maximum dry bulk weight (γd_{max}) of the original soil of 0.467 g/cm^3 and the optimum moisture content (OMC) of the original soil of 157%, can be seen in Fig. 6.

Fig. 5 UCS test

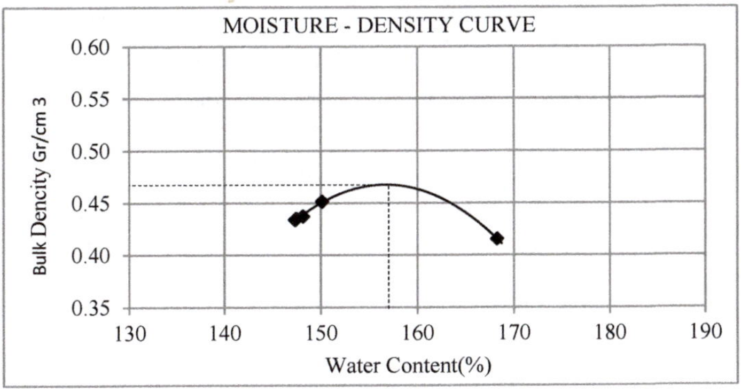

Fig. 6 Proctor standard result

Based on the tests that have been carried out, the physical properties of the peat soil in Siak Regency, Riau Province, were obtained in Table 1.

In this research, the free compressive strength test is carried out on the disturbed sample soil or in a state that has been mixed with other materials. However, for the test object that has been given a mixture of cementation solution, it must be stored for 14 days. This is to measure the ability of each sample of the specimen to free compressive strength.

The free compressive strength test value on the original soil sample +0% sand and +0% bacteria, then continued with the original soil test object +5% sand and with those that have been mixed the addition of cementation solution with variations in the percentage of the addition of 0, 5, 10, 15, 20, and 25% can be seen from Table 2 and Fig. 7.

From Fig. 7, there is an increase in the strength of the free compressive strength of the peat soil from the sample of the original soil sample +5% sand with that which has been mixed with bacterial cementation solution with variations in the addition of 0, 5, 10, 15, 20, and 25%. In this case, with the addition of bacteria, the bearing capacity of the peat soil occurs.

Table 1 Physical properties of the peat soil in Siak Regency

No	Description	Standard	Value	Unit
1	Specific gravity (GS)	1.2–1.95	1.30	–
2	Water content (%)	200–900%	407.5	%
3	γ_{dmax}	–	0.467	g/cm^3
4	Optimum water content (%)	–	157	%

Table 2 UCS test results

No	Strain (%)	Compressive strength (kg/cm^2)					
		0%	5%	10%	15%	20%	25%
1	0.00	0.00	0.00	0.00	0.00	0.00	0.00
2	0.175	0.06	0.06	0.06	0.03	0.06	0.07
3	0.35	0.09	0.10	0.10	0.06	0.09	0.10
4	0.70	0.16	0.16	0.16	0.12	0.14	0.16
5	1.05	0.22	0.22	0.22	0.19	0.19	0.23
6	1.40	0.29	0.27	0.26	0.23	0.24	0.29
7	1.75	0.30	0.30	0.30	0.29	0.30	0.33
8	2.10	0.29	0.31	0.31	0.36	0.36	0.33
9	2.45		0.28	0.28	0.41	0.41	0.28
10	2.80				0.42	0.38	
11	3.15				0.40		

Then from Fig. 7, the highest value of each sample of the test object is also obtained, namely the value of q_{Umax} (kg/cm^2) and can be seen in Table 3.

The results of Table 4 on the graph of the relationship between the value of the free compressive strength (q_{Umax}) of the sample of the original soil sample +5% sand with variations in the solution of bacterial cementation, there is an increase and decrease in the percentage of 5, 10, 15, 20 and 25%. The value of the free compressive strength of the original soil test object +5% sand is 0.30 kg/cm^2, while with the addition of variations in bacteria 5% = 0.31 kg/cm^2, 10% = 0.31 kg/cm^2, 15% = 0.42 kg/cm^2, 20% = 41 kg/cm^2, and 25% = 0.33 kg/cm^2. The highest value of free compressive strength is found in the addition of 15% cementation solution of 0.42 ^{kg}/cm^2 of the original soil sample + 5% sand, which has a free compressive strength value of 0.30 kg/cm^2.

From the results of the free compressive strength test sample of peat soil using the UCS tool with the addition of variations in the addition of +5% sand and bacteria 0, 5, 10, 15, 20, and 25% there is an increase and decrease in the strength of the peat

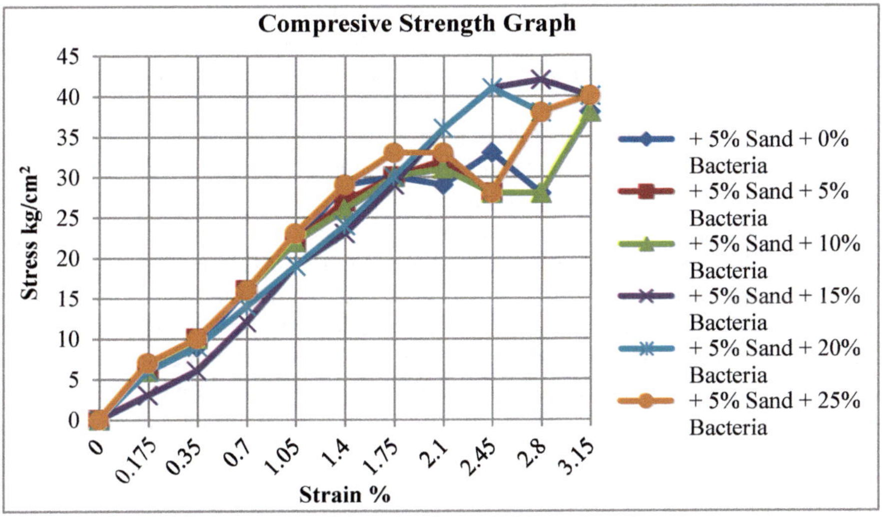

Fig. 7 Curve for UCS test

Table 3 q_{Umax} value of each sample specimen

Adding reagent (%)	Strain (%)	q_{Umax} (kg/cm^2)
0	2.10	0.30
5	2.45	0.31
10	2.45	0.31
15	3.15	0.42
20	2.80	0.41
25	2.45	0.33

Table 4 Comparison of the value of free compressive strength of this study of bio-grouting stabilization with other stabilizers

No	10% calcium carbonate + 8% Matos [17]		Sand and bio-grouting		Due to preloading [18]	
	Load (kg)	q_{Umax} (kg/cm^2)	Percentage (%)	q_{Umax} (kg/cm^2)	Time (kPa 1 day)	q_{Umax} (kg/cm^2)
1	1	1.1	0	0.30	10	0.09
2	2	1.6	5	0.31	20	0.11
3	3	1.8	10	0.31	10	0.12
4	–	–	15	0.42	20	0.14
5	–	–	20	0.41	–	–
6	–	–	25	0.33	–	–

soil. This is due to the addition of the subtilis bacteria cementation solution which can make the pores of the peat soil particles closed or filled with bacteria so that the peat soil does not easily decline. In [15, 16] the indication of this result is there are adding some amount of water on the reagent therefore the sample with reagent has a large amount of water and effect on the compression strength. There are also some investigations should be conducted to analyze this effect i.e., water effect, pH, and amount of carbonate. Organic soil had many amounts of water content and this effect should be analyzed in another research.

From the comparison of Table 4, the maximum free compressive strength (q_{Umax}) value is the highest on peat soil with stabilization of 10% lime $_{+}$8% matos while the lowest value (q_{Umax}) is due to preloading, and it turns out that in the addition of sand and bio-grouting with variations in the addition of bacteria 0, 5, 10, 15%, 20, and 25% there was an increase in the value of the free compressive strength of the peat soil. From the UCS test, the sample with 10% concentrate of reagent had a compression strength qu = 2.93 kg/cm^2 and the sample without the reagent had a compression strength qu = 4.97 kg/cm^2.

4 Conclusion

From the results of the research on the compressive strength of peat soil using a mixture of sand and bio-grouting techniques with the help of the bacterium Bacillus subtilis, the results obtained in accordance with the research objectives are as follows: the relationship between the compressive strength (q_{Umax}) of the original soil sample +5% sand with variations in the concentration of the bacterial cementation solution at a percentage of 0, 5, 10, 15, 20, and 25%. The value of the free compressive strength of the original soil test object +5% sand is 0.30 kg/cm^2, while with the addition of variations in bacteria 5% = 0.31 kg/cm^2, 10% = 0.31 kg/cm^2, 15% = 0.42 kg/cm^2, 20% = 41 kg/cm^2, and 25% = 0.33 kg/cm^2. The highest compressive strength value is found in the addition of 15% cementation solution of 0.42 $^{kg}/cm^2$ of the original soil sample +5% sand, which has a free compressive strength value of 0.30 kg/cm^2.

References

1. Mochtar NE (1998) Koefesien tekanan tanah ke samping at rest (ko) tanah gambut berserat serta pengaruh overconsolidation ratio (OCR) terhadap harga Ko. Jurnal Teknik Sipil ITB 5(4)
2. Mochtar NE, Yulianto FE (2017) Behaviour change in peat stabilized with fly ash and lime caco$_3$ due to water infiltration. ARPN J Eng Appl Sci 12(17):4967–4972
3. Moeyadi H, Kazemian S, Huat BBK (2013) Shear strength parameters of improve peat by chemical stabilizer. Geotech Geol Eng 31:1089–1106
4. Wibowo P, Suyatno N (1998) An overview of indonesian wetlands sites. In: Wetlands International—Indonesia programme (WI-IP)
5. Yulianto FE (2017) Perilaku tanah gambut berserat permasalahan dan solusinya. In: Konferensi Nasional Teknik Sipil dan Infrastruktur – I Jurusan Teknik Sipil, Universitas Jember
6. Parlan (2016) Pengaruh jumlah plat helical terhadap daya dukung pondasi tiang helical pada tanah gambut. Jurnal Online Mahasiswa Fakultas Teknik Universitas Riau 3(2)
7. Wardana IGN, Wardata I (2010) Korelasi strain rate dengan kadar organik pada test konsolidasi metode constant rate of strain. Jurnal Ilmiah Teknik Sipil 14(1):43–56
8. Fatnanta (2014) Pengaruh kadar lempung dan kadar air pada sisi basah terhadap nilai cbr pada tanah lempung kepasiran (sandy clay). Jurnal Online Mahasiswa Fakultas Teknik Universitas Riau 1(2)
9. Pangesti DR (2005) Pedoman grouting untuk bendungan. Depertemen Pekerjaan Umum, Derektoral Jendral Sumber Daya Air, Derektoral Sungai, Danau dan Waduk, Jakarta
10. Soesanto L (2008) Pengantar pengendalian hayati penyakit tanaman, PT Raja Grafindo Persada, Jakarta
11. Barkouki TH, Martinez BC, Mortensen BM, Weathers TS, De Jong JD, Ginn TR, Spycher NF, Smith RW, Fujita Y (2011) Forward and inverse bio-geochemical modeling of microbially induced calcite precipitation in half-meter column experiments. Transp Porous Med 90(23):23–39
12. Neupane D, Yasuhara H, Kinoshita N, Unno T (2013) Applicability of enzymatic calcium carbonate precipitation as a soil strengthening technique. J Geotech Geoenviron Eng 139(12):2201–2211
13. Putra H, Yasuhara H, Kinoshita N, Hirata A (2017) Optimization of enzyme-mediated calcite precipitation as a soil-improvement technique: The effect of aragonite and gypsum on the mechanical properties of treated sand. Curr Comput-Aided Drug Des 7(2):59
14. Yasuhara H, Putra H, Kinoshita N (2016) Enzyme-mediated carbonate precipitation (EMCP) for soil improvement: the effect of magnesium substitution. In: The 2016 world congress on advances in civil, environmental, and materials research (ACEM16), Jeju Island, Korea (2016)
15. Syarif F, Hardianto F, Mahardika G (2020) Penerapan teknik biocementation oleh bacillus subtilis dan pengaruhnya terhadap permeabilitas pada tanah organik. Jurnal Saintis 20(1):47–52
16. Syarif F, Hardianto F, Mahardika G (2019) Microbially induce calcite precipitation as bio grouting by bacillus subtilis on its shear strength parameter effects on organic soil (peat) from Siak Regency Riau Province Indonesia. In: The 7th International Conference of Euro Asia Civil Engineering Forum (EACEF) 2019—Green Engineering for Infrastructure and Safety against Hazards, The Universitat Stuttgart (2019).
17. Prabowo A, Fauziah M (2018) Pengaruh stabilitas tanah menggunakan kapur dan matos terhadap kuat geser dan konsolidasi tanah gambut. Universitas Islam Indonesia, Prosiding Sidang Program Studi Teknik Sipil (SPSTS) FTSP
18. Waruwu A (2013) Korelasi nilai kuat tekan dan cbr tanah lempung yang distabilisasi dengan abu batu dan semen. Jurnal Rancang Teknik Sipil 2(2):99–107

Sensitivity Analysis on the Effect of Reinforcement Materials Addition for Soil Stabilization

Ida Agustin Nomleni, Raden Harya Dananjaya, and Yusep Muslih Purwana

Abstract Sensitivity analysis is a widely used method to analyze uncertainty in the model. The sensitivity is calculated as the ratio of the output changes and parameters changes. This study was performed using a sensitivity index for sensitivity analysis to determine the effect of reinforcement material addition on clay soil. In this study, 54 samples of soil with additional material have been prepared for the unconfined compression test. The reinforcement materials used in this test are cement, tire waste strips, and water-cement ration for the hydration process. The unconfined compression test results were then analyzed for sensitivity analysis input. Based on this study, it can be concluded that the addition of reinforcement material, including cement, and tire waste strips, and water-cement ratio, can increase the unconfined compression strength of high plasticity clay. The maximum unconfined compressive strength value is 577.10 kN/m^2 with 3% of tire waste strips, 20% cement, and 35% water-cement ratio. The cement proportion is the most sensitive variable for unconfined compression strength from the sensitivity analysis. The sensitivity analysis results can help to become the baseline to help design improvement using tire waste strips on high plasticity clay.

Keywords Sensitivity analysis · Sensitivity index · Unconfined compression test

1 Introduction

Soil plays a vital role in civil construction because all buildings stand on soil. High plasticity clay is a problematic soil because it has a low bearing capacity, so it needs stabilization to increase its bearing capacity. Several studies have been performed to stabilize high plasticity clay, for example, using a mixture of tire waste, cement, and

I. A. Nomleni (✉)
Department of Civil Engineering, National Central University, Taoyuan, Taiwan

R. H. Dananjaya · Y. M. Purwana
Department of Civil Engineering, Universitas Sebelas Maret, Surakarta, Indonesia
e-mail: ymuslih@staff.uns.ac.id

S. A. Kristiawan et al. (eds.), *Proceedings of the 5th International Conference on Rehabilitation and Maintenance in Civil Engineering*, Lecture Notes in Civil Engineering 225,
https://doi.org/10.1007/978-981-16-9348-9_25

clay. Research to improve clay properties has been conducted, with the general aim is to improve the engineering properties of clay, especially to increase soil bearing capacity and reduce soil compressibility. Some soil improvement methods used are mechanical improvements by compacting, hydraulic improvements with dewatering, and physical and chemical improvements.

The increasing number of motorized vehicles in Indonesia has led to increased vehicle tire waste. Based on data reported by the Indonesian Central Bureau of Statistics, there was an increase in motorized vehicles from 2014 until 2017 [1–4]. The increasing number of vehicles has led to an increase in vehicle tire waste. Vehicle tires waste in large quantities can cause various health problems when piled up or burned. Therefore, the reuse of tire waste is needed to reduce environmental degradation. One effective way of utilizing tire waste is by using it as a mixture in geotechnical construction materials. Many researchers worldwide have implemented vehicle tire waste. It because tires developed rapidly throughout the world [5].

Research on low plasticity clay based on the Unified Soil Classification System (USCS) was performed [6]. In this study, crumb rubber and cement addition in clay mixture were investigated. Crumb rubber was cut into 0.8–2 mm. The study used crumb rubber of 2.5, 5, 7, 0.5, 10% of the soil's dry weight. It shows that the compressive strength and tensile strength increased in crumb rubber and cement addition. [7] conducted a study of soil reinforced by tire waste cut into fiber and cement. The study was performed on Perth sandy soil subjected to unconfined compressive strength and California bearing ratio test. From this study, it can be concluded that the addition of cement and or the addition of waste tires in Perth sandy soil can increase the value of the California bearing ratio and the unconfined compressive strength.

Iman et al. [8] stated that the model's sensitivity could be calculated as the ratio of the model's output and parameters changes. There are two reasons that cause problems in the sensitivity analysis study. The first problem arises when the model's response to parameter change is non-linear. In that case, the result will depend on how the initial value is chosen because small initial values produce slight variations, and larger initial values produce more significant variations. On the other hand, if the initial parameter is located near the upper or lower limit of a valid range of parameters, it can cause unacceptable values lying outside the range. Iman et al. [9] suggested several reasons for using sensitivity analysis, namely: (a) to determine parameters that require additional research to strengthen the theoretical basis so that it can reduce the uncertainty of output, (b) to determine parameters that do not affect the model so that it can be removed from the final model, (c) to determine parameters which give a significant influence on output, (d) to determine parameters which have a significant correlation with output, and (e) to know the change of output when input is changed.

Identification and representation of uncertainty are essential components in applying a model. One important approach often used to identify model uncertainty is the sensitivity analysis method. Many sensitivity analysis methods can be used, i.e., (a) the differential sensitivity analysis method, which can be analyzed based on sensitivity coefficient, (b) one-time sensitivity analysis or local sensitivity analysis, (c) factorial design method, which can be calculated by selecting several samples from each parameter and analyzing the model for all sample combinations, (d) sensitivity index which can be used to determine the sensitivity parameters by calculating the percentage difference in output, (e) important factors, (f) subjective sensitive analysis.

In [10], the sensitivity index method was used in the sensitivity analysis. This method is a simple method used to determine sensitivity parameters. The sensitivity index in the model is calculated based on the output % difference when the input parameter values vary from minimum to maximum. Lenhart et al. [11] used the ∂y/∂x differential approach in determining the sensitivity index value. Using the differential approach shows that changes in output parameters are affected by changes in input parameters. The sensitivity index for this calculation's input can be calculated with Eqs. 1 and 2 [11].

$$I' = \frac{y_2 - y_1}{2\Delta x} \tag{1}$$

$$I = \frac{(y_2 - y_1)/y_0}{2\Delta x/y_0} \tag{2}$$

In this research, sensitivity analysis was conducted using the sensitivity index method. The sensitivity index was obtained from the clay mixtures subjected to unconfined compression test (Fig. 1).

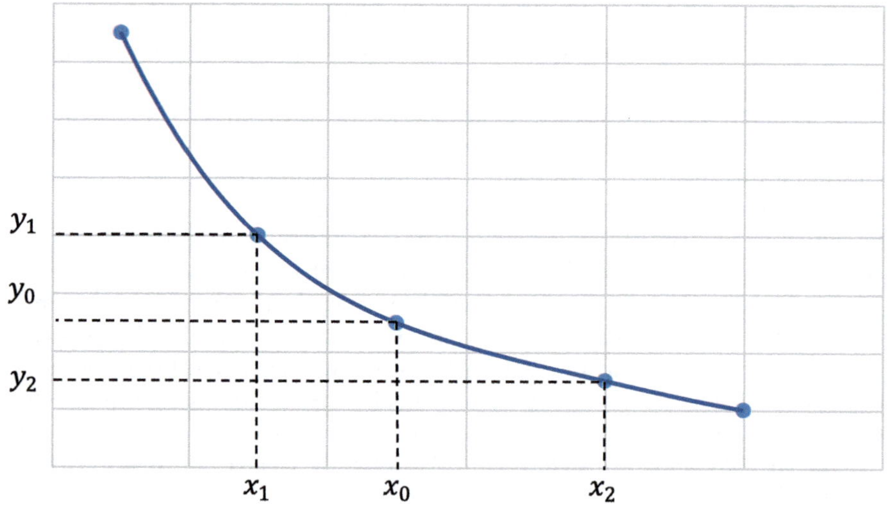

Fig. 1 Relationship between an output variable y and a parameter x proposed by Lenhart (2002)

2 Research Methodology

Primary data from unconfined compression test results have been used in this research. The unconfined compressive strength test was conducted on high plasticity clay. The reinforcement materials used in this test are cement, tire waste strips, and water-cement ration for the hydration process. Tire waste has been cut in a 1 mm × 1 mm × 30 mm dimension and mixed with soil, cement, and water in various weight conditions. The weight of tire mixtures varied from 0–7.5% of dry weight of the soil mixture, i.e. 0, 1.5, 3, 4.5, 6, and 7.5%. The cement's weight in the mixture is 10, 15, and 20% of the soil's dry weight. The water-cement ratios used in this study are 25, 35, and 45% of the cement's dry weight. Due to variations of the mixture compositions, a wide range of the data is obtained. The data obtained from the unconfined compression test then calculated and analyzed using sensitivity analysis. The sensitivity analysis used in this study is based on the hypothesis of a linear relationship between output and input parameters.

3 Results and Discussion

3.1 *Testing Material*

A preliminary test aims to determine the soil properties. In this study, soil was taken from Ploso village, Mojogedang, Karanganyar, at a depth of 0 to 0.5 m below the ground level. This soil's specific gravity is 2.66, with a liquid limit of 57.94% and a plastic limit of 29.31%. According to the Unified Soil Classification System (USCS), this soil can be classified as CH (inorganic clay or high plasticity clay). The summary of the preliminary test results can be seen in Table 1.

Table 1 Soil properties

Soil properties	Notation	Value
Specific gravity	G_s	2.66
Liquid limit (%)	LL	57.94
Plastic limit (%)	PL	29.31
Plasticity index	PI	28.64
Dry density (g/cm^3)	$\gamma_{d\ opt}$	1.404
Moisture content (%)	w_{opt}	29
Pass No. 200 sieve (%)		90.52
USCS classification		CH

3.2 Unconfined Compression Test Results

This research was conducted under Indonesian National Standard (SNI) 6887:2012, Standard Test Methods for Compressive Strength of Molded Soil–Cement Cylinders [12]. Figures 2 and 3 show that shredded tire waste strips addition can affect the unconfined compressive strength value. The value of unconfined compression strength without the addition of tire waste is 173.55 kN/m^2. At the addition of 1.5% tire waste strips, the value of unconfined compressive strength is increased up to

Fig. 2 Tire waste strips with a dimension of 1 mm × 1 mm × 30 mm

(a) 0% (b) 1.5% (c) 3%

(d) 4.5% (e) 6% (f) 7.5%

Fig. 3 Unconfined compression test results for soil sample with reinforcement material addition; 10% of cement and the water-cement ratio of 25%

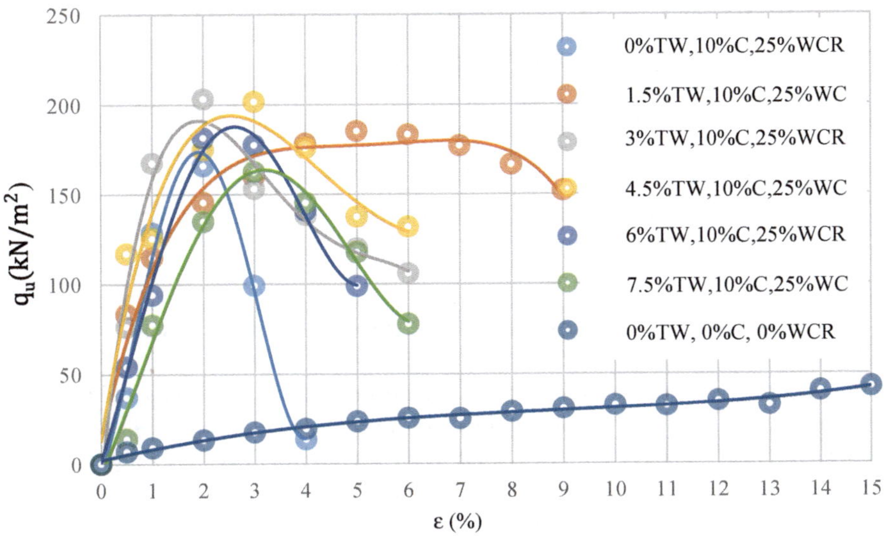

Fig. 4 Relationship between compression strength value and strain on the soil sample with reinforcement material addition

179.80 kN/m^2. When the addition of tire waste is 3%, the unconfined compression strength reaches its maximum value. The value of unconfined compressive strength decreases by adding 4.5, 6, and 7.5% of tire waste (Fig. 4).

3.3 *Sensitivity Analysis*

The sensitivity analysis can be done if the model has two or more dependent variables. In this analysis, the water-cement ratio of 45 and 20% cement is the dependent variable. Dependent variables are needed to get the simple 3D statistical model for predicting the unconfined compression strength of the model.

3.3.1 The Optimum Value of Reinforcement Material Addition

The first step of sensitivity analysis is determining the value of the addition of the optimum tire waste strips percentage for adding a water-cement ratio of 45% and 20% cement using 3D statistical models. Based on the 3D statistical model, we can get the percentage of tire waste strips that needs to add to the unconfined compression test sample. Based on this analysis, the percentage of tire waste strips can be found with Eq. 3. The maximum percentage of tire waste addition is 4.05%.

$$Z(X, Y) = 17,9419 - 1,7260\,Y + 0,0529\,Y^2 - 0,1617\,X + 0,0048\,XY + 0,0012\,X^2 \quad (3)$$

where:

X = water cement ratio (%) when X value is $25 \leq X \leq 45$.
Y = cement (%) when Y value is $10 \leq Y \leq 20$.
Z = tire waste strips percentage (%).

3.3.2 Sensitivity Analysis for Additional Reinforcement Material

The second step of sensitivity analysis is determining the sensitivity index for the addition of 3% tire waste for the addition of a water-cement ratio of 45% and 20% cement using 3D statistical models. The water-cement ratio, cement, and sensitivity index can be plotted with a 3D statistical model.

The sensitivity index value can be obtained using Fig. 5 and or Eq. 4. The index sensitivity for the addition of 3% tire waste strips for adding a water-cement ratio of 45 and 20% cement is 29.49.

$$Z(X, Y) = -22,8984 - 8,3648\,Y + 0,3147\,Y^2 + 4,7260\,X + 0,0360\,XY - 0,0747\,X^2 \quad (4)$$

where:

X = water cement ratio (%) when X value is $25 \leq X \leq 45$.

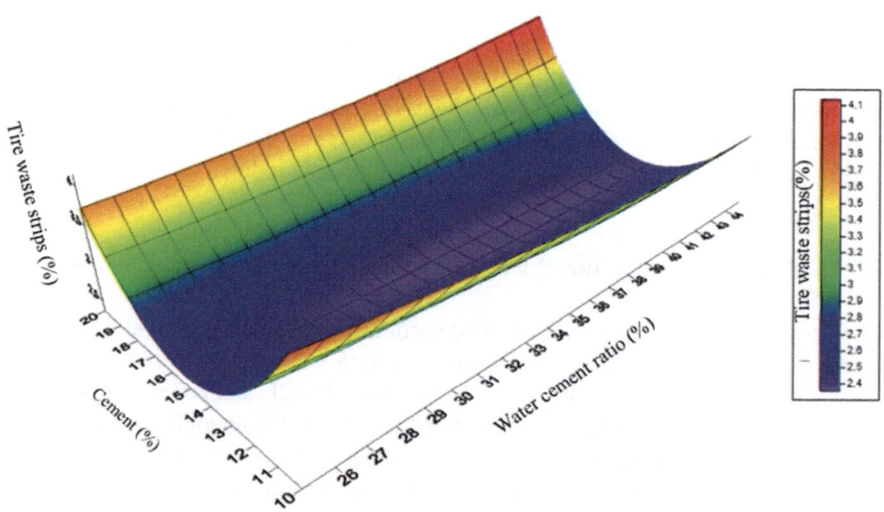

Fig. 5 3D statistical model of optimum value of reinforcement material addition

Y = cement (%) when Y value is $10 \le Y \le 20$.
Z = sensitivity index.

3.3.3 Unconfined Compression Strength

The third step of sensitivity analysis determines the optimum unconfined compressive strength for the addition of 3% tire waste strips for the addition of a water-cement ratio of 45 and 20% cement using 3D statistical models. The model can also be predicted using the following equation or Fig. 6. The predicted unconfined compression strength is 430 kN/m^2 can be obtained with the proposed equation as follow (Fig. 7):

$$q_u(X, Y) = -708,7262 - 19,4120Y + 1,1322Y^2 + 58,2342X + 0,2456XY - 0,8729X^2 \quad (5)$$

where:

X = water cement ratio (%) when X value is $25 \le X \le 45$.
Y = cement (%) when Y value is $10 \le Y \le 20$.
q_u = predicted unconfined compression strength of model.

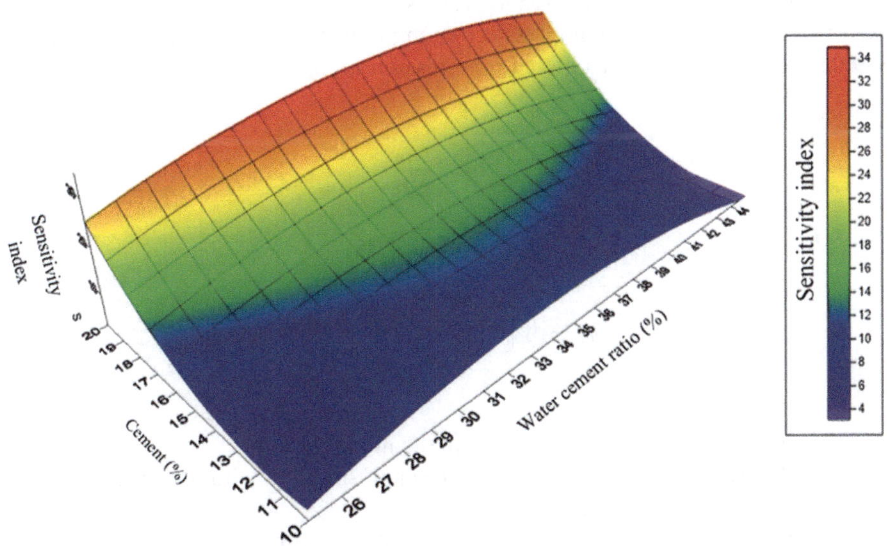

Fig. 6 3D statistical model of sensitivity index

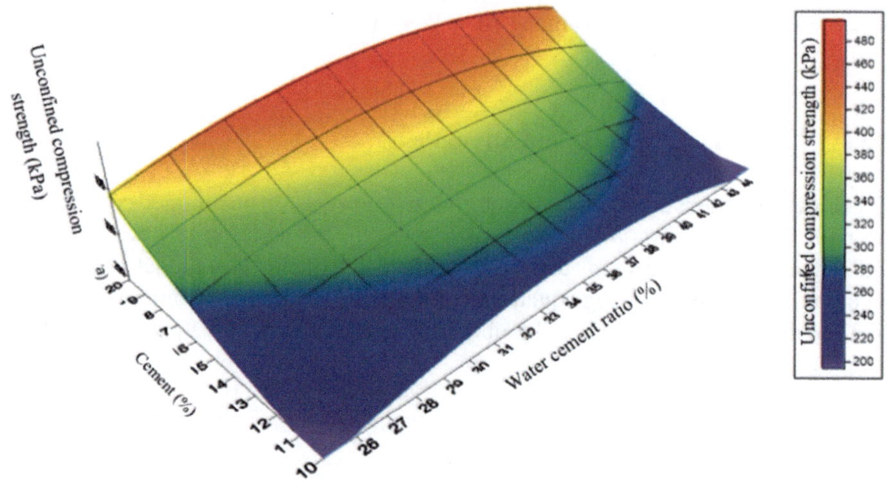

Fig. 7 3D statistical model for predicting the value of unconfined compression strength

3.3.4 Unconfined Compression Test Calculation in the Polynomial Model

The unconfined compressive strength values in the model can be predicted using Eq. (8). The unconfined compression strength value in the predicted model is 399.03 kN/m^2, while the unconfined compression strength value in the laboratory test is 412.40 kN/m^2.

$$\mathrm{IS} = \frac{\Delta q_u}{\Delta x} \tag{6}$$

$$\Delta \mathrm{q_u} = \Delta \mathrm{x} \times \mathrm{IS} \tag{7}$$

$$\mathrm{q_u} = \mathrm{q_{opt}} - \left(\left|x_{opt} - x_i\right| \times \mathrm{IS}\right) \tag{8}$$

where:

q_{u} = unconfined compression value in the model (kN/m^2).

q_{opt} = the value of unconfined compression strength with the reinforcement material addition (kN/m^2)

x_{opt} = independent variable at optimum conditions (the percentage of tire waste strips, cement, and the addition of water at optimum conditions)

x_{i} = the independent variable reviewed (percentage of the tire waste strips, cement, and water)

IS = sensitivity index (kN/m^2).

4 Conclusions

Based on this research, it can be concluded that the addition of reinforcement material, including cement, and tire waste strips, and water-cement ratio, can increase the unconfined compression strength of high plasticity clay. The maximum unconfined compressive strength value is 577.10 kN/m^2 with 3% of tire waste strips, 20% cement, and 35% water-cement ratio. The cement proportion is the most sensitive variable for unconfined compression strength from the sensitivity analysis. The sensitivity analysis results can help to become the baseline to help design improvement using tire waste strips on high plasticity clay.

References

1. Badan Pusat Statistik Indonesia (2014) Statistik Indonesia 2014. Badan Pusat Statistik ISSN: 0126-2912
2. Badan Pusat Statistik Indonesia (2015) Statistik Indonesia 2014. Badan Pusat Statistik ISSN: 0126-2912
3. Badan Pusat Statistik Indonesia (2016) Statistik Indonesia 2014. Badan Pusat Statistik ISSN: 0126-2912
4. Badan Pusat Statistik Indonesia (2017) Statistik Indonesia 2014. Badan Pusat Statistik ISSN: 0126-2912
5. Humphrey DN (1999) Civil engineering application of tire shreds. Presented at the tire industry conference, Hilton Head, South Carolina
6. Yadav JS, Tiwari SK (2017) Assessment of geotechnical properties of uncemented/cemented clayey soil incorporated with waste crumb rubber. J Mater Eng Struct 37–50
7. Bazazorde SM (2018) UCS and CBR behaviour of Perth sandy soil reinforced with waste tyre fibres and cement. Retrieved from https://ro.ecu.edu.au/theses/2076
8. Iman RL, Helton JC (1998) An investigation of uncertainty and sensitivity analysis techniques for computer models. Risk Anal 8:71–90
9. Iman RL, Helton JC, Campbell JE (1981) An approach to sensitivity analysis of computer models: part I-introduction, input variable selection and preliminary assessment. J Qual Technol 13:174–183
10. Hamby DM (1994) A review of techniques for parameter sensitivity analysis of environmental models. Environ Monit Assess 32:135–154
11. Lenhart T, Eckhardt K, Fohrer N, Frede H-G (2002) Comparison of two different approaches of sensitivity analysis. Phys Chem Earth 27:645–654
12. Badan Standarisasi Nasional. SNI 6887:2012: Metode uji kuat tekan silinder campuran tanah-semen. Badan Standarisasi Nasional, Jakarta (2012)

Numerical Simulation of Slope Stability for Soil Embankment Reinforced with Inclined Bamboo Piles

Ngudiyono and Tri Sulistyowati

Abstract Slope stability is an important parameter when designing for the embankment. Poor-quality and degradation properties of soil material became triggers slope failure. Bamboo culm (*Dendrocalamus asper*) provided a diameter of 6–15 cm and high tensile strength parallel to the grain, and it was suitable for slope stabilizing. Slope stabilization by bamboo pile is a passive technique in which soil displacements require to activate pile forces. In this manuscript, two-dimensional (2D) model slope reinforcement with bamboo pile with the shear strength reduction (SSR) finite element method (FEM) by using ABAQUS SE software has simulated. The bamboo piles with 5 m length and 6 cm, 10 cm, 15 cm for diameter were placed two in the edge, one in the middle, and inserted perpendicular to slope plane embankment with the inclination of 33.69°. The model of the slope was analyzed, its influence on bamboo piles soil interaction by evaluating the safety factor (FOS). The result has shown that reinforced embankment slope with inclined bamboo piles had increased slope stability, the safety of factor (FOS) increase to 1.515. The shear strength reduction (SSR) finite element method (FEM) can be used as an alternative for slope stability analysis.

Keywords Slope stability · SSR · FEM · Reinforcement · Bamboo pile

1 Introduction

Slope stability is an important parameter when designing embankment roads, highways, dams, and other civil engineering structures. The problem of poor-quality soil material and degradation properties of soil material became sometimes triggers slope failure. Several slope reinforcement techniques widely applied are geosynthetic, soil

Ngudiyono (✉) · T. Sulistyowati
Department of Civil Engineering, University of Mataram, Mataram, Indonesia
e-mail: ngudiyono@unram.ac.id

T. Sulistyowati
e-mail: trisulisyowati@unram.ac.id

S. A. Kristiawan et al. (eds.), *Proceedings of the 5th International Conference on Rehabilitation and Maintenance in Civil Engineering*, Lecture Notes in Civil Engineering 225,
https://doi.org/10.1007/978-981-16-9348-9_26

nailing, concrete or steel mini-pile, concrete block, etc. The technologies that require skilled labor are relatively expensive.

Bamboo is a natural material widely available globally, especially in Asia, and has been widely used as a traditional and modern building material. Bamboo also green building material, renewable, sustainable, typically be harvested in less than three years to four years, and relatively inexpensive. Bamboo culm (*Dendrocalamus asper*) provided with diameter 6–15 cm, tensile strength parallel to grain 228 MPa, flexural strength 134.97 MPa, and modulus of elasticity 12,888.48 MPa [1, 2] that make it suitable for slope stabilizing. Bamboo piles inserted in a soil mass can provide stability by adding the reactive portion to the resistant forces contrary to the ground. Slope stabilization by bamboo pile is a passive technique in which soil displacements require to activate pile forces [3, 4].

Using bamboo piles as material for reinforcement of slope sandy soil was carried out by Munawiret al. [5, 6]; the result of the experiment and numerical by using PLAXIS showed that pile reinforcement applied on the slope had improved the stability of the slope. It had indicated by the significant increase of safety factor and bearing capacity. Wahhab and Ramadhan were used bamboo as soil reinforcement to landslide in Kali Wungu–Boja road, Darupono Village, Kendal District, two varieties of bamboos with a diameter of 20 and 15 cm were analyzed. The safety factor was analyzed by using PLAXIS and SLOPE/W software. The result showed that of safety factor was give the illustration that bamboo could use as an alternative material to slope reinforcement [7]. Using the *Dendrocalamus Giganteus* species bamboo-pile for slope stability using the finite element method (FEM) analysis of bamboo-pile was developed PLAXIS software for 2D and 3D. The results presented that the capabilities of the bamboo piles are potential as an alternative for slope stability [8].

Generally, there are two methods for slope stability analysis, including limit equilibrium methods (LE), such as Ordinary, Bishop's, Janbu's, Spencer's, Morgenstern-Price, and numerical analysis methods such as finite element method (FEM). In the LE method for slope stability analysis, the critical slip surface has to be estimated. However, in FEM analysis, no potential slip surface was estimated in advance, and the stress–strain correlation in the slope can also be considered [8]. In this study, two-dimensional (2D) analyses evaluated bamboo piles using ABAQUS Student Edition (SE) software to simulate the slope model. The software is a personal finite element analysis tool for solving limited size problems (1000 nodes for Abaqus/Standard and Abaqus/Explicit. This numerical study aims to analyze the performance of the slope reinforced with inclined bamboo piles as an alternative material used in slope reinforcement. The variation of diameter bamboo piles was modeled to evaluate the influence of inclined bamboo piles on the safety factor (FOS).

2 Shear Strength Reduction (SSR) Finite Element Method (FEM)

Nowadays, shear strength reduction (SSR) has increasingly been used with the finite element method (FEM) for slope stability analyses. Previous researchers have widely used this method to analyze the slope stability of several cases [8–13].

In the SSR finite element method elastic–plastic strength is assumed for slope materials. The factor of safety (FOS) is applied to reduce the strength of soil until failure occurs. For Mohr–Coulomb material model shear strength reduced by a factor of safety can be determined from the equation below.

$$\frac{\tau}{FOS} = \frac{c\prime}{FOS}\frac{\tan \text{Ø}\prime}{FOS} \tag{1}$$

Equation (1) can be re-written as,

$$\frac{\tau}{FS} = c^{*} + \tan \text{Ø}^{*} \tag{2}$$

where

$$c^{*} = \frac{c'}{FOS} \tag{3}$$

$$\text{Ø}^{*} = arctan\left(\frac{\tan \text{Ø}'}{FOS}\right) \tag{4}$$

Equations (3) and (4) are reduced Mohr–Coulomb shear strength parameters, and these values can be input into the ABAQUS SE software and then analyzed. There are several types of criteria for determining the failure slope model: numerical iteration non-convergence, suddenly change displacement, plastic strain, or equivalent plastic strain [8–13].

In the computation process, the field variables change following increment step time. The process of reducing material strength be realized by adjusting the material parameters to change with field variables. Since the incremental time change is automatic and manual modification is no longer required, the strength is also automatically decreased. Eventually, according to specific failure criteria, a suitable increment step time is known. Then the safety of the factor and the critical sliding surface of the slope can be found.

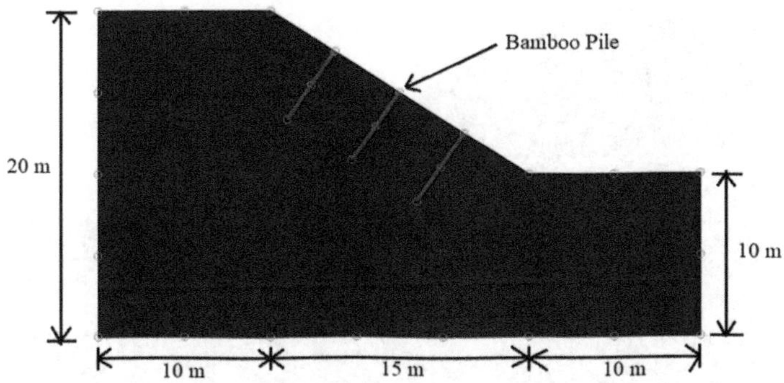

Fig. 1 The geometry of the slope model

3 Numerical Simulation

In the section describing step by step finite element analysis for slope model embankment reinforced with inclined bamboo piles using ABAQUS SE software, including model geometry, material properties, meshing, boundary condition, and loading as follows:

3.1 Model Geometry

According to the previous study [14], the geometry of the homogeneous embankment slope model was typically 20 m in height and 35 m long at the base with an inclination of 33.69°. The bamboo pile with 5 m length and 6, 10, 15 cm for diameter were placed two in the edge, one in the middle, and inserted perpendicular to the slope model plane. The geometry slope reinforced with bamboo piles is shown in Fig. 1.

3.2 Material Properties

The soil material properties of slope based previous study [14], the elastic–plastic Mohr–Coulomb failure criterion is used in the constitutive model of the soil in the analyses. Because clay assumed to be undrained, so the modulus of elasticity (E), Poisson's ratio (v), cohesion (c), friction angle (φ), and unit weight (γ) are 200 MPa, 0.25, 10°, 20 kPa, and 20 kN/m^3 respectively. In contrast the dilation angle (ψ) assumed zero because of the saturated clay. Meanwhile, the properties of the bamboo piles species *Dendrocalamus asper* are 5 m for length and 6, 10, 15 cm for diameter with the modulus of elasticity (E) 12,000 MPa and Poisson's ratio (v) 0.2.

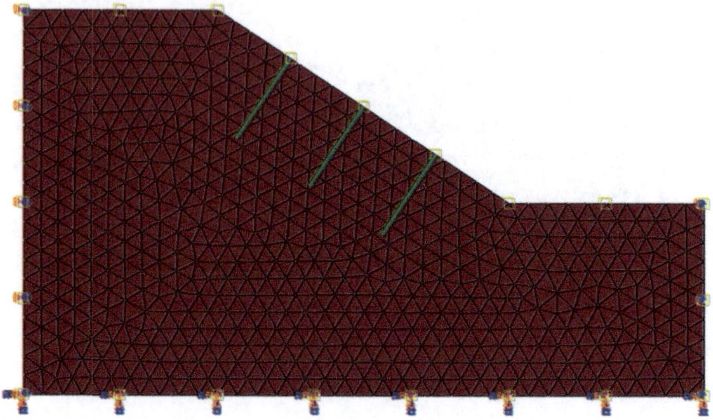

Fig. 2 Meshing and boundary condition of the slope model

3.3 Meshing, Boundary Condition, and Loading

Because the ABAQUS SE has having limitations, the 2D FEM technique has been adopted to solve the slope model in this study. There are 1185 element type CPE3 (3-node linear plane strain triangle) used for soil and type B21 (2-node linear beam in a plane) for the bamboo pile. The bamboo piles are assumed as an embedded elements beam in a set of solid elements. The boundary conditions (BC) on both sides are assigned to be roller which deformation on x-direction, and the fixed is used for the bottom. The meshing and BC of the model are shown in Fig. 2. Only the gravity load was applied in the slope model.

4 Result and Discussion

4.1 The Safety of Factor (FOS)

The slope stability analyses based on SSR using ABAQUS SE software, the factor of safety (FOS) has evaluated base criteria non-convergence of the numerical iteration, suddenly change the dis-placement node control on the top slope model. The displacement of the versus FOS has shown in Fig. 3. It illustrated that the displacement changes suddenly at FOS = 1.293 for clay embankment slope unreinforced. Meanwhile, the clay embankment slope reinforced inclined bamboo piles with diameters of 6, 10, 15 cm increase 1.503, 1.512, 1.515, respectively. The main principle of reinforcement in slopes is to provide stabilizing forces. When bamboo piles are inserted in the slope, additional pullout resistances are provided, so this structure changes safety of factor. The simulation results also indicate that the safety factor depends on the diameter of bamboo piles. With the larger diameter of bamboo piles, the pullout resistance was increased.

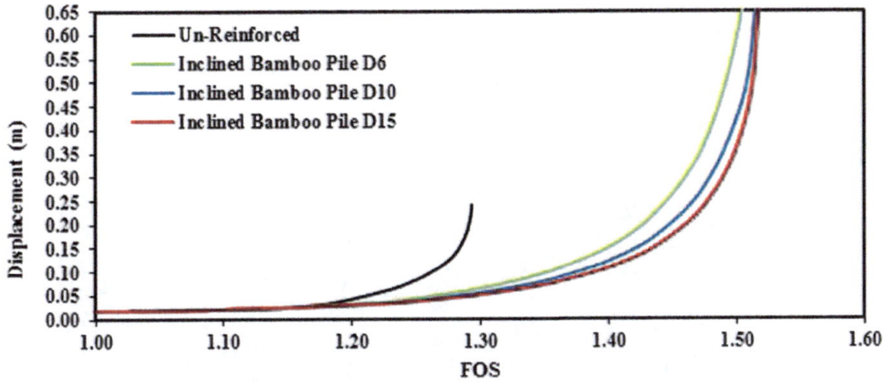

Fig. 3 Displacement versus FOS of the slope model

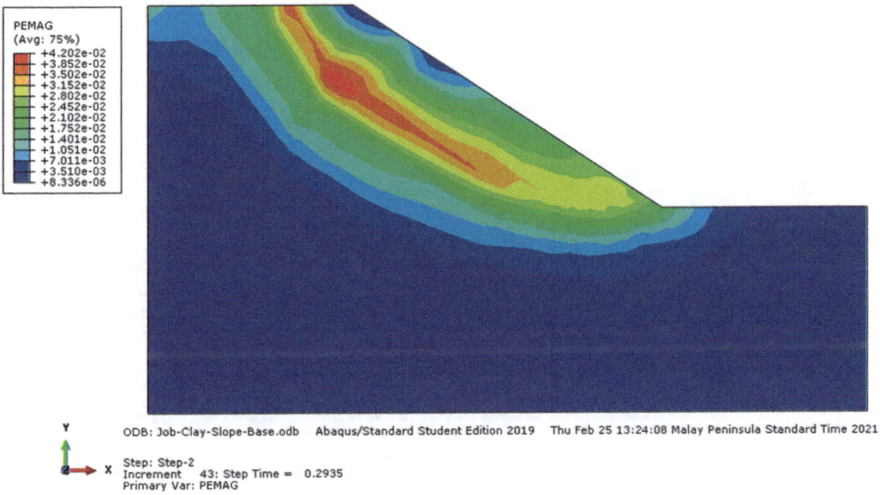

Fig. 4 The equivalent plastic strain (PEMAG) slope model unreinforced

4.2 The Equivalent Plastic Strain (PEMAG)

The bamboo piles are inserted in the slope, and it also changes the stress–strain and failure mechanism. Figure 4 until Fig. 7 shows the equivalent plastic strain (PEMAG) at increment 43rd for unreinforced and 158th for reinforced slope. In the figures, the contour indicates the location of the potential failure of the slope. The different contours shown in the figures indicate the location where the slope failure started (Figs. 5, 6 and 7).

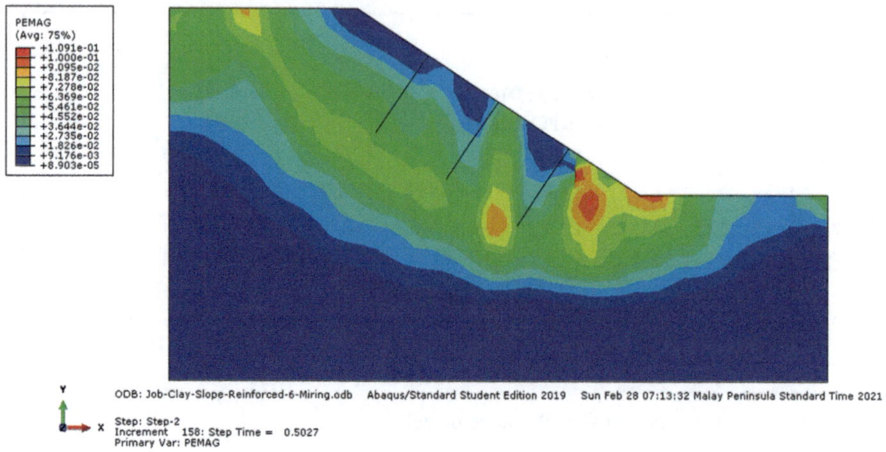

Fig. 5 The equivalent plastic strain (PEMAG) slope model reinforced with inclined bamboo piles diameter 6 cm

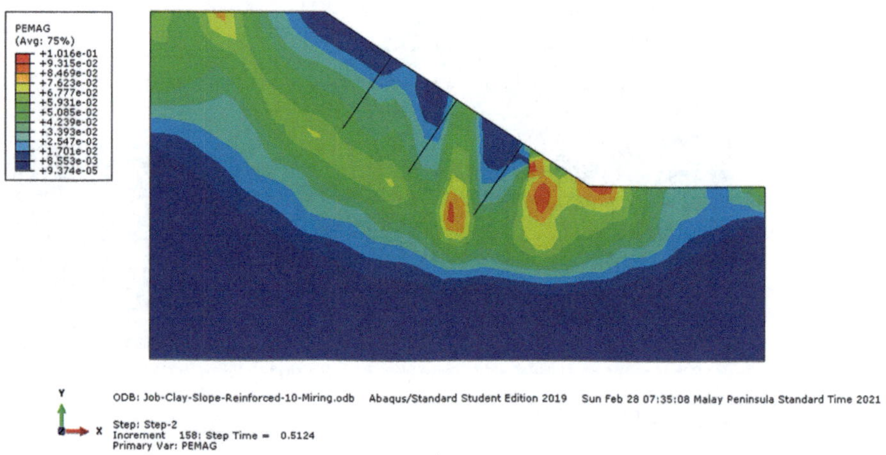

Fig. 6 The equivalent plastic strain (PEMAG) slope model reinforced with inclined bamboo piles diameter 10 cm

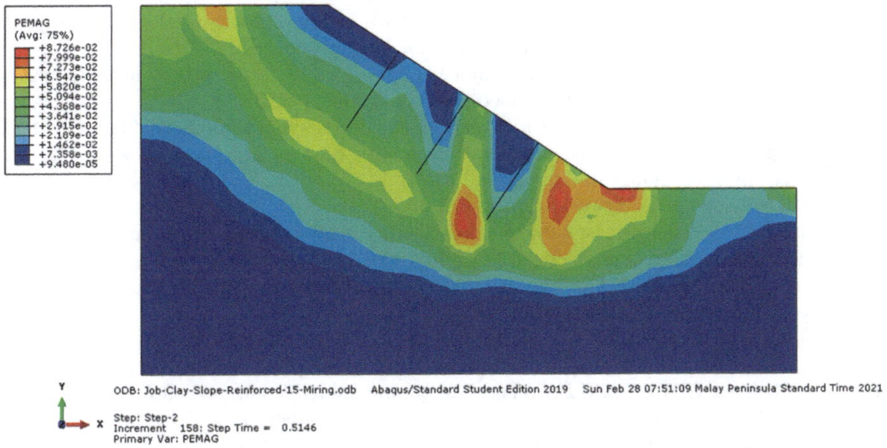

Fig. 7 The equivalent plastic strain (PEMAG) slope model reinforced with inclined bamboo piles diameter 15 cm

5 Conclusion

According to the 2D FEM simulation of soil embankment slope reinforced with inclined bamboo piles with the shear strength reduction (SSR) using ABAQUS SE software, the factor of safety (FOS), and the failure mechanism of the slope were obtained as follow:

1. The factor of safety (FOS) of the slope depends on the diameter of bamboo piles. The clay embankment slope reinforced with inclined bamboo piles with diameter 6, 10, 15 cm increase 1.503, 1.512, 1.515, respectively.
2. When the bamboo piles were inserted in the slope, it also changes the stress–strain and failure mechanism. The equivalent plastic strain (PEMAG) contour indicates the location of the potential failure of the slope.
3. The shear strength reduction (SSR) finite element method (FEM) can be used as an alternative for slope stability analysis of soil embankment slope model reinforced with inclined bamboo piles.

References

1. Morisco (1999) Rekayasa Bambu, Nafiri Offsett, Yogyakarta
2. Irawati IS, Saputra A (2012) Analisis Statistik Sifat Mekanik Bambu Petung. Proceeding Simposium Sinar Bambu I, Yogyakarta
3. Kumar P, Chakrabortty P (2018) A simple analytical model for bamboo reinforced slope using ordinary method of slices. Ind Geotech Conf
4. Davila AMJ (2019) Numerical Analysis of Bamboo Piles for Slope Stability, Geotechnical Engineering in the XXI Century. Pontificia Universidade Católica do Rio de PUC-Rio. Brazil

5. Munawir A, Dewi M, Zaika Y, Soehardjono A (2013) Bearing capacity on slope modeling with composite bamboo pile reinforcement. Int J Eng Adv Technol (IJEAT) 2(5)
6. Munawir A, Dewi M, Zaika Y, Soehardjono A (2013) Safety factor of continuous footing on slope modeling with composite bamboo pile reinforcement. EJGE 18
7. Wahhab MA, Ramadhan MR (2017) Analysis of slope stability using bamboo as an alternative slope reinforcement (Case Study In Kaliwungu—Boja Street, Darupono Village, Kendal District). In: Proceedings of international conference: problem, solution and development of coastal and delta areas. Semarang, Indonesia
8. Ho IH (2014) Parametric studies of slope stability analyses using three-dimensional finite element technique: geometric effect. J GeoEng 1:33–43
9. Dawson EM, Roth WH, Drescher A (1999) Slope stability analysis by strength reduction. Geotechnique 49(6):835–840
10. Griffiths DV, Lane PA (1999) Slope stability analysis by finite elements. Geotechnique 49(3):387–403
11. Yang X, Yang G, Yu T (2012) Comparison of strength reduction method for slope stability analysis based on ABAQUS FEM and FLAC. Appl Mech Mater 170–173:918–922
12. Dai w, Jiang P, Ding J, Fu B (2017) The influence of strength reduction method on slope stability under different instability criteria. Int Conf Arch Eng New Mater
13. Peng X, Li J, Li Y, Jiang Y (2020) Slope instability judgment criteria in FEM based on strength reduction method. IOP Conf Ser: Earth Environ Sci 571
14. Cai F, Ugai K (2000) Numerical analysis of the stability of a slope reinforced with piles. Soil Found 40(1):73–84

Analysis of Community Satisfaction Level on the Road Rehabilitation and Reconstruction Project (Learn from Palu Disasters Area)

Andri Irfan Rifai, Eko Prasetyo, and Rhismono

Abstract The rehabilitation and reconstruction project stages after the Palu disaster were almost completed. The massive earthquake disaster followed by the tsunami and the liquefaction required a relatively complex pre-during-past rehabilitation and reconstruction series. These stages will certainly be considered successful if the community affected by the disaster can feel performance satisfied. This research was conducted in the project area of rehabilitation and reconstruction of roads and bridges in Palu, Sigi, and Donggala. The research method was carried out through a questioner distributed to all communities around the project site. The analysis is carried out using a community profile and importance-performance analysis. The results showed that the community got satisfaction above the average (almost 4.00). The community is satisfied with the road's condition and the bridge's current state compared to the past. Meanwhile, essential indicators that must be improved are information and socialization and street lighting.

Keywords Community Satisfaction · Liquefaction · Rehabilitation · Reconstruction

1 Introduction

The earthquake that occurred in Palu—Indonesia, on September 28, 2018, caused severe damage. Until now, this incident still leaves quite a trauma. The earthquake that occurred has quite complete phenomena, including the movement of faults, tsunamis, landslides, and the devastating liquefaction period. Several incidents that took place simultaneously have paralyzed various community activities [1]. In particular, this liquefaction phenomenon has received attention from the world. The mudflow event

A. I. Rifai (✉)
Universitas Internasional Batam, Batam, Indonesia
e-mail: andri.irfan@uib.ac.id

E. Prasetyo · Rhismono
Directorate General Highway, Batam, Indonesia

S. A. Kristiawan et al. (eds.), *Proceedings of the 5th International Conference on Rehabilitation and Maintenance in Civil Engineering*, Lecture Notes in Civil Engineering 225,
https://doi.org/10.1007/978-981-16-9348-9_27

during the liquefaction had devastated infrastructure and housing on a massive scale and simultaneously during the incident.

Palu City and its surroundings based on topographic, geological, and seismological conditions can suffer damage due to earthquakes, including secondary disasters (tsunami, liquefaction, and cliff landslides). A similar earthquake with a magnitude of 7.6 on the Richter scale occurred in Palu in 1938 with many victims [2]. Several records of earthquake events in the last 100 years can be used as a lesson for all stakeholders to be better prepared for earthquakes.

Apart from being a natural occurrence that is destructive and has a devastating impact on human life and property loss, Earthquakes is also a natural laboratory on a full scale. This can provide many benefits; among them is to study earthquakes' characteristics to mitigate disasters in the future. Civil engineering experts have a fundamental relationship to earthquake disasters. Most of the Indonesia's territory is in earthquake-prone areas. Of course, the Palu earthquake that occurred should be used as a valuable lesson for practitioners, especially civil engineering, because it provides a complete picture of natural disasters. Apart from having to evaluate preparing a mitigation plan, we must immediately evaluate the rehabilitation and reconstruction that has been carried out within the past three years. Valuable for the continuation of humans to build an economy and a proper life order [3].

Evaluation of the rehabilitation and reconstruction process is needed to ensure that development direction is on the right track. One of the measuring tools that can be used is the community satisfaction level [4]. This can be used as a measure of accuracy in carrying out trauma healing. The community's existence and involvement can boost the success of the preparation, implementation, and implementation processes in implementing rehabilitation and reconstruction.

One method that is widely used to measure community satisfaction is the Importance Performance Analysis (IPA). The scientific method has proven to be a widely applicable tool that is relatively easy to manage and interpret. Thus, it is widely used among researchers and managers in various industries [5]. However, despite its widespread, this IPA model still has to be correctly ascertained about its method and implementation. The purpose of this study is to determine community satisfaction with rehabilitation and reconstruction using the IPA method.

2 Literature Review

Palu is an active earthquake area where according to the USGS Seismicity Map, it has an epicenter depth of less than 150 km [6]. The earthquake in the Palu region is included in the transform zone type, which is an earthquake caused by sliding two tectonic plates parallel to each other, but in opposite directions. The two of them neither gave each other nor supported each other. The earthquakes in this zone are generally shallow crustal earthquakes caused by the Palu-Koro Fault and the Matano Fault [7].

2.1 Disasters Vulnerability

Palu City is the capital city of Central Sulawesi Province below the equator with an altitude of 0–700 m above sea level. Palu's vulnerability to disasters is primarily the result of the Palu-Koro fault. This fault is the main fault on the island of Sulawesi and is classified as an active fault extending from central Sulawesi to the Karimata Strait [8]. The fault starts from the boundary of Sulawesi waters with the Makassar Strait to the bay of Bone. This fault crosses the city of Palu and reaches the Koro River area. Morphologically, the Palu-Koro Fault is a left-lateral slip [9].

The Palu-Koro fault is very active with a shift in the range of 33 mm/year. Because of its large dimensions, it is also called the Palu-Koro fault zone or system. This land fault with a length of 250 km is the second-longest fault in Indonesia, after the large-Sumatra fault [10]. Initially, this fault was called the Fossa Sarassina fault, and then it was called the Palu-koro fault, Lariang in the Koro River segment (Koro Valley). These geological and geophysical scientists familiar with the Palu-Koro fault agree that the fault is active [11].

2.2 Rehabilitation and Reconstruction

The scope of implementation of post-disaster rehabilitation and reconstruction is carried out through several activities. Among them are improving the environment in the disaster area, repairing public infrastructure and facilities, helping for community housing repairs, social-psychological recovery, health services, reconciliation and conflict resolution, social, economic, and cultural recovery, restoring security and order, restoring government functions, and restoration of public service functions. The main objective of rehabilitation and reconstruction is to restore disturbed community activities and public services [12].

In post-disaster management, especially in the handling of reconstruction, a proper reconstruction process is needed. Based on sound planning, so that it is right on target and is also orderly in the use of funds and can increase community resilience to the threat of future disasters [13]. A good post-disaster reconstruction process must recover the community's condition, both physically, mentally, socially, and economically, and reduce vulnerability to disasters, not exacerbate existing vulnerability conditions that lead to disasters.

For the reconstruction and rehabilitation process to run well, a particular standard is needed to manage disaster management. The government (central and regional) and non-governmental organizations and members of the public can carry out the rehabilitation and reconstruction process with planned, on time, on quality and budget, and under the target [14]. Every rehabilitation and reconstruction program must prepare a detailed technical plan, covering aspects of volume, network system, stages of work, the cost, the technical requirements for the implementation, and the actors who can do it.

Each rehabilitation and reconstruction program must meet specific achievement indicators, mainly so that each component of public infrastructure and facilities can adequately support the resumption of the communities' social and economic life in the disaster area. One of the critical indicators that must be considered is the disaster's social-psychological recovery to return to normal functions [15]. This can be achieved by activating community elements so that they can return to normal social functions.

2.3 Community Satisfaction

The main target of rehabilitation and reconstruction is the repair and restoration of all aspects of public services by normalizing all aspects of governance and community life in post-disaster areas [16]. The ability to provide essential services to the community cannot be adequately achieved without proper planning. The shift of authority from the central government to regional government is often not accompanied by the transfer of responsibility for services and protection to the public. As a result, regional government responses tend to be slow and often centrally dependent when disasters occur. This situation becomes even more complicated if the disaster affects more than one area.

As part of overall disaster management, the rehabilitation phase's implementation must be linked to other stages. In this sense, the rehabilitation phase activities relate to the pre-disaster and emergency stages and the reconstruction phase. The relationship and coordination between these stages dramatically determine the effectiveness and efficiency of disaster management. Therefore, disaster management staging should not be positioned as a goal but to achieve efficiency and effectiveness of disaster management as a whole [17]. Based on this definition, synchronization and coordination should be the keywords for disaster management that various parties must implement. The planning, preparation, implementation of post-rehabilitation and reconstruction processes are a series of interrelated activities.

In the implementation of rehabilitation and reconstruction, it is necessary to mobilize resources, including human resources, equipment, materials, and funds, by considering the available local resources. Human resources who understand and have professional skills are indispensable in all post-disaster rehabilitation processes and activities. Resources in the form of equipment, materials, and funds are provided and ready to be allocated to support the rehabilitation process [18]. Recovery activities involving the community are carried out by helping the community to revive and reactivate social, economic, and cultural activities. By actively involving the community, it is hoped that the rehabilitation and reconstruction activities carried out will get a high level of satisfaction.

A community satisfaction survey is needed in connection with some of the above, especially the community satisfaction level. This is a comprehensive measurement of the level of community satisfaction with the quality of rehabilitation and reconstruction services provided by public service providers [19]. The survey needs to

Table 1 Community satisfaction level

Code	Level	Weight
VD	Very dissatisfied	1
NS	Not satisfied	2
QS	Quite satisfied	3
S	Satisfied	4
VS	Very satisfied	5

be carried out to know each indicator's weaknesses or deficiencies in public service providers and determine the performance of the rehabilitation and reconstruction implementation being carried out.

3 Research Method

The research method used in this research is descriptive research. This research is used to answer questions about what or how an event or phenomenon. The intensive field observation of the phenomenon under study is required. This research was conducted on the community and stakeholders in the post-disaster area of Palu. Respondents in the community satisfaction survey were 125 respondents, but those who returned the questionnaire and filled it in completely were 110 respondents.

3.1 Survey Instrument

For demonstration purposes, community and stakeholder data around the rehabilitation and reconstruction sites were collected using a structured questionnaire. The questionnaire consists of two parts. In the first part, respondents are asked to provide information about the respondent's characteristics and information. The second part contains questions about the level of importance and respondent satisfaction with each attribute of the rehabilitation and reconstruction activity. The main attributes of implementing rehabilitation and reconstruction are taken from existing literature and measured by existing and tested scales adapted from several previous studies.

The assessment for the satisfaction level on the guideline using the Likert Scale method can be seen in Table 1.

3.2 Data Analysis

The application IPA begins with the identification of attributes that are relevant to the observed choice situation. Attribute lists can be developed using mean, median,

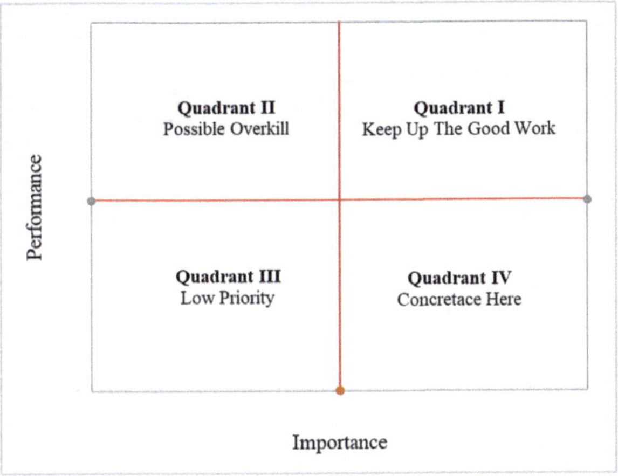

Fig. 1 Importance performance analysis

or ranking measures. The aggregated importance scores and performance attributes are classified into high or low categories; then, each attribute is assigned to one of the four predefined performance interest quadrants by pairing the two rank sets. IPA has been widely applied to evaluate importance and performance value in the market, identify opportunities for improvement, and guide strategic planning efforts. In IPA, service attributes are plotted in a two-dimensional matrix based on each attribute's importance and performance. The average or median importance and performance of all attributes divide the matrix into four quadrants [20]. The priority for improvement is then summarized based on the matrix points' location, which can be seen in Fig. 1.

Quadrant 1 (high importance and high performance) has the management scheme for this quadrant is continuing to work well. Quadrant 2 (low importance and high performance) have a management scheme for this quadrant that is likely to overdo. Quadrant 3 (low importance and low performance): having a management scheme for this quadrant is a low priority. Whereas Quadrant 4 (high importance and low performance) has a management scheme for this quadrant is concentrated here.

4 Data Analysis and Result

The policy for implementing rehabilitation and reconstruction in Palu is the central government's responsibility or local governments affected by the disaster. In carrying out rehabilitation and reconstruction activities in the strategic and national assets, it is directly carried out by the central government. One of them is the rehabilitation and reconstruction of national roads and other strategic roads carried out by the Directorate General of Highway-Indonesia, which loan from the World Bank. The series of stages of the rehabilitation and reconstruction process can be seen in Fig. 2.

Fig. 2 Pre-during-post rehabilitation and reconstruction stage

4.1 Damage Description

The damage that occurred in Palu after the disaster varied greatly but could be classified into several types of damage. Damage due to fault motion, fractures, and earthquake shocks caused by fault motion is an offset where the left side moves north, and the right side shifts south. The length of the most significant shear on the right side is about 4 m, while the left side shifts to the north along 3 m. This shift is visible on the map visible on googles map. Of course, buildings that are traversed by faults will suffer significant damage and soil fractures where fractures can be the impact of the movement of faults (or reactivated faults) with smaller offsets. Earthquake shocks are in the form of vibrations both horizontally and vertically. In Palu City, in general, the impact of damage due to shocks was not too much, except for buildings of low quality.

Furthermore, the damage caused by the tsunami is due to inundation (submerged buildings) and tsunami currents (the speed or force that acts to push or pull the building). The impact of current velocity is mainly the scouring of the subgrade. If it is loose sand, the erosion rate is very high. Generally, buildings with shallow foundations fail because the scour reaches the foundation's base, and the buildings are relatively light, so they are easily carried away by the flow of water. Another damage is due to the tsunami and at the same time carrying debris to cars and ships, so collisions with these objects often result in heavy damage.

Also, there is damage due to massive liquidation with high functionality. There are 4–5 locations that are prominent and wide, namely in Balaroa, Petobo, Jono Oge, Lolu village (also in Jono Oge), and Sibalaya. Although some spots also occur in the sand boil, they are not prominent and are not recorded.

Avalanche damage on several hills. Many locations on the hill suffer from local landslides, and a notable one is in the village of Poi near South Sibalaya. This landslide caused damage and threatened the people below it because the debris in that location was loose and ready to slide due to the rainfall that occurred (it could cause flash floods).

Damage due to liquefaction-induced landslides and landslides on the seabed. Avalanches of this kind are induced by liquefaction. The landslides in Balaroa and Sibalaya were a phenomenon of liquefaction-induced landslides. It is possible that the submarine landslides that occurred in Palu Bay which caused the tsunami impact had the exact mechanism as in Sibalaya.

Table 2 Sample profile

Measure	Items	Frequency	Percentage (%)
Gender	Male	82	74.55
	Female	28	25.45
Age	<21	2	1.82
	21–25	18	16.36
	26–30	15	13.64
	31–35	16	14.55
	36–40	44	40.00
	>41	15	13.64
Education	Diploma degree or lower	23	20.91
	Bachelor's degree	74	67.27
	Master's degree	13	11.82
Profession	Government employees	32	29.09
	Private employees	23	20.91
	Entrepreneur	42	38.18
	Unemployment	3	2.73
	Student	10	9.09

4.2 Sample Profile

Based on the results of the questionnaire, the respondent's profile can be seen in Table 2. Among the respondents, 74.55% were men, and 25.45% were women. Meanwhile, based on age, the respondents were dominated by the 36–40-year-old group, with most of them being undergraduate. Furthermore, based on the type of work, entrepreneurs occupy the most positions.

The questionnaire data obtained from 110 respondents have been tested for instrument reliability using the Cronbach's alpha value. The value of the reliability test results can be used if it exceeds the alpha value of 0.600. Based on the reliability test results, the value of Cronbach's alpha for reliability importance statistics is 0.944 and for reliability performance statistics is 0.909.

4.3 Importance Performance Analysis

IPA is carried out to determine the importance of each service quality attribute based on user perceptions of the importance of service quality. The analysis to determine the level of importance of service quality is carried out in 3 stages: pre-stage, during-stage, and post-stage.

Based on the questionnaire results regarding the importance level and performance attributes of the rehabilitation and reconstruction implementation, descriptive statistics are obtained on the data from all respondents in the study area, which can be seen in Table 3. The I is importance, P is performance, and G is the gap between importance and performance.

Based on the results of a survey of respondents in the rehabilitation and reconstruction after the Palu Disasters area, the average value of the rehabilitation and reconstruction series's satisfaction level was obtained. From these values, it is known that the indicator which has the highest level of importance according to the respondents is the quality road and bridge (4.34). Meanwhile, the indicator with the highest level of satisfaction is the current state of the road, and the bridge is compared to the past (3.99).

In the pre-reconstruction and rehabilitation stages, all performance indicators show an average number of 3.81. This shows that the community is quite satisfied with the pre reconstruction and rehabilitation performance. Likewise, the average performance during reconstruction and rehabilitation with an average performance of 3.81. Meanwhile, post-reconstruction and rehabilitation showed a higher number, namely 3.88. This shows that the community feels the overall performance satisfaction stages are above average.

The gap value is the ratio between importance value and performance value. When viewed from the value gap between importance and performance values, all attributes have a minus value, which means that no indicator can meet the community's expectations. The highest gap value is the current availability of street/environment lighting (−0.56). The performance indicators are considered to be still far below the value expected by the community. The indicator of the access road to residence current state is compared to the past, including attributes with a high enough importance and its performance is also relatively high. So, it can be concluded that respondents saw changes in access road to residence conditions better than before.

The IPA matrix of performance rehabilitation and reconstruction indicators is formed with an average value of importance as the X-axis and the average value of the performance level as the Y-axis, according to Table 3. Based on the matrix, indicators whose importance values are above the Y-axis are included in the high importance category (included in quadrants I and II). In Fig. 3, we can see the IPA matrix for each stage.

Based on the IPA matrix in Fig. 3, it can be seen that in the pre-reconstruction and rehabilitation stages. There are three indicators in quadrant I (keep up the good works) 1 indicator in quadrant II, three indicators in quadrant III, and one indicator in quadrant IV. From the questioners' results, the indicator in quadrant IV (concrete here) should get more attention, namely information and socialization about reconstruction and rehabilitation. While at the stage during rehabilitation and reconstruction, there are no indicators that require special attention. Furthermore, there is no indicator in quadrant IV at the post-reconstruction and rehabilitation stage, but the indicator of the current street/environment lighting availability is quite interesting. It can be seen that the community expects more on the availability of street lighting.

Table 3 The average level of community satisfaction

ID	Satisfaction indicator	I	P	G
A. Before the rehabilitation and reconstruction				
A1	Information and socialization	4.11	3.71	−0.40
A2	The time the reconstruction program began	4.06	3.85	−0.21
A3	Road and bridge damage identification process	4.22	3.85	−0.37
A4	Participation in the reconstruction and rehabilitation process	4.07	3.75	−0.32
A5	Collaboration between local communities in reconstruction and rehabilitation	4.07	3.78	−0.29
A6	The wishes of the people are fulfilled	4.10	3.90	−0.20
A7	Easy administration/disbursement process	3.84	3.62	−0.22
A8	The role of government in the reconstruction process	4.30	3.99	−0.31
Mean A		4.10	3.81	−0.29
B. During the rehabilitation and reconstruction				
B1	The Role of the Facilitator in the reconstruction and rehabilitation process	4.05	3.78	−0.27
B2	Labor availability	4.08	3.77	−0.31
B3	Work experience and skills	4.23	3.86	−0.37
B4	Availability of material for reconstruction and rehabilitation	4.28	3.85	−0.43
B5	Quality material available for reconstruction and rehabilitation	4.25	3.88	−0.37
B6	Quality of road and bridge	4.34	3.92	−0.42
B7	Community participation in the reconstruction and rehabilitation	3.93	3.62	−0.31
Mean B		4.17	3.81	−0.35
C. After the rehabilitation and reconstruction				
C1	With the results of existing assistance	3.98	3.69	−0.29
C2	The current state of the road and bridge is compared to the past	4.17	3.99	−0.18
C3	The road and bridge become more earthquake-resistant	4.25	3.94	−0.31
C4	The comfort of road and bridge compared to before	4.15	3.94	−0.21
C5	The quality of the road and bridge now compared to before	4.17	3.97	−0.20
C6	The road and bridge was had been as a community wish	4.22	3.96	−0.26

(continued)

Table 3 (continued)

ID	Satisfaction indicator	I	P	G
C7	Satisfaction with the current design	4.15	3.95	−0.20
C8	The access road to residence compared to before the reconstruction and rehabilitation	4.11	3.95	−0.16
C9	Current availability of street/environment lighting	4.11	3.55	−0.56
Mean C		4.15	3.88	−0.26
Grand mean		4.14	3.84	−0.30

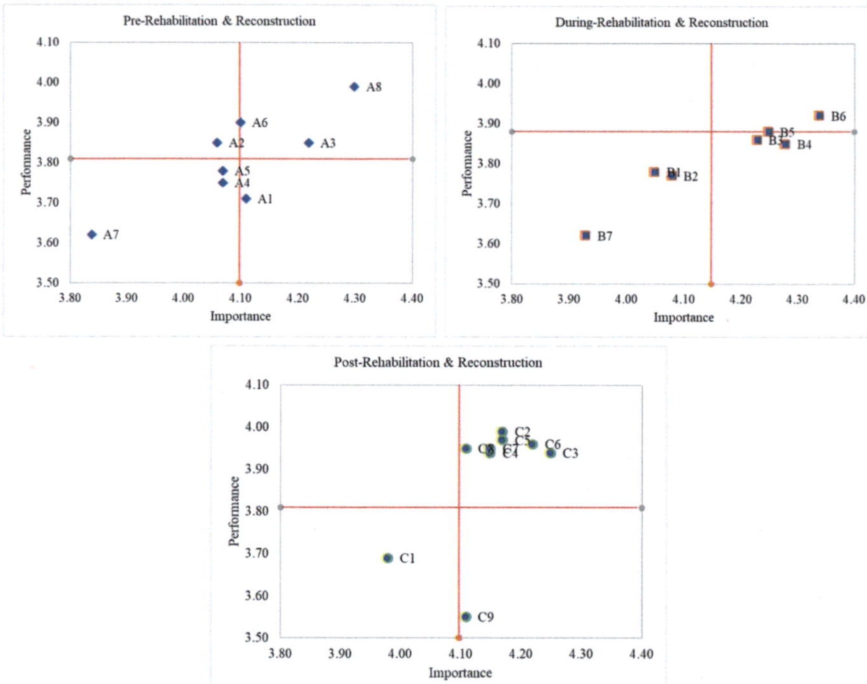

Fig. 3 IPA matrixes for each stage

5 Conclusion

From the data analysis that has been carried out, it can be concluded that the community gets satisfaction above the average for all stages. Consecutively starting from the pre-during-post reconstruction and rehabilitation, the average performance satisfaction level obtained was 3.81; 3.81; and 3.88, with an overall average of 3.84. The satisfaction performance almost met the satisfied category (almost 4.00). An interesting indicator is that the community is satisfied with the access road to residence current state compared to the past. This is indicated by the minimal gap between importance and performance. This study also found several important things that must be improved: information and socialization and street/environment lighting. This is consistent with several references that state that socialization is an important stage in rehabilitation and reconstruction.

Acknowledgements The authors are grateful to the editor and reviewers for constructive comments on the earlier version of the paper, the Rehabilitation and Reconstruction Project, which supported this research. We would also liketo thank those in Universitas Internasional Batam, Indonesia, who assisted in this study.

References

1. Yulianto E, Utari P, Satyawan IA (2020) Communication technology support in disaster-prone areas: Case study of earthquake, tsunami and liquefaction in Palu, Indonesia. Int. J Disaster Risk Redu 45:101457
2. Rajindra R, Suardi I, Sabara Z, Pushpalal D, Samad MA, Yani A, Umam R (2019) Diversity, resilience, and tragedy: three disasters in Palu of Indonesia. Int J Innovat Creat Change 5(2):1592–1607
3. Liu H, Zhang D, Wei Q, Guo Z (2017) Comparison study on two post-earthquake rehabilitation and reconstruction modes in China. Int J Disaster Risk Reduct 23:109–118
4. Wijegunarathna EE, Wedawatta G, Prasanna LJ, Ingirige B (2018) Long-term satisfaction of resettled communities: an assessment of physical performance of post-disaster housing. Proced Eng 212:1147–1154
5. Hosseini SY, Bideh AZ (2014) A data mining approach for segmentation-based importance-performance analysis (SOM–BPNN–IPA): a new framework for developing customer retention strategies. Serv Bus 8(2):295–312
6. Rusydi HE (2017) Vulnerability zoning of earthquake disaster of Palu. Int J Sci Appl Sci Conf Ser 1(2), 137–143
7. Jaya A, Nishikawa O, Jumadil S (2019) Distribution and morphology of the surface ruptures of the 2018 Donggala-Palu earthquake, Central Sulawesi, Indonesia. Earth, Planets Space 71(1):1–13
8. Rifai AI, Hendra H, Prasetyo E (2020) Data mining applied for liquefaction mapping and prediction learn from Palu Earthquakes. Civil Eng Arch 507–514
9. Patria A, Putra PS (2020) Development of the Palu-Koro Fault in NW Palu Valley Indonesia. Geosci Lett 7(1):1–11
10. Natawidjaja DH (2018) Updating active fault maps and sliprates along the Sumatran Fault Zone, Indonesia. IOP Conf Ser: Earth Environ Sci 118(1):012001
11. Massinai MF, Ariansyah MR, Trisartika R, Massinai MA (2020) Relative present-day motion on Palu-Koro Fault. Earth Environ Sci 575(1):012185
12. Moreno J, Shaw D (2019) Community resilience to power outages after disaster: a case study of the 2010 Chile earthquake and tsunami. Int J Disaster Risk Reduct 34:448–458
13. Lee DW (2020) The expertise of public officials and collaborative disaster management. Int J Disaster Risk Reduction 50:101711
14. Tagliacozzo S (2018) Government agency communication during post-disaster reconstruction: Insights from the Christchurch earthquakes recovery. Nat Hazard Rev 19(2):04018001
15. Asadzadeh A, Kötter T, Salehi P, Birkmann J (2017) Operationalizing a concept: The systematic review of composite indicator building for measuring community disaster resilience. Int J Disaster Risk Reduction 25:147–162

16. Nemeth DG, Kuriansky J, Onishi Y (2021) The 3 Es of psychosocial recovery after Disaster. The Intersection of Trauma and Disaster Behavioral Health Springer, Cham. pp 297–316
17. Yong Z, Zhuang L, Liu Y, Deng X, Xu D (2020) Differences in the disaster-preparedness behaviors of the general public and professionals: evidence from Sichuan Province, China. Int J Environ Res Public Health 17(14):5254
18. Santha SD (2018) Social interfaces in disaster situations: analyzing rehabilitation and recovery processes among the fisherfolk of Tamil Nadu after the Tsunami in India. In The Asian Tsunami and post-disaster aid. Springer, Singapore, pp 65–78
19. Sofyan M (2019) Community satisfaction of the Urban flood control system improvement project (UFCSI). Ilomata Int J Soc Sci 1(1):29–34
20. Chang RD, Zuo J, Soebarto V, Zhao ZY, Zillante G, Gan XL (2017) Discovering the transition pathways toward sustainability for construction enterprises: Importance-performance analysis. Journal of construction engineering and management, 04017013

Seepage and Piping Control of Earth Fill Dam

(Case Study of Pidekso Dam Indonesia)

Muhammad Zainal Arifin, Yusep Muslih Purwana, and Raden Harya Dananjaya

Abstract The Pidekso Dam is one of the major dams in the Central Java, situated in Wonogiri Indonesia. The stability analysis of the dam is required due to this dam is planned for national water and food security program covering the irrigation area of $15 \times 10^6\ m^2$, increasing the planting intensity from $20 \times 10^6\ m^2$ to $36 \times 10^6\ m^2$ and as a source of water around 300 L/s for Wonogiri, Sukoharjo, Solo and their surroundings. The control is needed for the safety of the dam from seepage and piping. Excessive seepage can cause piping and lead to dam failure. Seepage control is carried out by 3 alternatives, extended curtain grouting, cut off wall and upstream blanket respectively. Each alternative is divided into 4 models: (1) original length (L), (2) 2/3L, (3) 1/3L and (4) 2/3L with doubles in the upstream and downstream of the dam. The result indicates that, the installation curtain grouting with model 1 is the most effective alternative to reduce seepage up to 67.82% of dams.

Keywords Cut off wall · Curtain grouting · Seepage control · Upstream blanket

1 Introduction

The Pidekso Dam is one of the large dam in the Central Java area, located in Pidekso Village, Giriwoyo District, Wonogiri Regency. The map of the study location can be seen in Fig. 1. The stability analysis of the dam is required due to this dam is planned for national water and food security program covering the irrigation area of $15 \times 10^6\ m^2$, increasing the planting intensity from 133% ($20 \times 10^6\ m^2$) to 240% ($36 \times 10^6\ m^2$) and as a source of water 300 L/second for Wonogiri, Sukoharjo, Solo and their surroundings [1].

M. Z. Arifin · Y. M. Purwana (✉) · R. H. Dananjaya (✉)
Department of Civil Engineering, Universitas Sebelas Maret, Surakarta, Indonesia
e-mail: ymuslih@ft.uns.ac.id

M. Z. Arifin
e-mail: arifinmuzarif@student.uns.ac.id

S. A. Kristiawan et al. (eds.), *Proceedings of the 5th International Conference on Rehabilitation and Maintenance in Civil Engineering*, Lecture Notes in Civil Engineering 225,
https://doi.org/10.1007/978-981-16-9348-9_28

Fig. 1 Study location map

Dam stability studies need to be carried out in the construction and planning of seepage and piping. Excessive seepage can cause piping and lead to dam failure. For example, the dam failure that occurred in the Lempake-Indonesia dam and the Brumadinho-Brasil dam was due to the dam's condition which had seepage due to piping resulting in flooding which caused material and non-material loss and casualties [2, 3].

Embankment dams are constructed across rivers and valleys to retain water for flood protection, water storage and energy generation purposes. Embankment dams consist of homogeneous dams (constructed using one type of soil) or zonal dams (consisting of more than one soil type with high hydraulic permeability different) [4]. Seepage is a flow that continuously flows from upstream to downstream. This water flow constitutes the flow from the reservoir water through the material which passes the water (permeable), either through the body of the dam or the foundation [5]. Piping is a kind of movement or erosion of fine particles in soil along interspace channel between skeleton grains, by the action of seepage [6]. Costa (1985) reported that approximately 34% of dam failures were caused by overtopping, 30% by foundation defects, and 28% by piping [7]. If seepage is considered as a potential problem, countermeasures—such as filters, drains, clay blankets and flatter side slopes–introduced at the design stage can reduce any risks to a minimum [8].

Seeing the condition of the study area and the many objectives of the Pidekso Dam development, it is necessary to carry out detailed technical planning. Among other things, regarding the geological condition of the dam foundation, foundation repair, seepage capacity, the possibility of piping, and the stability of the dam body in various conditions. So that through this research entitled "seepage and piping control of earth fill dam". So that a dam that is safe against seepage and the piping can be planned.

1.1 Geotechnical Site and Construction Material

The body of the Pidekso Dam is a embankment dam with an upright core in the centre, a dam length of 387 m stretching from the Southeast to the Southwest, and a maximum height of 32.00 m from the riverbed or 44.00 m from the bottom of the excavation. The elevation of the top of the dam is designed at El.189.00 m with a freeboard height from the normal water level (El.185.0 m) [9]. Image of the cross-section of the dam can be seen in Fig. 2.

The embankment material used for the Pidekso Dam consists of a watertight, random, filter, transition and rip-rap core which is divided into zones as seen in Table 1.

2 Methodology

Seepage analysis that indicates the occurrence of piping, is determined based on the safety factor for the piping as follows [10]:

$$SF = \frac{Icr}{Ical} \tag{1}$$

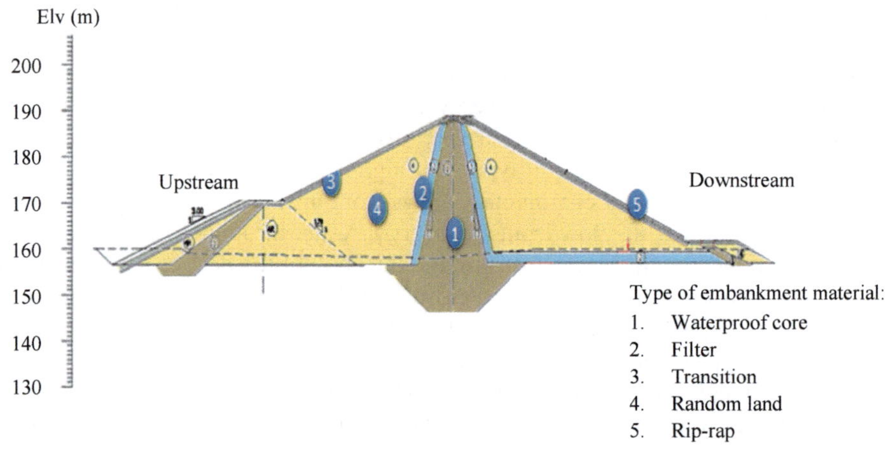

Fig. 2 Geometry section of the Pidekso Dam

Table 1 Dam material input data

Material	Earthfill	Randomfill	Rockfill	Filter	Transition	Foundation
Classification	ML	SC	GM	SM-SC	SM-SC	CH
Permeability (m/s)	10^{-7}	5.86×10^{-6}	–	10^{-5}	10^{-6}	5×10^{-6}

where:

Icr = Critical hydraulic gradient (dimensionless).

$Ical$ = Output gradient from seepage analysis results (dimensionless).

SF = Safety factor for symptoms piping > 4 (dimensionless).

$$Icr = \frac{Gs - 1}{1 + e} \quad (2)$$

where:

Gs = specific gravity.

e = porosity number.

The Laplace equation is the basis of mathematical equations for several models and methods used in seepage analysis and is used as a basis for modelling seepage flow in the case of steady flow conditions [11].

$$\frac{\partial^2 \mu}{\partial x^2} + \frac{\partial^2 \mu}{\partial \gamma^2} + \frac{\partial^2 \mu}{\partial z^2} = 0. \quad (3)$$

Fortunately, now that computers and numerical software tools are so readily available, it is no long necessary to rely on graphical solutions to the Laplacian equation. With a software tool like SEEP/W these types of situations can be readily considered in a seepage analysis [12]. Seepage analysis was carried out on 3 alternative installations of extended seepage controls as follows:

1. Alternative 1: installations of extended seepage control with cut off wall.
2. Alternative 2: installations of extended seepage control withurtain grouting.
3. Alternative 3: installations of extended seepage control with upstream blanket.

From each alternative is divided into 4 models: (1) original length (L), (2) 2/3L, (3) 1/3L and (4) 2/3L with doubles in the upstream and downstream of the dam. This analysis is calculated against normal water level conditions, a more detailed sketch can be seen Fig. 3.

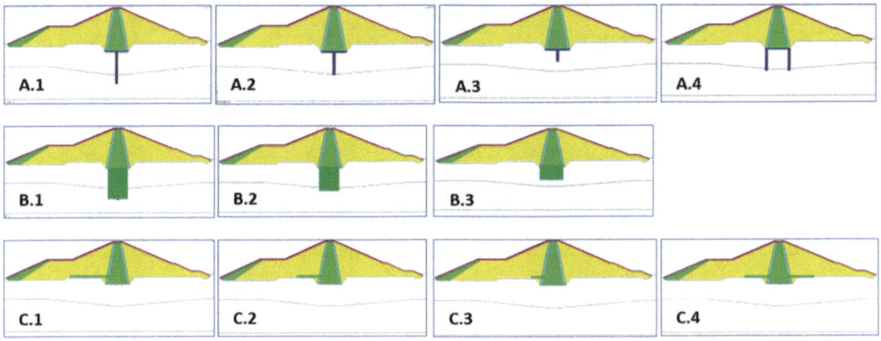

Fig. 3 **a** Extended cut off wall. **b** Extended curtain grouting. **c** Extended upstream blanket

3 Result and Discussion

The seepage discharge obtained from the modelling results with Seep/W was analyzed against the allowable seepage discharge. The permissible seepage discharge is less than 0.05% mean annual run-off [13]. Safety factor against piping analyzed based on Eqs. (3) and (4), while the hydraulic slope value of the flow and foundation material obtained from modelling and data from soil investigation and testing.

3.1 Without Extended Seepage Control

Seepage analysis in conditions without extended seepage control is carried out to determine the seepage discharge and piping. Seepage discharge analysis results and exit gradient by using the Seep/W program presented on Fig. 4. From Fig. 4, got exit gradient in the zone around the base of the core zone is 0.15. The critical hydraulic

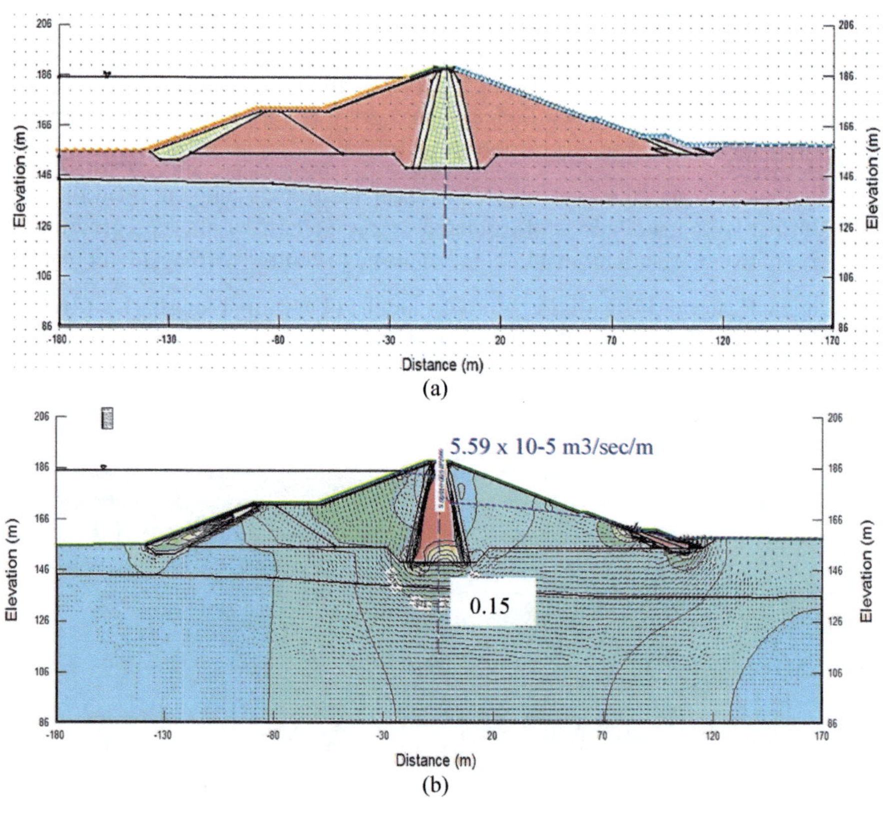

Fig. 4 **a** Dam sketches without extended seepage control. **b** Seepage and exit gradient analysis

magnitude of this embankment material is equal to 0.948, so the size of the Safety Factor is 6.32, which is greater than allowed. So it can be concluded that the dam safe against the dangers of piping.

The amount of discharge per time per length of the dam results from the analysis as amounting to $5.59 \times 10^{-5} m^3/s/m$. With the length of the dam section of 250 m, the amount of seepage discharge for the entire length of the dam is: $5.59 \times 10^{-5} \times 250 = 1.4 \times 10^{-2} m^3/s$ or 1206.60 m^3/day. This discharge is smaller from 0.05% of the effective volume of the dam per day, which is 8535 m^3/day, so it can be concluded that the dam without extended seepage control safe against volume seepage.

3.2 Curtain Grouting

Seepage analysis with extended seepage control of curtain grouting was carried out on the located at the base of the dam core. The determination of the curtain grouting depth model is based on empirical calculation according to Eq. 4 which is then tabulated in Table 2 [14].

$$D = 1/3h + c \tag{4}$$

Where:

D = depth of the curtain grouting.

H = height of the dam, EL.189 - EL.153 = 36 m.

C = coefficient that varies between 8 to 20, the value is 18.

The magnitude of the seepage discharge and the piping hazard is analyzed according to the appropriate model Table 2 using the Seep/W program which can be seen in Fig. 5 which produces the amount of seepage discharge and exit gradient. The seepage discharge value is then calculated based on the seepage permit discharge and the exit gradient value then calculated using Eq. (3). These results are more fully tabulated in Table 3.

Based on Table 3, Overall, the seepage discharge for the dam with extended seepage control of curtain grouting at the base of the foundation is still less than the allowable seepage permit discharge and safe against piping hazards. A complete comparison of the grouting curtain length and the seepage discharge can be seen in Fig. 5.

Table 2 Length model of curtain grouting extended seepage control

Model	Information	Length (m)
Model 1	Length (3/3) of empirical calculation of Eq. 4	30
Model 2	Length (2/3) of empirical calculation of Eq. 4	20
Model 3	Length (1/3) of empirical calculation of Eq. 4	10

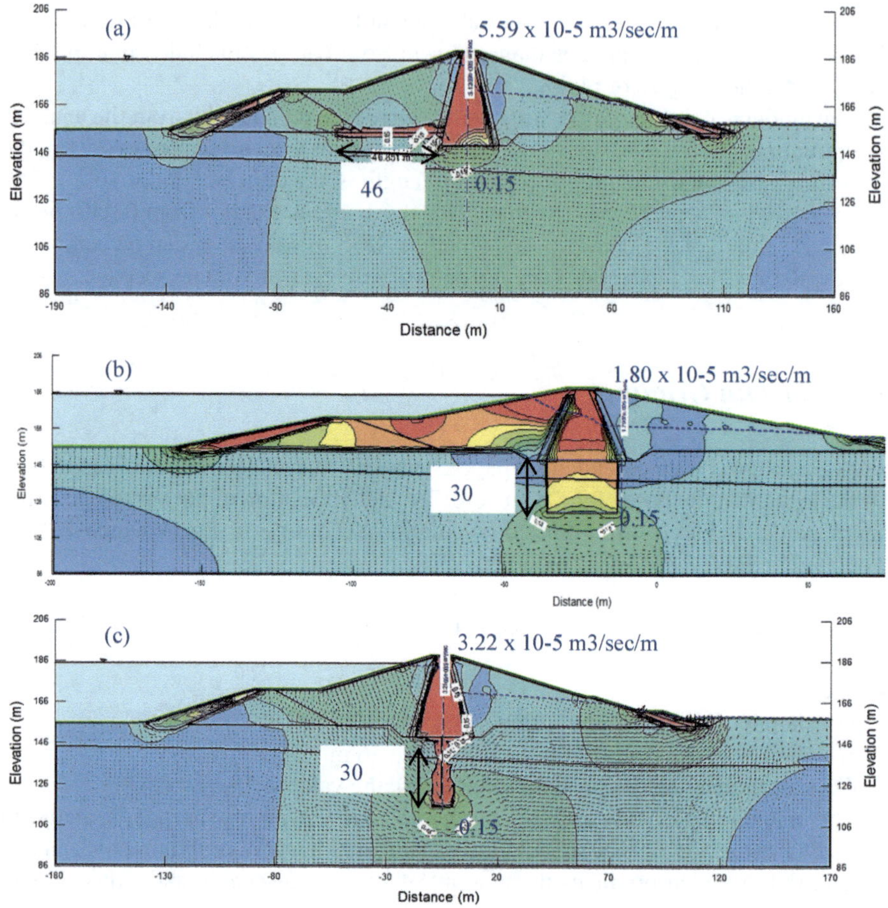

Fig. 5 Exit gradient analysis of **a** upstream blanket, $D = 46.75$ m. **b** Curtain grouting, $D = 30$ m. **c** Cut of wall, $D = 30$ m

Table 3 Result of curtain grouting extended seepage control

Model	Length (m)	Seepage (m^3/day)	*SF* Piping	Seepage reduction (%)
Model 1	30	388.3032	6.32	67.82
Model 2	20	472.4784	6.32	60.84
Model 3	10	647.5032	6.32	46.34

3.3 Cut off Wall

Seepage analysis with extended cut off wall seepage control was carried out on the cut off wall depth model which is located at the base of the dam core. The determination

Table 4 Result of cut of wall extended seepage control

Model	Length (m)	Seepage (m^3/day)	*SF* Piping	Seepage reduction (%)
Model 1	30	694.764	6.32	42.42
Model 2	20	985.3272	6.32	18.34
Model 3	10	1143.396	6.32	5.24
Model 4	20	801.2088	6.32	33.60

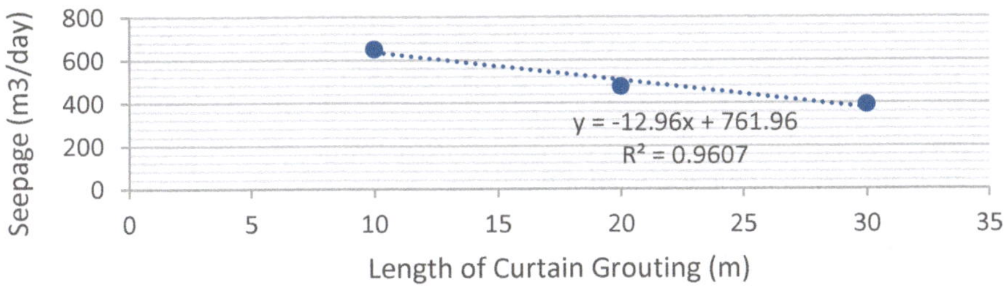

Fig. 6 Variation of curtain grouting verses seepage

of the cut off wall depth model is based on the empirical calculation of the cut off wall depth according to Eq. 4 which is then tabulated in Table 2.

The magnitude of the seepage discharge and the piping hazard is analyzed according to the appropriate model Table 4 using the Seep/W program which can be seen in Fig. 5 which produces the amount of seepage discharge and exit gradient. The seepage discharge value is then calculated based on the seepage permit discharge and the exit gradient value then calculated using Eq. (3). These results are more fully tabulated inward Table 4.

Based on Table 4, Overall, the seepage discharge for the dam with extended seepage control of cut off wall at the base of the foundation is still less than the allowable seepage discharge and safe against piping hazards. A complete comparison of the cut off wall length and the seepage discharge can be seen in Fig. 6.

3.4 Upstream Blanket

Seepage analysis with extended upstream blanket seepage control is carried out on the upstream blanket depth model which is located at the base of the dam core. Determination of the upstream blanket depth model based on empirical calculation of the upstream blanket depth according to Eq. 5 according to SNI 8065: 2016 which is then tabulated in Table 5 [15].

$$x = \frac{\sqrt{2}}{\sqrt{\frac{Kb}{Zb.Kf.Zf}}} \quad (5)$$

Where:

Zb = thickness of the waterproof base (m).
= 1 m.
Kb = K coefficient of waterproof base.
= 1×10^{-7} (m/s).
Kf = K coefficient of foundation (m/s).
= 1×10^{-5} (m/s).
Zf = depth of the water graduated layer on the foundation (m) = 9.93.

The magnitude of the seepage discharge and the piping hazard is analyzed according to the appropriate model Table 5 using the Seep/W program which can be seen in Fig. 5 which produces the amount of seepage discharge and exit gradient. The seepage discharge value is then calculated based on the seepage permit discharge and the exit gradient value then calculated using Eq. (3). These results are more fully tabulated in Table 6.

Table 5 Length model of upstream blanket extended seepage control

Model	Information	Length (m)
Model 1	Length (3/3) of empirical calculation of Eq. 5	46.75
Model 2	Length (2/3) of empirical calculation of Eq. 5	31.17
Model 3	Length (1/3) of empirical calculation of Eq. 5	15.58
Model 4	Length (2/3) of empirical calculation of Eq. 5 with double points upstream and downstream	31.17

Table 6 Result of upstream blanket extended seepage control

Model	Length (m)	Seepage (m^3/day)	*SF* piping	Seepage reduction (%)
Model 1	46.75	1109.5704	6.32	8.04
Model 2	31.17	1136.7216	6.32	5.79
Model 3	15.58	1170.2664	6.32	3.01
Model 4	31.17	1153.2888	6.32	4.42

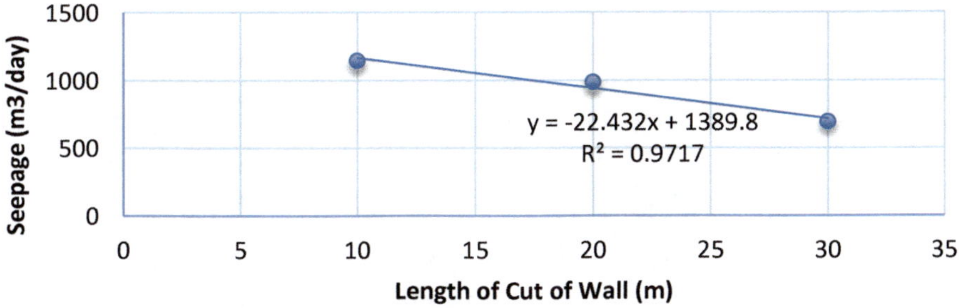

Fig. 7 Variation of cut off wall verses seepage

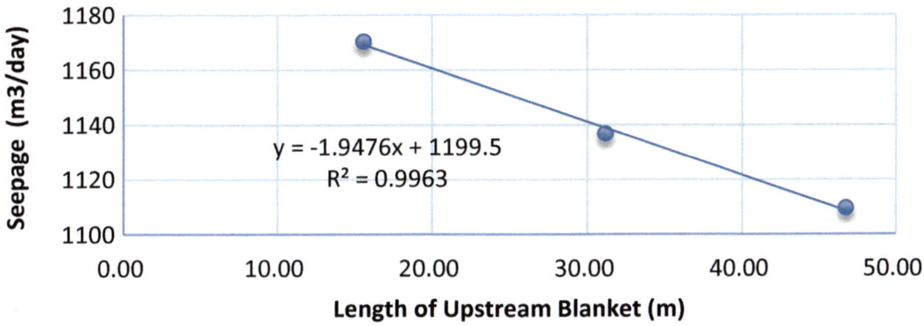

Fig. 8 Variation of upstream blanket verses seepage

Based on Table 6, Overall, the seepage discharge quantity for the dam with extended seepage control of upstream blanket at the base of the foundation is still less than the allowable seepage discharge quantity and safe against piping hazards. A complete comparison of the upstream blanket length and the seepage discharge can be seen in Fig. 7.

3.5 Comparison of Analysis Results

The following is a recapitulation of the results of various extended seepage control installations models Curtain Grouting, Cut Off Wall and Upstream Blanket with a variety of models Length which can be seen in Fig. 9.

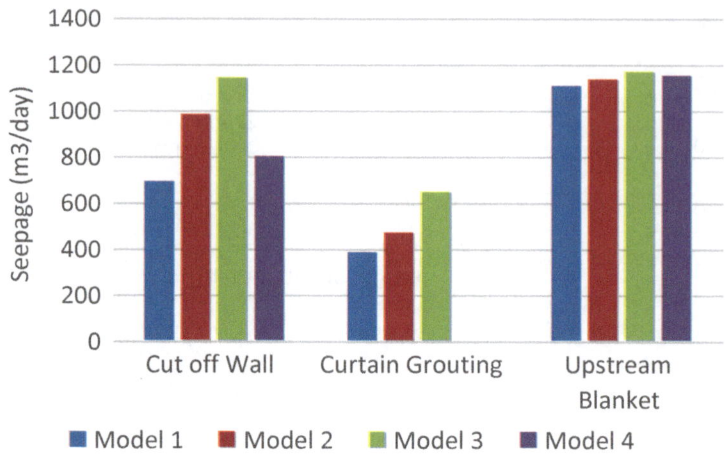

Model 1	original length (L)
Model 2	2/3L
Model 3	1/3L
Model 4	2/3L with doubles in the upstream and downstream of the dam

Fig. 9 Comparison of seepage discharge results from the installations of extended seepage control

4 Conclusion

Based on the results of the stability dam to the amount of seepage and safety against piping, the following conclusions can be made:

1. The stability of the Pidekso Dam without and with the installations of extended seepage controls to the safety of the seepage magnitude is safe, which is less than the allowable seepage discharge and safe against of the piping.
2. The longer the addition of curtain grouting, cut off walls and upstream blankets on the foundation of the dam, so the smaller the seepage discharge from the dam
3. Of the several alternatives extended seepage control, then alternative curtain grouting 30 m is the most effective alternative to reduce seepage up to 67.82% of dams.

References

1. Engineering Consultant (2011) Technical Specifications Report, PT. Mettana, Wonogiri
2. Nyoman N (2014) Hydraulic analysis of flood due to dam failure (Case Study of Lempake Dam, Samarinda City). Thesis, Gadjah Mada University. Yogyakarta.S

3. Raman (2019) An investigation of the Brumadinho Dam Break with HEC RAS simulation. Journal on arXiv:1911.05219v1, New Jersey. USA
4. Molla E (2019) Seepage through homogeneous earth dams provided with a vertical sheet pile and formed on impervious foundation. Ain Shams Eng J 10:529–539
5. Kanchanha (2015) Adequacy of seepage analysis in core section of the earthen dam with different mix proportions. In: International conference on water resources, coastal and ocean engineering (ICWRCOE 2015), Aquatic Procedia, vol 4. India, pp 868–875
6. Yongbiao (2012) Prediction methods to determine stability of dam if there is piping. In: International conference on mechanical, industrial, and manufacturing engineering, IERI Procedia, vol 1. China, pp 131–137
7. Sheng-shui (2019) Numerical modeling of earthen dam breach due to piping failure. Water Science and Engineering 12(3):169e178
8. Andreaa (2015) Unsaturated Slope Stability and Seepage Analysis of a Dam. Sustain Sol Energy Environ Energy Procedia 85:93–98
9. Planning Consultant (2011) Final Report of the FS Pidekso Dam Review in Wonogiri Regency. PT. Virama Karya, Wonogiri
10. Peng S (2020) Inverse analysis of laboratory data and observations for evaluation of backward erosion piping process. J Rock Mech Geotech Eng 12:1080e1092
11. Shivakumar (2015) Seepage and stability analyses of earth dam using finite element method. In: Proceedings of international conference on water resources, coastal and ocean engineering (ICWRCOE 2015), Aquatic Procedia, vol 4. India, pp 876–883
12. GEO-SLOPE International Ltd (2012) Seepage Modeling with SEEP/W. In: An engineering methodology July 2012 Edition. Canada
13. National Standardization Agency (2002) Procedure for dam body design. RSNI T-01–2002. Jakarta
14. Water Resources and Construction Education and Training Center (2005) Guidelines for dam grouting. Ministry of Public Works, Jakarta
15. National Standardization Agency (1992) Methods of analysis and control method of seepage water for urugan-type dam. SNI 8065: 2016. Jakarta

Maintenance Management of Buildings and Infrastructures

State-of-the-Art of Artificial Intelligence Methods in Structural Health Monitoring

I. G. E. A. Putra

Abstract Development of technology nowadays occur in every aspect of life, including in computational technology. Artificial Intelligence (AI) over the past years became trending topic in field related to computation as it is mimic the way human think. This is led into effective problem-solving execution for a complex problem. Solution based on AI method is proven to be time-efficient compare with traditional ways and require significantly less resource to conduct experiment. AI method is able to perform a decision making based on series of given data with relatively short time, less error and efficient in overall computation. The objective of this paper is to investigate what is the most popular AI method and with the intended use in Structural Health Monitoring (SHM) by referring to reliable publications over the past decade. This paper shows the Neural Network is the most popular AI method in terms of SHM with 38% from total research over the past decade which is used as a performance evaluation.

Keywords Structural health monitoring · Artificial intelligence · Machine learning · Neural network · Pattern recognition

1 Introduction

Structural health monitoring in structural engineering can be defined as a process of performing damage detection which involves observation of structural system over time. The structural system health is determined based on the analysis result of periodically measured dynamics response. It is also includes extraction of damage-sensitive feature and feature statistical analysis [1]. Damage that occurs on the structure can be detected either through parametric method (fundamental frequency, mode shape stiffness and damping) or non-parametric method [2]. In assessing the parametric performance of existing buildings, ASCE 41-17 provides three evaluation procedures; Tier 1: Screening, Tier 2: Deficiency-Based Evaluation and Retrofit and Tier

I. G. E. A. Putra (✉)
Department of Civil Engineering, Universitas Tarumanagara, Jl. S. Parman No. 1, Jakarta 11440, Indonesia

S. A. Kristiawan et al. (eds.), *Proceedings of the 5th International Conference on Rehabilitation and Maintenance in Civil Engineering*, Lecture Notes in Civil Engineering 225,
https://doi.org/10.1007/978-981-16-9348-9_29

3: Systematic Evaluation and Retrofit [3]. It is expected that the use of artificial intelligence methods in the field of structural health monitoring would make the work easier in making efficient decision [4].

Research on artificial intelligence has been conducted since the term "Artificial Intelligence" was founded at a meeting at Dartmouth College in 1956. Artificial intelligence consists of several disciplines such as computer science, cybernetics, information theory, psychology, linguistics and neuropsychology. The main purpose of artificial intelligence concept is to imitate the way of the human brain executing intelligence functions for a certain thing [5]. Recently, there has been numerous research on artificial intelligence in the world of structural health monitoring. Sun et al. [6] conducted a study on bridge structural health monitoring using neural networks and extreme machine learning. Neural networks are also used in predicting the effectiveness of the transfer length of precast concrete by considering the strands characteristics, concrete properties, the detailed geometry and the manufacturing methods [7]. Chun et al. [8] used the supervised machine learning to evaluate damage by measuring the velocity of the structure at several points. Evaluation of steel connections is also performed using a neural network to investigate the behavior of innovative C-shaped shear connectors or tilted angle connections [9]. Structural evaluation which involves a combination of artificial intelligence methods with commonly used methods. In this case the Monte Carlo method also occurs in the investigation of possible bridge failure due to the fatigue failure in overloaded traffic [10]. Hung et al. [11] estimated the ultimate load factor on a nonlinear inelastic steel truss with the help of a deep learning algorithm. While Liu et al. [12] developed a deep learning algorithm to automatically detected damage after a disaster on a concentrically braced frame structure with seismic loading.

This paper focuses on investigation of artificial intelligence method used in the field of structural health monitoring for the past decade. A general description regarding each artificial intelligence method is provided along its application in the structural health monitoring.

2 Research Method

In this research, several papers are used as references. They have been checked for the conformity to the research purpose. The papers that conform then are being investigated with the purposes, the artificial intelligence method used and the mechanism action in structural health monitoring. The reference used in this research has been reviewed and approved by the publisher. Because it has been peer-reviewed, the quality of the reference is proven. Reference articles are obtained from the reliable sources. The literature review process in this study is as follows:

- From academic sites on the internet, a search was made for Candidate papers. The sites used must have proven quality such as Scopus, Wiley Online Library, ASCE Library, Sage, Springer and Elsevier.

- Several keywords used for search candidate paper were "Structural health monitoring", "Artificial intelligence structural engineering", "Artificial intelligence structural health monitoring", "Machine learning structural health monitoring", "AI methods in structural health monitoring", "Deep learning structural health monitoring" dan "Neural network structural health monitoring". Papers that will be used as data are those published in the period 2010–2020.

In the process of assessing the contents of the artificial intelligence method, the quantitative and qualitative methods are used. By using quantitative data analysis methods, the popularity level of the artificial intelligence method in terms of structural health monitoring will be obtained. By using qualitative methods, the common utilization of AI methods in SHM will be obtained.

3 Result and Discussion

In this study, more than 256 samples of data were obtained from journals that had been indexed by Scopus and had proven quality, such as Scopus, Wiley Online Library, ASCE Library, Sage, Springer and Elsevier and 101 papers were selected. Samples selected from 2010 to 2020 obtained 38% using the Neural Network method, 27% Machine Learning, 17% Support Vector Machine, 8% Pattern Recognition, 5% Principal Component Analysis and the other 5% are Bayesian Network (see Fig. 1).

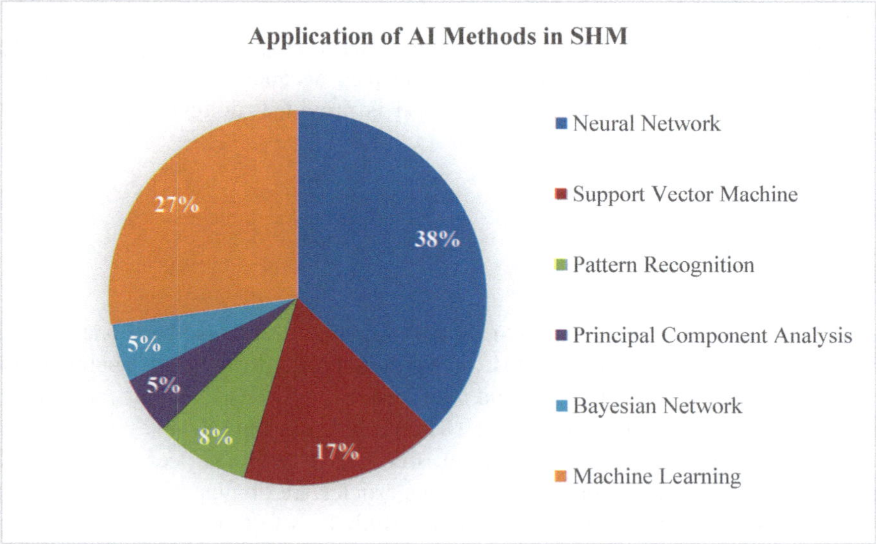

Fig. 1 Application of AI methods in SHM

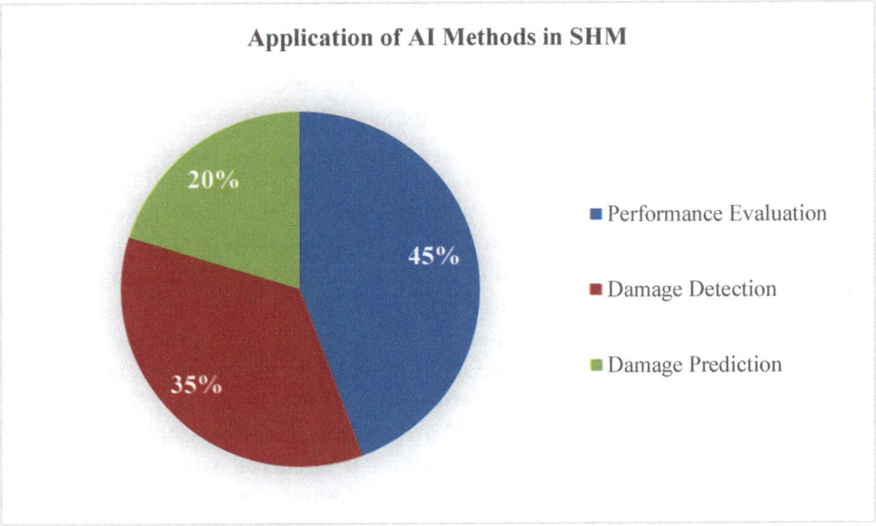

Fig. 2 Common usage of AI methods

The data above is then checked for contents to determine the purpose of using the artificial intelligence method. Generally categorized as; performance evaluation, damage detection and damage prediction. The data obtained are 45% for performance evaluation, 35% for damage detection and 20% for predicting damage (see Fig. 2).

3.1 Application of Artificial Intelligence Methods

Neural Network

Posenato et al. [13] implement Neural Network method to interpret anomalous behaviour of Riccolo viaduct located on Swizz. The result of analysis then compared to other statistical methods in which neural network superior toward others. An integrated multi-objective harmony search with artificial neural networks also proposed to minimize the run time the analysis in obtaining a cost optimal of a post-tensioned concrete box girder road bridge [14]. Several bridges performance evaluation also conducted using neural network methods [15–18]. Siddique et al. [19] performs a comparative performance evaluation of a model to predict the 28 days compressive strength of self-compacting concrete using neural network method while Saridemir et al. [20] evaluate its tensile strength. Dantas et al. [21] develop a neural network algorithm predict the compressive strength of concrete specimen at the age of 3, 7, 28 and 91 days which consist of demolition waste material. A concrete dam also identified to be analyzed using neural network method in order to evaluate its performance under thermal effect [22].

An early damage detection of International Bridge, a cable stayed bridge which connects Portugal and Spain, was performed using neural network [23]. Norman et al. [24] identify the performance of Creek Bridge in Canada, a long span bridge, and develop genetic algorithm to detect anomaly which often related with damage (cracks) on a certain part of the structure [25–27]. Concepcion II et al. [28] develop a thermal-vibration approach using hybrid soft computing using neural network algorithm to a reinforced concrete bridges experimental model to detect anomaly during its critical condition. A damage detection action also performed for a composite cantilever beam using signal which obtained from an optical fiber sensor supplemented using back propagation neural network [29, 30]. Ng et al. [31] perform a damage detection on an experimental model of 5 story steel structure acted as a benchmark structure using pattern recognition supplemented by Bayesian designed artificial neural network, which successfully identify its damage.

Mondoro et al. [32] develop an optimal risk-based management strategy of bridges in coastal region to predict required repair action to address the deterioration performance under traffic load and retrofit action by developing a neural network algorithm. By this study it is expected to minimize the potential failure during the occurrence of hurricanes. A framework of an autoencoder-based was developed to learning a mapping between the characteristic of vibrations and damages characteristics using Artificial Neural Network and Deep Learning method. Common usage of Neural Network can be seen on Fig. 3.

Support Vector Machine

Yan et al. [33] develop a support vector machine to predict the elastic modulus of normal and high strength concrete with the introduction of insensitive loss function based on learning theory. Similar research also developed to investigate the compressive strength of high-performance concrete using time-weighted fuzzy support vector machine inference model [34] and genetic weighted pyramid operation tree [35].

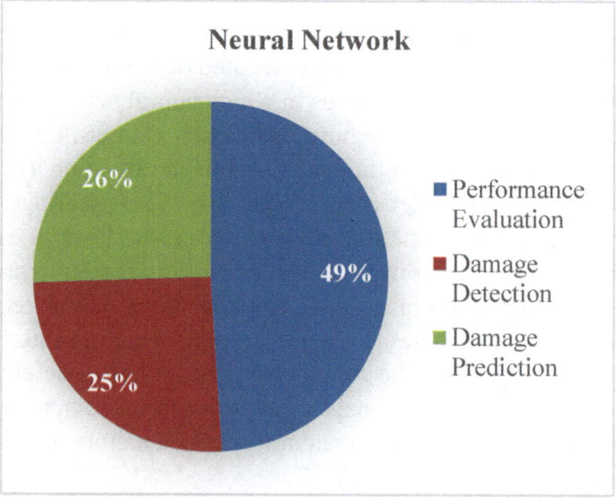

Fig. 3 Common usage of neural network

Support vector machine also used in diagnosis the damage [36] located in Wuhan [37]. Santos et al. [38] present a damage detection action by performing feature extraction, data normalization and statistical classification using a support vector machine method for Bridge in Portugal. A combination of support vector machine with the likes of pattern recognition method performed to investigate an experimental model for bridges [39, 40]. Radhika et al. [41] conduct a research of damage detection caused by cyclone by combining support vector machine method and pattern recognition technique. Several damage predictions also performed such as harnessing the structural data to recovery of a random missing vibration responses time-history by Yang et al. [42]. In bridge structure, calculation of damage in cable stayed bridge by analyzing its natural frequency and comparing the result of support vector machine analysis in the numerical example which shows its efficiency in the process [43]. Concrete dams also use support vector machine to predict the damage possibility.

SVM benefited the user who's working on a high dimensional space, in specifying a decision function it uses a subset of training data and it is versatile since it is able to use a different kernel function and make customization out of it. Common usage of support vector machine can be seen on Fig. 4.

Pattern Recognition

Abdollahzadeh et al. [44] use a pattern recognition technique to investigate the performance of beam column joint due to behavior of extreme non-elastic response, constant changes in roughness and resistance by using two models of the structure. Several simulation models also investigated using the same method such as image-based damage detection [45] and structural behavior based on learning algorithm. Lin et al. [46] perform a damage detection of a prestressed concrete using pattern recognition technique. Elwood et al. [47] also conduct similar research about damage detection in concrete structures using simulation model. Binary data is utilized to identify damage using an image-based pattern recognition which generated from self-powered sensor networks. A three stories bookshelves structure is analyzed

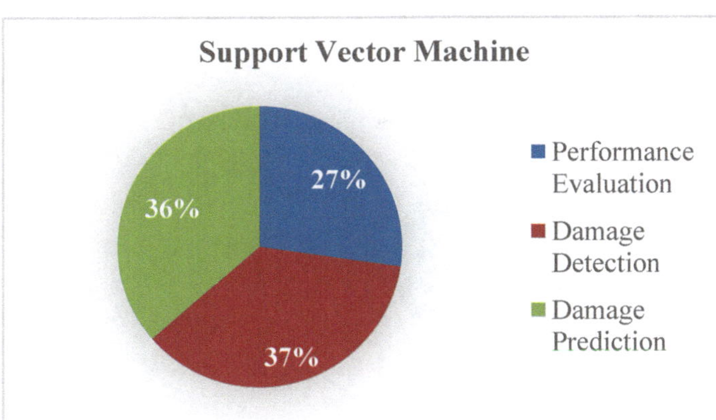

Fig. 4 Common usage of support vector machine

using experimental model to identify its damage by utilizing its frequency response spectrum [48]. Common usage of pattern recognition can be seen on Fig. 5.

Principal Component Analysis

Datteo et al. [49] present a study of structural health monitoring of G. Meazza stadium in the form of simulation model to detect any possibilities of damage in the future by applying principal component analysis. Damage prediction using principal component analysis also conducted by Balsamo et al. [50] in a simulation model using small training dataset. Reddy et al. [51] using principal component analysis to detect crack occurs in wind turbine blade based on data of wind turbine in Sweden in 1997–2005.

Principal component analysis was investigated on its usage purpose as performance evaluation method on structural health monitoring which often used together as a subset to other Artificial Intelligence method. Common usage of principal component analysis can be seen on Fig. 6.

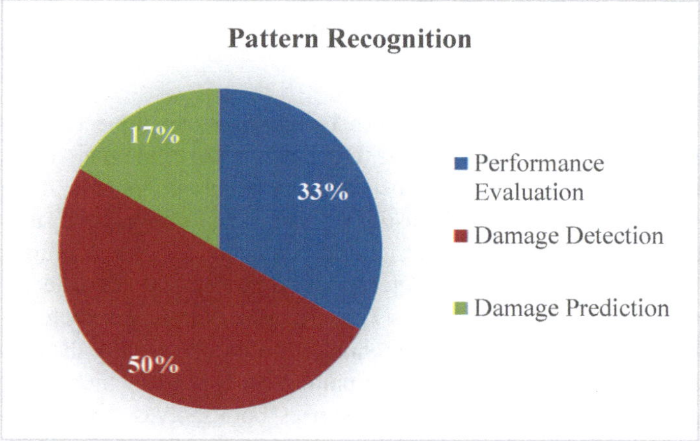

Fig. 5 Common usage of pattern recognition

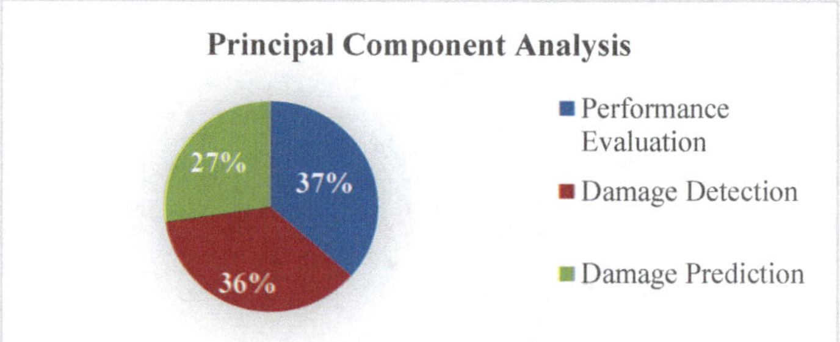

Fig. 6 Common usage of principal component analysis

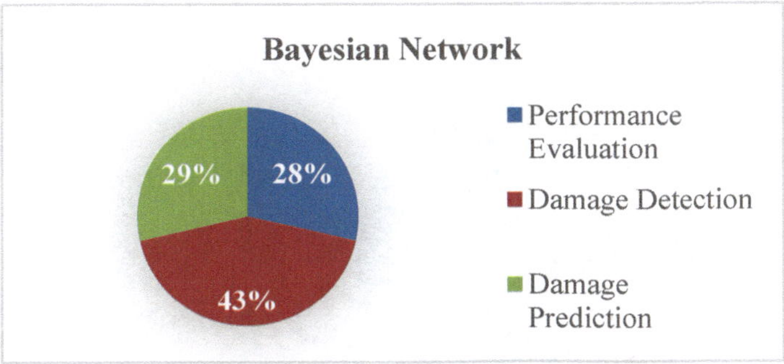

Fig. 7 Common usage of Bayesian network

Bayesian Network

Bayesian network method used as fault diagnosis on a rotary machinery in an industrial application since this kind of particular method has become an interesting topic for academia throughout the world [52]. A study of cyclone damage detection using satellite image and processed using Bayesian network. An optimization technique using Bayesian network as a subset of artificial network is conducted in order to obtain an optimal damage detection result. A methodology to diagnosis system level using Bayesian network was presented by Bartram et al. [53] in the presence of heterogeneous information. Common usage of Bayesian network can be seen on Fig. 7.

Machine Learning

Machine learning application in terms of performance evaluation was the most popular application among the others. An observation of prestressed concrete in terms of its initial yield and onset failure in bonded posttension concrete beams using machine learning method [54, 55]. A prediction of elastic modulus of high-performance concrete is conducted while Nedushan et al. [56] investigate similar object supplemented by ANFIS and optimal nonlinear regression models. A steel end plate was investigated using machine learning technique in order to obtain the most optimum type of connection [57]. Tamura et al. [58] use machine learning in his research to evaluate the optimal placement of steel brace on a steel frame. Several simulations of machine learning models was conducted to evaluate the performance of smart city concept [59], building performance prediction [60] and beam column joint in reinforced concrete structure [61].

Damage detection using machine learning technique in a composite material was presented by Sarkar et al. [62]. Machine learning in combination with several other artificial-based method to investigate the damage that occurs in a cable stayed bridge structure is. Several models based on machine learning framework was developed to detect damage in bridge using one-class machine learning [63], detect damage in a bridge under operational used and environmental variability [64] and detect damage with the most efficient energy in structure with delayed signal [65].

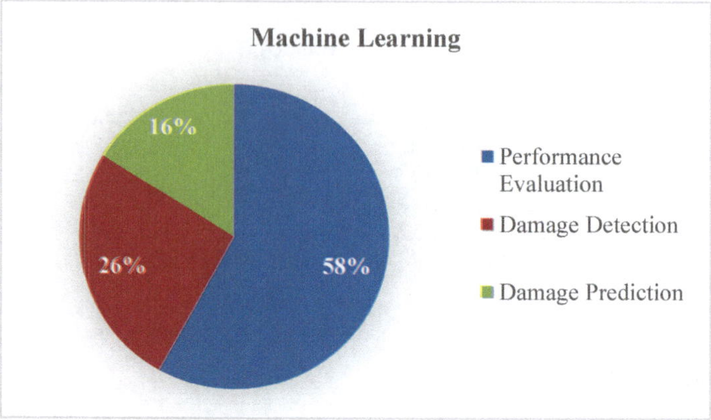

Fig. 8 Common usage of machine learning

In predicting damage that may occur, a machine learning framework in combination of Bayesian network is presented to evaluate Church in Portugal [66]. Zheng et al. [67] investigate the damage possibility by a collision in pier using probabilistic-based classification in machine learning. An experimental model to investigate damage location using comparison of sparse representation and Fourier discriminant method was presented by Wang et al. [68] and estimation of punching shear of FRP reinforced concrete slabs using hybrid machine learning presented by Vu et al. [69]. Common usage of machine learning can be seen on Fig. 8.

3.2 Artificial Intelligence Method as a Performance Evaluation, Damage Detection and Damage Prediction in SHM

Performance evaluation is defined as an ability of a certain AI algorithm in measuring how well is an object serve its function. Machine learning is mostly utilized as a performance evaluation (58%) followed by Neural Network (49%). Machine learning consist of component of task (T), performance measure (P) and improving from dataset by experience (E). The use machine learning in several research to evaluate the optimal placement of steel brace on a steel frame by extracting the characteristics using support vector machine to determine classes and binary decision tree are conducted. Characteristics are incorporated with maximum value of additional stress in existing beams and column and also the maximum interstory drift angle. Research regarding evaluation of anomalous behavior of Riccolo viaduct located on Swizz by using Neural Network and using several functions such as Auto regressive moving average (ARMA), Discrete Wavelet Transform (DWT), Robust Regression Analysis (RRA) and several other in defining the algorithm for dot product is conducted.

Damage detection is defined as an ability of a certain AI algorithm in detecting flaws of a system or object during a sudden change due to either natural disaster or human error. Pattern recognition and Bayesian network are mostly used as a damage detection by 50% and 43% respectively. A damage detection of a prestressed concrete is performed using pattern recognition technique by developing a discrete-time stochastic state space model and a standard form of matrix. These parameters then investigated using a damage index and the result is verified using several study cases. If the result from damage index is considerably acceptable then it can be concluded that the model is proven to be able to detect damage.

Damage prediction is defined as an ability of a certain AI algorithm in predicting a potential flaw of a certain system or object before any disaster or other dangerous event occurs. All the mentioned AI method above is shown to have a rather small portion in terms of predicting damage compare to performance and damage detection. Support vector machine has the relative highest percentage by 36% in its population. A support vector machine is developed to predict the elastic modulus of normal and high strength concrete with the introduction of insensitive loss function based on learning theory. The SVM is developed based on learning theory and regression technique toward loss function. The result then compared with the experimental result which shown SVM perform well in predicting elastic modulus of concrete.

4 Conclusion

In this research, it shows that the neural network is the most popular artificial intelligence method in the past decade in terms of structural health monitoring. Neural network mostly used for both performance evaluation and damage detection. Machine learning is the second most popular artificial intelligence method and just like neural network its intended use is mostly for either performance evaluation or damage detection. Third most popular artificial intelligence method is support vector machine and it is used as performance evaluation as well as damage detection. Other artificial intelligence methods that have almost equal proportion are pattern recognition, principal component analysis and Bayesian network with each majority intended used as a performance evaluation and damage detection. In the utilization of structural health monitoring, artificial intelligence is often used as a performance evaluation tool followed by damage detection and the least is for damage prediction.

References

1. Farrar CR, Worden K, Wiley J (2013) Structural health monitoring a machine learning perspective. Library of Congress Cataloging-in-Publication Data
2. Rafiei MH, Adeli H (2017) A novel machine learning-based algorithm to detect damage in high-rise building structures. Struct Des Tall Spec Build 26(18):1–11. https://doi.org/10.1002/tal.1400

3. Wijaya U, Putra IGEA, Tavio (2020) Assessment and evaluation procedure of existing industrial building structure based on ASCE 41–17. J Phys Conf Ser 1477(5). https://doi.org/10.1088/1742-6596/1477/5/052045
4. Smarsly K, Dragos K, Wiggenbrock J (2016) Machine learning techniques for structural health monitoring. In: 8th Eur Work Struct Heal Monit. EWSHM 2016, vol 2, pp 1522–1531
5. Modha DS, Ananthanarayanan R, Esser SK, Ndirango A, Sherbondy AJ, Singh R (2011) Cognitive computing. Commun ACM 54(8):62–71. https://doi.org/10.1145/1978542.1978559
6. Sun L, Shang Z, Xia Y, Bhowmick S, Nagarajaiah S (2020) Review of bridge structural health monitoring aided by big data and artificial intelligence: from condition assessment to damage detection. J Struct Eng (United States) 146(5) https://doi.org/10.1061/(ASCE)ST.1943-541X.0002535
7. Alhassan MA, Ababneh AN, Betoush NA (2020) Innovative model for accurate prediction of the transfer length of prestressing strands based on artificial neural networks: Case study. Case Stud Constr Mater 12:e00312. https://doi.org/10.1016/j.cscm.2019.e00312
8. Chun P-j, Yamane T, Izumi S, Kuramoto N (2020) Development of a machine learning-based damage identification method using multi-point simultaneous acceleration measurement results. Sensors 20(10):2780. https://doi.org/10.3390/s20102780
9. Shariati M et al (2020) A novel approach to predict shear strength of tilted angle connectors using artificial intelligence techniques. Eng Comput. https://doi.org/10.1007/s00366-019-00930-x
10. Yan W, Deng L, Zhang F, Li T, Li S (2019) Probabilistic machine learning approach to bridge fatigue failure analysis due to vehicular overloading. Eng Struct 193(May):91–99. https://doi.org/10.1016/j.engstruct.2019.05.028
11. Hung TV, Viet VQ, Van Thuat D (2019) A deep learning-based procedure for estimation of ultimate load carrying of steel trusses using advanced analysis. J Sci Technol Civ Eng NUCE 13(3):113–123. https://doi.org/10.31814/stce.nuce2019-13(3)-11
12. Liu H, Zhang Y (2019) Deep learning-based brace damage detection for concentrically braced frame structures under seismic loadings. Adv Struct Eng 22(16):3473–3486. https://doi.org/10.1177/1369433219859389
13. Posenato D, Kripakaran P, Inaudi D, Smith IFC (2010) Methodologies for model-free data interpretation of civil engineering structures. Comput Struct 88(7–8):467–482. https://doi.org/10.1016/j.compstruc.2010.01.001
14. García-Segura T, Yepes V, Frangopol DM (2017) Multi-objective design of post-tensioned concrete road bridges using artificial neural networks. Struct Multidiscip Optim 56(1):139–150. https://doi.org/10.1007/s00158-017-1653-0
15. Bao Y, Chen Z, Wei S, Xu Y, Tang Z, Li H (2019) The state of the art of data science and engineering in structural health monitoring. Engineering 5(2):234–242. https://doi.org/10.1016/j.eng.2018.11.027
16. Shrestha YR, Ben-menahem SM (2019) Organizational decision-making structures in the age of artificial intelligence, pp 1–18. https://doi.org/10.1177/0008125619862257
17. Sri C et al (2018) Structural damage identification based on autoencoder neural networks and deep learning. Eng Struct 172(April):13–28. https://doi.org/10.1016/j.engstruct.2018.05.109
18. Cury A, Crémona C, Diday E (2010) Application of symbolic data analysis for structural modification assessment. Eng Struct 32(3):762–775. https://doi.org/10.1016/j.engstruct.2009.12.004
19. Siddique R, Aggarwal P, Aggarwal Y (2011) Prediction of compressive strength of self-compacting concrete containing bottom ash using artificial neural networks. Adv Eng Softw 42(10):780–786. https://doi.org/10.1016/j.advengsoft.2011.05.016
20. Yan K, Xu H, Shen G, Liu P (2013) Prediction of splitting tensile strength from cylinder compressive strength of concrete by support vector machine. Adv Mater Sci Eng. https://doi.org/10.1155/2013/597257
21. Dantas ATA, Batista Leite M, De Jesus Nagahama K (2013) Prediction of compressive strength of concrete containing construction and demolition waste using artificial neural networks. Constr Build Mater 38:717–722. https://doi.org/10.1016/j.conbuildmat.2012.09.026

22. Kang F, Li J, Zhao S, Wang Y (2019) Structural health monitoring of concrete dams using long-term air temperature for thermal effect simulation. Eng Struct 180: 642–653. https://doi.org/10.1016/j.engstruct.2018.11.065
23. Santos JP, Cremona C, Orcesi AD, Silveira P (2017) Early damage detection based on pattern recognition and data fusion. J Struct Eng (United States) 143(2):1–11. https://doi.org/10.1061/(ASCE)ST.1943-541X.0001643
24. Noman AS, Deeba F, Bagchi A (2013) Health monitoring of structures using statistical pattern recognition techniques. J Perform Constr Facil 27(5):575–584. https://doi.org/10.1061/(ASCE)CF.1943-5509.0000346
25. Gui G, Pan H, Lin Z, Li Y, Yuan Z (2017) Data-driven support vector machine with optimization techniques for structural health monitoring and damage detection. KSCE J Civ Eng 21(2):523–534. https://doi.org/10.1007/s12205-017-1518-5
26. Dorafshan S, Thomas RJ, Maguire M (2018) Data in brief SDNET2018: an annotated image dataset for non-contact concrete crack detection using deep convolutional neural networks. Data Br 21:1664–1668. https://doi.org/10.1016/j.dib.2018.11.015
27. Heo G, Kim C, Lee C, Hur J, Seo S (2017) A damage assessment technique based on a revised Statistical Pattern-Recognition Technique (SPRT). KSCE J Civ Eng 21(3):882–888. https://doi.org/10.1007/s12205-016-0764-2
28. Ii RSC, Ilagan LC (2019) Application of hybrid soft computing for classification of reinforced concrete bridge structural health based on thermal-vibration intelligent system parameters. In: 2019 IEEE 15th Int Colloq Signal Process its Appl, March, pp 207–212
29. Mallik N, Wali AS, Kuri N (2016) Damage location identification through neural network learning from optical fiber signal for structural health monitoring. In: ACM international conference proceeding series, pp 157–161. https://doi.org/10.1145/3036932.3036937
30. Senniappan V, Subramanian J, Papageorgiou EI, Mohan S (2017) Application of fuzzy cognitive maps for crack categorization in columns of reinforced concrete structures. Neural Comput Appl 28:107–117. https://doi.org/10.1007/s00521-016-2313-9
31. Ng CT (2014) Application of Bayesian-designed artificial neural networks in phase ii structural health monitoring benchmark studies. Aust J Struct Eng 15(1):27–36. https://doi.org/10.7158/S12-042.2014.15.1
32. Mondoro A, Frangopol DM, Soliman M (2017) Optimal risk-based management of coastal bridges vulnerable to hurricanes. J Infrastruct Syst 23(3):1–12. https://doi.org/10.1061/(ASCE)IS.1943-555X.0000346
33. Yan K, Shi C (2010) Prediction of elastic modulus of normal and high strength concrete by support vector machine. Constr Build Mater 24(8):1479–1485. https://doi.org/10.1016/j.conbuildmat.2010.01.006
34. Cheng MY, Chou JS, Roy AFV, Wu YW (2012) High-performance concrete compressive strength prediction using time-weighted evolutionary fuzzy support vector machines inference model. Autom Constr 28:106–115. https://doi.org/10.1016/j.autcon.2012.07.004
35. Cheng MY, Firdausi PM, Prayogo D (2014) High-performance concrete compressive strength prediction using Genetic Weighted Pyramid Operation Tree (GWPOT). Eng Appl Artif Intell 29:104–113. https://doi.org/10.1016/j.engappai.2013.11.014
36. Çevik A, Kurtoğlu AE, Bilgehan M, Gülşan ME, Albegmprli HM (2015) Support vector machines in structural engineering: a review. J Civ Eng Manag 21(3):261–281. https://doi.org/10.3846/13923730.2015.1005021
37. Tan D, Qu W, Zhang J, Wei G, Liu J (2012) Damage diagnosis of cable of large span cable-stayed bridge based on the support vector machine. Appl Mech Mater 190–191:958–961. https://doi.org/10.4028/www.scientific.net/AMM.190-191.958
38. Santos JP, Crémona C, Orcesi AD, Silveira P (2013) Multivariate statistical analysis for early damage detection. Eng Struct 56:273–285. https://doi.org/10.1016/j.engstruct.2013.05.022
39. McGetrick PJ, Kim CW (2014) An indirect bridge inspection method incorporating a wavelet-based damage indicator and pattern recognition. In: Proceedings of international conference structural dynamics, EURODYN, vol 2014 January, no June 2014, pp 2605–2612

40. Bandara RP, Chan THT, Thambiratnam DP (2014) Frequency response function based damage identification using principal component analysis and pattern recognition technique. Eng Struct 66:116–128. https://doi.org/10.1016/j.engstruct.2014.01.044
41. Radhika S, Tamura Y, Matsui M (2015) Cyclone damage detection on building structures from pre- and post-satellite images using wavelet based pattern recognition. J Wind Eng Ind Aerodyn 136:23–33. https://doi.org/10.1016/j.jweia.2014.10.018
42. Yang Y, Sun P, Nagarajaiah S, Bachilo SM, Weisman RB (2017) Full-field, high-spatial-resolution detection of local structural damage from low-resolution random strain field measurements. J Sound Vib 399:75–85. https://doi.org/10.1016/j.jsv.2017.03.016
43. Kang F, Li J, Dai J (2019) Advances in engineering software prediction of long-term temperature effect in structural health monitoring of concrete dams using support vector machines with Jaya optimizer and salp swarm algorithms. Adv Eng Softw 131:60–76. https://doi.org/10.1016/j.advengsoft.2019.03.003
44. Abdollahzadeh G, Shabanian SM (2018) Experimental and numerical analysis of beam to column joints in steel structures. Front Struct Civ Eng 12(4):642–661. https://doi.org/10.1007/s11709-017-0457-z
45. Salehi H, Das S, Chakrabartty S, Biswas S, Burgueño R (2018) Structural damage identification using image-based pattern recognition on event-based binary data generated from self-powered sensor networks. Struct Control Heal Monit 25(4):1–21. https://doi.org/10.1002/stc.2135
46. Lin YQ, Ren WX, Fang SE (2011) Structural damage detection based on stochastic subspace identification and statistical pattern recognition: II. Experimental validation under varying temperature. Smart Mater Struct 20(11). https://doi.org/10.1088/0964-1726/20/11/115010
47. Elwood E, Corotis RB (2015) Application of fuzzy pattern recognition of seismic damage to concrete structures. ASCE-ASME J Risk Uncertain Eng Syst Part A Civ Eng 1(4):1–12. https://doi.org/10.1061/AJRUA6.0000831
48. Bandara RP, Chan TH, Thambiratnam DP (2014) Structural damage detection method using frequency response functions. Struct Heal Monit 13(4):418–429. https://doi.org/10.1177/1475921714522847
49. Datteo A, Lucà F, Busca G (2017) Statistical pattern recognition approach for long-time monitoring of the G. Meazza stadium by means of AR models and PCA. Eng Struct 153(July):317–333. https://doi.org/10.1016/j.engstruct.2017.10.022
50. Balsamo L, Betti R, Beigi H (2014) A structural health monitoring strategy using cepstral features. J Sound Vib 333(19):4526–4542. https://doi.org/10.1016/j.jsv.2014.04.062
51. Reddy A, Indragandhi V, Ravi L, Subramaniyaswamy V (2019) Detection of cracks and damage in wind turbine blades using artificial intelligence-based image analytics. Measurement 147:106823. https://doi.org/10.1016/j.measurement.2019.07.051
52. Liu R, Yang B, Zio E, Chen X (2018) Artificial intelligence for fault diagnosis of rotating machinery: a review. Mech Syst Signal Process 108:33–47. https://doi.org/10.1016/j.ymssp.2018.02.016
53. Bartram GW, Mahadevan S (2012) System modeling for SHM using dynamic Bayesian networks. AIAA Infotech Aerosp Conf Exhib 2012(June):1–12. https://doi.org/10.2514/6.2012-2423
54. Salamone S, Veletzos MJ, Lanza Di Scalea F, Restrepo JI (2012) Detection of initial yield and onset of failure in bonded posttensioned concrete beams. J Bridg Eng 17(6):966–974. https://doi.org/10.1061/(ASCE)BE.1943-5592.0000311
55. Chou JS, Tsai CF, Pham AD, Lu YH (2014) Machine learning in concrete strength simulations: multi-nation data analytics. Constr Build Mater 73:771–780. https://doi.org/10.1016/j.conbuildmat.2014.09.054
56. Ahmadi-Nedushan B (2012) Prediction of elastic modulus of normal and high strength concrete using ANFIS and optimal nonlinear regression models. Constr Build Mater 36:665–673. https://doi.org/10.1016/j.conbuildmat.2012.06.002
57. Greco L (2018) Machine learning and optimization techniques for steel connections, no. July, pp 1–8

58. Tamura T, Ohsaki M, Takagi J (2018) Machine learning for combinatorial optimization of brace placement of steel frames. Japan Archit Rev 1(4):419–430. https://doi.org/10.1002/2475-8876.12059
59. Chin J, Callaghan V, Lam I (2017) Understanding and personalising smart city services using machine learning, the Internet-of-Things and Big Data. In: IEEE international symposium Ind Electron, pp 2050–2055. https://doi.org/10.1109/ISIE.2017.8001570
60. Geyer P, Singaravel S (2018) Component-based machine learning for performance prediction in building design. Appl Energy 228(July):1439–1453. https://doi.org/10.1016/j.apenergy.2018.07.011
61. Baker JW (2007) Measuring bias in structural response caused by ground motion scaling. Pacific Conf Earthq Eng 056:1–6. https://doi.org/10.1002/eqe
62. Sarkar S, Reddy KK, Giering M, Gurvich MR (2016) Deep learning for structural health monitoring: a damage characterization application. In: Proceedings annual conference Progn Heal Manag Soc PHM, vol 2016 October, no. October, pp 176–182
63. Long J, Buyukozturk O (2014) Automated structural damage detection using one-class machine learning. Conf Proc Soc Exp Mech Ser 4:117–128. https://doi.org/10.1007/978-3-319-04546-7_14
64. Figueiredo E, Park G, Farrar CR, Worden K, Figueiras J (2011) Machine learning algorithms for damage detection under operational and environmental variability. Struct Heal Monit 10(6):559–572. https://doi.org/10.1177/1475921710388971
65. Salehi H, Das S, Biswas S, Burgueño R (2019) Data mining methodology employing artificial intelligence and a probabilistic approach for energy-efficient structural health monitoring with noisy and delayed signals. Expert Syst Appl 135:259–272. https://doi.org/10.1016/j.eswa.2019.05.051
66. Ramos LF, Miranda T, Mishra M, Fernandes FM, Manning E (2015) A Bayesian approach for NDT data fusion: the Saint Torcato church case study. Eng Struct 84:120–129. https://doi.org/10.1016/j.engstruct.2014.11.015
67. Zheng W, Qian F (2017) Promptly assessing probability of barge–bridge collision damage of piers through probabilistic-based classification of machine learning. J Civ Struct Heal Monit 7(1):57–78. https://doi.org/10.1007/s13349-017-0208-9
68. Wang Z et al (2013) Comparison of sparse representation and fourier discriminant methods: damage location classification in indirect lab-scale bridge structural health monitoring. In: 2013 Bridg Your Passion with Your Prof - Proc 2013 Struct Congr, pp 436–446. https://doi.org/10.1061/9780784412848.039
69. Vu DT, Hoang ND (2016) Punching shear capacity estimation of FRP-reinforced concrete slabs using a hybrid machine learning approach. Struct Infrastruct Eng 12(9):1153–1161. https://doi.org/10.1080/15732479.2015.1086386

Development of Preventive Maintenance Guidelines for Simple-Classification Government Buildings Based on Work Breakdown Structure Within the DKI Jakarta Provincial Government

Dyah Ayu Pangastuti and Yusuf Latief

Abstract DKI Jakarta, as the capital of the Republic of Indonesia, has the highest population density in Indonesia, with a population of 16,334 people/km^2. Therefore, community facilities and infrastructure within the DKI Jakarta Provincial Government play an important role so that community service can be carried out properly. As one of the main infrastructures, State Buildings must have building reliability as stated in the technical requirements stipulated in Presidential Regulation Number 73 of 2016. Building maintenance is an activity to maintain the building's reliability and infrastructure, and facilities so that the building always functions properly. Based on a survey conducted by the DKI Jakarta Provincial Office for The Creation of Works, Spatial Planning and Land Use in 2019, the building assets of Province DKI Jakarta are 9823. 60% of those buildings were classified as Simple-Building category. Due to numerous building assets, building maintenance activities need to be supported by an adequate maintenance guideline especially for the simple-classification building. This study aims to develop preventive maintenance guidelines based on Work Breakdown Structure. The Work Breakdown Structure (WBS) will be used as the basis for maintenance guidelines, including maintenance activities and frequency. The methodology used in this research is study literature, survey and expert validation for WBS for Simple-Building component and Preventive Maintenance Guidelines.

Keywords Simple-building maintenance · Maintenance guidelines · Preventive maintenance

1 Introduction

Based on the DKI Jakarta Provincial Office for The Creation of Works, Spatial Planning and Land Use survey results in 2019, the number of state-building assets in the DKI Jakarta province are 9823 consisting "in need of repair" condition are 2994 buildings and "severely damaged" condition are 796 buildings. It can be concluded

D. A. Pangastuti · Y. Latief (✉)
Department of Civil Engineering, Faculty of Engineering, University of Indonesia,
Depok 16424, Indonesia

S. A. Kristiawan et al. (eds.), *Proceedings of the 5th International Conference on Rehabilitation and Maintenance in Civil Engineering*, Lecture Notes in Civil Engineering 225,
https://doi.org/10.1007/978-981-16-9348-9_30

that 38% of the state buildings in the DKI Jakarta Provincial Government are in less reliable condition. Figure 1 shows that the building assets of the DKI Jakarta Provincial Government with Simple-Buildings classification are more numerous than buildings with non-Simple-Building classifications.

From interviews with the Head of DKI Jakarta Provincial Office for The Creation of Works, Spatial Planning and Land Use, there were several obstacles in maintaining state buildings DKI Jakarta Province. The technical incomprehension of those responsible for maintenance activities in planning building maintenance activities and other several factors that can be seen in Fig. 2.

Among the causes of building unreliability is the absence of an efficient system for data recording, paper-based forms and a lack of technical personnel to monitor and communicate information [1]. The influence of government policies, lack of clear job descriptions, and lack of technical experts are common problems in building maintenance [2]. Also, the lack of maintenance personnel, the lack of preventive maintenance [3], and the lack of IT use, the lack of adequate maintenance procedures are some of the main issues in the management of government assets [4].

For this reason, it is necessary to have a maintenance guidelines based on WBS to improve maintenance performance [5, 6].

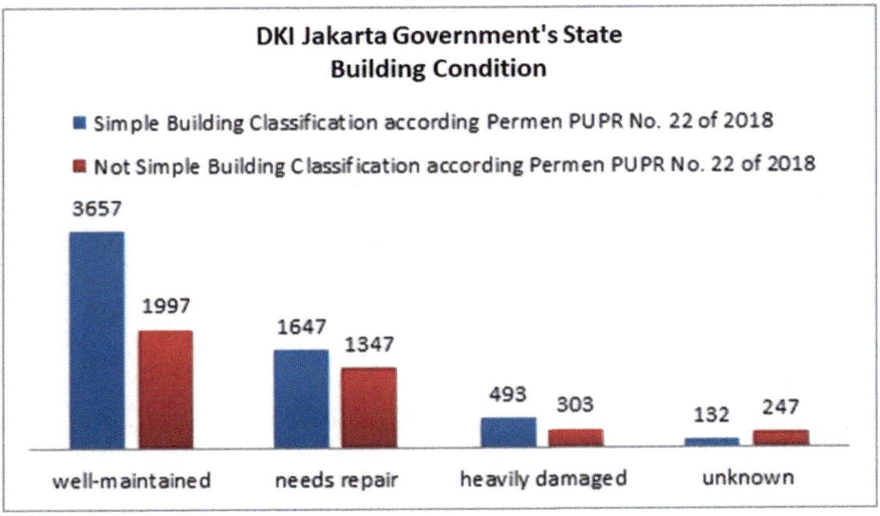

Fig. 1 DKI Jakarta provincial office for The creation of works, spatial planning and land use building asset survey data 2019

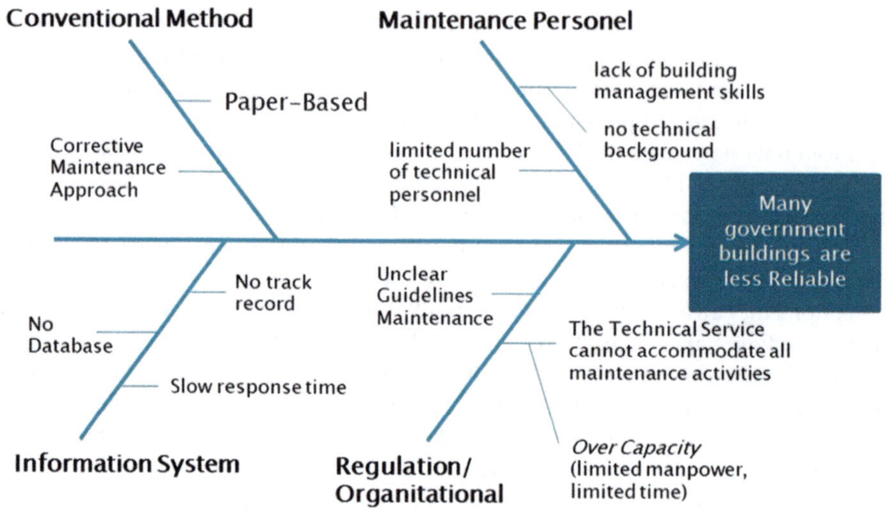

Fig. 2 Fishbone diagram about obstacles that occurs in building maintenance of state building within DKI Jakarta Province

2 Literature Review

2.1 Simple-Building Classification

Simple-Building classification for State buildings according to Ministerial regulation of Minister For Public Works and Human Settlements Number 22 in 2018 is described in article 14 paragraph (2), which reads State Buildings with simple classification as referred to in paragraph (1) letter a are buildings with simple technology and specifications including:

1. Office buildings and other state buildings with a number of floors up to 2 (two) floors;
2. Office buildings and other state buildings with an area of up to 500 m^2 (five hundred square meters); and
3. State houses include Type C, Type D, and Type E.

The technical specifications for simple classified State buildings are described in attachment II Table 1, including building layout, building materials, building structures, utilities, infrastructure & in buildings and safety facilities.

2.2 Work Breakdown Structure

Work Breakdown Structure (WBS) is a hierarchical decomposition oriented towards the work that the project team will carry out to achieve project objectives and create the required results [7]. In general, WBS can be defined as a result-oriented grouping (deliverable) to project elements, which organize and determine the project's overall structure [8]. Furthermore, standard practice for work breakdown structure explains that a more regularly made project describes individual work components, determining the entire scope of work on a project. This activity is called WBS [7].

WBS for this system is used for determines the building component so that no component are overlooked for maintenance. It is also used as a basis for preparing preventive maintenance guidelines for the system.

2.3 Preventive Maintenance

Building maintenance activities are classified into preventive maintenance, corrective maintenance [9]. Periodic maintenance is a preventive measure. its activities include regularly measuring, detecting, analyzing equipment problems to avoid failure [10]. The maintenance schedule is part of preventive maintenance. Efficiency and effectiveness in scheduling are also determined by preventive maintenance planning [11].

Preventive maintenance activities are attached to the guidelines. The guidelines will be containing: Preventive maintenance activities, frequency and interval each activity for every component of the building.

Table 1 Problem formulations and research strategy

No	Problem formulation	Question used	Strategy
1	How is the structure of a simple-building WBS within the DKI Jakarta provincial government?	How	Archive analysis, survey
2	How to develop maintenance guidelines for simple buildings within the DKI Jakarta Provincial government?	How	Archive analysis, survey

3 Research Method

3.1 Strategy and Process

After identifying the problems, some formulations of the problem were obtained, which were then determined by the research strategy. The research strategy used in this research is archive analysis (literature review and existing document for WBS and maintenance guidelines) and surveys (validation and deep interview).

The research process consists of two stages analysis. The first analysis is a simple-building WBS. The results of the validated WBS become basis for the identification of maintenance activities and the schedule of maintenance activities which are then validated. The second analysis is carried out to guide maintenance activities. Research process flow shown in Fig. 3

There are several stages of a questionnaire to determine WBS, preventive maintenance activities and schedules. This study uses the Delphi method, which involves expert respondent for every validation process: five experts in managerial positions within 10–25 years of experience and seven expert technical staffs within 5–15 years of experience in building maintenance activities.

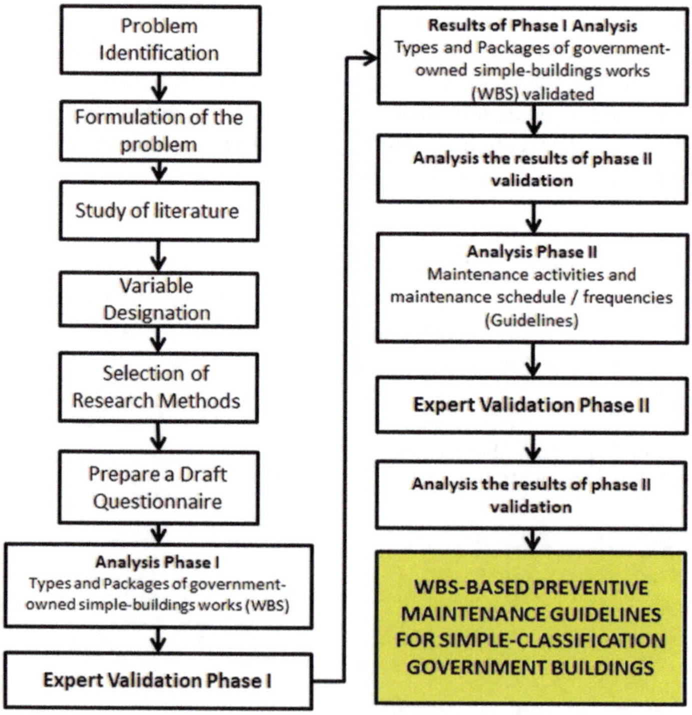

Fig. 3 Research method flow

3.2 Research Variables

Based on the results of the literature study, it is found that the activity variables and maintenance schedules form the maintenance guidelines. The list of activities and maintenance schedules are compiled based on identifying the building components to be maintained. This building component is defined by the type of work and the work package as variables of the work breakdown structure. The resulting maintenance guidelines are expected to improve the performance of building maintenance in accordance with applicable regulations. Table 2 shows a list of the variables and sub variables used in this research,

4 Result Discussion

4.1 Work Breakdown Structure

The first stage analysis is validation of the types and packages of simple building works. The questionnaire was compiled and obtained from 5 different references representing each construction work. The questionnaire was given to 7 experts with the answers "yes" or "no" to the types and packages of work deemed appropriate for simple-buildings. Validation of acceptance was carried out with a minimum of four "Yes" answers. Table 3 shows an example of expert data analysis, with the most answers in quantity being used as the acceptance or denial judgment.

The validation results of the WBS study of Simple-Building education functions are: Structural elements obtained 8 types of work and 16 work packages; Architectural elements obtained 12 types of work and 78 work packages; Mechanical elements obtained 4 types of work and 17 work packages; Electrical elements obtained 4 types of work, and 15 work packages; Landscape and Housekeeping element obtained 3 types of work and 50 work packages. Total 55 Type or work and 176 work package obtained. Example of WBS findings for electrical work are shown in Fig. 4.

4.2 Activity Maintenance and Schedule

After obtaining the WBS results for simple-buildings, then a literature study is carried out to identify the activities and schedule of preventive maintenance activities. The activities and periodic schedule of preventive activities are then compiled in a simple building maintenance guideline and validated using a questionnaire and deep interviews with 5 experts with 10–15 years of experience in managerial positions. The guidelines consist of preventive maintenance activity and periodic schedule for structural, architectural, mechanical, electrical, landscaping, and housekeeping

Table 2 Research variables and sub variables

	Variables		Sub variables			References
X.1	Work breakdown structure	X.1.1	Type of work maintenance of simple buildings	X.1.1.1	Types of structural work	Muhammad Naufal (2020)
				X.1.1.2	Type of architectural work	Luki Wijaya (2020)
				X.1.1.3	Type mechanical work	Yusuf Djonli (2019)
				X.1.1.4	Types of electrical work	Azhara Yudha (2019)
				X.1.1.5	Types of landscaping and housekeeping work	Rizkyana Fathoni (2019)
		X.1.2	Simple building maintenance work package	X.1.2.1	Structural work package	Muhammad Naufal (2020)
				X.1.2.2	Architectural work package	Luki Wijaya (2020)
				X.1.2.3	Mechanical work package	Yusuf Djonli (2019)
				X.1.2.4	Electrical work package	Azhara Yudha (2019)
				X.1.2.5	Landscape and housekeeping work package	Rizkyana Fathoni (2019)
X.2	Preventive maintenance guidelines	X.2.1	Preventive maintenance activity	X.2.1.1	Structural preventive maintenance activities	Muhammad Naufal (2020), Anita Handayani (2020)
				X.2.1.2	Architectural preventive maintenance activities	Luki Wijaya (2020), Anita Handayani (2020)

(continued)

Table 2 (continued)

	Variables		Sub variables			References
				X.2.1.3	Mechanical preventive maintenance activities	Yusuf Djonli (2019), Anita Handayani (2020)
				X.2.1.4	Electrical preventive maintenance activities	Azhara Yudha (2019), Anita Handayani (2020)
				X.2.1.5	Landscape and housekeeping Preventive maintenance activities	Rizkyana Fathoni (2019), Anita Handayani (2020)
		X.2.2	Schedule/frequency of preventive maintenance activities	X.2.2.1	Structural schedule/frequency of Preventive maintenance activities	Muhammad Naufal (2020), Anita Handayani (2020)
				X.2.2.2	Architectural schedule/frequency of Preventive maintenance activities	Luki Wijaya (2020), Anita Handayani (2020)
				X.2.2.3	Mechanical schedule/frequency of Preventive maintenance activities	Yusuf Djonli (2019), Anita Handayani (2020)
				X.2.2.4	Electrical schedule/frequency of Preventive maintenance activities	Azhara Yudha (2019), Anita Handayani (2020)

(continued)

Table 2 (continued)

	Variables		Sub variables			References
				X.2.2.5	Landscape and housekeeping Schedule/frequency of preventive Maintenance activities	Rizkyana Fathoni (2019), Anita Handayani (2020)
Y.1	Building Maintenance Performance	Y.1.1	Safety			Law of the Republic of Indonesia Number 28 of (2002), Regulation of the Minister of Public Works and Public Housing Number 24 of 2010
		Y.1.2	Health			
		Y.1.3	Convenience			
		Y.1.4	Easiness			

works. There are 1049 maintenance activities for 176 work package in the guidelines. Table 4 is an example of Tie Beam's preventive maintenance guidelines.

From deep interviews with the experts it was found differences of opinion about several activities or schedules. The differences of suggestions are collected in a list then analyzed and further discussed. Table 5 is an example of Expert's different opinions on periodic schedule activity in structural work.

From the results of the analysis and discussion, it was agreed that several work activities needed to be corrected, some if it was agreed to be eliminated. Examples of the results of corrections to preventive maintenance work activities can be seen in Table 6.

Total 6 activities were corrected, 4 of which were eliminated. The work that have been eliminated are painting the ACP walls, painting the acoustic ceiling, coating the acoustic ceiling and replacing the backup lamp battery for the emergency lights. Most of the corrected activities are changes in schedule status to incidental because there are conditions that require repairs, such as grouting the fractured part which was previously scheduled every 2 years, being corrected to incidental activities because its depending on the appearance of the crack. Further validation was carried out on the maintenance guidelines obtaining 1045 maintenance activities with each activity schedule.

Table 3 Validation analysis process example

WBS level 2		WBS level 3		WBS level 4			Does this type of work include WBS Level 4 simple-building maintenance activities?							Quantity		Modus	Status
		Work type		Work package													
		Reference: Luki Wijaya (2020)		Reference: Luki Wijaya (2020)			Expert							Yes	No		
Code	Family	Code	Type	Code	Package	Design alternatives	1	2	3	4	5	6	7				
X.2	Architecture Work	X.2.1	Parking and pedestrian facilities	X.2.1.1	Parking	Bicycle parking	Yes	Yes	Yes	No	Yes	Yes	Yes	6	1	Yes	Accept
						Motor parking	Yes	Yes	Yes	Yes	Yes	Yes	Yes	7	0	Yes	Accept
				X.2.1.2	Sidewalk	Sidewalk	No	No	Yes	No	No	No	Yes	2	5	No	Denied
		X.2.2	Means of Smoking	X.2.2.1	Smoking Room	Smoking room	No	No	No	No	No	No	No	0	7	No	Denied
		X.2.3	Means of exit	X.2.3.1	Exit	Exit	Yes	No	Yes	No	No	No	Yes	3	4	No	Denied
		X.2.4	Floor work	X.2.4.1	Floor	Floor h ardener	Yes	No	Yes	No	No	No	Yes	3	4	No	Denied
						Ceramic floor of the corridor area	Yes	Yes	Yes	Yes	Yes	Yes	Yes	7	0	Yes	Accept
						Ceramic floor dry area	Yes	Yes	Yes	Yes	Yes	Yes	Yes	7	0	Yes	Accept
						Tiled floor wet area	Yes	Yes	Yes	Yes	Yes	Yes	Yes	7	0	Yes	Accept
						Ceramic floor fire ladder	Yes	Yes	Yes	Yes	No	Yes	Yes	6	1	Yes	Accept
						Waterproof coating concrete floor	Yes	Yes	Yes	No	Yes	Yes	Yes	6	1	Yes	Accept

(continued)

Table 3 (continued)

WBS level 2		WBS level 3		WBS level 4			Does this type of work include WBS Level 4 simple-building maintenance activities?							Quantity		Modus	Status
		Work type		Work package													
		Reference: Luki Wijaya (2020)		Reference: Luki Wijaya (2020)			Expert							Yes	No		
Code	Family	Code	Type	Code	Package	Design alternatives	1	2	3	4	5	6	7				
						Waterproof membrane concrete floor	Yes	No	No	Yes	No	Yes	Yes	4	3	Yes	Accept
						integral waterproof concrete floor	Yes	No	No	Yes	No	Yes	Yes	4	3	Yes	Accept
						Granite floor	Yes	Yes	No	No	No	No	Yes	3	4	No	Denied
						Parquet floor	No	No	No	No	No	Yes	No	1	6	No	Denied

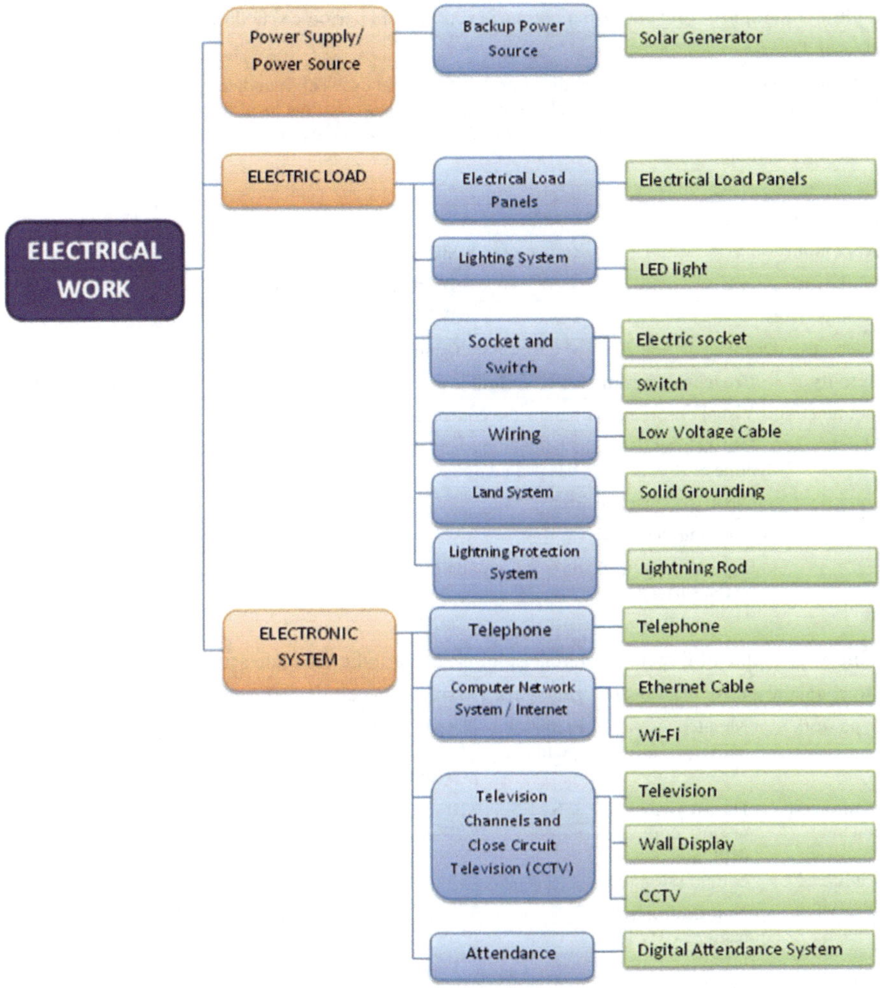

Fig. 4 WBS result for electrical work

As we can see in Table 7, the results of the validation of five experts obtained five "Yes" answers. That means, the maintenance guidelines can be used as a simple building maintenance guideline to improve maintenance performance. Suggestions are given by the respective experts for the purposes of developing this preventive maintenance guideline.

Table 4 Example of preventive maintenance activity and periodic schedule

Description		
Structural elements contained in buildings or structures that use deep foundations or local shallow foundations. This tie beam is located above the ground and on shallow local foundations such as footplat foundations or deep foundations. This tie beam is the same as the beam except that it is located in the lower structure		
Tie beam		
Description		
Work family	Structure	
Work type	Tie beam and sloof	
Package Work	Tie beam	
Alternative desain	Concrete tie beam	
Activity	**Preventive maintenance actions**	**Periodic schedule**
Examination	Visual observation of cracks in concrete	Annual
Maintenance	Use a chisel to tap the concrete around the crack to make sure that none of the concrete has peeled off	2 years
	Clean the chiseled surface	2 years
	Grouting the crack	2 years
	Repaint with emulsion paint or paint that is waterproof and acidic on the surface	2 years

Table 5 Example of Expert's different opinions on periodic schedule activity in structural work

Structural preventive maintenance guidelines			Type of suggestion/correction				
Type work	Maintenance activity	Frequency (years)	Expert 1 (years)	Expert 2 (years)	Expert 3	Expert 4 (years)	Expert 5
Reinforced concrete columns, reinforced concrete beams, conventional concrete floor slabs, conventional concrete stairs, conventional concrete plate roofs	Destructive testing: core drill, and compression test	3–5	10–15	3–5	Incidental	3–5	Incidental
	Non-destructive testing: UPVT, infrared thermograph, hammer test, loading test, covermeter test, carbonation test, impact echo test	3–5	10–15	3–5	Incidental	3–5	5 years
Steel column, steel beam	Destructive testing: radiography, ultrasonic test, and liquid Penetrant test	3–5	10–15	3–5	Eliminate	3–5	Incidental
Steel truss roof	Non-destructive testing	3–5	10–15	3–5	3–5	3–5	5 years

Table 6 Example of corrected preventive maintenance activity and schedule in electrical work

WBS level 3	WBS level 4	WBS level 5	Preventive maintenance activity	Schedule/frequency	Corrected
Work type	Work package	Work activity			
Electrical load	LED light	Treatment	Replace the back up battery on the emergency lights	Incidental	Eliminated
	Solid grounding	Service	Improve the resistance of detention if the resistance is above the standard	Monthly	Incidental
	Lightning rod	Maintenance	Improve grounding resistance if it is above standard	Monthly	Incidental

Table 7 Preventive maintenance guidelines validation analysis

Is this guidelines for the maintenance of simple buildings for the state building could improve maintenance performance in the DKI Jakarta Provincial environment?		
Respondents	Yes/No	Suggestion
Expert 1	Yes	According to the regulation of the Minister of Public Works and observes the real/real conditions regarding daily maintenance and maintenance (buildings)
Expert 2	Yes	Generator work can be eliminated for school functions building (no longer used in school buildings, replaced with Solar Panels (future school construction will use Solar Panels)
Expert 3	Yes	To be adjusted to the building object
Expert 4	Yes	Can be developed for use by other buildings, especially in the city hall complex
Expert 5	Yes	Agree if the waterproofing coating in question is to use a special waterproofing coating instead of re-coating it with the same waterproofing material

5 Conclusion

1. Preventive guidelines can be prepared based on the identification of building components using WBS based on relevant previous research and literature studies.

2. Preventive maintenance guidelines cannot be equated for all types of building classifications. In the same work package there can be different activities or schedules for preventive maintenance activities. For example, in a simple building, the structural test schedule is not necessary as often as a building with a non-simple classification.
3. WBS and preventive maintenance guidelines can improve maintenance performance, although not directly. Maintenance performance is very dependent on the activeness and concern of the building manager for maintenance activities. Further research can be developed to make this guideline implemented properly, such as using web-based system for maintenance activity.
4. These activities can be developed for the preparation of the budget plan by analyzing the budget requirements of WBS level 6, which are resources and equipment.

References

1. Ismail Z-A (2014) Improving maintenance management practices for building facility. J Constr Eng Proj Manag 4(3):21–32. https://doi.org/10.6106/jcepm.2014.4.3.021
2. Alshehri A, Motawa I, Ogunlana S (2015) The common problems facing the building maintenance departments. Int J Innov Manag Technol 6(3):234–237. https://doi.org/10.7763/ijimt.2015.v6.608
3. Waziri BS, Vanduhe BA (2013) Evaluation of factors affecting residential building maintenance in Nigeria: users ' perspective. Civ Environ Res 3(8):19–25
4. Abdullah S, Razak AA, Hanafi MH, Salleh MN (2011) Managing government property assets: the main issues from the malaysian perspective. J Techno-Social 3(1):35–52
5. Djonly Y (2019) Pengembangan Pedoman Pekerjaan Preventive Maintenance Komponen Mekanikal Pada Bangunan Gedung Pemerintah Berbasis Work Breakdown Structure (WBS) untuk Meningkatkan Kinerja Pemeliharaan dan Peawatan," Universitas Indonesia
6. Wijaya L (2020) Pengembangan Pedoman Pekerjaan Preventive Maintenance Komponen Arsitektur Pada Bangunan Gedung Pemerintah Berbasisi Work Breakdown Structure (WBS) Untuk Meningkatkan Kinerja Pemeliharaan dna Perawatan. Universitas Indonesia
7. PMI (2021) Work Breakdown Structure Practice Standard | PMI. https://www.pmi.org/pmbok-guide-standards/framework/practice-standard-work-breakdown-structures-2nd-edition. Accessed 12 Apr 2021
8. Jung Y, Woo S (2004) Flexible work breakdown structure for integrated cost and schedule control. J Constr Eng Manag 130(5):616–625. https://doi.org/10.1061/(asce)0733-9364(2004)130:5(616)
9. Ruparathna R, Hewage K, Sadiq R (2017) Developing a level of service (LOS) index for operational management of public buildings. Sustain Cities Soc 34:159–173. https://doi.org/10.1016/j.scs.2017.06.015
10. Lee PC, Wang Y, Lo TP, Long D (2018) An integrated system framework of building information modelling and geographical information system for utility tunnel maintenance management. Tunn Undergr Sp Technol 79:263–273. https://doi.org/10.1016/j.tust.2018.05.010
11. Basri EI, Razak IHA, Ab-Samat H, Kamaruddin S (2017) Preventive maintenance (PM) planning: a review. J Qual Maintenance Eng 23(2), 114–143. Emerald Group Publishing Ltd. https://doi.org/10.1108/JQME-04-2016-0014

Development of Preventive Maintenance Guidelines for Electrical Components on Government Building Based on Work Breakdown Structure

Azhara Yudha Pradipta, Yusuf Latief, and Rossy Armyn Machfudiyanto

Abstract This research aims to create preventive maintenance guidelines for electrical components in buildings, especially government buildings, by breaking down electrical components using work breakdown structures so that the identification of electrical components is complete and well-structured. The methodology used in this research uses archive analysis with validation by experts through a questionnaire based on the legislations and applicable research and previous research. The results of this research are guidelines for preventive maintenance for electrical components in government buildings that contain preventive maintenance steps for each component at an alternative design level with a time interval for each component. Preventive maintenance guidelines for electrical components in this building are formed using work breakdown structure described from level 1 to level 4 with their derivatives, namely alternative designs. From the results of the validation, there are 22 (twenty-two) work packages and 45 (forty-five) alternative designs for electrical components in buildings that need preventive maintenance with each activity, procedure and preventive action and time intervals.

Keywords Government building · Guideline · Preventive maintenance · Work breakdown structure

---

A. Y. Pradipta · Y. Latief (✉) · R. A. Machfudiyanto (✉)
Civil Engineering, Faculty of Engineering, Universitas Indonesia (UI), 16424 Depok, West Java, Indonesia
e-mail: latief73@eng.ui.ac.id

R. A. Machfudiyanto
e-mail: rossyarmyn@eng.ui.ac.id

A. Y. Pradipta
e-mail: azhara.yudha@ui.ac.id

S. A. Kristiawan et al. (eds.), *Proceedings of the 5th International Conference on Rehabilitation and Maintenance in Civil Engineering*, Lecture Notes in Civil Engineering 225,
https://doi.org/10.1007/978-981-16-9348-9_31

1 Introduction

Minister of Finance, through letter S-841/MK.02/2014 dated December 16, 2014, on the Postponement/Moratorium on Ministry/Institution Office Building Construction asking Ministry/Institution to postpone the construction of new government office buildings that began in 2015. Postponement/moratorium on new building construction is carried out to implement bureaucratic reform more effectively. Besides, it is also to improve the quality of state spending in the 2015 fiscal year [1]. Along with this, consequences arise in the operational phase of the building itself, namely for the building manager in maintaining the reliability of the building so that it is always worthy of function [2]. Furthermore, Wimala and Karim [3] stated that building maintenance is still considered less important in developing countries so that implementation priorities are not prioritized.

In Facility Management, two terms are known as Preventive Maintenance (PM) and Corrective Maintenance (CM). Both are part of the operational phase after the construction phase is completed and the difference is the time of implementation where maintenance is carried out at certain time intervals to reduce the possibility of malfunction while maintenance is carried out after a malfunction occurs [4, 5].

Electrical components in a building are a separate challenge to be described because they are many. WBS or Work Breakdown Structure is the process of dividing the results and project work into smaller ones so that it is easy to manage utilizing decomposition techniques [6]. The purpose of WBS is to organize project elements, known as project scope, all work is organized into smaller components so that the project can be better managed. The cluster hierarchy of WBS provides details of all levels, usually to the smallest level of resources consisting of materials, tools, labor and also methods of implementation. The decomposition technique in WBS can prevent the loss of an element in a component to be maintained. According to U.S. Government Administration Office [7], WBS provides a clear picture of what needs to be achieved and how the work will be done.

Without scheduling either routinely or periodically in building maintenance, which can result in maintenance costs because preventive maintenance is associated with costs to carry out maintenance/replacement (PM) activities and cost savings needed to increase the costs required by the system (CM) [8]. Based on this, research needs to be carried out on preventive maintenance of WBS-based electrical components in government buildings.

There are many cases of fire whose main cause is a short circuit. The U.S. Fire Administration noted that 119,000 fires were caused by malfunctioning electrical components in non-residential buildings between 2003 and 2016 and ranked third as the main cause of fires after cooking and intentional activities [9]. In Indonesia, cases of fire in buildings that occurred in DKI Jakarta as a result of damage to electrical components (electrical short circuit) as many as 494 events from a total of 692 events in 2018 [10, 32].

Maintenance and operations are the phases passed after the initiation phase, the planning phase and the construction phase are completed in a series of construction industry processes. Poor maintenance is associated with damage that occurs in a building [11]. Many factors cause damage to electrical components in building that have the impact as mentioned earlier. One of them is the poor performance of maintenance of electrical components in buildings [12].

The maintenance and upkeep of a building become important because it plays a role in maintaining the function of every component in the building and keeping the building in its original condition [13, 14] so that user satisfaction can be achieved. Lack of awareness regarding damage prevention and the mindset "If it hasn't been damaged, don't fix it" [15] must be eliminated.

2 Research Objective

This research will produce a preventive maintenance guideline for electrical components in buildings. In realizing it there are several stages so that the resulting guidelines are optimal. These stages are:

1. Identify policies and regulations that are in force today and are relevant to preventive maintenance
2. Identify the work package (level 4) electrical components of the building that need to be prevented from maintenance
3. Identify the procedures and guidelines used in preventive maintenance work so that a preventive maintenance guideline can be developed on a WBS-based

The above stages are carried out so that a guideline that is produced can cover all electrical components in buildings, especially government buildings or in other words comprehensive.

3 Literature Review

3.1 Government Buildings

Building is a physical form of the result of construction work which is integrated with the place of domicile, partly or wholly on top and/or in land and/or water, which functions as a place for humans to carry out their activities, both for residential or residence place, religious activities, activities business, social, cultural and special activities [2, 5, 16–18]. Furthermore, state buildings can simply be interpreted as buildings owned by the state or government or in other words, state buildings are buildings for official purposes that become/will become state property and are held with funding sources that come from state budget funds, and/or region budget funds, and/or other legitimate sources of financing [16, 18]. According to Wimala and Tamin [13], maintaining government buildings is a necessity coupled with inadequate policies and regulations related to the maintenance of buildings.

3.2 Preventive Maintenance

According to Beichelt and Fischer [19], there are two general maintenance policies, namely (the age of replacement) carried out when damage occurs or the replacement is done at a certain time and (minimal repair-replacement) replacement is always done at when certain time. Furthermore, [19] also stated, minimal repair in practice is not always possible to restore the function of a system after damage occurs. It can be concluded that it is important to prevent damage to a system. Damage to a component cannot be avoided but can be prevented by choosing the type of maintenance, namely preventive maintenance [20–22] because corrective maintenance is considered less effective in terms of frequent damage to a component that results in high maintenance costs [23].

Preventive maintenance is an activity to minimize the damage done at certain time intervals based on the experience or recommendations of the manufacturing company (OEM Recommendation) [21]. It is important to note these time intervals because too frequent time intervals will cause high maintenance costs [24]. Based on Peng et al. [25], one of the activities that must be carried out before preventive maintenance activities are inspection because inspection is a very effective activity to measure the performance of a system. Furthermore Ref. [5], states that preventive maintenance activities for building electrical components are inspection, service, and maintenance.

3.3 Work Breakdown Structure

WBS (Work Breakdown Structure) is a hierarchical decomposition of work-oriented results that will be carried out by the project team to achieve project objectives and create the required results [26]. WBS is created after the stage of determining the scope is completed. The predetermined scope of work is structured in WBS to make it easier to manage [27]. Every project activity, in this case, maintenance is important to be planned as well as possible. WBS is an effective tool for managing projects at the planning stage [28].

3.4 Guideline

Guidelines are defined as a framework, procedure, or process, or step by step used to plan or decide on something [21]. In a preventive maintenance work guide, usually presented several things, one of which is a table that lists maintenance activities with a daily, weekly, monthly or yearly schedule. The purpose of a guideline is as a reference for doing something that includes technical aspects and planning aspects [29].

If reviewed according to policies and regulations in Indonesia, the Guidelines are intended to meet the technical requirements of buildings [30]. Guidelines for building maintenance are a reference to maintaining the reliability of buildings and their infrastructure and facilities so that buildings are always worthy of function [5].

4 Research Methodology

The research methodology used is divided into several stages, namely archive analysis to obtain data and information related to preventive maintenance including relevant previous research, literature studies, and validation by experts in the field of building maintenance. The entire methodology is intended to identify policies and regulations that are applicable and relevant to preventive maintenance of buildings, identify WBS electrical components of buildings consisting of level 1, level 2, level 3, level 4 including alternative designs and identify applicable procedures and guidelines related to preventive maintenance with the time interval required to carry out each of its activities. After all the stages are carried out, a preventive maintenance guideline for electrical components of the building can be made based on WBS that has been identified previously, so that guidelines are obtained consisting of level 1 to level 4 of electrical components along with alternative designs with each preventive maintenance activity and intervals time for each component.

Validation carried out by experts in the field of maintenance includes validation of applicable policies and regulations, validation of the WBS that has been made, validation of the actions and time intervals of each electrical component and finally validation of guidelines for preventive maintenance of electrical components of the building as a whole.

5 Result and Discussion

5.1 Policies and Regulations

In Indonesia, the policies and legal regulations in the form of hierarchies are interconnected with one another. The law is the highest source of law followed by government regulations and ministerial regulations [31]. In connection with preventive maintenance, the underlying policies and regulations are Law of the Republic of Indonesia Number 28 of 2002 concerning Buildings, Government Regulation of the Republic of Indonesia Number 36 of 2005 concerning Implementing Law of Republic Indonesia Number 28 of 2002 concerning Building, Minister of Public Works Regulation number 24/prt/m/2008 concerning guidelines for building maintenance and Minister of Public Works Regulation number 16/prt/m/2010 concerning technical guidelines for periodic inspections of buildings.

5.2 WBS for Electrical Components of Buildings

Based on the results of archival analysis and literature studies on policies and regulations and previous research, WBS obtained for electrical components of buildings consisting of level 1, namely the name of the project, level 2, namely the work section, level 3, namely the type of work and level 4 in the form of work packages with derivatives, namely alternative designs. Alternative design is not a level in WBS because it cannot be measured [27] (see Fig. 1).

There are 4 (four) types of work (level 3) for WBS electrical components of buildings that consist of power supply, distribution systems, electricity loads and electronics. At level 4 (work packages) there are 22 (twenty-two) work packages consisting of transformers, UPS (Uninterruptible Power Supply), generators, solar panels, MVMDP, LVMDP, electricity loads, lighting systems, lighting control systems, switches, under floor duct/raised floor systems, cables, ground systems, lighting rod systems, fire alarm systems and detectors, telephones, sound systems, network systems, closed-circuit television (CCTV), signal amplifiers, building automation systems (BAS), presence machines.

Derivatives from level 4 or work packages are alternative designs that are not included in the WBS level but are still part of the WBS consisting of 45 (forty-five) components which are described in full in Table 1.

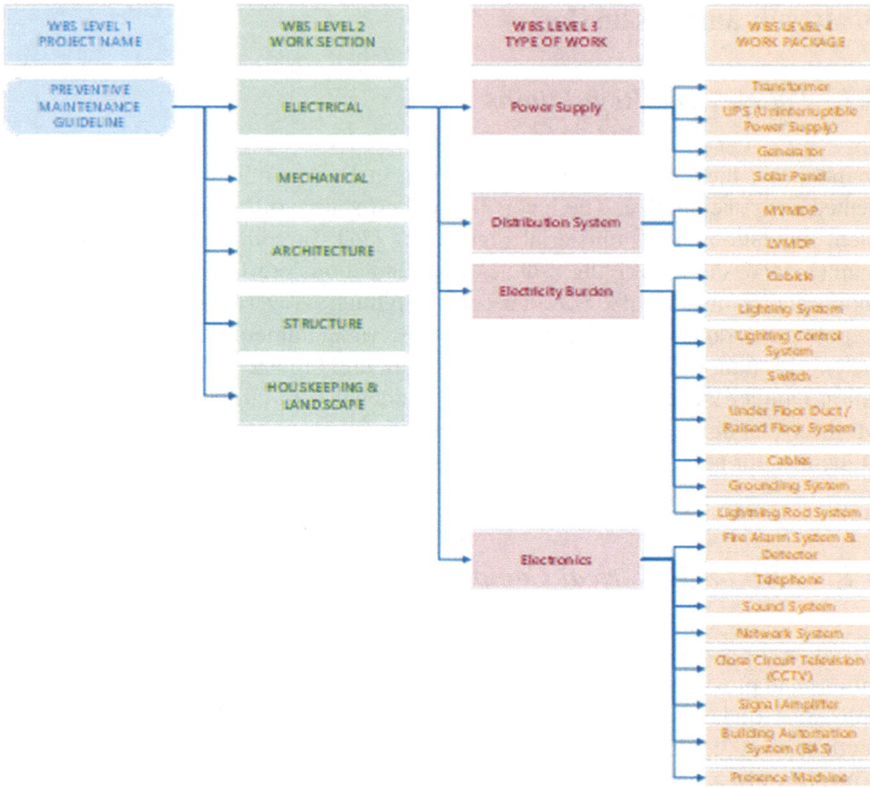

Fig. 1 WBS for electrical components of buildings

Table 1 Alternative design

WBS level 4 (work package)	Alternative design
Detail	Detail
Transformer	Dry transformer
	Liquid transformer
UPS (uninterruptible power supply)	Local UPS (uninterruptible power supply)
	Central UPS (uninterruptible power supply)
Reserved power supply	Generator

(continued)

Table 1 (continued)

WBS level 4 (work package)	Alternative design
Detail	Detail
	Gas generator
Renewable power supply	Solar panel
	Biodiesel generator
MVMDP	Oil circuit breaker
	Vacuum circuit breaker
LVMDP	Miniature circuit breaker
	Molded case circuit breaker
Cubicle	Cubicle
Lighting system	LED
	HID
	Fluorescent
Lighting control system	Dimmer
	Lighting censor
	Time censor
Switch	Electric socket
	Switch
Under floor duct/Raised floor system	Under floor duct
	Raised floor system
Cable	High voltage cable
	Medium voltage cable
	Low voltage cable
Grounding system	Solid grounding
Lightning rod	Lightning rod
	Lightning rod–tight wire
	Lightning rod—meshed cage
Fire alarm system and detector	Smoke detector
	Heat detector
	Fire panel
	Fire alarm
Telephone	Telephone
Sound system	Sound system
Network system	Ethernet cable
	Wi-Fi
CCTV	TV

(continued)

Table 1 (continued)

WBS level 4 (work package)	Alternative design
Detail	Detail
	Wall display
	Closed circuit television (CCTV)
Signal amplifier	Cellphone signal amplifier
Building automation system (BAS)	Censors
	Control panel
Presence machine	Digital presence system

5.3 Preventive Maintenance Guidelines for Building Electrical Components

Preventive maintenance activities in this guideline include inspection, service, and maintenance [5, 25]. These activities are attached to every electrical component included in the alternative design category. Each activity consists of preventive maintenance actions and procedures that have been validated by experts in the field of maintenance. Every preventive maintenance actions and procedure for each electrical component is equipped with a certain time interval. Generally, the time intervals for preventive maintenance are daily, monthly, weekly and yearly [29]. The time interval is also validated by experts in the field of maintenance.

For example, for liquid transformer components (Table 2), the procedures and preventive measures are checking safety relays, checking bushings, checking terminals, checking transformer stands, checking transformers' physical conditions, checking transformer temperatures, checking safety and measuring tools, checking temperatures and transformer chamber air condition, check cable connections in bushing terminals and grounding systems, clean the outside of the transformer, adjust the temperature and condition of the transformer chamber air and replace equipment and parts if damage occurs. All procedures and precautions are carried out at time intervals every 6 (six) months except for replacements done at time intervals every 1 (one) year.

In the end, all phases that have been carried out are grouped into a comprehensive format consisting of building electrical components, activities, procedures and preventive measures of building electrical components together with the time intervals of each of these components so that comprehensive guidelines for buildings are produced as shown in Table 3.

Table 2 Preventive maintenance for liquid transformer

Component	Activities	Preventive actions and procedures	Time interval
Liquid Transformer	Inspection	Check safety relays	Biannually
		Check bushing	Biannually
		Check the terminal	Biannually
		Check the transformer holder	Biannually
		Check the physical condition of the transformer	Biannually
		Check the temperature of the transformer	Biannually
		Check security and measurement equipment	Biannually
		Check the temperature and air condition of the transformer room	Biannually
		Check the cable connections on the bushing terminal and the grounding system	Biannually
	Service	Clean the outside of the transformer	Biannually
		Adjust the temperature and air condition of the transformer room	Biannually
	Repair	Replace equipment and spare parts if damage occurs	Annually

Table 3 Preventive maintenance guidelines

WBS		Detail	
WBS level 2		Work section	Electrical
WBS level 3		Type of work	Power supply
WBS level 4		Work package	Transformer
		Alt. design	Liquid transformer
Activities	Preventive actions and procedures	Time interval	
		Biannually	Annually
Inspection	Check safety relays	√	
	Check bushing	√	
	Check the terminal	√	
	Check the transformer holder	√	
	Check the physical condition of the transformer	√	
	Check the temperature of the transformer	√	
	Check security and measurement equipment	√	
	Check the temperature and air condition of the transformer room	√	
	Check the cable connections on the bushing terminal and the grounding system	√	
Service	Clean the outside of the transformer	√	

(continued)

Table 3 (continued)

WBS		Detail	
WBS level 2		Work section	Electrical
WBS level 3		Type of work	Power supply
WBS level 4		Work package	Transformer
		Alt. design	Liquid transformer
Activities	Preventive actions and procedures	Time interval	
		Biannually	Annually
	Adjust the temperature and air condition of the transformer room	√	
Repair	Replace equipment and spare parts if damage occurs		√

6 Conclusion

Some policies and regulations relating to government buildings, especially those relating to maintenance, are sufficient to be used as a basis. In preventive maintenance, there are 4 (four) policies and regulations that apply as a reference. The identification of electrical components obtained from the policies and regulations can be compiled using WBS so that all electrical components in buildings can be identified completely, thoroughly and well-structured. From the results of the validation, there are 22 (twenty-two) work packages and 45 (forty-five) alternative designs for electrical components in buildings that need preventive maintenance with each activity, procedure and preventive action and time intervals. The role of the WBS in developing Preventive Maintenance Guidelines is very large. The results obtained describe activities that are easy to understand and very neatly arranged. Each level in the WBS has its own role according to the existing theory. The greater the level shows the portion to break something down into more detail. Furthermore, Preventive Maintenance Guidelines can be developed again by identifying electrical components that have not been listed according to technological advances.

Acknowledgements The authors would like to thank the members of the Maintenance Building Team who have helped researchers to provide support in the form of both financial and moral support in completing this research.

References

1. Direktorat Jenderal Anggaran Kementerian Keuangan (2015) Penundaan/Moratorium Pembangunan Gedung Kantor Kementerian Negara/Lembaga, Januari 8. Dipetik September 24, 2019, dari Direktorat Jenderal Anggaran Kementerian Keuangan. http://www.anggaran.depkeu.go.id/dja/edef-konten-view.asp?id=1026

2. Pemerintah Indonesia (2002) Undang-Undang No. 28 Tahun 2002 Tentang Bangunan Gedung. Lembaran Negara Republik Indonesia Tahun 2002, No. 134, Desember 16. Jakarta, DKI Jakarta, Sekretariat Negara, Indonesia
3. Wimala M, Karim EKA, Adianto YLD, Setiawan TH, Roy AFV (2017) Pengembangan Sistem Pelaporan Online Kerusakan Bangunan Asrama Kampus di Malaysia. Raka Racana: Jurnal Online Institut Teknologi Nasional, Sept 2017
4. Chanter B, Swallow P (2007) Building maintenance management, 2nd edn. Blackwell Publishing, Oxford
5. Pemerintah Indonesia (2008) Peraturan Menteri Pekerjaan Umum No. 24/PRT/M/2008 Tentang Pedoman Pemeliharaan dan Perawatan Bangunan Gedung. Kementerian Pekerjaan Umum, Jakarta, DKI Jakarta, Indonesia , Desember 30
6. Project Management Institute (2017) A guide to the project management body of knowledge (PMBOK® Guide), 6th edn. Project Management Institute, Newtown Square
7. U.S. GAO (2009) GAO cost estimating and assessment guide: best practices for developing and managing capital program costs. USGAO, Washington, District Of Columbia, USA
8. Moghaddam KS (2008) Preventive maintenance and replacement scheduling: models and algorithms. Comput Ind Eng 1:85
9. U.S. Fire Administration (2016) Download fire data and data analysis tools. Dipetik Agustus 22, 2019, dari U.S. Fire Administration, Working for a fire-safe America: https://www.usfa.fema.gov/data/statistics/order_download_data.html
10. Setiyo B (2014) Korsleting Listrik Penyebab Kebakaran Pada Rumah Tinggal Atau Gedung. Edu Elektrika J 3(2).
11. Wardhana K, Hadipriono FC (2003) Analysis of recent bridge failures in the United States. J Perform Constr Facil 17(3):144–150. https://doi.org/10.1061/(ASCE)0887-3828(2003)17:3(144)
12. Ahzahar N, Karim NA, Hassan SH, Eman J (2011) A study of contribution factors to building failures and defects in construction industry. Procedia Eng 20:249–255. https://doi.org/10.1016/j.proeng.2011.11.162
13. Wimala M, Tamin RZ (2014) Pengembangan Konsep Perencanaan Biaya Pemeliharaan Rutin Gedung Pendidikan di Institut Teknologi Bandung. Jurnal Itenas Rekayasa, May 2014, 11. Retrieved from https://www.researchgate.net/profile/Mia_Wimala/publication/316282959_Pengembangan_Konsep_Perencanaan_Biaya_Pemeliharaan_Rutin_Gedung_Pendidikan_di_Institut_Teknologi_Bandung/links/5939f83e0f7e9b32b749bae3/Pengembangan-Konsep-Perencanaan-Biaya-Pemeliharaa
14. Yong CY, Sulieman MZ (2015) Assessment of building maintenance management practice and occupant satisfaction of School Buildings in PERAK, Malaysia. Jurnal Teknologi (Sciences & Engineering) 5:57–61
15. Njuangang S, Lasanthi Liyanage C, Akintoye A (2016) Performance measurement tool (PMT) to control maintenance-associated infections. Facilities 34(13–14):766–787. https://doi.org/10.1108/F-12-2014-0107
16. Pemerintah Indonesia (2005) Peraturan Pemerintah No. 36 Tahun 2005 Tentang Peraturan Pelaksanaan Undang-Undang No. 28 Tahun 2002 Tentang Bangunan Gedung, September 10. Lembaran Negara Republik Indonesia Tahun 2005, No. 83. Jakarta, DKI Jakarta, Indonesia: Menteri Hukum dan Hak Asasi Manusia
17. Pemerintah Indonesia (2010) Peraturan Menteri Pekerjaan Umum No. 16/PRT/M/2010 Tentang Pedoman Teknis Pemeriksaan Berkala Bangunan Gedung, Desember 31. Berita Negara Republik Indonesia Tahun 2010, No. 701. Jakarta, DKI Jakarta, Indonesia: Kementerian Hukum dan Hak Asasi Manusia
18. Pemerintah Indonesia (2011) Peraturan Presiden No. 73 Tahun 2011 Tentang Pembangunan Bangunan Gedung Negara. Jakarta, DKI Jakarta, Oktober 11. Sekretariat Kabinet, Indonesia
19. Beichelt F, Fischer K (1980) General failure model applied to preventive maintenance policies. IEEE Trans Reliabil R-29(1):39–41. https://doi.org/10.1109/TR.1980.5220704
20. Murthy DNP, Atrens A, Eccleston JA (2002) Strategic maintenance management. J Qual Mainten Eng 8(4):287–305. https://doi.org/10.1108/13552510210448504

21. Ahmad R, Kamaruddin S, Azid I, Almanar I (2011) Maintenance management decision model for preventive maintenance strategy on production equipment. J Ind Eng Int 7(13):22–34
22. Lavy S, Garcia JA (2012) KPIs for facility' s performance assessment Part II: identification of variables and deriving expressions for core indicators. Facilities 32(5):275–294. https://doi.org/10.1108/F-09-2012-0067
23. Chua SJL, Zubbir NB, Ali AS, Au-Yong CP (2018) Maintenance of highrise residential buildings. Int J Build Pathol Adapt 36(2):137–151. https://doi.org/10.1108/IJBPA-09-2017-0038
24. Vilarinho S, Lopes I, Oliveira JA (2017) Preventive maintenance decisions through maintenance optimization models: a case study. Procedia Manuf 11(June):1170–1177. https://doi.org/10.1016/j.promfg.2017.07.241
25. Peng R, Liu B, Zhai Q, Wang W (2019) Optimal maintenance strategy for systems with two failure modes. Reliabil Eng Syst Saf 188:624–632, July 2017. https://doi.org/10.1016/j.ress.2017.07.014
26. Project Management Institute (2006) Practice standard for work breakdown structures, 2nd edn. Project Management Institute, Newton Square
27. Hidayah DN, Latief Y, Riantini LS (2018) Development of work breakdown structure standard based on risk for safety planning on dam construction work. IOP Conf Ser Mater Sci Eng 420(1). https://doi.org/10.1088/1757-899X/420/1/012003
28. Burghate N (2018) Work breakdown structure: simplifying project management. Int J Commer Manag Stud (IJCAMS) 3(2):1–5. Retrieved from www.ijcams.com
29. Starova-Sulejman L, Mojsoski N, Lega NH (2015) Guidelines manual for maintenance of school buildings. Macedonian Civic Education Center (MCEC), Skopje, North Macedonia
30. Pemerintah Indonesia (2006) Peraturan Menteri Pekerjaan Umum No. 29/PRT/M/2006 Tentang Pedoman Persyaratan Teknis Bangunan Gedung. Jakarta, DKI Jakarta, Indonesia: Kementerian Pekerjaan Umum
31. Saraswati R (2009) Perkembangan Pengaturan Sumber Hukum Dan Tata Urutan Peraturan Perundang-Undangan Di Indonesia. Media Hukum IX(2):48–59
32. BPBD DKI Jakarta (2018) Data Rekapitulasi Kejadian Kebakaran di Provinsi DKI Jakarta Tahun 2018. Dipetik Agustus 22, 2019, dari JAKARTA open data, Berbagi data untuk transparansi. http://data.jakarta.go.id/dataset/0cb0984f-03a8-43b4-b481-7c55b24e947e/resource/7d595666-39bd-4803-aecc-801ac60014a9/download/REKAPITULASI-KEJADIAN-KEBAKARAN-TAHUN-2018.csv

Evaluation of the Implementation of Fire Safety Management Based on Work Breakdown Structure Affecting the Insurance Premium Costs of High-Rise Lecture Buildings

E. P. Mahardika, F. Muslim, Y. Latief, and P. S. Nugroho

Abstract High-rise buildings, especially lecture buildings, have the potential for fire hazards which can cause large losses. Fire risk is one of the factors that must be minimized in a building operation where insurance is a way of transferring the risk. However, the amount of fire insurance premiums in various insurance institutions varies greatly due to differences in the determinants of insurance premium costs. The purpose of this study is to evaluate the application of fire safety management in high-rise lecture buildings in Jakarta and determine the determinants of premium costs that are influenced by the implementation of fire safety management. This determining factor is a tool to determine the amount of the insurance company's fire insurance premiums. This research refers to literature studies, surveys, interviews, and data obtained using a questionnaire instrument. This study uses the work breakdown structure as a tool to detail indicators that meet the fire safety management criteria and the data are analyzed using the SmartPLS 3.0 software. Based on the survey results, the data shows that the lecture buildings in Jakarta haven't fully implemented fire safety management. In addition, there is no clear relation regarding the role of insurance in fire protection financing in lecture buildings. To address this, we propose to consider the extent to which high-rise lecture buildings implement fire safety management in determining insurance premium rates to improve the application of fire safety management in high-rise buildings, by processing the data using SmartPLS 3.0 and average analysis in order to obtain priority factors in fire safety management which can be used as a determinant of realistic premium costs. As the result, the most priority factor in fire safety management which can be used as a determinant of realistic premium costs is fire prevention in lecture buildings, and the least priority factor is safety of people in the event of fire.

Keywords Fire safety management · Work breakdown structure · Insurance premium costs · Lecture buildings

E. P. Mahardika (✉) · F. Muslim · Y. Latief · P. S. Nugroho
Civil Engineering Department, Faculty of Engineering, Universitas Indonesia, Depok, Indonesia
e-mail: ero.prahara@ui.ac.id

F. Muslim
e-mail: fadhilah@eng.ui.ac.id

S. A. Kristiawan et al. (eds.), *Proceedings of the 5th International Conference on Rehabilitation and Maintenance in Civil Engineering*, Lecture Notes in Civil Engineering 225,
https://doi.org/10.1007/978-981-16-9348-9_32

1 Introduction

Building fires can occur in any type of designation of buildings, both residential settlements, roles, and educational facilities. In Indonesia, several campus (educational) buildings occurred in the last ten years. The fire that occurred at the Campus of the Islamic University of Malang [1] was a campus fire incident in June 2016 which was suspected to have been caused by a short circuit (electric short circuit). The incident burned 2 of the 3 floors of the existing building and consumed several laboratory rooms and their contents.

Campus fires not only occur in Indonesia, but these incidents also often occur in several universities in America. It is recorded that from 2000 to 2015 there have been 85 campus fire incidents, or an average of 7 campus fires per year, totaling 118 lives. A total of 58% of fire incidents that claimed lives originated from the dormitory building, a student gathering place whose smoke detection equipment was not working (damaged), while 85% of fatalities occurred in buildings that were not equipped with sprinklers. And the cause of most events was power disruption (Fig. 1).

The causes of fires can vary but can be categorized into two aspects, namely human carelessness and failure of the building engineering system. These two aspects work both during planning, construction implementation, and building operations (post-occupancy). In Indonesia, the failure of engineering systems is higher, especially the failure of the electrical system in buildings, such as short-circuit electricity [2]. A good fire safety management implementation is the right effort in fire management, starting from the planning stage to the follow-up. This action needs to be done in Indonesia considering that Indonesia is one of the countries that have frequent fires [3].

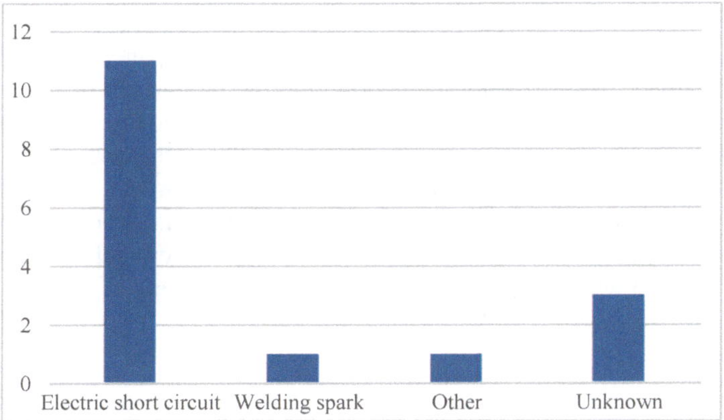

Fig. 1 Graph of the cases of fires in Jakarta. *Source* republika.com (2018), cnnindonesia.com (2019), and detik.com (2020)

To anticipate and mitigate the risk of fire, the owners of tall buildings cooperate with an insurance company as a form of risk transfer. The form of cooperation between individuals/legal entities and insurance companies is risk direction, either partially or completely [4].

Therefore, each building must have fire management that plays a role in risk prevention and countermeasures to achieve safety in fire events. The Provincial Government of DKI Jakarta issued a Regional Regulation of the Special Region of the Capital City of Jakarta Number 8 2008 on Fire Prevention and Prevention to overcome the fire problem in Jakarta. The government through the Ministry of Public Works and Public Housing also issued a provision referring to the Ministry of Public Works and Public Housing 20/PRT/M/2009 [5] on technical guidelines for fire protection management in urban areas, which among others state that every public building including apartments, which are inhabited by at least 500 people, or that have an area of at least 5000 m^2, or have a building height of more than 8 floors, is required to apply fire safety management.

The implementation of fire safety management in buildings can be divided into four points [6], namely:

1. Fire prevention in buildings
2. Safety of people in the event of a fire
3. Monitoring, auditing, and reviewing fire safety systems
4. Reactive monitoring (reporting, recording, and investigation)

Given the importance of fire safety, careful planning is required before the implementation of the project. According to the 6th edition of PMBOK [7], there are several stages in planning the scope of the project, such as collecting requirements, determining the scope of work, and creating a work breakdown structure.

The determination of fire premium rates in Jakarta is regulated in the Attachment to the Circular letter of the Otoritas Jasa Keuangan (OJK) Number 6/SEOJK05/2017 [8] Attachment I. The attachment determines the amount of premium tariff by the lower limit and upper limit, which is only divided based on the occupancy of the building and its Construction Class. The impact, according to the Indonesian Insurance Bureau, Ministry of Finance Decree, and Otoritas Jasa Keuangan said that in the last 21 years the growth rate of insurance premiums averaged 22% per year, while in the last 4 years (until 2017) growth according to up to only 8% per year.

This research was conducted in the province of DKI Jakarta and the object of research is a high-rise lecture building. The focus of this research is on mechanical electrical work in which there is fire safety management work.

2 Literature Review

2.1 High-Rise Lecture Buildings

According to Republic of Indonesia Government Regulations Number 36 of 2005 [9] and Regional Regulation of DKI Jakarta Province Number 7 of 2010 [10], a building is a physical form of construction work that integrates with its domicile, partly or wholly above and/or above. On land and/or in water, which functions as a place for humans to carry out their activities, whether for shelter or residence, religious activities, business activities, social, cultural, or special activities.

The development of science and technology must be accompanied by high formal education and adequate infrastructure. Therefore, a university was established which could provide knowledge widely and globally. Meanwhile, there is an increase in the need for lecture buildings that must be served, for that it is necessary to increase the usability of the building either in the form of adding new buildings or constructing new buildings in more appropriate and strategic places.

According to National Fire Protection Association 101 [11] or National Fire Protection Association 101A [12], one type of building based on the level of occupancy load is an educational building where this building has the objective of activities related to education through classrooms with 6 (six) occupants or more. for 4 h per day or more than 12 h per week.

2.2 Fire Safety Management Based on Work Breakdown Structure

Given the importance of fire safety, careful planning is required before the implementation of the project. According to the 6th edition of PMBOK [7] there are several stages in planning the scope of the project, such as collecting requirements, determining the scope of work, and creating a work breakdown structure.

Work breakdown structure is a multi-level framework for organizing and graphically representing the elements of work displayed to be completed in logical relationships [13]. Work breakdown structure is one of the tools available for a project manager to define and structure projects [14]. A well-designed work breakdown structure is critical to the success of the proposed cost estimation methodology [15]. WBS can also be used as a repetitive process in the form of templates useful for future projects [16].

2.3 *Insurance Premium Costs*

Insurance is a tool used in reducing the risk inherent in the economy where it combines a large enough number of units affected by the same risk to predict the potential that losses will occur and if these losses do occur, it will be shared by all parties that are joined together proportional [17].

The determination of fire premium rates in Jakarta is regulated in the Attachment to the Circular letter of the Otoritas Jasa Keuangan (OJK) Number 6/SEOJK05/2017 [8] attachment I. The attachment determines the amount of premium tariff by the lower limit and upper limit, which is divided based on the occupancy of the building and its Construction Class. The types of coverage according to Article 247 of Kitab Undang-Undang Hukum Dagang (KUHD):

1. Fire hazard
2. Dangers that threaten agricultural products that have not been harvested
3. The soul of one or more people
4. The danger of the sea and the danger of slavery
5. Transportation hazards on land, in rivers and inland waters

In Undang-Undang Number 40 of 2014 [18] on Insurance, paying premiums is the obligation of an insurance policy holder. Article 1 point 29 states that the premium is the amount of money determined by the insurance company or reinsurance company and approved by the policyholder to be paid under the insurance agreement or reinsurance agreement, or the amount of money stipulated under the provisions of the legislation underlying the insurance program is obliged to obtain liability benefits. The amount of premium for participation in the insurance to be paid has been determined by the insurance company about concerning to the circumstances of the insured [19].

3 Research Methodology

Research method is a scientific way to obtain data for specific purposes and uses [20]. The scientific way means that research activities are based on scientific characteristics, namely rational, empirical, and systematic. To answer the formulation of these research question, a literature review was conducted and validation was carried out using the Delphi method. The data was collected through a survey of respondents using a questionnaire instrument and a case study method with a simulation of the cost of fire insurance premiums for lecture buildings based on determined variables. The questionnaire used contains indicators of fire safety management from literature reviews and the results of leverage, clarification, and validation by experts. The questionnaire was filled out by the management of the high-rise lecture building. Each response item in the Likert sample scale has five responses, namely very influential, influential, moderately influential, not influential, and very unaffected. The results of this data are analyzed descriptively and processed using SmartPLS software, and the average analysis is to obtain priority factors in fire safety management which can be used as a realistic determinant of premium costs.

In this research, data were obtained as many as 31 respondents building management of the lecture building and 31 respondents are fire insurance underwriters. The questionnaire was filled in by the building management of high-rise lecture buildings and fire insurance agencies in Jakarta. Criteria for building management respondents include bachelor or diploma education and 5 years of building management experience, especially in the fire prevention department of lecture buildings. And the criteria for fire insurance underwriter include a minimum education of bachelor or diploma, have an underwriter position in an insurance company and at least 2 years of experience.

4 Results and Discussion

4.1 Mechanical Electrical Work Breakdown Structure on High-Rise Lecture Buildings to Plan Fire Safety Management

At this stage, interviews and discussions with experts are conducted to identify and validate the work breakdown structure related to the components of fire safety management for mechanical and electrical work in the high-rise lecture building.

The work breakdown structure of the literature review was then carried out by the experts to get the initial validation, clarification, and validation related to the fire safety management components of mechanical and electrical work in the high-rise lecture building. Here are the results (Table 1):

4.2 The Implementation of Fire Safety Management on High-Rise Lecture Buildings in Jakarta Based on Work Breakdown Structure

The work breakdown structure of mechanical electrical work in the study [21] was developed into a work breakdown structure for mechanical electrical work that meets fire safety criteria. The points contained in the work breakdown structure of mechanical electrical work are then carried out preliminary validation to several experts to obtain verification, clarification, and validation related to the components of fire safety management for mechanical and electrical in high-rise lecture buildings.

A book [6] entitled introduction to fire safety management explains and mentions indicators of fire safety management. The fire safety management indicator in the

Table 1 Work breakdown structure fire safety

WBS level 3	WBS level 4
Sub work section	Work package
Fire system	Fire pump
	Panel control
	Cable power
	Pipe riser
	Diesel tank
	Fire pipe
	Valve and accessories
	Hydrant box
	Sprinkler head
	Fire extinguisher
	Pillar hydrant
	Siamese connection
	Compartment
	Fire lift
	monitoring center
	Emergency lighting
	Fire system testing
Fire alarm	Main control panel
	Announciator
	grounding system
	Cable feeder
	Detector
	Manual break glass push button
	Alarm bell
	Firemen telephone
	Terminal box
	Cable tray
	Fire alarm testing
Sound system	Paging
Gas	–
Telephone	–
Data	–
MATV	–
CCTV	–
Security system	–
Radio communication	–
Audio videophone	–
Building autosystem	Monitoring center
Electrical	Emergency lighting
Air conditioning works	Pressurized fan
Transport equipment	Fire lift

books [6] are added by several additional indicators from the results of the development of work breakdown structures for mechanical electrical work that meet the fire safety criteria. Furthermore, these indicators are used to conduct this research.

The results of the implementation of fire safety management in the high-rise lecture building will be obtained using descriptive analysis of the data from the questionnaire. Descriptive analysis is done by describing or explaining the collected data. Based on data from samples of high-rise lecture buildings in Jakarta, data on the fire safety management indicator applied to the building were obtained. The following is a recapitulation of percentage data by high-rise lecture buildings which are presented in tables and diagrams:

Based on these four dimensions, fire safety management hasn't been fully implemented 100% in every indicator. The lowest average implementation is the dimension of safety people in the event of a fire which is 78%, while fire protection in a building is 83%, reactive monitoring (reporting, recording and investigation) 82%, and monitoring auditing and reviewing fire safety systems 80%. This percentage shows that high-rise lecture buildings in Jakarta have not fully implemented fire safety management, which is also in line with previous research [22].

4.3 *Indicators of Fire Safety Management Implementation that Affecting Insurance Premiums Cost*

Based on data testing through SmartPLS 3.0 software obtained by T-Statistic which shows that the four dimensions of fire safety management affect the cost of premium (Y) significantly Table 2. This is also in line with the results of the analysis that uses the average value, where the priority of dimension influence has the same order as the highest position, namely Fire Prevention in Building Buildings and the lowest is People Safety in Fire Incidents Fig. 2.

Based on Table 2 it is known that the order of variable implementation of fire safety management that affects premium costs are fire prevention, reactive monitoring, active monitoring and people safety. This is in accordance with the journal [22] which states that the implementation of fire safety management affects the calculation of premium costs.

In addition, it is also obtained through SmartPLS 3.0 software with path coefficient test as shown in Table 3, that the path coefficient of the calculation result is close to

Table 2 Significance indicators of fire safety management implementation

T-statistic		
X1 → Y	2.227	Significant
X2 → Y	1.951	Less significant
X3 → Y	2.009	Significant
X4 → Y	2.137	Significant

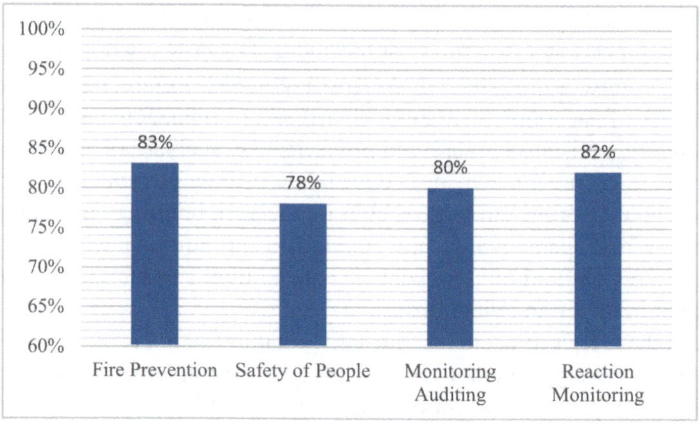

Fig. 2 Graph of the implementation of fire safety management in Jakarta

Table 3 Path coefficient

X	Y
X1	−0.291
X2	−0.069
X3	−0.194
X4	−0.280

−1. This indicates a strong negative relationship, and proves that the more applied fire safety management, the smaller the cost of insurance premium.

5 Conclusions

Based on the discussion of the research results that have been described in the previous chapter, it can be concluded:

1. Obtained work breakdown structure of mechanical electrical as shown in Table 1. in high-rise lecture buildings.
2. Safety management based on the work breakdown structure in the high-rise lecture buildings in Jakarta hasn‘t fully implemented. With an average value of 81%, it means that it is in the good category
3. Factors/indicators determining premium costs based on the quality of work breakdown structure based fire safety management in high-rise lecture buildings are derived from the four dimensions of fire safety management with the

order of the most priority is fire prevention in building buildings, reactive monitoring (reporting, recording, and investigation), monitoring, audit, and review safety system, then people safety in fire incidents.

Acknowledgements The Authors would like to thank the financial support provided by Ministry of Research and Technology/National Research and Innovation Agency through PDUPT Grant 2021 with contract number: NKB-220/UN2.RST/HKP.05.00/2021 managed by the Directorate for Research and Community Engagement (DRPM) Ministry of Research and Technology/National Research and Innovation Agency (BRIN).

References

1. LNCS Homepage. https://news.detik.com/berita-jawa-timur/d-3239846/kampus-unisma-kebakaran-diduga-konsleting-listrik. Last accessed 2016/06/22
2. Sufianto H, Green AR (2012) Urban fire situation in Indonesia. Fire Technol 48(2):367–387
3. Nugraha R (2018) Penerapan Sistem Manajemen Kebakaran Di PT. Application of fire management system in PT. Adiluhung. Indonesian J Occupat Saf Health 7(3):378–386. https://doi.org/10.20473/ijoshv7i3.(2018).378
4. Kusumoardi AN, Rinitami Njatrijani S (2016) Klausula Baku dalam Polis Asuransi Kebakaran Ditinjau dari Undang-undang Nomor 8 Tahun 1999 Tentang Perlindungan Konsumen (Studi pada PT. Asuransi Jasa Indonesia). Diponegoro Law J 5(3):1–13
5. Peraturan Menteri PUPR No. 20/PRT/M/2009. Pedoman Teknik Manajemen Proteksi Kebakaran di Perkotaan (2009)
6. Furness A, Muckett M (2007) Introduction to fire safety management. Routledge
7. PMBOK (2017) PMBOK Guide, 6th edn. Project Management Institute, pp 104–104. https://doi.org/10.1002/pmj.20125
8. Jasa Keuangan O (2017) Surat Edaran Otoritas Jasa Keuangan Nomor. 6/SEOJK.05/2017 Tentang Penetapan Tarif Premi Atau Kontribusi Pada Lini Usaha Asuransi Harta Benda dan Asuransi Kendaraan Bermotor Tahun 2017
9. Peraturan Pemerintah Nomor. 36 Tahun 2005. Peraturan Pemerintah Republik Indonesia Nomor 35 Tahun 2005 Tentang Peraturan Peaksanaan Undang-Undang Nomor 28 Tahun 2002 Tentang Bangunan Gedung (1), pp 1–5 (2005)
10. Peraturan Daerah Provinsi DKI Jakarta Nomor 7 Tahun 2010. Peraturan Daerah Provinsi DKI Jakarta Nomor 7 Tahun 2010 Tentang Bangunan Gedung (2010)
11. National Fire Protection Association. NFPA 101-Life safety code (2000)
12. National Fire Protection Association (2000) NFPA 101-Guide on Alternative Approaches to Life safety code
13. Burghate M (2018) Work breakdown structure: simplifying project management. Int J Commerce Manag 3
14. Sutrisna M, Ramanayaka CD, Goulding JS (2018) Developing work breakdown structure matrix for managing offsite construction projects. Architect Eng Des Manag 14:381–397
15. Sequeira S, Lopes E (2015) Simple Method Proposal for cost estimation from work breakdown structure. Procedia Comput Sci 64:537–544
16. Supriadi R, Sagita L, Latief Y, Susilo B, Rajasa M (2017) Development of risk-based standardized WBS (Work Breakdown Structure) for cost estimation of apartment's project. Int J Civil Eng Technol (IJCIET):822–833
17. Sunyoto D, Putri WH (2017) Manajemen Risiko dan Asuransi. CAPS, Yogyakarta
18. KUHD (1936) Kitab Undang-Undang Hukum Dagang 49(1):1–9. https://doi.org/10.1016/j.solener.2019. 02.027

19. Agustiranda W, Yuliani Y, Bakar SW (2018) Pengaruh Pendapatan Premi, Pembayaran Klaim, Dan Risk Based Capital Terhadap Pertumbuhan Laba Pada Perusahaan Asuransi Yang Terdaftar Di Bursa Efek Indonesia. Sriwijaya University
20. Darmadi H (2013) Metode Penelitian Pendidikan dan Sosial. Alfabeta, Bandung
21. Ajizah N (2018) Perencanaan Sumber Daya Pada Pekerjaan Mekanikal dan Elektrikal Bangunan Gedung Apartemen Berbasis Work Breakdown Structure. Universitas Indonesia, Fakultas Teknik
22. Winarningsih YR (2002) Pemodelan Pengaruh Sistem Proteksi Kebakaran terhadap Premi Asuransi Kebakaran pada Bangunan Tinggi di Jakarta. Universitas Indonesia, Fakultas Teknik

A Critical Review of Bridge Management System in Indonesia

Surya Dewi Puspitasari, Sabrina Harahap, and Pinta Astuti

Abstract The bridge is a crucial infrastructure in public mobilization. Recently, the utilization of old bridges has intensely come into an option during limited funds in many developing countries. Hence, the bridge management system plays an important role in maintaining and reducing bridge failure. In Indonesia, the bridge management system is called Interurban Bridge Management System (IBMS). However, the IBMS implementation in Indonesia is still poor. This system has been implemented throughout all provinces in Indonesia since 1992, yet many bridges were still found damaged, even collapsed. At this moment, Indonesia is still focusing on construction, not maintenance work. Some developed countries like the United States of America and Japan have expanded a complex system for a bridge management system, Pontis, and J-BMS. Thailand as a developing country is the same as Indonesia has also realized the importance of having a bridge management system. Thailand has established a Bridge Maintenance and Management System (BMMS) to guide the periodic bridge inspection and maintenance program. In this study, the authors intend to grow the awareness of the IBMS application in Indonesia which consist of the development of bridge management system, the causes of poor IBMS implementation, the improvement needs to enhance the IBMS implementation, and recommended aspects may extend to current IBMS implementation in Indonesia by seeking for and integrating the advantageous information the existing bridge management system (BMS) that has been established and implemented in the U.S., Japan, and Thailand.

Keywords Bridge · IBMS · Maintenance · Management system

S. D. Puspitasari (✉) · S. Harahap
Department of Civil Engineering, Faculty of Military Engineering, Universitas Pertahanan RI, Bogor 16810, Indonesia
e-mail: surya.puspitasari@idu.ac.id

S. Harahap
e-mail: sabrina.harahap@idu.ac.id

P. Astuti
Department of Civil Engineering, Faculty of Engineering, Universitas Muhammadiyah Yogyakarta, Yogyakarta 55138, Indonesia
e-mail: pinta.astuti@ft.umy.ac.id

S. A. Kristiawan et al. (eds.), *Proceedings of the 5th International Conference on Rehabilitation and Maintenance in Civil Engineering*, Lecture Notes in Civil Engineering 225,
https://doi.org/10.1007/978-981-16-9348-9_33

1 Introduction

Indonesia has been becoming the world's 10th largest economy and the most enormous one in Southeast Asia [1], encouraging Indonesia's infrastructure to grow rapidly. Indonesia has a total of 18.648 existing bridges in 2019 which were reported around 14% in unhealthy condition [2]. The increasing traffic volume and the weight of vehicle, that pass through the bridge indicates there has been overloading which led to being suspected as the main cause of bridges deteriorated [3]. Bridges have an indispensable role in the transportation system to support people and goods in their mobility and accessibility which affect social and economic development. It is becoming substantial to avoid severe damages. The Ministry of Public Works and Public Housing of the Republic of Indonesia is the executive of its Interurban Bridge Management System (IBMS) through its agency Directorate General of Bina Marga. Unfortunately, the implementation of IBMS in Indonesia is still poor. Consequently, it has caused some bridges to be damaged and collapse during their service life, such as the collapse of the Kutai Kartanegara bridge in 2011. An assessment of bridge maintenance practice in Indonesia is needed to enhance the practices of the IBMS whether the procedure, standard, bridge engineer, bridge maintenance engineer, or bridge inspector.

On the other hand, the U.S. and Japan as developed countries have expanded their bridge management system through the Federal Highway Administration (FHWA) and Road Bureau of the Ministry of Land, Infrastructure, Transport and Tourism (MLIT), respectively. Besides, our neighbour country, Thailand the same as a developing country as Indonesia, has come to realize the importance of having a bridge management system. This country has established a bridge maintenance management system (BMMS) to undergo a periodic bridge inspection and maintenance program. Therefore, to enhance the practices of the IBMS in Indonesia, the authors wrote this paper to increase the awareness-raising of the IBMS implementation in Indonesia which consist of the development of bridge management system, the causes of poor IBMS implementation, the improvement needs to enhance the IBMS implementation and recommended aspects may extend to current IBMS implementation in Indonesia by seeking for and integrating the advantageous information from the existing bridge management system (BMS) that has been established and implemented in the U. S., Japan, and Thailand.

2 Development of Bridge Management System

The bridge management system is a series of actions to assure the structural performance of a bridge structure in good condition during its service life without heavy maintenance [4]. The BMS encompasses all activities related to bridges' management, for example, inspections, planning, programming, design, construction, and maintenance [5–8]. Aiming to boost the efficacy and quality of maintaining the bridge structure, the Directorate General of Bina Marga constructed the IBMS associated with SMEC-KinHill in 1991. Since then, all the National and Provincial Roads in Indonesia, including bridge infrastructures, has been using the IBMS for maintainace guidance [5, 9].

The U.S. through the FHWA has started creating the National Bridge Inspection Standard (NBIS) and made a program to train their bridge inspectors in 1971, soon after the Silver Bridge collapsed in 1967. The NBIS was responsible for the national policy concerning bridge inspection frequentness, inspector eligibility, advisory form, and rating method. The revised NBIS was in 1988, and a few years later, developing PONTIS for the BMS to help the authority in decision-making and planning regarding bridge monitoring, maintenance, and rehabilitation. The updated and modified NBIS provide more comprehensive and useful information [10].

Japan evolves its J-BMS by using advanced technologies and techniques in information processing and health monitoring. A combined life-cycle management system was built as Doctor for Infrastructure Systems, called InfraDoctor. This system has three main features, i.e., basic features for GIS and 3D point group data; management and search features for GIS data; and further sophisticated management and advanced features [11]. From this system, we could possess real-time monitoring, structural performance, prediction of behaviour, and the type of maintenance that could be done [12].

In 1989, Thailand's Department of Highway (TDOH) collaborates with the Danish Road Directorate to establish the bridge management and maintenance system (BMMS) for more than 10,000 concrete highway bridges in Thailand [6, 13]. However, due to several constraints and limitations in utilizing them, BMMS is not thoroughly applied in the present system. Thus, the defined works on the inspection and evaluation of bridge performance conducted by TDOH ally with TESCO Ltd. in 2008 [6]. In 2013, Thailand, JICA and Japanese's company set a new manual for bridge maintenance in rural area [14].

3 Factors May Cause the IBMS Implementation in Indonesia is Still Poor

The IBMS has been employed in Indonesia since 1993 [5], yet the implementation of this system is still poor. There were at least 17 collapsed bridges were recorded have taken place in Indonesia from 2007 to 2011 [15]. Recently, a suspension bridge in West Java was also reported almost collapsed in January 2021. It is only five years after the bridge has been completely constructed. It was presumed that there is a shifting part on the bridge due to flooding [16]. The authors outlined three factors that may cause the poor implementation of IBMS in Indonesia as follows:

3.1 Scarcity of Funds

Insufficient financial budgeting in bridge management has been a problem for all countries, including Indonesia. In the period from 2013 to 2018, the Indonesian government entrusts about 42 billion USD[1] to The Ministry of Public Works and Public Housing of the Republic of Indonesia to enhance infrastructure development. This amount is not only for the bridge but also for water resources, highways, housing, etc. [17]. In 2018, maintaining the 46,812 km road and the 495,698 m length of the bridge requires 1.65 billion USD [18]. Dananjoyo has calculated one bridge maintenance cost for replacement works that sum up to 3,5 million USD with the IBMS method in 2020 [19]. In addition, about 2 thousand bridges with 71,549.6 km length were recorded in poor condition in 2019 [20]. Comparing the maintenance cost of one bridge and the repair work needs for all bridges to the available funding is inadequate. The high bridge maintenance cost cannot be covered all by the government, thus it is reasonable if the IBMS implementation in Indonesia is still unsatisfactory.

3.2 Insufficient Human Resources

The development of transportation networks in Indonesia does not equal human resources number with bridge inspection speciality [21]. Inspection activities are fundamental works in BMS to determine the effective treatment for the bridge at the right time [6]. Roughly 7,398 people from different backgrounds and level of educations work for the Directorate General of Bina Marga that liable for the 32,790,18 km provincial road, the 211,644,08 km district road, the 35,241.40 km city street, the 2100 km length of the existing toll road, and 18,648 bridges with 510,336 km length [2]. Looking at the numbers of officials compared to the length of highways and bridges, it does make sense if there are insufficiently qualified inspectors and analytical resources, particularly in a rural area. The bridge inspector or maintenance officer may not certify nor an expert in bridge management. Furthermore, currently, Indonesia still focuses on construction, not maintenance works [22].

3.3 Ignoring Standard, Procedure, Manual

The collapse of the Kutai Kartanegara Bridge on November 26, 2011, shocked people generally and the engineering community particularly. This bridge is supposed to last at least 40–100 years of service life [15]. However, in its 10th year of service life, this bridge collapsed unpredictably and rapidly after the maintenance works. This is evident that ignoring standard, procedure and manual are still becoming the biggest issue for Indonesia in terms of IBMS application. Kawai in 2014, mentioned that an

[1] 1 IDR = 0.000070 USD.

accumulation of design errors, misconduct in maintenance, and reconstruction work had driven the bridge to collapse [23]. Maintenance works in Indonesia is still weak and carried out poorly [24]. The bridge failure is due to the alteration of deflection by retightening the hangers clamp should have been gone in advance by the main cables of retightening. Analyzing joint conditions must come first before retightening was carried out [24, 25]. Ignoring standard might also happen due to human error, such as lack of knowledge and competence, the officer did not mean to neglect it, but it was pure owing to their lack of knowledge.

4 How to Improve the IBMS Implementation in Indonesia

Coping with some issues regarding the poor implementation of IBMS in Indonesia, the authors propose a 'bottom-up' approach as a fundamental philosophy. The standard helps in proposing a strategy. The proposed strategy's cost is compared to the budget to accommodate the standard and customized strategy [26]. Following the proposed concept, in the authors' opinion, the IBMS requires the optimization model of the bridge maintenance strategy. Some considerations are necessary to make an optimum bridge maintenance strategy, like the structural performance view (i.e., deteriorating structures) and economic view (i.e., the life-cycle costing and the net present value). The combination policy optimizes the structural performance, maximizes the net present value (NPV), and minimizes the life-cycle cost (LCC) that has been successfully investigated the application to the wharf [27–29], so does the bridge [19, 20, 30].

NPV describes more about the benefit, this extent maximizing NPV is the same as maximizing the benefit value. LCC got influence from the life-cycle period of the bridge structures, which usually between 50 and 100 years. Budget estimation and project priority poll are for 10 or 15 years budgeting period. Hence, the most presentable mix in planning and time scheduling on maintenance and improvement work under budget constraints could be achieved [31]. Following this approach, insufficient financial and ignoring standard issues could be tackled. According to the IBMS method, the bridge maintenance cost results are lower than using the LCC method, because the inspection and additional life-cycle cost did not exist on the IBMS method, which are important factors [19]. The success of the optimization model of bridge maintenance implementation depends on the data (structural condition and cost) that we have. The more available data, the more accurate analysis is, the better the optimization model we will obtain. Thus, collecting data activities are essential.

Insufficient human resources handling by recruiting more people, then train them to become qualified bridge engineers, bridge inspectors, or bridge maintenance engineers. Besides, changing the concern of the Indonesian government from constructing only to construction and maintenance is also needed.

Table 1 Proposed aspects to be added for IBMS from BMS in Thailand, U. S., and Japan

Aspects	Thailand	US	Japan	References
Safety	√	√	√	[7, 8, 32, 33]
Risk	√	√	√	[7, 8, 32, 33]
Degradation model	√	√	√	[7, 8, 33]
Effects of maintenance work	√	√	√	[7, 8, 33]
Life-cycle costing	√	×	√	[7, 8, 33]
Optimal maintenance strategies	√	√	×	[7, 8, 33]

5 Recommended Aspects from BMS Implementation in the U.S., Japan, and Thailand

Bridge management systems (BMS) employed in the United States, Japan, and Thailand for bridge maintenance practices, which could be applied to current IBMS procedures in Indonesia are provided in Table 1. Those aspects are such as safety, risk, degradation model, effects of maintenance work, and optimal maintenance strategy.

Safety is the assessment of structure level to know the collision possibility (probability of failure). The bridge could be assessed based on the design code. Risk, same as safety is also an assessment type to recognize the collapse's effect and possibility. Deterioration of a major member, deterioration of a minor member, pier failure because of scouring, road accident owing to sub-standard guardrails, and loss of life due to the earthquake are five risk factors to be considered. Both aspects are obtained from inspection activities and collected as data [33].

To forecast the structural condition (predict the degradation) of the bridge at any given time point in the forthcoming, thus the degradation should be modelled [31]. Effects of maintenance work are the improvement works after the bridge has already gotten the repair actions (i.e., change in physical condition and performance barometers). Generally, the degradation model and the effects of maintenance work following probabilistic model, for example, Markov Chain model [33]. The degradation model along with the cost model (e.g., LCC) are supposed to define the optimization model, at least for maintenance cost, repairing works, and rehabilitation works to obtain the optimal maintenance strategies [31]. By adding those mentioned aspects, the IBMS is expected to be able to define the most appropriate and optimum policies for the bridge management system in Indonesia.

6 Conclusions

In this paper, some factors that may cause the poor implementation of IBMS in Indonesia and some recommendations adopted from BMS implementation in the

U. S., Japan, and Thailand have been described. Finally, the following conclusions could be derived:

- Improvement of IBMS practice in Indonesia is highly required by making optimization of the bridge maintenance model with the structural performance and economic consideration. However, it is going to be a more perfect model if we also could consider the environmental issues in the optimization model for IBMS.
- Improving and updating the procedure and the standard quality as well as special training for bridge inspectors and maintenance officers. So that, the right treatment can be performed at the right time and on the right bridge.
- Having more detailed and comprehensive information for some aspects in IBMS, for example, safety, risk, degradation model, effects of maintenance work, life-cycle costing and optimal maintenance strategies are expected to be able to define and apply for optimum maintenance works within budget.

References

1. Worldbank Homepage. https://www.worldbank.org/en/country/indonesia/overview. Last accessed 2021/02/22
2. Data and Information Technology, General Secretary, The Ministry of Public Work and Public Housing of the Republic of Indonesia (2020) Infrastructure statistical information PUPR 2020. PUPR, Jakarta (In Bahasa Indonesia)
3. Marwoto S (2019) Damage evaluation application for concrete bridges based on expert systems with android technology. Jurnal Riset Rekayasa Sipil Universitas Sebelas Maret 3:34–41
4. Sriskandan K (1990) Bridge management—an overview. In: Harding JE, Parke GAR, Ryall MJ (1990) Bridge management—inspection, maintenance, assessment and repair. Springer Science+Business Media Dordrecht, London, pp 17–29
5. Directorate General of Bina Marga, Ministry of the Public Works Republic of Indonesia (1993) Bridge management system (BMS): bridge inspection manual. Directorate General of Bina Marga, Jakarta (In Bahasa Indonesia)
6. Suksuwan N (2010) Condition rating system for Thailand's concrete bridges. J Construct Dev Countries 15(1):1–27
7. TDOH (2012) BMMS: bridge management and maintenance system. Department of Highway, Bangkok
8. Suksuwan N, Hadikusumo BH (2009) Development of computer applications for bridge management system. In: IABSE symposium: sustainable infrastructure—environment friendly, safe and resource efficient. Sustainable infrastructure—environment friendly, safe and resource efficient, Bangkok, pp 130–139
9. Ginting DT (2001) Study on the efficiency of bridge maintenance at BMS Bina Marga and efforts to determine the pattern of maintenance cost allocation [Master Thesis]. Institut Teknologi Bandung, Bandung (In Bahasa Indonesia)
10. Lee S, Kalos N, Shin DH (2014) Non-destructive testing methods in the U.S. for bridge inspection and maintenance. KSCE J Civil Eng 1–10
11. MLIT Hompage. https://www.mlit.go.jp/road/road_e/s3_maintenance.html, Last accessed 2021/03/27
12. Miyamoto A (2012) Life-cycle performance of deteriorating structures. In: Proceedings of the 3rd IABMAS workshop on life-cycle cost analysis and design of civil infrastructures systems. American Society of Civil Engineers, Lausanne

13. Anders B, Sorensen FB (1990) Implementation of bridge management and maintenance systems (BMMS) in Europe and the Far East. In: Harding JE, Parke GAR, Ryall MJ (eds) Bridge management—inspection, maintenance, assessment and repair. Springer Science+Business Media Dordrecht, London, pp 28–38
14. Japan International Cooperation Agency (JICA) CHODAI Co., Ltd and Metropolitan Expressway Company Limited.: Kingdom of Thailand the Project for Bridge Master Plan and Bridge Maintenance Ability in Rural Area Report. Japan International Cooperation Agency, CHODAI Co., Ltd and Metropolitan Expressway Company Limited, Tokyo
15. Putra GA, Rio K, Hatmojo D (2012) Study on the importance of bridge management system (BMS) applications and standard operating procedures (POS) in maintaining the quality value of bridge infrastructure, case study bridge infrastructure—bridges in Indonesia. In: SCAN#3 National Proceeding Conference. Universitas Atma Jaya Yogyakarta, Yogyakarta (2012)(In Bahasa Indonesia)
16. Banten suara Homepage. https://banten.suara.com/read. Last accessed 2021/02/26 (In Bahasa Indonesia)
17. Data and Information Technology, General Secretary (2018) The ministry of public work and public housing of the Republic of Indonesia.: infrastructure statistical information PUPR 2018 (In Bahasa Indonesia). PUPR, Jakarta
18. Bisnis Homepage. https://ekonomi.bisnis.com/read/20180426/45/788964/dana-pemeliharaan-jalan-jembatan-rp2370-triliun. Last accessed 2021/05/27 (In Bahasa Indonesia)
19. Dananjoyo RA, Aminullah A, Nugroho ASB (2020) Application of life-cycle cost method in calculation of economic evaluation of bridges to determine the priority of bridge handling. Jurnal Teknosains 9(2):91–180 (in Bahasa Indonesia)
20. Liu Min AM, Frangopol DM (2005) Multiobjective maintenance planning optimization for deteriorating bridges considering condition, safety, and life-cycle cost. J Struct Eng 131(5):833–842
21. The Ministry of Public Work and Public Housing of the Republic of Indonesia.: Module 2 Bridge Management System. PUPR Internal Module, Bandung (2017) (In Bahasa Indonesia)
22. Sumargo, Hakiki R, Raafidiani R (2020) Condition inspection and handling of bridges in Nias Island, North Sumatra province by using the bridge management system method. Jurnal Sipil Politeknik 22(2):156–168 (in Bahasa Indonesia)
23. Kawai Y, Dionysius S, Yozo F (2014) Failure analysis of the hanger clamps of the Kutai-Kartanegara Bridge from the fracture mechanics viewpoint. J Jpn Soc Civil Eng 2:1–6
24. 逊 汪 Homepage. https://www.academia.edu/5832394/A_STUDY_ON_THE_CAUSE_OF_KUKAR_BRIDGE_COLLAPSE. Last accessed 2021/02/26
25. Luknanto Homepage. https://luk.staff.ugm.ac.id/kukar. Last accessed 2021/02/26 (in Bahasa Indonesia)
26. Small E, Philbin T, Fraher M, Romack G (1999) The current status of bridge management system implementation in the United States
27. Puspitasari SD, Kawabata Y, Yokota H (2019) Optimization of life-cycle management on port mooring facilities. In: Proceedings of the 21st International Summer Symposium in junction with the 74th JSCE Annual Conference. Japan Society of Civil Engineers, Takamatsu
28. Kawabata Y, Kato E, Tanaka Y (2018) A maintenance management plan for port mooring facilities based on cost-benefit analysis—a case study. In: Proceedings of life-cycle of engineering systems: towards an integrated vision. international association on life-cycle civil engineering, Ghent
29. Yokota H, Hashimoto K, Tani T (2016) Sensitivity analysis on repair prioritization for mooring facilities. In: Life-cycle of engineering systems: emphasis on sustainable civil infrastructure. International Association on Life-Cycle Civil Engineering, Delft
30. Xie HB, Wu WJ, Wang YF (2018) Life-time reliability based optimization of bridge maintenance strategy considering LCA and LCC. J Clean Prod 176:36–45
31. Jeong Y, Kim W, Lee I, Lee J (2018) Bridge inspection practices and bridge management programs in China, Japan, Korea, and US. J Struct Integr Maint 3(2):126–135
32. Ministry of Land, Infrastructure, Transportation, and Tourism (2014) Manual for bridge periodic inspection. ministry of land, infrastructure, transportation, and tourism, Tokyo

33. Mirzaei Z, Adey BT, Klatter L, Kong JS (2014) Overview of existing bridge management systems. The IABMAS Bridge Management Committee

Crack and Corrosion Inspections for Coastal and Marine Concrete Infrastructure: A Review

Sabrina Harahap, Surya Dewi Puspitasari, and Ahmad Aki Muhaimin

Abstract As one of the biggest archipelago countries, Indonesia has an extensive coastline where concrete was used as a construction material. Concrete can suffer cracks that may stimulate the ingress of chloride to the rebar then induced corrosion. Coastal and Marine Infrastructure (CMI) that are vulnerable to corrosion due to chloride attack should be considered for continual inspection or monitoring system. The inspection must use the optimal techniques to achieve the objectives of the investigation. However, some developing countries, such as Indonesia, have no standard related to corrosion investigation resulting in the lack of understanding and awareness of concrete deterioration due to corrosion. This paper compares some inspection techniques consist of visual observation of cracking and corrosion investigation. This paper provides the utilization condition and the limitation of the inspection technique. The paper also contributes to knowledge related to cracking and corrosion inspection in coastal and marine concrete structures by providing a readily available reference for researchers, practitioners, and society.

Keywords Coastal infrastructure · Concrete inspection · Crack · Corrosion

1 Introduction

Coastal and Marine Infrastructures (CMI) are crucial and strategic assets for the development, improvement, and defense of any country [1, 2]. Around half of the world's population live around the coastline which has increased many economic

S. Harahap (✉) · S. D. Puspitasari
Indonesia Defense University, Bogor 16810, Indonesia
e-mail: sabrina.harahap@idu.ac.id

S. D. Puspitasari
e-mail: surya.puspitasari@idu.ac.id

A. A. Muhaimin
The University of Tokyo, Tokyo 113-8654, Japan
e-mail: ahmadaki@iis.u-tokyo.ac.jp

S. A. Kristiawan et al. (eds.), *Proceedings of the 5th International Conference on Rehabilitation and Maintenance in Civil Engineering*, Lecture Notes in Civil Engineering 225,
https://doi.org/10.1007/978-981-16-9348-9_34

benefits [3]. Indonesia as the largest archipelago country has an extensive coastal area where concrete is used as a construction material [4]. Concrete is one of the popular materials for coastal and marine infrastructures as well as masonry and steel [5, 6]. In many developing countries, like Indonesia, the utilization of old coastal and marine infrastructures in port and coastal defense has intensely come into an option due to limited funds. Inevitably, some coastal and marine infrastructures suffered damage over time, one of the main damages is cracking.

Coastal and marine infrastructures should be a concern because they are susceptible to deterioration and corrosion due to chloride attacks [7]. Corrosion will be induced even by tiny cracks on the concrete surface. The presence of crack is the main factor of corrosion initiation. Cracks influence the durability of reinforced concrete structures in aggressive environments by accelerating the ingress of corrosive agents to the embedded steel [8–10]. In this case, the coastal area has a role as an aggressive environment while chloride ion has a role as corrosive agent.

Deterioration due to corrosion is an issue for concrete structures related to durability, safety, and sustainability [11, 12]. However, there is no standard nor manual of corrosion inspection for concrete in Indonesia. This situation leads to the lack of monitoring and the lack of inspection systems for coastal and marine infrastructure in Indonesia. This paper attempts to present the importance of corrosion monitoring in CMI that consists of visual monitoring and non-destructive testing.

2 Crack Inspection

Concrete structures suffer from cracks by several causes such as material, construction aging, environment, and external force that potentially decrease the durability of the concrete [13]. Hence, crack monitoring is a crucial maintenance process for civil infrastructure in general [14, 15] as well as coastal and marine infrastructure. Manual inspection, as the most common technique of visual observation, is normally inspected by the inspector by observing and monitoring the concrete surface. Manual inspection is also conducted for underwater structure. Diver is necessary for annual inspection to underwater structure such as breakwater and pier. However, divers have limited time to remain the certain depth of water [16]. This method has several disadvantages such as time consuming, costly, and some parts of the concrete structure might be dangerous to be observed manually [17].

Some researchers have tried to develop alternative crack inspection. Based on the literature, there are two classifications of crack inspection methods: the image-based method and the sensor-based method [18]. In image-based method, unmanned aerial vehicle (UAV) able to take the image of structure surface. UAV has been utilized in many civil engineering projects for mapping [19, 20] and monitoring [21, 22]. UAV utilization should be considered to use in the CMI monitoring system. The UAV was able to take the image of a crack on the concrete surface. Nonetheless, UAV-based application has a limitation on quantitative assessment of crack width and crack length [23]. Other type of the image-based method is a head-mounted display

(HMD). HMD could have 3D data of concrete structures that is more convenient to be checked in real time and onsite. Moreover, this type could show the cracks in which difficult to be seen by human eyes [24, 25]

In the sensor-based method, optical fiber sensor or capacitive dense sensor array installed in the infrastructure aim to detect the real time building condition or structural health monitoring by analyze the crack detection and propagation [26–28]. However, sensor-based methods are quite expensive, inconvenient, and need physical contact with structure in the application [18].

3 Corrosion Inspection

Corrosion is the consequence of the crack on concrete surface. Basically, corrosion inspection divides into electrochemical measurement method and non-destructive method [29]. Destructive method is avoided and become the last option in corrosion inspection. Half-cell potential (HCP) measurement is the simplest way to check the probability of corrosion on site. This method obtains the corrosion potential with respect to a standard reference electrode such as copper/copper sulfate electrode (CSE) [29]. This measurement based on ASTM standard C876-15 [30], which provides general guidelines for HCP evaluating corrosion in concrete structures as outlined in Table 1. Although, some studies have reported that HCP measurement may be unreliable [31, 32] because only provides the information and the estimation for corrosion probability.

Polarization resistance (Rp) is another way to estimate corrosion of reinforcement concrete. Rp is the ratio between applied voltage and the step of current [33]. The judgement of polarization resistance based on CEB [34] is shown in Table 2. Rp measurement able to conduct on cracked concrete, however, location with large crack (>1 mm) is avoided. Aside from that, wet concrete has to be avoided as well

Table 1 Corrosion probability based on ASTM standard C876-15

Half-cell potential (mV, CSE)	Corrosion estimation
$-200 < E$	90% no corrosion
$-350 < E < -200$	uncertainty
$E < -350$	90% corrosion

Table 2 The criteria of the polarization resistance

Corrosion current density, Icorr ($\mu A/cm^2$)	Rate of corrosion estimation
Icorr < 0.2	Low
$0.2 \leqq$ Icorr < 0.5	Low to medium
$0.5 \leqq$ Icorr < 1	Medium to high
1 < Icorr	High

Table 3 Grade of passivity (by anodic polarization curve)

Grade	Measurement result	Existence of passivity film
Grade 0	0.2 < E < 0.6 (V), Icorr > 100 μA/cm^2 at least one time	No passivity exist
Grade 1	0.2 < E < 0.6 (V), 10 < Icorr < 100 (μA/cm^2)	Certain degree of passivity exist
Grade 2	0.2 < E < 0.6 (V), Icorr > 10 μA/cm^2 at least once but not to qualified to Grade 1 and 3	
Grade 3	0.2 < E < 0.6 (V), 1 < Icorr < 10 (μA/cm^2)	
Grade 4	0.2 < E < 0.6 (V), Icorr > 1 μA/cm^2 at least once but not qualified to Grade 1, 2 and 3	
Grade 5	0.2 < E < 0.6 (V), Icorr < 1 μA/cm^2	Excellent passivity exist

because it may deviate the signal and resulting erroneous reading [35]. Hence, in CMI inspection, polarization resistance is considered to conduct on dry concrete surface.

Anodic polarization curve is related to quality of passivity film. The larger current density, the worse grade of passivity film is. Recently, anodic polarization measurement takes about 40 min [36]. There are two methods to obtain anodic polarization curve: immersion method and contact method. Immersion method is carried out by immersing the concrete specimen in a solution and usually conducted in laboratory. While contact method is carried out by using double layer counter electrode contacted on the concrete surface, so that, it is suitable for CMI inspection on site. The judgement of anodic polarization curve based on the finding of N. Otsuki [37] is shown in Table 3. Grade 0 represents no passivity film existence, Grade 1–4 represent certain degree of passivity film existence, and Grade 5 represents excellent passivity film existence. Table 4 illustrates the area of inspections and limitations for crack and corrosion inspection methods discussed above, respectively.

4 Proposed Framework of CMI Inspection

Indonesia, the developing country that mainly utilizes concrete as the main material of CMI, has develop several standards for inspection, design, implementation, maintenance and investigation in several public infrastructures such as bridge and road by Ministry of Public Work and Housing [38] even there are many standards in the Indonesian National Standard for concrete and steel application [39]. However, Indonesia has no standard code on corrosion monitoring and inspection. Corrosion is one of the main reasons for concrete structure failure. An early corrosion detection can reduce repairmen and rehabilitation costs [40]. Hence, it is necessary to develop the steps, rules, and standards to minimize corrosion damage. Figure 1 shows the proposed framework of CMI Inspection that can be conducted if the standard is

Table 4 Crack and corrosion inspection method

Method	Area of inspection	Limitations	Reference
Manual inspection	Reachable for inspectors (road, bridge, etc.)	Time consuming, costly, limited inspection	[17]
Image-based method	Unreachable area (offshore platform, breakwater, pier, port, etc.)	Limitation on quantitative assessment of crack width and crack length	[23]
Sensor-based method	Unreachable area or specified area to investigate real-time (dam, airport near coastal, tunnel, etc.)	Expensive, inconvenient, directly contact with the structure	[18]
Half-cell potential	Reachable for inspectors, susceptible of corrosion, (Pier, Port, etc. especially around tidal zone)	Some studies reported that HCP measurement may be unreliable	[31, 32]
Polarization resistance	Reachable for inspectors, susceptible of corrosion (Pier, Port, etc. especially around tidal zone)	Void (>1 mm) and wet concrete may deviate the signal and resulting erroneous reading	[35]
Anodic polarization curve	Reachable for inspectors, susceptible of corrosion (Pier, Port, etc. especially around tidal zone)	Time consuming: it takes 40 min to conduct one point	[36]

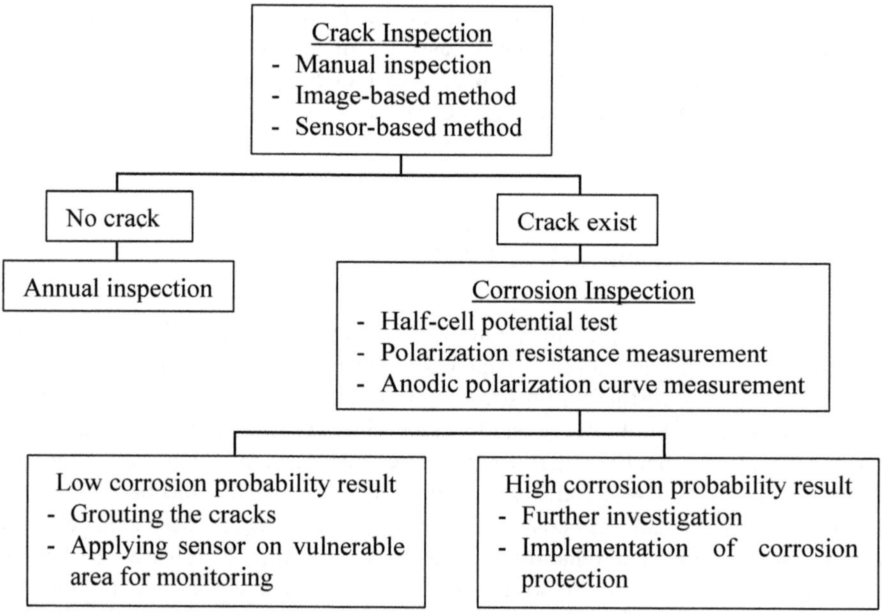

Fig. 1 Proposed framework of CMI inspection

unavailable in certain countries like Indonesia. The proposed framework emphasizes crack inspection and corrosion inspection. This proposed framework is not an official framework but it can illustrate the inspection plot in CMI.

At the first step, crack inspection can use or combine several methods such as manual inspection, image-based method and sensor-based method by UAV. If cracks are not found on the concrete surface, the inspection will be continued as an annual inspection. If the cracks exist on the concrete surface, it is highly recommended to following to corrosion inspection. The second step, corrosion inspection can be conducted by electrochemical measurements such as HCP, polarization resistance measurement, and anodic polarization curve measurement. These measurements estimate the corrosion probability so that it is strongly suggested to do at least two kinds of measurements to mutually validate the estimation results. Using one kind of measurement in corrosion inspection is not recommended.

Low corrosion probability result leads to various decisions such as grouting the cracks and applying sensor on vulnerable areas. These actions are taken to prevent corrosion initiation and to observe the vulnerable areas, respectively. The addition of corrosion inhibitor substances for example calcium nitrate in grouting paste can be an option to delay corrosion initiation in concrete inexpensively [41]. High corrosion probability result leads to conducting further investigation like destructive testing to confirm corrosion stage. Further investigation can be conducted by taking core drill samples of CMI structure then testing by advanced measurement in the laboratory. Furthermore, the implementation of corrosion protection like cathodic protection (CP) is needed to repair the concrete structure due to corrosion [42]. In general, CP installation has a high initial cost but it is considered as an economical long-term cost [41]. The probability corrosion results can be the fundamental data to decide the prevention or protection of CMI due to corrosion.

5 Conclusion

Costal and marine infrastructures, as the first line in the defense of the country, are valuable assets that need proper monitoring and inspection method. For that purpose, this paper reviewed some literature related to crack and corrosion inspection that enable to conduct in coastline area, but found that each method has limitations. By the review above, following conclusion could be derived.

- Each inspection has limitations that need to be developed to obtain the proper way of inspection. The inspectors are necessary to select the best inspection method based on location, cost, and condition.
- Even though crack and corrosion inspection standards are not available in certain countries like Indonesia, these inspections are highly considered to be conducted in CMI for monitoring the corrosion stage. An early crack and corrosion detections can reduce repair and rehabilitation costs.

- The corrosion inspections such as half-cell potential test, polarization resistance test, and anodic polarization resistance measurement, are estimation and probability only. It is highly recommended to conduct minimum two kinds of measurements to mutually validate the estimation results.
- Corrosion inspection results are precious data that help to decide the proper treatment in concrete structures, especially around coastline. These results can be a deliberation to implementation of corrosion prevention (low corrosion probability) or corrosion protection (high corrosion probability).

Future researches are necessary to develop the knowledge about proper way to inspect crack and corrosion on concrete structure in coastline area. This paper can be used as a tool to guide engineers and asset managers in decision-making about crack and corrosion inspection for concrete especially in coastal and marine infrastructure.

References

1. Bastidas-Arteaga E, Schoefs F (2015) Sustainable maintenance and repair of RC coastal structures. In: Proceedings of the institution of civil engineers-maritime engineering, vol 168, No 4, pp 162–173. Thomas Telford Ltd
2. Lucio D, Tomás A, Lara JL, Camus P, Losada IJ (2020) Stochastic modeling of long-term wave climate based on weather patterns for coastal structures applications. Coastal Eng 161
3. Creel L (2003) Ripple effects: population and coastal regions. Population Reference Bureau, Washington, DC
4. Wahyu S, Pramono GH, Purnawan B (2009) Establishment of marine and coastal spatial data infrastructure in Indonesia. In: Coastal and marine geospatial technologies, pp 97–103
5. Boéro J, Schoefs F, Capra B, Rouxel N (2019) Technical management of French Harbour structures, Part 1: description of built assets. Editions Paralia CFL, vol 6, pp 1–11. Revue Paralia, France, pp 1–7
6. McManus RS, Archibald N, Comber S, Knights AM, Thompson RC, Firth LB (2018) Partial replacement of cement for waste aggregates in concrete coastal and marine infrastructure: a foundation for ecological enhancement. Ecol Eng 120:655–667
7. Cai R, Yu M, Hu Y, Yang L, Ma H (2021) Influence of data acquisition and processing on surface chloride concentration of marine concrete. Construct Build Mater 273
8. Otieno MB, Alexander MG, Beushausen H-D (2010) Corrosion in cracked and uncracked concrete—influence of crack width, concrete quality and crack reopening. Mag Concr Res 62(6):393–404
9. Cao C, Cheung MM, Chan BY (2013) Modelling of interaction between corrosion-induced concrete cover crack and steel corrosion rate. Corros Sci 69:97–109
10. Shaikh FUA (2018) Effect of cracking on corrosion of steel in concrete. Int J Concr Struct Mater 12(1):1–12
11. James A et al (2019) Rebar corrosion detection, protection, and rehabilitation of reinforced concrete structures in coastal environments: a review. Constr Build Mater 224:1026–1039
12. Chaves IA, Melchers RE, Peng L, Stewart MG (2016) Probabilistic remaining life estimation for deteriorating steel marine infrastructure under global warming and nutrient pollution. Ocean Eng 126:129–137
13. Jung S, Lee S, Yu J (2021) Ontological approach for automatic inference of concrete crack cause. Appl Sci 11(1):252
14. Das Khan S, Topdar P, Datta AK (2020) Applicability of fuzzy-based visual inspection approach for condition assessment of bridges in developing countries: a state-of-the-art review. J Inst Eng India Ser A 101:835–846. https://doi.org/10.1007/s40030-020-00465-1

15. Fujita Y, Hamamoto Y (2011) A robust automatic crack detection method from noisy concrete surfaces. Mach Vis Appl 22:245–254. https://doi.org/10.1007/s00138-009-0244-5
16. Chen HH, Chuang WN, Wang CC (2015) Vision-based line detection for underwater inspection of breakwater construction using an ROV. Ocean Eng 109:20–33
17. Kim H, Sim SH, Cho S (2015) Unmanned aerial vehicle (UAV)-powered concrete crack detection based on digital image processing. In: 6th international conference on advances in experimental structural engineering, 11th international workshop on advanced smart materials and smart structures technology
18. Zhao X (2020) Automated rotational-and scaling-invariant image-based crack width monitoring with sub-millimeter accuracy and self-numbering label. Asian J Civil Eng 21(4):741–749
19. Remondino F, Barazzetti L, Nex F, Scaioni M, Sarazzi D (2011) UAV photogrammetry for mapping and 3d modeling–current status and future perspectives. Int Arch Photogramm Rem Sens Spat Inf Sci 38(1):25–31
20. Nex F, Remondino F (2014) UAV for 3D mapping applications: a review. Appl Geomat 6:1–15. https://doi.org/10.1007/s12518-013-0120-x
21. Ham Y, Han KK, Lin JJ et al (2016) Visual monitoring of civil infrastructure systems via camera-equipped Unmanned Aerial Vehicles (UAVs): a review of related works. Vis Eng 4:1. https://doi.org/10.1186/s40327-015-0029-z
22. Zhang C, Elaksher A (2012) An unmanned aerial vehicle-based imaging system for 3D measurement of unpaved road surface distresses 1. Comput Aided Civil Infrastruct Eng 27(2):118–129
23. Kim H, Lee J, Ahn E, Cho S, Shin M, Sim SH (2017) Concrete crack identification using a UAV incorporating hybrid image processing. Sensors 17(9):2052
24. Yamaguchi T, Shibuya T, Kanda M, Yasojima A (2019) Crack inspection support system for concrete structures using head mounted display in mixed reality space. In: 58th annual conference of the society of instrument and control engineers of Japan (SICE)
25. Yamaguchi T, Park T, Shibuya T, Yasojima A (2019) Development of crack inspection support system on concrete wall using MRHMD. In: The Proceedings of JSME annual conference on robotics and mechatronics (Robomec)
26. Antunes P, Rodrigues H, Travanca R, Ferreira L, Varum H, André P (2012) Structural health monitoring of different geometry structures with optical fiber sensors. Photon Sens 2(4):357–365
27. Yan J, Downey A, Cancelli A, Laflamme S, Chen A, Li J, Ubertini F (2019) Concrete crack detection and monitoring using a capacitive dense sensor array. Sensors 19(8):1843
28. Park S, Ahmad S, Yun CB, Roh Y (2006) Multiple crack detection of concrete structures using impedance-based structural health monitoring techniques. Exp Mech 46:609–618
29. Song HW, Saraswathy V (2007) Corrosion monitoring of reinforced concrete structures—A. Int J Electrochem Sci 2(1):1–28
30. ASTM C 876–15 (2015) Standard test method for half-cell potential of uncoated reinforcing steel in concrete
31. Leelalerkiet V, Kyung JW, Ohtsu M, Yokota M (2004) Analysis of half-cell potential measurement for corrosion of reinforced concrete. Constr Build Mater 18(3):155–162
32. Frølund T, Klinghoffer O, Sørensen HE, Denmark DD (2003) Pro's and con's of half-cell potentials and corrosion rate measurements. In: International Conference on Structural Faults
33. Stern M, Geary AL (1957) Electrochemical polarization: I. A theoretical analysis of the shape of polarization curves. J Electrochem Soc 104(1):56
34. CEB, Working Party V/4.1 (1997) Strategies for testing and assessment of concrete structures affected by reinforcement corrosion (Draft. 4), BRI-CSTC-WTBC, CEB Bulletin No. 243
35. Andrade C, Martínez I (2010) Techniques for measuring the corrosion rate (polarization resistance) and the corrosion potential of reinforced concrete structures. In: Non-destructive evaluation of reinforced concrete structures, pp. 284–316. Woodhead Publishing
36. Hamada H, Sagawa Y, Ikeda T, Morikawa R (2010) Several factors affecting the anodic polarization curve of steel bars embedded in mortar. In: 6th international conference on concrete under severe conditions-environment and loading, CONSEC'10, pp 201–208

37. Otsuki N (1985) A study of effectiveness of chloride on corrosion of steel bar in concrete. Report of Port and Harbor Research Institute, pp 127–134
38. Ministry of Public Work and Housing (PUPR). https://binamarga.pu.go.id/index.php/konten/standar-pedoman (in Bahasa). Last accessed 2021/04/25
39. National Standardization Agency of Indonesia (BSN). http://sispk.bsn.go.id/SNI/DaftarList (in Bahasa). Last accessed 2021/04/25
40. Sadowski L (2013) Non-destructive investigation of corrosion current density in steel reinforced concrete by artificial neural networks. Arch Civil Mech Eng 13:104–111
41. Goyal A, Pouya HS, Ganjian E et al (2018) A Review of corrosion and protection of steel in concrete. Arab J Sci Eng 43:5035–5055. https://doi.org/10.1007/s13369-018-3303-2
42. Byrne A, Holmes N, Norton B (2016) State-of-the-art review of cathodic protection for reinforced concrete structures. Mag Concr Res 68(13):664–677. https://doi.org/10.1680/jmacr.15.00083

e-Peralatan System as an Equipment Management for Disaster Mitigation on Indonesia National Roads

Adityo Budi Utomo, P. Gitaning, and S. Tosan Kunto

Abstract The e-Peralatan is an information system developed to optimize the cycle time of heavy equipment owned by National Road Construction Agencies of Central Java and Special Region of Yogyakarta. Heavy equipment is used to support the process of construction, maintenance, and disaster mitigation that impact the national road in the work area. The type of heavy equipment contained in the e-Peralatan is known as DRU (Disaster Reacting Unit) that consists of excavators, motor graders, wheel loaders, vibrator rollers, dump trucks, and dozers. This paper aims to explore how to create the e-Peralatan system as equipment management for disaster mitigation. Preliminary data were gathered by inventorying equipment, identifying existing business processes, and collecting documents related to equipment utilization. Then, the preparation of utilization framework involving National Road Construction Agencies of Central Java and The Special Region of Yogyakarta, local government, and publics. We found that there is no digital equipment management system for disaster mitigation. At the end of the dissemination, all parties have the hope of digital equipment management that can be utilized together. Our findings indicate that e-Peralatan provides many changes in the business process compared to conventional methods. The changes are including less standby time, faster equipment rental ordering time, cost certainty, time certainty, document administration certainty, and process certainty.

Keywords Disaster · Heavy equipment · Digital system · e-Peralatan

A. B. Utomo (✉) · P. Gitaning
Public Works Polytechnic, Semarang, Indonesia
e-mail: adityobudiutomo@pu.go.id

P. Gitaning
e-mail: gita.primaswari@pu.go.id

S. Tosan Kunto
National Road Construction Agency of Central Java and The Special Region of Yogyakarta, Semarang, Indonesia

S. A. Kristiawan et al. (eds.), *Proceedings of the 5th International Conference on Rehabilitation and Maintenance in Civil Engineering*, Lecture Notes in Civil Engineering 225,
https://doi.org/10.1007/978-981-16-9348-9_35

1 Introduction

DRU is a set of heavy equipment used for natural disaster emergency response activities that affect Indonesia's national roads and bridges. DRU consists of one set of work units consist of an excavator, motor grader, wheel loader, vibrator roller, bulldozer, and dump truck. DRU is alerted to handle natural disaster conditions such as landslides, floods, volcanic eruptions, and tsunamis. Floods and landslides are the dominant or common disaster occur on national roads and bridges.

Requirements that must be met in the management of the DRU include:

a. The equipment must always be in good condition or lightly damaged
b. The equipment must always be available whenever needed by any party
c. The equipment can reach the disaster location in less than 6 h so that the heavy equipment workshop must be placed in a strategic location prone to disaster.

To improve its productivity, it is essential to improve the performance of the construction systems. The desired production output is achieved through high equipment availability, which is influenced by equipment reliability and maintainability [1]. Idle time, downtime, poor equipment maintenance practices, improper determination of economic life and timing of replacement, poor training of equipment operators, equipment breakdown, over maintenance of equipment, huge capital investment during acquisition, the balance of interdependent equipment, misunderstanding the scope of work carried out, and incompatibility between the unit cost of production and equipment suitability for job condition were found to be the major problems that affect construction equipment planning and management [2].

The intended users are the most critical factor in the success of information systems and technologies; this also applies to disaster management. Moreover, as users' interactions with an information system play a significant role in determining whether or not it is useful, this suggests that understanding users' attitudes toward information systems is critical to their development [3]. The system designed includes being able to help command post commanders in making decisions based on the data displayed on the dashboard contained in the system [4].

National Road Construction Agencies of Central Java and The Special Region of Yogyakarta require equipment to support the construction and maintenance of national roads and bridges in Central Java and Special Region of Yogyakarta Provinces. Because the intensity of disaster occurrence in the national roads in Central Java and Special Region of Yogyakarta Provinces per year is not too much, so if The National Road Construction Agency of Central Java and Special Region of Yogyakarta invest funds in the form of heavy equipment, there will be a lot of idle time. Other options for the utilization of the DRU are rent to the public project on terms that they are always in good condition or slightly damaged also if a disaster occurs, the DRU can be mobilized at any time.

Problems arising from the management of DRU by National Road Construction Agencies of Central Java and The Special Region of Yogyakarta are:

a. DRU equipment rent services

1) Such as utilization of equipment by third parties through lending schemes on which between administration and realization are out of sync, no certainty of the rental fare, no certainty of the time of lending, the rental application process takes a long time, also disorderly administration.
2) Utilization of equipment by the internal unit of National Road Construction Agencies of Central Java and The Special Region of Yogyakarta is not accompanied by good administrative documents
3) The procedure for borrowing equipment has not been well-publicized.

b. Monitoring of DRU equipment assets

1) Many equipment assets in the Agencies are not monitored for their utilization
2) There is no periodical time of reporting system related to DRU equipment.

It is necessary to increase cooperation with local governments, disaster-affected communities for the continued dissemination of disaster mitigation information and to add disaster information and knowledge content from community experiences [5]. An information system is required in the management of DRU that accommodates all parties. Characteristics of information system which will be developed are:

a. The rent system, The large number of DRU in the National Road Construction Agencies of Central Java and The Special Region of Yogyakarta has a large country income potential but a system must be built with characteristic are:

1) Effective and Efficient. Certain rental fare, easy to monitor the process
2) Faster. There is certainty about the time of rental, the process for submitting a rent application must be faster
3) Better. A simple process that does not violate applicable Indonesian rules

b. Monitoring System, the system can be accessed easily by interested parties anywhere, anytime, and in real-time.

2 Methodology

The development of e-Peralatan consists of several stages as shown in Fig. 1. The first step is to conduct a regulatory study related to equipment operations and equipment utilization documents. From that documents, we made a framework of e-Peralatan involving National Road Construction Agencies of Central Java and The Special Region of Yogyakarta, local government, and IT expert in dissemination. Based on the e-Peralatan's framework can be determined about the data dan developing system that will be made. The method used in designing this management information system is the waterfall method, which consists of four stages, including the stages of analysis, design, coding, and testing. The stages of designing this information system include system requirements analysis consisting of two processes [6]:

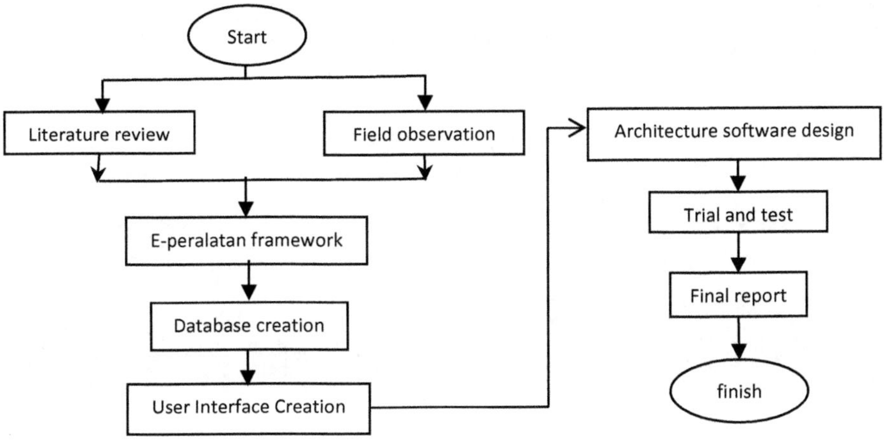

Fig. 1 Research methodology

a. Object-Oriented Analysis (OOA). Information system design is carried out in the Object-Oriented Analysis (OOA) stage, which is designing business processes from existing systems, then determining the parties involved in the system (users), designing use case diagrams¸ and designing use case descriptions.
b. Object-Oriented Design (OOD). OOD is designed by making a diagram that shows the design of the information system created. Diagrams used at this stage include activity diagrams, sequence diagrams, class diagrams, and Entity Relationship Diagrams (ERD). The system design described using the Unified Modelling Language.

Furthermore, the coding is done and followed by testing by the verification and validation process.

3 Results and Discussions

3.1 User Interface

The main page (home) is the user interface that can be accessed at www.palan.mapgeo.id (Fig. 2). This page displays simple information sections that can be accessed by the public, which contains header section, menu section, and banner section as well as search engine.

The new procedure for e-Peralatan by external users is designed more modern, simple, good lock, and easy to understand by anyone, which also can synchronizing with their email account as notifications. Users submits the letters to the equipment manager accompanied by a description of the type of equipment, location, period time, and purpose of use [7].

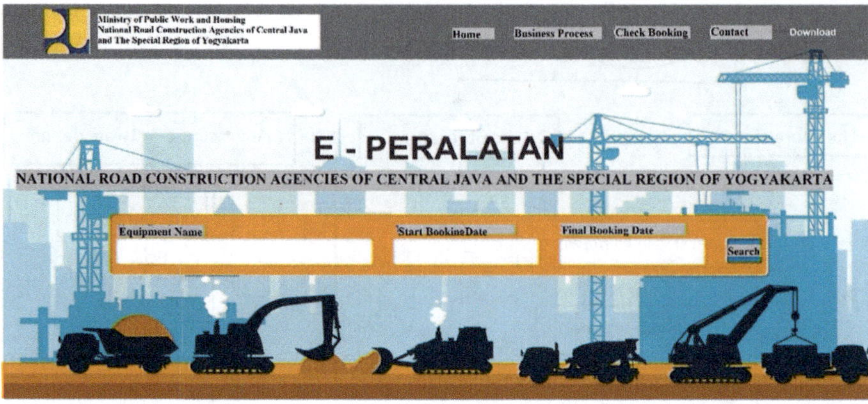

Fig. 2 Main page (home). *Source* www.palan

New procedure for e-Peralatan by external users can show in Fig. 3:

1. Open the website address www.palan.mapgeo.id [8] (Fig. 2)
2. Type the name of the equipment you want to rent for example "excavator" or "loader"
3. Enter the period time of the rent
4. Click the "search" icon

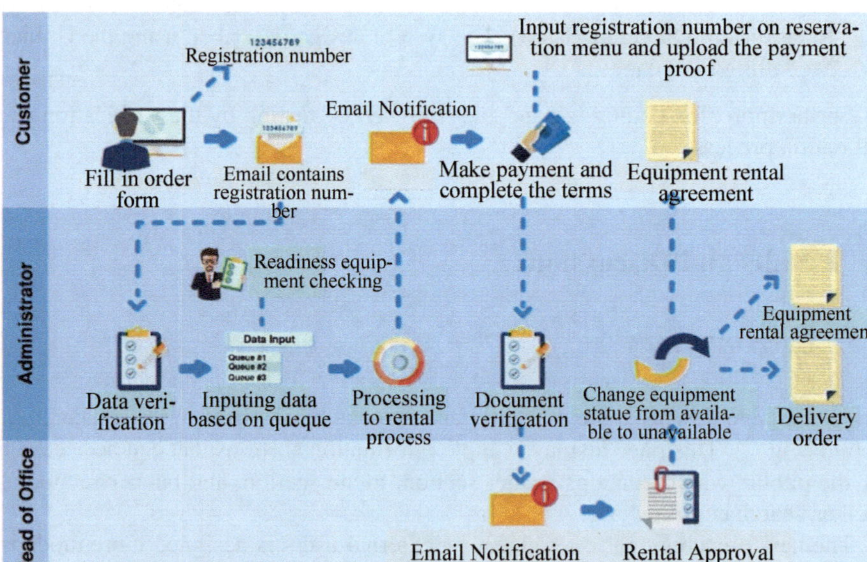

Fig. 3 Business process of new E-peralatan. *Source* www.palan.mapgeo.id

Fig. 4 Availability of DRU. *Source* www.palan.mapgeo.id

5. The website will guide you to the equipment availability page and location. Click on the details, to know the equipment identity (Fig. 4)
6. Click "booking" to proceed. Fill in data identity along with the residence card or passport and application letter. Application letter format can be found on the download menu.
7. When the file is complete the notification will be sent via email along with the attachments to the required administrative files
8. Make payment after email notification has been received and upload the proof of payment in the website
9. Complete the equipment retrieval file and the equipment is ready to operate
10. Perform equipment returns within the agreement period. If the time limit is exceeded, an additional fee will be charged according to the provisions

3.2 *Disaster Mitigation Management System*

The thing that often happens when disasters happened is that access roads to sites will be blocked due to landslides or floods. DRU management is required to open access for logistic road assistance and evacuation of victims. DRU support in disaster mitigation management that has been done by National Road Construction Agencies of Central Java and The Special Region of Yogyakarta are:

1. Conduct a study of potential risk disasters (floods, landslides, tsunamis, and eruptions) that impact national roads so that a map of potential risk disaster is obtained in the red zone, yellow zone, or green zone.
2. Relocations of all DRU workshops in a strategic red zone location so that it is easy and fastly to mobilize within a maximum of 6 h (Fig. 5)
3. Socialization about e-Peralatan's support system to Sub Project Manager, task force, Local Disaster Board, local government, and the community about disaster mitigation management.

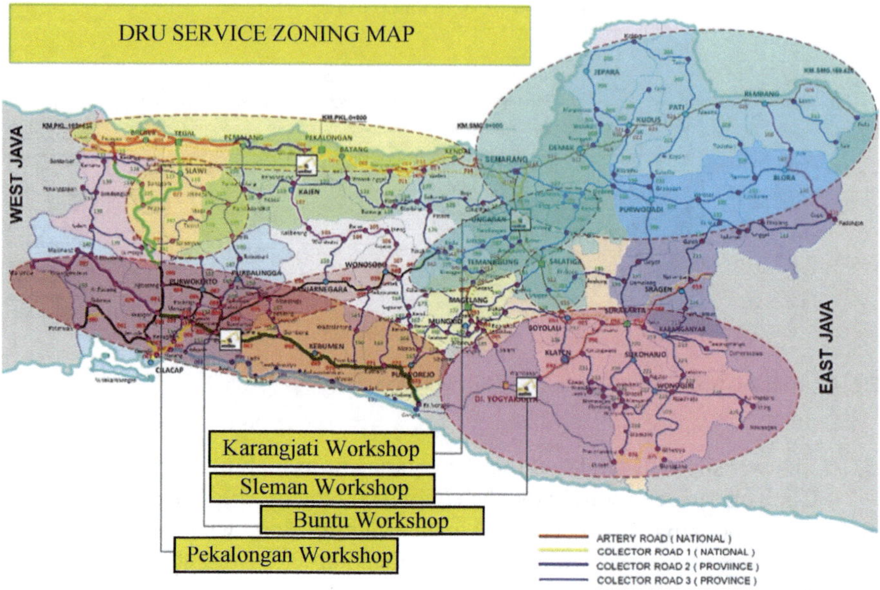

Fig. 5 DRU workshop location based on disaster risk management. *Source* National Road Construction Agencies of Central Java and Special Region of Yogyakarta

4. Delivery DRU to the disaster site in a clear, measurable, and effective way. An inappropriate choice of DRU will result in low productivity and will incur additional cost and time to open the access road [9].
5. Provides optimal DRU operation services until the disaster emergency response period is finished.

The development of the e-Peralatan gives many advantages compared to the conventional system. The comparative performance between e-Peralatan and the conventional system can be seen in Table 1.

The thing that is complained about this process is that there is no certainty how long it takes to get the permit and how much the total cost must be paid. Sometimes the equipment manager is so busy that the approval of the permit will take a long time. E-Peralatan is designed with a digital letter, digital document, and digital signature so that it can simplify the permit approval. we suggest that e-Peralatan management needs human resources for optimal services. Further studies are needed regarding the most suitable public service model for DRU operation in disaster mitigation management.

Table 1 Comparative performance between e-Peralatan and the conventional system

No.	Parameter	Conventional system	E-Peralatan	Description
1	Permit cost	There is a fee to create, duplicate, and send documents	There is no fee for document creation, copying, and shipping	With e-Peralatan all registration and document submissions are done digitally
2	Permit time	There is no certainty of completion time from 2 to 14 days	Permit time for the whole process will take 2 h	All processes and document can be monitored in real-time
3	Equipment rental fee	There is no certainty of equipment fee	Rates are stated in the application	All price are listed in the application
4	Mobilization for disaster mitigation	It is difficult to monitor existing equipment conditions and location	It is easy to monitor existing conditions and location	The equipment's location and conditions data recorded in the application in realtime
		Coordination takes up to 1 day for the equipment to arrive at the disaster site	Mobilization equipment to disaster area less than 6 h	Relocation in the workshop makes every disaster area more accessible
5	Equipment maintenance	Maintenance data is done manually so that spare parts planning is less effective	Maintenance is carried out periodically and digitally so that spare parts planning is more effective	Equipment maintenance data recorded in the application
6	Equipment usage documentation	Manually and untidy documentation	Digitally and tidy documentation	All documents are done digitally in the application

4 Conclusion

Based on the analysis, it was found that by using e-Peralatan, DRU operations have become more optimal. DRU standby time is utilized through equipment rental schemes for disaster mitigations or other needs. The combination of e-Peralatan and setting up the workshop location makes every disaster site more accessible so that it can be reachable in less than 6 h. Comparing to conventional systems, equipment data owned by National Road Construction Agencies of Central Java and Special Region of Yogyakarta can be better organized with e-Peralatan system information and the public can access it anywhere and anytime. E- Peralatan can make it easier to find out information about equipment and the condition of existing equipment.

DRU management includes purchasing (investment), procurement of spare parts, fuel-consuming, operating hours, maintenance, and removal when it is severely damaged in a DRU life cycle time. The life cycle time must be managed disaster mitigation that has an impact on the infrastructure. In this paper, an attempt is made to highlight the role of e-Peralatan in the DRU life cycle time management digitally to reduce cost and time for contingency.

References

1. Desai P, Darade PM, Khare PP, Engineering C, Pune DYPS (2017) Overall equipment effectiveness in construction equipments. Int Res J Eng Technol (IRJET) 4(1):1111–1115
2. Manikandan M, Adhiyaman M, Pazhani KC (2018) A Study and analysis of construction equipment management used in construction projects for improving. Int Res J Eng Technol (IRJET) 5(3):1297–1303
3. Sari KP, Kanegae H, Roychani M (2021) Local risk managers ' and disaster volunteers ' awareness of , attitudes toward , and intention to use a local government-developed disaster information system: a case study of SIKK Magelang. In: Universitas Andalas (ed.) 2nd International conference on disaster and management. IOP conference series: earth and environmental science, vol 708 (012066) (2021). https://doi.org/10.1088/1755-1315/708/1/012056
4. Patrisina R, Meilani D, Habib M (2021) An information management system for effective disaster relief operations. In: Andalas U (ed.) 2nd International Conference on disaster and management. IOP conference series: earth and environmental science, vol 708(012066), pp 1–11 (2021). https://doi.org/10.1088/1755-1315/708/1/012066
5. Siswanto T, Shofiati R, Hartini H (2018) Acceptance and utilization of technology (UTAUT) as a method of technology acceptance model of mitigation disaster website acceptance. In: The 4th international seminar on sustainable urban development, pp 1–6
6. Rajagopal D, Thilakavalli K (2017) A study UML for OOA and OOD. Int J Knowl Content Dev Technol 7(2):5–20
7. Kirmanto D (2014) Peraturan Menteri Pekerjaan Umum Republik Indonesia Nomor 09/PRT/M/2014 tentang Jenis dan Tatacara Penggunaan Peralatan Konstruksi di Kementerian Pekerjaan Umum (Issue 10 September 2014), pp 1–27
8. www.palan.mapgeo.id
9. Nugraha AS, Putranto L (2019) The effect of heavy equipment management on the performance of the construction project and the construction company. In: 1st International conference of construction, infrastructure, and materials, pp 7–14 (2019). https://doi.org/10.1088/1757-899X/650/1/012019

The Implementation of Functional Road Assessment on Pramuka Road Section in Klaten District

Amirotul Musthofiah H. Mahmudah, Syahrul Anggara Wuryatmaja, and Ary Setyawan

Abstract Roads as one of the important facilities in transportation have a role as a distributor of goods and services, as an economic driver, so it is necessary to improve the quality of the roads so as to improve driving safety. Based on the Minister of Public Works Regulation No. 11/PRT/M/2010 each regent/major must conduct an evaluation of functional road assessment on district/city roads. The district road that was reviewed for the implementation of the research was Pramuka Street with the function of being secondary local. The purpose of this study is to determine the functional assessment eligibility category of roads and to provide recommendations so that Pramuka Street can achieve eligible function. The difference with previous studies is a review of the location, types of roads based on their function, and an assessment of the administrative component. The data required is in the form of technical requirements and administrative requirements. Component assessment is carried out by means of direct measurement and assessment in the field and requires secondary data obtained from an agency. The analysis is done by calculating the deviation results for each component inspection. The results of this study concluded that the Pramuka Street was categorized as conditional function. The recommendations given still need improvements in geometric road techniques, pavement structure techniques, complementary building techniques, traffic engineering management techniques, and road equipment techniques. Road administration documents for those that do not yet exist are also immediately completed and the document procurement process also needs to be accelerated immediately. Recommendations are given so that roads can achieve an eligible function condition.

Keywords Functional road assessment · Secondary local · Functional assessment eligibility category · Recommendations

---

A. M. H. Mahmudah (✉) · S. A. Wuryatmaja · A. Setyawan (✉)
Department of Civil Engineering, Sebelas Maret University, Jl. Ir. Sutami 36 A, Surakarta, Indonesia
e-mail: amirotulmhm@staff.uns.ac.id

S. A. Kristiawan et al. (eds.), *Proceedings of the 5th International Conference on Rehabilitation and Maintenance in Civil Engineering*, Lecture Notes in Civil Engineering 225,
https://doi.org/10.1007/978-981-16-9348-9_36

1 Preliminary

The road is an important facility in transportation that plays a big role in people's lives. The road is also a connector between regions and a facility for distributing goods and services. Roads can also play a role in increasing the economic growth of a region. The number of residents in each region has increased every year, thus encouraging an increase in the number of private vehicles [1]. Based on these conditions, if the number of private vehicles and public vehicles has exceeded the road capacity, it will cause congestion. Based on data from the Central Statistics Agency of Klaten Regency, each year the population increases. In 2017 there was a population increase of 0.36%. The increasing number of vehicles also encourages an increase in traffic accidents. The cause is from humans themselves who are less focused when driving, road infrastructure namely roads, facilities or vehicles, and the environment [2]. We often encounter problems with road quality around us, causing a lack of safety and comfort in driving.

Based on data from the police, the number of traffic accidents in Klaten Regency in 2018–2020 was 2799 accident cases. Traffic accidents in the district during that year have claimed as many as 4957 victims, more specifically 389 victims died, 14 were seriously injured, and 4554 were slightly injured. These data indicate that the accident rate is still high in Klaten Regency, so there is a need for efforts to carry out road evaluations. From these various problems, it is necessary to hold a study aimed at realizing a safety road. One way to do this is the road function assessment. Road function assessment on national roads, provincial roads, and district roads must be carried out by the respective regional governments [3].

A study regarding the Road Function Assessment with the same objective was carried out by [4, 5], namely analyzing the level of functional assessment eligibility category. The research method refers to regulation of the Minister of Public Works Number: 11/PRT/M /2010. The results of the two studies show that the road that is evaluated is categorized as conditional function (CF).

Research on road infrastructure safety evaluation on Trans Kalimantan roads. The background of this research is suspected to have deficiencies in technical requirements. Data collection was carried out for two days in November 2016, calculating geometric parameters and observing road surface conditions and harmonizing signs, markings and other road attachments by referring to applicable regulations. The identification results in the conclusion that there has been a deficiency in the applicable geometric planning regulations and technical requirements [6].

Research to analyze road safety at bends based on the radius and transverse slope of the bend. The research was conducted on the Mataram-Senggigi-Pemenang road section. Sharp bends are very dangerous for riders because they can reduce the rider's visibility and centrifugal force. The magnitude of the radius is obtained from the image of the road situation as a result of the topography and the cross-slope of the road is obtained from the difference in road side elevation. The results of the study showed that the bend on the Mataram-Senggigi-Pemenang road, there were 71.43% of the radius of the bend which was included in the very dangerous category and only the Amarsvati bend (STA 21 + 500) was categorized as quite dangerous and the others were safe. The solution to this problem is the installation of bend mirrors, speed limiting signs, sharp bend signs, lighting lamps, guardrail, enlarge bend section [7].

In the above review, the types of roads selected for the inspection were national and provincial roads. From previous study that has not been carried out on district roads, it is necessary to conduct research on district roads. The difference in this study from previous studies is that the location review of this study is based on the function of roads, namely local and roads based on their network system, namely secondary roads. The research does not only examine the technical requirements but also the completeness of the administrative requirements owned by the roads. The section of road being examined was Pramuka Street in Klaten Regency because accidents were still found on the road that caused the victim to be injured and materially damaged. This research is primarily aims to determine functional assessment eligibility category and to determine recommendations for each assessment of road components so that the eligibility functional conditions on the roads can be realized.

2 Theoretical Basis

2.1 Road Function Assessment

Road function assessment according to the Regulation of the Minister of Public Works Number: 11/PRT/M/2010 is a condition of a road that meets the technical requirements of eligibility to provide safety for its users and administrative requirements that provide legal certainty for road operators and road users so that it can be operated for the public.

2.2 Technical Requirements for Road Function Assessment

The technical components of the road function assessment consist of road geometric techniques, pavement structure techniques, complementary building structure techniques, road section utilization techniques, traffic engineering management techniques, and road equipment engineering. It is important to pay attention to road geometric planning because it is very important to get uniformity in planning the geometrical roads and to produce smoothness and comfort for road users [8]. The assessment components that are included in the geometric road technicalities are horizontal alignment, vertical alignment, cross section of the road body, horizontal alignment coordination and vertical alignment. On road pavement structures, the results of the pavement structure testing can determine the function of the structure and the strength of the road construction which includes the suitability of the pavement structure with the class of road function, road surface conditions, and surface drainage. On the road complementary buildings the results of this study identified the structural functioning of each complementary building component and adjusted it according to the functional assessment eligibility category. Technical complementary buildings consist of bridges, pontoon, culverts, parking lots, retaining walls, and waterways. Assessment of the utilization of part of the road, which includes testing of the road benefit space, the road property space, and the road supervision. The components included in the Traffic Management and Engineering assessment are road equipment that supports traffic regulation which includes road marking, signs, separators, roundabout, sidewalks, traffic light and road crossings.

The road equipment technical assessment is almost the same as the traffic management and traffic engineering technical assessment of its components, but in this test it is only carried out on the road equipment specifications. Focus of the assessment is on the conditions and dimensions of the road equipment on a road segment being tested. The road equipment technical assessment component is road equipment that is directly related to road users and technical equipment that is not directly related to road users.

2.3 Administrative Requirements for Road Function Assessment

The administrative requirements for the road function acceptance test consist of documents determining instructions for orders and prohibitions, road status documents, road class documents, land ownership documents, road information documents, and road environmental documents. Each document must be checked for completeness so that it meets the road functional assessment eligibility category.

2.4 Functional Assessment Eligibility Category

The category of road function eligibility is the result of the final assessment of all components of the road segment and entire road segment tested both technically and administratively. Based on the Regulation of the Minister of Public Works Number. 11 PRT/M/2010, the determination of the status of road functional assessment eligibility is divided into three categories, namely Eligible Function (EF), Conditional Function (CF), and Ineligible Function (IF). If the result of an assessment of components or the entire road section is CF or IF, a recommendation must be given in order to achieve a eligible function condition (EF).

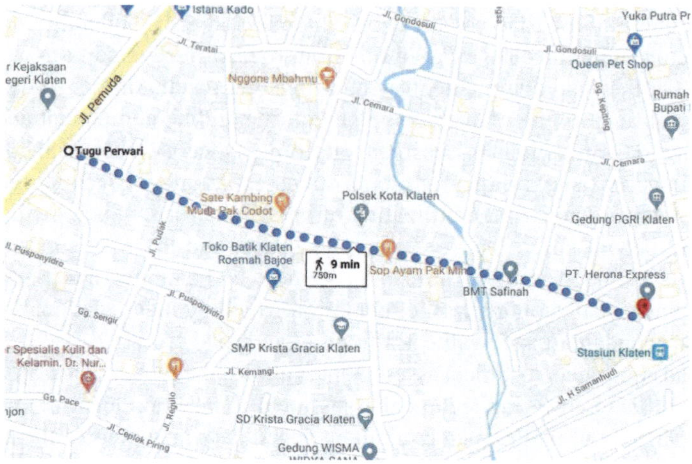

Fig. 1 Research location

3 Research Methods

3.1 Research Location

The location of the road function assessment research was carried out on Pramuka Street, Klaten Regency with a road length of 750 m. The results of the field identification of Pramuka Street are local roads, district roads, secondary roads, and small roads. The research location is located in the district center where there are important places such as police, schools, hospitals and stations. For a plan drawing of the research location, it can be seen in Fig. 1:

3.2 Data Collection Technique

The stages of data collection consisted of collecting primary data and collecting secondary data. The following will explain the primary data collection process:

Primary Data

The primary data inspection component consists of road geometric techniques, consisting of rider's visibility, pathways, median, and side gutter. Pavement structure techniques, namely pavement conditions using pavement condition index (PCI) method, road complementary building techniques, traffic engineering management techniques, and road equipment technical. Primary data is data that is obtained directly in the field by means of a survey according to approved procedures. The inspection was carried out in a comprehensive manner on Pramuka Street without dividing the segments. In the road function assessment when carrying out a survey, things that need to be prepared in advance are a survey form that is used to record the results of field research and tools such as roll meters, meters, pylox, cameras, wood, objects as high as 15 cm, stationery, and the road function assessment form.

Secondary Data

The secondary data inspection component consists of the road geometric technical inspection through road information documents, complementary building data, dimensions of the technical utilization of road sections through the road information document, LHRT data and all of the road administration requirements. Secondary data is data obtained from the results of inspections/surveys of certain people and agencies.

3.3 Data Analysis

Data analysis begins with the collection of primary data and secondary data. Deviation of each inspection result will be calculated and compared with the technical standard. To calculate each component deviation using the formula:

$$P = \frac{T - M}{T} \times 100\% \quad (1)$$

where P = Percentage Deviation, T = Technical Standard, M = Measurement Result.

Determination of the category/status of assessment component function eligibility criteria based on the deviation results. Determination of the functional eligibility category for the assessment component can be seen in Table 1.

Specifically for granting categories/status of road function eligibility to complementary building techniques using the analysis of the condition values in the Minister of Public Works Regulation Number: 11 of 2010 with a scoring number 1–5. The eligibility category for each weighted number will be explained in Table 2:

Assessment on the road administration documents, if the required documents are complete or exist, the document assessment is declared eligible functional, but if one of the documents does not exist, it will be declared conditional function,

Table 1 Determination of the functional eligibility category of the audit component

Functional category	Deviation	Information
Eligible function (EF)	0%	Can provide safety for riders
Conditional function (CF)	>0%	Can still provide safety for riders
Ineligible function (IF)	>0%	Cannot provide safety for the rider

Source [3]

Table 2 Determination of eligibility category for complementary building functions

Condition value	Functional category
0–1	Eligible function (EF)
2–4	Conditional function (CF)
5	Ineligible functional (IF)

Source [9]

Table 3 Matrix of how to determine functional assessment eligibility category/criteria technical, administrative, and a road section

Possibility	EF	CF	IF	Status/Category eligible functions for technical, administrative aspects, and a road
1	✓	✓	✓	IF
2	–	✓	✓	IF
3	✓	–	✓	IF
4	–	–	✓	IF
5	✓	✓	–	CF
6	–	✓	–	CF
7	✓	–	–	EF

Source [10]

but if the road status document is not exist, the road segment is declared ineligible functional. Providing recommendations only on components that are categorized as conditional functional (CF) and ineligible functional (IF). Determination of functional status/category is given for each technical component and administrative component as well as for all roads that are carried out by road functional assessment. Table 3 will explain the matrix for determining the functional assessment eligibility category for technical, administrative, and road segments:

After obtaining the functional eligibility category for each technical and administrative component, the percentage of each component functional eligibility category value can be determined using the formula:

$$P = \frac{TE}{TO} \times 100 \quad (2)$$

where P = Percentage of Each Category, TE = Total of Each Category, TO = Total Assessment of Overall Components.

4 Research Result

After analyzing the data on all assessment components, the following will be shown regarding the results of the categories/status of each assessment component:

4.1 Technical Geometric Roads

The geometric technical inspection of the road at the cross section of the road consists of road lanes, road shoulders, medians, safety thresholds, and traffic safety devices. Road shoulders and traffic safety devices are not inspected. The geometric technical inspection of roads on horizontal alignments consists of straight sections, bends, level intersections and parcel access. There is no inspection for access to parcels. A geometric technical inspection of vertical alignment consists of straight sections, climbing lanes and vertical arcs. For climbing lanes and vertical curves, no checks are made. The following will explain the results of inspection each technical geometric component in Tables 4, 5 and 6.

The overall result of the geometric technical inspection of the road is categorized as conditional functional, so there is still a need for recommendations for improvement. Recommendations given to geometric road technicalities on Pramuka Street are as follows:

1 The slope of the pavement should be further lowered to 3% according to technical standards so that water in the road body can quickly flow into the roadside drains.
2 It is necessary to add a rumble strip to each before entering the intersection.
3 Giving stop signs on minor roads.

Table 4 Cross section assessment results

Assessment Components	Pathways	Roadside	Road Markings	Threshold of Understanding	Side Gutter	Traffic Safety Tool
Functional eligibility category	CF	–	EF	EF	EF	–

Table 5 Horizontal alignment assessment results

Assessment components	Straight section	Bend section	Level intersection	Parcel access
Functional eligibility category	EF	EF	CF	–

Table 6 Vertical alignment assessment results

Assessment components	Straight section	Ascent lane	Vertical curve
Functional eligibility category	EF	–	–

Table 7 Results of Assessment Pavement Structures Technical

Assessment components	Types of pavement	Road pavement conditions	Road construction strength
Functional eligibility category	EF	CF	–

4.2 Pavement Structures Technical

In inspection the pavement structure, the strength of the road construction was not carried out. The results of the technical inspection of road pavement structures can be seen in Table 7.

The results of all technical assessment for the pavement structure of the Pramuka Street sections are categorized as conditional functional. The following recommendations are given so that the pavement structure can achieve eligible function:

1. Patching any damage.
2. Surface crack filling.

4.3 Road Complementary Building Technical

In the technical inspection of complementary buildings, what was not assessed were the pontoon, culverts, and retaining walls. The results of technical inspection of complementary buildings can be seen in Table 8.

The results of all technical inspections of complementary buildings for Pramuka Street are categorized as conditional functional. The following recommendations are given so that complementary buildings can achieve eligible function:

1. Marking parking spaces on the edge of Jalan Pramuka.
2. Addition of Parking Lot Signs on Jalan Pramuka.
3. Perform cleaning upstream of the roadside canal so as not to cause disruption of water flow to the downstream of the canal.

Table 8 Result of assessment complementary buildings technical

Assessment Components	Bridge	Pontoon	Water tunnel	Parking lot	Retaining Walls	Curbside Channels
Functional eligibility category	EF	–	–	CF	–	CF

Table 9 Results of assessment utilization of road parts technical

Assessment Components	Road Benefit Space	Road Property Space	Road Supervision Room
Functional Eligibility Category	EF	EF	EF

4.4 *Utilization of Road Parts Technical*

The technical assessment of the utilization of road parts as a whole component can be assessed which includes the road benefit space, the road property space, and the road supervision room. The following below will explain the results of the technical inspection of road sections utilization which can be seen in Table 9.

The results of all technical inspections on the use of parts of the road for the Pramuka Street section are categorized as eligible functional, so there is no need for recommendations.

4.5 *Traffic Engineering Management Technical*

The overall technical inspection of traffic engineering management, separators, traffic signal, zebra cross is not carried out. From the results of field inspection and analysis, the technical final results of traffic engineering management can be seen in Table 10.

The results of all technical assessment of road traffic management are categorized as conditional functional. The following recommendations are given in order to achieve eligible function:

1. Removal of the crossing warning signs by 50 m before the intersection.
2. Placement of signs of prohibited from turning right at the junction of Pramuka Street and Pemuda Street so that it is moved to the roundabout crossing so that it can be seen by drivers.
3. Signs that do not meet the standards for placing signs against the roadside must be adjusted to the standard.
4. Control of traders who are on the sidewalk.
5. Control of those who are still parked on the road sidewalks.

Table 10 Results of assessment traffic engineering management technical

Assessment components	Road markings	Sign	Separator	Kerb/Roundabout	Side-walk	APILL	Zebra cross	Clutter tape
Functional eligibility category	EF	CF	–	EF	CF	–	–	EF

4.6 Road Equipment Technical

The road equipment technical inspection consists of 2 components, namely the road equipment technicalities related to road users and the road equipment technicalities that are not related to the road users. Inspection of components equipment road technical related to road users is carried out on road markings, signs, kerb/roundabout, sidewalks, rumble strip, lighting lamps, traffic light, and separators. The inspection of traffic light, separators and road equipment technical that is not related to road users in this research was not carried out. The results of inspection road equipment technical can be seen in Table 11.

The results of all technical inspections of road equipment technical are categorized as conditional functional. The following recommendations are given in order to achieve eligible function:

1. Markings that are damaged/faded so that they can be repainted so that they can be seen by drivers.
2. The condition of the signs is obstructed by the leaves, so the flags must be handled immediately so that the signs can be seen clearly by the drivers.
3. The condition of the signs that have been filled with graffiti should be replaced with new ones.
4. The color of the signs is faded to be repaired/replaced immediately.
5. The dimensions of additional board signs that are not yet suitable, please be adjusted to technical standards.
6. Marking the edge of the roundabout at the intersection.

4.7 Road Administration

The assessment of the completeness of road administration for the scout roads is carried out at the Public Works and Spatial Planning Office of Klaten Regency by verifying whether each document is complete. The following in Table 12 below will be displayed regarding the results of road administration inspection.

The results of the examination of the administrative data for the Pramuka Street section at the Public Works and Public Housing Office of Klaten Regency were only the road status documents and road information documents. Road class documents, instructions and prohibitions documents, and environmental documents unavailable. Land ownership documents are still in the process of land certification. With

Table 11 Result of assessment road equipment technical

Examination components	Road markings	Sign	Kerb/Round-about	Sidewalk	Clutter tape	Lighting	APILL	Separator
Functional Eligibility Category	CF	EF	CF	EF	CF	CF	–	–

Table 12 Road administration assessment results

Completeness of road administrative documents and assessment results					
Road status	Establishment of instructions and don'ts	Street class	Land ownership	Environment	Leger Jalan
Available	Unavailable	Unavailable, only using reference to Law no. 22 of 2009 Article 19 paragraph 2C	Certification still Processing, Unavailable	Unavailable	Available
EF	CF	CF	CF	CF	EF

this result, the road administration can be categorized as conditional function. Recommendation given in order to achieve eligible function is:

1. Completing missing road administration documents and completing the document procurement process immediately in order to ensure legal certainty for road operators and road users.

4.8 Percentage of Functional Assessment Eligibility Category

Based on the results of the recapitulation above, it was found that there were 18 components that were eligible functional, 15 components had conditional functional values and there were no components that were ineligible functional, both technically and administratively. Based on Formula (2), it is obtained that the percentage of eligible function categories is 54.55%, conditional function is 45.45% and for ineligible categories is 0%. Based on previous research, the results obtained in the road functional assessment showed the same results, namely conditional function and some road components needed to be repaired in order to realize safe and proper road conditions for the driver. With these results based on Minister of Public Works Regulation No. 11.PRT/M/2010 it can be concluded that the Jalan Pramuka section has a status or has a conditional function category. The following will be displayed regarding the percentage of functional assessment eligibility categories of the Pramuka Street section in Fig. 2:

5 Conclusion

1. The status/category of functional assessment eligibility category for the Pramuka Street section is Conditional Function (CF) so that it still requires recommendations in order to achieve the Eligible Function Status

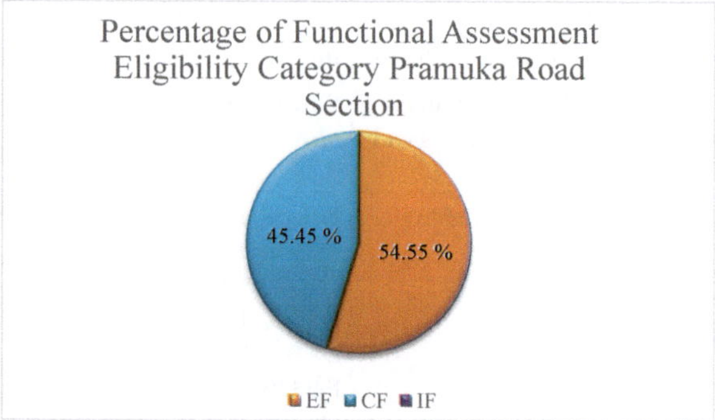

Fig. 2 Diagram of percentage functional assessment eligibility category

2. Recommendations for improvements in order to achieve an eligible functionally status on the Pramuka Street are:

   - Improvements to technical components which include road geometric techniques, pavement structure techniques, road complement building techniques, traffic engineering management techniques, and road equipment engineering.
   - Completing road administration documents consisting of documents stipulating instructions for orders and prohibitions, road class documents, land ownership documents, and environmental documents.

     - For further research, it is expected to carry out a road safety audit on the Pramuka Roads if the level of traffic accidents increases.
     - The role of related agencies is needed to complete supporting data to facilitate research on road functional assessment such as IRI data, as built drawings, DED documents, etc.

References

1. Amin CM (2017) Faktor-Faktor yang Mempengaruhi Pertumbuhan Kendaraan Bermotor Roda Dua Di Kota Pekanbaru. JOM Fekon 4:1106–1120
2. Priyanto S (1998) Accident analysis on Surabaya—Gempol toll road. Eng Media 20(4):50–59
3. Anonim (2010) Peraturan Menteri Pekerjaan Umum Nomor 11/PRT/M/2010 Tentang Tata Cara dan Persyaratan Laik Fungsi Jalan. Kementerian Pekerjaan Umum, Jakarta
4. Adwang J (2020) Analysis of road function assessment (ULFJ) on tumora national road (Bts. Kab. Parimo)—Tambrana section number 31 km. 157 + 800 to km. 168 + 000 Central Sulawesi Province. TEKNO, vol. 17, pp 122–138
5. Zachawerus J (2016) Uji Laik Fungsi Jalan Dalam Mewujudkan Jalan Yang Berkeselamatan (Studi Kasus Jalan Utama Di Pusat Kota Ternate). Prosiding Seminar Nasional Teknik Sipil 6, Surakarta, 25 October 2016, pp 169–176
6. Mayuni S, Widodo S, Sulandari E (2017) Evaluasi Keselamatan Infrastruktur Jalan (Studi Kasus Jalan Trans Kalimantan). Prosiding Konferensi Nasional Teknik Sipil dan Perencanaan (KN-TSP), Pekanbaru, 9 Feb 2017, pp 187–195

7. Widianty D, Rohani, Employees, IA (2019) Analisis Keselamatan Jalan Pada Tikungan Berdasarkan Jari-Jari Dan Kemiringan Melintang Tikungan. JRS-UNAND 15:103–114
8. Anonim (1997) Tata Cara Perencanaan Geometrik Jalan Antar Kota. Departemen Pekerjaan Umum Direktorat Jenderal Bina Marga, Jakarta
9. Anonymous (2014) Petunjuk Pelaksanaan Kelaikan Fungsi Jalan No. 09/P/BM/2014. Direktorat Jenderal Bina Marga Kementerian Pekerjaan Umum, Jakarta
10. Mulyono AT (2012) Formulir Laik Fungsi Jalan diwilayah BBPJN-VII. Yogyakarta

The Evaluation of Irrigation Maintenance in Pacal Irrigation Area at Bojonegoro Regency, East Java

Mahdika Putra Nanda, Rintis Hadiani, and Antonius Suryono

Abstract Pacal Irrigation Area, located at Bojonegoro regency and sourced from Pacal dam, was supporting 16.624 Ha agricultural field in 5 districts including Kapas, Balen, Sumberrejo, Kepoh Baru, and Baureno. Due to its vital function as agricultural support service, Pacal irrigation network should be able to function optimally to assist high crop yields. The maintenance of irrigation then became an important concern. However, the previous survey showed poor maintenance toward irrigation infrastructure, leading to non-optimal function to crop yields. This research aimed to identify the physical condition of the irrigation network and its infrastructure in Pacal irrigation area. The evaluation is based on the Minister of Public Works and Public Housing Regulation No. 12/2005 about exploitation and irrigation maintenance system. This particular regulation determined the appropriate maintenance for efficient irrigation operation and performance, leading to optimal agricultural productivity. The result showed that the primary canal was damaged in 12.55%, indicating a minor level of damage. Meanwhile, for the network and infrastructure surrounding the area, periodic maintenance was needed to ensure the irrigation performance running efficiently for supplying water.

Keywords Irrigation · Maintenance · Evaluation · Performance

M. P. Nanda · R. Hadiani (✉)
Department of Civil Engineering, Faculty of Engineering, Sebelas Maret University, Jl. Ir. Sutami 36 A, Surakarta, Central Java 57126, Indonesia
e-mail: rintis@ft.uns.ac.id

M. P. Nanda
e-mail: mahdika.nanda@student.uns.ac.id

R. Hadiani · A. Suryono
Procurement Officials of Water Resource Operation and Maintainance IV, Banaran, Pabelan, Kartasura District, Sukoharjo Regency, Central Java 57169, Indonesia

S. A. Kristiawan et al. (eds.), *Proceedings of the 5th International Conference on Rehabilitation and Maintenance in Civil Engineering*, Lecture Notes in Civil Engineering 225,
https://doi.org/10.1007/978-981-16-9348-9_37

1 Introduction

Water is a vital natural resource that supports a lot of sectors in life. This element is highly needed especially in the agricultural sector where the water supply supports most of the production. This sector then requires good water management to fulfill all the water need. That management is usually in the form of a dam or irrigation to accelerate and facilitate water distribution. In Indonesia, water sources are usually taken from a river, reservoir, groundwater, and tidal system. If the water availability is not adequate, another solution is made to still fulfill the water need. Water scarcity is a challenge due to increasing demand from various sources ranging from agriculture, mining, industry, hydroelectric power, and the increasing human population, especially in rural areas, the need for irrigation water is very important for agricultural land [1]. This stresses how important to calculate water need is because it is part of irrigation planning and maintenance. Irrigation management must consider the method of water use, the time of giving water, and how much water will be used [2]. Irrigation network performance is greatly affected by the physical condition of the building, function channels, and other factors related to the management of the irrigation network. Irrigation system performance is part of irrigation management and assessments are usually carried out once a year to monitor the function and performance of all aspects irrigation system. This assessment is carried out by Irrigation Area Manager by the location of the irrigation area [3]. Irrigation strategies and water demand have been hampered by a lack of understanding of the relationship between water scarcity, food production, food security, and sustainability [4]. The irrigation area is the basic territorial unit for the management of water resources, especially in rural areas. Irrigated areas are not only for the rural population and the amount of agricultural production on a large scale but also consume a lot of water resources [5]. This can be a factor that indirectly promotes the high productivity of rice crops. Irrigated agriculture appears to be a prime candidate for interventions aimed at reducing water demand and encouraging water savings [6].

Irrigation itself is the key to increase agricultural productivity [7]. In its principle, irrigation is an effort to distribute water through canals to agricultural fields and remove the excess water through wastegate [8]. The irrigation system includes infrastructures, water, operation management, and human resources [9]. Irrigation management is supported by five pillars including water availability, infrastructure, operation management, irrigation institution, and human resources [10].

By having this irrigation system, it is hoped that the need for water can be supplied to accelerate agricultural activities. It also can be a solution for water shortages in the dry season due to low rainfall. Maintenance of irrigation infrastructure then occurs to be important to promote agricultural activities. Besides that, the demand for irrigation's good performance guarantees people's food security, especially farmers. This is based on the fact that 23% of irrigation networks are damaged, only 11% of irrigation systems drain water regularly effective, and irrigation management is not efficient [11]. This problem causes water management to be ineffective and inefficient, leading to non-optimal agricultural production. The irrigation network will be customized with the water availability if the case shows that the water supply is lower than the water need. In an irrigation network, regular maintenance is important to support effective operation and performance. HIPPA, as the organization of agricultural irrigation management, can also participate in supervising and controlling the irrigation system.

Maintenance of damaged irrigation has become a priority. instead of constructing a new irrigation network. rehabilitation is the solution. Due to the large number of costs required for the rehabilitation of a large irrigation network, it is necessary to find an optimal rehabilitation scenario that can improve water management and hydraulic performance at the minimum possible cost [12]. The provincial government is responsible to restore the primary and secondary irrigation canals based on the irrigation maintenance level priority. Unfortunately, this priority data can be mistaken due to the limited knowledge and skill of farmers and government, leading to bad distribution of funding allocation. Therefore, technical training and data management are considered to be important in rehabilitating irrigation.

The irrigation area which covers an area of more than 3000 Ha is the government authority. Pacal irrigation area is under BBWS Bengawan Solo's supervision, a public river basin management organization. This area is located in Bojonegoro Regency with an agricultural field of 16.624 Ha. This irrigation supplies water for 5 districts, including Kapas, Balen, Sumberrejo, Kepoh Baru, and Baureno. However, an interview from Procurement Officials for Water Resource's Operation dan Maintainance IV stated that a water reservoir that conveys only 24 Million m^3 can not supply all the needs for these districts. Besides that, the canal and other irrigation facilities show some damage such as a crack on the canal network. This lead to a decrease in Pacal irrigation performance. This research then aims to investigate the physical condition of Pacal irrigation to function optimally and the kind of maintenance that can help increase the irrigation performance.

2 Research Method

This research used a descriptive method to explain the variables. In principle, descriptive research does not aim to form hypotheses or theory development. The characteristic of descriptive research is objectivity or neutrality. Descriptive research is about describes how reality is. In this case, descriptive research is different from prescriptive research which is primarily concerned with the question of what reality should be. Descriptive research is making an inventory prescriptive research is normative [13]. The data were taken from primary and secondary data. Primary data included traceability data and irrigation documentation. On the other hand, secondary data include irrigation network schematic data and irrigation map area. To know the physical condition of Pacal irrigation, evaluation is performed by investigating the existing condition, size, traits, the scope of work, and other parameters that have been set based on the standard and guidelines. In this research, primary data was the basis that determines the physical condition of Pacal irrigation. Then, after knowing

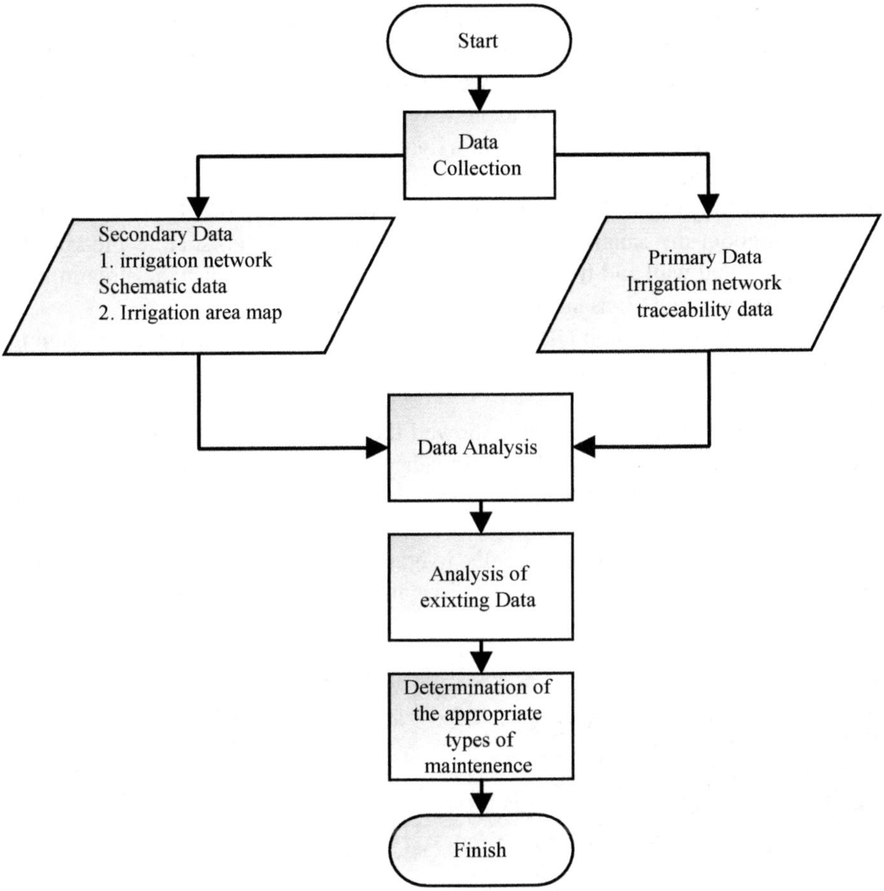

Fig. 1 Research flowchart

the data under existing condition, the type of maintenance and development for the irrigation can be determined to increase its function as the agricultural production support. The research data is presented in Fig. 1, as seen.

3 Result and Discussion

This research was conducted in Pacal Irrigation Area, Bojonegoro Regency, East Java. Existing condition evaluation was done to know Pacal Irrigation Area's situation in the part of physical infrastructure, the rice planting area, and growth, and also the fulfillment of irrigation need. From the research in 2017, the area of rice planting in the wet season used 15.096 Ha of agricultural field. It decreased in the first period of the dry season to become 14.528 Ha fields. On the other hand, in the second period of the dry season, it lowers to become 9.019 Ha. For the rice planting growth, Pacal irrigation area reached 90.46% in the wet season, 87.75% in the first dry season, and 90.59% in the second of the dry season.

Survey and documentation of Pacal irrigation were taken throughout the canals of 12.48 km of Klepek to Mengkuris weir. It also was taken from 8 km canals of Mengkuris to Kerjo weir. Pacal irrigation area consists of 16.624 Ha of agricultural field. Throughout the canals, maintenance was indeed needed such as clearing the weed on the canal wall and fixing the cracked wall that can disturb the streamline of water. It is seen in Fig. 2, as such.

The investigation through GPS and Google Earth is used to facilitate the damage tracking and inventory process. Figure 3 showed the map to the track location in Pacal irrigation, as seen.

After the investigation, the result showed the damage of Pacal irrigation infrastructure. The evaluation is based on Minister of Public Works and Public Housing Regulation No. 12 of 2015 where the physical condition index is presented in Table 1, as seen.

Based on the result above, Pacal's irrigation infrastructure condition can be determined. It is presented in Table 2 as irrigation damage index, as seen.

Fig. 2 The damage and weed on irrigation canals

Fig. 3 Map of track on irrigation network

Table 1 Irrigation condition index

Description	Condition index
	Maximum (%)
Infrastructure	
1. Main structure	13
2. Canals	10
3. Canal structure	9

Source Minister of Public Works and Public Housing Regulation No.12 of 2015

According to the table above, it can be seen that the damage of Pacal irrigation main infrastructure is 4.40%. It came from a lot of components such as a reservoir, sluice, sand trap, and wastegate. On the other hand, the damage of 7.40% happened in canals and 0.75% in canal structures. The total damage in Pacal irrigation infrastructure is 12.55%, showing a minor level of damage. For the damage, the major level is >40%, the moderate level is 21–40%, and the minor level is 10–20%. Compared to that, <10% level means the irrigation is in good condition. Pacal irrigation infrastructure damage is presented in a bar chart, as shown (Fig. 4).

Maintenance for irrigation infrastructure is important for agricultural sustainability [14]. Maintenance of the irrigation scheme for the whole Every year Indonesia requires a large amount of money, an allocation provided by the central government as well as local governments have not been able to meet these costs. requirements for the maintenance of the irrigation scheme of each The Irrigation Area must consider the performance of the Irrigation Area in the aspects of the physical infrastructure condition of the network and other supporting aspects [15]. Irrigation asset management, which includes planning, maintaining, and funding the network, is important to be done efficiently to ensure the optimal benefit for irrigation users. Operation and maintenance of irrigation network should be a responsibility of all, including water user farmers association as organization of agricultural irrigation management, provincial government, body corporate, social organization, people, and individuals that gain benefit from this irrigation [16]. The maintenance of an irrigation system is something that cannot be defined in technical, economic, or even institutional terms. Rather, it must be defined as a process that can be carried out within a technical, economic, and institutional framework and framework [17]. All should maintain irrigation to run optimally based on its function.

4 Conclusion

From the analysis above, it can be concluded that physical damage of Pacal's primary canal is 12.55%, indicating a minor level of damage. Meanwhile, for the network and infrastructure surrounding the area, regular maintenance was needed to ensure the irrigation performance run efficiently for watering the agricultural field. That

Table 2 Index percentage of irrigation damage

Details	Ratio	Value	Condition index	
	(%)	(%)	Inspection (%)	Maximum (%)
I. Infrastructure				
With the detail description as seen				
1. Main structure	4.40			**13.00**
1.1 Weir	1.62	100.00		**4.00**
a. Weir crest	0.36	20.00	20.00	**0.80**
b. Wing wall	0.00	15.00	0.00	**0.60**
c. Weir floor	0.00	20.00	0.00	**0.80**
d. Weir shutter	0.00	20.00	0.00	**0.80**
e. Bridgefoot	0.22	5.00	50.00	**0.20**
f. Information board	0.62	10.00	70.00	**0.40**
g. Head gauges	0.04	5.00	10.00	**0.20**
h. Safety bridge fence	0.38	5.00	85.00	**0.20**
1.2 Weir gate's gear that can be operated	2.78	100.00		**7.00**
a. Intake gate	0.68	55.00	8.00	**3.85**
b. Waste weir	2.10	45.00	30.00	**3.15**
1.3 Sand trap dan waste weir	0.00	100.00		**2.00**
a. Good condition of sand trap	0.00	50.00	0.00	**1.00**
b. Waste weir and gear of sand trap that can be operated	0.00	50.00	0.00	**0.50**
2. Canal network	7.40			**10.00**
a. Canal capacity	2.89	50.00	26.00	**5.00**
b. Levee canals	0.70	10.00	31.67	**1.00**
c. Leaked levee canals	0.75	10.00	33.64	**1.00**
d. Slided/Broken/Collapsed retaining wall of levee	3.06	30.00	46.00	**3.00**
3. Structure in canal network	0.75			**9.00**
3.1 Regulation structure (Conveyance/Distribution/Field application structure)	0.75	100.00		**3.50**
a. Conveyance/Distribution structure	0.75	100.00	9.60	**3.50**
3.2 Current meter	0.00	100.00		**3.00**
a. Current meter	0.00	30.00	0.00	**0.90**
3.3 Weir's ancillary works	0.00	100.00		**2.50**
a. Flumes, Syphon, Spillway, Drop Structure, Settling Basin, Inlet Drain	0.00	60.00	0.00	**1.50**
b. Bridgefoot, culverts	0.00	40.00	0.00	**1.00**
Total damage				**12.55**

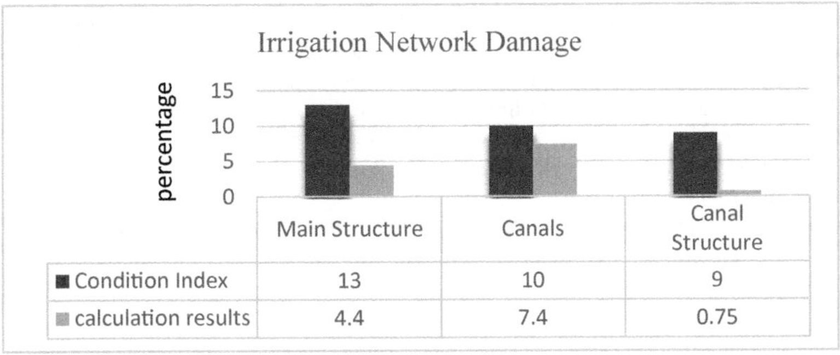

Fig. 4 Diagram for irrigation network damage

maintenance could be in the form of cleaning the weed that grew uncontrollably around the canals, plastering the cracks or the broken canal walls, lubricating the hinges of the sluice, coating the sluice gate with anti-corrosion paint, and change the lost or broken bolt.

References

1. Winters P, Salazar L, Maffioli A (2010) Designing impact evaluations for agricultural projects. Impact-Evaluation Guidelines, Technical Notes No. IDBTN-198 Strategy Development Division. Inter-American Development Bank
2. Allen LN, Adam JWM (2020) Irrigation and water management. The Science of Grassland. Wiley Online Library, Forages
3. Amrizal UL, Tohari B (2016) Performance evaluation of irrigation system in Cikeusik Irrigation Area, Cirebon Regency West Java Province. In: The 2en international conference on civil engineering research (ICCER). pp 12–18
4. Carruthers I, Rosegrant MW, Secler D (1997) Irrigation and food security in the 21st century. Irrig Drainage Syst 11:83–101
5. Pan H, Xu Q (2010) Quantitative analysis on the influence factors of the sustainable water resource management performance in irrigation areas: an empirical research from China. Econ Bus Aspects Sustain 10(1):264
6. Bekchanov MJ (2016) Water res and economic 13:30–45
7. Molden D, Orr S, Pittock J, Richter B, Riddell PJ, Scott CA, Philippe VJ, Vos J, Zwarteveen M (2020) A scale-based framework to understand the promises, pitfalls and paradoxes of irrigation efficiency to meet major water challenges. Glob Environ Change 65
8. Peraturan Menteri Pekerjaan Umum Dan Perumahan Rakyat (2015) Eksploitasi Dan Pemeliharaan Jaringan Irigasi
9. Kementerian Pekerjaan Umum dan Perumahan Rakyat (2019) Modul Pengenalan Sistem Irigasi
10. Direktorat Jenderal Sumber Daya Air (2019) Modul Pemeliharaan Dan Operasi Jaringan Irigasi
11. Waskito NT, Arif SS, Maksum M, Sutanto S (2010) Study on amortiztion in irrigation system management using knowledge management appoach. Paper on Asia Regional ICID
12. Elgamal T, Elfetyany M, Elkassar G (2019) Effect of irrigation network rehabilitation on water management-case study: Tanta navigation canal—Egypt. Alexandria Eng J 58(4)
13. Lans W, van der Voordt T (2002) Descriptive research. Delft University of Technology. Ways to study and research urban, architectural and technical design

14. Sharaunga S, Mudhara M (2018) Determinants of farmers' participation in collective maintenance of irrigation infrastructure. In: Kwazulu-Natal economic factors time ; Dependency on resource household characteristics. Physics and Chemistry of the Earth 105
15. Edhisono S, Hadihardaja IK, Suripin S (2016) Optimation budgeting distribution model for maintaining irrigation scheme. Int J Sci Eng 10(1):35–44
16. Peraturan Menteri Pekerjaan Umum (2007) Pedoman Pengembangan Pengelolaan Sistem Irigasi Partisipatif
17. Svendsen M, Huppert W (2003) Optimal maintenance in irrigation. Irrig Drainage Syst 17:109–128

Hazard Mitigation

Landslide Analysis Subject to Geological Uncertainty Using Monte Carlo Simulation (A Study Case in Taiwan)

Joni Fitra, Wen-Chao Huang, and Yusep Muslih Purwana

Abstract Landslide is the primary driver of the denudational process and sediment source dominantly onsite. Landslides are one of the most disastrous effects in Taiwan; groundwater or flood erosion is highly attributed to the landslide. Water induced to the slope increases driving force and decrease resisting force causing a slope landslide. This condition is generally affecting slope stability. In this study, we attempt to consider the uncertainty of the dip angle in slope stability analysis. In this research, the Monte Carlo simulation was used to quantify the effect of the geological uncertainty. Various sources of dip angles (with mean and standard deviation) were employed to generate 100,000 dip angle samples. All of the dip angles employed in this study were based on Highway no. 3 sliding events in Taiwan. Four different measurement sources, i.e., Central Geological Survey (CGS, Taiwan), Compass measurement before the sliding event, Surface measurement after the event, and LiDAR-derived data, were employed in this study. Further, the measured dip angles were converted to the projected dip angle based on the plane's strike. Simulation results show LiDAR Measurement Source provides the lowest failure probability of 16.9%, and Central Geological Survey (CGS, Taiwan) Measurement provides the highest failure probability of 78%. Therefore, based on the engineering design concept, if the design performed using the CGS data, the engineering design must be very conservative compared to the design using the LiDAR data.

Keywords Landslide · Geological uncertainty · Monte Carlo simulation

---

J. Fitra (✉) · Y. M. Purwana
Department of Civil Engineering, Sebelas Maret University, Surakarta, Central Java, Indonesia
e-mail: joni.fitra87@student.uns.ac.id

Y. M. Purwana
e-mail: ymuslih@ft.uns.ac.id

J. Fitra · W.-C. Huang
Department of Civil Engineering, National Central University, Taoyuan, Taiwan (R.O.C.)
e-mail: wenchaoh@g.ncu.edu.tw

S. A. Kristiawan et al. (eds.), *Proceedings of the 5th International Conference on Rehabilitation and Maintenance in Civil Engineering*, Lecture Notes in Civil Engineering 225,
https://doi.org/10.1007/978-981-16-9348-9_38

1 Introduction

Taiwan is an island formed by the active geological process on a complex of the convergent boundary. Landslide is the primary driver of the denudational process and sediment source dominantly onsite [1]. Landslides are one of the most geological disasters affecting Taiwan; groundwater or flood erosion is highly attributed to the landslide. Water induced to the slope increases driving force and decrease resisting force causing a slope landslide. This condition is generally affecting slope stability. The landslide has occurred widely globally, especially in the ring of fire' zone countries, i.e., Indonesia, Philippines, Taiwan, Japan, South Korea. A landslide occurred on Highway No. 3 near Xizhi, Taiwan (ROC) on 2010 April 25. A large amount of sediment buried both directions of the freeway. Four people were killed, and four cars were buried under debris [2]. For the above landslide, no rainfall or earthquakes have been recorded before the event. The landslide may be due to the local infiltration of groundwater between the weak planes or the weakening of the slope's stabilization facilities (Fig. 1).

The stability of rock slopes can be assumed as Mohr–Coulomb material expressed in terms of the cohesion c, friction angle ϕ and the factor of safety of the slope as follows [3–6]:

$$FS = \frac{cA + [W\cos(\beta) - \mathrm{Wp} - H\sin(\beta) + T\cos(\theta)]tan\Phi}{W\sin(\beta) + V\cos(\beta) - Tsin(\theta)} \quad (1)$$

c: the cohesion, ϕ: the friction angle (°); A: the slip area plane; θ: the angle of the rock anchor and normal vector (°); β: the dip angle (°), W: the weight of slipped, W_p: water pressure, H: horizontal force, T: rock anchors strength affecting to design force.

In this study, some assumption was applied for the equation. These are (1). No Cohesion was applied, (2). No water pressure effect, (3). No anchors were applied to stabilize the plane. If the sliding surface contains no water infiltration and zero

Fig. 1 Photograph of freeway No. 3 landslides in Taiwan [3]

cohesion, Eq. (1) reduces to

$$FS = \frac{\tan \Phi}{\tan (\beta)} \tag{2}$$

While a geologist does site investigation, one of the field's parameter measurements is strike and dip of the possible sliding planes. Site investigation and ground description are considered as one of geologic model uncertainty [7–9]. It is crucial to include a dip angle in slope stability analysis that best describes the potential sliding plane. However, it may not be realistic due to the intrinsic and extrinsic uncertainties. Therefore, in this study, we attempt to consider the dip angle's uncertainty in slope stability analysis. Various sets of dip angles that correspond to different measurement accuracy were applied in the plane failure analysis type of a simplified slope failure, hoping to shed some light on the influences of dip angle uncertainties on the slope's overall stability.

2 Monte Carlo Simulation Method

The Monte Carlo simulation was used to quantify the effect of the geological uncertainty on the safety factor and failure probability. Eight sources of dip angles (with mean and standard deviation) were employed to generate 100,000 dip angle samples. All of the dip angles employed in this study were based on the Highway no. 3 sliding events in Taiwan, i.e., the dip angles were measured or derived close to the sliding site. The sources include (a). Measured Central Geological Survey (M-CGS), (b). Projected Central Geological Survey (P-CGS), (c). Measured Compass (M-Com), (d). Projected Compass (P-Com), (e). Measured LiDAR (M-LiD), (f). Projected LiDAR (P-LiD), (g). Measured Surface (M-Sur) and (h). Projected Surface (P-Sur) as shown in Table 1.

Monte Carlo analysis randomly samples those parameters on the given probability distribution and its statistical parameters. Afterward, the generated random parameters were employed with the equation to calculate a corresponding safety factor several times (100,000 iterations in this study). We can obtain a series of safety factors. We can achieve a probability distribution of safety factors accounting for the dip angle's uncertainty through this process. We assumed that the distribution of all data sources of dip dingle and safety factors in this research was log-normal distribution.

2.1 Central Geological Survey (CGS) Taiwan

The mean value and standard deviations of the sliding plane dip angle are 22.5 and 9.57, while the friction angle is considered a constant of 15°. We addressed $\sigma^2_{\ln x}$

Table 1 Variation source of dip angle

Dip angle source	Mean value (degree)	Standard deviation (degree)
Measured CGS (M-CGS)	22.5	9.57
Projected CGS (P-CGS)	18.75	8.08
Measured Compass (M-Com)	17	4.37
Projected Compass (P-Com)	16.01	3.89
Measured LiDAR (M-LiD)	14.2	1.81
Projected LiDAR (P-LiD)	13.66	1.41
Measured surface (M-Sur)	14.91	2.55
Projected surface (P-Sur)	14.7	2.47

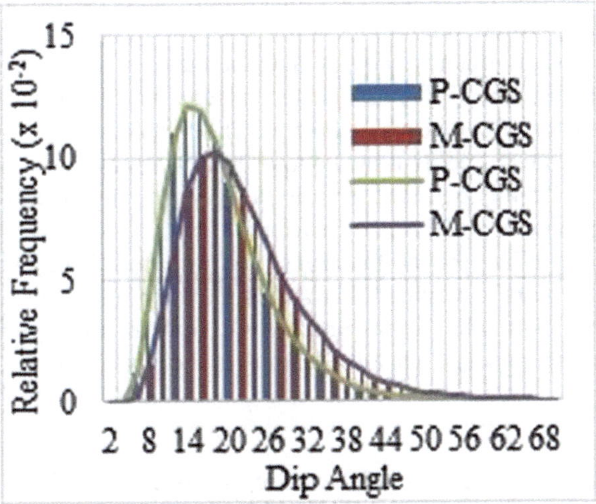

Fig. 2 Log-normal distribution data of central geological survey (CGS) dip angle

and $\mu_{\ln x}$ are 0.170 and 2.846. The log-normal distribution of the Central Geological Survey dip angle is shown in Fig. 2.

Figure 2 illustrated the probability density function of the measured Central Geological Survey (M-CGS) and projected Central Geological Survey (P-CGS) dip angle for the non-symmetric log-normal distribution. The measured Central Geological Survey (M-CGS) distribution data separate widely within a long-range between 4 and 60°. The distribution data of the projected Central Geological Survey (P-CGS) separate widely within range 4–50°. Based on this graphic, the measured Central

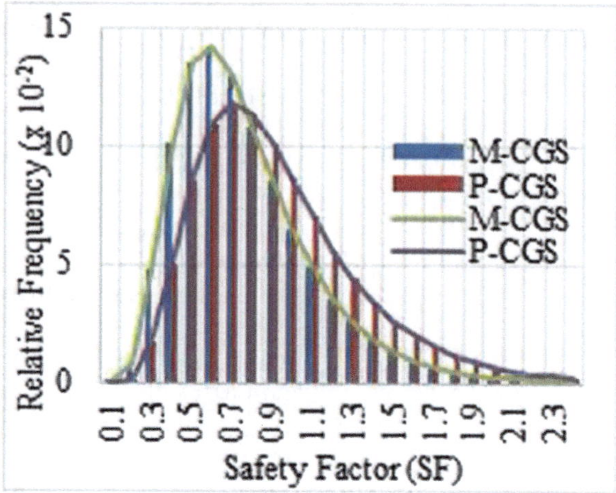

Fig. 3 Log-normal distribution data of central geological survey (CGS) safety factor

Geological Survey (M-CGS) dip angle has a broader range than the projected Central Geological Survey (P-CGS) dip angle. Furthermore, distribution data of central geological survey projected dip angle has a higher peak. The graph also shows a reasonable trend, as the projected dip angle indicates the effect of transferring from an apparent dip into the true dip.

Consequently, we calculate the factor of safety using the generated random variables. After 100,000 repeated calculations, the probability distribution curve for the factor of safety was obtained. The number of cases with a safety factor less than one and dividing it by the total cases were used to evaluated failure probability. The probability density function (PDF) of the safety factor is also shown in Fig. 3. Failure probability (P_f) of the measured Central Geological Survey (M-CGS) data is 78%, and failure probability (P_f) of the projected Central Geological Survey data is 63%. It can be concluded that the employment of the true dip angle in the analysis can yield a smaller failure probability. However, the P_f in the two analyzed cases showed an extremely high value leading to the slope's failure.

2.2 *Compass Measurement in the Field*

The second data source that was used in this research is the compass measurement dip angle. The probability density function of the compass measurement dip angle is shown in Fig. 4, in which one is the measured dip angle, and the other is the projected dip angle.

Figure 4 illustrates the probability density function of the compass measurement dip angle for the non-symmetric log-normal distribution. The distribution data of measured compass (M-Com) separate widely within range 6–32°, and the distribution data of projected compass (P-Com) separate widely within range 6–30°.

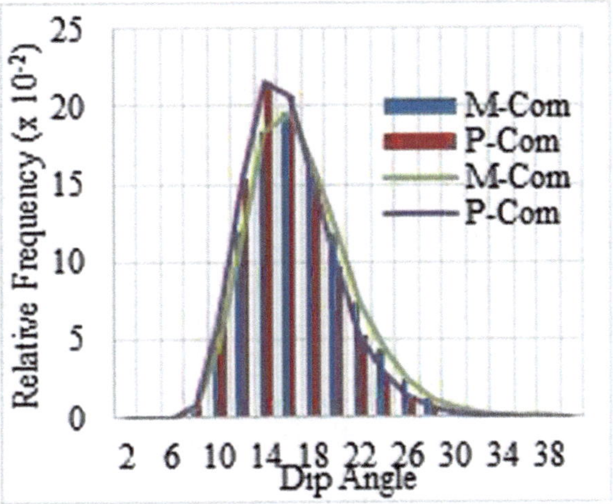

Fig. 4 Log-normal distribution data of compass measurement dip angle

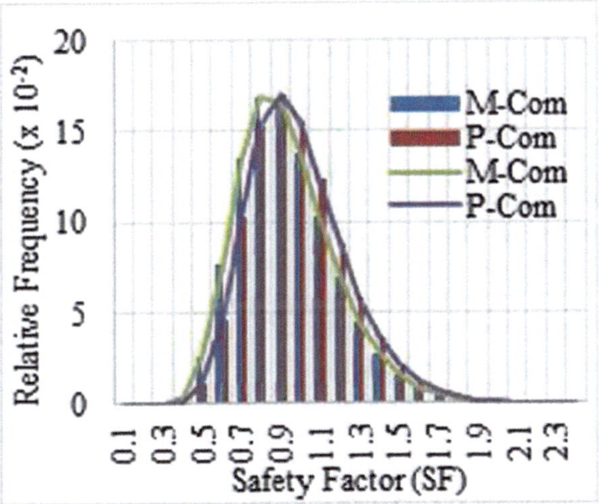

Fig. 5 Log-normal distribution data of compass measurement safety factor

Based on this graphic, the measured compass dip angle has a broader range than the projected compass dip angle. Therefore, distribution data of the projected compass dip angle has a higher peak. The probability density function (PDF) of the safety factor is shown in Fig. 5. Failure probability (P_f) of the measured compass data is 64.5%, and failure probability (P_f) of the projected compass data is 56.1% factor of safety <1 and evaluated as unstable.

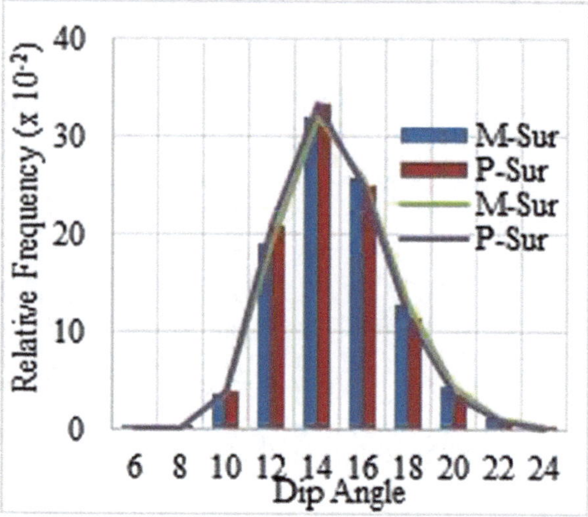

Fig. 6 Log-normal distribution data of surface measurement dip angle

2.3 *Surface Measurement of the Sliding Plane After the Event*

The third data source that was used in this research is the surface measurement dip angle. The surface measurement was carried out after the sliding event. The probability density function of the surface measured dip angle is shown in Fig. 6. Figure 6 illustrates the probability density function of the surface measurement dip angle for the non-symmetric log-normal distribution. The distribution data separate widely within the range of 8–24°. Based on this graph, the measured surface (M-Sur) dip angle and projected surface (P-Sur) dip angle have the same dip angle range.

Furthermore, distribution data of the projected surface dip angle has a higher peak. Finally, we can also observe that the measured and projected surface dip angles are close to each other, indicating that the field's measurement was indeed close to the true dip of the sliding plane. The probability density function (PDF) of the safety factor is shown in Fig. 7. Failure probability (P_f) of the measured surface data is 45.1%, and failure probability (P_f) of the projected surface data is 42.1%, given that factor of safety <1 is evaluated as unstable.

2.4 *LiDAR Measurement*

Traditional geological mapping full potential can be achieved using 3D rendering techniques on airborne Light Detection and Ranging (LiDAR) due to the exact three-dimensional (3D) Digital Terrain Model (DTM) information [10, 11]. The fourth

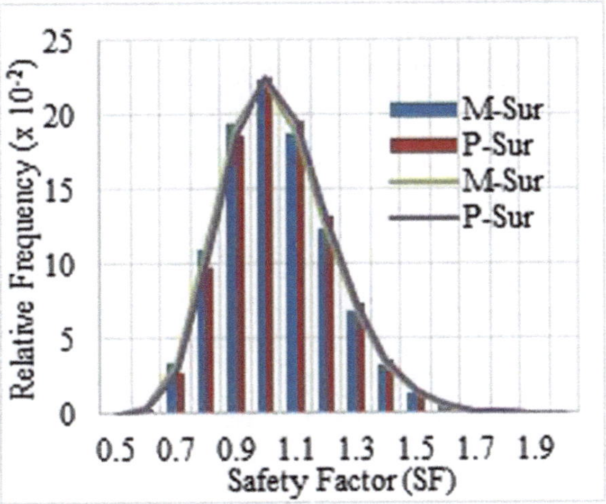

Fig. 7 Log-normal distribution data of surface measurement safety factor

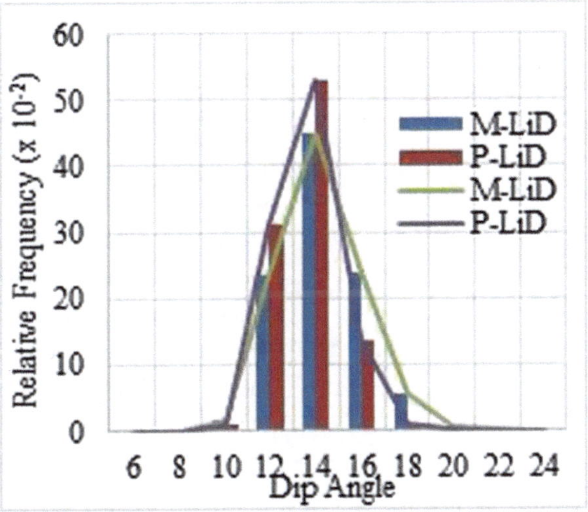

Fig. 8 Log-normal distribution data of LiDAR measurement dip angle

data source that was used in this research is the LiDAR measurement dip angle. The probability density function of the LiDAR measured dip angle is shown in Fig. 8.

Figure 8 illustrates the probability density function of the LiDAR measured dip angle for the non-symmetric log-normal distribution. The distribution data separate widely within the range of 8–20°. Based on this graphic, the measured LiDAR (M-LiD) dip angle has the same range as the projected LiDAR (P-LiD) dip angle. Therefore, distribution data of the projected LiDAR dip angle has a higher peak. We can also see that due to this source's high accuracy, the distribution of dip angles is more concentrated than other sources of dip angles.

The probability density function (PDF) of the safety factor is shown in Fig. 9.

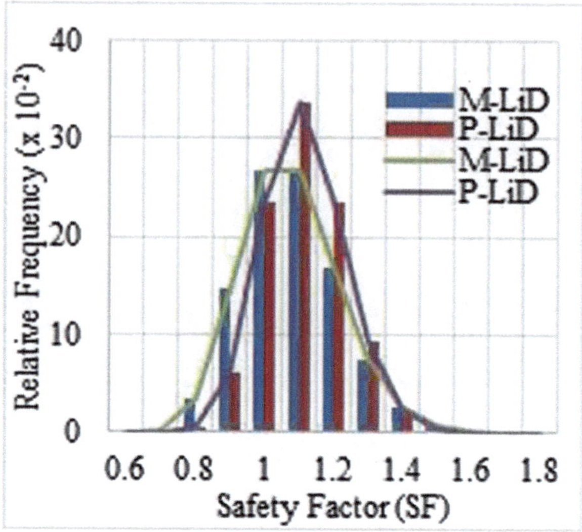

Fig. 9 Log-normal distribution data of LiDAR measurement safety factor

Table 2 Summary table of failure probability of the eight sources

Source	Failure probability (P_f)
Measured Central Geological Survey (M-CGS)	78%
Projected Central Geological Survey (P-CGS)	63%
Measured Compass (M-Com)	64.5%
Projected Compass (P-Com)	56.1%
Measured Surface (M-Sur)	45.1
Projected Surface (P-Sur)	42.1%
Measured LiDAR (M-LiD)	31%
Projected LiDAR (P-LiD)	16.9%

Failure probability (P_f) of the measured LiDAR data is 31%, and failure probability (P_f) of the projected LiDAR data is 16.9%, given that safety factor <1 is evaluated as unstable.

3 Discussions and Conclusions

Table 2 summarizes the failure probabilities of the eight sources of dip angles. Based on Table 2 results, we realize that the failure probability values are greater than 10%, indicating a high potential for slope failure if no stability measures were taken. In the current analysis, we have assumed no anchor applied in the design, while in the actual design, anchor forces were applied in the slope so that the factor of safety is acceptable [12]. Even the safety factor is acceptable when the anchor forces were applied, the failure probability is still not acceptable [13]. If the design is carried out only using the CGS data with a failure probability of around 60–80%, engineers would like to bring the P_f down to 0.01% (an acceptable failure probability for engineering design). For example, we can expect that the corresponding cost can be high. For example, the number of anchors can be more than the project's budget, or more cutting and filling may need to be performed to lower the slope geometry.

On the other hand, if the design is performed with the LiDAR data, it is evident that the required cost to bring down the original failure probability from 16% to 0.01% must be much lower than the case designed with CGS data. Hence, it is crucial to adopt more realistic dip angles of the analyzed slope to avoid unnecessary engineering designs. Although adopting the CGS data for design may cost less in the initial design phase, doing so could cost more in the actual design phase.

Acknowledgements The authors are grateful for the financial support provided by the Ministry of Science and Technology (MOST) Shackleton Program through Grant No. MOST108-2638-E-008-001-MY2 (Principal Investigator: Dr. Hsein Juang).

References

1. Broeckx J, Rossi M, Lijnen K, Campforts B, Poesen J, Vanmaercke M (2020) Landslide mobilization rates: a global analysis and model. Earth-Sci Rev 201:102972. https://doi.org/10.1016/j.earscirev.2019.102972
2. Chen L-K, Chen S-C, Ke M-C (2015) Investigation of the Freeway No. 3 landslide in Taiwan. In: Lollino G, Giordan D, Crosta GB, Corominas J, Azzam R, Wasowski J, Sciarra N (eds) Engineering geology for society and territory, vol 2. Springer International Publishing, pp 2093–2096. https://doi.org/10.1007/978-3-319-09057-3_374
3. Wang L, Hwang JH, Luo Z, Juang CH, Xiao J (2013) Probabilistic back analysis of slope failure—a case study in Taiwan. Comput Geotech 51:12–23
4. Hoek E, Bray J (1981) Rock slope engineering, 3rd edn. Institution of Mining and Metallurgy, London
5. Turner AK, Schuster RL (1996) Landslides—investigation, and mitigation: national research council. Transportation research board special report 247. National Academy Press, Washington (DC)
6. Wyllie DC, Mah CW (2004) Rock slope engineering, 4th edn. Spon Press, London
7. Juang CH, Zhang J, Shen M, Hu J (2019) Probabilistic methods for unified treatment of geotechnical and geological uncertainties in a geotechnical analysis. Eng Geol 249:148–161. https://doi.org/10.1016/j.enggeo.2018.12.010
8. Keaton JR (2013) Engineering geology: fundamental input or random variable? In: Proceeding of the geo-congress, 3–7 Mar 2013, San Diego, pp 232–253
9. Sullivan TD (2010) The geological model. In: Williams AL, Pinches GM, Chin CY, McMorran TJ, Massey CI (eds) Geologically active: proceeding of the 11th congress of the international association for engineering geology and the environment. CRC Press, Auckland, New Zealand, London, Sep 5–10, pp 155–170
10. Yeh C-H, Chan Y-C, Chang K-J, Lin M-L, Hsieh Y-C (2014) Derivation of strike and dip in sedimentary terrain using 3d image interpretation based on airborne LiDAR data. Terr Atmos Oceanic Sci 25(6):775. https://doi.org/10.3319/TAO.2014.07.02.01(TT)

11. Yeh C-H, Lin M-L, Chan Y-C, Chang K-J, Hsieh Y-C (2017) Dip-slope mapping of sedimentary terrain using polygon auto-tracing and airborne LiDAR topographic data. Eng Geol 222:236–249. https://doi.org/10.1016/j.enggeo.2017.04.009
12. TGS (2011) The investigation report of the landslide at Taiwan highway No.3, Ministry of Transportation and Communications investigation report. Taiwan Geotechnical Society, Taipei. (in Chinese)
13. Yeh C-H, Dong J-J, Khonevisan S, Juang CH, Huang W-C (2021) The role of the geological uncertainty in a geotechnical design—a retrospective view of Freeway No. 3 Landslide in Northern Taiwan (under review)

The Performance of Horizontal Drain as a Landslide Mitigation Strategy

Putu Tantri K. Sari

Abstract Landslides often occur during or shortly after heavy rains in several locations. One of the preventive efforts proposed is the removal of groundwater levels from the subsurface. A method proven to be effective in achieving this is the application of horizontal drain to increase the safety factor on a slope against landslides, especially during the rainy season. This research was, therefore, conducted to summarize the results of previous studies both on a field scale and numerical analysis to determine the effectiveness of horizontal drain on landslides. The results showed the horizontal drain has the ability to effectively lower the groundwater level at different degrees depending on soil parameters, rain conditions, the topography of the area, and the dimensions used. Moreover, the horizontal drain was also able to increase the slope safety factor by 1–4% for 1-m groundwater level decrease and al-most 20% for 9-m groundwater level decrease.

Keywords Landslide · Horizontal drain · Landslide due to rainfall

1 Introduction

Landslide, which is one of the impacts of the global climate change phenomenon, is increasing annually in Indonesia as reported by the National Disaster Management Agency (BNPB). It is also one of the top three disasters experienced between 2012 and 2014. Moreover, most of the landslide disasters recorded in Indonesia, especially in West Java, were reported by [1] to be caused by factors such as a steep inclination of the slope for more than 400, change of land usage from forest to farms and paddy fields, thick weathered rock soil from volcanic breccia, and high rainfall.

Several methods of slope reinforcement have been developed to mitigate the dangers of landslides and they include geotextile, soil retaining walls, sheet piles, ground anchors, soil nailing, and others [2, 3]. However, the use of reinforcement

P. T. K. Sari (✉)
Civil Engineering Department, Institute Technology of Sepuluh Nopember, Surabaya 60111, Indonesia

S. A. Kristiawan et al. (eds.), *Proceedings of the 5th International Conference on Rehabilitation and Maintenance in Civil Engineering*, Lecture Notes in Civil Engineering 225,
https://doi.org/10.1007/978-981-16-9348-9_39

is very uneconomical due to the occurrence of slope instability which triggers landslides during the peak-rainy season which lasts for only 1–3 months while stability is usually experienced in the remaining 9–11 months. Therefore, the control of rainwater infiltration into the soil is increasingly preferable and one mechanism for this is the use of a horizontal drain which is considered to be one of the most effective methods.

According to Tang et.al. [4], a horizontal drain is very effective in lowering the groundwater table. It was also reported by [5] to have the ability of increasing slope stability during rainfall, especially for dangerous slopes and embankments. Moreover, [6–8] showed it is the most economical method available in addition to its effectiveness and the same was also reported by other studies such as [9–16]. A horizontal drain is a hole made by drilling a manmade embankment or natural slope covered by a perforated metal or plastic pipe [17]. It is also defined by Barret [18] as a 3.7 cm or 1.5-inch diameter polyvinyl chloride (PVC) pipe mounted horizontally or inclined through drilling to form a permanently installed angle. Meanwhile, there are possible variation in drain dimensions such as the diameter and length of the installation depending on the topography of the area, climatic and rainy conditions, and subgrade parameters.

Several studies have been conducted through numerical modeling and field tests to obtain an effective horizontal drain design formulation. The results from these studies vary based on several variables and field conditions but almost all of them have positive findings. The differences in the previous studies make it important to have a reasonable summary and this is the focus of this research.

The advantages of using horizontal drain for landslide mitigation need to be summarized especially in Indonesia for construction designers. This is necessary considering their lack of information on the function, design, and advantages of this method. Moreover, the horizontal drain is generally installed for special occasions mostly based on previous experience without understanding the philosophy of its design. This research, therefore, intends to provide an explanation of the design and use of the horizontal drain.

The current state of horizontal drain practice through field applications and numerical analysis was summarized using results obtained from several locations with different conditions to determine the effective dimensions on landslides. Moreover, the factors of safety with and without horizontal drain installation were also compared based on the decrease in groundwater table elevation using the results from previous research. The impact of this method on slope stability based on different soil parameters, rainfall, and topographic conditions was also evaluated using information from previous studies.

2 Horizontal Drain Design Procedure

Martin et al. [13] showed that horizontal drain needs to satisfy certain basic requirements to perform successfully. These include the availability of sufficient size to carry the maximum water flow without any disturbance to the surrounding soil or the development of excessive outflow pressures and absence of significant loss of flow through infiltration along the drain. It also needs to be strong and sturdy enough to be easy to install according to the dimensions designed and able to support the borehole without collapsing. Moreover, the slotted or perforated length of each liner needs to be formed to prevent soil ingress or equipped with a suitable filter and the drainage is required to function properly without clogging and with minimum maintenance in the long run.

Some factors usually considered for effective design of horizontal drain include the shape, dimensions of the hole, length of the penetration into the ground as well as the installation spacing and configuration. In addition, the monitoring process, cleaning and inspection as well as design of the numerical modeling are also important.

Barret [18] and Santi [6] showed the formation process for horizontal drains. According to Barret, they are flat or low angle holes drilled with temporary or permanent case and are limited to small-diameter boreholes with a 3.7-cm or 1.5-in diameter slotted polyvinyl chloride (PVC) casing which is installed for permanent drains. Moreover, Santi also showed the hole of a horizontal drain is typically cased with a small diameter perforated polyvinyl chloride pipe. There is also a horizontal wick-drain model which is considered to be advantageous in resisting clogging, more affordable, has the ability to deform without rupture, and easy to install even by unskilled laborers with a minimal investment in equipment [6].

Another important aspect of horizontal drain design is the perforation size. These perforations or slots allow water in the soil or slope to enter the horizontal drain in order to be conveyed out of the slope. According to [19], the holes are usually at every 120° with the width varying between 0.25, 0.5, and 1.3 mm. Wide ones are generally applied on coarse sand or gravel soils and this allows fine sand and silt or smaller diameter materials to enter the horizontal drain. It is important to determine the hole to be used based on the analysis of soil in the installation area. Meanwhile, additional filtering can be conducted by surrounding the pipe with a fibrous filter material [20].

The horizontal penetration length and direction of the installation also need to be analyzed to ensure effectiveness. Several studies have been conducted to obtain the effective penetration length and the results vary depending on the topographic slope shape and subgrade parameters. For example, [19] reported that horizontal drains are most effective when placed in a location that holds water before reaching a zone of instability. It was also discovered that drains need to extend far enough into the slope to achieve the desired drawdown throughout the slope. This means they are to be installed beyond the area experiencing landslide due to the need to remove water from the landslide zone. Moreover, the determination of the penetration length is also influenced by the source and direction of groundwater movement in the area, the character of the strata or carrier layer, presence of open breccias or discontinuities, interconnected gaps in clay, and any perched water layers [21].

Pohll et al. [15] found the water table elevations to be dropping as the drain length increased. However, the horizontal drain becomes ineffective as it penetrates much longer than the predicted landslide zone. It was also discovered that the maximum reduction in subsurface water was not affected by the changes in drain length beyond a critical length [22]. Moreover, Royster [17] recommended that drains should not extend more than 3–5 m past the slip surface, and installations that significantly exceed this value is considered uneconomical. Lau and Kenney [9] also showed there is no additional benefit to installing a horizontal drain beyond the limit where the critical slip surface intersects the top of the slope. Furthermore, Cai et al. [5] indicated a smaller increase in safety factors as the drains extend beyond a critical length.

Several researchers have attempted to provide quantitative methods to estimate drain spacing. For example, Hunt [23] showed it depends on the type of material being drained with fine-grained soils reported to require 3–8 m while more permeable materials need up to 8–15 m. Huculak and Brawner [24] also recommended the installation of horizontal drains on an exploratory basis at a spacing of 9–12 m when the locations of high-permeability zones are unknown. Some researchers showed the need to place at least one additional drain between drains of the initial exploratory set in some cases based on site conditions and to tap the zones producing substantial amounts of water. Moreover, Cornforth [25] indicated parallel drains are usually 1–3 m apart and drilled normally into the slope in the direction of slide movement. These findings showed the effective spacing for horizontal drain installation varies depending on different conditions encountered in the field. The summary of all these studies is presented by [7] that the initial drains for parallel drains in high-permeability soils should be spaced at 8–15-m intervals while those in fine-grained soils should be at 1–8-m intervals.

There are 2 horizontal drain installation configuration models which are fan-shaped or array and parallel. There is, however, no significant difference in the maximum amount of water reduced between the two methods and they both have their respective advantages [22]. According to Mekechuk [26], the fan-shaped method has a faster installation process because it does not require a lot of time to set up the equipment. In addition, it also has some benefits related to cost and site access through an easier collection of water from several drains at once for conveyance off the slope but it requires the installation of sufficient drains to provide a mean distance equivalent to the guidelines provided for parallel horizontal drains. Meanwhile, parallel configurations provide more confidence in measuring the spacing of drains and in representing the slope with simplified models [7].

The changes in water conditions in the soil with a horizontal drain can be monitored using a piezometer which is a small-diameter well or tube used to measure the hydraulic head of groundwater. This instrument is usually applied to estimate the initial water heights in the slope for stability modeling purposes and also to evaluate the average drawdown over time to observe the effectiveness of the drainage system [21]. However, the use of piezometers to observe and determine the effectiveness of a drainage system requires understanding the possibility of a rapid change in water levels over short distances. This is a problem in areas with sparse piezometer coverage [6, 9].

The horizontal drain also requires some maintenance to keep it effective. This is necessary considering the possibility of clogging by root growth, sediments, or mineral deposits due to the time factor which needs to be cleaned. Several methods have been developed to clean these drains. For example, Cornforth [25] showed that high-pressure water systems are used to clean the slots and scour away sediments or mineral deposits. Moreover, the use of chemical treatments is sometimes necessary to remove stubborn mineral deposits. According to Santi et al. [6], horizontal wick drains tend to resist clogging but once clogged, they cannot be easily cleaned, thereby, requiring a new installation. The frequency of cleaning depends on the climate, local geology and vegetation, and several other factors [27]. Cornforth [25] recommended cleaning and inspection every 3 months in the first year of installation followed by once in a year and then after 4 years. An exception was, however, made when there is calcium carbonate blockage, in which case it is necessary to clean and inspect the drain every 2 years.

Another important factor to assess the effectiveness of horizontal drain is the method of conducting numerical modeling before the horizontal drain is installed in the field. Several studies have used different models as observed in the use of a hydrogeological software known as FEFLOW by [28] to analyze the transient hydrogeological and coupled hydromechanical modeling. Tacher (2005) also used the same software to calculate the groundwater pressure field in each landslide area. Meanwhile, Lin et al. [12] used SEEP/W and SLOPE/W software to determine the slope stability under rainfall seepage conditions and obtain safety factors with and without subsurface drainages. In another study, Lin et al. [29] used a 3-D finite element method program known as Plaxis 3D modeling to analyze groundwater flow and slope stability.

3 The Increase of Safety Factor of Slope After Horizontal Drain Installation

Horizontal drain has usually been applied to lower groundwater table and to ensure slope stability for a long time. It was initially developed for road construction in California in anticipation of landslides when the California Highway Division introduced the Hydrauger method in 1939 [17, 30]). A study was subsequently conducted on its use and the results were published by ([27, 31]). The application of horizontal drain to ensure slope stability then expanded to several countries such as Japan, England, Hong Kong, Germany, Czechoslovakia, Yugoslavia, Canada, France, and others as noted in previous studies published [31, 32].

The use of the horizontal drain for slope stability was observed to be continuously growing with a wide variety of conditions which were researched to determine its effectiveness. Those studies were generally divided into 3 major groups which include the analysis of the effectiveness of horizontal drain by observing water level fluctuations in the field before and after its installation. It is important to note that piezometer assay was commonly used in this research group. The second group includes the use of laboratory tests such as the centrifuge device and the results were later compared with those from the analysis conducted using a computer program or verification of the landslides in the field. The third group involved the use of mathematical modeling with the focus on the rate of movement of water in the porous media.

These studies were all conducted at several conditions including the presence and absence of the influence of rain intensity, soil parameters as well as the dimensions and elevations of the horizontal drain installed. The results of these studies showed that (i) the presence of horizontal drain was able to reduce the groundwater table to 3–7 m and, even in some cases, up to 15 m ([9, 11, 14, 27]), (ii) the presence of horizontal drain was able to increase slope stability by improving the slope safety factor ([4, 16, 33, 34]).

Nonveiller [33] also showed that horizontal drain increased the safety factor of slope value by 5–20% while [4] show the increment was up to 21.5%. The same trend was also observed in [16] and [34] even though the percentage increase was not stated. Moreover, the safety factor based on the groundwater table drop from previous studies was determined using 3 soil slope parameters which include the cohesive dominance, non-cohesive dominance, and cohesive and non-cohesive mixture. It is, however, important to note that the slope was considered to be a man-made embankment standing on a soft subgrade. The results presented in Fig. 1 showed the safety factor increases with a decrease in groundwater level in the three types of soil due to the installation of the horizontal drain. Meanwhile, the increase in safety factor varies from 1 to 4% for every 1-m drop in groundwater level while the value was observed to be up to approximately 20% for 9 m.

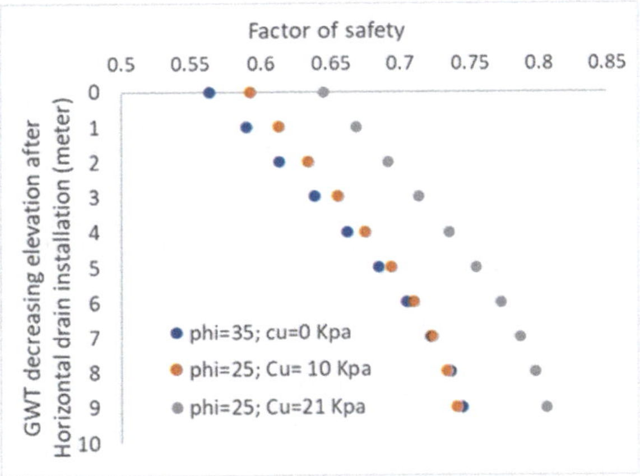

Fig. 1 The increase of safety factor after horizontal drain installation on slope

4 Conclusion

The summary of the findings from previous studies showed horizontal drain is an effective method to mitigate landslides, especially during the rainy season. It has been proven to have the ability of reducing groundwater level in order to increase the safety factor on a slope. This, however, depends on several factors such as the topography of the location, soil parameters, and rain conditions and this is the reason for the difference in the groundwater level recorded in the locations. Moreover, the horizontal drain was also able to increase the slope safety factor for every 1-m drop in groundwater level while the value was observed to be up to approximately 20% for 9 m.

Some other factors affecting the effectiveness of this method include dimensions, the shape of the horizontal drain material as well as the installation space and length. These factors influence the ability of the drain to hold water and drain rainwater out of the slope. Moreover, the cases summarized were observed to have different dimensions, shapes, and results after the installation but both the field scale and numerical analysis proved the effectiveness of this method. This means the horizontal drain is one of the best alternatives to mitigate landslides in Indonesia. It is, however, important to have very complete data on soil parameters especially those related to water infiltration rate and topography of the location for the preliminary design of the drain. In addition, the data on rain conditions at the location under review at a certain time is also very necessary.

References

1. Qodriyatun SN (2019) Landslide and disaster mitigation, in INFO singkat
2. Sari PTK, Lastiasih Y (2018) A General formulation to describe the empirical prediction of the critical area of a landslide. JESTEC 13(8):2379–2394
3. Sari PTK, Lastiasih Y (2020) Combined effects of vertical drain and pre-loading to reduce the number of geotextile reinforcement for road embankment design combined effects of vertical drain and pre-loading to reduce the number of geotextile reinforcement for road embankment design. J Phys Conf Ser 1569:1–13
4. Tang L, Tang X, Liu W (2011) A case study of effect of horizontal drains on rainfall-induced landslide. Adv Mater Res 253:1834–1837
5. Cai F, Ugai K, Wakai A, Li Q (1998) By three-dimensional finite element analysis effects of horizontal drains on slope stability under rainfall by three-dimensional ® nite element analysis. Comput Geotech 23(December):1998
6. Santi PM, Elifrits CD, Liljegren JA (2003) Design and installation of horizontal wick drains for landslide stabilization. 2(01)
7. Cook DI, Santi PM, Higgins JD (2008) Horizontal landslide drain design: state of the art and suggested improvements. Geol Soiety Am XIV(4):241–250
8. Rahardjo H (2012) Performance of horizontal drains in residual soil slopes. SOILS Found 51(3):437–447
9. Lau KC, Kenney TC (1983) Horizontal drains to stabilize clay slopes. Can Geotech J 21:241–249
10. Kopecký M, Ondrášik M, Antolová D (2013) Horizontal drains as effective measure for horizontal drains as effective measure. Stud Geotehnia Meh XXXV(March):129–141

11. Ahmed J, Ghazali MA, Mukhlisin M, Mukhlisin M (2012) Effectiveness of horizontal drains in improving slope stability : a case study of landslide event in Putra Jaya Precinct 9, Malaysia effectiveness of horizontal drains in improving slope stability : A case study of landslide event in drains Putra in Jaya, no. January, 2012
12. Lin D, Hung S, Ku C, Chan H (2016) Evaluating the efficiency of subsurface drainages for Li-Shan landslide in Taiwan. Nat Hazards Earth Syst Sci Discuss 1(January):1–22
13. Martin R (2013) Use of horizontal drains for slope stabilisation in Hong Kong use of horizontal drains for slope stabilisation in Hong Kong. HKIE Trans 3(April):37–41
14. Pfeiffer TJ (2014) Practical considerations for the design and construction of landslide mitigation using horizontal drains. In: 65th highway geology symposium
15. Pohll GM (2013) Design guidelines for horizontal drains used for slope stabilization, no. March 2013
16. Amatya SC, Mori M (2018) Horizontal drilling drainage as a preventive measure for water induced landslide risk reduction : a case study from Sindhuli Road, Section I, Nepal. J Nepal Geol Soc 55:109–122
17. Royster DL (1972) Horizontal drains and horizontal drilling: an overview. Transp Res Rec 16–20
18. Barrett RK (1980) Use of horizontal drains : Case histories from the colorado division of highways. Transp Res Rec 783(5)
19. Diana C, Paul S, Higgins J (2012) Prediction of piezometric surfaces and drain spacing for horizontal drain design. Landslides, no. June 2011
20. Choi (1983) Seepage around horizontal drains in hill slopes. J Hydraul Eng ASCE 109(10):1363–1368
21. Hoek E, Bray JW (1981) Rock slope engineering. In: Institution of Mining and Metallurgy, 3rd edn, p 23
22. Nakamura (1988) Landslide control works by horizontal drainage works. In: Landslides: proceedings of the 5th symposium, vol 2. Lausanne, Switzerland, pp 893–896
23. Hunt (2005) Geotechnical engineering investigation handbook, 2nd edn. Taylor and Francis Group
24. Huculak NA, Brawner CO (1961) The use of horizontal drains in landslide stabilization. In: Proceedings of the 42nd annual Canadian good roads conference, pp 383–400
25. Cornforth (2005) Landslides in practice: investigation, analysis, and remedial/Preventative options in soils
26. Mekechuk J (1992) Experience with horizontal drains as a means to improve slope stability. In: Proceedings of the 45th Canadian geotechnical conference. Toronto, Ontario, pp 1–7
27. Smith TW, Stafford GV (1957) Horizontal drains on California highway. Soil Mech Found Eng 83(SM-3):1300–1 to 1300–2
28. Matti B, Tacher L, Commend S (2012) Modelling the efficiency of a drainage gallery work for a large landslide with respect to hydrogeological heterogeneity. Can Geotech J 49(November):968–985
29. Lin D, Chang K, Ku C (2020) Applied sciences three-dimensional numerical investigation on the efficiency of subsurface drainage for large-scale landslides numerical models. Appl Sci 10:3346
30. Stanton TE (1948) California experience in stabilizing earth slopes through the installation of horizontal drains by the Hydrauger method. In: 2nd international conference on soil mechanics and foundation engineering. Rotterdam, the Netherlands, pp 256–260
31. Root (1958) Prevention of landslides. In Eckel EB (ed) special report 29: landslides and engineering practice
32. Hutchinson JN (1977) Assessment of the effectiveness of corrective measures in relation to Geolo_Gical conditions and types of slope Movem
33. Nonveiller (1981) International society for soil mechanics and 1981
34. Salmasi F, Dalir AH, Sarkarabad RN (2019) Investigation of the performance of horizontal drains in increasing slope stability in intense rainfall conditions by numerical simulation. Amirkabir J Civ Eng Amirkabir 51(3):2018–2020

The Analysis of Impact and Mitigation of Landslides Using Analytical Hierarchy Process (AHP) Method

A. Andriani, B. M. Adji, and S. Ramadhani

Abstract Landslides disaster is the third biggest disaster in the last ten years in Indonesia. The impact of landslides has resulted in many materials and non-material losses, such as casualties, physical impacts, environmental health, and socio-economy. Mitigation efforts are needed to reduce and even eliminate losses due to land-slides. In this study, the Analytical Hierarchy Process (AHP) method is used to assess the impact and mitigation of landslides so that the right decisions can be taken to deal with this problem. Questionnaires were conducted on 20 experts in the fields of disaster, geotechnical, geology and environment through direct distribution of questionnaires and google forms. The Likert scale is used as an additional method to obtain weights from the indicators contained in the last level of landslide mitigation. The results show that using the AHP method can identify the direct impacts felt by the community and the most dominant impacts occur due to landslides so that the most suitable mitigation efforts can be chosen based on the environmental conditions experiencing the disaster. Mitigation efforts before the disaster is one of the steps that must be taken to reduce the impact caused by landslides, structural and non-structural mitigation is carried out to mitigate the landslide hazard.

Keywords Landslide · Mitigation · AHP

1 Introduction

Indonesia was ranked first as the country with the most disaster victims in the world, and was ranked second after Haiti as the country with the most victims for the last 19 years [1]. This fact cannot be denied due to the geographical and geological

A. Andriani (✉) · B. M. Adji · S. Ramadhani
Civil Engineering Departement, Andalas University, Padang 25163, Indonesia
e-mail: andriani@eng.unand.ac.id

B. M. Adji
e-mail: bayu@eng.unand.ac.id

S. A. Kristiawan et al. (eds.), *Proceedings of the 5th International Conference on Rehabilitation and Maintenance in Civil Engineering*, Lecture Notes in Civil Engineering 225,
https://doi.org/10.1007/978-981-16-9348-9_40

conditions of Indonesia which are at the junction of the three major plates of Eurasia, Indoaustralia and the Pacific.

Indonesia's natural conditions such as topography, geology, and climatology are the dominant factors causing disasters, the movement of land masses. Land mass movement or landslide is a disaster that often occurs in hilly areas, valleys and volcanoes [2–5]. Landslides are the movement of slope-forming material in the form of rocks, scrap material or mixed material that moves down or out of the slope. The movement of the slope-forming material that moves out of the slope is due to the absorption of water into the soil until it penetrates the impermeable layer which acts as a sliding area, causing an increase in soil weight. Landslides will have a direct or indirect impact on the ecological condition of an area [6–8].

In the last ten years, landslides have occupied the third largest disaster with a total of 5022 events after a whirlwind disaster with a total of 6467 events and floods with a total of 7283 events [9]. According to calculations by the disaster management agency in Indonesia in 2019 from January to August, landslides are the biggest disasters after a tornado. The number of landslide disasters was 548 and the most areas that experienced landslides were Central Java with 242 incidents. Landslides are disasters that often occur in Indonesia. Especially in hilly areas and rivers that have steep cliff relief. One of the causes of frequent landslides in Indonesia is the change in land use. Land use change is caused by limited area. West Sumatra is one of the provinces that often experiences landslides disasters. The results of monitoring carried out by the Geological Agency, the Center for Volcanology and Geological Disaster Mitigation, in 2011 recorded 21 times the disasters of land movements in West Sumatra that are spread in several districts.

Landslide disasters have negative impacts, physically, socio-economically and on environmental health. In 2019, landslides in Indonesia caused 145 people to die and disappear, 125 injured victims, 5293 affected/displaced people, 2,298 damaged houses and 76 damaged facilities and infrastructure and displaced victims during landslides. in 2019, was ranked first with a score of 9473. Based on the impact of landslides, disaster mitigation efforts are needed to reduce these impacts. UU no. 24 (2007) concerning disaster management, mitigation is an effort or action that can reduce risks due to disasters either through physical efforts or awareness and enhancement of community capacity when facing disasters that occur. The steps taken to mitigate landslides can be grouped into three parts, namely mitigation before a landslide occurs, mitigation when a landslide occurs and mitigation after a landslide occurs [10–13].

To determine the priority of this mitigation action requires a lot of considerations, one of which is the most frequent impacts caused by the landslide disaster. The scientific approach can be used as a way to get the dominant impact of landslides and the right mitigation in reducing the impact of landslides. One method that can be used is the Analytic Hierarchy Process (AHP) method. The AHP method is a decision support model that uses a hierarchical arrangement and assessments derived from human perception. Opinions or perceptions used in this method are opinions that come from experts or people who understand the problem. AHP method is well known and widely used for decision making. This is because the AHP method has advantages, namely: make broad and unstructured problems into a flexible and easy-to-understand model through deductive integration. To get optimal results, this method requires people who are experts in their fields so that the assessment is subjective, so sometimes there are consistent answers or responses from respondents that are not always logical in the sense that they are in accordance with the existing problems. On the problem of landslides, many researchers use the AHP method. The use of the AHP method for mapping landslide-prone potential has been carried out by several researchers [14–19]. Besides that, disaster impact analysis can also be carried out using the AHP method [20–24].

Based on the results of previous research, it shows that AHP can be used as an alternative that can be used to determine appropriate mitigation efforts so that the impact of disasters can be reduced or eliminated.

2 Materials and Methods

AHP model uses human perception which is considered by experts as its main input. The criteria for "expert" here refers to people who understand the problem posed, feel the consequences of a problem or have an interest in the problem.

The main step taken in this study is a literature review, then a questionnaire is made to the impact and mitigation of landslides. The questionnaires were distributed to experts all over Indonesia who have experience in the fields of landslides, civil engineering and the environment. The experts used in the study were 20 people. From the results of the AHP analysis, only 14 experts met the criteria.

The following are the steps for the assessment using the AHP method:

2.1 Creating a Hierarchical Structure Begins with the Main Objective

The hierarchical structure begins with a general objective and then continues with the criteria criteria and alternative alternatives that will be sorted later. The network is made on the basis of a comprehensive structure related to the problems being developed.

2.2 Calculation Using the AHP Method

Assessment Matrix

Qualitative decisions are made by providing a perception of comparison that is scaled in pairs (pairwise comparison) and becomes a quantitative assessment by assessing

Table 1 Pairwise comparison matrix

	A_1	A_2		..	A_n
A_1	a_{11}	a_{12}		..	a_{1n}
A_2	a_{21}	a_{22}		..	a_{2n}
..	..	..		..	.
A_m	a_{m1}	a_{m2}		..	a_{mn}

Table 2 Random index value (RI)

No.	1	2	3	4	5	6	7	8	9	10
RI	0.00	0.00	0.58	0.90	1.12	1.24	1.32	1.41	1.45	1.48

the importance of an element compared to other elements. After the assessment is carried out using the appropriate scale according to Saaty, an assessment matrix or pairwise comparison matrix is made. Table 1 shows the paired matrix.

The values a_{11}, a_{22},..... a_{mn} are the comparison values for the elements of row A1 to column A1 which state the relationship of how far the importance of row A1 and column A1 is, or how far row A1 dominates over column A1.

Normalizing Data

Which normalizes this data is done by dividing the value of each element in the paired matrix by the total value of each column.

Calculating Priority Vector

In determining the priority value, a comparison of the total rows in the matrix is needed, where the total rows are obtained from the sum of the total row values of the calculated matrix that have been compared with the total value of the matrix in each column.

Testing of Hierarchy Consistency

The difference between the AHP method and other decision making methods is the requirement of absolute consistency. This consistency is also determined by the maximum eigen value. If the CI value is 0, the matrix is consistent (pairwise comparison). The inconsistency limit has been set by Saaty using the consistency ratio (CR) and the random index value (RI), following the formula below.

$$CR = CI/RI \tag{1}$$

where as

CI: Deviation ratio (deviation) consistency = $(\lambda max - n)/(n-1)$
λ_{max}: The largest eigenvalue of the n order matrix
n: Matrix order

RI: Random index (Table 2)

If the pairwise comparison matrix has a CR greater than 0.100, it can be said that the expert makes inconsistent decisions so that the judgment cannot be used.

Combining Opinions from Experts with Geomean

The geometric mean is the average obtained by multiplying all consistent data within a sample group, then rooted to the power of that sample size. To calculate the geometric mean the formula below is used.

$$\mathrm{GM} = \sqrt[n]{(X1)(X2)\ldots\ldots(Xn)} \tag{2}$$

where as

GM: *Geometric mean*
X_1: Expert-1
X_n: Expert-n.

This AHP method certainly has various disadvantages and advantages. The advantages of the AHP method are:

a. Unity. The model created by the AHP method is an understandable and flexible model for broad and unstructured problems.
b. Complexity. AHP can solve or solve complex or complex problems.
c. Hierarchy Structuring. AHP performs grouping by including each element or action into different levels. Each level contains elements that are similar.
d. Measurement. The assessment on the AHP method uses a scale to set priorities
e. Consistency, the assessment used in the AHP method must be consistent so that the level of consistency is tested for each assessment.

Meanwhile, the weakness of the AHP method is that this model is highly dependent on its main input, which is human perception, so that if the judgment is made wrong, the assessment will be meaningless.

2.3 Likert Scale

The Likert scale is a psychometric scale that has generally been widely used in questionnaires. In this study, the Likert scale was carried out for validation and reliability testing.

Validity Test

Validity is a measure that indicates that the measured variable is really the variable the researcher intends to study. Validation testing is carried out using the formula below.

$$r = \frac{n\Sigma XY - (\Sigma X)(\Sigma Y)}{\sqrt{(n\Sigma(X)^2 - (\Sigma X)^2)(n\Sigma(Y)^2 - (\Sigma Y)^2)}} \tag{3}$$

Reliability Test

In this study using a single test reliability with a split technique (odd–even). For the reliability test formula used the formula below.

$$r_{\text{calculted}} = 2r/(1 + r) \tag{4}$$

Calculation of Critical Score and Rating Scale

The criterion score is the score used to calculate the rating scale. The formula used to calculate the criterion score is as follows:

$$\text{Criterium Scale} = \text{Scale value} \times \text{Number of Experts} \tag{5}$$

In this study, the rating scale used is divided into four classes, namely strongly disagree (1), disagree (2), agree (3) and strongly agree (4).

Combining the Weights of All Experts

Combining the scores given by the expert on each question in the questionnaire using the formula below:

$$P(\%) = \sum f/z \times 100 \tag{6}$$

where as,

Σf: The sum of the scores of all experts on each question
P: Percentage of answers
Z: Total ideal score/highest criterion scale.

After obtaining the combined values of all the experts, the weight of each question is obtained using the formula below:

$$\text{Weight}(\%) = (P(\%))/(\text{Sum of } P(\%)) \times 100 \tag{7}$$

3 Results and Discussion

3.1 Assessment of Landslides Disaster Impact

The AHP method divides the level of impact into 3 parts where at level 2 the impact of landslides is classified into four parts, namely the impact of casualties, physical losses, environmental health and socio-economic. Furthermore, at the third level, each indicator is compared at level two. Table 3 shows the results of the expert assessment using AHP.

Based on the results of data processing using the AHP method, it was found that the most dominant impact occurred was the impact of casualties with a weight of 0.358, then the impact on environmental health was 0.272, the physical impact was 0.213 and the lowest weight was the socio-economic impact of 0.157. The weighting value obtained is in accordance with BNPB data from 2015–2020, which shows that the casualty impact is the most caused by landslides. In the data released by BNPB, the highest impact on casualties was in 2017, in which 163 people died, 185 were injured and 59,640 affected victims.

Table 3 Landslide impact assessment with AHP method

No.	Impact of Landslides	Weight	Score	Total
1	**Causalities impact**	0.358		
	Affected victims/relocation		42.84	15.35
	Injured victims		25.22	9.04
	The victim of died		31.93	11.44
	Total		100.00	
2	**Physical impact**	0.213		
	The transportation track was cut off		32.98	7.01
	Lost of the settlement		34.15	7.26
	The infrastructure damage		32.88	6.99
	Total		100.00	
3	**Environmental health impact**	0.272		
	Deteriorating environmental sanitation		31.96	8.71
	Disruption of ecosystem balance		32.23	8.78
	Loss of vegetation		35.81	9.76
4	**Socio-economic impact**	0.157		
	Incidence of crime		43.13	6.75
	Disruption of economic stability		56.87	8.90
	Total	1.00	100.00	100.00

3.2 Assessment of Landslides Disaster Mitigation

The magnitude of the impact caused by landslides needs to be anticipated, with mitigation efforts, the impact that occurs due to landslides can be reduced or eliminated. Based on the time of occurrence, mitigation efforts can be divided into mitigation before landslides, during landslides and after landslides. The method used can be done structurally or structurally. Based on the results of the expert assessment, the mitigation efforts that need to be done are shown in Table 4.

From Table 4, it can be seen that disaster mitigation measures can be carried out before having the greatest weight value compared to mitigation efforts during and after a disaster. It is hoped that with mitigation efforts before a landslide disaster occurs, the resulting impact can be reduced and can minimize casualties, so preventive

Table 4 Landslide mitigation assessment with AHP method

No.	Mitigations	Weight 1	Weight 2	Weight 3	Total
1	**Mitigation before landslides**	0.57			
	Structural mitigation		0.57		
	Making maps of areas prone to landslides and evacuation routes			34.53	11.29
	Establish an early warning system and install ground motion monitoring devices			34.53	11.29
	Construction of landslide retaining structures			20.94	10.11
	Non structural mitigation		0.43		
	The outreach to the community about the impact and mitigation of landslides			20.47	5.07
	The holding of landslide simulations to the affected communities			17.67	4.38
	Making posters about landslide disaster information			19.07	4.73
	Vegetation arrangement on slopes			20.93	5.19
	The Arrangement of RTRW (Regional Spatial Planning) in areas with potential for landslides			21.86	5.42
	Total		**1.00**		
2	**Mitigation during Landslides**	0.22			
	Structural mitigation		0.48		
	Making a special distribution track for aid			50.00	5.35
	Construction of a proper shelter			50.00	5.35
	Non structural mitigation		0.52		
	Coordination between the parties involved			48.00	5.55
	A responsive team of evaluators			52.00	6.02
	Total		**1.00**		

(continued)

Table 4 (continued)

No.	Mitigations	Weight 1	Weight 2	Weight 3	Total
3	**Mitigation after landslides**	0.21			
	Structural mitigation		0.58		
	Improvement of slope physical			33.33	3.88
	Replanting vegetation			34.78	4.05
	Improvement of residential areas and infrastructure buildings			31.88	3.71
	Non structural mitigation		0.42		
	Rehabilitation of communities affected by landslides			34.59	2.97
	Making of settlements in accordance with applicable standards			32.33	2.78
	Enforce a RTRW (regional spatial plan) policy that is suitable for landslide-prone areas			33.08	2.84
	Total	**1.00**	**1.00**		**100.00**

measures must be prepared as early as possible, preventive efforts are made to avoid a major impact on areas where disaster is indicated. Natural disasters can be local (micro) but can also be widespread, causing macro impacts. One way that can be done to mitigate disasters is through spatial planning, so as to reduce physical and economic losses due to disasters. Mitigation activities include risk assessment and determination of potential disaster.

4 Conclusions

Based on the results of the assessment using AHP, the impact of casualties/displacement is the impact with the highest weight of 15.35. For this reason, pre-disaster mitigation efforts need to be carried out to reduce the impact, such as by making maps of landslide disaster locations and easy-to-reach evacuation routes and secondly, creating a landslide early warning system and installing a rainfall monitoring device.

Acknowledgements Publication of this article is supported by Engineering Faculty of Andalas University, fiscal year 2022 and we would like to thank all those who have helped this research, especially all the experts who were involved in filling out the questionnaire.

References

1. BNPB (2018) Buku IRBI 2018 Diterbitkan Oleh : Badan Nasional Penanggulangan Bencana
2. Suriadi AB, Arsjad M, Hartini S (2014) Analisis Potensi Risiko Tanah Longsor di Kabupaten Ciamis dan Kota Banjar, Jawa Barat. Maj Ilm Globe 16:165–172
3. Nathania B, Muira F (2017) Remote sensing and GIS approach for landslide susceptibility mapping: a case study in Hofu City, Yamaguchi, Japan. Int J Environ Geosci 1(1):28–35
4. Wieczorek G (1996) Landslides: investigation and mitigation. Chapter 4—Landslide triggering mechanisms. Transp Res Board Spec Rep 247
5. Young R, Norby L, Wieczorek GF, Snyder JB (2019) Monitoring slope movements. Geol Monit 11:245–271
6. Polemio OP (2000) Landslides—Rainfall as a landslide triggering factor : an over view of recent international research. Landslides Res Theor Pract Thomas Telford, London 3
7. Berglar K, Gaedicke C, Lutz R, Franke D, Djajadihardja YS (2008) Neogene subsidence and stratigraphy of the Simeulue forearc basin, Northwest Sumatra. Marine Geol 253:1–13
8. Moradi S, Rezaei M (2014) A GIS-based comparative study of the analytic hierarchy process, bivariate statistics and frequency ratio methods for landslide susceptibility mapping in part of the Tehran metropolis, Iran. Geopersia 4(1):45–61
9. Yanuarto T, Pinuji S, Utomo AC (2019) Buku saku Tanggap Tangkas Tanguh Menghadapi Bencana 48
10. Tokmechi Z (2016) Landslide mitigation and its risk controlling landslide mitigation and its risk controlling. May 2016
11. Warburton J (2015) Peat landslides. Elsevier Inc.
12. Sultana N, Tan S (2021) Landslide mitigation strategies in southeast Bangladesh: lessons learned from the institutional responses. Int J Disaster Risk Reduct 62:1–18
13. Iinran BAM (2012) Malino- Manipi landslide, south Sulawesi Indonesia, vol. 2012
14. Oktorie O (2017) A study of landslide areas mitigation and adaptation in Palupuah Subdistrict, agam regency, West Sumatra Province, Indonesia. Geogr Geogr Educ 1(1):43–49
15. El Bchari F, Theilen-willige B, Malek HA (2019) Landslide hazard zonation assessment using GIS analysis at the coastal area of Safi (Morocco), no. July, pp 15–20
16. Roy J, Saha S (2019) Landslide susceptibility mapping using knowledge driven statistical models in Darjeeling district, West Bengal, India. Geoenvironmental Disasters 6(11):1–18
17. Moradi M, Bazyar MH, Mohammadi Z (2016) GIS-based landslide susceptibility mapping by AHP method , a case study, Dena GIS-based landslide susceptibility mapping by AHP method, a case study, Dena City, Iran. no. Dec 2016
18. El Jazouli A, Barakat A (2019) GIS-multicriteria evaluation using AHP for landslide susceptibility mapping in Oum Er Rbia high basin (Morocco). 7
19. Mondal S, Maiti R (2013) Integrating the analytical hierarchy process (AHP) and the frequency ratio (FR) model in landslide susceptibility mapping of Shiv-khola watershed, Darjeeling Himalaya. 4(4):200–212
20. Eddy AI, Dwi D, Putranto A, Affandi AK (2018) Evaluation of land subsidence impact at lowland areas using analitycal hierarchy process (AHP) Method. 7:1–7
21. Munthafa AE, Mubarok H, Teknik J, Universitas I (2017) Penerapan metode analytical hierarchy process dalam sistem kata Kunci : analytical hierarchy process, consistency index, Mahasiswa Berprestasi. keywords : analytical hierarchy process, consistency index, achievement student b. Kelebihan dan Kelemahan Metode AHP. 3(2):192–201
22. Whitaker R, Foundation CD (2017) The analytic hierarchy process—what it is and how it is used. 0255(October)
23. Szabó S, Gácsi Z, Balázs B (2016) Specific features of NDVI, NDWI and MNDWI as reflected in land cover categories. Landsc Environ 10(3–4):194–202
24. Studi P, Informatika T, Tinggi S, Ilmu M, Nusa K, Jakarta M (2015) Multi criteria analysis Menentukan point weight XI(1):11–19

The Stability of a Slope on Soft Soil Using the Hardening Soil Model

Yerry Kahaditu Firmansyah and Maharani Putri Dewanty

Abstract The development of reservoir construction does not always stand on with good carrying capacity soil, on the part of the existing soil layer with a soft to very soft constancy. About 10% or more of the total area in various regions in Indonesia or more precisely about 20 million hectares are soft-land which is composed of soft soil and moss soil. If we look at the problem of soft clay soil, which is very influential in the construction of the construction and its success, soil repair must be done first; therefore, the construction is not dented already the predetermined and intentional age. There is a soil repair solution that is minimal in cost but very good and effective for maintaining the stability of the landfill, namely by using a soil repair system with a retaining wall. The Finite Element Model is a method of research to solve a problem in a study or field of mathematics and physics. This method presents several parameter estimates for an amount of distinct scheme nodal constructions not known. The Hardening Soil Model Standard, or what can be abbreviated as (HS) is a continuation of the advanced soil behavior model. But the value or number of soil stiffness can be described, analyzed, and explained in more detail and accurately using stiffness. There is three dissimilar effort arduousness, called triaxial load stiffness (E_{50}), triaxial loading stiffness (E_{ur}), and one-way load stiffness (E_{oed}). In a convincing case, the Hardening Soil Model or (HS) method is proven to be more truthful and closer to field conditions. It can be seen and has been proven from numerous previous studies that show the Hardening Soil Model (HS) method is very distinctive and accurate by the outcomes of field testing. The charge of the safety factor or (SF) obtained after the investigation of slope constancy on soft soil using hardening soil modeling analysis is 1.59.

Keywords Reservoir construction · Soil improvement · Hardening soil model

Y. K. Firmansyah (✉) · M. P. Dewanty (✉)
Universitas Pembangunan "Veteran" Jawa Timur, Raya Rungkut Madya, Surabaya, East Java, Indonesia
e-mail: yerry.kahaditu.ts@upnjatim.ac.id

S. A. Kristiawan et al. (eds.), *Proceedings of the 5th International Conference on Rehabilitation and Maintenance in Civil Engineering*, Lecture Notes in Civil Engineering 225,
https://doi.org/10.1007/978-981-16-9348-9_41

1 Introduction

Temporary water or shelter is often referred to with the reservoir construction should be built as sturdy as possible so that it can function optimally [1]. most of the soft soils found in Indonesia are probably composed of Holocene clay deposits. This Holocene material is obtained from volcanic eruptions in which there is a potentially very significant volcanic ash content [2]. The nature of soft soil is fundamentally different from other jobs soils found on coastal plains or in inland areas, due to the leaching of freshwater [3]. And it can be identified that the existence of derivative volcanic soil has a very significant effect on the properties of the soil. In some cases, the deposits seen from a geological perspective are very young [4], where there is a relationship and a relationship between the rate of deposition and the degree of consolidation, which can be seen in Fig. 1 in the following. In terms of degrees of consolidation, the average for deltaic clays in the Southeast Asia region can range between and begin 20 to 100% reliant on where the location is located. Furthermore, for sediments in the Indonesian region, for example, the grade of the alliance is prophesied (on Fig.1) around 70 to 80%. It indicates that the soft clay in the area around the coast or the sea is experiencing very little consolidation.

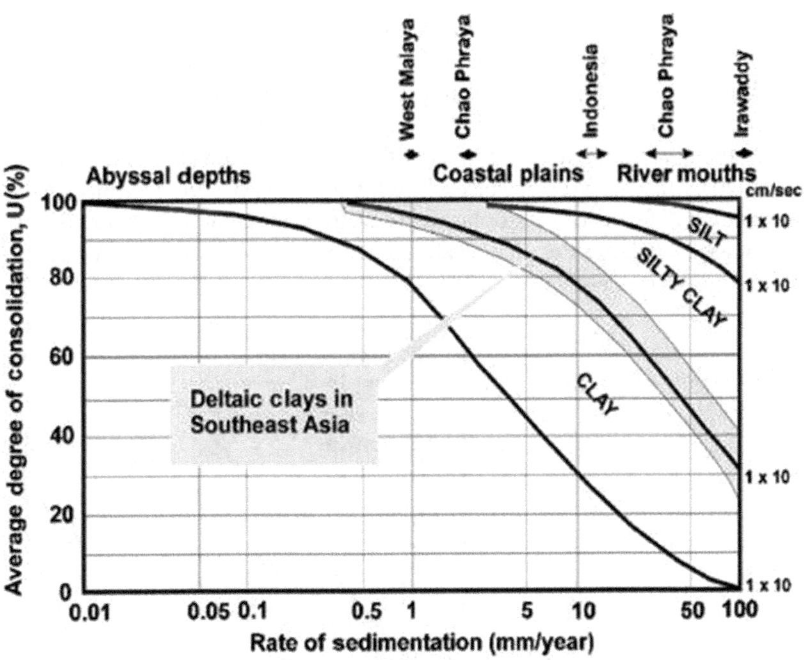

Fig. 1 The relationship between sedimentation rate and the consolidation degree [4]

Among a lot of obstacles, there is one of the obstacles in an embankment activity on soft soil is the number of landslides. Several other problems are frequently encountered when constructing roads in soft soils located around swamps, especially if there is no reinforcing material available for the road [5]. An embankment material sinks frequently into the softer soil because it is not sturdy enough to hold the ridge soil [6]. In general and basically, the general purpose of soil improvement is to surge the comportment volume and shave strength of soil, upsurge the modulus of soil, reduce the compressibility of soil, control the stability of the soil (reduction and inflammation), improve the quality of construction resources. Material and helps to diminish the effects of damage to work processes in a zone and its surroundings. A sheet pile is a structural tool where it is designed to be able to withstand horizontal forces when embedded in the ground. The sheet pile can usually also be used as a holding system. The pile of sheets is made of different materials, which play an important role in their use and application [7].

The above calculations that have been analyzed will then be carried out by an approach called the Hardening Soil or (HS) model method. The admiration of this model is based on very unpretentious restriction documentation of a predefined test and empirical formula. The HS Model is realized in most profitable FE ciphers which purpose has been designed to analyze geotechnical problems [8]. Developed by Schanz, Vermeer, and Bonnier, this rudimentary form of the Hardening Soil (HS) elastoplastic model is realized into the Plaxis FE cipher [10]. The key basis of the HS model can be further described and enlightened by illustrating the outlines of the resulting surface on the Roscoe p-q stress invariant plane, which can be seen in Fig. 2. The criteria of Mohr–Coulomb shear strength will later limit the level of shear stress. On the other hand, the hardening of the volumetric deviator and the pre-failure is controlled by the two yield surfaces. The HS model is entirely isotropic in terms of elastic range and terms of elastoplastic. Therefore, inherent anisotropy is absent or absent, and the available stress contradicts experimental evidence mainly or focused, e.g. in fine-grained deposits [11, 12].

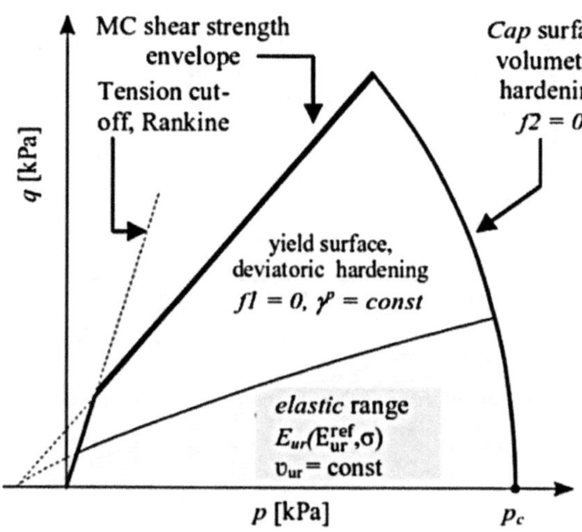

Fig. 2 Basic HS model in terms of surface yield and elastic range [8]

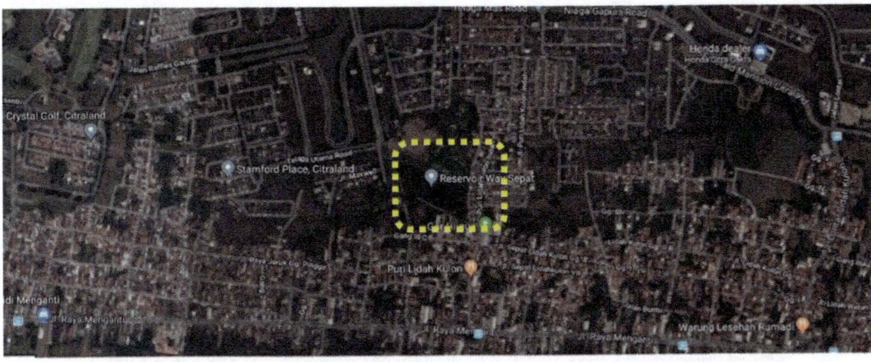

Fig. 3 Sepat reservoir location, Surabaya, East Java

In the previous HS version, the relationship of stress to hyperelastic strain was functional and applied in the elastic range over which Young's modulus was dependent on the stress. E_{ur} (E_{ur}^{ref}, σ) through a law called Poisson's law of power and constant ratio υ_{ur}. In an HSS version, increasing the small strain stiffness is based on and depends on the stress and the accumulated shear strain, whereas in the hypoelastic relationship, the actual reference shear modulus G_t^{ref} changes between one value and another G_{ur}^{ref} and G_0^{ref}.

2 Project Information in General

2.1 Description on a Experimental Test Site

The site that the experiment held is located at Sepat reservoir, Surabaya, East Java. The Sepat reservoir has a water holding capacity of 13,000 m^3. Furthermore, the alternative reinforcement in the talud area of the reservoir is used corrugated concrete sheet pile (CCSP) type W400A. Figure 3 shows the Sepat reservoir location.

2.2 Method of Construction

The CCSP installation method uses the jack-in method. This method is considered environmentally friendly, meaning that it does not cause damage to surrounding buildings. This material is in installed on the inside of the reservoir and surrounds the area with type of material used is W400A. Figure 4 shows the CCSP material properties.

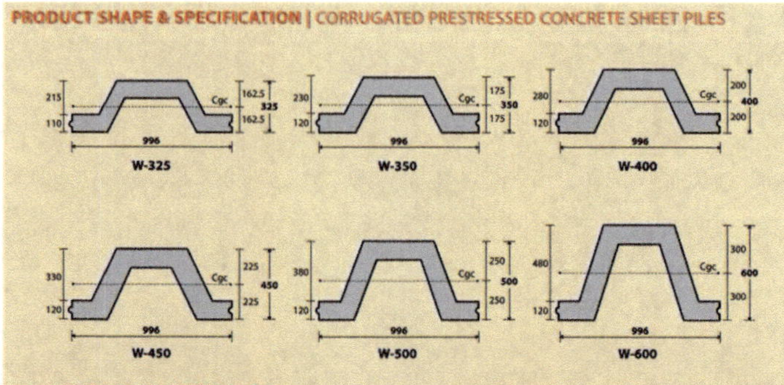

Type	Width (mm)	Cross Section (cm²)	Section Inertia (cm⁴)	Unit Weight (kg/m)	Class	Moment (ton.m) Crack	Moment (ton.m) Break	Allow. Service Moment (ton.m) Temporary	Allow. Service Moment (ton.m) Permanent	Length* (m)
W-325	996	1,315	134,264	329	A	11.40	22.80	10.07	6.74	8 - 15
					B	13.30	26.60	11.97	8.64	8 - 16
W-350	996	1,468	169,432	368	A	15.60	31.20	14.04	10.14	9 - 17
					B	17.00	34.00	15.44	11.54	10 - 18
W-400	996	1,598	248,691	400	A	20.10	40.20	18.10	13.08	10 - 18
					B	23.40	46.80	21.40	16.38	11 - 20
W-450	996	1,835	353,363	459	A	26.90	53.80	24.37	18.04	11 - 20
					B	30.70	61.40	28.17	21.84	12 - 21
W-500	996	1,818	462,373	455	A	35.20	70.40	32.22	24.76	12 - 22
					B	40.40	80.80	37.42	29.96	13 - 24
W-600	996	2,078	765,907	520	A	50.60	101.20	46.48	36.19	14 - 25
					B	59.60	119.20	55.48	45.19	15 - 27

Note : * Length of Concrete Corrugated Sheet Piles may exceed usual standard whenever lifted in certain position

Fig. 4 CCSP material properties

2.3 *Soil Field Investigation*

Results of soil investigations that have been carried out show that the condition of the soil layer has a very soft to soft consistency at a depth of 0 m to a depth of 5 m. The description on layer 1 is Inorganic clay with high plasticity, brown color, and layer 2 is Inorganic clay with high plasticity, grey color. Figure 5 shows the soil stratification.

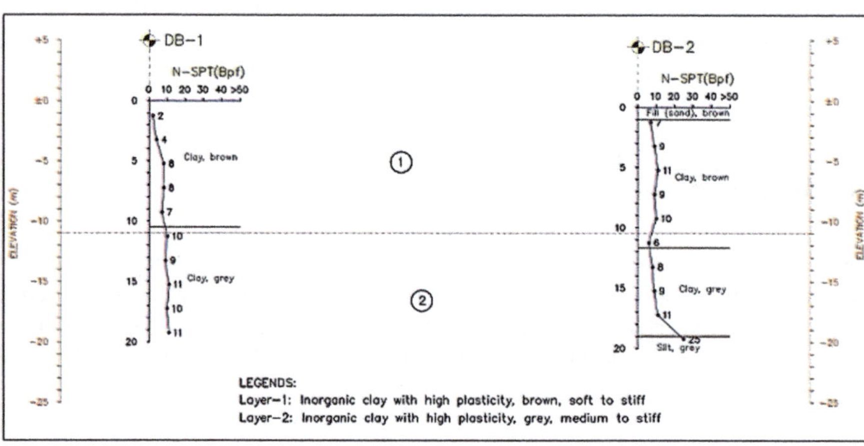

Fig. 5 Soil stratification

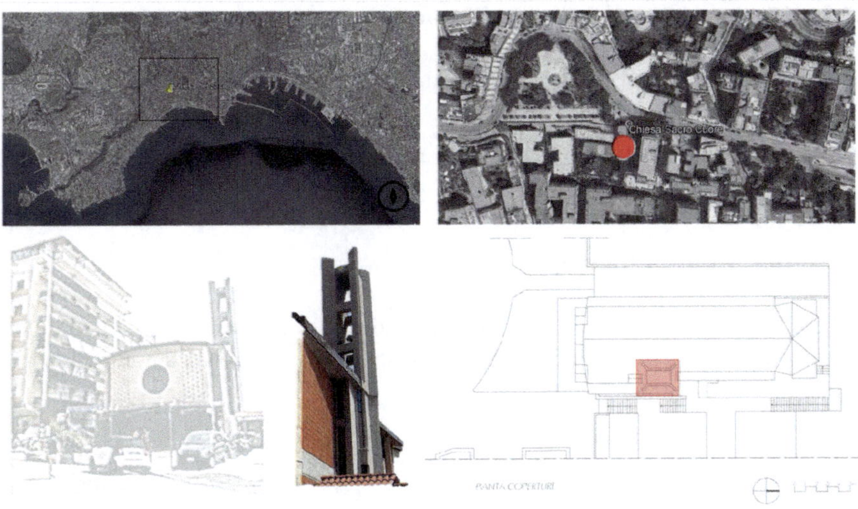

Fig. 6 An example of overshooting which is an error in modeling hysterical behavior during cyclic cutting [15]

3 Approaches for Analysis

3.1 *Review of the HS Base Model*

The HS construction of the extension of the small strain stiffness was first proposed by Benz and his colleagues [13, 14], whereupon the so-called original model was called the Hardening Soil Small or (HSS). This new construction presents and provides an unusual but specific algorithm to govern the present hypoelastic rigorousness in the flexible assortment of the basic HS model. There are further restrictions that are gotten on or after a test that is carried out by standard then extends or empirical formulas. This is depicted in the systematic schematic which can be seen in Fig. 6. Related to the problem of overshooting in the HSS model, it will be reviewed in a critical and detailed way [15]. Subsequent minor load reversals are accomplished by reloading or rebuilding, which can cause unexpected irregularities.

3.1.1 Mechanism of Deviatoric Plastic

The formula of the yield surface controlling the deviatoric hardening mechanism [9]:

$$f1 = \frac{q_a}{E_{50}} \frac{\sigma_{1-}\sigma_3}{q_a - \sigma_{1-}\sigma_3} - 2\frac{\sigma_{1-}\sigma_3}{E_{ur}} - \gamma^P = 0 \, \text{for} \, q < q_f, \quad (1)$$

3.1.2 Mechanism of Volumetric Plastic

The equation for the surface of the smooth cover:

$$f_2 = \frac{q^2}{M^2 r^2(\theta)} + p^2 - p_0^2 = 0, \tag{2}$$

3.1.3 Barotropy of Rigidity

In the HS model basic [9], loading and unloading E_{ur} and secant E_{50} The modulus of stiffness and stress depend on the following force rules and the corresponding empirical barotropy function f_σ:

$$E_{ur}(\sigma) = E_{ur}^{ref} f_\sigma(\sigma), \tag{3}$$

$$E_{50}(\sigma) = E_{50}^{ref} f_\sigma(\sigma), \tag{4}$$

$$f_\sigma(\sigma), = \left(\frac{\sigma_3 + c\cot\varnothing}{P_{ref} + c\cot\varnothing}\right)^m. \tag{5}$$

4 Results and Disscussions

The safety factor (SF) is generally used to measure stability on a slope in geotechnical engineering. In this study, finite element analysis was used to obtain deformation, bending moment at CCSP, and safety factors that occurred on slopes. Slope geometry data will be modeled according to field data. Figure 7 shows the input geometry model.

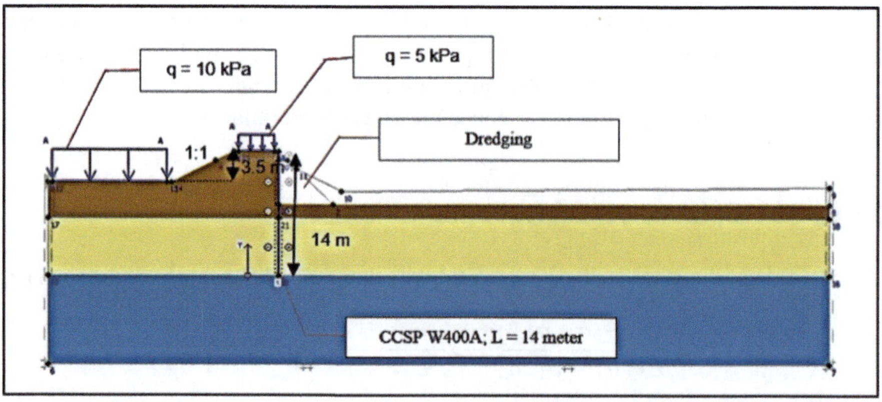

Fig. 7 Input geometry model

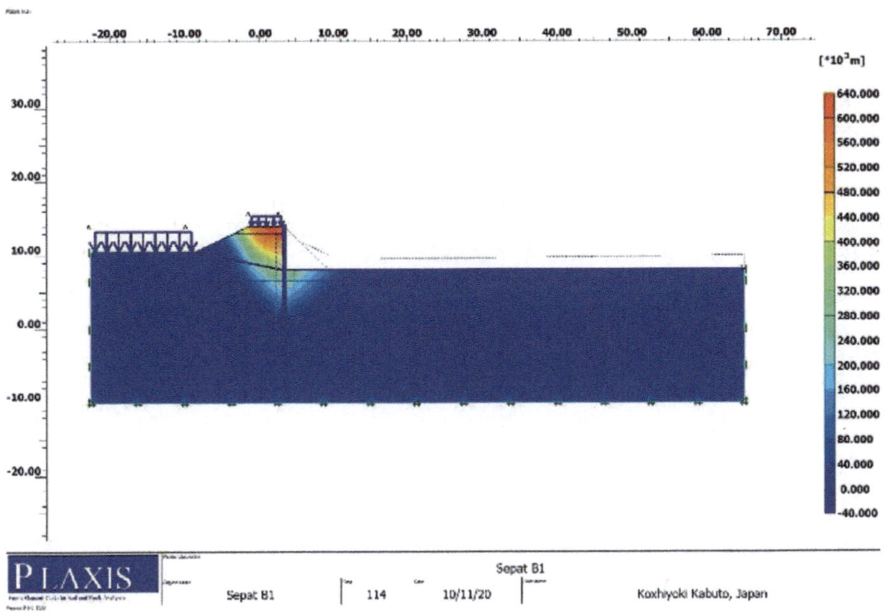

Fig. 8 Distribution of stress that occurs in the slope area

The use of CCSP material as deep as 14 m looks safe because the distribution of soil stress does not exceed the depth of the material (This can be seen in Fig. 8). The SF value obtained depends on the outcomes of analysis modeling using the finite element is 1.59. This means that the use of CCSP as an amplifier for the Sepat reservoir is safe. Based on the material strength control, the type of material W400A has obtained a bending moment value of 10.4 t.m while the permitted capacity based on the material brochure for type W400A is 20.1 t.m, meaning that the material is still safe to use. Furthermore, the permit deflection is 0.5% x the elevation difference is 25 mm, while the horizontal displacement that occurs is 13.06 mm, meaning that the deflection is still a safe limit. Figure 9 shows the force on CCSP material.

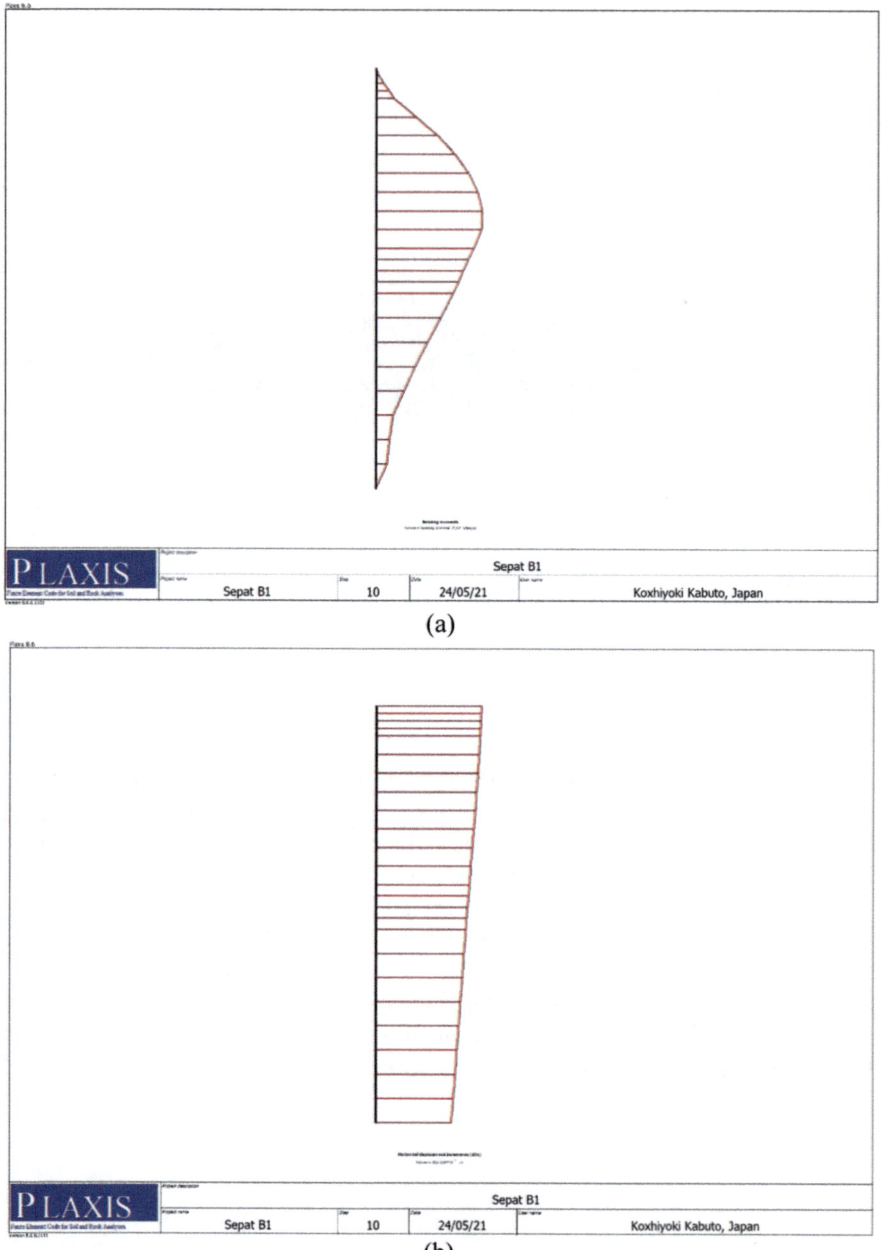

Fig. 9 Force distribution for **a**. Bending moment occured is 10.4 t.m and **b**. Deformation occured is 13.06 mm

5 Inference

This case focuses more on the study of an analytical approach using the Hardening Soil (HS) model. Furthermore, the conclusions obtained are:

a. From the talud stability modeling results in the Sepat reservoir, it is known that the critical condition when the reservoir is dry or without water filling.
b. Talud in the Sepat reservoir has the potential for landslides to occur if there is no reinforcement in the talud area. In order to avoid landslides, corrugated concrete sheet pile (CCSP) reinforcement type W400A with a total length of 14 m is used and installed on the inside of the reservoir.
c. The safety factor (SF) value of the use of corrugated concrete sheet pile (CCSP) type W400A with a total length of 14 m is 1.59. This is still within safe limits and in accordance with the recommendation of SNI 8460: 2017 "Geotechnical Design Requirements"
d. The used of the CCSP W400A material as reinforcement for the Sepat reservoir has taken into account several aspects, including: When the reservoir is full of water, the reservoir is empty, the load that may occur in the reservoir area, digging into the ground, material for permit capacity, and deflection limits for the use of these materials.

Acknowledgements We would like to thank *UPN "Veteran" Jatim* for backing and investigation funding. Next, the data for this paper is provided by Citraland's Developer Properti and Mr. Bangun.

References

1. Irmawan M, Firmansyah YK, Refani AN (2016) Modification of reinforced concrete retaining wall at Pilangbango reservoir construction project madiun, East Java. ICCER
2. Barry AJ, Rachlan A (2001) Embankment on soft soils in North Java. Embankments on soft soils. In: Proceedings of the preconference volume, international conference on in situ measurement of soil properties and case histories, Bali, Indonesia
3. Younger JS (1998) Ground improvement options for highways in Indonesia. In: Proceedings of the short course and seminar on ground improvement, application to Indonesia soft soils. Universitas Kristen Indonesia and Asian Institute of Technology in Cooperation with Himpunan Ahli Tanah Indonesia, Jakarta, pp S3/i–S3/32, 23–26 Aug 1988
4. Cox JB (1970) The distribution and formation of recent sediments in South East Asia. In: Proceedings of the second South East Asia conference on soil engineering, 1970, pp 30–47
5. Muhammad A, Indrasurya BM, Noor EM, Yulian FA (2020) Road embankment full- scale investigation on soft soil with geotextile stabilization. Int J GEOMATE 19(71):145–152
6. Hardiyatmo HC (2013) Geosynthetic for highway engineering-planning and application
7. Eskandari L, Kalantari B (2016) Basic types of sheet pile walls and their application in the construction industry—a Review. EJGE 16
8. Marcin C, Andrzej T (2020) Refinement of the hardening soil model within the small strain range. Acta Geotech. https://doi.org/10.1007/s11440-020-00945-5

9. Schanz T, Vermeer PA, Bonnier PG (1999) The hardening soil model: formulation and verification. In: Beyond 2000 in computational geotechnics—10 years of plaxis. Rotterdam, Balkema, pp 1–16
10. Brinkgreve RBJ, Kumarswamy, S, Swolfs WM, Zampich L, Ragi Manoj N (2019) Plaxis finite element code for soil and rock analyses. Plaxis BV , Bentley Systems, Incorporated, Philadelphia
11. Lings ML, Pennington DS, Nash DFT (2000) Anisotropic stiffness parameters and their measurement in a stiff natural clay. Géotechnique 50(2):109–125, 27
12. Masǐn D, Rott J (2014)Small strain stiffness anisotropy of natural sedimentary clays: review and a modelActa Geotech9(2):299–312
13. Benz T (2007) Small-strain stiffness of soils and its numerical consequences. Ph.D. thesis, University of Stuttgart
14. Benz T, Vermeer P, Schwab R (2009) A small-strain overlay model. Int J Numer Anal Methods Geomech 33:25–44
15. Niemunis A, Cudny M (2018) Discussion on "Dynamic soilstructure interaction: a three-dimensional numerical approach and its application to the Lotung case study". Poor performance of the HSS model. Comput Geotech 98:243–245

Software Performance of Risk-Targeted Maximum Considered Earthquake (MCE_R) Calculation

Windu Partono

Abstract Website software (WS) for Risk-targeted Maximum Considered Earthquake (MCE_R) acceleration calculation was announced in 2019. Single user non website software (stand-alone software/SAS) was also released in June 2020. The MCE_R acceleration is divided into three different values MCE_G (peak ground acceleration), MCE_R-S_S (short period) and MCE_R-S_1 (1-s period) spectral acceleration. The purpose of the SAS software development is to help civil engineers living in areas having no internet connection or internet connection problems. Both software packages should yield the same MCE_R result when the same building location or position or coordinate is entered as an input. This paper describes the evaluation of SAS in calculating the MCE_R compared to the result calculated using the WS. The evaluation was performed at three different islands in Indonesia; Sumatra, Java and Sulawesi islands and at 154, 115 and 81 coordinate positions on these islands, respectively. More than 10% difference in the data was observed when the MCE_R acceleration was calculated at the area having (MCE_G and MCE_R-S_1) greater than 0.6 g (g is gravitational acceleration). The minimum 10% difference in data was also observed when the MCE_R-S_S was greater than 1.5 g. The largest difference was also observed when the coordinate positions were located close to seismic sources (fault source traces).

Keywords MCE_R · Website software · Stand-alone software

1 Introduction

The Indonesian seismic code for buildings and other structures design was released in 2019 [1]. The code was implemented as part of the development of Indonesian seismic hazard maps 2017 [2–4]. Some of the information described in this code was developed following the American Standard Code for Building Design ASCE/SEI

W. Partono (✉)
Civil Engineering Department, Faculty of Engineering, Diponegoro University, Semarang 50275, Indonesia
e-mail: windupartono@lecturer.undip.ac.id

S. A. Kristiawan et al. (eds.), *Proceedings of the 5th International Conference on Rehabilitation and Maintenance in Civil Engineering*, Lecture Notes in Civil Engineering 225,
https://doi.org/10.1007/978-981-16-9348-9_42

7-16 [5, 6]. Compared to the previous 2012 Indonesian seismic code, two different software were announced as a complementary part of the 2019 code. However, only the website software was developed for the 2012 seismic code. All the software created during the development of the 2012 and 2019 seismic codes is related to the calculation of the Risk-targeted Maximum Considered Earthquake (MCE_R) and seismic Response Spectrum Acceleration (RSA) for seismic forces in the design of buildings and other structures.

The additional 2019 seismic code software was created to meet the requirement civil engineers or users living in areas having internet connection problems. The website software (WS) was first released in 2019 and after the improvement of the software performance, the final version of the website software was released in February 2021. A new and additional software, also known as stand-alone software (SAS) or binary software, was also created in 2020. The final stand-alone software was also released in February 2021. Both WS and SAS were developed using different programming languages. Although WS and SAS were developed using the same algorithm and procedures, the restrictions and differences in syntax and semantics that exist in these two programming languages sometimes lead to different results [7–10]. The main objective of the software application used by [7–10] related with the calculation of probabilistic seismic spectral acceleration [11, 12].

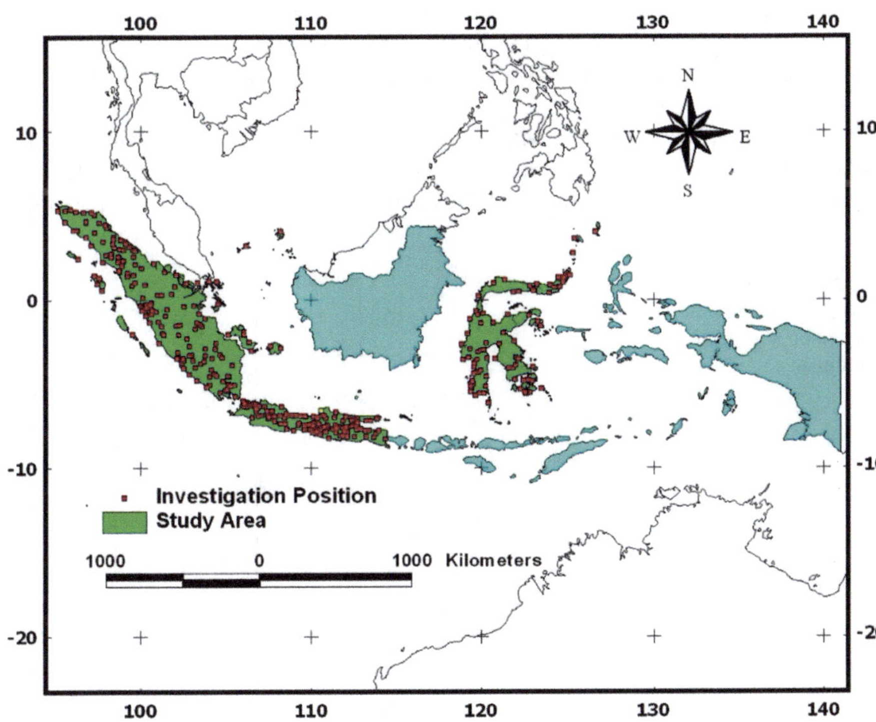

Fig. 1 Three different location of WS and SAS evaluation, Sumatra, Java and Sulawesi islands

This paper describes the evaluation of the MCE_R acceleration values calculated using WS and SAS software. The evaluation was conducted at three different locations: Sumatra, Java and Sulawesi islands and performed at 154 coordinate (longitude and latitude) positions on Sumatra Island and performed at 114 coordinate positions on Java Island and 81 coordinate positions on Sulawesi Island (Fig. 1). The evaluation was performed in terms of the mapped MCE_G peak ground acceleration and MCE_R at short period (MCE_R-S_S) and at 1-s period (MCE_R-S_1).

2 Methodology

The MCE_R acceleration (MCE_G, MCE_R-S_S and MCE_R-S_1) at a specific position or coordinate P can be calculated using linear interpolation of the MCE_R values calculated at four different points close to P. Figure 2 shows the basic concept of spectral acceleration calculation at point P. S1, S2, S3 and S4 represent four MCE_R accelerations (MCE_G or MCE_R-S_S or MCE_R-S_1). SP represents the MCE_R acceleration (MCE_G or MCE_R-S_S or MCE_R-S_1) calculated at point P. The distance "d" between two points is 0.1° or approximately 11.13 km. If the point P is located between two other points, the MCE_R acceleration can also be calculated using linear interpolation of the two points. Table 1 shows an example of spectral acceleration MCE_G calculation at point P. Table 2 shows an example of MCE_R-S_S calculation at point Q. The difference of the MCE_G or MCE_R-S_S results calculated using WS and SAS can also be seen in Tables 1 and 2.

The differences of MCE_G, MCE_R-S_S or MCE_R-S_1 calculated using the two different type of software WS and SAS were then evaluated and compared to determine the performance of the WS and SAS software. The evaluation was performed

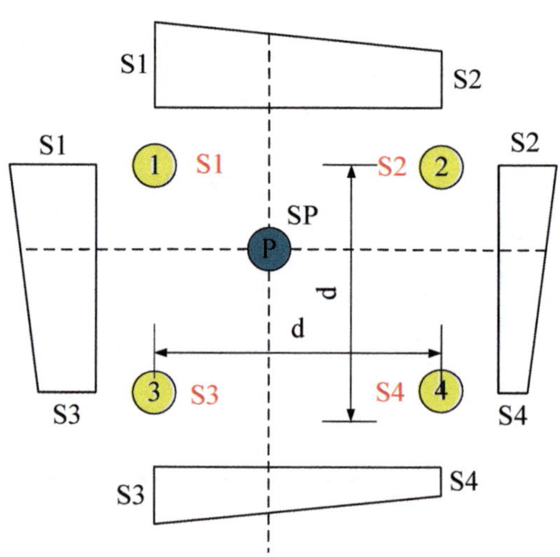

Fig. 2 Linear interpolation of MCE_R acceleration calculation at specific coordinate point P

Table 1 Sample spectral acceleration MCE_G calculation at P (110.57, −6.53)

Point no	Longitude (degree)	Latitude (degree)	MCE_G (g)	MCE_G at P (g)
1	110.5	−6.5	0.198	0.2032 (WS) 0.2031 (SAS)
2	110.5	−6.6	0.211	
3	110.6	−6.5	0.2	
4	110.6	−6.6	0.212	

Table 2 Sample spectral acceleration MCE_R-S_S at Q (99.73764, 1.05033)

Point no	Longitude (degree)	Latitude (degree)	MCE_R-S_S (g)	MCE_R-S_S at Q (g)
1	99.7	1.1	1.5	1.4762 (WS) 1.5 (SAS)
2	99.7	1.0	1.5	
3	99.8	1.1	1.176	
4	99.8	1.0	1.5	

based on the total data (percentage of the total points in one study area) having MCE_G or MCE_R-S_S or MCE_R-S_1 differences, as calculated using WS and SAS software, greater than 0.01 g.

3 Result and Discussions

Calculation of response spectrum acceleration design at specific building locations is an important step in seismic design for buildings. The purpose of this analysis is to obtain the MCE_R, S_{DS} (short period) and S_{D1} (1-s period) spectral acceleration design. According to [1], another parameter used for response spectrum acceleration design is TL (the long transition period). Only the MCE_R difference performance of the two software packages was evaluated in this study. Figure 3 shows the analysis results of the WS and SAS difference performance calculated at 154 points on Sumatra island. Comparative analysis of the MCE_R calculation between the WS and SAS software was also conducted on Java and Sulawesi islands. Figure 4 shows the analysis results of the WS and SAS difference performance conducted on Java island. Figure 5 shows the analysis results for Sulawesi island.

Based on the calculation results conducted using WS and SAS at Sulawesi island, approximately 10–11% of the total 81 MCE_G, MCE_R-S_S and MCE_R-S_1 data have a minimum 0.01 g difference. The largest MCE_G, MCE_R-S_S and MCE_R-S_1 differences between the WS and SAS calculation were observed on Sumatra island. At least 21–28% of the total 154 investigation data had a minimum 0.01 g difference. The largest MCE_G, MCE_R-S_S and MCE_R-S_1 differences for these two islands were observed at all points located close to seismic source traces. Figures 3 and 5 show the significant differences of MCE_G, MCE_R-S_S and MCE_R-S_1 calculated using the WS and SAS

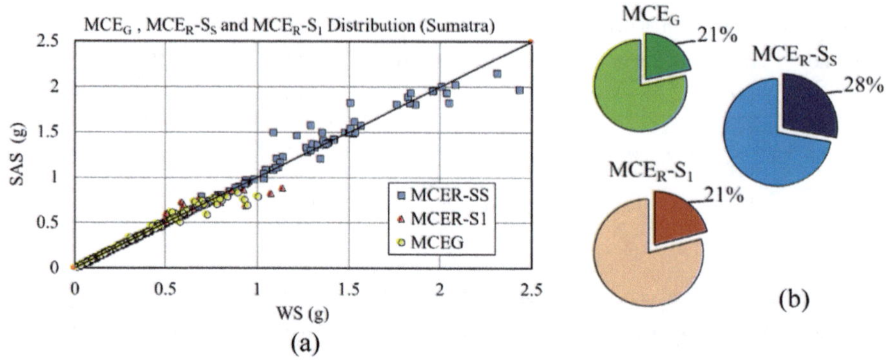

Fig. 3 MCEG, MCE$_R$-S$_S$ and MCE$_R$-S$_1$ difference performance calculated for Sumatra island (**a**) and the percentage of the total investigation data having at least 0.01 g differences (**b**)

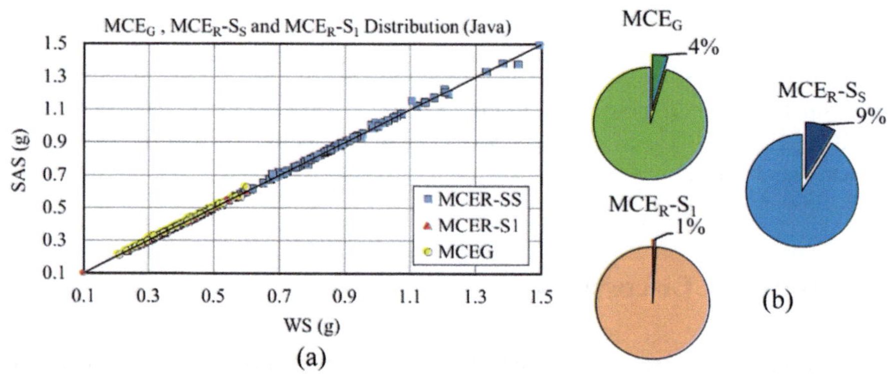

Fig. 4 MCEG, MCE$_R$-S$_S$ and MCE$_R$-S$_1$ difference performance calculated for Java island (**a**) and the percentage of the total investigation data having at least 0.01 g differences (**b**)

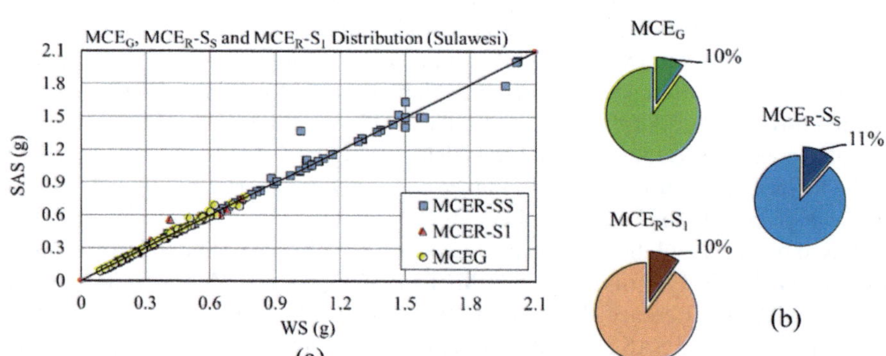

Fig. 5 MCEG, MCE$_R$-S$_S$ and MCE$_R$-S$_1$ difference performance calculated for Sulawesi island (**a**) and the percentage of the total investigation data having at least 0.01 **g** differences (**b**)

software for Sumatra and Sulawesi islands. Figure 4 shows the performance of the WS and SAS software used for Java island. Based on this figure, a minimum of 1% up to a maximum of 9% of the total 115 data were observed to have at least 0.01 g difference. The largest difference (9%) among the total of 115 data was observed for MCE_R-S_S calculation. However, only 1% of the total 115 MCE_R-S_1 results have at least 0.01 g difference.

To evaluate the performance of these two software packages especially at the position coordinates close to seismic sources, the analysis was performed at the area where the MCE_G and MCE_R-S_1 reached ±0.6 g and the MCE_R-S_S reached ±1.5 g. These two 0.6 g and 1.5 g values are two constant MCE_R acceleration values calculated based on the combination of deterministic, probabilistic and RTGM (Risk-targeted Ground Motion) seismic hazard analysis and following the same procedure proposed by [13]. Figures 6a, 7a and 8a show the predicted area where the MCE_G or MCE_R-S_S or MCE_R-S_1 reached constant acceleration values. The distribution area of the three constant acceleration values MCE_G, MCE_R-S_S and MCE_R-S_1 at one island is not always equal or in the same area. The performance of WS and SAS in calculating MCE_G, MCE_R-S_S and MCE_R-S_1 was also investigated at three different cross sections. The purpose of the analysis was to find the predicted position where significant MCE_G, MCE_R-S_S and MCE_R-S_1 differences were observed. Figure 6a, b, 7a, b and 8a, b show three different positions (cross-sections A, B and C) for MCE_G, MCE_R-S_S and MCE_R-S_1 evaluation. For each cross-section, the analysis was conducted at 21 coordinate position points: from point A1 (98.84, 1.31) to A21 (100.49, 2.21) at cross-section A (Sumatra island); from point B1 (106.75, −7.41) to B21 (107.01, −6.09) at cross-section B (Java island) and from point C1 (119.31, −1.48) to C21 (120.55, −1.06) at cross-section C (Sulawesi island).

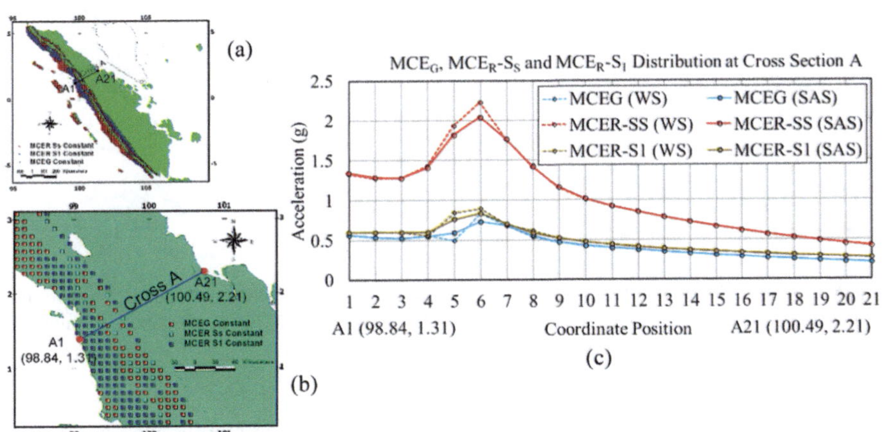

Fig. 6 Constant MCE_G, MCE_R-S_S and MCE_R-S_1 area on Sumatra island (**a**), detailed position of constant area and cross-section A (**b**), MCE_G, MCE_R-S_S and MCE_R-S_1 distribution calculated using WS and SAS at cross-section A (**c**)

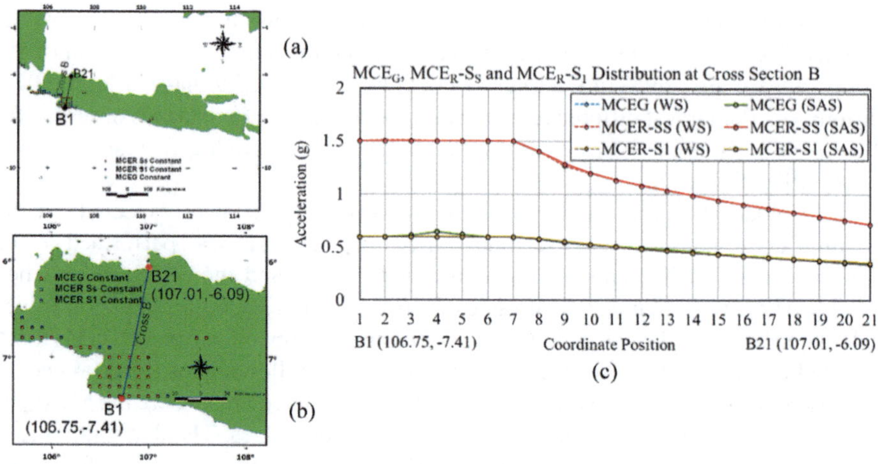

Fig. 7 Constant MCE_G, MCE_R-S_S and MCE_R-S_1 area on Java island (**a**), detailed position of constant area and cross-section B (**b**), MCE_G, MCE_R-S_S and MCE_R-S_1 distribution calculated using WS and SAS at cross-section B (**c**)

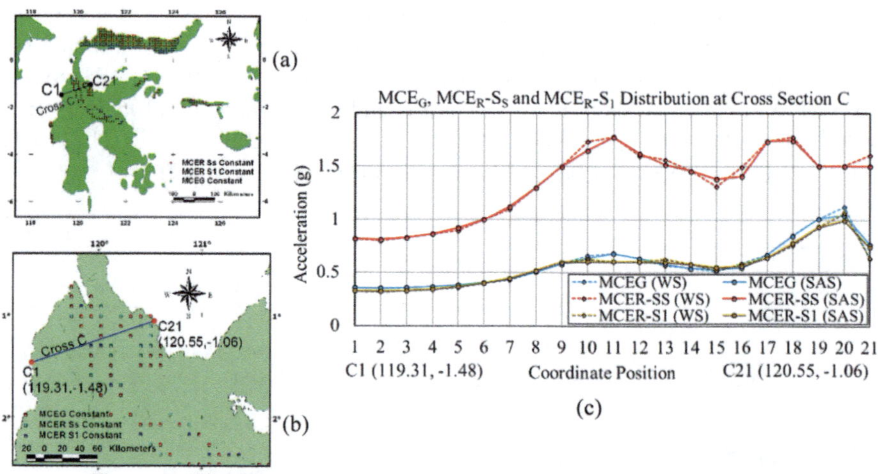

Fig. 8 Constant MCE_G, MCE_R-S_S and MCE_R-S_1 on Sulawesi island (**a**), detailed position of constant area and cross-section C (**b**), MCE_G, MCE_R-S_S and MCE_R-S_1 distribution calculated using WS and SAS at cross-section C (**c**)

The largest MCE_G, MCE_R-S_S and MCE_R-S_1 difference results calculated using the WS and SAS software were observed at the coordinate points close to seismic fault traces, shown in Fig. 6c and 8c. When the MCE_G and MCE_R-S_1 acceleration at one position coordinate was less than 0.6 g, the MCE_G and MCE_R-S_1 calculated using WS and SMS software were nearly equal. The same performance was also observed when the MCE_R-S_S calculated at one position coordinate was less than 1.5 g, when the MCE_R-S_S calculated using the WS and SAS software were almost equal. The MCE_G, MCE_R-S_S and MCE_R-S_1 calculated at all 21 position coordinates at Java island, (Fig. 7c), were almost equal due to the condition that there are no coordinate positions having MCE_G and MCE_R-S_1 greater than 0.6 g or MCE_R-S_S greater than 1.5 g.

The MCE_G, MCE_R-S_S and MCE_R-S_1 constant area at Sumatra island was observed in the western part, while the MCE_G, MCE_R-S_S and MCE_R-S_1 constant area was predicted at the small south-western area of Java island. However, there are two MCE_G, MCE_R-S_S and MCE_R-S_1 constant areas at Sulawesi island, at the middle and northern parts of the island. The seismic source traces were predicted close to the MCE_G, MCE_R-S_S and MCE_R-S_1 constant area [13].

4 Conclusions

Significant differences of MCE_G, MCE_R-S_S and MCE_R-S_1 calculated using WS and SAS software were observed at Sumatra and Sulawesi islands at coordinate positions close to the seismic sources area. Based on the calculation results performed at Sumatra island (154 points), the difference minimum of 0.01 g was observed at 21% (32 points) to 28% (43 points) of the data. However, for three spectral accelerations calculated at Sulawesi island (81 points), the difference minimum of 0.01 g was observed at 10–11% (approximately 9 points) of the data.

The MCE_G, MCE_R-S_S and MCE_R-S_1 acceleration values calculated at 151 coordinate positions on Java island using WS and SAS software were almost equal. Based on the calculation results conducted at 115 coordinate points, a maximum of 5% (5 points) of data with the minimum difference of 0.01 g were observed for the MCE_G and MCE_R-S_1 spectral acceleration. For MCE_R-S_S calculation, 9% (10 points) of the data with the minimum 0.01 g were observed on the Java island.

Acknowledgements Thank you to the Dean of the Faculty of Engineering, Diponegoro University, Semarang, Indonesia, for the approval of the research funding through the 2021 Strategic Research Scheme.

References

1. SNI 1726 (2019) Seismic resistance design codes for building and other structures. National Standardisation Agency of Indonesia
2. National Center for Earthquake Studies: Indonesian Seismic Sources and Seismic Hazard Maps (2017) Center for research and development of housing and resettlement. Ministry of Public Works and Human Settlements

3. Partono W, Irsyam M, Wayan SI, Asrurifak M, Kistiani F, Sari UC (2017) Development of design response spectrum acceleration for building based on new Indonesian Seismic Hazard maps 2017. In: AICCE conference, 21–22 August 2019, Penang Malaysia (2019), Lecture Notes in Civil Engineering, vol 53. Springer Nature Switzerland AG
4. Sengara IW, Irsyam M, Sidi ID, Mulia A, Asrurifak M, Hutabarat D, Partono W (2019) New 2019 risk-targeted ground motions for spectral design criteria in Indonesian seismic building code. In: 4th International conference on earthquake engineering & disaster mitigation (ICEEDM 2019), vol 156. Padang Sumatra Barat, E3S Web of Conference, 26–27 Sept 2019
5. ASCE/SEI 7–16 (2017) Minimum design loads for building and other structures. American Society of Engineers
6. Stewart JA, Seyhan E (2013) Semi-empirical nonlinear site amplification and its application in NEHRP site factors. Pacific Engineering Research Center (PEER) Report 2013/13, University of California, Berkeley, USA
7. Irwandi I, Umar M, Khaizal J, Asrurifak M, Usama F, Ridwan M (2020) The Neo-deterministic seismic hazard map (NDSHA) of Sumatra compared with official 2010 and 2017 derived from PSHA method. In: NCWE & ISSCE 2019, IOP conferences series materials science and engineering, vol 712. IOP Publishing
8. Danciu L, Monelli D, Pagani M, Wiemer S (2010) GEM1 hazard: review of PSHA software. GEM Technical Report 2010–2, GEM Foundation, Pavia, Italy
9. Syahbana AJ, Aditramulyadi DD, Irsyam M, Asrurifak M, Hendriyawan (2020) Application of modified PSHA USGS software in Java Island Bedrock Peak ground acceleration and hazard curve with 2475 years return period. Int J Adv Sci Technol 29(7):3138–3148
10. Pailoplee S, Palasri C (2014) CU-PSHA a Matlab software for probabilistic seismic hazard analysis. J Earthq Tsunami 08(04):1450008
11. Mulargia F, Stark PB, Geller R (2017) Why is probabilistic seismic hazard analysis (PSHA) still used. Physics of the Earth and Planetary Interiors, Elsevier, pp 63–75
12. Sianko I, Ozdemir Z, Khoshkholghi S, Garcia R, Hajirasouliha I, Yazgan U, Pilakoutas K (2020) A practical probabilistic earthquake hazard analysis tool: case study Marmara region. Bull Earthq Eng 18:2523–2555
13. Leyendecker EV, Hunt EJ, Frankel AD, Ruktales KS (2000) Development of maximum considered earthquake ground motion maps. Earthq Spectra 16(1):21–40

Seismic Microzonation of Yogyakarta Province Based on 2019 Risk-Targeted Maximum Considered Earthquake

Windu Partono, Ramli Nazir, Frida Kistiani, and Undayani Cita Sari

Abstract Yogyakarta Province is located close to two active seismic sources, Opak Fault which crosses the province area and the South Java subduction source which is located at the south of Java island. The province is located at the southern part of Java island. Based on the Indonesian earthquake database, from 1984 to 2015 at least seven earthquakes struck this province. The 2006 earthquake with 6.2 Mw magnitude was the largest earthquake to hit this area, causing approximately 88,249 buildings to be totally destroyed and 98,343 buildings to collapse. Most of the destroyed and collapsed buildings were constructed based on the old version of the seismic design code. Improvements in earthquake research have already been conducted in this area and the latest research was conducted at 2016. According to the New Indonesian seismic code 2019, improvements in seismic acceleration for building design need to be adjusted in this area. This paper described the seismic microzonation of Yogyakarta Province based on the New Indonesian seismic maps and 2019 seismic code in terms of the Risk-targeted Maximum Considered Earthquake. The analysis was performed by conducting a combination of three seismic hazard analysis, namely probabilistic, deterministic and Risk-targeted Ground Motion. Based on the Risk-targeted Maximum Considered Earthquake data calculated and its distribution in this area, an area with a maximum 10 km radius from the Opak Fault trace was detected as the largest acceleration area. This area can be used as an indicator of a dangerous area of the province when subject to earthquake ground motion.

Keywords Deterministic · Probabilistic · Risk-targeted maximum considered earthquake · Risk-targeted ground motion

W. Partono (✉) · F. Kistiani · U. C. Sari
Civil Engineering Department, Faculty of Engineering, Diponegoro University, Semarang 50275, Indonesia
e-mail: windupartono@lecturer.undip.ac.id

R. Nazir
Center of Tropical Geoengineering, School of Civil Engineering, Faculty of Engineering, Universiti Teknologi Malaysia, 81300 Skudai, Johor, Malaysia
e-mail: ramlinazir@utm.my

S. A. Kristiawan et al. (eds.), *Proceedings of the 5th International Conference on Rehabilitation and Maintenance in Civil Engineering*, Lecture Notes in Civil Engineering 225,
https://doi.org/10.1007/978-981-16-9348-9_43

1 Introduction

Yogyakarta Province is located on the Java island and close to Central Java Province. Yogyakarta Province is divided into four districts, Bantul, Gunung Kidul, Kulon Progo, Sleman, and one city, Yogyakarta the province's capital city. According to [1], the province is located close to two active seismic sources, the South Java Subduction Megathrust and Opak Fault trace. According to Sutiono et al. [2], the first predicted earthquake to strike Yogyakarta Province occurred in 1840 and by 2015 there were seven different earthquakes caused by two different seismic sources (the Subduction Megathrust and Opak Fault). The 2006 earthquake that occurred in this area was caused by the Opak Fault source and was declared as the largest earthquake to strike this area and the southern part of Central Java Province. Elnashai et al. [3] stated that 4659 people from Yogyakarta Province area died and 19,401 people were injured. However, this 2006 earthquake was also struck part of Central Java Province. According to [3], 1,057 people from Central Java Province were killed and another 18,526 people were injured (BAPPENAS/The Indonesian Ministry of National Development Planning). Elnashai et al. [3] also stated that 186,592 houses in Yogyakarta Province and another 172,104 in Central Java Province were totally destroyed or damaged by the 2006 Opak Fault earthquake. According to [3], there were 8 precisely known 2006 earthquake epicentre positions. Figure 1 shows these earthquake epicentre positions declared by 8 different institutions together with information on the buildings that collapsed or were destroyed by 2006 earthquake.

Based on the 2002 Indonesian seismic code [4], Yogyakarta Province is located in region 3 and has a peak horizontal ground acceleration of 0.3 g (g = gravitational

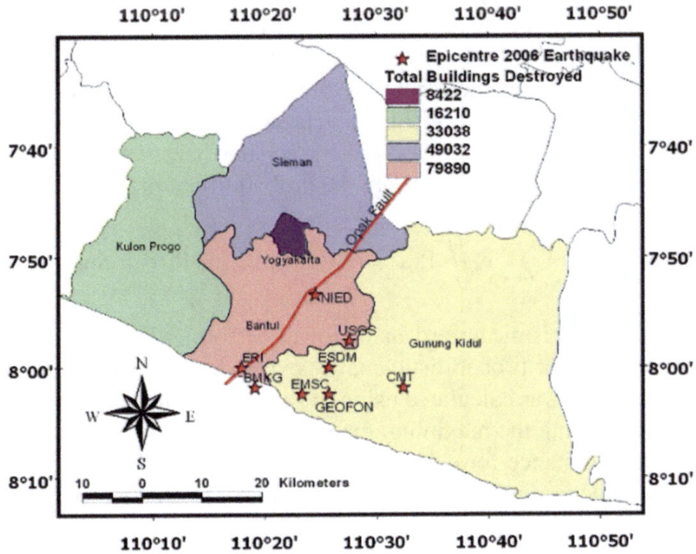

Fig. 1 Epicentre position and total buildings destroyed due to 2006 earthquake.

acceleration). During the 2006 earthquake, the peak horizontal ground acceleration in the study area was predicted at a maximum 0.5 g [3]. The seismograph equipment installed in the study area detected that the predicted peak horizontal ground acceleration maximum was 0.34 g and the maximum peak vertical ground acceleration was 0.3 g [3]. According to this seismic data analysis, the study area was not strong enough to resist the earthquake events having a minimum peak ground acceleration of 0.5 g. The existing buildings constructed before 2006 should be evaluated against earthquakes having a minimum peak ground acceleration of 0.5 g, especially in the areas close to the Opak fault trace (Bantul, Sleman districts and Yogyakarta City).

This paper describes the seismic microzonation analysis conducted for the whole of Yogyakarta Province. The analysis was performed by using the 2017 seismic hazard maps and following the 2019 Indonesian Seismic Code [5] for determining the Risk-targeted Maximum Considered Earthquake. Three basic seismic hazard analyses (probabilistic, deterministic and Risk-targeted Ground Motion) were combined together to obtain the Risk-targeted Maximum Considered Earthquake (MCE_R) [5–7]. The MCE_R acceleration values obtained at this study area were used as a parameter to evaluate the risk performance of the study area.

2 Methodology

The seismic hazard analysis used for calculating probabilistic spectral acceleration was conducted based on a 2% probability of occurrence in 50 years, or a 2500-year return period earthquake. Three different seismic sources (shallow crustal fault, subduction interface and deep intraplate) and eight attenuation functions [8–10] for a shallow crustal fault, [11–13] for a subduction interface and [12, 14] for the deep intraplate model were applied to calculate the probabilistic spectral acceleration. Uniform hazard curves obtained from the probabilistic analysis were then used to calculate the spectral acceleration (Uniform Hazard Ground Motion/UHGM). Equation (1) shows the basic model of probabilistic seismic hazard analysis, where "v" is the annual rate, and Pm(m) and Pr(r) are the probability of magnitude and distance.

$$\lambda(a) = \sum \nu \iint P(a > a^*)|m, r|\ P_m(m) P_r(r) dr\, dm \quad (1)$$

The deterministic seismic hazard analysis was performed by using the same attenuation function used for probabilistic analysis. The deterministic analysis for each seismic source model was calculated using the average of three different attenuation functions and extending the maximum magnitude used in the probabilistic analysis and the minimum distance for each investigation point to all seismic sources. The 84th percentile of deterministic spectral acceleration was applied for developing the MCE_R acceleration.

Risk-targeted Ground Motion (RTGM) analysis for each investigation point in the study area was performed following the same approach proposed by [15]. The purpose of this analysis is to obtain modified probabilistic analysis result by applying a probability of building collapse of 1% in 50 years. Equations (2) and (3) show the basic approach for calculation of the RTGM. The "β" value used in these two equations represents the logarithmic standard deviation, while "c" and "c10%" represent the spectral acceleration and the 10th percentile capacity of building collapse, respectively. For the 2019 seismic code, the value $\beta = 0.65$ was applied to calculate risk-targeted ground motion acceleration calculation.

Figure 2 shows three examples of risk-targeted ground motion calculated for three different positions A, B and C. Figure 3 shows the positions of the Risk targeted Ground Motion calculation. The analysis was performed for a short period (0.2 s) and a long period (1 s) of spectral acceleration. Tables 1 and 2 show the RTGM analysis results (RTGM 0.2 and 1 s) and the corresponding risk coefficient obtained from this analysis. The risk coefficient was calculated by comparing the RTGM value and the UHGM spectral acceleration from the probabilistic hazard analysis.

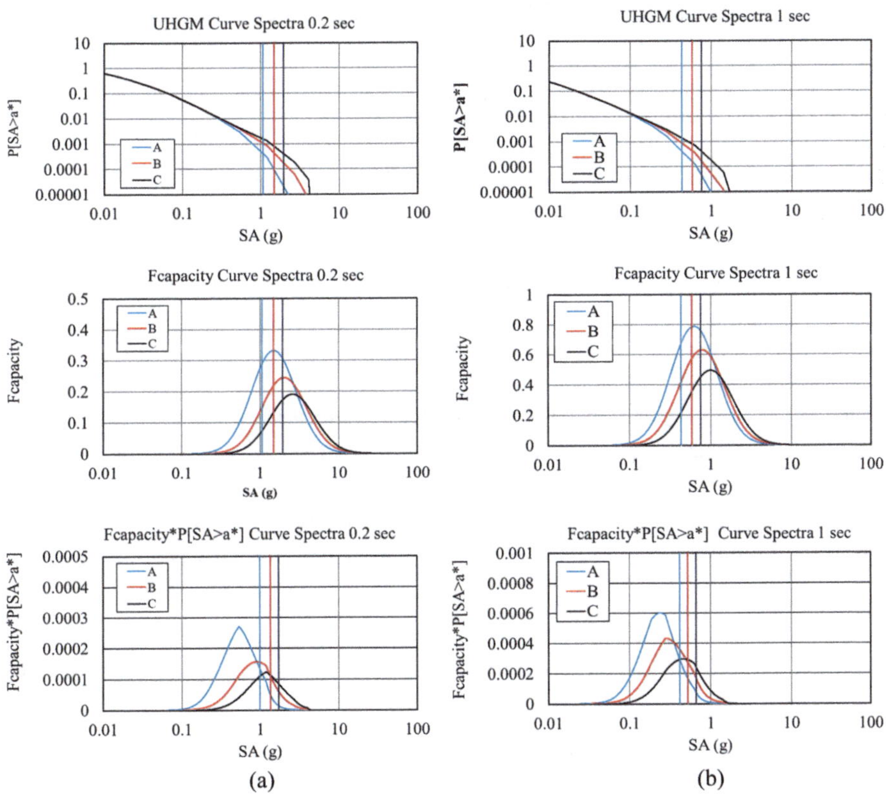

Fig. 2 Risk-targeted ground motion calculation at three position A, B and C for spectral acceleration 0.2 s (**a**) and 1 **s** (**b**).

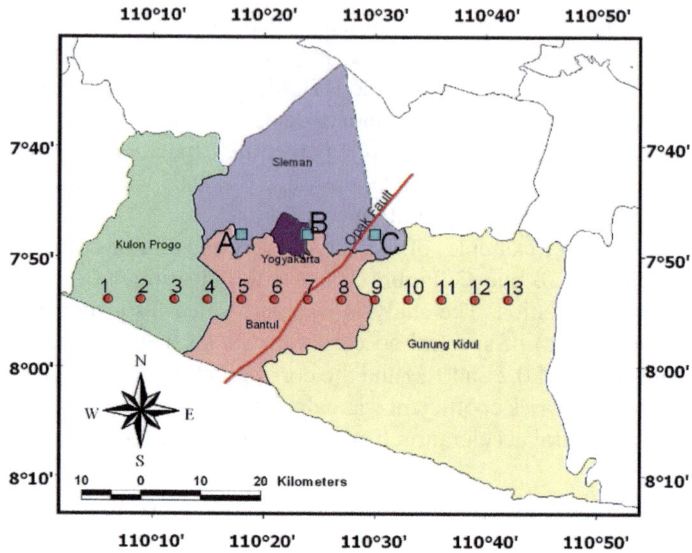

Fig. 3 Risk-targeted ground motion calculation position.

Table 1 Risk targeted ground motion calculation for 0.2 s spectral acceleration.

Point no	Longitude (degree)	Latitude (degree)	UHGM (g)	RTGM (g)	Cr
A	110.3	−7.9	1.1497	1.1070	0.9302
B	110.4	−7.9	1.3449	1.2378	0.9204
C	110.5	−7.9	1.7599	1.5815	0.8986

Table 2 Risk targeted ground motion calculation for 1 s spectral acceleration.

Point no	Longitude (degree)	Latitude (degree)	UHGM (g)	RTGM (g)	Cr
A	110.3	−7.9	0.5364	0.4931	0.9192
B	110.4	−7.9	0.6026	0.5379	0.8927
C	110.5	−7.9	0.7278	0.6446	0.8858

$$F(c) = \frac{1}{c\beta\sqrt{2\pi}} e^{-\left[\frac{lnc-(lnc_{10\%}-1.28\beta)^2}{2\beta^2}\right]} \tag{2}$$

$$\text{Pcollapse} = \int_0^{\infty} \text{F(c)P[S > c]dc} \tag{3}$$

The Risk-targeted Maximum Considered Earthquake (MCE_R) calculation was conducted by applying Risk-targeted Ground Motion and 84th percentile deterministic calculation results. To perform the calculation, direction factors 1.1 and 1.3 were applied for the spectral acceleration of 0.2 and 1 s respectively. The basic procedure for the Risk-targeted Maximum Considered Earthquake (MCE_R) acceleration calculation followed the method proposed by [16].

3 Result and Discussions

Risk-targeted Maximum Considered Earthquake distribution maps were developed for the whole of the study area. Three different spectral acceleration distributions (MCE_G, MCE_R-S_S and MCE_R-S_1) were performed by applying spatial analysis. The purpose of developing these three different seismic microzonation maps was to evaluate the distribution of the minimum to maximum spectral acceleration in the study area. Figure 4 shows the distribution of MCE_G, MCE_R-S_S and MCE_R-S_1. Based on the 2019 seismic code, the minimum and maximum MCE_G values were distributed between 0.35 and 0.6 g, the MCE_R-S_S values were distributed between 0.85 and

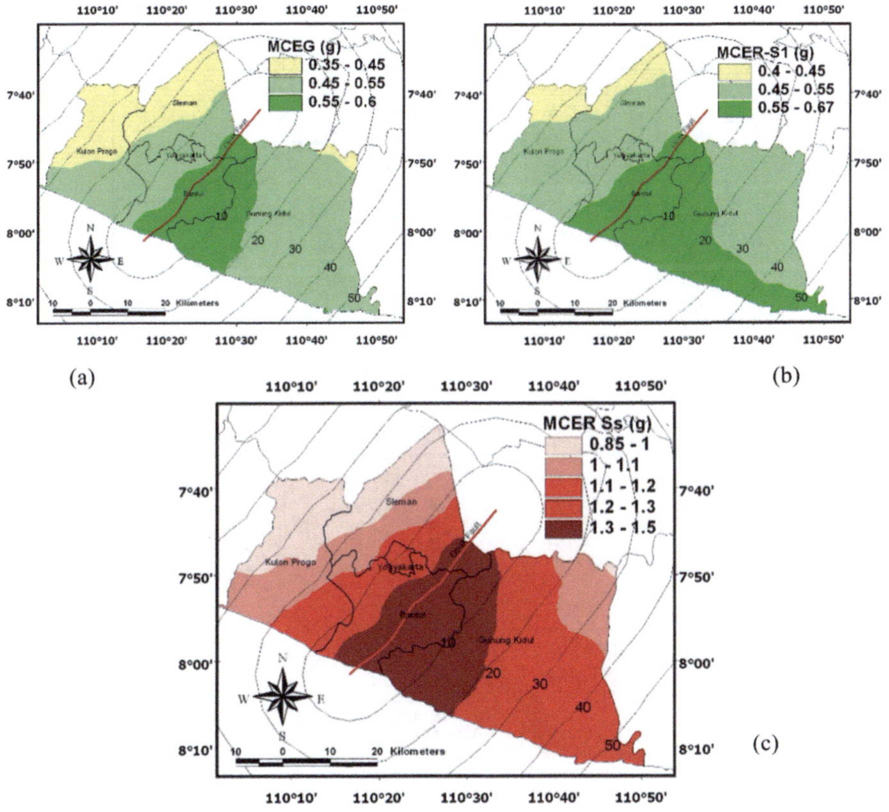

Fig. 4 MCE_G (**a**), MCE_R-S_1 (**b**) and MCE_R-S_S (**c**) distribution at the study area.

1.5 g and the MCE_R-S1 were distributed between 0.4 and 0.67 g. All the maximum Risk-targeted Maximum Considered Earthquake were distributed from 10 km north-west to 20 km south-east of the Opak Fault. According to these three Risk-targeted Maximum Considered Earthquake, the maximum ground acceleration at the study area was predicted to occur between 10 north-west to 20 km south-east of predicted Opak fault trace.

Evaluation of the MCE_R 0.2 s (MCE_R-S_S) and 1 s (MCE_R-S_1) distribution was also performed at a specific line (cross-section). The purpose of this evaluation was to obtain the information of deterministic, probabilistic and Risk-targeted Ground Motion analyses in developing a combined analysis of the Risk-targeted Maximum Considered Earthquake (MCE_R) spectral acceleration. The analysis was performed from coordinate positions 1 (110.1, −7.9) to 13 (110.7, −7.9) (see Fig. 3). All thirteen coordinate positions were distributed at the same interval or distance. Figure 5 shows the distribution of the three spectral accelerations and the related Risk-targeted Maximum Considered Earthquake (MCE_R) spectral acceleration developed from these three values. Only spectral accelerations of 0.2 and 1 s are displayed in these two figures because the PGA distribution is almost equal to spectrum of 1 s.

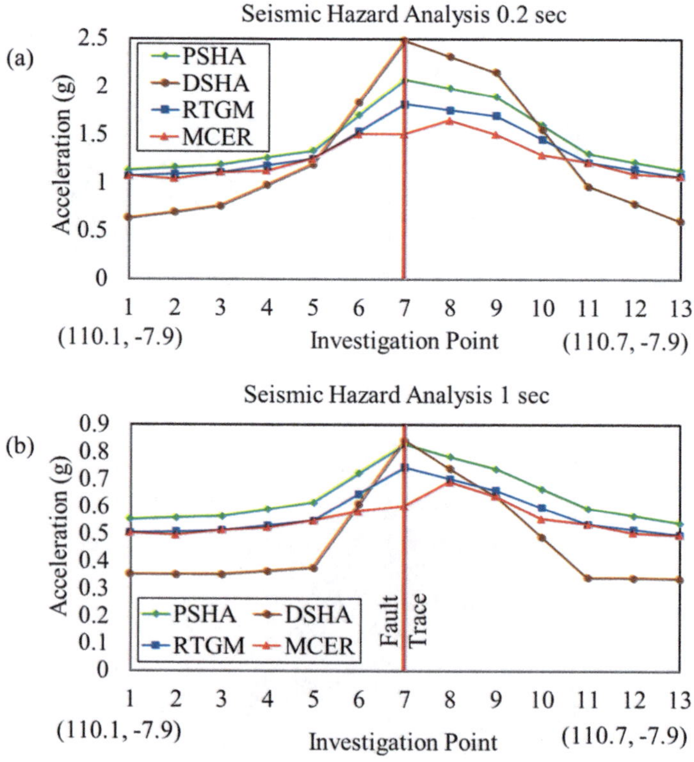

Fig. 5 PSHA, DSHA, RTGM and MCE_R performance at cross section 1–13 for spectra of 0.2 s (**a**) and 1 s (**b**).

Maximum values of 1.5 and 0.6 g capped for the 0.2 and 1 s spectral acceleration were applied following the same procedure as proposed by [5, 16]. The maximum spectral acceleration was detected at all points close to the predicted Opak Fault trace.

4 Conclusions

Seismic microzonation of Yogyakarta Province was conducted based on the Risk-targeted Maximum Considered Earthquake by combining three probabilistic, deterministic and Risk-targeted Ground Motion analyses. The analysis was developed based on all the earthquake data and earthquake sources used for developing the Indonesian Earthquake Maps 2017. The maximum spectral acceleration MCE_G, MCE_R-S_S (spectrum 0.2 s) and MCE_R-S_1 (spectrum 1 s) was detected within a 10 km north-western to 20 km south-east radius from Opak Fault trace.

Maximum values of 1.5 g for 0.2 s (MCE_R-Ss) period and 0.6 g for MCE_G and the 1 s (MCE_R-S_1) period spectral accelerations were obtained in the study area. The predicted maximum spectral acceleration and earthquake danger area associated with the Opak Fault was detected in a radius of 10 km from the Opak Fault trace.

Acknowledgements Thank you to the Dean of the Faculty of Engineering, Diponegoro University, Semarang, Indonesia, for the approval of the research funding through the 2021 Strategic Research Scheme.

References

1. National Center for Earthquake Studies: Indonesian Seismic Sources and Seismic Hazard Maps (2017) Center for research and development of housing and resettlement. Ministry of Public Works and Human Settlements
2. Sutiyono A, Prastistho B, Prasetyadi C, Supartoyo (2018) Opak fault a comparative review. In: IOP conference series, earth and environmental science 212, IOP Publishing
3. Elnashai AS, Sun JK, Gun JY, Sidarta D (2007) The Yogyakarta Earthquake of May 27 2006. Mid-America Earthquake Center, MAE Center Report 07-02
4. SNI 1726 (2002) Seismic resistance design for building structures. National Standardization Agency of Indonesia
5. SNI 1726 (2019) Seismic resistance design codes for building and other structures. National Standardization Agency of Indonesia
6. ASCE/SEI 7-16 (2017) Minimum design loads for building and other structures. American Society of Engineers
7. Sengara IW, Irsyam M, Sidi ID, Mulia A, Asrurifak M, Hutabarat D, Partono W (2020) New 2019 risk-targeted ground motions for spectral design criteria in Indonesian seismic building code. In: 4th international conference on Earthquake engineering and disaster mitigation (ICEEDM 2019) Padang Sumatera Barat 26–27 September (2019), E3S Web of Conference, vol 156

8. Boore DM, Stewart JP, Seyhan E, Atkinson GM (2013) NGA-West 2 equations for predicting PGA, PGV, and 5%-damped PSA for shallow crustal earthquakes. Earthq Spectra. Earthq Eng Res Inst 30(3):1057–1085
9. Campbell KW, Bozorgnia Y (2013) NGA-West2 Campbell-Bozorgnia, ground motion model for the horizontal components of PGA, PGV and 5%-damped elastic pseudo acceleration response spectra for periods ranging from 0.01 to 10 s. Pacific Earthquake Engineering Research Center, PEER Report 2013/6, pp xii+75
10. Chiou BSJ, Youngs RR (2013) Update of the Chiou and Youngs NGA ground motion model for average horizontal component of peak ground motion and response spectra. Pacific Earthquake Engineering Research Center
11. Abrahamson NA, Addo KO, Atkinson G, Chiou B, Gregor NJ, Silva W, Youngs RR (2014) Ground motion characterization for BC hydro SSHAC level 3 study. In: Tenth U.S. national conference on earthquake engineering, frontiers of earthquake engineering, IONCEE, 21–24 July 2014, Anchorage, Alaska
12. Atkinson GM, Boore DM (2003) Empirical ground motion relations for subduction zone earthquakes and their application to Cascadia and other regions. Bull Seismol Soc Am 93(4):1703–1729
13. Zhao JX, Irikura K, Zhang J, Fukushima Y, Somerville PG, Asano A, Ohno Y, Oouchi T, Takahashi T, Ogawa H (2006) An empirical site classification method for strong motion stations in Japan using H/V response spectral ratio. Bull Seismol Soc Am 96(3):914–925
14. Youngs RR, Chiou SJ, Silva WJ, Humphrey JR (1997) Strong ground motion attenuation relationships for subduction zone earthquakes. Seismol Res Lett 68(1):58–73
15. Luco N, Ellingwood BR, Hamburger RO, Hooper JD, Kimball JK, Kircher CA (2007) Risk-targeted versus current seismic design maps for the conterminous United States. In: Structural engineers association of California 2007 convention proceedings. California, pp 163–175
16. Leyendecker EV, Hunt EJ, Frankel AD, Ruktales KS (2000) Development of maximum considered earthquake ground motion maps. Earthq Spectra 16(1):21–40

Liquefaction Potential of Volcanic Deposits During Lombok Earthquake in 2018

Muhajirah

Abstract The Lombok earthquake and the Palu earthquake in 2018 were earthquakes that caused major damage in the two referenced areas. The Lombok earthquake occurred over a longer period than the Palu earthquake, but the Palu earthquake was able to cause enormous liquefaction. This study aimed to determine the behavior of the volcanic deposits of Mount Samalas during the corresponding earthquake. The research included a physical properties test, SEM photos, a shear strength test, a shaking table test, and grading tests before and after the soil samples were given cyclic loads. The results of the grading test after the shaking table test showed that there was the addition of fine grains before and after the shaking table test. This indicates that the pumice particles are scoured on the surface, resulting in the small grains becoming locked between the larger particles. This condition caused the shear strength to increase and the pore pressure to decrease significantly. The cavity in the pumice particles is thought to reduce the seismic waves during an earthquake. For the large particles, although the gradation curve has liquefaction potential, the pore water pressure did not increase significantly. However, pumice has a diameter of less than 0.075 mm because it does not have a cavity, therefore that it has the potential to experience liquefaction such as sand boiling.

Keywords Lombok earthquake · Pumice sand · Liquefaction

1 Introduction

In 2018, Lombok Island was shaken by a series of destructive earthquakes. The first earthquake with a magnitude of 6.4 M occurred on July 29, 2018, in the northern part of the island of Lombok. The intensity of the earthquake on the MMI VIII scale was recorded in the Loloan area. The epicenter was on land 47 km northeast of Mataram City at a depth of 24 km. A week later, on August 5, 2018, an earthquake again shook the island of Lombok that was 6.9 M_w. Seismologically, based on the

Muhajirah (✉)
Department of Civil Engineering, Mataram University, Mataram, Indonesia
e-mail: muhajirah@unram.ac.id

S. A. Kristiawan et al. (eds.), *Proceedings of the 5th International Conference on Rehabilitation and Maintenance in Civil Engineering*, Lecture Notes in Civil Engineering 225,
https://doi.org/10.1007/978-981-16-9348-9_44

distance between the two epicenters, magnitude, and the time of occurrence, the two earthquakes were related or were dependent events. The 6.4 M_w earthquake was a foreshock of the 6.9 M_w earthquake. Four days later, on August 9, 2018, strong tremors resulted from an earthquake that was 6.3 M_w. The epicenter of the earthquake was estimated to be to the southwest of the epicenter of the previous earthquake, closer to Mataram city [1].

The geological phenomena that emerged after the 6.4 M_w (July 29, 2018) and 6.9 M_w (August 5, 2018) earthquakes were dominated by land surface deformation in the form of landslides, lateral spreading, and liquefaction. The types of landslides that occurred included rockfalls and combined sliding debris. After the 7.0 M_w earthquake, several phenomena arose including the fractures that released water along with mud and sand. The fractures that occurred were in the form of longitudinal fractures and round holes. The elongated soil cracks due to liquefaction were in the form of longitudinal cracks with a width of 2–10 cm. From these cracks, sometime after the earthquake, water came out together with mud and/or sand. This phenomenon occurred in front of the village office of Selengan in North Lombok Regency [2].

The Palu earthquake occurred on Friday afternoon on September 28, 2018. It was 7.4 M_w centered 26 km north of Donggala Regency in Central Sulawesi Province. This earthquake caused a tsunami in Palu City and widespread liquefaction in the Petobo and Ballaroa areas. The series of large earthquakes that occurred on the island of Lombok for two months did not cause liquefaction effects like the one in Palu City.

The liquefaction events that occurred in Japan, New Zealand, and Indonesia indicate that most of the liquefaction occurs in sedimentary soils resulting from volcanic eruptions. Of the three countries, Japan suffered the most losses due to the impact of the earthquake. Several earthquakes resulted in liquefaction in parts of Japan. The liquefaction susceptibility of this area is thought to be because most of the land is covered by volcanic deposits that are easily destroyed. These deposits are widespread in the southern part of Kyushu known as Shirasu.

The Hokuseibu earthquake in 1997 resulted in heavy damage to the land reclamation which was covered with Masado soil. Masado soil is crushable granite residual soil. An investigation to determine the characteristics of the volcanic soil following the earthquake in Japan was carried out by Aramaki et al. [3] and Hyodo et al. [4].

Orense et al. [5] and Liu et al. [6] also researched the soil liquefaction that occurred in New Zealand due to the Edgecumbe earthquake in 1987 and the Canterbury earthquake in 2010–2011. Most of the New Zealand areas are built up by the volcanic eruptions concentrated in the Taupo and Rotorua regions, which are known as the Taupo volcanic zone.

On May 27, 2006, Yogyakarta and its surrounding areas experienced an earthquake with a magnitude of 6.3 on the Richter scale. The areas that were damaged by the earthquake included Bantul Regency and Klaten Regency. According to Soebowo et al. [7], most of Yogyakarta is made from the volcanic deposits of Mount Merapi. The phenomenon of liquefaction that occurred after the earthquake was in the form of sandblasting holes, wells drying up and being covered in fine sand, lateral ground movement, land subsidence, and the emergence of water on top of the ground shortly after the earthquake [7–9].

1.1 The Liquefaction Research in Volcanic Deposits

Kusumawardani et al. [8] conducted tests on soil samples from the UMY Yogyakarta campus. The cyclic triaxial test was carried out on the soil samples. The sample was prepared using the dry pluviation method with a relative density of 25%. In this test, the sample was loaded with a frequency of 0.05 and 0.1 Hz. The results showed that for the number of cycles equal to 100 at the same density ($D_r = 25\%$), the soil sample with a load frequency of 0.1 Hz reached a pore pressure ratio of 1.0 while the soil sample with a frequency of 0.05 Hz reached pore pressure ratio of 0.8.

Mase [10] conducted a shaking table test using Opak River sand from Imogiri District in Bantul Regency. He conducted an enlargement test on the Kali Opak sand particles. The test results, when examined through a stereo microscope, showed that the particles looked solid with no cavities (Fig. 1).

The sand samples had a field density of ($D_r = 28.40\%$). The amplitudes of acceleration ($a_{\max}$) used to move the shake table were 0.30, 0.35, and 0.40 g at frequencies of 1.4, 1.6, and 1.8 Hz. The duration of the shaking for each acceleration was 8, 16, and 32 s. The three amplitudes of acceleration applied to the sand model showed the same tendency, specifically an increase in the excess pore-water pressure ratio (r_u) to reach a value of one. The increase in the excess pore pressure at 32 s was the highest. This indicates that over a long duration of shaking time, the accumulated increase in excess pore pressure will be greater.

Kusumawardani [8] and Mase [10] evaluated the liquefaction potential of non-porous particles (Fig. 2). The pore water pressure increased after the soil sample was subjected to a cyclic load and reached the point of liquefaction [10]. Even though

Fig. 1 The shape of sand particles in the Opak River [10]

the cyclic triaxial test was carried out at low frequencies, the clean sand sample still reached a liquefaction condition [8].

This research was conducted in order to determine the behavior of the volcanic deposits of Mount Samalas following an earthquake. The research was carried out on the soil samples taken from the city of Mataram and the fine sand samples that came through the cracks during the Lombok earthquake.

1.2 Liquefaction Mechanism

Liquefaction is defined as the transformation of soil from a solid into a liquid. This occurs due to the increased pore water pressure and reduced effective stress. It generally occurs in less cohesive soils that are saturated with water. When the soil is subjected to rapid loads such as an earthquake load or other dynamic loads, the pore water cannot be drained in the short amount of time when the loading occurs. This loading condition is known as an undrained condition. If the soil is less dense, the less cohesive soil that receives cyclic loads tends to experience compaction. When the cavity between the soil grains contains water and the water cannot be drained, the excess pore water pressure will increase. Figure 2 shows the changes that occur in soil particles due to cyclic loads causing excess pore water pressure [11].

At a certain depth below the ground surface, Eq. (1) can explain the stress conditions when liquefaction occurs. The undrained condition causes the excess pore pressure to increase. At one point, the excess pore water pressure reaches the total stress value and the effective stress is close to or equal to zero [12].

$$\sigma' = \sigma - u \tag{1}$$

where:

σ' effective stress,

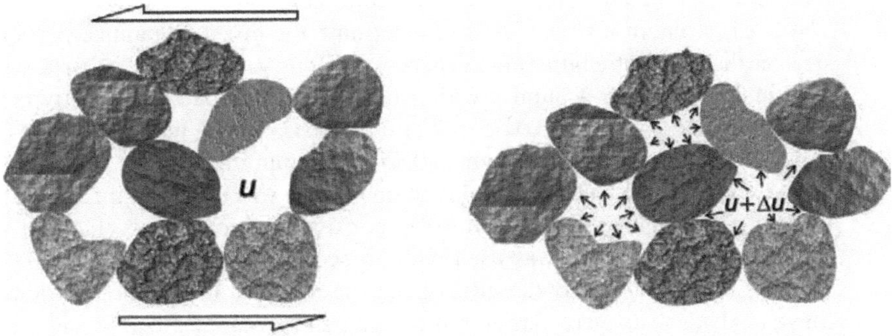

Fig. 2 Cyclic loads cause soil particles to move and excess pore pressure increase [9]

σ total stress,
u pore water pressure.

Not all soil deposits are susceptible to liquefaction. In general, liquefaction requires the following three conditions: (i) the cohesionless soil has a relative density of loose to medium dense, (ii) a shallow groundwater level to ensure that the soil is saturated, and (iii) a sufficiently strong ground motion or a cyclic load.

Several factors can influence the susceptibility of the soil to liquefaction. Based on the field investigations and laboratory test results, the liquefaction characteristics of cohesionless soils are influenced by the following factors: grain size distribution, soil type, relative density, earthquake loading characteristics, effective vertical stress and over-consolidation, the age of the soil, the origin of the soil, the seismic strain history, the degree of saturation and the thickness of the sand layer [13].

2 Materials and Method

The research was carried out in two stages: (1) a series of tests conducted on the soil samples from the City of Mataram, and (2) gradation tests and SEM photos conducted on the sand boil samples from Selengan village in North Lombok Regency.

Research procedures include:

(1) Pumice sand taken from Kebon Tallo in Mataram city. It was excavated to a depth of 2 m for the shaking table test. Furthermore, pumice sand was used in the analytical sieve test to obtain material with a potential liquefaction gradation curve.
(2) Testing the physical properties of the pumice samples using ASTM standards including a water content test, sieve test, relative density test, and dry weight density test.
(3) SEM (Scanning Electron Microscope), the particles used included the material retained from the No. 10 sieve (grain diameter $2.00 < D < 4.75$ mm) and No. 200 sieve (grain diameter $0.074 < D < 0.106$ mm).
(4) The shear strength test was used to determine the effect of pumice particle size on the shear strength value. There were four variations of particle size used in this test: type A pumice with a diameter of less than 4.75 mm, type B pumice with a diameter of 0.075–4.75 mm, type C pumice with a diameter of 0.106–4.75 mm, and type D pumice. 0.250–4.75 mm in diameter.
(5) In the shaking table test, the dry pluviation method was used to form a pumice sample with a relative density of 30%. For dynamic loads, a cyclic load in the form of a sine wave was used with an acceleration amplitude of 0.38 g for 20 s in order to study the effect of cyclic loads on the pumice particles. Sieve analysis tests were carried out on the pumice samples before and after the shaking test.

The evaluation of the liquefaction potential using a shaking table was conducted by measuring the pore water pressure increment of the sample. To determine the development of excess pore water pressure, three Pore Pressure Transducers, abbreviated as PPT, were placed on the side of the sample box at a distance of 0.10 m (Upper PPT), 0.25 m (Middle PPT), and 0.40 m (Base PPT) from the bottom of the sample box.

For the sand boil sample, only the sieve analysis test and SEM photos were carried out. The sieve analysis test was carried out to determine the particle size and percentage. The purpose of the SEM photos was to determine the shape of the particles.

3 Results and Discussion

3.1 Physical Properties of Mataram Pumice

The unique thing that can be seen in Table 1 is that the dry weight density of the pumice sand is very low (0.54 gr/cm^3). The void ratio of pumice (3.22) is greater than that of normal sand. This indicates that the low dry weight density of the pumice is due to the large number of cavities present in the pumice particles. The soil classification is SP, which refers to poorly graded sand with silt and gravel.

The shape of the Mataram pumice particles is subrounded to subangular [14]. The liquefaction susceptibility is influenced by grain characteristics such as particle

Table 1 Physical characteristics of Mataram pumice

No.	Properties		Mataram pumice
1	The specific gravity, G_s		2.28
2	Water content, w	%	114.37
3	Grain size		
	Gravel	mm	22.23
	Sand	mm	69.75
	Silt and clay	mm	8.02
	D_{50}	mm	1.20
4	Coefficient of uniformity, C_u		23.53
5	Coefficient of curvature,C_c		0.32
6	USCS classification		SP
7	Bulk density,ρ_s	gr/cm^3	1.16
8	Dry density,$\rho_{s-\text{dry}}$	gr/cm^3	0.54
9	Void ratio, e		3.22

size, gradation, and particle shape. Soil particle size is related to soil composition. Well-graded soils are generally less susceptible to liquefaction than poorly graded soils. Soil with a rounded particle shape will compact more easily when it undergoes shear stress than soil with an angular particle shape.

According to Youd and Poulos, when an earthquake occurs, spherical sand particles with a low relative density (loose) will more easily experience changes in the composition of the particles due to the shear stress and/or cyclic loads. Therefore, soils with a rounded particle shape are more susceptible to liquefaction compared to soils with an angular grain shape.

3.2 *Scanning Electron Microscope (SEM)*

Figure 3a shows the presence of cavities in the pumice particles. The cavities look like an elongated channel (arrow) and some of the cavities are not connected to other cavities (isolated cavities). Figure 3b shows a particle shape that is the result of the crushing of larger sand grains and the existence of small craters which are the end of isolated cavities (dotted white circles).

3.3 *Shear Strength Characteristic*

In Fig. 4, it can be seen that the shorter the range of soil particle sizes, the greater the friction angle of the soil or the increase in the soil shear strength. In this test, the size of the grain diameter contributed to the value of the shear strength of the soil. This can be seen in the figure where the type D particles have the largest internal friction angle.

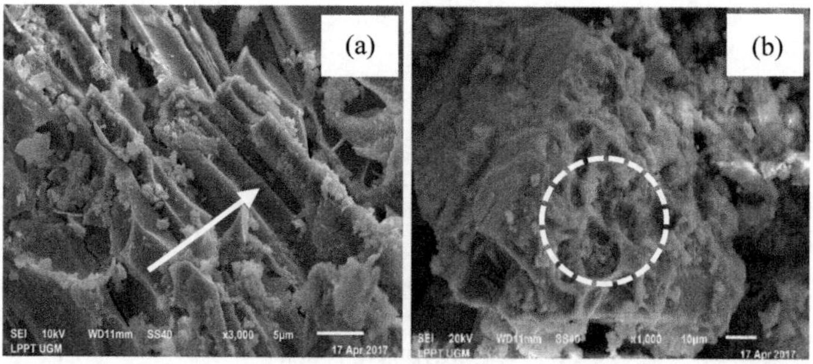

Fig. 3 SEM photograph of Mataram pumice: **a** retained No. 10 sieve and magnification of 3000×, **b** retained of No. 200 sieve and magnification 1000×

3.4 Shaking Table Test

Figure 5 shows the recorded Excess Pore Pressure (EPP) of a pumice sample at an acceleration of 0.38 g for 20 s. From this figure, the change in excess pore pressure is not very significant. The maximum pore pressure in Upper PPT and Middle PPT is only up to 0.10 kPa whereas the Base PPT records the maximum pore pressure at 0.20 kPa. Liquefaction condition did not occur in this sample. The maximum pore pressure recorded for the three transducers did not reach the effective stress value at

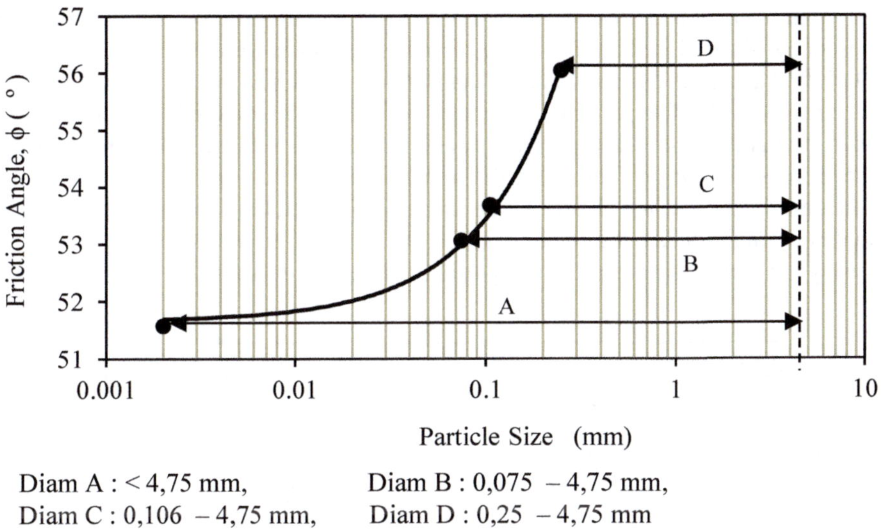

Fig. 4 Shear strength of Mataram pumice

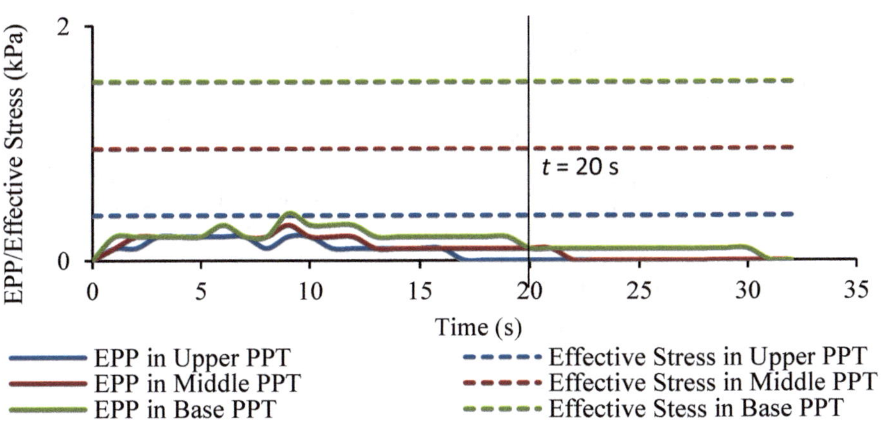

Fig. 5 Excess pore water pressure with time at an acceleration of 0.38 g and time 20 s

each level. The effective stresses in the Upper PPT, Middle PPT, and Base PPT were 0.38, 0.95, and 1.52 kPa respectively.

3.5 Degradation of Mataram Pumice

The gradation curve of the pumice sample was in the potential liquefaction zone (see Fig. 6). There were differences in the amount of the pumice sand before and after the shaking table test. The sieve analysis test showed that there was a reduction in the number of particles in the first sieve and the addition of particles in each subsequent sieve. Based on Fig. 6, the particle size ranges were 0.425–0.85 mm and 0.106–0.25 mm, and particles smaller than 0.074 mm with an increase of 3.0, 1.5 and 4.2% respectively. The particle size ranges of 2.00–4.75 mm, 0.85–2.00 mm, 0.25–0.425 mm and 0.074–0.106 mm decreased by 1.3, 5, 5, 1.6 and 0.4% respectively. The increase in the percentage of the number of particles in a certain size range indicates that there was a change in the shape of the particles from larger particles to smaller particles during the shaking table test. This condition was also confirmed by Wesley [15], indicating that pumice particles are easily destroyed in wet conditions.

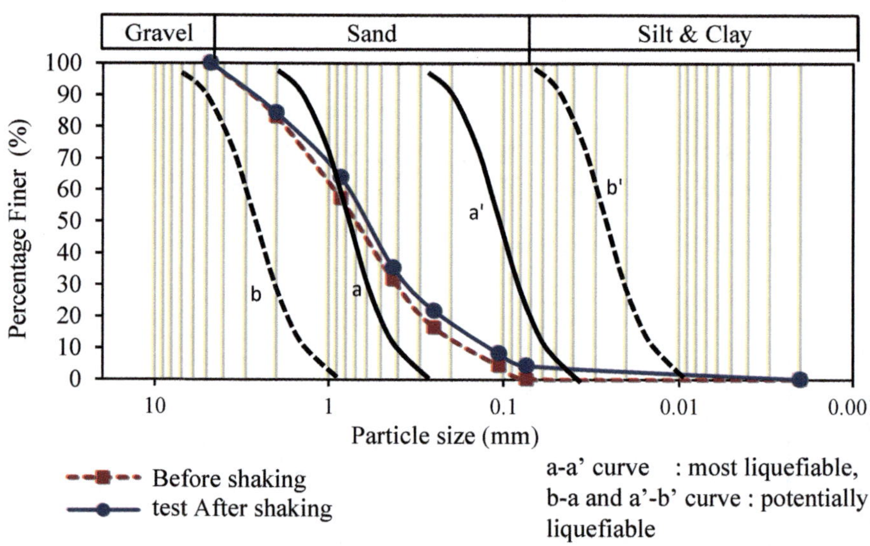

Fig. 6 Degradation of Mataram pumice

Fig. 7 SEM photograph of sand boil from Selengan village

3.6 *Sand Boil of Selengan Village*

Sand boil particles have a diameter of less than 2 mm and 65% passed the No. 200 sieve. The particles were made up of pumice sand, thought to be the result of the crushing of larger particles (Fig. 7). Based on the results of the Atterberg test, the particles were determined to be non-plastic. A similar gradation curve was obtained from the sand boil during the Padang earthquake in 2009 [16].

4 Conclusions

Based on the tests that have been done, it is concluded that:

1. The results of the grading test after the shaking table test show that there was an addition of fine grains before and after the shaking test. This indicates that the pumice particles were scoured on the surface, and that the small grains become locked between the larger particles. This condition causes the shear strength to increase and the pore pressure to decrease significantly.
2. The cavities in the pumice particles are thought to reduce the seismic waves during an earthquake. For large particles, although the gradation curve has liquefaction potential, the pore water pressure does not increase significantly. However, pumice has a diameter of less than 0.075 mm because it does not have cavities. This means that it has the potential to experience liquefaction like in a sand boil.

Acknowledgements The author wishes to thank Dr. es sc. tech. Ir. Ahmad Rifai, MT. and Ir. Agus Darmawan Adi, M.Sc. Ph.D. for their contributions to various aspects of the experimental work. The author would like to thank the Head and staff of the Soil Mechanic and the Structural Engineering Laboratories of Gadjah Mada University for providing the laboratory facilities and for their assistance conducting the shaking table test.

References

1. Maipark, di Lombok G, dan Donggala P, di Selat Sunda S. http://www.maipark.com/assets/uploads/publication/Gemuruh-di-Lombok-%20Palu-dan-Donggala-%20Senyap-di-Selat%20Sunda.pdf
2. Tim Pusat Studi Gempa Nasional (PusGen) (2018) Kajian Rangkaian Gempa Lombok Provinsi Nusa Tenggara Barat, Indonesia. Pusat Penelitian dan Pengembangan Perumahan dan Permukiman, Badan Penelitian dan Pengembangan, Kementerian PUPR, Cetakan Pertama, Bandung
3. Aramaki N, Okabayashi T, Hyodo M (2007) Liquefaction characteristics of crushable volcanic soil "Shirasu". In: 4th International conference on earthquake geotechnical engineering, June 25–28, Paper No. 1288
4. Hyodo M, Nakata Y, Aramaki N, Hyde AF, Inoue S (2000) Liquefaction and particle crushing of soil. In: Proceeding 12th world conference on earthquake engineering
5. Orense RP, Pender MJ, O'Sullivan AS (2012) Liquefaction characteristics of pumice sand, EQC Project 10/589
6. Liu L, Orense RP, Pender MJ (2015) Crushing-induced liquefaction characteristics of pumice sand. In: NZSEE Conference
7. Sebowo E, Tohari A, Sarah D (2007) Studi Potensi Likuifaksi di Daerah Zona Patahan Opak Patalan-Bantul Jogjakarta. Prosiding Seminar Geoteknologi Kontribusi Ilmu Kebumian Dalam Pembangunan Berkelanjutan, Bandung
8. Kusumawardani R, Suryolelono KB, Suhendro B, Rifa'i A (2014) The loading frequency effects of Yogyakarta's sand under cyclic Triaxial testing. Int J Civil Environ Eng 14(2)
9. Muntohar AS (2009) Evaluation of peak ground acceleration using CPT data for liquefaction potential. In: 4th annual international workshop & expo on Sumatera tsunami disaster and recovery
10. Mase LZ (2017) Experimental liquefaction study of Southern Yogyakarta using shaking table. Jurnal Teknik Sipil 24(1):11–18
11. Lenart L (2008) The response of saturated soils to a dynamic load. Acta Geotechnica Slovenia 1:37–49
12. Das BM (1993) Principles of soil dynamics. PWS-KENT Publishing Company, Boston
13. Geotechnical Design Procedure (2015) GDP-9: Liquefaction Potential of Cohesionless Soils. Revision #3, Geotechnical Engineering Bureau, Department of Transportation. State of New York
14. Hiden et al (2017) The Isopach mapping of volcanic deposits of Mount Samalas 1257 AD based on the values of resistivity and physical properties. Geosciences 7(67)
15. Wesley LD (2010) Geotechnical engineering in residual soils. Wiley, Hoboken
16. Hakam A, Suhelmidawati E (2013) Liquefaction due to September 30th 2009 earthquake in Padang. In: The 2nd international conference on rehabilitation and maintenance in civil engineering, Procedia Engineering, pp 140–146

Fault Structure Interpretation on the Western Part of East Java Using Second Vertical Derivative

Wien Lestari, Amien Widodo, Dwa Desa Warnana, Firman Syaifuddin, Rusba Saputra Rivensky, and Bagoes Idcha Mawardi

Abstract The National Earthquake Center (PusGeN) in 2017 stated that East Java was traversed by several active faults that extended towards Bali and Flores. The gravity research was carried out to map the fault structure in the western part of East Java. Gravity method is applied to identify subsurface geological structures by utilizing variations in Earth's gravity caused by differences in density of subsurface rocks. The processed data is satellite gravity data (TOPEX) as many as 110 data spread in Lamongan, Gresik, and Surabaya City. Gravity Topex is a Geodesy satellite altimeter launched by NASA to measure the altitude of the satellite above the closest sea surface point to a very high precision. Gravity data that collected from Topex/Poseidon satellite are already processed into free air anomaly, further, it is performed Bouguer Correction and Terrain Correction, resulting in Bouguer Anomaly data which is then filtered using Second Vertical Derivative (SVD). The results of SVD 2-D modeling have a high level of accuracy because the faults are well identified and highly correlated with the PuSGeN faults map (2017). The curve of the slice profiling results on the Surabaya Fault shows that the maximum value of SVD is 0.1 mGal/m^2 and the minimum value is 0.2 mGal/m^2 which is identified as a reverse fault. The curve of the slice profiling results on the Waru Fault shows that the maximum value of SVD is 0.2 mGal/m^2 and the minimum value is 0.3 mGal/m^2 which is identified as a reverse fault.

W. Lestari (✉) · A. Widodo · D. D. Warnana · F. Syaifuddin · R. S. Rivensky · B. I. Mawardi
Geophysical Engineering Department, Institut Teknologi Sepuluh Nopember, ITS Raya Street, Surabaya 60111, East Java, Indonesia
e-mail: wien@geofisika.its.ac.id

A. Widodo
e-mail: amienwidodo@geofisika.its.ac.id

D. D. Warnana
e-mail: dwa_desa@geofisika.its.ac.id

F. Syaifuddin
e-mail: firman@geofisika.its.ac.id

S. A. Kristiawan et al. (eds.), *Proceedings of the 5th International Conference on Rehabilitation and Maintenance in Civil Engineering*, Lecture Notes in Civil Engineering 225,
https://doi.org/10.1007/978-981-16-9348-9_45

Keywords Bouguer anomaly · TOPEX gravity · Fault · Second vertical derivative (SVD)

1 Introduction

1.1 Background

Indonesia occupied a very active tectonic zone because the world's three large plates and nine other smaller plates had encountered each other in the Indonesian territory and form complex plate meeting routes [1]. The existence of interaction between these plates located Indonesia as an area that is very prone to earthquakes. Victims of material and human lives because of the earthquake disaster were mostly caused by ignorance of a fault in an area, so there was no anticipatory measure against the movement of the fault. Therefore, active fault mapping is very necessary for the development of spatial planning in Indonesia because from now on, building construction must consider aspects of disasters, one of which is earthquakes.

Geodetic studies using the Global Positioning System (GPS) have been carried out on surface deformation measurements in Java [2]. Its GPS deployment identified an active extension or extension of Flores back-arc thrust westward along 300 km on land to East Java along the Kendeng Fault as described by Fig. 1 [2]. The results of this study denoted a new major earthquake threat for East Java and a tsunami threat to Bali, Lombok, Nusa Tenggara, and other shores along the Flores Sea. Information about near-surface faults forms the basis of studies of deformation studies in deeper areas. The western part of East Java, including Surabaya and its surroundings, is traversed by several geological structures including the active Lidah and

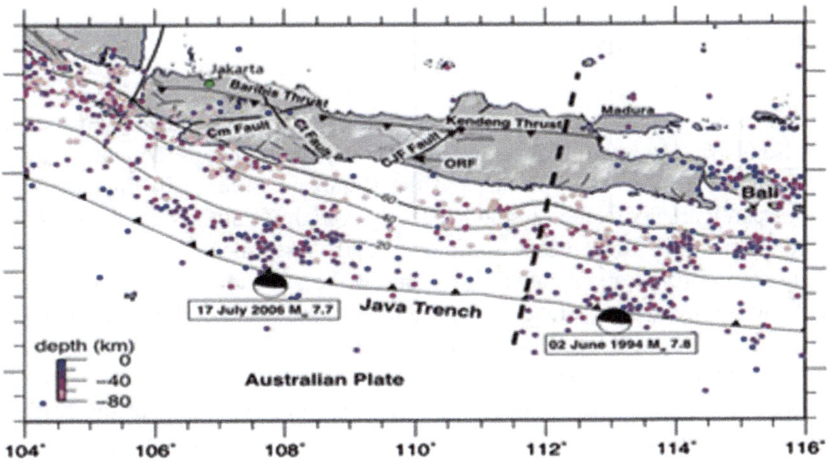

Fig. 1 Primary faults on Java Island [2]

Kedung Waru faults which are associated with the Lidah Anticline, the Gayungan Anticline, and the Kedung Waru Anticline, causing the western part of East Java and its surroundings to have the potential for earthquakes [3, 4]. The National Earthquake Center (PusGeN) published the 2017 national earthquake map with the discoveries of various new active faults in Indonesia as well as the Kendeng Zone of East Java which are located the Surabaya Fault, the Waru Fault, the Blumbang Fault, and the Cepu Fault. Their slip displacement of about 0.05 mm with a magnitude < 7 [5]. Earthquake potential sources are crucial in determining earthquake hazard, therefore geometric information and location of potential sources are very important [6]. Geodetic studies have been carried out, however, data or research on fault mapping on the western part of East Java and its surroundings is still incomplete, so further research elaboration is needed. The subduction paths that showed in Fig. 1 run through the Northeast–Southwest (NE-SW) to a relatively East–West (EW) which has occurred since the Oligocene to the present has took place in a very complicated Tertiary geological setting on Java Island [7]

Gravity is a passive method that can be implemented to identify subsurface geological structures by utilizing variations in Earth's gravity caused by differences in subsurface rock density [8–10]. Based on research related to the priority of potential earthquakes on the island of Java using multi-criteria analysis of decision-making, it is stated that the Kendeng Surabaya Fault is one of the main priorities for mapping so that this research will identify related subsurface structures covering the area. Surabaya, Gresik and Lamongan [11, 12]. Interpretation of the fault can be carried out using the Second Vertical Derivative (SVD) method [13] to indicate the type of fault by delineated the highest and lowest anomalies beneath the subsurface [14].

2 Basic Theory

2.1 Regional Geology

The Java Island physiography at the eastern part are divided into 5 zones, namely: the Southern Mountain Zone, the Solo Zone, the Kendeng Mountain Zone, the Randublatung Depression Zone, and the Rembang Zone [15]. Physiographically, the research area is located in the Rembang Anticlinorium, where the rock forming it consists of shallow marine sediments, clastic sediments, and carbonate rocks. Figure 2 informed about three directions of the structural pattern considering the tectonic setting of Java Island, namely the Northeast–Southwest (NE-SW) direction which is called the Meratus pattern, the North–South (NS) direction or the Sunda pattern and the East–West (EW) direction.

At the rest of the Cretaceous a subduction zone was built in the Karangsambung area continuously to the Meratus Mountains in Kalimantan. This zone created a geological structural framework that traverses Northeast-Southwest then during Tertiary the subduction zone is appeared at the southern of Java Island and the

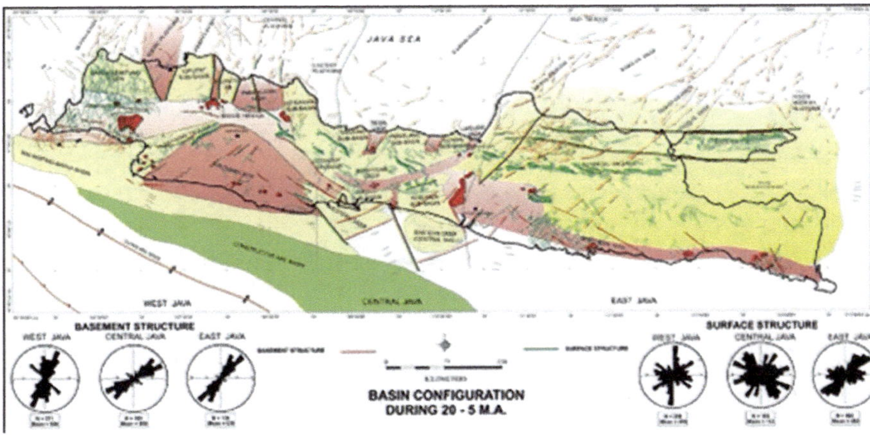

Fig. 2 Tectonic setting of Java Island [2]

structure formed is East–West trending. The Eurasian Plate and the Australian Plate produces the main force of north–south compression. This force forms an oblique wrench fault pattern in the Northwest-Southeast direction, which is more or less in line with the Late Cretacious ridge pattern [7, 16]. In the Pliocene–Pleistocene period the main stress direction is still similar, namely North–South, the tectonic activity of this period produces an upward and folded structure patterns in an East–West direction that can be recognized in the Kendeng Zone. In general, Java Island has an active geological structure which is dominated by shear and upward faults with downward faults which become minor structures. Based on The National Earthquake Hazard and Source Map of 2017 issued by the National Earthquake Center (PusGeN), it is declared that East Java is traversed by several active faults, including the Surabaya Fault, the Waru Fault, the Blumbang Fault, and the Cepu Fault [5].

2.2 *Gravity*

The Gravity Survey is a geophysical method which is based on measuring the variation of gravity values at a certain location [8, 17]. Figure 3 illustrated the Newton's Law as the basic formula of Gravity Method (Eq. 1) which is explained that the magnitude of the gravitational force between two substances or materials is proportional to the two masses of substances and inversely proportional to the square of the centre distance between them [18, 19].

$$\mathrm{Fg} = \mathrm{G}(\mathrm{m}_1 \cdot \mathrm{m}_2)/\mathrm{r}^2 \quad (1)$$

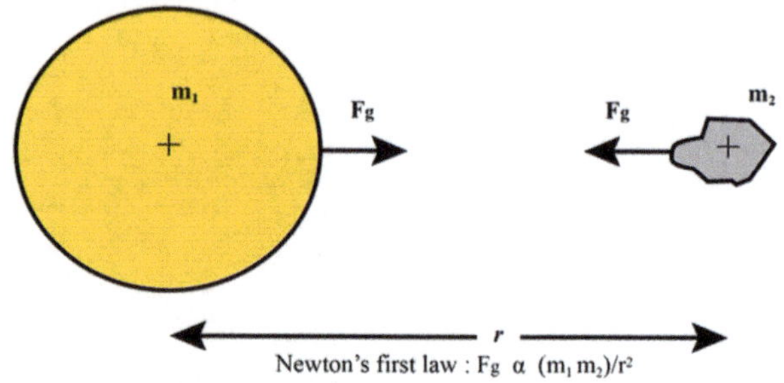

Fig. 3 The concept of the force of attraction between two modified bodies [18]

where Fg is the force of attraction (N) between the mass m_1 and m_2, G is the Universal Gravity Constant (6.674×10^{-11} Nm^2/kg^2), m_1 and m_2 are the mass of the object (kg), and r is the distance between the centers of the two objects (m).

2.3 *Second Vertical Derivative (SVD) of Gravity*

Second Vertical Derivative (SVD) was started popular in 1951 for gravity interpretation [20]. This method is performed by calculating the gradient to increase the subtle gravity data feature which is not visible visually from the original data [21]. The SVD method is based on the second derivative of the Bouguer Anomaly equal to zero contact of the density contrast between the two layers of lithologies, in this case the contact is a fracture or intrusion [22, 23]. The SVD of the vertical component of gravity, gz, can be calculated in the spatial domain from the horizontal gradients by using the Laplace's Equation. The SVD can be approximated by the second horizontal derivative of the gravity data in particular for an elongated anomaly along the y-axis [24]. The existence of the fault structure is identified by the change in the value of the second order vertical gradient which is very contrasting [17, 25]. In addition, the type of fault can be defined by Eq. 2 and Eq. 3 below:

$$\left|\frac{\partial^2 \Delta g}{\partial z^2}\right| \min < \left|\frac{\partial^2 \Delta g}{\partial z^2}\right| \max \quad \text{(Normal Fault)} \tag{2}$$

$$\left|\frac{\partial^2 \Delta g}{\partial z^2}\right| \min > \left|\frac{\partial^2 \Delta g}{\partial z^2}\right| \max \quad \text{(Thrust Fault)} \tag{3}$$

Where *g* is gravity acceleration (m/s^2), z is vertical component.

3 Methodology

3.1 Acquisition Data

This study used gravity data from the TOPEX satellite (https://topex.ucsd.edu/cgi-bin/get_data.cgi). Figure 4 exhibited the 110 acquisition points of gravity data that are scattered in Lamongan, Gresik, and Surabaya City districts. The area of data collection is 18 km × 17 km with a distance between stations of 1.8 km. The measuring station crosses the active structures are overlaid with the revised map 2017 of earthquake hazard and source published by The National Earthquake Center [5]. The gravity data obtained from TOPEX is in the form of Free Air Anomaly data, then it will performed Bouguer Correction (Bouguer Correction) and Terrain Correction, in consequence, Bouguer Anomaly data is generated and displayed through Kriging method.

Figure 5 displayed the complete processing where the Gravity Bouguer Anomaly were devided into regional and residual anomaly. Utilization the Second Vertical Derivative (SVD) method is to generate shallow sources of anomalies hence, the anomalies resulting from this process are residual anomalies. Second Vertical Derivative (SVD) is able to determine the boundaries of the existing structures in the research area.

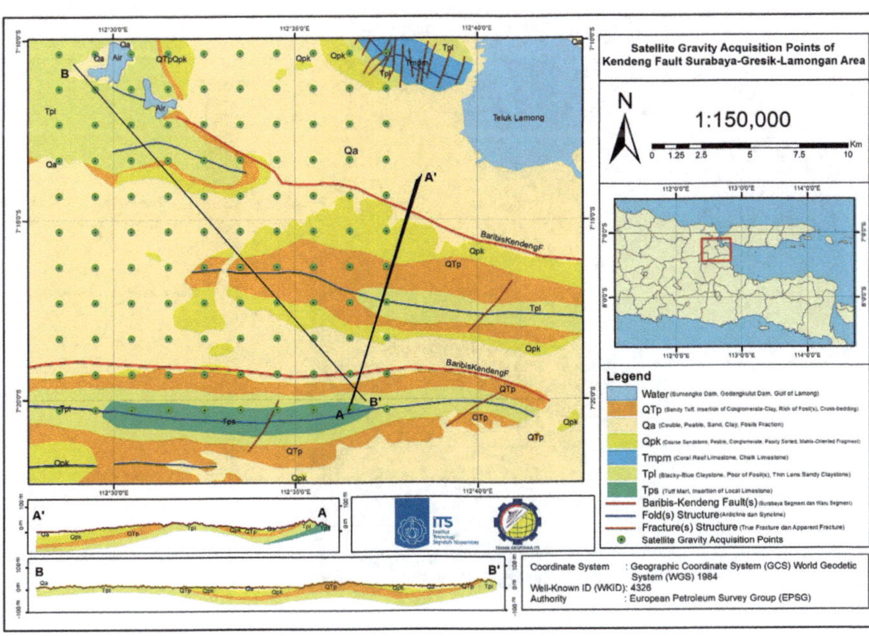

Fig. 4 Distribution of gravity data points

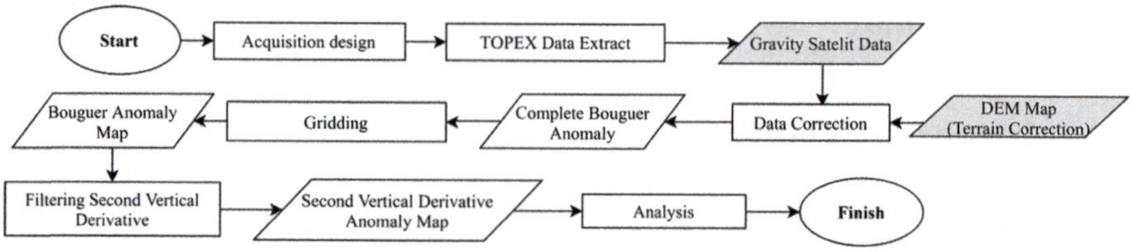

Fig. 5 Research flowchart

4 Result and Analysis

The gravity method is carried out by implementing variations in the earth's gravity caused by differences in the density of subsurface rocks. The gravity data extracted from the satellite then corrected and calculated to produce Bouguer Anomaly. The results of the calculation of gravity data carried out by the interpolation process (gridding) to produce a Bouguer Anomaly map with dimensions of 42 × 46 grid point. Figure 6 described the Bouguer Anomaly map which has a range of values from −20.8 to 10.3 mGal.

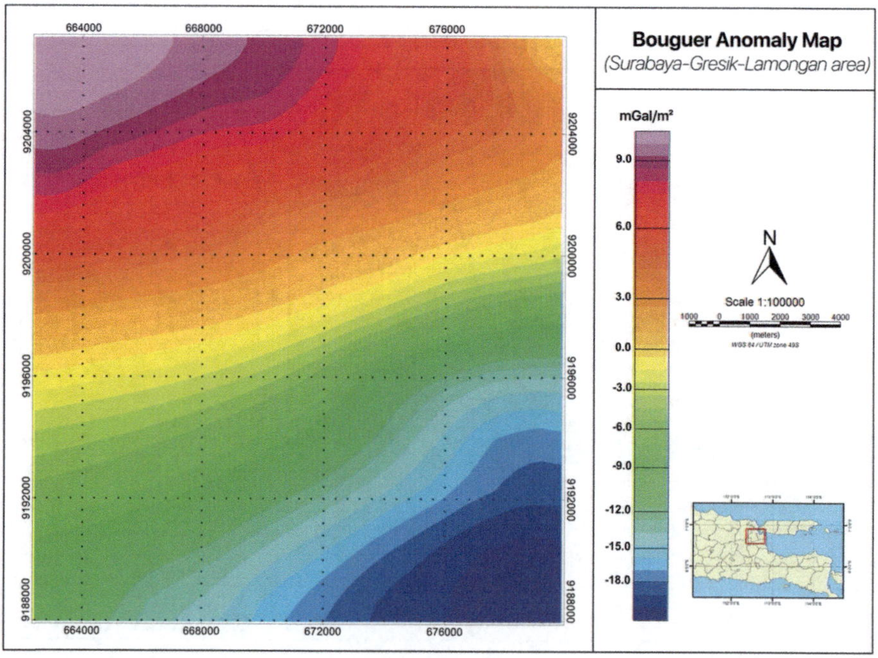

Fig. 6 Bouguer anomaly map

The distribution of anomalies are grouped into three parts, namely high anomalies with values ranging from 10.3 to −2.5 mGal marked in red to yellow. This high anomaly is distributed at the northern area of the study, high anomaly indicates the presence of rocks that have high density. Medium anomaly in the range −2.5 to −12.6 mGal, characterized by yellow to green color, this anomaly is deployed in the middle area of the study. Low anomalies are marked by light to dark blue with a value of −12.6 to −20.8 mGal, these low anomalies are at the western area, low anomalies predicted the existence of rocks with low density.

Second Vertical Derivative (SVD) analysis is used to clarify the location and identify the type of fault. Second Vertical Derivative is carried out to bring the shallow effect of its regional influence because the geological structure will have a maximum absolute value and a minimum absolute value which is limited to a value of zero (0) or close to zero as the limit of geological characteristics. The Bouguer anomaly map will be the input for calculating SVD (2-D) data.

The interpretation stage is carried out by integrating the results of SVD modeling with active fault structure mapping data published by the National Earthquake Center. The results of the SVD 2-D map are interpreted with five trends described by Fig. 7 (dotted black lines), the interpretation is based on the similarity of trends by considering the maximum and minimum values of global SVD. The results of this interpretation described that the interpretation of the existence of a fault using SVD has a high level of accuracy. This is shown in Fig. 8 where illustrated the highly correlated results between SVD-2D modeling and active fault structure mapping published by

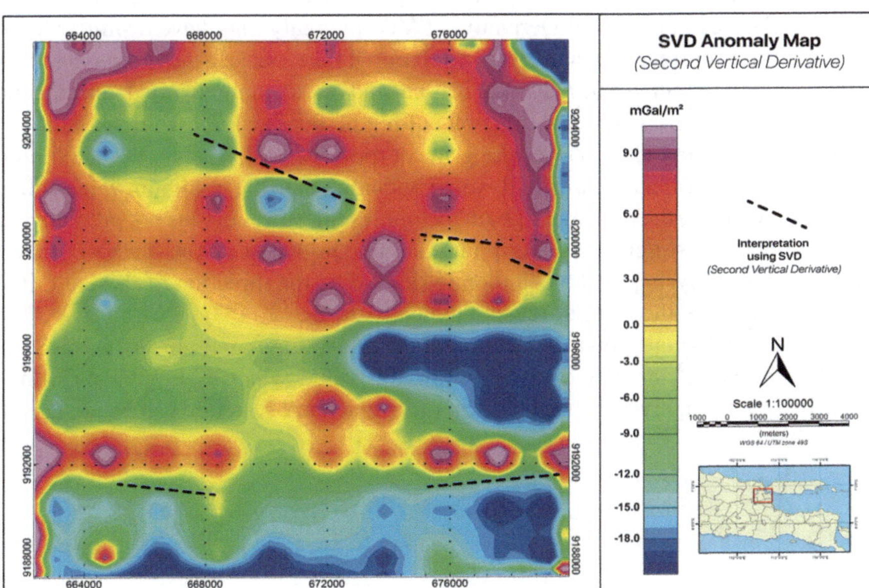

Fig. 7 Second Vertical Derivative (SVD) map with fault trend lines interpretation

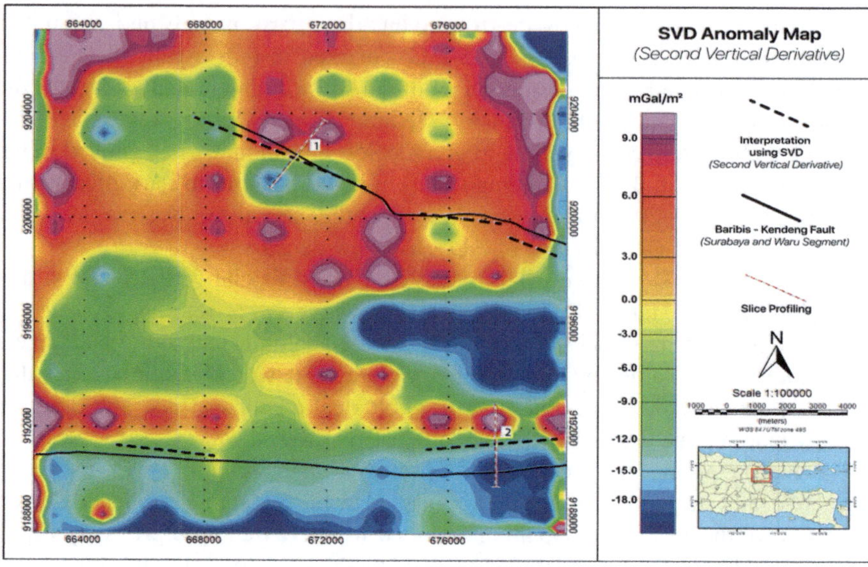

Fig. 8 Second vertical derivative (SVD) map with trend interpretation with fault correlation of the National Earthquake Study Center (PuSGen) and the slicing lines

PuSGeN (2017) [5] namely traversed by the Baribis–Kendeng Segment of Surabaya fault (top) and the Waru segment (bottom). It is detected that the five faults interpreted on the Second Vertical Derivative (SVD) anomaly map have a similar trend that is relatively close to the fault from the National Earthquake Study Center. Subsequently, slicing the SVD anomaly map was carried out to qualitatively determine the direction of the fault, which was done by cutting perpendicularly from the fault.

In the first slicing (Fig. 9), resulting in a profile of SVD values, the maximum value of SVD is 0.1435 mGal/m^2, this value is smaller than the minimum value which is 0.1678 mGal/m^2, the negative sign on the minimum value curve is not predicted as the value, but to distinguish the minimum and maximum values on the SVD curve. Based on this analysis and the equation to determine the direction of the fault, it is possible to identify the form of reverse fault structures.

In the second slicing that represented by Fig. 10, the maximum value of SVD is 0.2076 mGal/m^2 smaller than the minimum value which is 0.2668 mGal/m^2. According to this analysis hence the form of reverse fault structure can be identified.

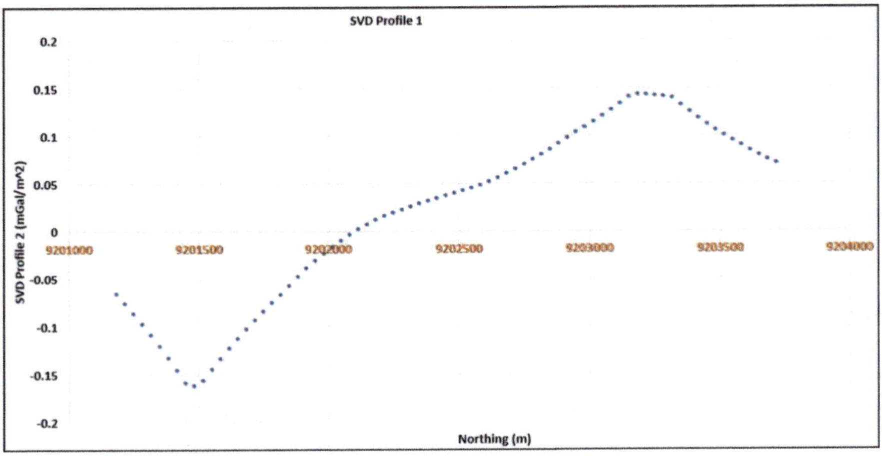

Fig. 9 Profile curve of the Second Vertical Derivative - 1

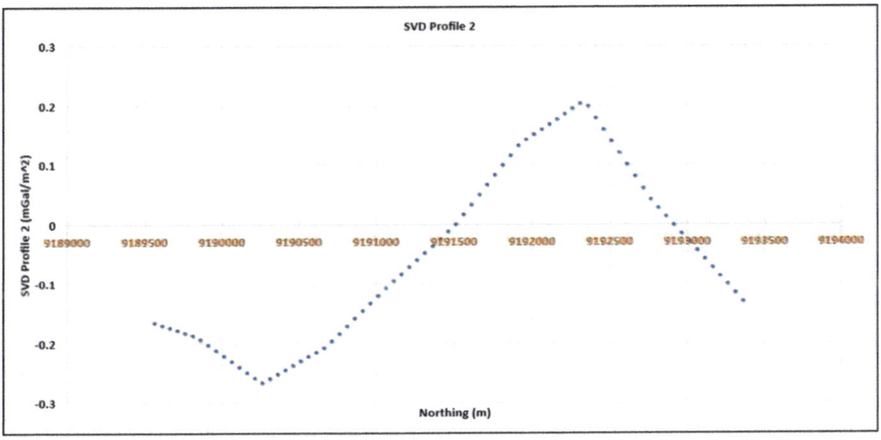

Fig. 10 Profile curve of the Second Vertical Derivative - 2

5 Conclusion

Based on the research that has been done, the following conclusions can be implied as:

1. The results of gravity data processing using SVD 2-D modeling have a high level of accuracy to interpret the fault, indicated by the highly correlated results between SVD-2D modeling and active fault structure mapping published by The National Earthquake Centre - PuSGeN (2017).

2. Slicing curve of the Surabaya Fault SVD profiling exhibited the SVD's maximum value is 0.1 $mGal/m^2$, this value is smaller than the minimum value of 0.2 $mGal/m^2$ which is identified as a reverse fault.
3. Slicing curve of the Waru Fault SVD profiling showed the maximum value of SVD is 0.2 $mGal/m^2$ smaller than the minimum value of 0.3 $mGal/m^2$ which is predicted as a reverse fault.

References

1. Bird P (2003) An updated digital model of plate boundaries. Geochem. Geophys Geosyst 4(3)
2. Koulali A, Susilo S, McClusky S, Meilano I, Cummins P, Tregoning P, Lister G, Efendi J, Syafii MA (2016) Crustal strain partitioning and the associated earthquake hazard in the eastern Sunda-Banda Arc. Geophys Res Lett 43:1943–1949
3. Gafoer S, Ratman N (1999) Peta Geologi Lembar Jawa Bagian Timur Skala 1:500.000
4. Supandjono JB, Hasan K, Panggabean H, Satria D, Sukardi (1992) Peta Geologi Lembar Surabaya & Sapulu, Jawa
5. National Earthquake Study Centre (Pusat Studi Gempa Bumi Nasional), Indonesian Earthquake Hazard and Source Map of 2017, Bandung
6. Widodo A, Syaifuddin F, Lestari W, Warnana DD (2020) Earthquake potential source identification using magnetotelluric data of Kendeng thrust Surabaya area. In: E3S Web Conference, vol 156, p 01002. https://doi.org/10.1051/e3sconf/202015601002
7. Sribudiyani IP, Muchsin N, Sapiie B, Ryacudu R, Asikin S, Kunto T, Harsolumakso AH, Astono P, Yulianto I (2003) The collision of the East Java microplate and its implication for hydrocarbon occurrences in the East Java Basin Proc. Indon Petrol. Assoc., 29th Ann. Conv. Indonesian Petroleum Association (IPA). https://doi.org/10.29118/IPA.1530.03.G.085
8. Kearey P, Brooks M, Hill I (2002) An introduction to geophysical exploration, p 281 (2002)
9. Firdaus S, Setianto A (2018) Interpretasi Struktur Geologi berdasarkan Citra Landsat 8, SRTM dan Anomali Medan Gravitasi Satelit di Cekungan Jawa Timur Utara, ISSN, p 12 (2018)
10. Sota I (2011) Estimation of fault structure using gravity method (Pendugaan Struktur Patahan Dengan Metode Gaya Berat). Positron **1**(1), (2011). https://doi.org/10.26418/positron.v1i1.1565
11. Lestari W, Widodo A, Syaifuddin F, Warnana DD (2020) Research priority of the potential earthquake on the java island using decision making analysis. In: E3S Web conference, vol 156, p 03003. https://doi.org/10.1051/e3sconf/202015603003
12. Rochman JPGN, Warnana DD, Widodo A, Syaifuddin F, Lestari W, Mahsa A (2018) Application of gravity method for local geological structures identification in Surabaya. In: EAGE-HAGI 1st Asia pacific meeting on near surface geoscience and engineering pp 1–5
13. Febriyansyah D, Haerudin N, Setiadi I, Study of hydrocarbon sub-basin patterns using spectral decomposition analysis (Studi Pola Sub-Cekungan Hidrokarbon menggunakan Analisis Spectral Decomposition, 2D Modeling and 3D Modeling based on Gravity Data for the Longiram Region, East Kalimantan (Pemodelan 2D dan Pemodelan 3D berdasarkan Data Gayaberat Daerah Longiram, Kalimantan Timur), p 16
14. Fitriastuti A, Aristo, Putri FF (2019) Identification of subsurface structures using gravity method analysis of first horizontal derivative (FHD) and second vertical derivative (SVD), for Earthquake disaster mitigation efforts in Wonosobo Regency, Central Java Province (Identifikasi Struktur Bawah Permukaan Menggunakan Metode Gaya Berat Analisis First Horizontal Derivative (FHD) dan Second Vertical Derivative (SVD), Guna Upaya Mitigasi Bencana Gempa Bumi di Kabupaten Wonosobo, Provinsi Jawa Tengah), The 12th Earth Science National Proceeding (SeminarProsiding Seminar Nasional Kebumian ke-12), p 11

15. Vanbemmelen RW (1949) The geology of Indonesia, Government Printing Office, The Hague, vol 1A
16. Hamilton W (1973) Tectonics of the Indonesian Region, pp 3–10
17. Parera AFT, Yusuf M (2015) Three dimensional modeling of gravity anomalies and identification of local faults in determining fault types in the Sidoarjo Region (Pemodelan Tiga Dimensi Anomali Gravitasi dan Identifikasi Sesar Lokal dalam Penentuan Jenis Sesar di Daerah Sidoarjo), p 10
18. Telford WM, Geldart LP, Sheriff RE (1990) Applied geophysics, 2nd edn. Cambridge University Press
19. Hinze WJ, Von Frese R, Saad AH (2013) Gravity and magnetic exploration: principles, practices, and applications. Cambridge University Press, New York
20. E. T. A. (1951) The second derivative method of gravity interpretation. Geophysics 23
21. Sumintadireja P, Dahrin D, Grandis H (2018) A note on the use of the second vertical derivative (SVD) of gravity data with reference to Indonesian cases. J Eng Technol Sci 50(1):127–139
22. Setiadi I (2018) Pattern of subsurface geological structure and configuration of the North West Java Basin based on gravity analysis (Pola Struktur dan Konfigurasi Geologi Bawah Permukaan Cekungan Jawa Barat Utara Berdasarkan Analisis Gayaberat), p 14
23. Sarkowi M (2014) Gravity exploration (Eksplorasi Gayaberat), Graha Ilmu
24. Sari L, Sehah, Hartono (2018) Application of second vertical derivative (SVD) on gravity data to identify the presence of faults along the South Serayu Mountains, Banyumas Regency (Penerapan Second Vertical Derivative (SVD) Pada Data GravitasiUntuk Mengidentifikasi Keberadaan Patahan di Sepanjang Pegunungan Serayu Selatan Kabupaten Banyumas). J Ter Fis **1**(2):8. https://doi.org/10.20884/1.jtf.2018.1.2.1127
25. Zain MA, Rozi MF, Septikasari AN, Nuruddianto M, Supriyanto, Zarkasyi A (2020) Study of application of derivative analysis methods on gravity potential data (Studi Penerapan Metode Analisis Derivatif pada Data Potensial Gravitasi). Open Sci Framework. https://doi.org/10.31219/osf.io/m9dg7

Lesson Learned from Weathering Clay Shale Residual Interface Shear Strength Testing Method

Fathiyah Hakim Sagitaningrum, Samira Albati Kamaruddin, Ramli Nazir, Budi Susilo Soepandji, and Idrus M. Alatas

Abstract Clay shale has been a problematic soil due to its weathering characteristic. Weathering is the degradation of shear strength due to exposure to water and air. As the depth of soil increases, the weathering effect would decrease, leading to different weathering grades. On a slope, the different weathering grades would lead to an interface slope failure. Until now, laboratory methods on finding the interface shear strength between the weathering grades seldom been investigated. This research will propose a method and give an evaluation of the results. An interface direct shear test has been conducted until its residual state between an undisturbed clay shale and weathering clay shale. The weathering condition of the bottom clay shale is achieved by soaking and drying the bottom of the sample using a two-days wetting-drying cycle. Results showed scattered values with different behavior for the soaked and unsoaked conditions for increasing weathering days. The soaked condition resulted in an increasing cohesion, increasing average stress ratio, and decreasing friction angle. The unsoaked condition resulted in decreasing cohesion, increasing friction angle, and increasing average stress ratio. Previous research shows that, the residual shear stress should have decreased as the weathering days increases. The cause for the contrary results is due to soil swelling and the absence of a separator when

F. H. Sagitaningrum (✉) · S. A. Kamaruddin
Razak Faculty of Technology and Informatics, University of Technology Malaysia, 54100 Kuala Lumpur, Malaysia

S. A. Kamaruddin
e-mail: samira.kl@utm.my

R. Nazir
Centre of Tropical Geoengineering, University of Technology Malaysia, 81310 Johor Bahru, Malaysia
e-mail: ramlinazir@utm.my

B. S. Soepandji
Civil Engineering Department, Faculty of Engineering, University of Indonesia, Depok 16424, Indonesia
e-mail: budisus@eng.ui.ac.id

I. M. Alatas
National Science and Technology Institute, Jakarta 12630, Indonesia

S. A. Kristiawan et al. (eds.), *Proceedings of the 5th International Conference on Rehabilitation and Maintenance in Civil Engineering*, Lecture Notes in Civil Engineering 225,
https://doi.org/10.1007/978-981-16-9348-9_46

weathering the bottom clay shale. In conclusion, the test conducted is not quite suitable to understand the interface shearing behavior of weathering clay shale. Thus, a modification is suggested for future research according to the identified causes.

Keywords Clay shale · Interface shear strength · Direct shear test · Residual shear stress · Weathering

1 Introduction

Landslide is known to be a common problem in civil engineering. One of the causes of landslide is the existence of problematic soil, such as clay shale within the soil profile. In Java island, clay shale had been a challenging case for engineers due to its rapid decreasing of shear strength. Some of the clay shale landslides recorded were the Cipularang tollway landslide [1], Hambalang landslide [2], and the Semarang-Bawen landslide [3].

Clay shale is known for its unique property of swelling and weathering, influenced by its minerals [3, 4]. Not only due to its minerals, but weathering might also happen due to exposure of clay shale to water and air. Due to weathering, the shear strength of clay shale would decrease. However, the effect of weathering decreased as the clay shale layer deepened. The different characteristics are defined as weathering grades. When the clay shale layer got closer to the surface, the weathering grade would increase, and the shear strength would decrease.

As the Semarang-Bawen slope, the different strengths of the soil layer would initiate a landslide at the interface of the two soil [3]. However, the behavior of the interface shear strength between the two soils are rarely discussed, especially for the different weathering grades.

Thus, this research tried to propose the interface weathering method. The method would describe the interface shear strength behavior of an undisturbed clay shale with a weathering clay shale. It would also evaluate and offer suggestions for further research.

2 Methodology

The proposed method involved an interface direct shear test on different weathering grades of clay shale soil. The samples of clay shale were taken in an undisturbed state. The samples were from the Semarang-Bawen toll road KM 483 + 200. It was classified as the Kerek and Kaligetas formation clay shale, dominated by claystone and shale classification [5].

From the undisturbed state, the samples were cut to fit the shear box. The samples were divided into two sections of the shear box where the lower part of the shear box would act as the weathering clay shale and the upper part would act as the undisturbed clay shale. The schematic of the sample can be seen in Fig. 1.

The interface shear test was conducted as a single-stage direct shear test until its residual state with three normal stresses: 35, 70, and 104 kN/m^2. Due to the limitation of the device, the effect of fissures and pore water pressure reading was ignored. Overall, the flowchart of the research can be seen in Fig. 2.

The consequence of using a single-stage direct shear test meant that the samples could only be used once for each test. Two conditions determined the number of samples: water at the interface and weathering days. The effect of water at the interface differentiated the samples into two states: testing the sample five minutes after the eight-days wetting–drying cycle and after one day of the eight-days wetting–drying cycle. For each condition, two samples were determined. The other condition represents several cycles of the eight-days wetting–drying cycle. From previous research in a similar location, the clay shale would completely slake after 48 days [6]. The 48 days were divided into an eight-day wetting–drying cycle with a two-time soaking period. Thus, a total of six sample series were needed. In total, twenty-four samples were needed.

Initially, four samples were tested beginning interface direct shear test until its residual state. Later, the remaining samples were air-dried for two days. The bottom part of the samples was then soaked for five minutes, followed by another drying for four days. After soaking the bottom part of the samples for five minutes, different treatments were used in the testing. Sample 1 and 2 were tested after air-drying the samples for five minutes, whereas samples 3 and 4 were tested one day after the soaking. Both tests were conducted until its residual state. The testing of the samples is terminated after the 48 days of slaking was reached.

Fig. 1 Sample schematic for the interface direct shear test

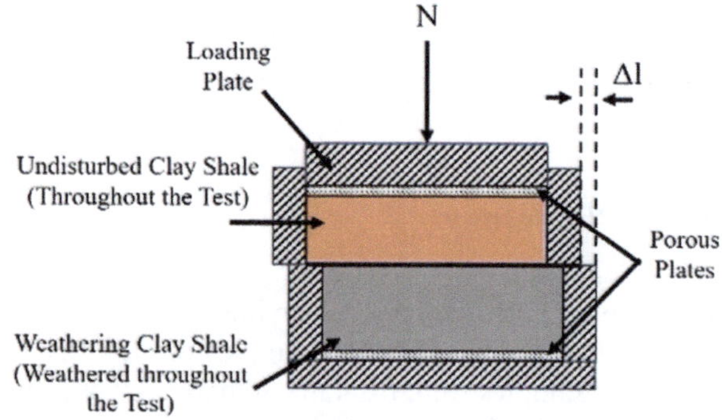

3 Result and Discussion

From the test, several results were extracted. Three parameters from the weathering interface shear strengths: interface cohesion, interface friction angle, and interface stress in the residual state. To understand the behavior of the parameters, a linear trendline was used to observe the trend of the values.

Additionally, the average stress ratio is also considered. The average stress ratio is the average ratio of shear stress and normal stress. The term was used to simplify the moving interface stress between one weathering day to another.

3.1 *Cohesion, Friction Angle, and Average Stress Ratio at Residual*

The interface cohesion results were shown in Fig. 3. From the graph, the values were

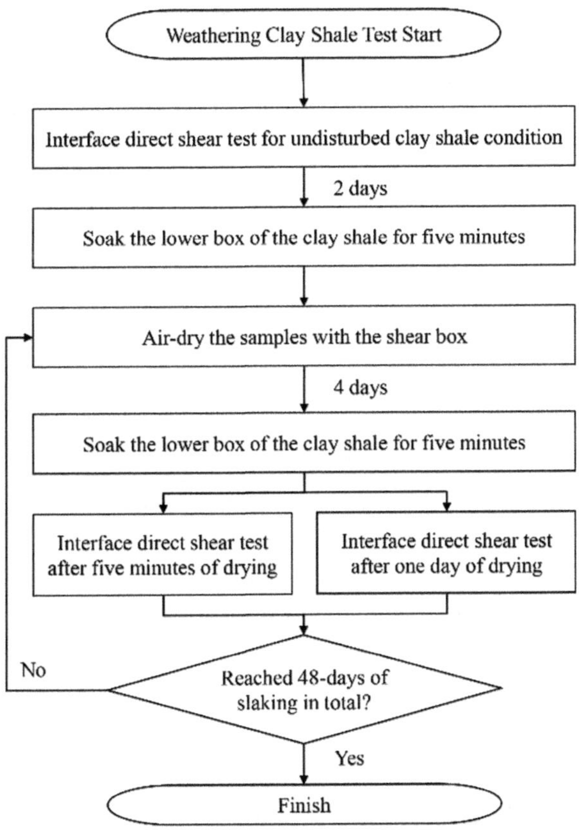

Fig. 2 Research schematic of the interface weathering direct shear test

generally scattered. The scattered values can be seen for both the soaked (Sample 1 and 2) and dried (Sample 3 and 4) conditions. The trend for the soaked conditions was not similar but had a positive value. However, the dried conditions had two different behavior. Sample 3 had a positive trend, whereas sample 4 had a negative trend. Overall, both at soaked and dried conditions, the trend gave a positive value. It means that as the weathering days increases, the interface cohesion increases.

The interface friction angle results are shown in Fig. 4. From the graph, the values were generally scattered. The scattered values can be seen for both the soaked (Sample 1 and 2) and dried (Sample 3 and 4) conditions. Contrary to the interface cohesion, the interface friction angle had a distinct behavior for the soaked and drying conditions. As seen in the graph, both the soaked samples exhibit a positive trend, whereas the dried samples exhibit a negative trend. It means that right after a wetting-drying cycle, the friction angle at the interface is higher than after the soil had some time to dry itself. The trend was consistent with the increasing weathering days, meaning that the testing condition of the sample is also crucial when portraying the field condition.

Direct shear used a linear equation to determine the cohesion and friction angle. Thus, the stress behavior might not be fully representing the shear strength condition. The average stress ratio was used to understand the stress behavior throughout the test comprehensively. The results for the average stress ratio were shown in Fig. 4.

From the graph, the trend shows similar trend for both the soaked and dried conditions. The values were generally scattered for all of the samples. The trend for all of the samples showed a positive value. It means that the average stress ratio of both the soaked and dried conditions when testing would increase as the weathering days increases.

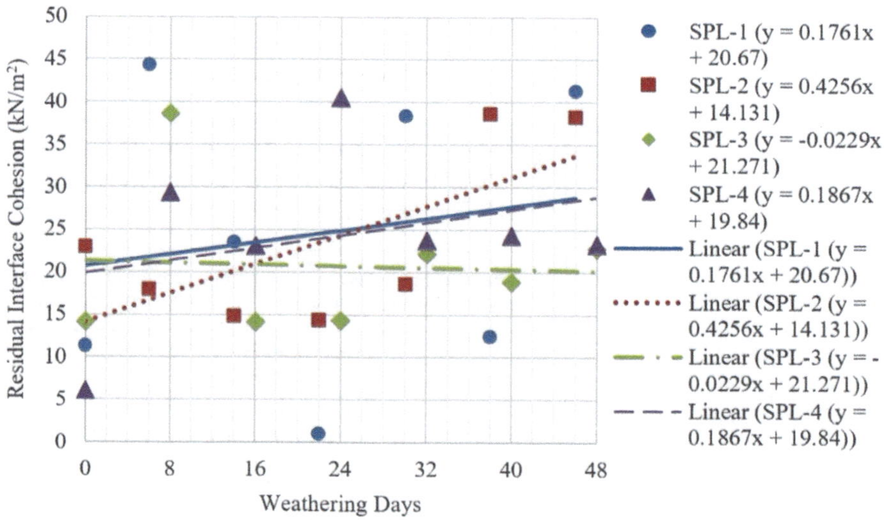

Fig. 3 Interface cohesion for increasing weathering days

3.2 Discussion

From the test results, the values were seen to be consistently scattered. Thus, evaluation for regression, R^2 values, were needed. The results were summarized in Table 1.

From the R^2 values, most of the samples were indeed very scattered. The precision of the data was not good. Identification of the reasons from the scattered values was done.

First of all, it was assumed that the samples were prepared similarly. Thus, the samples would have given similar results in return. The samples were indeed from the same location, same method, same undisturbed condition, and went through the same cutting method. However, even when the samples were similarly treated, the results did not behave similarly. A simple linear trendline was established, but the trend shows differently between one another.

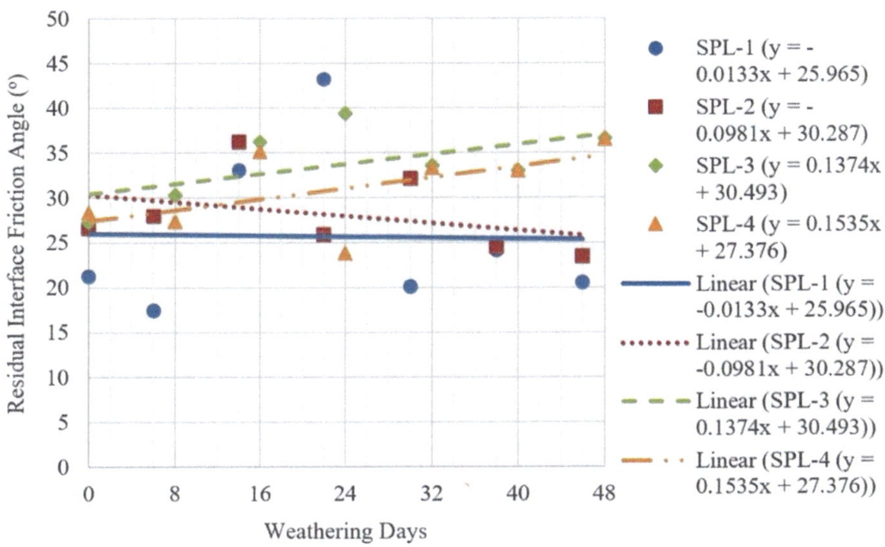

Fig. 4 Interface friction angle for increasing weathering days

Table 1 R^2 values for the tests

Samples	Interface cohesion	Interface friction angle	Interface average stress ratio
Sample 1	0.0320	0.0006	0.0452
Sample 2	0.4684	0.1326	0.4089
Sample 3	0.0020	0.3375	0.1007
Sample 4	0.1010	0.3255	0.4686

Second, the results indicated that using a single-stage method might not be the correct method. Both the values and trends were random. Theoretically, past research showed that with the increase of weathering days, the residual shear stress of clay shale should have decreased [3]. But, as shown in Fig. 5, the residual stress increases instead. There were several reasons for the behavior, but the significant factor might be due to the nature of clay shale as a swelling soil. As the soil swells when immersed in the water, the tested interface when shearing might have changed. Thus, the shear stress observed might have been the fresh clay shale, the weathered clay shale, or the accurate interface. Clay would naturally bind the particles, which effects seen at the opened interface after testing in Fig. 6.

Mesri pointed out that there should be a thin boundary that separates the two samples of direct shear before testing [7]. The partition would keep the interface as the shearing plane and act as a stopping point of the water effect. As seen in Fig. 6, the top part of the undisturbed clay shale was also affected by the water from the bottom weathered clay shale. Thus, the interface shear strength might not be precise.

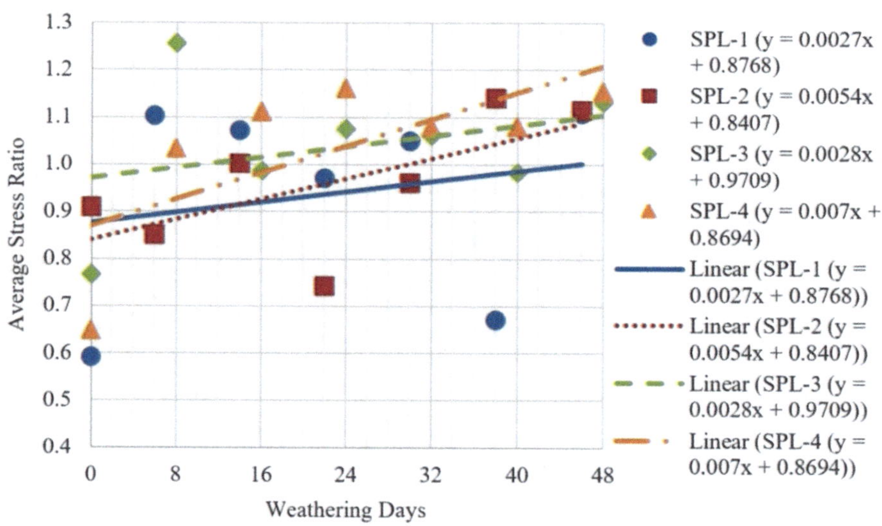

Fig. 5 Interface average stress ratio for increasing weathering days

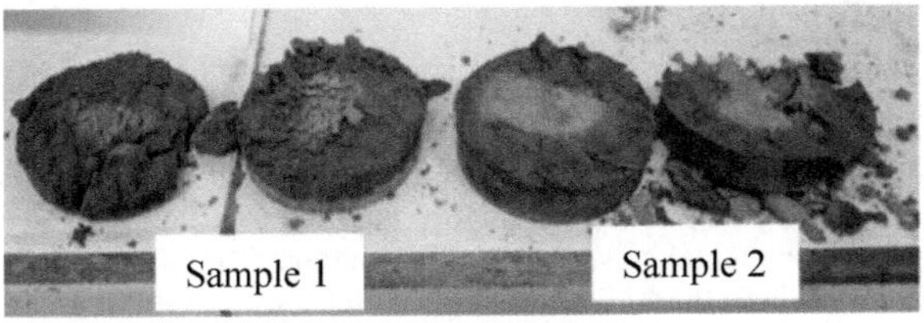

Fig. 6 Visual of interface for different weathering degree after testing in soaked condition

4 Conclusion

For determining the interface shear strength of different weathering grades of clay shale, a method was proposed. The results of interface cohesion, interface friction angle, and the average stress ratio from the interface direct shear test was evaluated. Several conclusions were determined from the analysis:

1. Values of the interface cohesion, interface friction angle, and average stress ratio was scattered. The trend of both the soaked and dried conditions was random. The R^2 values indicated that the results of the test had low precision.
2. As most of the trendline indicated a positive trend, the results were not in agreement with previous research [3] where the cohesion and friction angle should have decreased.
3. The adverse trends of the results might have been due to the swelling tendency of clay shale. The swelling would disrupt the shearing plane at the interface.
4. No partition was used in the interface direct shear test. Thus, water from the bottom part of the shear box was able to seep through the interface to the upper undisturbed clay shale. Water would reduce the interface shear strength and cause the undisturbed clay shale to deteriorate.
5. Finally, the single-stage shearing method is not be suitable for the weathering process.

It can be concluded that the interface direct shear test for the weathering clay shale was not suitable to obtain the objective. Future research should consider the lessons learned to enhance the proposed method. Future improvement should at least consider the use of the multistage direct shear method, include a pore-water pressure measurement on the test, use of thin partition to separate the upper and bottom part of the sample and use a separate weathering phase from the interface direct shear test.

The future research suggestions would give no water intrusion to the upper undisturbed part of the clay shale and confirm that the shearing would be at the interface of the different weathering grades.

Acknowledgements First Author is a PhD student funded with International Doctoral Fund (IDF) from University of Technology Malaysia.

References

1. Irsyam M, Susila E, Himawan A (2007) Slope failure of an embankment on clay shale at KM 97+500 of the Cipularang toll road and the selected solution. In: International symposium on geotechnical engineering, ground improvement, and geosynthetics for human security and environmental preservation, Thailand, pp 531–540
2. Widjaja B (2008) Engineering characteristics of Bukit Sentul clayshale based on laboratory and in situ tests. In: Huang, Mayne (eds) Geotechnical and geophysical site characterization, pp 1231–1237. Taylor & Francis Group, London
3. Alatas IM, Kamaruddin SA, Nazir R, Irsyam M (2016) Effect of weathering on disintegration and shear strength reduction of clay shale. Jurnal Teknologi 78(7–3):93–99
4. Sadisun IA, Shimada H, Ichinose M, Matsui K (2005) Study on the physical disintegration characteristics of Subang claystone subjected to a modified slaking index test. Geotech Geol Eng 23:199–218
5. Thaden RE, Sumadirdja H, Richards PW (1975) Geologic map of the Magelang and Semarang quadrangles, Java. In: Geological Survey of Indonesia, Ministry of Mines, Jakarta
6. Alatas IM (2017) Kesan luluhawa terhadap kekuatan ricih syal lempung dalam penentuan parameter kestabilan cerun. Universiti Teknologi Malaysia, Malaysia
7. Mesri G, Huvaj-Sharihan N (2012) Residual shear strength measured by laboratory test and mobilized in landslides. J Geotech Geoenviron Eng 138(5):585–593

Shear Strain Evaluation on Analysis of Additional Clay Liner Layer Modeling in Ngipik Landfill, Gresik

Siti Nurlita Fitri

Abstract The discovery of new faults in the East Java region has led to different perspectives of analyzing dynamic infrastructures including the landfills and one of the major factors in understanding the local site effect is the shear strain. The Ngipik landfill in Gresik, East Java has been using an open dumping method which involves direct disposal of wastes into the surface pit since 2002. There is, therefore, the need to implement strategies to avoid groundwater contamination and one of the appropriate methods is the addition of clay liner. Therefore, this research applied different forms of this method in four simulated modelings and four profiles to assess its efficiency in terms of the surface area dynamic properties using finite element analysis. The result showed the shear strain reduced by approximately 40% from the existing condition. This, therefore, shows the importance of adding a layer such as the clay liner in open dumping landfill.

Keywords Clay liner · Shear strain · Dynamic

1 Introduction

Several factors are included in the calculations associated with civil structures, especially the dynamic calculations. For example, the development of seismicity in the eastern region of Java significantly influences infrastructural analysis as indicated in the new faults recently identified by Kuncoro et al. [1]. This is in the form of the new active Kendeng fault located along Central-East Java with a strain rate of 0.99–1.93 mm/year. This further strengthens the importance of the local site effect in evaluating an infrastructure, including a landfill.

The local site effect is the reaction of soil to seismic motion on the ground surface. It has been reported that the most critical step in determining the possible characteristics of earthquakes is to assess the ground motion attributes on the surface when accounting for the conditions of local geological and geotechnical sites [2].

S. N. Fitri (✉)
Civil Engineering Department, Universitas Sebelas Maret, Surakarta, Indonesia
e-mail: sitinurlitafitri@staff.uns.ac.id

S. A. Kristiawan et al. (eds.), *Proceedings of the 5th International Conference on Rehabilitation and Maintenance in Civil Engineering*, Lecture Notes in Civil Engineering 225,
https://doi.org/10.1007/978-981-16-9348-9_47

Ngipik landfill is located in Gresik, East Java, and the waste treatment system applied is an open dumping system without a protective layer. Meanwhile, one of the layers recommended by the Environmental Protection Agency (EPA) is the clay liner. This is a layer of cohesive soil with certain parameters which is alternatively used to prevent leachate from contaminating the groundwater [3]. Moreover, Soemitro et al. [4] found the geological formation of the Lidah Structure with Quarterage where the landfill is located to be consisting of clay which is black and blackish, hard when warm, small-fossil content, and sandstone clay. It also has a fairly thick clay content on the layer and this makes the area vulnerable to earthquakes.

Dynamic analysis including simulation and evaluation has been conducted on the landfill in several countries and the results showed the landfill has a significant contribution to groundwater pollution due to the large strain on the protection layer of the solid waste. Moreover, the maximum shear strain produced in the landfill for low-frequency input movements is significantly higher due to the fact that the seismic inertial forces at all depths are in phase at lower frequencies. It is also possible to reduce the stiffness of the saturated soils through repetitive shear strain based on the consideration that soil stiffness modifications have the ability to drive long-term results [5, 6].

The preliminary data of the Ngipik landfill dynamic properties have been determined using microtremor analysis in previous studies. Fitri et al. [7] showed the shear strain is approximately 3.5×10^{-3} to 12.5×10^{-3} with high vulnerability to dynamic failure observed in the east-southwest area. The soil vulnerability index was also estimated to be 2.04–7.16 with the same prone area [8].

This research was, therefore, conducted to determine the effect of adding clay liner to the new model of municipal solid wastes in Ngipik Landfill through the use of dynamic analysis. Several models of clay liner were analyzed to determine the shear strain value on the surface zone despite the existence of values from previous research in order to determine the effectiveness of clay liner construction.

2 Methodology

2.1 Ngipik Landfill Description

The geotechnical properties were evaluated using the previous 3 borehole investigations [9]. The SPT results showed the soil mainly consists of 51% clay and 44% silt and a small amount of sand at 4%, high plasticity (mostly more than 50% with PI), and an average porosity estimated at 50%. The location of the borehole test and profile of soil are presented in Fig. 1.

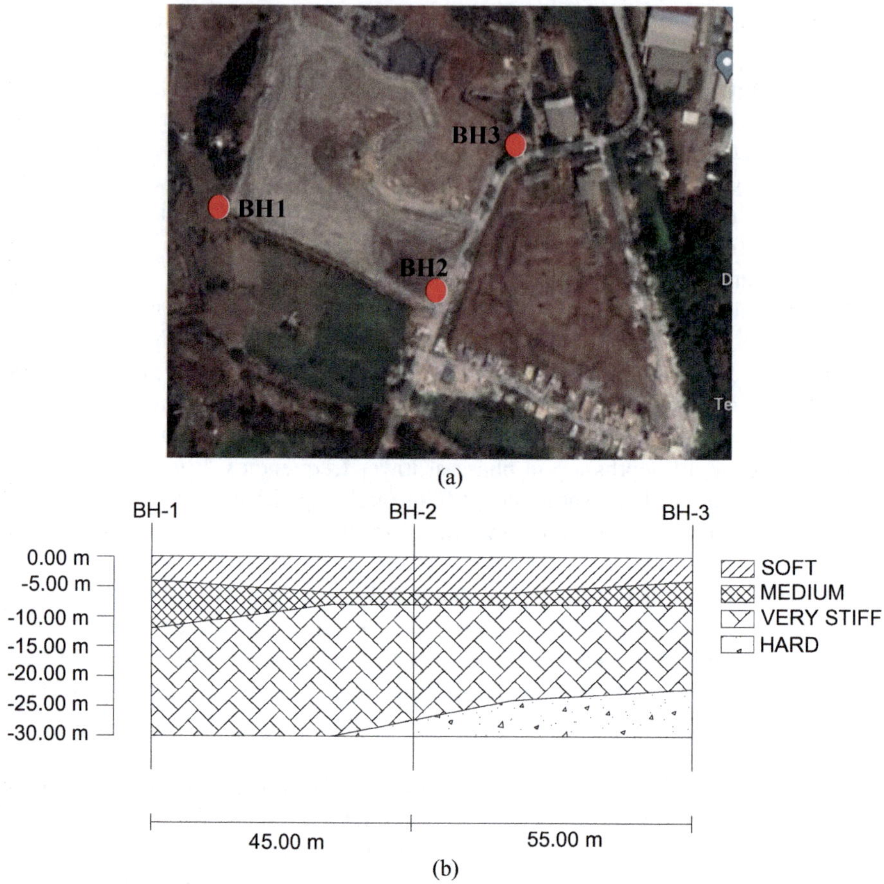

Fig. 1 **a** Location of SPT test, **b** Soil stratigraphy in Ngipik Landfill

Figure 1a shows the location of the SPT test while Fig. 1b describes the stratigraphy of the research site which was determined through statistical analysis of the 3 bored holes. The SPT investigation of BH1, BH2, and BH3 show the soil has clay with different consistencies and this involved the soft and medium at the top surface while those at 15 m depth are dominated by stiff clays. Meanwhile, the soil stratigraphy was used to determine the properties needed for the landfill simulation model as indicated in Table 1.

Table 1 The soil properties of the models

Soil properties	k (m/day)	C	φ	γ (kN/m^3)	μ	E
Clay liner	0.00864	30	0	13.1	0.35	4500
Solid waste	3.58×10^{-2}	10	15	3	0.25	1200

Ngipik Landfill is located in Gresik and has been functional since 2002. It has a total landmass of 6 ha but only 4 ha are being used for disposal with the landfill estimated to be 10–12 m high. Moreover, the municipal solid waste mound is placed in the clay soil deposited before the waste is dumped without treatment or compaction. Several models of the mound were, however, simulated according to the existing site as shown in Fig. 2.

Figure 2 shows the 5 models with different shapes and approximately 115–160 m widths to be analyzed. Moreover, the addition of clay liner divides the cross-section of the landfill into 5 parts as shown in Fig. 3.

Figure 3 shows the solid waste mound models with some variations of the clay liner. The first is the existing model without a clay liner layer, the second is the clay liner design with 1 m thickness around the solid waste mound while the third and last models were simulated with 2 m thickness of clay liner as recommended by EPA.

Finite Element Model was used to analyze the behavior of the site due to dynamic modeling while the dynamic properties were determined using Probabilistic Seismic Hazard Analysis. The results showed the Peak Ground Acceleration (PGA) in the landfill bedrock with a 2% chance of being exceeded in 50 years or a 2475-year earthquake return duration is 0.2 g [10]. The initial ground motion was used for the bedrock while the Nonlinear Earthquake site Response Analysis (NERA) software was applied to determine the ground motion to model the dynamic properties on the surface.

FEM model was simulated using Plaxis 2D at a constant-coefficient of $\alpha = 5.5$ and $\beta = 5$ while the average natural frequency on the surface ground was found to be 1.9 Hz based on a previous study [8]. The mound of landfills with dynamic properties which were determined based on the graph of acceleration vs time in NERA software was applied as an input model in Plaxis and the result produced is the shear strain value in the geometry models. Meanwhile, the final result was obtained from the section of shear strain in the surface layout using all the landfill models and profiles.

3 Result and Discussion

This research was conducted on the dynamic analysis of a finite element method with the profile variations of municipal solid mound waste in different clay liner models analyzed. The results are, therefore, presented in the following Figs. 4, 5, 6, 7 and 8.

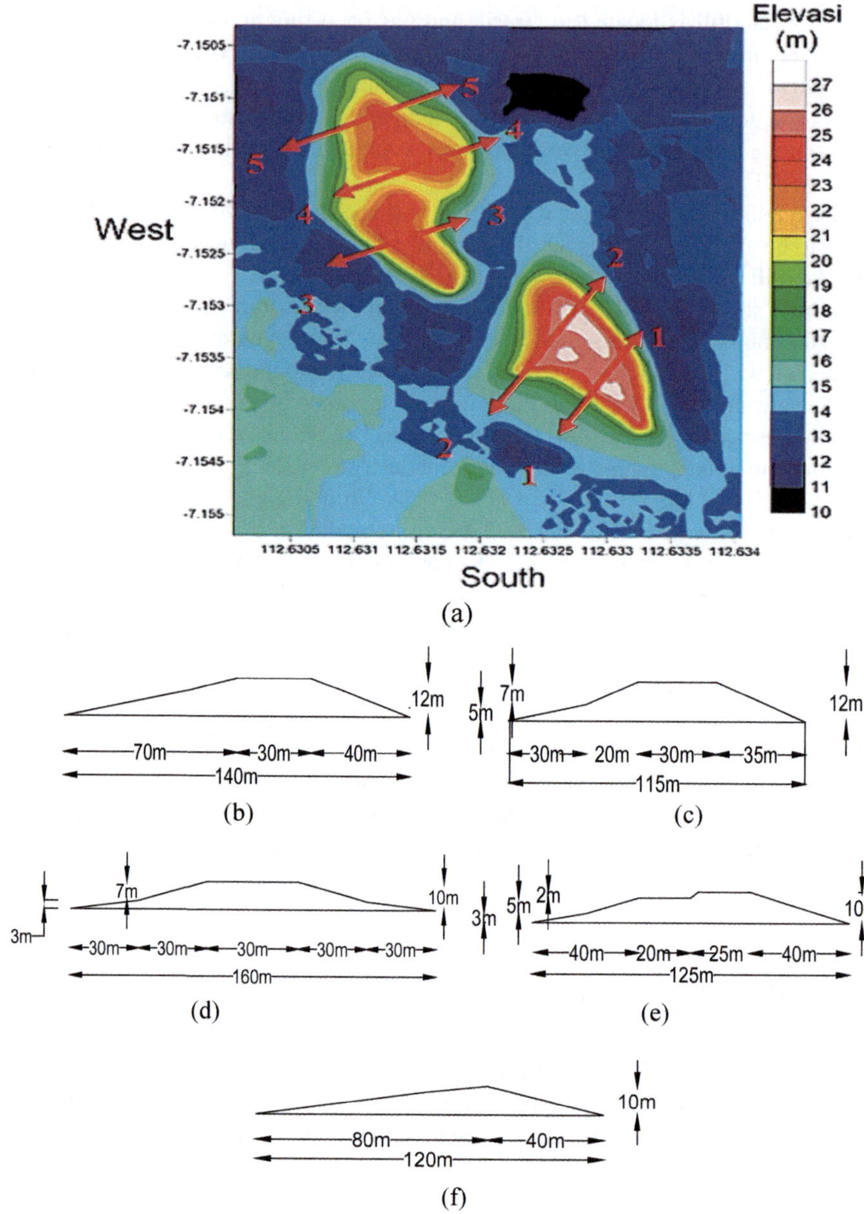

Fig. 2 Solid waste mound in Ngipik Landfill. **a** The sections of the landfill, **b** Profile 1, **c** Profile 2, **d** Profile 3, **e** Profile 4, **f** Profile 5

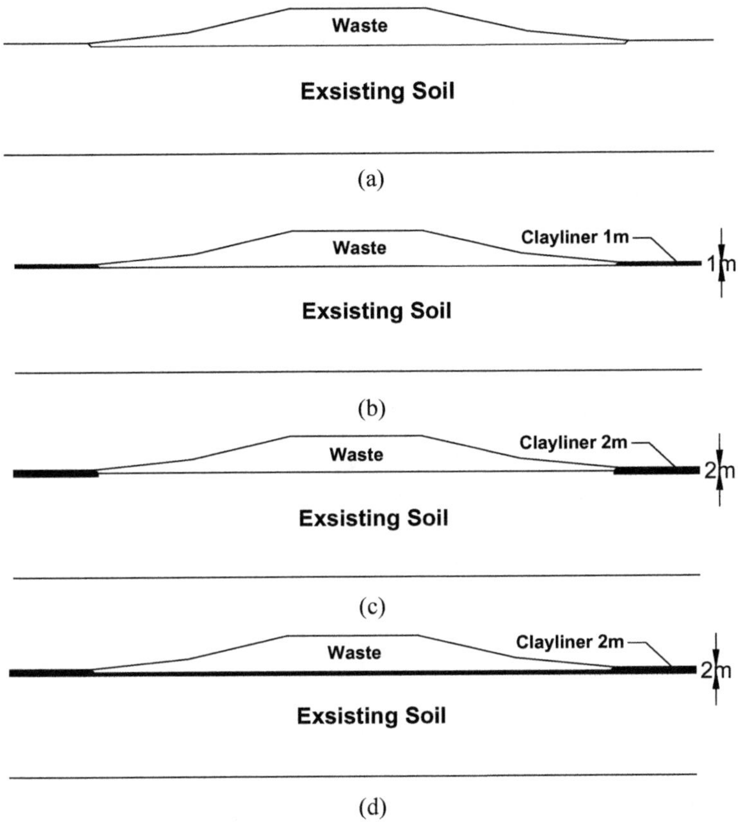

Fig. 3 Addition of clay liner, **a** Model 1, **b** Model 2, **c** Model 3, **d** Model 4

Figure 4 shows the shear strain parameter in the surface around profile 1 and the results of the dynamic parameters in several conditions were discovered to have different shear strain values as indicated in Fig. 1. The maximum shear strain was estimated to be from 6×10^{-3} up to 10^{-2} and, according to Nakamura et al. [11], 10^{-2} indicates the site is collapse while 10^{-3} means there is crack and settlement. A slight reduction in the maximum shear strain was, however, observed in Models 1 to 4 of profile 1 and this means the dynamic conditions do not have an influence on the site.

The summary of all the results is presented in Fig. 9 with the graph observed to illustrate the relationship between shear strain and clay liner modeling in several profiles of the solid waste mound. All the profiles have the same trend of falling slightly from Models 1–4. Moreover, the maximum change in shear strain was found to be 40% of the initial condition due to the addition of 2 m clay liner as indicated in Profile 4 while the minimum was 2% as observed in Profile 1. This was attributed to the difference in the shapes of mound and soil properties. Furthermore, the maximum result was discovered in Profile 2 due to its location as the highest area of amplification

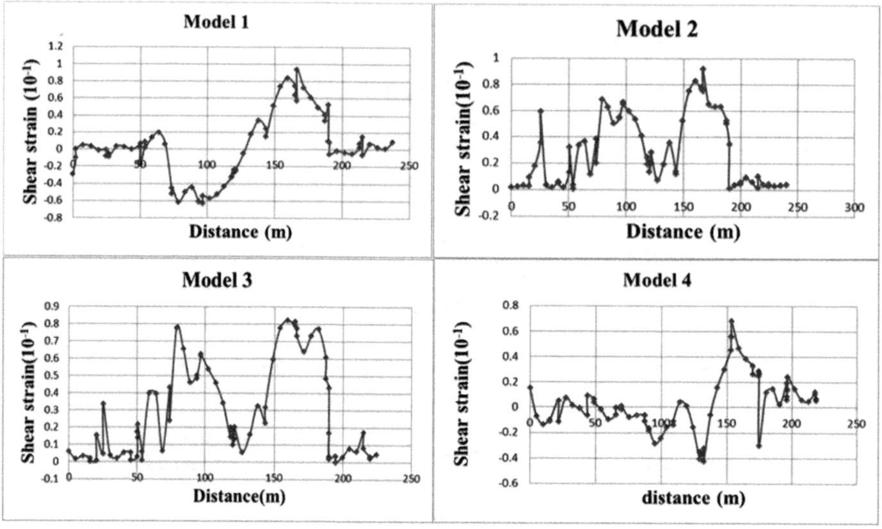

Fig. 4 Shear strain analysis of Profile 1

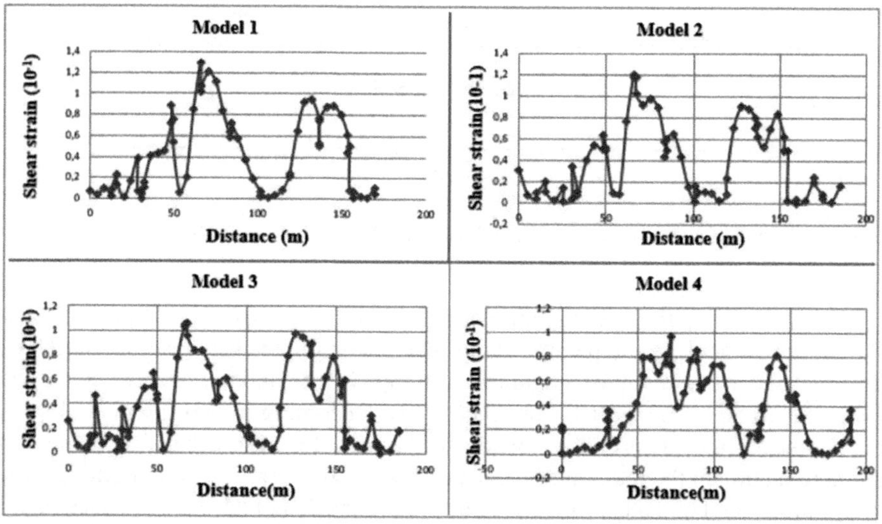

Fig. 5 Shear strain analysis of Profile 2

[8]. The increase in the shear strain indicates a better condition based on dynamic behavior. It is, however, important to note that a value ranging between 10^{-6} and 10^{-5} indicates an elastic characteristic of the soil and this usually leads to the vibration on the surface. Meanwhile, a higher shear strain value of 10^{-2}–10^{-1} indicates a collapsed structure due to dynamic conditions. This, therefore, means the site has

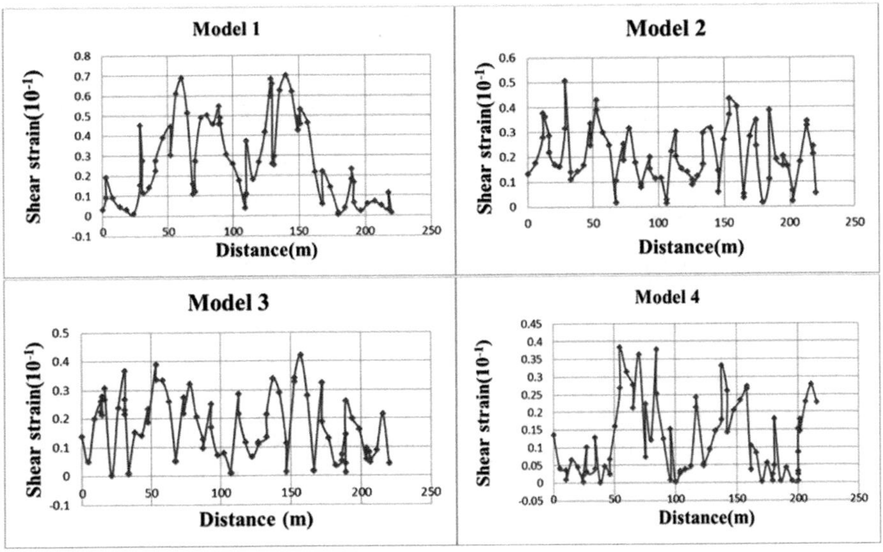

Fig. 6 Shear strain analysis of Profile 3

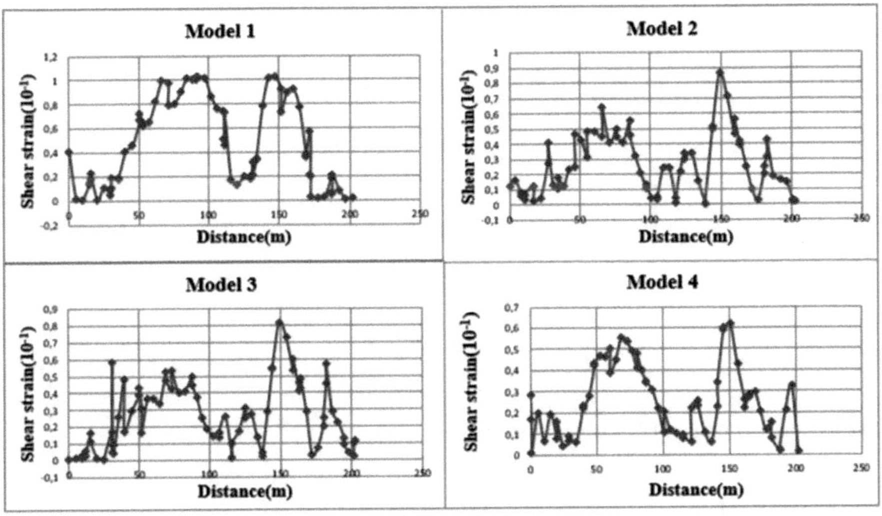

Fig. 7 Shear strain analysis of Profile 4

a significant effect in identifying seismic ground movements which significantly intensify or deamplify at the last moment just before impacting the surface of the ground or the basement of man-made structures [12].

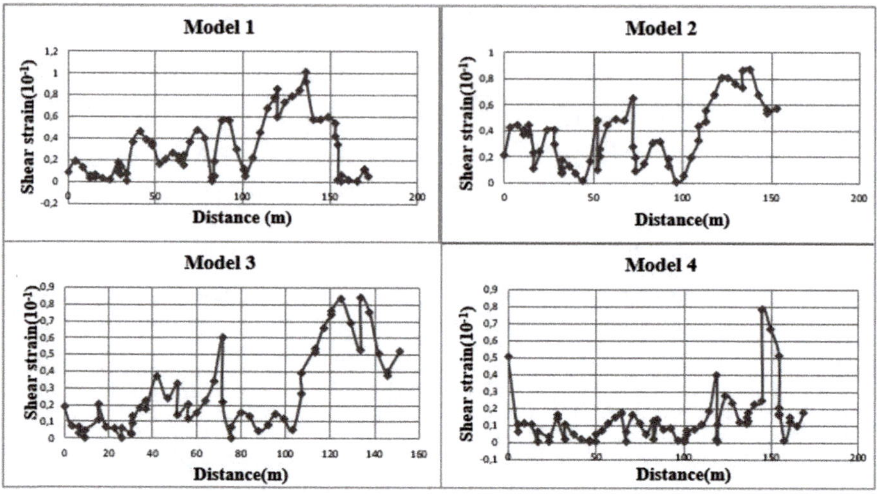

Fig. 8 Shear strain analysis of Profile 5

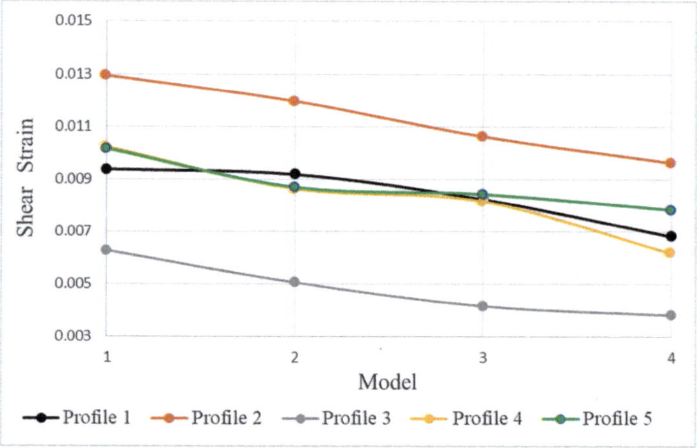

Fig. 9 Shear strain in all the models and profiles

4 Conclusion

Clay liner addition in Ngipik open dumping landfill was modeled in this study and the results showed similarity in all variations. The maximum difference was observed to be up to 40% of the existing condition and this was discovered in the 2 m clay liner model. This means it is important to consider the use of clay liner as a replacement material. It is, however, recommended that further study needs to analyze basic local sites in different geological areas.

References

1. Kuncoro H, Kartini GAJ, Meilano I, Susilo S (2019) Identifikasi Mekanisme Sesar Di Bagian Timur Pulau Jawa Dengan Menggunakan Data Gnss Kontinyu 2010–2016. Semin Nas Geomatika. https://doi.org/10.24895/sng.2018.3-0.1069
2. Parvez IA, Rosset P (2014) The role of microzonation in estimating earthquake risk. In: Earthquake hazard, risk and disasters
3. Agency EP (2000) Landfill manuals: landfill site design. Environmental Protection Agency, Ireland
4. Soemitro NSRAA, Warnana DD, Characterization of leachate distributions at Ngipik municipal solid waste disposal site
5. Pain A, Annapareddy VSR, Sarkar S (2019) Seismic stability analysis of municipal solid waste landfills using strain dependent dynamic properties. Indian Geotech J. https://doi.org/10.1007/s40098-018-0314-6
6. Zania V, Tsompanakis Y, Psarropoulos PN (2008) Seismic distress and slope instability of municipal solid waste landfills. J Earthq Eng. https://doi.org/10.1080/13632460701574605
7. Fitri SN, Soemitro RAA, Warnana DD (2020) Mitigasi Bencana Berdasarkan Nilai Shear Strain Tanah Akibat Gempa Bumi Pada Tempat Pembuangan Akhir (Tpa) Ngipik, Gresik. Rekayasa Sipil 14(3):227–231
8. Nurlita Fitri S, Asih Aryani Soemitro R, Dewa Warnana D, Sutra N (2018) Application of microtremor HVSR method for preliminary assessment of seismic site effect in Ngipik landfill, Gresik. In: MATEC web of conferences. https://doi.org/10.1051/matecconf/201819503017
9. Sutra N et al (2020) The impact of open dumping method in Ngipik landfill investigated with electrical resistivity tomography (ERT) and very low-frequency electromagnetic (VLF-EM). Int J GEOMATE. https://doi.org/10.21660/2020.73.52066
10. Irsyam M et al (2020) The new national seismic hazard maps for design of buildings and infrastructures in Indonesia. In: 16th Asian regional conference on soil mechanics and geotechnical engineering, ARC 2019
11. Nakamura Y, Sato T, Nishinaga M (2000) Local site effect of Kobe based on microtremor. In: Proceedings of sixth international conference Seism. Zo. EERI, 12–15 November 2000, Palm Springs, California
12. Chen Y, Wu Z (2004) International handbook of earthquake and engineering seismology. Acta Seismol Sin. https://doi.org/10.1007/bf03191403

Tsunami Hazard in Cilacap City Due to the Megathrust of West-Central Java Segment

Wahyu Widiyanto and Sanidhya Nika Purnomo

Abstract The source of the earthquake and tsunami in the Indian Ocean, particularly in South Java, is needed for mitigation purposes that are beneficial to life on the expanding south coast of Java. Cilacap City is one of the important cities in driving development in south coast of Java. This study aims to perform a tsunami simulation with the study area of Cilacap City. The simulations are carried out with software which is based on the shallow water equations. The epicenter varies in the form of coordinate points in the megathrust segment of west-central Java. The simulation results show that Cilacap City is prone to tsunami disasters because the tsunami amplitude on the coast can reach 10.7 m and the arrival time of the tsunami is between 10 and 20 min. These results may complement previous data assessing the potential for tsunamis along the Sumatran Java Ring of Fire.

Keywords Tsunami hazard · Megathrust · Java · Cilacap

1 Introduction

The megathrust that stretched from the northern part of Sumatra Island to the islands of Nusa Tenggara is the source of the earthquakes and tsunamis that are very intense. Earthquakes seem to occur regularly that is often felt by people on the west coast of Sumatra, the south coast of Java and Bali and Nusa Tenggara. Likewise, although with a lower frequency of occurrence, tsunamis have become the recognized type of disaster in the region. The greatest tsunami in history that occurred in modern era has certainly not disappeared from our memory, namely the 2004 Aceh Tsunami [1].

W. Widiyanto (✉) · S. N. Purnomo
Department of Civil Engineering, Universitas Jenderal Soedirman, Purwokerto, Indonesia
e-mail: wahyu.widiyanto@unsoed.ac.id

S. N. Purnomo
e-mail: sanidhyanika.purnomo@unsoed.ac.id

Department of Hydraulic and Ocean Engineering, National Cheng Kung University, Tainan, Taiwan

S. A. Kristiawan et al. (eds.), *Proceedings of the 5th International Conference on Rehabilitation and Maintenance in Civil Engineering*, Lecture Notes in Civil Engineering 225,
https://doi.org/10.1007/978-981-16-9348-9_48

Before and afterwards, tsunamis also occurred in the megathrust zone of the same ring of fire as the Aceh tsunami, i.e. the 1992 Maumere tsunami [2], the 1994 Banyuwangi tsunami [3, 4], the 2006 Pangandaran tsunami [5] and the 2010 Mentawai tsunami [6]. While studies and mitigation measures have been carried out, more extensive studies and follow-up are needed to raise the parties up prepare for possible disasters. Therefore, this study aims to provide information on the potential for a tsunami disaster in the city of Cilacap.

Cilacap is a densely populated city with a variety of vital economic activities on the south coast of Java. This city is administratively located in Central Java Province. The population of Cilacap region in 2020 which is living in the coastal area are 655,000. The Cilacap coastal area (Fig. 1) is the center of economic activity, containing a cargo port, a fishing port, two large power plants, a cement factory and an oil refinery. In addition, there is a Pendem Fort, which has cultural heritage status and is a valuable asset that cannot be easily valued in currency as well as the Indonesian naval base. With this in mind, it is important to evaluate the city of Cilacap against the risk of a tsunami.

The assessment of the potential tsunami height is carried out at critical facilities in Cilacap City. These facilities are described below. The first is Nusakambangan Prison. Nusakambangan Prison is the best super tight security prison in Indonesia. This prison is located on the island of Nusakambangan, one of the outer islands of Indonesia. The Nusakambangan Island in the north used to be a water area called Segara Anakan. Meanwhile, the southern part is directly facing the Indonesian Ocean. When it was first established, Nusakambangan Prison was a prison for political land. Currently, however, Nusakambangan Prison is used for big criminals and terrorist

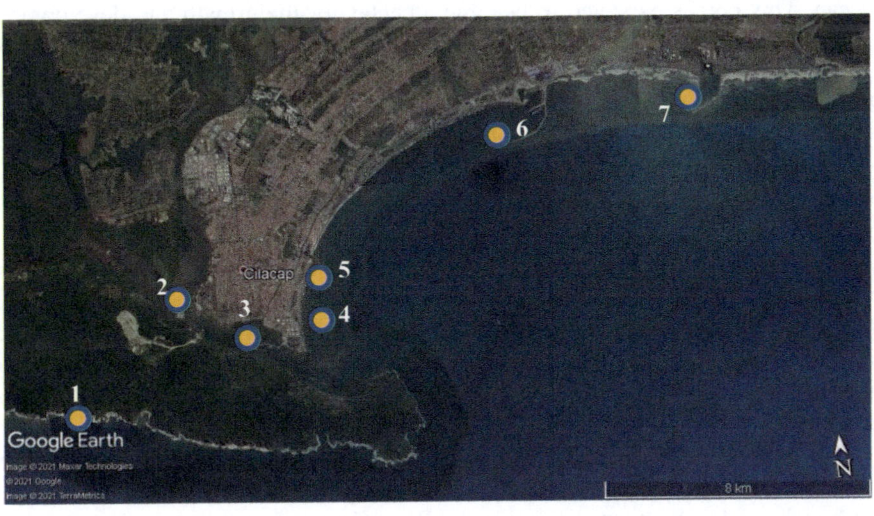

Fig. 1 Study area covers Cilacap City and its surroundings. Important points include (1) prison at Nusakambangan Prison, (2) Port of P.T. Pelindo III, (3) PT. Pertamina, (4) Fort Pendem, (5) Cilacap Ocean Fishing Port (PPSC), (6) Karangkandri PLTU, and (7) Adipala PLTU

prisoners. The oldest prison in Nusakambangan is Batu Prison, built in 1925. In the 2006 Pangandaran tsunami, one of the prisons near the south coast, Permisan Prison, recorded the highest increase of any other tsunami-hit area. Futhermore, the second assessed point is P.T. Pelindo III, a state-owned company (BUMN) operating in the transportation sector, having a role in the management and control of 43 ports in 7 Indonesian provinces. Established on December 1, 1992, P.T. Pelindo III has businesses and services in the form of increasing connectivity and competitiveness between regions across Indonesia, marine pilot and towing services, marine logistics, canal and basin dredging. The company also serves oil and natural gas raw materials, supplies terminals and energy (oil and gas), power plants, shore connections, fuel supply and waste management.

The evaluation has also been conducted on the third point, namely P.T. Pertamina (Persero) Refinery Unit IV Cilacap. This company is one of seven series of processing units in Indonesia with the largest production capacity of 348,000 barrel per day and the most comprehensive facilities. This refinery is a strategic value as it supplies 34% of the national fuel requirement or 60% of the fuel requirement of Java Island. The refinery, built in 1974, also produces asphalt and base oil for Indonesia's infrastructure needs. The fourth point is the fortress of Pendem. This object can be found in Teluk Penyu, Cilacap. This fort was built by the Dutch in 1861. At the time, Pendem Fort, built in stages over an area of 6.5 ha over 18 years, served as defense against attacks from the southern side of Cilacap. Currently, Pendem Fort is a cultural heritage as well as a historical and cultural tourism object.

The fifth point is the Cilacap Ocean Fishing Port (PPSC). This port complex is located in the village of Tegalkamulyan in the South Cilacap district. This port is the only large fishing port on the south coast of Java and directly faces the Indian Ocean. This port is expected to become a center for the growth and development of an integrated fishing industry. The sixth site is PLTU Karangkandri, also called PLTU 2 Central Java. This is a steam power plant that has been built since December 29, 2003 with an investment value of USD 510 million, with a capacity of up to 6,000 megawatts. The last point evaluated is PLTU Adipala, which is a power plant managed by P.T. Indonesia Power with a capacity of 1×660 MW. In contrast to other PPDE generators, this plant uses a Supercritical Boiler with a steam pressure of up to 25.4 MPa.

Many tsunami studies for the regions of Sumatra and Java were conducted from different research perspectives. The tsunami documentation in the form of field survey results was conducted by [1, 3, 7, 8] for the Banyuwangi tsunami. While [2, 5, 9, 10] provide field survey data for the Pangandaran tsunami. The recent tsunami triggered by flank collapse of anak Krakatau volcano occurred in Sunda Strait in 2018 was reported by [11].

This study is a brief analysis to assess whether Cilacap City is susceptible to tsunamis and how high a potential tsunami could be. The possible causes of the earthquake and tsunami are limited to the West-Central Java segment (Fig. 2).

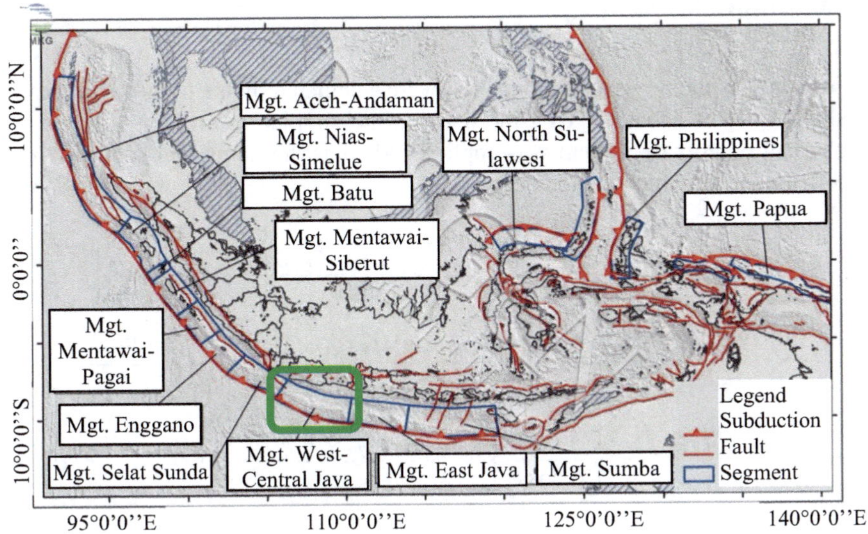

Fig. 2 Earthquake sources in Indonesia (Mgt. = Megathrust). This study uses megathrust of West-Central Java segment (see green box) as tsunamigenic to assess tsunami hazard in Cilacap city (Source of the figure: PuSGeN 2017)

Table 1 shows the coordinates of each assessed location. These coordinates are near the beach, but not directly on the coastline. The coordinate for location 1 is right on the edge of the Indonesian Ocean. Location 2 and 3 have a point at the mouth of the Donan River. Point 4 is directly across from Indonesian Ocean, but has north to south direction so that the coastline is in the same direction as the dominant wave propagation direction. Whereas the locations 5, 6 and 7 are harbors, the evaluation point is thus at the port mouth between the two breakwaters.

Table 1 Assessed location

No.	Site name	Long. (°)	Lat. (°)
1	Lapas Permisan Nusa Kambangan	108.882833	−7.745443
2	Pelabuhan P.T. Pelindo—TNI AL	108.987000	−7.735000
3	Dermaga Kilang Minyak P.T. Pertamina	109.004000	−7.748000
4	Benteng Pendem	109.022000	−7.750000
5	Pelabuhan Perikanan Samudera Cilacap	109.025000	−7.732000
6	PLTU Karangkandri	109.082642	−7.695936
7	PLTU Adipala	109.129000	−7.696000

2 Method

In this study, tsunami waves propagation were calculated by the long wave equation or shallow water equation. Considering that the simulation area is large enough, a spherical coordinate system was applied. Whereas the numerical solution uses a leap-frog scheme which can be expressed:

$$\frac{\eta_{i,j}^{n+1/2}-\eta_{i,j}^{n-1/2}}{\Delta t}+\left\{\frac{1}{R\cos\varphi}\right\}_{i,j}\frac{P_{i+1/2,j}^{n}-P_{i-1/2,j}^{n}}{\Delta\psi}+\left\{\frac{1}{R\cos\varphi}\right\}_{i,j}\frac{\left(\cos\varphi_{i,j+1/2}\right)Q_{i,j+1/2}^{n}-\left(\cos\varphi_{i,j-1/2}\right)Q_{i,j-1/2}^{n}}{\Delta\varphi}=-\frac{h_{i,j}^{n+1/2}-h_{i,j}^{n-1/2}}{\Delta t} \quad (1)$$

$$\frac{P_{i+1/2,j}^{n+1}-P_{i+1/2,j}^{n}}{\Delta t}+\left\{\frac{gh}{R\cos\varphi}\right\}_{i+1/2,j}\frac{\eta_{i+1,j}^{n+1/2}-\eta_{i,j}^{n+1/2}}{\Delta\psi}-fQ_{i+1/2,j}^{n}=0 \quad (2)$$

$$\frac{Q_{i,j+1/2}^{n+1}-Q_{i,j+1/2}^{n}}{\Delta t}+\left\{\frac{gh}{R}\right\}_{i,j+1/2}\frac{\eta_{i,j+1}^{n+1/2}-\eta_{i,j}^{n+1/2}}{\Delta\varphi}+fQ_{i,j+1/2}^{n}=0 \quad (3)$$

where η is the water surface elevation; (P, Q) denote the volume fluxes in X (West–East) direction and Y (South-North) direction, respectively; (φ, ψ) denote the latitude and longitude of the Earth; R is the radius of the Earth; g is the gravitational acceleration and h is the water depth. And the term $-\partial h/\partial t$ reflects the effect of transient seafloor motion, can be used to model landslide-generated tsunamis. F represents the Coriolis force coefficient due to the rotation of the Earth.

Equations (1)–(3) are then solved by the COMCOT (Cornell Multi-grid Coupled Tsunami model) [13]. This model is one of the most widely used tsunami models and has successfully modeled many tsunami cases. Tsunami generation originating from seabed deformation was modeled with elastic finite fault plane theory. This approach requires input in the form of epicenter, focal depth, fault length and width, strike angle, dip angle, slip angle, rake angle, and slip.

The tsunami propagation model from source to points assessed requires bathymetric data. This study implements Batnas (6 arcsec resolution) as bathymetry data. These data are sourced from the Geospatial Information Agency (Indonesian: BIG). Bathymetry data were used to form the simulation layer. In this study, three layers of simulation are used which form a nested grid system. COMCOT uses explicit staggered leap-frog finite difference schemes to solve numerical Eqs. (1)–(3). From this simulation a tsunami wave height will be obtained in the assessed area, in this study is Cilacap City.

3 Result

The results of the running model are processed to show the area affected by the tsunami. Figure 3 shows the beginning of the tsunami generation in the form of sea level changes due to the deformation of the seabed as a result of the Mw 8.7 earthquake. The red color represents the rise of the water level and the blue color represents the drop of the water level, so it is a wave that has a peak and a valley that represents the start of the tsunami wave. Furthermore, this wave propagates at high speed, so it slows down when it reaches the coastal area. Wave energy turns into rising waves and creeps upward in the form of attempts. Figure 4 shows the tsunami height of about 2–6 m approaching the south coast of Ciamis, Tasikmalaya, Garut and Cianjur districts as it spread about 10 min after the earthquake. The tsunami will slow down and reach the coast of Cilacap in about 36 min after it is generated (Fig. 5).

The detection of the water level provides the information that the wave height varies at the examined points. As can be seen in Fig. 6, the Permisan Nusa Kambangan Prison experienced a tsunami that was not too high, 2.1 m. At sites 2 and 3 in the river, no rise in the water level was detected. This perhaps because the numerical measurement point is not placed at the correct elevation, for example considered on

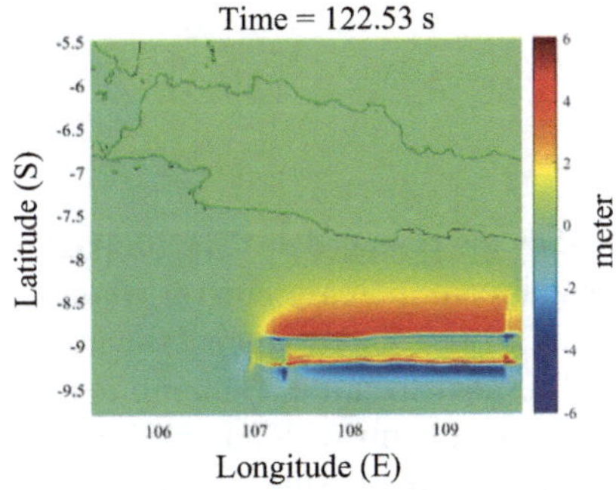

Fig. 3 The beginning of the simulation, a change in sea level due to the deformation of the sea floor by the earthquake Mw 8.7. The red color stands for the rise in the water level and the blue color for the fall of the water level from the MSL

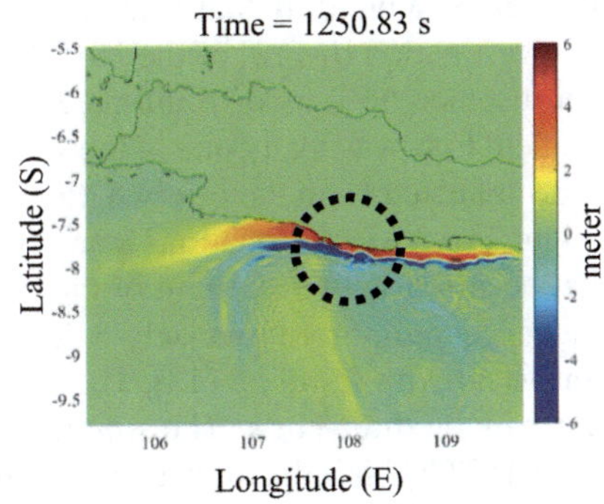

Fig. 4 Tsunami propagation process after 660 s from the earthquake. Dots circle indicates tsunami is reaching the shoreline

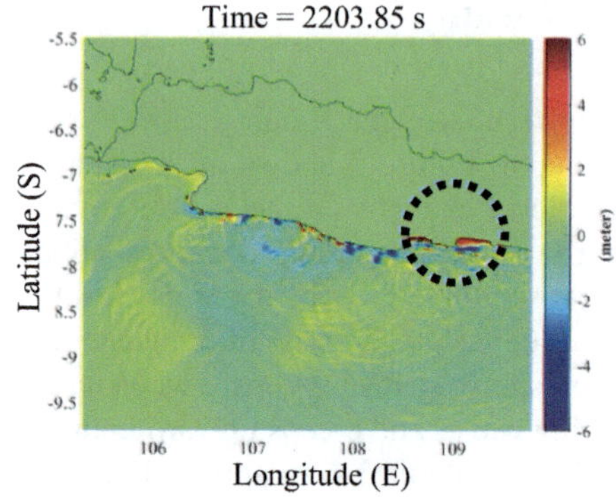

Fig. 5 The tsunami reached the city of Cilacap after traveling from the source of the earthquake in the Indonesian Ocean

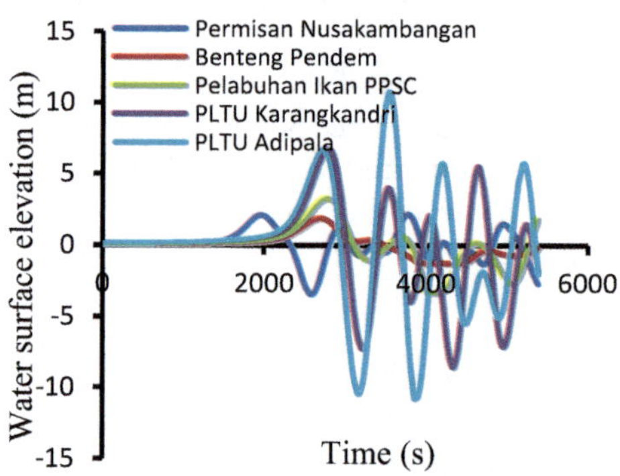

Fig. 6 The simulation results take the form of a rise in the water level due to the tsunami at several observed points in the city of Cilacap

land by the Batnas data. The propagation of the tsunami into the river actually causes a significant increase in river water level over a considerable distance as in the case of the 2011 Tohoku Japan Tsunami.

The next location, Benteng Pendem, based on simulations is expected to experience a tsunami with an amplitude of 1.8 m. This is not too high, but needs to be paid attention, considering that Pendem Fort is the location of an old building with a structure without reinforced concrete. A moderate run-up can not only collapse but also scatter a building that is predominantly masonry and brick. It needs to remark, an investigation by [12] of the 2018 Palu tsunami found that tsunamis 3.5–4 m high can displace 56 tons of reinforced concrete bridges. The low tsunami in front of Pendem Fort was likely due to the wave refraction, which caused the wave direction to be deflected from the original direction by almost 90°, reducing energy and wave height. For the last 3 locations, namely locations 5, 6 and 7, the height of the tsunami that occurred with a strength of more than 3 m was included in the category of the major tsunami, more precisely 3.2, 6.7 and 10.7 m, respectively. These number is consistent with result by [13]. The risk of a tsunami attack could damage the breakwater in the three ports. The simulation results at the seven locations show that the Adipala PLTU has the highest potential of tsunami wave, which in this computation produces ≈11 m. However, it seems that this has been eagerly awaited since the

PLTU was designed. For example, the breakwater chosen is quite heavy, 25 tons each of the protected layer stone [14]. However, this breakwater structure has yet to be prepared to be tested by a large wave, in this case a tsunami.

For damage control purposes, the arrival time of a tsunami is also a parameter that needs to be known. The results of this study show that a tsunami will hit this location within 45–60 min. Especially for the LP Permisan location, the arrival time will be half an hour earlier. Early warnings can be announced by the authorities and processed by the community for security reasons.

4 Conclusion and Recommendation

The results of the tsunami simulation show that the coast of Cilacap has significant potential for a tsunami. Medium height tsunamis can occur around Fort Pendem and the Cilacap Ocean Fishing Port (PPSC). Even at moderate height, the tsunami will hit a densely populated area with casualties. Higher tsunamis that can occur at the PLTU Karangkandri and the PLTU Adipala are 6.7 and 10.7 m high, respectively. The strong force could destroy the breakwater and pier, although the power house facilities on land are likely to remain safe. To find out more about the effects of the tsunami on land, a tsunami run-up must be simulated. High-resolution topographic data is required for this calculation. Tsunami arrival times between 45 and 60 min provide an opportunity for early warning measures that can save live.

This study is limited to the use of tsunami sources in the megathrust segment of west-central Java. In order to assess it more accurately, it is necessary to consider the source of the earthquake and the tsunami for other nearby megathrust segments, such as the Enggano segment in the western part and the East Java segment, and consider different scenarios. A more detailed analysis, covering many possibilities, can be done by making a probabilistic tsunami hazard assessment.

References

1. Borrero JC (2005) Field survey of northern Sumatra and Banda Aceh, Indonesia after the Tsunami and earthquake of 26 December 2004. Seismol Res Lett 76:312–320
2. Hidayat D, Barker JS, Satake K (1995) Modeling the seismic source and tsunami generation of the December 12, 1992 Flores Island, Indonesia, earthquake. Pure Appl Geophys PAGEOPH 144:537–554
3. Tsuji Y et al (1995) Field survey of the East Java earthquake and tsunami of June 3, 1994. Pure Appl Geophys PAGEOPH 144:839–854
4. Maramai A, Tinti ST (1994) The 3 June 1994 Java Tsunami: a post-event survey of the coastal effects. Nat Hazards 1997. https://doi.org/10.1023/A:1007957224367
5. Lavigne F et al (2007) Field observations of the 17 July 2006 Tsunami in Java. Nat Hazards Earth Syst Sci 7:177–183
6. Lay T et al (2011) The 25 October 2010 Mentawai tsunami earthquake (Mw 7.8) and the tsunami hazard presented by shallow megathrust ruptures. Geophys Res Lett 38:2–6

7. Borrero JC et al (2009) The tsunami of 2007 September 12, Bengkulu province, Sumatra, Indonesia: post-tsunami field survey and numerical modelling. Geophys J Int 178:180–194
8. Borrero JC et al (2011) Field survey of the March 28, 2005 Nias-Simeulue earthquake and Tsunami. Pure Appl Geophys 168:1075–1088
9. Fritz HM et al (2007) Extreme runup from the 17 July 2006 Java tsunami. Geophys Res Lett 34:1–5
10. Imamura F, Gica E, Takahashi T, Shuto N (1995) Numerical simulation of the 1992 Flores tsunami: Interpretation of tsunami phenomena in northeastern Flores Island and damage at Babi Island. Pure Appl Geophys PAGEOPH 144:555–568
11. Widiyanto W et al (2020) Run-up, inundation, and sediment characteristics of the 22 December 2018 Sunda Strait tsunami, Indonesia. Nat Hazards Earth Syst Sci 20:933–946
12. Widiyanto W, Santoso PB, Hsiao SC, Imananta RT (2019) Post-event field survey of 28 September 2018 Sulawesi earthquake and tsunami. Nat Earth Syst Sci 19:2781–2794
13. Widiyantoro S et al (2020) Implications for megathrust earthquakes and tsunamis from seismic gaps south of Java Indonesia. Sci Rep 10:1–11
14. Widiyanto W (2014) Assessment of armor layer stone weight of breakwater in ports of logending, PLTU Adipala and Cikidang. Civ Eng J. https://doi.org/10.24002/jts.v12i4.637
15. Pusat Studi Gempa Nasional (PuSGeN) (2017) Peta Sumber dan Bahaya Gempa Indonesia Tahun 2017 (Map of Indonesia Earthquake Sources and Hazards in 2017). Pusat Penelitian dan Pengembangan Perumahan dan Permukiman, Badan Penelitian dan Pengembangan, Kementerian Pekerjaan Umum dan Perumahan Rakyat, Bandung. (in Bahasa Indonesia)

Assessment of the Conditioning Factor for Flash Flood Susceptibility Potential Based on Bivariate Statistical Approach in the Wonoboyo Watershed in East Java, Indonesia

Entin Hidayah, Gusfan Halik, and Wiwik Yunarni Widiarti

Abstract Flash floods that occur suddenly, which cause damage to the weirs or embankments, immediately threaten human life. Identifying the causes of a flash flood is very important to reduce its negative impact. This paper examines changes in flash flood disasters in the Wonoboyo watershed based on estimates of flash flood hazard, land-use changes, and rainfall depth distribution patterns. The method of predicting susceptibility to flash flood hazards is based on various environmental factors that are integrated with GIS. Three bivariate statistics consisting of the Statistical Index (SI), Frequency Ratio (FR), and Predictor Rate (FP-PR) model are applied to select the best Flash Susceptibility Index (FFHSI) model. Changes in land use are then explored based on the conditioning factor for a flash flood. In the final stage, the estimation of areal rainfall uses Inverse Distance Weighting (IDW) to describe the position of rain and flash flood events. The best statistical bivariate statistical approach for the FFHSI is FR. Assessment of environmental factors using the FFHSI shows that 21% of the catchment area has moderate to high until very high vulnerability levels. Changes in land cover significantly affect flash floods, especially changes from forest to agricultural land or settlements. The distribution pattern and intensity of rainfall are closely related to the location of the flash flood. This study results can guide future flood mitigation measures.

Keywords Flash flood · FFHS · Land-use change · Rainfall distribution · Wonoboyo watershed

E. Hidayah (✉) · G. Halik · W. Y. Widiarti
University of Jember, Jember 68121, Indonesia
e-mail: entin.teknik@unej.ac.id

G. Halik
e-mail: gusfan.teknik@unej.ac.id

W. Y. Widiarti
e-mail: wiwik.teknik@unej.ac.id

S. A. Kristiawan et al. (eds.), *Proceedings of the 5th International Conference on Rehabilitation and Maintenance in Civil Engineering*, Lecture Notes in Civil Engineering 225,
https://doi.org/10.1007/978-981-16-9348-9_49

1 Introduction

Flash floods are local floods that occur suddenly with a high, difficult to predict peak flow [1, 2]. The occurrence of high rainfall intensity triggers flash floods. This kind of flood occurred in the rivers in the Wonoboyo watershed in East Java province. Flash floods in a short time can burst the embankments, damage the infrastructure, and generate runoff that flows over settlements and rice fields in the watershed area [3–7].

In addition to the rainfall, the natural terrain condition is a driving factor of the susceptibility to flash floods. A steep topography drives the water down to the estuary more quickly, endangering the river bed stability [8]. Especially during extreme rain-fall, flooding in this area is unavoidable. Other driving variables of the floods are often meteorological conditions in general, including low atmospheric pressure and land breezes that can cause storms and the entrance of air containing water vapor to the catchment area that produces heavy rains [9]. Furthermore, according to Pradhan [10], both the height and the slope of the hills significantly affect the floods occurrence, namely that steeper slopes resulting in faster flows. Accurate estimation of flood hazard susceptibility is essential to measure the relationship among the various spatial and temporal driving factors.

Furthermore, several studies suggest that changes in land use, distribution, and rain intensity are also the causes of floods. The land-use change in urban areas' growth contributes to increased flash floods [11]. The land-use change induces a hydrological response in the watershed in the form of a higher peak flow rate and a shorter concentration time, increasing the difficulty to avoid flash floods.

Several researchers have used various approaches and environmental factors that affect flash floods to identify the flash flood potentials. The commonly used factors are slope, convergence, Topographic Wetness Index (TWI), profile curvature, river networks, density, and land use. These factors have shown significant results to obtain potential susceptibility [12–15]. On the other hand, some researchers also used the NDVI factor.

In addition to determining the relevant factors, using the proper method is also very important in selecting a capable flash flood hazard model. Previous research has applied several methods including: Analytic Hierarchy Process (AHP) [16–18], decision tree analysis (DT) [15], weight of evidence [19, 20], Statistical Index [21], Entropy Shannon [21], and FR [20, 22]. The decision tree method provides better performance than AHP [12]; however, the decision tree method does not consider the temporal changes of various factors such as SPI, land use, and dynamic NDVI [15]. Compared to other methods, the relatively simple FR method applied in several flash flood studies has shown excellent training and validation results [9, 12, 20]. In addition, the FR method has provided good results for mapping the landslide hazard in the Wonoboyo watershed [23]. SI model gave the highest flood prediction result of 98.72% compared to weighting factors and the Shannon entropy model with predictive rates of 97.6 and 92.42% [21].

Based on the area and recording of flash flood events in the Wonoboyo watershed, this paper aims to develop a flash flood susceptibility map in the Wonoboyo watershed. The best model is obtained by comparing three methods: SI, FR, and FR-PR integration to find a better and more accurate model. Next, the impact of the causes of a flash flood is assessed with changes in land use and the areal rainfall events in the study area.

2 Study Area, Floods, and Flood Inventory Mapping

2.1 *Study Area*

Wonoboyo watershed, East Java Province, between 113°50′ E and 113°56′ E, and between 7°49′ N and 7°46′ N (Fig. 1) is the study area. This area is topographically hilly, flanked by two mountains, namely Mt. Malang and Mt. Rajekwesi, of which the heights are 1177 and 762 m above sea level, respectively. The watershed covers an area of 3968 ha.

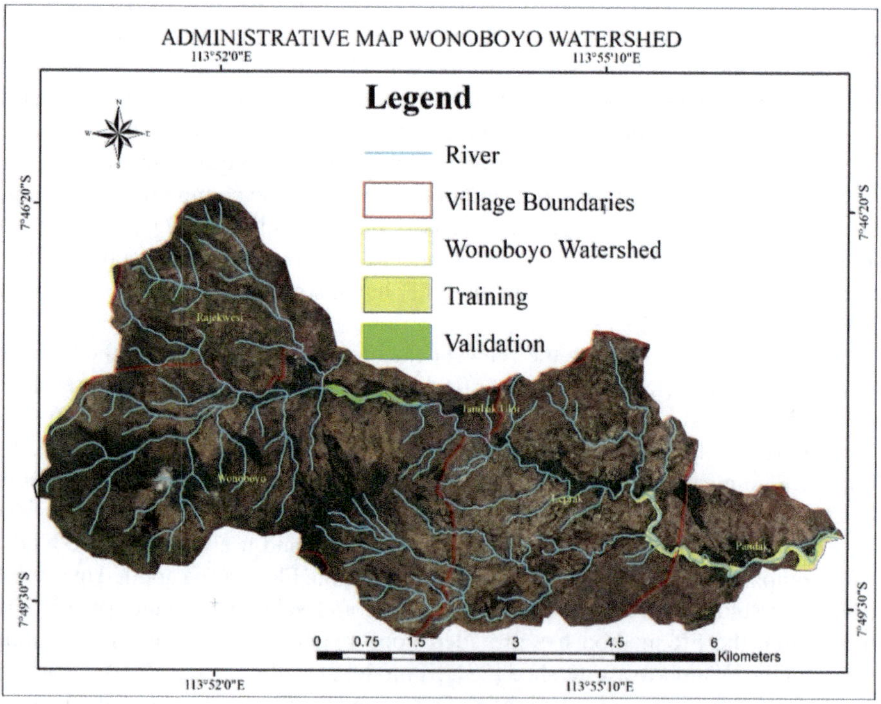

Fig. 1 Study area and flood inventory mapping at Wonoboyo Watershed

The climate has a distinct tropical monsoon with two seasons, namely rainy and dry seasons. The rainy season starts in October and ends in March. The dry season begins in April and ends in September. Based on the Agency of Civil Work of East Java Province, the mean annual rainfall in the study area varies from 460 to 2319 mm.

2.2 *Historical Records of Floods and Flood Inventory Mapping*

The historical records of the flood event were obtained from field surveys and various mass media reports. Base on the information gathered, 9 flash floods occurred in two decades. Typically, flash floods used to occur during heavy rain that lasted three days. The events and the impacts of the flash floods are described in Table 1.

The sampling of flash flood events using field observation data. The results in Fig. 1 show that 8859 pixels with a surface area of 44 ha are flood-affected areas with 78% for training and 22% for validation. The pixel size is 30 × 30 m.

3 Methodology

The methodology implemented in this study comprised of the following main steps: (i) determination of flood conditioning factors for flooding; (ii) estimating, evaluating, and mapping the best flash flood susceptibility index by comparing methods: SI, FR, and integrating FR-PR; (iii) statistical analysis of the changes in land covers and analyzing the distribution of rainfall on 6 flood events.

3.1 *Flood Condition Factor*

Flash flood parameterization is complicated because a flash flood has many causes generally interrelated. For an integrated analysis of the factors that influence a flash flood, it is crucial to determine data collection in essential variables to build a base map representing the conditioning factors for the flash flood in the area under study. Based on previous studies, this study uses 10 factors associated with the flash flood, which are shown in Table 2. [13–15, 19]. The 10 factors are slope, Topography Position Index (TPI), TWI, profile curvature, plan curvature, aspect, convergence index, the density of river networks, land cover, and NDVI. Of the ten factors, 7 geomorphological factors are derived from DEM: slope, TPI, TWI, profile curvature, plan curvature, aspect, and convergence index. Geomorphological data were obtained from the National DEM with a resolution of 8 m × 8 m. 2 other factors, namely land use and NDVI, are derived from Landsat 8 in 2018 and Landsat 7 in 2000, a resolution

Table 1 The historical record on the rainfall depths, the affected villages, and the flood hazards on several flash floods

Flood day evens (rain gauge)			18-Jan-08	31-Jan-16	29-Jan-17	27-Jan-18	25-Jan-19	31-Oct-20
Historical rainfall depth (mm)	Bluncong	D-2	75	0	23	7	0	0
		D-1	7	44	0	67	22	0
		D	27	15	96	190	74	55
		Total	109	59	119	264	96	55
	Wonoboyo	D-2	75	0	0	9	18	0
		D-1	7	50	0	18	17	0
		D	27	150	100	120	50	80
		Total	109	200	100	147	85	80
Exposure village			Wonoboyo and Rajekwesi,	Wonoboyo	Wonoboyo and Leprak	Pandak, Leprak, and Wonoboyo	Wonoboyo	Leprak, and Wonoboyo
Impact of flash flood			The breakdown of the leave, and the overflow of flood into roads and settlements	The breakdown of the leave, and overflow into settlements	The breakdown of the leave and damage to rice fields	The breakdown of the weir, and overflow into the bridge, mosque, and settlement	Small landslide	Overflow into settlement, road, rice field damage

Table 2 Identification driving factors of the flash flood susceptibility

Factor	The preparation procedure of preparation of each factor in Fig. 2 and its relationship with flood susceptibility
Slope (SL)	According to Pradhan [10], slope significantly affects flooding, where steeper slopes produce heavier flow. On the other hand, flat lowlands appear more prone to flooding. The Wonoboyo watershed ranges in slope from 0.008° to 52.602°
Topography position index (TPI)	TPI represents a morphometric index value that describes the difference in height between focused cells and neighboring cells [19]. The TPI value in the Wonoboyo watershed ranges from −11.46 to 10.38
Topographic wetness index (TWI)	The TWI value describes water accumulation's tendency to slope according to gravity, which controls flow direction. A high TWI value indicates that an area is increasingly vulnerable to saturated soil surfaces and areas that have the potential to produce surface runoff [24–26]. Equation (1) is the formula for TWI $TWI = \ln\frac{[As]}{\tan\beta}$ (1) Whereas represents the upstream contribution area, and β represents the slope angle. The TWI value in the Wonoboyo watershed ranges from 3.151 to 19.013
Profile curvature (Prof. C)	Curvature of the profile is a morphometric factor in the form of surface curvature in the vertical plan along the slope [27]. The profile curvature value at the Wonoboyo watershed ranges from −4.209 to 3.516
Plan curvature (Plan C.)	The curvature of the plan is another morphometric factor that affects runoff. This parameter is in the form of a contour line that is formed between the horizontal plane and the ground surface. The value is the difference between the size of the area and the runoff (affects the surface runoff) and those with convergent runoff (the area's infiltration capacity) [9, 10, 27]. The plan curvature value at the Wonoboyo watershed ranges from −4.209 to 4.737
Aspect (A)	Aspect is the direction of the slope. The output raster value is the compass direction of the aspect factor. The aspect factor in the Wonoboyo watershed is grouped into eight classes

(continued)

Table 2 (continued)

Factor	The preparation procedure of preparation of each factor in Fig. 2 and its relationship with flood susceptibility
Coverage index (CI)	CI is a terrain parameter that shows the relief structure as a set of convergence areas (channels) and divergent areas (ridges The CI value ranges from −93.116 to 92.441
River density (RD)	RD affects the level and intensity of flooding significantly. The denser the network and the area around the river, the more prone the area to flooding events [28, 29]. Network density is the division of the flow length (m) to the watershed area (km^2). The RD value ranges from −93.116 to 92.441
Land use (LU)	Land use contributes to infiltration rates in forests and vegetated areas that support infiltration. In contrast, residential areas and grasslands accumulate surface runoff due to impermeable layers that reduce the infiltration capacity. The five land cover classes identified for reclassification are scrubs, followed by plantation area, forest, settlement, and water
Normalized difference vegetation index (NDVI)	NDVI is an index used to measure vegetation characteristics in an area [30]. NDVI can sufficiently represent the decrease in land cover based on the image's local brightness [31]. Equation (2) is a formula to compute NDVI [32] $NDVI = \frac{(NIR-VIS)}{(NIR+VIS)}$ (2) VIS shows the spectral reflectance measurement obtained in the visible wave region in red (band 5). NIR shows the spectral reflectance measurements obtained in the infrared wave region (band 4). The NDVI value ranges from −1 to 1 [33]

of 30 m × 30 m. The river network map is obtained from the delineation of Indonesian Topographical Map data at a scale of 1:25,000. Hydro-climatological data in the form of rain data are obtained from the Bondowoso Technical Operating Unit of Water Resource Management (UPT PSDA) and the Agency of Public Works and Water Resources of Bondowoso Regency.

8 factors in the form of numerical variables are classified into several classes using statistical quantile classification according to the data grid's value [15]. 2 other factors, namely land use and aspect, are nominal variables and are classified based on the feature class.

Fig. 2 Maps of flood hazard susceptible conditioning factors, **a** S, **b** TPI, **c** TWI, **d** Prof. C, **e** Plan C. **f** A, **g** CI, **h** RD, **i** LU, **j** NDVI

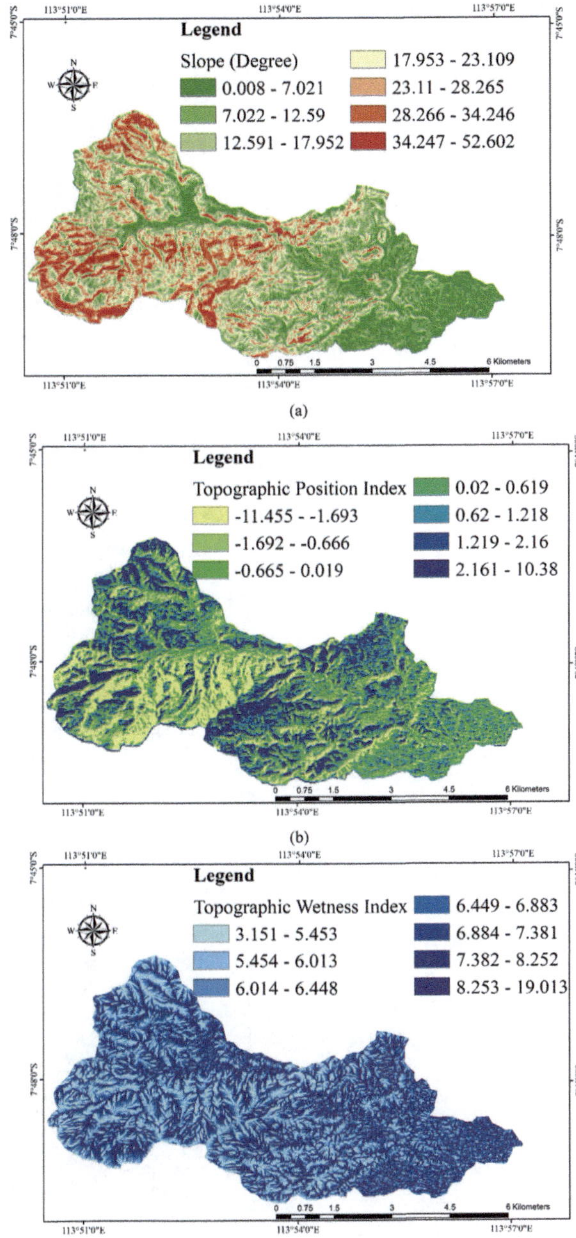

Fig. 2 (continued)

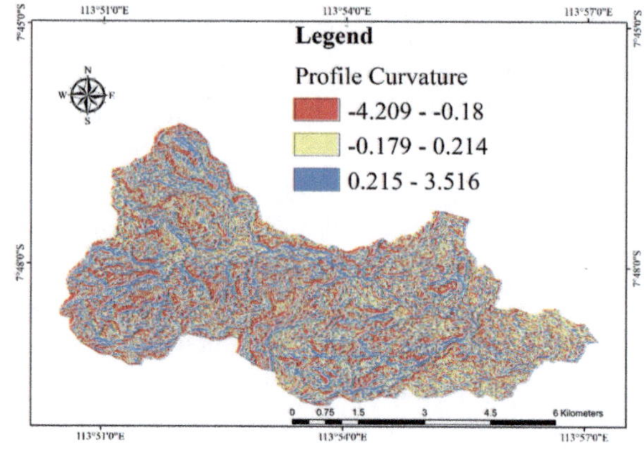

(d)

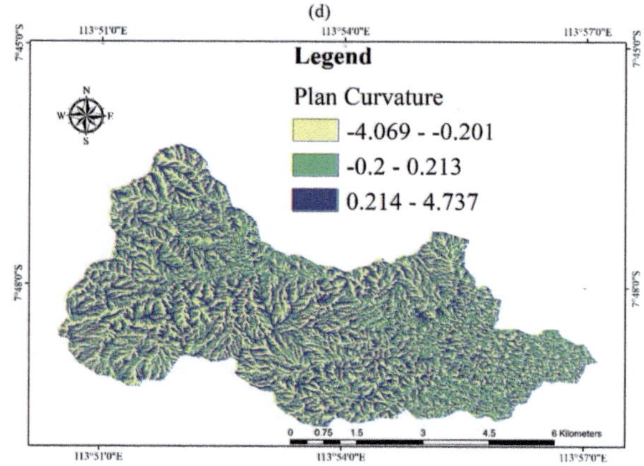

(e)

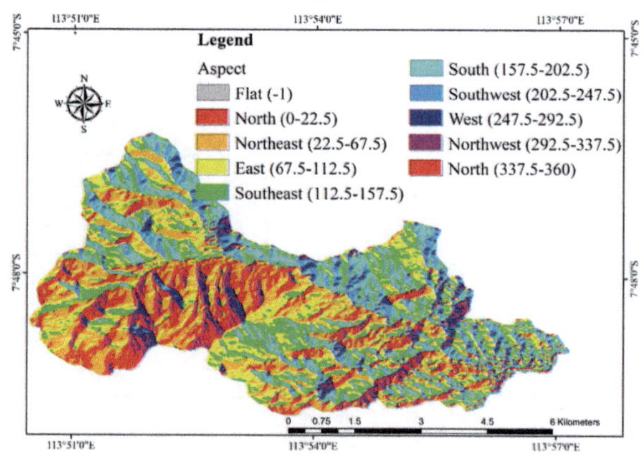

(f)

Fig. 2 (continued)

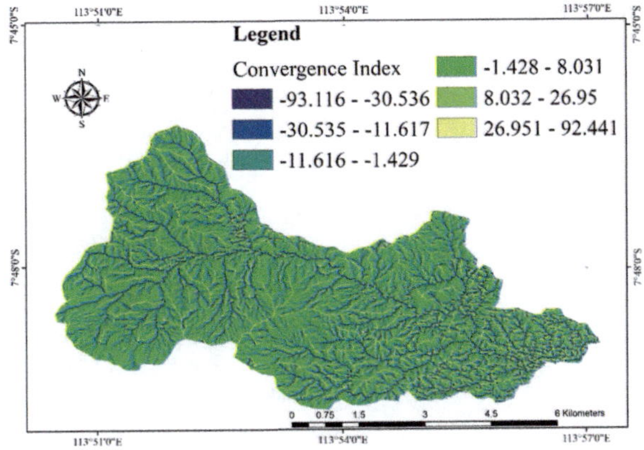

(g)

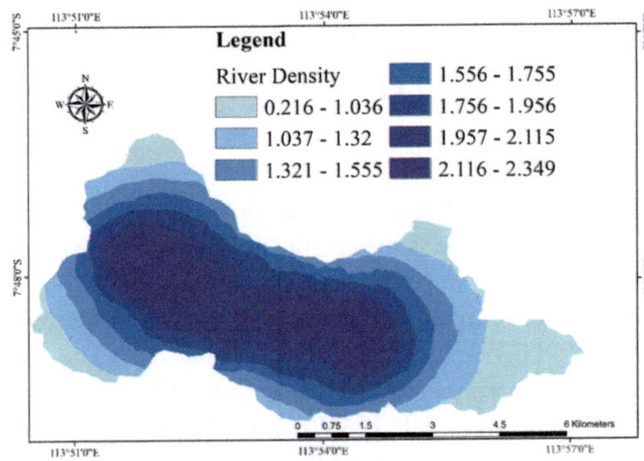

(h)

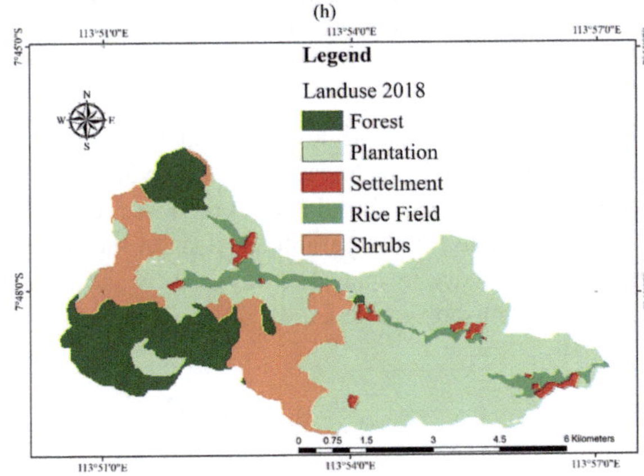

(i)

Fig. 2 (continued)

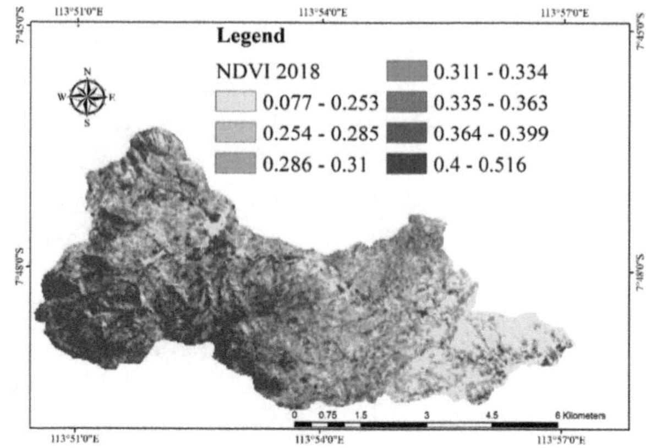

(j)

3.2 *Estimation of the Flash Floods Hazard Susceptibility Index*

The estimation of flash flood hazard is based on statistical observations between the inventory of flood events and the flood conditioning factors. This stage comprises four steps (1) Weighting each class's factors using various statistics bivariate methods. The weights for each class of geomorphological and environmental factors are computed by using GIS, resulting in FFHSI values. (2) The validation of the model is carried out by comparing the Area Under Curve (AUC) values between the flash flood hazard susceptibility predicted by the model with the actual flash flood locations from the field survey. The values of the indices were then grouped into six intensity classes by using natural break reclassification.

Statistical Index

The SI model is a bivariate statistical approach introduced by Van Westen [34] Initially, this model was to estimate landslide hazards. Recently, this model has been developed to be applied to flash flood hazard areas [9, 12]. The SI method describes the weight value for the parameter class in the form of the natural logarithm of the flash flood density in that class divided by the flash flood density across the map. The equation used is as in Eq. 3 [34]:

$$W_{ij} = \ln(\frac{D_{ij}}{D}) = \ln\frac{\left(\frac{L_{ij}}{L_T}\right)}{\left(\frac{P_{ij}}{P_L}\right)} \quad (3)$$

whereas W_{ij} is the weight assigned to class i given from factor *j; D_{ij} is the flash flood* density in class i from factor *j; D shows the totalflash flood density on the entire map; L_{ij} is the number of pixels with flash floo*ding in class i from factor j; L_T is the total number of flash floods on the entire map; P_{ij} is the number of pixels in class i of parameter j, and P_L is the total number of pixels on the entire map.

The results of the calculation can be interpreted that if the value of W_{ij} is positive, it indicates a viable and strong relationship between the class and the flash flooding distribution. The higher the score, the stronger the relationship. Conversely, if the W_{ij} value is negative, there is no correlation between the class and the occurrence of flooding. Furthermore, the FFHSI value can be calculated by adding all the weights for the number generated as in Eq. 4.

$$FFHSI_{SI} = \sum\nolimits_{R=1}^{R=n} W_{ij} \tag{4}$$

Frequency Ratio and Predictor Rate Method

The frequency ratio method was used in [9, 10] to estimate the probabilistic relationship between dependent variables and the independent variables and to determine the weighting coefficient values in each flood-related variable class Eq. 5 shows a formula used to estimate the values in Frequency Ratio.

$$FR = \left(F_i \Big/ \sum\nolimits_{i-1}^{m} F_i\right) / \left(A_j \Big/ \sum\nolimits_{j=1}^{n} A_j\right) \tag{5}$$

where F_i is the number of pixels with flash floods for each class of each factor; A_j is the number of pixels for each class of each factor; m is the number of classes in the F_i factor; n is the number of factors in the study area [35]. The FR value interpretation shows that if the value is more than 1, there is a correlation between the class and the occurrence of flooding. The higher the value, the stronger the relationship.

Based on the calculations for the FR sole conditioning factor, FFHSI can be calculated by aggregating the FR of all factors by using Eq. 6:

$$FFHSI_{FR} = \sum\nolimits_{j=1}^{n} FR \tag{6}$$

The next step is to calculate the relative frequency (RF) using Eq. 7 by normalizing the FR value (probability value range {0–1}).

$$RF = \frac{(FR_{ij})}{\left(\sum_{i=1}^{m} FR_{ij}\right)} \tag{7}$$

In RF, all factors have the same weight; FR-PR integration considers the disadvantages of these. Weights using the level predictor (PR) are calculated by assessing each flood conditioning factor with the training data set in Eq. 8 [34]

$$PR = \frac{(RF_{max} - RF_{min})}{(RF_{max} - RF_{min})_{Min}} \tag{8}$$

Furthermore, the FFHSI value is the sum of the results of each FP-PR factor with each RF class as in Eq. 9.

$$FFHSI_{PR} = \sum (PR \times RF) \tag{9}$$

3.3 Preparation of Datasets for Training and Validation

This stage evaluates and compares the training and prediction capability. The last step includes validating the model based on statistical metrics such as percentage of sensitivity, and specify, as shown in Eqs. 10 and 11.

The sensitivity indicates the proportion of incidents correctly classified as flash floods. Accuracy means the balance of incidents correctly classified as flash floods and correctly classified as non-flash-floods.

$$sensitivity(y) = \frac{TP}{TP + FN} \tag{10}$$

$$specifity(x) = \frac{TN}{FP + TN} \tag{11}$$

where TP (True Positive) and TN (True Negative) indicate correctly classified incidents and FP (False Positive) and FN (False Negative) indicate wrongly classified incidents [13]. The ROC plot is a sensitivity versus specificity graph, statistically unaffected by prevalence. It is calculated by estimating the parameters for the infinity limit. The Area Under the Curve (AUC) is estimated from the ROC curve as Eqs. (12) show the overall performance of the vulnerability model [36]. An AUC value that is close to 1.0 indicates that the model is very good and can classify the flash flood area correctly.

$$AUC = \sum_{i=1}^{n+1} \frac{1}{2} \sqrt{(x_i - x_{i+1})^2 \times (y_i - y_{i+1})} \tag{12}$$

3.4 Estimation of Rain Spatial Distribution by Using IDW

Rainfall is the main factor that causes flash floods. The higher the rainfall intensity, the higher the flooding potential will be. For a study on the causes of flash floods in an area, spatial rain distribution is essential for describing the locations where the flash floods occurred. The spatial rain distribution is estimated by using Inverse Distance Weighting (IDW) method [36]. IDW is a spatial interpolation method suitable for the prediction of rainfall possibility [37]. Besides, estimates of mean regional rainfall using the IDW method are slightly better than the forecast resulting in the other three ways (ANUDEM, Spline, and Kriging). The IDW method is also easier to implement in GIS [38]. IDW interpolation determines cell values using a weighted combination of a set of sample points. The weight is a function of the inverse distance. The data recorded by these devices are used to estimate mean regional rainfall. The study focused on determining flash flood causes. The estimate of spatial rainfall distribution in various times of occurrence (in Table 1) of flash floods occurrences is used to identify the effect of rainfall on the flash floods historically.

4 Result and Discussions

4.1 *Assessment of FFHS from SI, FR, and FR-PR*

In this paper, the potential of the flash floods on the Wonoboyo watershed has been assessed by using SI, FR, and FP-PR a bivariate statistical approach integrated with geospatial techniques. These methods are used to find a correlation between a flash flood in its causes.

In the SI method, all factors are calculated to obtain the resultant weight for each conditioning factor, as shown in Fig. 3. The slope factor in classes 1 and 2 has a positive value, and the lower the class, the stronger the value. TPI in classes 2–4 shows positive values. TWI has positive scores in grades 4–7, and the higher the class, the stronger the relationship. The curvature of the profile with the curvature of the plan has the opposite positive value strength. Aspects have positive values in classes 6, 7, 8, and 10. Convergent in class 1 and 2 has a positive and strong value in class 1. River density indicates that the lower the class, the stronger the relationship. The use of paddy fields and settlements has a strong positive value. The

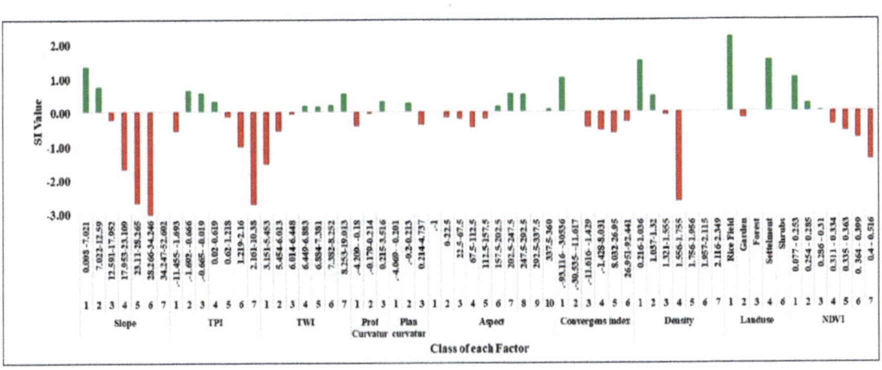

Fig. 3 The computed SI value variations for each class of the conditioning factors

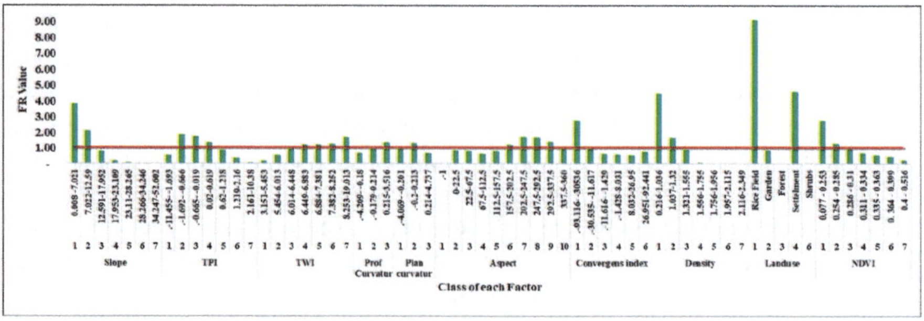

Fig. 4 The computed FR value variations for each class of the conditioning factors

NDVI in classes 1 and 3 had positive scores, and the lower the score, the stronger the relationship.

The FR method applied to measure the prediction capability from selected flash flood driving factors is shown in Fig. 4. Figure 4 shows that the dominant factors that drive flash floods are land use, followed by aspect. The other six other factors (river density, TWI, slope, NDVI, TPI, convergence index) have equal importance. The land-use factor is very closely related to flash flood occurrence. It mainly occurs in the rice fields and settlements class because it has the two highest FR values.

The map in Fig. 5 shows the estimation and identification of flash flood potentials. The map shows that the most susceptible areas to flash floods are the ones with the lowest slopes. The lower the slope, river density, NDVI, and convergence index, the higher the flash flooding potential.

Figure 5 shows a graph of the FR-PR method applied to measure the FP-PR of the selected flash flood driving factors. It shows the dominant factors that cause flash floods are density (3.76), land use (3.73), followed slope (3.26). The other six factors are NDVI (2.14), Convergence index (2.13), TPI (1.59), Prof Curvature (1.38), TWI (1.26), Plan Curvature (1.23), and Aspect. (1.00).

Figure 6 and 7 show a graphical comparison of the hazard levels for each model. The FR and FR-PR models have almost the same hazard pattern at moderate to high

Fig. 5 Flash flood conditioning factors from predictor rate

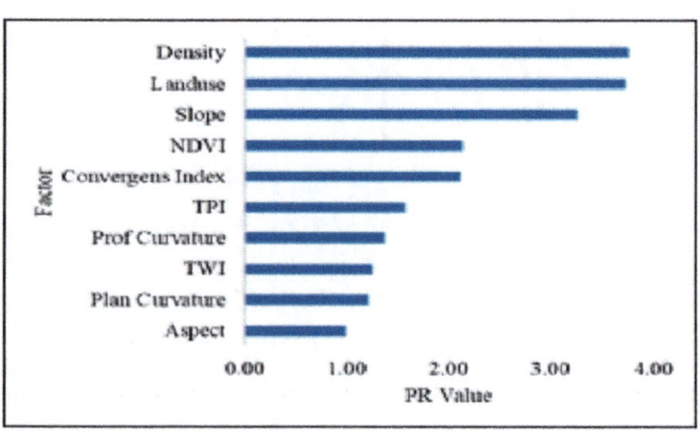

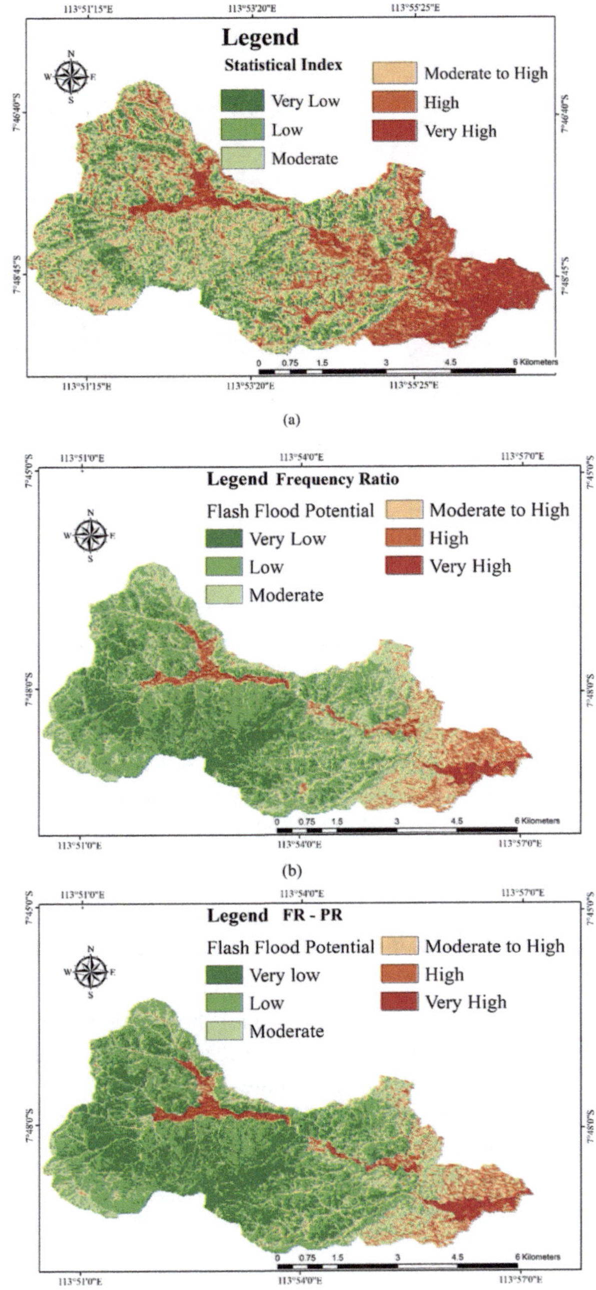

Fig. 6 Map of FFHSI at Wonoboyo Watershed with method: **a** SI, **b** FR, and **c** FR-PR

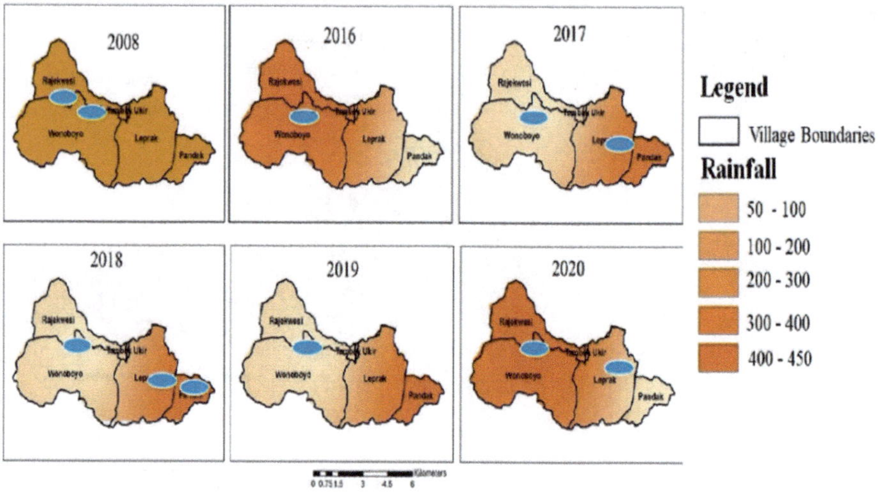

Fig. 7 Comparison of level of hazard between SI, FR, and FP-PR

to very high levels. These two models show that a very low to moderate hazard level has a high percentage value, but has decreased from a moderate to high to very high hazard level. On the other hand, the SI model has small percentage values at very low to moderate hazard levels but shows high percentage values at moderate to high until very high hazard levels downstream of the watershed. In general, from the modeling results, the highest percentage value is found in the FR method with a low hazard level with a value of 38%, while the lowest percentage value is found in the FR method with a very high hazard level with a value of 2%.

4.2 Validation of Flash Flood Hazard Susceptibility Map

The validation of the FFHSI model shows that the model has performed well for the SI model and very well for the FR and FR-PR. The AUC value represents this for success curve of 91.06%, and the AUC value for the prediction curve of SI, FR, and FR-PR is 79.38, 90.75, and 90.40, respectively. FR model is the best model from others. Therefore, the prediction of flash flood hazard susceptibility resulting in this the study can be used as a reference for flood disaster management plans for disaster management and decision makers.

Table 3 The change in land use

Land-use	Area (%)		% Change in land use from 2000 conditions
	2000	2018	
Rice field	4.6	4.7	2.84
Plantation	58.5	58.2	−0.5
Forest	17.4	17.3	−1.06
Settlement	1.6	2.0	19.84
Shrubs	17.8	17.8	0.13

4.3 Effect of Change in Land Uses to Flash Floods Occurrences

The change in land cover is analyzed by using a statistical approach. The land-use data of the year 2000 is used as the baseline, and the data of the year 2018 is used to show the change in land use (Table 3).

Of the total Area at Wonoboyo watershed of 4008 ha, the land uses most considerable proportion is for plantations (58%). The land use between 2000 and 2018 does not show any significant change in the total area ratio. The most significant change in land use was the 19.84% increase in settlement, followed by rice fields' rise and the decrease in plantations and forests. Based on Table 2, there is no significant evidence that the change in land use had conditionend the flash floods. Nevertheless, the rise of rice fields and settlements will increase the surface runoff and the infiltration capacity of an area, hence conditionend flooding.

4.4 The Effect of Rainfall on the Flash Floods Occurrence

Figure 7 shows the distribution pattern of rainfall depth and the distribution of six flash flood historical occurrences in 2008, 2016, 2017, 2018, 2019, and 2020. It shows that the distribution of flash flood occurrences is related to the distribution of the rainfall depths. Based on Table 1, the rainfall depth of 55 mm has conditioned the flash floods (Fig. 8).

The flash flood that was triggered by consecutive rainfall for 3 days, as shown in Table 1, would have been worse if one of the variables had become extreme. Based on this study's results, it is essential to consider factors that affect the occurrence of flash floods to determine the proper preparation and mitigation to minimize the risks of future flash floods.

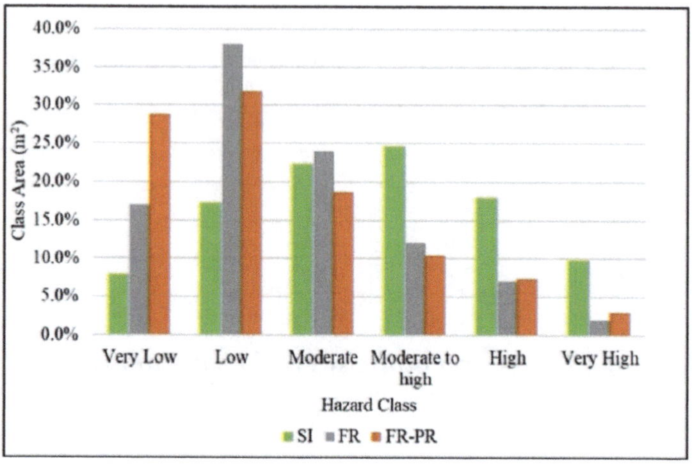

Fig. 8 The rainfall distribution pattern and the flash floods occurrences in Wonoboyo Watershed

5 Conclusions

Based on the comparative analysis of three statistical bivariate methods, the frequency ratio approach is the best method to describe the flash flood potentials, as implied from the AUC model value of 90.75. It is found that land use, the density of river networks, and slope are three dominant factors that conditionend the flash floods. Among the classes in the land-use factor, rice fields and settlement have the most significant effects. As for the factor of the density of the river network, it is found that the denser the river network is, the more susceptible it is to flash floods. Besides, the distribution pattern of rain and the intensity will determine the location of flash floods.

Rain intensity of more than 50 mm per day is the threshold in the occurrence of flash floods. The results of this study can be used as guidelines in the management of watershed spatial planning and steps to minimize the occurrence of flash flood.

References

1. Elkhrachy I (2015) Flash flood hazard mapping using satellite images and GIS tools: a case study of Najran City, Kingdom of Saudi Arabia (KSA). Egypt J Remote Sens Sp Sci 18(2):261–278. https://doi.org/10.1016/j.ejrs.2015.06.007
2. Hapuarachchi HAP, Wang QJ, Pagano TC (2011) A review of advances in flash flood forecasting. Hydrol Process 25(18):2771–2784. https://doi.org/10.1002/hyp.8040
3. Antarafoto (2008) Banjir Desa Wonoboyo, antarafoto
4. BPBD (2016) BPBD Jatim Bantu Korban Banjir dan Tanah Longsor di', BPBD. https://web.bpbd.jatimprov.go.id/2016/02/01/bpbd-jatim-bantu-korban-banjir-dan-tanah-longsor-di-bondowoso/. Accessed Feb 27 2021
5. Beritaekspress (2017) Akibat Banjir, Akses ke Desa Wonoboyo Jatim Lumpuh, beritaekspres

6. Terbitannews (2020) Akibat Hujan Deras, Banjir dan Longsor Terjadi di Desa Leprak Bondowoso, Terbitannews
7. Faktualnews (2018) 3 Kecamatan di Bondowoso Diterjang Banjir Bandang, faktualnews. https://faktualnews.co/2018/01/28/3-kecamatan-bondowoso-diterjang-banjir-bandang/60730/. Accessed 27 Feb 2021
8. Gaume E, Borga M (2008) Post-flood field investigations in upland catchments after major flash floods: proposal of a methodology and illustrations. J Flood Risk Manag 1(4):175–189. https://doi.org/10.1111/j.1753-318x.2008.00023.x
9. Cao C, Xu P, Wang Y, Chen J, Zheng L, Niu C (2016) Flash flood hazard susceptibility mapping using frequency ratio and statistical index methods in coalmine subsidence areas. Sustainability 8(9) (2016). https://doi.org/10.3390/su8090948
10. Pradhan (2009) Flood susceptible mapping and risk area delineation using logistic regression, GIS and remote sensing. J Spat Hydrol 9(2):1–18
11. Jodar-Abellan A, Valdes-Abellan J, Pla C, Gomariz-Castillo F (2019) Impact of land use changes on flash flood prediction using a sub-daily SWAT model in five Mediterranean ungauged watersheds (SE Spain). Sci Total Environ 657:1578–1591. https://doi.org/10.1016/j.scitotenv.2018.12.034
12. Costache R, Hong H, Pham (2020) QB (711) Comparative assessment of the flash-flood potential within small mountain catchments using bivariate statistics and their novel hybrid integration with machine learning models. Sci Total Environ 711:134514. https://doi.org/10.1016/j.scitotenv.2019.134514
13. Bui DT, Tsangaratos P, Ngo PTT, Pham TD, Pham BT (2019) Flash flood susceptibility modeling using an optimized fuzzy rule based feature selection technique and tree based ensemble methods. Sci Total Environ 668:1038–1054. https://doi.org/10.1016/j.scitotenv.2019.02.422
14. Popa MC et al (2020) Spatial assessment of flash-flood vulnerability in the Moldova river catchment (N Romania) using the FFPI. J Flood Risk Manag 2019:1–10. https://doi.org/10.1111/jfr3.12624
15. Khosravi K et al (2018) A comparative assessment of decision trees algorithms for flash flood susceptibility modeling at Haraz watershed, Northern Iran. Sci Total Environ 627:744–755. https://doi.org/10.1016/j.scitotenv.2018.01.266
16. Lappas I, Kallioras A (2019) Flood susceptibility assessment through GIS-based multi-criteria approach and analytical hierarchy process (AHP) in a River Basin in Central Greece, IRJET, pp 738–751
17. Stefanidis S, Stathis D (2013) Assessment of flood hazard based on natural and anthropogenic factors using analytic hierarchy process (AHP). Nat Hazards 68(2):569–585. https://doi.org/10.1007/s11069-013-0639-5
18. Kazakis N, Kougias I, Patsialis T (2015) Assessment of flood hazard areas at a regional scale using an index-based approach and analytical hierarchy process: application in Rhodope-Evros region, Greece. Sci Total Environ 538:555–563. https://doi.org/10.1016/j.scitotenv.2015.08.055
19. Costache R (2019) Flash-flood potential assessment in the upper and middle sector of Prahova river catchment (Romania). A comparative approach between four hybrid models. Sci Total Environ 659:1115–1134. https://doi.org/10.1016/j.scitotenv.2018.12.397
20. Khosravi K, Nohani E, Maroufinia E, Pourghasemi HR (2016) A GIS-based flood susceptibility assessment and its mapping in Iran: a comparison between frequency ratio and weights-of-evidence bivariate statistical models with multi-criteria decision-making technique. Nat Hazards 83(2):947–987. https://doi.org/10.1007/s11069-016-2357-2
21. Khosravi K, Pourghasemi HR (2016) Flash flood susceptibility analysis and its mapping using different bivariate models in Iran: a comparison between Shannon's entropy, statistical index, and weighting factor models. Environ Monit Assess. https://doi.org/10.1007/s10661-016-5665-9

22. Costache R, Zaharia L (2017) Flash-flood potential assessment and mapping by integrating the weights-of-evidence and frequency ratio statistical methods in GIS environment—case study. Bâsca Chiojdului River Catchment (Romania). J Earth Syst Sci 126(4):1–19. https://doi.org/10.1007/s12040-017-0828-9
23. Efendi D, Hidayah E, Hasanuddin A (2020) U KaRsT, Mapping of landslide susceptible zones by using frequency. Ratios Bluncong 5(1):126–141. https://doi.org/10.30737/ukarst.v3i2
24. Tehrany MS, Pradhan B, Jebur MN (2015) Flood susceptibility analysis and its verification using a novel ensemble support vector machine and frequency ratio method. Stoch Environ Res Risk Assess 29(4):1149–1165. https://doi.org/10.1007/s00477-015-1021-9
25. Beven K (1984) Infiltration into a class of vertically non-uniform soils, Hydrol Sci J 29(4):425–434. https://doi.org/10.1080/02626668409490960
26. Manfreda S, Di Leo M, Sole A (2011) Detection of flood-prone areas using digital elevation models. J Hydrol Eng 16(10):781–790. https://doi.org/10.1061/(asce)he.1943-5584.0000367
27. Duman TY, Can T, Gokceoglu C, Nefeslioglu HA, Sonmez H (2006) Application of logistic regression for landslide susceptibility zoning of Cekmece Area, Istanbul, Turkey. Environ Geol 51(2):241–256. https://doi.org/10.1007/s00254-006-0322-1
28. Fernández DS, Lutz MA (2010) Urban flood Hazard zoning In Tucumán Province, Argentina. Using GIS and multicriteria decision analysis. Eng Geol 111(1–4):90–98. https://doi.org/10.1016/j.enggeo.2009.12.006
29. Glenn EP, Morino K, Nagler PL, Murray RS, Pearlstein S, Hultine KR (2012) Roles of Saltcedar (Tamarix spp.) and capillary rise in salinizing a non-flooding terrace on a flow-regulated Desert River. J Arid Environ 79:56–65. https://doi.org/10.1016/j.jaridenv.2011.11.025
30. Falahnsia AR (2014) Vegetasi dengan metode skoring menggunakan citra satelit di, pp 400–416
31. Ullah K, Zhang J (2020) GIS-based flood hazard mapping using relative frequency ratio method: a case study of Panjkora River Basin, Eastern Hindu Kush, Pakistan. PLoS One 15(3):1–18. https://doi.org/10.1371/journal.pone.0229153
32. Ullah K, Zhang J, Id JZ (2020) GIS-based flood hazard mapping using relative frequency ratio method: a case study of Panjkora River Basin, Eastern Hindu Kush, Pakistan. PLoS 15(3):1–18. https://doi.org/10.1371/journal.pone.0229153
33. Abdulwahid WM, Pradhan B (2017) Landslide vulnerability and risk assessment for multi-hazard scenarios using airborne laser scanning data (LiDAR). Landslides 14(3):1057–1076. https://doi.org/10.1007/s10346-016-0744-0
34. Van Westen CJ, Rengers N, Terlien M (1997) Prediction of the occurrence of slope instability phenomena through GIS-based hazard zonation. https://doi.org/10.1007/s005310050149
35. Regmi AD et al (2014) Application of frequency ratio, statistical index, and weights-of-evidence models and their comparison in landslide susceptibility mapping in Central Nepal Himalaya. Arab J Geosci 7(2):725–742. https://doi.org/10.1007/s12517-012-0807-z
36. Bartier PM, Keller CP (1996) Multivariate interpolation to incorporate thematic surface data using inverse distance weighting (IDW). Comput Geosci 22(7):795–799. https://doi.org/10.1016/0098-3004(96)00021-0
37. Chen FW, Liu CW (2012) Estimation of the spatial rainfall distribution using inverse distance weighting (IDW) in the middle of Taiwan. Paddy Water Environ 10(3):209–222. https://doi.org/10.1007/s10333-012-0319-1
38. Yang X, Xie X, Liu DL, Ji F, Wang L (2015) Spatial interpolation of daily rainfall data for local climate impact assessment over Greater Sydney Region. Adv Meteorol. https://doi.org/10.1155/2015/563629

Flood Management Strategies in Indonesia: A Lesson Learned from Pepe River, Central Java

Rian Mantasa Salve Prastica and Amalia Wijayanti

Abstract Floods occur every year and threaten communities worldwide. Indonesia is among the countries faced with this problem and still struggling to seek practical solutions to date. Further, as the country's centers of trade, business, and technology, the island of Java cannot even perform better in dealing with floods. To further understand possible damage and losses, this research investigated structural mitigation measures as part of flood management in the Pepe River, notably river channelization to increase hydraulic efficiency. For the analysis, it used HEC-RAS to observe the river capacity. Existing conditions revealed that the river could not accommodate the predicted floods based on the return periods derived from the synthetic unit hydrograph of Nakayasu. However, after running the mitigation scenario wherein hydraulic modification was applied, flood risk could be reduced by lowering the river water level. Based on the results, it can be concluded that Indonesia needs to pay more attention to other aspects: ecology, green infrastructure, and water governance, to achieve more effective flood mitigation measures because structural mitigation only provides a temporary solution.

Keywords Flood risk · Disaster mitigation · Operation and maintenance

1 Introduction

Floods are natural disasters that often occur in Indonesia [1–3]. Some of its watersheds are degraded and, hence, experiencing critical problems [4]. In some areas

R. M. S. Prastica (✉) · A. Wijayanti
Civil Engineering Department, Gadjah Mada University, Yogyakarta 55281, Indonesia
e-mail: rian.mantasa.s.p@ugm.ac.id; rian.mantasa@alumni.ui.ac.id

A. Wijayanti
e-mail: amaliawijayanti@mail.ugm.ac.id

R. M. S. Prastica
Water Infrastructure Construction Technology Study Program, Public Works Polytechnic, Semarang 50275, Indonesia

S. A. Kristiawan et al. (eds.), *Proceedings of the 5th International Conference on Rehabilitation and Maintenance in Civil Engineering*, Lecture Notes in Civil Engineering 225,
https://doi.org/10.1007/978-981-16-9348-9_50

with low-lying and level terrains on Java Island, floods persistently become the main problem during the rainy season [5, 6]. Because of these issues, various water infrastructures must deal with diverse challenges, such as the decreasing performance of rivers, dams, and other flood-defense structures. Some scholars argue that the island's water management has shown signs of improvements [6–8] in various aspects. However, evidence indicates that some watersheds suffer hydrological and hydraulic problems [4, 8].

There are many mutually sustainable components in a watershed, including infrastructure, biotic and abiotic ecosystems, water quality [9], and humans. Therefore, watershed management must consider their safety, preservation, and sustainability. Further, characteristics and problems vary across watersheds, meaning that the decision-making process to address what appears to be the same problem needs to consider their respective hydrological principles, local wisdom, and ecosystems. Finally, infrastructure operations and maintenance in the watershed are crucial for reducing susceptibility to hazards. The infrastructure in question includes bridges, rivers, irrigation networks [10], drainage channels [11], reservoirs, and dams.

As an essential component of a watershed [12], the river is a medium that stores, retains, and transports water and serves as a drainage network connecting springs to estuaries within an area defined by topographic boundaries. The river is a complex ecosystem wherein water, sediments, and abiotic components exist and interact. Due to the complexity, river management must operate on a principle that sees the water body as a whole from the upper to the lower reaches and proceed in a sustainable manner. If met accordingly, these terms can help solve problems that are naturally inseparable from the river, e.g., flooding in level to gently sloping areas.

Flood is a natural phenomenon that can occur as a result of decreased infiltration capacity [13]. Increasing frequency and dimensions of flood occurrences are signs of watershed degradation, which entails infiltration capacity reduction and higher surface flow coefficient. Therefore, flood and watershed capacity analyses are necessary so as to capture trends based on historical data and predict the expected probability and frequency of occurrences according to the return period. For this reason, the research conducted rainfall data analysis, prediction of rain intensity at predefined return periods, frequency test, and hydrological analysis to calculate the river discharge according to the rainfall return period. Then, the calculated flood occurrences based on the river discharge were inputted to a model designed to determine predicted channel capacity. In Indonesia, floods mostly strike areas with small and flat slopes and inadequate river channel planning and management. Therefore, this study used HEC-RAS software to predict whether or not the channel could accommodate flood discharge at a certain rain return period. If the model shows that the existing channel cannot accommodate the calculated flood discharge, then it is necessary to redesign the observed channel [14].

This study was located in Pepe River, Pepe Subwatershed, Central Java, Indonesia, as depicted in Fig. 1. According to the above issues, the research aimed to investigate the river's hydraulic properties, particularly channel capacity. In addition, it evaluated existing mitigation measures to tackle floods in the Pepe River in particular and Indonesia in general. There are three specific questions answered in the research: Is

Fig. 1 The location of the Pepe Subwatershed

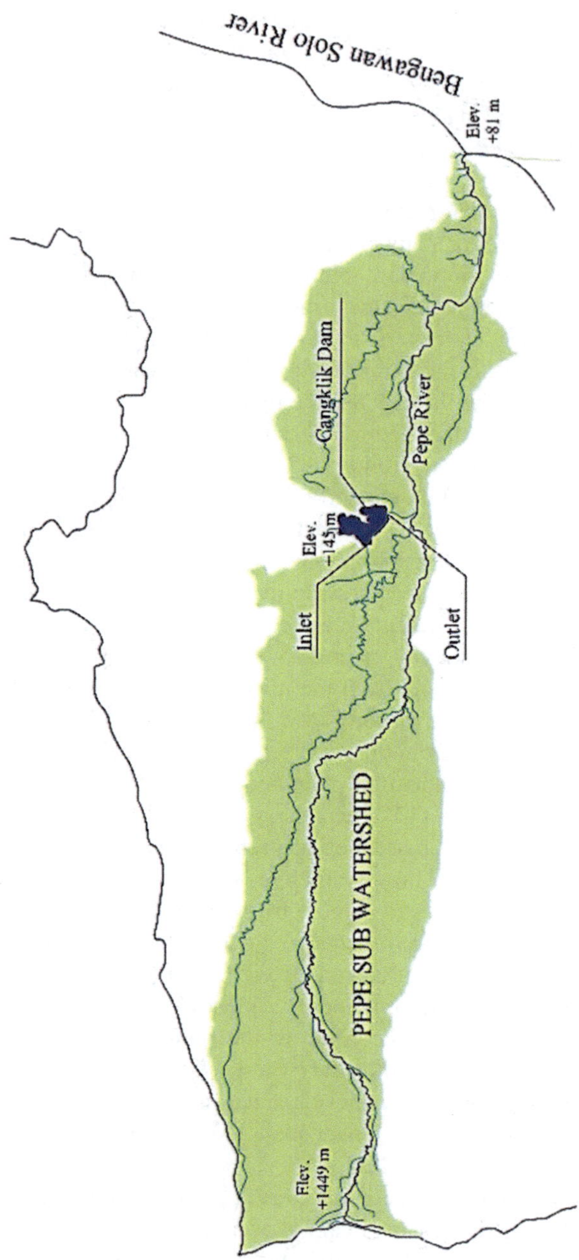

structural mitigation sufficient to prevent flooding? Are there other feasible solutions to sustaining the mitigation efforts? What causes the difficulties in flood mitigation in the current situation?

2 Method

The rainfall data were collected from the rain gauge stations distributed in the Pepe Subwatershed and then processed using the Thiessen polygon method [15] to obtain maximum daily rainfall. Afterward, the frequency analysis of maximum daily rainfall for various return periods (i.e., 5, 10, 25, 50, 100, 500, and 1000-years) was conducted using four statistical techniques: Normal Distribution, Gumbel Distribution, Log-Normal Distribution, and Log-Pearson Type III Distribution. The data frequency was tested for validity using the chi-square and Kolmogorov–Smirnov goodness-of-fit tests. Then, the Nakayasu method was employed to calculate the flood discharge at each predetermined return period [2].

Then, to model the existing river hydraulics, the limiting conditions for flood occurrences were set in HEC-RAS software, such as the flood events at the seven return periods. The flood discharges were denoted by Q_5, Q_{10}, Q_{25}, Q_{50}, Q_{100}, Q_{500}, and Q_{1000}.

Modeling with HEC-RAS involves several steps. First, the existing condition was analyzed and observed. If the results show an overflow, a new river channelization design (i.e., river channel dimensions) is analyzed as the selected structural mitigation scenario. The hydraulic design proposed in the research used the flood discharge at the 100-year return period (Q_{100}). Second, the model built in HEC-RAS was re-run with the new dimensions derived from the latest hydraulic analysis to observe the river flow condition at different return periods.

In HEC-RAS software [16], the flow simulation occurs in an open channel so that users can study the flow pattern along the channel. On a laboratory scale, simulations factor in actual conditions by flowing water into a channel in a physical model. Virtually, hydraulic numerical simulations can also be calculated using a computer application software with a mathematics-based model. Mathematical models mimic the physical phenomena of river flows in an actual or prototype channel using a mathematical equation describing the relationship between flow variables: geometric, kinematic, and dynamic variables. When the river flows are measured and observed from a physical model, the flow parameters can be obtained by calculating or solving the mathematical equation in the model.

3 Results and Discussion

The Cengklik Dam is located in Margorejo Village (Ngemplak District, Boyolali, Central Java). It was built in 1923–1931 during the Dutch colonization and designed and constructed under the Dutch East Indies Government to irrigate several subdistricts in Boyolali Regency. The Bengawan Solo River Basin Center currently manages the Cengklik Dam under the Directorate General of Water Resources, the Ministry of Public Works and Housing.

The three rain gauge stations closest to the Pepe Subwatershed were Sleman Geophysical Station, Semarang Meteorological Station, and Nganjuk Geophysical Station, as presented in Fig. 2. The rainfall data recorded at these stations were obtained from dataonline.bmkg.go.id and the Bengawan Solo River Basin Center. The Thiessen polygon was created based on the station's geographical position, dividing the into areas affected by the rainfall data at each station (in %). Flood occurrences predicted from the rain return period are depicted in Fig. 3. The analysis used the Nakayasu method to produce all predicted flood occurrences for further analysis.

According to the flow simulation in HEC-RAS software, all stations of the river cannot accommodate the design flood in all return period. Figure 4 describes the water level of the river by using several data of return period in STA 15 + 700 of the location samples.

Based on the model of the existing river conditions in HEC-RAS, it was found that the entire river overflowed and could not accommodate the 10, 25, 50, and 100-years flood events. Therefore, it is necessary to redesign the channel dimensions.

Structural modification is a popular flood mitigation measure in Indonesia. Government and policy-makers often decide to perform channelization by enlarging the river dimension to increase its capacity to accommodate floods of varying intensity and magnitude. Hence, this research factors in channelization strategies as the selected structural mitigation scenarios in the flood simulation. The new HEC-RAS simulation, as depicted in Fig. 5, revealed that the river could accommodate the predicted floods even without significant runoffs appearing in the simulation result.

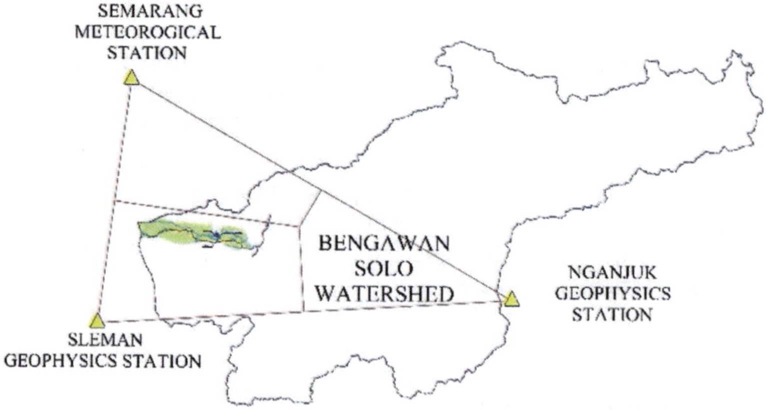

Fig. 2 Illustration of the Thiessen method used to calculate areal mean rainfall in Pepe Subwatershed

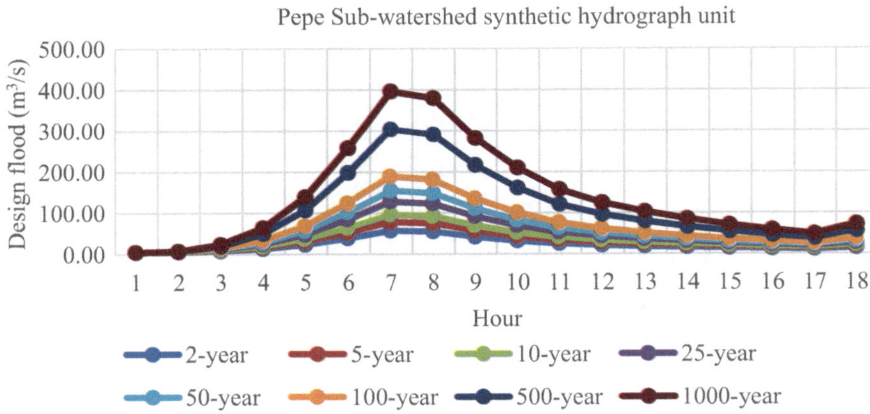

Fig. 3 Nakayasu's synthetic unit hydrograph for flood occurrences predicted for each rain return period in Pepe Subwatershed

However, some researchers argue that structural flood mitigation has many weaknesses, especially related to sustainability [17], ecological aspect [12], and prospects of economic benefits for the society [18]. For instance, water-sensitive urban design, or WSUD for short, is currently the leading topic discussed among urban management researchers and practitioners as a flood management strategy. It is also believed to be a potential solution to flood issues in Indonesia.

Over the last decades, water-related disasters have stolen the spotlights on various political, social, and scientific stages due to their impacts on many sectors and are often related to accelerating climate change dynamics and urban sprawl. As a leading center for urban water research, Australia has successfully established a new research field focusing on WSUD and waves a flag of new topic discovery worldwide. WSUD brings about improved environmental development in terms of water quantity and quality [19]. The United Kingdom, China, the Netherland, and several other countries prone to water-related disasters have adopted this concept and can offer their residents promising results to deal with urban stormwater management problems in the future. However, some countries claim that the concept cannot be implemented. Despite this negative result, the question remains: Can the WSUD concept be applied to solving flood issues in Indonesia? Some scholars argue that Indonesia's water system does not meet the conditions of the novel approach and that the WSUD concept does not encourage integrated water resource management. However, there are several reasons for Indonesia to implement WSUD effectively to manage water quantity and quality.

Green infrastructure, or GI, is one of the elements of WSUD. Many Indonesian regions have started introducing and applying the GI concept, especially since the infrastructures are widely available and easy to construct. In previous studies, it is mentioned that urban settings have the infrastructure arrangement but not proper installation, such as in the case of constructed wetlands [20], green roofs [21],

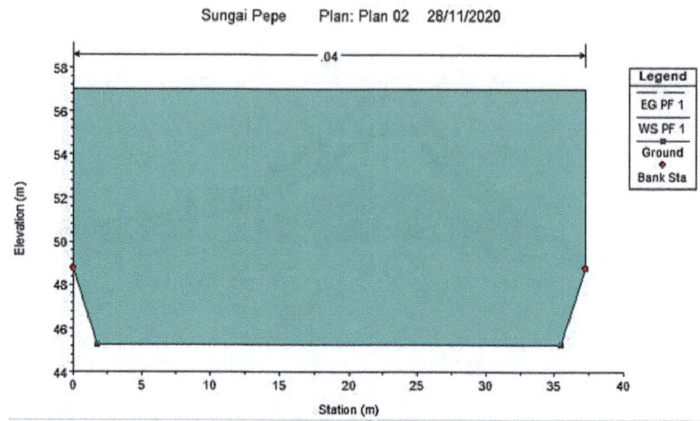

(a)

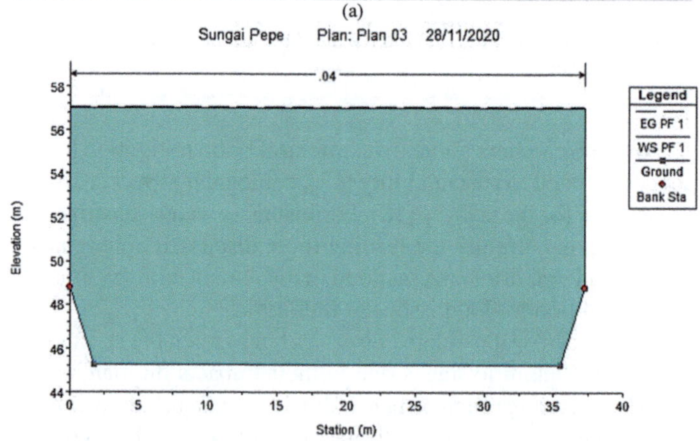

(b)

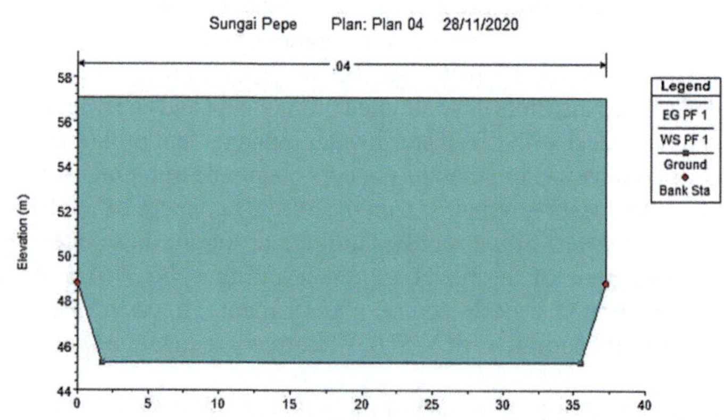

(c)

Fig. 4 Pepe River's flow simulation in HEC-RAS software for **a** 10-year, **b** 25-year, **c** 50-year and **d** 100-year floods

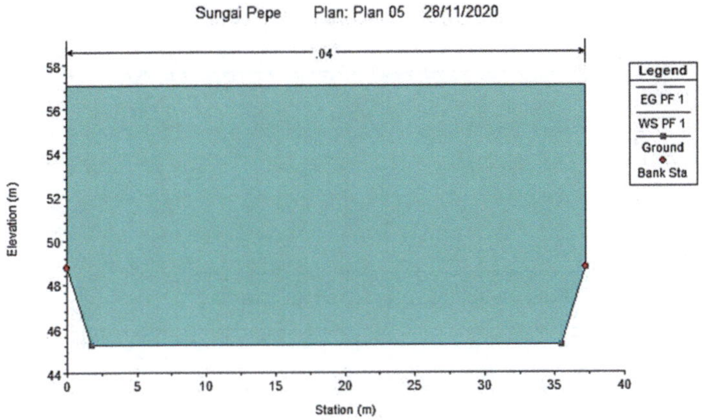

(d)

Fig. 4 (continued)

and rainwater harvesting [22]. Recently, society has developed awareness of the wide availability of GI in its surroundings and installed some systems personally without supervision from the experts. To tackle this issue, the government has legislated GI installation procedure standards as community guidelines. The GI installation benefits society in many ways. First, materials are easily found and locally sourced. Second, locally available green materials are less expensive than grey infrastructure. However, even though the installment procedures have been published, some green infrastructures are still built without following the measurement standards. In many cases, people decide to install large-sized rainwater harvesting tanks without factoring in mathematical concepts. Also, the product manuals or installation guidelines sometimes provide complicated measurements that should otherwise be translated to layman's terms and encourage time efficiency.

Besides structural mitigation and green infrastructure, river restoration has been a higher priority for most regional governments in today's urban areas [23, 24]. As a result of urban sprawl and rapid population growth [25], urban areas face a never-ending challenge: riverine floods. On the other hand, rivers play a pivotal role in rural and urban populations [26]. Farmers depend on river water for irrigation [10]; an unreliable river system can disturb water irrigation flows and decrease agricultural production. Rivers undoubtedly provide water for daily use, sustain aquatic habitats, and maintain the food chain in the environment. Without the nutrients provided by rivers, the population of the aquatic habitat will deteriorate [27]. This is possible when the river water quality is declining [28], partly due to human activities that adversely affect the water [29].

At the same time, Indonesia still lacks adequate water quantity management. Challenges often arise from inconsistent coordination between governmental agencies in conducting integrated flood management [3]. Similarly, joint efforts between the responsible ministries and local communities have yet to enforce strict rules and effective measures. These issues can lead to overlapping organizational functions

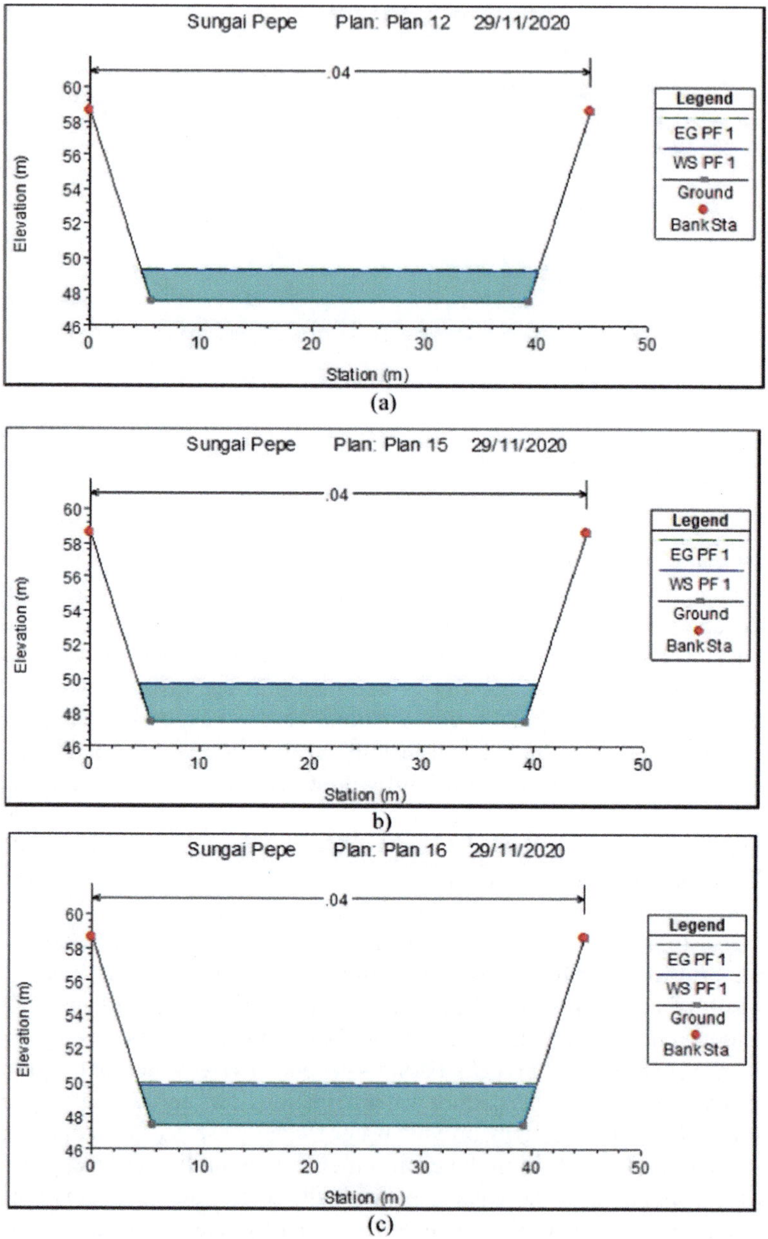

Fig. 5 Pepe River's flow simulation in HEC-RAS software (after redesigning the dimensions) for **a** 10-year, **b** 25-year, **c** 50-year and **d** 100-year floods

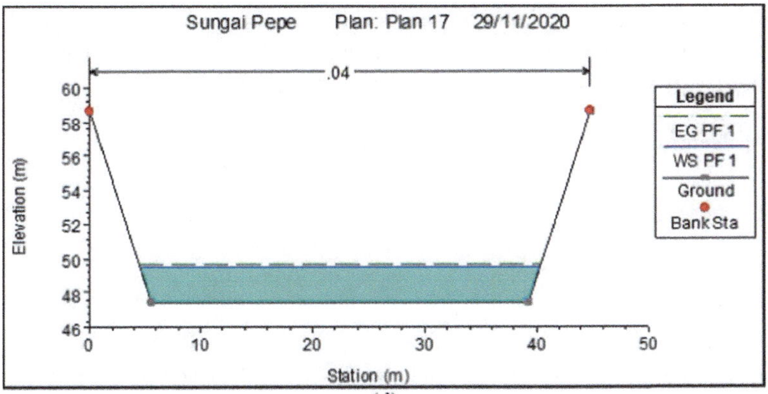

(d)

Fig. 5 (continued)

and programs and cause some measures to cease, receive no sufficient follow-ups, and be implemented partially [8]. An example includes a flood prevention infrastructure in Tegal (Central Java, Indonesia), which was destroyed by the society because, instead of protecting the affected areas, it worsened the flood impacts and material losses. This issue indicates the lack of well-formulated coordination between central and provincial governmental agencies. Even though poor responses to water-related disasters are the most pressing problem, the stakeholders involved (i.e., local, regional, and national authorities) still cannot agree on a well-synchronized communication during and after the disaster.

To sum up, structural mitigation strategies like channelization can entail an enormous change to flood mitigation, which may raise negative responses from varying parties. Another perspective might include providing rooms for other solutions, such as green infrastructure, river restoration, and legal coordination, and making them priorities in the management of the Pepe River and many rivers in Indonesia. Even though structural mitigation can tackle floods, this solution offers a temporary result, especially when it fails to integrate other aspects into its implementation.

4 Conclusion

Based on the flood and mitigation scenario in this research, structural mitigation proves effective in preventing floods. Before a selected measure (i.e., channelization) is introduced, the Pepe River's upper, middle, and lower reaches cannot accommodate the floods predicted from the rain return period. Redesigning channel dimensions leads to increased capacity that allows the river to accommodate said floods. However, the proposed mitigation appears to have negative effects, especially when other factors are not considered in the implementation. Therefore, other existing options like green infrastructure, community-based river restoration, and water governance need to be further developed to achieve sustainable flood mitigation in the Pepe River and other rivers with similar characteristics in Indonesia.

Acknowledgements The research grant funding this study was provided by the Directorate of Research, Universitas Gadjah Mada, Number 2458/UN1/DITLIT/DIT-LIT/PT/2021.

References

1. Prastica RMS, Apriatresnayanto R, Marthanty DR (2019) Structural and green infrastructure mitigation alternatives prevent Ciliwung River from water-related landslide. Int J Adv Sci Eng Inf Technol 9(6):1825–1832
2. Prastica RMS, Maitri C, Hermawan A, Nugroho PC, Sutjiningsih D, Anggraheni E (2018) Estimating design flood and HEC-RAS modelling approach for flood analysis in Bojonegoro city. In: IOP Conference series: materials science and engineering, vol 316(1)
3. Lin E, Shaad K, Girot C (2016) Developing river rehabilitation scenarios by integrating landscape and hydrodynamic modeling for the Ciliwung River in Jakarta, Indonesia. Sustain Cities Soc 20:180–198
4. Kefi M, Mishra BK, Masago Y, Fukushi K (2020) Analysis of flood damage and influencing factors in urban catchments: case studies in Manila, Philippines, and Jakarta, Indonesia. Nat Hazards 104(3):2461–2487
5. Wedawatta G, Kulatunga U, Amaratunga D, Parvez A (2016) Disaster risk reduction infrastructure requirements for South-Western Bangladesh: perspectives of local communities. Built Environ Proj Asset Manag 6(4):379–390
6. Kim J, Gim THT (2020) Assessment of social vulnerability to floods on Java, Indonesia. Nat Hazards 102(1):101–114
7. Marfai MA et al (2008) Natural hazards in Central Java Province, Indonesia: an overview. Environ Geol 56(2):335–351
8. Marfai MA, Sekaranom AB, Ward P (2015) Community responses and adaptation strategies toward flood hazard in Jakarta, Indonesia. Nat Hazards 75(2):1127–1144
9. Prastica RMS (2018) Employing Galerkin method and field study for model calibration to perform sediment transport modelling in Agathis Lake Universitas Indonesia. In: AIP conference proceedings, vol 2014
10. Lindenschmidt KE, Carr MK, Sadeghian A, Morales-Marin L (2019) CE-QUAL-W2 model of dam outflow elevation impact on temperature, dissolved oxygen and nutrients in a reservoir. Sci Data 6(1):312
11. Siqueira RC Moural PM, das Graças Silval TF (2019) Methodology for the construction of an urban flood hazard chart. Rev Bras Recur Hidricos 24:1–16
12. Talbot CJ et al (2018) The impact of flooding on aquatic ecosystem services. Biogeochemistry 141(3):439–461
13. Prastica RMS, Wicaksono D (2019) Integrated multimodal disaster mitigation management for urban areas: a preliminary study for 2-d flood modeling. In: IOP conference series: materials science and engineering, vol 650(1)
14. Prastica RMS, Maitri C, Nugroho PC, Hermawan A (2017) Analisis Banjir dan Perencanaan Desain Transportasi Sungai di Kota Bojonegoro. Media Komun Tek Sipil 23(2):91
15. Prastica RMS, Adi DAR, Famila N (2020) Mitigasi banjir dan alternatif pemeliharaan infrastruktur keairan pada sub-DAS code Yogyakarta. Tek J Sains dan Teknol 16(1):25–33
16. Geravand F, Hosseini, SM, Ataie-Ashtiani B (2020) Influence of river cross-section data resolution on flood inundation modeling: Case study of Kashkan river basin in western Iran. J Hydrol 584(Dec 2019):124743

17. Chan SL, Wey WM, Chang PH (2014) Establishing disaster resilience indicators for Tan-Sui river Basin in Taiwan 115(1)
18. De Vincenzo A, Jacopo Molino A, Molino B, Scorpio V (2017) Reservoir rehabilitation: the new methodological approach of economic environmental defence. Int J Sed Res 32(2):288–294
19. Kuller M, Bach PM, Ramirez-Lovering D, Deletic A (2017) Framing water sensitive urban design as part of the urban form: a critical review of tools for best planning practice. Environ Model Softw 96:265–282
20. Sheng X et al (2020) Management of rural domestic wastewater in a city of Yangtze delta region: performance and remaining challenges. Bioresour Technol Rep 11(Jul):100507
21. Liu C, Li Y (2016) Measuring eco-roof mitigation on flash floods via GIS simulation. Built Environ Proj Asset Manag 6(4):415–427
22. Zingiro A, Okello JJ, Guthiga PM (2014) Assessment of adoption and impact of rainwater harvesting technologies on rural farm household income: the case of rainwater harvesting ponds in Rwanda. Environ Dev Sustain 16(6):1281–1298
23. Setyoasri YP, Prastica RMS (2020) Rapid assessment of river watershed health and vulnerability level for restoration strategy: a study of river systems in Indramayu West Java, Indonesia. In: IOP conference series: earth and environmental science, vol 423(1)
24. Prastica RMS, Kurniawan RP, Apriatresnayanto R (2020) Environmental vulnerability assessment and stormwater management to enhance watershed performance. J Phys Conf Ser 1625(1)
25. Machado RAS, Oliveira AG, Lois-González RC (2019) Urban ecological infrastructure: the importance of vegetation cover in the control of floods and landslides in Salvador/Bahia, Brazil. Land use policy 89(Oct):104180
26. Kozlowski M, Yusof YM (2016) The role of urban planning and design in responding to climate change: the Brisbane experience. Int J Clim Chang Strateg Manag 8(1):80–95
27. Geist J, Hawkins SJ (2016) Habitat recovery and restoration in aquatic ecosystems: current progress and future challenges. Aquat Conserv Mar Freshw Ecosyst 26(5):942–962
28. Björklund K, Bondelind M, Karlsson A, Karlsson D, Sokolova E (2018) Hydrodynamic modelling of the influence of stormwater and combined sewer overflows on receiving water quality: benzo(a)pyrene and copper risks to recreational water. J Environ Manage 207:32–42
29. Tedford M, Ellison JC (2018) Analysis of river rehabilitation success pipers river Tasmania. Ecol Indic 91(Sep 2017):350–358

Comparison of Suitable Drought Indices for Over West Nusa Tenggara

Humairo Saidah, Heri Sulistiyono, and I Dewa Gede Jaya Negara

Abstract The drought has had a significant impact on West Nusa Tenggara. Drought experienced in some areas almost every year, causing crop failure and even causing famine in 1954 and 1966. This study aims to find the best method for obtaining the meteorological drought index because it is essential to understand the drought characteristics in the study area in guiding policymakers in early anticipation of this region's drought hazard. This study compares two drought index methods, SPI and PDSI, and tests their proximity to drought data in these locations. Evaluation is carried out to determine the best approach to choose for conducting a drought assessment. The best-fitted method was determined statistically by counting the number of dry months produced by the model and comparing it with the drought experienced over West Nusa Tenggara. The results obtained are the SPI method has an accuracy rate of 58% and a correlation coefficient r about 0.06. Meanwhile, the PDSI method has a better accuracy rate of 75% and a correlation coefficient r about 0.51. The PDSI method is also the most superior in accurately predicting the arrival of a dry month compared to SPI. So, it can be concluded that the PDSI method is the better method for evaluating and detecting dry periods in West Nusa Tenggara.

Keywords SPI · PDSI · Drought index

1 Introduction

Drought is one of the phenomena that occur due to seasonal circulation under the global climate's influence, characterized by water availability far from water

H. Saidah (✉) · H. Sulistiyono · I. D. G. J. Negara
Department of Civil Engineering, University of Mataram, Mataram, Indonesia
e-mail: h.saidah@unram.ac.id

H. Sulistiyono
e-mail: h.sulistiyono@unram.ac.id

I. D. G. J. Negara
e-mail: jayanegara@unram.ac.id

S. A. Kristiawan et al. (eds.), *Proceedings of the 5th International Conference on Rehabilitation and Maintenance in Civil Engineering*, Lecture Notes in Civil Engineering 225,
https://doi.org/10.1007/978-981-16-9348-9_51

needs for living, agriculture, and other activities. The impact of drought events is widely recognized. Drought is directly related to public health problems, usually resulted from poor sanitation to spread disease, and can widely lead to poverty and underdevelopment of an area.

Drought is a normal phenomenon, a recurrent feature of climate variability. It differs from aridity, which is described to low rainfall regions permanently as a feature of the site climate condition. Drought is a major natural hazard of nature in almost all climate zone types, although it varies significantly from one region to another spatially and temporally [1, 2]. Drought is an abnormal period of dry weather sufficiently prolonged caused by a lack of precipitation that causes hydrological imbalance and moisture deficiency in water use requirements [3].

Drought analysis is needed to support agricultural activities towards food security. Agriculture, especially rice, is the largest water consumer, which is very prone to drought. Calculation of drought analysis usually produces an index that describes the severity of the drought. Drought is classified into three categories, namely meteorological drought, agricultural drought, and hydrological drought. Meteorological drought is related to the amount of rainfall that occurs under normal conditions during a season. Agricultural dryness is associated with reduced water content in the soil (soil moisture), and hydrological dryness is related to reduced water supply from the earth.

Meteorological drought describes reduced rainfall which is the first indication of drought conditions. There are many analytical methods for detecting this meteorological drought with various advantages and disadvantages. Among the widely used methods are SPI (Standardized Precipitation Index), PDSI (Palmer Drought Severity Index), Decile Index, PNI (Precipitation Normal Index), WSI (surface water supply index), RDI (Reconnaissance Drought Index), Standardized Precipitation Evapotranspiration Index (SPEI) and others [4–9].

This study chooses two methods, namely SPI and PDSI, to be applied in the study location. The SPI model has the advantage of being simple. It only requires rainfall input and it can determine the severity of drought by measuring the rainfall deficit at various periods based on normal conditions [4, 5]. However, the use of SPI model in characterizing drought level should be done with caution [10]. The PDSI model is commonly used to evaluate droughts that have occurred, especially in semiarid areas and dry sub-humid climates [11, 12], where West Nusa Tenggara is a semiarid area. The PDSI model is also a standardization for all country's local climates to show relative drought or rainfall conditions [13]. The PDSI method can also assess and predict drought [14].

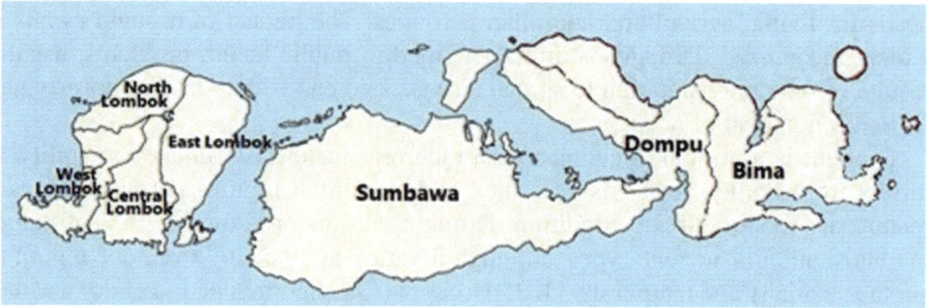

Fig. 1 West Nusa Tenggara region [15]

2 Materials and Method

2.1 Study Area

The research conducted on the two largest islands in West Nusa Tenggara Province, namely Lombok and Sumbawa islands, located at coordinates 8°10'–9°5' latitude and 115°46'–119°5' Longitude, which has a total area of around 20.164.84 km^2 (see Fig. 1) [15].

2.2 Data Availability

The data needed for the SPI model is rainfall, while the input data for the PDSI model is rainfall, evapotranspiration, soil type, soil structure, and land use. The primary data for this study are the monthly rainfall in the last 21 years from 1997 to 2018, which were collected from West Nusa Tenggara Regional Infrastructure Information Center. All available data used for drought index analysis, then the model's accuracy evaluation using only the indexes generated from 2012 to 2017. It is because the drought experienced data to be compared to has only been recorded since 2012.

The selection of rain stations used in this study is based on two mains of consideration: length of data records and data completeness (minimum blank data). In the calculation, SPI uses monthly rainfall data with a fairly long recording period as described in the WMO-No.1090 WMO SPI User Guide that the length of the data used ranges from 20 to 30 years, and it is better if it is more than 30 years [16].

2.3 Research Method

This study analyzed the drought index from SPI and PDSI then compared the results with experienced drought data collected from Regional Board for Disaster Management (BPBD). The month where the drought occurred will be an indexed score of −1, and when the drought has not happened, it will be an indexed score of 0. The next step is to provide a drought index score calculated by the SPI and PDSI methods. In the SPI and PDSI methods' indexing system, both have almost the same range of index values, where the index value of less than −1 indicates a drought. It makes the comparison of the two methods easier because they have the same indexing class. The next step is calculating the duration of drought by accumulating drought events every month started with a score of −1 and ended when the score is 0.

The accumulated drought duration score was obtained from SPI and PDSI methods then compared to the drought scores from the BPBD data. For the same month that shows the same score, for example, both show drought or no drought, it is interpreted as suitable and given a score of 1. And if it does not match, then the score given is 0. The suitability assessment is then carried out by summed up the number of match scores. The higher the number of corresponding scores, the better the method produced a drought index.

Method of completing the missing data

In some areas, the recording of rain data is sometimes incomplete. It could be due to administrative errors or recording equipment damaged. Then estimating data technique must be done to complete it. Ideally, estimating missing data is carried out by comparing the data from several close stations and correlated with the test station [17].

In this study, completing the missing rain data was carried out by a normal ratio method, a simple method to fill in the missing rainfall data based on rainfall data from several close stations at the same time and compared it with the annual rainfall data of each station. The formula in completing missing data at stations is [18]:

$$P_x = \frac{1}{m}\sum\nolimits_{i=1}^{m}\left[\frac{N_x}{N_i}\right]P_i \tag{1}$$

Where: P_x = estimated value of rainfall for the ungauged station (mm); P_i = rainfall values of other rain gauges used for estimation; N_x = Normal annual rainfall data for the ungauged station; N_i = normal annual precipitation of other surrounding stations; m is the number of surrounding stations.

Calculating of Drought Indices

SPI computation includes of matches probability density function of the distribution of Gamma is defined by:

$$g(x) = \frac{1}{\beta^{\alpha}(a)} \int_0^x t^{a-1} e^{\frac{-x}{\beta}} \quad (2)$$

Where α > 0 is the form parameter; β > 0 is a scale parameter; and x > 0 is the total monthly rainfall. While Γ (α) is the Gamma function defined as:

$$\Gamma(\alpha) = \int_0^{\infty} y^{\alpha-1} e^{-y} dy \quad (3)$$

The values of α and β are estimated for each rain station using the following formula:

$$\alpha = \frac{1}{4A}\left[1 + \sqrt{\left(1 + \frac{4A}{3}\right)}\right] \quad (4)$$

$$\beta = \frac{\overline{x}}{\alpha} \text{ and } A = ln(\overline{x}) - \frac{\sum \ln(x)}{n} \quad (5)$$

Where n is the amount of rainfall data.

Then calculating the cumulative chance then the Gamma spread integrated with x to give G (x):

$$G(x) = x \int_0^x g(x)\,dx = \frac{1}{\beta\alpha\Gamma(a)} \int_0^x t^{a-1} e^{\frac{-x}{\beta}} dx \quad (6)$$

Then substitution of $I = \frac{x}{\beta}$, so the formula (6) becomes:

$$G(x) = \frac{1}{\Gamma(a)} \int t^{a-1} e^{-t} dt \quad (7)$$

Since the gamma function is undefined for $x = 0$, the value of G (x) becomes:

$$H(x) = q + (1 - q)G(x) \quad (8)$$

Where: q = m/n; m is the number of rainfall events 0 mm in the rain data series. SPI value calculation:

$$Z = SPI = -\left(t - \frac{C_0 + C_1 t + C_2 t^2}{1 + d_1 t + d_2 t^2 + d_3 t^3}\right); \quad \text{for } 0 < \text{H(x)} \leq 0.5 \quad (9)$$

and transformation of gamma distribution: $t = \sqrt{ln\left[\frac{1}{(H(x))^2}\right]}$;

$$Z = SPI = +\left(t - \frac{C_0 + C_1 t + C_2 t^2}{1 + d_1 t + d_2 t^2 + d_3 t^3}\right); \quad \text{for } 0.5 < \text{H(x)} \leq 1.0 \tag{10}$$

and transformation of gamma distribution: $t = \sqrt{ln\left[\frac{1}{(1-H(x))^2}\right]}$;

Calculation of PDSI value uses the formula:

$$X = \left(\frac{Z}{3}\right)_{j-1} + \Delta X;\ and\ \Delta X = \left(\frac{Z}{3}\right)_j - 0.103\left(\frac{Z}{3}\right)_{j-1} \tag{11}$$

The number of $c_0 = 2.515517$; $c_1 = 0.802853$; $c_2 = 0.010328$; $d_1 = 1.432788$; $d_2 = 0.189269$; $d_3 = 0.001308$.

The index is calculated from water supply and demand computed using a complex water-balanced system based on recorded precipitation, air temperature, and the soil characteristics of the site. The calculation of potential evapotranspiration was proposed by Thornthwaite. This is required only daily averaged temperatures and latitude of the site to calculate the maximum amount of sunshine duration [19].

The SPI and PDSI classification index describe a relative moisture condition within 9 categories, as shown in Table 1.

The hydrologic accounting procedure of the water balance model involves the production of time series of the runoff values, recharge to the soils, moisture loss from the soils, and transpiration, as well as estimates of their potential values. From these time series, average monthly values of the potential and the actual values are calculated over the calibration interval. The coefficient of evapotranspiration for a particular month is the quotient of an average of the actual value ET and the average of the potential value PET for that month. Similar coefficients are b, giving the ratio of average recharge to the soils R and its potential value PR, g, giving the ratio of average runoff RO and its potential value PRO, and giving the ratio of average loss of moisture from the soils L to its potential value PL. Here the overbar denotes a long-term average. Palmer's Climatically Appropriate for Existing Conditions (CAFEC)

Table 1 The SPI and PDSI classification

SPI	Class	PDSI	Class
≥2	Extremely wet	≥4.00	Extremely wet
1.5 to1.99	Severely wet	3.00 to 3.99	Severely wet
1.00 to 1.49	Moderately wet	2.00 to 2.99	Moderately wet
0.5 to 0.99	Mild wet	1.00 to 1.99	Slightly wet
−0.49 to 0.49	Near normal	0.50–0.99	Incipient dry spell
−0.99 to -0.5	Mild drought	0.49 to −0.49	Slightly dry
−1.49 to -1.00	Moderately drought	−0.50 to −0.99	Moderately dry
−1.99 to -1.50	Severely drought	−1.00 to −1.99	Severely dry
≤ −2.0	Extremely drought	−2.00 to −2.99	Extremely dry

precipitation; ˆP is a function of the potential evapotranspiration, potential recharge, potential runoff and potential loss, weighted by their appropriate coefficients.

3 Result and Discussion

3.1 Data Preparation

The first step of this research is quality data control by checking its completeness. The requirements of drought index calculation, especially for the SPI model, must be filled in completely or without any missing data during that period of use. Then if there is blank data, it needs to be interpolated to predict it using formula 1.

This study calculated SPI and PDSI index, which are divided into several regional groups. Lombok Island was divided into four regions and two regions for Sumbawa Island. First, the calculation of the regional average rainfall is calculated using the Thiessen polygon method, then it is followed by the analysis of the monthly rainfall. Monthly rainfall is the primary input data in calculating the drought index. Regional averaged monthly rainfall for each region is presented in Fig. 2.

Other data used in PDSI calculations are air temperature. The air temperature data is needed to calculate the potential evapotranspiration value, which is then used to determine the criteria for wet and dry months based on rainfall data and potential evapotranspiration rates. Air temperature data throughout West Nusa Tenggara is presented in Fig. 3.

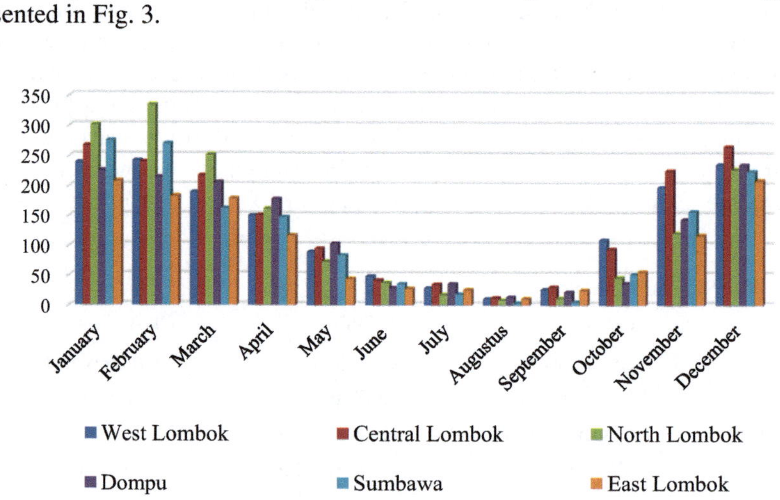

Fig. 2 Regional averaged monthly rainfall over West Nusa Tenggara

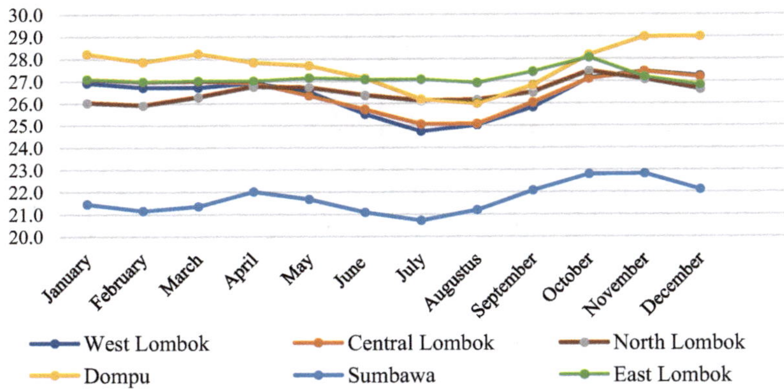

Fig. 3 Averaged monthly air temperature over West Nusa Tenggara

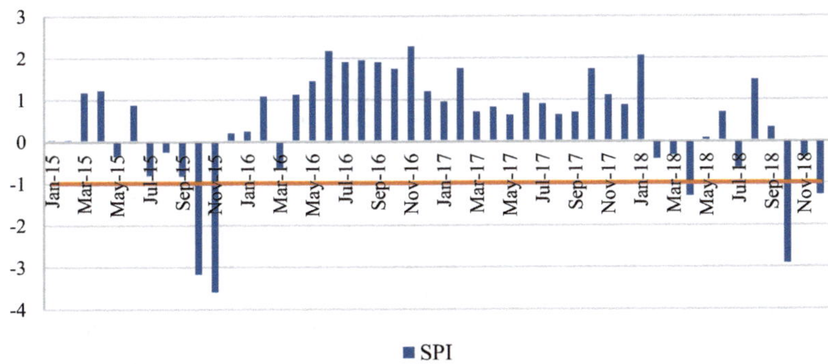

Fig. 4 The example of SPI index for West Lombok. The negative value shows the strength and drought duration. The orange line indicates the start of the dry periods

3.2 Calculation of Drought Indices

In the SPI and PDSI methods, drought occurs when the index value continuously gives a negative value and reaches a drought intensity with an index value of −1 or less. A positive index shows that the rain is greater than the median and a negative index shows that the rain is smaller than the median.

Firstly, the Indices of SPI and PDSI for any location were calculated based on the long-term precipitation record for the desired period. The SPI model is well known and becomes the most popularly used drought index [20] (Fig. 4).

The data input in the PDSI method is rainfall, groundwater capacity (WHC), and potential evapotranspiration. The advantage of this method is that in addition to producing index values, it also creates a coefficient of climate parameters, namely the evapotranspiration coefficient, the recharge coefficient, the runoff coefficient, and the moisture loss coefficient. From this coefficient, it can be calculated the

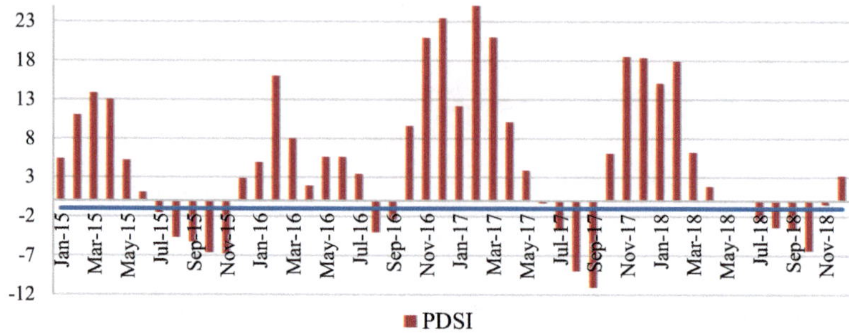

Fig. 5 The example of PDSI value for West Lombok. The negative value shows the strength and drought duration. The blue line indicates the start of the dry periods

Table 2 West Lombok Drought experienced data

BPBD	Jan	Feb	Mar	Apr	May	Jun	Jul	Aug	Sept	Oct	Nov	Dec
2014	0	0	0	0	0	0	−1	−1	−1	−1	−1	0
2015	0	0	0	0	0	0	0	0	−1	−1	−1	−1
2016	0	0	0	0	0	0	0	0	−1	−1	−1	−1
2017	0	0	0	0	0	0	0	−1	−1	−1	−1	−1
2018	0	0	0	0	0	0	0	−1	−1	−1	−1	−1

rainfall that occurs during a specific month support evapotranspiration, runoff, and soil moisture reserves which are considered normal conditions. Palmer's drought index classification is divided into 9 classes with zero indexes as a normal condition. Besides, the Palmer method requires land-use data, soil type, and soil texture (Fig. 5).

3.3 *Comparison of Drought Index Values*

The accuracy performance test of the model in evaluating the drought period was conducted by comparing the SPI and PDSI model's indices against drought data from Regional Board for Disaster Management (BPBD) West Nusa Tenggara. BPBD recorded a dry period in which the community did not have any access to water, both rainwater and surface water. The month when the drought recorded actually figured the water shortage that occurred on site. BPBD data for water deficiency were presented in numerical format, where the months with the drought occurrence are given a score of −1, while months without drought are provided 0 (see Tables 2 and 3).

The calculation of the model's suitability is started by first giving a dumb score on the index produced from both methods. The SPI and PDSI values of < -1 are given −1 score for that means the dry period has started, and a score of 0 for the

Table 3 Determination of both model's drought period scores and their matches to BPDB

Score of	Jan- 15	Feb- 15	Mar- 15	Apr- 15	May- 15	Jun- 15	Jul- 15	Aug- 15	Sep- 15	Oct- 15	Nov- 15	Sums (2015–2018)
SPI	0	0	0	0	0	0	0	0	0	−1	−1	
PDSI	0	0	0	0	0	0	−1	−1	−1	−1	−1	
BPBD	0	0	0	0	0	0	−1	−1	−1	−1	−1	
SPI matches to BPBD	1	1	1	1	1	1	0	0	0	1	1	32
PDSI matches to BPBD	1	1	1	1	1	1	1	1	1	1	1	37

Table 4 Correlation coefficients between SPI and PDSI to BPBD data drought period

Region	r SPI	r PDSI
West Lombok	0.282	0.512
Central Lombok	0.064	0.192
North Lombok	0.192	0.641
East Lombok	0.048	0.543
Sumbawa	−0.266	0.643
Dompu	0.076	0.529

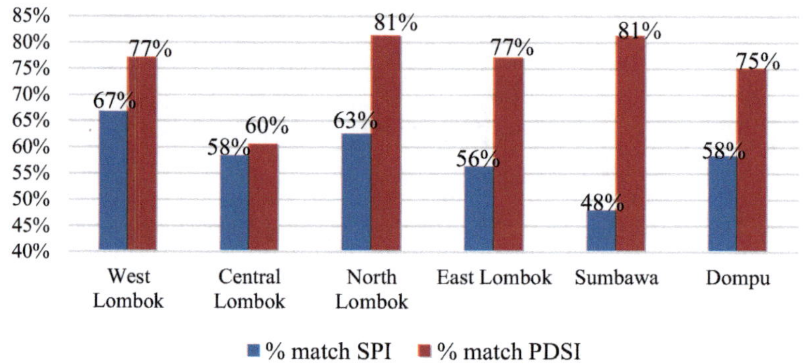

Fig. 6 The matches percentage of SPI and PDSI drought period to BPBD data

month with no drought indicated by the index values of both models are > −1. Next, check the matches of the score. Months with the same score will get a score of 1, and months with a different score will get 0. The correlation coefficients were calculated between the scores of the models and BPBD drought data (Table 4). The summary of the total match scores can then be added to find how suitable the model is. The higher the score, the higher the suitability drought index, especially in the dry month (Table 3).

The matches score of the drought period was then summed up and calculated the matches percentage (see Fig. 6).

Based on the calculation results presented in Fig. 6, it can be concluded that the PDSI method provides a better estimation of the drought period in the study area than the SPI method. The calculation of suitability produced the result that the PDSI method has an average suitability percentage of 75% and a mean correlation coefficient (r) of 0.510, and the SPI method has an average suitability percentage of 58% and a correlation rate (r) of 0.06.

The estimation of drought duration is done by accumulating a drought index score. The more scores we get, the longer the period of the drought. A technique in accumulating drought index score started when a score shows −1 that indicates a dry period has occurred and ends at a score of 0, which means there is no more drought. The same calculation is started again when a score of −1 is found.

The cumulative score obtained will show the length of the dry month. The higher scores mean the longer the duration of the drought. The calculated drought period results in a cumulative score for 2015–2018, presented in Fig. 7.

Figure 7 shows that the performance of the PDSI method appears to have a better ability to detect prolonged droughts in 2014 and 2016 than the SPI method. The SPI drought index often does not indicate the suitable occurred drought time. On the other hand, PDSI can detect the drought that happens relatively close to the BPBD data. The model accuracy is only reviewed for the presence or absence of the drought event, regardless of its depth, magnitude and severity. So, the results of drought index calculations that show specific criteria are ignored. Comparative data used is only for the drought period, while the wetness periods and the rank are ignored.

Besides requiring more data input, the PDSI method has a higher level of difficulty than the SPI method in the calculation process. So that in its application, the SPI method is still interesting to apply.

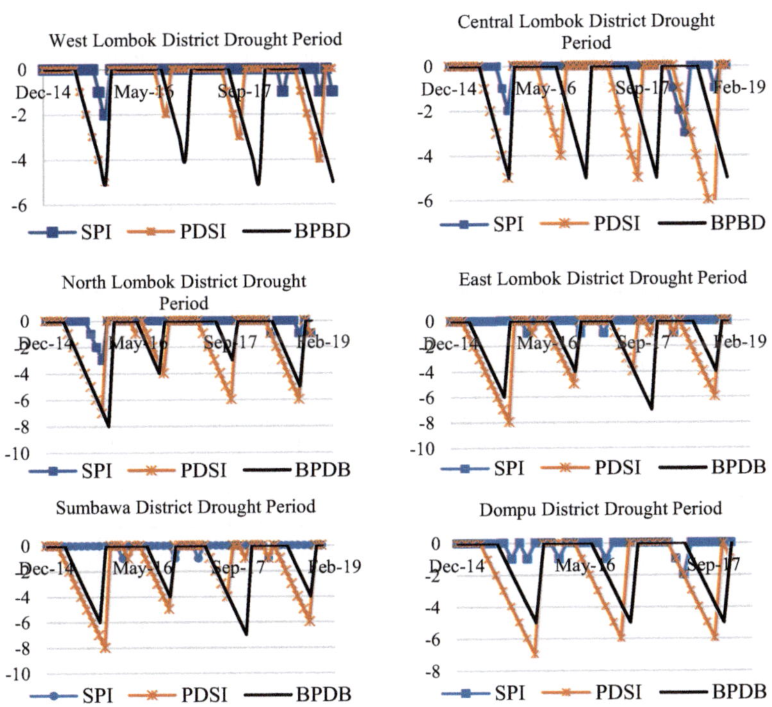

Fig. 7 Cumulative score of the SPI, PDSI, and BPBD drought period

4 Conclusions

The suitability of the drought indices calculating techniques compared to experienced drought data, the PDSI method can evaluate and determine the drought period properly for West Nusa Tenggara. The PDSI method gave an accuracy rate of 75% for dry months tested conditions, while SPI was only 58%. The correlation number of indexes correlated with drought occurrence is about 0.51 for PDSI and 0.06 for SPI.

References

1. Gao L, Zhang Y (2016) Spatio-temporal variation of hydrological drought under climate change during the period 1960–2013 in the Hexi Corridor. China J Arid Land 8:157–171. https://doi.org/10.1007/s40333-015-0022-3
2. Zhang J, Sun F, Lai W, Lim WH, Liu W, Wang T, Wang P (2019) Attributing changes in future extreme droughts based on PDSI in China. J Hydrol 573:607–615. https://doi.org/10.1016/j.jhydrol.2019.03.060
3. McMahon T (1982) Arenas A D. Methods of computation flows. Gustard 107
4. Pathak AA, Dodamani BM (2020) Comparison of meteorological drought indices for different climatic regions of an Indian river basin. Asia-Pac J Atmos Sci 56:563–576. https://doi.org/10.1007/s13143-019-00162-5
5. Chelton DB, Risien CM (2020) A hybrid precipitation index inspired by the SPI, PDSI, and MCDI. Part II: application to investigate precipitation variability along the West Coast of North America. J Hydrometeorol 21:1977–2002. https://doi.org/10.1175/JHM-D-19-0231.1
6. Zou L, Xia J, She D (2017) Drought characteristic analysis based on an improved PDSI in the Wei River Basin of China. Water 9:178. https://doi.org/10.3390/w9030178
7. Haied N, Foufou A, Chaab S, Azlaoui M, Khadri S, Benzahia K, Benzahia I (2017) Drought assessment and monitoring using meteorological indices in a semi-arid region. Energ Procedia 119:518–529. https://doi.org/10.1016/j.egypro.2017.07.064
8. Abbasian MS, Najafi MR, Abrishamchi A (2021) Increasing risk of meteorological drought in the Lake Urmia basin under climate change: introducing the precipitation–temperature deciles index. J Hydrol 592:125586. https://doi.org/10.1016/j.jhydrol.2020.125586
9. Swain S, Patel P, Nandi S (2017) Application of SPI, EDI and PNPI using MSWEP precipitation data over Marathwada, India In: 2017 IEEE International Geoscience and Remote Sensing Symposium (IGARSS) pp 5505–5507. https://doi.org/10.1109/IGARSS.2017.8128250
10. Tirivarombo S, Osupile D, Eliasson P (2018) Drought monitoring and analysis: Standardised precipitation evapotranspiration index (SPEI) and standardised precipitation index (SPI). Phys Chem Earth Parts A/B/C 106:1–10. https://doi.org/10.1016/j.pce.2018.07.001
11. Quiring SM, Papakryiakou TN (2003) An evaluation of agricultural drought indices for the Canadian prairies. Agric For Meteorol 118:49–62. https://doi.org/10.1016/S0168-1923(03)00072-8
12. Dehghan S, Salehnia N, Sayari N, Bakhtiari B (2020) Prediction of meteorological drought in arid and semi-arid regions using PDSI and SDSM: a case study in Fars Province. Iran J Arid Land 12:318–330. https://doi.org/10.1007/s40333-020-0095-5
13. Huang S, Li P, Huang Q, Leng G, Hou B, Ma L (2017) The propagation from meteorological to hydrological drought and its potential influence factors. J Hydrol 547:184–195. https://doi.org/10.1016/j.jhydrol.2017.01.041
14. Vasiliades L, Loukas A (2009) Hydrological response to meteorological drought using the Palmer drought indices in Thessaly. Greece Desalination 237:3–21. https://doi.org/10.1016/j.desal.2007.12.019

15. Badan Pusat Statistik Provinsi Nusa Tenggara Barat (2017) Luas Daerah Nusa Tenggara Barat 2015–2017. https://ntb.bps.go.id/indicator/153/56/1/luas-daerah-nusa-tenggara-barat.html. Last accessed 30 Mar 2021
16. World Meteorological Organization (2012) Standardized precipitation index user guide
17. Adhyani NL, June T, Sopaheluwakan A (2017) Exposure to drought: duration, severity and intensity (Java, Bali and Nusa Tenggara).In: IOP conference series: earth and environmental science, vol 58, p 012040. https://doi.org/10.1088/1755-1315/58/1/012040
18. De Silva RP, Dayawansa NDK, Ratnasiri MD (2007) A comparison of methods used in estimating missing rainfall data. J Agric Sci 3:101–108
19. Schrier van der G, Jones PD, Briffa KR (2011) The sensitivity of the PDSI to the Thornthwaite and Penman-Monteith parameterizations for potential evapotranspiration. J Geophys Res Atmos 116. https://doi.org/10.1029/2010JD015001
20. McKee TB, Doesken NJ, Kleist J (1993) The relationship of drought frequency and duration to time scales. 6

Small Debris Flow Simulation Using MORPHO2DH

Puji Harsanto, Dandy Darvin Septiandy, Berli Paripurna Kamiel, and Nursetiawan

Abstract Several studies based on experiments or field surveys reported the effect of de-bris flow from the eruption of volcanoes on river morphology. However, there has been a notable absence of scientific literature dealing with the prediction of river morphology at the middle or downstream after several weeks or months due to the remnants of material sediment from an eruption. The remaining material sediment at upstream will produce a small debris flow during the rainy season. This study analyzes the riverbed change due to small debris flow using the 2-dimensional numerical simulation in iRIC-Morpho2DH. The prediction of riverbed erosion on several segments shows that the erosion increases when the small debris occurs. The maximum erosion occurs when the sediment discharge ratio is 2 times, then decreases in the simulation of the sediment discharge ratio 3 times. Base on the field survey, the river segment with maximum erosion, the bank protection was failed.

Keywords Erosion · Sedimentation · Morpho2DH · Sediment discharge ratio · Gajah Wong River

P. Harsanto (✉) · D. D. Septiandy · Nursetiawan
Department of Civil Engineering, Faculty of Engineering, Universitas Muhammadiyah Yogyakarta, Yogyakarta, Indonesia
e-mail: puji_hr@umy.ac.id

D. D. Septiandy
e-mail: dandy.darvin.ft17@mail.umy.ac.id

Nursetiawan
e-mail: nursetiawan@umy.ac.id

B. P. Kamiel
Department of Mechanical Engineering, Faculty of Engineering, Universitas Muhammadiyah Yogyakarta, Yogyakarta, Indonesia
e-mail: berlikamiel@umy.ac.id

S. A. Kristiawan et al. (eds.), *Proceedings of the 5th International Conference on Rehabilitation and Maintenance in Civil Engineering*, Lecture Notes in Civil Engineering 225,
https://doi.org/10.1007/978-981-16-9348-9_52

1 Introduction

Mount Merapi is one of the most active volcanoes in Indonesia. The material sediment from its eruptions is an essential key for river morphology in rivers that originated from that volcano. Gajah Wong River, which flows through Yogyakarta City, is one of the rivers that originated on the slopes of Mount Merapi. A riverbank protection is built on the river to protect the bank erosion. However, some constructions failed because of the excessive erosion. Regarding [1], it is crucial to mitigate riverbank erosion in the river to get the proper design of riverbank protection.

Morpho2DH is a two-dimensional flow and bed open-source modeling system that is integrated into iRIC [2]. The purpose of this study was to analyze riverbed erosion due to the concentration of bedload transport. The advantage of this software is that the sediment transport concentration modeling can be set beyond the equilibrium condition to model the flow of debris. This is relevant because which would prevent instances such as the Gajah Wong River flood.

2 Study Area

This research was conducted in the segment of Gajah Wong River along ± 1 km due to the markers in the Prenggan area (−7.822571°, 110.399046°) as the upstream point to the Jagalan GW137 area (−7.828972°, 110.396000°) as the downstream point. The research location can be seen clearly in Fig. 1.

Fig. 1 Cross section GW157-GW137 of Gajah Wong River

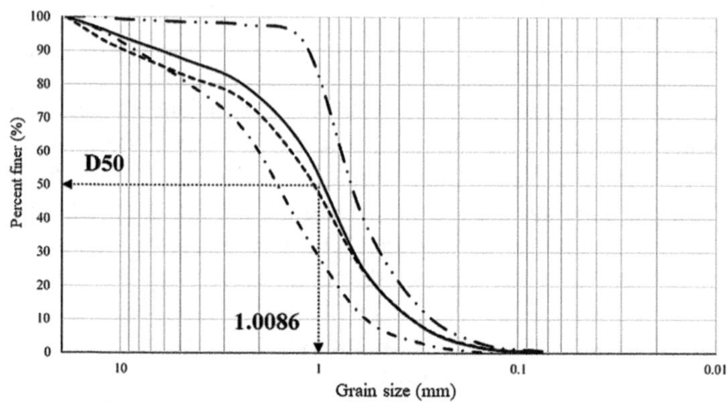

Fig. 2 Sediment grain size in Gajah Wong River

2.1 Discharge and Cross Section Data

In this study, two types of data are used, namely primary and secondary data. The primary data is in the form of sediment grain size based on laboratory test results from ± 2 kg of sediment samples taken directly from the Gajah Wong River with a total of 3 test samples (upstream, middle, downstream) which were tested for filter analysis in the laboratory. The secondary data in cross-section data and design flood discharge data is obtained from the Public Works Service of BBWS Serayu-Opak. The design flood discharge is 82.97 m^3/s [3].

2.2 Grain Size Data

Three samples of sediment were obtained from the upstream, middle, and downstream of the river field survey. As much as ± 2 kg was obtained from each section, then the sediment grain-sizes were determined through [4]. The results of the grading test are presented in Fig. 2.

3 Two Dimensional Flow and Bed Change Equations

Morpho2DH is a computational analysis program developed by Hiroshi Takebayashi of Kyoto University, which models the debris and sludge flow in rivers. In 2009, the model was added to RIC-Nays, which is free software developed by RIC. The simulation can predict the transport and deposition of debris flow and mud in the horizontal distribution of erosion depths. The equations of debris and mudflow consist of flowing water and sediment, which is treated as a single-phase continuous fluid

[5]. The mass conservation equation for the mixture of water and sediment is as follows:

$$\frac{\partial h}{\partial t}+\frac{\partial hu}{\partial x}+\frac{\partial hv}{\partial y}=\frac{E}{c_*} \tag{1}$$

where,

∂t	= time (s)
∂h	= flow depth (m)
u, v	= velocity in the x and y axis/direction (m/s)
c_*	= sediment deposition concentration in static deposition layer
E	= bottom erosion velocity

When the cohesive characteristics of riverbed layers can be neglected and the bed reaches equilibrium, where the rate of bed erosion is equal to the rate of sediment deposition, the rate of erosion of the bed can be estimated using the following equation:

$$\frac{E}{\sqrt{u^2+v^2}}=c_*\tan(\theta-\theta_e) \tag{2}$$

where,

θ = channel bed slope

$$\sin\theta=\frac{u\sin\theta_x+v\sin\theta_y}{\sqrt{u^2+v^2}} \tag{3}$$

where,

θ_x, θ_y	= bed slope in the x and y axis/direction
θ_e	= bed slope balance refers to the mean depth of sediment concentration c

If it is assumed that a laminar flow layer is formed near the bottom and a turbulent flow layer is formed in a laminar layer with a constant mean depth of sediment concentration c, the following relationship is obtained.

$$\tan\theta_e=\frac{(\sigma/\rho-1)\bar{c}}{(\sigma/\rho-1)\bar{c}+1}\frac{h_s}{h}\tan\varnothing_s \tag{4}$$

where $\varnothing_s$ is angle of repose, the momentum conservation equation is as follows [8].

$$\frac{\partial hu}{\partial t}+\frac{\partial huu}{\partial x}+\frac{\partial hvu}{\partial y}=-gh\frac{\partial z_b}{\partial x}-\frac{1}{\rho_m}\frac{\partial P}{\partial x}-\frac{\tau_{bx}}{\rho_m} \tag{5}$$

$$\frac{\partial hv}{\partial t}+\frac{\partial huv}{\partial x}+\frac{\partial hvv}{\partial y}=-gh\frac{\partial z_b}{\partial y}-\frac{1}{\rho_m}\frac{\partial P}{\partial y}-\frac{\tau_{by}}{\rho_m} \tag{6}$$

where,

g = gravitational acceleration
z_b = bed elevations
P = pressure (assumed as static pressure)

Then the debris flow density is as follows.

$$\rho_m=(\sigma-\rho)\overline{c}+\rho \tag{7}$$

where,

σ = sediment density
P = water density
τ_{bx}, τ_{by} = shear stress in the x and y axis/direction

When the turbulent flow area is dominant, the shear stress equation is as follows.

$$\tau_{bx}=\left\{\tau_y+\rho f_b\left(u^2+v^2\right)\right\}\frac{u}{\sqrt{u^2+v^2}} \tag{8}$$

$$\tau_{by}=\left\{\tau_y+\rho f_b\left(u^2+v^2\right)\right\}\frac{u}{\sqrt{u^2+v^2}} \tag{9}$$

If τ_y is the yield stress, use the following relationship for the non-cohesive material:

$$\tau_y=\left(\frac{\overline{c}}{c_*}\right)^{\frac{1}{5}}(\sigma-\rho)\overline{c}\,gh_s\cos\theta\tan\text{Ø}_s \tag{10}$$

where f_b = coefficient of friction, the equation is as follows.

$$f_b=\frac{C_{mu}}{8}a^2 \quad a=k/6 \tag{11}$$

$$f_b=\frac{4}{25}\left\{k_f\frac{(1-\overline{c})^{\frac{5}{3}}}{\overline{c}^{\frac{2}{3}}}+k_d\frac{\sigma}{\rho}\left(1-e^2\right)\overline{c}^{\frac{1}{3}}\right\}\left(\frac{h_s}{d_m}\right)^{-2}+9a^2\left(1-\frac{h_s}{h}\right) \tag{12}$$

where C_{mu} is the resistance coefficient of the sludge flow, $k_f = 0.16$, $k_d = 0.0828$, and e is the particle inelastic coefficient, and d_m is mean particle size of the sediment from the debris flow. So, the basic elevation equation becomes:

$$\frac{\partial z_b}{\partial t} = -\frac{E}{c_*} \tag{13}$$

4 Results and Discussion

The simulation on iRIC: Morpho2DH displays the results in the form of a scalar of flow velocity values. For example, a scalar display of the simulated flow velocity under 2 times sediment discharge ratio conditions can be seen in Fig. 3. The results indicate that under non-equilibrium sediment transport, more sediment supply, which is a small debris flow model, can produce the increase of velocity in some conditions. The velocity also increases the shear stress on the river bed. In these conditions, the erosion will increase. This condition is demonstrated in cross-sections around 151 and 152. This cross-section shows that the erosion maximum occurs under two times sediment supply, then erosion decrease consider to the increasing of sediment supply.

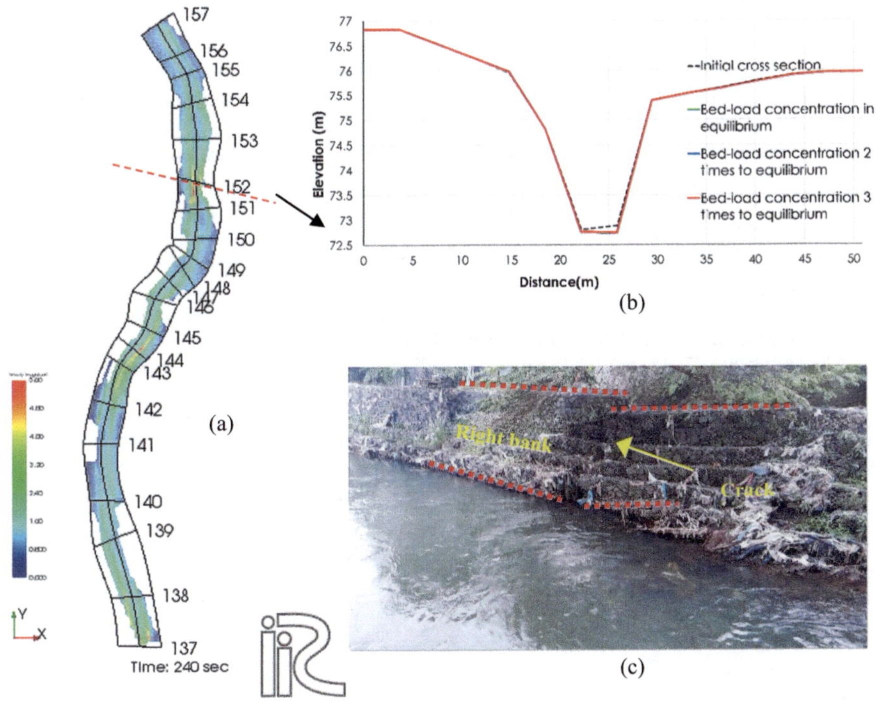

Fig. 3 The simulation result for velocity (**a**) and cross section elevation after 240 **s** (**b**), and condition in the field (**c**)

Regarding to the research findings, simulation of river bank erosion should be done before determining stream corridor reservations, especially along meandering rivers in urban area [6].

5 Conclusion

The two-dimensional model for flow and riverbed morphology was performed using iRIC: Morpho2DH. The model can simulate flow, bed degradation, aggradation, which are similar to real river conditions. In general, the maximum flow velocity in non-equilibrium conditions occurred when the sediment concentration was less than two times the equilibrium condition. When the sediment rate was more than two times the equilibrium condition, the aggradation will increase rapidly. The model should be utilized when designing riverbank protections.

References

1. Ikhsan J, Fujita M, Legono D, Rahardjo AP, Harsanto P (2020) Dynamics of lahar-affected river tributaries of the Progo river after the 2010 Mt. Merapi eruption, IOP Conf Ser Earth Environ Sci 437:012009
2. Takebayashi H, Fujita M (2020) Numerical simulation of a debris flow on the basis of a two-dimensional continuum body model. Geosciences 10(2):45
3. Nugroho AK (2015) Aplikasi Sig Untuk Evaluasi Sistem Jaringan Drainase Sub Das Gajahwong Kabupaten Bantul. SEMNASIF
4. Nasional BS (2012) SNI ASTM C136: 2012: Metode uji untuk analisis saringan agregat halus dan agregat kasar. BSN, Jakarta
5. Egashira S, Honda N, Itoh T (2001) Experimental study on the entrainment of bed material into debris flow. Phys Chem Earth Part C 26(9):645–650
6. Murniningsih S (2018) Study of erosion and sedimentation in urban areas for river with meander pattern. IOP Conf Ser Earth Environ Sci 191

Simulation of Debris Flow Using "SIMLAR" in the Watershed of Gendol River, Indonesia

Hendy Dwi Cahyo, Jazaul Ikhsan, and Ani Hairani

Abstract The eruption of Mount Merapi in 2010 released more than 140 million cubic meters of pyroclastic material, which still settles on cliffs and riverbeds. Gendol River is one of the rivers affected by the accumulation of sediment. When there is heavy rain in the river's upper reaches, rainwater will carry lava material at high speed called debris flow, which will cause damage to the area in its path. Sediment control buildings such as Sabo dams are built to reduce the impact of debris flows. To predict the impact of debris flows, it could be used using a software application. In this study, the debris flow simulation is used the SIMLAR V.2.1 application. The simulation uses data, namely rain data, to calculate the flood hydrograph using the Nakayasu method. The other data are grain size distribution data to determine the sediment material and a topographic map in the form of DEM to determine the length and area of the simulated watershed and modify the Sabo dam building's existing conditions in Gendol River. The simulation results show that at the beginning of the simulation hour to 0.5 with a discharge of 0.299 m^3/s produces a debris volume of 539.997 m^3 with a velocity of 0.198 m/s, at peak discharge with a discharge of 107.729 m^3/s has a debris volume of 452,129.320 m^3 with a velocity of 1.916 m/s. The simulation at hour of 15 with a discharge of 2.909 m^3/s, produces a volume of 75,038.150 m^3 with a velocity of 0.451 m/s.

Keywords SIMLAR V.2.1 · Sabo dam · Debris flow · Merapi

1 Introduction

Debris flow is a common occurrence in mountainous areas around the world, both volcanic and non-volcanic. These floods have caused major property damage and

H. D. Cahyo · J. Ikhsan (✉) · A. Hairani
Department of Civil Engineering, Universitas Muhammadiyah Yogyakarta, Yogyakarta, Indonesia
e-mail: jazaul.ikhsan@umy.ac.id

A. Hairani
e-mail: anihairani@umy.ac.id

S. A. Kristiawan et al. (eds.), *Proceedings of the 5th International Conference on Rehabilitation and Maintenance in Civil Engineering*, Lecture Notes in Civil Engineering 225,
https://doi.org/10.1007/978-981-16-9348-9_53

deaths, as well as significant morphological changes along riverbeds and mountain slopes [1–4]. Indonesia is one of the countries with the most active volcanoes in the world, which is more than 30% of the world's active volcanoes in Indonesia [5]. One of them is Mount Merapi, Mount Merapi is located on the island of Java and administratively belongs to the Sleman Regency, Yogyakarta Special Region. As an active volcano, of course, it has the potential to erupt at any time—it will happen again. It is recorded that Mount Merapi experienced its last eruption in 2010. According to [6]. The eruption in 2010 was the largest eruption of Mount Merapi since 1872. The eruption in 2010 released more than 140 million cubic meters of lava heading south, which resulted in the accumulation of lava deposits. It has the potential to become a debris flow, especially in the Gendol River. A system of sediment control was built, namely the Sabo dam, to anticipate debris flow impacts. It is hoped that the Sabo dam will reduce the impact of the disaster due to debris flow in Gendol River. The preparation of disaster mitigation is also one of the efforts to reduce the effects of debris flow. One of the ways to mitigate disasters is to simulate debris flows. In this case, the simulated lava flood was carried out using the SIMLAR application version 2.1. SIMLAR is a numerical modelling application to estimate the direction of the distribution of the lava flood that will occur. This application also has a function to analyze hydrology, analyze the lava flood propagation with 2D simulations, and display the simulation results in the form of a Geographical Information System (GIS) model to be used as an early warning system.

1.1 The Gendol River Watershed

Watershed (DAS) is defined as an area bounded by a ridge topography that receives and collects rainwater and other nutrients and then flows it to the river network and comes out at one point (outlet). The Gendol River Watershed is shown in Fig. 1.

1.2 Simulation Lahar (SIMLAR)

Simulation Lahar (SIMLAR) is a debris flow simulation application. The application consists of three sub-chapters of the programs; namely, the first is a flood hydrograph calculation program. The second is a hydrograph calculation program due to the collapse of a natural weir. The third is a 2D debris-flow simulation program. To estimate the debris flow direction, SIMLAR uses the debris flow modelling method with a numerical model in which sludge water is also considered a fluid unit. The debris flow simulation is based on a partial differential equation that regulates the debris flow as follows:

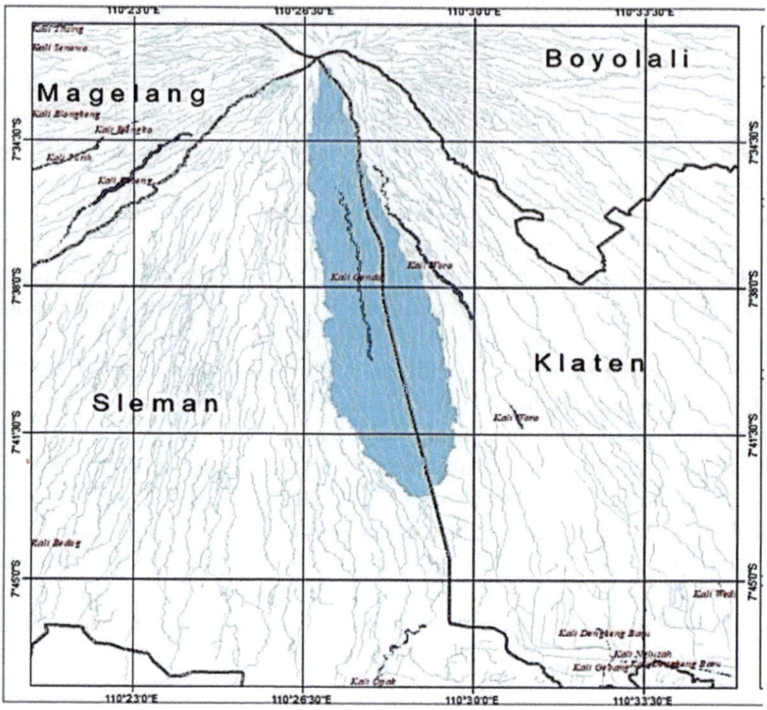

Fig. 1 Research locations in Gendol River watershed, Indonesia

1. Mass conservation equation

$$\frac{\partial h}{\partial t} + \frac{\partial M}{\partial X} + \frac{\partial N}{\partial y} = 0 \tag{1}$$

2. The momentum and force conservation equation (x direction)

$$\frac{\partial M}{\partial t} + \beta\frac{\partial (uM)}{\partial x} + \beta\frac{\partial (uM)}{\partial x}\frac{\partial (vM)}{\partial y} = -gh\frac{\partial H}{\partial x} - \frac{\tau bx}{\rho T} \tag{2}$$

The momentum and force conservation equation (y direction)

$$\frac{\partial M}{\partial t} + \beta\frac{\partial (uM)}{\partial x} + \beta\frac{\partial (vN)}{\partial y} = -gh\frac{\partial H}{\partial x} - \frac{\tau by}{\rho T} \tag{3}$$

where:

h = Flow height
t = Time (s)
M = Debris flow rate per unit width x direction (m^2/s)

N = Debris flow rate per unit width y direction (m^2/s)
β = Momentum correction
H = Flow depth (m)
g = Gravity
τ_{bx} = The component of the riverbed shear stress in the x direction
τ_{by} = The component of the riverbed shear stress in the y direction
ρ_T = Mass flow density
u = Average velocity in the x direction
v = Average velocity in the y direction

3. The equation for the conservation of sediment in the river

$$c_* \frac{\partial Z_b}{\partial t} + \left(\frac{\partial q_{Bx}}{\partial x} + \frac{\partial q_{By}}{\partial y} \right) = 0 \tag{4}$$

where:

c_* = Concentration of riverbed sediments
q_{Bx} = Sediment discharge in the x direction
q_{By} = Sediment discharge in the y direction
$q_{Bx} + q_{By}$ = The amount of base sediment (*bedload*)

4. Cliff erosion considered with the equations of Ashida, Egashira and Kamamoto

$$q_p = q_{p*} u_* d \tag{5}$$

$$q_{p*} = F_{0*} \tau_*^{\frac{1}{2}} \left(1 - \frac{1}{\varepsilon^2} \frac{\tau_{*sc}}{\tau_*} \right)^3 \left(\frac{h}{d} \frac{1}{\sin\theta} \right) \tag{6}$$

$$u_* = \sqrt{ghI} \tag{7}$$

$$\tau_* = \frac{u_*^2}{gd\left(\frac{\sigma}{\rho} - 1\right)} \tag{8}$$

$$\tau_{*sc} = \tau_{*c} \cos\theta_1 \sqrt{1 - \left(\frac{\tan\theta}{\mu} \right)^2} \tag{9}$$

where:

q_p = Cliff erosion per unit width (m^2/s)
q_{p^*} = Cliff erosion per unit width (dimensionless)
u_* = Basic shear speed (m/s)
I = The slope of the energy flow line

θ = The slope of the cliff
F_{0*} = Coefficient = 0.01
d = Sediment diameter

5. The discharge formula in 2D simulation using Meyer—Peter and Muller

$$q_{*_b} = 8\left(\tau_* - \tau_{*_c}\right)^{1,5} \tag{10}$$

where:

q_{*b} = Discharge volume per width (m^3/s)
τ_{*c} = The average value of the critical friction stress
τ_* = Shield's number

1.3 Sabo Dam

According to [7], the word "sabo" comes from Japanese, "sa" means sand, and "bo" means prevention or control. The broad understanding is the erosion control system, sediment, cold lava, and landslide countermeasures to protect human life from the threat of disaster. The sabo system is a synthesis technology from mechanical, civil engineering, and vegetative forestry construction with aspects of land and river channel conservation in river catchments. The sabo dam building is a building that controls the flow of debris or cold lava that occurs during heavy rains, and the sabo dam is built across the river channel [8, 9]. The Sabo dam building's working principle is sediment control by holding, accommodating and flowing the material or sand carried by the debris flow and bringing it downstream.

2 Research Method

2.1 Research Location

The research location that simulated debris flow was in Gendol River, which is administratively included in the Sleman Regency, Yogyakarta Special Region, Indonesia. The watershed area used in this study is 55.663 km^2 with a river length of 20.231 km.

2.2 Research Data

The research data used is secondary data obtained from Balai Sabo Yogyakarta. The data acquired will be inputted to perform simulations on the SIMLAR application.

The author does not collect data directly in the research area. The secondary data obtained are:

Rainfall Data

This study's Rainfall data were rainfall data from Ngandong, Sorasan, Plosokerep stations (Fig. 2). The rainfall data used are hourly rainfall data that occurred in 3 stations at the same hour with 3 h in April 2015. The rainfall selection is the most significant rainfall intensity in the 2015–2019 period.

Sediment Material Data

The grain size analysis from Balai Sabo Yogyakarta found that d50 was coarse sand with a grain size of 0.80 mm. The study of the characteristics of the lava sediment in Gendol River shows that the lava sediment is coarse sand. The analysis results are used to classify the type of debris flow, namely mudflow. Mudflow is a flow dominated by sedimentary material with an acceptable size of 0.3–1 mm, and grain below 0.1 mm is 10–30% (Fig. 3).

Topographic Data

The topographic data used were obtained from the website (http://tides.big.g.id/DEMNAS/), which has a spatial resolution of 0.27 or about 8 m × 8 m. The data is in the form of topographic information on DEM (Digital Elevation maps). The model with the .tif format is then converted into Asc format to be inputted into the SIMLAR software. The DEM map is then modified by increasing the river channel bed's elevation to be assumed as a Sabo dam building at the elevated point.

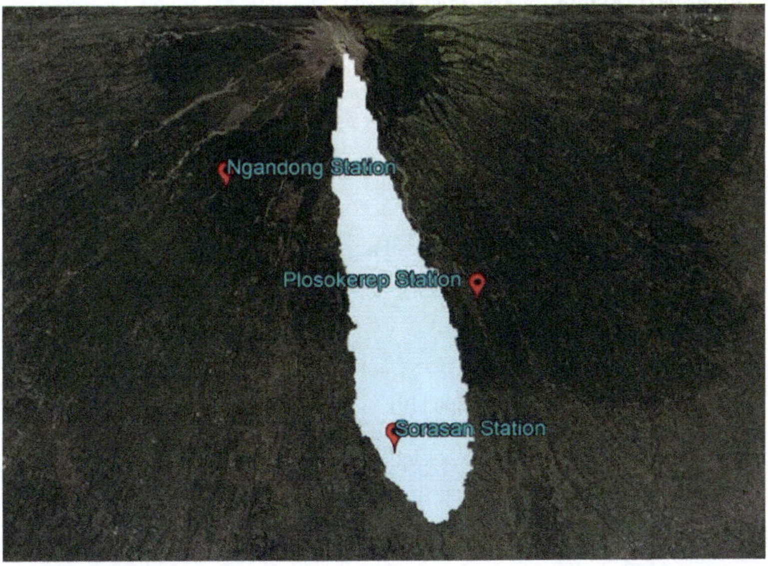

Fig. 2 Rain station location

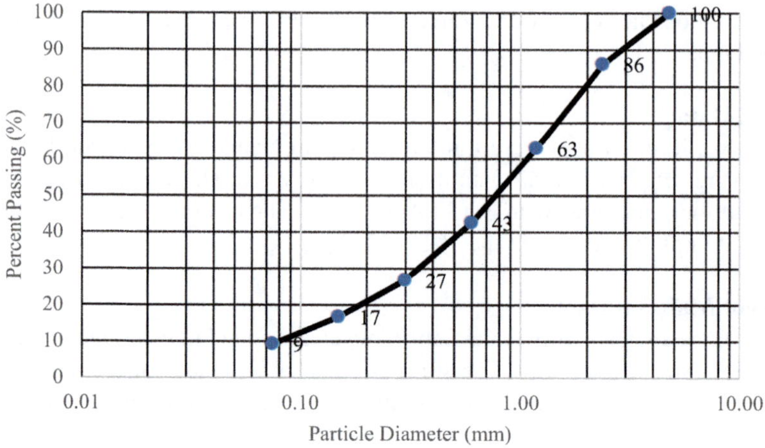

Fig. 3 Grain size distribution chart [11]

3 Results and Discussion

3.1 Flood Hydrograph

The hydrograph used in this research is the synthetic hydrograph of the Nakayasu method. The results of the calculation of the Nakayashu Synthesis Unit Hydrograph on the SIMLAR application obtained the discharge value starting at 0.5 h (30 min) to the 15th hour (900 min) as follows (Fig. 4).

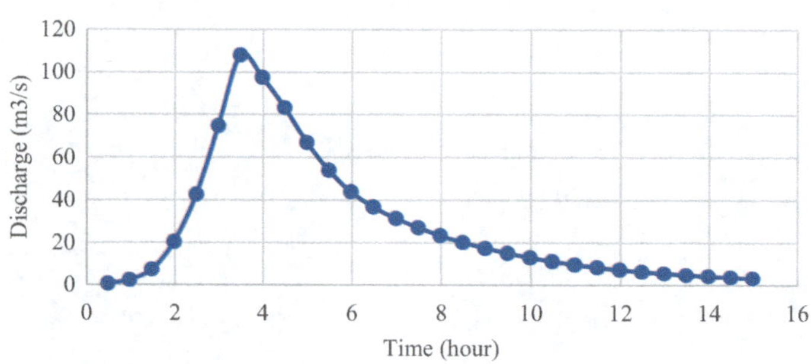

Fig. 4 Nakayasu method flood hydrograph [10]

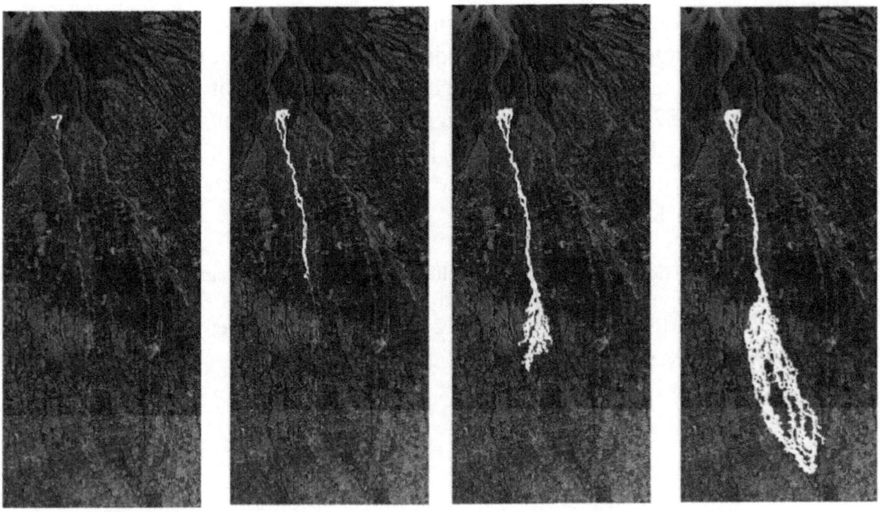

Fig. 5 Distribution map of simulation results

Table 1 SIMLAR simulation result

Hour	Discharge (m^3/s)	Volume (m^3)	Velocity (m/s)
0.5	0.29	539.99	0.19
5	66.58	581,309.16	1.58
7	31.04	341,761.91	1.16
15	2.0	75,038.15	0.45

3.2 Debris Flow Simulation

The simulation is carried out with an initial time of 0.5–15 h according to hydrograph calculations, and analysis is carried out every 30 min. The simulation results are then processed using the Arc-Map application to produce a raster map of the distribution of debris flow. The following is a map of the distribution of the simulation results in the SIMLAR application.

The simulation results is shown in Fig. 5, it can be explained that the simulation results on Gendol River are shown in Table 1.

4 Conclusion

The SIMLAR application program can model debris flows by producing a map of each hydrograph distribution that has been predetermined. The length and the velocity of debris flows are influenced by the river's cross-section and the Sabo dam building in the river. The simulation results show that at the beginning of the simulation hour to 0.5 with a discharge of 0.299 m^3/s produces a debris volume of 539.997 m^3 with a velocity of 0.198 m/s, at peak discharge with a discharge of 107.729 m^3/s has a debris volume of 452,129.320 m^3 with a velocity of 1.916 m/s. The simulation at hour of 15 with a discharge of 2.909 m^3/s, produces a volume of 75,038.150 m^3 with a velocity of 0.451 m/s. It means that the alert condition starts when a velocity of debris flow is about 1.916 m/s.

Acknowledgements The author is grateful for University of Muhammadiyah Yogyakarta, Indonesia, through the internal research grant 2021 and the Ministry of Education, and Culture, Indonesia through PDUPT Research Grand 2021 which have supported author to conduct the research and publicizing in this workshop.

References

1. Takahashi T (1991) Debris flow. IAHR monograph series, Rotterdam, Balkema
2. Hunt B (1994) Newtonian fluid mechanics treatment of debris flows and avalanches. J Hydraul Eng ASCE 120(12):1350–1363
3. Huang X, Garcia MH (1997) A perturbation solution for Bingham-plastic mudflows. J Hydraul Eng ASCE 123(11):986–994
4. Shrestha BB, Nakagawa H, Kawaike K, Baba Y (2008) Numerical simulation on debris-flow deposition and erosion processes upstream of a check dam with experimental verification. In: Annuals of Disaster Prevention Research Institute, Kyoto University, No. 51 B, pp 614–624
5. Pratomo I (2006) Klasifikasi gunung api aktif Indonesia, studi kasus dari beberapa letusan gunung api dalam sejarah. Indonesian J Geosci 1(4):209–227
6. Kholiq MA (2017) Simulasi aliran banjir lahar pasca erupsi gunung merapi 2010 terhadap keberadaan sabo dam di sungai gendol. J Teknisia 22(2):410–415
7. Munir MD (2019) Bangunan sabo dam, fungsi dan potensinya sebagai bagian dari geowisata gunung api merapi. J Lingkungan dan Bencana Geologi 10(2)
8. Ikhsan J, Legono D, Rahardjo AP, Harsanto P, Fujita M (2020) Dynamics of lahar-affected river tributaries of the Progo river after the 2010 Mt. Merapi eruption. In: Earth and Environmental Science
9. Khairi MAW, Rozainy MR, Ikhsan J (2020) Smoothed particle hydrodynamics simulation for debris flow: a review. IOP Conf. Ser. Mater. Sci. Eng. 864(1): https://doi.org/10.1088/1757-899X/864/1/012045
10. Pengembangan Sistem Prakiraan dan Peringatan Dini Aliran Debris di Kali Putih Kabupaten Jember
11. Yuniawan, R (2019) Hasil uji analisis saringan agregat halus dan kasar. Balai Sabo 1(1)

Wind-Generated Wave Simulation on Payangan Beach Utilizing DELFT3D

Enggar Setia Baresi, Retno Utami Agung Wiyono, and Wiwik Yunarni Widiarti

Abstract The Southern Coastal area of Jember has been damaged almost yearly by coastal flooding and high waves, especially Payangan Beach. A study on high waves is urgently needed for disaster risk preparedness. This study is conducted to model the high waves and simulate the effect on the beach. Delft3D-Wave model is used to provide a spatiotemporal characteristic of the event. The focus of the study is to analyze the high waves on Payangan Beach generated by wind forces. The domain model is a curvilinear grid with a grid size ranging from 130 m until 900 m. National bathymetric data and CDS Copernicus wind data were utilized as data sources. The simulation outcomes have shown that waves are generated far from the coastline across the ocean inside the fetch area. The simulations show the significant wave height increase during the high waves event according to specific meteorological characteristics from 1.6–2.0 m on normal days and exceeding 3.2 m when the high waves hit.

Keywords High waves · Delft3D · Wind-generated waves

1 Introduction

Shoreline is the point of intersection between land and sea. In this area, ocean energy reacted to the land vice versa. The system that works in this reaction mostly caused by the natural movement of the ocean which transferring energy into the system. Coastal zones are directly affected by the ocean forces especially beach and near-shore zone. Therefore, this particular area is the most dynamic in all coastal areas [1].

E. S. Baresi · R. U. A. Wiyono (✉) · W. Y. Widiarti
Civil Engineering Department, Universitas Jember, Jember, Indonesia
e-mail: retnoutami@unej.ac.id

W. Y. Widiarti
e-mail: wiwik.teknik@unej.ac.id

S. A. Kristiawan et al. (eds.), *Proceedings of the 5th International Conference on Rehabilitation and Maintenance in Civil Engineering*, Lecture Notes in Civil Engineering 225,
https://doi.org/10.1007/978-981-16-9348-9_54

The most common form of ocean energy is wave. The linear waves theory or small-scale wave amplitude theory is derived from Laplace equation for irrotational flow. The equation gives solution of periodic velocity for irrotational flow which later can be utilized to derive equations from various waves characteristic for example water level, velocity, particle acceleration, waves propagation speed, and others [2].

According to the previous study conducted in Southern Seas of Jember, this particular area has high amount of energy transfer about 190 MWh/m/year with significant wave height up to 4 m [3]. In Payangan Beach, a beach located in Southern Seas of Jember, high waves occurred annually e.g. 2018 [4]. There is no known study that simulate spatial variation of high waves in this area. Thus, this study is crucial for the various purpose related to high waves effect on the Payangan Beach coastal area.

This study aims to simulate high waves and analyze the significant wave height in Payangan Beach in the last five years. Every year, this particular location hit by high waves in late July when the atmosphere pressure shift in the Indian Ocean. But only in 2020, the high waves occurred in May, two months prior to the usual period of the event. This study focused to analyze the high waves on the study site on a small-scale simulation by using national bathymetric data.

2 Methodology

2.1 Study Area

Payangan Beach exposed directly to waves coming from Indian Ocean that can measure up roughly from 2 to 5 m in a bad weather condition. Average significant wave height in Payangan Beach during normal days is ranging from 1.6 to 2.0 m. This beach has natural defense against large waves and even massive tidal wave in a form of sand dunes and highest slope compared to other two beaches nearby (Fig. 1).

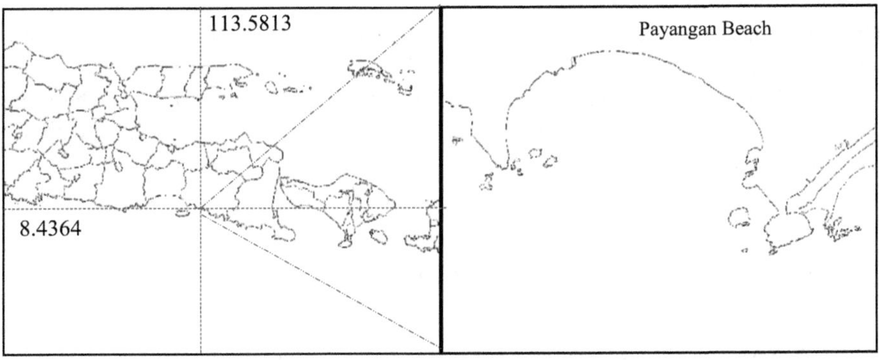

Fig. 1 Study location in east Java (left); the zoom-in map of study domain and study site at Payangan Beach (right)

Table 1 Wind data distribution

Wind speed (knot)	Wind direction (%)								Days (%)
	N	NE	E	SE	S	SW	W	NW	
0–3	0.00	0.00	0.00	0.00	0.14	0.00	0.00	0.00	0.14
3–6	1.74	2.90	1.88	0.39	21.11	0.24	3.77	0.43	32.46
6–9	3.38	4.11	2.71	0.24	16.91	0.24	5.89	0.97	34.44
9–12	2.42	2.46	1.21	0.14	13.91	1.01	6.76	0.97	28.89
12–15	0.14	0.05	0.10	0.00	1.59	0.29	0.24	0.05	2.46
15–18	0.00	0.00	0.10	0.00	0.77	0.29	0.14	0.00	1.30
18–21	0.00	0.00	0.00	0.00	0.10	0.05	0.00	0.00	0.14
21–24	0.00	0.00	0.00	0.00	0.05	0.05	0.00	0.00	0.10
30–33	0.00	0.00	0.00	0.00	0.05	0.00	0.00	0.00	0.05
Days	7.68	9.52	5.99	0.77	54.64	2.17	16.81	2.42	100.00

2.2 *Wind Data Analysis*

Wind data of this study was obtained from Class III Meteorological Station located in Banyuwangi which was the closest wind station to the study site. The data that was obtained from Class III Meteorological Station is the final wind data derived from u10 and v10 wind data. In the equation below [5, 6], the Ø means meteorology wind direction angle in radians.

$$u = -|V| \sin Ø \tag{1}$$

$$v = -|V| \cos Ø \tag{2}$$

$$|V| = \sqrt{(u^2 + v^2)} \tag{3}$$

In addition, spatial variation of wind data namely ERA5-Interim wind data was obtained from Climate Data Store Copernicus that was measured by Aeolus satellite. The data that was gathered classified within 3 speed interval, $0 < U < 3$, $3 < U < 6$, $6 < U < 9$, up to $30 < U < 33$.

Table 1 shows wind data distribution and Fig. 2 shows wind rose of the study location. Based on the data, the main direction of wind is South (225°) which is 54.64% from total wind data.

2.3 *Wave Simulation Using Delft3D*

Delft3D-WAVE is able to compute various geospatial even meteorological parameters, wave dispersion calculation, wave generation by wind force, non-linear interaction between the waves, and wave dissipation [7]. In this study, Delft3D-WAVE is utilized based on the Simulating Waves Nearshore (SWAN) spectral model.

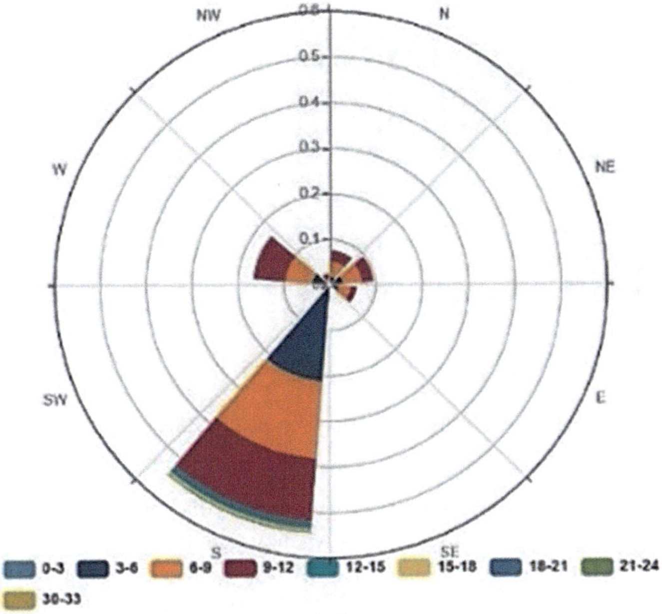

(a)

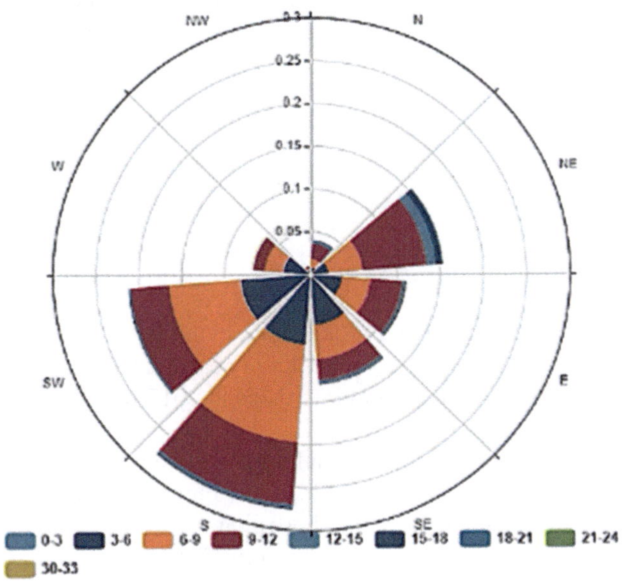

(b)

Fig. 2 Wind rose based on **a** peak wind direction; **b** mean wind direction

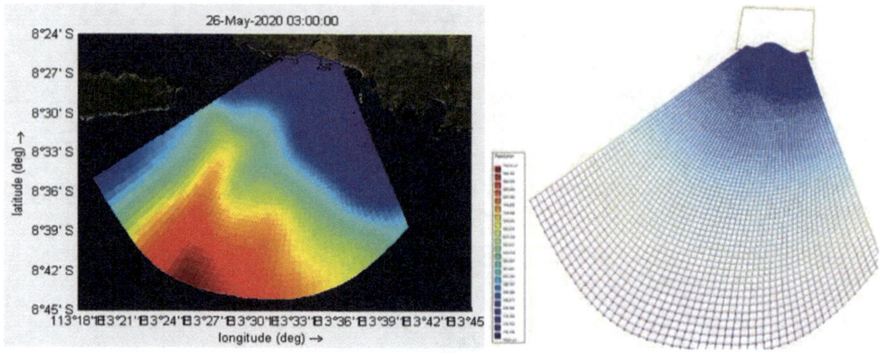

(a) (b)

Fig. 3 **a** Depth generated from bathymetric data projected on ESRI visual; **b** simulation grid

In this study, bathymetric data that was used is the bathymetric data obtained from Indonesian Geospatial Information Agency with 6-arcsecond resolution. Wind input used in this simulation is taken from Class III Meteorological Station which have temporal resolution of 24 h (daily). In addition, wind data from ERA5-Interim provided by CDS Copernicus which have temporal resolution of 1 h (hourly) is used to generate meteo file as input file in Delft3D simulation. The bathymetric data is mapped in Fig. 3a. The grid used within the study domain ranges from 138.346 m to 946.482 m shown in Fig. 3b.

3 Results

3.1 Simulation Results

Generally, results of Delft3D-WAVE simulations in 2017, 2018, 2019, and 2020 show significant wave height of 2.5 m to 4.5 m nearshore and around 5 m offshore. The high waves in 2018 almost similar to those in 2017 with the difference margin around 5–10%.

Figures 4 and 5 show simulation results based on SWAN modeling in Delft3D when high waves occurred on July 25th, 2017. Figure 4a shows significant wave height of 2.5–4.5 m at the coastal area. Those high waves may be the cause of coastal flooding in the coastal area as occurred on July 25th, 2017. Figure 4b shows peak wave period more than 11 s in offshore while in coastal area, peak wave period ranges from 6 to 10 s. Figure 5 shows that waves approaching the beach at angle of 80° and shifting to 70° then around 55° to 60° when the waves arrive at the beach.

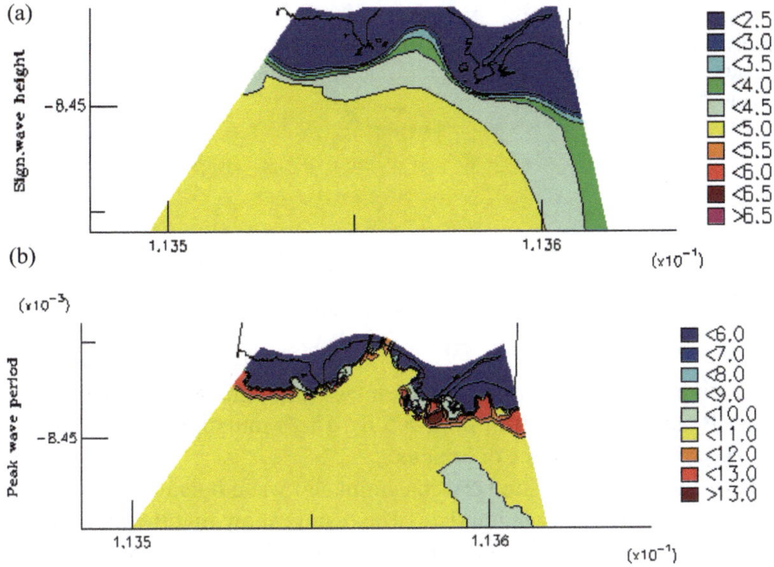

Fig. 4 Simulation result on July 25th, 2017. **a** Significant wave height; **b** peak wave period

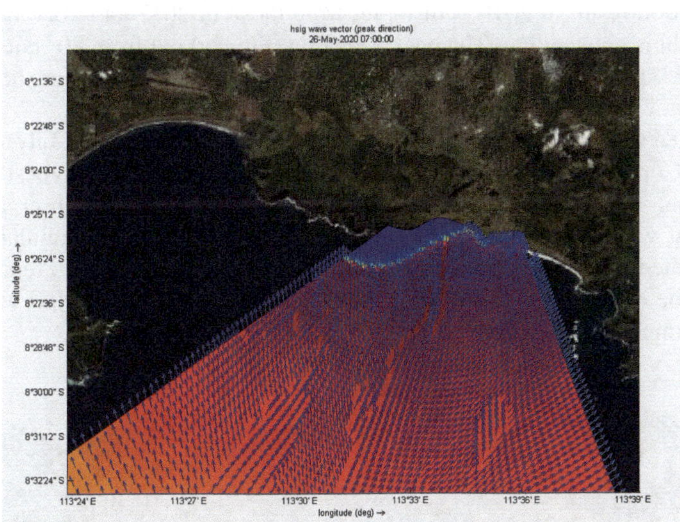

Fig. 5 Peak wave direction

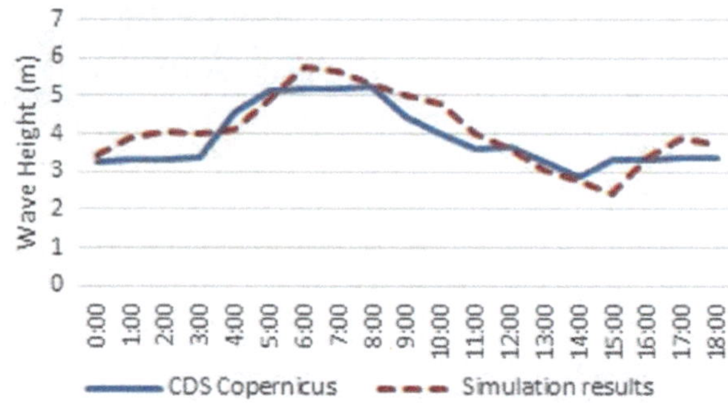

Fig. 6 Comparison between simulated significant wave heights and CDS Copernicus observation data on July 2017

When high waves occurred, maximum wave height at 20 km offshore from Payangan Beach reached 3.9 m until 5.5 m. The simulation results coincide with the observed data from CDS Copernicus.

Figure 6 shows comparison between simulated wave heights on July 25th, 2017 and CDS Copernicus observation data. The results show that the highest simulated wave height was 5.8 m at 6 am, while based on CDS Copernicus observation data, the highest wave height was 5.2 m at 7 am. The simulated results show close prediction of CDS Copernicus data. The waves were generated from 8.55 m/s wind speed.

Simulation result in 2019 is different from those in 2017 and 2020 because the wave height is outstanding. The peak direction (Fig. 7a) not from 80° but vary from 62° to 77.5°. The significant wave height reached 6 m to 7 m in nearshore of Payangan Beach (Fig. 7b).

Figure 8a shows comparison between simulated wave heights on July 26th, 2019 and CDS Copernicus observation data. The results show that the highest simulated wave height was 6.9 m at 8 am, while based on CDS Copernicus observation data, the highest wave height was 5.9 m at 8 am. Although there are simulated results overestimated the observed wave height, the simulation was able to reproduce the similar time of maximum wave height at 8 am. The outstanding high waves were also confirmed by the captured video shown in Fig. 8b.

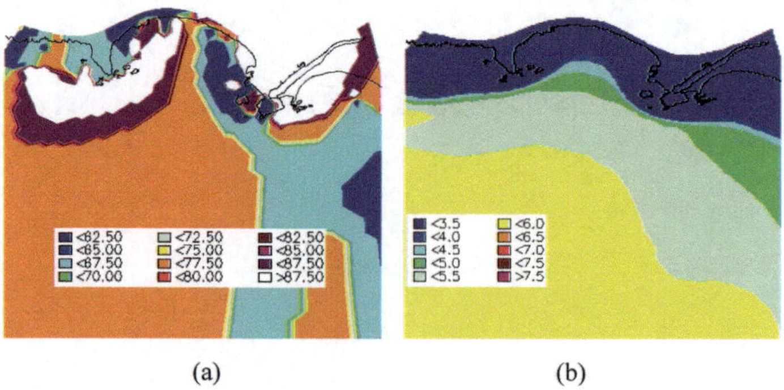

Fig. 7 Simulation result on July 26th, 2019. **a** Peak wave direction; **b** significant wave height

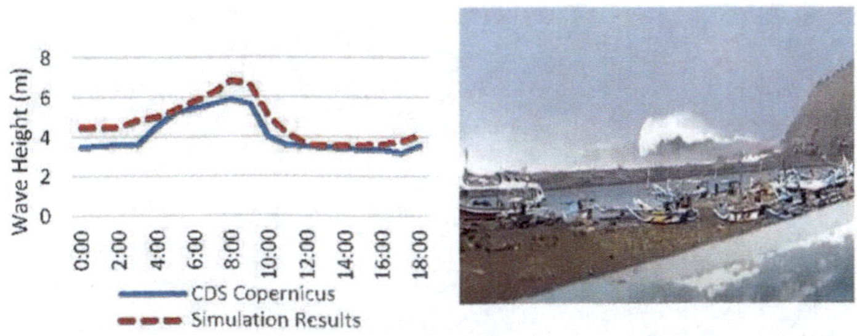

(a) (b)

Fig. 8 **a** Comparison between simulated significant wave heights and CDS Copernicus observation data on July 26th, 2019; **b** Payangan beach high waves documentation [8]

(a) (b)

Fig. 9 **a** Comparison between simulated significant wave heights and CDS Copernicus observation data on May 2020; **b** Payangan beach high waves documentation on May 27th 2020 high waves [9]

Different from 2017 and 2019, in 2020 the high waves occurred not in July but in May. Figure 9a shows comparison between simulated wave heights on May 27th 2020 and CDS Copernicus observation data. The results show that the highest simulated wave height was 5.6 m at 6 am, while based on CDS Copernicus observation data, the highest wave height was 5.9 m at 6 am. The waves were generated from 7.42 m/s wind speed. The simulated results slightly underestimated the observed wave height. However, the simulation was able to reproduce the similar time of maximum wave height at 6 am. The high waves were also confirmed by the photo shown in Fig. 9b.

4 Structural Measures

In order to protect the beach along with the community living within the area of study in Payangan Beach, one of the best solutions is to create safety measures based

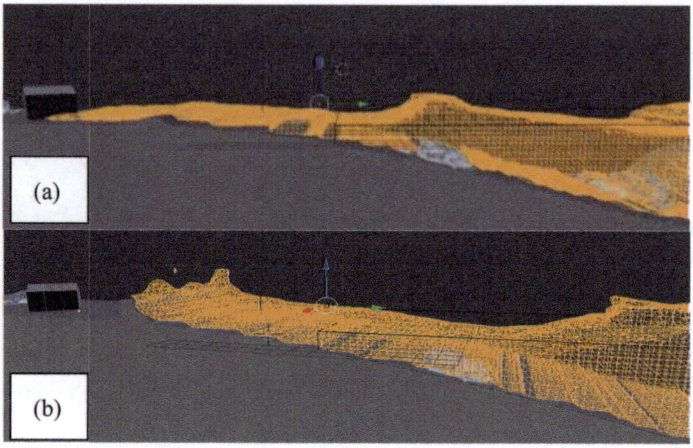

Fig. 10 High waves **a** without sea wall as protective measures; **b** with sea wall as protective measures

on the simulated wave height. Based on consideration and effectiveness according to the events, the best measures considered is to build a sea wall.

Sea wall is a coastal defense structure built parallel to the coastline to mitigate coastal flooding by blocking and or reflecting waves force to another direction so the wave will less affect the area protected by the sea wall. Sea wall vary in shape and type which based on the site condition. Payangan Beach condition require sea wall because of the direct wave effect on the beach during high waves.

Figure 10 show illustration of Payangan Beach without sea wall (Fig. 10a) and with sea wall (Fig. 10b). Without sea wall, waves will propagate to the coastal area and inundate residential area. By constructing sea wall, waves will be reflected and the residential area will be protected from severe inundation.

5 Conclusions

High waves occurred at Payangan Beach is significantly high because of the direct exposure of wind and wave from the open ocean. Wind meteorological direction is from the South with wind speed vary from 3.9 m/s to more than. When high waves occurred, the wind speed vary from 5 m/s to 16 m/s and generate waves with significant wave height of 3.4–5 m.

The high waves occurred in 2015, 2016, 2017, and 2018 show significant wave height of 2.5–4.5 m nearshore and around 5 m offshore with difference margin around 5–10%. However, in 2019 high waves occurred with significant wave height reached 6–7 m in nearshore of Payangan Beach. The significant wave height creating massive wave that visible from the distance. In 2020, high waves occurred two months prior than the other five years which always occurred in late July.

Preventive measures must be taken into consideration to protect the coastal area and residential area located in Payangan Beach. The most effective solution is to build coastal protection structure to cope with high waves and inundation that caused the coastal flooding. Sea wall has high efficiency to hold waves energy and to protect the residential area from inundation. However, with this solution, there may be sediment transport around the seawall. Further study is necessary to analyze coastal protection structure and the effect to the beach.

References

1. Coastal Engineering Research Center (1984) Shore Protection Manual, Vol 1. US Army Corps of Engineers, Washington DC
2. Triatmodjo B (1999) Teknik Pantai [Coastal Engineering]. Beta Offset, Yogyakarta
3. Warpindyasmoro HS (2018) Wave energy potency in east java coast. MATEC Web Conf 177:10–18
4. High Waves Attacked a Shop in Payangan Beach. https://www.youtube.com/watch?v=6yZlZBshfas. Last accessed 2020/12/11
5. Ostrenga D (2019) Derive wind speed and direction with MERRA-2 wind components. https://disc.gsfc.nasa.gov/information/data-in-action?title=Derive%20Wind%20Speed%20and%20Direction%20With%20MERRA-2%20Wind%20Components. Last accessed 2021/7/2
6. Yuwono N (1982) Teknik Pantai, Vol I. Biro Penerbit Keluarga Mahasiswa Teknik Sipil FT UGM, Yogyakarta
7. Deltares (2014) Delft3-Wave User Manual. Deltares, Delft
8. High waves in Payangan on July 2019. https://www.youtube.com/watch?v=iATXCUStZUQ. Last accessed 2021/6/1
9. High waves in Jember Payangan on May 2020. https://www.youtube.com/watch?v=rMFgTJDImaA. Last accessed 2021/6/1

GPSR Compliance

The European Union's (EU) General Product Safety Regulation (GPSR) is a set of rules that requires consumer products to be safe and our obligations to ensure this.

If you have any concerns about our products, you can contact us on ProductSafety@springernature.com

In case Publisher is established outside the EU, the EU authorized representative is:

Springer Nature Customer Service Center GmbH
Europaplatz 3
69115 Heidelberg, Germany

Batch number: 10165645

Printed by Printforce, the Netherlands

Lecture Notes in Civil Engineering

Volume 225

Lecture Notes in Civil Engineering (LNCE) publishes the latest developments in Civil Engineering - quickly, informally and in top quality. Though original research reported in proceedings and post-proceedings represents the core of LNCE, edited volumes of exceptionally high quality and interest may also be considered for publication. Volumes published in LNCE embrace all aspects and subfields of, as well as new challenges in, Civil Engineering. Topics in the series include:

- Construction and Structural Mechanics
- Building Materials
- Concrete, Steel and Timber Structures
- Geotechnical Engineering
- Earthquake Engineering
- Coastal Engineering
- Ocean and Offshore Engineering; Ships and Floating Structures
- Hydraulics, Hydrology and Water Resources Engineering
- Environmental Engineering and Sustainability
- Structural Health and Monitoring
- Surveying and Geographical Information Systems
- Indoor Environments
- Transportation and Traffic
- Risk Analysis
- Safety and Security

To submit a proposal or request further information, please contact the appropriate Springer Editor:

- Pierpaolo Riva at pierpaolo.riva@springer.com (Europe and Americas);
- Swati Meherishi at swati.meherishi@springer.com (Asia - except China, and Australia, New Zealand);
- Wayne Hu at wayne.hu@springer.com (China).

All books in the series now indexed by Scopus and EI Compendex database!

More information about this series at https://link.springer.com/bookseries/15087

Stefanus Adi Kristiawan ·
Buntara S. Gan · Mohamed Shahin ·
Akanshu Sharma
Editors

Proceedings of the 5th International Conference on Rehabilitation and Maintenance in Civil Engineering

ICRMCE 2021, July 8–9,
Surakarta, Indonesia

Set 2

Springer

Editors
Stefanus Adi Kristiawan
Department of Civil Engineering
Universitas Sebelas Maret
Surakarta, Indonesia

Buntara S. Gan
College of Engineering
Nihon University
Tokyo, Japan

Mohamed Shahin
Department of Civil Engineering
Curtin University
Perth, WA, Australia

Akanshu Sharma
Institute of Construction Materials
University of Stuttgart
Stuttgart, Germany

ISSN 2366-2557 ISSN 2366-2565 (electronic)
Lecture Notes in Civil Engineering
ISBN 978-981-16-9350-2 ISBN 978-981-16-9348-9 (eBook)
https://doi.org/10.1007/978-981-16-9348-9

This Springer imprint is published by the registered company Springer Nature Singapore Pte Ltd.
The registered company address is: 152 Beach Road, #21-01/04 Gateway East, Singapore 189721, Singapore

Foreword

The International Conference on Rehabilitation and Maintenance in Civil Engineering (ICRMCE) is a triennial conference that aims to provide a forum for researchers, academicians (professors, lecturers, and students), government agencies, consultants, and contractors to exchange experiences, technological advancement, and innovations in the world of civil engineering, specifically in the fields of rehabilitation and maintenance. The previous four ICRMCE conferences took place successfully in 2009, 2012, 2015, and 2018. Hundreds of researchers worldwide attended these events to present their scientific papers in various areas of civil engineering such as material engineering, structural engineering, geotechnical engineering, transportation engineering, and construction management.

This year's conference was organized by Sebelas Maret University in collaboration with Mataram University. The conference was initially scheduled offline in Mataram, Indonesia. However, due to the escalating coronavirus (COVID-19) outbreak and the need for social distancing, we decided to hold the conference online. Some reputable universities and institutions are participating in the current ICRMCE as partners. Among them are Nihon University, University of Stuttgart, National Taiwan University, TU Delft, Hiroshima University, Diponegoro University, Muhammadiyah University of Yogyakarta, Jenderal Soedirman University, University of Jember, UPN Veteran East Java, the National Center for Research on Earthquake Engineering (NCREE) Taiwan, Himpunan Ahli Konstruksi Indonesia (HAKI), and Himpunan Ahli Teknik Tanah Indonesia (HATTI).

The ICRMCE 2021 was successfully held on July 8–9. Presenters who joined this conference came from Japan, Singapore, Malaysia, China, Vietnam, Taiwan, England, the Netherlands, Kuwait, and Indonesia. Furthermore, several outstanding keynote speakers gave a presentation of the state-of-the-art findings in the field of civil engineering. Our esteemed speakers are Prof. Shyh-Jiann Hwang (National Taiwan University), Prof. Buntara Sthenly Gan (Nihon University), Dr. Edgar Bohner (VTT Technical Research Centre of Finland), and Prof. Mohamed Shahin (Curtin University).

In the process of organizing this conference, we received invaluable motivation, advice, and support from several individuals and institutions. I intend to express my gratitude and appreciation to all of them. First, my most profound appreciation goes to all organizing committee members who worked day and night preparing this conference. Special thanks to the conference and media partners for their generous support. We also express our gratitude to Prof. S.A. Kristiawan (Sebelas Maret University), Dr. Ing. Akanshu Sharma (University of Stuttgart), Prof. Mohamed Shahin (Curtin University), and Prof. Buntara Sthenly Gan (Nihon University) for their willingness to serve as the editors of the 5th ICRMCE proceedings.

Halwan Alfisa Saifullah
The 5th ICRMCE Chairman

Preface

Civil engineering infrastructures are the backbone for the continuous development of civilization. Managing these infrastructures is essential in keeping the quality of services they provide to the community. A decline in the performance of key infrastructure will have an impact on the quality of these services, which in turn can cause social and economic problems. A variety of factors affects the performance of infrastructure. In each case, the declining performance of infrastructure requires an appropriate and adaptive response to offer effective solutions. Protection, maintenance, repair, and retrofitting are part of the various solutions that can be implemented. All of these solutions are assisted by technological developments related to repair materials, methodologies, systems, management, and operational efficiency, as well as economic and social considerations.

Infrastructure performance is also inevitably affected by exposure to hazards originating from natural and environmental conditions such as earthquakes, landslides, and floods, among others. Therefore, hazard mitigation is also an interesting topic of discussion. In addition, risk reduction and safety are among the most important issues of infrastructure management. Finally, various perspectives on sustainability in civil engineering are also covered in this conference.

This book is a collection of papers presented at the 5th International Conference on Rehabilitation and Maintenance in Civil Engineering (ICRMCE) 2021 that deals with the issues stated above. The papers are grouped into sequential themes representing the structure of this book:

- Part I: Factors affecting performance of buildings and infrastructures
- Part II: Assessment, protection, maintenance, repair, and retrofitting of buildings and infrastructures
- Part III: Maintenance management of buildings and infrastructures
- Part IV: Hazard mitigation
- Part V: Risk reduction and safety management
- Part VI: Sustainability aspects in transportation engineering
- Part VII: Sustainability aspects in construction projects

- Part VIII: Sustainability aspects in water resources management
- Part IX: Construction materials for sustainable infrastructures

Postgraduate students, researchers, and practitioners who would like to update their knowledge on the topics above will find this book very useful.

Surakarta, Indonesia

Stefanus Adi Kristiawan
Chief Editor

Contents

Risk Reduction and Safety Management

Incorporating Cultural Attributes into Disaster Risk Reduction-Based Development Plans in Indonesia

Yusron Saadi

Abstract Frequent occurrences of natural disaster in Indonesia cause major problems in the form of financial losses and social disturbances. Disasters with tremendous impacts disrupt the administration of state and hinder the development processes, mainly because the events require expensive and long recovery process. The issuance of Law No. 24 of 2007 on Disaster Management has triggered a paradigm change in disaster management in Indonesia. The management of disaster which previously tended to be responsive and spontaneous has shifted toward preventive and more focused on disaster preparedness and risk reduction. This law explicitly states the responsibility of the government in the implementation of Disaster Management in the form of disaster risk reduction and it integration to development programs. In addition, in the implementation of Disaster Management the government has the authority in determining disaster management policies in line with national development policies. But it is undeniable that there were many cases where different approaches between government and community emerged and become sensitive issues in dealing with disasters management particularly in disaster risk reduction. In this paper a series of intensive community involvement in the form of focus group discussions has been proposed and should be carried out by government in the preparation of development plans. This includes early round of activities followed by formal round of consultation before a draft document of development plans is prepared. Accommodating local wisdom through this robust community engagement is essential in the governmental strategy preparation process. It is a valuable addition and an effective integration in the action strategies of integrated disaster risk reduction so that an appropriate planning decision including the resulting mitigation plan and disaster-resistant infrastructure accepted by the community can be implemented in the disaster prone areas.

Keywords Disaster management · Disaster risk reduction · Data and geospatial information · Cultural attributes · Disaster-resistant infrastructure

Y. Saadi (✉)
Center for Disaster Risk Management, Department of Civil Engineering, University of Mataram, Mataram, Indonesia
e-mail: y.saadi@unram.ac.id

S. A. Kristiawan et al. (eds.), *Proceedings of the 5th International Conference on Rehabilitation and Maintenance in Civil Engineering*, Lecture Notes in Civil Engineering 225,
https://doi.org/10.1007/978-981-16-9348-9_55

1 Introduction

Disaster events in Indonesia have a high intensity since this archipelago country is located in disaster-prone areas known as the Ring of Fire. Indonesian National Disaster Management Agency, known as BNPB, revealed that there were 2952 disaster events in Indonesia during 2020 [1]. The number is based on data recorded by BNPB between 1 January 2020 and 31 December 2020 at 3 pm of West Indonesian Time. The vulnerability to the disaster caused Indonesia to be referred to as the Disaster Laboratory. Multiple natural hazards in various regions have contributed to growing awareness that disasters can occur anytime. It is therefore necessary to make adequate efforts to anticipate and to avoid the devastating impacts of the disaster.

A disaster is an event or series of events that threaten and disrupt people's lives and livelihoods, caused either by natural factors and or non-natural factors or human factors resulting in the emergence of fatalities, environmental damage, property losses, and psychological impacts [2]. Disaster management is part of national development, i.e. a series of disaster management activities before, during and after a disaster. In Law No. 24 of 2007 on Disaster Management [2] it is stated that "the implementation of disaster management is a series of efforts which include the determination of development policies that risk the onset of disasters, disaster prevention activities, emergency response and rehabilitation". One of the responsibilities of the government in implementing disaster management is in the form of disaster risk reduction and integration of disaster risk reduction with development programs while the authority of the government in the implementation of disaster management includes the establishment of disaster management policies in line with national development policies [2]. At present there are many government policies or legal products that regulate the implementation of disaster management as stipulated in the Government Working Plan in 2007 which was promulced through Presidential Decree No. 19 of 2006 where disaster mitigation and management has become one of the 9 national development priorities.

It is recognized that the perspective of government is often considered as contradictive to the perspective of people who are affected by disaster and disrupting the recovery process and future mitigation plan. This is where the role of disaster management actors must have competence and a fundamental understanding of the concept of disaster to bridge these differences including readiness in the management of data and geospatial information which is indispensable to minimize losses and accelerate the rehabilitation and reconstruction process in the affected areas.

The impact of disaster on the community is strongly influenced by the level of vulnerability to the threat of such disasters. In order to understand the aspects that make people vulnerable to disasters, factors such as economic, socio-cultural, institutional, political and psychological factors that make up people's lives should be considered. Thus, the vulnerability of disasters is basically socially constructed [3].

According to Setianto [4], when dealing with disasters people used to embrace and adopt the belief of "fatalism" that nothing can be done against disasters, people must live with disasters and accept disasters as they are. Nowadays, people try to reduce loss of life and property with measurable pre-disaster preparation in line with disaster management. The problem is that academic studies of disasters in Indonesia are still very limited [5] and from the social aspect of disaster, the high frequency of disasters does not necessarily raise interest to conduct studies [6]. Furthermore, Abdullah [5] mentioned that a disaster is always considered as a new experience and something unprecedented. It is considered as something that has not become a collective knowledge and experience so that it has not been integrated in life and social policy.

2 Disaster and Disaster Management

Disaster Management is a science that studies disasters including all related aspects, especially disaster risk and prevention of the disaster. The cooperation of all parties involved in disaster management is important, therefore there must be equalizing perceptions about disaster and disaster management. This is conducted through an agreed system or rules of play, namely Disaster Management System. The parties involved must carry out comprehensive and continuous work programs or activities in each field. Periodic and momentary responses will lead to failure in the implementation of work programs since disasters will occur repeatedly [7].

Figure 1 shows the disaster management cycle based on the basic format put forward by Carter [8] and has been developed by Setianto [4] according to disaster management needs. This picture describes that if steps or activities have been taken during the Prevention and Mitigation and Preparedness phase and disaster occurs then the Emergency Response phase has started and continued with Post-Emergency phase before returning to the Prevention and Mitigation phase. The Prevention and Mitigation phase is necessary to deal with the possibility of future disasters and should be interpreted as the Prevention and Mitigation phase in the next Disaster Management Cycle.

Natural disaster events are irregular events disrupting normal pattern of life. The irregularity in frequency, location and intensity cause natural disasters are complicated to predict. Prevention, reduction, avoidance and self-recovery activities from disaster impacts require a series of activities, both before, during and after a disaster, referred as natural disaster management or disaster management.

Reflecting on various disasters in the past, disaster management actors must include elements of the government, non-governmental institutions and the community. Apparently, there is increasing recognition of the value of community-led initiatives that facilitate emergency management, risk reduction and community resilience [9]. In order for disaster management to run effectively and efficiently in accordance with the objectives of Law No. 24 of 2007 [2], disaster management actors must be competent and certified [10]. Therefore, the role of Disaster Management Professional Certification Body becomes very important and needs to be empowered. In

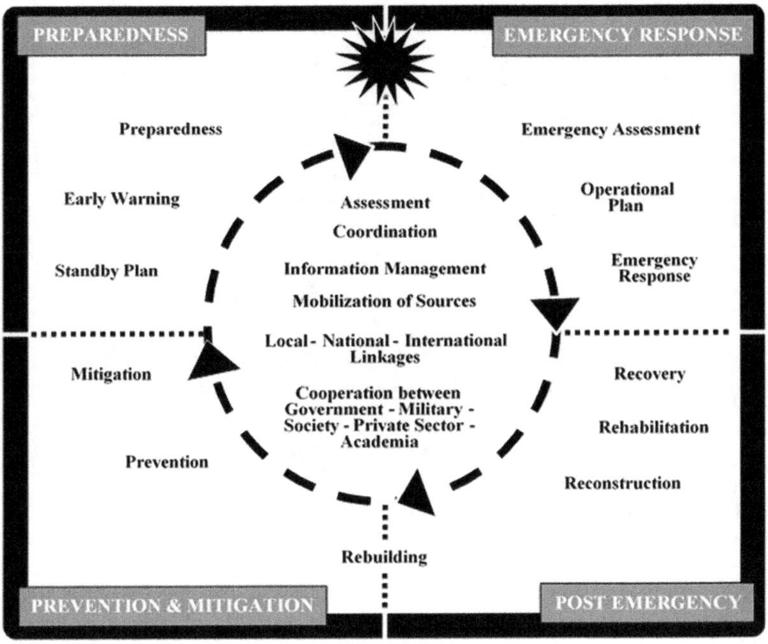

Fig. 1 Disaster management cycle (adopted and translated with permission from Setianto [4])

addition, the readiness of data management and geospatial information is very necessary to minimize losses and accelerate the rehabilitation and reconstruction process in the affected areas.

In recent years various parties have prepared disaster risk maps but the lack of standardization in the preparation resulted in the production of various types of disaster risk maps developed by each institution. Setianto [4] stated that fundamental understanding of the concept of disaster becomes a strong basis in mapping disaster risk that can be applied to a Geographic Information Systems (GIS) and presented spatially in the form of Threat Maps, Vulnerability Maps, Capacity Maps and Disaster Risk Maps respectively. Figure 2 describes methods containing stages required to produce a comprehensive Disaster Risk Map in accordance with disaster management objectives.

Figure 2 indicates the importance of data collection. The availability of data is essential and makes considerable contribution to analyzing and diagnosing broad scale and discrete problem. Understanding the larger scale problems mean that inappropriate areas for development can be avoided completely. Detailed data allow the authority to identify the potential hazards that are likely to be encountered.

In developed countries with a broad experience on disaster management, the development plans includes requirements relating to consultation with the community.

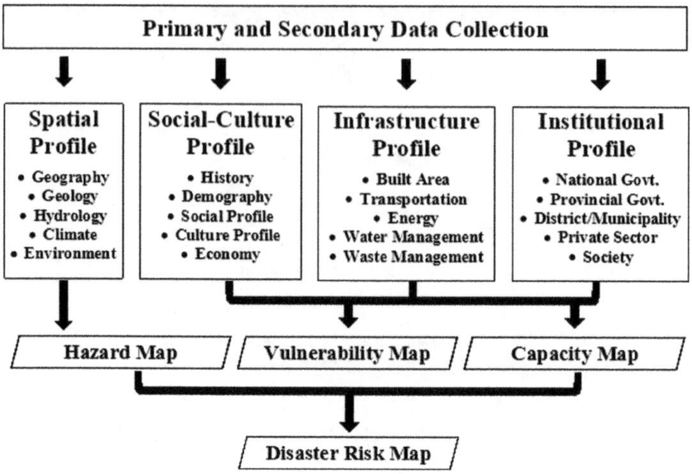

Fig. 2 Risk mapping methods (adopted and translated from StIRRRD [11])

Development plans, often referred to as strategic plans, are prepared by local councils with input from the community [12]. As such, any type of strategic plan produced by local council must consider the avoidance or mitigation of natural disasters.

3 Cultural Attributes in Disaster-Oriented Development Plans

The existence of Law No. 24 of 2007 on Disaster Management [2] shows that development policy in Indonesia that was previously less disaster-minded slowly but surely began to change. Previously disaster management tended to be responsive and spontaneous. Now disaster management has started to move towards preventive and more focused on disaster preparedness and risk reduction. Disaster risk reduction is one of a series of disaster management activities that take place in conditions before disasters occur (pre-disaster).

At a time when the role of the government is demanded more in disaster risk reduction, there are often differences in definitions and attitude towards disasters between the government and the community alike. The government's perspective in disaster management is often perceived the opposite way by the community in the location of disasters. Research conducted on one of the villages on the southwest slope of Mount Merapi in Central Java Province shows that there is a difference in perception between the government and the community on the threat of volcanic eruptions [6]. The government has been in a long persuasion to transmigrate villagers to one of the islands in Indonesia but the element of rejection appear within the community who decided to stay put on the slope of the volcano and adjusting themselves to the dangers based on local knowledge systems in the form of direct monitoring and story-telling of hereditary myths about volcanic activities. From economic aspect, when government set the exclusion zone as preparedness measures for volcanic hazard, the community often sees it as controversial. Recent study in other continent severely affected by a prolonged volcanic eruption suggests that the governmental decision to extent the exclusion zone because of volcano increasing activity met with large disagreement, particularly from people who lost income during the period of extension [13].

In the case of an earthquake in West Sumatra Province, research by Effendi [14] showed that local people believe natural disasters are not a symptom of nature. The indigenous Minangkabau view and impose natural hazards as well as all natural symptoms as natural or normal as possible within the framework of their diverse cultural cosmology. Some people view earthquakes as entirely natural events that are scientifically understandable but others view the hazard as divine interference with human life so that they can only be understood in religious discourse as a test or punishment [6].

The results of the above research show that socio-cultural diversity in Indonesia affects the perceptions toward natural disasters, the impacts or victims caused, and post-disaster management. Understanding the cultural system will help the government in trying to find answers of why a community acts on a particular pattern of natural disasters as well as their impacts. By exploring this potential, it is expected that the community will be capable to independently manage pre-disaster, emergency response, rehabilitation and reconstruction process.

The government should not feel right on its own and ignore the values and norms within the community in the form of indigenous knowledge that speaks of natural events and the potential consequences. Various efforts in the form of technocracy and partnerships with stakeholders provide opportunities for people to live better and not become victims in the event of a disaster. Technocratic actions supported by local wisdom will confirm that community development in the framework of mitigation and social development is based on high cohesion [6]. In the framework of adoption of local wisdom, mitigation that maintains harmony and local cultural-based will be adaptive and can be transformed into a modern mitigation system through partnerships with governments and non-governmental organizations to produce a more effective and successful mitigation systems. This will strengthen the mitigation strategies already attached to the community.

In planning infrastructure development, the government should accommodate local wisdom in the community. In addition, knowledge of the history of disasters obtained from local community will greatly help the development actors in planning the right infrastructure. The results based on theoretical calculation may not be suitable for different region and produce an over-estimated or under-estimated result. Marking flood water level, for instance, has been recognized for a long period of time. The evidence of flood levels with the year of occurrences can be seen in a wall or any surviving infrastructure.

The practice of engraving the flood level on tablets of stone, set in gardens or house walls seem to have started more than three centuries ago [15]. In more simple fashion

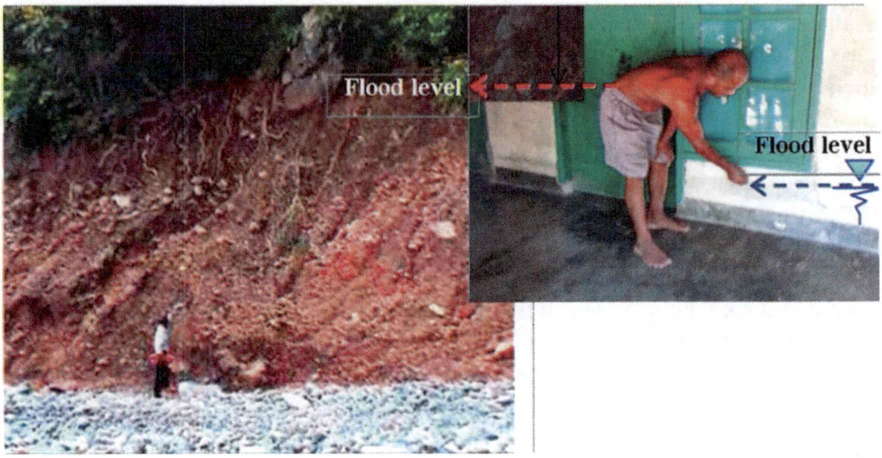

Fig. 3 Flood level-marking based on personal information (Courtesy of West Lombok public works and housing office, 2017)

the determination of flood elevation plans in the planning of river embankments pay attention to traces of flooding based on hereditary experience of the surrounding community (see Fig. 3).

In certain infrastructures such as reservoirs, operation and maintenance are based on standard operating procedure (SOP) that has been prepared. Other anticipation measures that Operational and Maintenance Officers must be equipped with knowledge of the Emergency Preparedness Plan (EPP) including warning systems and activities to increase public awareness of possible hazards [16]. It is important that the SOP and EPP accommodate the environmental consideration and the involvement of the community in deciding if the risk is acceptable for their particular area.

Countries with advanced knowledge in disaster management, prepare the development planning based on the need for reducing risk and emergency management plans. Experiences show that Indonesian are familiar with discrete hazardous events such as annual flooding and forest fires as well as ongoing adversity resulting from disasters. Adaptive strategies have been developed to ensure community wellbeing. These strategies may works well for small-scale disasters but the community requires a more effective strategy to anticipate larger-scale disasters. Effective strategies are perceived by community as a key component in ensuring well organized emergency management response. Government officials as disaster management actors are required to be in the forefront in implementing effective strategies and the community are eagerly expecting their part.

A schematic concept has been proposed to be applied in Indonesia within the context of disaster risk reduction. Figure 4 shows that consultation with the community is an important aspect in the process of preparing development plan. It should begin with engaging communities to allow them to forward opinion based on experience and local knowledge. During the early round of the processes, the community

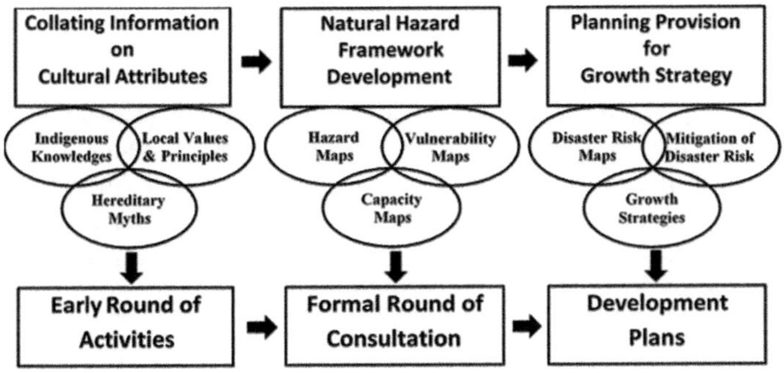

Fig. 4 Conceptualization of incorporating cultural attributes into development plans

has the opportunity to incorporate traditional knowledge, values and practices into disaster related development plans. The community can persuade government that the community perspective in disaster management and recovery initiatives are relevant to the government plan.

Under early round of the process, a large amount of information collection occurs. The relevant existing knowledge may have to be reviewed and the gaps with the existing information gathered by government may need to be identified. Once the required information has been obtained, the process continues to formal round of consultation with the involvement of the experts. The consequences of development planning associated with natural hazards require specialist reports. During this formal round of consultation, constraints and limitation have been taken into account in the preparation of development plans. The end result is the development plans which incorporate the risks of development associated with natural disasters in a manner that is aligned with the expectation of the community. In the event that developments are considered in areas subject to natural disasters, the primary focus is on the likelihood of the event with consideration of the potential consequences.

Disaster risk-based approach allow for a holistic consideration of both the consequences and the likelihood of a disaster, thereby allowing the government to make robust planning decision regarding natural disasters and future development. Development planning accompanied by a pre-event recovery planning [17], for example in the form of licensing and approval related to potential disasters, and also considers a "development moratorium" in which development decisions are halted for a certain time limit as well as zoning for evacuation infrastructure. For disaster risk reduction, land administration planning must be made in such a way, disaster risk-based and accommodate cultural attributes. It is an inevitable that the regional development planning should incorporate traditional knowledge, values and practices to address disaster-related risk.

4 Conclusions

Traditionally the planning approach for addressing natural disaster has been based on the likelihood of the occurrence of an event and disaster risk reduction have always been focused on the government. Recent developments require the need to increase community involvement in activities related to disaster risk reduction.

Based on the above description, some conclusions can be drawn as follows:

1. High frequency of disasters in Indonesia has led to the establishment of Law Number 24/2007 on Disaster Management as well as triggering a paradigm change in disaster management from responsive and spontaneous approaches to a preventive and more focused on preparedness and disaster risk reduction.
2. The existence of socio-cultural diversity in Indonesia which causes multi perceptions on natural disasters, the level of impacts or the number of casualties, and post-disaster management requires government to understand such cultural system in order to obtain information about community responses toward certain pattern of natural disasters along with their impacts.
3. Disaster management actors must be competent and certified, have a fundamental understanding of the concept of disaster risk reduction including readiness in the management of data and geospatial information that is indispensable to minimize losses and accelerate the rehabilitation and reconstruction process in the affected areas.
4. In the implementation of disaster risk reduction and infrastructure development planning, the government should accommodate local wisdom in the form of traditional attributes and values within the community including the absorption of information about the history of disasters from indigenous people.
5. It is important that there is a robust community engagement and communication process to enable community and the government to reach an agreement on risk acceptability as well as mitigation option. Consultation with the community is a key aspect in the process of preparing strategies for disaster risk reduction. A concept of development planning related to disaster risk reduction should take into account the risks of development associated with natural hazards.

Acknowledgements The author wishes to thank the GNS Science of New Zealand for their support during New Zealand Comparative Study Programme 2015 through the project of Strengthened Indonesian Resilience: Reducing Risk from Disaster (StIRRRD). The appreciations are also addressed to Dr. Agung Setianto for his permission to adopt and translate the picture of Disaster Management Cycle and to the West Lombok Public Works and Housing Office for providing photographs presented in this paper.

References

1. BNPB (2021) Indonesian disaster infographics 2020 in www.bnpb.go.id (in Bahasa Indonesia)

2. Indonesia State Gazette (2007) Law of the Republic of Indonesia number 24 year 2007 on disaster management. State Gazette of the Republic of Indonesia Year 2007 Number 66 (in Bahasa Indonesia)
3. Twigg J (2001) Sustainable livelihood and vulnerability to disaster. The Disaster Mitigation Institute
4. Setianto A (2015) Remote sensing application and geographic information system for mapping disaster prone areas. In: Proceeding of disaster risk reduction action plan workshop, strengthening Indonesia's resilience through disaster risk reduction, pp 33–44. Mataram (in Bahasa Indonesia)
5. Abdullah I (2008) Construction and social reproduction of natural disasters, working papers in interdiciplinary studies No. 01. Graduate School of UGM, Yogyakarta (in Bahasa Indonesia)
6. Anantasari E (2015) Living harmoniously with disasters: integration of socio-cultural aspects (local wisdom) in disaster risk reduction. In: Proceeding of disaster risk reduction action plan workshop, strengthening Indonesia's resilience through disaster risk reduction, pp 45–52. Mataram (in Bahasa Indonesia)
7. Nurjanah D, Kuswanda D, Siswanto A (2012) Disaster management. Alfabeta, Bandung (in Bahasa Indonesia)
8. Carter WN (2008) Disaster management: a disaster manager's handbook. Asian Development Bank, Manila
9. Kenney CM, Phibbs S (2015) A Maori love story: community-led disaster management in response to the Otautahi (Christchurch) earthquakes as a framework for action. Int J Disaster Risk Reduction
10. Saadi Y (2016) The role of universities in capacity building of disaster management actors. National Disaster Risk Reduction Month Commemoration, StIRRRD Technical Session (TS#3), Manado (in Bahasa Indonesia)
11. StIRRRD (2015) Handout of disaster risk reduction action plan workshop, strengthening Indonesia's resilience through disaster risk reduction, pp 49–57. Mataram (in Bahasa Indonesia)
12. Grace ES (2015) Incorporating natural hazards into development plans in New Zealand, strengthening Indonesia's resilience through disaster risk reduction. New Zealand Comparative Study Programme
13. Monteil C, Barclay J, Hicks A (2020) Remembering, forgetting, and absencing disasters in the post-disaster recovery process. Int J Disaster Risk Sci 11:287–299
14. Effendi N (2007) Disaster: experiences and cultural values of Minangkabau people. Indonesian Soc 33/2 (in Bahasa Indonesia)
15. Archer D (1992) Land of singing waters, rivers and great floods of Northumbria. Spredden Press, Stocksfield, England
16. Saadi Y (2013) Post-construction problems of Embung in Lombok Island and the operation and maintenance works. Proc Eng 54:648–660
17. Becker J (2015) Pre-event recovery planning, strengthened Indonesian resilience: reducing risk from disaster (StIRRRD), New Zealand Comparative Study Programme

Identification Risk Potential Hazard of Railway Project Based on the Work Breakdown Structure to Improve Safety Performance

D. V. Aryanto, L. S. Riantini, R. A. Machfudiyanto, and Y. Latief

Abstract Railway construction work is carried out in developing or developed countries around the world. The railway was built as a means of mass transportation to support connectivity between places. Railway construction work is one of the complex construction projects. Construction accidents can hamper many work activities. It is not uncommon for construction accidents to occur in construction that affects people, property, public safety, and the environment. For this reason, it is necessary to identify and analyze the risk of hazards from the beginning to prevent the occurrence of these construction accidents. This study aims to identify the potential hazards in each work package by reviewing work methods, work activities, and WBS-based resources for standardized railway construction using a qualitative approach. This research is expected to produce safety risk sources with the highest potential hazards in the railway construction works. The risk of construction accidents can be reduced or even zero risks of construction accidents (zero accident) during construction project activities.

Keywords Railway · Work breakdown structure · Risk · Safety performance

1 Introduction

In the past 5 years, Indonesia has participated in building infrastructure to become a developed country. One of the infrastructures currently under construction now is the Railway construction project. This is in line with the concept of the National Railway Master Plan [1], which was initiated by the Directorate General of Railways which aims to solve problems due to limited road service capacity so that the railways are ready for the implementation of national transportation capable of ensuring the movement of people and goods throughout Indonesia. Advanced technology, complex structure, quality, and high technical standards are required for long-term

D. V. Aryanto (✉) · L. S. Riantini · R. A. Machfudiyanto · Y. Latief
Universitas Indonesia, Jawa Barat 16424, Indonesia

S. A. Kristiawan et al. (eds.), *Proceedings of the 5th International Conference on Rehabilitation and Maintenance in Civil Engineering*, Lecture Notes in Civil Engineering 225,
https://doi.org/10.1007/978-981-16-9348-9_56

use [2]. Therefore, in planning the construction of a railway track, it must be done carefully with to obtain effective and efficient development results [3].

In general, the construction project building process is an activity that contains a lot of dangerous elements or substances [4]. Based on the study [3], Railway line construction has been investigated as one of the projects that causes the highest number of construction accidents. Particularly in the construction of the Burma-Siam Railway, the number of workers who fell victim to work accidents was 385.45 per 1000 workers [5]. The increase in cases of construction accidents that occur is due to a combination of unsafe acts and unsafe conditions [6].

In carrying out a railway construction project, the project plan must be appropriately considered, so that timeliness, quality, and cost can be achieved [7]. The planning process of the construction of the Railway Line construction needs to be considered carefully to achieve success in the rampant construction of the Railway Line construction. Correct and systematic planning and preparation of stages is the key to a successful project, for that WBS is needed as a basis for project planning.

WBS is very important in project planning. Planning that begins with the formation of the WBS needs to be explained as each construction activity as well as a structured work breakdown according to the scope of work and the standardized levels to produce good construction products. Each level of WBS that has been defined brings WBS to a level that is more and more complex namely activity. So that activities are significantly affected by risks, and that risks will impact work safety [9]. Thus, in this research the author will use the WBS method to identify the risks that occur in each activity in a work package to minimize all potential hazards.

2 Research Objectives

Identify sources of risk that are potentially dangerous and affect Construction Safety in Railway Line construction based on standardized WBS.

3 Literature Review

3.1 WBS of Railway

Railway Railroad tracks are a commercial service responsible for operating the rail transportation system that attracts passengers or goods [10]. Based on [11] as stated in Chapter 1 Article 1 paragraph 3, railway infrastructure is railway lines and stations, including the facilities necessary for the railway facility to be operated. Rail supporting facilities complement the operation of rail transportation, which can provide convenience and comfort to users of rail transport services. Railway infrastructure in more detail can be classified as Rail lines or roads, Station buildings, Bridges, Signals, and telecommunications.

Work Breakdown Structure Work Breakdown Structure (WBS) is a hierarchical arrangement of project components that breaks the project deliverable into smaller parts, making it easier for the project team to manage and implement [12]. Meanwhile, according to PMI [13], WBS is a foundation tool to help the planning team project overcome big project uncertainties by making it a project with a lower level of uncertainty. The creation of the WBS produces a foundational document that provides the basis for planning costs, resources, project schedules, and changes [14]. WBS has benefits that determine the success of the project; some of the benefits of WBS are to reduce the complexity of the project, create a schedule and cost estimation, compile a budget, plan risk management, and identify activities.

3.2 Construction Safety Performance

According to Endroyo et al. [15], the definition of construction safety is safety which includes 4 aspects, including worker safety (safe for labor), community safety (safe for people), construction safety during the project development process (safe for the property) and safe environment around the work location (safe for the environment). Safety performance is treated as a concept that revolves around maintaining a safe workplace. This can be achieved through the organization's safety culture and worker safety behaviors in the workplace environment or the ability to minimize work-related injuries in the workplace [16].

There are 6 indicators for construction safety performance: safety awareness, safety records, safety costs, accident costs, and the safety and productivity scores [17]. Safety performance has different indicators according to each researcher. According to [18], the work safety performance can be measured by 5 aspects: Injury severity rate, injury frequency rate, Project accident cost figures, average days change per disabling injury, and the number of accidents at work.

3.3 Risk

Risk is a measure of the probability or severity of an impact detrimental to life, health, property, or the environment [19]. Another definition of risk is an event or factors that are uncertain, and if they occur, it can negatively impact project objectives [12]. If these risk factors occur, it will increase the likelihood of not achieving the project objectives [20]. One way to achieve project objectives is to manage risk in the entire project implementation process [21].

According to Hamid et al. [6], the causes of construction safety risks are influenced by park behavior and conditions then are categorized as the important cause and

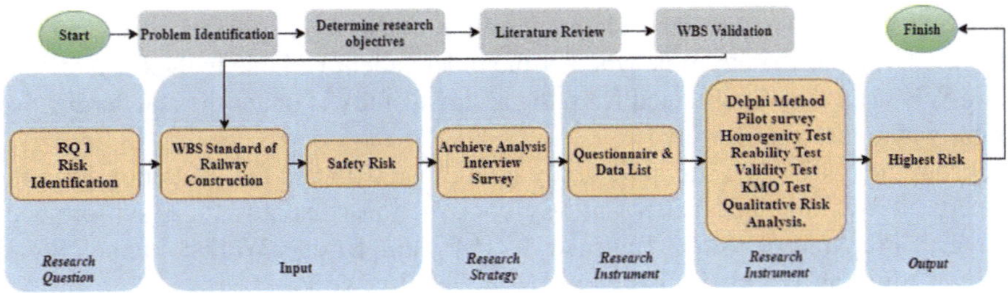

Fig. 1 Research flow chart

other causes. The among causes are (1) Use of defective or damaged equipment; (2) Failure to apply the use of PPE from its workers; (3) Human Factor—ex: under the influence of drugs; (4) Lack Of commitment from workers to policies safety; (5) Poor construction site management.

4 Methodology

This research was conducted with a qualitative approach to answer research objectives. The survey and discussion were conducted using a structured research instrument in the form of a questionnaire to experts from railway construction work with more than 10 years of experience. The research flow can be seen from Fig. 1.

5 Result and Discussion

5.1 WBS Standard of Railway Construction

In the WBS preparation process, the determination of the WBS level is from archival analysis with the benchmarking Bill of Quantity (BOQ) carried out 3 railway projects in Indonesia, the international railway regulatory standards, and refers to the Railways Law No. 23 of 2007. Then the WBS arranged into 6 levels/levels, which consists of WBS Level 1: Name of Sub-Project; WBS Level 2: Group of Work; WBS Level 3: Type of Work; WBS Level 4: Work Package; WBS Level 5: Work Activities; WBS Level 6: Resources (Labor, Materials, Equipment).

Experts in the field of railway construction need to validate the WBS to make standardized WBS. 3 experts were used as resource persons with at least 25 years of work experience in the railway construction field and understand the WBS concept in general construction work, especially in railway construction project work. The validation of WBS is to determine whether the independent variable in the form of standard WBS (level 1–6) obtained from the archival study is under the standardization of railway construction work.

From the results of a literature study of archival data and the validation form the experts, it was obtained WBS Railway construction standards for 6 sub-projects, namely: Preparatory Work; Railroad Works; Rail Bridge Works; Tunnel Work; Station Building Works; Railway Operation Facility Works.

5.2 *Identification of Risk Affecting Safety Performance*

The risk identification process is carried out from the WBS, which is viewed from the lowest level of WBS as at level 4: Work Packages, in which each work package is detailed per activity with some literature that looks at potential risks to 4 aspects, namely: Labor, Property (Material and Equipment), Environment, and Public Safety. The literature review produces as many as 985 risks that affect construction safety performance (229 risks for Railroad Works, 373 risks for Rail Bridge Works, and 383 risks for Tunnel Works).

Then on this potential risk, a validation process is carried out with an expert in the field of safety or an expert in railway projects by discussing whether if the risk is relevant to the work package.

By the validation results, those identified risks are admissible and there are 19 eliminated risks eliminated so that in total, there are 966 risks associated with construction safety performance (229 risks for Railroad Works and 365 risks for Rail Bridge Works, and 372 risks for Tunnel Works). The risk variables that experts have validated are then carried out by pilot survey testing to test whether all project workers easily understand the identified risks.

5.3 *Risks Assessment*

The validated risk variable is then calculated the risk value to determine the level of risk belonging to the 'high' category based on the assessment matrix in PMBOK and through respondents' opinions and expert validation. The number of respondents used in the study was 30 people with the criteria of respondents who filled out the frequency and impact questionnaires based on at least 5 years of experience working in railway construction. The risk assessment is obtained by multiplying the frequency rates and severity rates of each risk.

$$Risk = Probability \times Impact \tag{1}$$

Table 1 Probability rate scale indicators

Scale	Rating	Remarks	Weight
1	Very low	Very unlikely to occur, only under certain conditions	0.1
2	Low	Unlikely to occur	0.3
3	Moderate	Possible occur in every condition	0.5
4	High	Likely to occur	0.7
5	Very high	Very likely to occur	0.9

Table 2 Severity rate scale indicators

Scale	Rating	Remarks	Weight
1	Very low	Has no impact on construction safety performance	0.1
2	Low	Little impact on construction safety performance	0.3
3	Moderate	Sufficient impact on construction safety performance	0.5
4	High	Impact on construction safety performance	0.7
5	Very high	Very impact on construction safety performance	0.9

This risk assessment is carried out by distributing questionnaires using a Likert scale of 1–5 for frequency rate and severity rate with the following indicators scale (Tables 1 and 2).

Then, the multiplication result between the frequency level and the impact level is converted according to the risk assessment matrix referring to the PMBOK 6th edition and generate risk category, as follows (Table 3).

Based on the risk assessment results that has been carried out, there are 37 risks identified as high-risk levels. The method for assessing high risk is based on data processing results generated from 30 qualitatively analyzed respondents who refer to the PMBOK assessment matrix, with the following details (Table 4).

From the identified high risks, initial control can be carried out against high risks; the results of discussions with experts are that several preventive actions can prevent these risks: including: Completing workers' PPE, installing K3 signs, completing work safety equipment, conducting safety morning talks before work, Make Work Instructions for each job and tool operation, and Ensure safe worksite conditions.

Table 3 Risk category

Risk Score	Risk Category
0.18-0.72	High Risk
0.06-0.17	Moderate Risk
0.01-0.05	Low Risk

Table 4 Highest risks

Work package	Activity	Risk affect OHS performance	Risk score	Risk rank
Railroad works				
Soft rock excavation	Mechanical excavation	The fall of heavy equipment due to unstable soil	0.228	13
		A landslide occurred in the excavation	0.210	18
	Stony soil disposal	Traffic accident due to scattered land on the arterial road	0.229	12
Rock excavation	Stone blasting	Crushed or hit by blasting debris	0.249	3
		Noise resulting from explosion	0.228	13
Asphalt pavement excavation	Mechanical excavation	People buried by landslides	0.248	5
Structure excavation	Stony soil disposal	Lack of waste control/project material waste	0.195	24
Precast concrete drainage	Precast concrete drainage manufacturing	Precast concrete fell on the workers	0.186	27
Ballast work	Stone sowing	Heavy equipment crashed into nearby workers/facilities	0.182	36
Sub ballast work		There was a landslide (soil, rocks, etc.)	0.199	21
Rail bridge works				
Land clearing and stripping	Carry out land clearing and stripping	Accidents during equipment mobilization	0.190	25
Rock excavation	Stone blasting	Contaminated underground water	0.230	8
Precast concrete piles	Erection of precast concrete piles	Workers are hit by piles during lifting	0.216	16
Cast concrete drill piles	Wash drilling	Dirty/slippery roads due to scattered soil	0.203	20
Abutment work/column/pier head (concrete cast in situ)	Formwork work	Formwork collapse	0.213	17
	Concrete works	Broken sling crane	0.181	37
		Fall from a height	0.232	7
Elastomer bearing pad	Bearing pad installation	Fall from a height	0.233	6

(continued)

Table 4 (continued)

Work package	Activity	Risk affect OHS performance	Risk score	Risk rank
Post-tension prestressed steel girder	Setting girder	Crashed by precast material	0.184	30
	Stressing the bridge girder	Got electric shock	0.331	1
Transportation and storage of precast concrete	Precast concrete delivery units	Crashed by precast material	0.184	32
	Storage of precast concrete units	Crashed by precast material	0.184	32
Erection using the double crane method	The placement of the crane on two different sides of the bridge	A worker is hit by a crane	0.189	26
		Crane falls	0.182	35
	Girder placement on bearing pad	Girder fell and hit the worker	0.207	19
Diaphragm work	Stressing the bridge girder	Got electric shock	0.326	2
	Implementation of segmental diaphragm beams installation	Clamped beam segments	0.183	34
Floor plate work	Concrete works	Fall from a height	0.184	31
Tunnel works				
Regular excavation	Land disposal	Traffic accident due to scattered land on the arterial road	0.221	15
Rock excavation	Stone blasting	Ear disorders due to noise	0.230	10
	Stone removal	Material splattered on the road	0.185	29
Tunnel excavation	Stone blasting	Crushed or hit by blasting debris	0.249	3
		Ear disorders due to noise	0.230	10
		Contaminated underground water	0.230	8
Tunnel curb concrete	Formwork works	Formworks collapse	0.197	22
Longitudinal side drainage	Precast drainage installation	Precast concrete fell on the workers	0.186	27
Crossing drainage		Precast concrete is damaged/cannot be used again	0.196	23

6 Conclusions

From these results of the study, it can be concluded that 966 risks have been identified in railway construction projects. There are 37 highest risks in sub-project of railway, with details, 10 Railroad Works, 18 Rail Bridge Works, and 9 Tunnel Works, which are the most dangerous risks in the stressing process of bridge girders in Girder Installation Works. The identified risk factors can be used as a reference in preparing the safety plan document. With identifying these potential hazards, preparing a safety plan for contractors and other stakeholders to determine risks, and evaluating the project can be shortened because of the detailed explanation.

References

1. Kementrian Perhubungan DJP (2018) Reviu rencana induk perkeretaapian nasional (RIPNas). Kementrian Perhubungan 134
2. Li Q, Liu R, Zhang J, Sun Q (2014) Quality risk management model for railway construction projects. Proc Eng 84:195–203. https://doi.org/10.1016/j.proeng.2014.10.426
3. Yu QZ, Ding LY, Zhou C, Luo HB (2014) Analysis of factors influencing safety management for metro construction in China. Accid Anal Prev 68:131–138. https://doi.org/10.1016/j.aap.2013.07.016
4. Tsang YT, Fung IWH, Tam VWY, Sing CP, Lu CT (2017) Development of an accident modelling in the Hong Kong construction industry. Int J Constr Manage 17(2):124–131. https://doi.org/10.1080/15623599.2016.1222664
5. Migiro G (2018) The 10 deadliest construction projects in the world—WorldAtlas.com. [Online]. Available: https://www.worldatlas.com/articles/the-10-deadliest-construction-projects-in-the-world.html. Accessed 18 April 2020
6. Hamid ARA, Majid MZA, Singh B (2008) Causes of accidents at construction sites. Malays J Civ Eng 20(2):242–259
7. Hamzaoui F, Taillandier F, Mehdizadeh R, Breysse D, Allal A (2015) Evolutive risk breakdown structure for managing construction project risks: application to a railway project in Algeria. Eur J Environ Civ Eng 19(2):238–262. https://doi.org/10.1080/19648189.2014.939416
8. Elsye V, Latief Y, Sagita L (2018) Development of work breakdown structure (WBS) standard for producing the risk based structural work safety plan of building. MATEC Web Conf 147. https://doi.org/10.1051/matecconf/201814706003
9. Satrio G, Latief Y (2018) Development of risk-based standardized WBS (work breakdown structure) for safety planning of cable stayed bridge project. In: Proceedings of the international conference on industrial engineering and operations management, vol 2018, pp 2694–2710
10. AREMA (2003) Practical guide to railway engineering. AREMA COMMITEE
11. Undang-Undang Republik Indonesia Nomor 23 Tahun 2007. Perkeretaapian 3:108 (2007)
12. PMBOK (2017) A guide to the project management body of knowledge (PMBOK Guide), 6th edn. Project Management Institute, Newtown Square, Pennsylvania
13. PMI (2006) Practice standard for work breakdown structures. 2nd edn.
14. Wardahni NI, Latief Y, Machfudiyanto RA (2020) Development of safety plan to improve OHS (occupational health and safety) performance for construction of dam (supporting infrastructure) based on WBS (work breakdown structure). IOP Conf Ser Earth Environ Sci 426(1). https://doi.org/10.1088/1755-1315/426/1/012017
15. Endroyo B, Suraji A, Besari MS (2017) Model of the maturity of pre-construction safety planning. Proc Eng 171:413–418. https://doi.org/10.1016/j.proeng.2017.01.351

16. Sanni-Anibire MO, Mahmoud AS, Hassanain MA, Salami BA (2020) A risk assessment approach for enhancing construction safety performance. Saf Sci 121:15–29. https://doi.org/10.1016/j.ssci.2019.08.044
17. Lu M, Cheung CM, Li H, Hsu SC (2016) Understanding the relationship between safety investment and safety performance of construction projects through agent-based modeling. Accid Anal Prev 94:8–17. https://doi.org/10.1016/j.aap.2016.05.014
18. de la Garza JM, Hancher DE, Decker L (1998) Analysis of safety indicators in construction by Jesus M. de la Garza; Associate Member, ASCE, Donn E. Hancher, 2 and Lisa Decker. Construction 81(Cii):2–4
19. Kementrian PUPR (2019) Permen PUPR No. 21, no. 1 1–25
20. Yao Y, Peng Y, Li X, Zhang A (2018) Research on safety risk management of civil construction projects based on risk matrix method. IOP Conf Ser Mater Sci Eng 392(6). https://doi.org/10.1088/1757-899X/392/6/062080
21. Sharma SK, Swain N (2011) Risk management in construction projects. Asia Pacific Bus Rev 7(3):107–120. https://doi.org/10.1177/097324701100700310

Identification Factors of Safety Climate, Awareness, and Behaviors to Improve Safety Performance in Telecommunication Tower Construction at PT X

B. F. Athaya, L. S. Riantini, and R. A. Machfudiyanto

Abstract Telecommunication infrastructure has an important role in economic development in Indonesia. Along with the development of the times, the need for a network to access the internet increases in demand. This has resulted in increased demand for telecommunications tower infrastructure development as the infrastructure for telecommunications operators in Indonesia to develop their businesses. However, with the high development of infrastructure development, there is a possibility that there will be risks to worker safety. This study aims to describe the safety climate in the construction of PT X Telecommunication Tower which consists of, a form of work safety awareness among business owners and service providers, forms of work safety behavior from workers to business owners and service providers, as well as knowing, and overcoming obstacles or obstacles to the concept with the hope of reducing the number of work accidents. The research method used is literature review and case study review to obtain variables and indicators that show the relationship and influence of safety climate, safety behavior, and safety awareness to improve safety performance in PT X telecommunication tower construction. Then the selected variables and indicators will go through the expert validation process. The results of data collection from experts and interviews that have been conducted, obtained 6 indicators of safety climate, 5 indicators of safety behavior, 3 indicators of safety awareness, and 3 indicators of safety performance.

Keywords Telecommunication · Safety climate · Safety behavior · Safety awareness · Safety performance

B. F. Athaya (✉) · L. S. Riantini · R. A. Machfudiyanto
Civil Engineering Department, Faculty of Engineering, Universitas Indonesia, Depok, Indonesia
e-mail: barra.farras@ui.ac.id

L. S. Riantini
e-mail: lsagita@eng.ui.ac.id

R. A. Machfudiyanto
e-mail: rossyarmyn@eng.ui.ac.id

S. A. Kristiawan et al. (eds.), *Proceedings of the 5th International Conference on Rehabilitation and Maintenance in Civil Engineering*, Lecture Notes in Civil Engineering 225,
https://doi.org/10.1007/978-981-16-9348-9_57

1 Introduction

Cellular telecommunication tower/BTS tower (Base Transceiver Station) is a device that functions to place a signal transmitting antenna (access network) to provide services to subscribers around the tower. The need for telecommunication towers that are getting faster has encouraged people to always be creative by creating new technology [1].

In line with the high development of telecommunication tower construction in Indonesia, the level of work safety must also be increased to reduce the rate of work accidents in construction projects [2], especially at Telecommunication Towers. Statistics show that the construction industry still faces safety issues, which are worth studying. Most accidents are caused by humans, which means that the risk of accidents and injuries can be reduced with safety behaviors [3].

Meanwhile, individual safety behavior can be improved by increasing self-awareness of safety [4]. In other words, individual behavior may be the result of certain environmental factors in the organization, such as the safety climate [5]. Therefore, to reduce construction accidents, both individual and organizational factors must be considered. From an individual and organizational perspective, it is important to analyze the relationship between safety climate, safety awareness, and safety behavior [6].

This study aims to identify and analyze the work safety climate in the construction of PT X Telecommunication Tower which consists of forms of work safety awareness among business owners and service providers, forms of work safety behavior from workers to business owners and service providers, as well as forms of work safety behavior. From workers to business owners and service providers, as well as knowing, and overcoming conceptual barriers or obstacles in the hope of reducing the number of work accidents.

2 Theoretical Study

2.1 Safety Climate

The safety climate describes the perception of the value of safety in the work environment and can be distinguished by attitudes, namely individual beliefs and feelings about certain objects or activities [7]. Safety climate as a picture that is felt or related to workers' perceptions of the importance of safety and how this can be determined in the organization [8].

2.2 Safety Behavior

Safety behavior is an approach designed to directly change safety-related behavior through the application of behavioral principles and strategies such as peer observation of safe behavior, goal setting, and performance feedback [9]. Individual behavior may be the result of certain environmental factors in the organization, such as safety climate. Therefore, to reduce construction accidents, individual factors and organizational factors must be considered [6].

2.3 Safety Awareness

Safety awareness is a concept that contains several dimensions of work safety. Awareness is a process of internalizing the information obtained and becoming the values that are embraced and can be manifested every day [10]. Safety awareness is a reaction to work safety-related events that occur outside of the individual, including the experience of a condition that can affect worker activities [11].

2.4 Safety Performance

Safety performance shows the results of safe work records for future periods, where safety performance has a positive effect on project performance. Safety performance is a function of many factors including safety culture [12]. Safety performance is work behavior that is relevant to safety that can be conceptualized the same as other work behaviors that are the result of work, so it must be determined by the knowledge and skills needed to improve a safety performance [5].

3 Methodology

This research uses a qualitative approach. Through this approach, it is hoped that the development of a concept to improve safety performance in the construction of PT X Telecommunication Tower by using the parameters of Safety Climate, Awareness, and Behavior. This qualitative approach is also supported by quantitative data obtained from survey results through a questionnaire instrument to obtain data that is more relevant to related research. A qualitative approach is taken to explore further the existing conditions and the form of the relationship model of the climate safety variable, awareness, and behavior in the construction of the Telecommunication Tower PT X. Before conducting further research, expert validation will be carried out using the Delphi Method to validate the variables to be used in this study. The

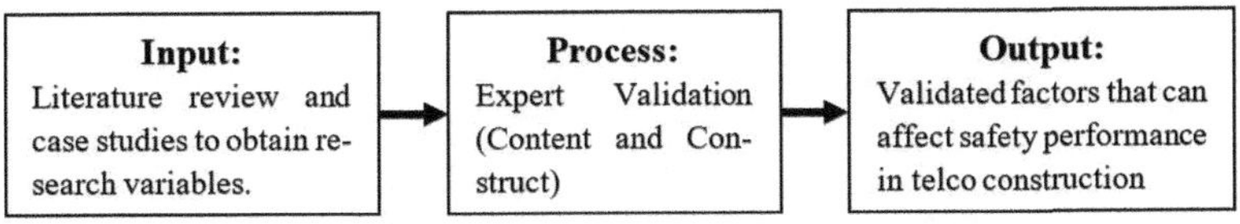

Fig. 1 Research flow line (input-process-output)

qualitative research strategy used in this research is a case study. That is, choosing an event or symptom to be investigated by applying various methods. The method used is in the form of interviews, observation and document analysis. The choice of strategy is related to the purpose of this study, namely explanative, this study aims to improve safety performance in the construction work environment of PT X Telecommunication Tower.

The survey questionnaire allows the researcher to answer questions related to the form of content analysis and the construct of this research. Thus, a validated questionnaire with several modifications was adopted for this study to collect and analyze data on Safety Climate, Safety Behavior, Safety Awareness, and Safety Performance. The questionnaire was also examined as a form of content validation and construct reliability by 1 academic and 4 industry practitioners as a form of validation by experts using the Delphi method.

Figure 1 shows that the research flow begins with literature review and case studies analysis to obtain research variables, then continues with expert validation (content and construct), and ends with the results of validated factors that can affect safety performance in Telco construction.

4 Result and Discussion

Research variables are used to deepen research methods based on factors related to research. In this case, the safety performance is the dependent variable and the independent variables are sought to facilitate research. A variable is defined as a symbol or concept that is assumed to be a set of values [13].

Based on previous studies and discussions with several experts, it has been determined and agreed that the three variables between safety climate, safety awareness and safety awareness are related to each other and play a role in improving safety performance. This linkage is supported by the existence of several additions to content and construction according to suggestions and input from experts to increase the range of research that will be carried out next.

Based on Fig. 2, it can be seen that there is a relationship between safety climate, safety awareness, safety behavior and safety performance. The safety climate is a predictor of accidents, although this relationship is bridged by the Safety Behavior, where in the safety climate there are two aspects, namely organizational policies and procedures as well as the work and implementation environment [14]. There is

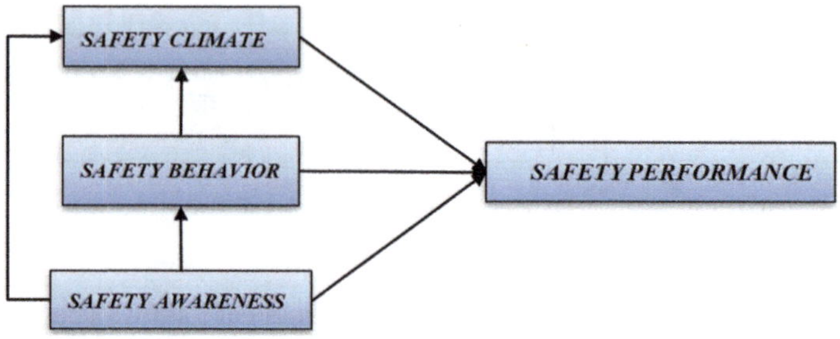

Fig. 2 Conceptual framework of this study

a relationship between safety climate and unsafe behavior which is a reference for measuring safety behavior and climate safety for work accidents [15]. The safety awareness element is about the causes of work accidents that most occur in construction work related to human errors, namely detecting and anticipating situations around when doing work. Safety behavior has been proven to prevent accidents from occurring in European and American countries [16].

After going through expert validation with 1 academic and 4 industry practitioners on this research instrument, the content and construct variables were obtained which will be used in the next research stage. Table 1 is the content and construct variables obtained based on expert validation as the result of this study.

Table 1 Variable and indicators that have been validated by experts

No. item	Variable	Indicators
X1	*Safety climate (SC)*	
X1.1	Communication support	1. Problems at work 2. Opinions of workers on policies 3. Discussion of the problem 4. The effect of changing procedures 5. Communication of changes 6. Concern between workers 7. Anticipate risks and consequences
X1.2	Job pressure	1. Time for planning 2. Time to do work 3. Time to get work done 4. Project schedule 5. Workload 6. External factor problem
X1.3	Work relationship	1. The future of workers 2. Work relationship 3. Company Morale

(continued)

Table 1 (continued)

No. item	Variable	Indicators
X1.4	Adequacy of procedures	1. Completeness of work procedures 2. Technically accurate job procedures 3. Work procedures are clearly available 4. Procedure compliance 5. Worker's understanding of procedures 6. Documentation management system
X1.5	PPE (Personal protective equipment)	1. PPE use is monitored 2. PPE users consulted 3. Findings from PPE monitoring 4. The use of PPE is enforced
X1.6	Safety rules	1. Understanding of safety rules 2. Implementation of safety rules 3. Safety rules constraints
X2	*Safety behavior (SB)*	
X2.1	Leadership	1. Superior support 2. Team engagement for a purpose 3. Superior involvement 4. Team involvement for decisions 5. OHS activity leadership 6. Responsibility 7. Role models in the team
X2.2	Management commitment	1. OHS organization 2. Making safety decisions 3. Safety policy 4. Check and investigate safety reports 5. Safety system audits 6. Information gathering mechanism 7. Safety management system
X2.3	Safety training	1. Safety training 2. OHS competency requirements 3. Efficiency of the safety training program 4. Job specific training 5. Worker monitoring 6. The relevance and effectiveness
X2.4	Allocation of resources	1. Availability of resources 2. Effectiveness of resource allocation 3. Adequacy of fund allocation
X2.5	Rewards	1. Recognition and praise 2. Worker performance 3. Incentive

(continued)

Table 1 (continued)

No. item	Variable	Indicators
X3	*Safety awareness (SA)*	
X3.1	Safety risk management	1. Possible danger 2. Engagement determines the likelihood of harm 3. Department of occupational safety and health 4. Hazard identification 5. Risk assessment procedures 6. Avoid or minimize the risk of harm 7. HIRADC guidelines 8. Management decision
X3.2	Safety promotion	1. Safety induction 2. Safety sign 3. Social security for workers 4. Work safety operational standards
X3.3	Workers safety awareness	1. Safety level awareness 2. Awareness of the use of PPE 3. Knowledge of safety 4. Employer assertiveness 5. Giving penalties to offenders 6. Worker culture 7. Worker safety culture
Y	*Safety performance (SP)*	
Y1	Safety compliance	1. Work safety procedures 2. Application of work safety procedures 3. The effectiveness of safety procedures
Y2	Safety participation	1. Importance of safety compliance 2. Safety participation of coworkers 3. Safety meeting
Y3	Number of accident and near-miss	1. Near-miss incident 2. Accident incident 3. Short-term lost time injury 4. Long-term lost time injury 5. Number of accident and near-miss

5 Conclusion

Safety climate, awareness, and behavior are the main factors that affect the work safety of a construction project. Aspects of safety, climate, awareness, and good behavior are expected to reduce and prevent disturbances to occupational health and safety, both on the part of the project owner and the service provider.

Based on literature review analysis and expert validation by five experts on safety climate, safety awareness, safety behavior, and safety performance to obtain research content and constructs, it is concluded that there are 6 content on climate safety:

Communication Support, Job Pressure, Work relationship, Adequacy of Procedures, PPE (Personal Protective Equipment), and Safety Rules. In the Safety climate there are 5 contents: Leadership, Management Commitment, Safety Training, Allocation of Resources, Rewards. There are 3 contents in Safety Awareness: Safety Risk Management, Safety Promotion, Workers Safety Awareness. Then the last one is the perception indicator of safety performance: Safety Compliance, Safety Participation, Number of Accident and Near-miss.

Acknowledgements The author is grateful to University of Indonesia for giving support in this research.

References

1. Ismail et al (2015) Analysis of BTS (base transceiver station) development planning based on earth curvature and Fresnel areas in the Southern Sumatra regional project
2. Asanka, Malik (2015) Study on the impact of accidents on construction projects
3. Hu K et al (2011) Factors influencing the risk of falls in the construction industry: a review of the evidence
4. Cooper (1994) Implementing the behavior based approach to safety: a practical guide
5. Neal et al (2000) The impact of organizational climate on safety climate and individual behavior
6. Mudan et al (2018) Relations between safety climate, awareness, and behavior in the Chinese construction industry: a hierarchical linear investigation
7. Guldenmund (2000) The nature of safety culture: a review of theory and research
8. Cooper et al (2004) Exploratory analysis of the safety climate and safety behavior relationship
9. Wirth, Sigurdsson (2018) WITHDRAWN: reprint of “when workplace safety depends on behavior change: topics for behavioral safety research
10. Yudhanto, Budiarto (2007) Education and behavior. Rineka Cipta, Jakarta, Indonesia
11. WHO—Regional Office for the Eastern Mediterranean (2001) Occupational health: a manual for primary health care workers. Cairo
12. Rossy, Yusuf (2017) A conceptual framework to development of construction safety culture in Indonesia
13. Goundar (2012) Chapter 3—Research methodology and research method
14. Neal, Griffin (2006) A study of the lagged relationships among safety climate, safety motivation, safety behavior, and accidents at the individual and group levels
15. Clarke (2006) The relationship between safety climate and safety performance: a meta-analytic review. J Occup Health Psychol 11(4):315–327
16. Ceren (2014) Examining safety behaviour with the safety climate and the theory of planned behaviour

Cost Structure of Construction Safety on High Residential Buildings in Indonesia

Ratih Fitriani, Yusuf Latief, and Rossy A. Machfudiyanto

Abstract The plan for the Cost implementation of Construction Safety Management System in the Public Works and Housing Ministry sector is part of the Construction Safety Plan, which is agreed upon and approved at the preparatory meeting for the implementation of construction work (Pre Construction Meeting). This study aims to identify and analyze the cost structure of safety in high residential buildings in Indonesia. The methodology used in this study is to use Bill of Quantity data from 16 projects in various regions in Indonesia and then analyzed using the Monte Carlo software to determine the optimal cost of safety. The results of this study indicate that the component of expert consultation related to construction safety is the least priority component, followed by the insurance and licensing component, then the construction safety personnel component which is not prioritized. For the construction safety personnel component, according to the construction cost budgeting, the personnel cost has been included, so that it is no longer budgeted for construction safety costs.

Keywords Cost structure · Construction safety · High residential buildings

1 Introduction

The high intensity of accidents in the construction sector in Indonesia is caused by a weak construction safety management system, namely lack of supervision [1], weak quality of human resources and a lack of understanding of external issues [2]. The Construction Safety Committee stated that the results of the investigation in general were that the contractor did not prioritize construction safety, in other

R. Fitriani · Y. Latief · R. A. Machfudiyanto (✉)
Department of Civil and Environmental Engineering, Faculty of Engineering, Universitas Indonesia, Depok, Indonesia

Y. Latief
e-mail: latief73@eng.ui.ac.id

S. A. Kristiawan et al. (eds.), *Proceedings of the 5th International Conference on Rehabilitation and Maintenance in Civil Engineering*, Lecture Notes in Civil Engineering 225,
https://doi.org/10.1007/978-981-16-9348-9_58

words the Construction Safety Management System had not been implemented optimally [3]. In connection with the target of building one million Flats listed in the National Medium Term Development Plan 2020–2024, the Indonesian Public Works and Housing Ministry stated that the implementation of the management system.

Construction safety is the key to infrastructure development [4]. However, most of the construction works in the public sector in many countries use the competitive tendering system. The implementation of competitive tendering which is carried out to select the contractor with the lowest price, this condition makes the contractor press down the price which results in a low budget for construction safety costs [5].

According to a survey conducted by the Public Works and Housing Ministry in 2017, the implementation of a construction safety management system in infrastructure development work in Indonesia is only 34% [3]. The main factor causing the low implementation of the construction safety management system is related to construction safety costs. This study aims to analyze the cost structure of safety in high residential buildings in Indonesia.

2 Theoretical Study

High Residential Buildings is a building built in an environment divided into sections that are functionally structured in horizontal and vertical directions and are units that each can be owned and used separately, which functions as a dwelling place, which is equipped with common parts, common objects and common ground. From the research results [5] the application of safety costs is one dimension in building a construction safety culture.

The Construction Expenses Index, especially in 2020, is a price index that describes the level of costly construction of a district/city compared to the reference city, namely Semarang City. The Construction Expenses Index data is obtained from the results of the Construction Expensive Price Survey especially for building/construction materials, heavy equipment rental, and construction service fees which are carried out in all districts/cities in Indonesia. The data is calculated based on quarterly price data for July 2019, October 2019, January 2020, and April 2020. The calculation of the Construction Expenses Index weigh diagram uses data from the Work Unit Price Analysis Book of the Public Works and Housing Ministry, Bill of Quantity (BoQ) and revenue budget realization data. And Regional Expenditure (APBD). The 2020 IKK is one of the main components used for calculating the General Allocation Fund (DAU) for the 2021 Fiscal Year [6].

In the Public Works and Housing Ministry Regulation 21/2019, it is stated that the cost components of the construction safety management system consist of at least [7]:

- preparation of a safety plan (comp. 1);
- outreach, promotion, and training (comp. 2);
- work protective equipment and personal protective equipment (comp. 3);

- insurance and licensing (comp. 4);
- construction Safety personnel (comp. 5);
- facilities, infrastructure and medical devices (comp. 6);
- signs and necessary traffic equipment or traffic management (comp. 7);
- consultation with experts related to Construction Safety (comp. 8), and
- activities and equipment related to the control of Construction Safety risk including the cost of environmental testing/inspection (comp. 9).

In the safety plan preparation component, there are subcomponents for preparing forms, work instructions as well as daily and monthly reporting. In the socialization, promotion and training component, there are subcomponents of socialization, communication and training related to construction safety. The third component, namely personal protective equipment and work protective equipment, consists of several subcomponents, namely safety helmet, safety shoes, safety harness, lifeline, safety deck, safety net, etc. The fourth component contains the construction insurance (CAR) subcomponent as well as heavy equipment testing. The fifth component consists of the subcomponent of construction safety experts, emergency responders, security officers, etc. Meanwhile, the sixth component consists of the initial health test sub-component for construction, fogging, clinic and ambulance workers if needed. The seventh component deals with the signs in the project such as warning signs, directions, prohibitions, etc. The eighth component is a facilitation related to consultation with construction safety experts, which can be structural experts, building engineers, geologists and others who ensure design standards have been met. The last component is activities and equipment related to construction safety risk control, including costs for environmental testing and inspection in accordance with the recommendations in environmental documents [5].

3 Methodology

In this study, a mapping of the construction safety cost structure will be carried out by the contractor implementing the construction of flats at the Public Works and Housing Ministry in 2020. The construction work of the flats in this study is a 3–6 storey building with a moderate construction safety hazard risk level. Meanwhile, the construction sites are grouped into 2 (two) parts, namely the western and eastern parts to be able to see the difference in the amount of costs in the two groups of locations, as shown in Fig. 1.

Furthermore, in the structure, the amount of construction safety costs will be seen and simulated using the Monte Carlo analysis. The analysis was carried out to obtain the optimal amount of the cost of implementing construction safety in the construction of flats in Indonesia.

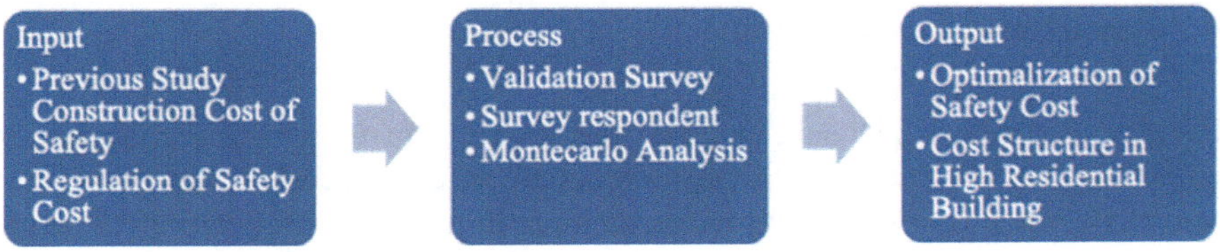

Fig. 1 Research methodology

Table 1 Project name

No.	Project name
West Indonesia	
1	High residential building in Batam
2	High residential building Indragiri Hulu
3	High residential building in Sarolangun
4	High residential building in BWS Kalimantan
5	High residential building in East Waringin City
6	High residential building in Surakarta City
7	High residential building in UGM student dormitory
8	High residential building in Pesisir Barat
9	High residential building in Brebes
10	High residential building in ASN PUPR
East Indonesia	
11	High residential building in ASN Region Sulawesi 1
12	High residential building in ASN Region Sulawesi 2
13	High residential building in TNI AL Sorong
14	High residential building in Maluku
15	High residential building in Tambrauw
16	High residential building in Tanimbar Island

The data used in the calculation of this simulation is the Bill of Quantity (BoQ) document on 16 (sixteen) construction works of the Public Works and Housing Ministry flats built in 2019–2020 as seen in Table 1. The location of the apartment construction is spread throughout Indonesia, which is grouped into 2 (two) namely west and east.

4 Result and Discussion

The Result of the study have mapped the structure of the existing construction safety costs for 16 flat projects. In the regulations, construction safety costs must be included in every infrastructure development in Indonesia [6] however, there are several components in construction safety costs that are left blank or not budgeted. In the expert validation process, information regarding the components of construction safety costs that are not budgeted can be obtained because there is no need for control using these components, or the items in these components are already owned by the contractor (obtained from previous projects), or the cost of these components has been obtained. Included in other costs. The following is an explanation of Structure of Construction Safety Costs in the High Residential Buildings as shown in Table 2.

Monte Carlo simulation

The calculation of construction safety costs in this study was carried out by simulating the data obtained from the BoQ of construction work at the Public Works and Housing Ministry in 2020. The simulation was carried out using the Monte Carlo method with Crystal Ball software. The data will be grouped into 2 (two) groups, namely based on the location of the western and eastern parts. This is done to be able to prove the difference in the amount of construction safety costs based on the construction location.

The first step is to compile a simulation design. For each project that will be simulated, construction safety costs are taken which consists of 9 components. The lowest and highest values for each component of construction safety costs are taken to calculate the estimated cost using the Monte Carlo analysis, as shown in Table 3.

Monte Carlo simulations are carried out by identifying the lowest and highest costs of each construction safety cost component as shown in the Table 4.

From the mapping results in Table 2, it is found that the components that are not budgeted for in many apartment constructions projects. These components are insurance and licensing, construction safety personnel and expert consultation related to construction safety. As for the three components, the component of consultation with construction safety experts was the component that was not budgeted at most, namely from 16 projects only 3 projects budgeted for this component. Next, the licensing insurance component was budgeted for only 7 projects, and the construction safety personnel component was budgeted for 8 projects.

There are two projects that do not budget for traffic signs and equipment components and each one project that does not budget for the safety plan preparation component, health infrastructure facility components and activity components and equipment related to construction safety control. All projects budget for socialization, promotion and training components and PPE and WPE components.

Table 2 Structure of construction safety costs in the high residential buildings

No.	Construction cost	Safety cost	Comp. 1	Comp. 2	Comp. 3	Comp. 4	Comp. 5	Comp. 6	Comp. 7	Comp. 8	Comp. 9
1	Rp. 17,542,751,000	Rp. 303,380,000	Rp. 3,250,000	Rp. 27,950,000	Rp. 74,180,000	Rp. 7,500,000	Rp. 176,000,000	Rp. 3,900,000	Rp. 6,300,000	Rp. –	Rp. 4,300,000
2	Rp. 18,835,689,900	Rp. 22,475,000	Rp. 800,000	Rp. 4,110,000	Rp. 11,055,000	Rp. –	Rp. –	Rp. 2,350,000	Rp. 250,000	Rp. –	Rp. 3,910,000
3	Rp. 18,600,195,000	Rp. 205,460,000	Rp. 3,500,000	Rp. 2,990,000	Rp. 80,870,000	Rp. 27,000,000	Rp. 80,000,000	Rp. 1,600,000	Rp. 5,200,000	Rp. –	Rp. 4,300,000
4	Rp. 45,515,773,600	Rp. 376,005,000	Rp. 5,000,000	Rp. 77,800,000	Rp. 251,105,000	Rp. –	Rp. –	Rp. 7,500,000	Rp. 10,000,000	Rp. –	Rp. 24,600,000
5	Rp. 21,553,900,000	Rp. 96,452,500	Rp. 2,700,000	Rp. 2,150,000	Rp. 57,325,000	Rp. 15,000,000	Rp. 187,500	Rp. 11,500,000	Rp. 850,000	Rp. –	Rp. 6,740,000
6	Rp. 16,344,700,000	Rp. 172,845,600	Rp. 2,896,000	Rp. 1,430,800	Rp. 31,879,800	Rp. 2,500,000	Rp. 114,000,000	Rp. 11,270,000	Rp. 5,390,000	Rp. –	Rp. 3,479,000
7	Rp. 13,455,651,000	Rp. 128,015,000	Rp. 5,500,000	Rp. 10,065,000	Rp. 37,165,000	Rp. 17,050,000	Rp. 5,000,000	Rp. 10,250,000	Rp. 5,000,000	Rp. 5,000,000	Rp. 21,550,000
8	Rp. 17,976,090,000	Rp. 107,028,000	Rp. 3,200,000	Rp. 12,840,000	Rp. 62,772,000	Rp. –	Rp. –	Rp. 4,800,000	Rp. 8,000,000	Rp. –	Rp. 15,416,000
9	Rp. 16,735,965,000	Rp. 48,500,000	Rp. 2,350,000	Rp. 1,400,000	Rp. 31,050,000	Rp. –	Rp. –	Rp. 11,500,000	Rp. –	Rp. –	Rp. 833,496
10	Rp. 27,853,207,271	Rp. 102,008,000	Rp. 3,000,000	Rp. 6,450,000	Rp. 54,940,000	Rp. –	Rp. –	Rp. 13,650,000	Rp. 4,500,000	Rp. –	Rp. 19,468,000
11	Rp. 42,514,842,000	Rp. 420,274,372	Rp. 7,221,500	Rp. 18,118,121	Rp. 351,778,921	Rp. –	Rp. –	Rp. 23,830,950	Rp. 4,101,955	Rp. –	Rp. 15,222,923
12	Rp. 38,334,670,000	Rp. 294,598,923	Rp. 5,050,000	Rp. 12,738,521	Rp. 246,630,801	Rp. –	Rp. –	Rp. 16,665,000	Rp. 2,868,500	Rp. –	Rp. 10,645,400
13	Rp. 23,664,697,622	Rp. 125,044,478	Rp. 2,200,000	Rp. 18,865,000	Rp. 31,460,000	Rp. 11,000,000	Rp. 38,500,000	Rp. 1,320,000	Rp. 1,039,500	Rp. 16,500,000	Rp. 4,159,978
14	Rp. 20,621,891,078	Rp. 67,012,000	Rp. –	Rp. 7,315,000	Rp. 5,137,000	Rp. –	Rp. 24,200,000	Rp. 22,825,000	Rp. –	Rp. –	Rp. 7,535,000
15	Rp. 25,775,233,000	Rp. 123,758,000	Rp. 4,800,000	Rp. 3,290,000	Rp. 25,410,000	Rp. 13,943,000	Rp. 45,500,000	Rp. 6,200,000	Rp. 1,750,000	Rp. 3,365,000	Rp. 19,500,000
16	Rp. 25,123,473,271	Rp. 104,243,100	Rp. 75,230,900	Rp. 1,126,250	Rp. 25,363,150	Rp. -	Rp. –	Rp. –	Rp. 2,522,800	Rp. –	Rp. –

Table 3 The lowest and highest costs of 9 (nine) components of construction safety costs for 16 flat projects by location

No.	Cost structure construction safety	Lowest cost	Highest cost
West Region			
1	Preparation component	800.000	5.500.000
2	Outreach and promotion	1.400.000	77.800.000
3	PPE	11.055.000	251.105.000
4	Insurance and licensing	0	27.000.000
5	Construction safety personnel	0	176.000.000
6	Health infrastructure facilities	2.350.000	13.650.000
7	Signs	0	10.000.000
8	Consultation with experts	0	5.000.000
9	Others related to hazard control	833.496	24.600.000
	Total cost of West SMKK	**22.475.000**	**376.005.000**
East Region			
1	Preparation component	0	75.230.900
2	Outreach and promotion	1.126.250	18.865.000
3	PPE	5.137.000	351.778.921
4	Insurance and licensing	0	13.943.000
5	Construction safety personnel	0	45.500.000
6	Health infrastructure facilities	0	23.830.950
7	Signs	0	4.101.955
8	Consultation with experts	0	16.500.000
9	Others related to hazard control	0	19.500.000
	Total cost of East SMKK	**67.012.000**	**420.274.370**

In the application of safety costs, in addition to paying attention to the components of safety costs, stakeholders must also be involved in its application so that it is optimal and effective [8]. The results of the Monte Carlo simulation can be seen in Table 4, where each project with a western and eastern location is analyzed and the amount is obtained according to 9 (nine) components of the construction safety cost structure in the ministerial regulation Public Works and Housing Ministry 21/2019.

Table 4 Simulations of high residential building construction safety costs in Indonesia by location

No.	Cost structure construction safety	Lowest cost	Highest cost	Optimal cost
West Region				
1	Preparation component	2.870.767	3.416.474	3.153.907
2	Outreach and promotion	35.793.495	43.311.460	39.542.540
3	PPE	119.309.513	143.488.280	131.182.954
4	Insurance and licensing	13.346.654	16.056.672	14.739.141
5	Construction safety personnel	7.299.301	8.705.269	7.984.271
6	Health infrastructure facilities	79.691.411	96.501.026	88.367.483
7	Signs	4.637.082	5.607.474	5.127.883
8	Consultation with experts	4.572.545	5.391.783	5.008.059
9	Others related to hazard control	11.488.262	13.875.758	12.720.852
	Total cost of construction safety	181.024.062	218.060.777	198.865.898
East Region				
1	Preparation component	35.179.322	42.389.405	38.675.792
2	Outreach and promotion	9.048.662	10.937.392	10.009.704
3	PPE	161.294.869	194.801.835	178.527.189
4	Insurance and licensing	11.454.381	13.448.432	12.479.709
5	Construction safety personnel	11.484.629	13.731.628	12.591.237
6	Health infrastructure facilities	31.817.964	37.524.153	34.870.913
7	Signs	2.348.325	2.789.993	2.571.673
8	Consultation with experts	9.080.267	10.794.800	9.952.653
9	Others related to hazard control	10.804.404	12.866.524	11.842.686
	Total cost of construction safety	222.439.408	263.697.106	243.536.924

5 Conclusion

From this research it can be concluded that in budgeting the construction safety costs for the construction of high-rise flats up to 6 floors or moderate risk, the component of expert consultation related to construction safety is the least priority component, followed by the insurance and licensing component, then the construction safety personnel component which is not prioritized.

For the construction safety personnel component, according to the construction cost budgeting, it has entered the personnel cost, so that it is no longer budgeted for construction safety costs. The Monte Carlo simulation clearly shows that the difference in the amount of construction safety costs for a flat construction project located in the west is smaller, compared to similar developments carried out in eastern Indonesia.

Acknowledgements The Authors would like to thank the financial support provided by Ministry of Research and Technology/National Research and Innovation Agency through PDUPT Grant 2021 with contract number: NKB-219/UN2.RST/HKP.05.00/2021 managed by the Directorate for Research and Community Engagement (DRPM) Ministry of Research and Technology/National Research and Innovation Agency (BRIN).

References

1. Rachman FF (2018) Menteri PUPR Buka Suara Penyebab Maraknya Kecelakaan Konstruksi
2. Bahfein S (2020) Kualitas SDM Rendah, Penyebab Kecelakaan Konstruksi
3. Fitriani R, Latief Y (2019) A conceptual framework of cost of safety model in infrastructure project based on work breakdown structure (wbs) to improve safety policy in indonesia. IOP Conf Ser Mater Sci Eng 650. https://doi.org/10.1088/1757-899X/650/1/012003
4. Choi TNY, Chan DWM, Chan APC (2011) Potential difficulties in applying the pay for safety scheme (PFSS) in construction projects. Accid Anal Prevent 48:10. https://doi.org/10.1016/j.aap.2011.04.015
5. Machfudiyanto RA, Latief Y, Arifuddin R, Yogiswara Y (2017) Identification of safety culture dimensions based on the implementation of OSH management system in construction company. Proc Eng 171:405–412
6. Peraturan Presiden Nomor 18 Tahun 2020 Tentang Rencana Pembangunan Jangka Menengah Nasional Tahun 2020–2024, Peraturan Presiden Nomor 18 Tahun 2020 C.F.R. (2020)
7. Pedoman Penerapan SMKK, 21 C.F.R. (2019)
8. Machfudiyanto RA, Latief Y, Suraji A, Sagita L (2020) Main task and functions stakeholders institutional to improve safety culture in construction sector. IOP Conf Ser Mater Sci Eng 771(1):012063

Adaptive Traffic Signal Control Using Fuzzy Logic Under Mixed Traffic Conditions

Budi Yulianto

Abstract The effectiveness of the intersection control performance is very influential on the occurrence of vehicular delays and traffic accidents. One of the systems for controlling vehicle movements at intersection is a traffic signal control. Traffic signal control widely used in Indonesia is Fixed Time Control (FTC). The deficiency of the FTC is that it cannot accommodate high traffic flow fluctuations, which causes excessive vehicle delays. The purpose of this study is to evaluate the Fuzzy Logic Traffic Signal Control (FLTSC) algorithm on the VISSIM simulation model for an isolated four-way intersection with two-stage signal type 42 and compare its performance with the optimized FTC under different traffic conditions. The simulation results showed the performance of the proposed FLTSC is generally better than the optimized FTC, especially in time-varying traffic cases.

Keywords Fuzzy logic · Signal control · Mixed traffic

1 Introduction

One of the most prominent problems in big cities of the world is traffic congestion, especially in developing countries where there is a significantly higher gap between growing vehicle ownership and new road construction. The traffic congestion is mainly associated with the use limited road space by high number of people at the same time. This is observed during peak periods in the road network of urban areas in most big cities in Indonesia, thereby, causing excessive delays and increment in air pollution. The inability to find a major solution is, however, projected to make the situation much worse in the years ahead due to the continuous growth of traffic. One of the efforts to overcome traffic congestion is by controlling the vehicles movement at intersections. Intersections are part of the road network which is a point of conflict for vehicle movements that have the potential to cause vehicular delays and traffic accidents as well as vehicle exhaust emissions that cause air pollution. In order to

B. Yulianto (✉)
Department of Civil Engineering, Sebelas Maret University, Surakarta, Indonesia
e-mail: budi.yulianto@ft.uns.ac.id

S. A. Kristiawan et al. (eds.), *Proceedings of the 5th International Conference on Rehabilitation and Maintenance in Civil Engineering*, Lecture Notes in Civil Engineering 225,
https://doi.org/10.1007/978-981-16-9348-9_59

minimize traffic congestion and accidents at the intersection, therefore, the vehicle movements at the intersection are regulated by using traffic signal control.

The performance of signalized intersection depends on type of traffic signal control in optimizing the signal setting parameters. The signalized traffic control at isolated intersections in Indonesia mostly uses FTC and this method allows the master controller to set the splits and cycle length based on predetermined values usually from historical traffic data which are kept constant during certain period. It was discovered that FTC produces good results when the condition of the traffic is normal but unable to deal the period when it is complex and time-varying [1–4]. The classic problem with the FTC is that updating of signal timing based on changing traffic conditions is rarely done so that signal timing is not optimized based on the latest conditions, because it is expensive and tiring works. In connection with this, to improve the performance of signalized intersections, it is necessary to use demand responsive traffic signal control, where signal timing based on the real time traffic data fluctuation according to information from the detector. Moreover, Adaptive Traffic Signal (ATS) control is discovered to have the ability of adjusting the green time length at a specific stage using the traffic demand associated with the stages having both right of way and red. A complex conventional mathematical model is, however, usually used to develop the controller algorithm.

The complexity associated with the determination of the appropriate signal setting parameters through the use of conventional mathematical models led scholars to apply artificial intelligence methods. An example of this is the fuzzy logic which was reported to be an interest mathematical method with prospects to model traffic and transportation processes described by subjectivity, uncertainty, and imprecision [5]. The first study observed to have applied fuzzy logic to the control of traffic signal was by Pappis and Mamdani [6] and this was followed by its wide usage in developing a theoretical and practical ATS controller [4, 7–12]. Yulianto [4] also applied this technique to design traffic signal to be used in mixed traffic conditions or when the traffic is untidy due to the fact that it does not follow lane-based operations. The fuzzy logic controller performs much better than optimized FTC in time varying traffic conditions. The purpose of this study is to evaluate the fuzzy logic traffic signal control algorithm proposed by Yulianto [4] to be applied in an isolated four-way intersection with two-stage signal type 42 [13] and compare its performance with the optimized FTC under two case studies namely constant traffic flow and fluctuated traffic flow during one-hour period.

2 Traffic Signal Control Strategies

2.1 Fixed Time Control

Webster [14] was reported to have developed the signal timing method which is mostly applied in FTC. The process includes the proposal of mathematical equations for the calculation of the signal setting plan to minimize the average vehicle delay for all vehicles through the use of an intersection. Meanwhile, the signal timing plan is usually designed in Indonesia using the methods provided by the Indonesian Highway Capacity Manual [13]. Moreover, the cycle length and green time were calculated using the equations adapted from the Webster model. It is important to note that the cycle length (seconds) was determined in this study based on the Indonesian Highway Capacity Manual (IHCM) using the following equation:

$$C = \frac{1.5LTI + 5}{1 - \sum FR_{CRIT}} \tag{1}$$

where LTI is the lost time per cycle (seconds), FR is the ratio of flow to saturation flow (Q/S), FR_{CRITn} is the maximum value of FR among the signal group for the same stage, and $\sum FR_{CRIT}$ is the summation of FR_{CRITn} for all N stages in the cycle which is written as

$$\sum FR_{CRIT} = \sum_{n=1}^{N} FR_{CRITn}. \tag{2}$$

The determination of the nth stage green time was through the following equation.

$$g_n = (C - LTI)\frac{FR_{CRITn}}{\sum FR_{CRIT}} \tag{3}$$

It is, therefore, possible for the intersection to perform optimally by setting the signal timing based on the traffic currently flowing through.

2.2 Adaptive Traffic Signal Control

The aim of the ATS control is to ensure the maintenance of the best desirable flow of traffic through the automatic adaptation of the individual controller characteristics including the cycle times, splits, and possibly stage sequences in order to satisfy the traffic conditions which are changing heavily [15]. Real-time traffic data on the traffic conditions are also set to the controller through the vehicular detectors installed on each approach lane upstream of the intersection. It is important to note that the ATS control logic is most times designed to be based on optimization process and this means it has the ability to optimize the flow of traffic using online method without confinement to particular cycle time. Meanwhile, it is possible for the signal timing plan to change at any time depending on the optimizing algorithm.

The major focus of recent research is on the application of artificial intelligence techniques such as fuzzy logic on intersection control. In general, the fuzzy logic traffic signal control structure consists of three main modules, namely fuzzification, fuzzy inference and defuzzification modules. Fuzzification is a process where the crisp input values for a fuzzy logic traffic signal control are converted into fuzzy values, which are represented by a degree of membership. The input data from vehicular detectors provide precise information on the current conditions of traffic in real time. Meanwhile, fuzzy inference has the ability to evaluate the control rules stored in the fuzzy rule base in order to served as the representation of the knowledge of experts through If–Then rule structure. The fuzzy rule base is, therefore, a set of fuzzy rules used in mapping the inputs to outputs. However, defuzzification involves the conversion of the fuzzy inference output values to crisp output values which are applied in determining the following cycle signal status through the signal controller program.

The fuzzy logic traffic signal control (FLTSC) strategies that have been developed can be categorized into three groups namely extension level, extension-termination, and urgency concept. Extension Level, the controller decides whether or not to give an extension time to the current green stage at the end of each intervention [6, 9]. Extension-Termination, the controller decides at regular time intervals to either extend or terminate the current green stage [3, 10]. Urgency Concept, the output of this controller is similar to the extension-termination strategy, but the urgency concept has the ability to determine what stage, among the possible ones, should be the next [4, 7, 8, 12].

3 Research Methods

This research was conducted using a simulation modeling approach. The FLTSC algorithm [4] which is designed for an isolated four-way without turning movements is applied to an isolated four-way intersection with two-stage signal type 42. Figure 1 shows the geometric intersection with its phase sequence. To determine effectiveness of the proposed FLTSC performance in minimizing delays, therefore, it is compared with the optimized FTC performance. An isolated four-way intersection with two-stage signal type 42 with both types of control is modeled using the VISSIM simulation program.

VISSIM simulation program has a traffic simulator to simulate traffic flow. FTC modeling uses the default signal controller program in the traffic simulator, while FLTSC uses the signal controller program created by using Vehicle Actuated Programme (VAP) language [4]. This signal controller program is then connected to the traffic simulator via the signal stage generator. The traffic simulator simulated the traffic operation during the simulation period and output values were obtained in the form of average delay of vehicles at the end.

Various case studies were carried out with different traffic conditions at the intersection under constant and fluctuating conditions to evaluate the effectiveness of the proposed FLTSC and optimized FTC performance. Traffic signal control that produces the smallest average delay of vehicles at intersection is the most effective traffic signal control in providing a good intersection level of service.

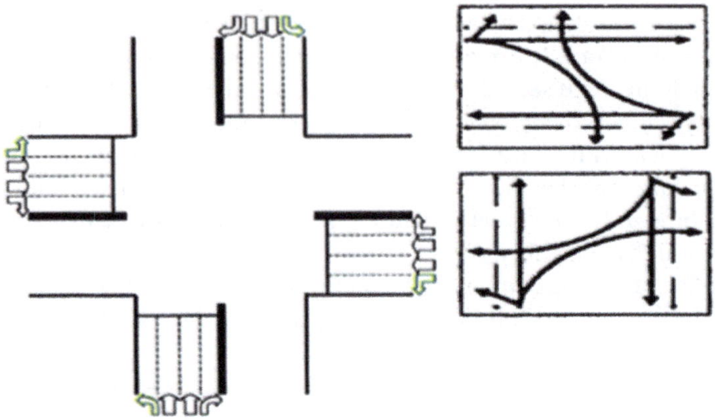

Fig. 1 Geometric intersection with its phase sequence

4 Proposed Fuzzy Logic Traffic Signal Control

4.1 Fuzzy Control Strategy

Development of the proposed FLTSC process is carried out in the VISSIM simulation program. The controller was developed by using Vehicle Actuated Programming [4]. The proposed FLTSC applied the urgency concept used in Sayers et al.'s research [7] which was selected as a suitable strategy due to its flexibility and computational simplicity. Moreover, traffic information on the maximum queue lengths and average occupancy rates retrieved from the previous cycle were also applied in estimating the green time length needed by each set of signal groups or stages in the next cycle based on the stage weight values. The process used in calculating the next cycle's new green time values is, therefore, presented in Fig. 2.

4.2 Fuzzification and Membership Function

There was an adjustment of the configuration for the input and output fuzzy variables membership functions through the analysis of the performance of controller at various traffic conditions to improve its performance. These membership functions including the maximum queue length and average occupancy rate were observed to have the linguistic labels of Low, Medium, High, and Very High while the weight has Very-Very Low, Very Low, Low, Medium, High, Very High, and Very-Very High respectively as indicated in Figs. 3, 4 and 5.

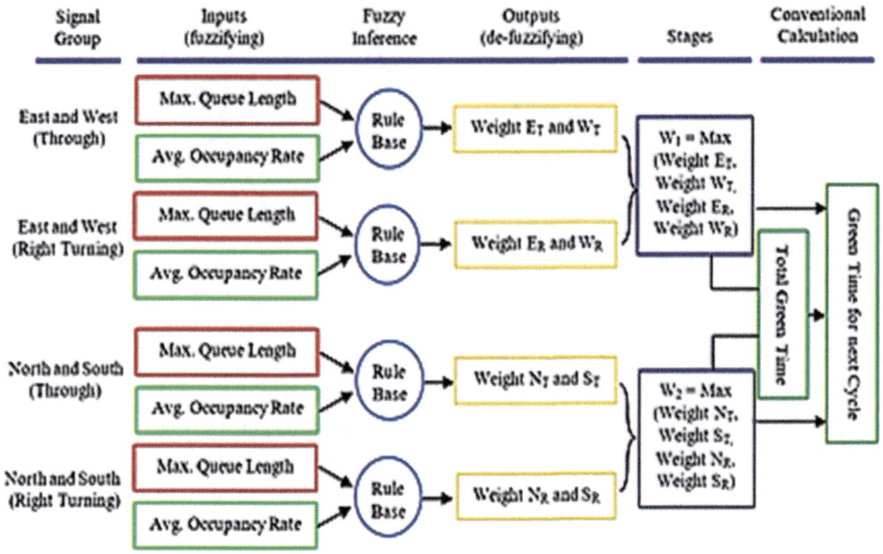

Fig. 2 Process of calculation of new green time values

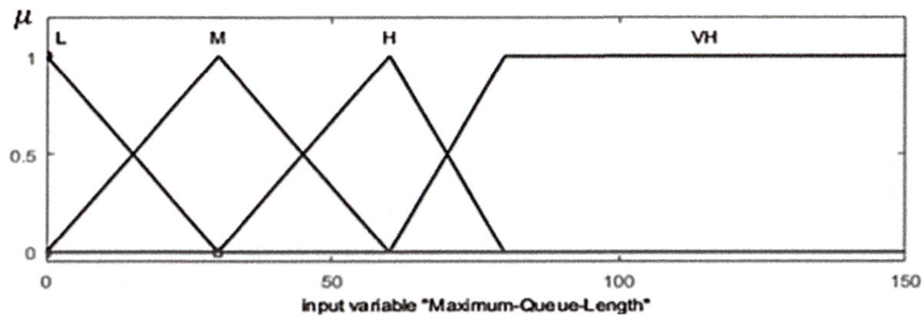

Fig. 3 Input maximum queue length

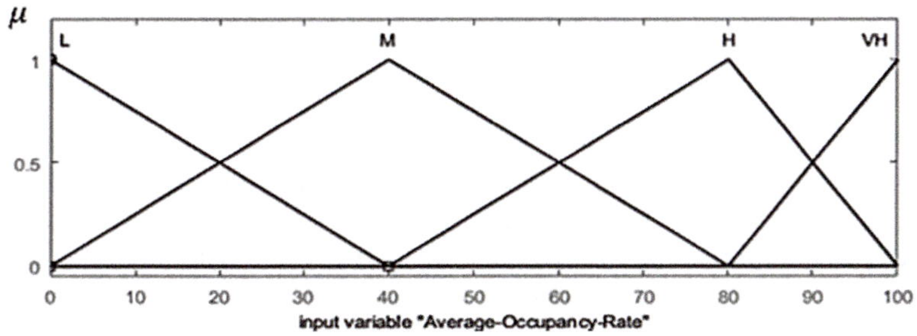

Fig. 4 Input average occupancy rate

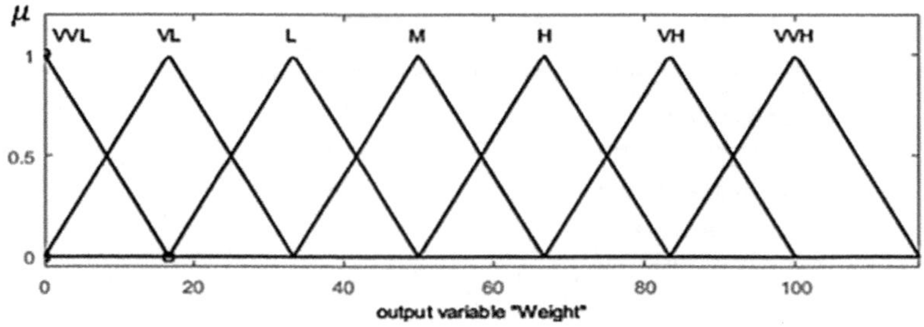

Fig. 5 Output weight

Table 1 Fuzzy rule base

		Maximum queue length			
		L	M	H	VH
Average Occupancy Rate	L	VVL	L	M	H
	M	VL	L	H	VH
	H	L	M	H	VVH
	VH	M	H	VH	VVH

4.3 Fuzzy Inference Rule Base

There are two inputs with each having four membership functions to produce sixteen fuzzy rules in the proposed FLTSC. Moreover, there was an adjustment of the fuzzy rule base through the use of the trial-and-error method [6] and the fuzzy rules configuration in the proposed FLTSC matrix structure is presented in Table 1 while the fuzzy inference system used was the Mamdani max–min composition method [16].

4.4 Defuzzification

This is the process of transforming the output value generated from the fuzzy inference system into a real crisp value. Moreover, the Centre-of-Sum method is usually applied in this process to determine the output variable Weight due to its ability to provide better performance in relation to disambiguate, continuity, and computer complexity.

5 Result and Discussion

This research used two case studies with different traffic volume to evaluate the effectiveness of controller. Case 1 shows the traffic flow is constant during one-hour period while Case 2 involves the variation of the traffic flow every 15 min and the set of traffic volumes for the case studies are indicated in Table 2. Traffic turning proportions for all cases is to turn right 10%, straight 75% and turn left 15%.

The IHCM method was used in calculating the optimized FTC signal timing using the cycle and green times presented in Table 3. The comparison between the proposed FLTSC and optimized FTC is only valid if the calculations of cycle time and green time using the IHCM method gives rise to optimum time settings. To find out whether the IHCM method produces the optimum signal timing, therefore, a sensitivity test is performed using the TRANSYT program. The results of the sensitivity test show that the IHCM method indeed produces the optimum signal timing [17].

There was imposition of the lowest and highest values of total green time at 10 s and 52 s respectively in the cycle of the proposed FLTSC to avoid the cycle length from being too long or too short. Meanwhile, the amber and all red periods were 3 s and 1 s respectively for both controllers. Identical simulations were run for approximately one-hour on both controllers to produce the output value which was the average delay at the intersection. The values for all cases were later compared for both controllers and the results are presented in Fig. 6.

Table 2 Set of traffic volumes (vehicle/hour)

Case	Traffic volume		Case	Traffic volume		Case	Traffic volume	
	E–W	N–S		E–W	N–S		E–W	N–S
1a, 2a	1.080	1.080	1e, 2e	2.520	1.080	1i, 2i	1.800	1.440
1b, 2b	1.440	1.080	1f, 2f	1.440	1.440	1j, 2j	2.160	1.440
1c, 2c	1.800	1.080	1g, 2g	1.800	1.440	1k, 2k	2.160	1.800
1d, 2d	2.160	1.080	1h, 2h	2.160	1.440			

Table 3 Cycle time and green time used for FTC

Case	Green time (s)		Cycle time (s)	Case	Green time (s)		Cycle time (s)
	E–W	N–S			E–W	N–S	
1a, 2a	14	15	45	1g, 2g	27	21	64
1b, 2b	19	15	50	1h, 2h	35	23	74
1c, 2c	25	15	56	1i, 2i	47	26	89
1d, 2d	32	16	64	1j, 2j	29	29	74
1e, 2e	42	17	75	1k, 2k	39	33	88
1f, 2f	20	20	56				

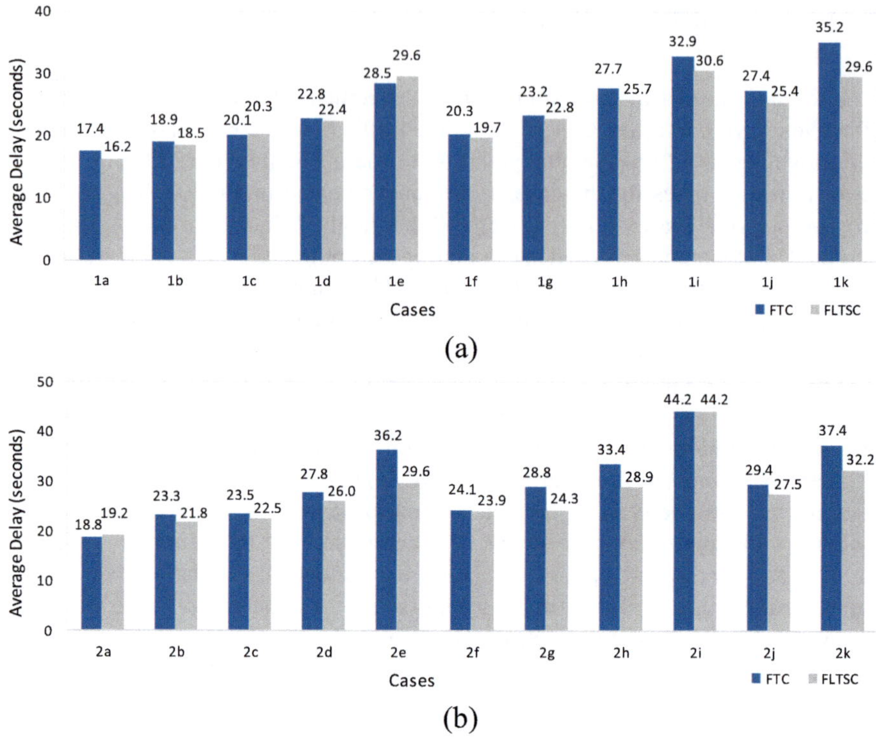

Fig. 6 Average delay of vehicles of the optimized FTC and proposed FLTSC, **a** case 1 **b** case 2

The simulation for Case 1 showed the optimized FTC generally produced an average delay higher than the proposed FLTSC with the proposed FLTSC observed to have reduced the average delay by 1.9–15.8% against the optimized FTC as indicated in Fig. 7. Meanwhile, the simulation for Case 2 indicated the proposed FLTSC is superior to the optimized FTC for all cases, except case 2a. The average delay saving

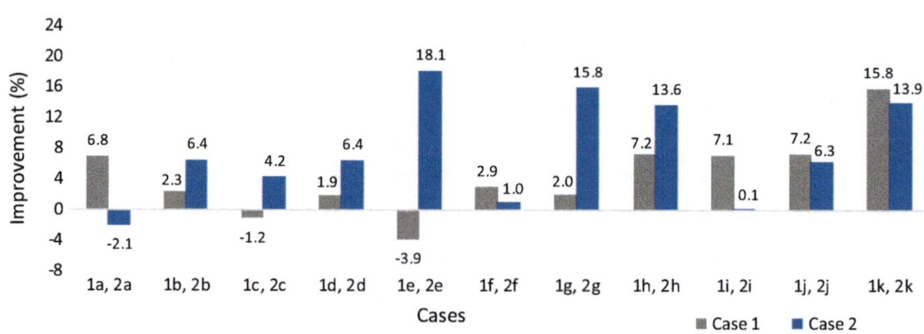

Fig. 7 Percentage improvement of the proposed FLTSC against the optimized FTC

by the proposed FLTSC is between 1.0% and 18.1% against the optimized FTC, see Fig. 7. Overall, the proposed FLTSC performs much better than optimized FTC. It is due to the proposed FLTSC's ability to cause an adjustment of the green time length due to the fluctuation of the traffic flow.

The main concern of this type of signal control is the driver behavior of right turning movements. For opposed approaches, the queue discharge is influenced by the fact that right turning vehicles do not respect the right of way of straight through traffic. This may reduce the proposed FLTSC's effectiveness, where the length of green time to discharge vehicles cannot be used efficiently. However, this behavior does not affect the performance of the controller significantly, since the number of right turning vehicles is low (less than 200 pcu/hr).

6 Conclusion

The development of an adaptive FLTSC for an isolated four-way intersection with two-stage signal type 42 suitable for mixed traffic has been presented. A wide range of traffic flow rates was applied to compare the controller's performance in terms of average delay at an intersection experienced with the optimized FTC and proposed FLTSC. The simulation showed the performance of the proposed FLTSC is generally better than the optimized FTC, especially in time-varying traffic cases.

References

1. Kell JH, Fullerton IJ (1991) Manual of traffic signal design—chapter 7: detectors. Institute of Transportation Engineers, Englewood Cliffs, New Jersey
2. Lee JH, Lee KM, Leekwang H (1994) Fuzzy controller for intersection group. In: IEEE conference on industrial automation and control emerging technology applications, pp 376–382. Taipei, Taiwan
3. Trabia MB, Kaseko MS (1996) A fuzzy logic controller for a traffic signal. In: IASTED international conference on applications of control and robotics, pp 117–122
4. Yulianto B (2003) Application of fuzzy logic to traffic signal control under mixed traffic conditions. Traffic Eng Control 44(9):332–336
5. Zadeh LA (1965) Fuzzy sets. Inf Control 8(3):338–353
6. Pappis CP, Mamdani EH (1977) A fuzzy controller for a traffic junction. IEEE Trans Syst Man Cybern 7(10):707–717
7. Sayers T, Bell MGH, Mieden T, Busch F (1996) Traffic responsive signal control using fuzzy logic—a practical modular approach. In: Proceedings of the fourth European congress on intelligent techniques and soft computing, pp 2159–2163. Aachen, Germany
8. Sayers T, Anderson J, Clement S (1999) The multi-objective optimisation of a traffic control system. In: Transportation and traffic theory. Papers presented at the abbreviated presentation sessions, pp 153–176
9. Niittymäki J, Pursula M (2000) Signal control using fuzzy logic. Fuzzy Sets Syst 116(1):11–22
10. Trabia MB, Kaseko MS, Ande M (1999) A two-stage fuzzy logic controller for traffic signals. Transp Res Part C Emerg Technol 7(6):353–367

11. Murat YS, Gedizlioglu E (2005) A fuzzy logic multi-phased signal control model for isolated junctions. Transp Res Part C Emerg Technol 13(1):19–36
12. Yulianto B (2012) Setiono: traffic signal controller for mixed traffic conditions. IOSR J Mech Civ Eng 4(1):18–26
13. Indonesian Highway Capacity Manual (1997) Directorate general Bina Marga, Indonesia
14. Webster FV (1958) Traffic signal settings. Road Research Technical Paper No 39
15. Strobel H (1982) Computer controller Urban transportation. John Wiley & Sons, UK
16. Mamdani EH, Assilian S (1975) An experiment in linguistic synthesis with a fuzzy controller. Int J Man–Mach Stud 7(1):1–13
17. Yulianto B (2019) Sensitivity test of IHCM 1997 traffic signal timings using TRANSYT program and VISSIM. AIP Conf Proc 2114(1):1–10

Application of Fuzzy Inference System Mamdani at Pelican Crossing

Salsabila Naura Putri

Abstract Fuzzy logic is one of the soft computing methodologies which aims to analyze the system that contains fuzziness. Fuzzy inference system (FIS) defines as a decision-making process based on fuzzy set theory, fuzzy rules, and fuzzy reasoning. In this research, we use FIS Mamdani which has been used in several applications, especially transportation. Pedestrians are the weakest mode of transportation that requires an infrastructure that is safe and comfortable when crossing the road. Pelican crossing is one of the pedestrian crossing types. Pelican crossing in Indonesia is based on fixed time where the pedestrian time is fixed. It is not effective because the pedestrian time is not based on the volume of traffic conditions. To overcome this problem, the FIS Mamdani method is used. Mamdani method is composed of three steps to obtain an output, namely pedestrian time. Fuzzification with two input variables, namely pedestrian and vehicle volumes, and the output variable is pedestrian time. Inference engine which executes all logic with the if–then rule. Apply defuzzification with the Centroid method to obtain the output result. The purpose of this research is to determine the pedestrian time on pelican crossing using FIS Mamdani. The result shows the pedestrian time change according to the volume of pedestrians and vehicles. FIS Mamdani can adjust the traffic light signal time to the traffic conditions that occur on a pelican crossing.

Keywords Fuzzy logic · Fuzzy inference system · Pelican crossing

1 Introduction

Problem-solving in the real world requires an intelligent system that has human-like skills, able to utilize knowledge, and learns to adapt to its environment. Soft computing is a collection of methodologies that aim to exploit the tolerance for imprecision and uncertainty to achieve tractability, robustness, and low solution cost [1]. Soft computing consists of several computing paradigms, including neural networks,

S. N. Putri (✉)
Department of Mathematics, Sebelas Maret University, Surakarta, Indonesia

S. A. Kristiawan et al. (eds.), *Proceedings of the 5th International Conference on Rehabilitation and Maintenance in Civil Engineering*, Lecture Notes in Civil Engineering 225,
https://doi.org/10.1007/978-981-16-9348-9_60

fuzzy set theory, approximate reasoning, genetic algorithm, and simulated annealing [2].

Fuzzy logic was first established by Lotfi A. Zadeh of California University in 1965 [3]. Fuzzy logic defines as a logic that has value fuzziness between true and false with a membership degree between 0 and 1. This logic proved to be a powerful and effective tool for dealing with uncertainty to fuzzy systems which could support decisions in several fields, from control theory to artificial intelligence.

Fuzzy Inference System (FIS) is a formulation process from input to output with fuzzy logic [4]. Generally, three methods may be used in FIS which are Mamdani method [5], Sugeno method [6], and Tsukamoto method [7]. Mamdani method is often applied in fuzzy models. Several applications have been used using FIS Mamdani, such as making decisions in transportations. The uses of FIS Mamdani in the transportation field help solve various complex traffic and transportation problems with many objectives, many constraints, unclear input information, and vague decision criteria. An example of application is the setting of traffic signal lights which can adjust to the density of traffic flow on the path arranged [8].

Pedestrians are an important form of transportation in urban areas. It is the weakest position if they are mix with vehicles, therefore pedestrian crossing is needed. The pedestrian crossing is a special area on the roadway that is allocated for pedestrians to cross the road [9]. Pelican crossing is a type of pedestrian crossing which provided traffic lights and push buttons to activate traffic lights.

The facility of pelican crossing provides safety and comfort for the pedestrians. However, the timing of pelican crossing signal lights should be concerned because pelican crossing determines the service level of the roads. The service level of the roads describes the traffic conditions in which the service level of a good road has a small delay for vehicles and pedestrians and vice versa. Poor timing of signal lights will have an impact on traffic congestion on the road and or traffic violations by pedestrians because it takes too long to wait for the time to cross [10].

Pelican crossings in Indonesia are based on pedestrian actuated control (PA-C) with fixed time operational. In this case, the pedestrian phase will activate when the pedestrian presses the button. The length of pedestrian time remains under the calculations established by several standards [11]. The existing pelican crossing is not effective because the pedestrian time is not based on the traffic conditions which may cause time-wasting by pedestrians and vehicles.

Turky et al. [12] have successfully simulated the performance between the control of pedestrian crossing using fixed-time and fuzzy logic. It found that the fuzzy logic pedestrian crossing controller performs better than the fixed-time controller. Hunsanon et al. [13] proposed a fuzzy logic control (FLC) strategy for analyzing the optimal traffic signal timing setting for pedestrian crossing signals in the urban area. The results showed that the operational performance of fuzzy logic at the crossing guard is better than PA-C. Makarova et al. [9] researched fuzzy logic to improve safety at road crossings. Vehicle queue length, pedestrian queue length, and time after the completion of the pedestrian phase were used as the input variable while waiting time for the pedestrian phase as the output variable. The results showed that the use of fuzzy logic can normalize the traffic parameters. In this research, we proposed FIS Mamdani to calculate pedestrian time based on traffic control, namely the pedestrian and vehicle volumes.

2 Fuzzy Inference System (FIS) Mamdani

Fuzzy inference system (FIS) is a decision-making process based on fuzzy sets, fuzzy rules, and fuzzy reasoning [14]. This system is compatible with solving real-world problems which are complicated to obtain a mathematical model, whereas it is uncomplicated to describe the solution linguistically with if–then rules [15].

FIS Mamdani provides understandable results with a clear structure, also for the basic nature of rules that are intuitive and explainable [16]. FIS Mamdani is composed of three parts namely fuzzification, fuzzy inference engine, and defuzzification [14].

3 Pelican Crossing

Pelican crossing is a type of pedestrian crossing which provided traffic signs and traffic lights [17]. It is generally featured with a push button to activate the traffic lights. On pelican crossing the pedestrian time is calculated using the following equation,

$$PT = \frac{L}{vt} + 1.7\left(\frac{N}{W} - 1\right) \tag{1}$$

where PT is pedestrian time (second), vt is pedestrian speed (1.2 m/s), L is crossing length (meter), N is the pedestrian volume per cycle, and W is crossing width (meter) [18].

The phasing sequence on pelican crossing can be shown in Fig. 1 where Y is yellow, FG is flashing green, and R is red.

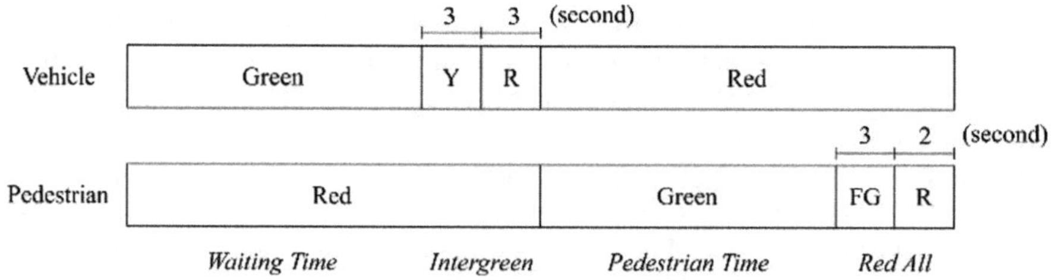

Fig. 1 Phasing sequence on a pelican crossing

The duration of flashing green is not always 3 s, but it depends on the crossing length and the average pedestrian speed (1.77 m/s) [11]. The flashing green can be written as

$$FG = \frac{L}{1.77} \tag{2}$$

where FG is flashing green (second) and L is crossing length (meter).

4 Research Methods

This research is applied research that uses pedestrians and vehicles data located on the pelican crossing in Jalan M. H. Thamrin, Jakarta [19]. The data were taken in April 2019 at 12.00–1.00 PM and 5.00–6.00 PM. The procedures that apply to this research are: (1) to explore related data, (2) describe the logic overview of pelican crossing with fixed-time and fuzzy logic, (3) describe the model FLC, (4) analyze the applied results of pedestrian time, (5) conclusions.

5 Result and Discussion

5.1 Data Description

The data of pedestrians and vehicles on the pelican crossing in Jalan M. H. Thamrin (Plaza Indonesia) are shown in Table 1. Pelican crossing Plaza Indonesia consists of two pelican crossings as presented in Fig. 2, where the traffic signal control is based on fixed-time.

Table 1 Data of pedestrians and vehicles

Time (PM)	Volume of pedestrian			Time (PM)	Volume of vehicle		
	East	West	Total		East	West	Total
12.00–12.15	146	158	304	12.00–12.15	915	751	1666
12.15–12.30	135	152	287	12.15–12.30	964	786	1750
12.30–12.45	111	123	234	12.30–12.45	1097	779	1876
12.45–1.00	154	199	353	12.45–1.00	1034	819	1853
5.00–5.15	232	197	429	5.00–5.15	928	1001	1929
5.15–5.30	184	255	429	5.15–5.30	856	935	1791
5.30–5.45	224	316	540	5.30–5.45	886	992	1878
5.45–6.00	190	312	502	5.45–6.00	712	720	1432

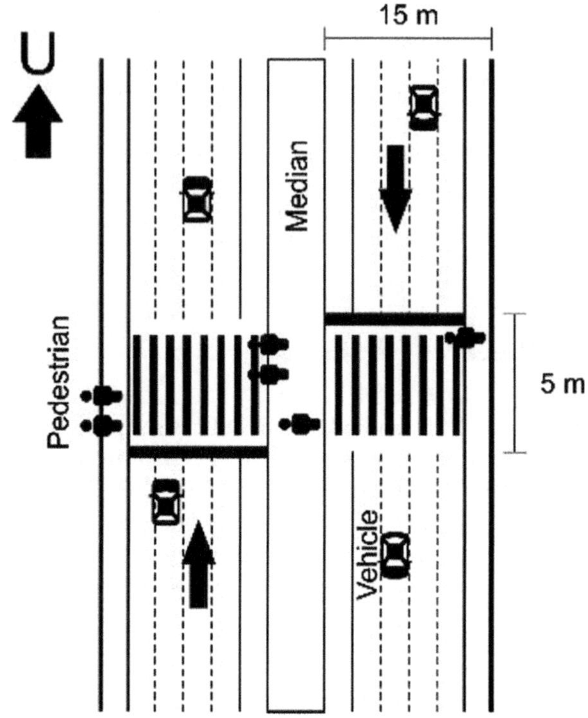

Fig. 2 The layout of the pelican crossing

5.2 *Logic Overview of Pelican Crossing*

Pelican Crossing Existing (Fixed Time) Pelican crossing has two-phase which are the pedestrian phase and vehicle phase. In this research, the cycle time on the pelican crossing Plaza Indonesia is 90 s. The minimum green time for pedestrian time with Eq. (1) during the day is 16 s and during the evening is 19 s. Pelican crossing signal diagram is shown in Fig. 3 with flashing green time using Eq. (2) is 8 s.

Fuzzy Logic Pelican Crossing In the fuzzy logic system, the pedestrian time changes according to the pedestrian and vehicle volumes. In this case, the cycle time is fixed but the vehicle time changes according to the changes in the pedestrian time.

Pedestrian time is calculated every cycle. Data from the volume of pedestrians and vehicles in the current cycle are calculated by FIS Mamdani to get the pedestrian time and vehicle time for the next cycle. The explanation is shown in Fig. 4.

In Fig. 4, vehicle time is obtained from the equation,

$$VT = CT - IG - RA - PT$$

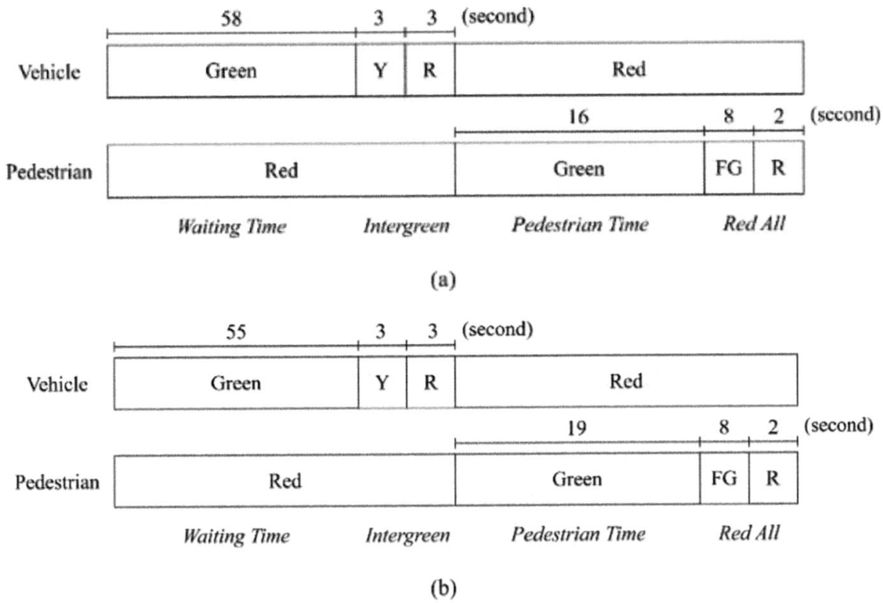

Fig. 3 Pelican crossing plaza Indonesia signal diagram during **a** day and **b** evening

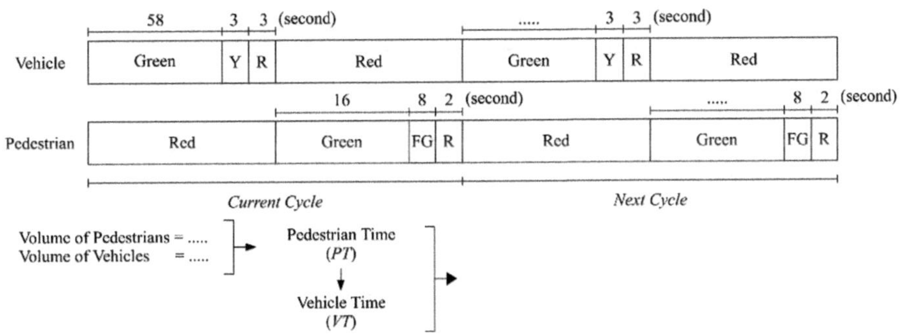

Fig. 4 Signal diagram fuzzy logic

where CT is cycle time (90 s), IG is intergreen (6 s), and RA is red all (10 s). Maximum green time for pedestrian time is calculated by using the equation $G_{max} = 1.5 \times PT$ [20]. During the day, the maximum green time is 24 s while evening the maximum green time is 28 s.

5.3 Fuzzy Logic Control

Fuzzification A FIS was constructed to output the pedestrian time with two input variables, namely the volume of pedestrians and vehicles. This research consists of two FIS which represent during the day and the evening. In fuzzification, the input and output variables are divided into three fuzzy sets. The plots of membership functions are shown in Figs. 5, 6, 7 and 8.

Inference System Based on Fuzzy Rules The general format of the rules is the following:

$$\text{IF } Vol\ Pedestrian\ is\ X_1 \text{ AND } Vol\ Vehicle\ is\ X_2 \text{ THEN } Ped\ Time\ is\ y$$

Where X_1, X_2, and y are natural language expressions of the condition of respective variables. The numbers of rules are combinations of fuzzy sets for X_1 and X_2. The sets of rules used are shown in Table 2.

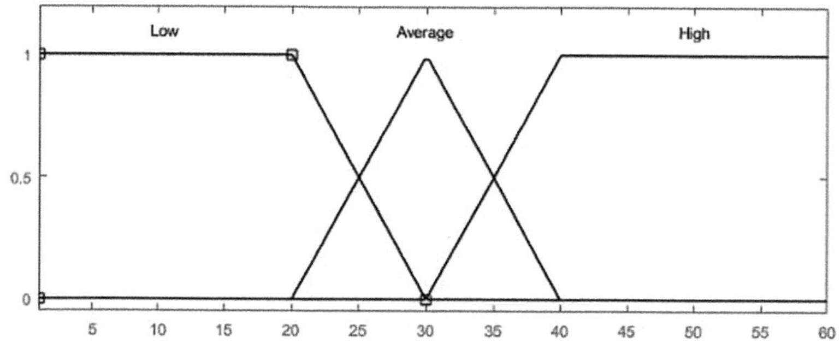

Fig. 5 Volume of pedestrians

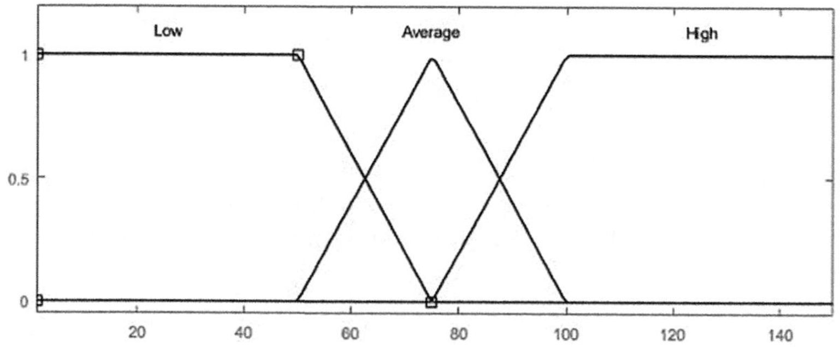

Fig. 6 Volume of vehicles

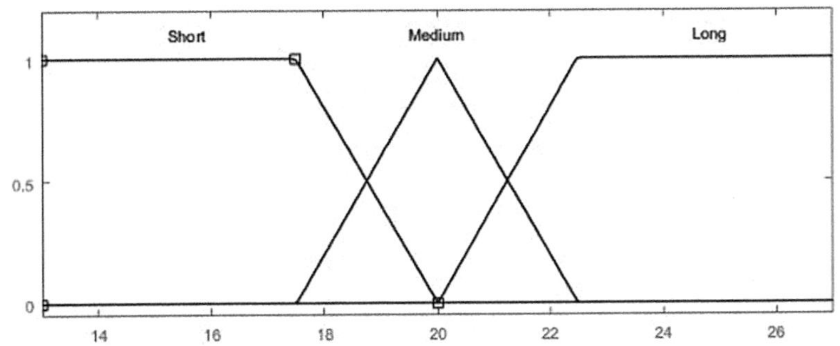

Fig. 7 Pedestrian time during the day

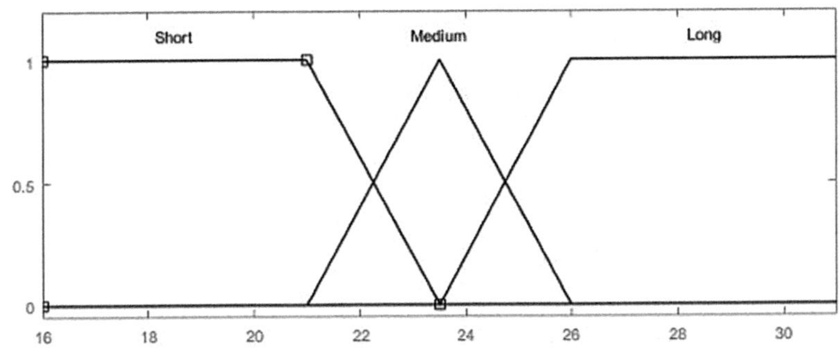

Fig. 8 Pedestrian time during the evening

Table 2 Fuzzy rules

		Volume of vehicles		
		Low	Average	High
Volume of pedestrians	Low	Short	Short	Short
	Average	Medium	Medium	Medium
	High	Long	Long	Medium

Defuzzification Defuzzification is defined as a transformation process in the output from FIS into a single numerical value. The centroid method is used in defuzzification to determine the pedestrian time.

5.4 *Result of Pedestrian Time*

The fuzzy sets are used to determine the pedestrian time, namely pedestrian and vehicle volumes as input and pedestrian time as output. An experiment of pelican

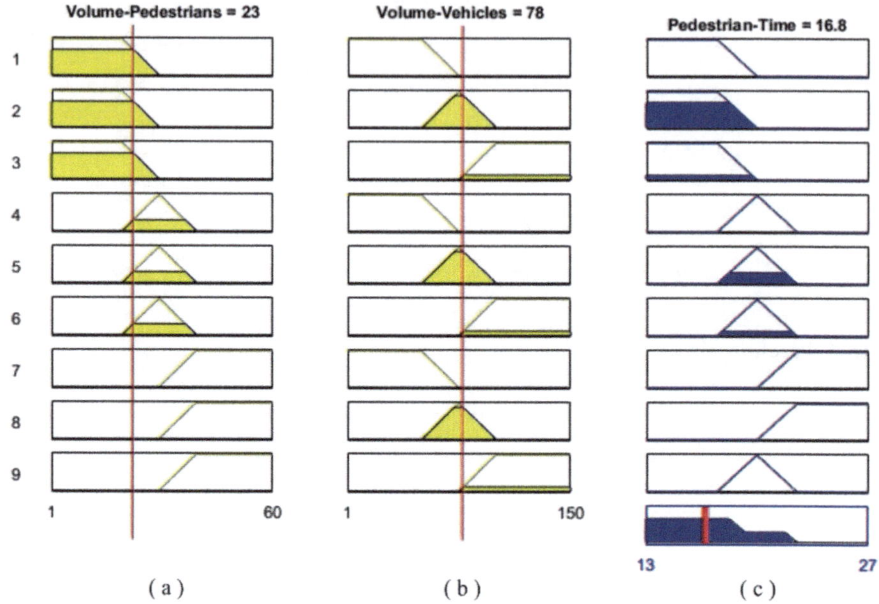

Fig. 9 Rule viewer of 23 pedestrians and 78 vehicles

crossing with FIS Mamdani based on the research data in Table 1 is given. The snapshot of the rule viewer with 23 pedestrians and 78 vehicles is shown in Fig. 9.

Figure 9a represent the membership degree of 23 pedestrians in the pedestrian volume variable, (b) represent the membership degree of 78 vehicles in the vehicle volume variable, and (c) represents the output results of the inference system according to the conditions and the last row is combined fuzzy solution area of each rule. The red line on the pedestrian time variable shows that the pedestrian time for 23 pedestrians and 78 vehicles is 17 s. Figure 10 represents the signal diagram of Fig. 9.

Other experiments of pelican crossing with FIS Mamdani during the day and evening are shown in Table 3. Based on Table 3, it appears that when the pedestrian volume is low, the pedestrian time is short. This shows that pedestrians have priority

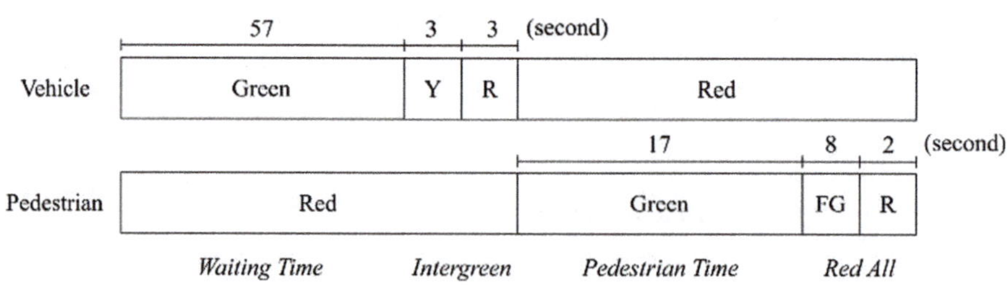

Fig. 10 Signal diagram of 23 pedestrians and 78 vehicles

Table 3 Pedestrian time experiment

No.	Vol. Ped.	Vol. Veh.	Pedestrian time		No.	Vol. Ped.	Vol. Veh.	Pedestrian time	
			D	E				D	E
1	8	35	16	19	9	32	42	21	25
2	12	59	16	19	10	34	74	22	26
3	14	87	16	19	11	37	80	23	27
4	19	110	16	19	12	38	113	20	24
5	22	30	17	20	13	44	24	24	28
6	25	60	17	21	14	48	64	24	28
7	28	90	19	22	15	50	83	23	27
8	29	108	19	23	16	53	115	20	24

on pelican crossing because pedestrians are a vulnerable mode of transportation which must be prioritized when crossing the road at the pedestrian crossing.

6 Conclusion

Pelican crossings in Indonesia are based on PA-C with fixed-time operational. It controls signal with pre-set timing however this controller has some limitations. For example, when many pedestrians are crossing the pelican crossing, the pedestrian time is terminated when reaches the maximum green time. Even when the pedestrian volume is low, the vehicles must wait until the pedestrian time ends which can cause congestions. To outcome this problem, FIS Mamdani is used to determining the pedestrian time of pelican crossing since it helps to control the traffic signal at the pelican crossing. In FIS Mamdani, the volume of pedestrians and vehicles was selected as input while the output is the pedestrian time. The results showed that when the pedestrian volume is low and vehicle volume is high, pedestrian time will be short. While, when the pedestrian volume is high and vehicle volume is low, pedestrian time will be long. This indicates that the pedestrian time changes according to the pedestrian and vehicle volumes. The use of FIS Mamdani for determining the pedestrian time able to adjust the traffic signal control to the traffic conditions that occur at the pelican crossing.

References

1. Zadeh LA (1994) Soft computing and fuzzy logic. IEEE Softw 48–56
2. Jang JSR, Sun CT, Mizutani E (1997) Neuro-fuzzy and soft computing: a computational approach to learning and machine intelligence. Prentice-Hall Inc., USA

3. Pamuji A (2017) Fuzzy logic inference system for determining the quality assessment of student's learning ICT. Sci J Inform 4(1):57–65
4. The MathWorks Inc (1998) Fuzzy logic toolbox user's guide, 4th edn. The MathWorks Inc., USA
5. Pourjavad E, Shahin A (2018) The application of Mamdani fuzzy inference system in evaluating green supply chain management performance. Int J Fuzzy Syst 20(3):901–912
6. Yulianto T, Komariyah S, Ulfaniyah N (2017) Application of fuzzy inference system by Sugeno method on estimating of salt production. AIP Conf Proc 1867(1):020039
7. Hastono T, Santoso AJ (2017) Honey yield prediction using Tsukamoto fuzzy inference system. In: 4th international conference on electrical engineering, computer science and informatics (EECSI), pp 1–6
8. Sumiati, Sigit HT, Kapuji A (2014) Mamdani fuzzy inference system application setting for traffic lights. Int J Fuzzy Logic Syst (IJFLS) 10(3):56–62
9. Makarova I, Shubenkova K, Mavrin V, Buyyol P (2018) Improving safety on the crosswalks with the use of fuzzy logic. Transp Prob 13(1):97–109
10. Yulianto B, Sugiyarto (2019) Evaluation of midblock pedestrian crossing facility in the Surakarta City, Indonesia. ICETsAS J Phys Conf Ser 1376(1):1–9 (2019).
11. Malkhamah S (2004) Evaluasi penetapan periode pada pelican di Indonesia. Media Teknik 26:74–81
12. Turky AM, Ahmad MS, Yusoff MZM, Sabar NR (2008) Fuzzy logic application for pedestrian crossing. Int Symp Inf Technol 1:1–6
13. Hunsanon T, Kronprasert N, Upayokin A, Songchitruksa P (2017) Control strategy for vehicular and pedestrian midblock crossing movements. Transp Res Proc 25:1672–1689
14. Putri SN, Saputro DRS (2021) Construction fuzzy logic with curve shoulder in inference system Mamdani. KNPMP V J Phys Conf Ser 1776:1–8
15. Castellano G, Castiello C, Pasquadibisceglie V, Zaza G (2017) FISDet: fuzzy inference system development tool. Int J Comput Intell Syst 10(1):13–22
16. Kovac P, Rodic D, Pucovsky V, Savkovic B, Gostimirovic M (2014) Multi-output fuzzy inference system for modeling cutting temperature and tool life in face milling. J Mech Sci Technol 28(10):4247–4256
17. Departemen Pekerjaan Umum (1999) Pedoman Perencanaan Jalur Pejalan Kaki pada Jalan Umum. PT Mediatama Septakarya, Jakarta
18. Dirjen Perhubungan Darat (1997) Keputusan Direktur Jenderal Perhubungan Darat Nomor: SK.43/AJ 007/DRJD/97, Perekayasaan Fasilitas Pejalan Kaki di Wilayah Kota. Departemen Perhubungan, Jakarta
19. Widyaningsih N, Daniel O (2019) Analisis karakteristik dan perilaku penyeberangan orang pada fasilitas penyeberangan zebra cross dan pelican cross (Studi Kasus Ruas Jalan M. H. Thamrin). J Pengembangan Rekayasa dan Teknologi 15(1):27–32
20. Roess RP, McShane WR, Prassas ES (1998) Traffic engineering, 4th edn. Prentice-Hall Inc., New Jersey

Determining the Maximum Speed Limit in Residential Area

Gito Sugiyanto, Fadli Wirawan, Eva Wahyu Indriyati, Yanto, and Mina Yumei Santi

Abstract The residential area which is located on the right or left side of the arterial or collector road is one of the areas with high potential to traffic accidents. An effort to reduce the number of traffic accidents is applying speed limits. This is important as most traffic accidents are caused by speeding factors. The aim of this research is to determine the value of the maximum speed limit in residential areas using the 85th percentile and stopping sight distance (SSD). The study locations are in three collector roads in Purbalingga Regency namely Padamara Street, Letnan Yusuf Street, and Letkol Isdiman Street. Based on the analysis result, the maximum speed limit of vehicles based on the 85th percentile method varies between 42 up to 44 kph. The maximum speed limit based on the SSD method is 39.32 kph for Padamara Street, 38.10 kph for Letnan Yusuf Street, and 38.40 kph for Letkol Isdiman Street. The maximum speed limit of vehicles proposed for collector roads with residential area functions on the right or left side is 40 kph.

Keywords Accident · 85th Percentile · Residential area · Speed limit · Stopping sight distance

G. Sugiyanto (✉) · F. Wirawan · E. W. Indriyati · Yanto
Department of Civil Engineering, Jenderal Soedirman University, Purbalingga, Central Java, Indonesia
e-mail: gito.sugiyanto@unsoed.ac.id

E. W. Indriyati
e-mail: eva.indriyati@unsoed.ac.id

Yanto
e-mail: yanto@unsoed.ac.id

M. Y. Santi
Health Polytechnic Ministry of Health of Yogyakarta, Mangkuyudan Street MJ III/304, Yogyakarta, Indonesia
e-mail: mina.yumei@poltekkesjogja.ac.id

S. A. Kristiawan et al. (eds.), *Proceedings of the 5th International Conference on Rehabilitation and Maintenance in Civil Engineering*, Lecture Notes in Civil Engineering 225,
https://doi.org/10.1007/978-981-16-9348-9_61

1 Introduction

Residential areas that are located on the right or left side of arterial or collector roads are areas that have high potential for traffic accidents. According to the Traffic Corps Republic of Indonesia, 75% of traffic accident locations are on roads with land use around the residential areas [1]. Traffic accidents are caused by several factors such as vehicles, negligence of road users (human error), road conditions and the environment [2–5]. From the four factors, human error such as an increase in vehicle speed is a factor that often results in traffic accidents [6, 7], which can even reach one-third of fatal accidents [8]. The highest human error as a cause of accidents is speeding [9, 10]. Speeding increases the chance of accidents occurring [8] and the severity of traffic accident victims [11, 12]. Speed limit was introduced to reduce the severity of crashes [7, 13], accident cost [14, 15], driving behavior [10, 16, 17], air pollution [18], travel time [19], generalized cost [20, 21], mobility [22], and increased efficiency in transportation cost [23–26]. Limiting the vehicle speed is one of strategies to improve traffic safety [27]. The purpose of speed limiting is to balance mobility and safety by ensuring safe speed limits that are suitable for road conditions [11, 28]. For this reason, a vehicle speed limit value is required in accordance with road and traffic conditions in residential areas.

Higher vehicle speeds will cause noise and increase the potential for traffic accidents. Based on the theory of movement, speed plays an important role as one of the causes of traffic accidents [29]. Higher vehicle speed will have an effect on the increase in reaction time and braking distance required for the vehicle to stop. Also, it can increase the likelihood of mistakes being made by the driver. The higher speed of the vehicle, the higher death rate due to accidents, both involving pedestrians and other vehicles [30]. The ideal and safe speed is less than 30 kph to avoid conflicts that could involve pedestrians and bicycles, less than 45 kph for conflicts with vehicles from the side-impact collision, and less than 70 kph to avoid possible conflicts with vehicles from the front (head-on collision). According to Baguley, traffic conflict is a condition in which one or more road users approach each other or approach another object at a certain time and place so that it causes a risk of collision if the movement cannot be stopped or changed [31]. Traffic conflicts occur because of various types of traffic flow movements. Xie et al. introduced an improved crash frequency model for signalized intersections in urban road networks using conditional autoregressive model [32]. Road network patterns signal density and access densities are used to analyze the safety impacts [33]. The other study, Tapp et al. analyze the lack of compliance with 20 mph limits in Great Britain and 30 kph limits in Europe is concerning [34].

Nowadays, the maximum speed limit setting is determined based on the vehicle speed in the 85th percentile method [35, 36]. The 85th percentile method has disadvantage because it only uses vehicle speed as the reference parameter. For this reason, another method is needed as a comparison in order to produce a more valid speed limit value. The method to be used is stopping sight distance (SSD). In this method, the determination of maximum speed limit refers to SSD, free-flow speed, and road friction coefficient between tire and road pavement surface. Kagaya et al. identified the traffic management specific to setting of speed limits implemented in Hokkaido. Based on the results of the questionnaire, it was found that more than half of the respondents indicated that the vehicle speed limit for roads in Hokkaido is too low and must be distinguished between summer and winter. The vehicle speed limit is proposed depending on the characteristics and road conditions. The free-flow speed was introduced to calculate the maximum speed limit of vehicles in summer and free-flow speed and SSD are used in winter [37]. The aim of this research is to determine the maximum speed limit in residential areas using the 85th percentile and stopping sight distance method.

2 Method

2.1 Study Location

The research was conducted on three collector roads in Purbalingga Regency, namely Padamara Street (access road to Abdi Negara Permai Housing), Letnan Yusuf Street (access road to Griya Perwira Asri Housing), and Letkol Isdiman Street (access road to Griya Abdi Kencana Housing). On the left or right side from three roads are residential areas as shown in Fig. 1.

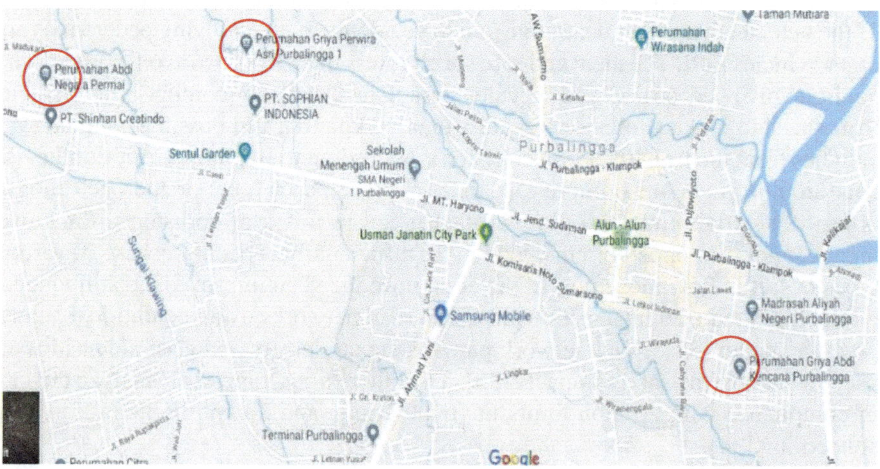

Fig. 1 Study location on three collector roads in Purbalingga Regency, Central Java [38]

2.2 Analysis Approach

There are several methods to determine the maximum speed limit value of a vehicle. The method that is most commonly used is the 85th percentile. Another method that can be used is the stopping sight distance method.

The 85th Percentile Method

In this method, the determination of the maximum speed limit refers to 85th percentile speed. The 85th percentile speed is a traffic speed where 85% of the driver driving the vehicle on the road without being affected by traffic speeds lower or bad weather [39]. The purpose of this method is to determine the ideal speed limit on the reviewed road sections based on the average speed of the vehicle. The 85th percentile method only uses one parameter namely the vehicle speed. Speed is one of the main risk factors in traffic safety [40].

Stopping Sight Distance Method

In this method, the determination of the maximum speed limit value refers to the vehicles stopping sight distance [41], free-flow speed, and road friction coefficient between tire and road pavement surface. The coefficient of friction between tire and road pavement surface is influenced by vehicle speed and roughness. The pavement type affects the road roughness value [42, 43]. Stopping sight distance is the distance the driver has to stop this vehicle. For comfort and safety in driving, every length of the road section should be filled with the minimum visibility along the minimum stopping sight distance [44]. Two parameters in stopping sight distance method are distance traveled during reaction time and braking distance. The sum of these two parameters becomes the minimum stopping sight distance. Formula for minimum stopping sight distance is:

$$d = d_1 + d_2 \tag{1}$$

To find the distance traveled during the reaction time (d_1), we must first know the total reaction time, which is the total time it takes the driver from seeing the obstacle to applying the brakes. The total reaction time is taken as 2.5 s which is the sum of 1.5 s of perception, identification, emotion, and volition or PIEV time (the time it takes the driver from the moment he realizes an obstacle until he makes a decision) and 1.0 s for the time it takes the driver to hit the brake pedal. So, d_1 is the distance traveled in 2.50 s. The equation for calculating the value of the distance traveled during the reaction time (d_1) is:

$$d_1 = 0.278 \cdot vt \tag{2}$$

With v is vehicle speed (kph) and t is the total reaction time (seconds).

The braking distance (d_2) is the distance a vehicle will travel from the point when its brakes are fully applied to when it comes to a complete stop. The braking distance is calculated due to the effect of friction between the tires and the road surface. The equation for calculating the braking distance (d_2) is:

$$d_2 = \frac{v^2}{2 \cdot g \cdot (f \pm G) \cdot (3.6)^2} \tag{3}$$

where v is the vehicle speed (kph), g is the acceleration due to gravity (m/s^2) and f is the factor of the road friction coefficient between tire and road pavement surface.

The value of road coefficient friction between tire and road surface is formulated:

$$f = -0.00065V + 0.192 \; for \; V \; value \; between \; 40 \; and \; 80 \text{ kph} \tag{4}$$

$$f = -0.00125V + 0.24 \; for \; V \; value \; between \; 80 \; and \; 112 \text{ kph} \tag{5}$$

Furthermore, the reaction time value of 2.50 s and the acceleration due to gravity of 9.81 m/s^2 are added to Eqs. (2) and (3) to become:

$$d = 0.694 \cdot v + \frac{v^2}{254.016 \cdot (f \pm G)} \tag{6}$$

To solve the polynomial function, Eqs. (7) and (8) are used. The equation for calculating maximum speed limit using the SSD method can be seen in Eq. (9).

$$ax^2 + bx + c = 0 \tag{7}$$

$$x = \frac{-b \pm \sqrt{b^2 - 4ac}}{2a} \tag{8}$$

$$v = \frac{-0.694 + \sqrt{0.694^2 + 4 \cdot \left(\frac{1}{254.016 \cdot (f \pm G)}\right) \cdot \left(0.694Vc + \frac{Vc^2}{254.016 \cdot (fc \pm G)}\right)}}{2 \cdot \left(\frac{1}{254.016 \cdot (f \pm G)}\right)} \tag{9}$$

where V is the maximum speed limit of the vehicle (kph), V_c is free-flow speed (kph), f is the value of road friction coefficient between tire and road pavement surface based on the design speed, f_c is the value of road friction coefficient between tire and road pavement surface based on free-flow speed and G is the road slope (%).

Table 1 Road geometric conditions and side friction

Name of road	Road type	Road width (m)	Left shoulder width (m)	Right shoulder width (m)	Side friction	Design speed (kph)	Terrain condition
Padamara street	2/2 UD	7.5	1.0	1.0	201.5 (low)	60	Flat
Letnan Yusuf	2/2 UD	7.0	1.2	1.2	299 (low)	60	Flat
Letkol Isdiman	2/2 UD	12.0	1.5	1.5	916.3 (very high)	60	Flat

3 Result and Discussion

3.1 The Geometric Conditions and Side Friction of the Road

The geometric data of the roads collected include the number of lanes, the lane width, the width of the shoulders and the conditions of the road terrain. The road geometric conditions include road type, road width, road shoulder width, design speed, and side friction on the three roads are shown in Table 1.

3.2 Traffic Flows During Peak Hour Time

The traffic flow data during peak hour time on the three collector roads in Purbalingga: Padamara Street, Letnan Yusuf Street, and Letkol Isdiman Street are shown in Table 2. Based on Table 2, the peak hours' time for Padamara Street and Letkol Isdiman Street are at 06.15–07.15 with a total traffic flow of 4049 vehicles/h and 3287 vehicles/h.

Table 2 Traffic flow during peak time

Name of road	Peak hour time	Vehicle type							Total (veh./h)
		Motor cycles	Passenger cars	Pick up, City trans	Trucks	Light trucks	Bus	Minibus	
Padamara street	06.15–07.15	3768	246	19	1	12	1	2	4049
Letnan Yusuf	06.30–07.30	2985	224	20	1	20	11	26	3287
Letkol Isdiman	06.15–07.15	2802	287	18	0	1	0	8	3116

The peak hour time for Letnan Yusuf Street at 06.30–07.30 with a total traffic flow of 3116 vehicles/h.

3.3 Determination of the Maximum Speed Limit Using 85th Percentile Method

S-Curve of Vehicle Speed

The S-curve of vehicle speed presents the relationship between vehicle speed and the percentage of the cumulative frequency of vehicle speeds. Vehicle speeds of motorcycle (MC), passenger car (PC), pick up and city transport (PU&CT), light truck (LT), minibus (MB), bus, truck (TR), obtained in the field using Speed Gun. S-curve of vehicle speed is used in the determination of the 85th percentile speed. The S-curves of vehicle speed on Padamara Street, Letnan Yusuf Street, and Letkol Isdiman Street are presented in Figs. 2, 3 and 4.

Maximum Speed Limit based on 85th Percentile Method

The maximum speed limit for each type of vehicle and all vehicles using 85th percentile method are shown in Table 3. The maximum speed limit of vehicles is 44 kph on Padamara Street and 42 kph to Letnan Yusuf and Letkol Isdiman.

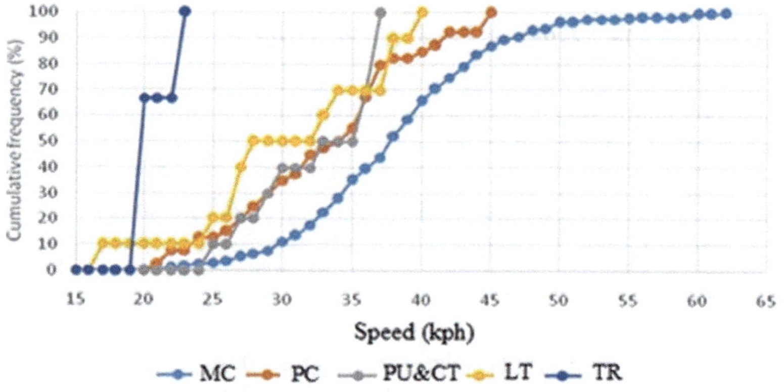

Fig. 2 S-curve of vehicle speed on Padamara Street, Purbalingga

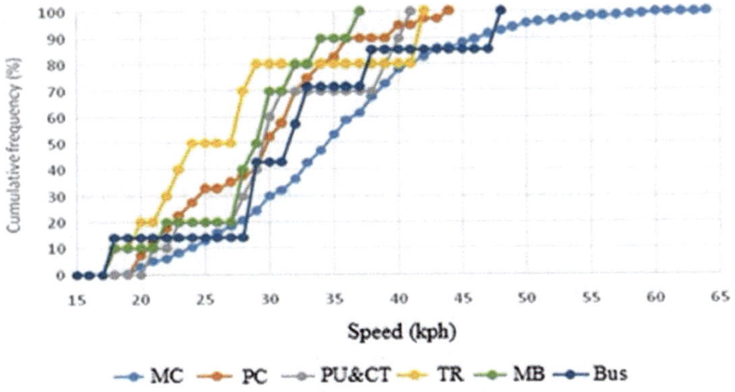

Fig. 3 S-curve of vehicle speed on Letnan Yusuf Street, Purbalingga

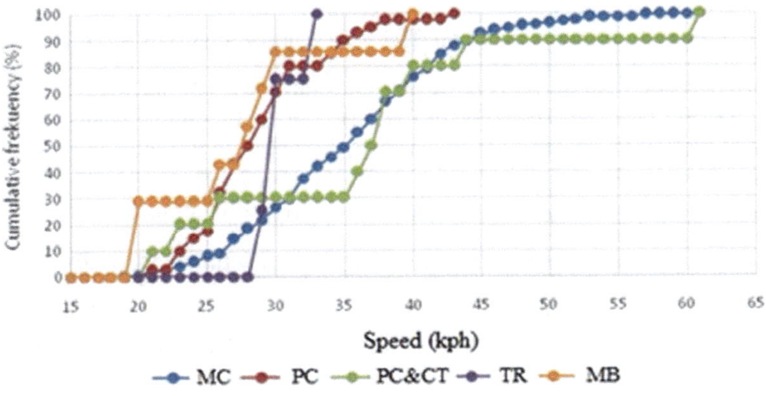

Fig. 4 S-curve of vehicle speed on Letkol Isdiman Street, Purbalingga

Table 3 Maximum speed limit of each vehicle using the 85th percentile method

Name of road	Maximum speed limit value of vehicles (veh/h)								
	Motor cycles	Passenger cars	Pick up, City trans	Truck	Light truck	Mini bus	Big buses	4/ more wheels	All vehicles
Padamara street	44	40	37	38	23	35	38	38	44
Letnan Yusuf	43	35	39	42	N.A	34	38	37	42
Letkol Isdiman	42	34	44	33	N.A	30	N.A	37	42

3.4 Determination of the Maximum Speed Limit Using SSD Method

Free-flow Speed

Free-flow speed is influenced by the value of basic current speed (C_o), correction factor of road width (FC_W), correction factor of side friction (FC_{sf}), correction factor of separation (FC_{sp}) and correction factor of citizen size (FC_{cs}). The value of the correction factor for speed refers to the Indonesian Highway Capacity Manual 1997 [45]. Data on road width and side friction in each road section refers to Table 1, while the correction factor of citizen size is determined based on the population in Purbalingga Regency in 2019 amounting to 933,989 peoples so the value of FC_{cs} is 0.95. The free-flow speed for Padamara Street at 40.50 kph, Letnan Yusuf Street at 39.26 kph, and Letkol Isdiman Street is 39.57 kph.

Road Friction Coefficient between Tire and Road Pavement Surface

Road friction coefficient between tire and road surface is obtained from calculations based on Eqs. (4) or (5). The value of road friction coefficient between tire and road pavement surface is calculated for two conditions, the first based on the design speed and second based on free-flow speed. Road friction coefficient for three collector roads in Purbalingga Regency with a design speed of 60 kph is 0.153.

Maximum Speed Limit based on Stopping Sight Distance Method

The maximum speed limit based on the SSD method is determined using Eq. (9). Vehicle speed limit for the three collector roads in Purbalingga are as follows:

Padamara Street, Purbalingga

The free-flow speed on Padamara Street is 40.5 kph. The road friction coefficient between tire and road pavement surface with a design speed of 60 kph is 0.153 and based on free-flow speed is 0.1657. Slope of the road is 0%. Obtained the maximum speed limit on Padamara Street is 39.32 kph.

Letnan Yusuf Street, Purbalingga

The free-flow speed on Letnan Yusuf Street is 39.26 kph. The road friction coefficient with a design speed of 60 kph is 0.153 and based on free-flow speed is 0.166. The maximum speed limit on Letnan Yusuf Street is 38.10 kph.

Letkol Isdiman Street, Purbalingga

The free-flow speed on Letkol Isdiman Street is 39.57 kph. The road friction coefficient with a design speed of 60 kph is 0.153 and based on free-flow speed is 0.166. The road slope is 0%. The maximum speed limit on Letkol Isdiman Street is 38.40 kph. The recapitulation of data from the analysis of the maximum speed limit using stopping sight distance method is shown in Table 4.

Road safety strategies potentially are improved by applying system theory and safety models [46]. Implementation of speed limit in arterial and collector roads

Table 4 Road friction coefficient and maximum speed limit using SSD method

Name of collector road	Design speed (kph)	Friction coefficient based on design speed (f)	Free-flow speed (kph)	Friction coefficient based on free-flow speed (fc)	Maximum speed limit (kph)
Padamara street	60	0.153	40.50	0.1657	39.32
Letnan Yusuf	60	0.153	39.26	0.1660	38.10
Letkol Isdiman	60	0.153	39.57	0.1660	38.40

is one of the strategies to increase road safety. Operating speed has a significant contribution to the total number of crashes [47]. The maximum speed limit of vehicles proposed for collector roads with residential land functions on the right or left side is 40 kph. The proposed maximum speed limit value is in line with Ministerial Decree PM 111/2015 [48] where the speed limit for vehicles in residential areas is 40 kph. The result of this study relevant with study by Soriguera et al. that a stable flow of 1942 veh/h/lane was measured with a 40 kph speed limit in force [49].

4 Conclusion

Based on the results of study to determine the maximum speed limit in residential area, the following conclusion can be drawn:

a. The maximum speed limit based on the 85th percentile method is greater than that obtained from stopping sight distance method.
b. The maximum speed limit of vehicles based on the 85th percentile method between 42 up to 44 kph and based on the stopping sight distance method between 38.10 up to 39.32 kph.
c. The maximum speed limit of vehicles proposed for collector roads with residential land functions on the right or left side is 40 kph.

Acknowledgements The authors fully acknowledged the Ministry of Education, Culture, Research and Technology and Jenderal Soedirman University for the approved fund through Research Grant Basic Research in the fiscal year 2021 which makes this important research viable and effective.

References

1. Traffic Corps Republic of Indonesia (*Korlantas Polri*) (2014) Kecelakaan Lalu Lintas dalam Angka 2013. Traffic Corps Republic of Indonesia. Korlantas Polri, Jakarta
2. Treat JR, Tumbas NS, McDonald ST, Shinar D, Hume RD, Meyer RE (1977) Tri-level study of the causes of traffic accidents. NTHSA, Washington

3. Sugiyanto G, Mulyono B, Santi MY (2014) Karakteristik kecelakaan lalu lintas dan lokasi black spot di Kabupaten Cilacap. Jurnal Teknik Sipil 12(4):259–266
4. Sugiyanto G, Fadli A, Santi MY (2017) Identification of black spot and equivalent accident number using upper control limit method. ARPN J Eng Appl Sci 12(2):528–535
5. Astarina L, Sugiyanto G, Indriyati EW (2018) Karakteristik kecelakaan lalu lintas dan analisis lokasi black spot di Kabupaten Bogor. Dinamika Rekayasa 14(2):65–76
6. Nilsson G (1991) Speed limits, enforcement and other factors influencing speed. Enforcement and rewarding: strategies and effects. Leidschendam, Netherlands, pp 46–50
7. Hoareau E, Newstead S, Oxley P, Cameron M (2002) An evaluation of the 50 kph speed limit in Southeast Queensland. Journal Report No.264, Monash Univ. Accident Research Centre
8. Organization for Economic Cooperation Development (OECD) (2006) Speed management. Transport Research Centre, OECD and European Conference of Ministers of Transport, Paris
9. Elvik R (2010) A restatement of the case for speed limits. Transp Pol 17(3):196–204
10. Schmid-Mast M, Sieverding M, Esslen M, Graber K, Jäncke L (2008) Masculinity causes speeding in young man. Accid Anal Prev 40:840–842
11. Haglund M, Aberg L (2000) Speed choice in relation to speed limit and influences from other drivers. Transp Res Part F: Traffic Psychol Behav 3(1):39–51
12. Sugiyanto G, Jajang, Santi MY (2019) The impact of lowering speed limit on mobility and the environment. AIP Conf Proc 2094(1):020019, 1–8
13. Russo BJ, Savolainen PT, Schneider WH, Anastasopoulos PC (2014) Comparison of factors affecting injury severity in angle collisions by fault status using random parameters bivariate ordered probit model. Anal Methods Accid Res 2:21–29
14. Sugiyanto G (2017) The cost of traffic accident and equivalent accident number in developing countries (case study in Indonesia). ARPN J Eng Appl Sci 12(2):389–397
15. Sugiyanto G, Santi MY (2017) Road traffic accident cost using human capital method (case study in Purbalingga, Central Java, Indonesia). J Teknol (Sci Eng) 79(2):107–116
16. Tarko AP (2009) Modeling drivers' speed selection as a trade-off behavior. Accid Anal Prev 41:608–616
17. Fuller R (2005) Towards a general theory of driver behavior. Accid Anal Prev 37(3):461–472
18. Delhomme P, Chappé J, Grenier K, Pinto M, Martha C (2010) Reducing air pollution: a new argument for getting drivers to abide by the speed limit? Accid Anal Prev 42:327–338
19. Fuller R, Gormley M, Stradling S, Broughton P, Kinnear N, O'Dolan C, Hannigan B (2009) Impact of speed change on estimated journey time: failure of drivers to appreciate relevance of initial speed. Accid Anal Prev 41(1):10–14
20. Sugiyanto G (2016) The impact of congestion pricing scheme on the generalized cost and speed of motorcycle to city of Yogyakarta, Indonesia. J Eng Appl Sci 11(8):1740–1746
21. Sugiyanto G (2018) The effect of congestion pricing scheme on the generalized cost and speed of motorcycle. Walailak J Sci Tech (WJST) 15(1):95–106
22. Schepers P, Hagenzieker M, Methorst R, Wee BV (2014) A conceptual framework for road safety and mobility applied to cycling safety. Accid Anal Prev 62:331–340
23. Sugiyanto G, Santosa PB, Wibowo A, Santi MY (2016) Hub and spoke airport networks in Indonesia based on Herfindahl-Hirschmann Index. J Eng Appl Sci 11(8):1804–1810
24. Sugiyanto G, Santosa PB, Jajang, Fadli A, Santi MY (2018) Evaluation of hub-spoke airport networks in Sumatra Island, Indonesia to increase efficiency of air transportation. MATEC Web Conf 195(04009):1–9
25. Sugiyanto G, Santosa PB, Wibowo A, Santi MY (2015) Analysis of hub-and-spoke airport networks in Java Island, Indonesia based on cargo volume and freight ratio. Procedia Eng 125:556–563. https://doi.org/10.1016/j.proeng.2015.11.061
26. Sugiyanto G, Santosa PB, Santi MY (2020) Hub and spoke airport networks in Sulawesi Island, Indonesia based on freight ratio. ARPN J Eng Appl. Sci. 15(10):1201–1209
27. Sugiyanto G, Suciningtyas R, Jajang, Santi MY (2020) Strategies for handling black spot area to increased road safety. ARPN J Eng Appl Sci 15(9):1101–1110
28. Wooley J (2005) Recent advantages of lower speed limits in Australia. Proc Eastern Asia Soc Transp 6:3562–3573

29. Global Road Safety Partnership (GRSP) (2008) Speed management: a road safety manual for decision-makers and practitioners, Geneva
30. Depart. of Planning, Transport, Infrastructure (2017) Speed limit guideline for South Australia
31. Baguley CJ (1984) The British traffic conflict technique. Transport and Road Research Laboratory, NATO ASI Series Vol. F5
32. Xie K, Wang X, Ozbay K, Yang H (2014) Crash frequency modeling for signalized intersections in a high-density urban road network. Anal Methods Accid Res 2:39–51
33. Wang X, Yuan J, Schultz G, Fang S (2018) Investigating the safety impact of roadway network features of suburban arterials in Shanghai. Accid Anal Prev 113:137–148
34. Tapp A, Nancarrow C, Davis A, Jones S (2016) Vicious or virtuous circles? Exploring the vulnerability of drivers to break low urban speed limits. Transp Res Part A 91:195–212
35. Sugiyanto G, Malkhamah S (2018) Determining the maximum speed limit in urban road to increase traffic safety. J Teknol (Sci Eng) 80(5):67–77
36. Sugiyanto G, Fadli A, Suciningtyas R, Indriyati EW, Santi MY (2094) Road safety audit at black spot area (Case study in Tlahab Lor, Karangreja, Purbalingga). AIP Conf Proc 2094(1):020020, 1–10
37. Kagaya S, Thanesuen S, Uchida K (2005) Traffic management in terms of speed limit in Hokkaido Area. J Eastern Asia Soc Transp Stud 6:2247–2262
38. https://www.google.com/maps/search/Perumahan+Griya+Abdi+Kencana (2021)
39. Abraham J (2001) Analysis of highway speed limits. Faculty of Applied Science and Engineering University Toronto, Canada
40. Pauw ED, Daniels S, Thierie M, Brijs T (2014) Safety effects of reducing the speed limit from 90 kph to 70 kph. Accid Anal Prev 62:426–431
41. Shaheed MS, Gkritza K, Carriquiry AL, Hallmark SL (2016) Analysis of occupant injury severity in winter weather crashes: a fully Bayesian multivariate approach. Anal Methods Accid Res 11:33–47
42. Sugiyanto G (2017) Characterization of asphalt concrete produced from scrapped tire rubber. Eng J 21(4):193–206
43. Sugiyanto G (2017) Marshall test characteristics of asphalt concrete mixture with scrapped tire rubber as a fine aggregate. J Teknol (Sci Eng) 79(2):55–64
44. Sukirman S (1994) Dasar-Dasar Perencanaan Geometrik Jalan. NOVA, Bandung
45. Ministry of Public Work (1997) Manual Kapasitas Jalan Indonesia (Indonesian highway capacity manual) 1997. Jakarta
46. Hughes BP, Anund A, Falkmer T (2015) System theory and safety models in Swedish, UK, Dutch and Australian road safety strategies. Accid Anal Prev 74:271–278
47. Vayalamkuzhi P, Amirthalingam V (2016) Influence of geometric design characteristics on safety under heterogeneous traffic flow. J. Traffic Transp Eng (Engl Ed) 3(6):559–570
48. Ministry of Transportation (2015) Peraturan Menteri Perhubungan Republik Indonesia Nomor PM 111 Tahun 2015 tentang Tata Cara Penetapan Batas Kecepatan. Jakarta
49. Soriguera F, Martinez I, Sala M, Menendez M (2017) Effects of low speed limits on freeway traffic flow. Transp Res Part C: Emerg Technol 77:257–274

Identification of Traffic Accident Hazardous Location and Cost of Accidents in Developing Country (Case Study of Tabanan Regency, Bali-Indonesia)

Putu Alit Suthanaya and **Made Oka Sugiana**

Abstract Road traffic accidents have long become a major issue in Indonesia. The number of accident fatalities has increased following the increase in population and motor vehicle ownership. The increase in traffic accidents will be followed by an increase in traffic accident costs. As limited funds are available, one important step required is to identify traffic accident hazardous locations. Previous studies on black spots in developing countries such as Indonesia mainly did not consider the frequency of each type of accident, the impact on mitigation measures such as shortcuts development in reducing accidents, and the cost of traffic accidents. The objectives of this study were to identify black sites and black spots, calculate economic loss due to road traffic accidents, and evaluate the impact on the shortcut's development. This study applied a z-score to identify black sites and CUSUM to identify black spots. The cost incurred due to the accident was analyzed by using the gross output method. Three road sections were found to be the locations of accident-prone areas (black sites). By comparing conditions before and after three shortcuts were built on black spot location, it was found that the existence of the shortcuts has reduced the number of traffic accidents by 70%. The refinement of the road infrastructure has reduced the number of traffic accidents significantly. The average annual cost of accidents in the Tabanan Regency was IDR 10,115,918,496/year.

Keywords Accident costs · Black-sites · Black-spots · CUSUM · Z-score

1 Introduction

Motor vehicle accidents cause more than 1.35 million deaths worldwide. The global rate of road traffic deaths is 18.2 per 100,000 population and regional rates of road traffic deaths in Africa and South-East Asia are 26.6 and 20.7 deaths per 100,000 population respectively [1].

P. A. Suthanaya (✉) · M. O. Sugiana
Civil Engineering Department, Engineering Faculty, Udayana University, Kampus Bukit Jimbaran Street-Bali, Kuta Selatan, Indonesia
e-mail: suthanaya@unud.ac.id

S. A. Kristiawan et al. (eds.), *Proceedings of the 5th International Conference on Rehabilitation and Maintenance in Civil Engineering*, Lecture Notes in Civil Engineering 225,
https://doi.org/10.1007/978-981-16-9348-9_62

In Asia, such as India, a study has analyzed traffic accidents at the national, state, and metropolitan city levels and found that road accidents vary according to age, gender, month, and time. It was also found that the age group of 30–59 years is the most vulnerable population group. Males tended to have higher fatality rates than females, and higher accidents occurred in extreme weather and during working hours [2].

Similar to India, Indonesia also experienced a high rate of traffic accidents. The traffic fatalities from traffic accidents in 2014 were predicted to be 28,000 with a rate of 12 per 100,000 population compared to only 4.8 in Singapore and 5.2 in Australia. It is estimated that traffic fatality in Indonesia will reach 65,000 in 2035. The Indonesian government has targeted to reduce these numbers by 50% in 2020 and 80% in 2035 [3]. The increase in the fatality rates in Indonesia was correlated with the increase in the number of vehicles on the road, which is dominated by motorcycles [4].

Road traffic accidents are a major issue in Indonesia that contribute to economic and social losses. Strategies need to be developed to reduce road crashes. One important field in road traffic accident studies is the identification of traffic accident hazardous locations. The accuracy of the black spots identification contributes to the effectiveness of the traffic management implemented [5]. Suitable criteria for identifying black spots are different from country to country and are also influenced by financial capabilities [6]. In developing countries like Indonesia, due to limited funds available, the treatments need to be prioritized. Therefore, it is required to develop an approach to identify the black spot and also a priority location to be treated so that it is economically effective.

Various research on black spots has been conducted. A study has employed Kernel Density through the GIS environment to identify black spot locations and also used the SI (severity index) method to generate a black site map to determine accident-prone zones and take appropriate measures to be implemented [7]. An empirical Bayes technique was used to identify black spots of two small-sized cities, in India, i.e. Patiala and Rajpura. It was found that the black spots mainly lie on the national highways, state highways, and major district roads passing through the cities [8].

Another study has identified the accident-prone zones within Cochin City using road accident data for three years (2014, 2015, and 2016) [9]. The Accident Severity Index (ASI) method was used to rank the accident locations, while other studies also applied the index [10, 11].

Much research on traffic accidents has been conducted in Indonesia to identify appropriate traffic management measures to be implemented [12]. A study focused on identification of the factors that influence fatal traffic accidents involving motorcycles [13]. Another study investigated types of and factors that influence traffic violations by young motorcyclists [14]. The cost of traffic accidents has been analyzed based on a gross output method [15]. A black spot analysis has been conducted on Siak highway, Riau Province, Indonesia [16]. There were some limitations on the previous studies such as only considering one section of the road, did not analyze black site areas in the region [16], and only focused on the traffic accident cost analysis [15] based on the gross output method developed by [17]. Mainly, previous studies on the black spots have not considered the frequency of each type of traffic accident, the impact on mitigation measures such as shortcuts development in reducing accidents, and the cost of traffic accidents. The objectives of this study were to identify the traffic accident hazardous locations (black sites and black spots), investigate the frequency of accidents at black spot locations, evaluate the impact on shortcut development, and analyze the cost of traffic accidents in developing countries with the case study of Tabanan regency in Bali, Indonesia, where the mode of transport was dominated by motorcycle with the average share of about 70% on the road.

This study applied a z-score to identify black sites and CUSUM to identify black spots. The cost incurred due to the accident was analyzed by using the gross output method. Unlike previous study [15], this study applied the gross output method developed by the Department of Public Works [18]. To analyze the impact on the shortcut development at the same time, this study used road traffic accident data over a 5-year time period from 2013 to 2017 as three shortcuts were developed in 2016. The period from 2013 to 2015 was analyzed as the time before the shortcuts developed and the period from 2016 to 2017 was the time after the development.

2 Theoretical Background

2.1 Black Sites and Black Spots

In an accident analysis, it is required to determine the highest rate of traffic accident locations based on the black-spot analysis [19]. Prioritization of hazardous locations can be done based on real-time data and traffic accident records from the police station. The black-spot is characterized by a high number of traffic accidents. Crash black site analysis is usually focused on the road segments or junctions. There is no general method to determine black sites. Thus, continued review of all methods is necessary to identify the most successful way of identifying black sites [20].

The definition of black spots has changed over time. Several statistical definitions of road traffic accident black spots in different studies have been discussed. The black spots has been defined as the "number of accidents (crashes), or accident consequences by kind and severity, expected to occur on the entity during the specified period on an entity (an intersection, a road segment, a driver, a bus fleet, etc.)" [21]. Another researcher defined black spots as the identification of transportation system locations bearing treatable safety problems which are observed through the risen frequency of crashes relative to similar locations [22].

2.2 CUSUM

CUSUM (cumulative sum) can be used to determine changes in a series of quantities over time. CUSUM method has been used in various studies, including in the medical and financial fields. CUSUM can also be used to identify black spots. CUSUM graph is a standard statistical procedure as quality control to detect changes in the mean value. The calculation to find the next year's CUSUM accident value is as follows [23]:

$$S = [S_0 + (X_1 - W)] \tag{1}$$

where S is the accident CUSUM value, S_0 is the first-year accident CUSUM value, X_1 is the number of accidents every year, and W is the mean value.

2.3 Z-Score

The Z-score is a widely used standardized score. Z-score and describes the position of a score from the mean. It can be used if the mean and standard deviation of the population is known. A positive z-value explains a standard score above the mean and, on the other hand, a negative z value describes a standard score below the mean. The Z-score is very useful in explaining the probability of a value occurring within a normal distribution. It can be calculated as follows [24]:

$$Z_i = X_i - \mu\sigma \tag{2}$$

where Z_i is the Z-score accident at i location, σ is the standard deviation, X_i is the number of accidents for location *i*, and μ is the mean value.

2.4 The Cost of Traffic Accident

To calculate the cost of traffic accidents, this study uses the Guidelines for Calculating the Cost of Traffic Accidents [18]. The approach used in this guideline is the gross output (human capital) approach. The cost of traffic accidents (CTA) in year n is calculated as follows [18]:

$$\text{CTA} = \sum_{i=1}^{k} (\text{NTA}_i \times \text{UCTA}_i(T_n)) \tag{3}$$

where CTA is the amount of traffic accident costs on the road or intersection or area in year n (IDR/year), NTA_i is the number of traffic accidents for each accident class (accidents/year), $UCTA_i(T_n)$ is the unit cost of traffic accidents in year 'n' for each accident class (IDR/accident), and i is the traffic accident class.

3 Research Methods

Figure 1 shows the study location in the Tabanan Regency, Bali Province, Indonesia. The population of the Tabanan Regency was 445,700 people with a total area of 839.33 km^2 and a total road length of 1,058.379 km [25]. The traffic accident data used were for a 5-year time period from 2013 to 2017 [26]. This study period was selected based on the development of three shortcuts in the study area. The period 2013–2015 was the time before the shortcuts were built and the period 2016–2017 was the time after the shortcuts were built. Therefore, the impact on the development of shortcuts to reduce traffic accidents could be measured by comparing conditions before and after they were built.

Fig. 1 Study location in the Tabanan Regency, Bali-Indonesia

The first stage of the research was analyzing the location of black sites. The Z-score values of each location were calculated to determine the growth of the number of accidents per year for a 5-year time period (2013–2017) and to determine the value of the Z-score based on the latest data (2017's data). The black site location was identified with the graph of the Z-score. Next, the second step was to analyze the black spot location. Black spot analysis was performed using the CUSUM (cumulative sum) method. The road was divided into stations every 1 km. The third stage was to analyze the frequency of accidents at each black spot location. Finally, the fourth stage was to analyze the costs incurred due to accidents that occurred in the Tabanan Regency using the Gross Output method [18].

4 Results and Discussion

4.1 Traffic Accident Characteristics

Table 1 shows that the number of traffic accidents on the roads in the Tabanan Regency from 2013 to 2017 varied and tended to decrease every year. The highest number of accidents occurred in 2013 with 299 events, while the lowest occurred in 2017 with 128 incidents.

Table 2 shows that during 2013–2017, the number of human victims due to accidents in the Tabanan Regency reached 1428 people, where the highest number of human victims occurred in 2013, as many as 445 people.

Table 1 The number of traffic accidents on the roads in Tabanan Regency (2013–2017)

No.	Year	Number of accidents
1	2013	299
2	2014	234
3	2015	167
4	2016	129
5	2017	128

Table 2 The number of traffic victims

No.	Year	Fatal	Serious	Minor	Total
1	2013	47	112	286	445
2	2014	37	81	243	361
3	2015	44	44	161	249
4	2016	40	20	108	168
5	2017	42	25	138	205
Total		210	282	936	1,428

Table 3 Traffic accident rates

No.	Road section	Number of accidents					Total
		2013	2014	2015	2016	2017	
1	Denpasar-Gilimanuk	2000	1649	1370	645	860	6524
2	Ir. Soekarno	702	401	281	273	234	1891
3	Ahmad Yani	344	253	244	76	189	1106
4	Perean Tengah	366	202	104	80	128	880
5	Batukaru	92	192	272	64	196	816
6	Baturiti-Mekarsari	236	177	88	171	73	745
7	Wisnu Marga	279	137	69	0	16	501
8	Marga-Apuan	48	324	52	20	16	460
9	Raya Baturiti	64	120	69	104	100	457
10	Raya Tanah Lot	136	132	97	32	32	429

4.2 *Traffic Accident Rates*

The top ten ranking results of the traffic accidents on the roads in the Tabanan Regency can be seen in Table 3. It was found that the Denpasar-Gilimanuk road segment had the highest accident rate compared to the accident rate for the other road sections in the Tabanan Regency. The total number of accidents on the Denpasar-Gilimanuk road during 2013–2017 was 6524. The highest number of accidents occurred in 2013 (2000 accidents).

4.3 *Analysis of Black Sites*

Figure 2 shows the graph of the Z-score. The three roads with the highest Z-score value were Denpasar-Gilimanuk, Ir. Soekarno, and the Ahmad Yani road section.

4.4 *Analysis of Black Spots on Denpasar-Gilimanuk Road Section*

Black spot analysis on the Denpasar-Gilimanuk road section was divided into two, namely black spot analysis before the shortcuts were built (during 2013–2015) and black spot analysis after the shortcuts were built (during 2016–2017). Before the three shortcuts were built, the Denpasar-Gilimanuk road section had a length of 34.84 km with 35 stations (station per 1 km). After the shortcuts were built, the Denpasar-Gilimanuk road section had a length of 33.10 km, with 34 stations. Figure 3 shows

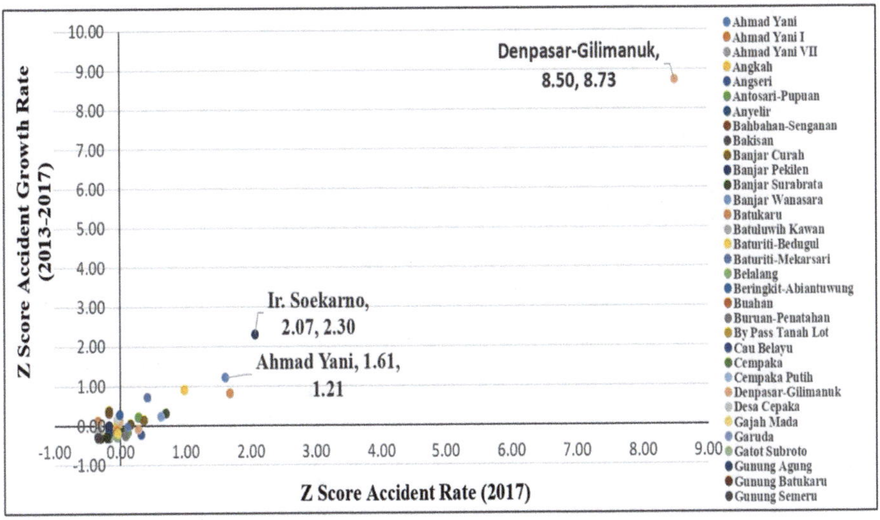

Fig. 2 Z-scores based on the accident growth rates and accident rate in 2017

the graph of the relationship between the year of the accident and the value of CUSUM.

From Fig. 3, it can be seen that the station identified as the location of the black spot was at Sta 27-Sta 28 (total 20 accidents) located in the Lalanglinggah Village area with the largest CUSUM value of 13,914. The accident happened in a flat area with slight corners in areas with several shops and schools. Several factors have caused the accidents such as a shaky vehicle, mistakes in overtaking, and crashing into the vehicle from behind. The type of accidents was dominated by collisions when overtaking (sideswipe) by 50%, head-on collisions, and out of control, each by 18.75%, and front-rear collisions (rear end) by 12.5%. The accident occurred in the morning 37.5% of the time during the day 25% of the time, and at night 18.75% of the time. Collisions were dominated by collisions between motorbikes and cars by 50%, motorbikes with motorbikes by 31.25%, and cars with cars by 18.75%.

After three shortcuts were built in 2016, the Denpasar-Gilimanuk road section had a length of 33.1 km. It was found that the station identified as the location of the accident or black spot was at Sta 10–Sta 11 (km 10,600 to km 10,900) which was located in the Megati area with the largest CUSUM value of 4,176. There were six accidents identified at this location. The cause of the accident, based on police data in the area, was a shaky vehicle, an error in overtaking and crashing into a vehicle from behind. The type of accidents was dominated by collisions when overtaking (sideswipe) and front-rear collision (rear end) each by 33.33%, head-on collision, and out of control by 16.67%. Traffic collisions were dominated by collisions between motorbikes-cars (83.33%) and motorbikes-motorbikes (16.67%).

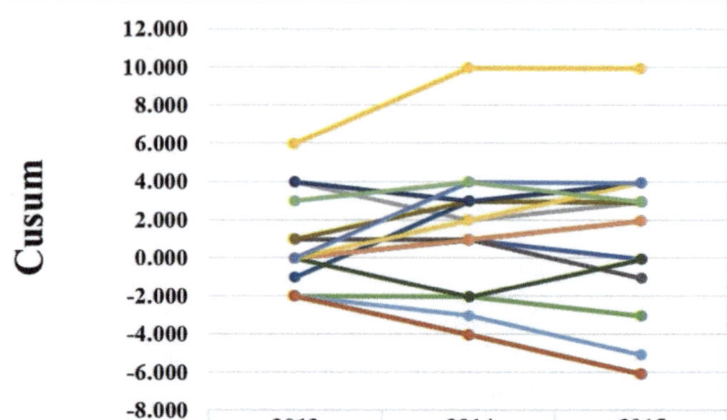

	2013	2014	2015
Sta. 0 - Sta. 1	3.971	2.943	3.914
Sta. 1 - Sta. 2	0.971	0.943	-0.086
Sta. 2 - Sta. 3	3.971	1.943	2.914
Sta. 3 - Sta. 4	5.971	9.943	9.914
Sta. 4 - Sta. 5	-0.029	0.943	-0.086
Sta. 5 - Sta. 6	-2.029	-2.057	-3.086
Sta. 6 - Sta. 7	-1.029	2.943	2.914
Sta. 7 - Sta. 8	-0.029	0.943	1.914
Sta. 8 - Sta. 9	0.971	0.943	-1.086
Sta. 9 - Sta. 10	0.971	2.943	2.914
Sta. 10 - Sta. 11	3.971	2.943	3.914
Sta. 11 - Sta. 12	-0.029	-2.057	-0.086
Sta. 12 - Sta. 13	-2.029	-3.057	-5.086
Sta. 13 - Sta. 14	-0.029	0.943	1.914
Sta. 14 - Sta. 15	-2.029	-4.057	-6.086
Sta. 15 - Sta. 16	-0.029	1.943	3.914
Sta. 16 - Sta. 17	-0.029	3.943	3.914
Sta. 17 - Sta. 18	2.971	3.943	2.914
Sta. 18 - Sta. 19	-2.029	-4.057	-6.086
Sta. 19 - Sta. 20	-2.029	-4.057	-6.086

Fig. 3 Black spot on Denpasar-Gilimanuk road section (2013–2015) (before the shortcut) based on CUSUM

4.5 Analysis of the Amount of Traffic Accident Costs (TAC)

Table 4 shows the total amount of traffic accident costs in the Tabanan Regency for the period of the year 2013–2017.

The highest traffic accident costs were obtained in 2013 where the total costs of fatal, severe, and minor accidents reached IDR 12,774,884,608/year. The region's gross domestic product (GDP) was IDR 6,452,645.72 million in 2013 [27]. Therefore, the total accident cost of the Tabanan regency was about 0.20% of the GDP, which was lower if compared to 0.38% in the Purbalingga Regency [15]. One way to reduce

Table 4 The total amount of traffic accident costs (2013–2017)

No.	Year	Accident	Cost of accidents on the road section (IDR/Year)	Cost of accidents at intersection (IDR/Year)	Economic loss (IDR/Year)	Total economic loss (IDR/Year)
1	2013	Fatal	7,833,677,574	–	7,833,677,574	12,774,884,608
		Serious	2,367,084,333	–	2,367,084,333	
		Minor	2,574,122,701	–	2,574,122,701	
2	2014	Fatal	5,960,338,731	–	5,960,338,731	9,845,880,419
		Serious	1,396,012,643	25,382,048	1,421,394,691	
		Minor	2,447,269,279	16,877,719	2,464,146,998	
3	2015	Fatal	6,857,404,771	–	6,857,404,771	9,378,780,377
		Serious	836,104,968	–	836,104,968	
		Minor	1,685,270,638	–	1,685,270,638	
4	2016	Fatal	7,244,775,957	–	7,244,775,957	8,971,414,324
		Serious	295,860,969	–	295,860,969	
		Minor	1,412,892,680	17,884,717	1,430,777,398	
5	2017	Fatal	7,840,222,236	–	7,840,222,236	9,608,632,751
		Serious	332,246,489	55,374,415	387,620,903	
		Minor	1,380,789,613	–	1,380,789,613	
					Average	10,115,918,496

the traffic accident costs was by improving the road geometry, such as developing shortcuts in the identified black spot location. As limited funds are available, the strategy to be implemented in reducing traffic accidents needs to be preceded by the identification of the black site and black spot. The unit cost of the traffic accident that has been developed by the Department of Public Work [18] needs to be updated to accommodate present conditions.

5 Conclusions

As limited funds are available, it is important to rank the location of accident-prone areas to prioritize the mitigation measures. This study applied a Z-score to rank black site location and CUSUM to determine black spot location. Three roads were found to be the locations of black sites. By comparing conditions before and after three shortcuts were built on the black spot location (Gilimanuk-Denpasar road section), it was found that the existence of the shortcuts has reduced the number of traffic accidents by 70%. This indicated that the refinement of road infrastructure can reduce the number of traffic accidents significantly. The highest total cost of accidents occurred in 2013 with the cost of fatalities and injuries reaching IDR 12,774,884,608/year or 0.20% of GDP. The average annual cost of accidents in the Tabanan Regency during the year 2013–2017 was IDR. 10,115,918,496/year. The unit cost of the traffic accident that has been developed by the Department of Public Work [18] needs to be updated to accommodate present conditions.

References

1. World Health Organization (2018) Global status report on road safety 2018 summary. Available at: https://www.who.int/publications-detail/global-status-report-on-road-safety-2018. Accessed 26 March 2019
2. Singh SK (2017) Road traffic accidents in India: issues and challenges. Transp Res Procedia 25:4708–4719
3. Directorate General of Land Transportation, Ministry of Transportation Republic of Indonesia (2011) National plan on road safety 2011–2035, Jakarta
4. Jusuf A, Nurprasetio IP, Prihutama A (2017) Macro data analysis of traffic accidents in Indonesia. J Eng Technol Sci 49(1):132–143
5. Elvik R (2006) New approach to accident analysis for hazardous road locations. Transp Res Record J Transp Res Board 1953:50–55
6. Jafri SF (2013) An effective and way forward approach for road safety acquisition of knowledge correlative study for Pakistan. Eng J 17:41–48
7. Shad R, Rahimi S (2017) Identification of road crash black-sites using geographic information system. Int J Traffic Transp Eng 7(3):368–380
8. Dhanoa KK, Tiwari G, Malayath M (2017) Method to identify black spots with minimal available police data. J Eastern Asia Soc Transp Stud 12:2054–2073
9. Fayaz MM, Mrudula SP, George SJ, Yoyak SP, Roy SS (2018) Black spot identification using accident severity index method. Int J Curr Eng Sci Res ((IJCESR) 5(3):2394–8374
10. Katre N, Pitale NH, Bobade S (2019) Analysis of black spots on NH-3 and its rectification. MAT J Transp Syst 4(2):16–32
11. Matthew NA, Silven A, Mathew EE, Dominic S, Joy T (2019) Identification of accident black spots along NH-85. Int J Sci Res Rev 8(6):222–226
12. Soehodho S (2017) Public transportation development and traffic accident prevention in Indonesia. IATSS Res 40:76–80
13. Suthanaya PA (2016) Analysis of fatal accidents involving motorcycles in low income region (case study of Karangasem Region, Bali-Indonesia). Int J Eng Res Afr 19:112–122
14. Joewono TB, Susilo YO (2017) Traffic violations by young motorcyclists on Indonesian urban roads. J Transp Saf Secur 9:236–261
15. Sugiyanto G (2017) The cost of traffic accident and equivalent accident number in developing countries (case study in Indonesia). ARPN J Eng Appl Sci 12(2):389–397
16. Sandhyavitria A, Zamrib, Wiyono S, Subiantoro (2017) Three strategies reducing accidents rates at black spots and black sites road in Riau Province, Indonesia. Transp Res Procedia 25:2153–2166
17. Silcock R, Transport Research Laboratory (TRL) (2003) Guidelines for estimating the cost of road crashes in developing countries. Department of International Development. Project R7780
18. Department of Public Works (2003) Calculation of the cost of traffic accidents using the gross output (human capital) method. Department of Public Work, Jakarta
19. Raut UM, Dhumal RK, Nagne SD, Kale KV (2015) GIS contribution for identification of accident black spots—a review. Int J Comput Sci Eng 3(7):75–80
20. Anderson T (2007) Comparison of spatial methods for measuring road accident ‘hotspots’: a case study of London. J Maps 3(1):55–63

21. Hauer E (1996) Identification of sites with promise. Transp Res Record J Transp Res Board 1542:54–60
22. Cheng W, Washington SP (2005) Experimental evaluation of hotspot identification methods. Accid Anal Prev 37(5):870–881
23. Austroads (1992) Road crashes, guide and traffic engineering practice part 4. Australia, Sydney
24. Base C (2018) Understanding basic statistics. Cengage Learning
25. Tabanan Bureau of Statistics (2020) Tabanan regency in figures 2019. Bureau of Statistics, Tabanan. Bali
26. Tabanan Police Department (2018) Annual traffic accident report. Tabanan
27. Tabanan Bureau of Statistics (2014) Tabanan regency in figures 2013. Bureau of Statistics, Tabanan, Bali

Characteristics of Foreign Motorcyclists in Tourism Areas in Bali

Cokorda Putra Wirasutama, Putu Alit Suthanaya, Dewa Made Priyantha Wedagama, and Anak Agung Gde Agung Yana

Abstract Along with the development of tourism, the number of foreign motorcyclists also increases. Meanwhile, it is widely known and empirically found that foreign riders are prone to greater risks than domestic riders in most countries around the world. This is due to different traffic cultures. Therefore, safety measures of foreign riders are highly important. This study examined the characteristics of foreign motorcyclists in tourism areas in Bali, as a preliminary study on the behaviour of foreign motorcyclists in tourism areas in Bali. This study was conducted through questionnaire survey on 30 foreign motorcyclists. The results obtained were analyzed using SPSS. Based on the survey results respondents' characteristics were categorized based on gender, age, education, marital status, and countries of origin, driver's license ownership, purpose of riding and the riding distance traveled.

Keywords Characteristics · Motorcyclist · Foreigner

1 Introduction

Foreigners coming to Indonesia come from many countries with varying characteristics, both in terms of age, purposes of arrival, education and cultural background. These differences in characteristics lead to differences in the choice of transportation modes in Indonesia. The mode of transportation used varied from large buses with

C. P. Wirasutama (✉)
Mahasaraswati University, Denpasar, Indonesia
e-mail: cokputra@unmas.ac.id

P. A. Suthanaya · D. M. P. Wedagama · A. A. G. A. Yana
Udayana University, Denpasar, Indonesia
e-mail: suthanaya@unud.ac.id

D. M. P. Wedagama
e-mail: priyantha@civil.unud.ac.id

A. A. G. A. Yana
e-mail: agungyana@unud.ac.id

S. A. Kristiawan et al. (eds.), *Proceedings of the 5th International Conference on Rehabilitation and Maintenance in Civil Engineering*, Lecture Notes in Civil Engineering 225,
https://doi.org/10.1007/978-981-16-9348-9_63

a capacity of 40–50 passengers, medium buses with a capacity of 20 passengers minibuses with a capacity of dozens of passengers, vehicles like MPV (multipurpose vehicle) with a capacity of 7 or 8 passengers, and some also prefer bikes as mode of transportation. Apart from using these modes, foreigners often use two-wheeled motorcycles as a means of transportation. The behaviour of foreigners who ride two-wheeled motorcycles will be examined in this study.

Foreign motorcyclists are considered to be prone to greater risks than domestic drivers in most countries around the world. Several empirical findings have been reported, including by [1]. It is said that the traffic cultures of different countries largely explain the differences of some specific technical parameters.

Previously, the risks faced by foreign motorcyclists were not thoroughly analyzed in detail, or at least not as thoroughly documented. The same applies, however, in most countries around the world [2]. Lack of knowledge of the road network, lack of understanding of local traffic regulations, and insufficient driving skills in unfamiliar conditions, can lead to an increase in the number of accidents, severity, and the risks faced by foreign motorcyclists.

Variations in general road safety attitudes expressed in riding behaviour might further lead to a greater risk of accidents for foreign motorcyclists compared to locals. A study on the characteristics of foreign motorcyclists in Bali tourist regions was done as a preliminary study on the behaviour of foreign motorcyclists in Bali tourist areas. The method used in this study is SPSS. The intended contribution of this study is to know the characteristics of foreign motorcyclists in tourist areas. Subsequently, it can be utilised as input for future study on foreign motorcyclists.

2 Research Method

See Fig. 1.

2.1 Research Location and Time

The research location was conducted in tourism destination areas in Bali, such as in Kuta, Ubud, Sanur, and Canggu. The research was conducted from September 2020 to December 2020.

2.2 Scope of Study

The study was carried out in tourism destination areas in Bali, such as in Kuta, Ubud, Sanur, and Canggu. Therefore, the questionnaires were made based on the characteristics that need to be observed from foreign motorcyclists in those areas.

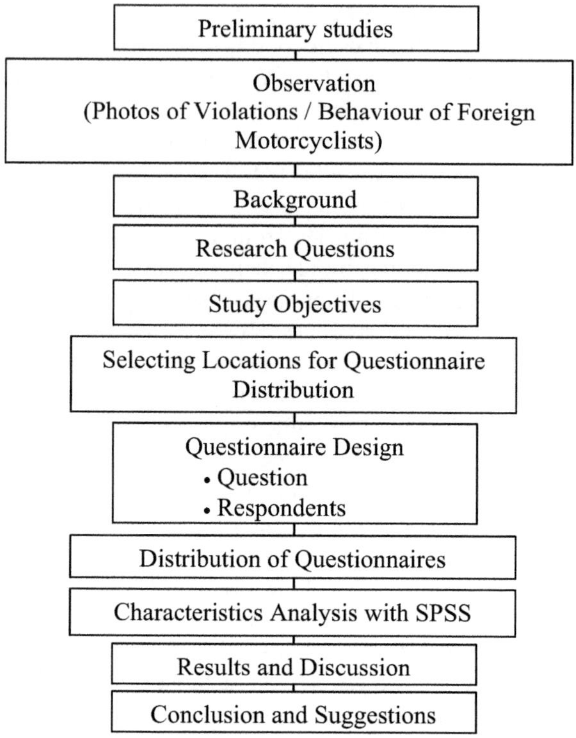

Fig. 1 Research design

2.3 Determination of Data Sources

Sources of data used are from the collection of observations or recordings in the field and the results of questionnaires about the characteristics of foreign motorcyclists in tourism areas in Bali. The questions given were designed in such a way as to get the desired results without being disturbance to the respondents. A structured questionnaire was developed to collect information on foreign motorcyclists' characteristics in tourism areas in Bali.

There are 30 respondents examined in this study, according to Bailey's statement [3]. The questionnaires were distributed as independent questionnaires. On this occasion, if the respondent does not want to write, the surveyor asks the question as a face-to-face interview. This implies that the sample selection process may not be completely random as respondents are approached based on respondents' availability. The questionnaires were given only to respondents who claimed to be motorcyclists. The questionnaire survey per respondent took about 20 min to complete. The data sources in this study are as follows:

1. Primary data is data obtained directly from primary data sources. The primary data source in this study is the result of direct field observations on foreign

Table 1 Data, data types and data sources

Data	Type of data	Data source	
		Primary	Secondary
Photos of foreign motorcyclists	Qualitative	Observation	
Characteristics of foreign motorcyclists	Quantitative	Questionnaire	Literature

motorcyclists' behaviour (field observation). Results of the questionnaires given to foreign motorcyclists are also included as primary data source.

2. Secondary data is obtained from secondary data sources which are not directly obtained by researchers, such as through relevant journals, books, and archives related to the discussions. These sources can be used as guidelines or references in this study, including data on Bali's foreign visitors (Tables 1 and 2).

3 Results

3.1 Respondents Characteristics Based on Gender

The results show that the questionnaires had been distributed to 30 respondents consist of 14 male respondents (46.7%), and 16 female respondents (53.3%) (Fig. 2 and Table 3).

3.2 Based on Age

The results showed that the questionnaire distributed to 30 respondents with the following age groups: 18–25 years (5 people) with a percentage of 16.7%, 26–35 years (7 people) a percentage of 23.3%, age 36–45 years (6 people) a percentage of 20.0%, age 46–55 years (8 people) a percentage of 26.7% and age > 50 years (4 people) a percentage of 13.3% (Fig. 3 and Table 4).

3.3 Based on Country of Origin

The results showed that the questionnaires distributed to 30 respondents from different country of origin, such as African countries (1 person) with a percentage of 3.3%, America (1 person) 3.3%, Australia (1 person) 3.3%, Canada (1 person) 3.3%, German (2 people) 6.7%, Holland (2 people) 6.7%, India (1 person) 3.3%, Italy (1 person) 3.3%, Japan (1 person) 3.3%, Malaysia (1 person) 3.3%, Poland (1 person) 3.3%, France (2 people) a percentage of 6.7%, Romania (1 person) a percentage of

Table 2 Number of foreign tourist visits to Indonesia in 2019 by entrance

Entrance	Number of visits					Changes in Dec 2019 from 2018 (%)	Changes in Des 2019 from Nov 2019 (%)	Changes in Jan-Des 2019 from 2018
	Dec 2018	Nov 2019	Dec 2019	Jan-Dec 2018	Jan-Dec 2019			
A. Air	825,635	777,244	838,978	10,088,781	9,834,706	1.62	7.94	−2.52
1. Ngurah Rai	495,641	492,904	544,726	6,025,760	6,239,543	9.9	10.51	3.55
2. Soekarno-Hatta	224,989	183,759	186,723	2,814,586	2,419,196	−17.01	1.61	−14.05
3. Juanda	26,609	20,780	20,546	320,529	243,899	−22.79	−1.13	−23.91
4. Kualanamu	19,966	20,798	22,431	229,586	244,530	12.35	7.85	6.51
5. Husein Sastranegara	17,311	14,616	14,951	155,566	157,833	−13.63	2.29	1.46
6. Adi Sucipto	10,813	9218	9795	138,822	113,028	−9.41	6.26	−18.58
7. Bandara Int. Lombok	3151	4635	5499	79,807	57,763	74.52	18.64	−27.62
8. Sam Ratulangi	8845	9261	10,743	122,104	130,285	21.46	16	6.7
9. Minangkabau	4801	5013	5180	54,383	61,131	7.89	3.33	12.41
10. Sultan Syarif Kasim II	3013	2917	4042	29,776	34,419	34.15	38.57	15.59
11. Sultan Iskandar Muda	1315	2417	4012	29,213	28,303	205.1	65.99	−3.12
12. Ahmad Yani	1930	2296	2627	23,592	23,977	36.11	14.42	1.63
13. Supadio	2033	1811	1527	23,050	22,500	−24.89	−15.68	−2.39
14. Hasanuddin	1406	1595	1821	14,126	17,771	29.52	14.17	25.8
15. Sultan Badaruddin II	1431	1218	762	13,862	15,846	−46.75	−37.44	14.31
16. Others	2381	4006	3593	14,019	24,682	50.9	−10.31	76.06

(continued)

Table 2 (continued)

Entrance	Number of visits					Changes in Dec 2019 from 2018 (%)	Changes in Des 2019 from Nov 2019 (%)	Changes in Jan-Des 2019 from 2018
	Dec 2018	Nov 2019	Dec 2019	Jan-Dec 2018	Jan-Dec 2019			
B. Sea	370,629	358,264	388,495	3,214,420	4,160,632	4.82	8.44	29.44
1. Batam	233,153	167,288	190,232	1,887,284	1,947,943	−18.41	13.72	3.21
2. Tanjung Uban	61,258	52,374	59,340	522,399	634,735	−3.13	13.3	21.5
3. Tanjung Pinang	16,020	12,378	16,426	140,603	169,364	2.53	32.7	20.46
4. Tanjung Balai Karimun	9020	9209	10,916	84,718	112,753	21.02	18.54	33.09
5. Tanjung Benoa	3178	5021	7677	31,062	29,456	141.57	52.9	−5.17
6. Tanjung Emas	2315	2202	5135	19,907	17,761	121.81	133.2	−10.78
7. Others	45,685	109,792	98,769	528,447	1,248,620	116.2	−10.04	136.28
C. Land	209,290	145,273	149,594	2,507,104	2,111,616	−28.52	2.97	-15.77
1. Jayapura	3202	5006	6155	104,075	56,287	92.22	22.95	-45.92
2. Atambua	13,915	9824	12,561	85,914	103,905	-9.73	27.86	20.94
3. Entikong	3641	1691	3177	23,213	23,284	−12.74	87.88	0.31
4. Aruk	1125	1009	1708	12,329	16,508	51.82	69.28	33.9
5. Nanga Badau	1826	954	1585	12,148	14,472	−13.2	66.14	19.13
6. Others	185,581	126,789	124,408	2,269,425	1,897,160	−32.96	−1.88	−16.4
Total (A + B + C)	1,405,554	1,280,781	1,377,067	15,810,305	16,106,954	−2.03	7.52	1.88

[a] Included as entrance via air (Hang Nadim airport)

Source BPS, 2016 [4]

Fig. 2 Percentage based on gender

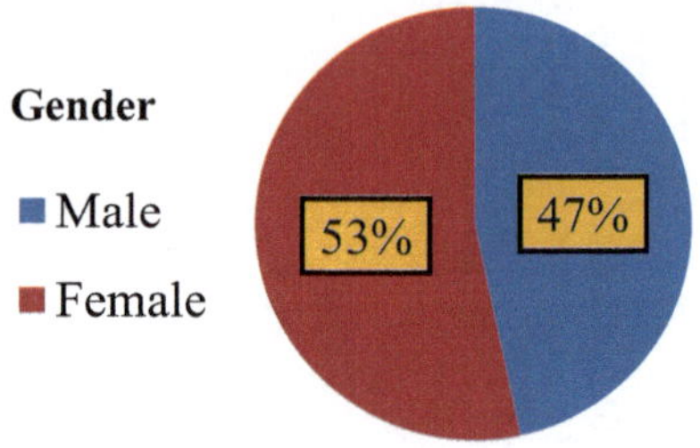

Table 3 Respondents characteristics based on gender

Gender	Total	Percentage (%)
Male	14	46.7
Female	16	53.3
Σ	30	100

Fig. 3 Percentage based on age

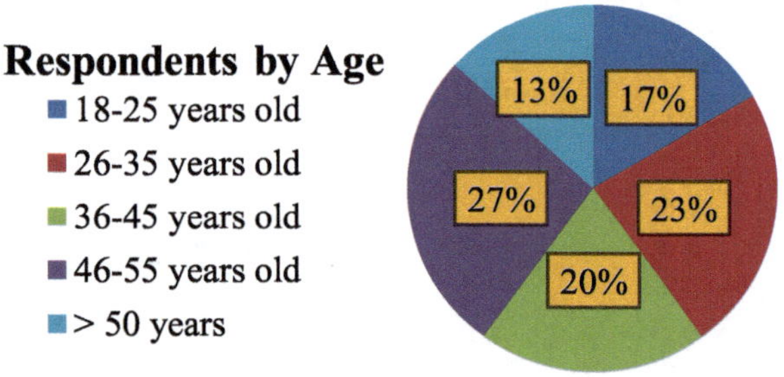

Table 4 Characteristics of respondents by age

Age	Total	Percentage (%)
18–25 years old	5	16.7
26–35 years old	7	23.3
36–45 years old	6	20
46–55 years old	8	26.7
> 50 years	4	13.3
Σ	30	100

3.3%, Russia (8 people) a percentage of 26.7%, Serbia (1 person) a percentage of 3.3%, Singapore (1 person) a percentage of 3.3%, Spain (1 person) a percentage of 3.3%, the UK (2 people) a percentage of 6.7%, USA (1 person) a percentage of 3.3%, and Welsh (1 person) a percentage of 3.3% (Fig. 4 and Table 5).

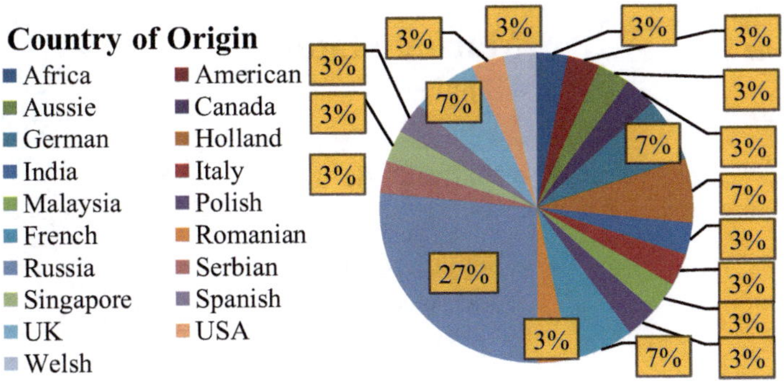

Fig. 4 Percentage based on country of origin

Table 5 Respondent character based on country of origin

Country name	Total	Percentage (%)
Africa	1	3.3
America	1	3.3
Australia	1	3.3
Canada	1	3.3
German	2	6.7
Holland	2	6.7
India	1	3.3
Italy	1	3.3
Malaysia	1	3.3
Poland	1	3.3
France	2	6.7
Romania	1	3.3
Russia	8	26.7
Serbia	1	3.3
Singapore	1	3.3
Spain	1	3.3
UK	2	6.7
USA	1	3.3
Welsh	1	3.3
Σ	30	100

3.4 *Based on Marital Status*

Based on marital status, there are single (12 people) with a percentage of 40.0%, married/de facto (13 people) with a percentage of 43.3%, and separated, divorced, widowed (5 people) with a percentage of 16.7% (Fig. 5 and Table 6).

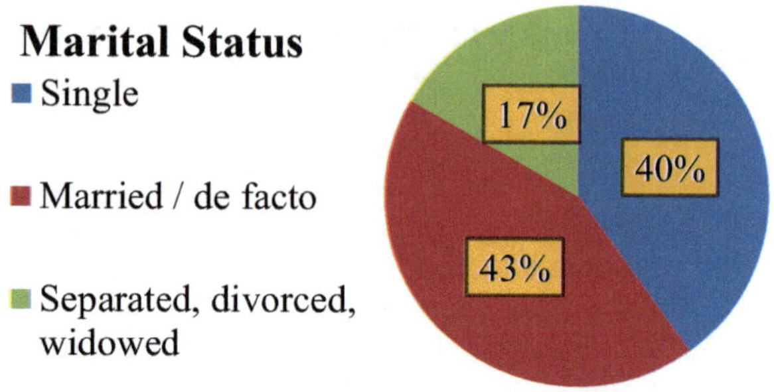

Fig. 5 Percentage based on marital status

Table 6 Respondent character based on status

Marital Status	Total	Percentage (%)
Single	12	40.0
Married/de facto	13	43.3
Separated, divorced, widowed	5	16.7
Σ	30	100

3.5 Based on Education

Based on education, there are Senior High School (4 people) with a percentage of 13.3%, completed tertiary education (26 people) with a percentage of 86.7%, while for Elementary school and Junior High School (0%) (Fig. 6 and Table 7).

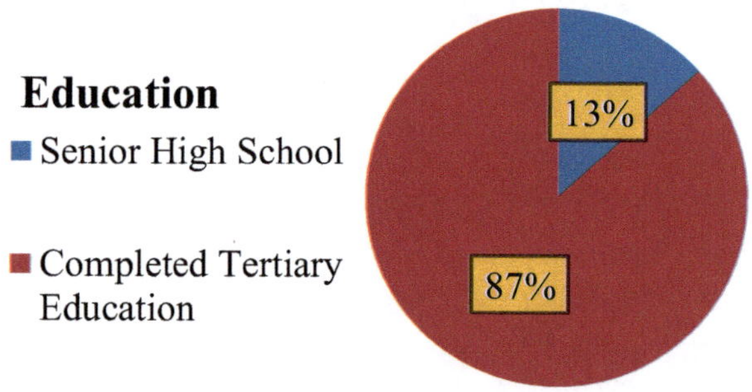

Fig. 6 Percentage based on education

Table 7 Characteristics based on education

Education	Total	Percentage (%)
Senior high school	4	13.3
Completed tertiary education	26	86.7
Σ	30	100

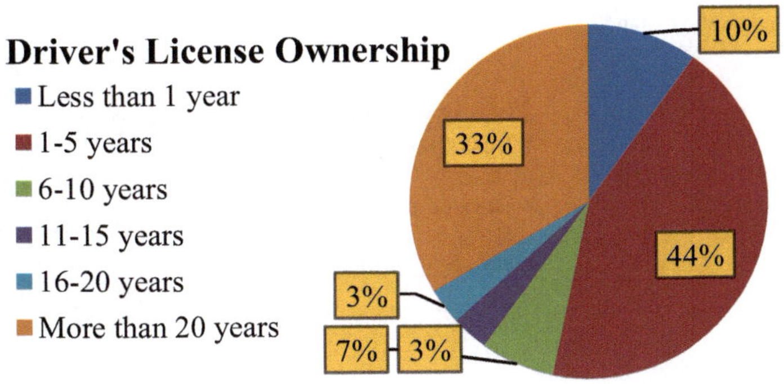

Fig. 7 Based on driver's license ownership

Table 8 Characteristics of respondents based on driver's license ownership

Driver's license ownership	Total	Percentage (%)
Less than 1 year	3	10.0
1–5 years	13	43.3
6–10 years	2	6.7
11–15 years	1	3.3
16–20 years	1	3.3
More than 20 years	10	33.3
Σ	30	100

3.6 *Based on Driver's License Ownership*

Based on Driver's License ownership, there are those with less than 1-year ownership (3 people) with a percentage of 10.0%, 1–5 years (13 people) 43.3%, 6–10 years (2 people) 6.7%, 11–15 years (1 person) 3.3%, 16–20 years (1 person) 3.3%, and more than 20 years (10 people) 33.3% (Fig. 7 and Table 8).

3.7 *Based on the Distance Traveled*

Based on the distance traveled, there are those who ride less than 20 km (18 people) with a percentage of 60.0%, 20–49 km (11 people) 36.7%, 50–99 km (1 person) 3.3%, and the riding distance beyond 100 km (0%) (Fig. 8 and Table 9).

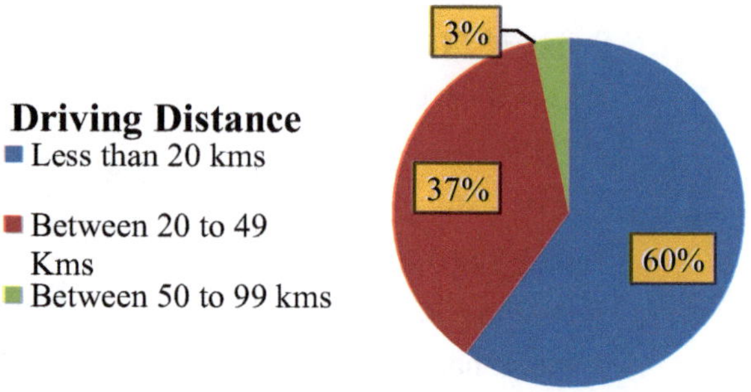

Fig. 8 Based on the distance traveled

Table 9 Characteristics of respondents based on riding distance traveled

Riding distance	Total	Percentage (%)
Less than 20 km	18	60
Between 20 to 49 km	11	36.7
Between 50 to 99 km	1	3.3
Σ	30	100

3.8 Based on the Purpose of Riding

Based on the purpose of riding, there are recreational purposes with 24 people or a percentage of 80.0%, while for commuting or chain trips were 6 people or a percentage of 20.0% (Fig. 9 and Table 10).

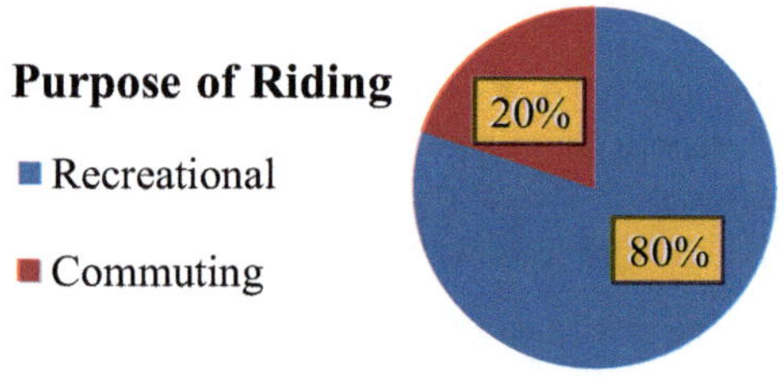

Fig. 9 Based on the purpose of riding

Table 10 Characteristics of respondents based on riding purpose

Riding purpose	Total	Percentage (%)
Recreational	24	80.0
Commuting	6	20
Σ	30	100

4 Conclusion

1. The results shows that the questionnaires distributed to 30 respondents consisted of 18–25 years (5 people) with a percentage of 16.7%, 26–35 years (7 people) a percentage of 23.3%, age 36–45 years (6 people) a percentage of 20.0%, age 46–55 years (8 people) a percentage of 26.7% and age > 50 years (4 people) a percentage of 13.3%.
2. Characteristics of respondents based on status are single (12 people) 40.0%, married/de facto (13 people) 43.3%, separated, divorced, and widowed (5 people) 16.7%.
3. Based on education, there are Senior High School (4 people) 13.3%, completed tertiary education (26 people) 86.7%, while for Elementary school and Junior High School (0%).
4. Based on driver's license ownership, there are those with less than 1-year ownership (3 people) 10.0%, 1–5 years (13 people) 43.3%, 6–10 years (2 people) 6.7%, 11–15 years (1 person) 3.3%, 16–20 years (1 person) 3.3%, and more than 20 years (10 people) 33.3%.
5. Based on the riding distance, there are those who drive less than 20 km (18 people) 60.0%, 20–49 km (11 people) 36.7%, 50 - 99 km (1 person) 3.3%, and the riding distance beyond 100 km (0%).
6. Based on the purpose of riding, there are recreational purposes with 24 people or a percentage of 80.0%, while to ride for commuting or chain trips, there are 6 people or a percentage of 20.0%.

References

1. Yannis G, Golias J, Papadimitriou E (2007) Accident risk of foreign drivers in various road environments. J Saf Res 38:471–480
2. Leviäkangas P (1997) Finnra's VIKING-programme. Finnra internal publications 24/1997. Finnish National Road Administration, Helsinki
3. Bailey KD (1994) Methods of social research, 4th ed. Free Press, Maxwell Macmillan, Canada
4. BPS (2016) Jumlah Kunjungan Wisatawan Mancanegara Tahun 2016 ke Indonesia menurut Pintu Masuk. Departemen Pariwisata, Badan Pusat Statistik, Jakarta

Analysis of the Behaviour Model of Foreign Motorcyclists in Tourism Areas in Bali

Cokorda Putra Wirasutama, Putu Alit Suthanaya, Dewa Made Priyantha Wedagama, and Anak Agung Gde Agung Yana

Abstract Foreign motorcyclists are prone to greater risks than the locals in most countries around the world. Some of the risk parameters that can be identified, such as lack of knowledge of traffic regulations, inadequate driving skills, different seasons, and general attitudes towards traffic safety that are reflected in driving behaviour. Along with the development of tourism, the number of foreign motorcyclists also increases. Therefore, the safety measures are essential. In this research, we study the influence of local road and traffic conditions, the influence of local motorcyclists and law enforcement on the behaviour of foreign motorcyclists around tourism areas in Bali. The method used is Structural Equation Modeling (SEM). The expected contribution is a model that finds the dominant factors influencing foreign motorcyclists' behaviour, which can significantly reduce the risk of accidents involving foreigners. The results are the local road and traffic factors affect foreign motorcyclists' behaviour with a bootstrapping value of 2.426. And local motorcyclists and law enforcement also affect with a bootstrapping number of 3.874. Both factors positively affect foreign motorcyclists' behaviour with a coefficient of 0.217 and 0.371. The research concluded that the behaviour of foreign motorcyclists in tourist areas in Bali is highly influenced by the behaviour of local motorcyclists and law enforcement by authorized officials. It shows that the more stringent enforcement of the law, the better the behaviour of foreign motorcyclists. Additionally, road and traffic conditions also affect the behaviour of foreign motorcyclists. The better foreign motorcyclists' behaviour will reduce the risk of accidents.

C. P. Wirasutama (✉)
Mahasaraswati University, Denpasar, Indonesia
e-mail: cokputra@unmas.ac.id

P. A. Suthanaya · D. M. P. Wedagama · A. A. G. A. Yana
Udayana University, Denpasar, Indonesia
e-mail: suthanaya@unud.ac.id

D. M. P. Wedagama
e-mail: priyantha@civil.unud.ac.id

A. A. G. A. Yana
e-mail: agungyana@unud.ac.id

S. A. Kristiawan et al. (eds.), *Proceedings of the 5th International Conference on Rehabilitation and Maintenance in Civil Engineering*, Lecture Notes in Civil Engineering 225,
https://doi.org/10.1007/978-981-16-9348-9_64

Keywords Motorcyclist · Foreigner · Behaviour · Model

1 Introduction

People who come to Indonesia come from many countries with varying characteristics, in terms of age, purposes of arrival, education, and cultural background. These differences in characteristics lead to differences in the choice of transportation modes in Indonesia. Some of them prefer motorcycle as a mode of transportation. The behaviour of foreigners who ride motorcycles will be examined in this study.

Foreign riders are prone to greater risks than domestic drivers in most countries around the world, proven by several empirical findings, including [1]. It is said that different countries' traffic cultures largely explain the differences in some specific technical parameters. Lack of knowledge of the road network, local traffic regulations and insufficient driving skills in unfamiliar conditions, can lead to an increase in the number of accidents, severity, and the risks imposed on foreign motorcyclists.

Variations in general attitudes towards road safety which are reflected in riding behaviour can further contribute to a higher accident risk for foreign motorcyclists than locals. In this research, we want to study the influence of local road and traffic conditions, the influence of local motorcyclists, and the influence of law enforcement on the behaviour of foreign motorcyclists around tourism areas in Bali. The method used is Structural Equation Modeling (SEM). Research on this field, especially on the effect of local motorcyclists and law enforcement on the behaviour of foreign motorcyclists has not been specifically addressed. The expected contribution from this study results is to provide a model that finds the dominant factors influencing foreign motorcyclists' behaviour, which can significantly reduce the risk of accidents involving foreigners.

The problems to be researched are:

- How the local road and traffic factors influence foreign motorcyclists' behaviour in Bali's tourism areas?
- How does the behaviour of local motorcyclists and law enforcement affect the foreigner motorcyclists' behaviour in tourism areas in Bali?

The benefit of this research is that it is expected to help develop corrective measures in foreign motorcyclists' behaviour to reduce the rate of motorcycle accidents involving foreign motorcyclists. Expectantly, the solution to reduce accidents against foreign motorcyclists will not only be 'hard' solutions such as infrastructure development or law enforcement but it is also hoped that the solution will be more directed at 'soft' improvements such as how to modify the motorcyclists' behaviour.

2 Theory of Planned Behaviour (TPB)

According to the Theory of Planned Behaviour (TPB), human behaviour is guided by three types of considerations: beliefs about possible consequences of behaviour (*behavioural beliefs*), beliefs about the normative expectations of others (*normative beliefs*), and beliefs about the existence of factors that can facilitate or hinder behavioural performance (*control beliefs*). In their respective proportions, beliefs result in behaviour that is favorable or unfavorable to that behaviour.

The intention is assumed to be the precursor to behaviour. However, since many behaviours present difficulties that may limit control over will, it is necessary to consider feeling control over behaviour other than the only intention. If the feeling that control over behaviour is sincere, it can serve indirectly as a solid control and can contribute to the prediction of the behaviour that is being examined [2].

The risk factors associated with injuries related to motor vehicles should be described and framed appropriately in the socio-cultural context. A thorough understanding of socio-cultural factors related to motorcycling is essential to develop targeted and efficient measures and interventions [3].

Road infrastructures, damages, and the maintenance condition apparently are partly responsible for the phenomenon of motorcyclists paying less attention to traffic regulations. According to statistics around the world, the number of traffic fatalities and injuries is estimated to have increased by about 65% between 2000 and 2020 [4, 5]. The outlook is devastating in low and middle-income countries where the number of traffic fatalities is estimated to increase by as much as 80% [6]. Road accidents have become big problems in many countries especially in urban areas [6, 7], and Indonesia did not escape this trend.

The characteristics and consequences of road accidents that result in death and injury in developing countries are different from those of high-income countries. In the United States, 80% of the victims who died from road collisions are motorcyclists and passengers of four-wheeled motor vehicles [8].

In low-income countries, most victims are pedestrians, cyclists, motorcyclists, and even passengers on public transportation systems [6].

What tourists do or not do contributes to many of these occurring accidents. Thus, the study of human behaviour is very important for road safety studies [9, 10]. Human behaviour factor is an important factor that should be considered when designing strategies to reduce traffic accidents and fatalities.

Furthermore, they also argued that motorcyclists' perceptions of the road and traffic environment were correlated with the rider's age, driving experience, fatigue, stress, and emotions. While there has been much research of this kind in developed countries, there has been little research into traffic violations committed by motorcyclists in developing countries, where the use of motorcycles has been growing rapidly recently. Motorcycles hold a large share of the mode of transportation choice in several Southeast Asian countries, such as Indonesia, Malaysia, Vietnam, and Thailand [11]. Meanwhile, motorcycles are also the main reason for traffic accidents pointed out by the researchers. For example, in Indonesia, there are 105.1 million motorcycles from 129.3 million units of motor vehicles in 2016 [12]. In Vietnam, 80–90% of households in Ho Chi Minh City have access to motorcycles [13].

The presence of motorcycles significantly impacts road performance and accident rates [14, 15]. It is widely believed that motorcyclists have relatively poor safety records compared to other groups of road users [16]. Motorcyclists are often seen as road users who trigger many collisions. Besides, they are seen to be more risk-taking and to have more speeding tendencies than car drivers [10]. In Indonesia, motorcyclists have the highest share of recorded accidents [17].

Most of the traffic accidents are related to human factors. Riders regularly deviate from safe driving conditions for various reasons [18]. Disobedience, driving errors, and road traffic violations are the main reasons for traffic accidents pointed out by many researchers such as by [18–22]. O'Connell [23] has stated that many undesirable aspects of rider behaviour happened due to intentional or unintentional actions, which means that the mistakes are committed by themselves and not being forced by others to make mistakes. A study by [24] in Australia, showed that a motorcyclist's intention to high-risk activities is primarily initiated by the desire to seek thrills. In contrast, the intention of safety is based on behavioural control in the form of personal discipline.

Many studies have shown that the attitudes of road users can be explored by studying noncompliance and violations. A detailed discussion related to this topic is provided by [21, 22, 25–27]. It is said that driver performance is influenced by the tendency to conduct driving errors and traffic violations, attitudes towards their own and other road users' driving style, attitudes towards the vehicles they control, and personal characteristics [26]. In other words, violations can occur due to the driver's attitude towards the behaviour, social norms regarding the action, behaviour control, and moral norms [27].

3 Hypothesis

The hypotheses of this study are:

- H1: The behaviour of foreign motorcyclists is influenced by local conditions, which are the road environment and local traffic conditions.
- H2: The behaviour of foreign motorcyclists is also influenced by the behaviour of local motorcyclists and law enforcement.
- H3: local motorcyclists behaviour factors and law enforcement moderate local condition factors
- H4: The behaviour factor of local motorcyclists is influenced by local conditions, which are the road environment and local traffic conditions

4 Methodology

Initial observations were made by observing the behaviour of foreign motorcyclists in tourist areas in Bali. In this observation, it was found that many foreign motorcyclists violated the traffic regulations. This phenomenon is what will be examined by initially determining the causal variables. Then a questionnaire was made to measure these variables. After the questionnaire was distributed and re-collected, the SEM method is used to measure reliability, validity and how much these variables affect the behaviour of foreign motorcyclists in tourist areas in Bali (Fig. 1).

The research will be carried out in tourism destination areas in Bali, which are in Kuta, Ubud, Sanur, and Canggu. The research was conducted from September 2020 to December 2020.

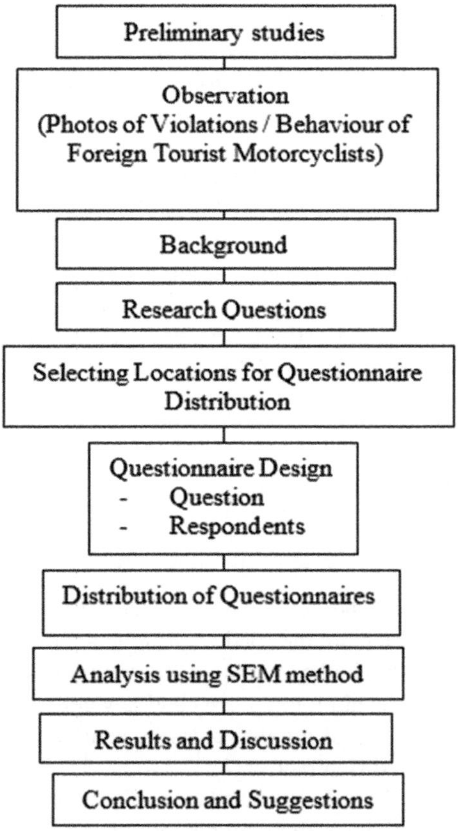

Fig. 1 Research design

5 Determination of Data Sources

Sources of data used are from collecting field observations or recording of foreign motorcyclists' behaviour. Then a survey was conducted using questionnaires to the respondents selected as samples of foreign motorcyclists. A structured questionnaire was developed to collect motorcyclists' information, motorcycle riding behaviour, and attitudes in tourism areas.

Three types of accidents were examined in this questionnaire, which are minor accidents, serious injury accidents and fatal accidents. The target number of respondents is based on [28], who recommends a sample size of at least 10 times the number of research variables for research using multivariate analysis. The variables in this study are 3, so the sample required is 30 respondents. If the respondent does not want to write, the surveyor asks the question as a face-to-face interview. This implies that the sample selection process may not be completely random as respondents are approached based on respondents' availability. The questionnaire was given only to respondents who claimed to be motorcyclists. The data sources in this study were as follows:

1. Primary data is data obtained directly from primary data sources. The primary data source in this study is the result of direct field observations on foreign motorcyclists' behaviour (field observation). And the results of questionnaires for foreign motorcyclists are also the primary data source.
2. Secondary data is obtained from secondary data sources, which are not directly obtained by researchers, namely through related journals, books, and archives related to the discussions. It can be used as guidelines or references in this study, including data on Bali's foreign visits (Table 1).

Table 1 Data, data types and data sources

Data	Type of data	Data source	
		Primary	Secondary
Photos of foreign motorcyclists	Qualitative	Observation	
Theory and conceptual basis – *Theory of planned behaviour* – *Structural equation modelling* (SEM)	Qualitative Quantitative		Literature
Research Question 1			
The influence of local road and traffic factors	Quantitative	Questionnaire	Literature
Research Question 3			
The influence of local motorcyclists and law enforcement factors	Quantitative	Questionnaire	Literature

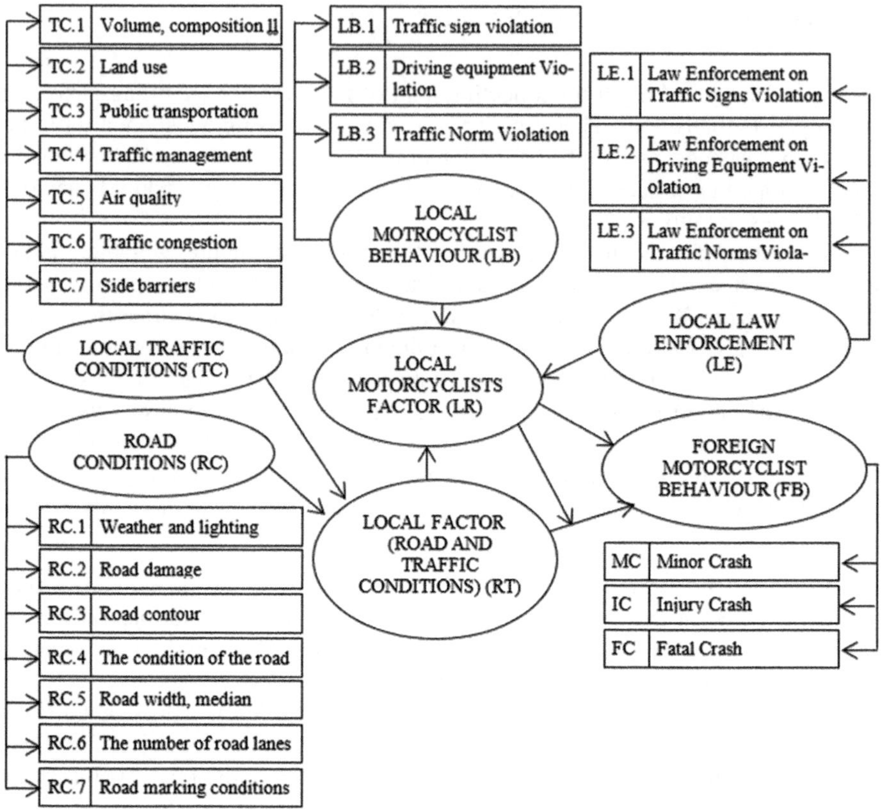

Fig. 2 Research variables

6 Research Variables

The variables used in this study were obtained from previous studies related to this research. The variables of Local Motorcyclist Behaviour (LB) and Local Law Enforcement (LE) are variables that are augmented by the hypothesis that the behaviour of foreign motorcyclists is influenced by local motorcyclists and local law enforcement (Fig. 2).

7 Findings

Outer Model Evaluation

The outer model's assessment is carried out to determine the validity and reliability of the indicators and research latent variables. Validity is known by using the values

Table 2 Convergent validity test

	Outer loading	AVE
LB	0.741	0.731
LE	0.956	
RC	0.921	0.804
TC	0.872	
MC	0.981	0.981
IC	0.996	
FC	0.994	

of *convergent validity* and *discriminant validity*. Meanwhile, reliability is known by using the indicator reliability value and value of *internal consistency reliability*.

7.1 Validity Test

The validity test can be seen from the results of convergent validity and discriminant validity.

Convergent Validity Test

The convergent validity of the model is known from the loading factor value and the AVE (average variance extracted) value. Table 2 shows that the loading factor value is between 0.741 and 1.172, which means that it exceeds 0.5 according to the recommendations of [29]. This figure shows that all measurement indicators meet the requirements of convergent validity, and each indicator is valid in measuring the variable in question.

Table 2 also shows the AVE construct values between 0.731 and 0.981. This value is greater than 0.5 [29] which means that the construct in this study can measure the corresponding latent variable.

Discriminant Validity

To calculate data validity, discriminant validity was used through the value of cross-loadings listed in Table 3. This indicator has a high correlation with its latent variable compared to other latent variables. This shows that all valid indicators and discriminant validity are met.

7.2 Reliability Test

The reliability value shows the consistency of the results even though the measurements are repeated. Therefore, reliability can be interpreted as a calculation that is free

Table 3 Cross loading

	LB	RT	Y
LB	0.741	0.053	0.134
LE	0.956	0.227	0.204
RC	0.161	0.921	0.249
TC	0.192	0.872	0.143
MC	0.160	0.228	0.981
IC	0.229	0.222	0.996
FC	0.217	0.220	0.994

LB Local Motorcyclist Behaviour, *LE* Local Law Enforcement, *RC* Road Conditions, *TC* Local Traffic Conditions, *MC* Minor Crash, *IC* Injury Crash, *FC* Fatal Crash

Table 4 Reliability Test

	Cronbach's alpha	Composite reliability
LR	0.776	0843
RT	0.759	0.891
FB	0.991	0.994

LR Local Motorcyclist Factor, *RT* Local Road and Traffic Conditions Factor, *FB* Foreign Motorcyclist's Behaviour

from random errors [30]. Testing reliability is calculated using PLS through internal consistency reliability. For internal consistency reliability, the value of composite reliability and Cronbach's alpha must be higher or equal to 0.7. Table 4 shows that the results of composite reliability and Cronbach's alpha are greater than 0.7, which means that these latent variables have consistent and reliable results.

7.3 Evaluation of the Inner Model (Structural Model)

This evaluation step is carried out to confirm the theoretical model outlined in the research structural model [31]. Here, the structural evaluation model uses two indicators: R2 and predictive relevance (Q2) [31].

R2 value

The R2 value in this PLS calculation represents the amount of variance contained in the model [31]. The R2 value in this study is listed in Table 5 and Fig. 3.

From Table 5 above it is known that:

- The LR variable (local motorcyclists behaviour) can be explained by the LB variable (behaviour of local motorcyclists), LE (local law enforcement), and RT (environmental condition factors for roads and local traffic) by 13.30% while the

Table 5 R square

Variable	R square
LB	
LE	
RT	
LR	0.133
FB	0.283

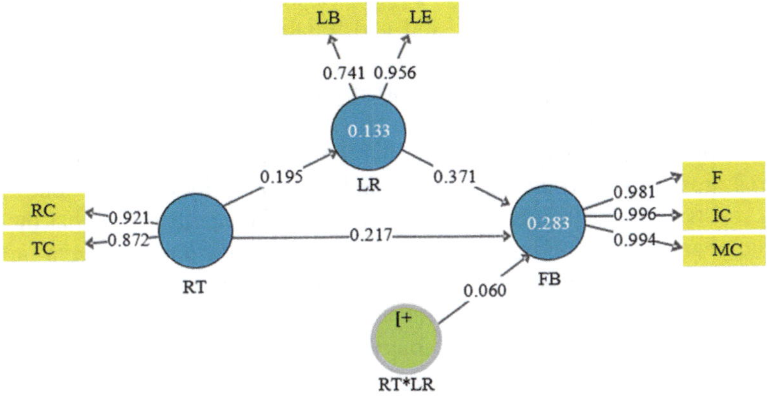

Fig. 3 PLS ALOG

remaining 86.70% is the contribution of other variables that are not included in this model.

- The FB variable (foreign motorcyclists behaviour) of 0.283 or 28.30% indicates that the behaviour of foreign motorcyclists (FB) is explained by the variables of RT (road conditions and local traffic factors) and LR (local motorcyclists) with 28.30%. While the remaining 71.70% is the contribution of other variables not included in this model.

Predictive Relevance

The Q2 predictive relevance value in a structural model is used to measure how well the observation value generated by the model and also its parameter estimation.

The value of Q2 in this study is calculated using the following formula:

$$\begin{aligned} Q2 &= 1 - (1 - R12)(1 - R22) \\ &= 1 - (1 - 0.133)(1 - 0.283) \\ &= 0.376 \end{aligned}$$

This value shows in general the contribution of the variables of road conditions and local traffic (RT), as well as the behaviour of local motorcyclists (LR), to the variable of the behaviour of foreign motorcyclists (FB) was 37.60%. The remaining 62.40% is the contribution of other variables that are not included in this model.

7.4 Summary of the Outer Model and Structural Model Evaluation

From the results of the outer model evaluation in this study, it is known that the constructs used in this study are valid and reliable so that the evaluation of the structural model can be continued. From the results of the structural model evaluation, it is also known that the proposed research model is appropriate, and hypothesis testing can be carried out.

The results of this analysis indicate that the behaviour of foreign motorcyclists in tourist areas in Bali is strongly influenced, especially by the behaviour of local motorcyclists and law enforcement by authorized officials. It shows that the better the behaviour of local motorcyclists, the better the behaviour of foreign motorcyclists, and vice versa. This also applies to law enforcement against traffic violations. Wherein the more stringent enforcement of the law, the better the behaviour of foreign motorcyclists. Road and traffic conditions also affect the traffic behaviour of foreign motorcyclists. The better the road conditions and the more organized traffic will result in better foreign motorcyclists' behaviour too.

8 Hypothesis Testing

Hypothesis testing is done to evaluate the relationship between latent variables, whether it is significant or not. Hypothesis testing can be calculated from the results of the path coefficient and the significance of the model based on the T-statistic value. The path coefficient value is significant if the T-statistic value is greater than 1.96.

The results of this study indicate that RT has a significant positive relation to FB (t-value 2.426). This significant result means that the better the local road and traffic conditions, the better the foreign motorcyclists' behaviour is. These results support Hypothesis 1.

The results of this study also indicate that LR has a significant positive relation with FB (t-value 3.874), which means that the better the behaviour of local motorcyclists, the better the foreign motorcyclists' behaviour is. Hypothesis 2 is proven.

In contrast to the RT and LR variables as latent variables, the results of this study indicate that the LR variable does not moderate the RT variable (t-value 0.455). This means that the behaviour of local motorbike riders (LR) does not play a role in improving the road conditions and local traffic (RT) factors on their effect on FB (behaviour of foreign motorbike riders). Thus hypothesis 3 is not proven.

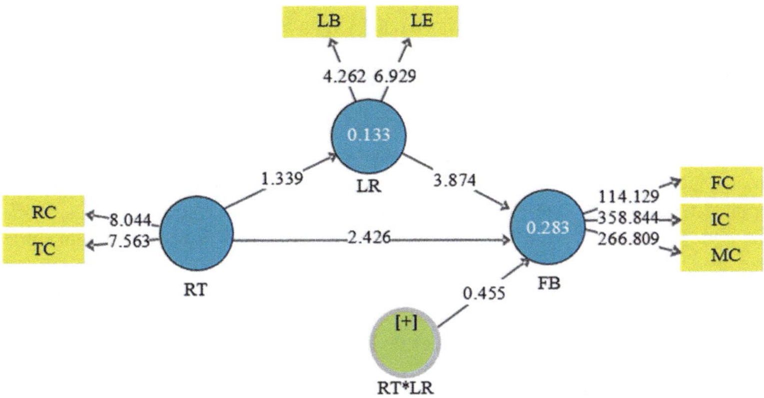

Fig. 4 Bootstrapping

Furthermore, this study found that the RT variable (road conditions and local traffic factors) did not have a significant relation with LR (local motorcyclists' behaviour), with a t-value of 1.339. Therefore, hypothesis 4 is not proven (Fig. 4).

9 Conclusion

From the results of the study above, it can be concluded that the local road and traffic factors influence the behaviour of foreign motorcyclists with a bootstrapping value of 2.426. Influence from local motorcyclists and law enforcement also affects foreign motorcyclists' behaviour with a bootstrapping number of 3.874. Both factors positively affect foreign motorcyclists' behaviour with a coefficient of 0.217 and 0.371. In conclusion, the behaviour of foreign motorcyclists in tourist areas in Bali is very much influenced, especially by the behaviour of local motorcyclists and law enforcement by authorized officials. This also applies to law enforcement against traffic violations. Wherein the more stringent enforcement of the law, the better the behaviour of foreign motorcyclists. Road and traffic conditions also affect the traffic behaviour of foreign motorcyclists. Where the better the road conditions and the more regular traffic will result in the better foreign motorcyclists' behaviour too. Better foreign motorcyclists' behaviour will reduce the risk of accidents for foreign motorcyclists, passengers, and other traffic users.

References

1. Yannis G, Golias J, Papadimitriou E (2007) Accident risk of foreign drivers in various road environments. J Safety Res 38:471–480. https://doi.org/10.1016/j.jsr.2007.01.014

2. Ajzen I (1991) The theory of planned behavior. Organ Behav Hum Decis Process 50:179–211. https://doi.org/10.1016/0749-5978(91)90020-t
3. Njå O, Nesvåg SM (2007) Traffic behaviour among adolescents using mopeds and light motorcycles. J Safety Res 38:481–492. https://doi.org/10.1016/j.jsr.2007.03.012
4. Murray CJL, Lopez AD (1996) The global burden of disease, a comprehensive assessment of mortality and disability from diseases, injuries, and risk factors in 1990 and projected to 2020. Harvard School of Public Health. https://doi.org/10.1016/s0140-6736(96)07495-8
5. Kopits E, Cropper M (2003) Traffic fatalities and economic growth. Policy Res. Work. Pap.
6. Peden M, Scurfield R, Sleet D, et al (2004) World report on road traffic injury prevention. Geneva
7. Zietlow DG (2006) The road safety cent-management and financing of road safety in low-income countries
8. TRB (2006) Improving road safety in developing countries. Opportunities for US Cooperation and Engagement
9. Fuller R, Santos JA (2002) Psychology and the highway engineer. Hum Factors Highw Eng, 1–10
10. Musselwhite C, Avineri E, Susilo Y (2013) Legitimising risk taking: articulating dangerous behaviour on the road. Transp Plan Technol 37:62–82. https://doi.org/10.1080/03081060.2013.844905
11. Kaltheier RM (2002) Urban transport and poverty in developing countries : analysis and options for transport policy and planning. Dtsch Gesellschaft für Tech Zusammenarbeit GmbH, Eschborn 49
12. BPS (2016) Number of motorized vehicle 1987–2009. Jakarta
13. PADECO (1986) Study on urban transport development, p 183
14. Department of the Environment T and the R (2000) Tomorrow's roads: safer for everyone, pp 1–7
15. Hossain M, Iamtrakul P (2007) Medical investigation of motorcycle accidents in. J East Asia Soc Transp Stud 7:2770–2785
16. Clarke DD, Ward P, Bartle C, Truman W (2004) Road safety research report No. 54 : In Depth Study of Motorcycle Accidents In-depth Study of Motorcycle Accidents, pp 1–67
17. Indriastuti AK, Sulistio H (2010) Influencing factors on motorcycle accident in urban area of Malang, Indonesia. Int J Acad Res 2:252–255
18. Rothengatter T (1998) Errors and violations as factors in accident causation, in Traffic and transport psychology. Theory and application. Rech - Transp - Sécurité 58:70–71. https://doi.org/10.1016/s0761-8980(98)80022-8
19. Yagil (1998) Gender and age-related differences in attitudes toward traffic laws and traffic violations. Transp Res Part F 1
20. Rimmö PA, Åberg L (1999) On the distinction between violations and errors: sensation seeking associations. Transp Res Part F Traffic Psychol Behav 2:151–166. https://doi.org/10.1016/S1369-8478(99)00013-3
21. Forward S (2006) The intention to commit driving violations—a qualitative study. Transportation. Transp Res Part F 9
22. Forward S (2009) An assessment of what motivates road violations. Transp Res Part F 12
23. O'Connell M (2008) Social psychological principles: the group inside the person. In: Fuller R, Santos JA (eds) Human factors for highway engineers. Emerald, Bingley
24. Watson B, Tunnicliff D, White K et al (2007) Psychological and social factors influencing motorcycle rider intentions and behaviour. ATSB Res Anal Rep 152
25. Rothengatter T (1997) Errors and violations as factors in accident causation. In Rothengatter T, Vaya EC (eds) Traffic and transport psychology: theory and application. Elsevier Science Ltd, Oxford
26. Underwood G, Chapman P, Wright S, Crundall D (1997) Estimating accident liability. In Rothengatter, T. and Vaya, E.C. (Eds.) Traffic and Transport Psychology: Theory and Application. Elsevier Science Ltd, Oxford (1997).

27. Rothengatter T, Manstead AR (1997) The role of subjective norm in predicting the intention to commit traffic violations. In: Rothengatter T, Vaya EC (eds) Traffic dan transport psychology: theory and application. Pergamon, Oxford
28. Roscoe (1982) Research methods for business. McGraw Hill, New York
29. Fornell C, Larcker DF (1981) Evaluating structural equation models with unobservable variables and measurement error. J Mark Res 18:39. https://doi.org/10.2307/3151312
30. Malhotra NK, Peterson M (2006) Basic marketing research: a decision-making approach. Prentice Hall, New Jersey
31. Chin WW (2010) How to write up and report PLS analyses

Knowledge and Practice of Helmet Usage Among Senior High School Students in Klaten Regency

Dewi Handayani, Putri Dewi Prasetianingrum, and H. M. Amirotul Musthofiah

Abstract Indonesia is one of the countries with the most significant number of motorcycle users in Southeast Asia. Because of the heavy traffic and high price of cars, most Indonesians choose to use two-wheeled vehicles. The accidents in Klaten Regency in 2019 to early 2020 increased by 30% from 2018 and were dominated by teenage riders (16–25 years old), with one of the leading causes, driving behavior. This research aims to obtain the correlation between the characteristics of respondents and their practice in using helmets—only four characteristics from 9 characteristics related to the habit of using a helmet. One of the characteristics is the sanctions if the driver does not use the helmet correctly while driving. The results of the data analysis obtained no correlation between knowledge about helmets and the practice of helmet usage while driving. This research has never been done in detail by linking the respondent's characteristics with the habit of respondents in using a helmet. Therefore, it is worth doing further research. This research applied to the proportion of respondents by 75% female & 25% male senior high school students in Klaten Regency.

Keywords Chi-square test · Cross tabulation · Helmet USAGE · Knowledge practice attitude (KPA)

1 Introduction

Indonesia is one of the countries with the most significant number of motorcycle users in Southeast Asia. This is due to heavy traffic and high car prices, making most Indonesians use two-wheeled vehicles. In 2018 the Central Bureau of Statistics

D. Handayani · P. D. Prasetianingrum (✉) · H. M. Amirotul Musthofiah
Civil Engineering, Sebelas Maret University, Surakarta, Indonesia

D. Handayani
e-mail: dewi@ft.uns.ac.id

H. M. Amirotul Musthofiah
e-mail: amirotulmhm@staff.uns.ac.id

S. A. Kristiawan et al. (eds.), *Proceedings of the 5th International Conference on Rehabilitation and Maintenance in Civil Engineering*, Lecture Notes in Civil Engineering 225,
https://doi.org/10.1007/978-981-16-9348-9_65

released data related to the number of motorcycles registered with the Indonesian police numbered 120 million units and data related to accidents in Indonesia that resulted in fatalities of 19.016 cases with a death toll of 4.115 people, severe injuries of 97 people, and minor injuries of 21.967 people with 74.49% involvement of motorcycles in accidents. A helmet is a mandatory attribute when driving a motorcycle. It is intended to protect the head from impacts in a traffic accident to avoid the user from minor to severe head injuries. In 2018 the World Health Organization (WHO) stated that proper use of helmets could reduce 42% of the risk of fatal traffic accidents and 69% can protect against head injuries. Therefore, research is needed on compliance studies for senior high school students in Klaten Regency to know the level of knowledge about helmets and the habits in using a helmet that is expected to minimize the risk in the traffic accident and know whether or not there is a correlation between the habits of using a helmet and the characteristics of respondents.

This research was conducted basically to see the correlation between the knowledge level of helmets and the habit of using helmets of senior high school students in the Klaten Regency. Similar to previous researches that used many interview methods and shared questionnaires in data collection as conducted by Rizky [1], Ambak and Hashim [2], Guntur [3], Pramono et al. [4], and Fadilah et al. [5], the data collection in this research was also conducted by sharing questionnaires online on google-form because the research was conducted when Indonesia and the world were affected by COVID-19 pandemic. To get the value of the correlation between knowledge about helmets and the habit of using helmets while driving cross-tabulation method is used. The specialty of the crosstab method is the ability of the method to analyze the correlation between rows and columns. This method has been used in previous research, but the previous research didn't link the characteristics of respondents and their habit in using helmets. So this research is expected can provide good results in transportation science.

Previous researchers have used cross-sectional methods and univariate-bivariate correlation analysis. Cross sectional methods conducted by Iskandar [6], Muryatma [7], Widiyawati [8], Dwinanda and Wijaya [9] and Fadilah et al. [5] as well as univariate-bivariate correlation methods have been conducted by Ambak and Hashim [2], Pramono et al. [4], Widyastuti and Trisnawati [10], and Putri et al. [11]. From previous researches with significant subjects in the adolescent age group, the majority only examined in general such as the relationship of knowledge with driving safety, the relationship of knowledge & attitude of using helmets with driving safety, driving behavior, compliance factors in driving, and the relationship of safety factors with safety riding. From this research, more in-depth research can be done on the characteristics of drivers as well as analysis of knowledge about helmets and their habit in using a helmet when driving a motorcycle, especially on senior high school students in Klaten Regency to know the relationship between knowledge about helmets and their habit in using a helmet that expected to minimize the risk of traffic accident.

2 Basic Theory

Safety riding means putting safety first, in personal safety or the safety of other road users. Drivers should never forget that the road is not our own private but rather a shared public facility. Thus, carelessness can cause others to be wretched [12]. Based on this idea, the drivers have to be aware of the importance of driving safety. The application of safety riding has been regulated in Law No. 22 of 2009 concerning Traffic and Road Transport in Chapter XI Article 203 Paragraph (2), which reads "To ensure Traffic Safety and Road Traffic Control as referred to in paragraph (1), a national general plan of Traffic Safety and Road Transport is established, including the preparation of a national program of Traffic Safety and Road Transport activities." Referring to the paragraph, the National Traffic Safety and Road Transport program is among others about the procedures of driving safety. Therefore, the government recommends the use of safety riding equipment, one of which is a helmet. The government made a regulation written in Law No. 22 of 2009 on Traffic and Road Transport Article 106 paragraph (8) to protect the public from potential head injuries in the accident. To discipline the public to use a helmet, the government also issued criminal laws contained in Law No. 22 of 2009 on Traffic and Road Transport Article 291 paragraphs (1) and (2), i.e.:

> Every person who drives a motorcycle that does not wear an Indonesian National Standard helmet as referred to Article 106 paragraph (8) shall be penalized with a maximum imprisonment of 1 (one) month or a maximum fine of Rp 250,000.- (two hundred and fifty thousand rupiahs).
>
> Every person who drives a motorcycle that allows his passengers not to wear a helmet as referred to Article 106 paragraph (8) shall be penalized with a maximum imprisonment of 1 (one) month or a maximum fine of Rp 250,000.- (two hundred and fifty thousand rupiahs).

KPA (Knowledge, Practice, Attitude) survey method is one of the methods to know precisely about education and knowledge, the tendency of attitudes and practices of respondents [13]. This research used the KPA survey method to determine the behavior of current drivers and set future targets by making policies that are expected to change the behavior of motorcyclists to be safer when driving.

3 Methods

Before conducting the research, problem identification & literature study were carried out to get an overview of the research. To support the validity of the data, secondary data was collected from local government that can be accessed through the website. Then the survey form and the pilot survey form were compiled. The survey results were analyzed using reliability and validity tests to obtain a suitable survey form for further research. Then the primary survey was conducted after determining the number of samples to be taken. Finally, the data obtained from the primary survey were analyzed using cross-tabulation analysis so that the conclusion could be taken from the research.

The research was conducted in several of Klaten Regency's senior high schools, especially SMAN 1 Jogonalan, SMAN 1 Klaten, SMAN 2 Klaten, and SMAN 3 Klaten. The samples count in this research was taken based on Krejcie and Morgan's table (1970) with the results of 266 respondents [14]. Online sampling techniques are assisted by google-form that contains questions related to respondent's characteristics, knowledge of helmets, and practices of helmet usage while driving. The data analysis method is done by cross-tabulation method and chi-square statistical test to see a correlation between variables with the Statistical Product and Service Solutions (SPSS) tool. Cross-tabulation is a method of analysis by presenting data with two or more variables into one matrix presented in rows and columns [15]. The Chi-square test used is as follows:

$$x^2 = \sum \frac{(f_0 - f_h)^2}{f_h} \tag{1}$$

where:

x^2 = Chi-square
f_0 = observed frequency
f_h = expected frequency.

Hypothesis tests conducted are

H_0 = no correlation between rows and columns
H_1 = there is a correlation between rows and columns.

Or

If $x^2_{count} < 0.05$, then the data does not support rejecting H_0.
If $x^2_{count} > 0.05$, then the data support to reject H_0.

4 Results and Discussion

Analysis of respondent's characteristics is one of the essential things in research. By analyzing the respondent's characteristics, researchers can see the characteristics that can affect the respondent's habit of using helmets while driving. In this research, respondent's characteristics include gender, age, ownership of the driving license, knowledge of the obligation to use Indonesian Standard Nasional helmet, knowledge of sanctions if the driver does not use the helmet while driving, knowledge of sanctions if the driver let the passenger not use the helmet properly while driving, knowledge of Indonesia Standard National helmets, frequency of correctly angled helmet, and ensuring passengers use helmet correctly. Then, all of these characteristics are correlated with the respondent's habits in using helmets while driving using cross-tabulation and chi-square tests. These correlations can be seen in Tables 1, 2, and 3.

Table 1 shows that the respondent's characteristics, namely gender, age, and ownership of the driving license, have a chi-square value of more than 0.05, which is no correlation or relationship with the respondent's habit in using helmets.

In the category of knowledge about helmets are made several categories to facilitate grouping the level of respondents' knowledge, such as sound, sufficient, and less, by summing the value of statements in questionnaires. As in the question of the obligation to use a helmet, respondents who had a score of 3 to 4 had "good" knowledge, a value of 2 had "sufficient" knowledge, and a value of only 1 had "less" knowledge. Table 2 shows that only 3 out of 4 characteristics of respondents who have a chi-square value below 0.05 which correlates with the habit of using helmets, namely knowledge about sanctions if not using a helmet when driving both drivers and passengers and knowledge about Indonesian Standard National helmets.

Table 3 shows that one of the characteristics of respondents has a chi-square value below 0.05 which is the habit of ensuring passengers use the helmet correctly, which means that characteristic correlates with the respondent's habit in using a helmet.

After analyzing the relationship between characteristics, an analysis of the relationship between knowledge about the helmet and the habit of using helmets while driving, in this variable, the categorization of variables becomes "good," "sufficient," and "less" derived from the recapitulation of the values of each segment in a single variable. The results of the analysis can be seen in Table 4.

Table 4 showed that the results of 20% of respondents who practiced "less" in using helmets while driving had "good" knowledge about helmets. Still, on the contrary, 1.74% of respondents who practiced "good" in using helmets while driving had "less" knowledge about helmets. The statistical chi-square test showed the value of 0,611

Table 1 Correlation between characteristics and respondent's habits of using helmet

Variables		Habits in using helmet								Chi-square
		Always		Sometimes		Never		Total		
		n	%	n	%	n	%	n	%	
Gender	Girls	138	69.0	60	30.0	2	1.0	200	100.0	0.373
	Boys	41	62.1	25	37.9	0	0.0	66	100.0	
Age	14	3	50.0	3	50.0	0	0.0	6	100.0	0.248
	15	35	60.3	21	36.2	2	3.5	58	100.0	
	16	55	66.3	28	33.7	0	0.0	83	100.0	
	17	73	73.7	26	26.3	0	0.0	99	100.0	
	18	10	58.8	7	41.2	0	0.0	17	100.0	
	19	3	100.0	0	0.0	0	0.0	3	100.0	
Ownership of driving license	Have	46	74.2	16	25.8	0	0.0	62	100.0	0.344
	Don't have	133	65.2	69	33.8	2	1.0	204	100.0	

Table 2 Correlation between knowledge regarding helmet and respondent's habit in using helmet

Variable		Habits in using helmet								Chi-square
		Always		Sometimes		Never		Total		
		n	%	n	%	n	%	n	%	
Knowledge of the obligation to use Indonesian Standard National Helmet	Good	82	64.5	34	28.8	2	1.7	118	100.0	0.387
	Sufficient	42	61.7	26	38.2	0	0.0	68	100.0	
	Less	55	68.8	25	31.3	0	0.0	80	100.0	
Knowledge of sanctions if the driver doesn't use helmet	Know	109	75.2	35	24.1	1	0.7	145	100.0	0.011
	Don't know	70	57.9	50	41.3	1	0.8	121	100.0	
Knowledge of sanctions if the driver allows the passenger not to use the helmet properly	Know	83	79.1	22	20.9	0	0.0	105	100.0	0.003
	Don't know	96	59.6	63	39.1	2	1.2	161	100.0	
Knowledge of Indonesian Standard National helmet	Good	165	69.6	71	29.3	1	0.4	237	100.0	0.000
	Sufficient	12	50.0	12	50.0	0	0.0	24	100.0	
	Less	2	40.0	2	40.0	1	20.0	5	100.0	

Table 3 Correlation between practice of helmet usage and respondent's habits of using helmet

Variable		Habits in using helmet								Chi-square
		Always		Sometimes		Never		Total		
		n	%	n	%	n	%	n	%	
Frequency of angled helmet correctly	Always	129	73.3	47	26.7	0	0.0	176	100.0	0.179
	Sometimes	18	50.0	18	50.0	0	0.0	36	100.0	
	Never	32	59.3	20	37.0	2	3.7	54	100.0	
Ensuring passengers using helmets correctly	Always	134	80.7	32	19.3	0	0.0	166	100.0	0.000
	Sometimes	19	48.7	20	51.3	0	0.0	39	100.0	
	Never	26	42.6	33	54.1	2	3.3	61	100.0	

Table 4 Cross-Tabulation between respondent's knowledge regarding helmet and respondent's practice of using helmet while driving

Variable		Respondent's practice of using helmet						Chi-square
		Good		Sufficient		Less		
		n	%	n	%	n	%	
Respondent's knowledge regarding helmet	Good	69	40.12	28	31.46	1	20.00	0.611
	Sufficient	100	58.14	59	66.29	4	80.00	
	Less	3	1.74	2	2.25	0	0.00	
	Total	172	100.0	89	100.0	5	100.0	

(> 0.05) so that H0 was received, which means there is no correlation between the respondent's knowledge of helmets and the respondent's practice of using helmets while driving.

This research is in line with the research of Nova (2017), which states that there is no correlation between knowledge and driving behavior in 2nd-grade students of Vocational High School 4 Surabaya with a p-value of 1.921 and research by Novita (2018), which states that there is no correlation between the level of knowledge and safety riding behavior in online motorcycle and taxi drivers in Surakarta with a p-value of 0.507.

5 Conclusion

Based on the results of research on the study of helmet use compliance among Senior High School students in Klaten Regency with the cross-tabulation method and chi-square test, the following conclusions can be drawn:

1. The results of the analysis of the relationship between habits in using a helmet and the characteristics of respondents were found that only 4 out of 9 characteristics correlated, namely knowledge about sanctions if not using helmet both the driver and passenger, knowledge of Indonesian Standard National helmets, and the habits of the driver in ensuring his passengers use the helmet correctly.
2. From the results of the analysis of the relationship between knowledge regarding helmets and the practice of helmet usage, there is no correlation between those variables.

Recommendations that can be given so the further research can give better results are:

1. From the conclusions can be recommended the material for further socialization on driving safety guidance is emphasized more on age, ownership of the driving license, the reason why it is mandatory to use a helmet while driving, and how to use helmets correctly.

2. This research was conducted in the COVID-19 pandemic, and data collection was collected with the google-form tool. However, it is possible if the data collection results will be ineffective if respondents do not understand the intent of the questions asked. Therefore, it is recommended to conduct further research under normal conditions with a questionnaire/live interview method.
3. In the following research, some respondents who know helmet regulation while driving can be further followed up or analyzed in more detail.

References

1. Rizky EDE (2019) Analisa Perilaku Pengendara Sepeda Motor Tanpa Helm pad Ruas Palembang-Banyuasin
2. Ambak K, Hashim H, Yousff I, David B (2006) An evaluation on the compliance to safety helmet usage among motorcyclists in Batu: an evaluation on the compliance to safety helmet usage among motorcyclists In Batu Pahat, Johor. Int J Integr Eng (Issue on Civil and Environ Eng) 2(October):45–51
3. Guntur M (2015) Studi Perilaku Pengendara Sepeda Motor Di Kota Makassar. Retrieved from https://core.ac.uk/download/pdf/77620427.pdf
4. Pramono TD, Subekti AT, Atmoko D (2017) Hubungan pengetahuan dan sikap dengan praktik keselamatan berkendara sepeda motor pada siswa smk semesta bumiayu. Jurnal Ilmu Kesehatan (JIK) Bhamada 9(1):8
5. Fadilah D, Supriyanto, Ginanjar R (2018) Faktor-Faktor Yang Berhubungan Dengan Perilaku Pemakaian Helm Pengendara Sepeda Motor Pada Pelajar Kelas X (Sepuluh). Promotor: Jurnal Mahasiswa Kesehatan Masyarakat 1(1):29–36
6. Iskandar NM (2015) Keselamatan Berkendara Sepeda Motor Siswa Smkn 3 Pekalongan Tahun Ajaran 2015/2016
7. Muryatma NM (2017) Hubungan Antara Faktor Keselamatan Berkendara Dengan Perilaku Keselamatan Berkendara. Jurnal PROMKES 5(2):155
8. Widiyawati N (2018) Hubungan Tingkat Pengetahuan dengan Perilaku Safety Riding pada Pengendara Ojek Online di Surakarta
9. Dwinanda D, Wijaya O (2019) Hubungan Antara Pengetahuan Milenial Road Safety, Sikap Berkendara Dan Fasilitas Berkendara Dengan Perilaku Berkendara Pada Mahasiswa FKM UAD Kota Yogyakarta
10. Widyastuti SD, Trisnawati N (2017) Gambaran pengetahuan dan sikap tentang keselamatan berkendara (safety riding) pada mahasiswa STIKES Indramayu. J Chem Inf Model 53(9):1689–1699
11. Putri EPP, Nurhidayat S, Sukamto FI (2019) Hubungan pengetahuan dengan sikap remaja dalam keselamatan berkendara. In: 1st seminar nasional and call for paper, pp 176–183
12. Kusmagi MA (2019) Pengertian safety riding. Retrieved from https://pengertianmenurutparaahli.org/pengertian-safety-riding/. Last accessed 3–4 June 2020
13. Yapeka R (2018) Pelaksanaan Survei KAP Desa Braja Yekti Tahun 2018. Retrieved from https://yapeka.or.id/survei-kap-desa-braja-yekti-tahun-2018/. Last accessed 5 June 2020
14. Basri S (2012) Cara Menentukan Jumlah Sampel Penelitian Menggunakan Tabel dan Berdasarkan Taraf Keyakinan Penelitian. Retrieved from https://setabasri01.blogspot.com/2012/04/cara-menentukan-sampel-penelitian.html?m=1. Last accessed 9 June 2020
15. Savi (2018) Analisis Crosstab. Retrieved from https://savinotes.wordpress.com/2018/05/22/analisis-crosstab/. Last accessed 3 June 2020

Sustainability Aspects in Transportation Engineering

Investigation of CO_2 Emissions on Two Local Streets by Means of IPCC and Direct Method

Florentina Pungky Pramesti, Annisa Tri Utami, Iva Yenis Septiariva, and Fajar Sri Handayani

Abstract One of the most widely used modes of transportation is passenger cars. The increasing use of passenger cars can have an impact on the environment. The residue from combustion in a vehicle engine can cause CO_2 emissions. Excessive CO_2 Emissions leads to greenhouse gas emissions. The purpose of this study is to calculate the CO_2 emissions produced by passenger cars using the Tier 1 of IPCC method and a so-called direct method, then comparing the two results afterwards. The case study in this research is two local roads with an average annual daily traffic of 3305 and 2673 vehicles/day. The IPCC method analyzes emissions based on the vehicle's fuel consumption, while the direct method identifies and analyzes emissions based on the measurement of actual gases emitted by the passenger cars based on its production year. Exhaust gas emission data is obtained from an emission survey conducted by the local Environmental Agency. The calculation of emissions using the direct method on both roads gave higher results than the IPCC method. This is because the IPCC method uses the tier 1 approach while the direct method is closer to the tier 3 approach.

Keywords CO_2 emissions · Passenger cars · IPCC method · Direct method

F. P. Pramesti (✉) · A. T. Utami · I. Y. Septiariva · F. S. Handayani
Civil Engineering Department, Faculty of Enginering, Universitas Sebelas Maret, Surakarta, Indonesia
e-mail: f.p.pramesti@ft.uns.ac.id

I. Y. Septiariva
e-mail: ivayenis@staff.uns.ac.id

F. S. Handayani
e-mail: fajarhani@ft.uns.ac.id

S. A. Kristiawan et al. (eds.), *Proceedings of the 5th International Conference on Rehabilitation and Maintenance in Civil Engineering*, Lecture Notes in Civil Engineering 225,
https://doi.org/10.1007/978-981-16-9348-9_66

1 Introduction

As the economy grows, it stimulates the opportunity to invest more in many other sectors including transportation infrastructure. Vice versa, investments in infrastructure are prescribed to trigger economical growth. In the short term, it creates construction work employment and in the long term it generates a more efficient and reliable transportation system, which later will drive the growth of the economy.

Road infrastructure has developed very well in Indonesia. It has lowered the cost of moving people and goods, which in turn, increases economic productivity. Improved productivity leads to higher prosperity, marked by the growing ownership of a private car. Nowadays it can be seen that road transportation is dominated by the use of private vehicles. Typical light passenger vehicles nowadays are commonly seen operating in city street networks as people enjoy driving within and between cities.

On the other hand, an increase in the number of light vehicles may have a negative impact on the environment. The combustion engine emits exhaust gases such as carbon monoxide (CO), carbon dioxide (CO_2), nitrogen oxides (NOx), sulphur dioxides (SOx) and hydrocarbons (HC) that may also lead to public health issues.

Nurdjanah [1] estimated that the average CO_2 emission on 14 streets of Denpasar, Bali, in 2012 was as much as 245.08 kg/hour.km [1]. This amount was calculated using 2011 annual average daily traffic (AADT) and based on a 2012 fuel usage survey as well as local vehicle emission factors. CO_2 emissions were predominantly contributed by light passenger cars and motorbikes. Muziansyah [2] reported that the largest annual emission of CO_2 on the 0.2 km long Pasar Bawah Ramayana Street terminal section in the city of Bandar Lampung reached 12.46 tons [2]. Due to the enclosed terminal building located at the ground floor of the central business district, the pollutant may affect the people working in this packed environment.

This ever-increasing greenhouse gas (GHG) may hamper public health directly. On a larger scale, this situation may contribute to the global warming and climate change. Yuliana [3] investigated the level of GHG emitted in Indramayu [3], where the impact attributed to climate change was profound, such as the local fishermen's inability to catch fish consistently due to disrupted seasonal patterns, the difficulty of local traditional farmers in maintaining rice production as unusual pests and other disturbing organisms attack their paddy field, and the rising sea level which disrupts socio-economic activity through floods.

Identification and estimation of the amount of GHG emissions is one of the first step to take in order to mitigate the problem and propose strategies for sustainable development.

One of the common methods used to identify and estimate the GHG emissions is the IPCC method [4–7]. The IPCC method was developed by the Intergovernmental Panel on Climate Change more than 10 years ago by using a database of emission factors on each activity. Although frequently updated, many reports have raised questions about the accuracy of this method, especially when it is used in certain conditions where traffic and vehicle engine maintenance are not ideal [5]. This non-ideal situation may lead to identification errors, which in turn can lead to inaccurate emission analysis results.

Furthermore, recent development in science has led environmental researchers to develop new methods for measuring and estimating emissions [5, 8, 9]. More data has been collected from field observation. This data collection is stored and maintained by local municipal environmental agencies in Indonesia. However, data analysis has not yet been carried out to provide broader dan deeper insight that in turn will provide the authority with science base policy making. This situation leads to some question; Can a large number of vehicle emission data points be used as actual shift factors for determining vehicle emissions?

This paper aims to identify and estimate the emission of GHGs due to light passenger cars by means of two methods: the Intergovernmental Panel on Climate Change (IPCC) method [4, 10, 11] and the so-called Direct method. In this case, a passenger car is a light vehicle defined as a two-axle motorized vehicle with four wheels and an axle distance of 2.0–3.0 m.

2 IPCC Method and Direct Method

Greenhouse Gases may be defined as gases contained in the atmosphere, both due to natural and anthropogenic activities, which absorb and re-emit infrared radiation. Some of the solar radiation in the form of short waves that is received by the earth's surface is reflected to the atmosphere in the form of long-wave radiation. This long-wave radiation is absorbed (entrapped) by GHGs in the lower atmosphere (near the earth's surface) and causes a heating effect known as the 'Greenhouse Effect'. The larger the volume of GHGs in the atmosphere, the more intense the warming effect of earth's surface can be observed on the whole world. This phenomenon is known as global warming.

Two methods of calculating GHGs will be employed in this paper; (1) IPCC method, (2) Direct method. IPCC method is the approach developed by the Intergovernmental Panel on Climate Change by using a database of emission factors on each activity. While direct method uses direct measurement data to estimate total emission.

2.1 IPCC Method

In its 2006 report, the IPCC suggested three categories for calculating the GHG emission based on data, approach, and accuracy. These are what are known as tier 1, tier 2, and tier 3. Tier 1 estimates GHG emission based on activity in general and default emission factors proposed by IPCC. In tier 2, estimation is based on more accurate activity data and default IPCC emission factors or country/plant specific factors. Meanwhile in tier 3, estimation of GHG emission is based on specific methods established in the country with accurate data of activities (direct measurement at country/plant specific emission factors) [4].

Tier determination in the GHG inventory is largely determined by the availability of data and the level of progress country or factory has made in terms of researching country/factory specific factors that are specific and applicable to that. In Indonesia and non-Annex 1 countries, the emission sources for key sectors/activities in the GHG inventory use Tier-1, which is based on activity data and IPCC default emission factors [11].

This paper analyses GHG emission on 2 streets, namely RA Kartini which will later be referred to as street 1, and HOS Cokroaminoto which will later be referred to as street 2, in Sragen, a regency with nearly 900 thousand inhabitants [12].

Based on the IPCC method, the amount of emissions, in this case CO_2 emissions, is calculated using the following formula (1) [11]

$$GHG_{emission}\left(\frac{\text{kg}}{day}\right) = Energy\,consumption\left(\frac{TJ}{day}\right) \times Emmision\,factor\left(\frac{\text{kg}}{TJ}\right) \quad (1)$$

which is adopted from Eq. (2) below [5]:

$$GHG_{emission} = Data\,activity \times Emmision\,factor \quad (2)$$

whereas emission factor is the emission shift factor of each activity. It has been modified and updated by the Ministry of the Environment from the IPCC database, taking into account the types of fuel available in Indonesia [10]. This database contains emission factors for calculating the GHG emission emitted by equipment and/or activities consuming certain types of fuel [11].

The energy consumption, as an independent variable in Eq. 1, is calculated based on Eq. 3

$$Energy\ consumption\left(\frac{TJ}{day}\right) = Fuel\,consumption\left(\frac{liter}{day}\right) \times Calorific\,value\left(\frac{TJ}{liter}\right) \quad (3)$$

where the calorific value of premium fuel in gasoline type combustion is 0.000033TJ/Liter [11]. The fuel consumption ($KBBM_i$ in liter/km) in Eq. 3 is calculated through Eq. 4 [13].

$$KBBM_i = \frac{\left(\alpha + \frac{\beta_1}{V_R} + \beta_2 \cdot V_R^2 + \beta_3 \cdot R_R + \beta_4 \cdot F_R + \beta_5 \cdot F_R^2 + \beta_6 \cdot DT_R + \beta_7 \cdot A_R + \beta_8 \cdot SA + \beta_9 \cdot BK + \beta_{10} \cdot BK \cdot A_R + \beta_{11} \cdot BK \cdot SA\right)}{1000} \quad (4)$$

where:

α	= Constant
$\beta 1$-$\beta 11$	= Parameter coefficients
VR	= Average speed
RR	= Average road inclination (uphill)
FR	= Average road declination (downhill)
DTR	= Average degree of road curve
AR	= Average acceleration
SA	= Standard deviation of acceleration
BK	= Vehicle weight.

Equation (4) shows that said data (average speed, road inclination and declination, acceleration, standard deviation of acceleration and vehicle weight) needs to be collected prior to emission determination.

2.2 *Direct Method*

Meanwhile, in determining the amount of vehicle greenhouse gas emissions using the direct method (actual multiplier), a direct vehicle emission measurement survey was conducted using the Emission Analyzer tool. This measurement was carried out in collaboration with the local district's environmental office. The measurement produces data on CO_2 emission levels which are classified by years of vehicle production. The parameter required is the average emission, which can be calculated using Eq. 5.

$$Average\,CO_2\,emission = \frac{\sum CO_2\,emission}{Number\ of\ vehicle} \tag{5}$$

After obtaining the average CO_2 emission, the multiplier factor of CO_2 per year of vehicle manufacture can be determined by multiplying said average by the emission sampling discharge being tested. The discharge is obtained from the specification of the Emission Analyzer tool.

$$Multiplier\ factor\ of\ CO_2\ Emission = Average\ CO_2\,emission \times Discharge \tag{6}$$

Afterwards, the actual CO_2 emission can be calculated using Eq. 7:

$$CO_2\,emission = multiplier\ factor\ of\ CO_2\ emission \times Traffic\ volume \times Traveling\ time \tag{7}$$

3 Results and Discussion

The emission of CO_2 on the two streets were carried out using two methods; the IPCC method and the direct method. Table 1, 2, and 3, below recapitulates the results of these methodologies.

For the direct method, it is necessary to first identify the emission data per year of vehicle production. See Fig. 1.

It was decided to classify emissions for every 5 years of production, with the consideration that in those 5 years the types of vehicle engines commonly used in Indonesia did not experience significant design changes, and the emissions produced by vehicles in that range were also almost the same. The value is transformed into Kg/L and shows in Table 2 column 3. These values can be understood as the shift factor to estimate the daily CO_2 emission in the street having the average daily traffic.

Table 1 Recapitulation of CO_2 emission calculated using the IPCC method for light vehicle (fuel type: gasoline)

Street	Fuel consumption l/km	Fuel consumed (Liter)	Average daily traffic Veh./day	Calorific value TJ/Liter	Energy consumed (TJ/day)	Emission factor of CO_2 Kg/TJ	CO_2 Emission (kg/day)
(1)	(2)	(3) = (2)× street length	(4)	(5)	(6) = (3)× (4)x(5)	(7)	(8) = (6)×(7)
1	0,0949	0,1044	3305	0,000,033	0,0114	69,300	789,1792
2	0,0937	0,0515	2673	0,000,033	0,0045	69,300	315,0219

Table 2 Recapitulation of CO_2 emission calculated using the direct method on street 1

Vehicle production year range	Average daily traffic (Veh./day)	CO_2 emission (Kg/L)	Discharge (L/m)	Emission multiplier CO_2 (Kg/hour)	Travelling time (hour)	CO_2 emission (Kg/day)
(1)	(2)	(3)	(4)	(5) = (3)×(4) × 60	(6)	(7) = (2)× (5)×(6)
>1989	0	0,1066	5	31,9714	0,0342	0,0000
1989–1993	20	0,1077	5	32,3000	0,0342	22,0907
1994–1998	109	0,1003	5	30,1033	0,0342	112,2065
1999–2003	237	0,1220	5	36,5850	0,0342	296,5026
2004–2008	663	0,1343	5	40,2872	0,0342	913,3925
2009–2013	1266	0,1410	5	42,3086	0,0342	1831,6383
2014–2018	1010	0,1419	5	42,5823	0,0342	1469,9831
Total						4645,8138

Table 3 Recapitulation of CO_2 emission calculated using the Direct method on street 2

Vehicle production year range	Average daily traffic (Veh./day)	CO_2 emission (Kg/L)	Discharge (L/m)	Emission multiplier CO_2 (Kg/hour)	Travelling time (hour)	CO_2 emission (Kg/day)
(1)	(2)	(3)	(4)	(5) = (3)×(4) × 60	(6)	(7) = (2)× (5)×(6)
> 1989	0	0.1066	5	31.9714	0.0164	0.0000
1989–1993	0	0.1077	5	32.3000	0.0164	0.0000
1994–1998	1	0.1003	5	30.1033	0.0164	0.2462
1999–2003	13	0.1220	5	36.5850	0.0164	7.4798
2004–2008	128	0.1343	5	40.2872	0.0164	84.3438
2009–2013	659	0.1410	5	42.3086	0.0164	456.0271
2014–2018	1873	0.1419	5	42.5823	0.0164	1304.1487
Total						1852.2456

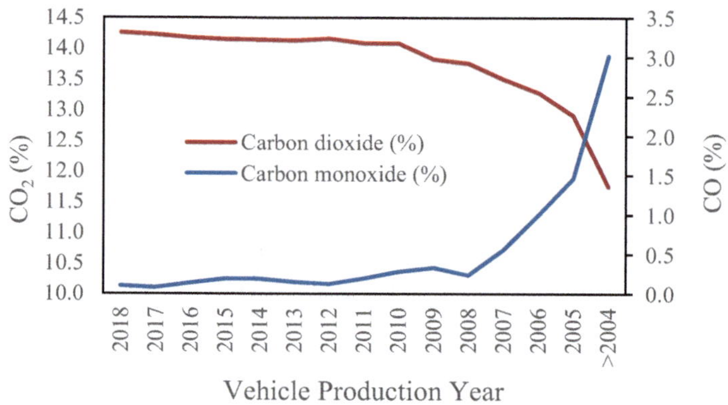

Fig. 1 Diagram of average CO_2 emission per vehicle production year tested in the beginning of 2019

Table 1 shows the amount of CO2 emission for street 1 and street 2 being 789.2 kg/day and 315 kg/day, respectively. This IPCC method uses emission factors issued by the Ministry of the Environment which are the average number of vehicles per day and energy consumption. The emission factors of CO2 used for gasoline-fueled passenger cars is 69300 kg/TJ. As seen from the tables above, CO2 emissions on street 1 are greater than street 2. This can be easily explained, as street 1 is longer and has a higher average number of vehicles per day than street 2.

From Table 2, the amount of CO_2 emission calculated using the direct method give values of 4645.8 kg/day for street 1 and 1852.2 kg/day for street 2. The direct method is carried out by identifying the average amount of CO_2 emitted by passenger cars which are categorized according to the year of vehicle's manufacture. It can be observed from Table 2 and Table 3 that the number of vehicles on street 1 is greater than street 2, hence the amount of CO_2 emission calculated on street 1 is greater than that of street 2.

By observing Fig. 1, we can summarize that the older the engine's production year, the lower the average value of CO_2 emission of said engine is. This is because vehicles with the latter manufacturing year usually have good engine performance so that complete combustion may occur as often as possible. This can be attributed to the chemical reaction in the internal combustion engine in which complete combustion will yield more CO_2 and less Carbon monoxide (CO), Meanwhile incomplete combustion engine produces more CO and less CO_2. The newer engine production performs complete combustion therefore yield more CO_2. Also confirmed by Dineen is the fact that the average amount of CO_2 emissions of all new private cars purchased in 2013 in Ireland was 162.4 g/km which is lower than the average of cars purchased in 2014 which shows an average value of 170.3 g/km [14, 15]. It should be noted that CO_2 gas is the result of complete combustion of fuel, hence why the engine's production year may influence the amount of CO_2 it produces. This is different from the results of Kusumawati's research [16], which states that the amount of CO_2 emissions is not closely related to the year of engine production but strongly influenced by the engine's performance. The performance of the engine itself is influenced by its level of maintenance. In other words, instead of the production year, the level of machine maintenance has more influence on the amount of CO_2 emissions produced [16].

From Fig. 2, we can see that the amount of CO_2 emissions calculated by the direct method are approximately 6 times higher than that of the IPCC method. We can see that, as the calculation of the IPCC method carried out in this paper is classified as tier 1 which only uses activity data in the form of fuel consumption and IPCC default emission factors, the IPCC method's level of accuracy is the lowest. All the while, the direct method belongs to tier 3 wherein the CO_2 emission test is carried out directly on the vehicle so that its accuracy is higher than that of the IPCC method.

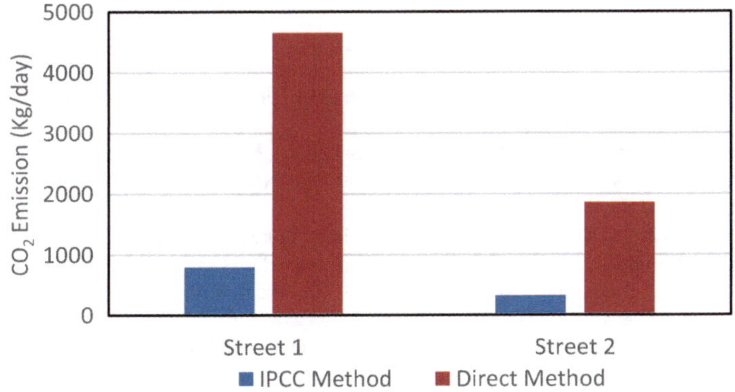

Fig. 2 Diagram of CO_2 emissions on street 1 and street 2

Considering the equations used by the IPCC method and the direct method, there are several variables that greatly affect the calculation's results. These variables are travel distance, average speed, and traffic volume. These variables are directly proportional to the amount of emissions produced. The travel distance and the average speed affect the required energy consumption, thus the greater each variable is, the greater the emissions produced. The difference between the two methods is the emission factor used. The IPCC method uses emission factors issued by the Ministry of the Environment, which are sourced from IPCC 2006, while the direct method uses data from emission tests conducted by the Environment Agency.

4 Conclusion

Investigation of CO_2 emission on two local streets has been carried out and presented in this paper by means of the IPCC tier 1 method and the direct method. It appears that the results of calculating CO_2 emissions using the direct method give 6 times larger results than the IPCC method. This is because the calculated emissions in the direct method are basically the result of the traffic volume times a CO_2 emission multiplier. This multiplier is determined from the measurement of passenger cars' exhaust emissions in the field. Meanwhile, in the IPCC method, the CO_2 emission factor used is atmospheric CO_2 gas that has either accumulated or interacted with other gases. In addition to this, the IPCC method used in this research is a tier 1 approach, while the direct method is more likely a tier 3 approach.

Acknowledgements The authors express their gratitude and high appreciation to the Institute for Research and Community Service (LPPM) UNS which has provided research grants from non-APBN UNS funds to carry out this research.

References

1. Nurdjanah N (2014) Emisi CO_2 akibat kendaraan bermotor di Kota Denpasar. J Penelitian Transp Darat 16(4):189–202
2. Muziansyah D, Sulistyorini R, Sebayang S (2015) Model emisi gas buangan kendaraan bermotor akibat aktivitas transportasi (studi kasus: terminal pasar bawah ramayana kota bandar lampung). JRSDD 3(1):57–70
3. Yuliana DK (2018) Tingkat emisi gas rumah kaca di kabupaten indramayu. Jurnal Sains dan Teknologi Mitigasi Bencana 12(2):1. https://doi.org/10.29122/jstmb.v12i2.2098
4. Eggleston HS, Buendia L, Miwa K, Ngara T, Tanabe K (2006) 2006 IPCC guidelines for national greenhouse gas inventories.
5. Borgen SK, Grønlund A, Andrén O, Kätterer T, Tveito OE, Bakken LR, Paustian K (2012) CO_2 emissions from cropland in Norway estimated by IPCC default and Tier 2 methods. GHG Measure Manage 2(1):5–21

6. Ramírez-Melgarejo M, Reyes-Figueroa AD, Gassó-Domingo S, Güereca LP (2020) Analysis of empirical methods for the quantification of N_2O emissions in wastewater treatment plants: comparison of emission results obtained from the IPCC Tier 1 methodology and the methodologies that integrate operational data. Sci Total Environ 747:141288
7. Kim KD, Ko HK, Lee TJ, Kim DS (2011) Comparison of greenhouse gas emissions from road transportation of local government by calculation methods. J Korean Soc Atmos Environ 27(4):405–415
8. Oh Y, Park J, Lee JT, Seo J, Park S (2017) Estimation of CO_2 reduction by parallel hard-type power hybridization for gasoline and diesel vehicles. Sci Total Environ 595:2–12
9. Seo J, Park J, Park J, Park S (2021) Emission factor development for light-duty vehicles based on real-world emissions using emission map-based simulation. Environ Pollut 270:116081
10. Hidup KL (2012) Pedoman Penyelenggaraan Inventarisasi Gas Rumah Kaca Nasional [Buku I Pedoman Umum]. KLH, Jakarta
11. Hidup KL (2012) Pedoman Penyelenggaraan Inventarisasi Gas Rumah Kaca Nasional [Buku II Volume 1, Metodologi Penghitungan Tingkat Emisi Gas Rumah Kaca]. KLH, Jakarta
12. BPS Statistics of Sragen Regency (2020) Sragen regency in figures, bilingual report, pp 29–31
13. DPU (2005) 'Pedoman Perhitungan Biaya Operasi Kendaraan (Pd T-15–2005- B)', pp 1–32
14. Dineen D, Howley M, Holland M (2014) Energy in transport 2014 report, Sustainable Energy Authority of Ireland
15. Alam MS, Duffy P, Hyde B, McNabola A (2017) Improvement in the estimation and back-extrapolation of CO_2 emissions from the Irish road transport sector using a bottom-up data modelling approach. Transp Res Part D: Transp Environ 56:18–32
16. Kusumawati PS, Tang UM, Nurhidayah T (2013) Hubungan Jumlah Kendaraan Bermotor, Odometer Kendaraan dan Tahun Pembuatan Kendaraan dengan Emisi CO_2 di Kota Pekanbaru. Ilmu Lingkungan 7(1):49–59

Reducing the Release of Greenhouse Gases in the Rigid Pavement Material Transportation Process Unit

Fajar Sri Handayani, Florentina Pungky Pramesty, and Ary Setyawan

Abstract Material transportation of the road construction project has had some negative impact on the environment, one of which is the release of greenhouse gas emissions. In an attempt to develop sustainable strategy, there is a need to identify and quantitatively evaluate the GHG emission reduction opportunities that exist in material transportation of the rigid pavement life cycle. The aims of this study is to estimate and evaluate the reduction of greenhouse gas emissions in the rigid pavement material transportation process unit. The life cycle assessment was used to estimate the greenhouse gases emissions. The evaluation of greenhouse gas emission reductions is not only on the percentage of the reduction, but also on the cost effectiveness of the reduction. The cost effectiveness analysis (CEA) was used to calculate cost effectiveness. The result shows that a reduction in greenhouse gas emissions in the variation of the use of dump trucks in the rigid pavement material transportation process unit ranged from 2.89% to 6.24%. Each alternative to reducing emissions has the potential to reduce greenhouse gas emissions at a cost comparable to reduced emission.

Keywords Reducing · Greenhouse gases · Rigid pavement

1 Introduction

Greenhouse gas (GHG) emissions due to construction projects in Indonesia in the period of 2009–2019 was estimated at 29,941,737 tons of carbon dioxide equivalent (CO_2e) of which 20% was contributed by toll road construction projects, 39% was issued by national road construction projects, 17% was issued by projects of

F. S. Handayani (✉) · F. P. Pramesty · A. Setyawan
Civil Engineering Department, Faculty of Engineering, Sebelas Maret University, Surakarta, Indonesia
e-mail: fajarhani@ft.uns.ac.id

F. P. Pramesty
e-mail: f.p.pramesti@ft.uns.ac.id

S. A. Kristiawan et al. (eds.), *Proceedings of the 5th International Conference on Rehabilitation and Maintenance in Civil Engineering*, Lecture Notes in Civil Engineering 225,
https://doi.org/10.1007/978-981-16-9348-9_67

provincial road construction, and 14% was generated from regional road construction projects [1]. Road construction activities also have a significant impact on the environment. To anticipate this, the government responds by issuing important strategies in infrastructure development, namely sustainable development strategies or development that applies the principles of sustainability.

Sustainable development is defined as the development of meeting the needs of the present without limiting the ability of future generations to meet their needs. This form of development is a stable relationship between activity and nature so that future generations obtain a good quality of life. Sustainable road development is development that is designed and built with reference to the concept of sustainability, which is a system that reflects the capacity to support nature and human values (ecology, economy, and equity) [2].

One strategic method to reduce the potential for global warming on pavement is development based on the concept of increasing environmental performance most effectively by focusing on components with high potential impacts. This system takes advantage of the fact that small changes in high-impact components will have a greater effect of change than low-impact components [3].

In the rigid pavement activities, the transportation process contributes 26.5% of greenhouse gas emissions from the total emissions. The transportation of aggregate to the Batching Plant contributes 9.48% of greenhouse gas emissions from the total emissions [4]. Therefore, the efforts to reduce greenhouse gas emissions in this article focus on opportunities for reduction in the pavement construction material transportation process unit.

Reducing greenhouse gas emissions on pavement requires a complete understanding of how the life cycle of a construction project impacts the environment. Like other products, road pavement also produces greenhouse gas emissions throughout its service life, which starts from the material production, construction, operational and maintenance stages.

Life cycle assessment (LCA) is a method used to estimate the environmental impact resulting from the use of materials, distribution/transportation, construction, maintenance and congestion, during the service life of the road construction [5].

Cost-effectiveness analysis (CEA) is a form of economic analysis that compares the relative costs and outcomes (effects) of different courses of action. In several references, efforts to reduce emissions have not yet carried out cost-effectiveness analysis. A comparison of the additional costs incurred for emission reduction activities with a number of reduced greenhouse gas emissions is necessary [6]. In this paper, applying the concept of cost effectiveness to the opportunities for reducing greenhouse gas emissions at the road pavement construction stage, alternative reductions can be evaluated not only on the potential for reduction, but also on the relative costs incurred for efforts to reduce these emissions.

An analysis of greenhouse gas emissions in variation dump truck (DT) needs to be done to determine the percentage of emissions reductions that can be done. Determination of alternative options for reducing greenhouse gas emissions is not only in the percentage of the reduction, but also in the relative costs of the reduction. Therefore, the purpose of this article is to estimate the percentage of reduction in greenhouse gas emissions in the material transport process unit and an analysis of its cost effectiveness, on rigid pavement.

2 Methods

Stages of calculation of greenhouse gas emissions in this research are the determination of the activities of each stage of construction carried out, the determination of emission factors per unit of activity, and the calculation of greenhouse gas emissions which are the results of multiplication of activity data and emission factors per unit of activity. Emission factors per unit of activity are based on the Eco-Invent data base and the results of previous research [7–10]. Emission factors of transportation equipment and construction equipment in units kilogram of carbon dioxide equivalent (kgC02e) per hour use of the tool is based on an Eco-Invent database [11]. Greenhouse gases conversion into equivalent carbon dioxide equivalent (CO2e) units used TRACI 2.1 data base. TRACI is an environmental impact assessment tool. It provides characterization factors for Life Cycle Impact Assessment (LCIA), industrial ecology, and sustainability metrics. Characterization factors quantify the potential impacts that inputs and releases have on specific impact categories in common equivalence units.

The emissions analysis stages reviewed in this research are the material production stage and the construction stage. The material production stage consists of cement, aggregate, sand, and steel production processes. The construction phase consists of pavement material transportation activities to the Batching Plant (BP), the process of the Batching Plant, transportation of concrete to the project site, and the spread of concrete at the project site.

Calculation of GHG emission reduction and cost was done on 3 alternative variations in the capacity of the used dump truck bouquet, alternative 1 used dump truck DT 10, alternative 2 used dump truck DT 20, and alternative 3 used dump truck DT 24. The existing mode of transportation of the material uses DT 7. The calculation of the CE was done by comparing the magnitude of the cost difference for reducing greenhouse gas emissions with a reduction in greenhouse gas emissions in each alternative.

3 Result and Discussion

Material production process unit is a material production process that exists in nature. This process looks at the CO2e gas produced from each material used in the production process. Material transportation process unit is a process of transportation of a number of pavement materials needed to reach BP. Greenhouse gas emissions calculated in this process are vehicle emissions resulting from the transportation of material to BP. The rigid pavement construction which is the case study in this research is the construction of rigid pavement with the volume of material usage

Table 1 The material usage

Material	Usage (kg)
Cement	884,502
Gravel	2,137,602
Sand	1,499,699
Steel	13,538

Table 2 GHG emission in the transportation material to BP

Process unit	Distance to BP (km)	Repetition	GHG emission factor ($kgCO_2e$/t.km)	GHG emission ($kgCO_2e$)
The transportation of gravel	31.2	286	0.353	47,257
The transportation of sand	90.3	200	0.353	95,643
The transportation of cement	238.1	169	0.119	36,046

presented in Table 1. The mixing process unit is the process of mixing raw materials at BP. In this activity, it was found that the amount of greenhouse gas emitted due to the mixing process at BP.

In the material transportation process unit to BP, the calculation of emissions is done by multiplying the time of use of the transportation mode (hours) with the emission factor ($kgCO_2e$/hour). The time of use of the transportation mode depends on the volume of material, the distance of quarry to the BP, the volume of the bouquet of transportation modes (dump truck), and the average speed of the transportation mode. Calculation of greenhouse gas emissions in the process unit for transporting material to BP is presented in Table 2.

The recapitulation of the calculation of greenhouse gas emissions released in the rigid pavement construction process case study using material dump truck DT 7 means of transportation in the material transport process unit is presented in Table 3.

In this case study of rigid pavement construction, the LCA process calculation results in the scope from raw material to the gate before the operation process (cradle to gate), the material production process unit releases the largest greenhouse gas emissions of 1,395,089.58 $kgCO_2e$ or 86.56%, while the transportation process unit releases greenhouse gas emissions by 178,945.62 $kgCO_2e$ or 10.97% of total emissions. However, when viewed from the LCA process in the scope of construction (gate to gate), 75.99% of emission releases is generated by the process of transporting pavement material.

The strategy to reduce greenhouse gas emissions in this case study was carried out by looking at how much greenhouse gas emissions will be reduced by using a dump truck transportation with a larger bouquet capacity, that is DT 10, 20, and 24. The assumption was established that with a larger dump truck bouquet capacity,

Table 3 Recapitulation of GHG emissions in rigid pavement

Stage	Process unit	GHG emission ($kgCO_2e$)
Material production	The production of cement	943,763.46
	The production of gravel	6,840.34
	The production of sand	4,799.04
	The production of steel	439,686.74
Transportation to BP	The transportation of cement	36,046.40
	The transportation of gravel	47,256.01
	The transportation of sand	95,643.21
Mixing process	Mixing process in BP	19,454.01
Transportation to site	The transportation of concrete	22,616.57
	The transportation of steel	12,157.66
	Concrete spreading	2,326.83
Sum of emission		1,630,590.26

the number of repetitions and time of dump truck usage will be reduced, so that greenhouse gas emissions can be reduced.

The reduction effort was evaluated not only on the percentage of the reduction, but also on the relative costs incurred for the emission reduction effort. The simulation results of the calculation of greenhouse gas emissions in the process of transporting materials to BP with alternative means of transportation 1, 2, and 3, the difference in emissions, the percentage of emission reduction, cost difference, and effective cost effectiveness, are presented in Table 4.

The results of the calculation of Table 4 show that the variation of dump trucks usage in the aggregate and sand transportation unit causes a reduction in greenhouse gas emissions by 2.89%, 4.73%, and 6.24%, respectively. Alternative 3 results in the highest reduction in greenhouse gas emissions. With the same volume, the use of large capacity means of transportation will be able to reduce repetition and reduce the time of its use, so that the resulting emissions will be reduced. The emission reduction strategy with the use of large capacity dump trucks will reduce the biggest emissions as well, but this strategy requires the biggest effective cost. Therefore, when viewed from the value of cost effectiveness (CE), alternatives 2 and 3 have a lower cost effectiveness value (CE) compared to alternative CE value 1. In this case study, when evaluated from its CE value, alternative 1 is a greenhouse gas emission

Table 4 Recapitulation of GHG emission, percentage of reductions and cost effectiveness

Process unit	Case study	GHG emission Alt. 1	($kgCO_2e$) Alt. 2	Alt. 3
The production of cement	943,764	943,764	943,764	943,764
The production of gravel	6840	6840	6840	6840
The production of sand	4799	4799	4799	4799
The production of steel	439,687	439,687	439,687	439,687
The transportation of cement	36,046	36,046	36,046	36,046
The transportation of gravel	47,256	45,890	45,021	44,307
The transportation of sand	95,643	92,879	91,119	89,675
The transportation of concrete	22,617	22,617	22,617	22,617
The transportation of steel	12,158	12,158	12,158	12,158
Concrete spreading	2327	2327	2327	2327
GHG emission (kgCO2e)	1,630,590	1,626,460	1,623,831	1,528,841
Δ emission (kgCO2e)		4130	6759	8917
Percentage of reductions (%)		2.89	4.73	6.24
Δ cost (IDR*1000)		7000	13,000	21,000
CE (IDR*1000/kgCO2e)		1.69	1.92	2.36

reduction strategy in the most effective pavement material transportation process unit.

4 Conclusion

The reduction of greenhouse gas emissions in the variation of the use of dump trucks in the material transportation process unit to BP ranged from 2.89% to 6.24%. Based on the value of cost effectiveness, the use of dump truck DT 10 is the most effective strategy to reduce greenhouse gas emissions in the material transportation process unit. Each alternative to reducing emissions has the potential to reduce greenhouse gas emissions and requires a comparable cost for emissions that can be reduced.

References

1. The Word Bank Group (2011) Green gas emission mitigation in road construction and rehabilitation. A toolkit for developing countries, Washington DC
2. Muench ST, Anderson JL, Hatfield JP, Koester JR, Soderlund M (2011) Greenroads manual. V1.5. University of Washington
3. Santero NJ, Horvath A (2009) Global warming potential of pavements. Environ Res Lett 4(034011):1–7
4. Handayani FS, Pramesti FP, Wibowo MA, Setyawan A (2019) Estimating and reducing the release of greenhouse gases in local road pavement contructions. Int J Adv Sci Eng Inf Technol 9(5):1709–1715
5. Yu B, Qin L (2012) Life cycle assessment of pavement: methodology and case study. Transp Res Part D 17(5):380–388
6. Santero NJ, Loijos A, Ochsendorf J (2013) Greenhouse gas emissions reduction opportunities for concrete pavements. J Ind Ecol 17(6):1–10
7. Marceau ML, Nisbet MA, VanGeem MG (2007) Life cycle inventory of portland cement concrete SN3011. Portland Cement Association Skokie Illinois
8. Zapata P, Gambatese JA (2005) Energy consumption of asphalt and reinforced concrete pavement materials and construction. J Infrastruct Syst 11(1):9–20
9. Hughes B, Hare W (2012) LCI Data for steel product report produced. World Steel Association
10. Loijos A (2011) Life cycle assessment of concrete pavements: impacts and opportunities. National Ready Mixed Concrete Association (NRMCA) San Diego
11. Wernet G, Bauer C, Steubing B, Reinhard J, Moreno-Ruiz E, Weidema B (2016) The ecoinvent database version 3 (part I): overview and methodology. Int J Life Cycle Assess 21(9):1218–1230

Cost Effectiveness Analysis of Greenhouse Gas Emissions Reduction in the Flexible Pavement Material Transportation Process Unit

Fajar Sri Handayani, Florentina Pungky Pramesti, and Ary Setyawan

Abstract The activity of material transportation regarding the project of road construction indicated several negative impacts on the environment including the greenhouse gas emissions release. In developing sustainable development strategy especially in transportation infrastructure, it is necessary to find a strategy in reducing greenhouse gas emissions of road construction cycle. This study aimed to estimate and evaluate the greenhouse gas emissions reduction in the flexible pavement material transportation process unit. The life cycle assessment was used to estimate the emissions produced. The greenhouse gas emissions evaluation did not only underline the reduction percentage, but also on the relative cost of the reduction. The approach of cost effectiveness analysis concept was implemented to estimate the cost effectiveness. The results showed that the greenhouse gas emissions in the dump trucks use variation in the flexible pavement material transportation process unit ranged from 2.99% to 6.23%. Every alternative in emissions reduction showed the potential to decrease the greenhouse gas emissions at a cost comparable to emissions that could be reduced.

Keywords Cost effectiveness · Greenhouse gases · Flexible pavement

1 Introduction

Pavement construction activities have mostly used 22% of global energy consumption, 25% of fossil fuels and 30% of global air pollution, namely greenhouse gases [1]. Greenhouse gas emissions due to the projects of construction in Indonesia in the period of 2009–2019 showed an estimate of 29,941,737 ton CO_2e of which 20% was donated by the project of toll road construction, 39% was issued by the project of

F. S. Handayani (✉) · F. P. Pramesti · A. Setyawan
Civil Engineering Department, Faculty of Engineering, Sebelas Maret University, Surakarta, Indonesia
e-mail: fajarhani@ft.uns.ac.id

F. P. Pramesti
e-mail: f.p.pramesti@ft.uns.ac.id

S. A. Kristiawan et al. (eds.), *Proceedings of the 5th International Conference on Rehabilitation and Maintenance in Civil Engineering*, Lecture Notes in Civil Engineering 225,
https://doi.org/10.1007/978-981-16-9348-9_68

national road construction, 17% was issued by provincial road construction project, and 14% was resulted from the project of local road construction [2].

A significant impact on the environment has emerged regarding road construction and improvement activities. In anticipation, the government has given responses by issuing a strategy in infrastructure development that is necessary as a strategy of sustainable development or development implementing the principles of sustainability.

The definition of sustainable development refers to meeting the present needs development without restricting the future generations' ability in fulfilling their needs. This development form is a stable relationship between activity and nature and therefore there is a good quality of life provided for the future generations. Sustainable development is designed and built referring to the sustainability concept which is a system indicating the capacity in order to support nature and the values of human (ecology, economy, and equity) [3]. A sustainable road construction is a road construction starting from the design, construction, maintenance, and rehabilitation stages using environmentally friendly, environmentally friendly measures and applying sustainable principles [4].

One of the strategic methods in reducing the global warming potential on pavement includes the development based on the environmental performance increase most effectively by underlining the high potential impacts components. The benefit taken in this system is that the fact showing small changes in high-impact components will get a greater effect than low-impact components [5].

In the flexible pavement material production stage, the material extraction process contributed 92.11% of greenhouse gas emissions from the total emissions during the material production stage. In the flexible pavement construction phase, the material transportation process to the project site accounted for 70.09% of the total emissions generated during the construction phase [6]. Thus, the approach used to reduce greenhouse gas emission in this article underlined the reduction opportunities in the process unit of construction material transportation.

In order to reduce the greenhouse gas emissions on pavement, it needs a deep understanding about how the cycle of life of a construction project affects the environment. Similar to other products, road pavement also results in greenhouse gas emissions throughout its service life starting from the stages of material production, construction, operational and maintenance.

Life cycle assessment (LCA) is a method used to estimate the environmental effects resulting from the application of materials, distribution/transportation, construction, maintenance and congestion, during the road construction service life [7].

In general, cost effectiveness analysis is applied in the fields of health and medicine. It compares the costs and health effects of an intervention to assess the extent to which the business can give economic value. In implementing the cost effectiveness concept aiming to decrease the emissions of greenhouse gas at the construction stage of road pavement, reductions alternative can be assessed not only on the reduction potential but also on the relative costs acquired for the emissions reduction efforts. Cost Effectiveness (CE) becomes a comparison of the additional costs incurred for the activities of emission reduction with a number of reduced greenhouse gas emissions [8].

An analysis of greenhouse gas emissions on several material transportation alternative is necessary to be conducted to determine the emissions reductions percentage that can be done. The determination of alternative choices of greenhouse gas emissions is not only in the reduction percentage but also in the reduction relative costs. Thus, this article aimed to estimate the percentage of greenhouse gas emissions reduction in the material transportation process unit and its cost effectiveness analysis, on the road pavement construction project.

2 Methods

This research took a case study of regional road construction in Sukoharjo district. The road construction was flexible pavement. The data used in the life cycle assessment method consisted of primary data obtained from direct observations in the field and secondary data obtained from the database of related agencies and the official website of the database provider. Data on quarry locations for pavement material materials, asphalt mixing plant locations, types of material transportation equipment, types of heavy equipment and construction methods used were primary data, while secondary data required were general project data, TRACI 2.1 conversion database, and life cycle inventory database from previous research and the Ecoinvent database [9].

The estimate stages of greenhouse gas emission in this research were the determination of the activities of each construction stage which was conducted, the factors of emission per unit of activity, and the estimate of greenhouse gas emissions as the results of multiplication of activity data and emission factors per unit of activity. The flexible pavement construction activities of this case study included the production process of flexible pavement material (asphalt, filler, sand, aggregate), the transportation of pavement material to the Asphalt Mixing Plant, the process at the asphalt mixing plant, the transportation to the project site, and the laying process at the project site. The emission data for pavement material factors in units of $kgCO_2e$ per kilogram-material in this calculation was based on the results of research by Chappat [10], Blomberg [11] and Korre [12]. Energy use and specifications for asphalt mixing plant were based on the results of Zapata's research [13]. The emissions factor of transportation and construction equipment in units of $kgCO_2e$ per hour of tool use was based on the ecoinvent database [14].

The estimate of greenhouse gas emissions reduction and costs was conducted on three alternative variations in the capacity of dump trucks that were used, alternative 1 applied dump truck DT 10, alternative 2 applied dump truck DT 20, and alternative 3 applied dump truck DT 24. The Cost Efficiency estimate (CE) was calculated by comparing the cost difference magnitude to reducing the greenhouse gas emissions with greenhouse gas emissions reduction in each alternative.

3 Result and Discussion

The road pavement construction which is the case study in this research is the flexible pavement construction of regency road sections with a road length of 1446 m, a road width of 15.6 m and a pavement thickness of 0.4486 m consisting of 0.1486 m thick base course, sub-base thickness course of 0.2 m and the thickness of the surface course of 0.10 m.

The emission analysis stages reviewed in this study were the material production stage and the construction stage. The material production and construction stages were divided into several process units, namely the material production process unit, material transportation to asphalt mixing plant (AMP), the mixing process at AMP, transportation to the project site, and the on-site laying process.

The unit of the material production process is a material process occurring in nature. This process sees the CO2e gas resulted from every material applied in the process of production. The unit of material transportation process is the transportation process of a number of pavement material required to towards AMP. Greenhouse gas emissions calculated in this process were vehicle emissions resulting from the material transportation process to AMP. Material transportation in the form of aggregate, sand and filler in this case study used a 7.5 ton dump truck with a transportation effectiveness of 90%, which means a truck capacity of 6.75 tonnes. Transportation of materials in the form of asphalt and binder used a tank truck with a capacity of 8.75 tons and the effectiveness of transporting was 90% meaning the capacity of the truck was 7.785 tons. All trucks used are assumed to use a speed of 40 km/hour.

The mixing process unit is the process of mixing raw materials. In this mixing activity, the amount of greenhouse gases emission that was released was due to the asphalt mixing process in AMP. The transportation process unit to the project site was the transportation activity from AMP to the project site. The distance from AMP to the work site required a certain amount of fuel to be used for truck operations. The use of this fuel resulted in the release of greenhouse gas emissions in the process of moving asphalt from AMP to the project site.

The unit of the spreading process at the project site was the activity of spreading the asphalt mixture at the project site. Heavy equipment used in the sub-base course construction process included an asphalt finisher, motor grader, water tank and dump truck. Heavy equipment used in the base course construction process included asphalt finisher, pneumatic tire rollers, tandem rollers and dump trucks. Heavy equipment used in the surface course construction process included asphalt finisher, pneumatic tire rollers, tandem rollers and dump trucks.

The material production process in the form of aggregate, sand, asphalt, filler and binder was calculated by multiplying the volume of material required by the material emission factor. Material emission factors in units of kgCO_2e per ton and calculation of emissions in the material production process can be seen in Table 1.

In the material transportation process unit to AMP, the emission calculation is done by multiplying the time used for the transportation mode (hours) by the emissions factor (kgCO_2e/hour). The time to use the transportation mode depends on the volume

Table 1 Greenhouse gas emissions generated in the material production process unit

Material	Volume	Unit	Emissions factor ($kgCO_2e$/ton)	Usage (kg)
Aggregate	1997.83	Ton	10.0	19,978.32
Sand	459.89	Ton	2.5	19,978.32
Asphalt	60.68	Ton	285.0	1149.71
Filler	9.35	Ton	10.0	93.51
Asphalt binder	2.13	Litre	221.0	470.60

of material, the quarry distance to AMP, the volume of the dump truck, and the average speed of the mode of transportation. The calculation of greenhouse gas emissions in the material transportation process unit to AMP is presented in Table 2.

Material processing at AMP is in the form of mixing asphalt, aggregate, sand and filler into a flexible pavement material mixture. The calculation of emissions in the processing unit at AMP was done by multiplying the amount of time spent in AMP by the AMP emission factor per hour. Based on the Ecoinvent 3.3 and TRACI 2.1 database, the emission factor in the processing unit at AMP was 119.0141 kg CO_2e/hour. At an AMP capacity of 60 tonnes/hour with the use time in AMP in this case study of 16.6 $_h$, and therefore the greenhouse gas emissions produced in the processing unit at AMP were 1975.63 $kgCO_2e$.

The calculation of greenhouse gas emissions in the asphalt spread process unit at the project site is divided into 3 sub-process units, namely the base course, sub-base course, and surface course. The calculation of greenhouse gas emissions at the

Table 2 Greenhouse gas emissions in the material transportation process to AMP

The unit of material transportation process to AMP	Distance to AMP (km)	The time length (hour)	GHG emission factor ($kgCO_2e$/hour)	GHG emission ($kgCO_2e$)
The transportation of aggregate	33	488.4	5.06	463.99
The transportation of sand	33	113.85	5,06	2471.30
The transportation of asphalt	238	91.70	24.83	576.08
The transportation of filler	33	3.30	24.83	60.21
The transportation of a. binder	238	11.9	5.06	16.70

Table 3 Greenhouse gas emissions in the pavement construction process unit

	Tool capacity (ton/hour)-(l/hour)	Volume (ton)	Tool use (hour)	Emissions factor ($kgCO_2e$/hour)	Greenhouse gas emission ($kgCO_2e$)
Motor grader	234.77	1015.20	4.32	24.80	107.25
Tandem roller	73.94	1015.20	13.73	24.80	340.54
Water tank	71.14	1015.20	14.27	5.06	72.20
Asphalt distributor	4800.00	2248.13	0.47	3.58	1.68
Compressor	4800.00	2248.13	0.47	5.06	2.37
Asphalt finisher	72.79	994.80	13.67	24.80	338.98
Tandem roller	73.94	994.80	13.45	24.80	333.70
Pneumatic tire roller	172.34	994.80	5.77	24.80	143.17

construction stage was carried out by multiplying the total time used by the heavy equipment at each construction stage by the heavy equipment emission factor. Heavy equipment used in the sub-base course construction process includes an asphalt finisher, motor grader, water tank and dump truck. Heavy equipment used in the base course construction process includes asphalt finisher, pneumatic tire rollers, tandem rollers and dump trucks. Heavy equipment used in the surface course construction process includes asphalt finisher, pneumatic tire rollers, tandem rollers and dump trucks. The results of calculating the emission of construction heavy equipment in the flexible pavement construction process unit are presented in Table 3.

The recapitulation of the greenhouse gas emissions calculation released in the flexible pavement construction process using the DT 8 dump truck material transportation unit in the material transportation process unit is presented in Table 4.

In this case study of flexible pavement construction, the estimate results of LCA process in the scope from raw material to gate before the process of operation (cradle to gate), the unit of material production process released greenhouse gas emissions of 38,986.74 $kgCO_2e$ or 79.52%, while the unit of transportation processes released greenhouse gas emissions of 6727.51 $kgCO_2e$ or 13.72% of total emissions. However, when it is viewed from the process of LCA in the construction scope (gate to gate), 66.99% of emissions release was produced by the pavement material transportation process towards the site location.

The greenhouse gas emissions reduction strategy in this case study was conducted by observing how much greenhouse gas emissions would be reduced by applying dump truck transportation tool with bigger bucket capacity, such as DT 10, 20, and 24. The assumption built that the bigger the dump truck bucket capacity, the lower the repetitions number and the dump truck use time so the greenhouse gas emissions can be reduced.

The efforts of reduction evaluated not only on the reduction percentage but also on the relative costs incurred for the efforts of emissions production. The estimate

Table 4 The recapitulation of greenhouse gas emissions in flexible pavement

Process unit	Process sub unit	GHG emission ($kgCO_2e$)
Material production	Asphalt production	17,294.60
	Aggregate production	19,978.32
	Sand production	1149.71
	Filler production	93.51
	Asphalt binder production	470.60
Transportation to AMP	Asphalt transportation	463.99
	Aggregate transportation	2471.30
	Sand transportation	576.08
	Filler transportation	60.21
	Asphalt binder transportation	16.70
Mixing process	Mixing process in AMP	1975.63
Transportation to site	Mixed asphalt transportation	3130.88
	Asphalt binder transportation	8.35
Pavement construction	Base course	520.01
	Sub-base course	4.05
	Surface course	815.85
Sum of emission		49,029.29

simulation results of greenhouse gas emissions in the material transportation process unit to AMP by using means of transportation of alternatives 1, 2, and 3, the difference in emissions, the emissions decrease percentage, the cost difference, and the effective cost effectiveness are presented in Table 5.

The estimate results presented in Table 5 indicated that the dump trucks use variation in the unit of sand and aggregate transportation resulted in greenhouse gas emissions reduction by 2.99%, 4.74%, and 6.23% respectively. Alternative 3 produced the highest greenhouse gas emissions reduction. Having similar volume, the use of means of transportation in large capacity will be able to lower repetition and reduce the time of its use, and therefore the emissions that are produced will be reduced.

Table 5 Recapitulation of GHG emission, percentage of reductions and cost effectiveness

Process unit	Existing	GHG emission Alt. 1	($kgCO_2e$) Alt. 2	Alt. 3
Asphalt production	17,294.60	17,294.60	17,294.60	17,294.60
Aggregate production	19,978.32	19,978.32	19,978.32	19,978.32
Sand production	1149.71	1149.71	1149.71	1149.71
Filler production	93.51	93.51	93.51	93.51
Asphalt binder production	470.60	470.60	470.60	470.60
Asphalt transportation	463.99	463.99	463.99	463.99
Aggregate transportation	2471.30	2229.183	2087.25	1962.02
Sand transportation	576.08	517.64	484.242	459.20
Filler transportation	60.21	60.21	60.21	60.21
Asphalt binder transportation	16.70	16.70	16.70	16.70
Mixing process in AMP	1975.63	1975.63	1975.63	1975.63
Mixed asphalt transportation	3130.88	3130.88	3130.88	3130.88
Asphalt binder transportation	8.35	8.35	8.35	8.35
Base course	520.01	520.01	520.01	520.01
Sub-base course	4.05	4.05	4.05	4.05
Surface course	815.85	815.85	815.85	815.85
GHG emission ($kgCO_2e$)	49,029.79	48,729.23	48,553.90	48,403.63
Δ emission ($kgCO_2e$)		300.56	475.89	626.16
Percentage of reductions (%)		2.99	4.74	6.23
Δ cost (IDR*1000)		7000	13,000	21,000
CE (IDR*1000/$kgCO_2e$)		23.29	27.32	33.54

The strategy of emissions reduction by using large capacity dump trucks will also reduce the biggest emissions. However, this strategy needs the biggest effective cost. Therefore, when it is seen from the cost effectiveness value (CE), alternative 3 has highest cost effectiveness value. In this case study, if evaluated from its CE value, alternative 1 is the most effective alternative strategy for reducing greenhouse gas emissions in the pavement material transportation process unit.

4 Conclusion

The greenhouse gas emissions in the dump trucks use variation in the unit of material transportation process towards AMP ranged from 2.99% to 6.23%. Based on the cost effectiveness value, the dump trucks DT 10 use was the most effective strategy alternative of greenhouse gas emissions reduction in the unit of pavement material transportation. Every emissions reduction alternative has the potential in lowering greenhouse gas emissions and need a cost which is comparable with the emissions reduced.

References

1. Resmi CS, Satyakumar M (2011) Sustainable pavement: the green pavement for future. National Technological Congress, Kerala, College of Engineering Trivandrum
2. The Word Bank Group (2011) Green gas emission mitigation in road construction and rehabilitation. A toolkit for developing countries, Washington DC
3. Muench ST, Anderson JL, Hatfield JP, Koester JR, Soderlund M (2011) Greenroads manual. V1.5. University of Washington
4. Santero NJ, Horvath A (2009) Global warming potential of pavements. Environ Res Lett 4(034011):1–7
5. Handayani FS, Pramesti FP, Wibowo MA, Setyawan A (2017) Strategy toward sustainable local road network infrastructure. MATEC Web of Conf EACEF 138(07007):1–8
6. Handayani FS, Pramesti FP, Wibowo MA, Setyawan A (2019) Estimating and reducing the release of greenhouse gases in local road pavement contructions. Int J Adv Sci Eng Inf Technol 9(5):1709–1715
7. Yu B, Qin L (2012) Life cycle assessment of pavement: Methodology and case study. Transp Res Part D 17(5):380–388
8. Santero NJ, Loijos A, Ochsendorf J (2013) Greenhouse gas emissions reduction opportunities for concrete pavements. J Ind Ecol 17(6):1–10
9. USEPA (2012) Tool for the reduction and assessment of chemical and other environmental impacts (TRACI) users guide tool for the reduction and assessment of chemical and ther environmental impacts
10. Chappat M, Bilal J (2003) The environmental road of the future: life cycle analysis, Energy Consumption and Greenhouse Gas Emissions. Colas Group
11. Blomberg T (2011) Life cycle inventory: bitumen. (Second Edi). Brussels, European Bitumen Association
12. Korre A, Durucan S (2009) Life cycle assessment of aggregates. EVA025 Final Report
13. Zapata P, Gambatese JA (2005) Energy consumption of asphalt and reinforced concrete pavement materials and construction. J Infrastruct Syst 11(1):9–20
14. Wernet G, Bauer C, Steubing B, Reinhard J, Moreno-Ruiz E, Weidema B (2016) The ecoinvent database version 3 (part I): overview and methodology. Int J Life Cycle Assess 21(9):1218–1230

A Probabilistic Model of Container Port Demand in Java Concerning the Port Hinterland Connectivity

Lydia Novitriana Nur Hidayati, Gerard de Jong, Anthony Whiteing, and Munajat Tri Nugroho

Abstract The port demand could be modelled from the port choice of the ports' actors. From the actors' perspectives, factors influencing port choice do not solely depend on the port performance itself. As a port is an interchange location of land and sea transport, the port selection is closely associated with the inland mode choice. This study applies a multinomial logit (MNL) model to predict the probability of a joint port and inland mode choice in Java, Indonesia. Six factors classified into port performance and port hinterland connectivity are applied to estimate the probability of port demand. The model also allows us to estimates the port hinterland boundaries as a corollary of port choice made by the actors. Further from the simulation on several transport strategies, introducing Patimban port as a new container port in Java greatly affects the adjacent port demand. Reducing rail transport time and improving port performance also leads to substantial port and mode shifting for container transport in Java.

Keywords MNL model · Port choice · Inland mode choice · Port hinterland connectivity

L. N. N. Hidayati (✉) · G. de Jong · A. Whiteing
Institute for Transport Studies, University of Leeds, Leeds, UK
e-mail: tslnnh@leeds.ac.uk

G. de Jong
e-mail: g.c.dejong@its.leeds.ac.uk

A. Whiteing
e-mail: a.e.whiteing@its.leeds.ac.uk

L. N. N. Hidayati
Civil Engineering Department, Universitas Sebelas Maret Surakarta, Surakarta, Indonesia

G. de Jong
Significance, Amsterdam, The Netherlands

M. T. Nugroho
Industrial Engineering Department, Universitas Muhammadiyah Surakarta, Surakarta, Indonesia
e-mail: munajat3n@ums.ac.id

S. A. Kristiawan et al. (eds.), *Proceedings of the 5th International Conference on Rehabilitation and Maintenance in Civil Engineering*, Lecture Notes in Civil Engineering 225,
https://doi.org/10.1007/978-981-16-9348-9_69

1 Introduction

This paper presents a study on forecasting probability of container port demand using a discrete choice model. The demand of a port could be estimated through the choice of port made by the ports' actors (e.g., shippers and carriers), in which their loyalty in choosing the port depends on many factors. In this paper, six variables classified into two main categories, port-hinterland connection, and port performance, are examined. The model simulation results can be used for predicting the port hinterland boundaries and predict the demand volatility of port and inland mode if particular transport strategies are applied.

Port hinterland connectivity provides port authorities information about factors that may affect their port market share. Wang et al. argue that efficiently managing transport connectivity of the port and its hinterland areas could build customer devotion and potentially attracting more customers hence expand the port market share [1]. For example, Nugroho et al. [2] demonstrated that a reduction of 20% in hinterland mode time could raise port market share by 2% [2]. Secondly, Wang et al. stated that the merit of analyzing port hinterland is to identify competitiveness with surrounding ports, thus knowing the major port rivals or opening up a chance to collaborate with other ports to develop transport strategies to guarantee port sustainability [3].

A surge of research effort has been directed to utilize discrete choice models to analyze port choice. Malchow and Kanafani [4] discussed applying the logit-based disaggregate model to predict choice made by shipper over the route and port choices. Zondag et al. [5] demonstrate the MNL model to estimate the probability of using a specific transport logistics chain which included port within the transport chain determined by carriers. Tang et al. [6] offered an approach to improve the MNL model's use to predict shipping companies' port choice by integrating a transport network attributes into the choice explanatory variables. Tapia et al. [7] perform an advanced discrete choice model, a Multiple Discrete Extreme Value Model (MDCEV), to estimate the mode and port choice in Argentina and further use the model to evaluate impact of rail transport investment.

The model discussed in this paper is the multinomial logit (MNL) model on port and inland mode choice. This model combines both the port choice and inland mode choice into one choice alternative and examines the actor decision over the possible alternatives. Nugroho et al. [2] stated that not many literatures investigated these merged choices compared to abundance of literature analyzing the choices separately. Port is a midway interchange hub connecting land and sea transport; further, joint estimation of port and inland mode seems complicated. It may generate a massive number of possible alternatives that create difficulties in data collection. Thus, some studies either estimate port choice mode choice separately or treat port choice as part of intermodal transport routes to transport goods from origin to final destination location [4, 5, 8]. Applying the joint method, particularly for Java island, Indonesia, is possible as there are only three major container ports. Moreover, by this joint choice model, the impact of managing port hinterland connectivity to port demand could be identified. In the model, factors affecting port and inland mode choice categories as port performance variables are port cost and port ship calls. On the other hand, inland transport cost, transport time, reliability and GHG emission are the relevant variables for the port hinterland connection.

The rest of this paper is arranged as follows. The following section elaborates on the modelling framework applied. It starts with the intermodal transport builder to generate the value of explanatory variables of port hinterland connectivity and then continues with the MNL model's explanation. The third section summarizes the model application, and the final conclusions are in section four.

2 Modelling Framework

2.1 Intermodal Transport Builder

We begin this section by introducing the model of an intermodal transportation network in Indonesia, which we will denote as the transport chain builder (TC builder), to generate the explanatory values for the variables of port hinterland connectivity (i.e., inland mode cost, time, and GHG emission). The TC builder model portrays the interaction between transportation network and possible transport chain[1] combinations and the level of service connecting origin and destination location. In the TC builder, a node could represent a TC zone (i.e., the centroid of a group of subdistrict areas), transport terminal, and road junction. The link acts as a connector between nodes that depict the road, railways, sea routes, and flight routes. For this paper, only road and rail transport network within Java considered.

The mode characteristics applied in TC builder are provided in Table 1. The transport cost function used here is derived from the cost function applied in Zukhruf et al. [9], while the speed values are obtained from the data of the ministry of transport and Nugroho [10]. The transport chain's determination for only truck mode and truck-train mode combination follows the shortest path algorithm based on the total transport cost. From this path, then the total transport time and total GHG emission will be calculated. Apart from the mode cost function and speed variable, the total

Table 1 Mode characteristics

Mode	Cost function (Rp/TEU per km)[a]	Time function (hour)[b]	Emission function (KgCO_2e)
Truck	distance*846.96*11.6	distance/37.00	distance*62*11.6
Rail	distance*686.33*11.6	distance/36.24	distance*22*11.6

[a]In this paper, the cost value of the TC builder will be converted into TEU/Ton by multiply the initial value of 1 TEU = 11.6 Ton

[b]Time function is the calculation of distance divided by speed in km/hour

[1] A transport chain defines as a sequence of modes employed in the process of shipping goods from the point of origin to the destination location.

transport cost and total transport time also consider the handling cost and waiting time at the mode interchange location (i.e., railway station/terminal).

2.2 Probabilistic Model in Port and Inland Mode Choice in Java

Probabilistic model using discrete choice modelling usually roots on observed behavioral data that better reflect the reality of decision-making. An individual (in this case a shipper or exporter firm) chooses the alternative based on its attractiveness, characterized by a utility $\left(V_{pm}\right)$, in comparison to other alternatives. In this paper, the utility of each alternative is derived from the research performed in Nugroho [10] which estimates its model based on the stated preference data. The utility function of port and inland mode choice in Java can be expressed by Eq. (1).

$$
\begin{aligned}
V_{pm} = ASC_{pm} - 0.257\left(PC_p\right) - 0.817\left(PSC_p\right) + 1.93\left(IMC_{opm}\right) \\
- 0.863\left(IMT_{opm}\right) - 0.406\left(IMR_{opm}\right) + 0.683\left(IMG_{opm}\right) \qquad (1)
\end{aligned}
$$

V_{pm} The observed utility of the alternative p using mode m
ASC_{pm} Alternative specific constant for alternative port p using inland mode m
PC_p Port cost for 1 TEU in port p (thousand Rp).
PSC_p Ship calls of international container vessel per week in port p.
IMC_{opm} Inland mode cost for transporting 1 TEU FCL from origin o to port p using inland mode m (thousand Rp).
IMT_{opm} Inland mode time for transporting 1 TEU FCL from origin o to port p using inland mode m (hours).
IMR_{opm} Inland mode reliability for transporting 1 TEU FCL from origin o to port p using inland mode m (%).
IMG_{opm} Inland mode GHG emission for transporting 1 TEU FCL container from origin o to port p using inland mode m (Kg CO_2e).

As inferred from Eq. (1), the alternative i defined as alternative choosing port p using inland mode m, consequently the choice probability of alternative i for individual n, for a Multinomial Logit (MNL) model (P_{in}) can be calculated as in Eq. (2):

$$
P_{in} = \frac{e^{V_{in}}}{\sum_{j=1}^{J} e^{V_{jn}}} \qquad (2)
$$

3 A Case Study

In a case study of port hinterland connectivity in Java island, the proposed models will be used to calculate the probability of port and inland mode choice from surrounding zones. Java island has three main container port: Tanjung Priok, Tanjung Emas, and Tanjung Perak, which hold about 60% of total container transport in Indonesia [8]. In an attempt to meet the growth of container transport, the government has a plan to build the biggest container Port in Java, namely Patimban Port. This port expected to support the container traffic existing in Tanjung Priok port which currently faces a capacity problem and reduces the logistics cost due to its strategic location in between industrial areas of Bekasi, Karawang, and Purwakarta [2]. Referring to the Eq. (1) and Eq. (2), the probability of port and inland mode choice is depended on some variables related to the port and inland mode characteristics. The port hinterland connectivity variables except for inland mode reliability (no data available, thus it is assumed to be 80% for all alternatives) have a specific value for each zone area. The port performance variables applied are refer to Munajat [10] as provided in Table 2.

3.1 Port Hinterland Boundaries

From the probability of port and inland mode choice value, we can also infer the information about the port hinterland boundaries. A high probability of a particular port indicates that the respective area is the hinterland area of the port. The hinterland boundaries of the Tanjung Priok Port presumed by the probability value can be seen as in Fig. 1. Further, by the operation of Patimban Port, Tanjung Priok Port and Patimban port's hinterland boundaries are quite overlapping due to the proximity between these two ports that reduce the demand of Tanjung Priok (see Fig. 2).

Table 2 Port performance

Port	Port cost (thousand Rp)	Port ship call
Tanjung Priok	1127	82
Tanjung Emas	1098	12
Tanjung Perak	1154	23
Patimban[a]	1126	82

[a]Patimban port is not yet operated. However, as Patimban port is expected to be the biggest port in Java, hence all of the port performance values are assumed to be the same as Tanjung Priok Port which currently is the biggest port.

Fig. 1 Hinterland boundaries of Tanjung Priok port prior to Patimban port operation

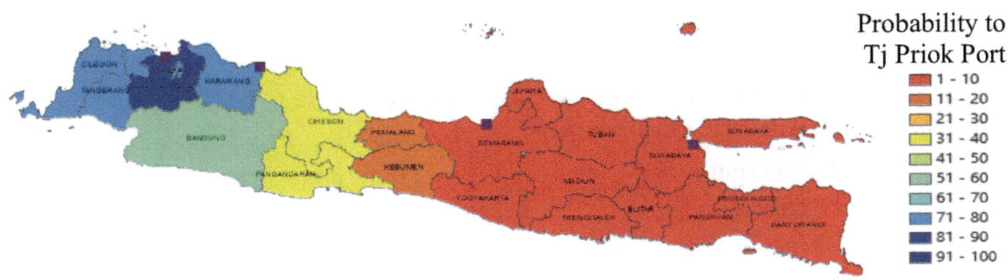

Fig. 2 Hinterland boundaries of Tanjung Priok port after Patimban port operated

3.2 *Elasticities*

The effect of changes in each variable on the probability of alternative could be expressed through the elasticity value. Elasticities can be calculated as percent changes of probability in response to one percent changes in the observed variables. Table 4 depicts the elasticities of probability choosing Tanjung Priok port as a service port and truck as the port hinterland transport mode from three zones, namely Jakarta, Cirebon, and Surabaya. Those three zones are selected to represent zones with high, medium, and low probability. Probability value close to 1 is assumed as high probability, and close to 0 is assumed as low probability, while medium probability is around 0.5.

Table 3 shows that the probability of choosing Tanjung Priok port by using truck mode for Jakarta area is inelastic demand (10% reduction of inland mode time cause only a 0.76% increase in probability). It means almost all containers from Jakarta will be choosing Tanjung Priok via road transport even if there are big changes in the observed variables. Unlike Jakarta, which has inelastic probability, the probability form Cirebon area is elastic (i.e., 10% increase in inland mode time could drop the probability of choosing Tanjung Priok port using truck by 9.30%). It means that the container demand from this area is potentially shifting to other ports if the port hinterland connectivity variables or the port performance are getting worse. For Surabaya, even though its probability elasticity is high, as the initial probability is low; hence the probability changes will not be very substantial. This result is in line with the derivative theory of MNL modelling. The impact of observed variable changes

Table 3 Elasticities of probability choosing Tanjung Priok port using Truck as inland mode respect to the one percent increase of observed variables in Jakarta, Cirebon, and Surabaya

	Jakarta	Cirebon	Surabaya
The probability to Tanjung Priok port using Truck	93.67%	43.78%	0.33%
Inland mode cost	−0.003	−0.055	−0.195
Inland mode time	−0.076	−0.930	−2.772
Inland mode GHG emission	−0.005	−0.693	−4.058
Inland mode reliability	0.098	0.868	1.539
Port cost	−0.029	−0.251	−0.467
Port ship call	0.035	0.046	0.157

Table 4 MNL simulation result: variation of probability demand as respond to transport policy scenario in Cirebon, Kebumen, and Trenggalek

Port	Inland mode	Probability (%)											
		Cirebon				Kebumen				Trenggalek			
		Scenario				Scenario				Scenario			
		0	1	2	3	0	1	2	3	0	1	2	3
Tj Priok	Truck	43.8	38.9	43.5	31.7	15.6	13.8	15.3	13.4	1.4	1.3	1.3	1.4
	Train	11.2	15.5	11.1	8.1	7.5	11.4	7.3	6.4	0.9	1.7	0.8	0.9
Tj Emas	Truck	33.5	29.8	33.2	24.2	59.0	52.1	57.7	50.4	32.8	29.9	29.7	32.3
	Train	7.4	10.4	7.4	5.4	5.0	7.6	4.9	4.3	2.5	4.0	2.2	2.4
Tj Perak	Truck	1.8	1.6	2.1	1.3	6.9	6.1	7.9	5.9	50.4	45.9	53.3	49.5
	Train	2.3	3.7	2.7	1.7	6.0	9.1	6.8	5.1	11.9	17.2	12.6	11.7
Patimban	Truck	0.0	0.0	0.0	22.9	0.0	0.0	0.0	9.7	0.0	0.0	0.0	1.0
	Train	0.0	0.0	0.0	4.7	0.0	0.0	0.0	4.9	0.0	0.0	0.0	0.7

is higher when the degree of uncertainty from the probability is high (P around 0.5). Meanwhile, the impact of changes will be diminished when the probability is almost certain (P close to 1 or 0) [11].

3.3 Transport Policies and Estimation Result

In this section, the multinomial logit (MNL) model will be used to estimate the probability of port and inland mode demand in Java. For this simulation, three zones with probability of around 50%, namely Cirebon, Kebumen, and Trenggalek will be examined. Further, the effect of two transport policiy scenarios and the operation of Patimban port on the port and inland mode demand will also be discussed. Scenario 0 represents the existing condition without transport policy implementation. The transport policy scenarios that will be simulated are:

- Scenario 1: improving the transshipment process of rail transport is expected to reduce train transport time by 20%.
- Scenario 2: investment in Tanjung Perak port is expected to double the port ship call.
- Scenario 3: Patimban port starts to operate.

The effect of these transport policy scenarios on port and inland mode demand is presented in Table 4. Improvement of rail transport as applied in scenario 1 is expected to encourage mode shifting from road to rail transport. Based on the simulation, reducing rail transport time by 20% has a high impact on boosting the demand for rail inland mode in all zones. However, the probability of using a train as the inland mode is still lower than road transport, indicating the low initial market share of rail transport. Comparing the base mode cost of road and rail transport; train is cheaper than truck. However, unlike trucks that offer cost efficiency as its cost only consider the cost from origin to destination (port), the cost component in rail transport requires an additional cost of loading and unloading process and the short-distance trucking cost that could scale up the total cost considerably. Moreover, inadequate law enforcement of overloading trucks also impacted the low market share of rail transport.

The second scenario is expanding the port performance, for example, by planning investment on Tanjung Perak port. By developing the Tanjung Perak port's infrastructure, it is expected that enlarging the pier will pull the number ship calls from 23 to 46. Referring to Table 5, the effect of increasing port performance positively impacts the port choice. The probability to Tanjung Perak port increases in all zones for both road and rail inland modes. It is quite explicit that by providing better performance on Tanjung Perak port, the shipper will be more attracted to shipping their containers through this port.

The last scenario is the operation of Patimban port which is expected to be fully operated in 2027. Cirebon has the highest port demand shifting from Tanjung Priok port to Patimban port, followed by Kebumen. In contrast, Trenggalek has slight port shifting from Tanjung Perak port to Patimban (about 1%). The high impact of Patimban port on Cirebon, which is located in West Java area is apparently due to close proximity to this port, and the assumption that both ports, Patimban and Tanjung Priok, are similar in port performance make the hinterland boundaries hazier (see Figs. 1 and 2).

4 Conclusion

This paper has demonstrated the use of a multinomial logit model to identify the port hinterland boundaries and the impact of changes in observed factors on the probability demand of port and inland mode. The transport time and reliability of inland mode are variables from the port hinterland connectivity category that greatly affects the joint demand of port and inland mode. At the same time, the number of port ship calls has the highest impact from the port performance category. From the simulation on several transport strategies, introducing Patimban port as a new container port in Java will significantly impact Tanjung Priok port's demand. Reducing rail transport time and improving port performance also impacted in encouraging port and mode shifting for container transport in Java.

Acknowledgements The authors wish to thank the Directorate of Higher Education, Ministry of Education and Culture, the Republic of Indonesia for financial support for this research. The authors also wish to thank the Universitas Sebelas Maret Surakarta for the support on this publication. The authors are also grateful to anonymous reviewers for their suggested improvements to this paper.

References

1. Wang X, Meng Q, Miao L (2016) Delimiting port hinterlands based on intermodal network flows: Model and algorithm. Transp Res Part E: Logistics Transpo Rev 88:32–51
2. Nugroho MT, Whiteing A, de Jong G (2016) Port and inland mode choice from the exporters' and forwarders' perspectives: case study—Java, Indonesia. Res Transp Bus Manag 19:73–82
3. Wang G et al (2016) Port connectivity in a logistic network: The case of Bohai Bay, China. Transp Res Part E: Logistics Transp Rev 95(C):341–354
4. Malchow MB, Kanafani A (2004) A disaggregate analysis of port selection. Transp Res Part E: Logistics Transp Rev 40(4):317–337
5. Zondag B et al (2010) Port competition modeling including maritime, port, and hinterland characteristics. Marit Policy Manag 37:179–194
6. Tang LC, Low J, Lam S-W (2011) Understanding port choice behavior—a network perspective. Netw Spat Econ 11:65–82
7. Tapia RJ et al (2020) Application of MDCEV to infrastructure planning in regional freight transport. Transp Res Part A: Policy Practice 133:255–271
8. Frazila R, Zukhruf F, Burhani J (2018) Developing a probabilistic model for constructing seaport hinterland boundaries. IOP Conf Series: Earth Environ Sci 158:012023
9. Zukhruf F, Frazila R, Burhani J (2017) A stochastic discrete optimization model for designing container terminal facilities 1903:060007.
10. Nugroho MT (2015) Stated preference study of port and inland mode choice for containerized export from java, in institute for transport studies. The University of Leeds, United Kingdom
11. Train K (2009) Discrete Choice Methods with Simulation. Cambridge University Press

Evaluation of Logistics System Performance-Based on Indonesian Government Policy

Muhammad Rizky Prakoso, Mohammed Ali Berawi, and Gunawan

Abstract Currently, Indonesia's logistics system has not been maximized, as evidenced by the high cost of logistics that account for about 23.5% of GDP nationally and still far from developed countries that are only 10% of GDP even less. This paper evaluates the performance of logistics systems in Indonesia based on government policy. The cost component that is a problem in Indonesia's logistics system will be mentioned in this paper based on journals, reports and related regulations. The data is taken by distributing questionnaires to the logistics delivery service providers. The data that has been obtained is analyzed using matrix paired comparison. This method is applied to rank the cost components presented. The result was that the logistics cost component that most affects the cost of logistics in Indonesia are the cost of sea and land transportation, due to too much tax and levy that needs to be spent and inadequate infrastructure.

Keywords Logistics system · Government policy · Logistics · Indonesia

1 Introduction

Indonesia is the largest island nation in the world, with more than 17,000 islands. Logistics distribution is a challenge for Indonesia, which is mostly territorial waters. The constraints of the national logistics system are the high cost of logistics in Indonesia. Currently, Indonesia's logistics system is under neighbouring countries such as Singapore, Malaysia and Vietnam. According to Logistic Parameter Index (LPI), data in 2018 shows that Indonesia is ranked 46th with logistics costs in Indonesia currently at 23.5% of GDP nationally [1, 2], while in developed countries, logistics costs are only 10% of GDP and even less [3].

M. R. Prakoso (✉) · M. A. Berawi · Gunawan
Civil Engineering Department, Faculty of Engineering, Universitas Indonesia, Depok, Indonesia
e-mail: muhammad.rizky98@ui.ac.id

M. A. Berawi
e-mail: maberawi@eng.ui.ac.id

S. A. Kristiawan et al. (eds.), *Proceedings of the 5th International Conference on Rehabilitation and Maintenance in Civil Engineering*, Lecture Notes in Civil Engineering 225,
https://doi.org/10.1007/978-981-16-9348-9_70

Logistics is defined as a process of planning, implementing, and monitoring efficiently and effectively the flow and storage of goods, services and related information from the point of origin to the destination point of goods to meet the user's need [4]. Logistics aims is to ensure the proper flow of finished goods and various materials in usable circumstances to the location where they are needed and at the lowest total cost [5]. High logistics costs impact the competitiveness of the domestic industrial sector with other countries' product and service. Moreover, it also causes people's purchasing power to decrease.

Inter-island logistics delivery has been mainly through ship due to its large size. It can accommodate a larger volume than other transportation, and shipping intra-island still relies on truck because of its advantages that can do door to door delivery from warehouse/factory to consumers. According to previous research, the performance of logistics systems at the port is still not optimal, high logistic costs also occur as long as the truck enters the port until it is loaded onto the ship and vice versa [6–10]. While the performance of logistics systems on land is still low, based on previous studies it is due to inadequate road infrastructure, causing increased logistics delivery costs [11].

The price disparity between regions is still relatively high, indicating that there are problems in the logistics system. Based on previous studies, improved connectivity is needed to reduce logistics costs. Although connectivity is closely related to infrastructure, government policy is necessary so that various relevant stakeholders can use existing infrastructure to be more optimal [12]. Therefore, this paper aims to evaluate the performance of the logistics system in Indonesia based on government policy.

2 Methodology

This paper will use a paired comparison analysis method. The Paired comparison method selects one of the two available options or which objects are more suitable or preferred according to the specified characteristics [13]. This method is applied to rank the cost components presented. The cost component as a variable in this study is taken from relevant reports, regulations and journals and obtained as follows: goods transportation operation permit cost (Biaya Izin Penyelenggaraan Angkutan Barang) [14, 15], Cross-district/city permit cost (Biaya Izin Lintas Antar Kabupaten) [16], port entry fees [16, 17], trucking costs [11], port service costs [11, 16, 18–21] and shipping costs [11, 12, 16]. The data was taken by sharing questionnaires to observe the highest cost components that led to Indonesia's high logistics cost. There was 30 sample in this study with a population of employees in the transportation and warehousing sector according to Central Bureau of Statistics, as much 2,296,109 people. The steps taken in this study are: first, conducting literature studies related to logistics systems in Indonesia based on reports, regulations and journals; second, determine the variables of the literature studies that have been conducted; third, the preparation of questionnaires; fourth, distribute questionnaires of expedition service

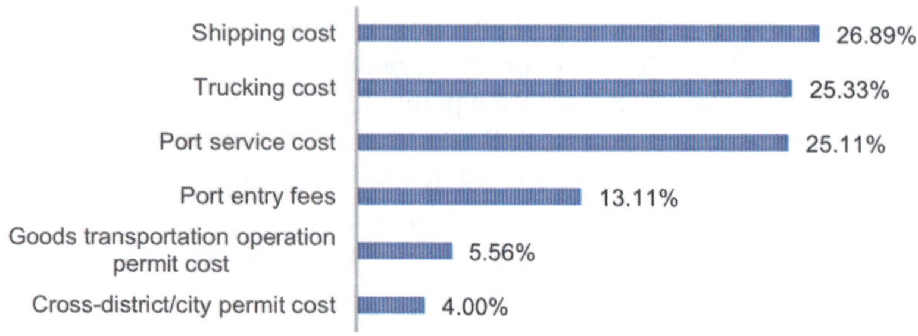

Fig. 1 Cost component ranking

companies/Freight forwarders and the like; lastly analyzed the data using matrix paired comparison and generated a ranking order for logistics costs.

3 Results and Discussion

3.1 Identification of Logistics Cost Components

Based on the variables that have been provided in the previous section, given six (6) cost components, namely: (1) goods transportation operation permit cost; (2) Cross-district/city permit cost; (3) port entry fees; (4) trucking cost; (5) port service cost; (6) shipping cost. Thirty respondents filled out this questionnaire, consisting of expedition services, freight forwarders and shipping freight expeditions that send logistics to all regions in Indonesia; they were asked to choose what costs are the most expensive of the two cost options presented. The result is provided in Fig. 1.

The picture above shows that the highest cost component ratings were the shipping cost after the analysis method paired comparison obtained. Shipping costs are an essential part of the shipping cost by sea or river. The high cost of shipping is caused by several factors that cause shipping companies to charge high fees, such as service fees to be paid by ships at the port and ship operating costs, followed by trucking cost, port service cost, port entry fees, goods transportation operation permit cost and Cross-district/city permit cost. Each cost component will be described in the next section.

3.2 Factors Affecting Logistics System Performance Based on Logistics Cost

The government plays a vital role in the performance of the logistics system through its policies and regulations. Logistics systems performance is still lacking attention from the government with proven high logistics costs. The previous section's data, shipping costs, trucking costs and dock service fee-fees are considered by logistics delivery service companies as the most expensive costs, followed by port entry fees, goods transportation operation permit cost and Cross-district/city permit cost. The costs are interrelated with each other. Here is a discussion on the logistics cost component based on government policy.

Shipping costs

Shipping costs are considered expensive due to some things such as service fees to be paid by ships at the port and ship operating costs. The results align with the previous study that shipping costs are the most expensive logistics delivery [22]. Services to be paid at the port by shipping companies are port dues, tug services, wharfage and pilotage services included in the Non-Tax State Revenue (PNBP) [17]. The longer the ship is in the port, the more handling done by the port, the cost of paying for services at the port will be expensive. So the cost of shipping is considered the most expensive cost, among others.

Trucking costs

Trucking costs become expensive due to several factors, including inadequate infrastructure, the existence of some areas that impose a ban on passing for container trucks so that it is necessary to move logistics transported to smaller trucks, as well as the presence of local governments that issue various permits and impose various levies [11, 23]. According to LPI 2018 data, Indonesia's infrastructure performance is still low with a score of 2.89, still below 3.00 where the infrastructure performance score in Indonesia is below other neighbouring countries such as Vietnam's infrastructure performance with a score of 3.01, Malaysia with a score of 3.15, and Singapore with a score of 4.06 [1]. Additionally, currently in Indonesia, most logistics shipments within the island use truck mode. This is because there is still no complete regulation on the concept of multimodal transportation. The regulation on multimodal transportation in Government Regulation No. 8 of 2011 has not regulated the procedures and documents for goods that switch modes of transportation.

Port service cost

Port service cost consists of services for trucks at the port and services for ships. The service cost for the ship has been explained in advance regarding the high cost of the ship. According to the Ministry of Transportation Directorate General of Sea Transportation, port services for trucks consist of three groups of costs; namely, the cost of Loading Port Fee (LPF)/Destination Port Fee (DPF) Stevedoring, LPF/DPF Cargodoring cost, and LPF/DPF Receiving/Delivery fee. LPF/DPF Stevedoring costs are costs incurred for loading and unloading services of goods/containers from the hatch/deck to the dock/truck/barge or loading from the dock/barge/truck on board by using a ship crane. If interference with the ship's crane is not used, it uses a mechanical device owned by the port and is charged a certain fee above the stevedoring tariff. LPF/DPF Cargodoring costs are costs incurred for the work of removing cargo from the network (sling /ex tackle) above the dock, transporting from the dock, transporting to the warehouse and arranging in the warehouse or vice versa. LPF/DPF Receiving/Delivery fee is the cost charged for taking cargo from the warehouse and delivering it to the top of the vehicle docked in the warehouse or vice versa until making its preparation in the warehouse. The number of services to be paid at the port makes the cost of port services considered expensive. However, the Republic of Indonesia has just issued Presidential Instruction No. 5 of 2020 on Structuring the National Logistics Ecosystem. The Existence of this Presidential Instruction will simplify the process of activities at the port that will shorten the time at the port, but whether it can lower the costs that service users must incur.

Port entry fee

Trucks entering the port must pay the entrance fee as Non-Tax State Revenue (PNBP) following Government Regulation No. 15 of 2016. This rate is paid each time service user enter the port. If the service user sends several logistics and requires in and out of the port, the costs incurred to pay the port entrance fee will increase.

Goods transportation operation permit cost

According to the regulation of Law No. 22 of 2009 concerning Traffic and Road Transport, the permit for the implementation of freight transport is issued by the Minister responsible for traffic facilities and infrastructure and road transport. This permit must be owned by the public transport company that organizes people's transportation and goods.

Cross-district/city permit cost

Cross-District permit fees are charged for the delivery of goods whose functional area exceeds provincial limits. An example is the cost of escorting at a particular location to ensure that the goods transported are avoided from criminality and the cost of passing through the City district that applies the rules prohibited crossing at certain hours [11, 14].

If logistics costs can be lowered, it can reduce the disparity in the price of products/services between regions. The purchasing power of the community will be the price of products/services offered is higher. Therefore, the Indonesian government must make logistics costs fall for long-term economic development purposes. There is an evaluation of factors causing high logistics costs, it is expected that further research can model the development of existing national logistics systems to lower logistics costs.

4 Conclusion

This paper aims to provide input to the Government of Indonesia on logistics systems in Indonesia based on government policy. The logistics system's evaluation improves Indonesia's logistics system, especially by paying attention to the existing logistics costs. The questionnaire results with paired comparison analysis method can be seen as the ranking of logistics cost components considered expensive by logistics delivery businesses in Indonesia. From the study literature and the results of questionnaires with paired comparison analysis methods can be seen that the ranking of logistics cost components that are considered expensive by logistics delivery business in Indonesia, namely shipping costs, related to expensive port services such as port dues, tug services, wharfage and pilotage services; and trucking costs due to several factors such as inadequate infrastructure, prohibition of crossing by local regulations and the number of permits and levies, followed by port service cost; port entry fee; Goods transportation operation permit cost, and Cross-district/city permit cost. The paper's results provide direction for the Indonesian government to pay more attention to the logistics cost components described above to improve the logistics system in Indonesia.

Acknowledgements The authors would like to thank Universitas Indonesia for the support given to this research.

References

1. The World Bank (2018) LPI aggregated rankings. In: World Bank. https://lpi.worldbank.org/international/global. Accessed 14 Sep 2020
2. Kemenkeu (2020) National logistic ecosystem, Upaya Indonesia Perbaiki Daya Saing Logistik. In: Kementeri. Keuang. Republik Indones
3. Zhao X, Tang Q (2009) Analysis and strategy of the chinese logistics cost reduction. Int J Bus Manag 4:188–191. https://doi.org/10.5539/ijbm.v4n4p188
4. Council of Logistics Management (1991) Careers in logistics. Oak Brook, IL 60521, 2803 Butterfield Road, Suite 380
5. Agustini ED, Yarlina L, Murtadho A (2015) Effectiveness and efficiency of air transport logistic in soekarno-hatta international airport. War Ardhia 41:97. https://doi.org/10.25104/wa.v41i2.148.97-110
6. Iswanto (2015) Simplifikasi Sistem dan Prosedur di Pelabuhan dapat mengurangi celah penyimpangan dan Ekonomi Biaya Tinggi. XIII:1–20
7. Kadarisman M, Suharto Y, Majid A (2016) Policy formulation of sea transportation system. J Manaj Transp Logistik 03:161–183
8. Rumambi CC, Sendouw TK, Manoppo MRE (2016) Evaluasi kinerja operasional pelabuhan bitung. 14:0215–9617
9. Prasetyo SA, Wicaksono A, Anwar MR et al (2014) Evaluasi Sistem Logistik Di Pelabuhan Tanjung Perak Surabaya 8:113–120
10. Sirajuddin, Sunaryo, Zagloel TY (2018) Effect of concession model and deregulation logistics policy for increasing port customer satisfaction in Indonesia. MATEC Web Conf 204. https://doi.org/10.1051/matecconf/201820406008
11. Kemendag (2014) Kajian Kinerja Logistik Perdagangan Antar Pulau: Studi Kasus Baja
12. Sandee H (2016) Improving connectivity in Indonesia: the challenges of better infrastructure, better regulations, and better coordination. Asian Econ Policy Rev 11:222–238. https://doi.org/10.1111/aepr.12138
13. Lawless HT, Heymann H (1999) Sensory evaluation of food: principles and praactices, 2nd edn. Kluwer Academic/Plenum Publisher, New York

14. Safitri RD (2020) Implementation of samarinda mayor regulation No. 23 of 2010 concerning The determination of freight transportation tracks in Samarinda city area against containers. J Law 1:86–101
15. Undang-Undang RI (2009) Undang-Undang Republik Indonesia Nomor 22 Tahun 2009 Tentang Lalu Lintas dan Angkutan Jalan
16. Kemendag (2013) Kajian Pengembangan Indikator Kinerja Logistik Indonesia
17. Peraturan Pemerintah RI (2016) Peraturan Pemerintah Republik Indonesia Nomor 15 Tahun 2016 Tentang Jenis Dan Tarif Atas Jenis Penerimaan Negara Bukan Pajak Yang Berlaku Pada Kementerian Perhubungan
18. Brooks MR, Frost JD (2004) Short sea shipping: a Canadian perspective. Marit Policy Manag 31:393–407. https://doi.org/10.1080/0308883042000304881
19. Miftach Z, Pasek P (2016) Analysis of Port Costs and Loading and Unloading Costs on Income PT. Abdi Nusantara Indonesia Line branch of Gresik. Surabaya
20. Direktorat Jenderal Perhubungan Laut K (2020) Penyelenggaraan dan Pengusahaan Bongkar Muat Barang di Pelabuhan. Jakarta, Indonesia
21. Peraturan Pemerintah RI (2010) Peraturan Pemerintah Republik Indonesia Nomor 20 Tahun 2010 Tentang Angkutan di Perairan
22. Nascimento DLM, Alencastro V, Quelhas OLG et al (2019) Exploring Industry 4.0 technologies to enable circular economy practices in a manufacturing context: a business model proposal. J Manuf Technol Manag 30:607–627. https://doi.org/10.1108/JMTM-03-2018-0071
23. Mulyono AT (2020) Implementation of national transportation infrastructure. The role and challenges of future development. Jakarta

Analysis of the Influence of Region Development Factors, Individual and Activity, Internal Operator and External Operator on the Demand for the Jakarta Bandung High-Speed Rail

Samijan, Mohammed Ali Berawi, and Andyka Kusuma

Abstract For the last few decades the High-Speed Rail has been growing quite rapidly in various parts of the world. However, many High-Speed Rail Projects were not as successful as expected. Most of the contributing factors were over-estimate demand and cost-overrun. This study aims to formulate a strategy in order to increase the demand for the Jakarta-Bandung High-Speed Rail by investigating the factors of region development, individuals and activities, internal operators and external operators. In conducting this research, the primary data collection was done by using a stated preference questionnaire and the data analysis used the structural equation modeling (SEM) method with the help of Lisrel software. The analysis results concluded that the strong factors influencing the demand for the Jakarta-Bandung High-Speed Rail are region development and external operators. This result is useful as a reference for planners in planning the next similar project.

Keywords High-speed rail · Demand · Region development · Individuals and activities · Internal operators · External operators · Structural equation modeling (SEM)

1 Introduction

During the period of half a century, the development of the High-Speed Rail in the world was quite rapid. Since Japan introduced the world's first High-Speed Rail in 1964 with its Shinkansen [1, 2], many European countries such as France in 1981

Samijan (✉) · M. A. Berawi · A. Kusuma
Department of Civil and Environmental Engineering, Faculty of Engineering, Universitas Indonesia, Depok, Indonesia
e-mail: samijan@ui.ac.id

A. Kusuma
e-mail: andyka.k@ui.ac.id

M. A. Berawi
Faculty of Engineering, Center for Sustainable Infrastructure Development, Universitas Indonesia, Depok, Indonesia

S. A. Kristiawan et al. (eds.), *Proceedings of the 5th International Conference on Rehabilitation and Maintenance in Civil Engineering*, Lecture Notes in Civil Engineering 225,
https://doi.org/10.1007/978-981-16-9348-9_71

[1, 2] and Germany (1988–1998) [2] had followed suit. The next countries were Italy, Spain, the Netherlands, Belgium, South Korea, Taiwan, China, England, Turkey and Saudi Arabia [2]. Whereas in Southeast Asia the first High-Speed Rail to be built was in Indonesia, namely the Jakarta-Bandung corridor, whose construction began in 2019 [3].

However, not all of the High-Speed Rail Projects were successful, even many of which were just wishful in their success, due to cost overruns and demand overestimation [4]. Demand is the key factor in the High-Speed Rail Project so that a very large investment is economically feasible [4–6]. This paper aims to investigate four factors that have the potential to affect the demand for the Jakarta-Bandung Indonesian High-Speed Rail. The four factors are region development, individuals and activities, internal operators, and external operators. Those factors were selected due to refer to the literature review below. The Jakarta-Bandung High-Speed Rail was chosen as the case because there were several controversies related to investment break even and demand [7].

2 Literature Review

The construction cost of the High-Speed Rail infrastructure is very high because there are many underground constructions [8], bridges, overpasses, and underpasses to avoid level meeting with other modes of transportation. Thus, high cost of construction needs a large income in order to achieve a breakeven and be economically viable. Passenger volume is the key factor for investment in High-Speed Rail [6, 8]. Demand levels below 10 million per year, such as Madrid-Seville and Madrid-Barcelona in Spain, shows a surprising fact in terms of return of capital or break even [6].

There is a relationship between region development factors such as transit-oriented development with High-Speed Rail passenger demand [5, 9–13]. Meanwhile, high-rise buildings and mixed-use are also the key factors to improve demand for transportation passengers [13].

Individual factors such as income [14, 15], activity [16], the purpose of the trip [14, 17] and the number of members or cohort [14, 17, 18] can influence the choice of mode and transportation demand decisions.

Internal operator factors that affect the demand for High-Speed Rail are usually related to service levels [19]. The variables cost, travel time and distance have a significant effect on the choice of transportation modes [9]. The fare per mile and the time spent between the origin and the destination also influence demand [20]. The utility function is defined as the attribute level of service: travel cost, the different components of total travel time and service frequency [21]. The selection of alternative time and cost options is very significant for transport users [22]. The security factors of the station also turns out to be a barrier for people to use the train [5, 23]. Other factors that affect demand are safety [5], comfort [21, 23], frequency [15, 24] and reliability [5].

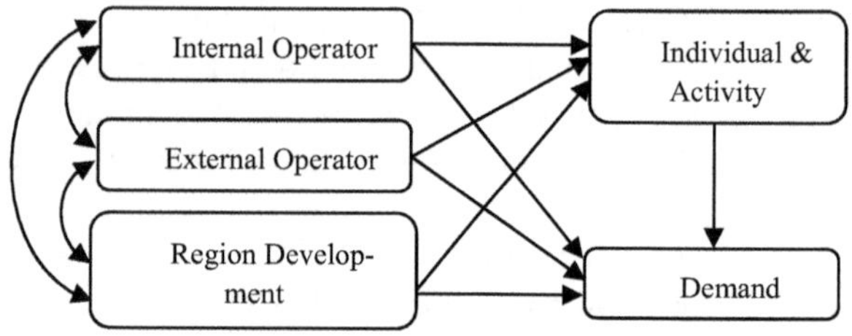

Fig. 1 The design SEM model

Accessibility was found as a significant factor that can induce higher mobility [18, 25]. The time consumed to get to and out of stations and the lack of integration of local transportation also greatly reduce the number of passengers on the High-Speed Rail [26]. People using trains depend on the quality of integration into local transport, and the number of passengers is predominantly constrained by accessibility [27]. The availability of parking facilities and shuttle transportation can also affect people in using public transportation modes.

In England, a friendly environment is one of the variables and strategies to attract passengers [23]. But this conclusion did not explain what the environmental specifications were meant to be.

Based on the literature review above, the author made a *structural equation model* (SEM) hypothesis as shown in the following Fig. 1. Of the five latent variables, three of them are independent latent variables (exogenous), while the other two are dependent variables (endogenous). Internal operator, external operator and region development are exogenous latent variables. Whereas, individual and activity and demand are endogenous latent variables.

3 Methodology

The primary data collection was carried out by means of a questionnaire survey, both offline and online. Selected respondents were the regular users of several modes of transportation to the Jakarta-Bandung route and vice versa, namely the users of the Argo Parahyangan executive train, executive travel cars, executive buses and private car users. These respondents are potential because they use these modes routinely so that they can be encouraged to shift to High-Speed Rail. The primary data collection was carried out from November 24 to December 18, 2020, on weekdays and weekends.

The data analysis was performed by structural equation modeling (SEM) and using Lisrel 8.7 software. The reason for using SEM is because SEM has several advantages, such as using confirmatory factor analysis (CFA), capability for test several dependent variables, model modification feature, etc. [28].

4 Results and Discussion

The data were collected from the questionnaire survey with a total number of 212 respondents. From the data collected, many respondents did not fill in questions about personal data such as occupation, work location and especially income. This perhaps because the questions were too personal. As a result, Lisrel could not execute those data. So those indicators were excluded from the model. Then the individual and activity latent variable was renamed into group activity to match the indicators that compose them, namely the number of members in group activities. Table 1 shows the indicators (manifest variables) of each latent variable.

The final result of running the above model on Lisrel 8.7 is shown in Fig. 2. Then performed tests the validity of the instrument, construct reliability (CR) and variance extracted (AVE). The validity of the instrument was measured by the value of λ (standardized loading factor) > 0.5, CR ≥ 0.5 and AVE ≥ 0.5 (Hair et al., 2010 in Refs. [29, 30]).

Table 2 shows the results of these measurements, in which CR and AVE are calculated using the formulas (1) and (2) below [29]. The formula (1) was also stated by Bagozzi and Yi [30]. Where, λ is the loading factor of each indicator for each latent construct, n is the number of indicators for each latent construct, and ε is the error of each indicator. So it can be concluded that all instruments are valid.

$$CR = \frac{\left(\sum_{i=1}^{n} \lambda_{yi}\right)^2}{\left(\sum_{i=1}^{n} \lambda_{yi}\right)^2 + \left(\sum_{i=1}^{n} \epsilon_i\right)} \tag{1}$$

$$AVE = \frac{\sum_{i=1}^{n} \lambda_i^2}{n} \tag{2}$$

The next test is goodness of fit (GOF) to the whole model. The result can be seen within the output of Lisrel, written in Table 3. Most of the tests met the thresholds so that the model was concluded fit.

Table 1 Latent variables and indicators

Latents variables	Indicators
Internal operator (INT)	Security, safety, amenity, cleanliness, friendliness, frequency, simplicity and reliability
External operator (EXT)	Accessibility, availability of parking facilities (P&R), availability of shuttle transportation, Integration of the transportation and the environment
Region development (REG)	New city, Transit-Oriented Development (TOD)
Group activity (GRO)	Group 1 (with 2 members or more), Group 2 (with 4 members or more)
Demand (DEM)	New region, TOD1, TOD2, TOD3, TOD4

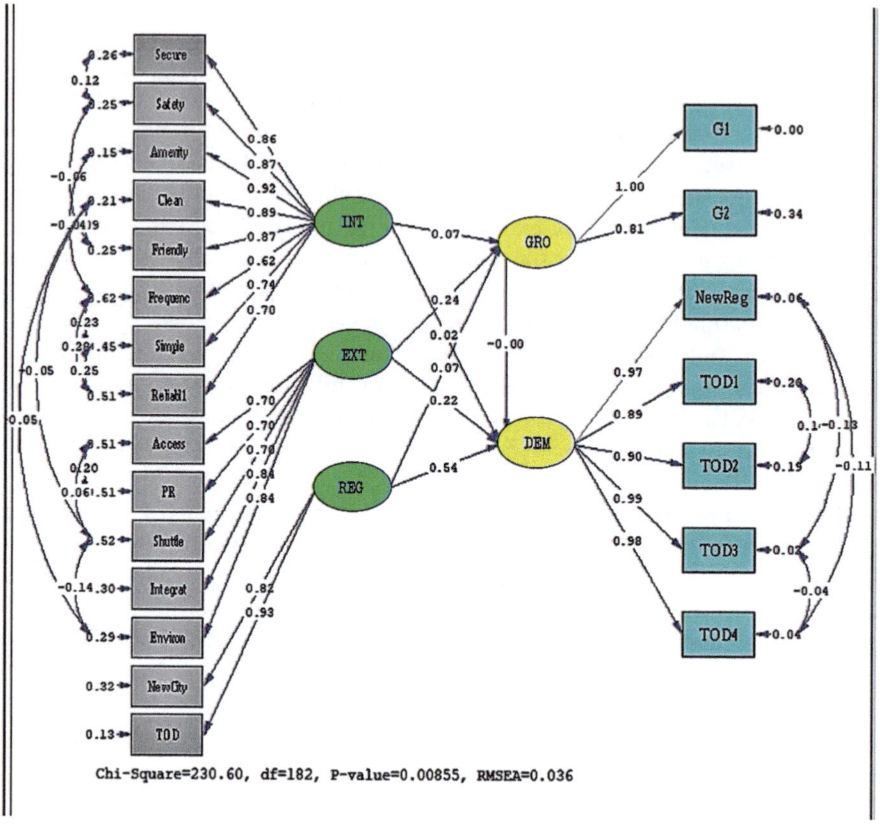

Fig. 2 The result of running Lisrel 8.7

Table 2 λ, CR and AVE

Latent variable	λ		CR		AVE		Conclusion
	Criteria	Result	Criteria	Result	Criteria	Result	
INT	>0.5	min 0.62	≥0.5	0.939	≥0.5	0.664	Reliable
EXT	>0.5	min 0.70	≥0.5	0.870	≥0.5	0.603	Reliable
REG	>0.5	min 0.82	≥0.5	0.872	≥0.5	0.769	Reliable
GRO	>0.5	min 0.81	≥0.5	0.906	≥0.5	0.828	Reliable
DEM	>0.5	min 0.89	≥0.5	0.978	≥0.5	0.897	Reliable

The significance test of the structural relationship among latent variables is measured by the T-value. It is considered significant if the T-value is more than 1.96 [29]. T-value describes the value of γ which is the level of relationship between exogenous latent variables and endogenous latent variables or shows the value of β namely the relationship between endogenous latent variable with one another. The

Table 3 Goodness of fit test

Test	Threshold	Result	Conclusion
Chi-square/df	≤2 [29]	1.267	Fit
P-value	>0.05 [31]	0.00855	Less fit
RMSEA	≤0.05 [31]	0.036	Fit
NFI	>0.9 [29]	0.98	Fit
NNFI	>0.9 [29]	0.99	Fit
PNFI	>0.5 [29]	0.77	Fit
CFI	≥0.95 [30]	0.99	Fit
IFI	>0.9 (Bollen in [29])	0.99	Fit
RFI	>0.9 [29]	0.97	Fit
CN	>200 (Hoetler in [29])	200.12	Fit

output structural model of Lisrel 8.7 can be seen in Fig. 3. It appears that there are three significant structural relationships, namely the relationship between EXT to GRO, EXT to DEM and REG to DEM, with T-values are 2.11, 2.56 and 7.95, respectively.

The structural equation coefficient can be seen in Figs. 4 and 5. The significant direct effect coefficients, according to the T-value test above, are 0.24 (EXT against GRO) with the standard error is 0.11, the error variance is 0.89 and the determination coefficient R^2 is 0.11; 0.22 (EXT against DEM) with the standard error is 0.086, the error variance is 0.54 and the determination coefficient R^2 is 0.46; and 0.54 (REG against DEM) with the standard error is 0.068, the error variance is 0.54 and the determination coefficient R^2 is 0.46. Meanwhile, the other coefficients are not significant. Based on the result, there is no significant indirect effect. The direct effect, the indirect effect and the total effect among latent variables can be seen more easily in Table 4.

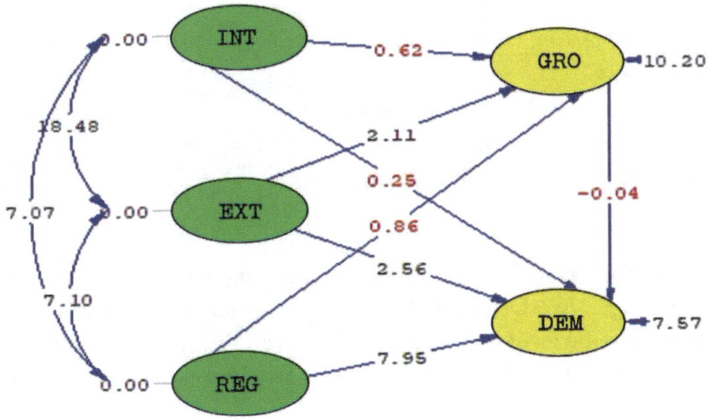

Fig. 3 "T-value" mode, output of Lisrel 8.7

```
Structural Equations


GRO = 0.067*INT + 0.24*EXT + 0.069*REG, Errorvar.= 0.89  , R² = 0.11
      (0.11)      (0.11)     (0.080)                (0.088)
      0.62        2.11       0.86                   10.20

DEM =  - 0.0019*GRO + 0.020*INT + 0.22*EXT + 0.54*REG, Errorvar.= 0.54  , R² = 0.46
         (0.052)      (0.081)     (0.086)    (0.068)              (0.071)

         -0.037       0.25        2.56       7.95                 7.57
```

Fig. 4 Structural equations, output of Lisrel 8.7

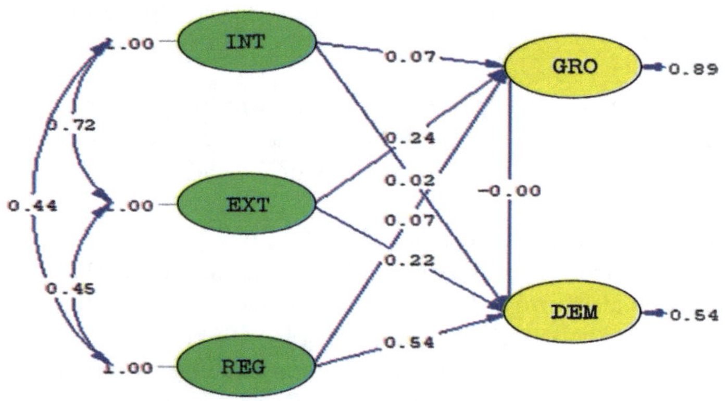

Fig. 5 The "structural model" mode, output of Lisrel 8.7

Table 4 Direct effect, indirect effect and total effect

Path	Direct effect	Indirect effect through GRO	Total effect	Remark
INT → GRO	0.07	0	0.07	Not significant
INT → DEM	0.02	$0.07 \times (-0.00) = 0$	0.02	Not significant
EXT → GRO	0.24	0	0.24	Significant
EXT → DEM	0.22	$0.24 \times (-0.00) = 0$	0.22	Significant
REG → GRO	0.07	0	0.07	Not significant
REG → DEM	0.54	$0.07 \times (-0.00) = 0$	0.54	Significant

Table 5 shows the significant influences among the latent variables, their indicators, and their respective statistical attributes. The consecutively of influence coefficients starting from the most powerful variables are TOD3, TOD4, new region, TOD, TOD2, TOD1, environment, integration, new city, accessibility, P&R and shuttle transit.

Table 5 Significant variables and statistical attributes

Variables	Coefficient	Standard error	T-value	Error variance	R^2
External	0.22	0.086	2.56	0.540	0.46
Accessibility	0.70	0.028	11.11	0.100	0.49
P&R	0.70	0.029	11.23	0.110	0.49
Shuttle transit	0.70	0.029	10.95	0.110	0.48
Integration	0.84	0.029	14.57	0.077	0.70
Environment	0.84	0.031	14.42	0.081	0.71
Region development	0.54	0.068	7.95	0.540	0.46
New city	0.82	0.034	13.52	0.100	0.68
TOD	0.93	0.035	15.91	0.046	0.87
Demand					
New region	0.97			0.036	0.94
TOD1	0.89	0.035	19.32	0.120	0.80
TOD2	0.90	0.033	19.57	0.100	0.81
TOD3	0.99	0.033	22.27	0.012	0.98
TOD4	0.98	0.030	23.46	0.020	0.96

According to the questionnaire, TOD1 describes respondents living at TOD Karawang and working in Jakarta and no shuttle transit facility, and TOD2 is vice versa. And then TOD3 describes respondents living at TOD Karawang and working in Jakarta and there is shuttle transit facility, and TOD4 is vice versa. While TOD shows a residential and business area that is built with a transit-oriented concept, as referred to in the definition of TOD in general. Although EXT has a significant effect on GRO (see Table 4), but GRO has no effect on DEM (see Fig. 5), then in Table 5 the relationship between the influence of the EXT variable on the GRO variable is not included.

In practice in the real world, it makes sense that a fast train-based TOD which has shuttle facilities as a means of transportation between stations and workplaces has the high influence on the demand because of the many advantages such as saving time, fast and convenience. Meanwhile, TODs without shuttle facilities have lower influence. The concept of region development such as new city and TOD in general also has a strong direct effect on demand. And the external operator factors namely environment, integrated transportation, shuttle transportation, parking facilities and accessibility have a strong influence on the demand and on the group activity. Yet, because the group activity has a very little effect on the demand so the indirect effect of the external operator on the demand through the group activity factor can be ignored. In general, the influence of the external operator factors on the demand is lower than the region development factors.

Thus, the accepted hypothesis is that the external operator factors (with its indicators) and the region development factors (with its indicators) strongly influence the demand. The influence of internal operator on demand is very small (coefficient is

0.02). With this result, to increase the demand of the Jakarta-Bandung High-Speed Rail, the first strategy that must be taken is to promote region development around the High-Speed Rail stations by building TODs and or new cities and then equipping shuttle facilities. The second strategy is building access infrastructure to the High-Speed Rail station from the surrounding areas, providing park & ride facility, providing shuttle transit and integrating transportation system with fast, safe, cheap and comfortable facilities. The road to the station must smooth and free from congestion, with the result that reduces the resistance for people to drive to the station and park their vehicles at the parking lot close the station and then take the High Speed Train to their destination.

5 Conclusion

The analysis result shows that there are two factors that are strong enough to influence the demand for the Jakarta-Bandung High-Speed Rail, namely external operator factors and region development factors. The external operator factors consist of access to and from the High-Speed Rail Station, the availability of parking facilities close to the station, the availability of shuttle transportation, the integration of the High-Speed Rail with other transportation systems, and the environment (minimal CO_2 emissions). And region development factors include TODs and new cities or new areas. TODs which have shuttle transit from/to High-Speed Rail Station have very powerful impact.

The strategies that must be taken in order to increase the demand for the Jakarta-Bandung High-Speed Rail are by promoting the development of region around the fast train stations namelyTODs, new cities, new business centers and equipping its with shuttle transit; building access infrastructure to and out of stations, providing parking facilities close to the stations, providing shuttle transportation between residential and activity centers and stations, and connecting other transportation modes with the fast train mode.

Acknowledgements We will appreciate editors and reviewers very much for positive and constructive comments and suggestions for our paper.

References

1. Givoni M, Dobruszkes F (2013) A review of ex-post evidence for mode substitution and induced demand following the introduction of high-speed rail. Transp Rev 33(6):720–742. https://doi.org/10.1080/01441647.2013.853707
2. Palacin R, Raif L, Deniz Ö, Yan N (2014) High speed rail trends, technologies and operational patterns: a comparison of established and emerging networks. Transp Probl 9:123–129
3. Anwar MC (2019) Akhir 2019, Mimpi RI Punya Kereta Cepat Bisa Separuh Terwujud. www.cnbcindonesia.com

4. Beria P, Albalate D, Bel G (2018) Delusions of success: costs and demand of high-speed rail in Italy and Spain. Transp Policy 68:63–79. https://doi.org/10.1016/j.tranpol.2018.03.011
5. Blainey S, Hickford A, Preston J (2012) Barriers to passenger rail use: a review of the evidence. Transp Rev 32(6):675–696. https://doi.org/10.1080/01441647.2012.743489
6. Preston J, Albalate D, Bel G (2013) The economics of investment in high speed rail. Public Adm Rev 2013–30(3):336–349. https://doi.org/10.1111/j.1540-6210.2011.02492.x
7. Afrianto D (2016) Pembangunan Kereta Cepat Diprediksi Timbulkan Kerugian. economy.okezone.com. https://economy.okezone.com/read/2016/02/02/320/1302855/pembangunan-kereta-cepat-diprediksi-timbulkan-kerugian. Accessed 02 Feb 2016
8. Nash C (2014) When to invest in high-speed rail, pp 45–72. https://doi.org/10.1787/9789282107751-3-en
9. Asensio J (2002) Barcelona's CBD. Urban Stud 39(10):1881–1895. https://doi.org/10.1080/0042098022000000300
10. Loo BPY, Chen C, Chan ETH (2010) Rail-based transit-oriented development: lessons from New York City and Hong Kong. Landsc Urban Plan 97(3):202–212. https://doi.org/10.1016/j.landurbplan.2010.06.002
11. Pan H, Li J, Shen Q, Shi C (2017) What determines rail transit passenger volume? Implications for transit oriented development planning. Transp Res Part D Transp Environ 57(September):52–63. https://doi.org/10.1016/j.trd.2017.09.016
12. Duncan M (2019) Would the replacement of park-and-ride facilities with transit-oriented development reduce vehicle kilometers traveled in an auto-oriented US region? Transp Policy 81(August 2017):293–301. https://doi.org/10.1016/j.tranpol.2017.12.005
13. Nyunt KTK, Wongchavalidkul N (2020) Evaluation of relationships between ridership demand and transit-oriented development (TOD) indicators focused on land use density, diversity, and accessibility: a case study of existing metro stations in Bangkok. Urban Rail Transit 6(1):56–70. https://doi.org/10.1007/s40864-019-00122-2
14. Ranjbari A, Chiu YC, Hickman M (2017) Exploring factors affecting demand for possible future intercity transit options. Public Transp 9(1–2):463–481. https://doi.org/10.1007/s12469-017-0161-3
15. Yunpeng Z, Zhuojun W, Xuemeng G (2016) Study on the influencing factors of passengers' demand for high-speed rail. Proc Third Int Symp Manage Innov Dev Bks One Two, 29:648–653. https://doi.org/10.5503/J.ME.2017.29.013
16. Rui F, Wei W, Min Y (2010) Analysis of individuals' activity-travel behavior based on a structural equation model. In: Proceedings of 2010 International conference on optoelectronics and image processing, ICOIP 2010, vol. 1, pp. 247–250. https://doi.org/10.1109/ICOIP.2010.100
17. Tyrinopoulos Y, Antoniou C (2013) Factors affecting modal choice in urban mobility. Eur Transp Res Rev 5(1):27–39. https://doi.org/10.1007/s12544-012-0088-3
18. Weis C, Axhausen KW (2009) Induced travel demand: evidence from a pseudo panel data based structural equations model. Res Transp Econ 25(1):8–18. https://doi.org/10.1016/j.retrec.2009.08.007
19. Zhou Z, Wang W, Hu Q (2015) An application of hierarchical structure model for trip mode choice forecasting in China. Math Probl Eng 2015. https://doi.org/10.1155/2015/925963
20. Iseki H, Liu C, Knaap G (2018) The determinants of travel demand between rail stations: a direct transit demand model using multilevel analysis for the Washington D.C. Metrorail system. Transp Res Part A Policy Pract 116(March):635–649. https://doi.org/10.1016/j.tra.2018.06.011
21. Román C, Martín JC (2010) Potential demand for new high speed rail services in high dense air transport corridors. Int J Sustain Dev Plan 5(2):114–129. https://doi.org/10.2495/SDP-V5-N2-114-129
22. Liu R, Li A (2012) Forecasting high-speed rail ridership using a simultaneous modeling approach. Transp Plan Technol 35(5):577–590. https://doi.org/10.1080/03081060.2012.701816

23. Harvey J, Thorpe N, Caygill M, Namdeo A (2014) Public attitudes to and perceptions of high speed rail in the UK. Transp Policy 36:70–78. https://doi.org/10.1016/j.tranpol.2014.07.008
24. Diao M, Zhu Y, Zhu J (2017) Intra-city access to inter-city transport nodes: the implications of high-speed-rail station locations for the urban development of Chinese cities. Urban Stud 54(10):2249–2267. https://doi.org/10.1177/0042098016646686
25. Chang J, Lee JH (2008) Accessibility analysis of Korean high-speed rail: a case study of the Seoul metropolitan area. Transp Rev 28(1):87–103. https://doi.org/10.1080/01441640701421495
26. Wong B, Habib KMN (2015) Effects of accessibility to the transit stations on intercity travel mode choices in contexts of high speed rail in the Windsor-Quebec corridor in Canada. Can J Civ Eng 42(11):930–939. https://doi.org/10.1139/cjce-2014-0493
27. Shen Y, Zhao J (2017) Capacity constrained accessibility of high-speed rail. Transportation (Amst) 44(2):395–422. https://doi.org/10.1007/s11116-015-9660-8
28. Sarwono Y (2010) Pengertian Dasar structural equation modeling (SEM). J Ilm Manaj Bisnis Ukrida 10(3):98528
29. Riadi E (2018) Statistik SEM structural equation modeling. ANDI
30. Bagozzi RP, Yi Y (2012) Specification, evaluation, and interpretation of structural equation models. J Acad Mark Sci 40(1):8–34. https://doi.org/10.1007/s11747-011-0278-x
31. Barrett P (2007) Structural equation modelling: adjudging model fit. Pers Individ Dif 42(5):815–824. https://doi.org/10.1016/j.paid.2006.09.018

Investigating the Factors Influencing the Demand of School Bus

Willy Kriswardhana, Syamsul Arifin, and Ainal Akbar

Abstract Mode choice to school is frequently influenced by both student and parent responses. However, the school bus is not a popular mode of school trip in Indonesia. Escalating the demand for the school bus is essential as reducing the use of the private motorised vehicle is a must. A stated preference survey and binary logistic regression method were performed to understand which factor influencing the willingness to ride the school bus. This study aims to investigate which characteristic of both children and parents significantly influence the school bus ridership. Factors that significantly influence students' willingness to ride school buses are gender, car ownership, and travel time to school. Furthermore, in terms of parents' permission, household factors, namely car ownership and income are significant. Direction for future study is presented.

Keywords School bus · Demand · Influencing factors · Stated preference

1 Introduction

The school bus is not a popular mode of school trip in Indonesia. Escalating the demand for the school bus is essential as reducing the use of the private motorised vehicle (for example, 2-wheeled scooter, which is very popular in Indonesia) is a must. In the United States, children coming from no vehicles owned household tend to use the school bus, while those from households with two vehicles are associated with a lower likelihood of riding a school bus for school travel [1]. Furthermore, a study mentioned that a good service scheduled bus could provide a proper service for school travels [2]. A study explained that if the walking conditions are unsafe, the

---

W. Kriswardhana (✉) · S. Arifin · A. Akbar
Universitas Jember, Jember 68121, Indonesia
e-mail: willy.teknik@unej.ac.id

S. Arifin
e-mail: syamsul.teknik@unej.ac.id

S. A. Kristiawan et al. (eds.), *Proceedings of the 5th International Conference on Rehabilitation and Maintenance in Civil Engineering*, Lecture Notes in Civil Engineering 225,
https://doi.org/10.1007/978-981-16-9348-9_72

use of the bus is essential. Nevertheless, the initial cost of operating the school bus is not cheap [3]. Moreover, the demand for the school bus in Indonesia is declining.

Mode choice to school is frequently influenced by both student and parent responses. Some studies have identified several significant factors related to travel modes for school trips [4]. Socio-demographic factors included age and gender, influence the school mode choice [5]. Meanwhile, parent's attitude is the strongest factor influencing the transport mode of school preferences for children [6]. The leading factor considered important to parents in choosing school trips is safety and security [7]. Furthermore, distance also plays a role as a consideration in modal choice for school. A study in Helsinki, Finland, implied that walking and cycling to school are favourable if the travel distance is less than 1 km. On the contrary, students choose to use or ride motorised vehicles when the distance exceeds 2 or 3 km [3, 8]. However, some studies stated that travel distance does not influence the mode choice of the school trip. For instance, variables of the built environment could affect the willingness to walk [4, 9]. When the distance exceeds the students' willingness to walk or ride bicycles, and the need to minimise the use of private motorised vehicles is essential, a school bus could be the possible alternative mode to school.

In Surabaya, Indonesia, the number of students riding school bus declined continuously (approximately 45% in the period of 2010–2013). In 2010, there was 25,200 students/year using a bus for their school travel. However, it decreased to 14,400 students/year in 2013 [10]. Whereas, the potential number of students who were willing to ride a school bus from 5 schools of High School Complex Area in Surabaya reached 4195 students [11]. Meanwhile, the vehicle's ownership is found to be a significant variable in influencing the use of school bus in Surabaya. The higher level of private vehicle ownership will affect students to use private vehicles as school transportation. Therefore, it needs to improve school bus transportation services as a strategy to increase students' willingness to use school buses as transportation to schools in the city of Surabaya [12]. A study on improving the services of school buses in Malang, Indonesia, found that two factors, namely the availability of information boards in each bus stop and the CCTV in buses, are important to increase the performance [13]. It can be concluded that the declining number of students using school bus for their school travel is mainly caused by the household and bus services factors.

Therefore, this study aims to investigate which characteristic of both children and parents significantly influence the school bus ridership. Previous studies in Indonesia mainly used descriptive and cross-classification analysis to obtain the demand for school bus. Furthermore, more studies concerned with the school bus level of services. A little attention has been paid to investigate the relevant factors influencing students' use of school bus. Logistic regression is accommodated to understand factors related to and probability of demand of school bus ridership.

2 Methodology

The population of this study is 3209 students. A total of 550 respondents from Senior and Junior High Schools in Jember, Indonesia, were asked to fill the questionnaire. A Slovin method was performed to calculate the number of samples.

2.1 Stated Preference Survey

The most leading use of stated preference (SP) methods is people's preference and willingness to pay (WTP) measurement for non-market goods. SP valuation has two main approaches: contingent valuation (CV) and choice experiment (CE). Generally, SP methods have several advantages, especially when dealing with sets of goods that do not currently exist [14].

This study used CV approach to understand the influence of household background characteristics on school bus ridership. The characteristics of the respondents in this study were divided into 2 (two), namely the socio-economic characteristics of students and the characteristics of student trips. Socio-economic characteristics consist of gender, school origin, class, age, parent's occupation, parents' income and vehicles owned by the students' parents. The characteristics of the trip consist of the student's home address, the distance from the house to the school, the travel time, the time students leave for school, and the mode of transportation that is often used (Fig. 1).

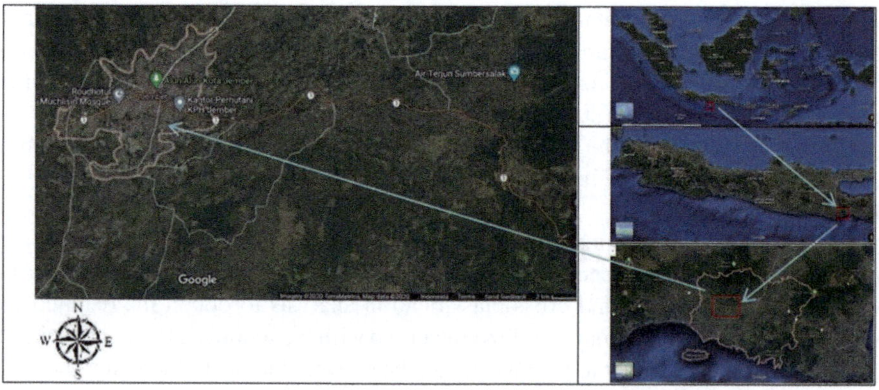

Fig. 1 Location of the study

2.2 Data Analysis

A binary logistic regression was performed to understand significant factors and the probability of each factor influencing the willingness to ride the school bus. Various modelling techniques have been applied for investigating household travel data. This study used the binary logistic regression to deal with binomial choices of whether students are willing to use the school bus or not. Statistical analysis software was performed to understand the model estimation.

3 Results and Discussions

Table 1 shows the general socio-demographic characteristics of the respondents.

Generally, students are willing to ride school bus when they travel form home to school. Furthermore, 70% of parents allows their children to ride school bus. This finding could be promising in implementing the school bus service in Jember.

3.1 Model Estimation and Goodness of Fit Test

Two estimation modelling were performed to understand which socio-demographic factors influence students' willingness to ride school bus and parents' allowing their children to ride school bus. Because of the fact that parent's attitude is the strongest factor influencing the mode school preferences for children [6] (Table 2).

The significance test shows that explanatory variables which were statistically significant at the 0.1 level are gender and car ownership. Meanwhile, an explanatory variable called travel time to school was found to be significant at the 0.05 level. Therefore, the general estimated equation was described as follows:

$$\begin{aligned} logit(\rho) &= \ln\left(\frac{\rho}{1-\rho}\right) \\ &= -0.286 + 0.330\, gender - 0.233\, car\ ownership \\ &\quad + 0.036\, travel\ time\ to\ school \end{aligned}$$

The model indicates that women students are more likely to ride a school bus. Furthermore, a group of people with the characteristic of having no car would be likely to choose school bus than people who have one or more car. This finding confirmed earlier studies [1] which reported children coming from no vehicles owned household tend to use the school bus, while those from the household with two vehicles are associated with a lower likelihood of riding a school bus for school travel. In terms of travel time, students experiencing a longer travel time to school also found to be more likely to use school bus.

Table 1 Socio-demographic characteristics

Characteristics		Percentage
Gender	Male	36
	Female	64
Parents' income	<IDR 1000,000	17
	IDR 1000,000–1999,999	19
	IDR 2000,000–2999,999	18
	IDR 3000,000–3999,999	14
	IDR 4000,000–4999,999	16
	>IDR 5000,000	16
Parents' motorcycle ownership	0	2
(2-wheeled scooter)	1	32
	2	44
	3	17
	4	3
	5	2
Parents' car ownership	0	47
	1	43
	2	8
	3	2
	4	0
Frequently used transport modes	Motorcycle	80
	Car	8
	Online taxibike/taxi	4
	Bicycle	4
	Walking	4
Students' willingness to ride school bus	Yes	67
	No	33
Parents' allowing their children to ride school bus	Yes	70
	No	30

Meanwhile, based on the parameter of estimation on parents' characteristics, the significance test shows that an explanatory variable which was statistically significant at the 0.1 level is parents' income. Meanwhile, an explanatory variable called car ownership found to be significant at the 0.05 level. Therefore, the general estimated equation was described as follows:

$$Logit(\rho) = \ln\left(\frac{\rho}{1-\rho}\right) = 0.505 - 0.295\, car\ ownership + 0.121\, parents'\ earning$$

Table 2 Parameter estimation

Variables	Student			Parent		
	Coeff.	S.E.	Sig.	Coeff.	S.E.	Sig.
Gender*	0.330	0.192	0.085*	0.233	0.197	0.236
Age	0.008	0.018	0.678	−0.044	0.071	0.535
Motorcycle ownership	0.042	0.096	0.665	0.076	0.103	0.462
Car ownership*	−0.233	0.132	0.077*	−0.295	0.134	0.027**
Parents' earning	0.030	0.061	0.619	0.121	0.063	0.054*
Distance to school	0.000	0.000	0.171	0.000	0.000	0.250
Travel time to school**	0.036	0.014	0.009**	0.015	0.013	0.239
Constant	−0.286	0.509	0.574	0.505	0.843	0.5
H–L test	Sig. = 0.455			Sig. = 0.152		
Classification plot	67.3%			69.8%		

Coeff. is coefficients, S.E. is the standard error, *variables were statistically significant at the 0.1 level, **variables were statistically significant at the 0.05 level. H–L Test is Hosmer Lemeshow Test (sig = 0.455 > 0.10), the model is valid

Among individual socio-demographic characteristics, car ownership and parents' income was found to be significant and associated with parents' willingness to allow their children riding school bus.

3.2 Interpretation of the Model and Discussion

The probability value of the willingness to ride a school bus is obtained by changing the odds ratio to the probability prediction. The probability of a student who is willing to ride a school bus can be formulated using the model. If a student is male (1), has 1 car, and spending 10 min to school, then the probability model is formulated as:

$$logit(p) = -0.286 + 0.330\,gender - 0.233\,car\ ownership + 0.036\,travel\ time\ to\ school$$

$$logit(p) = -0.286 + 0.330(1) - 0.233(1) + 0.036(10) = 0.171$$

And the probability:

$$p = 1 - \frac{e^{0.171}}{1 + e^{0.171}} = 0.457$$

The interpretation of the model is that a male student whose parent has a car and spends 10 min to school would be willing to ride the school bus by 45.7%.

Meanwhile, in terms of parents' willingness to allow their children to ride school bus, the probability could be formulated using the model. Car ownership and parents' income are two socio-demographic variables found to be statistically significant; therefore, those variables were used to obtain the probability. If the parents have 1 car and earn IDR 3000,000–3999,999. Then the probability is presented as follows:

$$logit(p) = 0.505 - 0.295\, car\ ownership \\ + 0.121\, parents'\, earning$$

$$logit(p) = 0.505 - 0.295(1) + 0.121(4) = 0.694$$

And the probability:

$$p = 1 - \frac{e^{0.694}}{1 + e^{0.694}} = 0.333$$

The result indicates that the parents whose income is IDR 3000,000–3,999,999 and have one car would be allowing their children to ride school bus by 33%.

This study shows that travel time to school significantly influences students' willingness to ride a school bus. This finding is aligned with previous studies [2, 4]. Students with longer travel time are more likely to ride the school bus. However, it should be noted that the distance variable does not show a significant influence on this behaviour. Furthermore, male students are found to be less likely to ride school bus. They prefer riding a motorcycle to the school bus.

Meanwhile, parents' attitude also has a strong role in influencing school travel mode. Household factors found to affect students' mode choice to school. This study found that household income significantly influences parents' permission to ride a school bus. Unlike the previous studies that stated that higher income led to less intention to ride the school bus, this study showed that parents with higher income are more likely to permit their children to ride a school bus. Car ownership found to be statistically significant in both students' and parents' categories. It indicates that having more car could decrease the willingness to ride a school bus.

By understanding the factors influencing the willingness to ride the school bus, the strategies regarding this issue could be prepared and implemented. This concept would undoubtedly support the sustainable transportation concept. Decreasing the use of private motorised vehicles could decline the level of Green House Gases (GHG) emitted to the air. Furthermore, safer transportation to school could decrease the number of accidents involving students.

4 Conclusion

Travel mode to school is an essential issue because factors influencing this behaviour could range from psychological, socio-demographic characteristics, built-environment, congestion issues, and road safety. This study performed a stated preference survey to understand the household background and the existing journey to school characteristics influencing school bus demand ridership.

Factors significantly influence students' willingness to ride a school bus are gender, car ownership, and travel time to school. Furthermore, in terms of parents' permission, household factors, namely car ownership and income are significant. This study provides a valuable baseline for future studies exploring the demand for other travel modes to school. Policymakers should consider the changing behaviour of parents and students' attitude and behaviour. The finding of this study could be useful to make an improvement to the travel mode to school.

Future studies could explore the attitude of students to active transportation; because school bus can only be implemented when students are willing to walk. Because of the pandemic situation, we performed an online-based survey. Future studies should do a direct survey to bring a better understanding of respondents.

References

1. Lidbe A, Li X, Adanu EK, Nambisan S, Jones S (2020) Exploratory analysis of recent trends in school travel mode choices in the U.S. Transp Res Interdiscip Perspect
2. Stark J, Beyer Bartana I, Fritz A, Unbehaun W, Hössinger R (2018) The influence of external factors on children's travel mode: a comparison of school trips and non-school trips. J Transp Geogr
3. McDonald NC, Steiner RL, Palmer WM, Bullock AN, Sisiopiku VP, Lytle BF Costs of school transportation: quantifying the fiscal impacts of encouraging walking and bicycling for school travel. Transportation (Amst)
4. Ewing R, Schroeer W, Greene W (2004) School location and student travel analysis of factors affecting mode choice. Transp Res Rec J Transp Res Board 1895:55–63
5. Noland RB, Park H, Von Hagen LA, Chatman DG (2014) A mode choice analysis of school trips in New Jersey. J Transp Land Use 7(2):111–133
6. Ahern SM, Arnott B, Chatterton T, de Nazelle A, Kellar I, McEachan RRC (2017) Understanding parents' school travel choices: a qualitative study using the theoretical domains framework. J Transp Heal 4:278–293
7. He SY, Giuliano G (2017) Factors affecting children's journeys to school: a joint escort-mode choice model. Transportation (Amst) 44(1):199–224
8. Ito K, Reardon TG, Arcaya MC, Shamsuddin S, Gute DM, Srinivasan S (2017) Built environment and walking to school: findings from a student travel behavior survey in Massachusetts. Transp Res Rec
9. Mitra R, Buliung RN (2014) The influence of neighborhood environment and household travel interactions on school travel behavior: an exploration using geographically-weighted models. J Transp Geogr
10. Kusuma OWA (2015) Evaluasi Program Bus Sekolah Di Kota Surabaya. Kebijak. Dan Manaj. Publik

11. Sekartadji R, Widyastuti H, Herijanto W (2015) Study demand and supply bus Sekolah Rute Dukuh Menanggal—SMA Kompleks Surabaya. In: Seminar Nasional Sains dan Teknologi Terapan III, pp 781–788
12. Ermirasari S, Handayeni KDME (2017) Keterkaitan Karakteristik Pelajar & Perjalanannya Terhadap Kesediaan Menggunakan Bus Sekolah Di Surabaya. J Tek ITS
13. Hariyani S (2017) School bus's level of service in Malang City. IOP Conf Ser: Earth Environ Sci
14. Hanley N, Czajkowski M (2019) The role of stated preference valuation methods in understanding choices and informing policy. Rev Environ Econ Policy

Sustainability Aspects in Construction Projects

Key Performance Indicator for Analytical Hierarchy Process Used for Determining the Effect of Reverse Supply Chain Toward Green Building Projects

Hermawan, Jati Utomo Dwi Hatmoko, and Jovana Neilkelvin

Abstract The Indonesian state also assesses that construction is an important part of the development of a country. Concern about the negative impact of construction waste is increasing every year and encourages the development of green or sustainable buildings. Reverse supply chain aims to obtain a production output from an end-of-life process that is environmentally friendly as much as possible. The calculation of the final performance of the supply chain can be done using the Key Performance Indicator weights obtained from the Analytical Hierarchy Process method. Key performance indicator (KPI) is determined to measure the level of achievement of objectives in the supply chain.

Keywords Green building · Reverse supply chain · Analytical hierarchy process · Key performance indicator

1 Introduction

Indonesia as a country valuing construction as an important part to the development of countries. The growth of Gross Domestic Product (GDP) in the construction sector is 5.4% according to data on Indonesia's economic indicators in 2017 at the Asia Construct Conference [19]. GDP value of the country of Indonesia estimate is

---

Reference to this paper should be made as follows: Sangwan, K.S., (2017) 'key activities, decision variables and performance indicators of reverse logistics' and Chinda, T., and Ammarapala, V. (2016) 'Decision making on reverse logistics in the construction industry'.

Hermawan (✉) · J. Neilkelvin
Soegijapranata Chatolic University, Semarang, Indonesia

J. Neilkelvin
e-mail: 17b10006@student.unika.ac.id

J. U. D. Hatmoko
Diponegoro University, Semarang, Indonesia
e-mail: jati.hatmoko@ft.undip.ac.id

S. A. Kristiawan et al. (eds.), *Proceedings of the 5th International Conference on Rehabilitation and Maintenance in Civil Engineering*, Lecture Notes in Civil Engineering 225,
https://doi.org/10.1007/978-981-16-9348-9_73

equivalent to USD 1015.54 million US in 2017 which represents 1.64% of the world economy [19].

Widhiawati et al. [29] and Tazi et al. [28] state the same thing, that along with the increasing number of construction projects, the need for materials will also increase. At the same time, more construction waste is generated during the construction process. Concerns about the negative impact of construction waste are increasing every year and encouraging the development of green or sustainable building both from the public and private sides [1].

Management of construction waste has four important aspects in waste management, namely reduction, reuse, recycling and landfilling. According to Rogers and Tibben-Lemke [21] Reverse supply chain is a process of planning, implementing and controlling in reverse from raw materials, work in progress, finished goods to consumer points to recycling points to proper disposal. Based on Rogers and Tibben-Lemke's explanation, it can be concluded that the reverse supply chain includes four important things in waste management [21].

Setiawan [26] states that key performance indicators (KPI) are determined to measure the level of achievement of objectives in the supply chain. Setiawan [26] also states that these measures can be financial and non-financial which can be used to measure the strategic performance of the organization. As a tool for measuring the strategic performance of an organization, KPI shows the health and development of the organization, the success of activities, programs or service delivery to realize organizational targets or goals [26].

According to Kuswandi et al. [16] stated that the calculation of the final performance of the supply chain can be done using the KPI weights obtained from the AHP method. The purpose of weighting the KPI is to see the priority interests according to the questionnaire filling stakeholders [16].

2 Related Literature

2.1 Supply Chain

Kusumastuti and Sugiama [15] state that in the supply chain process there are also planned logistics activities, implementation and control of effective flow and storage of goods, as well as services and information related from the starting point of manufacture to the point of use to fulfill customer desires. Kusumastuti and Sugiama [15] also state that logistics is an effort to manage goods needed for jobs related to the supply of raw materials, equipment and equipment for the production process.

Govinand and Soleimani [8] state that in supply chain management there are two types. The first type of supply chain management is the classic type which is commonly used and is commonly called the forward supply chain [8]. The second type of supply chain management is a reverse supply chain [8]. The purpose of a reverse supply chain is to obtain a production result from an end-of-life process that is environmentally friendly as much as possible [8].

2.2 Green Building

Green building define as a planning and management of construction projects (in accordance with contract documents) to minimize the impact of the construction process on the environment so that there is a balance between environmental capabilities and the needs of human life for present and future generations [6].

Green building or green construction is a sustainable building that leads to the structure and use of processes that are environmentally responsible and resource-efficient throughout the life cycle of the building, starting from site selection to design, construction, operation, maintenance, renovation and demolition. This practice extends and complements classic building designs in terms of economy, utility, durability and comfort [20].

The Green Building Council Indonesia [9] stipulates that a building in Indonesia can only be considered a green building if it meets the assessment set by the GBCI. Greenship assessment categories are appropriate site development (ASD), energy efficiency and conservation (EEC), water conservation (WAC), material resource and cycle (MRC), indoor health and comfort (IHC), and building environment management (BEM) [9].

A summary of the greenship assessment criteria in the source category and material cycle (Material Resources and Cycle-MRC) in the final assessment (FA) can be shown in Table 1.

Table 1 MRC Greenship assessment criteria

Material resource and cycle			
MRC P	Fundamental refrigerant	P	1 prerequisite criteria; 6 credit criteria
MRC 1	Building and material reuse	2	
MRC 2	Environmentally friendly material	3	
MRC 3	Non ODS usage	2	
MRC 4	Certified wood	2	
MRC 5	Prefab material	3	
MRC 6	Regional material	2	
Total		14	13.9%

Source GBCI [9]

Table 2 Alternative options for RSC criteria

Citation	Criteria alternative
Thierry et al. [29]	Repair, refurbish, remanufacture, cannibalize and recycle
Johnson and Wang [13]	Combination of remanufacture, reuse, and recycle
Rose and Ishii [22]	Reuse, service, remanufacture, recycle and disposal
Guide et al. [10]	Repair, remanufacturing and recycling
Ferguson and Browne [7]	Reuse, remanufacture and recycle
Lee et al. [18]	Reuse, remanufacture, recycle, landfill, and incineration
King et al. [14]	Repair, recondition, remanufacture, and recycle
Sasikumar and Kannan [24]	Direct reuse, recycling, remanufacturing, and repair
Skinner et al. [25]	Destroying, recycling, refurbishing, remanufacturing, and repackaging
Srivastava [27]	Repair & refurbish, remanufacturing, and secondary market
Lambert et al. [17]	Repair, reuse, remanufacture, upgrade, repackage, recycle, reconfigure, and revaluation
Jindal and Sangwan [12]	Repair, refurbishing, remanufacturing, cannibalizing, and recycling

Source Sangwan [23]

2.3 Key Performance Indicator

Key performance indicators (KPIs) are determined to measure the level of achievement of objectives in the supply chain [26]. Setiawan [26] also states that these measures can be financial and non-financial which can be used to measure the strategic performance of the organization. As a tool for measuring the strategic performance of an organization, KPI shows the health and development of the organization, the success of activities, programs or service delivery to realize organizational targets or goals.

Some of the product improvement processes include repair, reuse, remanufacture, re-consumption, recycle and disposal. Sangwan [23] has summarized alternative options in the process of repairing a product from various citation which can be shown in Table 2.

2.4 Analytical Hierarchy Process

The Analytical Hierarchy Process (AHP) method is used to find a priority supply chain process. The supply chain needs to be measured to ensure the success of planning [2]. The main tool of the AHP method is the hierarchical model of the problem. The hierarchy is created using a tree diagram which can be shown in Fig. 1.

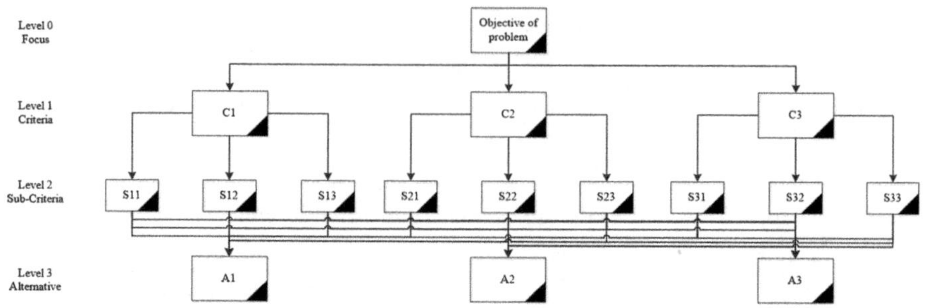

Fig. 1 Tree diagram of analytical hierarchy process

3 Research Identification

It is necessary to identify reverse supply chain criteria related to KPI determination. The KPI determination on the project comes from the criteria used. The criteria used are based on the literature study that has been conducted, namely economic and environmental criteria. Economic factors are a concern because if there is no economic value, companies will find it more difficult to engage in reverse logistics [5]. Meanwhile, environmental factors were chosen because they relate to green buildings.

3.1 Key Performance Indicator for Analytical Hierarchy Process

There are several KPIs that can be used for AHP in reverse supply chain. KPI choices from several citations can be summarized in Table 3.

3.2 Key Performance Indicator for Green Building Research

There are fifteen key performance indicators for reverse supply chain [4]. The fifteen KPIs were eliminated based on their relationship to the environment and economy. The elimination of KPIs can be shown in Table 4.

Table 3 Types of reverse supply chain KPI's

Citation	Criteria	KPI	Alternative
Barker and Zabinsky [3]	1. Cost 2. Business relations	1. Recycled product 2. Test Costs 3. Scrap shipped 4. Original facility 5. Proprietary knowledge 6. Customer interactions	1. Collection 2. Sort-test 3. Processing
Chinda [4]	1. Economic 2. Site constraint	1. Labor cost 2. Inventory cost 3. Transportation cost 4. Processing cost 5. Specific sorting machine 6. Specific technology 7. Matured market 8. Landfill charge 9. Availability of landfill 10. Site space 11. Social image 12. Requirement of virgin material 13. Limited project time 14. Environmental concern 15. Knowledge of sorting	1. Direct reuse 2. Remanufacturing 3. Recycle 4. Landfill
Hosseini et al. [11]	1. Economic gains 2. Environmental consideration	1. Direct reuse or resale 2. Repairing 3. Refurbishing 4. Remanufacturing 5. Cannibalization 6. Recycling 7. Landfilling	
Chinda and Ammarapala [5]	1. Economic 2. Site constraint	1. Labor cost 2. Inventory cost 3. Transportation cost 4. Processing cost 5. Specific sorting machine 6. Specific technology 7. Matured market 8. Landfill charge 9. Availability of landfill 10. Site space 11. Green image site 12. Replacement of virgin material 13. Limited project time 14. Legislative pressure 15. Knowledge of sorting	1. Direct reuse 2. Remanufacturing 3. Recycle 4. Landfill

(continued)

Table 3 (continued)

Citation	Criteria	KPI	Alternative
Kuswandi et al. [16]	1. Reliability 2. Responsiveness 3. Agility 4. Cost 5. Assets management	1. On time pick up 2. % utility of truckload 3. Deliver return cycle time 4. Return cycle time 5. Return order reestablished and sustained 6. Current deliver return volume 7. Recovery rate 8. Scrap rate	

Table 4 Justification of selection of RSC research KPI's

KPI RSC Chinda (2014)	Used	Justification
Labor cost	☐	Unrelated with environment
Inventory cost	☐	Unrelated with environment
Transportation cost	☒	Related with waste
Processing cost	☒	Related with waste
Specific sorting machine Specific technology	☒ ☒	Related with waste Related with waste
Matured market	☐	Unrelated with environment
Landfill charge	☒	Related with environment
Availability of landfill	☒	Related with environment
Site space	☒	Related with waste
Social image	☐	Unrelated with waste
Requirement of virgin material	☒	Related with environment
Limited project time	☐	Unrelated with environment
Environmental concern	☒	Related with environment
Knowledge of sorting	☒	Related with environment

4 Conclusion

This journal on the use of key performance indicators for analytical hierarchy process becomes an additional material in knowledge about supply chain management, green building and reverse supply chain itself. This research suggested a reverse supply chain approach with help from key performance indicators for analytical hierarchy process as one of many ways to determined reverse supply chain impact toward green building.

References

1. Agyekum K, Kissi E, Andku JC (2020) Professionals' views of vernacular building materials and techniques for green building delivery in Ghana. Sci. Afr. 8
2. Al Farisi MS (2018) Perancangan sistem pengukuran kinerja green supply chain management dengan metode *green supply chain operation* (GSCOR) and *anathycal hierarchy process* (AHP). Undergraduate (S1) Thesis, University of Muhammadiyah Malang
3. Barker TJ, Zabinsky ZB (2011) A multicriteria decision making model for reverse logistics using analytical hierarchy process. Omega 39:558–573
4. Chinda T (2014) Decision making on reverse logistics in the construction industry. Int J Civ Struct Eng IJCSE 1(3):135–138
5. Chinda T, Ammarapala V (2016) Decision-making on reverse logistics in the construction industry. Songklanakarin J Sci Technol 38(1):7–14
6. Ervianto WI (2012) Selamatkan bumi melalui konstruksi hijau, CV. Andi Offset, Yogyakarta
7. Ferguson N, Browne J (2001) Issues in end-of-life product recovery and reverse logistics. Prod Plann Control 12(5):534–547
8. Govinand K, Soleimani H (2016) A review of reverse logistics and closed-loop supply chain. J Clean Prod 142:371–384
9. Green Building Council Indonesia (2013) Greenship untuk bangunan baru versi 1.2. GBC Indonesia
10. Guide Jr. VDR (2000) Production planning and control for remanufacturing: industry practice and research needs. J Oper Manag 18(4):467–483
11. Hosseini MR, Chileshe N, Rameezdeen R, Lehmann S (2014) Reverse logistics for the construction industry lesson from the manufacturing context. Int J Constr Eng Manage 3(3):75–90
12. Jindal A, Sangwan KS (2016) A fuzzy based decision support framework for product recovery process selection in reverse logistics. Int J Serv Oper Manage. https://doi.org/10.1504/IJSOM.2016.10000346
13. Johnson MR, Wang MH (1995) Planning product disassembly for material recovery opportunities. Int J Prod Res 33(11):3119–3142
14. King AM, Burgess SC, Ijomah W, McMahon CA (2006) Reducing waste: repair, recondition, remanufacture or recycle? Sustain Dev 14(4):257–267
15. Kusumastuti D, Sugiama AG (2017) Manajemen logistik organisasi publik. Universitas Terbuka, Tangerang Selatan
16. Kuswandi RY, Ridwan AY, El Hadi RM (2018) Perancangan sistem monitoring reverse logistic untuk industri penyamakan kulit dengan model SCOR. E-Proc Eng 5(3):6912–6919
17. Lambert S, Riopel D, Abdul-Kader W (2011) A reverse logistics decisions conceptual framework. Comput Ind Eng 61(3):561–581
18. Lee SG, Lye SW, Khoo MK (2001) A multi-objective methodology for evaluating product end-of-life options and disassembly. Int J Adv Manuf Technol 18(2):148–156

19. Pribadi KS, Soemardi BW (2018) The construction sector of Indonesia. In: 23rd Asia construct conference 2018, Malaysia
20. Purnawan PY (2017) Pengelolaan limbah untuk pekerjaan strukturpada proyek konstruksi di Daerah Istimewa Yogyakarta. S1 Thesis Atma Jaya Yogyakarta University
21. Roger D, Tibben-lembke RS (2002) Differences between forward and reverse logistics in a retail environment. Suply Chain Manage Int J 7:271–282
22. Rose, CM, Ishii K (1999) Product end-of-life strategy categorization design tool. J Electron Manuf 9(01):41–51
23. Sangwan KS (2017) Key activities, decision variables and performance indicators of reverse logistics. Procedia CIRP 61:257–262
24. Sasikumar P, Kannan G (2008) Issues in reverse supply chains, part I: end of life product recovery and inventory management – an overview. Int J Sustain Eng 1(3):154–172
25. Skinner LR, Bryant PT, Richey RG (2008) Examining the impact of reverse logistics disposition strategies. Int J Phys Distrib Logistics Manage 38(7):518–539
26. Setiawan NR (2018) Perancangan sistem pengukuran kinerja green supply chain dengan metode green SCOR berbasis AHP and OMAX (studi kasus: CV.cool clean). Undergraduate (S1) Thesis, University of Muhammadiyah Malang
27. Srivastava SK (2008) Network design for reverse logistics. Omega 36(4):535–548
28. Tazi N, Idir R, Fraj AB (2020) Sustainable reverse logistic of construction and demolition wastes in French regions: toward sustainable practices. Procedia CIRP 90:712–717
29. Thierry MC, Salomon M, Nunen JAEEv, Wassenhove LNv (1995) Strategic issues inproduct recovery management. Calif Manage Rev 37(2):114–135
30. Widhiawati IAR, Astana NY, Indrayani NLA (2019) Kajian pengelolaan limbah konstruksi pada proyek pembangunan gedung di Bali. Jurnal Ilmiah Teknik Sipil 23(1)

Development of Blockchain and Machine Learning System in the Process of Construction Planning Method of the Smart Building to Save Cost and Time

Christiantono Tedjo, Mohammed Ali Berawi, and Mustika Sari

Abstract Achieving project objectives that are effective, efficient, and safe requires careful planning of construction methods. The construction method of a project may vary depending on the description and design of the project. Planning for construction methods generally takes a long time to review, involves many parties, and requires consideration in terms of costs, resources, and safety. And the need for construction planning characteristics for smart buildings recently. Machine learning combined with the blockchain system is an instrument that can accommodate processes in the construction method. Start from planning to integration-evaluation with a historical database. This instrument is expected to be able to provide suggestions and evaluation of methods, to be able to assist the planning process of construction methods so that the application of work methods will be more effective and efficient.

Keywords Construction method · Smart building · Machine learning · Blockchain

1 Introduction

The construction method of a project may vary depending on the description and design of the project. Construction method planning is generally not brief. The construction process requires integration between disciplines and divisions. To obtain efficient results in time, cost, ease of achievement, and a sustainable process [1]. While the more advanced and developing technology, the amount of information in the construction process will increase, and the more dynamic it is, the larger the knowledge network will be [2]. In interdisciplinary integration, directly or indirectly,

C. Tedjo (✉) · M. A. Berawi · M. Sari
University of Indonesia, Jakarta, Indonesia
e-mail: christiantono.tedjo@ui.ac.id

M. A. Berawi
e-mail: maberawi@eng.ui.ac.id

M. Sari
e-mail: mustika.sari01@ui.ac.id

S. A. Kristiawan et al. (eds.), *Proceedings of the 5th International Conference on Rehabilitation and Maintenance in Civil Engineering*, Lecture Notes in Civil Engineering 225,
https://doi.org/10.1007/978-981-16-9348-9_74

a process of transferring information is required. The knowledge transfer process, especially by manual method, tends to have several impacts that can disrupt the project deliverable process. One of them is the process of knowledge transfer manual or between individuals. There is a huge potential for the information required to experience delays [3]. In addition to delays in information needs, information reduction is also very possible because the knowledge transfer process requires good relationships between individuals or stakeholders so that communication can run smoothly. Also, the increasing intensity of work will make it difficult for each individual to transfer information accurately and quickly [3].

Construction project clients recently need increasingly complex facilities in their buildings, requiring a fast construction process, and low costs [4]. This is shown by the presence of various building design concepts in the world such as green building, healthy building, and smart building, this of course requires technological development and an increasingly high level of complexity. Apart from the consideration of the information transfer process, the construction project requires consideration in terms of costs, resources, process handling, and safety. One of the criteria for a good work method is that it can accommodate the transportation of tools, goods, and people. Minimizing narrow areas of transportation access and accommodating the creation of good HSE management [5]. Demand in the building sector is also increasing with collaboration on technological advances and energy consumption [6]. For each building to be able to manage its energy, smart building technology needs to become a new adoption, especially in Indonesia, where construction is currently starting to lead to the development of the smart building sector. Machine learning combined with the blockchain system is an instrument that can accommodate processes in the construction method. By reducing bureaucracy and an orderly and integrated database storage process. The study results show the potential cost savings from using blockchain is 8.3% of the total cost of building a house, with a standard deviation of 1.26% [4]. So that this instrument is expected to be able to provide suggestions and evaluation of methods, to be able to assist the planning process of construction methods so that the application of work methods will be more effective and efficient.

2 Blockchain and Machine Learning

2.1 Blockchain

Construction is a process of collaboration and integration in every element and stage. Collaboration and integration in transaction activity, for example in its current application, namely economic activity, whatever the form of the transaction requires a system that has a high level of "trust". Most people don't give trust free of charge, they want verification and assurance that the other party is worthy of their trust, and in commercial situations, they often buy that guarantee from intermediaries. There are many intermediaries in all sectors [7].

Intermediaries need time to process some information (for example, in the process of credit card payments, it will take a few days for the payment to arrive in our account) and the cost of the transaction is borne by us as consumers (the vendor only accepts part of the money charged to the card account. Credit). In general, this process applies, intermediaries collect information about our credit cards—to verify agreement data, to be able to convince other parties, or other intermediaries, that our data is valid.

The process takes time, effort and money, but provides no value for the customer. To reduce this process with the potential of blockchain as described in the previous section, blockchain is expected to be an alternative instrument to reduce this process. Blockchain is considered to have the potential to change many global industries including construction. Blockchain was originally developed as a supporting technology for the world's first cryptocurrency, Bitcoin, which was introduced in 2008 as a verification tool for the transaction process [8]. One of the breakthroughs developed by using blockchain for delivery systems in the field of construction includes the use of smart contracts. The application of smart contracts is used by utilizing 4 characteristics of the blockchain system, namely distributed, which means that data and functionality are distributed or not centralized (located on different servers or users), replicated which means data can be backed up, immutable which means data cannot be modified and non-repudiation, which means that every person who does it can be proven validity.

2.2 *Machine Learning*

The application of machine learning in the world of construction has been widely applied, one of which is the planning mapping of needs and formwork areas as done [9] in his research journal which explains how machine learning can develop the planning and procurement of the amount of formwork for an atypical building area. This process saves time and resources, the identification results of machine learning are more detailed and accurate than the identification of formwork requirements that is done manually. There is also a journal discussed by Ahmed Gondia et al. [10] about how machine learning is used to identify the causes of project delays. In addition to safety identification. The application and need for machine learning are examined for the level of need in a journal written by Ahna et al. [11]. What are the needs and what aspects can apply machine learning in its applied in the construction world, such as tower cranes area identification, an inspection of structural joints, prediction of construction safety accidents, operation of construction lifts, an arrangement of tower crane layouts? So that the areas that can be accommodated by machine learning include mitigating risks, identifying safety, and construction planning.

3 Blockchain Structure in Study

Blockchain in this study is used as data storage and records the approval of the work method planning that has been approved by several related parties. In its use as data storage to be uploaded or downloaded will have a hash. Hash on the data will serve as a process of identifying the authenticity of the data.

3.1 Relations Between Blocks

As supervised data, the relationships between blocks are only additions. This means that any data recorded and inputted will increase the number of data libraries. The input data will receive verification and identification with an identity in the form of a hash which will later function to see the approval time and data input time.

3.2 Data Record Process

Blockchain data recording is done with the IPFS platform. In this case, IPFS will provide an Identity Document (ID) to the block. IPFS can give ID to any type of file, but in this research, the file used is a file with the .xlsx extension, because the resource data and project specifications are inputted in table form to facilitate grouping and sorting data if needed. The ID given by the IPFS platform will be used by machine learning to pull new data and download historical data which will be used as a data train in the form of supervised data.

3.3 Use and Submit Data

Prediction data that has been evaluated and approved will be verified by machine-learning, which will then be put back into the block as new historical data. The new data, which already has an ID or hash, will be the train data that machine-learning will download automatically when machine-learning will predict work methods on the new project. To be uploaded to the blockchain platform, this data needs to be evaluated and approved by the relevant parties. In this process, it is not necessarily mandatory to use blockchain prediction results, but the prediction results can still be adjusted according to productivity assumptions or other needs. Once approved, machine learning can upload this data into the blockchain platform to obtain an ID.

4 Development of Machine-Learning Structures

In this research, machine learning acts as artificial intelligence that can identify or predict based on existing construction data. The process carried out by machine-learning includes data collection, identification of inputted data, prediction using the supervised method, and giving suggestions of prediction results to engineers.

4.1 *Data Identification*

The data table is divided into three parts, namely the test, train, and prediction sections. The test section functions to test the accuracy of the supervised learning method. The data used are all historical data, both building, and resource specification data. Data train is the development of a test data model that is used as a reference for machine learning to predict input data. In this research, the modeling requires an input of building and volume specifications to be able to predict the required resources. Because the data is predictive, evaluation can be carried out by engineers, to adjust to detailed project requirements that are not contained in the machine-learning input process.

4.2 *Supervised Learning*

The data table is arranged in the form of data that can be used by machine-learning as supervised data, meaning that the data has been grouped into columns that will become input data and which will become predictive data. In this process, building specification data is inputted, such as building layers, basement depth, area, and volume of material. The input data becomes raw data which will be entered in the historical data table, which will calculate the value of the resource requirements which will be input data for engineers. Due to the input data structure that has been adapted to the historical data table structure, machine-learning identifies data according to a given table pattern without changing the format. With an appropriate format, the .xlsx file can be connected to other tables in the file, so that the results of the predicted data will be processed automatically in a visual display so that it is easier for users to understand.

4.3 *Manage Prediction Result Data*

After the prediction results and the report are issued, the raw data of the report file will automatically be entered into the platform block and obtain an ID, to obtain authentication and facilitate data tracking. Uploading data to the platform is done with the .xlsx file extension, also without changing the structure of the files and tables, because the data is needed to make machine-learning predictions in the next process. To make it easier to retrieve and group data, machine learning with the flask platform directly downloads all uploaded data and stores it in a folder in the form of a pdf report, with the name and time of the creation of the method data.

5 Research Process

This research examines the process of defining work methods and resources in preparation for the implementation of construction projects. Based on previous data and designs contained in literature studies, machine-learning enrichment will be carried out to predict the characteristics of the construction planning model both in terms of resources and construction methods.

5.1 Data Identification

In research, problem identification is made more conical on the machine learning model plan. What algorithms will be used, as well as how machine learning processes data on the blockchain.

5.2 Machine Learning Algorithm Selection

Algorithm selection is carried out after the data cleaning process. Data training is carried out, to obtain training results whose value is close to one. This value shows the suitability of using the algorithm for the data.

5.3 Machine Learning Model

Machine learning begins to be modeled after the appropriate algorithm is obtained. Input definition based on the dependent variable is carried out. Defined the output plan that will be generated by machine learning. Machine-learning development process show in Fig. 1.

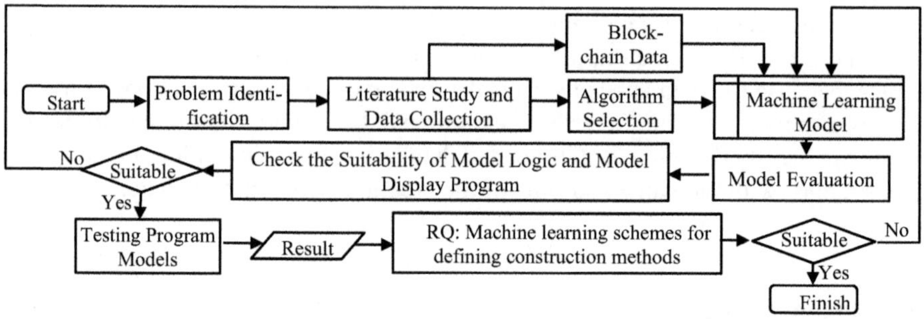

Fig. 1 Research process

Table 1 Project identification data form

No.	Site visit form		
	Data	Grade	Remark
1	Building type	…	1 = Apartment 2 = Office 3 = School 4 = Commercial area 5 = Smart building
2	Workforce monitoring	…	Schedule and list of workers
3	Equipment monitoring	…	Schedule and equipment list
4	Construction duration	…	Month period
5	Contract value	…	Amount of money

Evaluation and Model Testing. The machine learning output is evaluated, and it is validated whether the results can be used to develop predictive resources and tools in the construction process. All machine-learning data variables are inputted into the form shown in Table 1.

6 Result

Based on the research stages described in the previous section, the prediction results of machine learning are obtained with the following process.

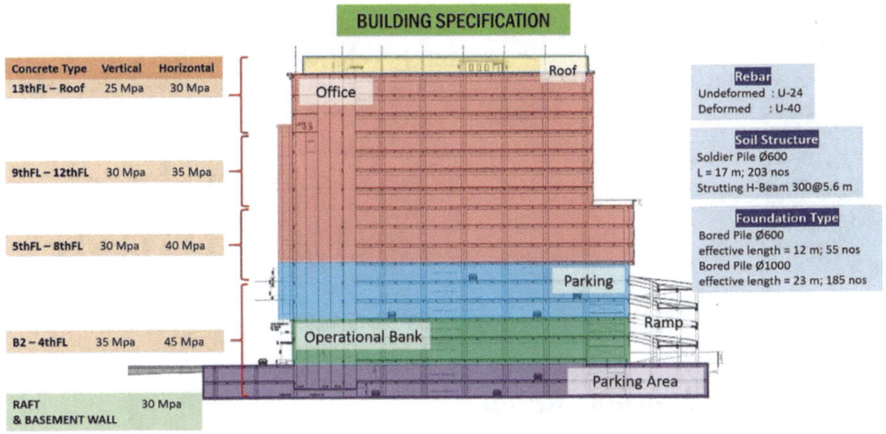

Fig. 2 Building specification

6.1 Data Identification

The data collected based on the results of the project visits were included in several tables, including building specification tables, manpower tables, and equipment usage tables, which are defined in the form of a schedule. Figure 2 explains the definition of building structure specifications per floor.

6.2 Machine Learning Algorithm Selection

To confirm the accuracy of the prediction results, machine learning needs to process the data training. Machine learning data training with a supervised learning model is generally carried out using three analysis methods, namely linear logistic, random forest, and decision tree, The selection algorithm is shown in the Fig. 3.

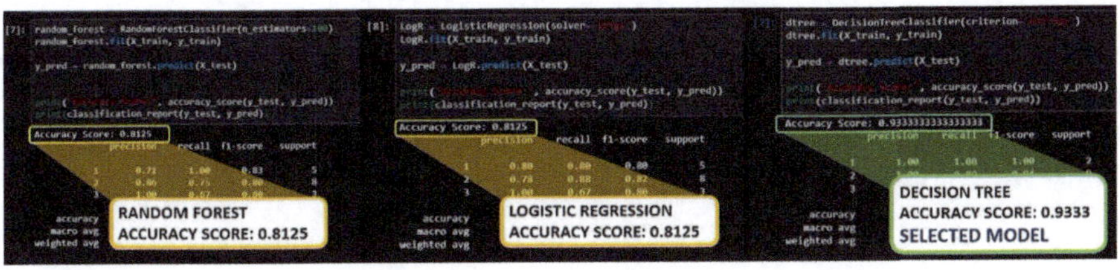

Fig. 3 Algorithm selection—Random Forrest, Logistic Regression, Decision Tree (selected)

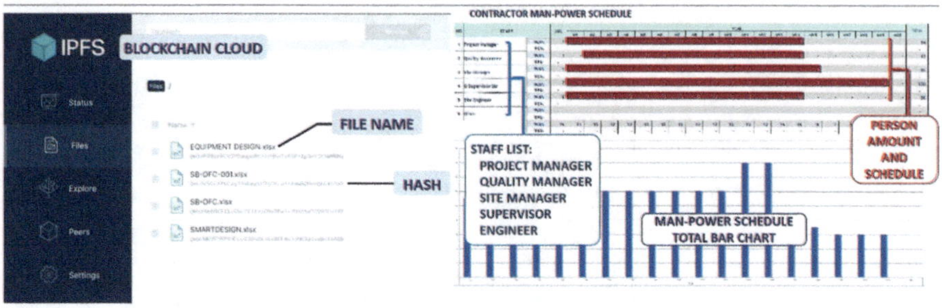

Fig. 4 Storing, collecting, and interacting data in IPFS and machine learning result

6.3 Machine Learning Result

The process of managing input data and predictive data from machine learning is carried out in blockchain-based clouds using the IPFS platform. Data that has been managed and predicted is displayed in the form of a schedule table as follows Fig. 4.

7 Conclusion

Based on the formulation of the problem, research objectives, and the results of machine learning enrichment trials based on python programming and block-chain cloud IPFS, a research conclusion is obtained where if a new management system is developed using machine-learning technology, which has a detailed database and is well verified with the system blockchain, it can produce a construction method that is more efficient and faster and can be applied, so that faster construction method designs can be produced. By using a supervised machine-learning model, with a data management system using blockchain, the results of machine learning are even applied in one of the building construction projects, because they match the project overview and evaluation of experts. This development can be particularly useful in implementing smart building construction in Indonesia. However, this research cannot manage the waste from the construction method used. It is hoped that later on, resource management can also be done using machine learning. Moreover, to support further research, it is necessary to try to also examine the process of enriching machine learning in infrastructure projects and the natural resource industry in Indonesia.

References

1. Oti AH, Tah JH, Abanda FH (2018) Integration of lessons learned knowledge in building information modeling

2. Jian Sun P, Xu Ren P, Chimay J, Anumba PD (2019) Analysis of knowledge-transfer mechanisms in construction project cooperation networks
3. Dakhli Z, Lafhaj Z, Mossman A (2019) The Potential of Blockchain in Building Construction - Civil Engineering, Ecole Centrale de Lille, 59651 Villeneuve d'Ascq, France
4. Dakhli Z, Lafhaj Z, Mossman A (2019) The potential of blockchain in building construction
5. Ellis RD, Thomas HR (2017) Construction site management and labor productivity improvement
6. Joud Al Dakheel CD (2020) Smart buildings features and key performance indicators: a review
7. Keiser J (2012) Leadership and cultural change: necessary components of a lean transformation
8. Rasoloharijaona S, Rakotosamimanana B, Randrianambinina B, Zimmermann E (2003) Pair-specific usage of sleeping sites and their implications for social organization in a nocturnal Malagasy primate, the Milne Edwards' sportive lemur
9. Lee D, Lim H, Kim T, Cho H, Kang K-I (2018) Advanced planning model of formwork layout for productivity improvement
10. Ahmed Gondia S, Siam A, Wael El-Dakhakhni F, Nassar AA (2019) Machine learning algorithms for construction projects delay risk prediction
11. Ahn H, Lee D, Lee S, Kim T, Cho H, Kang K-I (2018) Application of machine learning technology for construction site

The Development of Blockchain Based Knowledge Management System Model at EPC Projects to Improve Project Time Performance

D. Y. Priyambodo, M. A. Berawi, and M. Sari

Abstract EPC project is one of complex construction project types which differs from ordinary construction projects. It owns more frequent delayed occurrence than ordinary construction projects. EPC projects tends to be labor and knowledge intensive. Inputs and outputs at EPC project management involve various knowledge and intellectual creative activities. Thus, the EPC project management process is a knowledge management process. Knowledge Management System (KMS) aims to improve the quality of human resources in an organization by improving communication among all parts of the organization and increasing knowledge mastery by transferring knowledge. The application of KMS with a blockchain-based system is expected to be a solution for the needs of distributed knowledge transfer, high-speed transfer of knowledge and information, fast access to knowledge and high information security. This study uses the Tiwana model framework, and for organizational knowledge mapping uses the Zack KM model cycle, while the SECI model is used for knowledge formation. The quality of the resulting blockchain-based KMS platform is tested by using expert validation. The research result shows that blockchain-based KMS model can be well implemented with excellent validation so it becomes a solution to increase knowledge and to improve project time performance in EPC-X project.

Keywords Knowledge management system · EPC project · Blockchain

D. Y. Priyambodo (✉) · M. A. Berawi · M. Sari
Civil Engineering, University of Indonesia, Jakarta, Indonesia
e-mail: dimas.yossi@ui.ac.id

M. A. Berawi
e-mail: maberawi@eng.ui.ac.id

M. Sari
e-mail: mustika.sari01@ui.ac.id

S. A. Kristiawan et al. (eds.), *Proceedings of the 5th International Conference on Rehabilitation and Maintenance in Civil Engineering*, Lecture Notes in Civil Engineering 225,
https://doi.org/10.1007/978-981-16-9348-9_75

1 Introduction

The era of globalization era is indicated by rapid development of science and technology as well as a high and dynamic level of competition. This condition requires companies to innovate and to improve their performance. Due to the high-competitive competition, it is necessary to change the company's paradigm from being resource-based to knowledge-based, which is based on managing data, information, and knowledge. The condition also affects the construction industry [1]. Knowledge management is a way to improve company capabilities by managing the knowledge they have. EPC projects tend to be labor-intensive and knowledge-intensive. Inputs and outputs in EPC project management involve various knowledge and intellectual creative activities. The requirement for knowledge to be accessed collectively among different project phases, stakeholders, and management functions still poses as a challenge to be solved [1–3].

At the present era, the implementation of Knowledge Management processes at both organizational and national levels require an innovation. Many researches have been done to implement knowledge management in organizations, but from the technological point of view, it is still possible to discuss the processes of knowledge management in a fully transparent and high level style throughout the organization. The problems of organizational knowledge storage, information, and knowledge safety, sharing knowledge, knowledge access, intellectual property rights of scholars are some of the challenges faced by a traditional knowledge management. Blockchain technology is one of the IT-based tools which has a high level of innovation, which can be used to implement knowledge management at the enterprise level or even at the national level. It enables high speed of knowledge and information transfer, fast access to knowledge, and high information security. When information is stored in Blockchain, it is immutable. This design allows Blockchain to be capable of having a permanent historical record [4].

2 Theoretical Study

2.1 Knowledge Management

Tiwana defines knowledge management broadly in the sense of managing knowledge as "… management of organizational knowledge for creating business value and generating a competitive advantage". Knowledge Management provides the ability to create, communicate and apply knowledge necessary and useful for the achievement of all types of business goals. Tiwana also explains "Knowledge management is the ability to create and retain greater value from core business competencies" [5]. Effective knowledge management must be started by using a strategic look at what knowledge the organization needs. This is to ensure that the KMS will be developed is in accordance with what is needed and the core competency of the organization.

This alignment can be based on the needs and the core by determining what strategies are or will be implemented and what the organization has implemented [6, 7].

2.2 *Blockchain Technology*

Blockchain can act as a distributed, programmable, and encrypted database to transfer, protect, store, and quickly access knowledge from one location to another by using high security [8]. Blockchain also provides the potential for scientists to share their unique knowledge with other users. These guaranteed transfers are safe, and nothing can challenge their legality. The implementation of Blockchain's unique features include high reliability, privacy, accountability and data sharing, information disorganization, responding to traditional knowledge management challenges [4].

Blockchain has become a revolutionary paradigm in record keeping systems and has been seen as an emerging technology in the industry and research community. Through its immutable property, this design allows blockchain to have permanent historical records. Knowledge Management is based on management trust. As such, blockchain trust properties can be used to manage trust in generating, storing, sharing, protecting, and applying various knowledge [9].

3 Research Method

This research uses a qualitative descriptive approach through case studies. The method of data collection is done by doings of interviews, historical data, and documentation. Amrit Tiwana's 4 phased roadmap for knowledge management method is employed for the data analysis process in this study, accompanied with some adjustments. The infrastructure evaluation stage is supported by observation data, interviews, and document review. The system analysis and design stage include database design, system architecture, and blockchain-based knowledge management system interfaces using an object-oriented approach with expert validation. The expert validation related to the blockchain-based Knowledge Management System model was carried out by 2 blockchain technology experts and 1 knowledge management expert on the EPC project (Fig. 1).

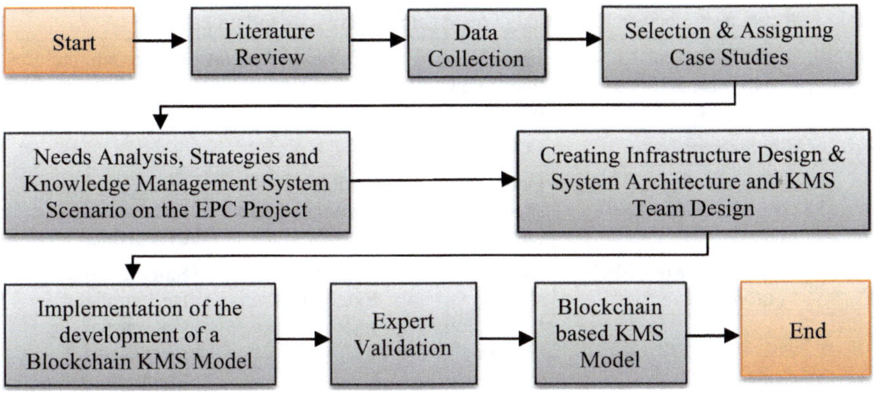

Fig. 1 Research stages

4 Result and Analysis

4.1 Knowledge Management System Scenario Analysis

The process of creating organizational knowledge occurs because of the interaction or conversion between tacit knowledge and explicit knowledge through the processes of socialization, externalization, combination, and internalization, which are SECI Nonaka's model. Every activity in the EPC-X Project to create and formulate knowledge needs to be carried out in a knowledge management process cycle which consists of 4 stages in accordance with the SECI Nonaka model. Displayed below is the relationship between the order to create and formulate knowledge (Table 1).

4.2 Infrastructure Design and Architecture System

In developing a knowledge management system prototype, a knowledge management system architecture design is carried out. The prototype was developed by using a blockchain-based system. The prototype runs on the existing infrastructure and becomes a separate module from other existing information systems at EPC-X Project (Fig. 2).

The knowledge storage mechanism on the blockchain enables to store knowledge, change the knowledge stored within, and get knowledge from inside the blockchain. The knowledge structure on the blockchain will be adjusted to the knowledge management structure. The implemented blockchain system has a feature to record every process of the storage mechanism, add knowledge management to the blockchain, and distribute knowledge to all blockchain members.

Table 1 Relationship between knowledge management activities and processes

No	EPC-X project activities	Knowledge management process	SECI model stages	Blockchain features
Project Management				
1	Project contract	Document sharing, Discussion	Socialization, Externalization	Smart contract, Decentralized, Immutable
2	Scope of work EPC project	Document sharing, Discussion, Sharing experience, Workshop	Socialization, Externalization, Internalization	Smart contract, Decentralized, Consensus mechanism, Transparency
3	Change order	Document sharing, Discussion, Sharing experience, Workshop, Case study	Socialization, Externalization, Combination, Internalization	Smart contract, Decentralized, Consensus mechanism, Immutable
Engineering				
4	Basic engineering	Document sharing, Discussion, Sharing experience, Sharing idea, Workshop	Socialization, Externalization, Internalization	Smart contract, Decentralized, Consensus mechanism, Immutable
5	Detail engineering	Document sharing, Discussion, Sharing experience, Workshop	Socialization, Externalization, Combination, Internalization	Smart contract, Decentralized, Consensus mechanism
6	MTO and BQ	Document sharing, Discussion, Sharing experience, Workshop	Socialization, Externalization, Internalization	Smart contract, Decentralized, Consensus mechanism, High Security
7	Construction drawing and specification	Document sharing, Discussion, Sharing experience, Workshop	Socialization, Externalization, Internalization	Smart contract, Decentralized, Consensus mechanism, Immutable

(continued)

Table 1 (continued)

No	EPC-X project activities	Knowledge management process	SECI model stages	Blockchain features
Procurement				
8	Procurement plan	Discussion, Sharing experience, Internet browsing	Combination, Internalization	Smart contract, Decentralized, Transparency, High security
Vendor				
9	Vendor databased—Product Information	Document sharing, Sharing experience, Internet browsing	Combination, Internalization	Decentralized, Transparency, High security
Material Control				
10	Material tracking	Discussion, Workshop, Sharing experience	Socialization, Combination, Internalization	Decentralized, Transparency, High security, Immutable
11	Material storage	Discussion, Workshop, Sharing experience	Socialization, Combination, Internalization	Decentralized, Transparency, High security, Immutable
Fabrication—Construction				
12	Fabrication procedure	Workshop, Sharing experience	Socialization, Combination, Internalization	Smart contract, Decentralized, Consensus mechanism
13	Construction procedure	Workshop, Sharing experience	Socialization, Combination, Internalization	Smart contract, Decentralized, Consensus mechanism
14	QA-QC procedure	Workshop, Sharing experience	Socialization, Combination, Internalization	Smart contract, Decentralized, Consensus mechanism
15	Completion certificate	Document sharing	Combination	Smart contract, Decentralized, Immutable
Commissioning				
16	Commissioning procedure	Workshop, Sharing experience	Socialization, Combination, Internalization	Smart contract, Decentralized, Consensus mechanism

(continued)

Table 1 (continued)

No	EPC-X project activities	Knowledge management process	SECI model stages	Blockchain features
17	Performance test	Workshop, Sharing experience	Socialization, Combination, Internalization	Consensus mechanism, Transparency, Decentralized, Immutable

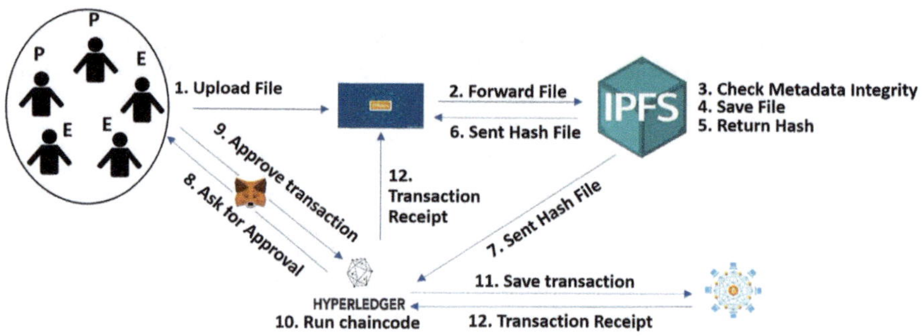

Fig. 2 Blockchain based knowledge management architecture

4.3 Knowledge Management Team Design

The design of the knowledge management team involved in the KMS application for the EPC project is tailored to the existing users. Blockchain based KMS managers consist of experts and other peers.

4.4 Knowledge Management Model

From a series of analyses based on the Tiwana framework, a blockchain based knowledge management system model can be created as follows in Fig. 3.

The blockchain based Knowledge Management System model for the EPC-X project consists of a data repository containing knowledge, files, and discussion summaries. Then this data is verified by peers comprised of experts, content management, consensus, and search modules.

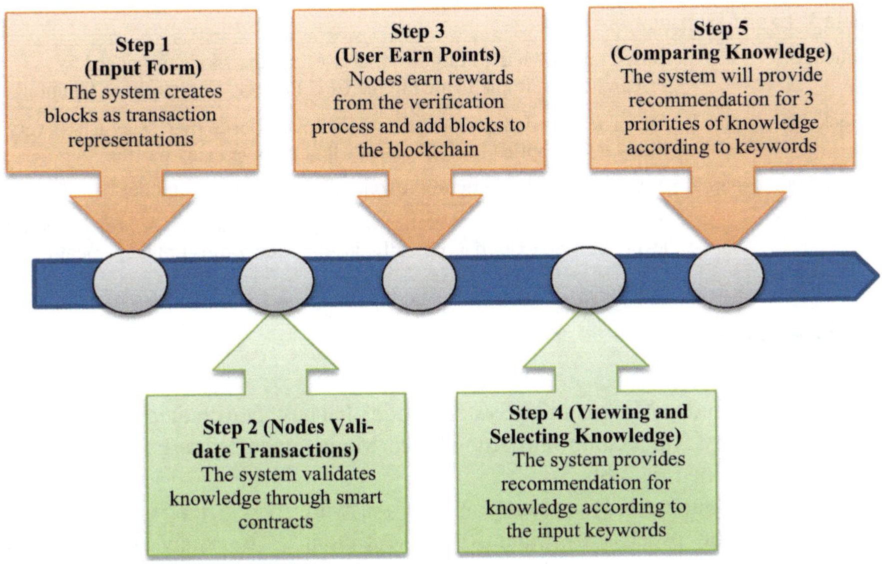

Fig. 3 Blockchain based KMS model

Table 2 Entity relationship diagram blockchain-based KMS

<table>
<tr><td>Key partners and customer segments</td><td>Key activities</td><td>Key resources</td><td>Cost structure</td></tr>
<tr><td rowspan="4">1. EPC project contractor
2 EPC project owner
3 EPC project planning and supervisory consultant
4 Cryptoasset service provider</td><td>1. Knowledge gathering and processing
2. Platform development</td><td>1. Blockchain
2. Big data
3. Knowledge verification</td><td>1. Platform development
2. Production cost
3. Peers acquisition cost
4. Overhead cost</td></tr>
<tr><td>Value propositions</td><td>Customer relationships</td><td>Channels</td></tr>
<tr><td>Fast, secure and distributed solution for storing project data</td><td>1. User friendly website
2. Points as gifts</td><td>1. Service excellence
2. Advertising
3. Media release</td></tr>
</table>

4.5 *Implementation*

Based on the needs of the system, the design of the entity relationship diagram (ERD) for the KMS prototype. The EPC project consists of entities that are interrelated within the system, the ERD diagram is as follows in Table 2.

Table 3 Expert validation

Validation	Based on the Model Framework above, can the Model proposed by researchers be implemented in the EPC Project? Explain Your Opinion
Blockchain expert 1	Yes. The knowledge management model proposed by researchers is very likely to be applied in EPC projects that involve many parties
Blockchain expert 2	Yes, but it needs a little improvement on the business model
KM expert	Theoretically correct, overall it will depend on the type of project and the level of complexity but basically has the same project management flow

5 Discussion

From the Knowledge Management System model that has been created by researchers and validated by Blockchain and Knowledge Management experts, the following results are obtained in Table 3.

Qualitative data Analysis using Nvivo software is carried out during the expert validation process by "identifying sentiment". The analysis reveals that the Blockchain based Knowledge Management System Model in the EPC Project returns "Moderately Positive" and "Very Positive" results of 44.19% and 55.81% respectively.

6 Conclusion

Blockchain technology implemented in the Knowledge Management System of the EPC Project returns a "Moderately Positive" and "Very Positive" results of 44.19 and 55.81%. Analysis employing the Tiwana framework reveals designing and implementing blockchain-based KMS models is able to produce prototypes that can be applied to support the implementation of EPC projects, thereby improving knowledge management and sharing among EPC project stakeholders. The developed model carries the potential to be applied to several other projects outside of the EPC project. In addition, Blockchain's unique features are expected to improve project time performance.

Acknowledgements The authors would like to thank University of Indonesia for the support given to this research.

References

1. Zhu F, Sun X, Xu Z, Haider Z (2014) A knowledge integration framework of EPC project based on KBS and stakeholder networks. Int J Innov Comput Inf Control 10(2)

2. Mee San K, Choy CF, Fung WP (2019) The potential and impacts of blockchain technology in construction industry: a literature review. IOP Conf Ser: Mater Sci Eng
3. Turk Z, Klinc R (2017) Potentials of blockchain technology for construction management. In: Creative construction conference 2017, CCC 2017, 19–22 June 2017
4. Akhavan P, Philsoophian M, Rajabion L, Namvar M (2018) Developing a blockchain knowledge management model (BCKMM) beyond traditional knowledge management. In: 19th European conference on knowledge management ECKM 2018
5. Tiwana A (1999) The knowledge management toolkit: practical techniques for building a knowledge management system, 1st edn. Prentice Hall, New Jersey
6. Nazori A, Samdi, Wirda, Hilwa (2014) Prototipe knowledge management system (KMS) Praktikum Perbankan Syariah Berbasis Web dan Android dengan Kerangka Kerja "Tiwana" pada Perguruan Tinggi: Studi Kasus STEI SEBI. Jurnal Telematika MKOM 6(2)
7. Zack MH (1999) Developing a knowledge strategy, vol 41, no 3. California Management Review. Nyame G, Qin Z, Kwame OBOA, Emmanuel BS (2020) An ECDSA approach to access control in knowledge management systems using blockchain. MDPI J
8. Tian F (2016) An agri-food supply chain traceability system for China based on RFID & blockchain technology. In: 13th International conference on service systems and service management (ICSSSM), pp 1–6
9. Zheng Z, Xie S, Dai HN (2018) Blockchain challenges and opportunities: a survey. Int J Web Grid Serv 14(4):352–375

Risk Allocation Implementation Analysis of Public-Private Partnership for Infrastructure Project (Case Study of the Solo-Yogyakarta-NYIA Kulon Progo Highway Project)

Widi Hartono, Aloysia Putri Hastari Purnomo, and Sunarmasto

Abstract Indonesia is now concentrating on infrastructure development; the construction of roads, bridges, airports, dams, and toll roads are seemingly given more attention in comparison to others. As the infrastructure sector rapidly grows, it opens wider opportunities for investors. This study aims to identify and monitor the risk allocation process and preferences along with risk response(s) applied by related parties due to critical risk variables that might have happened in highway projects—especially Public–Private Partnership based infrastructure projects situated in Central Java, Indonesia; the Solo—Yogyakarta—NYIA Kulon Progo Highway Project. The result from collected questionaries—processed with descriptive analysis—shows that: (1) 15% of the risk variables are allocated to the Ministry of Public Works and Housing (the Government), (2) 50% are allocated to private parties, and (3) 35% of the variables are shared between both public and private parties. Using the Risk Failure Mode and Effect Analysis (RFMEA), Risk Score and Risk Priority Number of each risk variable is compared, resulting in the knowledge that 7 out of 60 risk variables related to the project completion experienced: delays, budgetary and work schedule amendment, project supervision and control, and risk events that occur due to force majeure. Those are 7 of critical risks that would need to be responded or mitigated.

Keywords Risk · Public-Private Partnership · Highway · Risk Failure Mode and Effect

---

W. Hartono (✉) · A. P. H. Purnomo · Sunarmasto
Civil Engineering Department, Sebelas Maret University, Surakarta, Indonesia
e-mail: wieds_ts@ft.uns.ac.id

Sunarmasto
e-mail: sunarmasto@ft.uns.ac.id

S. A. Kristiawan et al. (eds.), *Proceedings of the 5th International Conference on Rehabilitation and Maintenance in Civil Engineering*, Lecture Notes in Civil Engineering 225,
https://doi.org/10.1007/978-981-16-9348-9_76

1 Introduction

Every construction project faces triple constraints (cost, time, and quality) as its main project success indicators. Besides, there is no risk-free project; there are plenty of risk possibilities that might affect project success. Well-planned risk management is much needed for anticipating harmful risks and also to predict various risks that significantly affect or are supposed to inhibit project success as soon as possible. Avoidance may seem the answer to all risks, but avoiding risks also means missing the potential gain that by accepting or retaining the risk may have allowed (Sichone 2002) [1].

Risk allocation is one of the tipping points to start risk management, especially for Public-Private Partnership (PPP) infrastructure projects, which are conducted not only by the government or private parties alone but both parties involved. Risks should be allocated as precisely as possible, as PPP projects serve a wide range of goals [2]. The Public–Private Partnership process demands not only carefulness but also extra transparency in allocating risks, as it shows not only the project's performance but the government's performance as well.

This study aims to get to know the risk allocation scheme and process in a highway project based on its Public–Private Partnership and to understand responses taken by related parties due to critical risk variables which might occur in highway projects, especially the Public–Private Partnership-based infrastructure project located in Central Java, Indonesia; the Solo—Yogyakarta—NYIA Kulon Progo Highway Project.

2 Review of Literature

In Architectural Construction and Engineering (ACE) projects, the risk allocation of related parties has specific mechanisms, clause types, payment methods, and responsibilities [1]. The risk allocation process in a PPP project always has a significant impact on the whole project; risk allocation implementation can only be successful when it is optimally implemented between related parties [3]. Maharani et al. [4] pointed out that in a PPP project, risk allocation strategies do not only define which parties are most capable of accepting risks, they also define the best timing, and the most suitable risk solution—all of which need to be put in consideration to achieve optimal risk allocation. To optimize the risk allocation process, Wang et al. [5] suggested that the government has a significant role in PPP projects; he claimed that a better government environment leads to better and more effective regulation, thus attracting more powerful investors; posing less risk, increasing profitable potentials.

In technical terms, Nasirzadeh et al. [6] stated that if risk allocation in a construction project is not done properly, the contractor party will suffer huge losses. This may result in a project cost overrun, as contractors have to use defensive strategies to keep work schedules and goals on track and against one-sided risk allocation. In 2011 Rosadin [7] researched PPP Optimization, while in 2015, Nurfarida [8] researched PPP for Performance-Based Contract (PBC) infrastructure projects.

To guide all parties related to risk allocation—especially for the PPP projects in all sectors. The World Bank has issued a Public–Private Partnership Reference Guide 3.0 [9]. This guidebook was adapted by the Indonesian Government; they published PPP Risk Allocation Tool 2019 Edition [9], and *Acuan Alokasi Risiko 2020* [10]. These handbooks are part of the references used for this research.

3 Methodology

3.1 Data

Collected data from questionnaires are processed with descriptive statistics to find the percentage of risk allocation preference from (1) the government as the owner of the project, and (2) private parties as project initiators and contractors; a risk is allocated to certain parties based on popular votes by participants.

Risk Failure Mode and Effect Analysis (RFMEA) compares the Risk Score and Risk Priority Number with the critical value. Risk Score (RS) and Risk Priority Number (RPN) values are processed by multiplying the probability, severity, and detection score of every determined risk variable. The calculation is based on the following equations [11]:

$$RPN = O \times S \times D \tag{1}$$

$$Risk\ Score = O \times S \tag{2}$$

where O represents the Occurrence (Probability) value, S represents the Severity (Impact) value and D represents the Detection value.

And to define critical value for RPN, the calculation is based on the following equation:

$$\text{Critical Value RPN} = \frac{\varepsilon\, RPN}{N\ cases} \tag{3}$$

To add, to determine the RS's critical value, this following risk level matrix is used (Table 1).

Table 1 Risk level matrix

Occurrence	Severity				
	1	2	3	4	5
5	Medium	Medium	High	High	High
4	Low	Medium	Medium	High	High
3	Low	Medium	Medium	Medium	High
2	Low	Low	Medium	Medium	Medium
1	Low	Low	Low	Medium	Medium

On Risk Failure Mode and Effect Analysis, critical risks are determined by comparing Risk Score and Risk Priority Number with the critical value. If the RS and or RPN value of a particular risk variable is equal to or more than the critical value, that indicates that the risk variable is categorized as a critical risk that needs to be responded at the early stage of the project. This method focuses on examining risk variables based on its detection mode effectiveness; one is stated critical as it has a less effective detection mode [11].

3.2 Risk Variables Identification

In this research, risk variables are determined based on risk factors from supporting research and risk matrix compiled by PT. Penjaminan Infrastruktur Indonesia (Persero)—which are adapted into research objectives. Risk variables used in this research are identified to analyze how risk allocation between related parties is done concerning government regulation, work scheme, construction process, operational policy, and force majeure issues. There are 60 risk variables chosen, classified into 6 groups;

- V1-V12 (Risk variables related to contract and regulation)
- V13-V23 (Risk variables related to location and land acquisition)
- V24-V30 (Risk variables related to design issues)
- V31-V46 (Risk variables related to construction process)
- V46-56 (Risk variables related to operational policy)
- V57-V60 (Risk variables related to *force majeure* issues)

4 Case Study Background

Solo-Yogyakarta-New Yogyakarta International Airport (NYIA) Kulon Progo Highway Project is considered as one of the National Strategic Projects (*Proyek Strategis Nasional*) that would be part of Trans Java Highways. It is planned to be the fastest route to connect three international airports from Yogyakarta to Central Java. This highway sets out to be the connector between 3 big cities, the golden triangle of Yogyakarta and Central Java; Semarang, Solo, and Yogyakarta.

This project's construction is set out to be built in gradual steps and should be fully operated by the end of 2024. There will be three location markers to its construction; starting from Kartasura Region (as a follow-up route from the Solo-Semarang Highway) to Purwomartani, Yogyakarta as the first marker, then from Purwomartani to Gamping, Yogyakarta as the second marker (where most of the construction designed to be an elevated road), and the last section would start from Gamping and lead to Kulon Progo, Bantul Regency, Yogyakarta.

As a Public-Private Partnership-based Infrastructure Project, this project plans to have a 40 years long concession time, and costs nearly Rp 15.81 trillion (USD 1 billion) for its construction, and Rp 26,637 trillion (USD 1850 billion) for investments value. Financially, this project is estimated to have USD 2.104 million of NPV (Net Project Value) and 12% of IRR (Internal Rate of Return) (simpulkpbu.com). This PPP project was proposed by PT. Gama Group; PT. Daya Mulia Turangga as a consortium, together with PT. Adhi Karya (Persero) Tbk. and PT. Jasa Marga (Persero) who appointed PT. Jogjasolo Marga Makmur as construction and operation provider (Fig. 1).

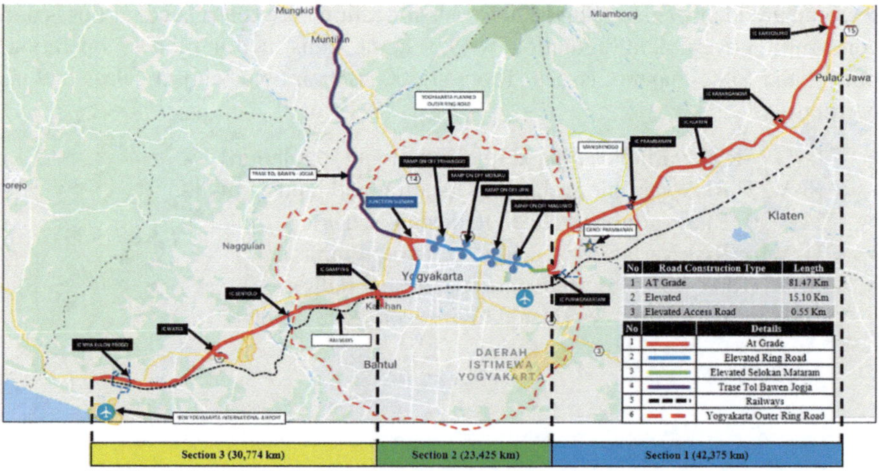

Fig. 1 Planned route of Solo-Yogyakarta-NYIA Kulon Progo Highways (courtesy of PT. Jogjasolo Marga Makmur)

5 Analysis

There were 30 respondents who have given their perspectives; respondents chosen were from the Ministry of Public Works and Public Housing (of involved regions) and private parties. Detailed information of respondents is presented in Table 2.

To process risk allocation implementation of the Solo-Yogyakarta-NYIA Kulon Progo Highway Project, collected data are processed with descriptive statistics to figure out the related parties' preferences on how risks have or should be allocated. Risk allocation analysis can be observed in the following table.

As shown in Table 3, 15% of risk or 9 out of 60 risk variables are allocated to the public party, 50% of risks or 30 of them are allocated to private parties in partnership with the Government (investors), and 35% of risks or about 21 risk variables are shared between related parties (public and private).

To figure out which variables included in critical risk need to be responded to, Risk Failure Mode and Effect Analysis was used. Participants was asked to score risk variables Occurrence, Severity and Detection Methods from 1 to 5 (mild to very serious). Analyzed data is shown in Table 4.

As shown in Table 4, 7 out of 60 risk variables are critical. Those are the ones that need to be responded to as soon as possible. Those variables are classified as critical because their RPN and RS values are higher than their critical values; those determined as High-Risk Variables are RPN $\geq$ 26, RS $\geq$ 10, which is an occurrence where high probability and impact severity is unsupported by suitable nor applicable detection method.

Those critical variables found are related to the occurrence of delays (V40, V35, V12), quality controls (V39), budgetary and work schedule changes (V27, V28), and natural disaster issues (V57). Responds taken by related parties are risk transfer, avoidance, risk sharing, and acceptance.

Table 2 Detailed information of respondents

Respondents	Quantity
Public Works Department of Central Java	3
Public Works Department of Klaten Regency	6
PT. Jogja Solo Marga Makmur	3
PT. Trans Marga Jateng	1
Public Works Department of Boyolali Regency	6
Public Works Department of Karanganyar Regency	6
PT. Jasamarga Solo-Ngawi	5

Table 3 Risk allocation analysis

Code	Risk variables	Risk allocation (%)			Risk allocation
		Public	Private	Shared	
V1	Regulation and tax changes (Generally or specifically)	60.00	20.00	20.00	Public
V2	Design approval delay	10.00	60,00	30.00	Private
V3	Project permission and approval delay	13.30	50.00	36.70	Private
V4	Delays in location access permission	13.30	63.40	23.30	Private
V5	Parastatal risk	16.60	56.70	26.70	Private
V6	Competitor risk	16.70	63.30	20.00	Private
V7	Non-toll network management	50.00	33.30	16.70	Public
V8	Road network and transportation connectivity	50.00	16.70	33.30	Public
V9	Government support delay (subsidies, incentive, tax)	30.00	33.30	36.70	Public
V10	Unorganized decision-making process	26.70	23.30	50.00	Shared
V11	Expropriation risk	16.60	26.70	56.70	Shared
V12	Land acquisition delay and land price increase	50.00	10.00	40.00	Public
V13	Land couldn't be acquisitioned	53.30	10,00	36.70	Public
V14	Land couldn't be used after acquisition process	50.00	23.30	26.70	Public
V15	Obscurity of land status	66.70	13.30	20.00	Public
V16	Unanticipated location condition	23.30	33.30	43.40	Shared
V17	Complex resettlement process	46.70	16.60	36.70	Public
V18	Affected artifacts and antiquities	23.30	20.00	56.70	Shared
V19	Soil structure risk	20.00	43.30	36,70	Private
V20	Contaminated environment	10.00	43.30	46.70	Private
V21	Damaged conservational area	20.00	20.00	60.00	Shared
V22	Hindered access of public transportation	20.00	23.30	56.70	Shared
V23	Disturbed public security and convenience	13.30	36.70	50,00	Shared
V24	Project scope accuracy	10.00	80.00	10.00	Private
V25	Resources' specification (tools, materials, human resources)	6.70	80.00	13.30	Private
V26	Technology implementation	6.60	76.70	16.70	Private
V27	Project budgetary changes	10.00	43.30	46.70	Private

(continued)

Table 3 (continued)

Code	Risk variables	Risk allocation (%)			Risk allocation
		Public	Private	Shared	
V28	Work schedule changes	0.00	60.00	40.00	Private
V29	Undefined output specification	3.30	40.00	56.70	Shared
V30	Design correction or adjustment	3.30	33.30	63.40	Shared
V31	Project scope change after contract signing	0.00	30.00	70.00	Shared
V32	Construction cost increase	10.00	56.70	33.30	Private
V33	contractor/sub-contractor bad performance	13.30	80.00	6.70	Private
V34	Resources' availability	0.00	90.00	10.00	Private
V35	Resources' stock delays	0.00	90.00	10.00	Private
V36	Work methods discrepancy or defective tool	0.00	90.00	10.00	Private
V37	Unpredicted geotechnical condition	3.30	60.00	36.70	Private
V38	Unapplicable design	1000	40.00	50.00	Shared
V39	Quality control and work target control	13.30	60.00	26.70	Private
V40	Inhibition of completion	0.00	70.00	30.00	Private
V41	Unfinished project by contractor/subcontractor error	3.30	76.70	20.00	Private
V42	Security and safety issues	3.30	63.40	3330	Private
V43	Testing and commissioning risk	330	63.40	33.30	Private
V44	Divergent standards	3.30	60.00	36.70	Private
V45	Missed communication	3.30	40.00	56.70	Shared
V46	Project management failure	6.70	53.30	40.00	Private
V47	Facilities availability	6.60	66.70	26.70	Private
V48	Industrial action	60.70	43.30	50.00	Shared
V49	Socio-culture conflict	60.70	20.00	73.30	Shared
V50	Project control and monitoring default	30.30	43.30	53.40	Shared
V51	Life cycle estimation error	60.60	46.70	46.70	Private
V52	Inefficient operation performance	16.60	56.70	26.70	Private
V53	Irregular facilities availability	13.30	60.00	26.70	Private
V54	Toll technology and information system failure	26.70	46.70	26.60	Private)
V55	Accidents due to design error	20.00	30.00	50.00	Shared

(continued)

Table 3 (continued)

Code	Risk variables	Risk allocation (%)			Risk allocation
		Public	Private	Shared	
V56	Failure by accident	13.30	33.30	53.40	Shared
V57	Natural disasters	16.70	13.30	70.00	Shared
V58	Politic force majeure (war, demonstration)	36.70	16.60	46.70	Shared
V59	Extreme weather	33.30	23.30	43.40	Shared
V60	Long drawn force majeure	20.00	16.70	63.30	Shared

Table 4 Critical risk analyzed with RFMEA

Code	Risk variables	O	S	D	RPN	RS
V40	Inhibition of completion	4.87	3.87	2,60	48.93	18.82
V39	Quality control and work target control	4.97	3.57	3.53	62.59	17.71
V28	Work schedule changes	3.57	4.53	2.53	40.96	16.17
V35	Resources' stock delays	3.17	4.63	2.83	41.57	14.67
V57	Natural disasters	3.63	3.57	3.73	48.38	12.96
V12	Land acquisition delay and land price increase	3.77	3.67	2.73	37.75	13.81
V27	Project budgetary changes	3.57	3.67	2.77	36.18	13.08

6 Conclusion

It is known from the research that in an unsolicited PPP infrastructure project, private parties are being responsible for most of the risk as their role as investor, designer, contractor, and operational agent. As it is a joint cooperation of private and Ministry of Public Works and Housing, Ministry of Public Works and Housing (government) is giving their support as they processed supportive regulation, working on land acquisition, and general public conditioning.

Those critical variables found are related to the occurrence of delays (V40, V35, V12), quality controls (V39), budgetary and work schedule changes (V27, V28), and natural disaster issues (V57). The responses taken by related parties are risk transfer, avoidance, risk-sharing, and acceptance.

References

1. Tembo-Silungwe CK, Khatleli N (2017) A Conceptual model for risk allocation in the construction industry. Am J Appl Sci 14(7):690–700. https://doi.org/10.3844/ajassp.2017.690.700

2. Subprasom K, Chen A (2005) Analysis of policy and regulation on build-operatetransfer scheme: a case study of the ban Pongkanchanaburi motorway in Thailand. J Eastern Asia Soc Trans Stud 6(January):3883–3898
3. Valipour A, Mohammadi F, Yahaya N, Sawvari H, Noor NM (2014) Identification and evaluation of risk allocation criteria and barriers: a Malaysian public private partnership project case study. J Appl Sci 14(18):2023–2031.
4. Maharani Y, Palupie R, Yuniarto HA (2016) Alokasi Risiko Proyek Infrastruktur Dengan Skema Kerjasama Pemerintah dan Badan Usaha (KPBU): Suatu Tinjauan Literatur. Seminar Nasional Teknik Industri Universitas Gajah Mada 96–103
5. Wang Y, Gao HO, Liu (2019) Incentive game of investor speculation in PPP highway projects based on the government minimum revenue guarantee. Trans Res Part A Policy Pract 125(15):20–34. https://doi.org/10.1016/j.tra.2019.05.006
6. Nasirzadeh F, Khanzadi M, Rezaie M (2014) Dynamic modeling of the quantitative risk allocation in construction projects. Int J Project Manage 32(3):442–451. https://doi.org/10.1016/j.ijproman.2013.06.002
7. Rosadin IM (2011) Optimalisasi Alokasi Risiko dalam Pembangunan Jalan Tol Studi Kasus: Jalan Tol Bandara Juanda -Tanjung Perak. Universitas Indonesia, Jakarta
8. Nurfarida S (2015) Alokasi risiko proyek pembangunan jalan dengan sistem Performance Based Contract. Institut Teknologi Sepuluh November, Surabaya
9. The World Bank (2017) Public-Private partnerships reference guide. PPP Knowl Lab Library 3:238. http://doi.wiley.com/10.1002/tie.5060290205
10. PT (2020) Penjaminan Infrastruktur Indonesia. Acuan PT. Penjaminan Infrastruktur Indonesia. (2020) Acuan Alokasi Risiko 2020. JakartaAlokasi Risiko 2020. Jakarta
11. Carbone TA, Corporation FS (2004) Project risk management using the project risk FMEA. Eng Manage J 16(4)

Risk Assessment of Construction Project Scheduling

Zetta Rasullia Kamandang

Abstract To perform a successful construction project, accuracy in managing project risks has to be fulfilled by project parties. Regarding time-related risk, failure to estimate project construction duration is one of the common problems that can affect the entire project performance during the execution. To avoid the impact of these risks, a scheduling method using probabilistic duration has been developed, namely Monte Carlo simulation. In this study, optimistic, most likely, and pessimistic duration of each task and task relationship information of a real case study were collected and were examined in several steps, (1) determining the number of iterations, (2) Monte Carlo simulation, (3) scheduling analysis, and (4) probabilistic of success analysis. Based on the calculation, the success probabilities of as-planned, optimistic, most likely, pessimistic, and simulated durations are 0.479%, 0.054%, 2.108%, 57.912%, and 52.575% respectively. By only have 0.479% of success probability, the as-planned duration is unsuited to be submitted as project completion duration in the contract. Furthermore, to avoid risk in scheduling, the simulation suggests the project parties to allocate at least 332 days of project duration which has a 100% of success probability.

Keywords Project scheduling · Monte Carlo simulation · Success probability

1 Introduction

Construction project is a part of the country's development which is executed to expand the infrastructure and building facilities. In its implementation, a construction project may be associated with various potential risks that can harm the project parties and lead to loss of the project's values [1, 2]. In literature, construction project risks are divided into several categories. International Construction Risk Assessment Model (ICRAM) [3] proposed three major categories of risk; risk at a

Z. R. Kamandang (✉)
University of Pembangunan Nasional 'Veteran' Jawa Timur, Jl. Rungkut Madya No.1, Gunung Anyar, Surabaya, Jawa Timur 60294, Indonesia
e-mail: zerasullia.ts@upnjatim.ac.id

S. A. Kristiawan et al. (eds.), *Proceedings of the 5th International Conference on Rehabilitation and Maintenance in Civil Engineering*, Lecture Notes in Civil Engineering 225,
https://doi.org/10.1007/978-981-16-9348-9_77

macro level, risk in construction market, and risk at project level. Skorupka [4] also divided risk that is linked to the stage of the construction industry investment process including preliminary design, tender, detailed design, construction, and completion of payment. Another research [5] provided risk approaches into cost-related, time-related, quality-related, environment-related, and safety-related aspects. For example, the risks associated with time are unsuitable construction program planning, inadequate program scheduling, tight project schedule, design variations, excessive approval procedures in administrative government departments, variations by the client, and high performance or quality expectations.

To perform a successful construction project, accuracy in managing project risks has to be fulfilled by project parties. Risk management (RM) can be described as a process that helps project parties to control arising risk in the project without affecting time, cost, duration, and scope of the project. RM assists to obtain better control over project scheduling, estimation, and quality [6], it also gives project parties beforehand ideas about future situations that may occur and make them prepare a decision as early as possible to avoid any risk later.

Regarding time-related risk, failure to estimate project construction duration is one of the common problems that can affect the entire project performance during its execution. The total project duration is one of the essential indicators representing the efficiency of resources management to accomplish the results with the proper quality standards [7]. Therefore, an assessment of construction project scheduling has to be done precisely to reduce any loss possibilities.

2 Case Study

This study uses a real case project in Indonesia with more than Rp. 81 billion of the project's contract to assess its risk of duration estimate at the planning stage as a case study. The project is a six-floors building constructed in Bengkulu, Indonesia, consisting of several works: structure work, architecture work, finishing work, interior work, façade work, supporting building work, and landscape work with 294 days of execution.

The collected data from this case is an as-planned schedule with duration and tasks relationship information. The total duration of this project later will be used as a decision-making baseline while the relationship of the tasks plays a significant role in the calculation. Table 1 shows the information of case study data for structure work.

Table 1 The information of structure work data

ID	Task Name		Dur. (day)	Predecessors		
7	Structure work	Foundation	30	3	SS	
8		Semi-basement floor	28	7	FS−	14
9		1st floor	14	8	FS−	20
10		2nd floor	14	9	FS	
11		3rd floor	14	10	FS	
12		4th floor	14	11	FS	
13		5th floor	14	12	FS	
14		6th floor	14	13	FS	
15		Rooftop floor and roof	14	14	FS	
16		Ramp and front gardens	7	7	FS+	13
17		Pit lift & core lift	105	7	FS	
18		Stairs	63	10	SS+	10

3 Material and Methodology

To conduct a risk assessment of construction project scheduling, this study implements Monte Carlo simulation to a case study that illustrates as a flowchart in Fig. 1. Monte Carlo simulation itself is a mathematical technique that uses a randomizing device to assign a value to each random variable, in accordance with its probability distribution [8]. It helps to explain the impact of risk and uncertainty in prediction and forecasting models.

After collecting the total duration and tasks relationship of the case study, a questionnaire was organized to collect information about the pessimistic, most likely, and optimistic duration of each task and was distributed to the project contractor of the case study. The main data that used in this simulation is the pessimistic, most likely, and optimistic duration of each task and its relationship to each other. By building models using Crystal Ball software, a Monte Carlo simulation is conducted to find out the ideal project duration that able to cover risk in project scheduling and is expected to calculate the probability of success of construction project duration. Table 2 shows the questionnaire results of structure work.

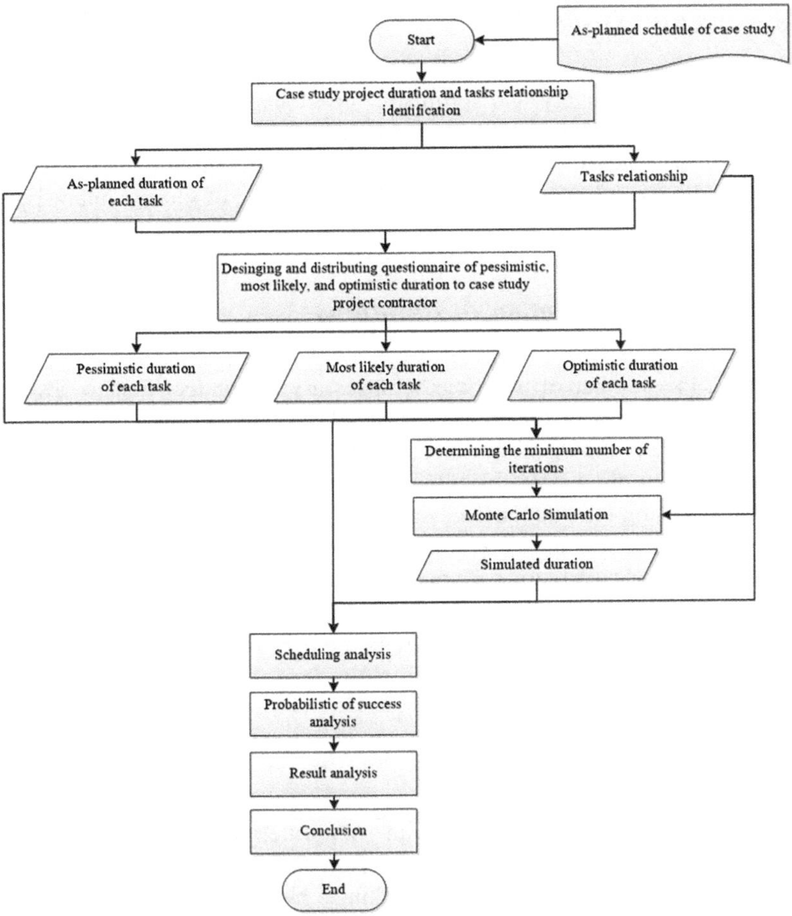

Fig. 1 The flowchart of this study

Table 2 Questionnaire result sample for structure work

ID	Task name		Duration (day)		
			Optimistic	Most likely	Pessimistic
7	Structure work	Foundation	30	36	43
8		Semi-basement floor	28	31	38
9		1st floor	14	20	27
10		2nd floor	14	14	21
11		3rd floor	14	14	21
12		4th floor	14	14	21
13		5th floor	14	14	21
14		6th floor	14	14	21
15		Rooftop floor and roof	14	16	23

4 Results and Discussion

The pessimistic, most likely, and optimistic of each task duration of the project case study are the main data that will be processed using Monte Carlo simulation by conducting several steps; (1) determining the number of iterations, (2) Monte Carlo simulation, (3) scheduling analysis, and (4) probabilistic of success analysis.

4.1 Determining Number of Iterations

Determining the number of iterations is necessary to conduct first to observe the minimum number of iterations that are needed to complete the whole simulation. Several steps in determining the number of Monte Carlo simulation are (1) calculating the standard deviation of optimistic, most likely, and pessimistic duration (population), (2) calculating the population average, (3) calculating the absolute error with the expected error is less than 2%, and (4) calculating the number of iterations [9]. The results of those calculations are presented in Table 3 below.

Based on the calculation above, the main data of this study will be processed using Monte Carlo simulation for 16500 iterations.

4.2 Monte Carlo Simulation

The main data that will be processed with 16,500 iterations of Monte Carlo simulation is the optimistic, most likely, and pessimistic duration of each task and task relationship. Starting the simulation by clicking a blank cell among the start and the end cell of each task, this blank cell later will be categorized as define assumption with the triangular distribution. For the first task, the start duration is 0 and the end duration is the summation of the start duration and the blank cell. For the second task, the start duration is the end duration of the previous task by also considering their relationship, while the second task end duration is obtained by summing the start duration with the blank cell of the second task. The same calculation is also applied for the remaining tasks. The end duration of the last task later will be assumed as define forecast and then the simulation can be conducted.

Table 3 The results of determining iterations number

Item	Calculation results	
Standard deviation	37.07595807	
Absolute error	0.872	
Average	43.6	
Iteration number	16270.25	≈16500

Table 4 Statistical results of the simulated project duration

Statistic	Fit: beta	Forecast values
Trials	–	16500
Base case	–	−97.00
Mean	311.71	311.71
Median	311.65	311.63
Mode	311.50	–
Standard deviation	6.89	6.89
Variance	47.53	47.53
Skewness	0.0582	0.0582
Kurtosis	2.94	2.94
Coeff. of variations	0.0221	0.0221
Minimum	255.23	285.82
Maximum	386.16	337.27
Mean Std. error	–	0.05

Monte Carlo simulation provides some information including the frequency, cumulative frequency, and statistical data (that consist of minimum, mean, median, skewness, kurtosis, etc.). Based on the calculation result, Monte Carlo simulation suggests 311.71 or 312 days as the project duration (mean result). Table 4 shows the statistical results of the simulated project duration.

4.3 Scheduling Analysis

Optimistic, most likely, and pessimistic duration of each task are not only used in Monte Carlo simulation to achieve the simulated project duration, but they also were processed by using Microsoft Project to calculate the total duration of the project. There are five total durations based on the schedule types that were obtained in scheduling analysis; total duration based on as-planned schedule, optimistic schedule, most likely schedule, pessimistic schedule, and simulation result schedule. The project total duration for each type schedule is presented in Table 5 below.

Based on the scheduling analysis results above, as-planned schedule duration is between optimistic and most likely duration. It indicates that the contractor has considered the ideal duration of the project.

Table 5 The total duration of the project based on schedule types

Schedule type	As-planned	Optimistic	Most likely	Pessimistic	Simulation result
Days	294	290	298	313	312

4.4 Probabilistic of Success Analysis

The probabilistic of success analysis for each type schedule is done by Monte Carlo simulation. Figure 2 and 3 showed the probability analysis for as-planned duration and for the simulated duration. As seen in Fig. 2, 294 days of as-planned duration only has 0.479% of success probability, while simulated duration has 52.575% (Fig. 3).

Table 6 presents the probability of success of other schedule types. For optimistic, most likely, and pessimistic duration with 290 days, 298 days, and 313 days of project completion duration, the probabilities are 0.054%, 2.108%, and 57.912% respectively. Furthermore, this simulation also presents its prediction of total project completion duration with expected probability which is summarized in Table 7. For example, to achieve 50% of success probability, the project completion duration is 312 days.

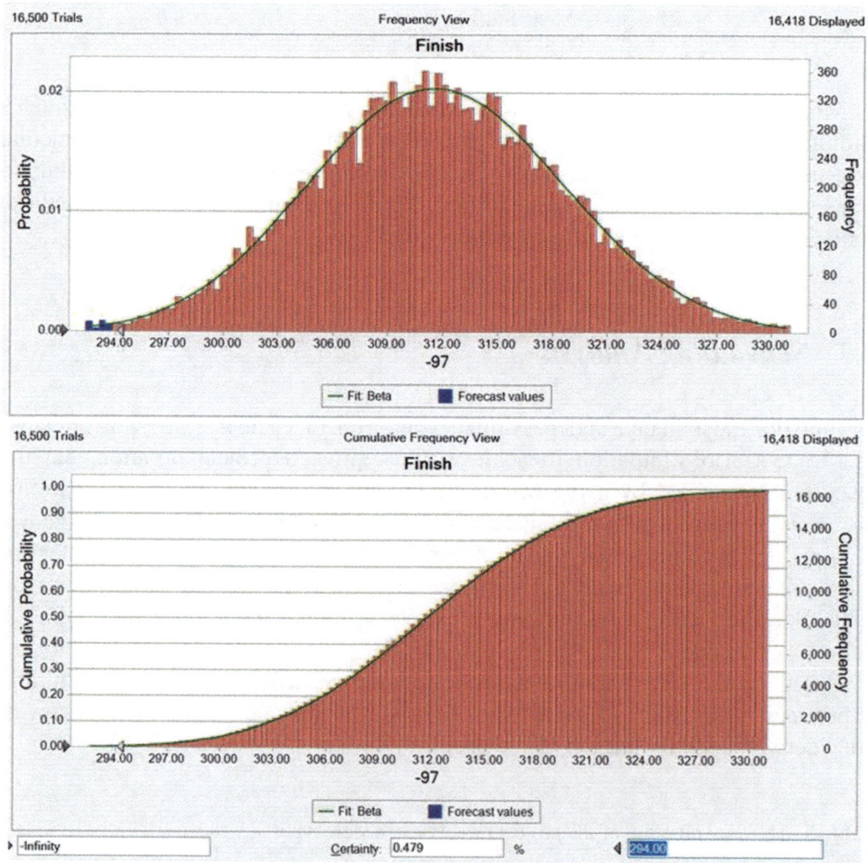

Fig. 2 Probability of success for as-planned duration

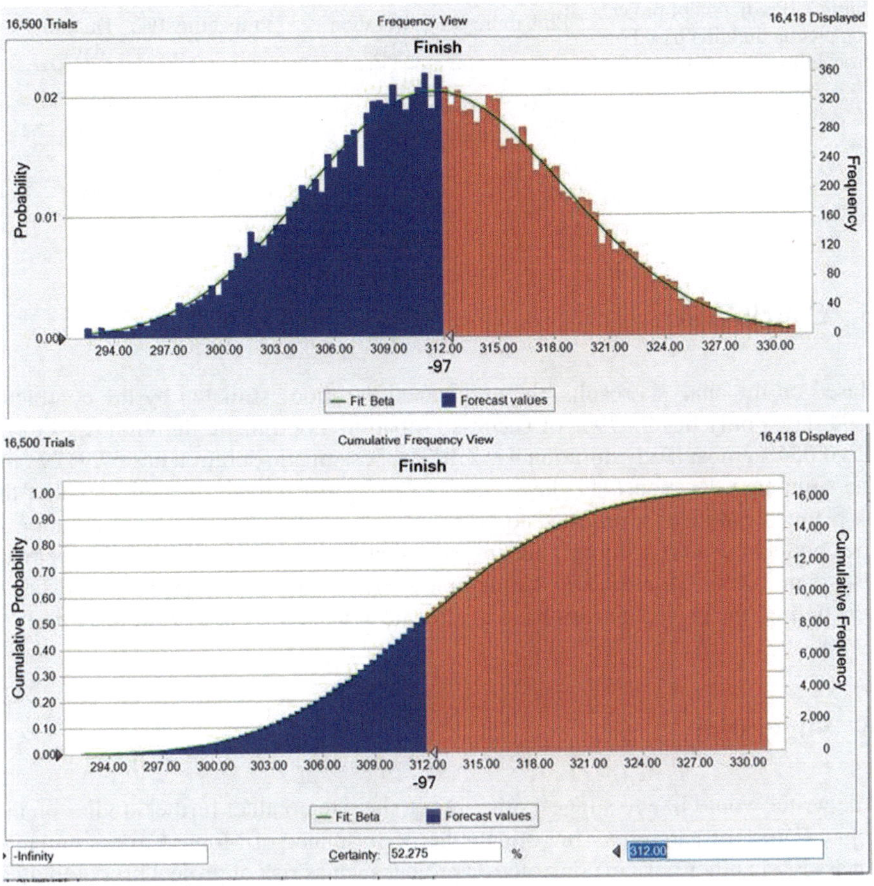

Fig. 3 Probability of success for the simulated duration

Table 6 Probability of success of project completion duration based on schedule types

Schedule type	Duration (day)	Probability of success (%)
As-planned	294	0.479
Optimistic	290	0.054
Most likely	298	2.108
Pessimistic	313	57.912
Simulation results	312	52.275

Table 7 Prediction of project completion duration based on probability of success

Probability (%)	Duration (day)	Probability (%)	Duration (day)
10	303	60	314
20	306	70	316
30	309	80	318
40	310	90	321
50	312	100	332

5 Conclusion

Based on the analysis results, the as-planned duration estimated by the contractor (294 days) only has 0.479% of success probability, optimistic duration (290 days) has 0.054%, most likely duration has 2.108%, pessimistic duration has 57.912% and the simulated duration (312 days) has 52.275%, respectively. It indicates that the as-planned duration is unsuited to be submitted as project completion duration in the contract because according to simulation results, its success probability is much less than 100%. Monte Carlo simulation analysis concludes that to avoid risk in scheduling, the project parties have to allocate at least 332 days of project duration.

6 Suggestion

The writer would like to suggest other researchers to conduct further studies on this topic. Future research may investigate the performance of Monte Carlo simulation on assessing other risks in construction project such as risk of project productivity or construction project emissions. The emissions topic itself is a rising one in research regarding the sustainability construction issue. Assessing risk using Monte Carlo could be used as a decision-making tool for construction project parties, therefore they can avoid and reduce the risks in the future.

References

1. Gładysz B, Skorupka D, Kuchta D, Duchaczek A (2015) Project risk time management—a proposed model and a case study in the construction industry. https://doi.org/10.1016/j.procs.2015.08.459
2. Pant D, Srinivas K (2019) Cost estimation of construction project using monte Carlo simulation. 16(2):13–19. https://doi.org/10.9790/1684-1602031319
3. Hastak M, Shaked A (2000) ICRAM-1: model for international construction risk assessment. J Manag Eng. https://doi.org/10.1061/(asce)0742-597x(2000)16:1(59)
4. Skorupka D (2005) The method of identification and quantification of construction projects risk. Arch Civ Eng 51(4):647–662

5. Zou P, Zhang G, Wang J-Y (2006) Identifying key risks in construction projects: life cycle and stakeholder perspectives
6. Ganame P, Chaudhari P (2015) Construction building schedule risk analysis using Monte-Carlo simulation. Int Res J Eng Technol 2(4):1402–1406
7. Koulinas GK, Xanthopoulos AS, Tsilipiras TT, Koulouriotis DE (2020) Schedule delay risk analysis in construction projects with a simulation-based expert system. Buildings. https://doi.org/10.3390/buildings10080134
8. Meerschaert MM (2013) Simulation of probability models. In: Mathematical modeling. Elsevier, pp 301–358
9. Fadjar A (2008) Aplikasi Simulasi Monte Carlo dalam Estimasi Biaya Proyek. SMARTek

Investigating Construction Project Delay Using Fault Tree Analysis Based on Its Dominant Risk on Private Project

Jojok Widodo Soetjipto, Amalia Martha Sukmana, and Syamsul Arifin

Abstract The team project has already arranged the project planning and scheduling, but so many projects have deviations in the schedule. Simultaneously, the project is highly dependent on the existence of all project participants and their environment. The previous study has resulted in many variables that cause project delays but has no ability to predict them in the future. In this research, the author did the survey and interviewed to solve this problem by knowing the causative factors and predicting the probability of their occurrence. Therefore, this study uses the FTA model to predict project delay events based on their variables and mitigate them. Due to so many risk variables, this study used the Pareto diagram to filter them. This diagram can reduce 16 variables into seven variables with the highest risk. The simulation results state that the probability rate of construction project delay on all variables is 0.1000. Still, each variable's contribution, namely environment, worker variables, contractor performance, owner roles, and Construction Management (CM) consultant roles, is 0.0766 0.0138, 0.0074, and 0.0064, respectively. This model has been validated based on existing project data and produces relatively the same occurrence, i.e., the project has behind schedule. Finally, this research also provides mitigation to minimize the risk of project delays.

Keywords Project delay · FTA · Simulation

1 Introduction

One of the indicators of project success is measured from the time of completion of the work in a cost-efficient manner and meets the owner's satisfaction. Private

J. W. Soetjipto (✉) · A. M. Sukmana · S. Arifin
Universitas Jember, Kalimantan 37, Jember 68122, Indonesia
e-mail: jojok.teknik@unej.ac.id

S. Arifin
e-mail: syamsul.teknik@unej.ac.id

S. A. Kristiawan et al. (eds.), *Proceedings of the 5th International Conference on Rehabilitation and Maintenance in Civil Engineering*, Lecture Notes in Civil Engineering 225,
https://doi.org/10.1007/978-981-16-9348-9_78

projects in Indonesia generally determine the completion time as the primary indicator in evaluating the project's success. They will give a penalty if the contractor is unable to complete the work within the defined time. Several previous studies on the benchmarks of construction project success, namely: (i) a construction project is declared successful if it can finish on time with a budget and quality that matches the project owner's expectations [1]; (ii) criteria for a successful construction project based on time, cost, quality, safety, client's satisfaction, employees' satisfaction, cash-flow management, profitability, environmental performance and learning, and development [2]; (iii) Critical Success Factors (CSFs) in project success are the cost (budget), time, quality and management [3]; etc.

Many construction projects are still experiencing delays even though planning and scheduling have been prepared. Problems will arise due to a discrepancy in the process with the actual reality [4]. The several random events caused the delay, either intermediate or primary. These events should be able to be identified, understood, and mitigated by the project manager.

This study aims to determine the probability of project delay based on the variables causing the delay that have dominant risks. The detailed mitigation used only the most dominant variable and ignoring others. This study uses a Fault Tree Analysis to identify delay factors and reveal the cause-and-effect relationship between them and the Pareto method to determine the variables with dominant risk.

2 Literature Review

2.1 Variables Causing Delays in Construction Projects

Many previous studies have resulted in various variables causing delays in construction projects. They used questionnaires and interviews with construction project experts and several literature studies to obtain the variables [5–8]. Based on the theoretical framework, the causes of construction projects' delays are grouped into two main factors, namely excusable delay and non-executable delay [9]. However, the variables that cause project delays depend very much on each case studied. The identified factors that can be the reason for delay are project identification (impractical duration, poor contract, scope, etc.); the contractor delay (team member, lower tender cost, improper planning, etc.); the owner delay (owner's inference, lack of decision power, payment, etc.); consultant delay (lack of decision power, suspension of work, late in approving, change in material, etc.); environment delay (bureaucracy of government, delay in issuing permits, change in regulation, criminal activity, conflict, etc.) [5]. The project delay variables are contractor delay, owner-related delay (payment progressively, meet-up to improve management, etc.), consultant-related delay (communication, coordination, etc.), material-related delay, labor-related delays, equipment-related delays, and external factor-related delays [6]. The most significant delay causes according [7]: internal environment (internal contractor), exogenous factors (according to the perception of all project parties), and input factors (resources project). Result of the investigation the causes of delays facing the Norwegian construction industry [8]: poor planning and scheduling; slow/poor decision-making process; internal administrative procedures and bureaucracy within project organizations; resources shortage (human resources, machinery, equipment); poor communication and coordination between parties; slow quality inspection process of the completed work; design changes during construction/change orders; sponsor/owner/client lack of commitment/clear demand; late improper design, office issues, and user's issues.

2.2 Risk Dominant Analysis of Variables Causing Delays in Construction Projects

The risk variable that causes project delay was so many variables so that need analyze the risk variable dominant. This dominant risk determination aims to obtain the most influential project delay variables for mitigation in more detail. The stages of dominant risk analysis are as follows [10]:

1. Interview the experts/practitioners to respond to the event's probability (P) and the impact (I) on each variable.
2. Analyzing the data's validity and reliability, they eliminated if there are invalid or unreliable variables.
3. Calculating the risk value using the equation below:

$$R = P * I \tag{1}$$

4. Determining the dominant risk uses the Pareto method based on the risk value from the largest to the cumulative 80% of the data.

2.3 Fault Tree Analysis

FTA is a method that uses a deductive approach to describe the process of system failure. FTA is build based on the symbols and gates to show the system behavior. The analysis of FTA uses the rules of Boolean Algebra. FTA is useful to describe the root cause of an accident logically. The FTA has already applied model delays in construction projects [11], investigation on the grounds of project problem [4], root-case analysis of construction cost over-run [6], determine delay in construction [7], model prediction deterioration of bridge [12], etc.

The FTA can find the root cause of potential failure systems and control them. The FTA can be modeled through the following steps:

1. Define system activity behavior
2. Define the top event
3. Define the tree of the structure
4. Expand each branch in successive levels of detail
5. Solve the fault tree for the combination of events contributing to the top event
6. Calculate the probability of failure.

3 Research Method

3.1 Background the Case Study

The case study taken in this research is the Suncity Sidoarjo Apartment Development Project. This project is a private project which is a high-rise building located on Jalan Pahlawan No. 01 Sidoarjo. Some of the obstacles that often occur in implementing this project are: (i) the location of this project which is very close to community settlement; (ii) coordination between the parties involved in this project because there are several participants, namely the contractor, the management construction consultant, and the owner; and (iii) a very complex scope of work.

3.2 Data Collection

The short project data was obtained from a construction company and included written records, i.e., detailed engineering design (DED) projects, planning-scheduling projects, a problem report and event logs, and an interview with the project manager. DED has been used to describe the work scope; the planning and scheduling could help determine the time frame of project work. The project's report and event logs give all of the project's milestones.

To obtain response data from the contractor, CM consultant, and owner, the author uses questionnaires and interviews. This questionnaire data is needed to: (i) confirm the variables that occur in the project. The author determines the variables based on both the previous study and the survey and data collecting that might occur in the project case, then confirm whether these variables occur or not (see Table 1); (ii) obtain risk data on each variable. The author makes questions about the probability and impact of each confirmed variable (see Table 2); and (iii) determines the probability and dependence logic between variables that cause the project to be late. The author interviewed all respondents to make FTA logic.

Table 1 The confirmed delay variables

Code	Risk variable	Yes		No		Relevant/no
A	Owner					
A1	Delay in contractors payments and related parties by the owner	9	75%	3	25%	Relevant
A2	Delay in approving change order by the owner	7	58%	5	42%	Relevant
A3	Delay in preparing the project land by the owner	9	75%	3	25%	Relevant
A4	Lack of communication between the owner and related parties	7	58%	5	42%	Relevant
A5	Delay in permit by the owner	8	67%	4	33%	Relevant
B	CM Consultant					
B1	Lack of oversight of design	8	67%	4	33%	Relevant
B2	Lack of coordination between CM and contractors	9	75%	3	25%	Relevant
B3	Lack of control over every job	10	83%	2	17%	Relevant
C	Contractor					
C1	Damage and effectiveness of equipment	11	92%	1	8%	Relevant
C2	Planning and control has no well carried out by contractors	7	58%	5	42%	Relevant
C3	A production system is no well	8	67%	4	33%	Relevant
C4	Lack of construction materials	7	58%	5	42%	Relevant
D	Environmental and worker conditions					
D1	The occurrence of work accidents	9	75%	3	25%	Relevant
D2	Problems with residents around the project	7	58%	5	42%	Relevant
D3	Late work due to environmental conditions	10	83%	2	17%	Relevant
D4	Soil conditions on construction projects	9	75%	3	25%	Relevant

Table 2 Percentage cumulative of risk value from largest to smallest

Code	Risk variable	P	I	R	R cum.	Percent cum.	Pareto diagram
A1	Delay in contractors payments and related parties by the owner	4	3	12	12	14%	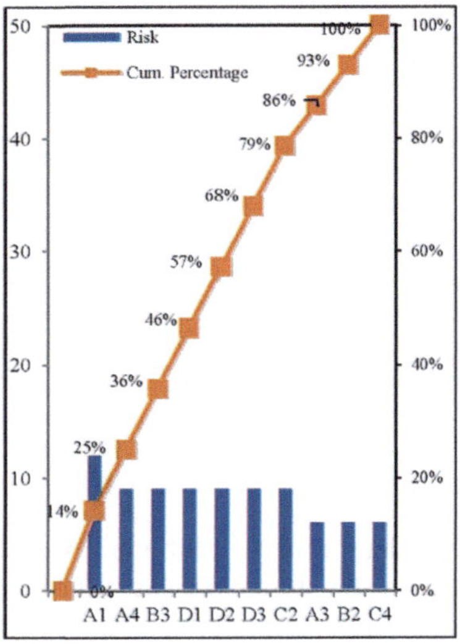
A4	Lack of communication between the owner and related parties	3	3	9	21	25%	
B3	Lack of control over every job	3	3	9	30	36%	
D1	The occurrence of work accidents	3	3	9	39	46%	
D2	Problems with residents around the project	3	3	9	48	57%	
D3	Late work due to environmental conditions	3	3	9	57	68%	
C2	Planning and control has no well carried out by contractors	3	3	9	66	79%	
A3	Delay in preparing the project land by the owner	3	2	6	72	86%	
B2	Lack of coordination between CM and contractors	2	3	6	78	93%	
C4	Lack of construction materials	3	2	6	84	100%	

3.3 The Research Stages

This study can show the research's road map and achieve objectives step by step (see flow chart in Fig. 1). First step: starting with identification of delay variable of the construction project, second step: analyze the risk of project delay variable and

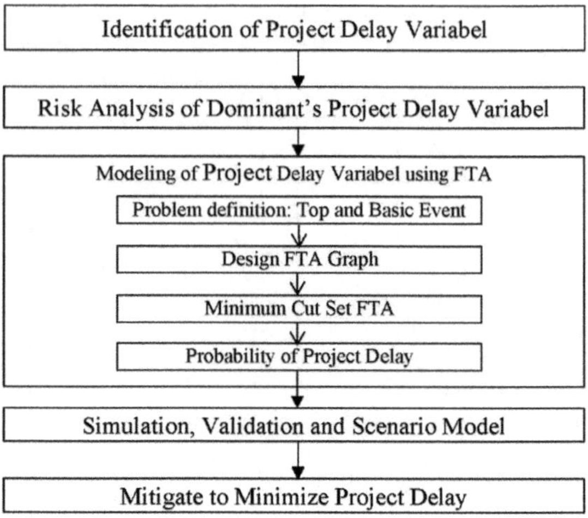

Fig. 1 Flowchart for the research stage

then determine the dominant variable based on the highest risk until 80% of total risk, the third step: construct the project delay model using FTA, fourth step: running the model to simulate and make a scenario. Finally: this research result mitigate to minimize the risk of project delay.

4 Results and Discussion

4.1 Identification Project Delay Variable

This study confirmed the literature study variables with the conditions of private projects in Indonesia, especially in the case study of the Suncity Apartment's construction in Sidoarjo. These variables are subject to confirmation interviews with contractors, CM consultants, and owners involved in the project. Confirmation of this variable was carried out on 12 respondents with different responses. These variables are relevant if more than 50% of respondents agreed, so that the author will analyze further. The variables are grouped based on project participants and environmental and worker conditions. The distribution data of respondents on the confirmation survey is 58–92% agree these variables causing the project delays (see Table 1).

4.2 Risk Analysis of Dominant's Project Delay Variable

This research analyzed the dominant risk of a variable to determine the risk's primary cause. It is done to simplify the risk calculation process but produce the right variables for mitigation. This study obtained the variable's dominant risk causing project delays by giving questionnaires and interviews to respondents/experts. The author tested the responses, both probability, and impact, to determine the variables' validity and reliability. If the variables are invalid or unreliable, then the variables were ignored. The validity and reliability analysis results showed that there were 10 out of 16 valid and reliable variables. Ten variables are analyzed, both the risk value using Eq. (1) and the Pareto diagram (see Table 2).

The Pareto diagram is used to choose variables that have a dominant risk by taking 80% of the cumulative value of risk. The Pareto diagram can reduce the risk variable that causes project delays from 10 to 7 dominant variables, i.e.: (1) Delay in contractors payments and related parties by the owner; (2) Lack of communication between the owner and related parties; (3) Lack of control over every job; (4) The occurrence of work accidents; (5) Problems with residents around the project; (6) Late work due to environmental conditions; and (7) Planning and control has no well carried out by contractors.

4.3 *Modeling of Project Delay Variable Using FTA*

This paper constructs the model of project delay using FTA to find the contributing variable. This model uses the dominant variable determined in the previous process. The FTA model was generated by several stages, i.e.:

Step 1: Define System Activity Behaviour

In this study, the causes of project delays were classified based on participants' roles and the environmental and worker condition. The number of dominant variables causing project delays due to owner-related delay, contractor-related delay, CM consultant-related delay, and environment and worker condition-related delay is 2 variables, 1 variable, 1 variable, and 3 variables, respectively.

Step 2: Define the Top Event

The top event is the main objective seeking a solution, namely the probability of project delays.

Step 3: Define the Tree of the Structure

The tree of FTA structure on project delay is built based on the project delay variable dominant. The diagram of the tree of the structure can be shown in Fig. 2.

Step 4: Expand Each Branch in Successive Levels of Detail

To expand each branch, we must search and explore as many causes as possible to meet each fault. Each event has a gate symbol based on the relationship between each event and successive levels of detail. In this study, consecutive levels have got by in-depth-interview with contractor project managers and CM Consultant team leaders. The expansion of each branch of the FTA until to find the basic event. Figure 3 is the

Fig. 2 FTA diagram of the top event of the project delay

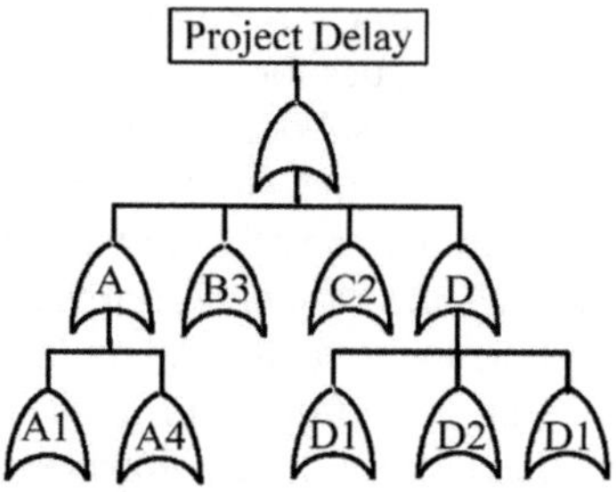

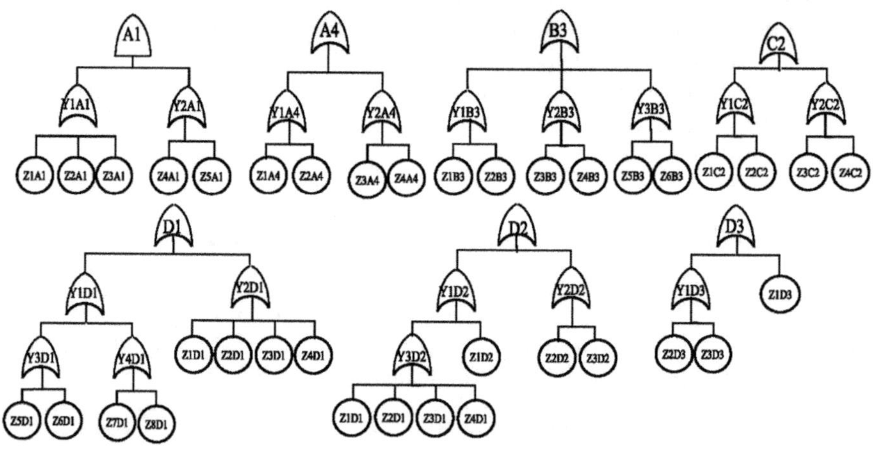

Fig. 3 FTA diagram of branch in successive levels of detail

detailed extension of the FTA Diagram.

Step 5: Solve the Fault Tree for the Combination of Events Contributing to the Top Event

Problem-solving of the FTA uses the Boolean Algebra formula. Each event contributes to the upper event based on its logic gate. An AND gate represents an independent event relationship and unaffected by any other event to the same gate. The probability formula for AND gate is given in Eq. 2.

$$P(A\ and\ B) = P(A \cap B) = P(A)\,P(B) \tag{2}$$

On the other hand, an OR gate has represented a dependent event or corresponds to a set union. The probability of OR gate is given in Eq. 3.

$$P(A\ or\ B) = P(A \cup B) = P(A) + P(B) - P(A \cap B) \tag{3}$$

4.4 Simulation, Validation and Scenario Model

Simulation Model

The model simulation requires data from interviews with experts/team leaders/project managers to predict the basic event probability value. Meanwhile, to indicate each basic event's delay rate using data from a previous study, the delayed payment's research results affected the project completion [13] and revisiting critical delay factors for construction [10]. The probability of each top event can be calculated using Eqs. (2) and (3) using a minimum cut set in Boolean Algebra. The minimum cut set produces the project delay variable (see Table 3).

This study obtained the model simulation results by calculating a minimum cut based on the probability value of the basic event, giving probability rate for each in-term mediated event, namely A1, A4, B3, D1, D2, D3, and C2 of 0.0019, 0.0074, 0.0065, 0.0020, 0.0498, 0.0102 and 0.0138 respectively. Meanwhile, this project's probability of experiencing delays if all basic events occur is 0.1000 (see Table 4).

Validation Model

This study validates the FTA model using data that occurred in 2020. The work plan and milestones that the contractor has set are as follows:

1. The event log of the project obtained the following data:
   - There has been a termination of work in March due to the public's complaints regarding the contractor's negligence, which caused the material to fall in the housing area.
   - Many houses in the area were damaged, and the charges filed were ignored by the contractors, so the community complained to the government to help solve this problem.
   - There is a problem of land acquisition for the expansion of the apartment area with the community.
   - The government requests the contractor to solve the public complaints.
   - The use of heavy equipment has experienced many obstacles due to the surrounding community's complain.
   - Topping off will be conducted in June 2020 but realize in September 2020.
   - Project completion delayed
2. Based on these data, this research runs the FTA model (see Table 5).
3. The FTA model can calculate the probability of delays in a construction project based on the project event log data in Table 5. The results of the running model show that the probability of the project experiencing delays is 0.3407. This shows that the probability of the project experiencing delays is 34%.
4. Data obtained from the project schedule shows that in the 7th month, the project experienced a delay of 15%.
5. This shows that the FTA Model can estimate the probability of project delays even though it has not been able to determine the percentage of the delay.

Table 3 Basic event and the minimum cut set of an intermediate event

Code	Basic event	Probability
A1	*Delay in contractors payments and related parties by the owner*	
Z1A1	Regulations from banking that complicate lending	0.0019
Z2A1	There is an increase in foreign currency exchange rates	0.0019
Z3A1	The bank refused the loan from the owner	0.0019
Z4A1	The owner is unable to pay off the remaining funds	0.0019
Z5A1	The allocation of funds is not used to pay contractors	0.0074
	The result from the minimum cut set	0.0019
A4	*Lack of communication between the owner and related parties*	
Z1A4	There is a conflict between the owner and related parties	0.0019
Z2A4	Lack of understanding of contract documents	0.0019
Z3A4	The results of the meeting were not carried out properly	0.0019
Z4A4	Unclear mechanisms for providing requests for information	0.0019
	The result from the minimum cut set	0.0074
B3	*Lack of control over every job*	
Z1B3	Frequent billing of shop drawing approval by contractors	0.0009
Z2B3	Slow administration handling by CM consultant staff	0.0009
Z3B3	The results of the meeting were not held	0.0019
Z4B3	Often some disagreements cause miscommunication	0.0009
Z5B3	Slow job review by CM consultant staff	0.0009
Z6B3	Evaluation results cannot be done/executed	0.0009
	The result from the minimum cut set	0.0065
D1	*The occurrence of work accidents*	
Z1D1	A competent officer does not carry out material identification	0.0025
Z2D1	There are no proper procedures for moving and storing materials	0.0025
Z3D1	The use of heavy equipment exceeds the maximum capacity	0.0025
Z4D1	The work done is not up to standard	0.0025
Z5D1	Workers fall from a height	0.0025
Z6D1	Workers are punctured by the material and crushed by the material	0.0050
Z7D1	Workers smoke in a prohibited/dangerous place	0.0025
Z8D1	Workers are in a hurry to do work	0.0050
	The result from the minimum cut set	0.0200
D2	*Problems with residents around the project*	
Z1D2	There was a crack due to the installation of the bored pile	0.0116
Z2D2	Heavy equipment activities carried out simultaneously	0.0116
Z3D2	There is work activity in overtime hours	0.0039
Z4D2	The load exceeds the strength limit of the sling rope	0.0039

(continued)

Table 3 (continued)

Code	Basic event	Probability
Z5D2	The scaffolding ladder is not strong enough to withstand the weight of the material	0.0010
Z6D2	Worker's negligence in carrying materials	0.0077
Z7D2	Sling rope is not strong enough to withstand the weight of the material	0.0116
	The result from the minimum cut set	0.0498
D3	*Late work due to environmental conditions*	
Z1D3	There is a natural disaster	0.0026
Z2D3	Rain	0.0026
Z3D3	Hot	0.0051
Z4D3	Wind	0.0026
	The result from the minimum cut set	0.0102
C2	*Planning and control has no well carried out by contractors*	
Z1C2	The specified time duration does not match	0.0046
Z2C2	The action plan doesn't work	0.0046
Z3C2	The method of doing work is not appropriate	0.0023
Z4C2	Incomplete job identification	0.0023
	Hasil minimal cut set	0.0138

Note Probability = rate of the event probability per month

Table 4 The minimum cut set (probability) of project delay

Code	Variable	Probability
A1	Delay in contractors payments and related parties by the owner	0.0001
A4	Lack of communication between the owner and related parties	0.0074
B3	Lack of control over every job	0.0065
D1	The occurrence of work accidents	0.0200
D2	Problems with residents around the project	0.0498
D3	Late work due to environmental conditions	0.0102
C2	Planning and control has no well carried out by contractors	0.0138
	The Result Minimal Cut Set (Probability Project Delay)	0.1000

Note Probability = rate of the event probability per month

6. The FTA model can also be used to predict probability with the following requirements: (i) monthly event prediction data is required as shown in Table 5; (ii) the event data is entered at the rate of probability to determine the basic event probability value; (iii) The prediction of project delay probability can be determined by the FTA Model by calculating the basic event recapitulation.

Table 5 Recapitulation of data variable based on project event log

Code	Risk variable	Event on month to							Occurrence log
		1	2	3	4	5	6	7	
A1	Delay in contractors payments and related parties by the owner	N	N	N	N	N	N	N	The contractor makes payment on time
A4	Lack of communication between the owner and related parties	Y	Y	Y	Y	Y	Y	Y	Many changes in project implementation: drawing, material, contract, etc.
B3	Lack of control over every job	Y	Y	Y	Y	Y	Y	Y	Delay in approval, the meeting didn't been implemented well, etc.
D1	The occurrence of work accidents	N	N	N	N	Y	Y	Y	The worker has an accident, the equipment didn't well, the material was improper, etc.
D2	Problems with residents around the project	Y	Y	Y	Y	N	N	N	So many complaints due to project: community property damage, loading activities, equipment activities, etc.
D3	Late work due to environmental conditions	Y	Y	Y	Y	N	N	N	Land acquisition and weather problem
C2	Planning and control has no well carried out by contractors	Y	Y	Y	Y	Y	Y	Y	The construction method can't be implemented, incomplete in scope of work, etc.

Note Y = Yes (occurrence); N = No (No occurrence)

Table 6 The scenario of project delay probability rate

Code	Simulation	Scenario 1	Scenario 2	Scenario 3	Scenario 4
A1	0.0001	0.0001	0.0000	0.0000	0.0000
A4	0.0074	0.0074	0.0000	0.0000	0.0000
B3	0.0065	0.0000	0.0065	0.0000	0.0000
D1	0.0200	0.0000	0.0000	0.0200	0.0000
D2	0.0498	0.0000	0.0000	0.0498	0.0000
D3	0.0102	0.0000	0.0000	0.0102	0.0000
C2	0.0138	0.0000	0.0000	0.0000	0.0138
Total	0.1000	0.0074	0.0064	0.0766	0.0138

Scenario Model

This research developed the scenario to determine the impact of each intermediate event on project delays by designing some basic events that occur together or not. This scenario can be used as a priority scale determination to control events, especially those that significantly impact project delays. The author arranges 4 scenarios based on the main variable that caused project delay: (1) Scenario 1: the project delay due to owner related-delay; (2) Scenario 2: due to the CM Consultant related-delay; (3) Scenario 3: because of the contractor related-delay; and (4) Scenario 4: caused by the environmental and worker condition. The results of these scenarios are shown in Table 6.

From the scenario above, this study concluded that the project environment and worker variables significantly affect the construction project delay, namely 0.0766. The 2nd ranking of the variable that affects delay is the contractor's performance with a probability of 0.0138. The 3rd is the owner's role, which is 0. 0074, and the least effective is the performance of CM consultant with a contribution of 0.0064. Based on this scenario, managing the project should maintain the project environment and worker properly to reduce the possibility of project delays. Furthermore, the contractor's performance must also be maintained, especially in conducting internal planning, scheduling, and monitoring. Meanwhile, the owner and the CM consultant contributed to the relatively small causes of delay because of their role in implementing the project as supporting only.

4.5 Mitigate to Minimize Project Delay

Based on the scenario above, this research mitigates reducing the risk of project delays. This mitigation is carried out based on the priority of the variable due to project delays that have been identified in the scenario of the FTA model. This mitigation results from discussions with the project manager and the team whose concepts have been drafted based on the debate results in this paper. This mitigation can be an input for owners, contractors, and CM consultants to improve their future performance. This study resulted in the mitigation in Table 7.

Table 7 The risk mitigation due to project delay

No	Variable	Risk response
1	*Environment and worker variable*	
1.1	The occurrence of work accidents	Carry out risk control by compiling the JSA and implementing the safety control hierarchy, namely Elimination, Substitution, design, and PPE (Protect Personal Equipment) and WPE (Work Protect Equipment)
1.2	Problems with residents around the project	Outreach and coordinate with the community around the project
1.3	Late work due to environmental conditions	Develop and evaluate work methods to suit environmental conditions and stop work when the weather is extreme
2	*Contractor variable*	
2.1	Planning and control of work carried out by contractors is not going well	• Arranging planning and scheduling using historical data/previous project data as a reference in project work • Always monitor and evaluate every activity regularly. If there is a deviation, immediately take corrective action
3	*Owner variable*	
3.1	Delay in the owner in making payments to contractors	• There is a need for mutual agreement and commitment in carrying out the work contract, especially in carrying out each party's rights and obligations • Conduct regular meetings with all project parties, namely the owner, contractors, and MC consultant • The owner always maintains a financial presence so as not to interfere with physical progress on the project, which has an impact on project delays • Contractors are expected to be able to find alternative solutions through additional working capital
4	*CM consultant variable*	
4.1	Lack of control over each employee	• CM consultant should reference in conducting supervision based on the work plan that the contractor has prepared • Always coordinate/meet so that any problems are be resolved quickly

5 Conclusion

This study has identified the variables that cause delays in construction projects based on literature studies fitted in project conditions. The main variables obtained are quite a lot caused difficulty in in-depth analysis. Therefore, to determine the dominant variable, use a risk analysis and Pareto diagram.

The FTA model established is proven to assess the probability of project delays well. In addition, with the framework arranged, this model can also predict the likelihood of project delays based on the prediction of basic events in the scheme project time. This model can also prioritize the main variables that cause project delays and make set mitigation to control and reduce project delays.

While this study has several contributions to predict the probability of project delays, it is only an estimate of the event that has not determined the behind schedule. It needs to study further to afford to determine the delay. In addition, this model uses a one-way hierarchical assumption, whereas the actual event of each variable can influence each other. Therefore, this model needs to be improved by the System Dynamics to solve interrelationships and looping between variables to make it as close as possible to the actual response.

References

1. Gündüz M, Nielsen Y, Özdemir M (2013) Quantification of delay factors using the relative importance index method for construction projects in Turkey. J Manag Eng 29(2):133–139
2. Silva GASK, Warnakulasuriya BNF, Arachchilge BJH (2016) Criteria for construction project success: a literature review. In: 13th International conference on business management. Faculty of Management Studies and Commerce, Sri Lanka, pp 697–717
3. Ramlee N, Tammy NJ, Mohd RNH, Musir A, Karim AKN, Chan HB, Nasir SRM (2016) Critical success factors for construction project. In: International conference on advanced science. In: 2015, AIP conference proceeding engineering and technology (ICASET), pp 030011–1–030011–6
4. Silvianitaa, Mahandekaa DS, Rosyida DM (2014) Fault tree analysis for investigation on the causes of project problems. In: 2nd International seminar on ocean and coastal engineering, environment and natural disaster management, ISOCEEN. Science Direct, Indonesia, pp 213–219
5. Bhatti IA, Abdullah AH, Nagapan S, Sohu S, Lohar MA, Bhatti NB (2018) Delay in high-rise building construction projects of dubai: a review. Eng Sci Tech Int Res J 2(2):7–15
6. Fashina AA, Omar MA, Shikh AA, Fakunle FF (2021) Exploring the significant factors that influence delays in construction project in Hargeisa. Heliyon 7(2021):e06826
7. Sweis G, Sweis R, Hammad AA, Shboul A (2008) Delays in construction projects: the case of Jordan. Int J Proj Manag 26(6):665–674
8. Zidane YJT, Andersen B (2018) The top 10 universal delay factors in construction projects. Int J Proj Manag Bus 11(3):650–672
9. Hamzah N, Khoiry MA, Arshad I, Tawil NM, Che Ani AI (2011) Cause of construction delay—theoretical framework. In: The 2nd international building control conference 2011. Science Direct, Malaysia, pp 490–495
10. Jeffrey BHY, Pei LG, Yoke BW, Martin S (2021) Revisiting critical delay factors for construction: analysing projects in Malaysia. Alex Eng J 2021(60):1717–1729

11. Al-Humaidi HM, Hadipriono FT (2010) A fuzzy logic approach to model delays in construction projects using rotational fuzzy fault tree models. Civil Eng Environ Syst 27(4):329–351
12. Soetjipto JW, Adi T, Anwar N (2016) System dynamics approach for bridge deterioration monitoring system. Int J Eng Technol Innov 6(4)
13. Maurice PO, Charles MR, Paul AO (2015) Effects of delayed payment of contractors on the completion of infrastructural projects: a case of Sondu-Miriu hydropower project. Chin Bus Rev 14(7):325–336

Implementing a Relational Database in Processing Construction Project Documents

Mik Wanul Khosiin and Ardian Umam

Abstract Construction project documents are extensive and complex, and most of those are still carried out with traditional systems. Consequently, it takes much time to process, low accuracy, cannot be stored for long periods, and make it difficult for future engineers to access historical data project for the next project planning. The current researchers have responded to this situation by implementing database management. However, it has not had a significant impact on the industrial field. Thus, the study of the application of relational databases through SQL with a different approach, namely integrating project tables such as resource tables, cost items, work items, and project drawings, which are currently being worked on separately. As the output, it would be demonstrated into a supply chain management table (table_3), construction cash flow (table_4), and the quantity take-off management table (table_5). The total project price from the result of the SQL program shown in supply chain and project cash flow tables is IDR 1.42 billion, with a project duration of 109 days. The SCM table describes the types of resources used in the project, including the quantity and price values. Then, the table of CSF contains a list of cost items and the volume and price of work for the budget plan and cash flow plan. Moreover, the QTO table describes the association between cost items and project drawings intended to project surveyors for the quantity take-off management process. Project managers are always required to make decisions for project control appropriately, such as cost estimation, resource planning, and project schedule effectively. Therefore, through those three tables, the authors hope they can help construction practitioners to improve their construction projects efficiently in the future.

Keywords Relational database · Construction · SQL

M. W. Khosiin
Civil Engineering, East Java Branch of Veterans National Development University, Surabaya, Indonesia
e-mail: mik.wanul.ts@upnjatim.ac.id

A. Umam (✉)
School of Electrical Engineering and Informatics, Bandung Institute of Technology, Bandung, Indonesia
e-mail: ardianumam@itb.ac.id

S. A. Kristiawan et al. (eds.), *Proceedings of the 5th International Conference on Rehabilitation and Maintenance in Civil Engineering*, Lecture Notes in Civil Engineering 225,
https://doi.org/10.1007/978-981-16-9348-9_79

1 Introduction

The construction industry has a large market and will be stable for the next few years because all countries are competing to build monumental buildings as the identity of the nation's progress [1]. Therefore, all construction companies also compete in the use of information technology as part of project management [2]. Most of the current construction projects have implemented building information modelling (BIM), artificial intelligence (AI), virtual reality (VR), as well as data science [3]. Scientists and practitioners agree that in the future, the construction sector will be dominated by the automation system in construction by answering various problems and challenges currently faced by project managers [4].

During the construction phase, the project will be faced with various kinds of documents with a large amount of data [5]. These data are interrelated, which are always reviewed, revised and controlled during the project. However, in practice, in the construction field today, they still use conventional methods so that the project database process is very long and the level of accuracy is quite low. Thus, the project manager is difficult to make decisions in real time, and then it will have an impact on project implementation performance [6].

This study was carried out to solve problems regarding complexity of project databases, by totally using SQL programming, all project documents will be integrated and simplified to make it easier for engineers in creating supply chain management table, cash flow construction, and quantity take-off management table.

2 Database Management System (DBMS)

Today databases are very important for every industry, including the construction sector. There is always a database behind the scenes on every data or document we request, because all companies always maintain all important data for future use. The database management system is a powerful tool for creating and managing large areas of data efficiently and can maintain it for very long periods in a secure manner. Basically a database is nothing more than a collection of information that has existed for a very long time, then users can create a new database with a different scheme, dig up the desired data, be able to accommodate a very large amount of data, and can be done with access control for a sufficient period of time long. The DBMS can be designed through several programming languages such as structured query language (SQL), R language, python, unified modelling language (UML), object definition language (ODL), Application Programming Interface (JDBC), hypertext pre-processor (PHP), and etc. [7, 8].

Several previous studies on the role and application of relational databases in the construction industry have been carried out by many researchers, starting from simple management in the form of table management in MS. Excel to involve building information modelling (BIM) and the internet in data management. In addition, the programming language that used in the current studies are different according to the their research objectives.

Database in Construction Project

Project database management in Indonesia has not been well managed. The hundreds or even thousands of cost items and work items which are the core of all project documents are still done traditionally and have not been integrated. Through the use of data analytics (Pivot Table) the correlation between the cost breakdown structure (CBS) and the work breakdown structure (WBS) can be done better, thus helping contractors to estimate costs and schedule better and more efficiently [9]. Project documents that require integration between CBS and WBS are very diverse, including S-curve, bill of quantity, budget plan, resources plan, project scheduling, progress reports, payment progress, and others. Research collaborations between countries such as Indonesia and Taiwan have been carried out to develop the best scheme for integrating CBS and WBS and demonstrated in the form of a construction progress curve [14].

Internet-Based Database Management

The advanced database management system can be generated by the internet, all work packages, data history, and formulas are integrated to develop the project database stores. Therefore, this system can help the fund contractor to track and control their construction projects in real time and accurately [10].

Building Information Modelling and DBMS

Information about historical project data is very difficult to trace, because it is developed separately, so that the engineers worked hard to estimate the project costs. The application of a relational database that contains tables of project history connected to BIM models is proven to be able to carry out costs tracking to determine the final costs required for future projects [11]. The BIM system can also be used to maintain the sustainability of project procurement by considering life-cycle cost analysis (LCCA) through a database management system, this is very helpful for project managers to manage supply chains in construction projects [12].

DBMS for Current Construction Issues

Construction projects have a high risk of delays due to natural conditions or mismanagement. Through the use of historical data for certain project types, database management is carried out which can identify what jobs are often affected by delays, as well as the costs and duration of work due to delays. By using a relational database, contractors can anticipate delays or can submit claims supported by accurate data through the results of studies that have been conducted [13]. In another study, the application of the DB2 database management system as a database design can also improve management and information levels. By using UML class diagrams to find a logical structure in order to accelerate the development of project costs to make it faster and more accurate [14].

The review that has been carried out from previous studies (Table 1) can be concluded that DBMS is proven to be able to help the construction sector to manage

Table 1 Comparison model of the previous studies

Year	Researchers	Model	Brief concept	Output
2006	Ji Li et al.	Internet-based	A relational database in storing all project data through internet based	Developed system to track an control the construction project in real time
2016	Nannan Zhang et al.	UML, DB2	Integrating the database management system to accelerate the construction cost system	Index design on construction project cost system
2015	Gozde Bilgin et al.	MS. Access	Relational database in processing previous projects data to predict construction delay	Delayed activities and response party
2020	Hang Thu Thi Le et al.	BIM, MS. Access	BIM integrated with database management to compute building cycle	Building life-cycle costs for sustainable procurement
2020	Wijatmiko et al.	MS. Excel	Using pivot table and unique code to integrate WBS and CBS	The association table of cost item and work item
2021	Khosiin et al.	SQL	Assigning entity relationship on complex project data through SQL script	Supply chain, cash flow and take-off tables
2021	Khosiin et al.	SQL	Assigning entity relationship on complex project data through SQL script	Supply chain, cash flow and take-off tables

data efficiently. An advanced scheme using the internet and a combination with project controlling software such as BIM is also able to overcome the latest problems in the construction industry such as mitigating the risk of delay, cost estimation, data mining, etc. However, due to the complexity of the problem, and the scope of the data to be faced, there are still many challenges that must be resolved immediately. Therefore, this indicates that there is a need for further study on project database management through one of the programming languages like SQL to correlate data so that it can be realized in a simpler and more accurate table form to help project managers make real-time and efficient decisions (Table 1).

3 Case Study

A simple one-story residential building and all its construction data are plotted to demonstrate the function of a relational database. The project document is formed in several stages, starting with a detailed engineering design (DED) which contains a project drawing, then the contractor will list the work items and their work order and end with the creation of a project schedule. At the same time the engineer will also compile a cost item for making a project budget plan. Each of those has a unit item, quantity, and unit price that previously analysed according to the price of the resources used (labour, material, equipment and subcontractor). All of these data categories are interrelated, project drawings are used as a reference for making project schedules and budgets, while cost items are integrated with work items with one to many or one to one relationship model as in Fig. 1. These project documents will always be reviewed by engineers during the project.

Step One, Relationship Design

To implement a relational database begins with creating an entity relationship diagram (ERD) through the MySQL programming language such as Fig. 2, this diagram consists of several tables that represent the project database that has been mentioned earlier. Then, the tables are integrated through the ID numbers that have been assigned to each database category, the details of the table used as an ERD compiler are,

1. Drawing table, contains drawing_id, description and number of drawing pages
2. The cost_item table has a cost_id, cost breakdown structure (CBS), units, work volume, and project overhead
3. Table work_item, contains the work_id, work breakdown structure (WBS), project schedule
4. The resource table, consisting of the resource_id, resource name, unit, and resource unit price

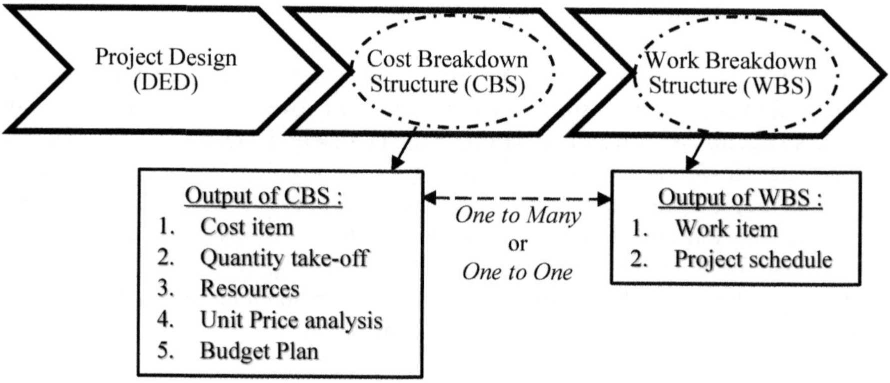

Fig. 1 The overview of construction database

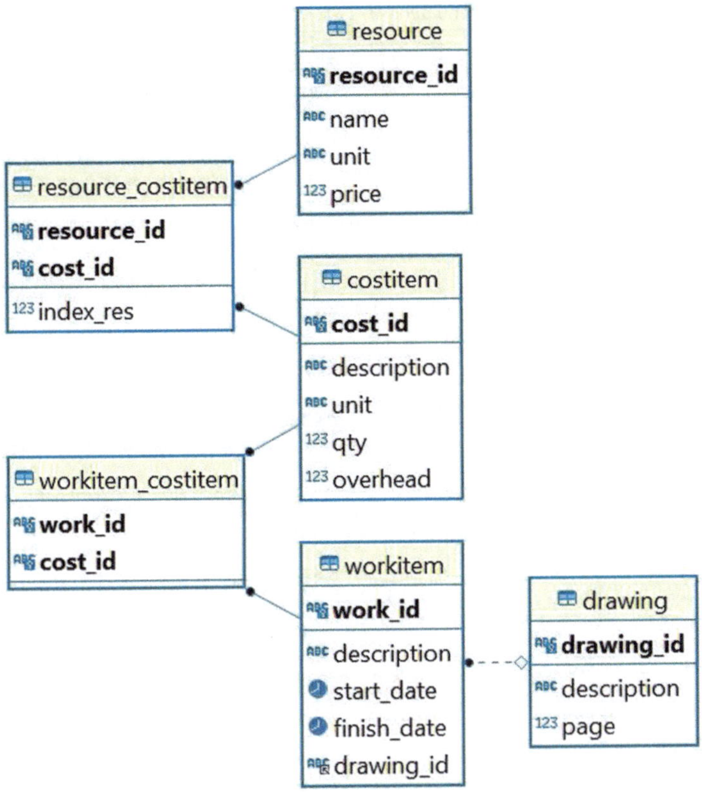

Fig. 2 ER diagram for construction database

5. Table workitem_costitem, describes the relationship between work_id and cost_id
6. Table resource_costitem, is the relation between resource_id (along with index_resource) and cost id

Step Two, SQL Script

The process of joining between tables is the second step in the project database management process. SQL scripts are arranged according to the required output table form. This study will present three project management tables as a demonstration of the application of a database management system, namely the supply chain management table, project financial plans, and quantity take-off management. Table 2 SQL script to prove that a relational database can help project managers in processing construction documents better through the three tables previously mentioned.

To obtain the Table_3 with columns of resources name, unit, quantity (qty), start_date, finish_date, duration, price and cost, we can join the tables of resource, resource_costitem, costitem, workitem_costitem and workitem using attributes of

Table 2 SQL script for construction database

```
#Table 1
WITH temp(name, cost_id, unit, qty, start_date, finish_date, duration, price, cost) as
(SELECT r.name, c.cost_id, r.unit, c.qty, MIN(w.start_date), MAX(w.finish_date),
DATEDIFF(MAX(w.finish_date), MIN(w.start_date)) as duration, r.price,
CAST(((rc.index_res*c.qty*r.price)) AS DECIMAL(14,2)) as cost
FROM (((resource r JOIN resource_costitem rc USING(resource_id)) JOIN costitem c
USING(cost_id)) JOIN workitem_costitem wc USING(cost_id)) JOIN workitem w USING(work_id)
GROUP BY r.name, c.cost_id, rc.index_res, r.price, r.unit)
SELECT name, cost_id, unit, qty, start_date, finish_date, duration, price, cost
FROM temp
UNION
SELECT "~TOTAL", "-", "-", "-", MIN(start_date), MAX(finish_date),
DATEDIFF(MAX(finish_date),MIN(start_date)), "-", SUM(cost)
FROM temp
ORDER BY name
```

```
#Table 2
WITH temp1(cost_id, cost_item, unit, qty, price, total_price, start_date, finish_date, duration) as
(WITH temp2(cost_id, cost_item, resource, unit, qty, price, start_date, finish_date) as
(SELECT c.cost_id, c.description, r.name, c.unit, c.qty, (rc.index_res*r.price), MIN(w.start_date),
MAX(w.finish_date)
FROM (((costitem c JOIN resource_costitem rc USING (cost_id)) JOIN resource r USING (re-
source_id)) JOIN workitem_costitem wc USING(cost_id)) JOIN workitem w USING(work_id)
GROUP BY c.cost_id, r.name, rc.index_res, r.price
ORDER BY c.cost_id)
SELECT cost_id, cost_item, unit, qty, CAST(SUM(price) AS DECIMAL(14,2)) as price,
CAST((qty*SUM(price)) AS DECIMAL(14,2)) as total_price, MIN(start_date) as start_date,
MAX(finish_date) as finish_date, DATEDIFF(MAX(finish_date), MIN(start_date)) AS duration
FROM temp2
GROUP BY cost_id)
SELECT cost_id, cost_item, unit, qty, price, total_price, CAST((total_price/(SELECT
SUM(total_price) FROM temp1)*100) AS DECIMAL(3,2)) as "weight (%)", start_date, finish_date, dura-
tion
FROM temp1
UNION
SELECT "TOTAL", "-", "-", "-", "-", SUM(total_price), "100", MIN(start_date), MAX(finish_date),
DATEDIFF(MAX(finish_date),MIN(start_date))
FROM temp1
```

```
#Tabel 3
SELECT DISTINCT c.cost_id, c.description as cost_item, c.unit, c.qty, w.drawing_id, d.description
FROM (costitem c JOIN workitem_costitem wc ON (c.cost_id = wc.cost_id)) JOIN workitem w ON
(wc.work_id = w.work_id) JOIN drawing d ON(w.drawing_id = d.drawing_id)
ORDER BY c.cost_id, w.drawing_id
```

resource_id, cost_id, cost_id and work_id, respectively. The price is calculated by multiplying qty with price in 'SELECT' clause, meanwhile for total cost is aggregated using 'sum function' toward the 'cost column', and in order to do that, we implement a temporary table called 'temp' using 'WITH' clause. Table_4 is composed of several columns, namely cost_id, cost_item, unit, qty, price, start_date, and finish_date. Thus, it is necessary to combine the resource_costitem, resource, workitem_costitem, and workitem tables through the attribute id of each table. To retrieve the unit price for each cost item, it is necessary to analyse the resources consisting of human resources, equipment, and material as detailed items for unit price compilers through SELECT clause, DATEIFF, and WITH clause as the formula in Table_2. In the last table is quantity take-off management, the column cost_id, cost_item, column unit, qty, drawing_id and drawing description column are presented. This is to show

Table 3 Supply chain management

Resources	Cost ID	Unit	Qty.	Start	Finish	Dur.	Unit price	Cost
Bar bender & Cutter	CS001	Ls	2 K	21-01-13	21-02-23	41	5 K	10 M
Bared steel	CS009	Kg	250	21-03-14	21-03-23	9	70 K	14.9 M
Baseplate & stiffener	CS008	Kg	200	21-03-14	21-03-23	9	14 K	3 M
Wooden glue	CA004	Kg	4.2	21-03-10	21-03-13	3	13 K	55 K
Wooden rafters	CS004	M3	500	21-02-02	21-02-23	3	3.6 M	72.5 M
Etc.	..	..	..	..	..	..	..	..
Total				21-01-01	21-04-20	109		1.42B

the relationship between each volume of work on the cost item having an association with a particular project drawing, so that the quantity surveyors can more easily calculate the quantity of buildings. Table_5 is obtained from the merging of the workitem_costitem, workitem, and drawing tables by using ORDER BY and SELECT clause.

Step Three, SQL Statement Execution

The rendering process of the SQL script for the three demo tables cannot be generated at same time, but must be executed one by one script, so that there is no failure in the execution process. Table 3 contains the types of resources used in the project along with detailed data including the volume of work, resource usage schedules and cost analysis to determine the total project value. Based on the results of the analysis in Table 3, it is also obtained 36 types of resources involved or 146 resource items after the repetition process, this shows that the construction project has a large enough order for material, equipment, and labour, so it's no wonder this will be a huge data when the project is scaled up. The next table is the project cash flow, which explains the project cost budget plan through a cost breakdown structure along with a time schedule to estimate the project cost flow during the construction process. Table 4 can also be transformed into an S-curve, which is the correlation between the accumulated costs and the project implementation time (weeks or months). Therefore, the project manager can easily control the amount of costs that must be spent within a certain time and can find out what kind of jobs and resources involved in it, [9]. Table 4 shows that of the 23 cost items, the total project value is IDR 1,420,522,875.06 with a project duration of 109 days starting from January 1 to April 20, 2021. Table 5 provides information that those cost items whose work volume is calculated based on the particular project drawing as its association. This process is part of the quantity take-off management that is usually managed by an engineer or quantity surveyor in a construction project.

Table 4 Project cash flow

Cost ID	Cost item	Unit	Qty.	Price	Total price	%	Start	Finish	Dur.
CA001	Brick laying	M2	500	149 K	74 M	5	21-03-02	2103-09	7
CA002	Stucco work	M2	500	90 K	44 M	3	21-03-10	21-03-09	7
CA003	Wooden Door and windows	M2	14.4	1 M	14 M	1	21-03-14	21-03-13	3
CS009	Steel bracing	Kg	250	630 K	15 M	1	21-03-14	21-03-23	9
CS010	Metal roof	M2	750	167 K	125 M	9	21-01-01	21-03-23	9
Etc.	..	..	..	..	..	..	..	..	
Total					1.42B	100	21-01-01	21-04-20	109

Table 5 Quantity take-off management

Cost ID	Cost item	Unit	Qty.	Drawing ID	Description
CA001	Brick laying	M2	500	AR001	Walls and stucco
CA002	Stucco work	M2	500	AR001	Walls and stucco
CA003	Wooden Door and windows	M2	14,4	AR002	Doors and window
CS009	Steel bracing	Kg	250	ST005	Roof plan
CS010	Metal roof	M2	750	ST005	Roof plan
Etc.	..	..	..	..	..

4 Conclusion

A large market in the construction industry sector must be accompanied by professional project database management. Project data including project design tables, work items, cost items, and resource lists are interrelated and generated separately. The database management system (DBMS) is a tool for managing a broad set of information effectively so that it can be accessed easily and can last a long time. Previous researchers have initiated the application of relational databases to address construction issues but their role is not significant yet, so a new approach is needed in this case. A case study through a simple residential project database was fully demonstrated in SQL language in the form of three strategic tables in the construction industry including supply chain table, the table of construction cash flow, and quantity take-off management table.

There are three stages that must be made in designing a database in SQL language. The first step is integration design, by compiling the project tables such as cost_item, work_item, resource, etc. through entity relationship diagram (ERD) with the unique

ID in each data. The next step is arranging the data analytical script to combine, classify, and analyze all of tables automatically and obtain the purposed tables. Finally, the tables in question are obtained by running the script (final step) shows that the total project price is IDR 1.42 billion with a project duration of 109 days. Furthermore, resource types and project scheduling can also be identified properly.

Acknowledgements The authors would like to thank the scientific committee and the organizing committee especially the editor, the paper reviewer for their evaluation and supports to this paper. The authors thank the Department of Civil Engineering, East Java Branch of Veterans National Development University for financially supporting this 5th International Conference on Rehabilitation and Maintenance in Civil Engineering (ICRMCE). In addition, The authors are also grateful for the research collaboration between universities (UPNVJT-ITB) in Indonesia, and hopefully this collaboration can be upgraded to a top research level in the future.

References

1. T. B. Research Comp (2020) Global construction market, the business research company, Philadelphia, USA
2. Halpin DW, Senior BA (2011) Construction management fourth edition, Colorado. Wiley, USA
3. Haynes ME (2002) Project management practical tools for success. Axzo Press, Arkansas
4. Fryer B (1996) The practice of construction management, 3rd edn. Blackwell Science, London
5. Institute PM (2013) A guide to the project management body of knowledge (PMBOK Guide)-fifth edition. Project Management Institute Inc., Pennsylvania
6. Oxley R, Poskitt J (1971) Management techniques applied to the construction industry, 2nd edn. Crosby Lockwood Staples, London
7. Molina HG, Ulman JD, Widom J (2009) Database systems the complete book, 2nd edn. Pearson Prentice Hall, New Jersey
8. Sumanthi S, Esakkirajan S (2007) Fundamentals of relational database management systems. Springer, Tamil Nadu
9. Khosiin MW, Wang KC, Wijatmiko I, Wang HH, Wang WC (2019) S-Curve development model by integrating Taiwanese and Indonesia Practices. In: The 23rd Symposium on construction engineering and management, Taichung, Taiwan
10. Li J, Moselhi O, Alkass S (2006) Internet-based database management system for project control. Eng Constr Archit Manag 13(3):242–253
11. He X, Liu R (2020) A relational database scheme of construction cost tracking from BIM model to final costs. In: American association of civil engineering, Florida, USA
12. Thi Le HT, Likhitruangsilp V, Yabuki N (2020) A BIM-Integrated relational database management system for evaluating building life-cycle costs. Eng J 75–86
13. Bilgin G, Eken G, Dikmen I, Birgonul MT (2015) A relational database for construction delay. In: Eighth international conference in the 21st Century, Thessaloniki, Greece
14. Zhang N, Song W (2016) Database design on construction project cost system. In: 3rd International conference on materials engineering, manufacturing technology and control (ICMEMTC), Jinzhou, China

Evaluation of the Public Procurement Principles Implementation in Surabaya Construction Projects

Patrisius Valdoni Sandi and Mohammad Arif Rohman

Abstract Public procurement is essential to the success of infrastructure development as through this process the best service providers can be selected. Within Indonesian context, there are seven basic principles of good public procurement, namely: efficient, effective, transparent, open, competitive, fair, and accountable. This study aims to investigate and assess seven principles implementation based on the providers' perceptions in the public construction procurement in Surabaya as the second biggest city in Indonesia. Importance Performance Analysis (IPA) was employed to analyze the level of importance and performance of each principle's indicator. The result shows that the implementation of the construction procurement principles in Surabaya obtained score of 87.55%, which can be considered quite good as the score was relatively high and close to the maximum (100%). In more details, from each procurement's principle, the implementation of the 'efficient' principle had score of 87.91%, 'effective' was 86.31%, and 'transparent' was 86.18%. Meanwhile 'open' principle was 89.87%, 'competitive' was 89.11%, 88.90% for 'fair', and 85.61% for 'accountable'. Meanwhile, there were some gaps between the importance and the performance levels. This means the Surabaya City Government still needs to improve or at least to maintain the existing performance to promote a successful infrastructure development in their region.

Keywords Public procurement principles · Construction procurement · Importance performance analysis

1 Introduction

Construction procurement leads to a system of policies that can be used to acquire new buildings, either through purchasing or leasing, or by designing and constructing infrastructure facilities [1]. Public procurement, which is a government policy

P. V. Sandi · M. A. Rohman (✉)
Construction Management Laboratory, Department of Civil Engineering, Sepuluh Nopember Institute of Technology, Surabaya, Indonesia
e-mail: arif@ce.its.ac.id

S. A. Kristiawan et al. (eds.), *Proceedings of the 5th International Conference on Rehabilitation and Maintenance in Civil Engineering*, Lecture Notes in Civil Engineering 225,
https://doi.org/10.1007/978-981-16-9348-9_80

system, is closely related to issues of culture, leadership, management, economy, environment, and even politics [2]. On the other hand, construction must ensure the integration of the interests of all stakeholders, including the community, in meeting its objectives. This is also related to the issue of the vulnerability of public trust due to the use of public funds as a source of system funding.

Surabaya as the second biggest city in Indonesia has implemented e-procurement system since 2004, which is considered successful in the Indonesian national scale. The success of Surabaya in implementing this system in the midst of various global issues related to the application of this system is an interesting thing to research. This will inform how the Surabaya City Government implements the system and becomes a national pilot, especially in terms of implementing the principles of procurement, especially in the construction sector. As part of Indonesia which also implements national procurement regulation, namely the Presidential Regulation 16/2018, Surabaya have to apply the principles of efficiency, effectiveness, transparency, openness, competition, fairness, and accountability in its procurement system.

However, even though the procurement principles have been implemented, there is still limited study which investigates these principles within Indonesian context. According to that, this study aims to evaluate the implementation of the principles of procurement public construction works in Surabaya. The data in this study were based on a questionnaire which was distributed to construction sector involving contractors, consultants, and suppliers who have been involved in government construction procurement in Surabaya.

2 Literature Review

2.1 Public Construction Procurement

Public procurement has been consistently used to advance public policy in a variety of areas, including industrial policy, reducing unemployment, supporting small businesses, local development, and equalizing wages [3]. In its implementation, public procurement must guarantee the effectiveness of the implementation of work contracts for those involved [4]. Meanwhile, construction procurement can be interpreted as a procurement system related to the purchase of construction services to build new buildings or structures. This includes all construction works such as repairing, maintaining, expanding, or demolishing a building or structure. Several previous studies focused on irregularities, problems, or issues hindering the implementation of procurement, or the impact of the application of electronic technology on the procurement system. For example, Vaidya and Campbell examine public procurement in terms of information systems, e-commerce/e-government, supply chain management, and their impact on procurement efficiency [5]. Detkova et al.

examined the relationship between procurement competition and the problem of corruption in the form of violations [6].

2.2 *Procurement Principles in Indonesia*

Indonesia has seven principles of public procurement that should be implemented, namely: efficient, effective, transparent, open, competitive, fair, and accountable. Based on the principles, (i) Efficient is a condition that is closely related to efforts to optimize the use of resources [7]; (ii) Effective can be defined as the use of available resources to obtain goods and services that have the highest benefits [5]; (iii) Transparent means that this principle requires that information related to the implementation of public procurement to be obtained through sources that are easily accessible and freely available [8]; (iv) Open emphasizes providing all competent providers of goods and services the opportunity to be involved in procurement [8]; (v) Competitive emphasizes that public procurement should avoid having a single bidder and prevent restrictions on competition [9]; (vi) Fair is related to equal treatment that applies to every party or individual involved in the public procurement process [7]; and (vii) Accountable emphasizes the existence of an effective control mechanism in public procurement accompanied by a clear and structured chain of responsibility based on their respective authorities at each stage of public procurement [10].

3 Research Methodology

Importance Performance Analysis (IPA) was used to analyze performance by measuring customer satisfaction with a product or service. IPA is a simple tool, but it has strong and strategic benefits in identifying the performance of product or service attributes, so it can be seen that the attributes most need to be improved under conditions of cost savings but not significant quality loss [11].

Research indicators were obtained from relevant previous studies which were then validated using preliminary survey to understand their relevancy to Indonesian context. There are 63 indicators of the 7 principles developed from various literatures which were validated using preliminary survey involving 4 experts in the field with academic and practitioner backgrounds and work experience ranges of at least 11–15 years. From the survey, 48 indicators were considered relevant for this study. Due to space limitation, the detail of the indicators is inserted in Table 1 together with their performance assessment.

The 48 indicators were then used as criteria to evaluate the procurement principles. The main survey was conducted involving 30 respondents through questionnaire distribution involving respondents from 18 contractors (60%), 10 consultants (33%), and 2 suppliers (7%) who had been participating in public procurement in Surabaya

Table 1 The result of importance (I) and performance (P) analysis

No	Indicators	Source (s)	Mean		Gap score	Conformity level (%)
			P	I		
A *Efficient*						
1	Use of electronic based systems	[13]	3.967	4.633	−0.667	85.61
2	Short processing time	[5, 7]	3.733	4.500	−0.767	82.96
3	Decreased number of objections or claims	[5]	3.967	4.100	−0.133	96.75
4	Centralized access	[13]	3.933	4.400	−0.467	89.39
5	Less processing cost	[5]	3.833	4.467	−0.633	85.82
6	Reducing the number of errors due to human error	[6]	3.767	4.533	−0.767	83.09
7	A classification of the type of work or service	[13]	4.200	4.533	−0.333	92.65
	Mean A		3.914	4.452	−0.600	87.91
B *Effective*						
1	The achievement of the objectives of the procurement implementation	[14, 15]	4.100	4.500	−0.400	91.11
2	Completion of the procurement process on time	[16]	3.933	4.500	−0.567	87.41
3	Procurement is in accordance with community needs	[17]	3.800	4.600	−0.800	82.61
4	Integration into procurement programs	[14]	3.600	4.367	−0.767	82.44
5	There is a control mechanism for the implementation of procurement	[13]	4.033	4.667	−0.633	86.43
6	Quality system	[15]	3.867	4.467	−0.600	86.57
7	A communication process that minimizes face-to-face systems	[18]	3.767	4.300	−0.533	87.60
	Mean B		3.871	4.486	−0.614	86.31
C *Transparent*						
1	Simple rules and procedures	[19]	3.833	4.633	−0.800	82.73

(continued)

Table 1 (continued)

No	Indicators	Source (s)	Mean		Gap score	Conformity level (%)
			P	I		
2	Access to procurement information	[18, 20]	4.033	4.033	−0.367	91.67
3	Availability of complete information	[4, 9]	4.033	4.633	−0.600	87.05
4	Easily accessible system	[8]	4.067	4.533	−0.467	89.71
5	Visibility of the procurement budget	[21]	3.867	4.567	−0.700	84.67
6	The existence of a control mechanism and sanctions	[22]	3.733	4.333	−0.600	86.15
7	Archiving procurement documents	[21]	3.667	4.233	−0.567	86.61
8	Mass media involvement in the procurement process	[21]	3.533	4.367	−0.833	80.92
	Mean C		3.846	4.463	−0.617	86.18
D *Open*						
1	Acceptance of participants based on qualifications	[4]	3.967	4.367	−0.400	90.84
2	Blacklisted business entity publications	[23]	4.067	4.500	−0.433	90.37
3	Disseminate information on procurement implementation	[24]	3.867	4.400	−0.533	87.88
4	There are no restrictions on participation due to the participant's origin, location, religion, ethnicity, or political orientation	[8]	4.067	4.500	−0.433	90.37
	Mean D		3.992	4.442	−0.450	89.87
E *Competitive*						
1	There is no intervention from any party	[4]	3.967	4.700	−0.733	84.40
2	There are clear rules regarding the procurement competition cycle	[9, 25]	3.867	4.500	−0.633	84.40

(continued)

Table 1 (continued)

No	Indicators	Source (s)	Mean		Gap score	Conformity level (%)
			P	I		
3	Participants have the same opportunity to compete	[25]	4.000	4.567	−0.567	87.59
4	The procurement agency is not affiliated with or cooperating with any particular party	[26]	4.133	4.367	−0.233	94.66
5	Completeness of the required documents	[25]	4.200	4.500	−0.300	93.33
6	Involvement of three or more bidders as participants	[9, 25]	3.967	4.467	−0.500	88.81
7	Compliance with procurement ethics	[8]	3.967	4.433	−0.467	89.47
	Mean E		4.014	4.505	−0.490	89.11
F *Fair*						
1	Winners are determined based on eligibility for qualifications	[4]	4.033	4.733	−0.700	85.21
2	Avoiding conflicts of interest	[24]	4.100	4.400	−0.300	93.18
3	Equal distribution of rights and obligations among participants	[7]	3.967	4.667	−0.700	85.00
4	There is a code of ethics that regulates the integrity of the procurement	[27]	4.067	4.467	−0.400	91.04
5	Each participant is treated equally	[7, 20, 25]	4.100	4.433	−0.333	92.48
6	The required technical specifications must not refer to certain brands	[13]	3.967	4.367	−0.400	90.84
7	The number of complaints against procurement implementation has decreased	[28]	3.833	4.400	−0.567	87.12
8	Bid consideration is based on compliance with bid documents	[20]	3.967	4.567	−0.600	86.86

(continued)

Table 1 (continued)

No	Indicators	Source (s)	Mean		Gap score	Conformity level (%)
			P	I		
	Mean F		4.004	4.504	−0.500	88.90
G *Accountable*						
1	There is a strict monitoring mechanism	[4, 21]	3.733	4.633	−0.900	80.58
2	There are internal and external audits	[18, 21]	3.700	4.500	−0.800	82.22
3	Reporting System which is open to the public	[13]	3.833	4.400	−0.567	87.12
4	Professionalism standard	[29]	3.967	4.600	−0.633	86.23
5	There are rewards for those who have good performance and sanctions	[4, 21, 25]	4.100	4.567	−0.467	89.78
6	Clear organizational framework	[21, 29]	3.933	4.667	−0.733	84.29
7	Conflict resolution	[21]	4.100	4.600	−0.500	89.13
	Mean G		3.910	4.567	−0.657	85.61
	Total mean for overall principles		3.936	4.488	−0.552	87.55

construction projects. The respondents had background of 5–20 years working experience and involved in at least 5 procurement processes in Surabaya. The questionnaire employed 5-point Likert scale, ranging from 1 (= not at all important) to 5 (= extremely important) for the importance, and from 1 (= poor) to 5 (= excellent) for the performance part.

4 Analysis Result

The data acquired from the main survey were analyzed using Importance Performance Analysis (IPA). Table 1 shows the respondents' perceptions regarding the performance of the overall indicators. According to the results, it was found that the indicators' performance is below the expectation (level of importance). This is indicated by negative gaps on the overall indicators.

The overall mean score of the indicators' performance (P) was 3.936 while the mean score of the indicators' importance (I) was 4.488. As the gap between them was −0.552 (smaller than 0), it means it still has not met the user's expectation [12]. This also can be seen from the parameter of suitability level between the performance and interests (%), which is still below the ideal (<100%).

In addition, the overall implementation of the procurement principles reached 87.35%. In more detailed, the implementation of the principle of 'efficient' has a score of 87.91%, 'effective' of 86.31%, 'transparent' of 86.18%, 'open' of 89.87%, 'competitive' of 89.11%, 'fair' of 88.90%, and 'accountable' at 85.61%. Even though the result can be considered good as the value of the conformity level is still less than 100%, it can be concluded that the quality of the services provided still has not met the criteria or expectation [30].

5 Findings and Discussion

The analysis results can be discussed with regard to the indicators which had low performance. For the principle of efficient, A2 (Short processing time) is the indicator with the lowest score. The public procurement service system should be conducted in short period of time [5, 7]. As the indicator had low score, it means this indicator performance has not met the providers' expectation, so it should be improved.

With regard to the principle of effective, indicator B3 (Procurement is in accordance with community needs) is the indicator with the lowest score. The procurement must be suitable with the community needs of the users. Non-conformity with community needs will lead to waste of resources, including the budget. This can occur due to lack of community involvement or lack of monitoring of the absorption of priority community needs.

Indicator C8 (Mass media involvement in the procurement process) is the lowest ranking in the 'transparent' principle. This means media has not been effectively involved in the process. Indeed, the media is an important component for controlling the implementation of public procurement to ensure 'transparent' principle as this can be source and channel of information regarding procurement planning and implementation. The media also has a role in educating public regarding the public procurement process [21].

Meanwhile, indicator D3 (Disseminate information on procurement implementation) is the lowest on the 'open' principle. Information dissemination is very important to increase market responsiveness (potential bidders) to participate with competitive bidding [24]. This low score could happen due to the impact of the relationship with media involvement. As previously explained, media involvement is very important as a source of information.

Indicator E2 (There are clear rules regarding the procurement competition cycle) is the lowest on the principle of competition. This indicator relates to the clarity of regulations, including those related to competition restrictions such as experience, technical resources, origin, and other documentation requirements [25, 31]. The low score of this indicator is probably due to the relatively complicated regulations. As explained in previous studies, procurement procedures should be simple and flexible [31].

In the 'fair' principle, the indicator with the lowest score is F3 (Equal distribution of rights and obligations among participants). The low score of this indicator is probably due to some parties who feel they are not being treated fairly in meeting the qualification requirements. As explained by Khan, this indicator relates to the equal treatment in meeting the qualification requirements, including the provision of equal time and the process of obtaining information [7].

Furthermore, on the 'accountable' principle, indicator G1 (There is a strict monitoring mechanism) is the lowest indicator. This indicator relates to a clear mechanism related to the responsibilities and roles of each party in each stage of procurement, so that the implementation of procurement is directed and in accordance with the planned [4]. The low score of this indicator is probably due to the current mechanism which has not provided understanding to all parties, especially to the participants who are involved in the procurement process.

6 Conclusion and Recommendation

This paper presents several attributes or indicators to measure the implementation of the procurement principles in Surabaya construction project as well as to assess their performance. Based on the analysis, the overall implementations of the public procurement principles in Surabaya construction work can be said quite good. This can be seen from the achievement of the conformity level which reached 87.55%. However, this achievement is still not satisfactory as the gap between the importance and performance implementation was −0.552. Based on these results, the implementation of these principles should be improved to achieve successful procurement process.

Acknowledgements The authors thank the Indonesian Ministry of Research and Technology/National Agency for Research and Innovation for providing grants to have conducted this research in 2021.

References

1. Mohsini R, Davidson CH (1991) Building procurement-key to improve performance. Build Res Inf 19(2):106–113
2. Noor MA, Khalfan MMA, Maqsood T (2013) The role of procurement practices in effective implementation of infrastructure projects in Pakistan. Int J Manag Proj Bus 6(4):802–826
3. Hawking P, Stein A, Forsters S (2004) E-Procurement: is the ugly duckling actually a swan down under. Asia Pac J Mark Logist 16(1):1–26
4. Lynch J (2013) Public procurement: principle, categories and methods. Leanpub
5. Vaidya K, Campbell J (2016) Multidisciplinary approach to defining public e-procurement and evaluating its impact on procurement efficiency. Inf Syst Front 18(2):333–348
6. Detkova P, Podkolzina E, Tkanchenko A (2018) Corruption, centralization and competition: evidence from Russian public procurement. Int J Public Adm 41(5–6):414–434
7. Khan N (2018) What is public procurement: introduction. Public Procurement Fundamentals, Emerald Publishing Limited

8. Hardjowijono B, Muhammad H (2009) Prinsip dasar dan kerangka hukum pengadaan barang dan jasa pemerintah. Indonesian Procurement Watch, Jakarta
9. Lembaga Kebijakan Pengadaan Barang/Jasa Pemerintah and Millennium Challenge Corporation United States of America (2016) Modul 1: Menerapkan Prinsip-Prinsip Dasar Pengadaan, Pelatihan Pengadaan—Tingkat Dasar, Proyek Modernisasi Pengadaan LKPP/MCA-I. LKPP/MCA-I, Bogor
10. Organisation for Economic Co-operation and Development (2009) OECD principles for integrity Public Procurement. OECD, Paris
11. Abalo J, Varela J, Manzano V (2007) Importance values for importance performance analysis: a formula for spreading out values derived from preference rankings. J Bus Res 60(2):115–121
12. Shia BC, Chen M, Ramdansyah AD, Wang S (2016) Measuring customer satisfaction towards localiation website by WebQual and Importance Performance analysis (case study on Aliex Press Site in Indonesia. Am J Ind Bus Manag 6:117–128
13. Institute for Development of Freedom of Information (2018) Recommendations for the public procurement system of the Republic of Indonesia. IDFI, Jakarta
14. Siagian SP (1982) Manajemen modern. Gunung Agung, Jakarta
15. Joesbury P (2016) Improving the effectiveness of procurement identification and improvement of key determinant factors—the PEPPs project. Doctorate Thesis, Aston University, Birmingham
16. Panayiotou NA, Gayialis SP, Tatsiopoulos IP (2004) An E-Procurement system for governmental purchasing. Int J Prod Econ 79–102
17. Andriani S (2012) Analisa efektivitas hukum dalam penerapan pengadaan barang dan jasa secara elektronik (e-procurement) serta peranan lembaga pengawas terhadap pengadaan barang dan jasa pemerintah. Tesis Universitas Indonesia, Jakarta
18. Neupane A (2014) The potential of public e-procurement technology to reduce corruption in public procurement. Doctorate Dissertation, University of Southern Queensland, Australia
19. Vicente RS (2008) Evaluation of efficiency and effectiveness of the regional procurement service depots in the Philippines: a case of region 1. Master Thesis Massey University, New Zealand
20. Krivinsh A, Vilks A (2013) Prevention of corruption in public procurement: Importance of general legal principles. Jurisprudence 20(1):235–247
21. Organisation for Economic Co-operation and Development (2016) High-level principle for integrity, transparency, and effective control of major events and related infrastructures. OECD, Paris
22. Asian Development Bank (1999) International survey of government procurement system. International Government Solution Ltd
23. Thai KV (2017) Global public procurement theories and practices. Springer International Publishing, Switzerland
24. Eady R, Parera S, Heany G, Carlisle J (2007) Construction contractor's perspective on web-based advertising of public sector construction contracts. In: Boyd D (ed) Proceedings 23rd annual ARCOM/association of researchers in construction management. Belfast, 767–776
25. Komakech RA (2016) Public procurement in developing countries: objectives, principles and required professional skills. Public Policy Adm Res 6(8)
26. Racca GM, Perrin RC, Albano GL (2016) Competition in the execution phase of public procurement. Public Contract Law J 41(1):89–108
27. Byung GS, Chang HL, Byoung CH, Hyunjeonget N (2019) Investigating the fair treatment of suppliers and its trust fostering role and performance benefits. Int J Prod Econ 216:54–66
28. Jourdain R, Balgobin N (2003) Controlling corruption in Asia and pacific. In: 4th Regional anti-corruption conference of the ADB/OECD anti-corruption initiative for Asia and the Pacific, Kuala Lumpur
29. Diggs SN, Roman AV (2012) Understanding and tracing accountability in the public procurement process. Public Perform Manage Rev 36(2):290–315

30. Supranto J (2006) Pengukuran tingkat kepuasan pelanggan untuk menaikkan pangsa pasar. Rineka Cipta, Jakarta
31. Khorana S, Bouher KF, Kerr WA (2015) Governance Issues in EU's e-Procurement framework. J Common Market Stud 53(2):292–310

Relationship Model Between Conceptual Cost Estimation Process of Flyover Development in the Provincial Government of DKI Jakarta with the Accuracy Level

Putika Yussi, Yusuf Latief, and Rossy Armyn Machfudiyanto

Abstract The conceptual cost estimate for flyover development is the cost estimate when the flyover construction does not yet have complete planning data. In the DKI Jakarta Provincial Government, the forecast is used for the proposed flyover construction budget in the 5 Year Strategic Plans and Multi-Year Activities. In the estimation process, there is a level of accuracy that still needs to be improved, which causes the regional development planning to be less effective because other allocations have shifted in priority. Various factors can cause this in the estimation process that affects the level of accuracy. This study aims to identify the factors or variables in flyover construction's conceptual cost estimation process in the DKI Jakarta Provincial Government that affect accuracy. And then, to analyze the model of the relationship between variables of conceptual cost estimation process to the level of estimation accuracy, using the SEM-PLS method. 8 variables in the conceptual cost estimation process, which affect the estimation accuracy, have been identified through literature studies and expert validation. The variables are information quality, scope quality, estimator performance, estimating procedures, project definition, cost information, project characteristics, and government regulations. Based on respondents' results and the relationship model's analysis with the SEM-PLS method, the model produces 12 significant relationships between the variables of the conceptual cost estimation process for flyover development, with an R-square value of 0, 279.

Keywords Conceptual cost estimation process · Flyover development · Accuracy · Structural equation modeling

1 Introduction

Conceptual cost is calculated without complete information from design details and does not have a clear work scope [1]. Conceptual cost is also part of the feasibility

P. Yussi · Y. Latief (✉) · R. A. Machfudiyanto (✉)
Civil Engineering Department, Faculty of Engineering, University of Indonesia, Depok, Indonesia

P. Yussi
e-mail: putika.yussi@ui.ac.id

S. A. Kristiawan et al. (eds.), *Proceedings of the 5th International Conference on Rehabilitation and Maintenance in Civil Engineering*, Lecture Notes in Civil Engineering 225,
https://doi.org/10.1007/978-981-16-9348-9_81

analysis at the start of a project, with limited information on project scope, design, and technical data [2]. Meanwhile, a flyover is an intersection with the unique construction of a bridge built over a level intersection to allow two ways free passage on one of the main roads. Its function is to increase traffic flow capacity and reduce traffic congestion at intersections. And the traffic under the existing bridge intersection is still usable [3]. For flyover construction budgeting in the DKI Provincial Government that has not had complete planning data, the DKI Jakarta Provincial Government uses the conceptual cost estimate for the proposed Multi-Year Activities proposal. By definition, accuracy means close to the actual value, minimal error [4]. The conceptual cost estimation accuracy is also compared with the cost of completion, adjusted for agreed changes [5]. According to True (1998), accuracy can be expressed as the percent difference between the estimated value and the actual value [6]. Table 1 gives a range of conceptual cost estimates accuracy based on literature.

Based on Dinas Bina Marga DKI Provincial Government's data in 2016–2021, 3 flyovers use conceptual cost estimate for the proposed Multi-Year Activities Proposal. Each of them has 104.21; 23.67 and 12.47% for conceptual cost estimate accuracy. The flyover with 23.67 and 12.47% conceptual cost estimates accuracy has not yet reached optimal accuracy based on [5] and [7]. And the flyover with 104.21% accuracy, beyond the conceptual cost estimate accuracy range based on [5, 7], and [8]. It describes that the Provincial Government of DKI Jakarta's problem is the accuracy of the conceptual cost estimation of flyover construction, which still needs improvement.

Based on interviews with conceptual cost estimators at the Dinas Bina Marga of the DKI Jakarta Provincial Government, there are several causes for the accuracy of the conceptual cost estimation of flyover construction that needs to be improved. Figure 1 uses a fishbone diagram to describe those causes.

In actual condition, this problem causes regional development planning to be less effective or wasteful and less beneficial because other shifting allocations prioritize. Meanwhile, from a literature review, the impact is that if the conceptual stage cost estimate is too low, it will experience difficulties in the contract. If the forecast is too high, it will reduce funds for other projects [9]. Based on this, the Provincial Government of DKI Jakarta's accuracy of the conceptual cost estimation of flyover construction needs improvement.

This research aims to identify the factors or variables in flyover construction's conceptual cost estimation process in the DKI Jakarta Provincial Government that affect the accuracy and analyze the relationship between the conceptual cost estimation process variables with accuracy through a model. Therefore, this research can

Table 1 Conceptual cost estimate accuracy range based on literature

Source	Conceptual cost estimate accuracy range		
Ballard and Pennanen [5]	−10%	–	+10%
Asal [7]	+10%	–	+20%
AACE [8]	+20%	–	+30%

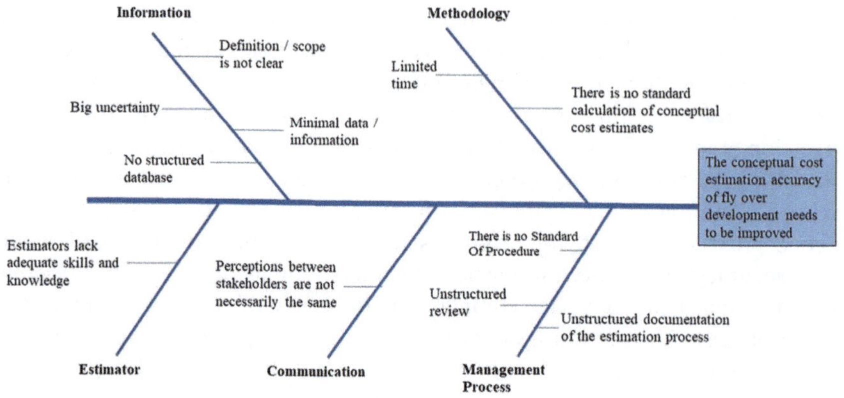

Fig. 1 Fishbone diagram of causes for the accuracy of the conceptual cost estimation of flyover development needs to be improved

be used as a base to improve the accuracy of the conceptual cost estimation of flyover development.

2 Literature Study

2.1 Conceptual Cost Estimation Process

The conceptual cost estimation process is a process to estimate project costs before there are significant design information and clear scope of work. Due to the lack of knowledge and the element of uncertainty, the conceptual cost estimation process is carried out based on deep consideration and experience [1]. Bley (1990) states that conceptual estimation is a process that is based mainly on judgment and knowledge because of the lack of information and uncertainty that always exists at the conceptual stage [2]. This statement is supported by Serpell [1], who stated that the conceptual cost estimation process is a process to estimate project costs before there are significant design information and clear scope of work. Due to the lack of knowledge and uncertainty, the conceptual cost estimation process is carried out based on deep consideration and experience. According to Honn An et al. [10], the conceptual cost estimation process is one of the critical stages in the early stages of the project cycle because it relates to several project uncertainties. It needs an estimator who must have specific expertise and knowledge to reduce the level of risk caused by uncertainty to be an acceptable level of risk in estimating the conceptual cost.

Conceptual cost estimates are generated at the conceptual stage where there is little information, no design details are available, and the project definition is unclear.

The estimation approach looks at the project in outline and is calculated on a top-down basis. This estimate is the basis for decision-making for funding so that if this estimate is inaccurate, it will result in an incorrect funding decision [11]. At the conceptual cost estimation stage, the project definition has not been fully defined. Most of the estimators will calculate the conceptual cost based on the characteristics of the project, which is an outline of the project [12]. In preparing the project budget, the estimated performance is influenced, among other things, by cost data, which comes from the Government and factors of regulatory conditions [13].

This research concludes that the conceptual cost estimation process is a series of stages carried out in the early phases of the project when there is no clear project information, scope definition, and project design. The process uses an approach to analyze the project in general and calculate with top-down estimates using cost information related to similar projects and government regulations. The process needs estimators who have the expertise and knowledge to reduce the level of risk caused by uncertainty to an acceptable level of risk to produce conceptual cost estimation.

2.2 *Factors in Conceptual Cost Estimation Process that Affect the Level of Accuracy*

There is a need to identify the factors and determine what factors may affect the level of accuracy in the conceptual cost estimation process. Several factors affect the accuracy of conceptual cost estimates in the estimation process, which these factors are essential to identify [6]. As mentioned by [10], which states that various factors influence conceptual costs, it is necessary to identify the influencing factors to determine its degree [4] also indicated that it is essential to identify critical factors for the practical estimation of each project.

Based on literature studies, some factors affect accuracy at each stage of the conceptual cost estimation process. The research performed by Serpell [6] identified 5 factors that affect the level of accuracy, namely scope quality, information quality, uncertainty level, estimator's performance, and estimating procedure. Serpell draws on these factors from expert interviews. Then a qualitative causal model is created using these factors to assess the accuracy and how this model is used to improve the scoring system. The journal provides a further explanation of each factor's details that helps further understand and clarifies the meaning of each of these factors.

According to [10], the research uses information quality, uncertainty level, estimator's performance, estimating procedure, and project definition. This study aims to detect and assess the factors that affect the cost estimation of a project. The author seeks to provide other solutions to evaluate the factors that influence conceptual cost estimates and evaluate conceptual cost estimates with a method that is easier to understand than the method offered by Serpell [6].

The study performed by L Liu and Kai, who stated that output measurability is likely to be low at the conceptual stage because the estimation is based on inaccurate data, identified the critical factors for practical estimation at various stages of typical construction. From 9 categories, there are essential factors that affect conceptual cost estimates, namely: project information, team experience and cost data [14].

Meanwhile, [15], which conducted research based on the data analysis, produces 11 groups of factors that affect the accuracy of estimation, in the following order: basic process design, team experience and cost information, the time allowed to prepare the estimate, site requirement, bidding and labour climate, team alignment, owner's cost, contingency and reviews, formal estimating process, money issues, technology issues. Then the five most dominant are taken, namely basic process design, team experience and cost information, time allowed to prepare the estimate, site requirements, and bidding and labour climate.

In [16], the research identifies factors that influence cost estimation in 5 construction phases, states that conceptual cost estimation is a rough estimate whose accuracy is effected by 12 factors. Among them are the clarity and adequacy of the client brief, the volume of the consultant's workload during estimation, project size and complexity and estimation method. This study describes clearly through a detailed table the factors and their influence on each phase. It can be seen what factors have an impact and do not have an effect in that phase.

Thomas Pasco [17] explored project characteristics that affect the variances between conceptual cost estimates and tender estimates to investigate the most effective solution to increase early-stage forecast accuracy. With the initial hypothesis, Ballard [5] states that the conceptual cost accuracy rate can reach 10%, saying that information quality, project definition, and target cost can affect accuracy. The results of this study answer the initial hypothesis that conceptual cost accuracy can reach ±10% with a 95% confidence level with the Haahtela method.

According to [13], there are factors used to evaluate project cost proposals and propose quantitative project costing procedures. These factors consist of project conditions, project duration, environmental conditions, regulation conditions, planning conditions, estimation conditions.

Eventually, this research uses factors that affect accuracy in the conceptual cost estimation process based on those existing literature studies. These factors are Information Quality, Project Definition, Estimator's Performance, Scope Quality, Estimating Procedure, Cost Information, Project Characteristics, and Government Regulations.

2.3 Factor's Definition in Conceptual Cost Estimation Process that Affect the Level of Accuracy

After identifying the factors in the conceptual cost estimation process that affect accuracy, each factor's definition needs to be described.

X.1 Information Quality. Gustavsson and Wanstrom (2009) explain that information quality is the ability of information to meet user needs. Information quality is the embodiment of workers' knowledge and customer expectations, which consists of 2 elements: the level of truth and usefulness of the information [18]. The minor difference between the information required and the information obtained increases the quality of the information [19].

X.2 Project Definition. Project definition is the result of the process of exploring all aspects of a proposed project [20]. Project definition can also be understood as a process by which stakeholder needs are identified [21, 22].

X.3 Estimator's Performance. An estimator is a person who is responsible for organising and analysing all information into estimates based on the expertise and experience he has. With expertise, an estimator will better understand new projects based on previous projects, collect data by making assumptions and justifying existing information, and visualize the scope of work [2].

X.4 Scope Quality. Quality of scope is the suitability of investment returns or results that have been delivered by the performance of an activity or project, with the requirements set by project stakeholders [20].

X.5 Estimating Procedure. Estimating procedure is an activity stage or step-by-step method determined to estimate the number and costs of uncertain resources required for the project's scope or activity to minimize the level of uncertainty from the quality of the scope definition and maximize the opportunity for costs [20].

X.6 Cost information. From a cost management perspective, the cost is an investment of resources in strategic assets, including time, finance, human resources, and physical resources [20]. Meanwhile, concerning estimation, information is a reference estimator for generating costs [2].

X.7 Project Characteristic. According to Skitmore (1991), project characteristics are project features, which can be physical or non-physical characteristics, which explain the nature of the target. The factors are objective and measurable [23].

X.8 Government Regulations. Regulations can take many forms and can be enforced by governments at various levels. At the local level, regulations can take permit fees, installation and implementation of construction standards that directly increase construction costs or may cause delays resulting in increased costs higher [24].

3 Conceptual Framework

The relationship between variables in the conceptual cost estimation process and the level of accuracy based on literature consists of:

- Information Quality and Accuracy Relationship. The accuracy of conceptual cost estimation is primarily determined by the level of information provided by both current information and historical information [2, 6].
- Information Quality and Estimating Procedures Relationship. According to [25], current cost data and historical cost data will affect the estimation method.

- Information Quality and Cost Information Relationship. The cost information consists of historical data related to similar projects and historical data quality [14]. Similar project data can be a source for conceptual estimation cost data and can be the basis for calculating the bridge's average cost per m^2 [12].
- Project Definition and Accuracy Relationship. Project definition is one factor that affects the accuracy of conceptual cost estimates. It increases the level of detail definition to produce a more credible conceptual cost estimate [12].
- Project Definition and Estimating Procedure Relationship. The estimating procedure used during the initial project design stage depends on its completeness; a certain design percentage is complete [26].
- Project Definition and Project Characteristic Relationship. According to [12], with an incomplete project definition, the estimator must see the project as a whole. Thus, the estimation will focus on "bigger picture" characteristics such as project location, potential environmental problems, and utility impacts.
- Project Definition and Cost Information Relationship. A project's definition can influence cost information in design quality, construction complexity, and the project components' condition [27].
- Estimator's Performance and Accuracy Relationship. According to [2, 6, 10, 25], estimator performance is one factor that affects the accuracy of conceptual cost estimates [2, 6, 10, 25].
- Estimator's Performance and Information Quality Relationship. Oberlender and Trost (2001) state that quantity surveyor's competence and experience determine and affect information quality [25].
- Estimator's Performance and Estimating Procedure Relationship. According to Bley (1990), the estimator's ability to visualize the scope of work from incomplete scope definitions is essential in the conceptual cost estimation process. The estimator is a process coordinator and brings his expertise and experience in estimating costs. Guidance is sometimes needed to assess the inputs that enter the process [2].
- Scope Quality and Accuracy Relationship. The scope quality is one factor that affects the accuracy of conceptual cost estimates [6].
- Scope Quality and Estimator's Performance Relationship. According to Bley, the availability of a complete scope definition is the essential factor in the conceptual cost estimation process. An estimator's first thing is learning and interpreting the project scope and producing an estimated plan [2].
- Estimating Procedure and Accuracy Relationship. In its effect on the accuracy level, the estimation procedure's quality is influenced by the time available to calculate conceptual costs [6, 10, 25].
- Cost Information and Accuracy Relationship. Cost information is one factor that affects the accuracy of conceptual cost estimates [14, 15].
- Cost Information and Estimating Procedure Relationship. Cost data taken from current data and historical data can affect the estimation method [25].
- Characteristic Project and Accuracy Relationship. According to several studies, project characteristics are one of the factors that affect the accuracy of conceptual cost estimates [2, 12].

- Characteristic Project and Estimator's Performance Relationship. The estimator prepares estimates based on project characteristics related to other factors, and some are subjective [25].
- Characteristic Project and Cost Information Relationship. The use of previous actual project cost data following the project's physical data allows the formation of cost factors to form conceptual cost estimates, with different types, for example, the cost per meter or the cost per square meter [12].
- Government Regulation and Accuracy Relationships. According to [13, 28], regulations are a factor that affects the quality of proposed costs and is an external factor to the accuracy of estimates.

The relationship between variables and accuracy based on the literature study was then used to compile 19 hypotheses. These hypotheses are shown in Table 2. The arrows indicate that one variable has a significant effect on the other variable. These hypotheses also form a relationship model and are considered as a model's component of SEM. The relationship model is described in Fig. 2.

Table 2 Hypothesis significance relationship table

No.	Hypothesis significance relationship	No.	Hypothesis significance relationship	No.	Hypothesis significance relationship
1.	Information quality → estimating procedure	8.	Estimator's performance → information quality	15.	Cost information → accuracy
2.	Information quality → cost information	9.	Estimator's performance → estimating procedure	16.	Project characteristic → estimator's performance
3.	Information quality → accuracy	10.	Estimator's performance → accuracy	17.	Project characteristic → cost information
4.	Definition of project → estimating procedure	11.	Scope quality → estimator's performance	18.	Project characteristic → accuracy
5.	Definition of project → cost information	12.	Scope quality → accuracy	19.	Government regulation → accuracy
6.	Definition of project → project characteristic	13.	Estimating procedure → accuracy		
7.	Definition of project → accuracy	14.	Cost information → estimating procedure		

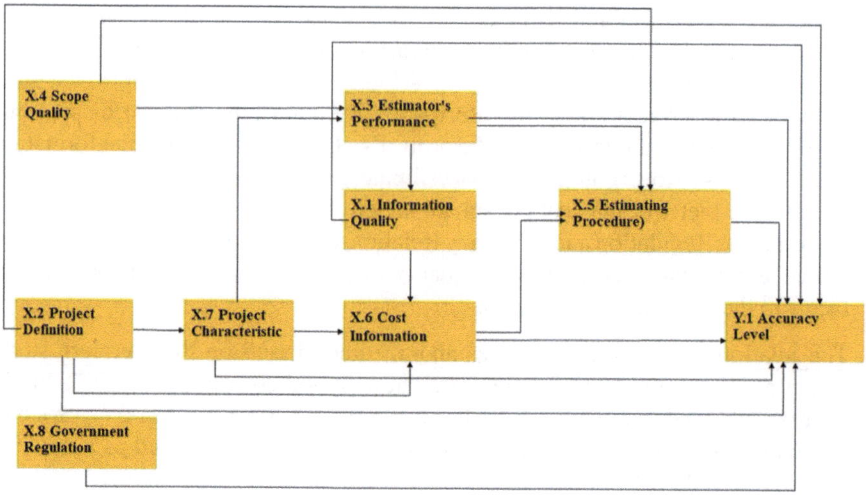

Fig. 2 Relationship model

4 Research Methodology

4.1 Research Methodology Stages

The research methodology in this study consists of several stages. In the early stage, this study uses the literature study method to obtain variables and indicators in the conceptual cost estimation process that affect estimation accuracy at the conceptual stage. A review of several literature studies resulted in a conceptual definition of the variables. Then based on the literature study, the conceptual definition is broken down into indicators that can measure variables.

And then, 5 experts in relevant fields will validate the variables and indicators with the criteria of having a minimum education of bachelor degree and having work experience of at least ten years in construction or flyover planning. The result of expert validation is analysed using the Delphi method. Delphi method is a method to reach agreement from experts to deal with problems [29]. In this study, the questionnaire was given to five experts in several rounds. The first round is carried out to obtain opinions from experts, which will be summarised to serve as the basis for the second round until we reach a mutual agreement. This expert validation impacts mutually agreed variables and indicators. Also, variables and indicators that experience improvements in the form of additions, subtractions, or changes.

The results from expert validation are used for a pilot survey. The pilot survey was conducted on 7 respondents with a minimum bachelor in engineering education qualifications in this study. The purpose of this pilot survey was to find out whether the delivery in the questionnaire was easy to understand or not. If there is a questionnaire that is not easy to understand, it will be corrected.

After a pilot survey, the survey was conducted with respondents using a questionnaire with corrected questions based on input from the pilot survey. This survey was conducted on 151 respondents with the criteria they had or are currently working in the DKI Jakarta Provincial Government and have a background or work relationship with flyover construction or planning. In this research, the number of respondents has exceeded the reference. With the maximum number of arrows pointing to the latent variable in the model, the minimum sample size needed is 84 [30].

Respondent survey results are compiled and applied to the relationship model between variables in the conceptual cost estimation process to accuracy. Then the model is analyzed using the SEM-PLS method with Smart-PLS software. The model in SEM is described in Fig. 3.

Structural Equation Modeling (SEM) is a multivariate analysis technique that combines factor analysis and regression methods, allowing researchers to perform simultaneous analyses and examine a series of dependency relationships simultaneously [31]. This analysis determines the influence level of each indicator in the variable on the level of accuracy, based on the hypothesized relationship model that has been made previously. After the SEM analysis is carried out, then validation will be carried out by experts.

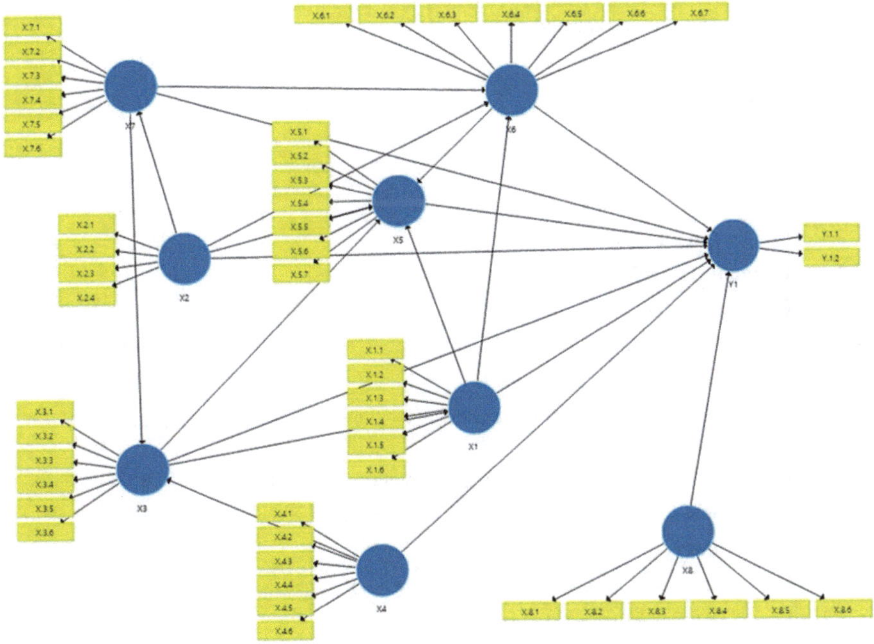

Fig. 3 Relationship model in SEM

4.2 Research Variables and Indicators

Based on literature studies and validation experts, this research identifies 8 variables in the conceptual cost estimation process that affect the accuracy. The variables are information quality consisting of 6 indicators, project definition composed of 4 indicators, estimator's performance composed of 6 indicators, scope quality composed of 6 indicators, estimating procedure composed of 7 indicators, cost information composed of 7 indicators, project characteristics composed of 6 indicators, and government regulations composed of 6 indicators. For accuracy, the variable has 2 indicators. Table 3 gives a list of variables and their indicators.

Table 3 Research variables and indicators

No	Variables and indicator's of variable	References
X.1	**Information quality**	
X.1.1	Quality of historical information	[2, 6, 10, 25]
X.1.2	Quality of current information	[2, 6, 25]
X.1.3	Accuracy of available information	[6, 32]
X.1.4	Information accessibility	[6, 32]
X.1.5	Fulfillment of user expectations	[19, 32]
X.1.6	Meeting stakeholder needs	[33]
X.2	**Project definition**	
X.2.1	Planning definition level	[15]
X.2.2	Plan drawing quality level	[10]
X.2.3	The level of project complexity	[10, 12, 26, 34]
X.2.4	Project communication level	[21]
X.3	**Estimator's performance**	
X.3.1	Estimator's experience and knowledge level	[6, 10, 25, 35, 36]
X.3.2	Estimator's commitment and perceptions level	[6, 10, 25, 34]
X.3.3	Estimator's ability to communicate	[25, 36]
X.3.4	Estimator's ability to see cost components and analyse project scope	[36, 37]
X.3.5	Estimator's ability to make assumptions about project details	[28, 37, 38]
X.3.6	Estimator's ability to gain access to cost information	[28]
X.4	**Scope quality**	
X.4.1	Completeness of the scope relating to the level of consistency	[6]
X.4.2	Project definition that is filled in for execution	[2, 39, 40]
X.4.3	Stability or change in project scope related to owner commitment level	[6]

(continued)

Table 3 (continued)

No	Variables and indicator's of variable	References
X.4.4	Stability or changes in the project scope related to the level of complexity of the project	[2, 6, 16]
X.4.5	Level of scope control	[26, 39]
X.4.6	Compliance with stakeholder understanding or requirements	[39–41]
X.5	**Estimating procedure**	
X.5.1	Estimated time	[6, 10, 25, 42]
X.5.2	The expected level of accuracy	[6, 25, 43]
X.5.3	Standard estimation process procedure	[10, 26, 42]
X.5.4	Documentation	[26, 42]
X.5.5	Communication in the estimation process	[26]
X.5.6	Management support	[26]
X.5.7	Estimation review	[26]
X.6	**Cost information**	
X.6.1	Accuracy, reliability and completeness of cost information	[15, 25, 34]
X.6.2	Formulation of construction component costs, in the form of costs per m^2 based on similar previous projects	[12, 26]
X.6.3	Formulation of construction component costs, in the form of costs per m^2 typical project section based on similar previous projects	[12, 26]
X.6.4	Formulation of the cost of the non-construction component to calculate contingency costs	[12, 15, 27, 43]
X.6.5	Formulation of the cost of the non-construction component is in the form of inflation	[12, 26]
X.6.6	Price regulation	[13, 44]
X.6.7	Economic conditions, market and price level	[6]
X.7	**Project characteristic**	
X.7.1	The physical characteristics of the project are the dimensions of the project	[25, 45]
X.7.2	The physical characteristics of the project include the type of structure used	[42, 46]
X.7.3	The non-physical characteristics of the project are the project location	[47]
X.7.4	The non-physical characteristics of the project are the type and duration of the contract	[25, 42]
X.7.5	Non-physical characteristics of the project in the form of project type	[25]
X.7.6	The uniqueness of the project conditions and the availability of information on the related data references	[12]
X.8	**Government regulation**	
X.8.1	Application of applicable regulations regarding licensing	[47]
X.8.2	Application of applicable regulations regarding technical road requirements, general requirements of bridges, construction standards	[12, 40, 44]

(continued)

Table 3 (continued)

No	Variables and indicator's of variable	References
X.8.3	Implementation of regulations containing the budget proposal and budget phasing for strategic planning and multiple years	[48]
X.8.4	Establishment of regulations regarding fees	[44]
X.8.5	Application of regulations regarding fulfillment of project requirements relating to environmental policies and public engagement	[47, 49]
X.8.6	Application of building requirements within the DKI Jakarta area and accessibility technical requirements for buildings and the environment	[50]
Y.1	**Accuracy**	
Y.1.1	The percentage represents the difference between the conceptual cost estimate and the actual cost of building the flyover	[10, 51]
Y.1.2	The Percentage represents the average difference between the conceptual cost estimate and the actual cost of building flyovers	[25, 52]

Table 4 Analysis results on construct reliability and validity based on results from SEM-PLS

Variable	Construct reliability and validity		
	Cronbach's Alpha	Composite reliability	AVE
X.1 Information quality	0.704	0.818	0.529
X.2 Project definition	0.621	0.839	0.723
X.3 Estaimtor's performance	0.851	0.889	0.574
X.4 Scope quality	0.854	0.892	0.579
X.5 Estimating procedure	0.785	0.853	0.538
X.6 Cost information	0.837	0.877	0.507
X.7 Project characteristic	0.825	0.873	0.536
8. Government regulations	0.875	0.901	0.606
Y.1 Accuracy	0.813	0.914	0.842

5 Result and Discussion

Based on SEM-PLS analyzed, after testing the validity and reliability, the model meets the condition that AVE is higher than 0,5 and composite reliability is higher than 0,7 [53], Cronbach's alpha is higher than 0,6 [54], as shown in Table 4. AVE for

Table 5 Analysis results on discriminant validity—Fornell-Larcker criterion based on results from SEM-PLS

Variable	Discriminant validity—Fornell Larcker criterion								
	X1	X2	X3	X4	X5	X6	X7	X8	Y1
X.1. Information quality	0.727								
X.2. Project Definition	0.314	0.850							
X.3. Estaimtor's Performance	0.517	0.565	0.757						
X.4 Scope Quality	0.519	0.550	0.726	0.761					
X.5 Estimating Procedure	0.442	0.516	0.685	0.626	0.734				
X.6 Cost Information	0.402	0.417	0.636	0.574	0.666	0.712			
X.7 Project Characteristic	0.329	0.384	0.561	0.520	0.553	0.566	0.732		
X8. Government Regulations	0.190	0.318	0.376	0.470	0.519	0.522	0.474	0.778	
Y.1 Accuracy	0.241	0.293	0.311	0.367	0.259	0.436	0.245	0.401	0.918

each latent variable is higher than R^2 with all other latent variables [53], as shown in Table 5. Based on the bootstrapping results of the SEM-PLS analysis, the relationship model's output is as shown in Table 6.

Based on the SEM-PLS analysis results, from the 19 relationship hypotheses, there are 12 hypotheses with a T-statistic value (significance value) above 1.96, which indicates that the relationship between these variables has a significant relationship. The significant relationships are information quality and cost information relationship, project definition and estimating procedure relationship, project definition and cost information relationship, project definition and project characteristics relationship, estimator's performance and information quality relationship, estimator's performance and estimating procedure relationship, scope quality and estimator performance relationship, information cost and estimation procedures relationship, cost information and accuracy relationship, project characteristics and estimator performance relationship, project characteristics and cost information relationship, and government regulations and accuracy relationship. Besides that, there is 1 hypothesis (X.5 Estimating procedure and Y.1 Accuracy level relationship) with a T-statistic value of 1.627, which is considered to have a significant relationship even though it is less significant. Based on the T-statistic, the value of X.5 → Y.1 can increase (although not powerful) as the respondents' number in analyzing increases.

Table 6 Analysis results of relationship between variables based on SEM-PLS

No	Hypothesis relationship	T Statistics	The relationship between variable	No	Hypothesis relationship	T Statistics	The relationship between variable
1	Information quality → estimating procedure	1.042	Insignificant	11	Scope quality → estimator's performance	9.728	Significant
2	Information quality → cost information	2.904	Significant	12	Scope quality → accuracy	1.029	Insignificant
3	Information quality → accuracy	0.836	Insignificant	13	Estimating procedure → accuracy	1.627	Less significant
4	Definition of project → estimating procedure	2.097	Significant	14	Cost information → estimating procedure	3.680	Significant
5	Definition of project → cost information	2.942	Significant	15	Cost information → accuracy	2.972	Significant
6	Definition of project → project characteristic	4.294	Significant	16	Project characteristic → estimator's performance	3.160	Significant
7	Definition of project → accuracy	1.066	Insignificant	17	Project characteristic → cost information	4.870	Significant
8	Estimator's performance → information quality	7.884	Significant	18	Project characteristic → accuracy	0.743	Insignificant
9	Estimator's performance → estimating procedure	3.318	Significant	19	Government regulation → accuracy	2.536	Significant
10	Estimator's performance → accuracy	0.014	Insignificant				

Based on the outer loading results, the model has an R-square of 0.279, categorized as weak [53]. The constructed variable, the accuracy of conceptual cost estimation, is affected by information quality, project definition, estimator performance, scope quality, estimating procedure, cost information, project characteristics, and government regulations amounted to 27.9%. And this model has an NFI value of 0, 545, which means that this model can represent the field's actual conditions amounted to 54.5%.

After the SEM-PLS analysis is carried out, 3 experts validate the relationship model. Expert validations show that they agree with significant variable relationships based on SEM-PLS analysis, and there are 3 additional significant relationships. The additional significant variables relationships are information quality and accuracy relationship (from insignificant become significant relationship), the estimator's performance and accuracy relationship (from insignificant become significant relationship), and the estimating procedure and accuracy relationship (from less significant become significant relationship). There are differences in the relationship model based on the results of SEM-PLS analysis and expert validation. These differences can be illustrated by the relationship model's result based on SEM-PLS analysis in Fig. 4 and the relationship model's result based on expert validation in Fig. 5.

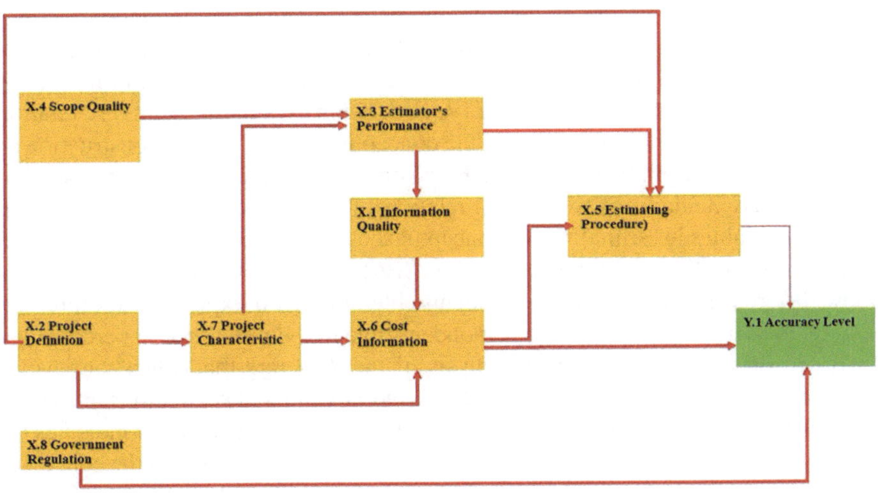

Fig. 4 The relationship model's result based on SEM-PLS analysis

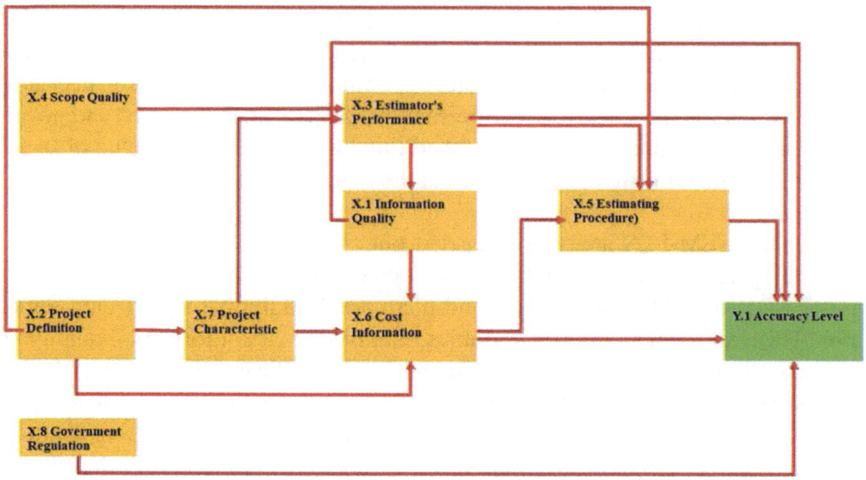

Fig. 5 The relationship model's result based on expert validation

6 Conclusions

Based on literature studies and expert validation, 8 factors affect the accuracy of conceptual cost estimates: information quality, scope quality, estimator performance, estimating procedures, project definition, cost information, project characteristics, and government regulations. And from SEM-PLS analysis, those factors affect the accuracy as big as 27.9%. Other factors outside of this study still determine the rest.

Based on the SEM-PLS analysis of 19 relationship hypothesis results, 12 relationships between variables in estimating the conceptual cost of flyover construction are considered significant. In comparison, based on expert validation results to the SEM-PLS, there are 3 additional significant relationships. So it becomes 15 relationships between variables in estimating the conceptual cost of flyover construction is considered significant. These differences between SEM-PLS analysis and expert validation show different perceptions of respondents, literature, and experts. Therefore, it is suggested to do more outreach to respondents due to the difference in perceptions.

Based on the results of SEM analysis of the 8 factors that affect the level of accuracy, it turns out that cost information and government regulations are factors that have a significant direct effect on the level of accuracy. This result indicates that these two factors may be the main cause of the accuracy in estimating the conceptual cost of flyover development in DKI Jakarta Province needs to be improved. Suppose they need to increase accuracy; first, they need to improve government regulations and cost information. In addition, it is also required to enhance factors that have a significant effect, either directly or indirectly, to cost information, such as information quality factors, project definition, project characteristics, and estimator performance.

References

1. Serpell AF (2005) Improving conceptual cost estimating performance. 2005 AACE Int Trans
2. Phaobunjong K (2002) Parametric cost estimating model for conceptual cost estimating of building construction projects, 2002
3. Slatoom N, Taneerananon P (2015) A study of a flyover-bridge—improved intersection. Eng J 19(1):1–12. https://doi.org/10.4186/ej.2015.19.1.1
4. Jumas DY, Zainon N, Rahim FAM (2015) The significance of cost variables and their influences on the conceptual cost estimation accuracy 2015(December)
5. Ballard G, Pennanen A (2013) Conceptual estimating and target costing. In: 21st Annual conference of the international group for lean construction, IGLC 2013, January 2013:1–10
6. Serpell AF (2004) Towards a knowledge-based assessment of conceptual cost estimates. Build Res Inf 32(2):157–164. https://doi.org/10.1080/0961321032000172373
7. Asal EM (2014) Factors affecting building construction projects' cost estimating
8. AACE International (2020) Cost estimate classification system-as applied in engineering, procurement, and construction for the building and general construction industries
9. Azman MA, Abdul Samad Z, Ismail S (2013) The accuracy of preliminary cost estimates in Public Works Department (PWD) of Peninsular Malaysia. Int J Proj Manag 31:994–1005. https://doi.org/10.1016/j.ijproman.2012.11.008
10. Honn An S, Cho H, Kyun Lee U (2011) Reliability assessment of conceptual cost estimates for building. Int J Civ Eng, vol 9
11. Gardner BJ (2015) Applying artificial neural networks to top-down construction cost estimating of highway projects at the conceptual stage
12. AASHTO (2013) Practical guide to cost estimating
13. Lai Y, Wang W, Wang H (2008) AHP- and simulation-based budget determination procedure for public building construction projects. Autom Constr 17:623–632. https://doi.org/10.1016/j.autcon.2007.10.007
14. Liu L, Zhu K (2007) Improving cost estimates of construction projects using phased cost factors. J Constr Eng Manag 133(1):91–95. https://doi.org/10.1061/(ASCE)0733-9364(2007)133:1(91)
15. Trost SM, Oberlender GD (2001) Predicting accuracy of early cost estimates based on estimate quality. J Constr Eng Manag 127(June):173–182
16. Dandan TH, Sweis G (2019) Factors affecting the accuracy of cost estimate during various design stages cost estimate. https://doi.org/10.1108/JEDT-08-2019-0202
17. Pasco T, Aibinu Ayodeji A (2008) Project factors influencing the accuracy of early stage estimates. In: Proceedings from international conference on building education and research (BEAR), 2008, pp 102–112
18. Woodall P, Parlikad AK (2013) Approaches to information quality management: state of the practice of UK asset intensive organizations (January):0–19. https://doi.org/10.1007/978-1-4471-2924-0
19. Eppler MJ, Christ DW (2000) Conceptualizing information quality: a review of information quality frameworks from the last ten years. In: Proceedings of the 2000 conference on information quality, 2000, January
20. AACE International (2019) Cost engineering terminology
21. Kuchta D, Chojnacka JM (2020) An approach to increase the sustainability of projects and their outcomes in public sector through improving project definition
22. Cano JL, Lidón I (2011) Guided reflection on project definition. Int J Proj Manag 29:525–536. https://doi.org/10.1016/j.ijproman.2010.04.008
23. Azman MA (2012) The accuracy of preliminary cost estimates in Public Works Department (PWD) of Peninsular Malaysia dissertation submitted in fulfilment of the requirements for the Degree of Master of Science Faculty of Built Environment University of Malaya. University of Malaya, Kuala Lumpur
24. Emrath P (2011) How government regulation affects the price of a new home. HousingEconomics.com

25. Azman MA (2012) The accuracy of preliminary cost estimates in Public Works Department (PWD) of Peninsular Malaysia. University of Malaya, Kuala Lumpur
26. Anderson S, Molenaar K, Schexnayder C (2006) Final report for NCHRP Report 574: guidance for cost estimation and management for highway projects during planning, programming, and preconstruction
27. GSA Public Building Service (2007) Project estimating requirements. Washington
28. Enshassi A, Mohamed S, Abdel-Hadi M (2013) Factors affecting the accuracy of pre-tender cost estimates in the Gaza Strip. J Constr Dev Ctries 18(1):73–94
29. Grime MM, Wright G (2016) Delphi method (2016). https://doi.org/10.1002/9781118445112.stat07879
30. Kay W (2013) Partial least squares structural equation modeling (PLS-SEM) techniques using SmartPLS. Mark Bull 24(November)
31. Hair JF, Black WC, Babin BJ, Anderson RE (2014) Multivariate data analysis
32. Alshikhi OA, Abdullah BM (2018) Information quality: definitions, measurement, dimensions, and relationship with decision making. Eur J Bus Innov Res 6(5):36–42
33. Azemi NA, Zaidi H, Hussin N (2018) Information quality in organization for better decision-making. Int J Acad Res Bus Soc Sci (January). https://doi.org/10.6007/IJARBSS/v7-i12/3624
34. Hatamleh MT, Hiyassat M, Sweis GJ, Sweis RJ (2018) Factors affecting the accuracy of cost estimate: case of Jordan Engineer. Eng Constr Archit Manag (January). https://doi.org/10.1108/ECAM-10-2016-0232
35. Lim B, Nepal M, Skitmore M, Xiong B (2015) Drivers of the accuracy of developers' early stage cost estimates in residential construction. J Financ Manag Prop Constr 21:4–20. https://doi.org/10.1108/JFMPC-01-2015-0002
36. Alroomi A, Seok Jeong DH, Oberlender GD (2012) Analysis of cost-estimating competencies using criticality matrix and factor analysis. J Constr Eng Manag 138(November):1270–1280. https://doi.org/10.1061/(ASCE)CO.1943-7862.0000351
37. Carr BRI (1990) Cost-estimating principles. 115(4):545–551
38. Hwang S (2016) A Bayesian approach for forecasting errors of budget cost estimates. J Civ Eng Manag 22(2):178–186. https://doi.org/10.3846/13923730.2014.897981
39. Kraus W, Kenneth R (1992) Project scope definition: a practical approach: a publication of the .... Cost Eng 34(12):15–19
40. Project Management Institute (2017) A guide to the project management body of knowledge (PMBOK guide), 6th edn. Project Management Institute, Inc.
41. AACE International (2019) Cost estimate classification system-as applied in engineering, procurement, and construction for the process industries
42. Arif F, Lodi S, Azhar N (2015) Factors influencing accuracy of construction project cost estimates in Pakistan: perception and reality. Int J Constr Manag (July). https://doi.org/10.1080/15623599.2015.1012141
43. Dysert LR (2006) Is 'estimate accuracy' an oxymoron? AACE Int Trans, pp 1–5
44. Khemani R, Shapiro DM (1999) Glossary of industrial organisation economics and competition law. Organization for Economics Co-Operation And Development
45. Elbeltagi E, Hosny O, Abdel-Razek R, El-Fitory A (2014) Conceptual cost estimate of Libyan highway projects using artificial neural network. J Eng Res Appl 4(8):56–66. www.ijera.com
46. Fayek AR, Ruben J, Flores R (2010) Application of fuzzy logic to quality assessment of infrastructure projects at conceptual cost estimating stage. 1147(December):1137–1147. https://doi.org/10.1139/L10-036
47. Washington State Department of Transportation (WSDOT) (2015) Cost estimating manual for projects
48. Jakarta PPD (2013) Peraturan Gubernur Provinsi DKI Jakarta Nomor 100 Tahun 2013
49. Singh RR, Chugh G (2016) Various risks involved in highway projects. Int J Innov Res Sci Technol 3(06):131–135
50. Jakarta PPD (1991) Peraturan Daerah Daerah Khusus Ibukota Jakarta Tentang Bangunan Dalam Wilayah Daerah Khusus Ibukota Jakarta

51. Aibinu AA, Pasco T (2008) The accuracy of pre-tender building cost estimates in Australia. Costr Manag Econ 6193. https://doi.org/10.1080/01446190802527514
52. Skitmore M (1991) Early stage construction price forecasting: a review of performance
53. Hair JF, Ringle CM, Sarstedt M (2011) PLS-SEM: indeed a silver bullet. J Mark Theory Pract 19(2):139–151. https://doi.org/10.2753/MTP1069-6679190202
54. Memon AH, Badullah NH, Rahman IA (2013) Using structural equation modelling to assess effects of construction resource related factors on cost overrun. World Appl Sci J (January). https://doi.org/10.5829/idosi.wasj.2013.21.mae.9995

The Sustainability Aspect of the Consulting Firm in Terms of Its Competitiveness in Indonesia

Nofita Harwin, Mairizal, Zelmi Sriyolja, Abd Rahman Mohd Sam, and Mohd Zaimi Abd Majid

Abstract Infrastructure development aims to improve services and accessibility. In terms of realizing the ideals of independence, infrastructure development for economic growth is a must. The government increased the infrastructure budget in the 2020 State Revenue and Expenditure Budget or RAPBN by 4.9%. To achieve this economic growth scenario, we need quality human resources, thus contributing to Indonesia's competitiveness in the 4.0 industrial era. Reliable, competent, and quality resources to compete will also encourage the growth of business entities in all areas, including consulting firms. The term competitiveness comes from the word power which means strength, and the word competitiveness which means to achieve more than others, or differ from others in terms of quality, or have certain advantages. The determinants of competitiveness are determined by factor conditions, demand conditions, relevant supporting industries, and company strategy & competitive structure. In this study, a literature review methodology is used to see what competitiveness factors are needed by a consulting firm to sustain the sustainability of getting a project every year. The results of this study obtain 13 competitiveness factors and two determinant factors that impact the sustainability of consulting companies validated by an expert.

Keywords Sustainability aspect · Consulting firm · Competitiveness

N. Harwin · Mairizal · Z. Sriyolja · A. R. M. Sam · M. Z. A. Majid
School of Civil Engineering, Faculty of Engineering, Universiti Teknologi Malaysia, 81310 Skudai, Johor, Malaysia
e-mail: harwin.1971@graduate.utm.my

A. R. M. Sam
e-mail: abdrahman@utm.my; dosen01742@iunpam.ac.id.com

M. Z. A. Majid
e-mail: mzaimi@utm.my

Mairizal (✉)
Program of Industrial Engineering, Faculty of Engineering, Universitas Pamulang, 15417 Tangerang Selatan, Banten, Indonesia
e-mail: dosen01742@unpam.ac.id

S. A. Kristiawan et al. (eds.), *Proceedings of the 5th International Conference on Rehabilitation and Maintenance in Civil Engineering*, Lecture Notes in Civil Engineering 225,
https://doi.org/10.1007/978-981-16-9348-9_82

1 Introduction

1.1 Infrastructure in Indonesia

Infrastructure development aims to improve services and accessibility. In order to realize the ideals of Indonesian independence, infrastructure development built for economic growth is a must. One of the important factors in infrastructure development is the quality of human resources [1]. As reported by the McKinsey Global Institute in September 2012, Indonesia will become the 7th largest economy in the world by 2050. In a report entitled "Archipelago Economy: Unleashing Indonesia", it is stated that Indonesia's economic growth is very fast and fast. Indonesia will become the 4th largest country in the world for the largest GDP, which is equivalent to 7275 USD billion by 2050 [2].

In this regard, the government of Indonesia has increased the infrastructure budget allocation in the 2020 year State Revenue and Expenditure Budget *(Rancangan Anggaran Pendapatan dan Belanja Negara)* posture by 4.9% to IDR 419.2 trillion. The 2020 infrastructure budget allocation consists of Rp. 405.1 trillion for economic infrastructure, Rp. 8.7 trillion for social infrastructure, and Rp. 5.3 trillion for supporting infrastructure (Fig. 1).

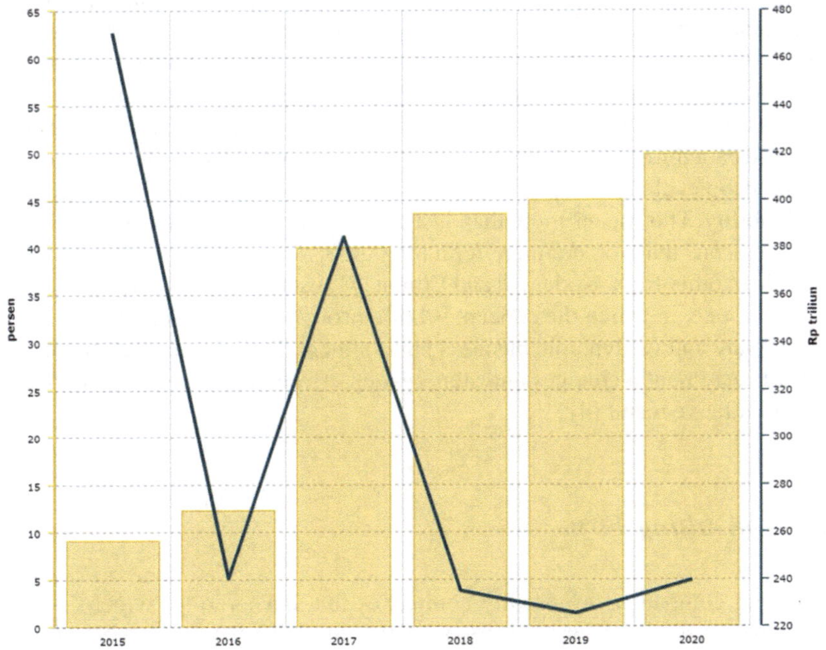

Fig. 1 2015–2020 Infrastructure growth and budget. *Sources* https://databoks.katadata.co.id/datapublish

1.2 *Infrastructure in Indonesia*

To achieve this economic growth scenario, Indonesia needs the support of quality human resources, thereby contributing to Indonesia's competitiveness in the industrial era 4.0. Sources that can be trusted, competent and qualified to compete will also promote the growth of commercial entities in all fields, including negotiating companies [3].

The need for infrastructure investment in Indonesia amounts to USD 1.7 trillion until 2045. If on average every year he needs USD 77 billion, together with IDR 1078 trillion (Global Infrastructure 2017). The need for large infrastructure, of course, the need for a negotiation agreement is also very large. In 2018 the number of negotiators accompanying Inkindo was 4776 companies, a decrease from the previous year's 5053 companies.

2 Literature Review

2.1 *Sustainability*

In research conducted by Mairizal [4] it has been explained that what is meant by sustainability according to Kotler and Keller [5] is an ability to meet human needs without harming the future of the current generation which is the top of many company agendas. Large companies have described in detail how they try to increase the long-term impact of all their actions on society and the environment. As one sustainability consulting company has also said, "There are three very basic things—people, planet and profit—and also which parts of society should come first. Sustainability. This means more than just being friendly to people, environment, and that also means that you are there for a long time.

While in other Refs. Kotler [6] and Ermon [7] sustainability is a dynamic concept, not a static one, in which the concept is built through intensive interactions between the company and its dynamic business environment. Sustainability is the result of a series of incremental changes and step-by-step efforts which collectively become a large cumulative result [4].

2.2 *Consulting Firm*

Consulting companies are usually companies that have various experts or professionals who have experience in certain fields and also have extensive knowledge about these various problems [8]. There is also another opinion that gives meaning to professional consultants or consultants according to Shenson [9], which states that individuals or companies have special skills and talents available to those who need them (clients), by receiving a certain amount of compensation (wages). The job of a professional consultant is to advise and often assist others in carrying out that advice both on the same side and for the client.

According to Darmadi's opinion, written in Diversity of Indonesia [10], stated that the first consultant was Arthur D. Little who founded his business in 1886 in Cambridge, Massachusetts. He also provides technical (technical) assistance to his clients. A man named Booz Allen Hamilton later founded a company with a similar structure in the early twentieth century, after the first company went bankrupt. After that in 1926, James McKinsey a professor from the University of Chicago also founded a company with the field of "accounting advisor and engineer" which introduced the concept of a different approach and framework. Mc Kinsey did not recruit traditional engineers, but instead employed experienced executives who were trained in the various collections of contemporary analysis and knowledge concepts that existed at that time, which included strategy, policies, objectives, organization, procedures, facilities and personnel. In historical facts, several extraordinary innovations have been made by the Boston Consulting Group (BCG). They used a different approach, and BCG developed the concept of the growth division matrix which is a tool for assessing the attractiveness of a company in various industries. Finally, the framework they created was later adopted by many other consulting firms in understanding the problems and business opportunities that could be exploited. So since then, various consulting companies in the world have shown themselves to be growing and developing rapidly. Some of them have merged with others, and some have changed their brands and changed their organizational structures. Likewise, the approach taken, the methodology and framework used and developed are also increasingly complex and more comprehensive.

2.3 Competitiveness

The history of the scientific field notes that the term competitiveness comes from the word "power" which means strength, and the word "competitiveness" which means achieving more than others, or something that is very different from others in terms of certain qualities, or has certain advantages too. Competitiveness can Competitiveness can also be interpreted as a strength that is used to try to be superior in certain matters and is carried out by certain groups or institutions [4, 7, 11–13].

According to the opinion of Porter [14] in research by Mairizal [4], it is said that the meaning of competitiveness or competition is the essence of the success or failure of a company, where competition is very similar to an activity in a company that will be able to contribute to its performance. such as innovation, a cohesive culture, or good practice. In addition, the meaning of the word competitive strategy is an effort to find a competitive position in an industry that is the basis of competition, which aims to form a profitable and sustainable position. Sustainable prosperity is governed by an underlying economic competitiveness which is largely determined by the level of productivity achieved by a company.

Table 1 Competitiveness factors

No.	Author's	Competitiveness factors
1	Porter [13], Ermon [7], Mairizal [17]	Human resources
2	Porter [13], Ermon [7], Mairizal [17]	Physical resources
3	Porter [13], Ermon [7], Mairizal [17]	Knowledge resources
4	Porter [13], Ermon [7], Mairizal [17]	Capital resources
5	Porter [13], Ermon [7], Mairizal [17]	Infrastructure
6	Porter [13], Ermon [7], Mairizal [17]	Demand segment
7	Porter [13], Ermon [7], Mairizal [17]	Sophistication
8	Porter [13], Ermon [7], Mairizal [17]	Anticipation
9	Porter [13], Ermon [7], Mairizal [17]	Material supply
10	Porter [13], Ermon [7], Mairizal [17]	Supporting equipment
11	Porter [13], Ermon [7], Mairizal [17]	Product innovation
12	Porter [13], Ermon [7], Mairizal [17]	Supply of experts
13	Porter [13], Ermon [7], Mairizal [17]	Company qualification
14	Porter [13], Ermon [7], Mairizal [17]	Strategy focused
15	Porter [13], Ermon [7], Mairizal [17]	Management structure
16	Porter [13], Ermon [7], Mairizal [17]	Future orientation

Meanwhile, according to a competitiveness expert, Porter [13] in the discussion of his book The Competitive Advantage of Nations, states that the potential competitiveness of a country or company comes from several sources, namely 4 groups of factors, namely: Factor Condition, Demand Condition, Related & Supporting and Firm Strategy. There are consist of 16 sub-factors of competitiveness (Table 1).

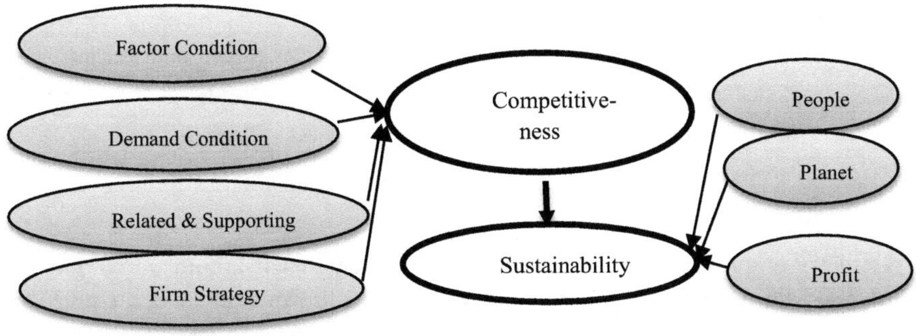

Fig. 2 Relationship model between competitiveness and sustainability. *Source* Porter [13] and developed by researcher

2.4 Relationship Model Between Competitiveness and Sustainability

In this study it will be seen to what extent the relationship between aspects of sustainability with competitiveness factors as illustrated in Fig. 2.

3 Research Methodology

3.1 Data Collecting

This paper uses two types of statistical approaches that can be used in research, namely quantitative and qualitative approaches [15, 16]. The two statistical approaches differ in terms of the methods and techniques used such as objectives, concepts, research approaches, data collection methods, data analysis and sampling [4, 15]. In the early stages of this study only collected all forms of relevant literature on issues of sustainability and competitiveness as mentioned by the various researchers reviewed. Literature review is not only limited to articles published in peer-reviewed and prestige journals, but also covers various theses and books. The literature search dates ranged from 1990 and 2018 to cover a wide range and at the same time capture new findings. The next stage, all factors or aspects related to sustainability and competitiveness are collected, and then used as a questionnaire which is distributed to respondents using the purposive sampling method. All respondents involved had experience in technical consulting activities in various projects in Indonesia. The questionnaire format consists of some general information and the respondent's background, followed by data on the importance of competitiveness factors identified through the literature as shown in Tables 1 and 2.

Table 2 Sustainability factors

No.	Author's	Sustainability factors
1	Kotler and Keller [5], Ermon [7], Mairizal [17]	People
2	Kotler and Keller [5], Ermon [7], Mairizal [17]	Planet
3	Kotler and Keller [5], Ermon [7], Mairizal [17]	Profit

Delivery of questionnaires to experts is also through self-administered questionnaires. A total of 20 responses have been received in full. Respondents consist of contractors, consultants and project owners. The Likert scale consists of five points, which is used to measure the importance of competitiveness for the sustainability of a consulting company. All data obtained were analyzed using descriptive statistical methods. All data from the questionnaire answers were analyzed using the Statistical Package for Social Sciences (SPSS). The results of the reliability value were calculated and found to be above 0.70 (Cronbach's Alpha) while the validity value must exceed the "r" value in the reference table. The highest factor value will be a determining factor for competitiveness in the sustainability of consulting companies in Indonesia.

3.2 Data Analysis

Data processing uses SPSS version 24.0 and according to Supriyadi [18], the testing techniques that are often used to test validity are Pearson Bivariate Correlation (Pearson Moment Product) and Item-Total Correlation Correction. In this paper using Corrected Item-Total with the results for the amount of data (n) = 16 obtained r table = 0.4259 (Table r for Pearson Product Moment) for a significance level of 0.05. The results of the SPSS data analysis can be seen in Table 3.

Referring to Table 3 of the Corrected Item-Total Correlation value, the calculated r value (Pearson Correlation) is smaller than r table 0.4259, namely X1 (Human Resources), X2 (Physical Resources) and X5 (Infrastructure). In the reliability testing method is to use the Cronbach's Alpha method whose value is >0.7 which is used to determine reliability and its shown the Cronbach's Alpha of all factors >0.700.

Table 3 Corrected item-total correlation results

	Scale mean if item deleted	Scale variance if item deleted	Corrected item-total correlation	Cronbach's Alpha if item deleted
X1	60.9000	67.463	0.135	0.904
X2	61.2500	65.987	0.174	0.906
X3	61.5000	59.000	0.728	0.888
X4	61.3500	58.871	0.775	0.887
X5	61.5000	63.421	0.392	0.899
X6	62.8000	56.484	0.777	0.886
X7	62.0500	59.418	0.618	0.892
X8	61.9000	57.358	0.673	0.890
X9	61.9000	57.779	0.643	0.891
X10	61.8000	57.432	0.659	0.891
X11	61.3500	61.397	0.545	0.895
X12	61.7500	61.250	0.525	0.895
X13	61.3500	61.608	0.527	0.895
X14	61.6500	58.976	0.808	0.887
X15	61.7000	60.116	0.505	0.897
X16	61.5000	61.842	0.481	0.897

4 Result and Discussion

4.1 Result

The first step has to ratify the Competitiveness Factors of sixteen (16) Porter's Competitiveness Factors in 3 Concepts of Competency Theory. The analysis of thirteen (13) factors what has reliable and valid in this research such as follows:

1. Knowledge Resources
2. Capital Resources
3. Demand Segment
4. Sophistication
5. Anticipation
6. Material Supply
7. Supporting Equipment
8. Product Innovation
9. Supply of Experts
10. Company Qualification
11. Strategy Focused
12. Management Structure
13. Future Orientation.

4.2 Discussion

Seen from the highest means of the 13 competitiveness factors, there are 2 competitiveness factors with the highest value, namely X6 (Demand Segment) with means = 62.8000 and X7 (Sophistication) with means = 62.0500. This means that these two factors are determining factors in the competitiveness of a consulting firm in Indonesia. All 13 factors mentioned above are mentioned to 3 Competitiveness experts and quite understanding about the competitiveness concept and they agree that all of these factors apply to the competitiveness of consultants in Indonesia.

It is argued that all marketing strategies involve the search for competitive advantage (Bharadwaj and Varadarajan 1993; Day and Wensley 1988; Varadarajan and Cunningham 1995). For market segmentation strategy factors are based on the achievement of competitive advantage and thus superior financial performance is generated by the company by: (1) identifying demand segments, (2) targeting specific segments, and (3) developing specific marketing "mix" for each segment. targeted market (Dibb, Simkin, Pride, and Ferrell 1994; Hunt 2002b) [19, 20].

On the other hand, in research Hunt and Arnett [19] say as confirmed by Shawney [21], p. 54, that "Customers become very sophisticated and demand products and services that are tailored to individual preferences and tastes." Likewise with Lancaster (1990) who argues that product variation is the result of consumers looking for variations for their own consumption and/or different consumers will want different variants of different flavors. From this stated perspective, a company that uses the concept of a market segmentation strategy will actually benefit both consumers and society by providing market offerings that better satisfy individual consumers' own wants and needs.

5 Conclusion

From all parts of the discussion above, the competitiveness of consulting firms in Indonesia is determined by 13 competitiveness factors referring to Porter's Competitiveness and previous research. From the 13 competitiveness factors, the results of data analysis using SPSS, obtained 2 competitiveness factors with the highest means value, namely Demand Segment and Sophistication. This means that consulting firms in Indonesia must take these two competitiveness factors into account in planning a competitive strategy for projects in increasingly fierce competition.

References

1. Bappenas (2017) Bahan Paparan Rapat Kerja Bappenas - Kementerian Lembaga dan Temu Konsultasi Triwulanan I Bappenas - Bappeda Provinsi Seluruh Indonesia pada Senin, 22 Februari 2016. Kementerian PPN/Bappenas

2. Price Waterhouse Coopers. Indonesia Akan Menjadi Negara Dengan Perekonomian Terbesar ke-4 di 2050. https://www.pwc.com/id/en/media-centre/pwc-in-news/2017/indonesian/pwc--indonesia-akan-menjadi-negara-dengan-perekonomian-terbesar-.html
3. Inkindo (2018) Roadmap Menuju INKINDO Emas 2030: "INKINDO Mandiri Demim Keunggulan Negeri". Ikatan Nasional Konsultan Indonesia (INKINDO), Jakarta, Indonesia
4. Mairizal (2020) Kompetensi Pengurus Projek Dalam Penilaian Prestasi Projek Pembinaan Bersepadu di Indonesia. Universiti Teknologi Malaysia (UTM), Johor, Malaysia
5. Kotler P, Keller KL (2012) Marketing management, 14th edn. Prentice Hall, New Jersey
6. Kotler P (2003) Marketing Management, 11th edn. Prentice Hall. Inc., New Jersey
7. Ermon DHN (2011) Strategi peningkatan daya saing untuk meningkatkan profit perusahaan jasa konstruksi, UI, Depok, Indonesia
8. Tordoir PP (1995) The Professional Knowledge Economy. The Management and Integration of Professional Services in Business Organizations. Springer Science+Business Media Dordrecht, Springer Nature Switzerland AG
9. Shenson HL (1990) The Contract and Fee-Setting Guide for Consultants and Professionals. Penerbit Wiley, USA
10. Darmadi S (2011) Board diversity and firm performance: the Indonesian evidence. Diversity of Indonesia. Corporate Ownership Control J 8:1–38
11. Durand M, Giorno C (1987) Indicators of international competitiveness: conceptual aspects and evaluation. OECD J: Econ Stud, ISSN 1995-2848, ZDB-ID 858174-5. 9:147–182
12. Hudori M (2013) analysis of competitiveness of the agribusiness sector companies using Porter's five forces. In: Proceedings: 2nd international conference on adaptive and intelligent Agroindustry (ICAIA) IPB International Convention Center Bogor—Indonesia, Sept 16–17 (2013)
13. Porter ME (1990) Competitive advantage-creating and sustaining superior performance. The Free Press, New York
14. Porter ME (2004) Global competitiveness report 2004–2005: findings from the business competitiveness index, Press Conference, New York
15. Chua YP (2001) Kaedah dan statistik penyelidikan: kaedah penyelidikan. Mcgraw-Hill Education, Kuala Lumpur, Malaysia
16. Fielding N, Schreier M (2001) Qualitative and quantitative research: conjunctions and divergences. Forum: qualitative social research. Center for Digital Systems, Freie Universität Berlin
17. Mairizal (2016) Perancangan Strategi Keunggulan Bersaing Perusahaan EPC. Institut Sains & Teknologi Nasional, Jakarta, Indonesia
18. Supriyadi (2014) Edy.SPSS+AMOS, Perangkat lunak statistic, In Media, Jakarta
19. Hunt SD, Arnett DB (2004) Market Segmentation Strategy, Competitive Advantage, and Public Policy: Grounding Segmentation Strategy in Resource-Advantage Theory. Australasian Marketing Journal (AMJ) 12(1):7–25
20. Hunt SD, Arnett DB (2006) The Explanatory Foundations of Relationship Marketing Theory. Journal of Business & Industrial Marketing 21(2):72–87
21. Shawney MS (1998) Leveraged high-variety strategies: from portfolio thinking to platform thinking 26(1)54–61

Sustainability Aspects in Water Resources Management

Sustainability Analysis of Minimization of Spills from a Reservoir

Syamsul Hidayat, Ery Setiawan, Ida Bagus Giri Putra, M. Bagus Budianto, and Salehudin

Abstract Reservoir operation in general is planned towards either maximization of benefit or minimization of loss. The operation plans may take form in optimization, simulation or combination of the two. Spills from reservoir or system of reservoirs can be considered as a loss, so that it must be minimized. In order to get a better perception of a study reservoir's behavior to which minimization of spills is implemented, sustainability of the plan needs to be assessed. The present work is aimed to evaluate the sustainability of an optimization-simulation model applied to the Batujai reservoir in Lombok Island, Central Indonesia. The model is set to minimize the spills from the study reservoir system, after which the sustainability of spill minimization is appraised in terms of reliability, resilience and vulnerability. The model was run at three consecutive years (2014–2017) so that sensitivity of the model to hydrologic variability can be evaluated. Results show that reliability, resilience and Sustainability Index (SI) clearly have same trends while vulnerability is in inversely trend with these three measures. In contrast, no conclusion can be made about the trade-offs between SI and irrigable area.

Keywords Sustainability · Minimization · Spills

1 Introduction

Reservoir management and operations deal with minimizing the loss caused by water shortages, flooding, adverse environmental impacts; or optimizing the benefit of

---

S. Hidayat (✉) · E. Setiawan · I. B. G. Putra · M. B. Budianto · Salehudin
Civil Engineering Department, Faculty of Engineering, University of Mataram, Mataram, Indonesia
e-mail: syamsul.hidayat@unram.ac.id

E. Setiawan
e-mail: ery.setiawan@unram.ac.id

I. B. G. Putra
e-mail: giriputra@unram.ac.id

S. A. Kristiawan et al. (eds.), *Proceedings of the 5th International Conference on Rehabilitation and Maintenance in Civil Engineering*, Lecture Notes in Civil Engineering 225,
https://doi.org/10.1007/978-981-16-9348-9_83

using natural resources such as land, water and energy, allocating water between multiple users and uses, and developing reservoir regulation strategies and operating rules [1]. Systems Approach is a concept underlying associated models and analysis methods in reservoir management and operations, which is aimed to improve the decision making process by applying optimization, simulation, multi-objective or combination of these previous three techniques to a whole picture of the reservoir problems [2].

Existing operating rules in multi-reservoir systems which have been formulated based on simulation or experience can be found, for instance, in [3, 4]. One of the purposes of these rules summarized in [3] is to gain a better understanding of the study system's behavior. In spite of their usefulness for quite a long time, alterations to these rules may be required due to changes in objectives and demands. Meanwhile, optimization refers to mathematical formulation which aimed to seek the maximum or minimum values of objective functions subjected to constraints [5]. While it is often said that simulation is descriptive and optimization is prescriptive, sometimes simulation cannot be distinguished clearly from optimization as some optimization models also simulate the study water system [1].

Objective functions in optimization may vary from irrigation, flood control, hydropower production, and environmental preservation [6], recreation and navigation [7]. A huge number of studies have been accomplished in optimizing these objectives, employing various techniques such as linear programming e.g. [8], non-linear programming e.g. [9], dynamic programming e.g. [10], network flow programming e.g. [11], neural network e.g. [12], fuzzy logic e.g. [13] and heuristic techniques e.g. [11]. Little to no attention, however, has been paid to sustainability analysis of the optimization results in spite of the importance of such an analysis as a means to gain a better understanding of a study water system's behavior.

The present work is aimed to a better understanding of the behavior of a reservoir to which an optimization plan is implemented. The objectives of the present work are twofold: firstly, to develop an optimization model to minimize spills from the study reservoir, and secondly to analyze the sustainability of the optimization results. The sustainability is measured through the most often used terms which are reliability, resilience and vulnerability.

2 Methods

2.1 Study Reservoir System and Data

Batujai reservoir, built in 1982, is located in Central Lombok Regency which is about 25 km from Mataram, the capital of West Nusa Tenggara Province. The specification and the location of the reservoir system is summarized and depicted in Table 1 and Fig. 1, respectively.

Table 1 Specification of the study reservoir system [14]

Specification	Batujai reservoir
Dead storage (MCM)	23.00
Maximum storage (MCM)	1.40
Purposes	Irrigation, flood control, raw water supply
Irrigation command area	3350 ha

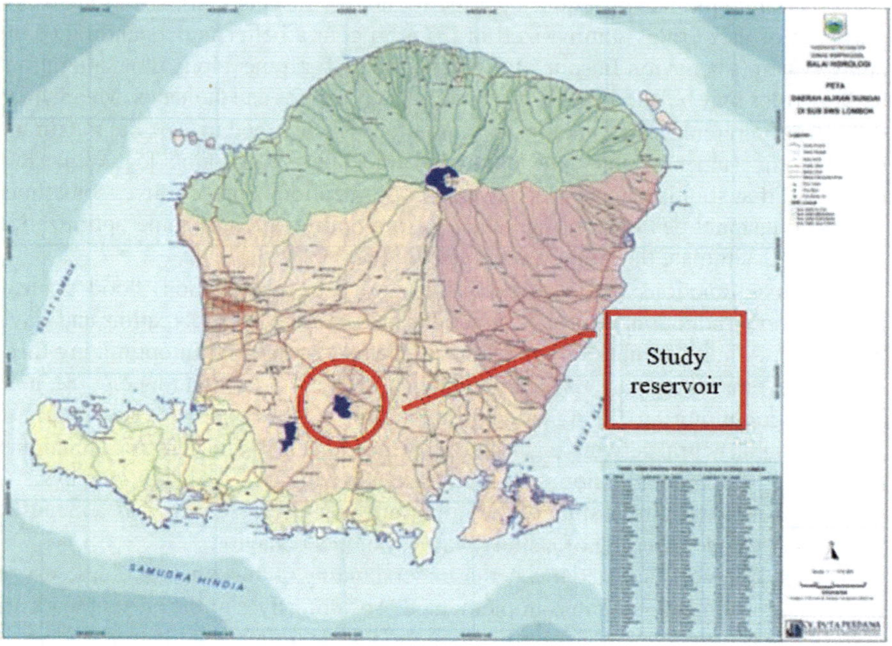

Fig. 1 Study water system [14]

The optimization was conducted for three years from November 2014 to October 2017, which bi-monthly was taken as optimization time step. One year consists of three cropping seasons lasting for four months each with the first season commences on Nov and cropping patterns paddy-paddy-corn. The evaporation rate is expressed in mm/day and is obtained from local evaporation pan observations. Trade-offs between reservoir's water elevation and reservoir's storage is represented by a quadratic form (Eq. 1) with a coefficient of regression (R^2) of 0.9957, whereas that of reservoir's storage and reservoir's free water surface area is interpolated based on Table 2.

$$V_t = 10^{-12} * A_t^2 + 0.0004 * A_t + 616.02 \quad (1)$$

Table 2 Study reservoir's storage and corresponding surface area

Storage (MCM)	Surface area (ha)	Storage (MCM)	Surface area (ha)
23.000	8900	12.575	5112
21.517	7869	9.567	4276
18.434	6908	4.650	2222
14.941	5989	1.400	1042

2.2 Optimization Model

The objective function is to minimize total spills from the study reservoir (Eq. 2), subject to constraints expressed in water balance (Eq. 3), storage bounds (Eq. 4) and steady state condition (Eq. 5)

$$\text{Minimize } Z = \sum\nolimits_{t=1}^{12} S_t, t = 1, 2, \ldots, 12 \tag{2}$$

Subject to:

$$V_{t+1} = V_t + I_t - IR_t - RW_t - E_t - S_t \tag{3}$$

$$V_{min} \leq V_t \leq V_{max} \tag{4}$$

$$V_1 = V_{24} \tag{5}$$

where:

V_t: volume at time *t* (MCM),
I_t: inflow at time *t* (MCM),
IR_t: irrigation releases at time *t* (MCM),
RW_t: raw water supply at time *t* (MCM),
E_t: evaporation at time *t* (MCM),
S_t: spill at time *t* (MCM),
V_{min}: minimum volume (1.400 MCM),
V_{max}: maximum volume (23.00 MCM),
V_1: volume at the beginning of Nov-1 (MCM),
V_{24}: volume at the end of Oct-2 (MCM).

2.3 Sustainability Analysis

The terms reliability, resiliency and vulnerability were first introduced by [15] and were later used in evaluating water management sustainability by, among others [16–20]. The definition of each term and the formulas here are taken from [17] as follows and are modified accordingly to spill minimization instead of hydropower maximization. Reliability (α) is defined as the probability of the system being in satisfactory state, which in the present work is represented by the number of occasions whenever no spill occurs at any time period.

$$\alpha = \frac{\sum_{t=1}^{24} S_t = 0}{T} \tag{6}$$

The transition of a system from being unsatisfactory to satisfactory is denoted by W_t, whose value is either 1 if the transition exists or 0 elsewhere [15]. The speed of recovery from an unsatisfactory condition is defined as resiliency (β); where in the present work is represented by a time period when spill occurs followed by another period without spill and can be expressed by Eq. 7. Both α and β values are ranging from 0 to 1.

$$\beta = \frac{\sum_{t=1}^{24} W_t}{T - \sum_{t=1}^{24} S_t = 0} \tag{7}$$

The extent or duration of a failure can be measured by expected extent-vulnerability expressed by Eq. 8. In the present work, deficit at period t (D_t) is defined as an event when the reservoir volume is below the dead storage. Further, a dimensionless vulnerability measure is proposed by [21] as expressed in Eq. 9.

$$Vul_{avg} = \frac{\sum_{t=1}^{24} D_t}{T - \sum_{t=1}^{24}(\text{Vmin} \leq \text{Vt} \leq \text{Vmax})} \tag{8}$$

$$\varphi = \frac{Vul_{avg}}{Max \sum D_t} \tag{9}$$

Finally, Sustainability Index (SI) is a geometric mean of reliability, resiliency and vulnerability and can be computed as follows:

$$SI = \sqrt[3]{\alpha\beta(1 - \varphi)} \tag{10}$$

3 Results and Discussion

The optimization model was run three times for one agricultural year each run commencing on November at any year and concluding on October the following year. Irrigable area, reliability, resiliency, vulnerability and SI for each period are summarized in Table 3 and depicted in Figs. 2 and 3.

Figure 2 shows that there is no clear trend between SI and irrigable area. While the irrigable area increases from 1551 hectares at 2014–2015 to 2290 hectares at 2015–2016, the values of SI decrease from 0.368 to 0.329. Meanwhile, at 2015–2016 period both irrigable area and SI increase.

In contrast to trade-offs between SI and irrigable area, Fig. 3 shows that there exists clear trend between reliability, resilience and SI. The more reliable the study reservoir, the more resilience and the more sustainable it is and so forth. Conversely, the more reliable the study reservoir, the less vulnerable it is and so forth.

Table 3 Results of sustainability analysis at each period

Period	Irrigable area (ha)	A	B	Φ	SI
Nov 2014–Oct 2015	1551.10	0.708	0.143	0.490	0.368
Nov 2015–Oct 2016	2289.82	0.625	0.111	0.510	0.329
Nov 2016–Oct 2017	2388.05	0.792	0.400	0.153	0.535

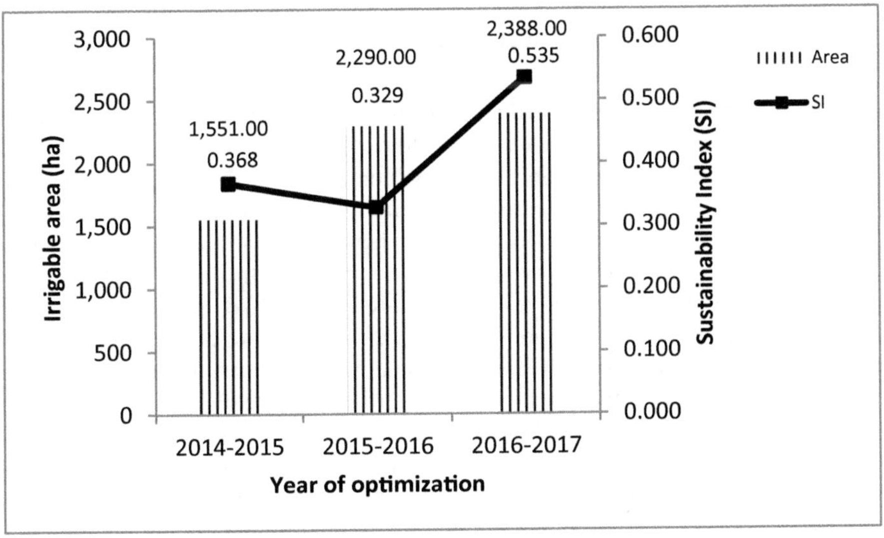

Fig. 2 Irrigable area (ha) and Sustainability index (SI)

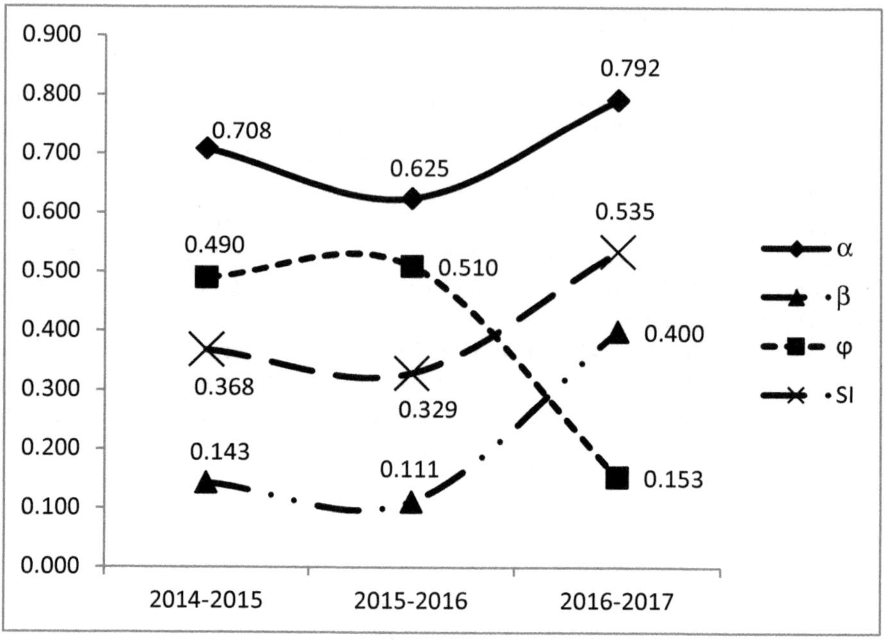

Fig. 3 Reliability (α), resilience (β), vulnerability (φ) and Sustainability index (SI) of the study reservoir during 3-year optimization horizon

4 Suggestions and Recommendation

By comparing the irrigable area to the designated irrigation area of 3350 ha (see Table 1), it can be seen that the former is much lower to the latter; this requires further research to investigate the possible causes which may include sediment deposition in the reservoir, or significant decreasing of inflows. Another result obtained in the present work is that there exist similar trends between reliability, resilience and SI whereas these measures' trends are inversely compared to that of vulnerability.

The formulas of reliability, resilience, vulnerability and Sustainability Index (SI) in the present work were adopted from a study whose research objective is hydropower maximization. While the mathematical formulas remain exactly the same, their physical meaning are modified accordingly to the objective of the present work i.e. spill minimization. This is believed to be true for future studies with different research objectives such as, for example, mitigation of droughts and reservoir hedging. The measures of sustainability mentioned above may need to be defined appropriately in relation to the future studies' objective functions.

Acknowledgements The authors would like to express their thankfulness to the Rector of University of Mataram as well as the Dean of the Faculty of Engineering for providing international publication research grant. The authors also wish to extend their gratitude to the Head of Civil

Engineering Department for his encouragement to the authors to conduct and to publish research in international conferences and journals.

References

1. Wurbs RA (1993) Reservoir-system simulation and optimization models. J Water Resour Plan Manag 119(4):455–472
2. Simonovic SP (1992) Reservoir systems analysis: closing gap between theory and practice. J Water Resour Plan Manag 118(3):262–280
3. Oliveira R, Loucks DP (1997) Operating rules for multireservoir systems. Water Resour Res 33(4):839–852
4. Smith J (1993) Conditional-yield operating rules for low-flow management. J Water Resour Plan Manag 119(3):324–338
5. Fayaed SS, El-Shafie A, Jaafar O (2013) Reservoir-system simulation and optimization techniques. Stoch Env Res Risk Assess 27(7):1751–1772
6. Rani D, Moreira MM (2010) Simulation-optimization modeling: a survey and potential application in reservoir systems operation. Water Resour Manage 24(6):1107–1138
7. Ahmad A, El-Shafie A, Razali SM, ZS Mohamad (2014) Reservoir optimization in water resources: a review. Water Resour Manag 28:3391–3405
8. Yoo J-H (2009) Maximization of hydropower generation through the application of a linear programming model. J Hydrol 376(1–2):182–187
9. Abdulrahman MKA, Ismail AAMd, Kek SL (2013) Determination of normal operating heights of reservoirs in a network using non-linear reservoir method. Br J Appl Sci Technol **3**(1):34
10. Zhang Z et al (2013) Use of parallel deterministic dynamic programming and hierarchical adaptive genetic algorithm for reservoir operation optimization. Comput Ind Eng 65(2):310–321
11. Lerma N et al (2013) Development of operating rules for a complex multi-reservoir system by coupling genetic algorithms and network optimization. Hydrol Sci J 58(4):797–812
12. Cancelliere A et al (2002) A neural networks approach for deriving irrigation reservoir operating rules. Water Resour Manage 16(1):71–88
13. Kumar ARS et al (2013) Application of ANN, fuzzy logic and decision tree algorithms for the development of reservoir operating rules. Water Resour Manag 27(3):911–925
14. Perdana C (2015) Map of catchments within Lombok Greater Basin (in Indonesia language)
15. Hashimoto T, Stedinger JR, Loucks DP (1982) Reliability, resiliency and vulnerability criteria for water resources system performance evaluation. Water Resour Res 18(1):14–20
16. Moy WS, Cohon JL, ReVelle CS (1986) A programming model for analysis of the reliability, resilience, and vulnerability of a water supply reservoir. Water Resour Res 22(4):489–498
17. Ghimire BNS, Reddy MJ (2013) Optimal reservoir operation for hydropower production using particle swarm optimization and sustainability analysis of hydropower. ISH J Hydraul Eng 19(3):196–210
18. Shin S et al (2018) A systematic review of quantitative resilience measures for water infrastructure systems. Water 10:1–25
19. Sule BF, Surajudeen M (2017) Evaluation of the reliability, resilience and vulnerability of Jebba hydropower reservoir operation, Nigeria. Saudi J Eng Technol 2(9):315–323
20. Jain SK, Bhunya PK (2008) Reliability, resilience and vulnerability of a multipurpose storage reservoir. Hydrol Sci J 53(2):434–447
21. Kjeldsen TR, Rosbjerg D (2001) A frame work for assessing the sustainability of a water resources system. In: Regional Management of Water Resources, 2001. Maastricht, the Netherlands

Infiltration Wells as an Alternative Eco Drainage System a Case Study in Mangkubumen Surakarta

Retnayu Molya, R. R. Rintis Hadiani, and Adi Yusuf Muttaqien

Abstract The city of Surakarta is a lowland area located in the slopes of Mount Lawu and Mount Merapi. Therefore, the city of Surakarta is prone to inundation and flooding. Mangkubumen Village is one of the urban villages that is quite dense and located in the middle of the city, often inundation and flooding occurs during the rainy season. This is due to the possibility of runoff flooding from the Pepe River that crosses the edge of the Mangkubumen Village. Besides that, this problem occurs because the conventional drainage system cannot accommodate runoff water. Urban drainage planning must take into account the environmental drainage function (Eco drainage). One of the eco drainage methods is to use infiltration wells. This research is descriptive quantitative. The main objective of this research is to determine the reduction of runoff by implementing infiltration wells in Mangkubumen Village so that the channel can accommodate water according to its normal capacity. Comparative analysis shows the return peak discharge of 2 and 5 years of Kali Pepe is 15.4848 and 19.5255 m^3/s. The peak discharge of 5 years resulted in inundation of 224,322.08 m^3 while the drainage channel capacity of 223,415.7 m^3 was not sufficient to accommodate the discharge. There are two alternatives in designing infiltration wells. The first alternative can re-duce the burden of drainage channels existing by 5.4% and the second alternative can reduce the burden of drainage channels existing by 2.8% so that the channel can accommodate runoff discharge by its capacity.

Keywords Eco drainage · Infiltration well · Urban drainage · Conventional drainage

---

R. Molya (✉) · R. R. Rintis Hadiani · A. Y. Muttaqien
Sebelas Maret University, Ir. Sutami 36A Street, Surakarta 57126, Indonesia
e-mail: retnayumolya@student.uns.ac.id

A. Y. Muttaqien
e-mail: adiyusufmuttaqien@staff.uns.ac.id

S. A. Kristiawan et al. (eds.), *Proceedings of the 5th International Conference on Rehabilitation and Maintenance in Civil Engineering*, Lecture Notes in Civil Engineering 225,
https://doi.org/10.1007/978-981-16-9348-9_84

1 Introduction

Population growth increases every year and has an impact on increasing the level of population density in an area. This has resulted in a change in land use from open land to developed land [1]. The development of land is being developed rapidly and uncontrollably resulting in the built-up land becoming a dense residential area. This results in an increase in runoff flow and a decrease in the quantity of water that is absorbed into the soil out of balance with the ability of the drainage channels, and this is what causes inundation and even flooding during the rainy season [2]. Mangkubumen Village which is a village in the middle of the city, also experiencing the same symptoms as indicated by population growth and urban development. In the end, there was very little open land in this village. Every rainy season there are always inundation that interfere with the activities of local residents.

Drainage is the most feasible and economical solution for managing surface runoff by draining it as quickly as possible [3]. However, the conventional drainage concept is considered ineffective for long-term use because it has to be expanded gradually over time and requires a large amount of money and a design that does not pay attention to water quality [4]. Eco drainage is an option to replace conventional drainage that has been used in urban areas. Environmentally friendly drainage has the principle of controlling excess water by absorbing it into the ground rather than discharging it directly into drains. One of the eco drainage methods is to use infiltration wells to support drainage channels [5]. Starting from this background the authors conducted a study "Infiltration Wells as an Alternative to Eco Drainage Systems, a Case Study in Mangkubumen Surakarta". The main objective of this research is to determine the reduction of runoff by implementing infiltration wells in Mangkubumen Village.

2 Research Method

The location of this research is located in Mangkubumen Village, Banjarsari District, Surakarta City which is geographically located at 7° 33′ 34,83″ South Latitude and 110° 48′ 43,48″ East Longitude and is administratively located surrounded by Manahan Village, Kestalan Village, Gilingan Village and Punggawan Village. The border between Mangkubumen Village and Kestalan Village is bordered by Kali Pepe.

The research stages are first, primary data collection, namely measuring the dimensions of the existing drainage channel. Then, rainfall data processing at Pabelan station to determine the rainfall intensity. After that, calculate the runoff discharge. The next is calculating the need for infiltration wells. The last, is calculate runoff reduction and compare with existing drainage capacity.

2.1 Data

The data is divided into two, namely primary data and secondary data. Primary data is data taken directly in the field by researchers. In this study, the primary data is the dimension of existing drainage channels. Secondary data is data obtained from various existing sources. This study uses secondary data from related agencies. The data are daily rainfall data, climatological data, topographic maps, dan DEM map.

2.2 Data Durability Test

The data durability test was conducted to determine whether the data used was consistent or not. If after being tested the results are inconsistent, it is necessary to replace or fill in the missing data. The method in testing the robustness of the data in this study is the RAPS (Rescaled Adjusted Partial Sums) method.

2.3 Eco Drainage

A drainage is a form of effort to drain or divert water from rainwater, seepage, or excess irrigation water from an area that does not interfere with the function of the area [6]. Conventional drainage is a drainage system that has the principle of draining excess water into drains as quickly as possible so as to allow the channel to be overloaded. Meanwhile, the environmentally drainage or ecodrainage has the principle of controlling excess water by absorbing it into the soil so that it has a role in contributing to groundwater reserves during the dry season [7].

2.4 Mononobe Method

The analysis of rainfall intensity using the Mononobe method is carried out to estimate the peak discharge in small catchments such as urban drainage system planning, culverts, and bridges [8]. The rainfall intensity is determined as

$$I = \frac{R_{24}}{24}\left(\frac{24}{tc}\right)^{\frac{2}{3}} \quad (1)$$

where I = the rainfall intensity in mm per hour; R24 = the maximum annual daily rainfall for Return Period in mm; and t = concentration time in hours.

2.5 Rational Method

The Rational Method is the development of a rational method where the concentration of rainfall occurs over time. The rational method is intended to consider the effect of discharge capacity in estimating peak runoff [9]. The modified formula for the rational method in determining the peak discharge is as follows

$$Q = 0.,278 \text{ x } C \text{ x } I \text{ x } A \tag{2}$$

where Q = the peak discharge with a certain period of time in m^3 per second; 0.278 = the conversion factor; C = runoff coefficient; I = the rainfall intensity in mm per hours; and A = catchment area in km^2.

2.6 Soil Permeability

According to the SNI 8256:2017 about ditch and infiltration wells, soil structures that can be used for infiltration wells must have a permeability coefficient of more than 2 cm per hour [10].

2.7 Number of Infiltration Wells

The number of infiltration wells that need to be calculated using the following formula:

$$n = \frac{L + s}{D + s} \tag{3}$$

where L = the length of drainage channel; s = the space between wells and D = diameter of wells.

2.8 Dimension of Infiltration Well

The depth of the infiltration wells is generally designed not exceeding the groundwater depth in the area. The formula that can be used to calculate the depth of infiltration wells is Sunjoto formula [11]. Calculations using the Sunjoto formula get a shallower well depth (H) than other formulas [12].

$$H = \frac{Q}{\omega \pi r K}\left(1 - e^{-\frac{F \times K \times T}{\pi \times R^2}}\right) \tag{4}$$

where H = infiltration well depth in m; Q = runoff discharge m^3 per hours; r = the width of the infiltration well in m; K = soil permeability coefficient in m per hours; $\omega = 2$, for empty infiltration wells with impermeable walls or infiltration wells without walls with fill stones; and $\omega = 5$, for empty infiltration wells with porous walls.

3 Analysis and Discussion

3.1 Type of Soil

Most of Surakarta has sandy clay types, including alluvial and gray regosol soil types [13]. In the western region leading to the south of the Surakarta area, including in Mangkubumen Village, is a gray regosol soil type. This gray regosol soil is a type of soil with quick rather-fast infiltration rate. This permeability coefficient is about at 36 cm per hours.

3.2 Soil Permeability

The permeability value that I use comes from previous research in the area around the study with the same soil type, gray regosol soil type, namely "Analisis Hidrologi dan Kapasitas Sistem Drainase Kota Surakarta". The results of the laboratory test in this study showed that the permeability coefficient was $k = 1.1 \times 10^{-4}$ m per second [14], so according to the permeability coefficient value, it could be categorized as soil with fast absorption. This fast absorption soil type is qualified to make infiltration wells.

3.3 Ground Water Level

The existence of the Surakarta free aquifer is at a depth of 4–10 m [15] which means that it meets the requirements for making infiltration wells according to SNI 8456: 2017 the depth of the groundwater level is more than 2 m. According to SNI 8456: 2017, the depth of infiltration wells should not exceed the depth of free aquifer (groundwater) so the dimension depth of infiltration wells is about less than 4 m. In this research, infiltration wells are designed with a depth of 1 m.

Table 1 Rainfall intensity

Repeat period (year)	Kt	Concentration time (h)	Maximum rainfall (R24) (mm)	Rainfall intensity (mm/h)
2	−0.1484	3.46	102.08	15.4848
5	0.9113	3.46	128.718	19.5255

Source Calculation

3.4 Catchment Area

The area of the rainwater catchment is the area of Kali Pepe watershed. Based on the Surakarta City Profile book, the area of watershed is 5.65 km^2. According to the area, the peak discharge of the Pepe River is calculated using the Rational Method. Before that, we should calculate the rainfall intensity.

3.5 Rain Data Processing

The rainfall intensity is determined by rainfall data at Pabelan Station in 21 years (2000–2020). The rainfall data obtained need to be calculated for the probability of recurrence in the future, namely at a return period. This is necessary to plan the peak discharge at a certain return period. The type of distribution used in this study is the Gumbel distribution. The distribution method needs to be tested to determine whether the distribution is carried out to determine whether the distribution method for determining design rain can represent the statistical distribution of the analyzed data. Distribution checking using the Chi Square method. Then, the rainfall intensity is calculated using the Mononobe method as shown in Table 1.

Based on Fig. 1. IDF curves above, it can be seen by the time the rainfall intensity equation. The equation for various 2-year and 5-year is I_2 return $= 542.38x^{-0.667}$ and $I_5 = 683.92x^{-0.667}$.

3.6 Peak Discharge

Furthermore, the return period of peak discharge in this study was calculated using the rational method. In this study, the return period used is a return period of 2 years and 5 years because the research area is an urban residential area with the most channels being tertiary channels. Here is Table 2 the result of peak discharge of Kali Pepe and Fig. 2 is the hydrograph of peak discharge of Kali Pepe.

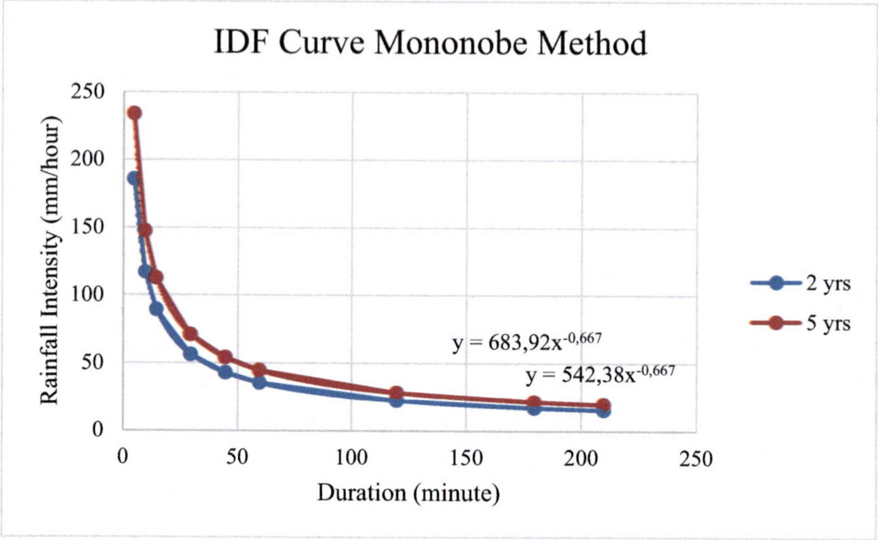

Fig. 1 IDF curve of rainfall intensity using Mononobe method. *Source* Calculation

Table 2 Peak discharge of Kali Pepe

Repeat period (year)	Rain intensity (mm/h)	Q (m^3/s)
2	15.4848	18.2089
3	19.5255	22.9937

Source Calculation

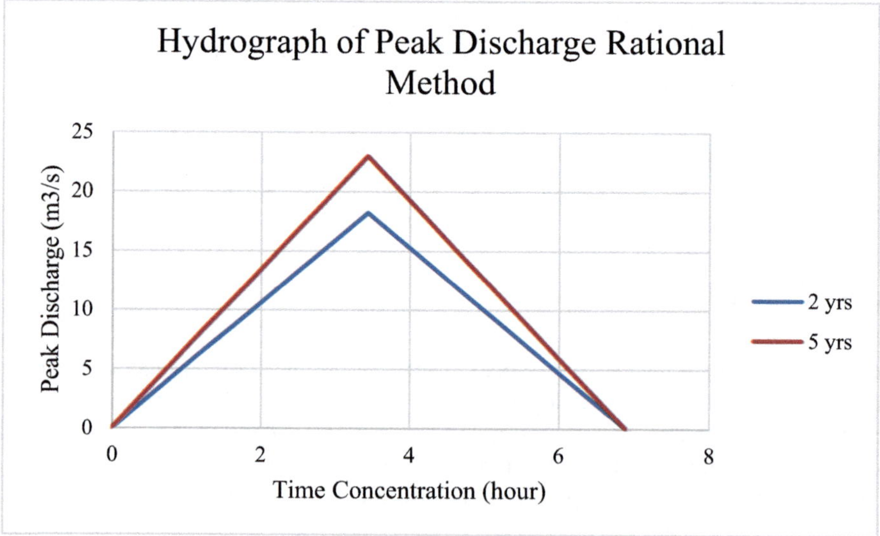

Fig. 2 Hydrograph of peak discharge of Kali Pepe using rational method *Source* Calculation

3.7 Inundation Resulting to Peak Discharge

After obtaining the peak discharge of Kali Pepe with a return period of 2 and 5 years, it was then modeled using Hecras to determine the inundation that occurred [16]. The results of the analysis using Hecras produce an inundation area of 0.25 km^2 when the peak discharge is applied for a 2-year return period and 0.28 km^2 for a 5-year return period as shown in Fig. 3. Existing channel's capacity is able to accommodate runoff due to return period discharge Q_2 years but it can't accommodate the runoff due to period discharge Q_5 years at 906.42 m^3.

Then to calculate the volume of inundation that occurs, the assumption is that the inundation is in the shape of a rectangular pyramid so that the volume of inundation due to Q_5 can be calculated by the following formula:

$$V = \frac{1}{3} \times \text{area} \times \text{elevation} \tag{5}$$

$$V = \frac{1}{3} \times 280402.6 \times (2.7 - 0.3)$$

V = 224,322,08 m^3 and so Q = 224,322,08 m^3/3600 = 62.3889 m^3/s.

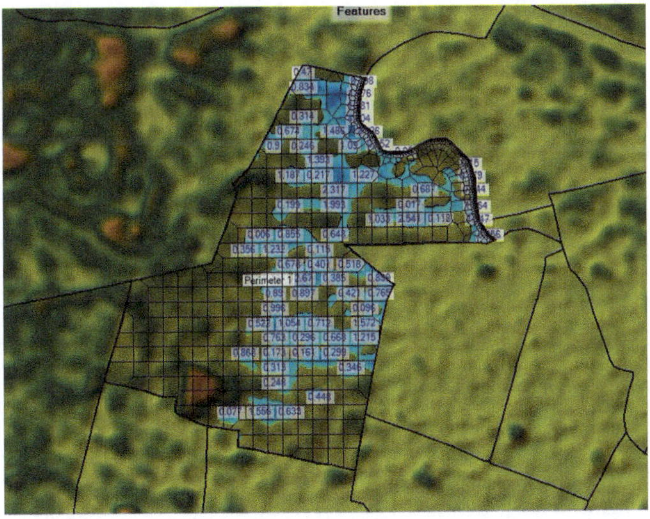

Fig. 3 Inundation area of Q_5 years from running Hecras. *Source* Author document

3.8 Existing Drainage Capacity

Based on the drainage channel tracing survey, the length of drainage channel in Mangkubumen is 10.127 km and the capacity of the channel can accommodate the discharge of 62.0599 m^3/s.

3.9 Number and Dimension of Infiltration Wells

The steps in designing an infiltration well include determining the geometric factor, the depth of the infiltration well, and the number and distance of the infiltration well. The infiltration well is planned to be located at the bottom of the channel existing with the same diameter as the smallest tertiary channel width, which is 0.4 m. In this research, there are two alternative of infiltration wells as shown in Fig. 4. The different between those alternatives is about the construction and cost.

The designed infiltration wells are cylindrical with a flat bottom. Here is the example of calculating infiltration wells alternative 1. The first step is to determine the geometric factor of the infiltration wells [17].

$$F = 5 \times 3.14 \times 0.2$$
$$F = 3.14\,\mathrm{m}$$

where $\omega = 2$, for empty holes with waterproof walls or holes without walls with fill stone and $\omega = 5$, for an empty hole with a porous wall. Then, look for the required total depth of infiltration hole with the following Sunjoto method:

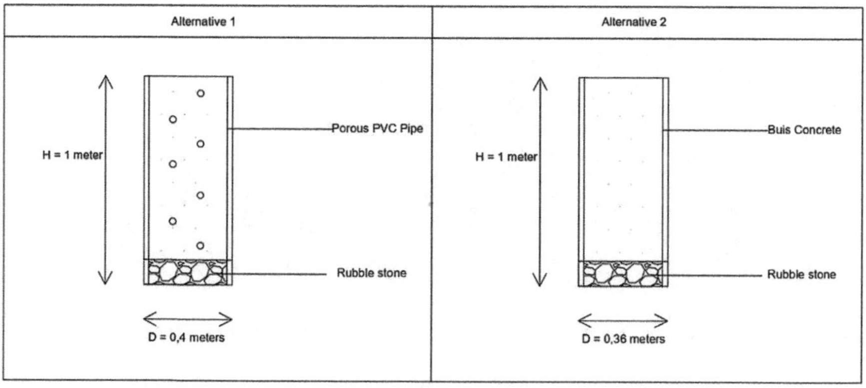

Fig. 4 Alternatives of infiltration wells design. *Source* Author document

$$H = \frac{Q}{\omega \pi r K}\left(1 - e^{-\frac{F \times K \times T}{\pi \times R^2}}\right)$$

$$H = \frac{62.3889}{3.14 \times 1.1 \times 10^{-4}}\left(1 - e^{-\frac{3.14 \times 1.1 \times 10^{-4} \times 207.3}{\pi \times 0.2^2}}\right)$$

$$\mathrm{H} = 77956\,\mathrm{m}$$

where H = total depth of infiltration well (m), Q = runoff discharge (m^3/s), F = geometric factor (m), K = coefficient of permeability (m/s), T = flow time (min), and R = radius of infiltration well (m).

After obtaining the required total depth, then look for the number of infiltration wells with a designed depth of 1 m for each hole.

$$n = \frac{H}{1\,\mathrm{m}}$$

$$\mathrm{n} = 77,956\,\mathrm{wells.}$$

From the calculation, the number of wells is 77956 while the length of the existing channel is 10.127 km or 10,127 m. Of course, this cannot be done because there is no space for the infiltration wells. Therefore, it is necessary to recalculate so that the space in the channel can accommodate the infiltration hole. If the depth of each hole is designed at 1 m and the distance between holes is planned 2 m, then the number of holes that can be made are as follows:

$$n = \frac{L + s}{D + s}$$

$$n = \frac{10127 + 2}{0.4 + 2}$$

$$\mathrm{n} = 4221\,\mathrm{wells.}$$

The same calculation above to calculate the second alternative. Table 3 is a recapitulation of the dimensions of the two alternatives.

Table 3 The dimension of infiltration wells Alternative 1 and 2

Alternative	D (m)	H (m)	n	S (m)	Q (m^3/s)
Alternative 1	0.4	1	4421	2	3.3739
Alternative 2	0.3	1	4404	2	1.0560

Source Calculation

Table 4 Construction and cost of Alternative 1 and 2

Alternative	D (m)	Q (m^3/s)	Reduction (%)	Construction	Cost (IDR)
Alternative 1	0.4	3.3739	5.4	PVC pipe	800.000
Alternative 2	0.3	1.0560	2.8	Concrete	89.000

Source Calculation

3.10 Runoff Reduction

After obtaining the dimensions of the infiltration wells, the following is the discharge that is absorbed into the well alternative 1.

$$Q = \frac{F \times H \times}{\left(1 - e^{-\frac{F \times K \times T}{\pi \times R^2}}\right)}$$

$$Q = 3.3739\,\text{m}^3/\text{s}.$$

While the discharge capacity of channels is 62.0599 m^3/s, the drainage channel load reductions are:

$$\begin{aligned} Q_{\text{capacity}} &= Q_{\text{existing}} - Q_{\text{absorp}} \\ Q_{\text{capacity}} &= 62.0599 - 3.3739 \\ &= 58.6860\,\text{m}^3/\text{s} \\ \% &= 100 - (58.6860/62.0599) \\ &= 5.4\% \end{aligned}$$

From the calculations above, a total of 4221 pieces infiltration wells placed in the Village Mangkubumen able to reduce the burden of existing drainage channels by 5.4%. The same calculations above to calculate the second alternative. Table 4 is a recapitulation of the construction and cost of the two alternatives.

4 Conclusion

The conclusions of this study are as follows:

There are two alternatives on infiltration wells. The first one is by applying 4221 infiltration wells that was designed with a radius of 0.4 m, a depth of 1 m and the space between each hole is 2 m can reduce the burden of drainage channels existing by 5.4%. And the second is by applying 4401 infiltration wells that was designed with a radius of 0.3 m, a depth of 1 m and the space between each hole is 2 m can reduce the burden of drainage channels existing by 2.8%. The different between those alternatives is the construction. The first one using PVC pipe construction and the second using concrete cylinder construction.

References

1. Karya J, Sipil T (2014) Perencanaan Eko-Drainase Kawasan Perumahan Tembalang Pesona Asri-Semarang. In: ejournal3.undip.ac.id 3(2). Halaman. http://ejournal-s1.undip.ac.id/index.php/jkts
2. Kusumastuti C, Chandra H, Wibisono K, Hartono A (2019) Eco drainage system for surface runoff reduction in Indonesia. http://repository.petra.ac.id/18281/
3. Idfi G, Yulistyorini A, T A-M W, (2018), undefined. (n.d.). Injection well as an eco-drainage solution to reduce surface run-off at the State University of Malang. Matec-Conferences.Org. Retrieved March 17, 2020, from https://www.matec-conferences.org/articles/matecconf/abs/2018/63/matecconf_imiec2018_03017/matecconf_imiec2018_03017.html
4. Mguni P, Herslund L, Marina Jensen B (2016) Sustainable urban drainage systems: examining the potential for green infrastructure-based stormwater management for Sub-Saharan cities. Springer 82:241–257
5. Wahyuningtyas A, Hariyani S, Sutikno FR (2011) Strategi Penerapan Sumur Resapan sebagai Teknologi Ekkodrainase di Kota Malang (Studi Kasus : Sub DAS Metro). Jurnal Tata Kota Dan Daerah 3(1):25–31
6. Suripin (2004) Sistem Drainase Perkotaan yang Berkelanjutan. ANDI, Yogyakarta
7. Hamzah Syahruddin M, Halide H (2019) To cite this article: Muhammad Hamzah Syahruddin et al. IOP Conf Ser: Earth Environ Sci 279:12021. https://doi.org/10.1088/1755-1315/279/1/012021
8. Suleman AR, Yusuf H, Nabi A, Erdiansa A, Aulia N (2020) Determining biopore infiltration hole as catchment flood in inundation area Eastern of Makassar City Based on Geographic Information System (GIS). J Eng Appl Sci 15(2)
9. Prayitno G, Hariyani S (2013) The effectiveness of bio-pore as an alternative eco drainage technology to control flooding in Malang City (case study: metro sub- watershed). In researchgate.net
10. Indonesia BSN (2017) Sumur dan Parit Resapan Air Hujan. Sni 8256(2017):12–13
11. Sunjoto (2011) Teknik Drainase Pro-Air. Universitas Gadjah Mada, Yogyakarta
12. Fachrurazie C, Firman Arifin Y, Sri Susanti D (2002) Analisa Drainase Sumur Resapan Pada Kampus UNLAM Banjarbaru. Info-Teknik 3(1):24–34
13. Pemerintah Kota Surakarta (2016). Badan Perencanaan, Penelitian, dan Pengembangan Daerah. Rencana Program Investasi Jangka Menengah (RPIJM) Bidang Cipta Karya Tahun. Bab ii profil wilayah kota surakarta 2.1:1–28
14. Sataloff RT, Johns MM, Kost KM (n.d.) Buku II Informasi Kinerja Pengelolaan Lingkungan Hidup Daerah Kota Surakarta. Pemerintah Kota Surakarta Dinas Lingkungan Hidup, Surakarta (2017)
15. Sipil JT, Teknik F (2016) Analisis hidrologi dan kapasitas sistem drainase kota surakarta
16. Amdan Y, Studi P, Sipil T, Teknik F, Maret US (2019) Drainage channel capacity analysis for flood controlling solution in Sewu Surakarta drainage channel capacity analysis for flood Yuliarini Amdan
17. Kamila N, Wardhana IW, Sutrisno E (2016) Perencanaan Sistem Drainase Berwawasan Lingkungan (Ecodrainage) Di Kelurahan Jatisari, Kecamatan Mijen, Kota Semarang. Jurnal Tehnik Lingkungan 22(2):63–72. https://doi.org/10.5614/j.tl.2016.22.2.7

Analysis of Leb Irrigation Patterns of Pipe System in Sorgum Plants in Sand Dry Lands Akar Akar Village

I Dewa Gede Jaya Negara, Sasmito Soekarno, Suwardji, Humairo Saidah, and Atha Adi

Abstract To take advantage of the dry land potential in North Lombok district which reaches 38,000 ha, the government has facilitated groundwater irrigation networks (JIAT). The most visible use of groundwater irrigation networks so far has been in the community for maize farming with wasteful water use, and it is necessary to develop crops such as sorghum in order to obtain more feed availability that supports the provision of animal feed. An adequate irrigation system is necessary so that it can provide economic improvement or reduce costs used. This study aims to examine the pattern of leb pipe irrigation water use in sorghum to until harvest on a land area 500 m^2. Leb irrigation network using 2 inches pvc pipe on 6 blocks of sorghum crop land, with irrigation duration of 20 and 30 min. The irrigation test is carried out on the uniformity, depth irrigation and soil water content. The test results show that the irrigation uniformity produced is 94% very good, reaching a soil depth of about 20–30 cm with a water content of about 35–40% so that this irrigation system has the potential to support irrigation of sorghum plants.

Keywords Uniformity · Pipe width · Irrigation pattern

I. D. G. J. Negara (✉) · S. Soekarno · H. Saidah · A. Adi
Department of Civil Engineering, Mataram University, Mataram, Indonesia
e-mail: jayanegara@unram.ac.id

S. Soekarno
e-mail: sasmitosoekarno@unram.ac.id

H. Saidah
e-mail: h.saidah@unram.ac.id

Suwardji
Department of Soil Science, Faculty of Agriculture, University of Mataram, Mataram, Indonesia
e-mail: suwardji@unram.ac.id

S. A. Kristiawan et al. (eds.), *Proceedings of the 5th International Conference on Rehabilitation and Maintenance in Civil Engineering*, Lecture Notes in Civil Engineering 225,
https://doi.org/10.1007/978-981-16-9348-9_85

1 Introduction

The Akar-Akar area of Bayan District, North Lombok Regency is one of the dry land potentials in the province of West Nusa Tenggara with an area of about 38,000 ha. The obstacle to providing plant irrigation water that is often faced in the field is the provision of irrigation water on land using open boxes, which causes a lot of water loss due to evaporation of the land surface during the trip and uneven infiltration into the soil resulting in uneven plant growth and crop yields. not good. The results of the optimization survey on the use of deep groundwater wells show that only 10–15% of the total deep groundwater wells built by the government are utilized by farmers while the rest is optimal due to high operational costs [1]. The practice of inundation irrigation systems has been carried out so far. 1 ha of land requires irrigation for about 14 h, the results of the study [2] show that the inundation irrigation system that has been carried out is a system that is very wasteful of water and is not profitable for farming. irrigation on dry land so it is necessary to develop an irrigation system to support agricultural development on dry land which is more water efficient.

To increase the productivity of agricultural land in this area, it has been tried to apply water-efficient irrigation techniques such as JIAT, sprinkler and drip systems to increase agricultural yields on dry land so that it can work well, especially in maize and horticultural crops. Based on field experience, it is not enough to improve the welfare of people in dry land, but other innovations are still needed, especially in the selection of plant types so that the community can benefit from multiple economic effects. Besides, the harvest can produce fruit multiple times, but it is also able to provide a continuous reserve of animal feed, so that livestock production and sales can contribute together in helping the economy of dry land communities.

To increase the productivity of agricultural land in this area, water-efficient irrigation techniques such as the JIAT system, big gun and drip sprinklers have been tried to increase agricultural yields on dry land. Based on field experience, in order to improve the welfare of the community on dry land, the irrigation system requires new investment which is expensive, so that it is not necessarily possible for the general public in JIAT to implement it. For this reason, it is necessary to develop an existing irrigation system so that the irrigation method becomes more efficient and agricultural crops can provide a higher economic effect. Sorghum plants are considered suitable, so it needs to be tested on land with over-pipe irrigation so that water use is known according to its growth phase. With this effort, it is hoped that later sorghum can be cultivated so that it can help livestock businesses towards improving the economy of the community on dry land. According [3], an irrigation system over a pipe with a duration of 10 min in dry land with finely graded soil can provide irrigation up to 30 cm with an average humidity of 25% and at 30 min it can provide about 32% moisture. After four days of testing, the soil moisture is obtained, the remaining moisture is above 20%. So the irrigation system may be more suitable to be applied to sandy soil such as in this study location, to support future development of sorghum crops.

Test of over-irrigation pipe system at the Salut location, North Lombok, shows that the over-irrigation test system on an area of about 1 acre requires a shorter irrigation time than the inundation system, with a uniformity of more than 91%. At irrigation duration of 10 min, 20 min and 30 min, the irrigation duration efficiency was 0.42 h, 0.83 h and 0.63 h, respectively, with cost efficiency of 25%, 33% and 25%, respectively [4]. Based on this test, irrigation provides water more efficiently than the inundation system, so that the puddle irrigation method which is also applied to JIAT needs to be developed with a pipeline system.

Soil moisture is water in the soil in the root area/root zone that fills part of the soil pore space expressed in units of weight percent or volume percent. According to [5], the most accurate measure of water content is the gravimetric method, namely by weighing the soil sample, drying it in an oven at 100–110 °C for 24 h and weighing it again.

High output uniformity is very important in the development of irrigation systems. The goal is to achieve a uniform irrigation water output level at each pipe outlet that can meet the needs of plants, especially in the root zone. Irrigation uniformity can be calculated by the formulation of Christianes [6].

To overcome this, it is necessary to innovate dry land plants in order to obtain a double economic effect in future farming. Tests need to be carried out on types of sorghum plants that have the ability to grow long with a long harvest period. The stover from the sorghum harvest can be used as animal feed, and can grow more and more so that animal feed reserves will be obtained due to repeated harvesting. Plant testing needs to be applied with a drip irrigation system and a wide pipe so that the distribution of water to plants can be evenly distributed to support optimal sorghum planting results.

To overcome the above problems, it is necessary to innovate dry land plants in order to obtain a double economic effect in future farming. Plant tests need to be carried out on types of sorghum plants that have the ability to grow longer than corn plants with a harvest period of more than once. The stover from the sorghum harvest can be used as animal feed, and can grow more and more so that animal feed reserves will be obtained due to repeated harvests. Plant testing needs to be applied with an over-pipe irrigation system, so that the water distribution pattern to the land is known and the discharge size and duration required for the depth of sorghum roots can be known. With this effort, sorghum plants will later be developed in a certain area with more water efficiency, and provide better economic benefits to farming communities.

This study aims to determine the amount of irrigation water demand for sorghum farming tests, pipeline and drip irrigation systems including irrigation time and irrigation duration. It also includes how the water services are carried out by the two irrigation systems until the sorghum plants can be harvested and what are the advantages and disadvantages of the irrigation system used during the growing season.

2 Materials and Methods

The equipment used during the study was a water meter, sample pipe, hoe, oven, gravimetric, hammer and stop time. For the research material consists of 2 "pvc pipe, 2" pvc dop, L 2 inches pvc, T 2 inches pvc and pvc 2 inches timer. This research was carried out by field test research where drip irrigation and over-irrigation systems with sorghum plants were tested directly in the dry land of Akar Akar, North Lombok district Fig. 1. The stages of conducting field research were carried out as follows.

The preparation stage consists of preparing references, tools and materials for the location of sorghum plants including clearing 4 acres of land for two irrigation systems, making 6 blocks of more irrigation test fields. In 6 irrigation blocks, block 1 consists of 9 outlet holes, land 2 has 6 outlet holes, test area 3 has 9 outlet holes, test area 4 with 4 outlet holes, test area 5 has 6 outlet holes and test area 6 there are also 6 irrigation outlet holes.

Testing the uniformity of irrigation over pipes, taking soil samples before and after irrigation at 3 (three) different location points in one plot of land and testing the initial soil moisture content. Provision of irrigation water for over-pipe irrigation systems is carried out in 20 and 30 min. The recording of the discharge of water used is carried out at each different irrigation time. Take a soil sample before and after irrigation, and test soil moisture.

Data analysis and discussion. Data analysis of test results was carried out on used discharge data, soil moisture data and irrigation uniformity. The need for irrigation water at each stage of plant growth is calculated as the volume of used water at each provision of irrigation water and the amount of water used until the initial harvest is used as data on the use of early harvest irrigation water. The results of the analysis were presented in tables and graphs and concluded descriptively.

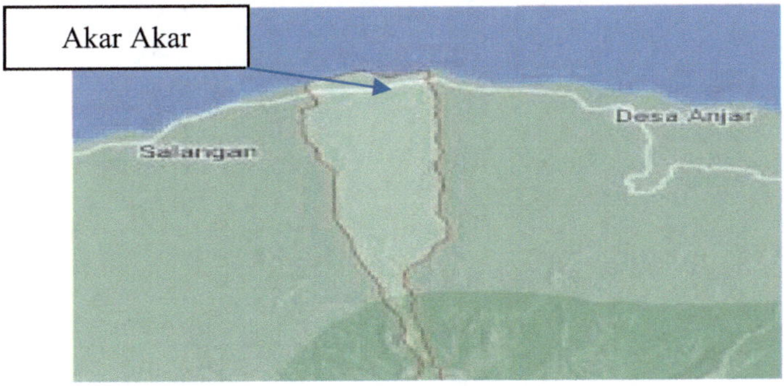

Fig. 1 Akar Akar, North Lombok district

3 Discussion

Based on the results of the analysis of the test data using the Chrietiansen's formula, the irrigation uniformity value is 94% so that it is considered good, so that this irrigation system is thought to be applicable for farming such as sorghum. The results of the analysis are shown in Table 1.

This indicates that over-piped irrigation will be able to provide irrigation services as well as drip irrigation, but the application of over-irrigation on a wider plot of land and on a large availability of discharge.

The amount of water flowing through the cross section for each unit of time is called the flow rate and is given the notation Q. for the calculation of the discharge, the following equation can be used [7] Table 2.

For irrigation, the discharge given is around 0.00017 m^3/s.

Soil moisture is water in the soil of the root area/root zone that fills part of the soil pore space expressed in units of weight percent or volume percent. According [5], the most accurate measure of water content is the gravimetric method, namely by weighing the soil sample, drying it in an oven at 100–110 °C for 24 h and weighing it again. The results of soil moisture analysis from wide irrigation can be seen in Figs. 1 and 2.

The result of providing irrigation to land is in the form of soil moisture, which is the amount of water stored in the soil layer which will later be used by plant roots for plant growth. Based on the existing soil conditions, soil moisture that can be given from wide irrigation is shown in Fig. 1. Based on this figure, the irrigation is given with a duration of 20 min, at three sample points a depth of 10–15 cm where soil

Table 1 Uniformity of pipe Leb irrigation system

Outlet pipe	CU (%)
1	94.02
2	93.89
3	93.44
4	97.24
5	93.80
6	93.38

Table 2 Irrigation hole discharge over

Outlet pipe	leb (m^3/dt)
1	0,00,016
2	0,00,017
3	0,00,016
4	0,00,018
5	0,00,017
6	0,00,017

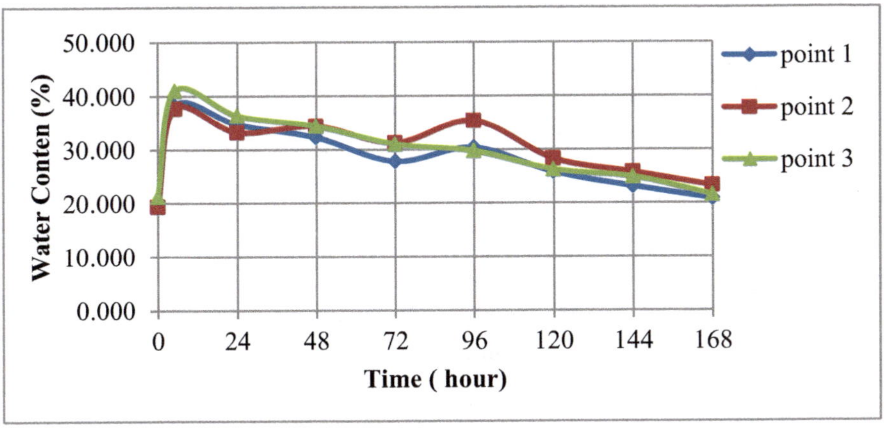

Fig. 2 Soil moisture and daily changes at a depth of 10–15 cm

moisture after irrigation is obtained at 38.88%, and after 7 days of irrigation, the remaining moisture is 21.77% approaching the initial soil moisture.

The result of providing irrigation to land is in the form of soil moisture which will later be used by plant roots at a depth of 30 cm for plant growth shown in Fig. 2. Based on this figure, irrigation is given with a duration of 30 min, at three sample points a depth of 20–30 cm where Soil moisture after irrigation was obtained by 40.66% and after 7 days of irrigation, the available moisture in the soil depth was 25.15%. This moisture is still higher than the initial soil moisture.

Based on the moisture condition of the soil obtained from excessive irrigation, the provision of irrigation in its application can still follow the JIAT irrigation pattern in general, where irrigation is given in 7 turns so that it does not change the irrigation schedule of the existing pumper. However, the provision of irrigation will be shorter than the existing JIAT irrigation system. With more irrigation it is still possible to adjust the uniformity of irrigation into the soil and this can shorten the irrigation time, reduce irregularity in soil and plant growth critical down that will be detrimental to plant growth and irrigation can be done sooner than 7 days.

The graph of the results of pipe irrigation for sorghum plants on sandy dry land can be seen in Fig. 4. So the amount of irrigation water used is irregular and follows the conditions of the plant so that it will follow the growth phase of the sorghum plant.

Based on Fig. 3, it is known that the provision of irrigation water on average in one irrigation in the initial growth phase is 1.242 m^3 and at the planting age of 20 days to 40 days is given irrigation water of 1.242 m^3, at the age of the plant 41 days to 67 days it is given irrigation water. an average of 1.446 m^3 and after the plants are 70 ^{days} old until harvesting, the average amount of irrigation water given by sorghum plants is 1.86 m^3. Increasing the provision of irrigation water from the initial phase to harvest to 1.242–1.86 m^3 when the sorghum plants are ready to be harvested at 88 days.

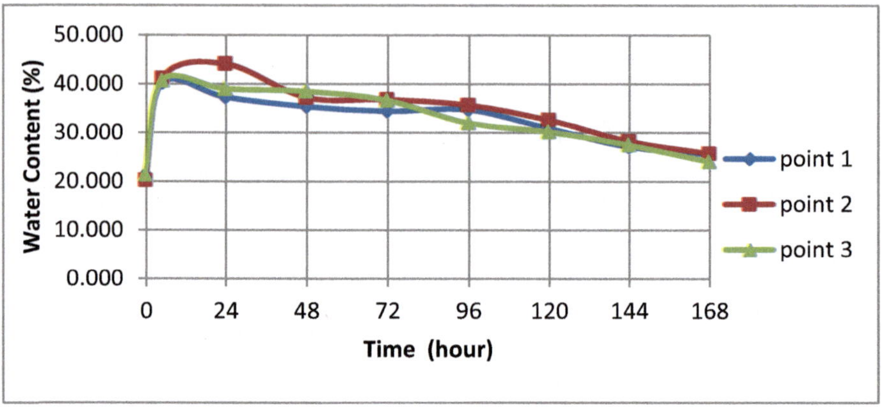

Fig. 3 Soil moisture and daily changes at a depth of 20–30 cm

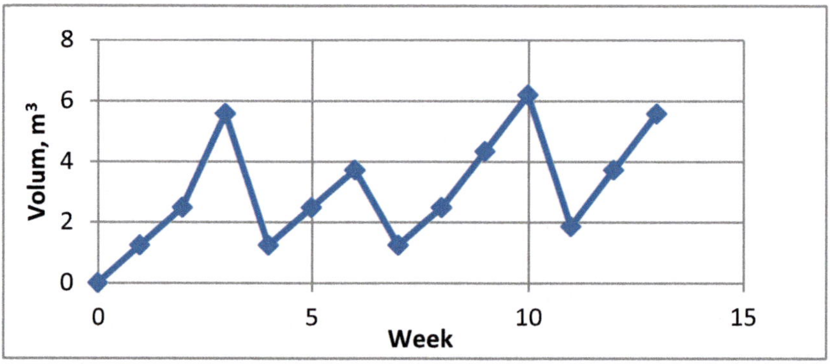

Fig. 4 Provision of low irrigation water for each planting phase

The provision of irrigation water during the growing season for each growth phase of sorghum is certainly not the same, because it depends on plant development. When compared with the provision of more irrigation water, it can be seen in Fig. 5.

Leb irrigation water is given every 1 week with the amount of water used for initial irrigation of 1.242 m^3, plant age up to 20 days, plant age 21 days to 43 days given irrigation water of 1.242 m^3, at 44 days to 67 days of 1.46 m^3 and at the age of 70 days of sorghum until harvest 1 is given 1.86 m^3 of water in one irrigation.

When compared to the irrigation water distribution pattern of the two irrigation systems above, it is very different, in drip irrigation the variation of the discharge from the irrigation system is still very high, whereas in irrigation the pattern is very regular.

Based on Fig. 6, it is known that, for the over-pipe irrigation system in the 5-acre sorghum area, it is irrigated once every 7 days with the amount of irrigation 13

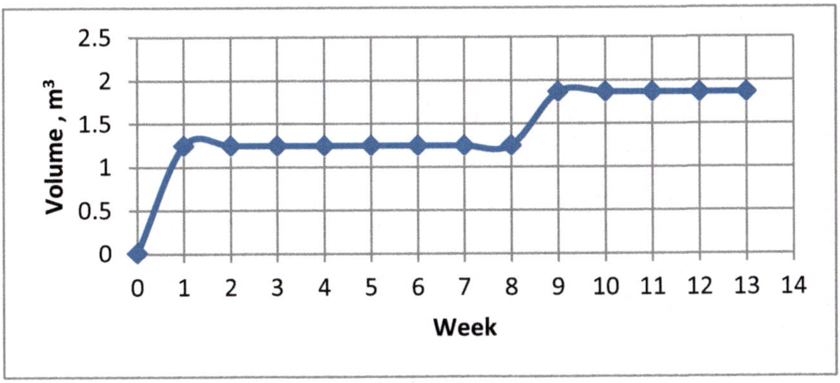

Fig. 5 Distribution of Leb irrigation water supply to harvest 1

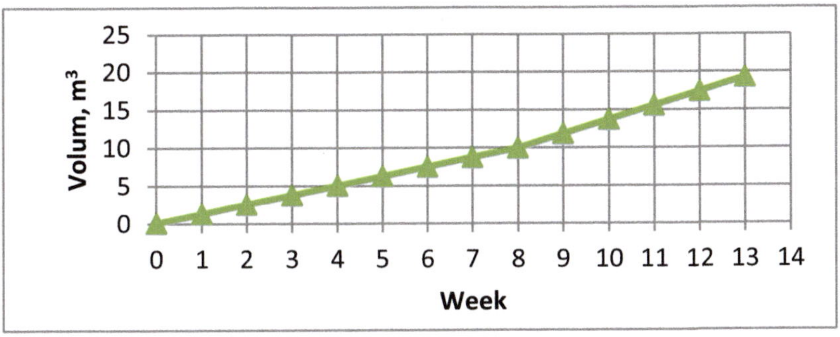

Fig. 6 Accumulation of irrigation water Leb

times, and the water used is 21.8 m^3. So the use of water by irrigation is lower than by pipeline irrigation, by a difference of about 4.75 m^3.

In irrigation, plants are given water with irrigation duration, namely the duration of irrigation is 20 and 30 min with the aim of achieving water at the roots with a depth of 10–15 cm and 20–30 cm. Irrigation is given once every 6 days with 8 times the duration of the irrigation is 20 min and 5 times the duration of the irrigation is 30 min. The total duration of irrigation over pipes needed to harvest crops is 26 h, with a used water volume of 21.80 m^3.

4 Conclusions

The test results show that the irrigation uniformity produced is 94% very good, reaching a soil depth of about 20–30 cm with a water content of about 35–40% so that this irrigation system has the potential to support irrigation of sorghum plants.

References

1. Eterna and Suwardji (2003) Survey of deep groundwater pump utilization throughout Lombok Islan, Collaborative research report on the groundwater development project section of the NTB Provincial Public Works Office And The Unram Dryland Research and Development Center. 76h
2. Suwardji (2005) Assessment of irrigation water needs for inundation systems and sprinkle big gun systems. Research Report of the Mataram University Research Institute
3. Zulvarisandy I (2017) Analysis of providing combined water for irrigation LEB pipes and mini sprinklers against changes in daily soil moisture in Pringgabaya District, Mataram
4. Muta'al Atha Hadi (2018) Analysis of time and costs of irrigation in the Leb pipe irrigation system in the dry land of Arungan Bali Hamlet, Akar-Akar Village, Bayan District, North Lombok, FT Unram Thesis, Mataram
5. Soemarto CD (1986) Engineering hydrology. National Business, Surabaya
6. Triatmodjo B (2003) Hidraulika II, Beta Offset, Yogyakarta
7. Negara JDG, Supriyadi A (2016) Analysis of the design of integrated water-saving irrigation systems based on groundwater irrigation network (JIAT) on dry land, smoothly degraded soils in Pringgabaya, East Lombok Regency, Mataram

Hourly Rainfall Simulation Using Daily Data

Suroso, Fatimatus Sholihah Marush, Purwanto Bekti Santoso, Irfan Sudono, Edvin Aldrian, and Nelly Florida Riama

Abstract Rainfall is one of the main inputs in the analysis of the hydrological system. However, rainfall data are often found to be inadequate in length and completeness. The purpose of this study is to overcome the problem of the limited availability of high resolution of rainfall data. Rainfall data on a daily scale generally have a better quantity and quality of data available to be used as a basis or reference for deriving hourly rainfall data. The research location is limited to Java Island, with 25 rain gauge stations used. Hourly rainfall has a diurnal cycle, where some rainfall is concentrated in the afternoon or evening. The result of the evaluation using Root Mean Square Error, Mean Absolute Error, Pearson Correlation, and Spearman Correlation shows that the hourly rain simulation model using daily rainfall data can be used for design, especially in the field of water structures due to producing similar statistical characteristics.

Suroso (✉) · F. S. Marush · P. B. Santoso
Department of Civil Engineering, Faculty of Engineering, Jenderal Soedirman University, Purbalingga 53371, Indonesia
e-mail: surosotsipil@unsoed.ac.id

P. B. Santoso
e-mail: purwanto1409@unsoed.ac.id

I. Sudono
Research Centre for Water Resources, Agency of Research and Development, Ministry of Public Works and Housing, Jakarta, Indonesia

E. Aldrian
Agency for the Assessment and Application of Technology (BPPT), Jakarta, Indonesia

N. F. Riama
Meteorology, Climatology, and Geophysical Agency (BMKG), Jakarta, Indonesia
e-mail: nelly.florida@bmkg.go.id

S. A. Kristiawan et al. (eds.), *Proceedings of the 5th International Conference on Rehabilitation and Maintenance in Civil Engineering*, Lecture Notes in Civil Engineering 225,
https://doi.org/10.1007/978-981-16-9348-9_86

1 Introduction

A flood is a pool of water that causes a detrimental impact on human life. A flood is a flow of water whose height exceeds the normal water level. Among all the disasters, floods are the most destructive because they can cause serious damage to human life, infrastructure, agriculture, and social-economic systems [1]. In Indonesia, almost all areas have a high potential for flooding such as West Java [2], Central Java [3], and East Java [4]. Flooding can take in many forms, including slow river flooding, rapid flash floods, accumulation of rainwater in poorly drained environments, and coastal flooding caused by tides and extreme waves.

Rainfall is one of the main inputs in the analysis of the hydrological system. However, rainfall data are often inadequate in length and completeness due to the small number of observations or instrument errors. Insufficient rainfall data series can cause severe problems in hydrological analysis. Accessibility of high-resolution rainfall is one of the main problems for hydrologists due to various reasons such as cost and government policies [5]. Therefore, scientists have attempted to produce accurate high-resolution rainfall data with many computer programs. The simulation of hourly rainfall is an important area of hydrological research, for example for designing hydraulic structures in urban areas or urban drainage systems, and for evaluating the frequency of flooding in small natural catchments characterized by a rapid hydrological response [6]. Hourly rainfall data or even smaller by temporal scale are critical for applications in many sectors, such as urban hydrology, infrastructure design, and risk assessment [7].

Daily rainfall is generally a difficult thing in rainfall modelling because daily rainfall has unique characteristics and is difficult to make a model. Unlike the annual and monthly rainfall, which can be modelled by the simple autoregressive moving average (ARMA) process, the incidence and amount of daily rainfall are challenging to model with the ARMA model [8]. There are many problems in hydrology, agricultural science and engineering that require extensive rainfall records [9], especially for flood risk assessment, water resources management or hydraulic structure design [10]. However, hourly or higher resolution rainfall data tend to have lower temporal and spatial coverage due to high instrument costs and technical limitations.

The purpose of this study is to overcome the problem of the limited availability of high-resolution rain data. High-resolution rainfall data in the hourly or higher (minute) is necessary for the hydrological aspect. Rainfall data on a daily scale generally have a better quantity and quality of data availability. Hence, daily rainfall can be used as a basis or reference for deriving hourly rainfall data.

2 Material and Method

2.1 Data

The data used were obtained from the Meteorology, Climatology, and Geophysical Agency (BMKG). The data required is hourly and daily rainfall data in Java. The location of selected rainfall stations is presented in Fig. 1, and the name of that can be seen in Table 1. The rainfall data used in this study are selected with having relatively complete data from all stations in Java. By knowing the number of NA (Not

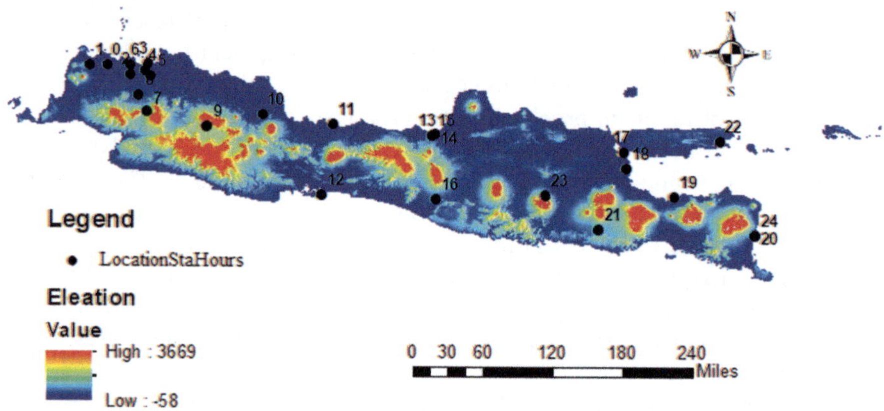

Fig. 1 Map of 25 rain station in Java Island

Table 1 The name of rain station in Java Island

Code	Rain station	ID rain station	Code	Rain station	ID rain station
0	Tangerang	96,735	13	Semarang_Klimatologi	96,835
1	Serang	96,737	14	Semarang_Maritim	96,837
2	Bawean	96,925	15	Ahmad Yani_Semarang	96,839
3	Curug	96,739	16	Surabaya Tg. Perak	96,937
4	Mataram	96,742	17	Surabaya_Perak1	96,933
5	Jakarta BMKG	96,745	18	Juanda	96,935
6	Cengkareng	96,749	19	Karang Ploso	96,943
7	Citeko	96,751	20	Tretes	96,945
8	Dermaga	96,753	21	KarangKates	96,949
9	Bandung	96,783	22	Kalianget_Madura	96,973
10	Jatiwangi_Cirebon	96,791	23	Sawahan	96,975
11	Tegal	96,797	24	Banyuwangi	96,987
12	Cilacap	96,805			

Available) or the countless data, there are three rain recording stations are obtained. The relatively complete rainfall recording stations are Citeko (96,751) and Dermaga (96,753).

2.2 Research Steps

Data Preparation
The following steps in preparing the data are the hourly rainfall recording stations that have a relatively long measurement period, complete and continuous are identified and selected for further analysis. Then at each hourly rainfall recording station that has been selected, the cumulative rainfall calculation is carried out into a daily period. Two data sets of rainfall are collected, so that hourly and daily rainfall data will be obtained at the same time.

Model Development
There are two types of modelling approaches were taken, namely the approach that the rainfall data series for both zero and non-zero rainfall was calculated and another approach that only non-zero rainfall data series were carried out in the modelling stage.

Model Evaluation
After obtaining the hourly rain simulation modelling using daily rainfall data, it is necessary to evaluate it to obtain the performance of the model. There are several measuring tools to find out how well the model developed consists of RMSE, MAE, Pearson Correlation, Spearman Correlation.

RMSE (Root Mean Square Error)
RMSE is a way to evaluate linear regression models by measuring the accuracy of the estimated results of a model. RMSE is calculated by squaring the error (prediction - observation) divided by the amount of data, then rooted. The RMSE formula is shown in Eq. (1). This prediction model is said to be the best if the value is 0 (zero).

$$\mathrm{RMSE} = \left(\frac{\sum (y_i - \hat{y}_i)}{n}\right)^{1/2} \tag{1}$$

where y = the value of the observation result, ŷ = predicted value, i = sequence of data in the database, N = amount of data.

MAE (Mean Absolute Error)
MAE is one of the methods used to measure the accuracy of the forecasting model. The MAE value shows the absolute average error between the prediction results with the real value. In general, the MAE formula is shown in Eq. (2). This prediction model is said to be the best if the MAE value is 0 (zero).

Table 2 Class of correlation coefficient value

Class	Correlation coefficient value	Information
1	0–0.25	Very weak
2	0.25–0.50	Weak
3	0.50–0.75	Strong
4	0.75–1	Very strong

$$\text{MAE} = \frac{1}{n}\sum_{i=1}^{n}\left|y_i - \hat{y}_i\right| \tag{2}$$

where n = amount of data, $\left|y_i - \hat{y}_i\right|$ = absolute error, y_i = predicted value, $\hat{y}_i$ = real value.

Pearson and Spearman Correlation

Pearson correlation is used to determine whether there is a relationship between two variables on an interval or ratio scale. The Pearson correlation denoted by r is shown in Eq. (3).

$$r = \frac{\sum xy - \frac{(\sum x)(\sum y)}{n}}{\sqrt{\left(\sum x^2 - \frac{(\sum x)^2}{n}\right)\left(\sum y^2 - \frac{(\sum y)^2}{n}\right)}} \tag{3}$$

where r = correlation value, x = variable x, y = variable y. The strength of the correlation relationship classified into 4 classes that shown in Table 2.

The Spearman-rho rank correlation coefficient was used for the measurement of correlation in nonparametric statistics.

3 Result and Discussion

3.1 Analysis of the Quality of Hourly and Daily Rainfall Data

Completeness of rain data is an important aspect in knowing the quality of hourly and daily rainfall data. Rainfall data on the hourly and daily time scales used in this study were taken from 1981 to 2019. However, the available rain data, especially on the hourly time scale, is inadequate so that a lot of rain data is not recorded or is called NA (Not Available). From the 25 stations, the percentage of missing values (NA) that is less than 31% is the Dermaga station (96,753) and Citeko (96,751), that are 24.57 and 30.44%. Meanwhile, there are 4 stations that have rainfall data between 30–50%. The remaining 19 stations with more than 70% missing values. For the daily rainfall scale, the percentage of missing values (NA) for the 25 stations is not

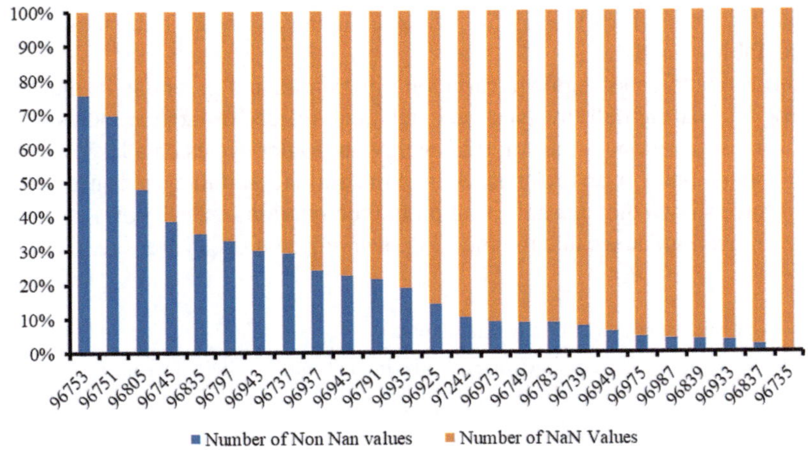

Fig. 2 Percentage of number of non-NA and NA at each station

much different from the hourly rain scale. The following graph shows the percentage of NA numbers at each station as in Fig. 2.

In another study, it was stated that in Malaysia, for the daily rainfall scale, the number of missing values was less than 10% for the period 1975–2004. The number of missing values is estimated using various types of weighting methods. Meanwhile, in Singapore, the daily time scale analyzed from 30 stations for the period 1980 to 2010 shows very few missing values data, namely less than 1%. However, the area used in Singapore is relatively small, around 716 km^2 compared to Java, which is 128.297 km^2.

3.2 *Statistical Analysis of Hourly and Daily Rainfall Data*

From 25 rain stations for the hourly rain scale, the largest average value of rainfall occurred at Semarang Maritim Rain Station (96,837) of 1.17 mm. Meanwhile, for the daily rain scale, there is only one rain station that has an average value below 1 mm, namely Banyuwangi Rain Station (96,987) of 0.06 mm. Then for the hourly scale of rain, the largest maximum amount of rain occurs at the Jatiwangi Cirebon Rain Station (96,791) of 147 mm which occurs precisely at 17.00 on 25 December 2013. For the daily rain scale, the Cilacap rain station (96,805) has the largest maximum amount of rain, which is 302.7 mm. For the minimum amount of rain in the hourly and daily rain scale at each rain station, it is 0 mm, which means that each rain station has dry hours. Furthermore, the standard deviation value for the largest hourly rain scale occurred at Semarang Maritim Rain Station (96,837) of 4.08. Meanwhile, the largest standard deviation value for the daily rain scale occurred at the Cilacap Rain

Station (96,805), which was 19.56. Statistical analysis for 25 rain stations can be seen in Tables 3 and 4.

Diurnal cycles and regional variations in rainfall are essential in the tropics. Because tropical rainfall involves the uptake and release of large amounts of latent heat from evaporation, in the study of energy and water cycles on a large scale such as the Asian monsoon, an understanding of the diurnal rainfall cycle is needed. Most studies have shown that diurnal variation in precipitation in the Tropics peaks in the afternoon overland areas and the morning over adjacent ocean areas (e.g. [11–14]).

Table 3 Statistical analysis of hourly rainfall data

Rain station	Average (mm)	Maximum (mm)	Minimum (mm)	Standard deviation
Tangerang	0.69	55	0	3.16
Serang	0.18	85	0	1.61
Curug	0.31	80	0	2.57
Jakarta BMKG	0.24	82	0	2.15
Cengkareng	0.15	73.9	0	1.69
Citeko	0.37	100	0	2.21
Dermaga	0.41	107	0	2.96
Bandung	0.3	67	0	2.09
Jatiwangi_Cirebon	0.27	147	0	2.46
Tegal	0.18	99.7	0	1.86
Cilacap	0.35	111.2	0	2.56
Semarang_Klimatologi	0.37	94	0	2.71
Semarang_Maritim	1.17	63	0	4.08
Ahmad Yani_Semarang	0.64	85	0	3.37
Surabaya_Perak1	0.23	99	0	2.3
Juanda	0.26	80	0	2.31
Karang Ploso	0.21	77	0	1.89
Tretes	0.8	99	0	4.03
KarangKates	0.62	61.5	0	3.17
Kalianget_Madura	0.18	100	0	1.84
Sawahan	0.53	77.2	0	3.21
Banyuwangi	0.36	74	0	2.5
Mataram	0.18	70	0	1.86
Bawean	0.32	99	0	2.51
Surabaya Tg. Perak	0.21	91.1	0	2.01

Table 4 Statistical analysis of daily rainfall data

Rain station	Average (mm)	Maximum (mm)	Minimum (mm)	Standard deviation
Tangerang	2.13	67.8	0	8.64
Serang	4.33	101.4	0	10.41
Curug	7.55	187.8	0	17.07
Jakarta BMKG	5.68	277.5	0	15.47
Cengkareng	3.58	125.2	0	11.72
Citeko	8.91	245.3	0	15.98
Dermaga	9.7	240	0	18.58
Bandung	7.28	108.4	0	13.39
Jatiwangi_Cirebon	6.6	157.5	0	15.44
Tegal	4.38	140	0	12.15
Cilacap	8.27	302.7	0	19.56
Semarang_Klimatologi	8.78	245.2	0	17.86
Semarang_Maritim	7.14	100	0	16.26
Ahmad Yani_Semarang	1.99	54.5	0	6.78
Surabaya_Perak1	4.31	187	0	14.78
Juanda	5.73	118.2	0	14.15
Karang Ploso	5.11	152	0	12.63
Tretes	7.27	187.4	0	18.35
KarangKates	5.26	91.9	0	11.6
Kalianget_Madura	4.31	157.9	0	11.79
Sawahan	9.96	176	0	18.67
Banyuwangi	0.06	13.3	0	0.71
Mataram	4.4	164	0	12.78
Bawean	7.09	221.2	0	16.53
Surabaya Tg. Perak	4.72	130	0	12.36

3.3 *Daily Cycle Pattern of Hourly Rainfall*

In this study, two seasons were used to determine the daily rain pattern at selected stations, namely the wet and dry seasons. Based on the rain event in Java, it is chosen for the wet season to occur from October to March and the dry season occurs from April to September. Furthermore, a rain station that has the least amount of NA (Non-Available) is selected to determine the daily rainfall pattern based on hourly rainfall data. The selected rain stations are the Dermaga rain station (96,753) and Citeko rain station (96,751). The results of the daily rainfall patterns for the two stations in the two predetermined seasons can be seen in Fig. 3. The boxplot for the daily rain pattern at the Citeko rain station and Dermaga in the wet season and dry season have

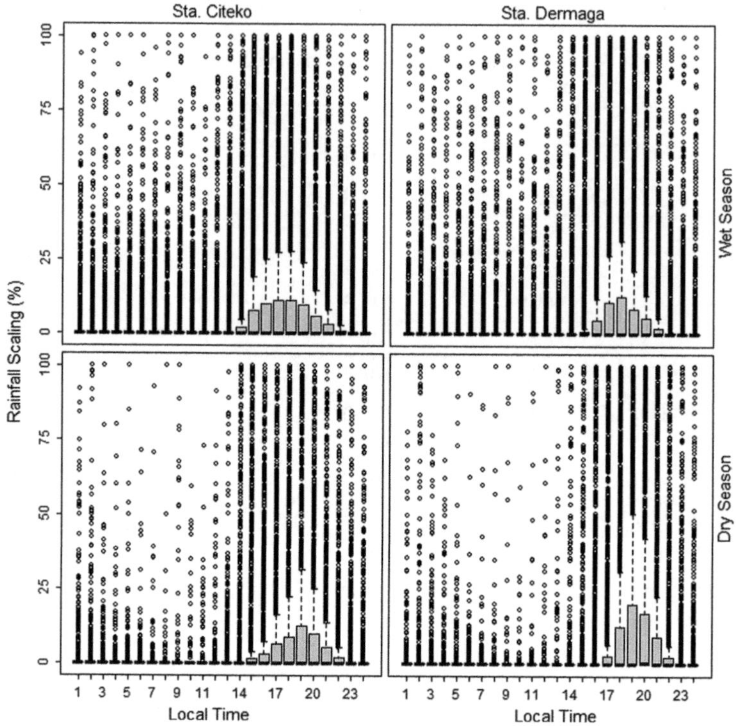

Fig. 3 Boxplot for the daily rain pattern at Citeko and Dermaga rain station

many outlier values, which means that there is a lot of hourly rain data which has very high values differ greatly from other data.

3.4 Simulation Model of Hourly Rainfall

In this research, to develop a simulation model, the assumption is that for rain below 0.1 mm in an hourly period, it is considered 0 mm. The results of the daily rain pattern in Fig. 4 shows that the daily rain has the same pattern for both the Citeko and Dermaga rain stations in two seasons, the wet and dry seasons. Furthermore, a daily rainfall model is made by considering the mean hourly rainfall as shown in Fig. 4. The results of the development of a simulation model from known patterns show that intermittent rainfall has a daily cycle pattern; that is, some concentrated rain occurs in the afternoon or at night. This is because morning to noon has evaporated and formed clouds. So that in the afternoon or evening, the clouds begin to condense so that it rains.

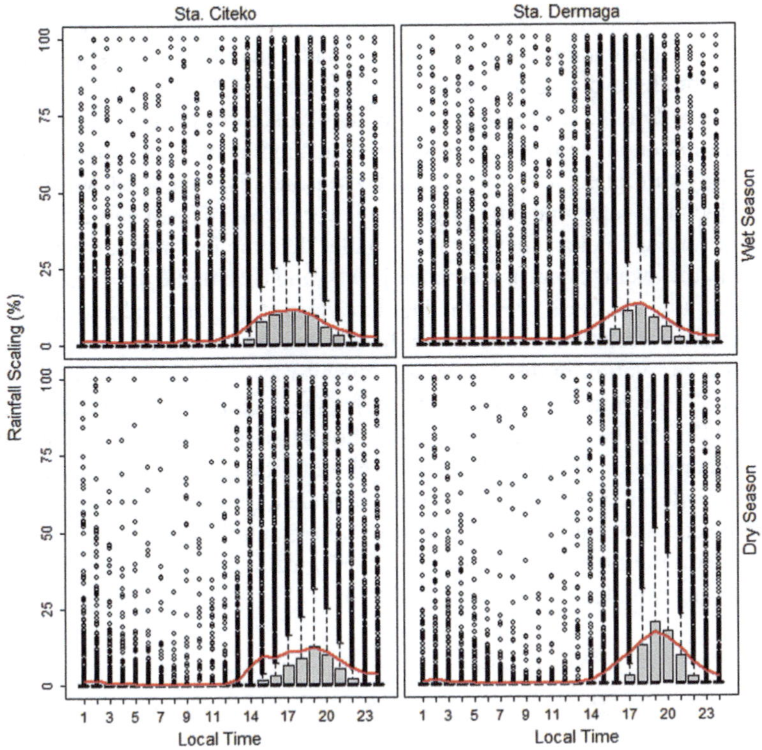

Fig. 4 Daily rain modeling for Citeko and Dermaga rain station

3.5 Model Evaluation

After obtaining the hourly rain simulation modelling using the daily rainfall data, it is necessary to evaluate it to obtain the performance of the model. Evaluation based on the mean value for simulated and observed rain can be seen in Fig. 5. The mean value for the simulated and observed rain has quite a small difference. For the Citeko rain station, the average difference in the mean value for simulated rain and observations in the wet season is 0.193 mm and for the dry season 0.185 mm. Then for Dermaga rain station, in the wet season the average difference in the mean value is 0.214 mm and for the dry season is 0.317 mm.

The next model evaluation is MAE and RMSE. The two prediction models are said to be the best if the value is 0 (zero). For the Citeko rain station in the wet season, the average of MAE value is 1.05 mm, and for the dry season 0.85 mm. Furthermore, the average of RMSE value in the wet season is 2.56 mm and for the dry season is 2.12 mm. For the Dermaga rain station in the wet season, the average of MAE value is 1.17 mm and for the dry is 1.25 mm. Furthermore, the average of RMSE value in the wet season is 2.96 mm and for the dry season is 2.98 mm. From Fig. 6 shows that the pattern of MAE and RMSE values is close to the rain pattern simulations that have been made. The average error value is both above 0 but not too big. The average value for the Citeko rain station tends to be smaller than Dermaga rain station.

The last model evaluation is seen from the value of the Pearson and Spearman correlation coefficient. The average value of the Pearson correlation coefficient for the Citeko rain station in the wet season is 0.29 which indicates that entering into class 2 is sufficiently correlated. Meanwhile, the dry season falls into class 1 with an average of 0.21 which indicates that the correlation is very weak. The Spearman correlation coefficient for the Citeko rain station is smaller than the Pearson correlation. In the wet season, the average coefficient value is 0.24 and 0.15 in the dry season indicates that it is classified as class 1. For the Dermaga rain station, the value of the Pearson and Spearman correlation coefficient in the wet and dry season falls into class 1. For the Pearson correlation in the wet season, the average value is 0.21 and 0.17 in the dry season. Furthermore, the average Spearman correlation value in the wet season is 0.14 and 0.12 in the dry season. Figure 7 shows that the pattern of the Pearson and Spearman correlation coefficient values is close to the rain pattern simulations that have been made. However, the correlation coefficient for both of them is quite small, so it can be concluded that the evaluation of rain simulations and observations has a very weak correlation. For this reason, further modelling is needed if it is to be used for forecasting rain data.

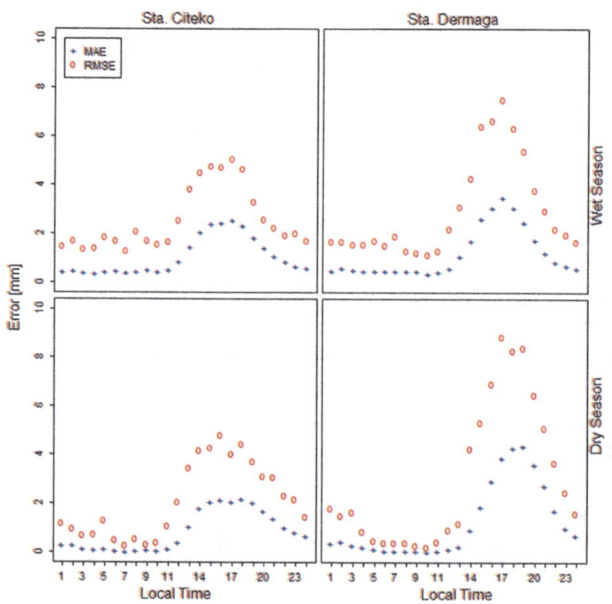

Fig. 5 Comparison of mean value of simulation and observation rain

Fig. 6 Comparison of MAE and RMSE values

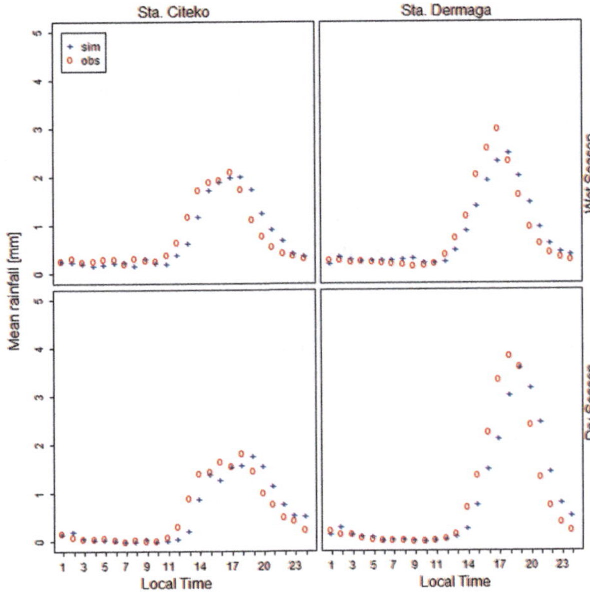

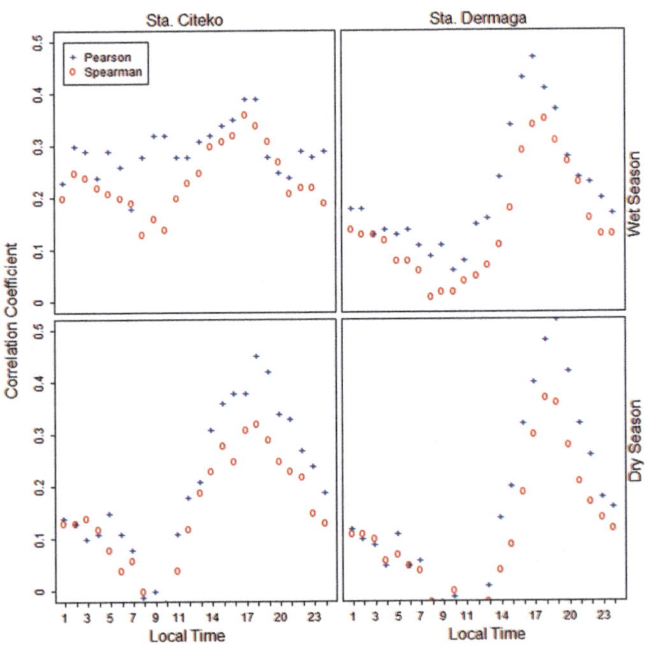

Fig. 7 Comparison of Pearson and spearman correlation coefficient value

4 Conclusion

Based on the quality of hourly and daily rainfall data, it shows that the quality of rain at 25 rain stations in Java Island is not good so that only two rain stations were selected with the least amount of NA (Not Available) data, that are Citeko and Dermaga. The results of the development of a simulation model show that the hourly rainfall has a daily cycle pattern, that is some concentrated rain occurs in the afternoon or at night. The mean values for simulated and observed rain at the Citeko and Dermaga rain stations in both the wet and dry seasons have quite small differences. The average error (MAE and RMSE) value of the two rain stations is above 0 but not too big. The average error value for the Citeko rain station smaller than Dermaga. The value of the Pearson and Spearman correlation coefficient for the two rain stations is quite small. The evaluation of simulated and observed rain has a very weak correlation. Hourly rain simulation modelling using daily rainfall data can be used for design, especially in the field of water construction. However, for forecasting purposes, further modelling is required.

Acknowledgements This research is part of a strategic research collaboration between the Indonesian government and the UK government with the title "Extreme Rainfall and Its Impact on Flood Risk in Indonesia".

References

1. Mosavi A, Ozturk P, Chau KW (2018) Flood prediction using machine learning models: literature review. Water 10(11):1–40
2. Jati MIH, Suroso S, Santoso PB (2019) Prediction of flood areas using the logistic regression method (a case study of the provinces Banten, DKI Jakarta, and West Java). J Phys Conf Ser 1367(1):012087 (IOP Publishing)
3. Pratidina G, Suroso S, Santoso PB (2019) Detection of satellite data-based flood-prone areas using logistic regression in the central part of Java Island. J Phys Conf Ser 1367(1):012086 (IOP Publishing)
4. Suroso S, Ghifari IA, Santoso PB (2021) Detection of trend behaviour of extreme rainfall over Java using Mann-Kendall. In: IOP conference series: earth and environmental science, vol 698, issue No. 1. IOP Publishing, p 012010
5. Bakhshi M, Al Janabi F (2019) Disaggregation the Daily rainfall dataset into sub-daily resolution in the temperate oceanic climate region. Int J Mar Environ Sci 13(1):11–16
6. Brigandì G, Aronica GT (2019) Generation of sub-hourly rainfall events through a point stochastic rainfall model. Geosciences 9(5):11–14
7. Diez-Sierra J, del Jesus M (2019) Subdaily rainfall estimation through daily rainfall downscaling using random forests in Spain. Water 11(1).https://doi.org/10.3390/w11010125
8. Suroso S, Bárdossy A (2018) Multisite daily precipitation simulation in Singapore. MATEC Web of Conferences 195:05007 (EDP Sciences)
9. Mehrotra R, Li J, Westra S, Sharma A (2015) A programming tool to generate multi-site daily rainfall using a two-stage semi-parametric model. Environ Model Softw 63:230–239
10. Vandenberghe S, Verhoest NEC, Onof C, De Baets B (2011) A comparative copula-based bivariate frequency analysis of observed and simulated storm events: a case study on Bartlett-Lewis modeled rainfall. Water Resour Res 47(7)

11. Gray WM, Jacobson RW Jr (1977) Diurnal variation of deep cumulus convection. Mon Weather Rev 105(9):1171–1188
12. Murakami M (1983) Analysis of the deep convective activity over the western Pacific and Southeast Asia Part I: Diurnal variation. J Meteorol Soc Jpn Ser II 61(1):60–76
13. Mori S, Jun-Ichi H, Tauhid YI, Yamanaka MD, Okamoto N, Murata F, Sribimawati T (2004) Diurnal land-sea rainfall peak migration over Sumatera Island, Indonesian Maritime Continent, Observed by TRMM satellite and intensive rawinsonde soundings. Mon Weather Rev 132(8):2021–2039
14. Sakurai N, Murata F, Yamanaka MD, Mori S, Hamada JI, Hashiguchi H, Suhardi B (2005) Diurnal cycle of cloud system migration over Sumatera Island. J Meteorol Soc Jpn Ser II 83(5):835–850

Water Quality Mapping on the Coast of Bangkalan Madura Based on the Acidity Value from Aqua MODIS Satellite Imagery

Hendrata Wibisana, Zetta Rasullia Kamandang, and Kartini

Abstract This study aims to present a map of the acidity values distribution along the Bangkalan, Madura coastline to illustrate the water quality based on its acidity value by implementing remote sensing technology that used the reflectance value at wavelengths of visible light, which is correlated with acidity value (pH). This information could be essential to people who live on the coast, so they could be more aware of the water quality changing because most of them depending on their livelihood from fishery activities. Based on the calculation, the mathematical models representing the pH value are exponential and linear algorithms due to their number of R^2 values compared to other models, with 0.779 and 0.773 values, respectively. Furthermore, the study result indicates that acidity in the Bangkalan coast is in the range of 6.4 to 6.98, obtained by an exponential algorithm, and 6.57 to 6.98, obtained by a linear algorithm, respectively. This study concludes that the distribution of acidity in the Bangkalan coast is still within the safe pH range for aquatic ecosystems on the coast.

Keywords Acidity value · Aqua MODIS satellite imagery · Madura

1 Introduction

Good water quality affects the healthiness of the ecosystem. It has a significant influence on the growing process of living creatures because it is contained of compounds, minerals, and ions. In terms of monitoring the quality of water, various existing technologies and also parameters are used such as total suspended solids [1, 2], sea

H. Wibisana (✉) · Z. R. Kamandang · Kartini
Universitas Pembangunan Nasional "Veteran" Jawa Timur, Jl.Raya Rungkut Madya, Gunung Anyar, East Java 60294 Surabaya, Indonesia

Z. R. Kamandang
e-mail: zerasullia.ts@upnjatim.ac.id

Kartini
e-mail: kartini.if@upnjatim.ac.id

S. A. Kristiawan et al. (eds.), *Proceedings of the 5th International Conference on Rehabilitation and Maintenance in Civil Engineering*, Lecture Notes in Civil Engineering 225,
https://doi.org/10.1007/978-981-16-9348-9_87

surface temperature [3, 4], salinity, chlorophyll-a, and acidity of waters. One of the considered technologies that able to map parameters of water quality determining is remote sensing. Remote sensing utilizes satellite imagery in a monitoring process that is currently widely used by researchers in various parts of the world. The satellite imagery has a variety of characteristics in its use, some use low resolution such as Meris satellite imagery [5], Aqua MODIS satellite imagery [6–8], Terra MODIS [9–11], and Suomi VIIRS [12, 13]. Some experts also prefer intermediate resolution such as ASTER satellite imagery [14], Landsat imagery [9, 15, 16], and imagery SPOT [17]. Overall, this satellite image relies on sunlight to record data in the form of objects on the earth's surface, so that experts who have examined the parameters of water quality many use wavelengths of visible light. In this case, the wavelength is represented by red light, green light, and blue light or ordinary given RGB notation (red, green, blue).

Bangkalan, Madura is a coastal area that is surrounded by fish ponds, shrimp farms, and mangrove areas. The coastal area is a home for people who depends on the water creatures such as fish and shrimp for their livelihood. Therefore, the ecosystem problems that might be caused by water quality issues, such as water pollution could affect their fishery products.

The purpose of this study is to analyze water quality in the Bangkalan coast by mapping the acidity distribution of the coastal area using the Aqua MODIS satellite imagery. The expected results could be essential to people who live on the coast, so they could be more aware of the water quality changing, regarding its acidity.

2 Material and Method

2.1 Satellite Image Data

To collect Aqua MODIS satellite imagery data, a process including downloading the data is conducted by accessing http://oceancolor.glfc.nasa.com [18] for May 2020 with name A2020128061500.L2_LAC_OC.nc, where A is an in initial for Aqua MODIS and the next four-digit means year 2020, three-digit following it is 128 indicates the 128th of days in a year according to the 'Julien Days' rule, the rest of the next digit 061,500 is the satellite recording time wherein recorded at 6 a.m. past 15 min. This satellite image is a Level-2 image that has been corrected geometrically. The Aqua MODIS imagery in full view can be seen in Fig. 1.

Figure 2 above illustrates the scene of the Aqua MODIS satellite image of Madura island. The figure was obtained by cropping the wider image (Fig. 1) using SeaDAS 7.5 software.

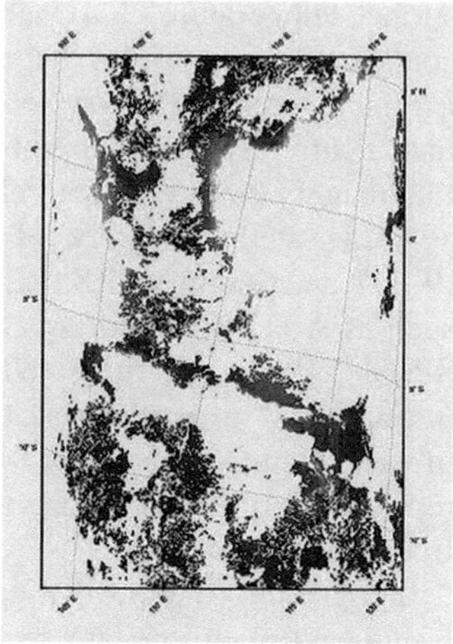

Fig. 1 A full scene of Indonesia Aqua MODIS satellite images on 8 May 2020

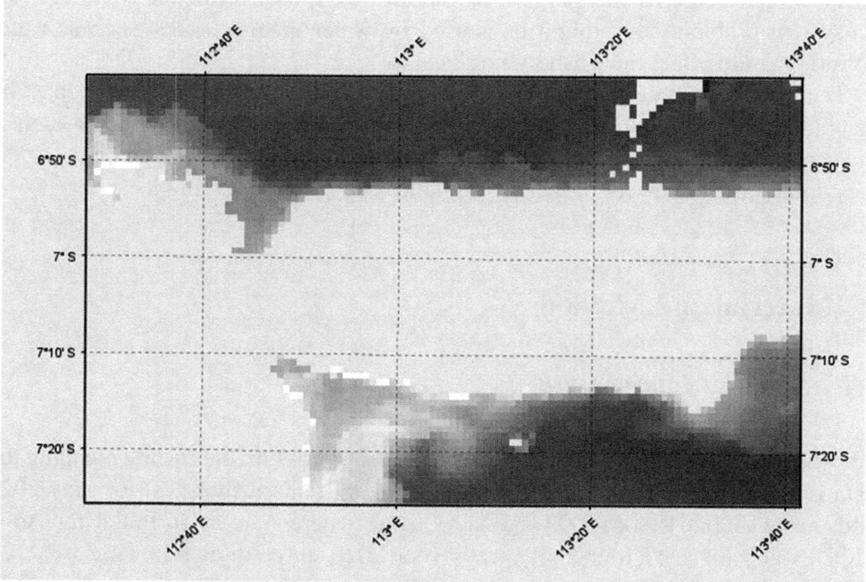

Fig. 2 A cropped scene from Fig. 1 of research location, Madura Island

Table 1 Acidity data (pH) with the field coordinates

Point	Longitude	Latitude	pH
1	112.8911	−7.2262	7.5
2	112.9138	−7.2261	7.8
3	112.9365	−7.2260	7.3
4	112.9478	−7.2259	7.2
5	112.9705	−7.2372	7.1
6	112.9819	−7.2371	6.7
7	113.0045	−7.2370	6.8
8	113.0159	−7.2370	6.5
9	113.0385	−7.2369	6.6
10	113.0386	−7.2482	6.5
11	113.0159	−7.2483	6.7
12	113.0046	−7.2483	6.8
13	112.9819	−7.2484	6.8
14	112.9706	−7.2485	6.7
15	112.9592	−7.2485	6.9
16	112.9478	−7.2373	7.1
17	112.9366	−7.2486	7.3
18	112.9139	−7.2487	7.6
19	112.8912	−7.2488	7.5
20	112.8798	−7.2375	7.6

2.2 Sea Water Acidity Data (pH)

Acidity value data denoted by the pH value is obtained directly in the field by measuring sea level at a depth of 10 to 20 cm using a digital pH meter and directly reading the pH value generated at the corresponding coordinates. The coordinates that accompany the pH measurements were performed using GPS Navigation and at each stop, the measurements of coordinates Latitude and longitude were conducted before measuring the pH value. The results of the field measurements can be seen in Table 1, where the total number of measurements is 20 points.

2.3 Algorithm of Mathematical Models

After resulting the Aqua Modis satellite image reflected data generated by extracting these values in the SeaDAS 7.5 software by inserting the coordinate values, the next step was to analyze the mathematical model with trend analysis to obtain mathematical equations and the existing R^2 correlation values. From the obtained R^2-value, it can be seen which the most optimal model that can describe the acidity distribution conditions in the Bangkalan coast.

2.4 Thematic Map of pH Distribution on the Bangkalan, Madura Coast

To illustrate the distribution of pH value on the coast, this study implemented the two most optimal mathematical model algorithms as indicated by the accepted R^2-value of other models using SeaDAS 7.5 software. It concluded that acidity distribution which has processed using a wavelength of 667 nm has the best results compared to 412 nm and 531 nm.

3 Result and Discussion

The most optimum mathematical model results using the trend analysis method presented in the scatter diagram form are shown in Table 2. It presents the resume of all analyzed mathematical models with the existing R^2 correlation value and the initial of variable x that represents a reflectance of satellite images.

Table 2 shows that the linear model and the exponent model have the highest R^2 correlation value with 0.773 for the linear and 0.779 for the exponent, respectively. The corresponding wavelength is 667 nm which is in the red spectrum, so it can be concluded that the red wavelength can provide a strong enough signal to detect changes in the acidity value in coastal waters.

Furthermore, to see how far the chosen mathematical models could provide the same response to others coordinate points, a validation test was conducted using 5

Table 2 Resume the algorithm of a mathematical model from 667 nm wavelength

No	Algorithm	Mathematical models	R^2
1	Linear	$pH = 83.531x + 6.3879$	0.773
2	Exponent	$pH = 6.4089e^{11,834x}$	0.779
3	Logarithmic	$pH = 0.4845\ln(x) + 9.4612$	0.692
4	Power	$pH = 9.9189x^{0,0689}$	0.703

Source Calculation results

remaining field data that has not been used in the previous calculation. The result of the validation test is presented in Fig. 3.

As seen in Fig. 3, the pH value of each mathematical model has a slight difference compared to each other. After the pH value of the field (in-situ) data has processed using algorithm calculation of satellite image data, the mathematical models that can represent the situation of Bangkalan, Madura coast water were obtained, however, an advanced correlation test is still required to identify their correlation to each other. The results are seen in Table 3.

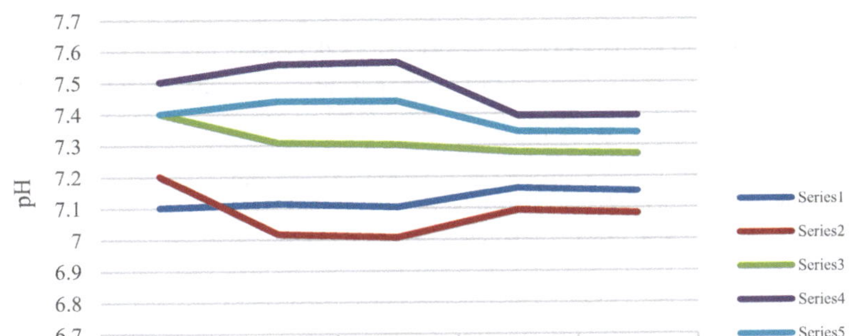

	In-situ (pH)	Linear (pH)	Exponent (pH)	Logarithmic (pH)	Power (pH)
Series1	7.1	7.112	7.101	7.161	7.151
Series2	7.2	7.015	7.004	7.091	7.08
Series3	7.4	7.307	7.3	7.276	7.27
Series4	7.5	7.556	7.562	7.392	7.391
Series5	7.4	7.44	7.439	7.342	7.338

Fig. 3 Validation test result for each mathematical model

Table 3 Comparison of the correlation coefficient of mathematical model

	In-situ	Linear	Exponent	Logarithmic	Power
In-situ	1				
Linear	0.9056	1			
Exponent	0.9059	0.99995	1		
Logarithmic	0.8970	0.99676	0.99590	1	
Power	0.8979	0.99719	0.99638	0.99998	1

Source Calculation results

Table 4 shows that the in-situ pH value has a large correlation with the linear model and the exponent model with a 0.9056 correlation value or around 90.56% of the available data. It can be represented as a description of the actual field conditions in comparison with the logarithmic model and the power model that only shows 0.8979 of value, but indeed these values are also able to describe the situation well.

The final result is shown by depicting acidity mapping in Bangkalan, Madura coast, where Fig. 4 shows the results of the thematic map for linear algorithms at 667 nm wavelength, and Fig. 5 shows the results of thematic maps for the exponent algorithm at the same wavelength. Those figures show that the distribution of acidity ranges is in the value of 6.4–6.98.

Table 4 Comparison of the correlation of mathematical model

	In-situ	Linear	Exponent	Logarithmic	Power
In-situ	1				
Linear	0.9056	1			
Exponent	0.9059	0.99995	1		
Logarithmic	0.8970	0.99676	0.99590	1	
Power	0.8979	0.99719	0.99638	0.99998	1

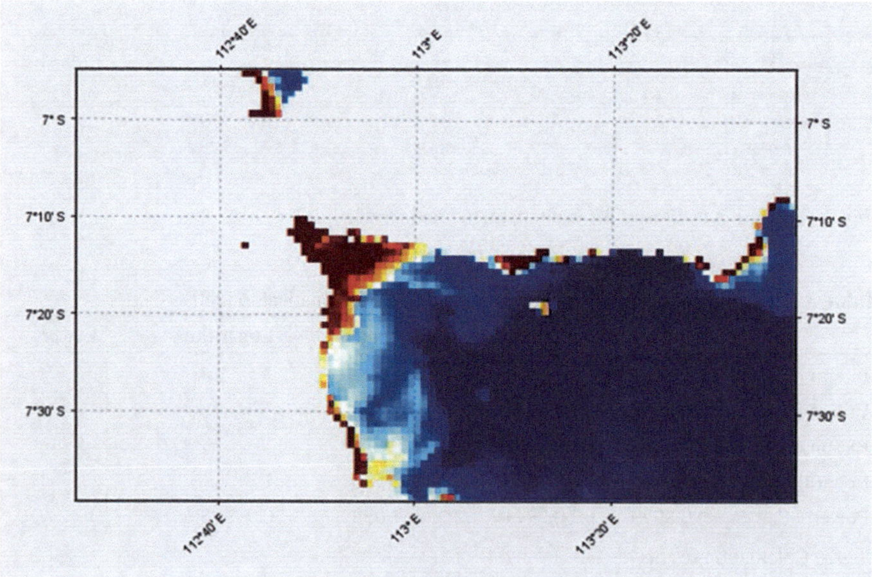

Fig. 4 Map of acidity (pH) using the linear algorithm at 667 nm wavelength

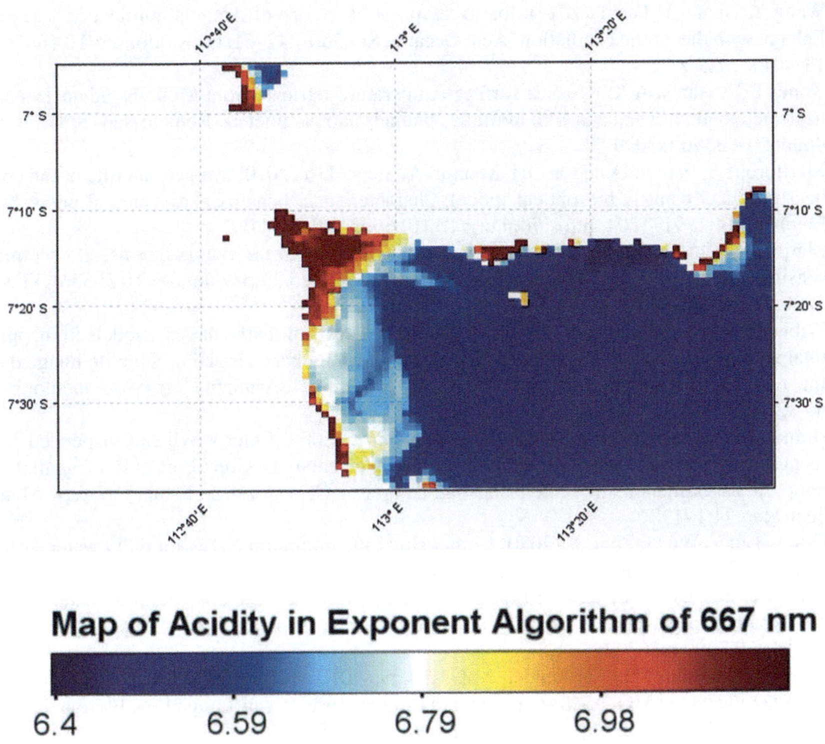

Fig. 5 Map of acidity (pH) using the exponent algorithm at 667 nm wavelength

4 Conclusion

Aqua MODIS image data can be used as a reference to identify the mathematical model algorithms that able to calculate any parameters such as the acidity or pH value of the coastal water environment. This study concludes that the distribution of acidity in the Bangkalan coast is still within the safe pH range for aquatic ecosystems on the coast.

References

1. Wibisana H, Soekotjo BM, Lasminto U (2019) Preliminary study of total suspended solid distribution in coastal ujung Pangkah Gresik based reflectance value of Landsat satellite imagery. Indones J Geogr 51(1):. https://doi.org/10.22146/ijg.38967
2. Sirjacobs D et al (2011) Cloud filling of ocean colour and sea surface temperature remote sensing products over the Southern North Sea by the data interpolating empirical orthogonal functions methodology. J Sea Res. https://doi.org/10.1016/j.seares.2010.08.002

3. Wang Y, Huang F, Fan T (2017) Spatio-temporal variations of Arctic amplification and their linkage with the Arctic oscillation. Acta Oceanol Sin 36(8):42–51. https://doi.org/10.1007/s13131-017-1025-z
4. Koner PK , Harris A (2016) Sea surface temperature retrieval from MODIS radiances using truncated total least squares with multiple channels and parameters. Remote Sens 8(9). https://doi.org/10.3390/rs8090725
5. Maritorena S, Fanton D'andon OH, Mangin A, Siegel DA (2010) Merged satellite ocean color data products using a bio-optical model: characteristics, benefits and issues. Remote Sens Environ 114:1791–1804. https://doi.org/10.1016/j.rse.2010.04.002
6. Marpaung S et al (2020) Analysis of water productivity in the Banda Sea based on remote sensing satellite data. Int J Remote Sens Earth Sci 17(1):25. https://doi.org/10.30536/j.ijreses.2020.v17.a3280
7. Wibisana H, Kamandang ZR, Zaenab S (2020) Analysis of mathematical models in mapping total suspended solids (TSS) on the coast of Banyuwangi beach based on satellite image data. Int. J Sci Eng Investig 9. Accessed: 23 Nov 2020 [Online]. Available https://oceancolor.gsfc.nasa.gov/
8. Brando V, Dekker A, Marks A, Qin Y, Oubelkheir K (2006) Chlorophyll and suspended sediment assessment in a macrotidal tropical estuary adjacent to the Great Barrier Reef: spatial and temporal assessment using remote sensing. Coop Res Cent Coast Zo Estuary Waterw Manag Tech Rep 74:1–128
9. Fan X, Liu Y, Wu G, Zhao X (2020) Compositing the minimum NDVI for daily water surface mapping. Remote Sens 12(4):700. https://doi.org/10.3390/rs12040700
10. Ody A et al (2016)"Potential of high spatial and temporal ocean color satellite data to study the dynamics of suspended particles in a micro-tidal river plume. Remote Sens 8(3). https://doi.org/10.3390/rs8030245
11. Moreno-Madrinan MJ, Al-Hamdan MZ, Rickman DL, Muller-Karger FE (2010) Using the surface reflectance MODIS Terra product to estimate turbidity in Tampa Bay, Florida. Remote Sens 2(12):2713–2728. https://doi.org/10.3390/rs2122713
12. Shi W, Zhang Y, Wang M (2018) Deriving total suspended matter concentration from the near-infrared-based inherent optical properties over turbid waters: a case study in Lake Taihu. Remote Sens 10(2):1–17. https://doi.org/10.3390/rs10020333
13. Platnick S et al (2020) The NASA MODIS-VIIRS continuity cloud optical properties products. Remote Sens 13(1):2. https://doi.org/10.3390/rs13010002
14. Wang C, Chen S, Li D, Wang D, Liu W, Yang J (2017) Engineering technology center for remote sensing big data application. Geosci Model Dev 10:4347–4365. https://doi.org/10.5194/gmd-10-4347-2017
15. Miller RL, Liu CC, Buonassissi CJ, Wu AM (2011) A multi-sensor approach to examining the distribution of total suspended matter (TSM) in the Albemarle-Pamlico estuarine system, NC, USA. Remote Sens 3(5):962–974. https://doi.org/10.3390/rs3050962
16. Nazirova K et al (2021) Comparison of in situ and remote-sensing methods to determine turbidity and concentration of suspended matter in the estuary zone of the Mzymta River, Black Sea. Remote Sens 13(1):143. https://doi.org/10.3390/rs13010143
17. Rocchini D (2013) Seeing the unseen by remote sensing: satellite imagery applied to species distribution modelling. J Veg Sci 24(2):209–210
18. Feldman GC (2020) NASA Goddard Space Flight Center, Ocean Ecology Laboratory, Ocean Biology Processing Group. Moderate-resolution Imaging Spectroradiometer (MODIS) Aqua A2020128061500.L2_LAC_OC.nc Data; NASA OB.DAAC, Greenbelt, D, USA, 2021. https://oceancolor.gsfc.nasa.gov/data/aqua/

Multi-attribute Analysis of Raw Water Treatment from Deep Wells at PDAM Tirta Mahottama, Klungkung Regency, Bali

Ni Kadek Dian Utami Kartini, Nurulbaiti Listyendah Zahra, Ariyanti Sarwono, Intan Rahmalia, Almira Davina Nastiti, Iva Yenis Septiariva, and I. Wayan Koko Suryawan

Abstract PDAM Tirta Mahottama is a regional drinking water company located in Klungkung Regency, Bali Province. The community uses 21 deep wells to meet their water needs for daily activities. This study aims to determine the reliability characteristics of raw water from deep wells and the quality of raw water used in the Klungkung Regency. This study was conducted to determine what processing can be done to improve raw water quality using the Multi-Attribute Utility Theory (MAUT). The deep well used has a production capacity of 1–9 L/s. In terms of raw water quality, the parameter that does not meet the quality standard is aluminum, followed by color and TDS. The selected processing is adsorption with consideration of its high ability to remove color and aluminum. Based on MAUT analysis, the best adsorbs are using GAC (granular activated carbon).

Keywords Sludge treatment · Water treatment · Decision analysis

1 Background

PDAM Tirta Mahottama is a PDAM located in Klungkung Regency, Bali Province. PDAM Tirta Mahottama serves four sub-districts and uses water sources in groundwater, springs, and seawater. PDAM Tirta Mahottama has 21 deep wells, seven

N. K. D. U. Kartini · N. L. Zahra · A. Sarwono · I. Rahmalia · A. D. Nastiti · I. W. K. Suryawan (✉)
Department of Environmental Engineering, Faculty of Infrastructure Planning, Universitas Pertamina, Jakarta 12220, Indonesia
e-mail: i.suryawan@universitaspertamina.ac.id

N. L. Zahra
e-mail: nurulbaiti.lz@universitaspertamina.ac.id

A. Sarwono
e-mail: ariyanti.sarwono@universitaspertamina.ac.id

I. Y. Septiariva
Department of Civil Engineering, Faculty of Engineering, Universitas Sebelas Maret, Surakarta 57126, Indonesia

S. A. Kristiawan et al. (eds.), *Proceedings of the 5th International Conference on Rehabilitation and Maintenance in Civil Engineering*, Lecture Notes in Civil Engineering 225,
https://doi.org/10.1007/978-981-16-9348-9_88

springs, and one reverse seawater osmosis (SWRO) treatment unit. According to the 2019 PDAM Performance Book, PDAM Tirta Mahottama had received a healthy category from 2017 to 2019 with customers' water quality aspect value of 88% in 2019. This value shows the quality of water in customers has not been maximal or has not fully met quality standards. Because it has not reached 100%, this value is still included in the three lowest scores compared to other PDAMs in Bali Province. Improving the quality of the environment in Klungkung Regency will be in line with water resources management. Water resources in Klungkung Regency have not been utilized until 100%, with most deep wells [1]. Deep wells pollution depends on groundwater infiltration, which carries wastewater [2].

The quality and quantity of deep well water in the Klungkung Regency need to be studied further to determine community consumption feasibility. Besides, the technology required to improve quality needs to be studied further. This study aims to develop the right technology selection based on the quantity and quality of deep wells located in Klungkung Regency.

2 Method

2.1 Data Collection

The data used in this research is from monitoring the deep well quality of PDAM Tirta Mahottama. Brucine measured inorganic chemical parameters Nitrite was measured by colorimetric method and Nitrate. The physical parameters measured were odor, color, TDS, taste, and temperature tested by organoleptic, calorimetric, organoleptic, and thermometer methods, respectively. The chemical parameters measured were aluminum with titrimetry, iron with phenanthroline, hardness with titrimetry, chloride with titrimetry, manganese with trichloride, pH with pH meter, sulfate with turbidimetry, and ammonia with Nessler.

2.2 Selection of Alternative Processing

The selection of alternatives is carried out using the Multi-Attribute Utility Theory (MAUT) method. This method is a method of decision-making by considering more than one criterion. According to Kiker et al., in applying multicriteria decision analysis in environmental decision making, this method has often been used in various fields to assist decision-making, including management, environmental management, economics, and engineering [3].

3 Result and Discussion

3.1 Reliability of Raw Water Sources

PDAM Tirta Mahottama is one of the PDAMs that uses deep wells as a source of raw water. There are 21 deep wells utilized, with a total production discharge of 164.5 L/s (Table 1). In the archipelago area, namely Nusa Penida Island, the water sources in Kutampi I, II, and III deep wells are only installed half of the mainstay discharge. Development of groundwater sources has been carried out in Nusa Penida by drilling wells on Nusa Penida's island. The availability of deep well water drops dramatically during the dry season, so it is necessary to think about other ways to meet the water needs in the Klungkung Region archipelago [4].

The production process in the deep well of PDAM Tirta Mahottama only consists of taking water from the well through a submersible pump and disinfection using chlorine gas. Then the water is directly distributed to customers. The disinfection process is carried out by injecting chlorine gas directly into the distribution pipe. However, some deep wells still have some parameters that exceed the quality standard, so that these parameters are not excluded.

3.2 Water Quality

Table 2 showed the predominant parameter is aluminum, followed by color and TDS. The excess of TDS occurs in deep wells located in the Nusa Penida area. The Nusa Penida area is composed of limestone, which causes seawater intrusion to occur quickly and affects the quality of deep well water to become brackish. It is planned that the advantages of TDS in the area will be processed using a distillation process. Water treatment in the Nusa Penida area is well planned [1], so that in this study, no further discussion was carried out. Also, nutrient parameters are seen under normal conditions, so they will not negatively impact streams [5, 6].

Aluminum metal is the metal most found in nature. Aluminum can be in the soil, water, and air and makes up about 8% of the earth's crust. Aluminum is a very reactive element that is easily oxidized. Next source of intake of aluminum in drinking water. According to Permenkes No. 492 of 2010 concerning Drinking Water Quality Requirements, aluminum's full drinking water content is 0.2 mg/L. The source of aluminum in water can come from chemical residues used during the drinking water treatment process, such as aluminum, commonly used as a coagulant. Besides, aluminum can naturally be contained in groundwater. The color in water indicates organic content and ions metals such as iron, manganese, and turbidity [5]. Color is a component of physical water quality. The presence of color will interfere with the water's aesthetics and indicate polluted water [6, 7]. True color was measured in water samples filtered using a filter with a size of 0.45 μm. In simple terms, the true color is the color that is measured when the turbidity

Table 1 Characteristics utilization of raw water from deep wells in Klungkung Regency

Location	The reliability of clean water (L/s)	Installed (L/s)	Production (L/s)	Not yet Be used (L/s)	Sub-district
Sema Agung	5	5	4	1	Banjarangkan
Desa Pikat	4	2.5	2	2	Dawan
Paksebali	10	10	7	3	Dawan
RSU Klungkung	10	10	4	6	Klungkung
Tegalinggah	5	5	5	0	Klungkung
Kutampi (I)	5	2.5	2	3	Nusa Penida
Kutampi (II)	5	2.5	1	4	Nusa Penida
Kutampi (III)	5	2.5	2	3	Nusa Penida
Dawan Kaler	10	10	5	5	Dawan
Kusamba	8	4	2,5	5,5	Dawan
Buayang	18	15	8	10	Dawan
Mincidan	18	15	8	10	Dawan
Swecapura	10	5	3	7	Klungkung
Sedap Malam	7	5	5	2	Klungkung
Akah	20	12	12	8	Klungkung
Koripan	5	5	5	0	Banjarangkan
Pura Kentelgumi	2,5	2,5	0	2,5	Banjarangkan
Balai Budaya	12	12	9	3	Klungkung
Cucukan	0	0	0	0	Klungkung
Besan	0	0	0	0	Dawan
Pesinggahan	5	5	5	0	Dawan

in the water has been removed. True color comes from the breakdown of organic compounds, such as hummus, lignin's, tannins, and other organic acids. The presence of this compound makes the color in water challenging to remove. Water that contains true color is generally characteristic of yellow to reddish-brown or relatively clear.

3.3 *Water Treatment Decision Analysis*

The design unit's determination is done through a literature study by looking at previous research on aluminum removal and color. Several removal methods can be used to remove aluminum and color, including adsorption, precipitation, membranes,

Table 2 The water quality of deep wells in Klungkung Regency

Location	NO_2	NO_3	Color	TDS	Al	Fe	Hardness	Chloride	Mn	pH	Sulfate	NH_3
	mg/L	mg/L	TCU	mg/L	mg/L	mg/L	mg/L	mg/L	mg/L		mg/L	mg/L
Sema Agung	nd	1.78	3	217	**0.26**	0.0004	194	16.08	nd	7.7	11.64	0.10
Desa Pikat	nd	0.99	0	414	**0.29**	0.047	388	38.32	0.0008	8	5.27	0.19
Paksebali	nd	1.01	7	291	**0.28**	nd	291	16.66	nd	7.9	11.88	0.11
RSU Klungkung	nd	2.79	1	264	**0.26**	0.012	232.8	22.49	nd	8.2	19.84	0.08
Tegalinggah	nd	2.7	**40**	368	**0.25**	0.247	291	20.9	0.0012	7.4	17.63	0.12
Kutampi (I)	nd	4.55	1	**2740**	**0.23**	nd	**853.6**	**1037.16**	0.0187	7	58.35	1.95
Kutampi (II)	nd	4.23	4	**1979**	**0.28**	0.026	**659.6**	**755.76**	0.021	7.3	45.72	0.98
Kutampi (III)	nd	3.41	7	**2650**	**0.33**	nd	**756.6**	**1157.76**	0.0184	7.1	52.43	1.27
Dawan Kaler	nd	1.43	7	**339**	**0.27**	0.007	426.8	17.49	nd	8.1	0.69	0.14
Kusamba	nd	1.32	7	**558**	**0.26**	nd	388	141.62	nd	8.3	24.12	0.13
Buayang	nd	4.42	**20**	300	**0.33**	0.018	388	27.44	0.0021	7.5	7.71	0.26
Mincidan	nd	1.89	11	327	**0.25**	nd	291	16.08	0.0005	7.4	12.90	0.14
Swecapura	0.0004	3.15	9	335	**0.25**	0.002	291	27.49	nd	8.4	27.23	0.12
Akah	0.0015	1.85	2	274	**0.28**	0.039	252.2	18.32	nd	7.9	15.06	0.11
Koripan	nd	3.46	4	224	**0.26**	0.018	213.4	17.68	0.0005	7.1	12.49	0.10
Pura Kentelgumi	nd	1.33	9	246	**0.22**	nd	194	15.27	0.0009	7.9	11.32	0.10
Balai Budaya	nd	3.01	2	309	**0.28**	0.66	213.4	27.49	nd	8.1	25.68	0.11
Cucukan	0.0715	0.68	**20**	194	**0.27**	0.334	174.6	17.68	0.0012	7.8	20.56	0.18
Besan	0.0269	0.88	**40**	251	**0.24**	nd	194	38.59	0.0022	10.1	19.45	0.29
Pesinggahan	0.442	0.61	**40**	**621**	**0.2**	nd	465.6	60.3	0.0011	7.4	16.45	0.59
Standard	3	50	15	500	**0.2**	0.3	500	250	0.4	6.5–8.5	250	1.5

*nd: not detected

granular filtration, and ion exchange. However, the PDAM wants processing that is as simple as possible and does not require high costs. Therefore, several considerations were made to choose a design alternative. The choice of alternatives at this stage is early selection. At this stage, the selection unit will be varied in processing methods, materials, or configurations for further evaluation using the Multi-Attribute Utility Theory method. Table 3 shows the considerations used in the selection process.

Table 3 Various considerations for suitable treatment methods to be applied in water treatment

Method	Can be remove aluminum and color all at once	Installation fee less than Rp 250,000,000	Ease of operation and treatment
Adsorption	Adsorption can be used to set aside the color and aluminum metal [5]	40.000.000–65.000.000 [8]	The adsorption operation is not required materials or specific treatment. The only backwash is necessary and desorption periodic [9] for treatment
Granular Filtration	The pure filtration process can only remove a specific size of material, but aluminum with a minimal size can only be removed through membrane filtration [9]	5.000.000–10.000.000 [10]	Operational filtration is not required materials or specific treatment, and only backwash is necessary periodically [5, 11, 12]
Precipitation	Precipitation can be used to remove aluminum by forming a precipitate Al (OH) 3+ [9] but cannot be used for color removal	50.000.000–65.000.000 [10]	Operations and care relatively easy precipitation, but precipitation need tub sediment or filter tub to put aside residue [9]

In this method, six steps must be taken are describing the alternatives and criteria, determining the criteria's importance, determining the qualitative criteria weights, calculating the utility value, calculating the relative weight, and calculating the final evaluation value. The description of alternatives has been discussed in the previous discussion that will be proposed three alternatives processing, namely adsorption using GAC (granular activated carbon) adsorbent with the adsorption column (fixed-bed adsorption column) processing method, adsorption with silica adsorbent with the adsorption column processing method (fixed-bed adsorption column), and adsorption with PAC (powdered activated carbon). After the weight value for the qualitative criteria is determined, a summary is made of all the measures values for each alternative as in Table 4.

Table 4 Considerations in choosing alternative media in the adsorption process

Alternatives	Criteria					
	Aluminum removal	Color removal	Cost	Area required (m^2)	Operational	Maintenacne
GAC	99.2%	99.67%	63,792,200.00	3.82	Very easy	Kinda easy
Silica	94.6%	63%	63,792,200.00	5.73	Very easy	Kinda easy
PAC	99%	80.1%	65,262,304.40	25,22	Kinda easy	Very easy

The utility value is calculated on each criterion for each alternative. This utility requires quantified criterion value data. Relative weight calculation is carried out on each measure based on the value of the level of importance. The calculation results can be seen in Table 5. In calculating the final evaluation, v(x), the utility value data and the weight normalization of each criterion are needed. Based on the analysis of the multi-attribute utility theory (MAUT) method, an alternative ranking was obtained for the Deep Well as listed in Table 6. Based on the analysis using the multi-attribute utility theory (MAUT) method, adsorption with GAC adsorbent was chosen as the best alternative in the well.

Table 5 Utility value is calculated on each criterion for each alternative

Alternatives	Criteria					
	Aluminum removal	Color removal	Cost	Area required (m^2)	Operational	Maintenacne
GAC	1	1	1	1	1	0
Silica	0	0	1	0,91	1	0
PAC	0.956	0.466	0	0	0	1
Relative Weights	0.2083	0.2083	0.1667	0.1667	0.125	0.125

Table 6 Alternate ranking order

No	Alternative	Final evaluation results	Ranking
1	**GAC**	**0.875**	**1**
2	Silica	0.443	2
3	PAC	0.421	3

The bold indicate the selected water treatment method through the MAUT analysis.

The adsorption process using granular adsorbents usually uses the batch method column or fluidized-bed operation. The contact system usually applied is a fixed bed

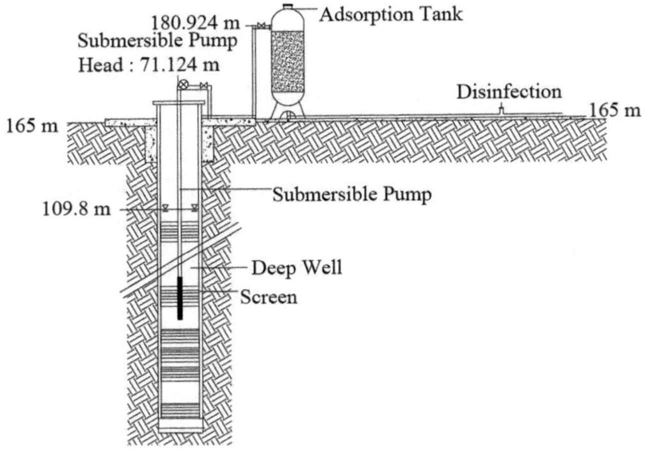

Fig. 1 Diagram of deep well processing in Klungkung Regency with fluidized-bed granular activated carbon

contacting system or countercurrent moving bed. This application is because these two methods have low operational costs and streamline the adsorbent's absorption ability [13]. Fixed bed columns usually have an up-flow or downflow flow. Figure 1 shows an example of a circuit used from a fixed bed adsorption column in a deep well in Klungkung Regency. The water will be entered through the influent distributor and will undergo adsorption in the adsorption zone (carbon bed) and flowed through the underdrain system as an effluent. The spent water drain tank tube helps store carbon during backwash to be cleaned, and then after cleaning the carbon, it will be regenerated using a furnace. GAC can be used in water or air treatment. GAC can be divided into three types based on the pores' size: micropores, mesopores, and macropores [14]. According to IUPAC, micropores have a pore size of <2 nm, mesopores have a pore size of 2 nm-50 nm, while macropores have a pore size of >50 nm. The choice of pore size depends on the type of adsorbate to be adsorbed. Using GAC as an adsorbent, pore size and pore structure are the most important things that need to be considered. These pores' size and design depend on the level of carbon activation and the relative carbon area. The activation rate is determined by the raw material's ratio and dehydrating agent and heating time and temperature.

After selecting the best alternative, detailed planning should be carried out regarding the GAC processing implemented in Klungkung Regency. Good environmental infrastructure planning is supported by considering the needs of the community. Currently, the COVID19 pandemic outbreak has caused the need for drinking water to increase [15]. In Bali Province, the environmental impact of the COVID19 pandemic affects the amount of waste generation [16, 17]. If it is linearly predicted that a pandemic, COVID19 will also impact the need for clean water needed, which is immensely increasing. Therefore, future water treatment planning to improve public health in Klungkung Regency.

4 Conclusion

The measurement of water discharge produced is 1–9 L/s for each deep well in Klungkung Regency. The parameters that need to be considered in water treatment are aluminum and color. Based on evaluation using MAUT on several considerations, namely the efficiency of aluminum removal and color, area required, cost, ease of operation and maintenance, processing of environmental parameters can use fluidized-bed granular activated carbon.

References

1. Septiariva IY, Suryawan IWK, Suhardono S, Sofiyah ES, Zahra NL (2021) Strategy and master plan for water supply and domestic wastewater in Klungkung Regency, Bali. In: IOP conference series: materials science and engineering, vol 1096, issue No. 1. IOP Publishing, p 012058
2. Septiariva IY, Suryawan IWK (2021) Development of water quality index (WQI) and hydrogen sulfide (H2s) for assessment around Suwung Landfill, Bali Island. J Sustain Sci Manag 16(4):137–148
3. Kiker GA, Bridges TS, Varghese A, Seager TP, Linkov I (2005) Application of multicriteria decision analysis in environmental decision making. Integr Environ Assess Manag Int J 1(2):95–108
4. Sudipa N, Mahendra MS, Adnyana WS, Pujaastawa IB (2020) Daya Dukung Air di Kawasan Pariwisata Nusa Penida, Bali, Jurnal Sumberdaya Alam dan Lingkungan 7(3):117–123
5. Apritama MR, Suryawan I, Afifah AS, Septiariva IY (2020) Phytoremediation of effluent textile wwtp for NH_3-N and Cu reduction using pistia stratiotes. Plant Arch 20(1):2384–2388
6. Afifah AS, Suryawan IWK, Sarwono A (2020) Microalgae production using photo-bioreactor with intermittent aeration for municipal wastewater substrate and nutrient removal. Commun Sci Technol 5(2):107–111
7. Davis ML (2010) Water and wastewater treatment engineering: design principle and practice. McGraw-Hill. Companies, Inc.
8. Suryawan IWK, Prajati G, Afifah AS, Apritama MR (2021) NH_3-N and COD reduction in Endek (Balinese textile) wastewater by activated sludge under different DO condition with ozone pretreatment. Walailak J Sci Technol 18(6):1–11
9. Suryawan IWK, Septiariva IY, Helmy Q, Notodarmojo S, Wulandari M, Sari NK, Sarwono A, Pratiwi R, Jun-Wei L (2021) Comparison of ozone pre-treatment and post-treatment hybrid with moving bed biofilm reactor in removal of remazol black 5. Int J Technol 12(2)
10. Rini ADAM, Ali F, Suwartha N (2013) Desain Kolom Adsorpsi granular activated carbon. Jurusan Teknik Lingkungan Fakultas Teknik Universitas Indonesia, Jakarta
11. Crittenden JC, Trussell RR, Hand DW, Howe KJ, Tchobanoglous G (2012) Water treatment principle and design, 3rd edn. Wiley, USA
12. Andini DRSDJ, Ainun S (2017) Perencanaan Unit Pengolahan Air Bersih di Kecamatan Sumedang Selatan. Jurnal Reka Lingkungan 5(2):503–514
13. ReynoldsT D, Richards PA (1996) Unit operation and processes in environmental engineering. PWS Publishing Company, California
14. Water Quality Association (2016) Granular activated carbon (GAC) fact sheet. Water Granular Activated Carbon (GAC) Fact Sheet, Illinois
15. Abu-Bakar H, Williams L, Hallett SH (2021) Quantifying the impact of the COVID-19 lockdown on household water consumption patterns in England. NPJ Clean Water 4(1):1–9

16. Suryawan IWK, Rahman A, Septiariva IY, Suhardono S, Wijaya IMW (2021) Life cycle assessment of solid waste generation during and before pandemic of Covid-19 in Bali Province. J Sustain Sci Manag 16(1)
17. Sari MM et al (2022) Identification of face mask waste generation and processing in tourist areas with thermo-chemical process. Arch Environ Prot (48)

Determination of Produced Wastewater Treatment Systems for Reclaim Water in the Oil and Gas Industry

Novena Lany Pangestu, Nurulbaiti Listyendah Zahra, Ariyanti Sarwono, Intan Rahmalia, Iva Yenis Septiariva, and I. Wayan Koko Suryawan

Abstract Reclamation of water from the oil and gas industry activities from the produced water can reduce the environmental impact and increase its proactive efforts. This study was conducted to assess the appropriate treatment system for produced water. This study was conducted by taking a case study at the Lindai Gathering Station (GS). The highest quality parameters that do not meet the quality standards for class III water reclamation are COD and oil and grease. Some of the recommended processing units are the American Petroleum Institute (API) separators, Parallel Plate Interceptors (PPI), and Corrugated Plate Interceptors (CPI). Each alternative is given the addition of ultra-filtration technology to improve the quality of the effluent produced water. The results of selecting the best technology using the Analytical Hierarchy Process (AHP) method show the use of the most suitable CPI to be applied for water reclamation.

Keywords Produced water · CPI · Reclamation water

1 Background

The activities of the oil and gas industry in the upstream sector are one of the activities that can cause pollution and environmental disturbance. In the oil and gas industry,

N. L. Pangestu · N. L. Zahra · A. Sarwono · I. Rahmalia · I. W. K. Suryawan (✉)
Department of Environmental Engineering, Faculty of Infrastructure Planning, Universitas Pertamina, Jakarta 12220, Indonesia
e-mail: i.suryawan@universitaspertamina.ac.id

N. L. Zahra
e-mail: nurulbaiti.lz@universitaspertamina.ac.id

A. Sarwono
e-mail: ariyanti.sarwono@universitaspertamina.ac.id

I. Y. Septiariva
Department of Civil Engineering, Faculty of Engineering, Universitas Sebelas Maret, Surakarta 57126, Indonesia

S. A. Kristiawan et al. (eds.), *Proceedings of the 5th International Conference on Rehabilitation and Maintenance in Civil Engineering*, Lecture Notes in Civil Engineering 225,
https://doi.org/10.1007/978-981-16-9348-9_89

the exploration activities are considered as the primary key of upstream activities. Exploration activities carried out by the oil–gas industry can produce fluid containing oil and produce gas, mud, and water. Based on the Ministry of Environments and Forestry Regulation (PermenLH) Number 19 of 2010, the produced water is defined as the water carried out from the strata that contain hydrocarbon to the surface during the oil and gas extraction activities. The quantity of produced water is often more significant than the quantity of oil produced, especially in old fields. According to Hayes, 80% of the total volume of wastewater produced by the oil and gas industry is produced water [1].

Produced water, often referred to as brine, is groundwater present with hydrocarbons in oil and gas rock formations and becomes waste produced by drilling [2]. Tiana also states that produced water is a by product of the processing of petroleum, natural gas, or Coal Bed Methane (CBM) [3]. Produced water is always found with petroleum. This water is slightly acidic and resides under hydrocarbons in formations or reservoirs. According to Ivory, produced water can be used as clean water to meet water needs [4]. The use of produced water as clean water requires paying attention to its quality because the government has regulated the quality standard of raw water according to its designation in Government Regulation (PP) of the Republic of Indonesia Number 82 of 2001. Produced water can be used for irrigation, wildlife consumption, industrial use, or processed into drinking water [5].

The primary treatment of produced water is used to remove oil and grease. Gravity-based separation technology is a technology that companies generally choose in removing oil and grease in produced water. According to Forero et al., gravity-based separation technologies that are usually used in industry are American Petroleum Institute (API) separators, Parallel Plate Interceptors (PPI), and Corrugated Plate Interceptors (CPI) [6]. This type of separator generally consists of an initial separator that functions to separate large particles and the primary separator used to separate oil from water. The produced water's primary treatment does not separate the hydrocarbons contained in produced water 100% [7]. Secondary treatment is also often required if the desired effluent has not reached the target. The technologies commonly used for secondary processing of produced water are membrane technology, coagulation, and biological processing. Some companies choose to use biological processing because the nutrient concentration is still very high. Biological processing that several companies commonly use is the Biological Aerated Filter (BAF) and activated sludge [8]. However, the physic-chemical processing method has been preferred by companies because of the need for a smaller area and operational simplicity [9].

Various alternatives can be used in water reclamation from oil and gas-produced water; it is necessary to select the best alternative according to the company's priority scale. This research aims to determine the most appropriate alternative in water reclamation from oil and gas industry activities.

2 Method

2.1 Location

The design of produced water treatment is carried out in the onshore facility oil and gas industry, namely PT Pertamina Hulu Energi Kampar-Siak, to be precise, at the Lindai Gathering Station (GS). Administratively, Lindai GS is in Lintas Suram, Senama Nenek Village, Tapung Hulu District, Kampar Regency, Riau.

2.2 Produced Wastewater Quality

Secondary data includes the produced water characteristics data in Lindai GS, which can be found in the Environmental Management Plan and Environmental Monitoring Plan (RKL-RPL) document belonging to PHE Kampar-Siak. The main parameters that are seen, are the parameters regulated in PermenLH No. 19 of 2010, specifically for produced water. These parameters include Chemical Oxygen Demand (COD), oil and grease, temperature, pH, and Total Dissolved Solid (TDS) measured according to the standard method. The data taken as a design reference is data from 2016 to 2019.

2.3 Decision Method

Treatment alternatives were selected based on literature studies from journals, proceedings, and books on produced wastewater treatment. The selection of alternative processing units is carried out using the Analytical Hierarchy Process (AHP) method. The AHP method is quite commonly used to determine the best wastewater treatment unit, which similar as in research conducted by Ramadan et al. [10]. There are five criteria for selecting alternative production water treatment units.

The assessment criteria include the removal of oil and grease, the effectiveness of eliminates the COD, the operation and maintenance (O&M), the land requirements, and the initial cost of processing unit materials. These criteria will then be assessed and compared in pairs to determine which standards are prioritized. The rating scale 1–9 was chosen because it is the best scale in formulating opinions for various problems [11].

3 Result and Discussion

3.1 Existing Produced Wastewater Treatment System

The production water treatment unit is carried out in the oil and gas industry, namely at the Lindai Gathering Station on PT Pertamina Hulu Energi (PHE) Kampar-Siak. In treating produced water from Gathering Station (GS) Lindai used a storage pool called a pit. There are four pits in Lindai GS, which function to reduce the temperature and oil levels in the produced water before being channeled into the canal and forwarded to the Lindai River. The industry's proactive efforts to reduce environmental impact are water reclamation [12, 13]. For this reason, an additional treatment unit is required in the existing wastewater treatment system.

3.2 Produced Water Quality

Produced water meets the wastewater quality standards of PermenLH No. 19 of 2010 regarding wastewater quality standards for oil, gas and thermal power businesses and activities. The processing load of producing water into class III raw water is more easier and more possible to do. Table 1 shows the parameters of produced water that compared to the effluent quality standard. Produced water at Lindai GS can be reused to watering plant. However, there is a need for further processing of produced water so that its quality can be accepted following applicable regulations. The quality standard reference used is PP. 82 of 2001, specifically on third-class clean water quality. Third-class water quality can be used as water for freshwater fish cultivation, animal husbandry, watering plants, and similar uses [4]. It can be seen in Table 1 that the parameters of concern are COD and oil and grease. Nutrient parameters in the form of NH_3-N are not a problem in water production. Usually, nutrients parameters are the main factors in the pollution flow in streams [14, 15].

Table 1 Comparison of produced water quality in effluent quality standards

Parameters	Example	Maximum produce water quality	Average produce water quality	Standard PP No. 82 2001 Class III
COD	mg/L	287.8(*)	61.29(*)	50
Oil and Grease	mg/L	16.28(*)	5.9(*)	1
TDS	mg/L	989	725.87	1000
pH	-	8.8	8.04	6–9
NH_3-N	mg/L	2.09	0.78	Not required
Phenol total	mg/L	<0.001	<0.001	0.001
H_2S	mg/L	<0.002	<0.002	0.002

(*) The parameter being concern because the value that exceeding the quality standard

3.3 Decision Analysis

The produced water treatment unit's design aims to obtain water quality following class III water quality which then to be used as a new water source for watering plants. Parameters that exceed the quality standard are COD and oil and grease, so the processing process focuses on setting these parameters aside. In the process of designing a produced water treatment unit, there are three alternative treatment series. According to Kundu and Mishra, the stages of produced water treatment consist of primary processing and secondary processing [12]. In water processing production at PHE Kampar Siak, primary processing is used to remove oil & grease. Secondary processing is used to remove COD content to comply with quality standards. The alternative manufacture for the series of produced water treatment units at Lindai GS has differences in primary processing, namely oil and grease removal technology. This is because removing oil and grease is crucial process and can affect the subsequent processing unit. The technology to be designed to remove oil and grease must be considered carefully. The most effective and efficient processing unit is obtained so that the oil and grease content will not interfere with the following unit's performance. Three alternative series of produced water treatment units with variations in primary treatment are shown in Fig. 1.

The selection of alternative processing units is carried out using the Analytical Hierarchical Process (AHP) method. The AHP method is quite commonly used to determine the best wastewater treatment unit. There are five criteria for selecting alternative production water treatment units. The evaluation criteria include oil and grease removal efficiency, COD removal efficiency, operation and maintenance (O&M), land requirements, and the initial cost of processing unit materials. According to PP, if the produced water is will be used as plant water, the produced water must meet the Class III water quality standard. 82 of 2001, which regulates the parameters of oil, grease, and COD. Therefore, the efficiency of removing oil, grease, and COD is used as a criterion for consideration because it has not met the specified quality standards. It is hoped that the processing unit will set aside these two parameters. Operation and maintenance (O&M) is directly proportional to the human resources required. The frequency of cleaning and maintaining the unit and the costs to pay

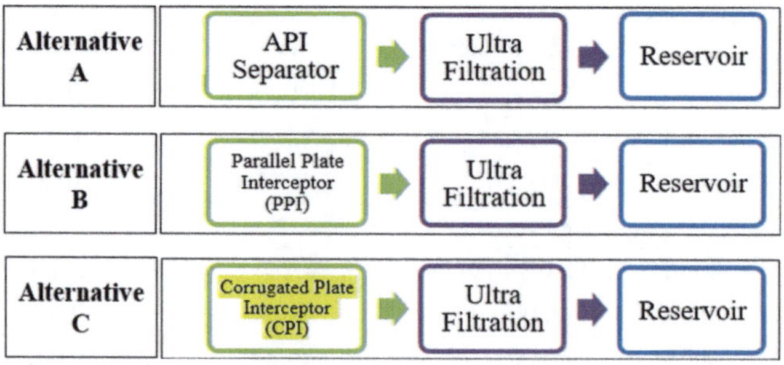

Fig. 1 Schematic diagram of produced wastewater treatment

for staff and experts can be reduced. Land requirements are considered so that the unit wastewater treatment plant can be built in a predetermined vacant area and minimize construction costs. Cost criteria also need to be considered so that costs can be reduced. The smaller the unit material cost, the more profitable the company will be.

The efficiency criteria for removing oil and grease were compared with the removal efficiency of COD. The efficiency of removing oil and grease is considered very important compared to removing COD so that it is given a value of 7. COD concentration can be removed in the subsequent processing unit, namely ultrafiltration membrane technology. Membrane technology can only remove emulsified oil, because free-oil type can damage membranes and interfere with the processing process. Therefore, the concentration of oil and grease needs to be released with as best as possible to not interfere with the membrane process and cause clogging. Based on these reasons, the comparison value of COD efficiency criteria to oil and grease efficiency is the opposite of 7, 1/7. Oil and grease removal efficiency criteria were compared with O&M criteria. The efficiency of oil and grease removal is more important than operation and maintenance. It has a score of 5. the efficiency of removing oil and grease affects the unit's effluent quality. If the removal's efficiency is small, the oil and grease concentrations cannot meet the class III water quality standards, so that the produced water cannot be used as plant flush. On the other hand, O&M can still be overcome by assigning experts to handle it if maintenance and operations are difficult or quite complicated. The value for O&M criteria compared to oil and grease efficiency is the opposite of 5, which is 1/5. The criteria for the efficiency of oil and grease removal compared to land requirements criteria. Compared to land requirements, the removal efficiency of oil and grease is more critical so that the value is 5. The removal efficiency of oil and grease is mandatory because the main function of the unit is intended to set aside oil and grease. The available vacant land at the Lindai GS location is still sufficient to build a processing unit with a large land area. Based on this, the land requirement criteria' comparison value against the oil and grease removal efficiency criteria is 1/5.

The criteria for oil and grease removal efficiency are compared with unit material costs criteria. The costs incurred to build a treatment unit will be replaced by the cost of using raw water from produced water. Thus, compared with the cost criteria, oil and grease removal efficiency is critical, because the concentration of oil and grease have to meet class III water quality standard, which limited to a maximum of 1 mg/L. The treatment unit selected must ensure that it can produce effluent less than 1 mg/L. Therefore, the comparison value of the oil and grease efficiency criteria against the unit material cost is given a value of 7. The ratio of the cost criteria to oil and grease efficiency is the opposite, which is 1/7. The COD removal efficiency criteria were compared with the O&M criteria. COD efficiency does affect the quality of produced water effluent to be used as plant flush. Still, COD concentration is not a priority in selecting the primary processing unit because it can be removed in the subsequent processing unit. The comparison value of COD removal efficiency criteria against O&M criteria is given a value, 1/5. The O&M criterion is crucial compared to COD efficiency so that the comparison value of O&M to COD efficiency is 5.

Efficiency criteria for COD removal compared to the requirements for land requirements the criteria for the land requirement are moderate or slightly more critical than COD removal efficiency requirements. COD removal will be focused on the subsequent processing unit, namely the ultrafiltration membrane. Ultrafiltration membrane technology is well-known as a technology with high removal efficiency, so the value of the COD removal efficiency criterion against the land requirement criteria is 1/3. The COD removal efficiency criteria were compared with the unit material cost criteria. Compared to the cost criteria, the efficiency of COD removal is assigned a score of 3 because it is slightly more critical. This is because the company prioritizes the quality of produced water to be used later than the processing unit's cost. Besides, the water used for watering the plants will be replaced with water from this processing to make savings in the future. Therefore, the comparison value of cost criteria to COD removal efficiency is 1/3. O&M criteria compared to land requirements. Operation and maintenance are slightly more important than land because operation and maintenance activities are carried out continuously as long as the processing unit is used. The comparison value of the O&M criteria to land requirements is 3. The comparison value of the land requirement criteria against O&M is 1/3. O&M criteria compared to unit material costs. Cost is not a problem for the company. Still, the availability of human resources is quite limited. The company needs to increase workers' time or recruit operators or experts to operate and maintain the processing unit. Therefore, the O&M criterion was slightly more critical than the cost criterion and got a score of Criteria for land requirements compared to unit material costs. The criteria for land needs are considered to be slightly more critical than unit costs and get a value of 3. This is because the large land area will also affect costs, such as clearing the area. Therefore, a unit that requires an area that is not too large will save on company costs. The comparison value between the material cost criteria and the land requirement criteria is 1/3. After conducting a comparative assessment, calculations are carried out according to the AHP method to determine the eigenvector value. The eigenvector value is directly proportional to the priority level of the criteria. Eigenvector values for each measure are listed in Table 2. The comparison assessment of each alternative with each criterion is based on the criteria data for each alternative listed in Table 3.

The assessment of the comparison of each alternative with each criterion is based on the criteria data for each alternative listed in Table 3. The selected processing alternative is the processing unit with the highest weight value than other alternatives,

Table 2 Eigen vector each criteria

No	Criteria	Eigenvector
1	Efficiency of removal of oil and grease	0.54
2	COD removal efficiency	0.04
3	Operations and maintenance (O&M)	0.21
4	Land requirements	0.13
5	Material cost unit	0.08

Table 3 Alternative calculation results against criteria

Alternative	Criteria					Total
	Efficiency of removal of oil and grease	COD removal efficiency	Operations and maintenance (O&M)	Land requirements	Material cost unit	
API	0.07	0.09	0.65	0.07	0.08	0.19
CPI	**0.24**	**0.22**	**0.23**	**0.77**	**0.19**	**0.51**
PPI	0.69	0.69	0.12	0.16	0.72	0.30

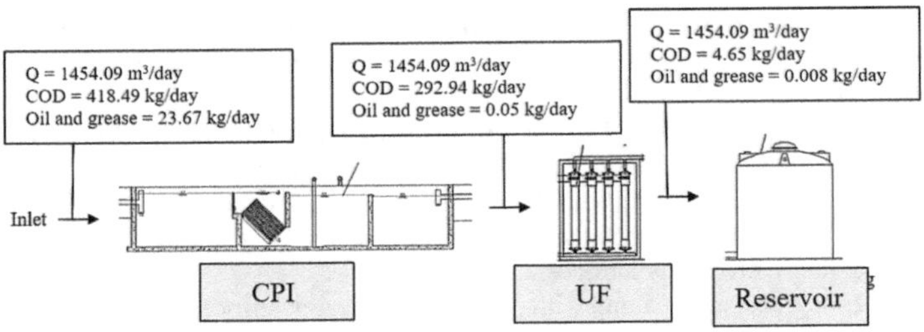

Fig. 2 Schematic diagram of the mass balance of selected produced wastewater treatment

namely CPI, with a weight of 0.51 or 51%. Thus, based on the AHP analysis results, a CPI unit will be built at the Lindai GS location to treat produced water into plant water. After processing in the CPI unit, the produced water will be reprocessed using ultrafiltration membrane technology (Fig. 2).

4 Conclusion

Parameters that must be eliminated in water reclamation are COD along with high oil and grease. A process hierarchy analysis for a suitable treatment system is prescribed consisting of a CPI unit, UF, and reservoirs. Reclamation of produced water from upstream activities in the oil and gas industry is carried out using a CPI unit and Ultrafiltration that selected based on the Analytical Hierarchy Process (AHP) method to eliminate high COD and oil and grease content from the water.

References

1. Hayes (2010) Produced water management: challenges and solutions. E&P Center Gas Technology Institute.

2. Astono W, Suswantoro E (2016) The analysis of oil and grease distribution and the evaluation of produced water process unit on Kf star energy platform (Kakap) LTD. Indonesian J Urban Environ Technol 6(5):123–131
3. Tiana AN (2015) Air Terproduksi: Karakteristik dan Dampaknya Terhadap Lingkungan. Fakultas Teknologi Industri, Institut Teknologi Bandung 1–11
4. Ivory D (2015) The prospect of utilization produced water. Teknik Kimia Institut Teknologi Bandung, Bandung
5. Arthur JD, Langhus BG, Patel C (2005) Technical summary of oil and gas produced water treatment technologies. Gulf Professional Publishing, Tulsa
6. Forero JE, Ortíz OP, Nariño FA, Díaz J, Peña H (2008) Design and development of a high efficiency tank for crude oil dehydration (I), CT&F-Ciencia. Tecnología y Futuro 3(4):185–199
7. Arnold K, Stewart M (2011) Produced water treatment field manual (1st edn.) Gulf Professional Publishing
8. Tellez GT, Nirmalakhandan N, Gardea-Torresdey JL (2005) Kinetic evaluation of a field-scale activated sludge system for removing petroleum hydrocarbons from oilfield-produced water. Environ Prog 24(1):96–104
9. Camarillo MK, Stringfellow WT (2018) Biological treatment of oil and gas produced water: a review and meta-analysis. Clean Technol Environ Policy 20(6):1127–1146
10. Ramadan HS, Moersidik SS, Priadi CR (2014) Selection of wastewater treatment plant effluent recycling unit at Muara Karang steam power plant with analytical hierarchy process method. Fakultas Teknik, Universitas Indonesia, Jakarta
11. Saaty TL (2008) Decision making with the analytic hierarchy process. Int J Serv Sci 1(1):83–98
12. Suryawan IWK, Helmy Q, Notodarmojo S (2020) Laboratory scale ozone-based post-treatment from textile wastewater treatment plant effluent for water reuse. J Phys Conf Ser 1456(1):012002 (IOP Publishing)
13. Suryawan IWK, Septiariva IY, Helmy Q, Notodarmojo S, Wulandari M, Sari NK, Sarwono A, Pratiwi R, Jun-Wei L (2021) Comparison of ozone pre-treatment and post-treatment hybrid with moving bed biofilm reactor in removal of Remazol Black 5. Int J Technol 12(2)
14. Apritama MR, Suryawan I, Afifah AS, Septiariva IY (2020) Phytoremediation of effluent textile wwtp for NH3-N and Cu reduction using pistia stratiotes. Plant Arch 20(1):2384–2388
15. Suryawan IWK, Prajati G, Afifah AS, Apritama MR (2021) NH_3-N and COD reduction in Endek (Balinese textile) wastewater by activated sludge under different DO condition with ozone pretreatment. Walailak J Sci Technol 18(6):1–11

Preference of Sludge Treatment Plan in IPA II Pejompongan Water Treatment Plant

Nailatul Fadhilah, Nurulbaiti Listyendah Zahra, Fatimah Dinan Qonitan, Imroatus Sholikhah, Intan Rahmalia, Iva Yenis Septiariva, and I. Wayan Koko Suryawan

Abstract Water treatment is inseparable from the processing of sludge resulting from the deposition process. IPA II Pejompongan currently has no sludge treatment, where the treated sludge is discharged directly into the surrounding water bodies. This research was conducted to determine the suitable sludge treatment system for existing conditions. This study was conducted using data on the quantity of sludge and the quality of the resulting sludge, which was then discussed descriptively. This study also uses a literature review to find various appropriate treatments for the Pejompongan II IPA. The alternative is selected with the Technique for Order Preference by Similarity to Ideal Solution (TOPSIS) method. The highest mud generation in 2019 is during the rainy season, namely January-June. In terms of quality, the parameters of suspended solids, iron, nickel, lead, manganese, organic ($KMnO_4$), BOD, and COD do not meet Jakarta City's quality standards government. There are two alternatives: the belt filter press and the screw press, where the decision analysis results show a better screw press.

Keywords Sludge treatment · Water treatment · Decision analysis

1 Background

The sludge from the IPA II Pejompongan, the water treatment plant, is currently directly discharged into surface water, namely the Krukut River, without any treatment process. Discharge of mud into rivers can damage the river's aesthetics and endanger the aquatic biota in it. The release of sludge into rivers can increase water bodies' turbidity, accumulate sludge, and ultimately lead to rivers' silting. Increased turbidity in water bodies will interfere with sunlight's penetration, which will then

N. Fadhilah · N. L. Zahra · F. D. Qonitan · I. Sholikhah · I. Rahmalia · I. W. K. Suryawan (✉)
Department of Environmental Engineering, Faculty of Infrastructure Planning, Universitas Pertamina, Jakarta 12220, Indonesia

I. Y. Septiariva
Department of Civil Engineering, Faculty of Engineering, Universitas Sebelas Maret, Surakarta 57126, Indonesia

S. A. Kristiawan et al. (eds.), *Proceedings of the 5th International Conference on Rehabilitation and Maintenance in Civil Engineering*, Lecture Notes in Civil Engineering 225,
https://doi.org/10.1007/978-981-16-9348-9_90

block the photosynthesis process and disrupt aquatic biota's life [1]. Research conducted by Pradel et al., states that direct disposal of sludge into water bodies has the potential to cause eutrophication due to the nitrogen content in the sludge [2]. The Krukut River is a river of concern to the DKI Jakarta Provincial Government because the current depth of the Krukut River is less than 3 m, while the ideal depth for the river is around 8 m. From this explanation, it can be stated that IPA II Pejompongan has an urgency in building a sludge treatment plant to comply with applicable regulations and maintain environmental quality so as not to be disturbed.

Preventing environmental pollution can be done by adequately treating waste by considering various existing technologies even to waste to energy [3]. The benefit of the study of alternative design is to provide input to IPA II Pejompongan regarding the sludge treatment plant, superannuation management, and solid cake that can be carried out in the future. Suppose the design of the sludge treatment plant is applied. In that case, it can comply with Government Regulation Number 16 of 2005, which states that the water treatment plant's residue must be treated first before being discharged into the environment. The purpose of this study was to determine the existing conditions of sludge treatment at IPA II Pejompongan and to determine the appropriate alternative for sludge treatment based on the characteristics of the sludge in IPA II Pejompongan. By carrying out a study on selecting the best alternative for sludge treatment in IPA II Pejompongan, it is hoped that it will become the basis of reference in determining the sludge classification unit to then design the sludge treatment plant.

2 Method

2.1 *Location and Sampling Period*

This study was conducted at IPA II Pejompongan, located in Central Jakarta. IPA II Pejompongan can implement the drinking water supply system in the western part of DKI Jakarta. The data used in this study is secondary data during January–December 2019.

2.2 *Data Collection*

In the design of the sludge treatment plant, data on the quantity of sludge is needed to determine the sludge treatment plant's design capacity. The sludge quantity data comes from the sludge discharge calculation at the sludge discharge valve. The sludge discharge value will depend on the interval and duration of the sludge discharge valve opening. Determination of the density value of mud and sludge solids content was carried out by sampling in the field and continued testing at the IPA II Pejompongan Science Laboratory.

2.3 *Selection of Alternative Processing*

Determination of design alternatives is carried out using the Technique for Order Preference by Similarity to Ideal Solution (TOPSIS) method. The TOPSIS method can determine the best alternative from several alternative solutions with various criteria [4]. The selection of alternatives to the TOPSIS method is carried out using the Euclidean distance associated with the Pythagorean Theorem. Determining the ideal solution matrix is carried out to determine each alternative's distance with a positive ideal solution (Si+) and a negative ideal solution (Si−). The following are the equations used to determine positive and negative ideal solutions:

$$\mathrm{Si}+ = \left(\sum \mathrm{j} = \mathrm{n}1(\mathrm{vij} - \mathrm{vj}+)2\right)1/2 \quad (1)$$

$$\mathrm{Si}- = \left(\sum \mathrm{j} = \mathrm{n}1(\mathrm{vij}-\mathrm{vj}-)2\right)1/2 \quad (2)$$

With: Si+ = distance to positive ideal solution; Si− = distance to negative ideal solution; vj+ = best weighted normalized data for each criterion; vj− = worst weighted normalized data for each criterion.

The preference value states the alternative distance to the positive and negative ideal solution. The greater the preference value, the smaller the distance to the positive ideal solution and the further away distance to the negative ideal solution. The value of greatest preference will be the chosen alternative. The following is the equation in determining the preference value:

$$\mathrm{P} = (\mathrm{Si}- /(\mathrm{Si}- + \mathrm{Si}+) \quad (3)$$

With P = preference value; Si+ = positive ideal solution; Si− = negative ideal solution.

3 Result and Discussion

3.1 *Sludge Generation*

The sludge produced from IPA II Pejompongan comes from the pre-sedimentation unit, pulsator, and backwash filter processes. The sludge that will be treated in the sludge treatment plant is only the sludge from the pre-sedimentation unit and the

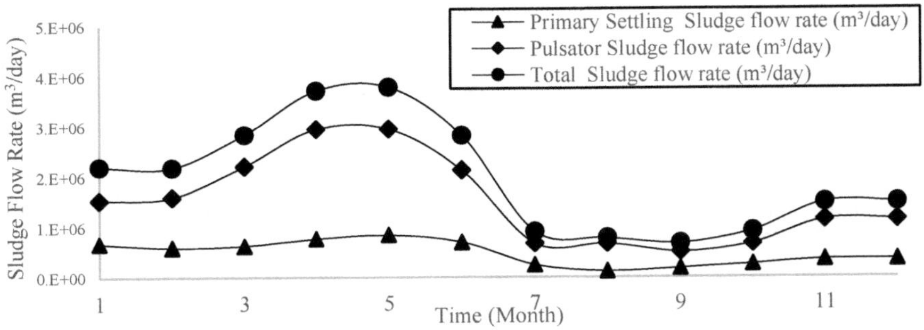

Fig. 1 Sludge flowrate produced at IPA II Pejompongan from January (1)–December (12) 2019

pulsator, while the sludge from the backwash filter process will be directly flowed to the pre-sedimentation unit to be reprocessed. The sludge from the pre-sedimentation unit and pulsator will flow to the sludge drain through the sludge discharge valve. The longer the duration of sludge disposal, the greater the daily sludge discharge produced. Interval, period, and monthly average sludge discharge data through the sludge discharge valve are shown in Fig. 1.

From Fig. 1, the most significant sludge discharge comes from the pulsator unit because the most significant allowance for suspended solids is in the pulsator unit. In January–June, the resulting mudflow is more important than in July–December. This was due to an increase in suspended solids in January–June. The increase in suspended solids was caused by the rainy season during January–June. Rainwater carries a high content of suspended solids due to soil erosion. Thus, there is an increase in turbidity during the rainy season [5].

3.2 Sludge Quality

The sludge sampling method is area-integrated, that is, by combining the samples taken from the pre-sedimentation valve and the pulsator. This sampling is carried out to determine the samples' physical quality that enter the processing plant and calculate the dimensions of the sludge treatment plant. The quality of the combined sludge of the pre-sedimentation unit and the pulsator, testing the sludge's quality is also carried out in each unit. The sludge sampling method is also carried out by using the area-integrated sampling method, which is by combining the samples taken from each MPA pre-sedimentation unit or pulsator. Sampling was carried out to calculate the mass balance of the resulting sludge. Sludge density data, sludge solids content, and contribution to total sludge are shown in Table 1.

The sludge tested came from the sludge drainage channel, a combined channel between the sludge from the pre-sedimentation unit and the pulsator. Routine sludge quality testing is carried out to compare the quality of the IPA II Pejompongan sludge with the quality standards stipulated in the Regulation of the Governor of

Table 1 Characteristics of sludge produced in the primary settling unit and pulsator at IPA II Pejompongan

Parameters	Primary settling	Pulsator	Combined
Density of sludge (kg/m^3)	1004	1013	1011
Solid level	0.30%	2.42%	1.92%
Contribution to total sludge at IPA II Pejompongan	24%	76%	100%

the Special Capital Region of Jakarta Number 69 of 2013. The average value of the results of testing the IPA II Pejompongan sludge parameters is shown in Table 2. Several parameters exceed the quality standards of the Regulation of the Governor of the Special Capital Region of Jakarta Province Number 69 of 2013 as effluent standards, namely parameters of suspended solids, iron, nickel, lead, manganese, organic (KMnO4), BOD, and COD. The value of suspended solids (TSS) in the sludge exceeds the quality standard because the pre-sedimentation and sedimentation processes aim to remove TSS from raw water. Thus, the resulting sludge will contain high TSS. The high TSS content contributes to increasing the BOD and COD content in waste [6]. The high levels of iron, nickel, lead, and manganese in the sludge are caused by the raw water's quality containing these metals. The metal contained in sludge is generally in the form of metal hydroxide (MOH), insoluble in water. In the process of clarification, these metals will experience precipitation and residue together with the floc. Thus, the resulting sludge contains high metal content. Based on Table 1, the value of sludge-specific gravity produced by IPA II Pejompongan is greater than the value of water-specific gravity. Thus, the separation of solids and water in the sludge by gravity can be carried out. Based on Table 2. The comparison of BOD and COD values in the sludge is very low, namely 0.135. The value of the BOD/COD ratio for processing with a biological treatment unit more than 0.31 [7–10]. Thus, the sludge produced by IPA II Pejompongan cannot be treated biologically and cannot be discharged to a wastewater treatment plant that uses a biological treatment unit.

3.3 *Existing Mass Balance*

The sludge volume balance describes the volume flow of sludge into and out of the treatment unit in units of m^3/day. Solid mass balance states the amount of solid that enters and leaves a processing unit in units of kg/day. The mass flow in the pre-sedimentation unit and pulsator represents the number of solids suspended in the sludge flow. The sludge mass and volume flow diagrams are prepared to determine which units produce the most significant sludge volume. The flow of the average volume of existing sludge at IPA II Pejompongan is shown in Fig. 2.

In general, the sludge treatment plant consists of collecting, thickening, conditioning, dewatering, and final disposal units. The thickening unit for all alternatives

Table 2 Comparison of sludge quality with standards in Jakarta city

No	Parameters	Unit	Standard*	Concentration (Average)
1	Temperature	°C	38	25.933
2	TDS	mg/L	1000	186
3	TSS	mg/L	100	8337.5*
4	Hg	mg/L	0.02	0.003
5	NH_3	mg/L	5	0.242
6	As	mg/L	0.1	0
7	Fe	mg/L	5	154.71*
8	F–	mg/L	2	0.315
9	Cd	mg/L	0.05	<0.01
10	Free chlorine	mg/L	1	0.142
11	Chrome (total)	mg/L	0.5	0.075
12	Hexavalent chrome	mg/L	0.1	<0.003
13	Ni	mg/L	0.1	0.158*
14	Nitrate	mg/L	10	8.072
15	Nitrite	mg/L	1	0.0425
16	pH	–	6–9	7.033
17	Zn	mg/L	2	6.475*
18	Sulfide	mg/L	0.05	0.0097
19	Copper	mg/L	1	0.35
20	Pb	mg/L	0.1	0.457*
21	Mn	mg/L	2	25.857*
22	Phenol	mg/L	0.5	0.0132
23	Oil and Grace	mg/L	5	1.58
24	Methylene blue active compound	mg/L	1	0.13
26	Organic ($KMnO_4$)	mg/L	85	241.77*
27	BOD_5	mg/L	75	92.417*
28	COD	mg/L	100	682.5*

* Regulation of the Governor of the special capital region of Jakarta province Number 69 of 2013

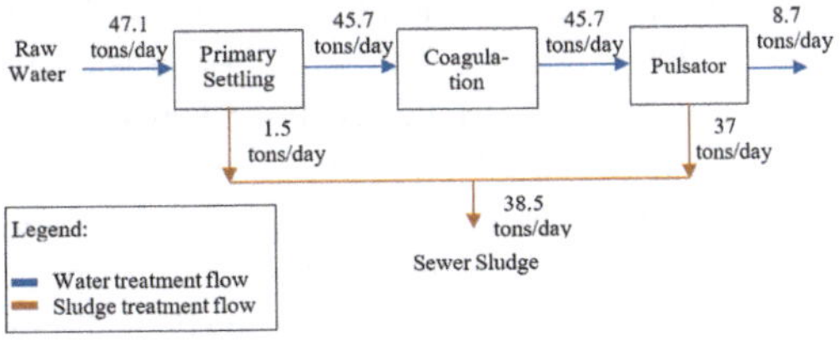

Fig. 2 Mass balance of sludge production from existing water treatment at IPA II Pejompongan

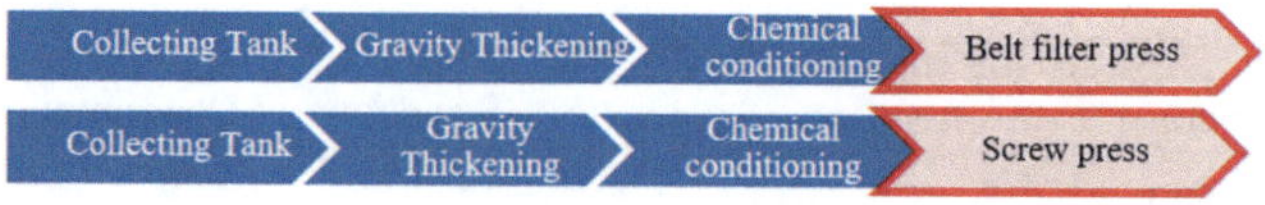

Fig. 3 Schematic diagram of alternative sludge processing with belt filter press and screw press

is gravity thickening because the sludge's quality is still possible to use this unit. Thus, alum sludge is easier to settle with a gravity thickening unit than to float with a dissolved air flotation thickening unit. Besides, mechanical thickening, such as gravity belt thickening, is rarely used in water treatment plant sludge because it requires much energy and chemicals to treat sludge in water treatment plants with large discharges. Conditioning is necessary when using a mechanical dewatering treatment unit to increase the compressibility of the sludge. The type of conditioning commonly used in water treatment plants is chemical conditioning with polyelectrolytes or polymers because it does not produce additional sludge. The use of inorganic chemical conditioning reagents produces additional sludge, namely metal hihydroxides and sludge from lime [11]. Dewatering is the final process before the waste is discharged into the environment. The dewatering unit consists of mechanical and non-mechanical units. The mechanical units commonly used are the belt filter press, centrifuge, screw press, and pressure filter press. In selection alternative sludge treatment plants, the parameters that must be met are land area and electrical energy requirements. The available land area is 1300 m^2 and the maximum electrical energy that can be used for a sludge treatment plant is 100 kW. Based on these four mechanical units, the units that require more than 100 kW of energy are the centrifugal and pressure filter press unit, so that the alternative options to be considered are the belt filter press and screw press (Fig. 3).

The land area criterion has a weight value of 1 because the land area will not be related to future operations and costs. In the land area criteria, the smaller land area required by an alternative, the closer the alternative is to a positive ideal solution. The alternative screw press requires a smaller space than the alternative belt filter press. The TSS value will affect compliance with quality standards. Based on the Regulation of the Governor of the Special Capital Region of Jakarta Number 69 of 2013 concerning Wastewater Quality Standards for Activities and/or Businesses, the maximum total suspended solids that can be discharged into water bodies is 100 mg/L. Supernatant TSS values obtained for alternatives 1 and 3 have met the quality standard. The smaller the supernatant TSS value of an alternative, the closer an option is to the positive ideal solution. Alternative belt filter press produces smaller supernatant TSS than screw press. The belt filter press unit requires regular maintenance, especially for belt cleaning. The belt filter press unit has an automatic belt cleaning system using water with a minimum pressure of 6 kg/cm^2 [12]. The belt on the belt filter press unit needs to be replaced periodically [13]. In the screw press unit, the fixed and moving rings' gap has an automatic cleaning system using a nozzle located above the ring. Things that need to be considered in cleaning are the interval and duration of cleaning [14]. The easier the treatment of an alternative, the closer the alternative is to a positive ideal solution.

The alternative screw press is easier to maintain than alternative belt filter press. The potential odor and noise criteria have a weighted value of 1 because they will not be related to future operations and costs. The belt filter press unit's noise level without a cover is ±80 dB [15]. Meanwhile, the screw press unit's noise level without a cover is ±70 dB [16]. The belt filter press can have high odor and noise because it is designed to be open [12]. Alternative screw press has a lower odor and noise potential than alternative belt filter press. From the criteria data for each alternative and the weighted value of each criterion, the normalized and weighted data are determined using Eqs. 1 and 2. Normalized and weighted data are shown in Table 3. The value of V+ represents the best value of normalized and weighted data, while V− represents the worst value of normalized and weighted data. From the weighted normalization data, then a positive ideal solution (Si+) and a negative ideal solution (Si−) are determined using Eqs. 1 and 2. A preference value (Pi) is selected, which states the proximity of the alternative to ideal conditions using Eq. 3.

The alternative with the highest preference value (Pi) will be the best design alternative applied in IPA II Pejompongan is screw press. Screw press was chosen because it is closer to the positive ideal solution than the belt filter press. In addition, screw press is also further away from negative ideal solution than belt filter press. With the sludge treatment at IPA II Pejempongan, a continuous process is needed to achieve zero waste. Although the screw press requires higher energy, the energy recovery of the mud can help supply energy to the IPA II Pejempongan. Determination

Table 3 Comparison of sludge quality with standards in Jakarta city

Criteria	Belt filter press	Screw press	(V+)	(V−)
Land requirements	0.0322	0.0320	0.0320	0.0322
Dryness cake solid	0.0650	0.0636	0.0650	0.0636
Volume cake solid (m^3/day)	0.0955	0.0973	0.0955	0.0973
Massa cake solid (kg/day)	0.0645	0.0641	0.0641	0.0645
TSS supernatant	0.0169	0.0422	0.0169	0.0422
Energy requirements	0.1237	0.0575	0.0575	0.1237
Chemical requirements	0.1169	0.0702	0.0702	0.1169
Operation and maintenance	0.0407	0.0813	0.0813	0.0407
Order impact	0.0407	0.0203	0.0203	0.0407
Noise	0.0407	0.0203	0.0203	0.0407
Budget	0.0410	0.1301	0.0410	0.1301
Si+	0.0951	0.0926		
Si−	0.0926	0.0951		
Pi	**0.4935**	**0.5065**		
Ranking	2	1		

of the energy potential of solid waste can be done by knowing the application that will be applied in the field [16–18].

4 Conclusion

The best alternative selection analysis results show that the screw press can be used as sludge treatment in IPA II Pejompongan. Before using the screw press, pretreatment in the form of gravity thickening and chemical conditioning is required.

References

1. Bilotta GS, Brazier RE (2008) Understanding the influence of suspended solids on water quality and aquatic biota. Water Res 42(12):2849–2861
2. Pradel M, Reverdy AL, Richard M, Chabat L (2014) Environmental impacts of sewage sludge treatment and disposal routes: a life cycle assessment perspective. In: 4th European conference on sludge management, vol 8
3. Afifah AS, Suryawan IWK, Sarwono A (2020) Microalgae production using photo-bioreactor with intermittent aeration for municipal wastewater substrate and nutrient removal. Commun Sci Technol 5(2):107–111
4. Roszkowska E (2011) Multi-criteria decision-making models by applying the TOPSIS method to crisp and interval data. Multiple Criteria Decis Mak/Univ Econ Katowice 6(1):200–230
5. Saputra DGTB, Arthana IW, Pratiwi MA (2016) Analisis Kualitas Fisika Perairan Berdasarkan Nilai Padatan Tersuspensi dan Kekeruhan Perairan di Bendungan Telaga Tunjung Desa Timpag, Kecamatan Kerambitan, Kabupaten Tabanan-Bali. ECOTROPHIC 10(2):130–136
6. Aygun A, Nas B, Berktay A (2008) Influence of high organic loading rates on COD removal and sludge production in moving bed biofilm reactor. Environ Eng Sci 25(9):1311–1316
7. Wang XJ, Xia SQ, Chen L, Zhao JF, Renault NJ, Chovelon JM (2006) Nutrients removal from municipal wastewater by chemical precipitation in a moving bed biofilm reactor. Process Biochem 41(4):824–828
8. Suryawan I, Koko W, Rahman A, Lim JW, Helmy Q (2021) Environmental impact of municipal wastewater management based on analysis of life cycle assessment in Denpasar City. Desalin. Water Treat. 244:55–62
9. Suryawan IWK, Septiariva IY, Helmy Q, Notodarmojo S, Wulandari M, Sari NK, Sarwono A, Pratiwi R, Jun-Wei L (2021) Comparison of ozone pre-treatment and post-treatment hybrid with moving bed biofilm reactor in removal of Remazol Black 5. Int J Technol 12(2)
10. Suryawan IWK, Prajati G, Afifah AS, Apritama MR (2021) NH_3-N and COD reduction in Endek (Balinese textile) wastewater by activated sludge under different DO condition with ozone pretreatment. Walailak J Sci Technol (WJST) 18(6)
11. Turovskiy IS, Mathai PK (2006) Wastewater sludge processing. Wiley
12. Andreoli CV, von Sperling M, Fernandes F (2007) Sludge treatment and disposal. IWA Publishing
13. EPA U (2000) Biosolids technology fact sheet–belt filter press. EPA, Washington D.C.
14. Mumbi AW, Fengting L, Mwarania F, Uuganchimeg B (2017) An assessment of multiplate screw press in Dewatering process of sludge treatment. Int J Adv Res 740–747
15. Aventura (2011) Drum thickener belt-type filter press. Aventura, New Delhi
16. Macomber S (2016) Huber's New Q-Press. Huber, Huntersville

17. Suryawan IWK, Sarwono A, Septiariva IY, Lee CH (2021) Evaluating marine debris trends and the potential of incineration in the context of the COVID-19 Pandemic in Southern Bali, Indonesia. Jurnal Ilmiah Perikanan dan Kelautan 13(2)
18. Qonitan FD, Suryawan IWK, Rahman A (2021) Overview of municipal solid waste generation and energy utilization potential in major cities of Indonesia. J Phys Conf Ser 1858
19. Sarwono A, Septiariva IY, Qonitan FD, Zahra NL, Sari NK, Fauziah EN, Ummatin KK, Amoa Q, Faria N, Lim JW, Suryawan IWK (2021) Municipal solid waste treatment for energy recovery through thermal waste-to-energy in Depok City, Indonesia. J Adv Res Fluid Mech Thermal Sci 85

Design of Typical Rainwater Harvesting Storage Tanks Based on Housing Type (Case Study in Indonesia)

Lina Indawati, Setyo Budi Kurniawan, Siti Rozaimah Sheikh Abdullah, and Raden Harya Dananjaya

Abstract Indonesia, a tropical country, has high rainfall intensity and has an opportunity to harvest rainwater as a daily water supply. However, most Indonesians still rely on groundwater or tap water as their primary water source. Rainwater harvesting is a feasible technology to be applied to decrease the dependency on the current water source. A simple rainwater harvesting using rooftop is considered to be applicable in a residential area. This study calculated the typical storage tank for rainwater harvesting based on the house type in Indonesia. Two scenarios were used during the calculation. In the first scenario, storage was designed to collect water during rainfall to be used during the rainy season only. In the second scenario, the tank was designed to collect rainwater to be used for the whole year. Calculation results showed the number of typical storage tank (∅ 1.8 m and height of 1.9 m) required for the tiny house type was 1 tank, the medium house type was 1–2 tanks, and the large house type was 3–6 tanks. In the second scenario, the number of storage tanks for each type of house was 25 tanks. The second scenario is suggested to be used in areas where water scarcity is the main issue during the dry season, while the first scenario is suggested to be applied for a user with no land availability and construction cost issues.

Keywords Design · Rainwater harvesting · Storage tank

1 Introduction

Some areas in Indonesia still utilize groundwater as their primary water supply [1]. The use of groundwater for a long time can result in the depletion of groundwater,

L. Indawati (✉) · R. H. Dananjaya
Department of Civil Engineering, Faculty of Engineering, Universitas Sebelas Maret, 57126 Surakarta, Indonesia
e-mail: linainda@staff.uns.ac.id

S. B. Kurniawan · S. R. S. Abdullah
Department of Chemical and Process Engineering, Faculty of Engineering and Built Environment, Universiti Kebangsaan Malaysia, UKM, 43600 Bangi, Selangor, Malaysia
e-mail: rozaimah@ukm.edu.my

S. A. Kristiawan et al. (eds.), *Proceedings of the 5th International Conference on Rehabilitation and Maintenance in Civil Engineering*, Lecture Notes in Civil Engineering 225,
https://doi.org/10.1007/978-981-16-9348-9_91

while it can also cause land subsidence due to the empty aquifer [2]. Meanwhile, a catchment of rainwater is also reduced due to the development of housing and industry. The reduced catchment of rainwater may cause the increase of runoff and the decrease of water infiltration, which contribute to the groundwater recharge [3]. The increase of rainwater runoff is the leading cause of flood in many areas, especially in the densely populated area in Indonesia.

Reducing the rainwater runoff can become a solution to prevent a flood. Flood reduction in urban area by rainwater harvesting can reach up to 28.66% [4]. One technology to utilize rainwater further, while also reducing runoff is rainwater harvesting using a storage tank [5]. Rainwater harvesting is a technology to catch rainwater, store it, and utilize it for daily needs [6]. A simple rainwater harvesting technology involves house roofs, soil surface, or rock as catchment unit, while storage might be carried out in reservoirs or ponds [7]. The utilization of harvested rainwater might go to the households need and watering/irrigational purposes to reduce the dependence on groundwater.

Most of areas in Indonesia have medium to high rainfall intensity, which become very suitable for applying rainwater harvesting technology. The harvested rainwater may be used during the dry season, which usually caused drought in several areas of Indonesia. Rainwater harvesting can be applied in the rainy and dry season. Moreover, the difference in rainfall intensity can have an impact on the harvested rainwater [8–10]. Harvesting can be considered effective if the rainfall is over the minimum threshold to hold reliable water consumption [11–13]. However, rainwater harvesting is currently not familiarized in Indonesia. Most housing in the residential areas still rely on groundwater and or tap water supply, while the application of simple rainwater harvesting using the rooftop is currently very limited. This phenomenon is occurred due to the limitation of the knowledge, especially in terms of the technical part for applying rainwater harvesting technology, including the minimum requirement for a storage tank. Based on the problems as mentioned above, the aim of this paper is to design a typical rainwater harvesting tank, which focused based on the house type commonly built in Indonesia. The result of this study is expected to contribute as a basic understanding to provide a typical tank design for the application of simple rainwater harvesting in a residential area. The obtained result may also be adapted outside Indonesia, especially in tropical countries with the similar season and rain intensity characteristics.

2 Research Methodology

In this study, an approach was carried out using 2 scenarios. The first scenario assumes that rainwater harvesting was used to fulfil the water supply in the rainy season only. The second scenario assumes that rainwater harvesting was used to fulfil the water supply in the rainy season and dry seasons. All data used in this study were collected from 131 rain station of 33 provinces in Indonesia. This research method started from literature study, data collection, data processing, and calculation of typical rainwater storage tanks.

2.1 Literature Study

Literature studies focused on scientific journals, which also sourced from websites that discuss rainwater harvesting design criteria, the application of rainwater harvesting technology in tropical and developing countries, rainwater intensity in Indonesia, and the typical water consumption in Indonesia.

2.2 Data Collection and Analysis

The data used in this research were secondary data such as statistical data on housing types in Indonesia, rainfall data, and water consumption data. The rainfall data in Indonesia was obtained from the Meteorology, Climatology, and Geophysics Council (BMKG). Housing data was obtained from Central Bureau of Statistics (BPS). The water consumption data was obtained from Local drinking water company (PDAM) and Minister for Public Works and Human Settlements (PUPR). This study observed rainfall data in February 2021 and rainfall data started from 1981 to 2010 (the average, minimum, and maximum). Then, those rainfall data were visualized and mapped with GIS.

The volume of the storage tank depends on runoff and water consumption. Determination of the storage tank volume was carried out based on the rainwater precipitation and rainwater as a water supply. Moreover, the period of collection was calculated based on the ordinarily rainy season in Indonesia. The following equations were used to calculate the required water balance.

Calculation of the roof area as harvesting area

$$A = \frac{Floor\ area}{cos(the\ degree\ of\ roof)} \quad (1)$$

$$A = \frac{Total\ volume\ of\ water\ consumption\ (V)}{R} \times 1000 \quad (2)$$

Calculation of water consumption

$$The\ total\ volume\ of\ water\ consumption\ (V) = Va + Vb \quad (3)$$

$$Va = \frac{n \times hh \times h}{1000} \quad (4)$$

$$Vb = \frac{n \times hk \times k}{1000} \quad (5)$$

where, n is the number of water users (person), hk is the number of days during the dry season (day), hh is the number of days during the wet season (day), k is the water consumption in the dry season (L/person/day), h is the water consumption in the rainy season (L/person/day), A is the roof area (m^2), Va is the volume of accumulated water consumed during the wet season (m^3), Vb is the volume of accumulated water consumed during the dry season (m^3), and R is the precipitation during the rainy season (mm).

Calculation of runoff flowrate, rainwater harvesting quantity, and spill on tth day [14].

Runoff flow rate

$$Qt = I_{eff} x\, Cr\, x A \quad (6)$$

Rainwater harvesting quantity

$$St = S_{t+1} + Qt - Dt, 0 \leq S_{t+1} \leq C \quad (7)$$

Spill on t^{th} day

$$Spill\ on\ t^{th}\ \text{day} = S_{t+1} - C \quad (8)$$

where, I_{eff} is the effective daily rainfall intensity–first flush; with the first flush value of 0.33 mm, S_{t+1} is the storage volume at the end of the rainy period, St is the storage value at the beginning of the rainy period, Qt is a runoff from the roof, Dt is the total demand for water supply in the rainy period, C is the trawling capacity that can be assumed based on the available size in the market, and Cr is a runoff coefficient [15].

3 Result and Discussion

3.1 Typical Houses in Indonesia Based on Floor Area

A typical house in Indonesia can be seen from the floor area of the building. According to BPS 2010 statistical data, 56.98% of Indonesia's population owns a house with a floor area of more than 13 m^2 [16]. Most areas in Indonesia, such as Yogyakarta, classify the types of houses based on the middle to upper economic class. Upper-class house type has an area of 169–216 m^2. Middle-class house types have a floor area of

73–120 m^2 [17]. Other city such as Palembang City has 4 types of houses, namely type A with a floor area of 72 m^2, type B with a floor area of 65 m^2, type C with a floor area of 85 m^2 and type D with a floor area of 49 m^2 [18]. In addition, houses on stilts in urban areas have 3 types, including small, medium and large types. The tiny type has a floor size of 30 m^2, the medium type has a medium size of 56 m^2, and the large type has a size of 359 m^2 [7]. In general, minimalist home designs in Indonesia have a floor area of 30–125 m^2 [19]. Based on the type of the floor area, the approximate roof area will be known. Based on typical house data, a typical rainwater harvesting storage tank will be classified based on its size. Based on the several obtained data, the house type was classified into three types, namely tiny, medium, and large types. A tiny rainwater harvesting storage tank was estimated from a floor area of 30–72 m^2, the medium size rainwater tank was determined from the floor area of 73–120 m^2, while the large size was calculated from the floor area of 169–359 m^2.

3.2 *Rainwater Quality in Indonesia*

Rainwater quality in 2015 in several areas of Indonesia such as Java, Sulawesi and Sumatra still met the quality standards (Table 1). However, based on BMKG data in March 2021, the rainwater quality in most parts of Indonesia tended to be acidic, with a pH range of 4.11–5.55 [20]. Only in some areas such as Beto Ambari, Bone, Branti, Gawpalu, Jalaludin, Lasiana, Baai Island, Siantan, Simpang Tiga, Sorong,

Table 1 Rainwater quality in Indonesia

Parameter	Unit	Rainwater quality (source: 22–25)					
		Malang	Residential area (Lampung)	Industrial area (Lampung)	Bagan Siapi (Riau)	Sulawesi	Standard (minister of health regulation no. 32 of 2017)
pH	–	7.4	7.31	6,72	8	5.3	6.5–9.0
Iron	mg/L	<0.26	0.392	0.157	0.04	–	0.3
Fluoride	mg/L	0.34	0.494	0.332	0.003	–	1.5
$CaCO_3$	mg/L	39.6	26	32	15.29	–	500
Mangan	mg/L	0.01	11	0.308	0.02	–	0.5
Nitrate	mg/L	0.53	0.095	0.359	0.376	0.41	10
Nitrite	mg/L	<0.001	0.005	0.011	0.0074	–	1
Cyanide	mg/L	–	–	–	–	–	0.1
Zinc	mg/L	–	4	1.174	–	–	15
Organic	mg/L	0.9	–	–	7.79	–	10
Sulphate	mg/L	–	–	–	–	1.34	400

Supadio had an ideal pH (5.6 ≤ pH ≤ 6.21). According to the Ministry of Health No. 32 of 2017, the water pH that can be used for sanitation purposes was around 6.5–8.5 [21]. Therefore, the operation of rainwater harvesting must be considered to maintain the quality of rainwater. Rainwater for the first 15 min is channeled as a daily flush so that the rainwater only harvested after the first 15 min of rain to maintain the rainwater quality.

3.3 Rainfall in Indonesia

Most area in Indonesia has a high peak rainfall, especially in Java, Southeast Nusa, Sulawesi, Papua, which the maximum rainfall exceed 500 mm/year (Fig. 1a). The minimum rainfall was around 50–100 mm/year in most area in Indonesia (Fig. 1b). Moreover, North Sumatra, North Sulawesi, Northern Papua, North Kalimantan, and a small part of East Kalimantan had a minimum rainfall of less than 50 mm/year. However, most area in Indonesia was still in the range of moderate category, and rainwater harvesting technology can still be used in minimum rainfall condition.

The average rainfall in some parts of Indonesia was classified as moderate to high. (Fig. 2a). In addition to data for a period of 10 years from 1987 to 2010 which was used to determine rainfall trends in Indonesia, the latest rainfall data was used as a comparison. In February 2021, most areas in Indonesia had moderate to high rainfall intensity except Central to North Sumatra which had low rainfall intensity (Fig. 2b).

3.4 Estimated Roof Area

There are three types of roofs based on the angle of inclination. First, the flat roof has a slope of <10°, second, sloping roofs have an angle of inclination between 10° and 30°, and the sloping roof has a slope above 30°. Most housings in Indonesia used the high inclination; thus a 30° was used to calculate the roof area in this study. The estimated roof area based on determined house type is summarized in Table 2.

3.5 Estimated Rainwater Flowrate for Harvesting

The flow rate of rainwater was estimated from the runoff that flows on the roof, so that, it can be obtained from processing rainfall data, Cr and roof area. The average monthly rainfall data in Indonesia is 246 mm/month, with the daily rainfall data is 8.2 mm/day [26]. Effective daily rainfall data was obtained from daily rainfall data minus 0.33 mm, where this value was assumed as rainwater mixed with pollutants

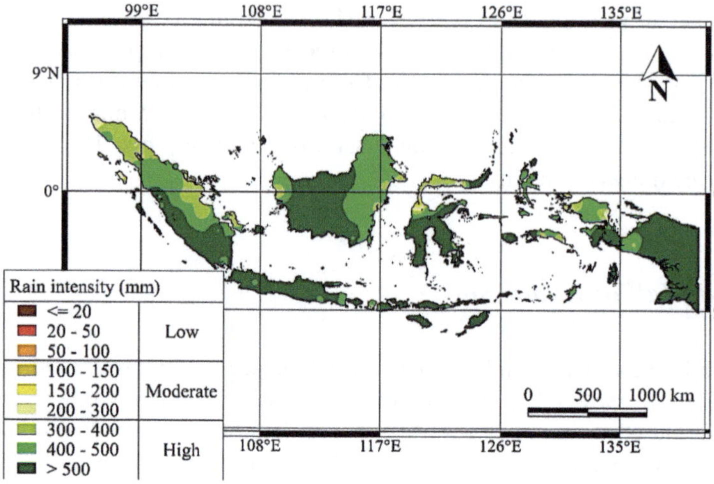

(a)

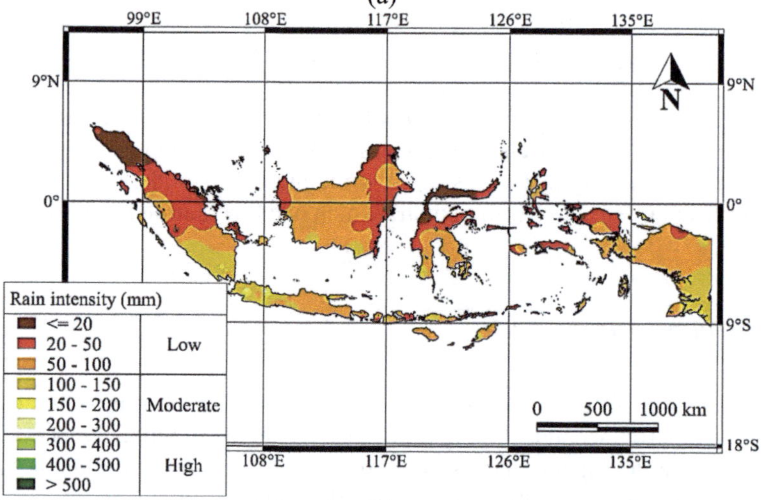

(b)

Fig. 1 Rainfall intensity in Indonesia mapped using GIS **a** maximum rainfall; **b** minimum rainfall [26]

(first rainfall till 15 min' rainfall) which will not be collected [27]. The runoff coefficient for the roof of the house is 75–95% [28]. Therefore, 80% was defined as the average runoff coefficient. Table 3 shows the calculation of rainwater flow rate.

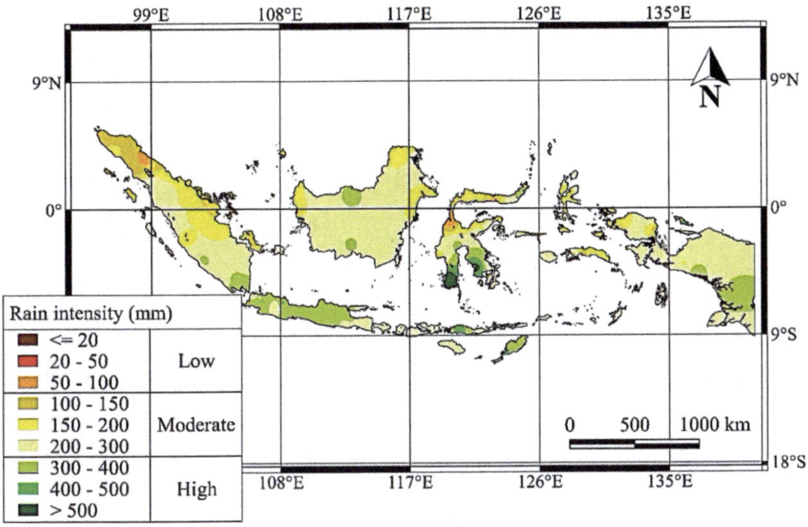

(a)

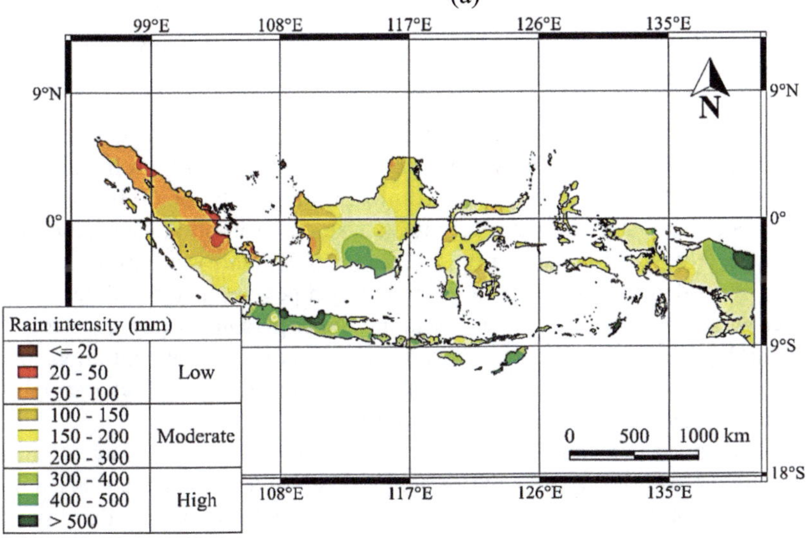

(b)

Fig. 2 Rainfall intensity in Indonesia mapped using GIS **a** average rainfall from 1987 to 2010; **b** average rainfall in 2021 [26]

3.6 Analysis of the Volume and Requirements of Rainwater Storage Tanks

The volume and requirement of rainwater storage tanks were calculated from the runoff flow rate, the water consumption in the rainy and the dry seasons, and

Table 2 Roof area each type of house

Type	Floor area (m^2)	Roof area (m^2)
Tiny	30	194.49
	73	473.25
Medium	73	473.25
	120	777.95
Large	169	1095.61
	359	2327.37

Table 3 Rainwater flowrate for harvesting

Type	Floor area (m^2)	Roof area (m^2)	Ieff (mm/day)	Cr	Q (m^3/day)
Tiny	30	194.49	7.88	0.8	1.23
	72	466.77	7.88	0.8	2.94
Medium	73	473.25	7.88	0.8	2.98
	120	777.95	7.88	0.8	4.91
Large	169	1095.61	7.88	0.8	6.91
	359	2327.37	7.88	0.8	14.68

the storage volume at the end of the rainy period (S_{t+1}). The final volume of the storage period was obtained from the difference between runoff discharge and water consumption in the rainy season. The results of the calculation (Table 3) show that the rainwater harvesting flowrate still meets water consumption for each unit of the house type. The water consumption for each housing unit was estimated from the number of residents (assumed to be occupied by 5 people), and the water consumption per person per day. The water consumption per person per day was set to 126.9 L/person/day [29–32]. In the first scenario, it was assumed that continuous use and continuous disposal of excess runoff or meeting the needs of water supply with rainwater harvesting during rainy season only.

The quantity of rainwater storage tanks was obtained from the volume of the tanks divided by the size of the tanks with a height of 1.9 m and a diameter of 1.8 m. The determination of the storage tank size considered with the limited availability of land in Indonesia. The number of tank requirements for the tiny house type was set to 1 tank, the medium house type was 1–2 tanks, and the large house type was 3–6 tanks (Table 4). The difference in the number of tanks for each type of house that has different roof sizes was considered to be not much different, but the difference in the size of the roof of the house as a rainwater catchment area provides a quite visible difference in the spilled rainwater.

Table 5 shows the number of the storage tank in the second scenario, in which there was no rainfall for the next 6 months in the dry season. Therefore, the fulfillment of clean water needs in the dry season was obtained from rainwater harvesting carried out in 6 months of the rainy season. As a result, the number of rainwater harvesting

Table 4 The number of rainwater harvesting tank for the first scenario

Type	Floor area (m^2)	Roof area (m^2)	Q (m^3/day)	Va (m^3/day)	St+1 (m^3)	Tank volume (St) (m^3)	Number of tanks (unit)
Tiny	30	194.49	1.23	0.635	0.59	1.18	1
	72	466.77	2.94	0.635	2.31	4.62	1
Medium	73	473.25	2.98	0.635	2.35	4.70	1
	120	777.95	4.91	0.635	4.27	8.54	2
Large	169	1095.61	6.91	0.635	6.28	12.55	3
	359	2327.37	14.68	0.635	14.04	28.09	6

Table 5 The number of rainwater harvesting tank for the second scenario

Type	Floor area (m^2)	Roof area (m^2)	Number of tanks for the first scenario (unit)	Vb (m^3/6 month)	Number of tanks in both season (unit)	Additional tank (unit)
Tiny	30	194.49	1	115.80	25	24
	72	466.77	1	115.80	25	24
Medium	73	473.25	1	115.80	25	24
	120	777.95	2	115.80	25	23
Large	169	1095.61	3	115.80	25	22
	359	2327.37	6	115.80	25	19

storage tanks at the beginning of the rainy season was quite a lot compared to the first scenario.

3.7 *Analysis of Spilled Rainwater*

In rainwater harvesting with the provided storage tank capacity, it was impossible for all runoff to be harvested. Some runoffs need to be discharged due to the limited storage tank capacity. In the first scenario, the discharged runoff was greater for each day than in the second scenario because the number of tanks in the second scenario was larger. The first scenario was recommended if there is no issue related to the land availability and the cost of constructing a rainwater storage tank. Table 6 shows the amount of daily spilled rainwater.

In the second scenario, rainwater runoff spilled was less than the first scenario (Table 7) during the rainy season. It is also able to meet the water consumption for both seasons. The type of tiny house with a roof area of 194.49 in scenario 2 even has no spillage due to the available volume of the tank. However, it would be

Table 6 Spilled rainwater in the first scenario

Type	Floor area (m^2)	Roof area (m^2)	V (m^3)	The number of tank (unit)	Spilled rainwater (m^3/month)	Spilled rainwater (m^3/day)
Tiny	30	194.49	1.18	1	30.95	1.03
	72	466.77	4.62	1	133.98	4.47
Medium	73	473.25	4.70	1	136.43	4.55
	120	777.95	8.54	2	247.15	8.24
Large	169	1095.61	12.55	3	362.78	12.09
	359	2327.37	28.09	6	815.14	27.17

Table 7 Spilled rainwater in the second scenario

Type	Floor area (m^2)	Roof area (m^2)	V (m^3)	The number of tank (unit)	Spilled rainwater (m^3/month)	Spilled rainwater (m^3/day)
Tiny	30	194.49	4.99	25	−1.27	−0.04
	72	466.77	8.43	25	50.3	1.67
Medium	73	473.25	8.51	25	51.5	1.72
	120	777.95	12.35	25	109	3.64
Large	169	1095.61	16.36	25	110.3	3.68
	359	2327.37	31.89	25	402.3	13.41

difficult to apply this second scenario if there are issues related to land availability and construction costs.

3.8 *Operational Procedure and Maintenance*

Operational procedures and maintenance include preparation of rainwater harvesting, collecting rainwater and maintenance including the cleaning process, rainwater quality monitoring and checking the installation. Following are the operation and maintenance procedures for rainwater storage tanks and rainwater quality:

1. Check if there is a blockage in the inlet or outlet pipe, if any, clean it immediately.
2. Check the roof area every week, if there is dirt, it must be cleaned.
3. Rainwater for the first 15 min is channeled as a daily flush so that the rainwater only harvested after the first 15 min of rain because rainwater quality after 15 min of rain is less pollutant.
4. Close the faucet when the rain stops, or the tank is full
5. Rainwater storage tanks, inlet, and outlet pipes should be cleaned if there is sediment, moss or dirt.

6. Check the pH, turbidity, TDS, temperature, taste and odor. If those do not meet the standard (Minister of Health Regulation No. 32 of 2017), further treatment will be needed.

4 Conclusion

Rainwater harvesting is an applicable technology with many benefits for a tropical country like Indonesia. This study shows the number of typical storage tank (∅ 1.8 m and height of 1.9 m) requirements in the first scenario (rainfall occurred for the whole year) for the tiny house type was 1 tank, the medium house type was 1–2 tanks, and the large house type was 3–6 tanks. In the first scenario, rainwater harvesting only meets water consumption during the rainy season and at the beginning of the dry season unless the area has rain throughout the year. In the second scenario, the number of storage tanks for each type of house is quite a lot, namely 25 tanks. The first scenario is considered to be very suitable for a user with no land availability and construction costs issues, while the second scenario is suggested to be used in areas where water scarcity in the dry season is the main issue. Based on the results of this study, further research is needed regarding the design of rainwater harvesting systems that pay attention to land use, geography, rainwater quality, and innovation in rainwater harvesting infrastructure that considers the limited availability of land.

References

1. Yudhistira MH, Sastiono P, Meliyawati M (2020) Sustainable water resources through harvesting rainwater and the effectiveness of a low-cost water treatment. Water Resour Econ 32(100161):1–15
2. Patunru AA (2015) Access to safe drinking water and sanitation in Indonesia. Asia Pac Pol Stud 2(2):234–244
3. Krishnaswamy J, Bonell M, Venkatesh B, Purandara BK, Rakesh KN, Lele S, Kiran MC, Reddy V, Badiger S (2013) The groundwater recharge response and hydrologic services of tropical humid forest ecosystems to use and reforestation: support for the "“infiltration-evapotranspiration trade-off hypothesis.”' J Hydrol 49(8):191–209
4. Akter A, Tanim AH, Islam MK (2020) Possibilities of urban flood reduction through distributed-scale rainwater harvesting. Water Sci Eng 13(2):95–120
5. Helmreich B, Horn H (2009) Opportunities in rainwater harvesting. Desalination J 248(1–3):118–124
6. Richards S, Rao R, Connelly S, Raj A, Raveendran L, Shirin S, Jamwal P, Helliwell R (2021) Sustainable water resources through harvesting rainwater and the effectiveness of a low-cost water treatment. J Environ Manage 286(112223):1–8
7. Angkasa Z (2017) Application of architecture concepts stage house in urban area. Arsir J 1(2):175–183
8. Ali S, Zhang S, Chandio FA (2021) Impacts of rainfall change on stormwater control and water saving performance of rainwater harvesting systems. J Environ Manage 280:111850
9. Wang R, Dijk SV, Lounsbury AW, Hoekstra AY (2020) Strategic design and finance of rainwater harvesting to cost-effectively meet large-scale urban water infrastructure needs. Water Res 184:116063

10. Zhang S, Zhang J, Jing X, Wang Y, Wang Y, Yue T (2018) Water saving efficiency and reliability of rainwater harvesting systems in the context of climate change. J Clean Prod 196:1341–1355
11. Che-Ain AI, Shaari N, Sairi A, Zain MFM, Tahir MM (2009) Rainwater harvesting as an alternative water supply in the future. Eur J Sci Res 34:132–140
12. Lee KE, Mokhtar M, Hanafiah MM, Halim AA, Badusah J (2016) Impacts of rainfall change on stormwater control and water saving performance of rainwater harvesting systems. J Clean Prod 126:218–222
13. Shaaban AJ (2009) Research and development of rainwater harvesting. IRCS 14:1–56
14. Khastagir A, Jayasuriya N (2007) Parameters influencing the selection of an optimal rainwater tank size: a case study for Melbourne. In: Proceedings of the rain water and urban design conference 2007, Sydney 21–23 Aug 2007
15. Khastagir A, Jayasuriya N (2010) Optimal sizing of rain water tanks for domestic water conservation. J Hydrol 381(3–4):181–188
16. Central Bureau of Statistics (BPS) (2010) Indonesian housing statistics. Central Bureau of Statistics, Jakarta
17. Leo KSD (2020) Preferences of selecting house type and housing locations of middle and upper communities near Yogyakarta city. Reka Ruang J 2(2):75–85
18. Triyuly W (2013) Residential development identification of house type in the Palembang city. Sriwijaya Eng J 22(1):1–7
19. Anditya, Afriansyah H (2016) Minimalist home design creative inspiration. Elex Media Komputindo, Jakarta
20. BMKG (2021) Rainwater quality information. BMKG, Jakarta
21. Minister of Health (2017) Regulation of The Minister of Health of the Republic of Indonesia No 32 of 2017: environmental health quality standards and water health requirements for hygiene sanitation, swimming pools, solus per aqua, and public baths. The Minister of Health of the Republic of Indonesia, Jakarta
22. Joleha J, Mulyadi A, Wawan W, Suprayogi I (2019) Assessment of rainwater quality in coastal area for domestic water supply. Civ Eng Nat Conf 13
23. Environmental Agency of South Sulawesi: Regional Environmental Status. Environ-mental Agency of South Sulawesi, South Sulawesi (2015)
24. Rahmayanti AE, Soewondo P (2015) Provision of drinking water in the coastal areas of Bandar Lampung city through rainwater harvesting. Environ Eng J 21(2):115–126
25. Untari T, Kusnadi J (2015) Utilization of rainwater as water consumption in Malang city using simple filtration modification method. J Food Agroindustry 3(4p):1492–1502
26. BMKG (2021) Rainfall analysis for February 2021 and rainfall forecast for April, May and June 2021. BMKG, Jakarta
27. Li Z, Boyle F, Reynold A (2010) Rainwater harvesting and greywater treatment systems for domestic application in Ireland. Desalination 260(1–3):1–8
28. Agustianto DA (2014) Rainfall and runoff relationship model. Civ Environ Eng J 2(2):214–224
29. PDAM (2014) Rispam of PDAM of Gorontalo city. PDAM of Gorontalo City, Gorontalo
30. PUPR Ministry (2010) Urban drinking water technical instructions and manuals. PUPR Ministry
31. Rustan FR, Sriyani R, Talanipa R (2019) Water consumption analysis of Bumi Graha Asri residents, Kendari city. Stab J 7(2):151–159
32. Salilama J (2018) Analysis of water consumption (PDAM) in Gorontalo city. J Sci Civ Eng Technol 6(2):102–114

Risk Analysis of Shared Marine Space in the View of Traditional Fishermen Perceptions in the National Tourism Strategic Area of Lombok, Indonesia

Ida Ayu Oka Suwati Sideman, R. M. Nyoman Budiartha, Ida Bagus Putu Adnyana, and Ngakan Ketut Acwin Dwijendra

Abstract This research is in Gili Air, North Lombok Regency. This island is one of the three islands, which is the National Tourism Strategic Area. The dominant livelihoods of residents are traditional fishermen, self-employment and tourism support activities. It is a natural thing if the port in the region is designed as a tourist port. Likewise, various risks are reviewed when submitting a plan for operating a tourist port, but they often do not involve the economic value of fishermen's failure to catch fish. This is a gap that is considered important in this study. This study is the result of preliminary research from the Doctor Program thesis; the purpose of this study is a risk analysis that places fishermen in the common use of marine space as part of the risk forming components of a tourist port. The research method used "Cause and Effect Analysis", which is a structured method to identify possible causes of risk. This method classifies the causes of risk into categories to facilitate identification. This study proves that the causes of risk formation from the low income of fishermen are water traffic, equipment and catch marketing, respectively. The detailed analysis shows that the synergy of the organization, schedules and services of the wharf and anchor grooves are the sub-cause of the risk.

Keywords Risk analysis · Traditional fisherman · Cause and effect analysis · Perceptions

I. A. O. S. Sideman (✉)
Engineering Science, Faculty of Engineering, Udayana University, Bali, Indonesia
e-mail: suwatisideman@unram.ac.id

Department of Civil Engineering, University of Mataram, Mataram, NTB, Indonesia

R. M. N. Budiartha · I. B. P. Adnyana
Deparment of Civil Engineering, Faculty of Engineering, Udayana University, Bali, Indonesia
e-mail: budiartharm@unud.ac.id

I. B. P. Adnyana
e-mail: bagusadnyana@unud.ac.id

N. K. A. Dwijendra
Engineering Science, Faculty of Engineering, Udayana University, Bali, Indonesia
e-mail: acwin@unud.ac.id

S. A. Kristiawan et al. (eds.), *Proceedings of the 5th International Conference on Rehabilitation and Maintenance in Civil Engineering*, Lecture Notes in Civil Engineering 225,
https://doi.org/10.1007/978-981-16-9348-9_92

1 Introduction

1.1 Research Background

Several countries, including Indonesia, This tourism industry is used by businesses providing services to visitors as a source of taxes and revenue [1, 2]. Tourism is defined as a temporary trip, carried out from one place to another to enjoy sightseeing and recreational trips [3] Unfortunately, there are several weaknesses of tourism factors such as weak human resources, long distances, lack of transportation access and less of community creativity [4]. Gili Indah area is one of the Strategic National Tourism Areas (KSPN) that located in West Nusa Tenggara Province. This area has 3 small islands that use sea transportation as the only access to the island of Lombok. Sea lanes are used as shared spaces in tourism transportation, fishing, and domestic movements. Shared space projects have been shown to greatly improve the public's view of an intersection, corridor, or locale through the use of enhanced architectural elements [5]. So that various ways must be done to fix this problem, one of this is by analyzing sea transportation through the perceptions of local communities such as fishermen. That is the reason why this research is important to be held.

All ship movements require port services in various classes and completeness. A port is a place that is deliberately built to be a place for ships to dock. This area is used as a stopover for ships before finally anchoring or continuing their journey [6]. Through this research, it is hoped that the risks borne by fishermen in the movement in the common room can be identified. The risk referred to is the risk based on the fishermen's perception as one of the actors in the movement. Risk is defined as the possibility of a potentially detrimental event or an uncertain situation faced which can have an adverse impact [7] in case of traffic, traffic congestion is a global trend that is expected to intensify in the future [8]. So that, traffic risks and the solving problems can be work as a universal recommendation in future.

The characteristic of traditional fishermen is to use a navigation system that combines nature, feeling, and instinct [9] with the characteristic of moving at a slow speed and stopping for fishing, while the characteristic of local logistics movement is fast moving because commodities are easily damaged [10]. On the other hand, the two movements are moving in an area designated as a tourist movement, where the characteristic is fast movement with higher speed [11]. Due to the condition above, it is predicted that there will be problems due to differences in the characteristics of the movement in the same roar from the beginning, namely from the port.

Many ports in tourism areas are built based on the perception of policy makers (the government) but these ports are not well utilized, the reason is that they are not in accordance with the needs of their users so they do not participate in supporting the port [12]. The most common cause of fear over the operation of an activity is risk [9]. The views or perceptions of different interests on the risk of an activity are often different [13].

Based on this, it is necessary to carry out research that examines from the point of view of fishermen who also receive the impact of the operation of the port in the tourism area.

The area used by the three movements with different characteristics certainly creates various risks [14]. The most common study to assess risk is the assessment of incidence [15]. The data ignores the feeling of one of the important components involved, namely fishermen. Research on risk is rare by looking at fishermen as a subject who has a sensitivity to risk assessment. Meanwhile, the sensitivity of the assessment is expressed in perceptions by the fishermen. A risk assessment relying solely on the number of occurrences eliminates the opportunity for fishermen to provide an opinion on future potential.

By involving fishermen's perceptions, neglect of traditional fishermen in potential risks can be eliminated. Discrimination in assessment in the formation of risk components can be more fair and humane, considering that fishermen in this area have existed before the stipulation of tourism areas is determined.

1.2 Literature Review and Novelty

One of the internal factors causing fishermen's poverty is the high dependence on fishing occupations, while one of the external factors is the geographical isolation of fishing villages which does not support the mobility of goods, services, capital and people [16]. When it is compared with research in Sambas district, where fishermen have a risk of unpredictable fishing season which is 41%. This number is relative high for fishermen's profit perception of 70% [7]. So, it can be said that fishermen have high optimism about the life expectancy of the traditional fishing profession. Both studies can be interpreted as strengthening the need for a research on risk factors, as fishermen still have a high dependence on their work and the sea. So that a research that based on fishermen's economic risk is absolutely necessary to provide the justice for fishermen. The new of this research is that the study area is designed as a tourism area, not a fishing area. So that, fishermen's perceptions will be found towards other movements that have different characteristics from the movements of fishermen, but work in the same sea space.

The three main characteristics of standard procedures from a typical port project evaluation are: diversity of port size and operations, diversity of port destinations and missions, and private sector participation in the tourism industry. A program is formulated in the framework of the importance of broad community participation in the success of the program [17]. Thus, the perception of fishermen as an affected community is important to research. This research took the point of fishermen's perception as the new of this research.

1.3 Preliminary Hypothesis

Economic problems are the main risk for fishermen in the Gili Indah region in the case of sharing marine space in the tourism area. This economic problem is supported by ship problems, especially the age of the ship's engine. Engine condition has the potential to affect safety and comfort, especially in terms of vehicle speed. Similar to land vehicles, water vehicles also have speed, where the journey speed is the effective speed of a vehicle traveling between two places [18, 19].

2 Research Method

2.1 Research Method

Various problems that occurred during the pandemic period caused many things to be done to continue to carry out research but in the health novel. This stimulates this research to be held by using the 'cause and effect' analysis method.

Cause and Effect Analysis is a structured method to identify possible causes of a risk [20]. This method classifies the causes of risk into categories to facilitate identification. This cause information is displayed in a fishbone diagram or Ishikawa diagram or in a tree form. The basic steps for Cause and Effect Analysis are as follows: (1) determining the risk or effect to be analyzed, this risk can have a positive or negative impact depending on the circumstances; (2) Determine the main categories of risk causes to be the main branches in the diagram; (3) Identify each possible cause covered in each category to the extent possible. The causes are described as twigs from each branch; (4) check each category branch to ensure completeness and consistency. 5. Identify the most likely cause based on opinion and evidence [21].

2.2 Research Design

Samples are the share of the population taken in a certain way and in number smaller than the population [22]. Then the sample is built from the accident sampling method. Accidental sampling is sampling accidental by taking the respondent who happened to be there is somewhere in accordance with the place of research [23].

2.3 Characteristic of Sample and Its Validation

Even though the method used was accident sampling, for data accuracy, samples were grouped based on characteristics related to their work.

Law of Republic of Indonesia No. 13 of 2003 about manpower, appealed that children can do some light works as long as they do not interfere with their physical development, are supervised by adults and work a maximum of 3 h a day [24]. Furthermore, in the same law it is stated also if the limit of children is 15 years. In Indonesia, the age limit for having various official personal documents such as identity cards and driving licenses is 17 years, so this research places a minimum of 17 years of workforce as fishermen. Furthermore, this limitation will apply in determining the eligibility requirements for the sample. Further characteristics analyzed from the sample were gender, work experience (in years), skills (training, courses, and other supporting skill) and side jobs. This is to obtain the feasibility of using samples to represent the fishing population in the Gili Indah area.

As the work experience is not clearly stated, a detail was made on the minister of manpower regulation No. 12 of 2016 [25] which follows the Indonesian National Work Competency Standards (SKKNI) where the workforce is determined to be skilled with a minimum work experience of 2 years.

The reason for including the competency component of fishermen is because the competency standard is expected to be a benchmark for fishermen's income, where fishermen with more competence will get more income [26]. Table 1 shows the characteristics of the respondent's data in this study.

Table 1 is a requirement to choose if the respondent's research methods are to determine the sample using the accidental sampling method, which is to capture a sample from a number of populations while doing activities [27]. In accident sampling which is also referred to as captive sample, the population must have high uniformity so that the sampling does not produce data bias [28].

From these data, the characteristic of respondents that stands out is the dominance of the productive age with an almost even distribution from the ages of 22–51 years. The gender is entirely male, very few have received training and it is concentrated on training on disasters. Although there are 12.5% of respondents who have a side job, they have a dependency on marine life. Work experience is dominated by a work period of 6–10 years. The characteristics of these **respondents are valid to be continued as a research** sample because they approach the results of previous research in selecting the sample model. The conclusion of the sample test, the research can be continued with the sample.

3 Results and Discussion

3.1 *Risk Analysis of Shared Marine Space*

The aim of any risk assessment tool is to ensure a consistent, knowledge-based decision-making process that represents the shared understanding of stakeholders. When the implementation of risk matrices is checked, it becomes evident that there are some notable weaknesses in this simple tool [29]. Risk evaluation at sea is a critical

Table 1 Characteristics of respondents

Number	Description	% of 16 respondents	Sample test standards
1	Age group 17–21 22–26 27–31 32–36 36–41 42–46 46–51 51–56 More than 56	0 6.25 6.25 18.75 31.25 18.75 18.75 0 0	Meet the requirements in accordance with law of Republic of Indonesia No. 13 of 2003
2	Gender Male Female	100	Meet the requirements in accordance with Law of Republic of Indonesia No. 13 of 2003
3	Residence Gili Indah Pemenang	80 20	Meet the requirements in research problem boundaries made by researchers based on references
4	Side job Yes No	12.5 (in sea tourism field before pandemic) 87.50	Meet the requirements in research problem boundaries made by researchers based on references
5	Work experiences 0–5 years 6–10 years More than 10 years	12.5 62.5 25	Meet the requirements in the minister of manpower regulation No. 12 of 2016

and time-consuming process. In order to determine the rules and strategies by which the team can conduct the risk assessment, the selection of suitable risk assessment tools is critical [30, 31]. Protection is characterized as the state of safety from the risk of pain, injury, threat, loss, injury prevention or hazard prevention devices. Security may also be described as the role of ensuring that accidents are prevented. Protection, however, cannot be strengthened by looking at the past and taking precautions against injuries [32]. Three key attributes are required by a resilient system: (1) anticipation: avoidance of disturbance; (2) perception: avoidance of aggravation of the effects of disturbance; (3) response: recovery and survival after disturbance [33]. The accident ratio will be an important issue if the respondent's perception points to sea accidents as a risk. The risk ratio is the potential for accidents per incident [34].

Of the three keys, the first key cannot be used because fishermen cannot avoid the risk of using the shared space. Meanwhile, the third key is the key that can be used by the government as the policy maker. Thus, it is increasingly important to conduct research on perceptions because it is a key that can be used by fishermen to overcome risks.

3.2 The Traditional Fishermen Perceptions

Perception in general is a process of obtaining, interpreting, selecting and organizing sensory information [33]. Fisherman is a community who lives in a coastal area with the main livelihood utilizes the natural resources contained in it sea, whether in the form of fish, shrimp, seaweed, shellfish, coral reefs and other products other marine wealth [35]. The life of coastal communities is different from aspects of life in agrarian societies (residents living in rural areas on generally).

This is due to natural environmental factors, due to coastal communities more related to the dominant sea, whereas agrarian society by natural environment in the form of rice fields, moor or fields. With that condition different, allowing them to have a culture and knowledge system different in meeting their daily needs [16]. Small island challenges can involve numerous creative approaches, as they all turn out to be wicked issues because of size, isolation, geography, tradition and culture, governance and education [36].

Based on the Table 2, the priority sub case handling is on traffic cooperation between related components in the movement of the sea space together. The first handling is on the movement schedule information, especially on tourist movements that have different movement characteristics from fishermen movements. The movements of these different characteristics can influence each other, such as when a fishing boat is at the anchor stage, the movement of a tourist boat can be a nuisance, as well as for tourist boats, the existence of a fishermen's boat can become an obstacle to movement.

The next risk treatment is information, especially about sea and weather conditions. Fishermen need up-to-date information as a reference to go to sea. Port services and synergy between organizations can be corrected by implementing cooperation between the organizations involved in it. The rating for this matter is that it can be

Table 2 Causes and effect

Impact	Categories of causes	Causes	Sub causes	Fishermen perception (%)
	Equipment	Machine	Machine life	7
			Organizational synergy	13
Low income	Water traffic	Traffic management	Time schedule	19
			Services of the wharf and anchor grooves	14
			Information	25
	Market	Types of fish	Not the kind of tourist taste, so it must be marketed locally and to Lombok	10

Table 3 Risk respond

Level of risk		Respond
1–3	Acceptable	Sufficient controlled
4–6	Monitored	Sufficient controlled
6–9	Control of management	Sufficient controlled
10–14	Urgent	Acceptable only with excellent controlled
15–25	Unacceptable	Acceptable only with excellent controlled

Source [19, 37]

done in a category that is not very urgent but requires handling. Meanwhile, the risk of vessel conditions and cooperation in selling fish is cooperation with outside parties which can be carried out in a non-urgent status and if it is to be implemented it requires more detailed research because the score is 7 that due to Table 3 it is on category of control management. For this category the respond of the risk must be sufficient controlled.

Verification of the data obtained from fishermen's perceptions is to use source triangulation. Triangulation is a validation and verification method that can be relied on to obtain valid data [38]. One of them is by triangulating data sources; this method uses other sources to test the data that has been obtained [39]. In this study, data from fishermen's perceptions were tested by cross check on the port operator. The data that is cross checked is schedule data and other information, which according to fishermen's perception is the cause of risk. From the cross check results, it was found that the port operator did not provide services to fishermen, because the schedule was informed at the port for the benefit of passenger movement. Other related information such as weather and height, waves are also carried out at the port for the benefit of passenger transportation operators. Thus it can be denied that the fishermen's perception of traffic management risk is correct and acceptable.

Furthermore, to support the results of the research, it resulted in two sub causes as things that must be managed with excellence; the following table of interests was created by conducting detailed interviews with fishermen.

Based on the results of in-depth interviews of the schedule and information needed by fishermen to reduce risk, the data obtained are as shown in Table 4.

Table 4 Forms of risk handling

Needs	Form
Time schedule	Cooperation agency/organization
Information Weather Wave Wind Accident	Regularly sharing information

Table 5 Fishermen's perceptions of digital information systems

Fishermen's perceptions of SWOT	Level
Weakness: ability	25
Strengths; relationship	18
Opportunity: information	10
Threat: decreasing local wisdom	3

Based on Table 4, it can be seen that fishermen representatives expect the involvement of fishermen in compiling the schedule of tourist movements in the Gili Indah national tourism strategic area. Fishermen also expect to share information regularly and always be updated as part of reducing risks. This will be risk mitigation for fishermen in the fishermen's perception.

To bridge the need for space to receive up-to-date movement schedules and shipping information, it is again to explore opinions or perceptions from fishermen about digital-based information systems. To comply with the causes and effects method that has been used with the results in Table 2, point 25 is set as the highest value to determine fishermen's perceptions of digital-based information. Perception-forming factors were extracted from the Strength Weakness Opportunity and Threat (SWOT). The data obtained are as shown in Table 5.

Furthermore, to facilitate the reading of the SWOT, the mapping was carried out into a spider web diagram. Thus, the focus of the various organizations involved can lead to dominant SWOT formers.

The results obtained are illustrated in Fig. 1.

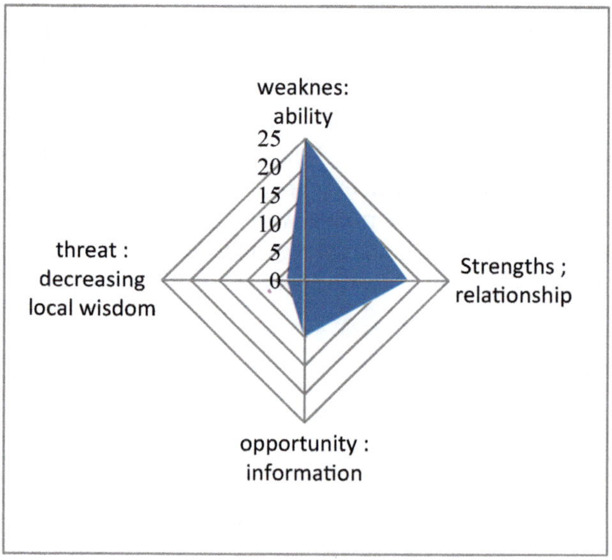

Fig. 1 SWOT analysis

Based on this figure, the direction of the use of digital systems will clash with media ownership (gadgets) but if the problem of gadget ownership can be resolved, the digital information system program will be successful because it has the support of "good relations" between port organizations in the region.

4 Conclusion

Thus, the existing recommendations, the preliminary hypothesis are refuted in case of risk causes. Even though the main risk is about economic problem, but the problem of ship's engine is not a priority problem in the perception of Gili Indah fishermen, because the priority issues are capacity institutions and networking port management. The recommendation for further research is the development of an institutional system based on fishermen's culture and local wisdom.

Funding Acknowledgement This research is a preliminary study for a dissertation of a Ph.D. student, so it uses the full cost of the student as the main author.

References

1. Pariwisata S (2009) Pitana dan Gayatri, Sosiologi Pariwisata, 2005, hal 45, pp 1–15
2. Ketut N, Dwijendra A (2020) Design model innovations for tourism villages in Bangli, Bali Indonesia: debate between environmental and cultural protection versus community economic development, vol 29, no 7, pp 1428–1443
3. Primadany S (2013) Analisis Strategi Pengembangan Pariwisata Daerah (Studi Pada Dinas Kebudayaan Dan Pariwisata Daerah Kabupaten Nganjuk). J Adm Publik Mhs Univ Brawijaya 1(4):135–143
4. Nugraha YE, Siti YE (2020) Pengembangan Ekowisata Berbasis Masyarakat Di Desa Pemo Taman Nasional Kelimutu Kabupaten Ende. J Destin Pariwisata 8(2):169. https://doi.org/10.24843/jdepar.2020.v08.i02.p01
5. Frosch C, Martinelli D, Unnikrishnan A (2019) Evaluation of shared space to reduce traffic congestion. J Adv Transp 2019. https://doi.org/10.1155/2019/6510396
6. Adnyana I, Dwijendra (2011) Arsitektur dan tata ruang pelabuhan di Bali, January 2012, 2011 [online]. Available http://erepo.unud.ac.id/id/eprint/26973/1/a01638a63e692dd76bbe4d4c1ca42df0.pdf
7. Lindawati L, Rahadian R (2016) Identifikasi Faktor Dan Penilaian Risiko Pada Usaha Perikanan Tangkap Di Kabupaten Sambas. J Sos Ekon Kelaut dan Perikan 11(1):99. https://doi.org/10.15578/jsekp.v11i1.3175
8. Kesuma PA, Rohman MA, Prastyanto CA (2019) Risk analysis of traffic congestion due to problem in heavy vehicles: a concept. IOP Conf Ser Mater Sci Eng 650(1). https://doi.org/10.1088/1757-899X/650/1/012011
9. Profesi AK, Navigasi S, Pratomo AH, Rustamaji HC (2020) Analisis Karakteristik Nelayan Tradisional Berdasar Jenis dan Klasifikasi Nelayan , Kelompok Kerja , Jenis Perairan , Teknologi 2(1):29–34
10. Chandra A (2013) Analisis kinerja distribusi logistik pada pasokan barang dari pusat distribusi ke gerai indomaret di kota semarang

11. Ketut N, Dwijendra A, Dani AA, Sjuchro DW, Klochko EN, Patriani Y (2020) Political and social infrastructure towards high quality tourism sites 7(11):1764–1773
12. Situbondo K (2013) pelabuhan panarukan, pp 1–8
13. Bilondatu MR, Kepercayaan DAN, Terhadap P (2016) motivasi, persepsi, kepercayaan, dan keputusan pembelian 1(3):710–720
14. Sideman I (2021) Proc ICST 2:1–9
15. Sideman I (2021) Traffic management of Gunung Sari intersection base on problem solving hierarchy traffic management of Gunung Sari intersection base on problem solving hierarchy. https://doi.org/10.1088/1742-6596/1779/1/012082
16. Wijaya AB, Fauzie A (2020) Pemaknaan Hidup Nelayan (Analisis Makro dan Mikro pada Kemiskinan Nelayan). Indonesia Psychol Res 2(2):96–108. https://doi.org/10.29080/ipr.v2i2.259
17. Arnatha M, Budiartha NRM (2011) Prioritas Program Terminal Kapal Pesiar Tanah Ampo: Suatu Penerapan Benefit-Cost Analysis (Bca) Untuk Penilaian Proyek. J Ilm Tek Sipil 15(1)
18. Karyawan IDMA, Widianty D (2014) Analisis Jarak Pandangan Henti sebagai Elemen Geometrik pada Beberapa Tikungan Ruas Jalan Mataram-Lembar 18(2):40–48
19. Ketut N, Dwijendra A (2020) Expanding the oil and gas sector of Indonesia through better marketing performance: analysing the role of online engagement, interactivity and brand attachment, Aug 2020
20. Kurniawan Z, Aknuranda I, Setiawan NY (2018) Analisis dan Penentuan Prioritas Pencegahan Masalah Menggunakan failure mode and effect analysis dan cause and effect analysis Pada Proses Bisnis Pengajuan proposal program Kreativitas Mahasiswa (PKM) (Studi Kasus: Universitas Brawijaya Malang). J Pengemb Teknol Inf dan Ilmu Komput Univ Brawijaya 2(12):7105–7112
21. Sebab-dan-akibat A. Cause-and-effect analysis
22. Iii BAB (1992) Jenis Penelitian Lokasi Penelitian, pp 55–77
23. Sunarta (2018) Metode Penelitian 53(9):1689–1699
24. Undang-Undang Republik Indonesia No. 13 Tahun (2003) Undang-Undang Republik Indonesia No. 13 Tahun 2003 tentang Ketenagakerjaan. Undang-Undang 1:1–34 [online]. Available http://www.kemenperin.go.id/kompetensi/UU_13_2003.pdf
25. Ketenagakerjaan K (2016) Peraturan Menteri Ketenagakerjaan Republik Indonesia Nomor 2 tentang Sistem Standarisasi Kompetensi Kerja Nasional. Lembaran Negara RI
26. Anugerah Y, Nurani TW, Sondita MFA (2016) Kompetensi Nelayan Rawai Tuna Ditinjau Dari Standar Kompetensi Kerja Nasional Indonesia (Skkni) Di Ppn Palabuhanratu. J Sos Ekon Kelaut dan Perikan 11(2):251. https://doi.org/10.15578/jsekp.v11i2.3693
27. Ekonomi DR, Perbankan DAN (2020) Menentukan Populasi dan Sampel Dalam Riset-Riset Ekonomi dan Perbankan Islam, Apr 2020
28. Iii BAB (2006) Vinna Indahtianti, 2013 Hubungan Pelayanan Sirkulasi Dengan Pembentukan Citra Perpustakaan (Studi Deskriptif di Perpustakaan Universitas Pendidikan Indonesia). Universitas Pendidikan Indonesia|repository.upi.edu, pp 25–41
29. Duijm NJ (2015) Recommendations on the use and design of risk matrices. Saf Sci 76:21–31. https://doi.org/10.1016/j.ssci.2015.02.014
30. Parra NM, Nagi A, Kersten W (2018) Risk assessment methods in seaports—a literature review, May 2018
31. Ketut N, Dwijendra A (2018) Quality of affordable housing projects by public and private developers in Indonesia: the case of Sarbagita Metropolitan Bali, May 2013. https://doi.org/10.5897/JGRP12.050
32. Raduan S, Dayang RAJ, Assis K, Diana H (2017) Passengers perception on safety level of ferry transport: a case study in Labuan Island, Malaysia. J Adv Res Soc Behav Sci 1(2017):7–17
33. Langard B, Morel G, Chauvin C (2015) Collision risk management in passenger transportation: a study of the conditions for success in a safe shipping company. Psychol Fr 60(2):111–127. https://doi.org/10.1016/j.psfr.2014.11.001
34. Zeisel J (2015) Publication preview source observing environmental behavior—Chapter 8 in inquiry by design: tools for environment-behavior, Jan 1981

35. Sasmita S, Martasuganda A, Purbayanto S, Hestirianoto T (2013) Keselamatan Kerja pada Operasi Penangkapan Ikan Cantrang Nelayan Tanjung Sari, Kabupaten Rembang (Safety Assessment of Cantrang Operation in Tanjung Sari, Regency Rembang Central Java). Bul PSP 21(1):11–17
36. Mijts E, Arens P (2019) Capacity building for sustainable development in small island states through science and technology research and education. https://ic-sd.org/wp-content/uploads/2019/11/eric_mijts, Nov 2019
37. Moeller J (2015) A word on standardization in longitudinal studies, Dec 2015. https://doi.org/10.3389/fpsyg.2015.01389
38. Ikasari AC (2017) Konsep Kebijakan Alih Kelola Pelabuhan Perikanan di Jawa Barat. J. Ilm. Magister Ilmu Adm. No. 2 Tahun XI Juni, no 2, pp 44–55
39. Hadi S (2010) Pemeriksaan Keabsahan, pp 21–22

Construction Materials for Sustainable Infrastructures

Residual Stress Evaluation on Cold-Formed Steel C-Section by X-Ray Diffraction

T. Widya Swastika, Heru Purnomo, M. R. Muslih, R. Apriansyah, Henki W. Ashadi, and Mulia Orientilize

Abstract It is known that the existence of residual stress in cold-formed steel profile plays an important role in determining the nominal strength of the profile in resisting the external forces. Ignoring the residual stress will be very risky due to the serviceability and ultimate strength requirements of the structure constructed from these profiles. This study focuses on the determination of residual stress from local manufacture cold-formed steel lip channel C-sections which are widely used today for the construction of residential roof truss. Two types of cold-formed steel from two different manufactures have been investigated from the residual stresses in the cold-formed lip channel C-sections and the flat steel sheets before being formed into the lip channel C-section by X-Ray Diffraction apparatus. The results show that the tension stress is measured on the flanges surface, while the compression stress is found in the web surface. The residual stress in the bent area of lip channel C-section changes significantly, while the residual stress in flat steel sheet experiences compression. In addition, the residual stress in PC-065-1 specimen is around 200 MPa, while it is about 150 MPa in LB-065-1, whereas for the flat steel sheet specimen of PC-065-1 it is around 100 MPa, while in LB-065-1 it is as high as 240 MPa.

Keywords Residual stress · Cold-formed steel · X-Ray diffraction

T. Widya Swastika · H. Purnomo (✉) · H. W. Ashadi · M. Orientilize
Civil Engineering Department, Universitas Indonesia, Depok, Indonesia
e-mail: heru.purnomo@ui.ac.id

T. Widya Swastika
e-mail: tri.widya@ui.ac.id

H. W. Ashadi
e-mail: henki@eng.ui.ac.id

M. Orientilize
e-mail: mulia@eng.ui.ac.id

M. R. Muslih · R. Apriansyah
Neutron Scattering Laboratory, PSTBM-BATAN Serpong, South Tangerang, Indonesia
e-mail: rifai@batan.go.id

S. A. Kristiawan et al. (eds.), *Proceedings of the 5th International Conference on Rehabilitation and Maintenance in Civil Engineering*, Lecture Notes in Civil Engineering 225,
https://doi.org/10.1007/978-981-16-9348-9_93

1 Introduction

Residual stress is the stress that exists within a steel profile after it has been formed. With the increasing scarcity of wood and the vulnerability of wood to termites as a roof construction material in Indonesia, cold-formed steel has been commonly used for roof constructions in the last fifteen years. Cold-formed steel lip channel C-section is widely used for the roof construction in Indonesia because of the ease of fabrication and installation. However, it is often the case that builders or contractors perceive cold-formed steel to behave in the same way as hot rolled steel, where in the calculation of its strength uses hot rolled steel. On one hand, the literature regarding the residual stress of cold-formed steels is rare; on the other hand, the builders often ignore the presence of residual stress in the cold-formed steel strength as there are no data related to this, although efforts to investigate residual stresses in cold-formed and hot rolled steels have been carried out overseas [1–7]. In general, the residual stress measurement method in steel can be selected by using one of the three method groups, namely non-destructive test, semi-destructive test, and destructive test [8–15]. The non-destructive test consists of tests based on Ultrasonic, Diffraction, Magnetic, Raman Spectroscopy, and Nano Indentation technique. Meanwhile, semi-destructive tests are based on Hole Drilling, Ring Core, and Deep Hole. The destructive test is based on Sectioning or Contour. The choice of the method will depend on the availability of equipment in the laboratory and the level of accuracy desired.

This study presents an experimental investigation on cold-formed steel lip channel C-section to evaluate the residual stress from two local manufacture products having the same thickness but with a different flat width of the largest sub-element on the web stiffener. Subsequently, the flat steel sheets before being formed into the lip channel C-section will also be evaluated to have insight on the history of the residual stress. Later, the coating of these specimens will be removed by water cooled sanding on the center of the section and then the residual stress will have been re-evaluated.

2 Method

In the X-Ray Diffraction method, the strain is measured in the crystal lattice. This method is applied to find grainy crystalline materials that produce a diffraction pattern and appropriate intensity peaks, and free of interference in high back reflection area for any orientation sample surface. Surface residual stress measurements are non-destructive. Good macroscopic residual stress and line widening caused by micro stresses and crystal damage can be determined independently. The X-Ray Diffraction method is claimed to have high accuracy and simple operation [15–17].

This study investigated the cold-formed steel lip channel C-sections from two different local manufactures which have a similar dimension in terms of height and thickness; however, each will have its associated peculiarities of web stiffeners. In the initial stage, two types of lip channel C-section were measured for the residual stress.

The next stage was removing the surface coating using water cooled sanding on the center of the section. The residual stress in the two flat steel sheets has also been evaluated by the same procedure as the lip channel C-section specimens. Evaluating the residual stress in flat steel sheets is necessary to see the difference of the residual stress state due to the fabrication process. The residual stress from all the specimens was measured by the X-Ray Diffraction cos-α method. The residual stress from the same coordinates located at the specimens will be compared.

2.1 Residual Stress Evaluation by X-Ray Diffraction

The relationship between normal stress components σ_{ij} and the strain $\varepsilon_{\phi\Psi}$ along the scattering vectors ϕ and Ψ in the case of elastically isotropic specimens is shown in Eq. 1 and Fig. 1, as follows [18]:

$$\begin{aligned}\varepsilon_{\phi\psi} = {} & (1/2)S_2\{(\sigma_{11}\cos^2\emptyset + \sigma_{12}\sin2\emptyset + \sigma_{22}\sin^2\emptyset)\sin^2\Psi \\ & +(\sigma_{13}\cos\emptyset + \sigma_{23}\sin\emptyset)\sin2\Psi + \sigma_{33}\cos^2\Psi\} \\ & + S_1(\sigma_{11} + \sigma_{22} + \sigma_{33})\end{aligned} \quad (1)$$

where:

$\sigma_{11}, \sigma_{22}, \sigma_{33}$ = Normal stress in the sample coordinate axes in X, Y, and Z directions.

σ_{23} = Shear stress in the Z axis direction normal to the Y axis.

σ_{12} = Shear stress in the Y axis direction normal to the X axis.

σ_{13} = Shear stress in the Z axis direction normal to the X axis.

S_1 and S_2 = The X-Ray elastic compliance constants. They can be related to Young's modulus E and Poisson's ratio v of the body. $S_1 = -\ v/E$ and $S_2 = (1 - v)/E$.

The strain $\varepsilon_{\phi\Psi}$ is defined in the following [18]:

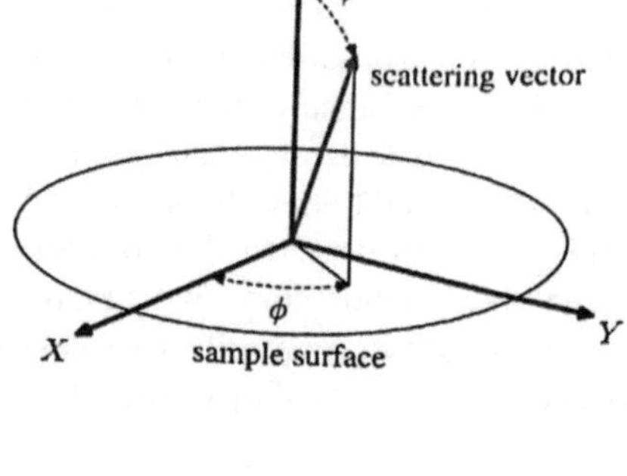

Fig. 1 A coordinate system for stress analysis using X-Ray diffraction [18, 19]

$$\varepsilon_{\phi\psi} = \frac{d_{\o\Psi} - d_0}{d_0} \tag{2}$$

where:

$d_{\o\Psi}$ = Lattice spacing along a scattering vector (the directions will be ϕ and Ψ).
d_0 = Strain-free lattice spacing.

The strain-free lattice spacing varies easily depending on the crystalline state, and thus it is difficult to know the accurate value for d_0. However, d_0 can be determined by an optimization procedure.

2.2 Experimental Investigation

The X-Ray Diffraction method was carried out at the Neutron Scattering Laboratory of BATAN, located in Serpong, South Tangerang. This method is to determine the residual stress in the work area on the surface (Plane Stress) with a very thin specimen thickness (100 μm – 17 mm). The accuracy produced by this technique is very good as it can reach ±20 MPa. The X-Ray Diffraction measurement is useful for residual stress analysis if the following conditions are met in the material specimen to be measured, namely the material must have a crystal structure, the material must have small grains, and the elastic constant of the material is known.

Figure 2 shows the X-Ray Diffraction residual stress method equipment. The operation of the tools is simple. In its operation, every time it is used, the tool must be set in advance with the conditions for using the tool, where the sample distance is set at 45 mm with ψ angle (in Fig. 1). Then it is calibrated by firing an X-Ray beam for the stress-free steel specimen. Seeing that in this diffractometer there are no moving parts (rotating), the center point of its axis, therefore, must be determined using a standard sample. The standard sample used is a stress-free sample in the form of iron powder moulded with resin. With the correct settings, the residual stress in this sample is in the range of 10 MPa (close to zero), and the tool is ready to use for other specimens.

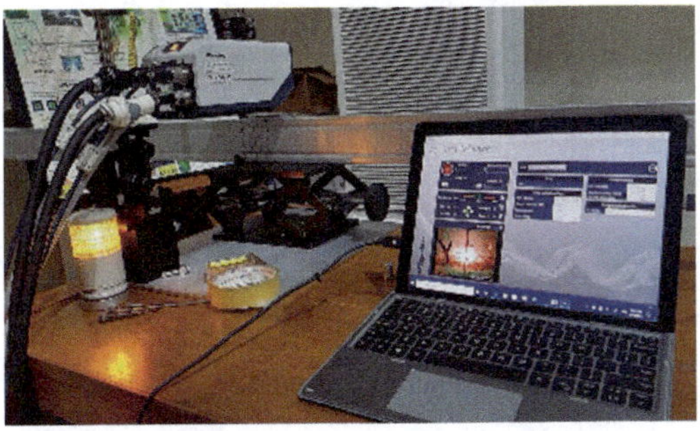

Fig. 2 The X-Ray diffraction residual stress method equipment

The data measured by the X-Ray Diffraction method are classified into 3 classes with a diffraction pattern of Debye-Scherer circle: very good, good, and bad. Only the very good and good data were taken into consideration.

The measurement results show the residual stress in the direction 11 (σ_{11}) and direction 12 (σ_{12}), where σ_{11} is the perpendicular direction of the cross section, while σ_{12} is the parallel lengthwise direction of the cross section.

2.3 *Test Specimens*

The test specimens were made from grade G550 AZ100. The specimens used are cold-formed steel lip channel C-section with dimensions of 75 × 35 of 0.65 mm thick and with a length of 2 m but with different rib positions from two local manufactures: LB-065 and PC-065, illustrated on Fig. 3. Meanwhile, the flat steel sheet (SS-LB-065 and SS-PC-065) is 0.65 mm thick with 151 mm of width (3U and 6U for C-section and SS for flat steel sheet are the label of measurement location configuration). Figure 4 shows the specimens of cold-formed steel lip channel C-section and flat steel sheet, with coating and uncoating.

3 Results and Discussion

The results of the residual stress measurement are shown in Figs. 5, 6 and 7.

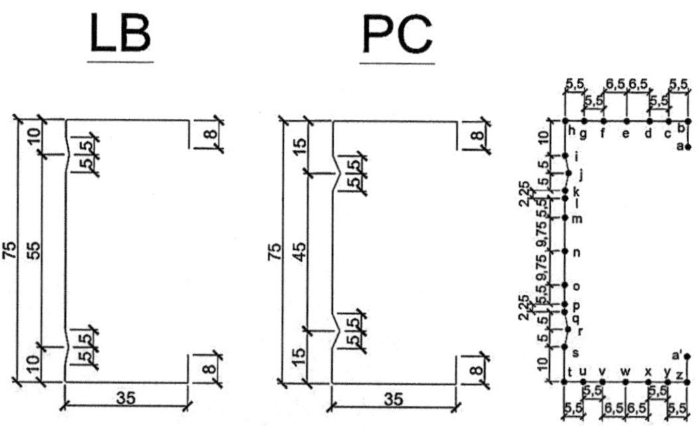

Fig. 3 Dimension detail and position of measurement nodes

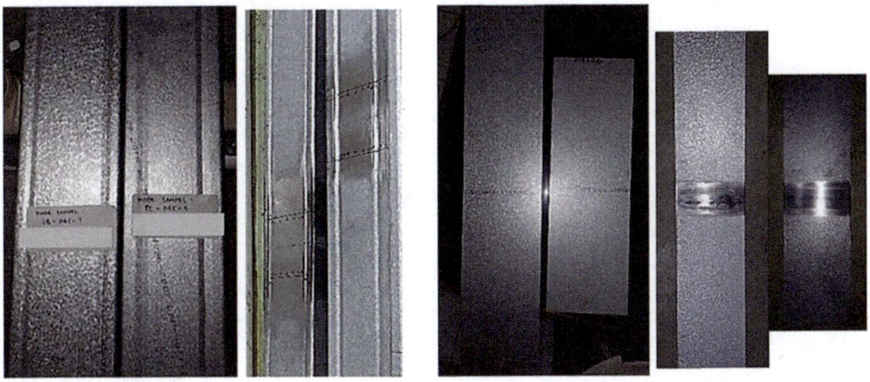

Fig. 4 Cold-formed steel specimens with coating and uncoating; lip channel C-section (left) and flat steel sheet (right)

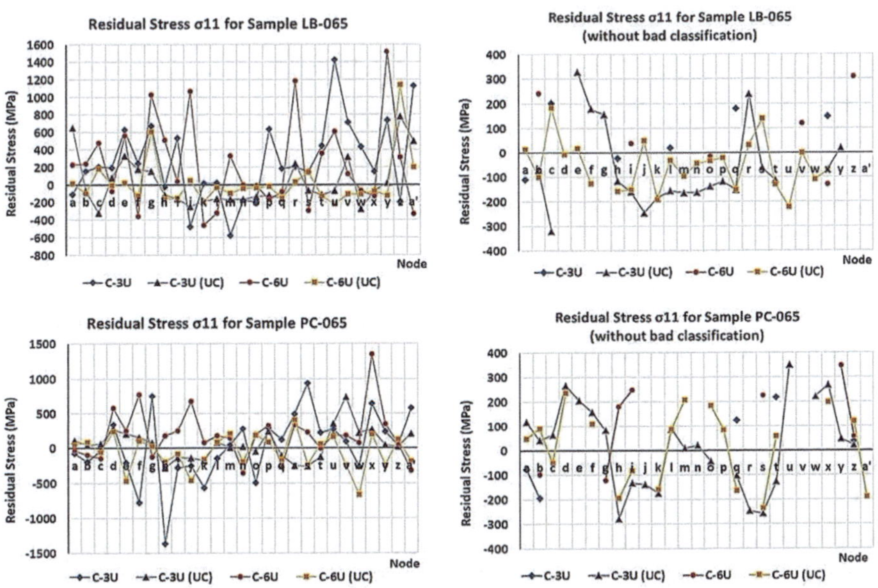

Fig. 5 Residual stress comparison for lip channel C-section specimens LB-065 and PC-065

3.1 *Specimens for Cold-Formed Lip Channel C-section*

Figure 5 shows the residual stress σ_{11} of the 2 different cold-formed lip channel C-section specimens which are taken before and after removing the coating. Lower residual stresses are given after removing the coating. We can see that the sample LB-065 gives a better classification than specimen PC-065. Moreover, all specimens show

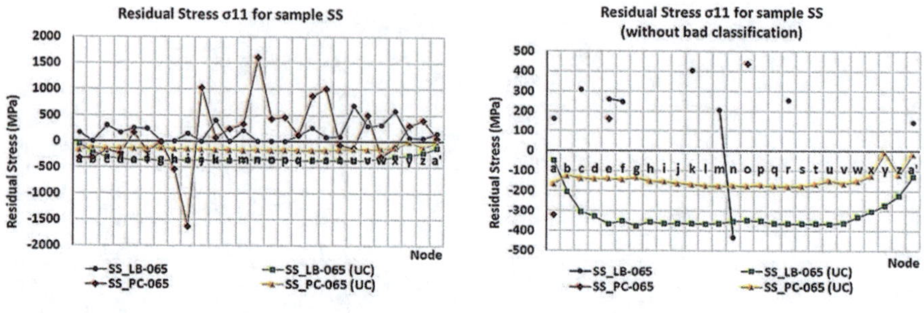

Fig. 6 Residual stress comparison for steel sheet specimens LB-065 and PC-065

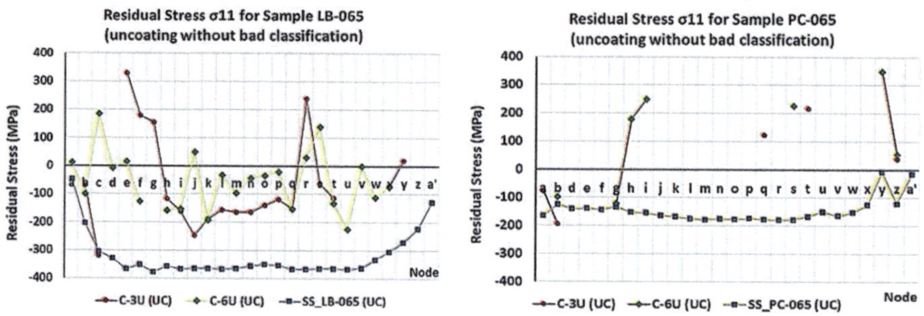

Fig. 7 Residual stress comparison between steel sheet and lip channel C-section specimens (LB-065 and PC-065)

that the residual stress represent the forming method of the specimens, compression in web, and tension in flanges.

3.2 *Specimens for Steel Sheet Cold-Formed*

We can see in Fig. 6 that the uncoated sample gives a lower value of residual stress than the coated sample. After uncoating the surface, the value of residual stress is all in compression. In addition, the uncoated surface also shows a better classification, indicated by the number of nodes read in the X-Ray Diffraction measurement result. We also see that the sample LB-065 in general gives a better result than the PC-065.

3.3 Comparison of Cold-Formed Steel Sheet and Lip Channel C-section

Figure 7 indicates that the two specimens of cold-formed lip channel C-sections at several location have almost similar pick values with the residual stress σ_{11} for flat steel sheet. After the coating removal, the residual stress σ_{11} decreased, and the result showed that the residual stresses represent the forming method of the specimens; in other words, the sample lip channel C-section experienced compression in the web and tension in the flanges, while the steel sheet was in compression. We can see that the sample LB-065 gives a better classification than specimen PC-065. Furthermore, the sample LB-065 gives a better result than the sample PC-065.

3.4 Residual Stress Contour

Figures 8 and 9 show the contours of the residual stress σ_{11} for cold-formed lip channel C-sections after coating removal and without considering the bad classification.

From Figs. 8 and 9, all specimens show that the residual stress represents the forming method of the specimens, compression in web, and tension in flanges.

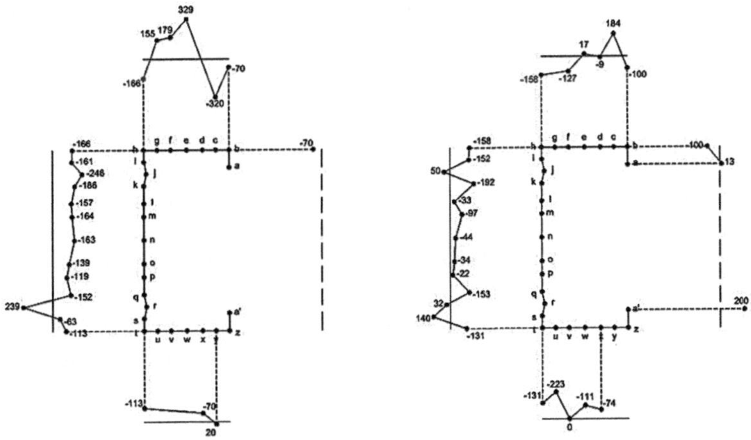

Fig. 8 Residual stress σ_{11} of 3U (left) and 6U (right) for specimen LB-065

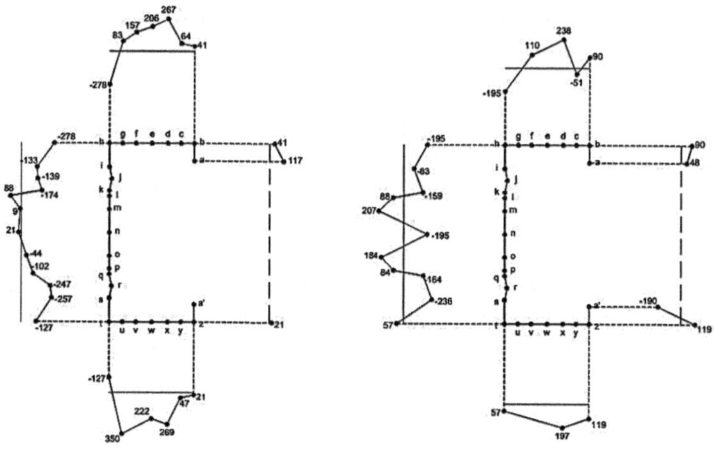

Fig. 9 Residual stress σ_{11} of 3U (left) and 6U (right) for specimen PC-065

4 Conclusions

Based on the experimental test results of residual stress measurement with X-Ray Diffraction cos-α method, conclusions could be taken:

Removing the coating from the specimen's surface is the key point to have a success in the X-Ray Diffraction method applied at a cold formed section. After experiencing forming, the compressive residual stress at the flat steel sheet has been transformed at both mid flanges to residual tensile stress and at the web; the values of compressive residual stress have been reduced by almost 50% for the LB-065 specimens. However, for the PC-065 specimens, the same phenomenon happens at both mid flanges, but at the web, the values of compressive residual stress do not change too much from its flat steel sheet residual stress state. It might come from the different forming process from both manufactures. It is interesting to note that ribs at the web from both specimens LB-065 and PC-65 have caused the residual stress to shift from compression to tensile stress at some nodes near the rib's location. In the near future, another investigation method to obtain residual stress will be conducted to confirm the average compressive and tensile residual stress resulted from this study.

Acknowledgements This research was funded by a research grant: "Publikasi Terindeks Internasional (PUTI) Universitas Indonesia Tahun 2020", NKB-2015/UN2.RST/HKP.05.00/2020 on November 6th, 2020. The experiment was conducted at Neutron Scattering Laboratory of PSTBM-BATAN at Serpong, Indonesia. The authors gratefully acknowledge the support.

References

1. Somodi B, Kövesdi B (2017) Residual stress measurements on cold-formed HSS hollow section columns. J Constr Steel Res 128:706–720
2. Mehner T et al (2018) Residual-stress evolution of cold-rolled DC04 steel sheets for different initial stress states. Finite Elem Anal Des 144:76–83
3. Moen CD, Igusa T, Schafer BW (2008) Prediction of residual stresses and strains in cold-formed steel members. Thin-Wall Struct 46(11):1274–1289
4. Weng CC, Pekoz T (1990) Residual stresses in cold-formed steel members. J Struct Eng 116(6):1611–1625
5. Kabir MZ, Parvizi M, Askari A. determination of residual stresses and strains in cold-formed steel members due to roll-forming process
6. Chinnaraj K, Prasad MS, Lakshmana RC (2014) Investigation of manufacturing residual stresses in cold formed truck frame rail sections. Int J Eng Res Appl 4:162–169
7. Akinlabi SA et al (2018) Investigating resulting residual stresses during mechanical forming process. In: IOP conference series: materials science and engineering. IOP Publishing
8. Queen H et al. Novel electromagnetic sensors for stress measurement and materials testing
9. Civín A et al (2011) Ring-core residual stress measurement: analysis of depth increment distribution for integral equation method. In: International conference engineering mechanics 2011
10. Epp J, Zoch H-W, Bobrov I (2018) Micromagnetic analysis of residual stress distribution in 42CrMo4 steel after thermal and mechanical surface treatment. Mater Res Proc 6
11. Petrucci G, Scafidi M (2008) A new procedure for the evaluation of residual stresses by the hole drilling method based on Newton-Raphson technique. In: Proceedings of ICRS-8 the 8th international conference on residual stress—Denver, CO (USA). Citeseer
12. Lestari S (2004) Residual stress measurements of unblasted and sandblasted mild steel specimens using X-ray diffraction, strain-gage hole drilling, and electronic speckle pattern interferometry (ESPI) hole drilling methods
13. Yerman J, Kroenke W, Long W (1996) Accuracy evaluation of residual stress measurements. Westinghouse Electric Corp., West Mifflin, PA (United States)
14. Schajer GS (2013) Practical residual stress measurement methods. Wiley
15. Haiyang GJAF (2018) Methods of residual stress measurement
16. Deveci M. Types of residual stresses
17. Engstler M (2021) Nondestructive micromagnetic materials characterization at high measuring speed [cited 2021]. Available from https://www.pm-forum.info/nondestructive-micromagnetic-materials-characterization-at-high-measuring-speed
18. Rigaku (2020) Smart site RS portable stress analyzer: instruction manual. Rigaku
19. Yasukawa S (2016) X-ray stress analysis technique using the optimization of d0 with error term—direct refinement solution (DRS) method. Rigaku J 32(2):6–12

Mechanical Properties of Fine-Grained Concrete Using Fine-Red Sand and Fly Ash for Road Construction: A Case Study in Vietnam

Nguyen Thanh Sang, Thai Minh Quan, May Huu Nguyen, and Lanh Si Ho

Abstract Nowadays, in order to decrease environmental impacts such as exhaustion of natural resources and CO_2 discharge, the partial replacement of cement by fly ash and coarse aggregate by crushed fine aggregate in concrete have been attracted many researchers. This study concentrated on an experimental examination of the mechanical properties of fine-grained concrete using fine-red sand placement for crushed sand. Experimental tests were performed to investigate compressive, splitting tensile, and flexural strength. The results reveal that the compressive, splitting tensile, and flexural strength increased with curing time for all mixtures. The increase in the amount of fine-red sand replacement caused a decrease in all mixtures. It was also found that there is a strong correlation between compressive and splitting tensile strength with a high coefficient of determination. Besides, the results of compressive, splitting tensile, and flexural strength satisfied the requirement of concrete used for pavement.

Keywords Compressive strength · Fine-grained concrete · Splitting tensile strength · Flexural strength

N. T. Sang · T. M. Quan
University of Transport and Communications, Hanoi, Vietnam
e-mail: nguyenthansang@utc.edu.vn

T. M. Quan
e-mail: minhquan.thai@utc.edu.vn

M. H. Nguyen · L. S. Ho (✉)
University of Transport Technology, 54 Trieu Khuc, Thanh Xuan, Hanoi 100000, Vietnam
e-mail: lanhhs@utt.edu.vn

M. H. Nguyen
e-mail: maynh@utt.edu.vn

Civil and Environmental Engineering Program, Graduate School of Advanced Science and Engineering, 1-4-1, Kagamiyama, Higashi-Hiroshima, Hiroshima 739-527, Japan

S. A. Kristiawan et al. (eds.), *Proceedings of the 5th International Conference on Rehabilitation and Maintenance in Civil Engineering*, Lecture Notes in Civil Engineering 225,
https://doi.org/10.1007/978-981-16-9348-9_94

1 Introduction

In the construction field, the utilization of Portland cement and coarse aggregate as the main ingredient is a common manner to manufacture conventional concrete. However, overexploitation has led to the depletion of these resources and caused numerous environmental problems [1, 2]. Thus, the need for friendly replacement material is crucial in adapting to environmental requests and sustainable development [3, 4].

In line with the need for friendly construction material, the utilization of fine materials as a substitute replacement solution for coarse aggregate is becoming a typical topic [5, 6]. Theoretically, a fine-grained concrete either does not consist of any coarse aggregate at all or only a small proportion ensuring the fine/coarse ratio keeps higher than 1.0 [7, 8]. In this kind of concrete, the local materials (e.g., seawater, saline sand) and the various industrial wastes (e.g., ground blast furnace slag (GGBFS), fly ash) can be effectively used and also further brings many environmental and economic benefits [9, 10].

The reuse of local sand to utilize fine-grained concrete has been broadly investigated [11, 12]. Bédérina et al. [11] studied the applicability of local sand available in Algeria as a partial effort to substitute coarse aggregate in concrete. Three types of local sand, including dune, river, and a mixture of dune and river sands were employed to produce three corresponding sand concretes that were used to examine the workability and mechanical strength. The obtained results indicated that the local sands could be used to manufacture a more workable, more compact, and more resistant sand concrete while still meeting the requirement in compressive strength compared to conventional concrete. Li et al. [13] presented an investigation to investigate the application of 100% replacement of river and manufactured sand to produce san concrete. A series of low and medium strengths (25–60 MPa) was then achieved, which indicated the application of these sands in producing sand concretes.

The utilization of fly ash (FA) in a concrete mixture can lead to changes in the hydration process and hence influencing concrete properties such as workability, mechanical, and durability characteristics [14, 15]. For example, Agrawal et al. [14] investigated the influence of replacement of natural sand with FA on hardened properties of concrete including mechanical and durability characteristics. The results indicated that the FA fine-grained concrete could obtain a comparable mechanical strength compared to the conventional concrete. The influence of fly ash on the mechanical properties and other properties such as shrinkage, chloride permeability, and carbonation of coral sand concretes was also carried out by Cheng et al. [15]. They indicated that the compressive strength development of fine-grained concrete was higher than conventional concrete at 3 and 7 days, but then lower at 28 days.

In Vietnam, as a developing country, the development of fine-grained concrete utilizing fine sand and fly ash is a promising approach [8, 16]. Le et al. [8] experimentally studied the performance of fine-grained concrete utilizing sea sand and fly ash. Several properties of the concrete including workability, compressive strength under influence of curing conditions were examined. The obtained results indicated that it is feasible to manufacture fine-grained concrete with high strength using sea sand and fly ash. Nevertheless, an investigation of mechanical properties of fine-grained concrete using red-fine sand and fly ash for road construction has not been carried out so far in a condition of Vietnam. Thus, the purpose of this study is, therefore, to investigate mechanical properties of fine-grained concrete using local fine-red sand as a partial replacement for road-pavement construction in a case study in Vietnam.

2 Materials and Methods

2.1 Materials and Specimen Preparation

Materials

To produce fine-grained concrete, ordinary Portland cement (PC 40), crushed sand, and local fine-red sand (dune sand) were used. To reduce the amount of cement used, cement was partially replaced with FA, and the content of fly ash was fixed by 90 kg/m^3. The crushed sand (hereafter, CS) was taken from Binh Thuan province, which can satisfy ASTM C33 standard on grain size distribution. The local fine-red sand (hereafter, RS) was obtained from a red fine sand mine in Binh Thuan province. The RS used in this study partially replaces CS to utilize local materials. The chemical compositions of cement, CS, RS, and FA are introduced in Table 1. The grain size distribution of CS, RS, and four combined mixtures of two sands is shown in Fig. 1. The specific density of cement, crushed sand, red sand, and fly ash is 3.10, 2.69, 2.57, and 2.50 g/cm^3, respectively.

Mixture Proportions

The mixtures of fine-grained concrete were calculated based on the absolute volume of materials. This study used an admixture to reduce the quantity of free water; as a result, the water/binder (w/b) of 0.37 was prepared for entire mixtures to reduce the

Table 1 Chemical compositions of materials used

Chemical composition (%)	PC40	Fly ash (FA)	Red sand	Crushed sand
SiO_2	21.49	49.60	92,69	62.48
Al_2O_3	5.4	21.97	1.61	16.82
Fe_2O_3	3.49	4.90	2.1	1.04
CaO	63.56	0.66	0.50	1.17
MgO	1.40	0.78	1.08	0.45
Na_2O	0.12	0.16	0.27	4.33
K_2O	0.3	3.52	0.46	2.19
LOI	0.19	12.63	–	1.46

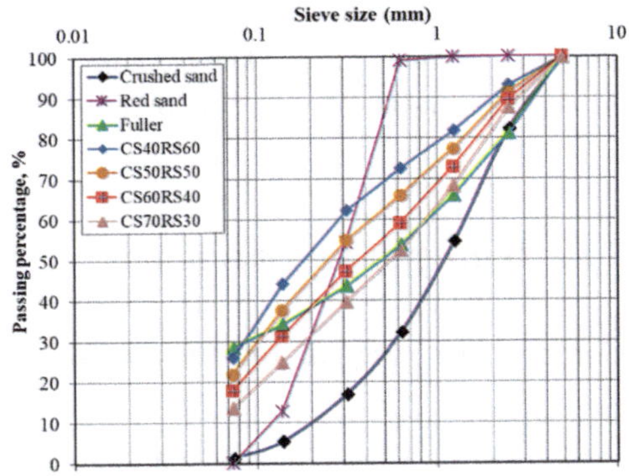

Fig. 1 The grain size distribution of crushed sand, red sand, and different mixtures

Table 2 Mixing proportion of fine-grained concrete

Mixture	w/b	Water (kg/m^3)	Cement (kg/m^3)	Fly ash (kg/m^3)	Crushed sand (kg/m^3)	Red sand (kg/m^3)	Admixture (kg/m^3)
70CS30RS	0.37	165	360	90	1259	540	4.50
60CS40RS	0.37	165	360	90	1075	717	4.50
50CS50RS	0.37	165	360	90	892	892	4.50
40CS60RS	0.37	165	360	90	711	1066	4.50

Note mixture name 70CS30RS indicates that this mixture containing 70% crushed sand and 30% red sand. This is also applied to all remaining mixtures

porosity of this concrete. Four different mixtures were prepared, in which the amount of crushed sand was partially replaced with red sand at 30, 40, 50, and 60%. The amount of fly ash (FA) was kept constant for all mixtures (90 kg/m^3). The amount of admixture (superplasticizer—a water reduction agent with the name RoadCon-SPR3000) was designed as 1.25% by mass of cement. The designed mix proportions of different mixtures are shown in Table 2.

Preparation and Casting of Specimens

All concrete mixtures were produced using a mechanical mixer and the total mixing time was 8 min. First, aggregates (crushed sand and red sand) including fly ash, and cement were mixed together in 2 min under dry conditions. Then, approximately 80% of the water was supplied and mixed for 2 min. Finally, the remaining water and admixture were added, and the mixtures were mixed again for 4 min., The cylindrical specimens having a size of 100 × 200 mm were used to determine compressive, tensile strength and elastic modulus. Regarding the flexural strength, the beam specimens with a size of 400 × 100 × 100 mm were adopted. The cylindrical specimens were prepared with three layers, each layer was compacted 25 times using a steel rod. For beam specimens, they were cast with two layers, and they were compacted using a vibration table. After completing compaction, the top surfaces of the mold were sealed with polyethylene sheets and stored in room condition at 20 °C. After one day, all specimens were demolded and cured in water at 20 ± 2 °C until designated ages. All tests in this study were conducted in triplicate and the mean value was used.

2.2 *Testing of the Specimens*

Compressive Strength and Elastic Modulus Tests

The compressive strength of this concrete was determined via the compression test according to ASTM C39. The compression tests were conducted for the specimens at 3, 7, and 28 days. The value of compressive strength is the mean value of three specimens. The elastic modulus of concrete was calculated based on ASTM C469 standard for the specimen at 28 days. The elastic modulus of each specimen was computed using the following Eq. (1).

$$E_0 = \frac{\sigma_2 - \sigma_1}{\varepsilon_2 - 0.000050} \tag{1}$$

where:

E: elastic modulus, Mpa,
σ_2: stress corresponding to 40% of ultimate load, Mpa,
σ_1: stress with a longitudinal strain, ε_1 of 50 millions, MPa,
ε_2: longitudinal strain caused by stress σ_2.

Splitting Tensile Strength, and Flexural Strength Tests

It is known that the splitting tensile strength is an important property that influences many characteristics of concrete such as controlling cracks, stiffness, bonding capacity to reinforcement, and durability. The splitting tensile strength was carried out based on ASTM C496 on specimens at 3, 7, and 28 days of curing. To determine the flexural behavior of this fine-grained concrete, the flexural strength test was then conducted for the beam specimens at 28 days in accordance with ASTM C293/C293M–16 standard.

3 Results and Discussion

3.1 *Compressive Strength and Elastic Modulus*

Figure 2 shows the compressive strength development of different mixtures. It can be observed that the compressive strength increased with curing time for all mixtures as

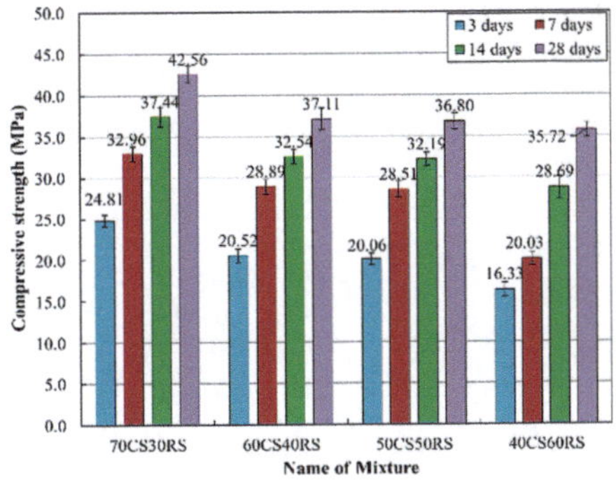

Fig. 2 Compressive strength of different mixtures

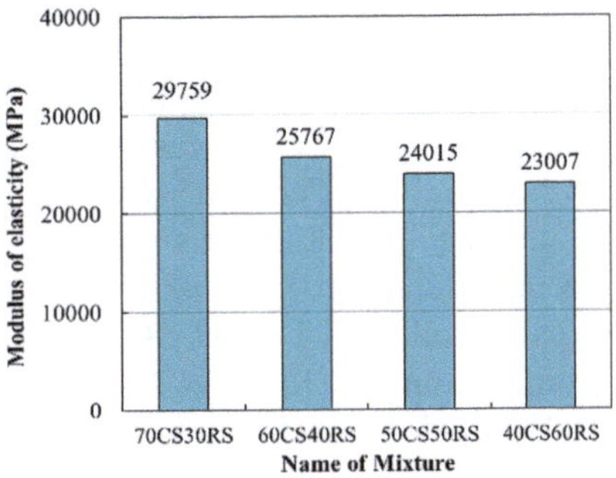

Fig. 3 Elastic modulus of different mixtures at 28 days

found in previous studies [8, 16]. The increase in compressive strength, in this case, is mainly contributed to the hydration of cement. For all curing ages, the compressive strength decreased with an increase in the amount of RS. This decrease can be caused by the low strength and frictionl of red sand compared to that of crushed sand. The rate of decrease in the compressive strength of all curing ages is approximately the same for all mixtures, this can also be explained by the decrease in the amount of crushed sand.

The elastic modulus of four mixtures at 28 days is presented in Fig. 3. Similar to the case of the compressive strength, the elastic modulus decreases with increase in the amount of red sand (i.e. from 30 to 60%). This was also caused by the increase in the amount of RS that causes the reduction of compressive strength and modulus. The highest elastic modulus was observed in mixture 70CS30RS (the lowest amount of RS), while the lowest value was found in mixture 40CS60RS.

3.2 *Splitting Tensile and Flexural Strength*

The splitting tensile strength of different mixtures is presented in Fig. 4. It can be seen that the splitting tensile strength increased with curing time for all mixtures,

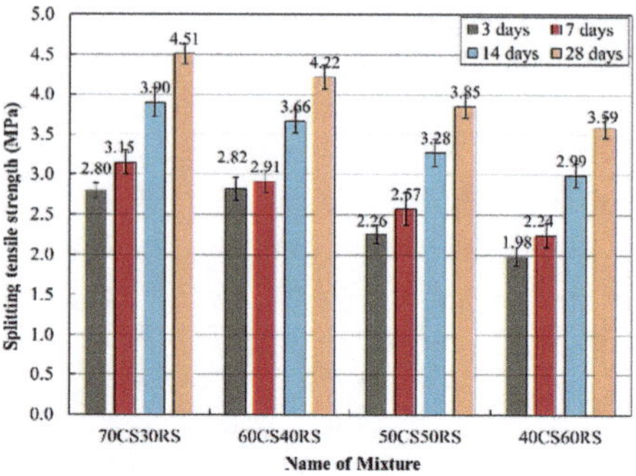

Fig. 4 The splitting tensile strength of different mixtures

as found in previous studies [8, 16]. Comparable to the compressive strength, the increase in the splitting tensile with time is mainly attributed to the cement hydration. Comparing with different mixtures, in general, it can be observed that the splitting tensile strength also decreased with increasing in the amount of RS, it is confirmed with the result of compressive strength. At the age of 3 days, the splitting tensile of mixture 70CS30RS was also almost equal to that of mixture 60CS40RS, while it decreased rapidly from mixture 60CS40RS to mixture 50CS50RS. For other curing ages, including 7, 14, and 28 days, the splitting tensile strength decreased gradually from mixture 70CS30RS.

Figure 5 shows the flexural strength of four mixtures at the age of 28 days. It was found that the flexural strength generally decreased with decreasing amount of crushed sand (CS) (i.e. increasing amount of RS) and it also agreed well with the results of the compression test. The flexural strength gradually decreased from mixture 70CS30RS to mixture 50CS50RS, and then it decreased rapidly from mixture 50CS50RS to mixture 40CS60RS. The highest value (7.70 MPa) of flexural strength was observed in mixture 70CS30RS, and the lowest value (5.18 MPa) was found in mixture 40CS60RS. It indicates that the increase in the amount of red sand leads to a decrease in flexural strength of this type of concrete.

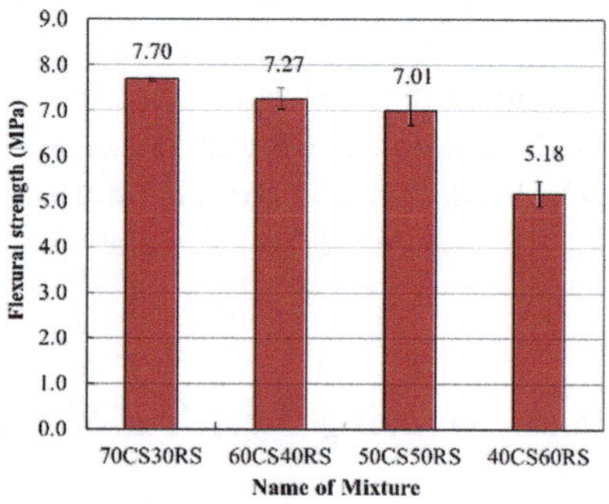

Fig. 5 The flexural strength of different mixtures at 28 days

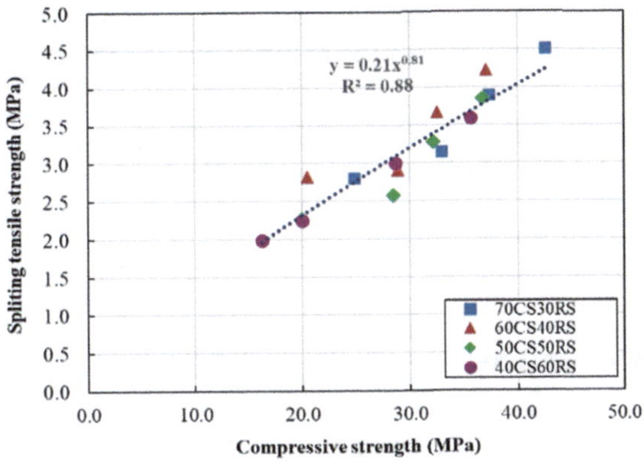

Fig. 6 Relation between splitting tensile strength and compressive strength

3.3 *Correlation Between Compressive Strength and Splitting Tensile Strength*

The correlation between compressive and splitting tensile strength is described in Fig. 6. Generally, this relationship is presented in a power-type equation as being seen in many previous studies of conventional concrete [17, 18] and fine-grained concrete [16]. According to the figure, it can be seen that the splitting tensile strength grew up with an increment of compressive strength as found in normal concrete and fine-grained concrete in previous publications [16, 19]. In addition, it is also found that this relationship has a high coefficient of determination, as found in the previous study of fine-grain concrete [16]. It indicates that this fine-grained concrete using red sand has similar behavior compared to normal concreted and fine-grained concrete using saline sand in the previous study.

4 Conclusions

In this study, the mechanical properties such as compressive, tensile, and flexural strength of fine-grained concrete using local fine red sand as a partial crushed sand replacement and fly ash. Some main concluding remarks can be derived from the experiment results as follows.

- The compressive, tensile, and flexural strength increased with curing time. When the amount of fine red sand increased, those strengths decreased.
- The increase in the amount of fine-red sand replacement caused a decrease in all mixtures.
- There is a strong correlation between compressive strength and splitting tensile strength with a high coefficient of determination, as found in previous studies.

- Compressive, tensile, and flexural strength results satisfied the requirement of concrete used for the requirement of concrete for pavement.

Acknowledgements This research is funded by the Project of Vietnam Ministry of Science and Technology, "Study on using saline sand to build transport work" under grant number DTDL.CN-23/19."

References

1. El-Fadel M, Zeinati M, El-Jisr K, Jamali D (2001) Industrial-waste management in developing countries: the case of Lebanon. J Environ Manage 61:281–300
2. Henry RK, Yongsheng Z, Jun D (2006) Municipal solid waste management challenges in developing countries–Kenyan case study. Waste Manage 26:92–100
3. Troschinetz AM, Mihelcic JR (2009) Sustainable recycling of municipal solid waste in developing countries. Waste Manage 29:915–923
4. Dash MK, Patro SK, Rath AK (2016) Sustainable use of industrial-waste as partial replacement of fine aggregate for preparation of concrete—a review. Int J Sustain Built Environ 5:484–516
5. Akono A-T, Zhan M, Chen J, Shah SP (2021) Nanostructure of calcium-silicate-hydrates in fine recycled aggregate concrete. Cem Concr Compos 115:103827
6. Li J, Yang E-H (2017) Macroscopic and microstructural properties of engineered cementitious composites incorporating recycled concrete fines. Cem Concr Compos 78:33–42
7. Bederina M, Marmoret L, Mezreb K, Khenfer MM, Bali A, Quéneudec M (2007) Effect of the addition of wood shavings on thermal conductivity of sand concretes: experimental study and modelling. Constr Build Mater 21:662–668
8. Le HT, Nguyen ST, Ludwig H-M (2014) A study on high performance fine-grained concrete containing rice husk ash. Int J Concr Struct Mater 8:301–307
9. Bai Y, Darcy F, Basheer PAM (2005) Strength and drying shrinkage properties of concrete containing furnace bottom ash as fine aggregate. Constr Build Mater 19:691–697
10. Xiao J, Qiang C, Nanni A, Zhang K (2017) Use of sea-sand and seawater in concrete construction: current status and future opportunities. Constr Build Mater 155:1101–1111
11. Bédérina M, Khenfer MM, Dheilly RM, Quéneudec M (2005) Reuse of local sand: effect of limestone filler proportion on the rheological and mechanical properties of different sand concretes. Cem Concr Res 35:1172–1179
12. Gurumoorthy N, Arunachalam K (2016) Micro and mechanical behaviour of treated used foundry sand concrete. Constr Build Mater 123:184–190
13. Nanthagopalan P, Santhanam M (2011) Fresh and hardened properties of self-compacting concrete produced with manufactured sand. Cem Concr Compos 33:353–358
14. Agrawal US, Wanjari SP, Naresh DN (2019) Impact of replacement of natural river sand with geopolymer fly ash sand on hardened properties of concrete. Constr Build Mater 209:499–507
15. Cheng S, Shui Z, Sun T, Yu R, Zhang G, Ding S (2017) Effects of fly ash, blast furnace slag and metakaolin on mechanical properties and durability of coral sand concrete. Appl Clay Sci 141:111–117
16. Sang NT, Quan TM, Nguyen MH, Ho LS (2021) Performances of eco-fine-grained concrete containing saline sand as partial fine aggregate replacement. J Appl Sci Eng 24:527–539
17. Behnood A, Verian KP, Gharehveran MM (2015) Evaluation of the splitting tensile strength in plain and steel fiber-reinforced concrete based on the compressive strength. Constr Build Mater 98:519–529

18. Legeron F, Paultre P (2000) Prediction of modulus of rupture of concrete. Materials Journal 97:193–200
19. Chhorn C, Hong SJ, Lee SW (2018) Relationship between compressive and tensile strengths of roller-compacted concrete. J Traff Transp Eng (English Edn) 5:215–223

A Systematic Review of Concrete Material for Noise Reduction of Transportation Sectors

Ecky Ferry Ferdyan, Dewi Handayani, Sholihin As'ad, and Ubaidillah

Abstract Noise is a sound whose presence is considered to interfere with hearing which can cause health problems and is caused by noise sources. The most common source of moving noise is transportation. Therefore it is necessary to have a noise absorber to reduce noise levels. The most used noise-reducing material is concrete. This paper using a systematic review method and aims to investigate the development of concrete materials as a noise barrier wall. Concrete for noise absorbers is divided into two, conventional and non-conventional. Non-conventional concrete is a mixture of concrete with added material and is generally a waste material. Non-conventional concrete with different added materials produces other characteristics and different noise attenuation results. Some examples of added materials such as waste materials from burning coal, fishery waste, waste ceramics, and polymeric materials provide significant results. The absorption coefficient was able to reach 0.9, and the NRC value earned 0.61. However, a higher ability to absorb noise, smaller compressive strength of the concrete produced. It is necessary to understand the characteristics of the material for use in applications in transportation.

Keywords Reducing noise · Transportation · Material concrete

1 Introduction

Noise is a sound whose presence is considered to interfere with hearing [1]. Noise is unwanted sound from a business or activity at a certain level and time that can

E. F. Ferdyan (✉) · D. Handayani · S. As'ad
Department of Civil Engineering, Faculty of Engineering, Sebelas Maret University, Surakarta 57126, Indonesia

D. Handayani
e-mail: dewi@ft.uns.ac.id

Ubaidillah
Department of Mechanical Engineering, Faculty of Engineering, Sebelas Maret University, Surakarta 57126, Indonesia

S. A. Kristiawan et al. (eds.), *Proceedings of the 5th International Conference on Rehabilitation and Maintenance in Civil Engineering*, Lecture Notes in Civil Engineering 225,
https://doi.org/10.1007/978-981-16-9348-9_95

cause disturbances to human health and environmental comfort. Based on the Decree of the State Minister for the Environment No. Kep-48/MENLH/11/1996 concerning noise level quality standards, there are noise thresholds determined and depend on the area's designation. The area with the lowest noise limit is green open space, while the site with the highest noise limit is the trade, industry, recreation, and seaport area, which is 70 dB [2].

A noise source causes noise. Noise sources are sound sources whose presence is considered to interfere with hearing both from moving and immovable sources [1]. A movable source is a moving or non-permanent emission source in a place originating from a motorized vehicle. Meanwhile, an immovable source is a source of emission which is fixed in one place [3].

Highway traffic noise is a source of environmental noise and comes from moving sources, namely motor vehicles. Road traffic noise is influenced by the type of vehicle and vehicle speed [4]. Apart from the vehicle engine factor, road noise occurs due to the interaction between tires and pavement. Noise that results from the interaction between tires and the dominant pavement at speeds above 40 km/h and influenced by four main factors, namely:

- Pavement surface characteristics (surface texture, friction, aggregate properties, porosity, pavement materials, and their properties, pavement life, and durability).
- Tire characteristics (tire dimensions, rubber hardness, trend pattern, tire age and wear, poles, and tire number).
- Environmental factors (temperature, humidity, wind, water layer thickness, and dust).
- Driver-controlled factors (driving speed, tire load, and pressure, tangential force, and acceleration) [5].

There are two methods for reducing noise: the noise isolation method and the noise absorption method. Noise isolation methods reduce noise by interfering with and blocking the propagation of sound. In contrast, the noise absorption method reduces noise by absorbing and weakening sound energy [6]. The most frequently used noise isolation method is a reflective noise barrier. Reflective noise barriers are designed to reflect most of the traffic noise, which can be problematic when minimizing sound reflection to noise-sensitive areas adjacent to roadways is required. Simultaneously, noise reduction can reduce the reflected sound's intensity on the wall surface [7].

This paper aims to determine how the development of concrete material as a noise barrier wall in the transportation sector can be helpful for research or use of noise barrier walls in the future as a reference in determining the material to be selected. So that later, it can be seen how the properties, characteristics, and effectiveness of various materials for noise barrier walls.

2 Method

The method used is a literature review. Literature Review is a description of the theory, findings, and other research materials obtained from reference materials to be used as a basis for research activities to formulate a clear frame of mind from the formulation of the problem to be investigated. This journal uses 32 references consisting of government regulations, books, journals, and proceedings. There are four references to government regulations, two books, five proceedings, and 21 Scopus indexed journals. From 21 journals, there are 14 journals rated Q1 and seven journals rated Q2.

3 Concrete

3.1 Conventional

In Indonesia, the required noise barrier walls are used ALWA material. ALWA (Artificial Light Weight Aggregate) is a kind of artificial aggregate processed at high temperatures and relatively lighter than ordinary aggregate or other artificial aggregates. ALWA can be a by-product of the iron mill as a result/process in a blast furnace. Composition = Cement:Sand:ALWA = 1:4:4 [8]. Recent studies have shown that an effective noise barrier wall is used as porous concrete. Porous concrete is a rigid acoustic absorbent material containing large cavities (pores) intentionally made for acoustic absorption or permeation. Although ordinary concrete is also an acoustic absorbent material, Porous concrete is a material that is more widely used because of its excellent mechanical properties and durability compared to ordinary concrete [9]. Porous concrete is made using coarse aggregate with little or no fine aggregate. So that porous concrete has a continuous cavity and has high permeability. As a result, the pores in the porous concrete absorb sound energy through internal friction. The type of aggregate also affects the properties of the porous concrete itself [10].

3.2 Non-conventional

Coal Bottom Ash. Fly ash, coal bottom ash (CBA), boiler slag, and flue gas desulphurization (FGD) sludge are waste generated during the combustion process at coal-fired power plants [11]. Coal waste is hazardous to health and the environment, so there is a need for innovation to use this waste not to endanger the environment but can be helpful, one of which is the use of CBA. CBA is gradually introduced into the construction industry as a base layer for roads, pavements, structural stockpiles, drainage media, aggregates for concrete, and prefabricated soil products to reduce environmental pollution [12]. CBA has been used as a partial substitute for natural aggregate (fine or coarse aggregate) [13]. CBA as a substitute for fine aggregate with a percentage of 80, 90, 100% can absorb sound by 35% [11]. CBA as a substitute for aggregate coarse with a ratio of 100% in concrete can absorb noise up to a coefficient of 0.43; however, CBA concrete has poor resistance to acids and sulfuric [10]. Apart from using laboratory testing, testing has also been carried out using the software. The use of software to determine the absorption coefficient of sound is adequate. The highest sound absorption coefficient that the material can absorb is more than 0.9. For semi-applicative use, concrete with coal bottom ash and made semi-industrial with multilayer can produce a sound reduction of 4 dB [14].

Steel Slag. Steel slag is a by-product of steel production, accounting for about 15% of steel output [15]. The porosity and NRC of the steel slag's sound-absorbing material can reach more than 60.0 and 0.50%, respectively, and the compressive strength of the sintered specimen is above 6.5 MPa [16]. One product that is similar to steel slag is Air-Cooled Blast Furnace Slag. Air-Cooled Blast Furnace Slag (ACBFS) is a by-product of steel and iron manufacturing. ACBFS is produced by compacting the liquid slag formed on top of the cast iron in a blast furnace under environmental conditions. It is a slow process that allows the liquid slag molecules to be oriented to a crystalline state [17]. The larger the ACBFS aggregate size, the greater the absorption coefficient value, and it can even reach 0.9 at a frequency of 1000 Hz. However, the higher the absorption value, the smaller the concrete's compressive strength [18]. The proportion of steel slag that reaches 50% can produce NRC 0.47, and the compressive strength is 3.0 MPa [19].

Mollusk Shell. The amount of waste produced from industry increases every year and can cause environmental and public health problems. In the water sector, about 6–8 million tonnes of waste a year are produced worldwide, and only a quarter of the waste generated is reused [20]. One of the water wastes is shellfish. Shell waste with a proportion of 80% as a substitute for aggregate, an aggregate size of 2–7 mm, can produce NRC reaching 0.45 while ordinary porous concrete only reaches 0.24 [21]. Noise barriers made with shell waste have a higher impact between 32 and 267% than usual porous concrete noise barriers [22].

Ceramic Waste. It is estimated that about 40% of the daily production in the ceramic industry is stockpiled. Therefore, the need to find new applications in other sectors where ceramic waste can be recycled is significant [23]. The proportion of ceramic waste is 80% as a substitute for coarse aggregate, and the particle sizes are divided into fine, medium, and coarse. Thickness 40, 80, 120 mm. Ceramic waste with a percentage of 80% and a large particle size, 120 mm thick, resulted in NRC 0.61 [24].

Fly Ash. Fly ash is a material that pollutes the environment and has the potential to damage human health. Research on the application of the waste is being intensified, a mixture for making concrete. Fly Ash has a proportion of 20, 30, 40, 50% used as a substitute for cement. Fly Ash's result with a percentage of 50% produces an NRC of 0.5 [25]. The proportion of fly ash and construction waste is 20, 50, 80%

as a substitute for coarse aggregate. 80% construction waste and 20% fly ash yield a maximum absorption coefficient of 0.9 [16]. With an average volume density of 0.93 g/cm^3.

Polymeric Material. Polymer foam includes several cells in the polymer matrix. Foam can have an open, partially open, or closed-cell structure. Polymer foams require less material than solid polymers of the same volume, so the material costs are reduced in many applications [26]. The use of polyuteran type polymer foam with a proportion of 60% and textile waste of 40% produced an NRC of 0.593 [27]. Mixing 70% metakaolin and 30% blast furnace slag powders can produce inorganic polymer foam with densities ranging from 0.4 to 1.0 g/cm^3. The most significant sound absorption coefficient is at a density of 0.4 g/cm^3, with a maximum value reaching 0.9 [28].

Table 1 has summarized the studies using the materials described above. We can see that each material has a different sound absorption coefficient and NRC value. There are also materials here that can replace coarse aggregate, fine aggregate, or cement. It depends on the type of material itself. The maximum absorption coefficient reaches 0.9, and the highest NRC can earn 0.61.

Table 1 NRC values of the use of unconventional concrete from several studies

References	Year	Type of material	As a replacement	Percentage (%)	Absorption coefficient (Max)	NRC
[7]	2017	Coal bottom ash	Fine aggregates and coarse aggregates	100	0.9	–
[10]	2017	Coal bottom ash	Aggregate coarse	100	–	0.43
[11]	2020	Coal bottom ash	Fine aggregates	80, 90, 100	–	0.35
[16]	2015	Steel slag	Fine aggregates	60	–	0.5
[16]	2015	Fly ash	Aggregate coarse	80	0.9	–
[19]	2014	Steel slag	Cement	50	–	0.5
[21]	2019	Shell waste	Fine aggregates and coarse aggregates	80	–	0.45
[24]	2016	Ceramic waste	Aggregate coarse	80	–	0.61
[25]	2016	Fly ash	Cement	50	–	0.5
[27]	2016	Polymeric material	Aggregate	60	–	0.593
[28]	2014	Polymeric material	Aggregate	70	0.9	–

4 Use for Transportation

A conventional concrete barrier to reduce noise indicates that the noise-absorbing material is generally uneven and has a porous cavity to reduce transportation noise [29]. Panel Material Paragon 23-T produces a noise suppression effect on the highway due to the most effective transportation, namely 13.58 dB, then precast concrete is second, and the last is steel. It is probably because the 23 T paragon panel is the thickest material than the others [30]. The tree material, Concrete Hollow, Concrete Panel, shows that those that meet the requirements of an effective noise barrier that can reduce noise at least 5 dB are hollow concrete and panel concrete, while vegetation is not. Concrete Hollow shows the results of noise reduction above 5 dB but at specific frequencies below 5 dB. It is inconsistent in reducing the impact of noise compared to concrete panels. It may be because hollow concrete requires many joints, so leaking sound from these joints [31].

5 Conclusion

One of the efforts to prevent noise in transportation is to use a noise barrier wall. Research on concrete materials for noise barrier walls has been widely carried out, and the type of material significantly affects the effectiveness in reducing noise. Various studies have shown that non-conventional materials such as waste materials from burning coal, fishery waste, ceramic waste, and polymeric materials can absorb more than 35% noise. However, because the structure of noise-reducing concrete is concrete that has many cavities, the strength of the concrete is also tiny.

References

1. Construction and Building Guidelines (Pd T-16-2005-B). Mitigation of noise impacts due to road traffic. Public Works Department (2005)
2. Minister of Environment Decree no. Kep-48/MENLH/11/1996 (1996)
3. Government Regulation of the Republic of Indonesia Number 41. Concerning air pollution control (1999)
4. Dubey AK, Shukla AK, Kumar S (2019) Assessment of noise variation on different pavements with four wheeler vehicle. Int J Sci Res Rev 7
5. Kleiziene R, Šernas O, Vaitkus A, Simanavi R (2019) Asphalt pavement acoustic performance model. Sustainability 11
6. Ling S, Yu F, Sun D, Sun G, Xu L (2021) A comprehensive review of tire-pavement noise: generation mechanism, measurement methods, and quiet asphalt pavement. J Clean Prod 287
7. Arenas C, Leiva C, Vilches LF, Ganso J (2017) Approaching a methodology for the development of a multilayer sound absorbing device recycling coal bottom ash. Appl Acoust 115:81–87
8. Keputusan Direktur Jenderal Bina Marga No. 076/Kpts/Db/1999. Pedoman Perencanaan Teknik Bangunan Peredam Bising (1999)
9. Kim HK, Lee HK (2010) Acoustic absorption modeling of porous concrete considering the gradation and shape of aggregates and void ratio. J Sound Vibr 329:866–879

10. Carlos L, Celia A, Luis F, Fatima A, Yolanda L (2017) Assessing durability properties of noise barriers made of concrete incorporating bottom ash as aggregates. Euro J Environ Civ Eng 10
11. Hannan R, Shahidan S, Ali N, Bunnori N, Zuki S, Ibrahim M (2020) Acoustic and non-acoustic performance of coal bottom ash concrete as sound absorber for wall concrete. Case Stud Constr Mater 399
12. Kurama H, Kaya M (2008) Usage of coal combustion bottom ash in a concrete mixture. Constr Build Mater 9:1922–1928
13. Singh M, Siddique R (2015) Properties of concrete containing high volumes of coal bottom ash as fine aggregate. J Clean Prod 91:269–278
14. Arenas C, Leiva C, Vilches LF, Cifuentes H, Galán M (2015) Technical specifications for highway noise barriers made of coal bottom ash-based sound absorbing concrete. Constr Build Mater 95:585–591
15. Guo WB, Cang DQ, Yang ZJ, Li Y, Wei CZ (2011) Study on preparation of glass-ceramics from reduced slag after iron melt-reduction. Bull Chin Silica Soc 30(5):1189–1192
16. Sun P, Guo Z (2015) Preparation of steel slag porous sound-absorbing material using coal powder as pore former. J Environ Sci 536:67–75
17. WS Association: Steel Statistical Yearbook (2014) World steel committee on economic studies, Brussels
18. Rios JD, Arenas C, Cifuentes H, Peceño B, Leiva C (2019) Porous structure by X-Ray computed tomography and sound absorption in pervious concretes with air-cooled blast furnace slag as coarse aggregate. Acoustics, Australia
19. Sun P, Zhan-cheng G (2014) Sintering preparation of porous sound-absorbing materials from steel slag. Trans Nonferrous Metals Soc China 25:2230–2240
20. Yan N, Chen X (2015) Sustainability: don't waste seafood waste. Nature 524(7564):155–157
21. Peceno B, Arenas C, Farin B, Carlos (2019) Substitution of coarse aggregates with mollusk-shell waste in acoustic-absorbing concrete. J Mater Civ Eng 31:6
22. Peceño B, Leiva C, Fariñas B, Schmid A (2020) Is recycling always the best option? Environmental assessment of recycling of seashell as aggregates in noise barriers. Processes 8:776
23. Torgal F, Jalalib S (2010) Reusing ceramic wastes in concrete. Constr Build Mater 24(5):832–838
24. Arenas C, Vilches LF, Leiva C, Fariñas B, Galán M (2016) Recycling ceramic industry wastes in sound-absorbing materials. Mater Constr 66:324
25. Arenas C, Luna-Galiano Y, Leiva C, Vilches LF, Arroyo F, Villegas R, Fernández-Pereira C (2017) Development of a fly ash-based geopolymer concrete with construction and demolition wastes as aggregates in acoustic barriers. Constr Build Mater 134:433–442
26. Niaounakis M (2015) Biopolymers: processing and products. Chapter 9—foaming and foamed products. Elsevier
27. Tiuca A, Vermeşan H, Gabora T, Vasile O (2016) Improved sound absorption properties of polyurethane foam mixed with textile waste. Energy Proc 85:559–565
28. Hung T, Huang J, Wang Y, Lin K (2014) Inorganic polymeric foam as a sound-absorbing and insulating material. Constr Build Mater 50:328–334
29. Christopher W, Menge BD (2011) Reflections from highway noise barriers and use of absorptive materials in the United States. J Transp Res Board:161–166
30. Clavijo P, Yang JJ (2020) A comparative study of insertion loss of traffic noise barriers in Georgia. In: International conference on transportation and development
31. Halima H, Abdullah R, Abdullah A, Ali A, Nor M (2015) Effectiveness of existing noise barriers: comparison between vegetation, hollow block concrete, and concrete panel. Proc Environ Sci 30:217–221

Mapping Literature of Reclaimed Asphalt Pavement Using Bibliometric Analysis by VOSviewer

Mochammad Qomaruddin, Han Ay Lie, Widayat, Bagus Hario Setiadji, and Mochamad Agung Wibowo

Abstract The purpose of this research is to map the literature on reclaimed asphalt pavement (RAP) and to determine future related potential topics. Data were collected from the Scopus meta-data and analyzed using the bibliometric analysis technique with the VOSviewer tool. The result showed that there are several relationships between RAP and the topics related to aggregate, mixing, hot-mix asphalt, and compressive strength. However, some topics such as extraction, microstructural properties, furnaces, and interfacial transition zone in the RAP have not been widely researched. Numerous studies have been carried out on RAP by authors such as Xiao F, Amirkhanian S N, Daniel J S, Canestrari F, Huang B, Zaumanis M, and Arulrajah. Some of the countries that have contributed significantly to RAP research are the United States of America, China, Italy, and India.

Keywords Reclaimed asphalt pavement · Bibliometric · VOSviewer

M. Qomaruddin (✉) · H. A. Lie · B. H. Setiadji · M. A. Wibowo
Department of Civil Engineering, Universitas Diponegoro, Semarang, Indonesia
e-mail: qomar@unisnu.ac.id

H. A. Lie
e-mail: hanaylie@live.undip.ac.id

B. H. Setiadji
e-mail: bhsetiadji@ft.undip.ac.id

M. A. Wibowo
e-mail: agung.wibowo@ft.undip.ac.id

M. Qomaruddin
Department of Civil Engineering, Universitas Islam Nahdlatul Ulama, Jepara, Indonesia

Widayat
Department of Chemical Engineering, Universitas Diponegoro, Semarang, Indonesia
e-mail: widayat@live.undip.ac.id

Advanced Material Laboratory Central of Research and Service Unit, Universitas Diponegoro, Semarang, Indonesia

S. A. Kristiawan et al. (eds.), *Proceedings of the 5th International Conference on Rehabilitation and Maintenance in Civil Engineering*, Lecture Notes in Civil Engineering 225,
https://doi.org/10.1007/978-981-16-9348-9_96

1 Introduction

Reclaimed Asphalt Pavement (RAP) is a waste material obtained from dredging flexural pavement using a Cold Milling Machine [1]. Numerous studies have been conducted [2] to determine the more efficient and economical use of RAP [3]. Topic distribution is very broad in mapping the RAP because it has various purposes. During data collection, the word "Reclaimed Asphalt Pavement" was entered into a search engine, and the majority of the information that appeared was on materials, hot mix, and concrete. However, multidisciplinary RAP research was found from the Scopus database. The problem associated with RAP is the difficulty in mapping the scientific development cluster, based on the relationship between the authors and countries in this research's citation and correspondence section. Generally, studies on the mapping of publications related to asphalt and bitumen [3] have been carried out, although it is not specific to RAP waste. Therefore, this research aims to explore the topic of RAP using the bibliometric VOSviewer method for authors to identify and acquire research gaps from its phenomena globally easily. This quantitative research comprises systematic articles, transparently and synthetically collected, to assess study findings on topics related to RAP [4, 5]. Bibliometric, which is a statistical analysis method that connects text or keywords between papers [6], including the citation relationship between authors [7], the co-occurrence of keywords with author institutions [8, 9], and analysis was used to carry out this research [10]. An author from Indonesia also wrote on RAP.

2 Methodology

This research analyzes the SCOPUS indexed literature on RAP using the bibliometric method with the help of VOSviewer. This software has the ability to map keyword, author, and citation relationships. Scopus has a large database and complete metadata analysis, which was downloaded on February 7, 2021. This method is used to determine the information that contributes to answers to research questions on RAP. Systematic Review is carried out through various stages of data collection [11, 12], Filtering, Identifying and Interpreting on VOSviewer. The process flow is shown in Fig. 1. The data collection stage is carried out by entering the keyword "Reclaimed Asphalt Pavement" to obtain the associated articles. The data collection process using the keywords "Reclaimed Asphalt Pavement" is shown in Fig. 2.

The second process in filtering needs to focus on the subject area by increasing the relevance between the articles. This filtering is limited to journals published from

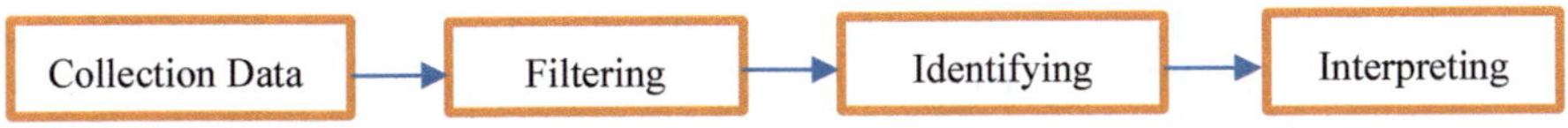

Fig. 1 Scopus bibliometric data interpretation flow

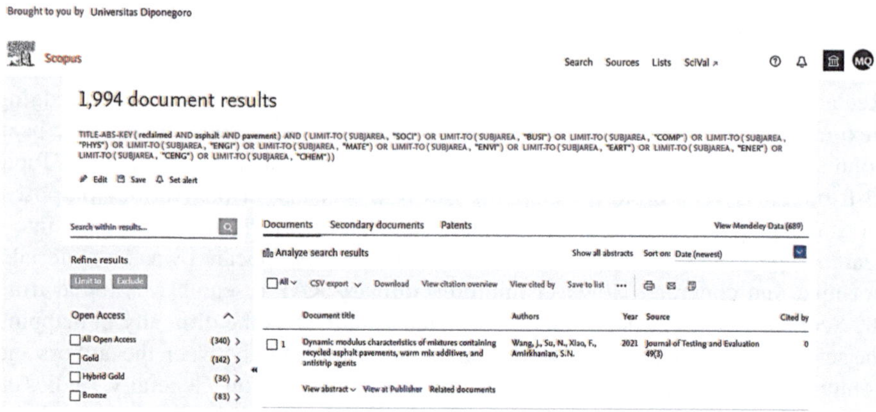

Fig. 2 Scopus document results (scopus.com)

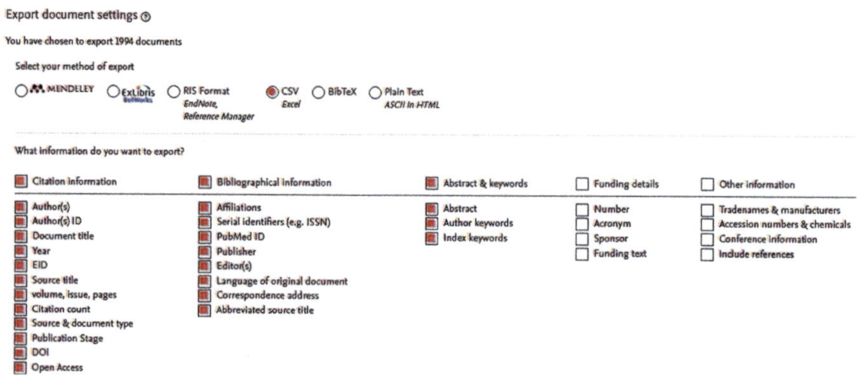

Fig. 3 Export Scopus document to excel CSV (scopus.com)

1999 to 2021. Subject areas selected are Engineering, Material science, Environmental science, Earth and Planetary Sciences, Energy, Chemical Engineering, and Chemistry. Types of documents include articles, conference papers, book chapters, reviews, and conference reviews, while the choice of sources is taken from journals, conference proceedings, book series, trade journals, and books. The publication stage was selected for the final document and article in press with the filtering process shown in Fig. 3. A total of 1994 articles were obtained from the search results using the relevant keywords in accordance with the citation, bibliographical, and abstract. The selected results are saved in a CSV file format, which is then used to further analyze VOSviewer.

The identification process using Co-occurrence was summarized in CSV and created on VOSviewer with the following steps: Create a map based on bibliographic data → Read data from bibliographic database files → Choose the type of analysis and counting method (Co-occurrence or Co-authorship) → choose threshold Co-occurrence with a minimum of 5 keywords → choose the number of keywords to be selected (500), then run analysis Visualization VOSviewer and interpretation.

3 Result and Discussion

3.1 RAP Publication Development

The Scopus data that is mined in relation to the publication on the topic of RAP increases every year, as shown in Fig. 4. However, from 1999 to 2003, there was no increase in publications on RAP by looking for asphalt content using the centrifugal method [13] and mixed research variants [14]. RAP was explored in concrete [15, 16] after 2011 and continued to increase irrespective of the slight decrease from 2015 to 2018.

The authors that conducted the reclaimed asphalt pavement were recorded in a total of 44 countries divided into 10 clusters, and 17 citations and corresponding links are shown in Fig. 5. Research on RAP has been carried out in various countries, which indicates that this topic is quite interesting and of a global nature. However, this study is dominated in America, China, Italy, and India.

The United States of America has the largest citation and contributors on RAP research. In addition, a total of 8 countries reinforce each other in citation between the authors, as shown in Table 1.

The citation analysis in Fig. 6 visualizes the network between one author and another through digitalization. The total number of authors on the topic of "reclaimed asphalt pavement" is 239, grouped into 19 clusters and interconnected with citation and relationship. They are further marked by differences in color and interconnected

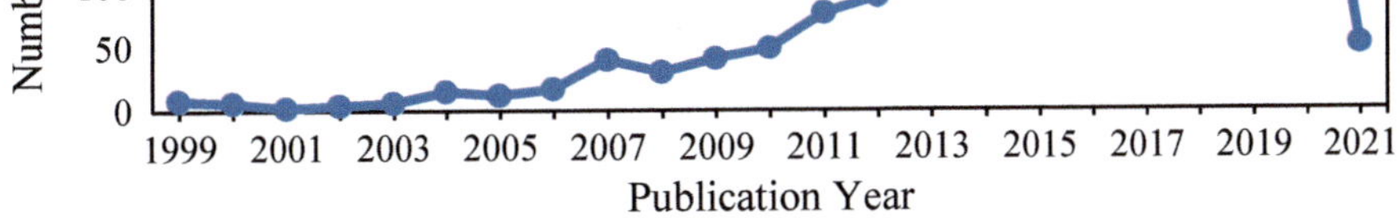

Fig. 4 Graph of the number of RAP publications each year

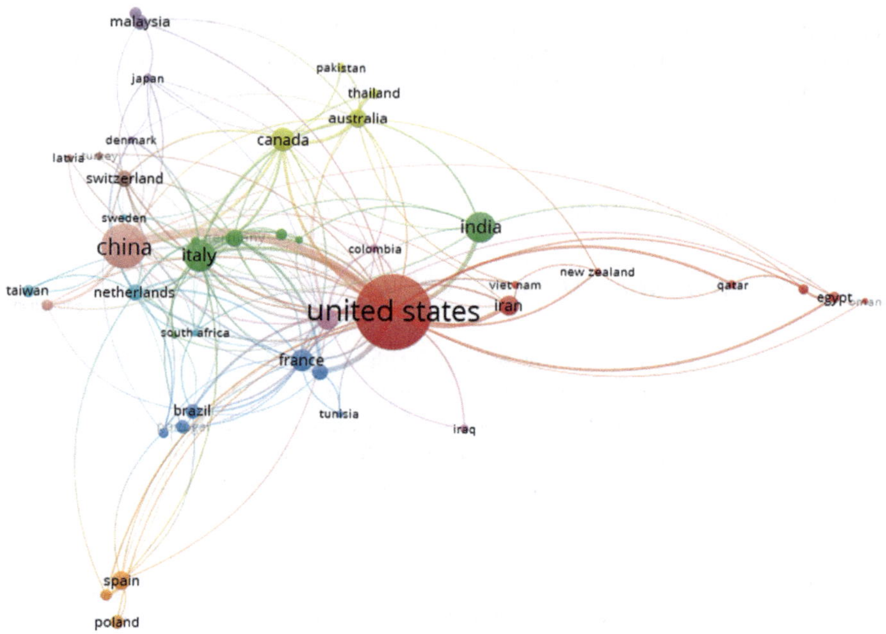

Fig. 5 Network visualization country

Table 1 Number of country clusters

Cluster country	Number of countries	Country name
Cluster 1	8	Egypt, Iran, New Zealand, Oman, Qatar, Saudi Arabia, United State, Vietnam
Cluster 2	7	Czech Republic, Germany, Greece, India, Italy, San Marino, Serbia
Cluster 3	6	Belgium, Brazil, France, Portugal, South Korea, Tunisia
Cluster 4	4	Australia, Canada, Pakistan, Thailand
Cluster 5	4	Denmark, Japan, Malaysia, Nigeria
Cluster 6	4	Netherlands, South Africa, Sweden, Taiwan
Cluster 7	3	Chile, Poland, Spain
Cluster 8	3	Latvia, Switzerland, Turkey
Cluster 9	3	Colombo, Iraq, United Kingdom
Cluster 10	2	China, Hong Kong

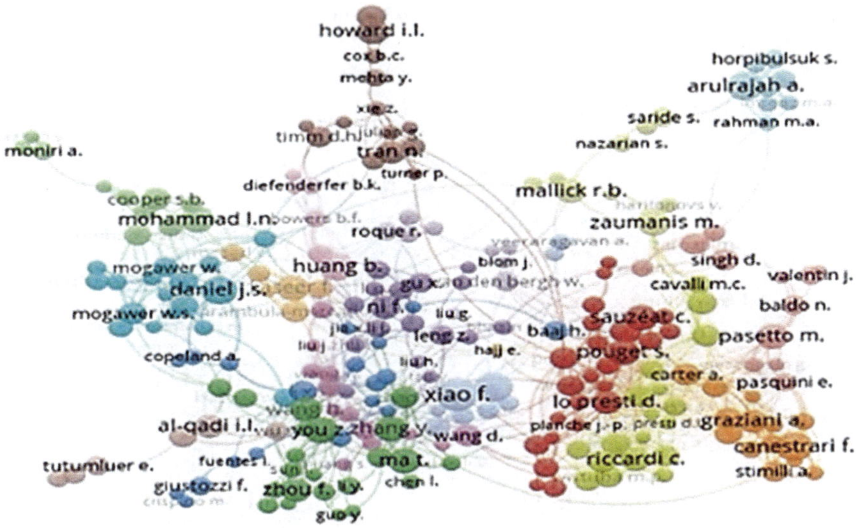

Fig. 6 Overlay visualization author

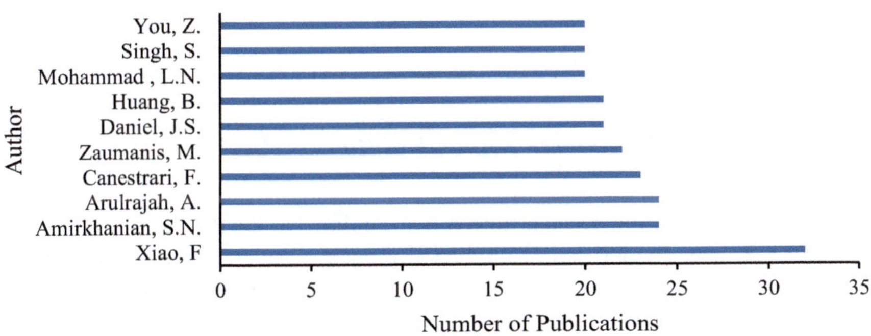

Fig. 7 Graph paper per Scopus data author

by 848 networks. Cluster 1 is light blue and consists of an author named Xiao F with a total of 32 papers.

Author contributions with research topics on RAP are identified in the network and overlay of Figs. 6 and 7. The author with the most citations is Huang B, with a total of 1325 citations from 21 articles. The year 2021 is expected to have a significant increase in the number of new authors interested in the topic of RAP around the world.

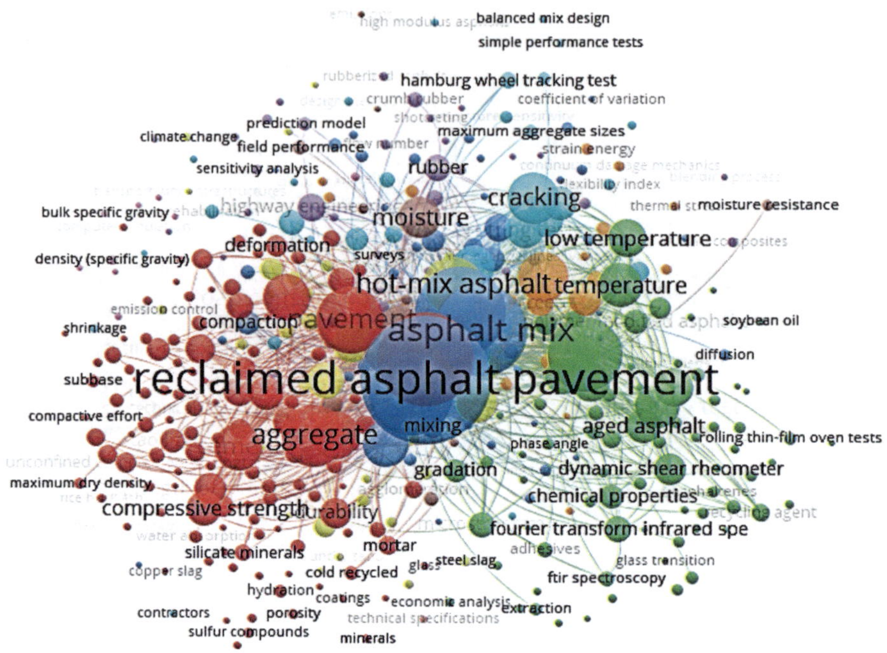

Fig. 8 Network visualization RAP

3.2 *Analysis of Research Topics Related to RAP*

Figure 8 is a visualization of the article network related to RAP which consists of 396 relevant keywords divided into 10 clusters (color differences) interconnected with 18,569 links. Based on the observations, related studies on RAP have been widely conducted. However, few topics intersect with RAP, such as extraction, chemical properties, and microstructure. The blue cluster was more significant in this research with 123 related keywords on aggregate, mixing, hot-mix asphalt, and compressive strength relatively high.

The special network on the topic of "reclaimed asphalt pavement" has 3 clusters, with 395 links in several 1629 papers, as shown in Fig. 8. Meanwhile, the number of related papers accumulated in the Scopus database is 1994. This shows that there is a relationship close to the word "asphalt" and "pavement," which is associated with "reclaimed asphalt pavement" or "recycled asphalt pavement" with a total of 1629 papers.

The authors were able to find research gaps in linking networks that had few links between keywords such as extraction, microstructural properties, furnaces, and interfacial transition zone. For example, the word "extraction" in Fig. 9 used to visualize the relationship between the words "reclaimed asphalt pavement" and "extraction" is found in 16 papers out of 1994. The next step is to download these 16

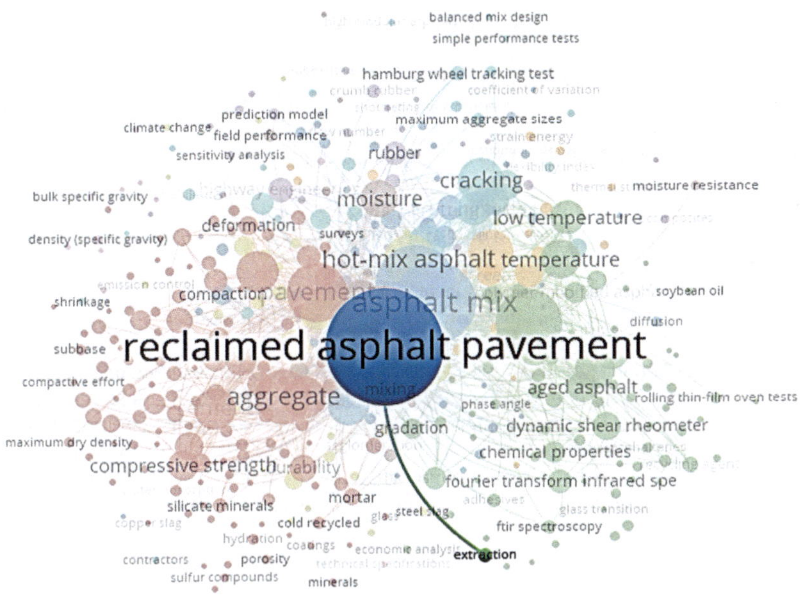

Fig. 9 Network visualization RAP with extraction

papers by breaking and technically describing their state of the art to fill the search gaps easily.

4 Conclusion

In conclusion, Scopus meta data and VOS Viewer can be used to map the bibliometric network on RAP. These tools are also used to describe the relationship between authors' distribution, countries, and related keywords.

Acknowledgements This research was financially supported by The Faculty of Engineering, Diponegoro University, Indonesia Through Strategic Research Grant 2021.

References

1. Widayanti A, Soemitro Ria AA, Ekaputri JJ, Suprayitno H (2018) Characterization of reclaimed asphalt pavement (RAP) as a road pavement material (National Road Waru, Sidoarjo). MATEC Web Conf 181. https://doi.org/10.1051/matecconf/201818105001
2. Al-Qadi IL, Elseifi M, Carpenter SH (2014) Reclaimed asphalt pavement—a literature review

3. Chang X, Zhang R, Xiao Y, Chen X, Zhang X, Liu G (2020) Mapping of publications on asphalt pavement and bitumen materials: a bibliometric review. Constr Build Mater 234:117370. https://doi.org/10.1016/j.conbuildmat.2019.117370
4. Callahan JL (2014) Writing literature reviews: a reprise and update. Hum Resour Dev Rev 13(3):271–275. https://doi.org/10.1177/1534484314536705
5. Sweet M, Moynihan R (2007) Improving population health: the uses of systematic reviews, Dec 2007
6. Tsay M, Shu Z (2011) Journal bibliometric analysis: a case study on the journal of documentation. J Doc 67(5):806–822 [online]. Available http://dblp.uni-trier.de/db/journals/jd/jd67.html#TsayS11
7. Franceschini F, Maisano D, Mastrogiacomo L (2015) Influence of omitted citations on the bibliometric statistics of the major Manufacturing journals. Scientometrics 103(3):1083–1122. https://doi.org/10.1007/s11192-015-1583-9
8. Ravikumar S, Agrahari A, Singh SN (2015) Mapping the intellectual structure of scientometrics: a co-word analysis of the journal scientometrics (2005–2010). Scientometrics 102(1):929–955 [online]. Available https://econpapers.repec.org/RePEc:spr:scient:v:102:y:2015:i:1:d:10.1007_s11192-014-1402-8
9. Cainelli G, Maggioni MA, Uberti TE, de Felice A (2015) The strength of strong ties: how co-authorship affect productivity of academic economists? Scientometrics 102(1):673–699. https://doi.org/10.1007/s11192-014-1421-5
10. Zhai L, Yan X, Shibchurn J, Song X (2014) Evolutionary analysis of international collaboration network of Chinese scholars in management research. Scientometrics 98(2):1435–1454. https://doi.org/10.1007/s11192-013-1040-6
11. Petticrew M, Roberts H (2006) Starting the review: refining the question and defining the boundaries. In: Systematic reviews in the social sciences, pp 27–56, Jan 01 2006. https://doi.org/10.1002/9780470754887.ch2
12. Huang C, Yang C, Wang S, Wu W, Su J, Liang C (2019) Evolution of topics in education research : a systematic review using bibliometric analysis. Educ Rev:1–17.https://doi.org/10.1080/00131911.2019.1566212
13. SNI-03-6894 (2002) Metode pengujian kadar aspal dari campuran beraspal dengan cara sentrifus 1, pp 1–6
14. Taha R, Al-Harthy A, Al-Shamsi K, Al-Zubeidi M (2002) Cement stabilization of reclaimed asphalt pavement aggregate for road base and subbases. J Mater Civ Eng 14(3):239–245
15. Qiang W, Peiyu Y, Ruhan A, Jinbo Y, Xiangming K (2011) Strength mechanism of cement-asphalt mortar. J Mater Civ Eng 23(9):1353–1359. https://doi.org/10.1061/(ASCE)MT.1943-5533.0000301
16. Miller SR, Hartmann T, Dorée AG (2011) Measuring and visualizing hot mix asphalt concrete paving operations. Autom Constr 20(4):474–481. https://doi.org/10.1016/j.autcon.2010.11.015

Literature Study: Alternative Materials for Hot Rolled Sheet-Wearing Course (HRS-WC) Pavement

Elsa Eka Putri, Purnawan, Bayu Martanto Adji, and Bobby Herman

Abstract HRS mixture is a concrete asphalt mixture using gaps that have a high bitumen content so that it has high flexibility and is resistant to deformation. Moreover, this pavement is suitable for application in Indonesia because Indonesia is a tropical country with high enough heat. HRS has two types of pavement layers, namely HRS-Base and HRS-WC. As a non-structural layer containing more fine aggregate and bitumen, the HRS-WC Mixture's structural strength is prone to plastic deformation with rutting on the asphalt surface. Therefore, the quality of HRS-WC needs to be improved through new specifications and the type of material used must have good physical and mechanical properties. Efforts have been made to improve the quality of the existing asphalt. Modifying the physical and chemical properties of asphalt with various added materials was to obtain cheap and good quality asphalt. This literature study aims to identify the alternative materials that can improve the HRS-WC Mixture performance based on previous studies in terms of the Marshall test results of related research. The collection and analysis of research data from books, documents, journals, and research results have obtained eight alternative materials to improve the HRS-WC Mixture's performance, namely used tire powder, concrete waste, tras, bagasse ash, rice husk ash, coconut shells, palm oil, natural rubber (latex) and coal ash. On the other hand, alternative use has advantages from economic and environmental aspects. They are replacing/reducing the use of new aggregates and reducing the amount of waste that pollutes the environment.

Keywords HRS-WC · Alternative materials · Marshall parameter · Waste

---

E. E. Putri (✉) · Purnawan · B. M. Adji · B. Herman
Department of Civil Engineering, Faculty of Engineering, University of Andalas, Padang, Indonesia
e-mail: elsaeka@eng.unand.ac.id

Purnawan
e-mail: purnawan@eng.unand.ac.id

B. M. Adji
e-mail: bayu@eng.unand.ac.id

S. A. Kristiawan et al. (eds.), *Proceedings of the 5th International Conference on Rehabilitation and Maintenance in Civil Engineering*, Lecture Notes in Civil Engineering 225,
https://doi.org/10.1007/978-981-16-9348-9_97

1 Introductions

Hot Rolled Sheet (HRS) is a hot asphalt mixture developed by Bina Marga around 1980, which has durability and flexibility [1]. HRS or Lataston is a concrete asphalt mixture that uses gaps-graded aggregates and has a high bitumen content [2]. The HRS-WC layer is non-structural which contains fine aggregate and high bitumen content. Therefore, the HRS-WC structure is prone to plastic deformation, which is indicated by the appearance of grooves on the pavement surface [3].

Modification of the physical and chemical properties of the asphalt mixture is one of the efforts that have been implemented to improve the quality of an asphalt mixture. Modifications include adding additives or replacing ordinary materials with better materials to obtain excellent and cheap quality and performance of the asphalt mixture. As a result, the government's funds for the manufacture and repair of the asphalt mixture are becoming more economical [4].

This literature study will discuss alternative materials for the HRS-WC Mixture based on previous related research to improve the asphalt mixture's performance. Such as a substitute for coarse aggregate, fine aggregate and a substitute for asphalt as a binder.

2 Literature Review

2.1 Hot Rolled Sheet

Hot Rolled Sheet (HRS) or a thin layer of asphalt concrete (Lataston) is a mixture of road pavements, forming material consisting of asphalt, gaps graded aggregate, and filler spread and compacted at a specific temperature. The thickness of the layer ranges from 2.5 to 3 cm [5]. According to the Transportation Infrastructure Research and Development Center, 2005 there are two HRS mixtures, namely HRS-Base and HRS-WC.

HRS-WC is a type of HRS pavement layer that is at the top or on the surface. HRS-WC has water-resistant coating properties, is resistant to the formation of rutting, has a good surface smoothness, can transmit loads, and is resistant to slipping. The HRS-WC layer is prone to wear because it directly contacts the vehicle's wheels and the weather's effects. The appearance of cracks on the surface is caused by evaporation and ageing of asphalt due to temperature and climate. Therefore, the HRS-WC layer must be designed to have a good level of stability, flexibility, durability and resistance to fatigue [6].

There are some characteristics after modification of the HRS-WC asphalt mixture. Sufficient levels of asphalt to produce a durable pavement, high stability to withstand traffic loads without changing shape, sufficient void in total pavement mixture after compacting [1]. It can provide space for extra compaction and the development of the asphalt due to increasing the temperature. Moreover, maximum percentage of air voids to limit permeability because water can cause damage if it gets into the pavement, as well as having high workability so that becomes more efficient without shifting between pavement layers and without reducing the stability of the Mixture during spreading.

2.2 *Marshall Parameters*

Asphalt concrete performance is determined through Marshall testing, should fulfill the following result, stability is the strength of the Mixture to withstand deformation due to traffic loads. Stability can be obtained through inter-aggregate friction resistance, interlocking aggregates, and asphalt's cohesion power [7].

Flow is the magnitude of the change in the plastic shape of an asphalt mixture test object that occurs due to a load to the limit of collapse, expressed in units of length (Department of Public Works 2010) [8].

Marshall Quotient is the quotient of stability to flow, which is used to approach the level of stiffness or flexibility of the Mixture. A high Marshall Quotient value indicates a high hard layer stiffness value. Hard layers with too high a Marshall Quotient value will easily crack due to repeated traffic loads. On the other hand, the Marshall Quotient value, which is too low, indicates the Mixture is too flexible (plastic), which results in the hard layer being easily deformed if it withstands traffic loads [9].

Void in the Mineral Aggregate (VMA) is the voids space between the aggregate particles on pavement, including the air voids and the effective asphalt volume (excluding the volume of asphalt absorbed by the aggregate). Void Filled with Asphalt (VFA) is the per cent of the voids between the aggregate particles (VMA) filled with asphalt, excluding asphalt, which is absorbed by the aggregate. Void in Mix (VIM) is the air voids in the asphalt pavement mixture consisting of air spaces between aggregate particles covered with asphalt.

3 Methodology

This literature study focuses on finding various ideas, theories, laws, propositions, principles, or ideas used to analyze, complete and formulate questions in according to the research object. Through literature research or literature review, knowledge, ideas and findings contained in academic-oriented literature are summarized and theoretical and methodological contributions to the topics are formulated [10].

As well as in a descriptive analysis, where the data obtained is presented and described regularly and given explanations and insights, the research can be easily understood by readers.

The data collection method used in this research is the documentation method. The documentation method is a data collection method by finding or extracting data from literature related to the formulated problem [10]. The data obtained from various literatures are collected as a single document and used to answer the problems that have been formulated.

The four procedures used in this research are;

1. *Organize*, namely organizing the literature to be reviewed. The literature must be relevant and in accordance with the problem statement. It was conducted by looking for ideas, general goals, and conclusions from the literature by reading the abstract, introductory paragraphs and conclusions, and classifying the literature based on certain categories.
2. *Synthesize*, which is to combine the results of the literature organization into a summary so that they become a whole, by looking for linkages between the contents.
3. *Identify*, namely identifying controversial issues in the literature. The controversial issue in question is an issue that is considered very important to be explored or analyzed, in order to get an interesting article to read.
4. *Formulate,* namely formulating questions that require further research.

4 Results and Discussions

4.1 Used Tire Powder

Used tire powder is one of the added materials (additives) that is expected to improve the asphalt mixture's quality to fulfil the characteristics of asphalt as a binder and increase the stability, flexibility, and durability of the asphalt concrete mixture [11]. The use of used tire powder as an added material in improving the quality of asphalt is motivated by the abundant availability of used tire waste in Indonesia. So, in addition to improving the quality of the asphalt mixture, the use of used tire waste can also minimize environmental pollution due to waste [12].

The stability value of the HRS-WC Mixture without used tire powder is higher than the stability value with the addition of used tire powder [13]. HRS-WC stability value without using used tire powder is higher than HRS WC with the addition of used tire rubber but still meets specifications [14].

The flow value of HRS-WC with the addition of used tire powder is higher than the typical HRS-WC Mixture. In contrast to the research, it stated that the flow value at HRS-WC without the addition of used tire powder is higher when compared to HRS-WC with the addition of used tire powder [14]. The variation in the size used in Hardwiyono's study was 20, 21 and 22%. The flow value will increase to the maximum flow value limit, and the flow value will decrease with increasing levels of used tire powder [11]. The MQ value tends to decrease along with the increase in the content of the used powder used. It is confirmed that the HRS-WC Mixture with the addition of used tire powder shows a lower MQ value compared to the HRS-WC Mixture without the addition of used tire rubber powder waste [15].

The VIM value tends to decrease with increasing levels of used tire powder. The Voids in the Mixture (VMA) will decrease along with the increasing levels of used tire powder. The VMA value in all used tire powder mixtures has decreased [16]. The VFA value tends to increase with increasing levels of biofuels powder used. The research of Faisal (2014) supported that there was an increase in the VFA value and the increase in the number of biofuels powder, although the increase was not too significant [17].

It can be concluded that used tire powder can be used as additional alternative material to improve the performance of the HRS-WC Mixture [13]. It is also supported by Jatmiko's research (2019) that the used tire powder is suitable for use in the HRS-WC asphalt mixture. The asphalt mixture is more elastic, flexible and the stability will decrease with increasing levels of used tire powder [18].

4.2 Concrete Waste

Concrete waste is a waste product of unused construction materials, which is very easy to find in Indonesia. Concrete waste originates from the ruins of buildings due to earthquakes, demolition of buildings and concrete originating from precast concrete factories, which are no longer used and, if left without handling, can cause problems for the environment [19].

The physical properties of Concrete Waste are almost as good as, in essence, that of new aggregates (stones). Based on this equation, the rest of the buildings with concrete waste can be used as an alternative to coarse aggregates in the construction of road pavement mixtures [20].

In doing so, the use of new aggregates might be reduced by using aggregates from recycled concrete waste [21].

In the research of Suwastika et al. (2019), the variations of the desired level used to obtain the Optimum Asphalt Content (OAC) were 6, 6.5, 7, 7.5, 8%, from the OAC test results obtained at 7.13% in Mixture using natural aggregate and aggregate 7.08% in the Mixture with concrete waste as coarse aggregate. Then a mixture is made with variations of 25% natural aggregate and 75% concrete waste, 50% natural aggregate, 50% concrete waste and the last variation is 75% natural aggregate 25% concrete waste [19]. The stability value tends to decrease in the use of high levels of concrete waste. Moreover, the stabilization value is decreasing with each addition of variations in the bio-concrete waste, which is currently large. The worse ability than new aggregate is the cause of the decrease in the value of stability in using concrete waste in the asphalt mixture [19, 21].

The test results show that the flow value has increased from 3.3 to 3.75 mm in the variation of Concrete waste content from 50 to 100%. Then there is a decrease in the variation of the concrete waste content of 0–25%, namely 3.8–3.5 mm, that the value of Flow tends to increase in higher levels of waste. The reason is that concrete waste has the greatest absorption properties, so mixing with more waste concrete will easily absorb more asphalt than the mixture with less concrete waste content [20].

The VIM value has increased as the use of concrete waste content in the mixture increases. Utomo et al. (2017) confirmed that the higher the level of concrete waste used, the higher the VIM value obtained will be as well [21].

The test results show that the VMA value tends to increase when the waste concrete is higher and with the higher variation in the amount of concrete waste, the VMA value tends to increase due to the high abrasion value of concrete waste [21].

The MQ value decreases along with the increase in the waste content used. As well as at the higher level of concrete waste, the results of the Marshall test show the MQ value tends to decrease. The higher the level of concrete waste used, the VFA value will decrease. The greater the variation of concrete waste, the VFA value tended to decrease. Asphalt, which is supposed to fill the voids in the mixture, is absorbed by waste which has a higher absorption rate than new aggregate, which causes a decrease in the VFA value [21].

Based on the Marshall parameter value obtained from the resolved test results, it can be concluded that the use of concrete waste aggregates in the HRS-WC Mixture can still be done because it still meets the General Specifications of Bina Marga 2018 [19].

4.3 Tras

Volcanic rocks that have undergone changes in chemical composition as a result of the weathering process and affect the water conditions of the soil are called tras [22].

Tras or commonly referred to as natural polypozzolans, is a natural material that occurs due to the weathering of volcanic rocks, which have a high content of silica (SiO_2) and aluminium and alkaline iron compounds chalk, and others that are not in the inconsiderable amount [23].

It was investigated by varying the coarse, medium and fine aggregates including sand and tras with a composition according to specifications. The mixture composition is the proportion of 10% coarse aggregate, 20% of dry aggregate, 50% of fine aggregate, 10% of sand and 10% of tras and 6% of asphalt percentage [24].

It concluded that with a composition of 10% coarse aggregate, 20% aggregate, 50% fine aggregate, 10% sand and 10% Tras, they are in ideal limits because they are still within limits required in the specification [22]. Then, at the maximum asphalt level of 6%, the stabilization value and VIM are far from the minimum limit, the values of the Marshall test results can still be optimized by trying to re-vary the use of Tras and asphalt content that are still sufficient so that they can support the performance of the HRS-WC Mixture. Moreover, the use of raw materials affects the amount of use of aggregates where the percentage of use of medium and fine aggregates is reduced by 10–20%. So, there are material and cost savings for the manufacture of the HRS-WC Mixture.

Hence, Tras can be used as an additive material in the smooth aggregate diffraction in the hot HRS-WC semi-heated mixture because it can produce a good and economical mixture while still meeting the specifications [22].

4.4 Bagasse Ash

The sugar factory produces waste in the form of sugarcane waste. In the sugar factory, sugar cane is used for burning material in the factory's steam boiler. The results of burning sugarcane waste will produce ash which can be used as a filler in a mixture of asphalt concrete [24].

The main content of bagasse ash is Silica (SiO_2) 70.97% and Na_2O of 22.27%, and a few other elements such as K_2O of 4.82%, $C_5H_8O_4$ of 0.9%, Magnesium (MgO) of 0, 82%, Al_2O_3 and Al_2O_3 each of 0.36%. The content contained in bagasse ash, when mixed with asphalt, will form a compound reaction that makes the structure of the Mixture hard and stiff [25].

In the field of road engineering, the use of bagasse ash continues to be optimized in order to reduce the impact of environmental pollution, improve mixed performance and at the same time increase the economic activities of the community in an area [26].

A study on the use of bagasse ash on HRS-WC mixtures' performance by using sugar cane ash as a filler and compared it to rock dust filler. From comparing the Marshall HRS WC Mixture Test with Rock Ash Filler and Sugarcane Ash Filler, conclusions can be drawn as follows, the optimum asphalt content in the rock ash filler is 6.75%, while bagasse ash is 6.95%. It indicates that the filler composition affects the optimum asphalt content. This difference indicates that the absorption with bagasse ash filler is greater than that with rock ash filler. The effect of using bagasse ash filler on the VMA value at the optimum asphalt content, the VMA value in the Mixture with rock ash filler was 20.64%, smaller than the bagasse ash filler of 21.36% [24].

Using bagasse ash filler on the VIM value on the optimum asphalt content shows that the VIM value of rock ash is 5958% and bagasse ash is 5960%, resulting in slightly large voids so that the durability of the pavement layer is reduced. The effect of using bagasse ash filler on the VFB value at the optimum asphalt content shows that the VFB value of rock ash filler is 71.68%, and bagasse ash is 72.12%, resulting in small air voids, the greater stiffness of pavement the higher the durability.

Using bagasse ash filler on the stability value of the optimum asphalt content shows that the stability value of rock ash filler is 1280.98 kg and bagasse ash is 1231.07 kg, resulting in the asphalt mixture able to withstand traffic loads. The effect of using bagasse ash filler on the flow value at the optimum asphalt content shows that the flow value in rock ash filler is 4.76 mm and bagasse ash is 4.72 mm, resulting in the pavement layer not easily cracked.

The effect of using bagasse ash filler on the Marshall Quotient value on the optimum asphalt content shows that the MQ value of rock ash filler is 268.97 kg/mm and bagasse ash is 260.58 kg/mm.

The results of the analysis indicate that the use of bagasse ash filler with HRS-WC mixture meets the requirements as per General Specifications of Bina Marga 2010 as an alternative material to replace rock ash filler. Hence, bagasse ash waste as well rock ash filler can be used for HRS-WC asphalt concrete mixture [24].

4.5 Rice Husk Ash

Rice husk ash is a local material that is available in large quantities and relatively cheap. The high silica content makes rice ashes suitable for use as a substitute filler in asphalt pavement mixtures [26].

Rice husk ash contains more than 70% silica thus it categorized as the pozzolanic material. It is a popular material used as an additive in the concrete making, especially in increasing concrete strength. Silica will react with cement and water to form calcium silicate hydrate as an adhesive [27]. It is possible to use rice husk ash waste to replace rock ash for the asphalt mixture on the surface layer. The materials used are 60/70 penetration bitumen, coarse aggregate, fine aggregate and filler. The use of husk ash filler can increase the density of the mixture so that the contact area between the aggregates has increased and the voids in the Mixture are reduced. The increase in interlocking between aggregates causes the stability value tends to increase. At 8.6% asphalt content, the stability value reaches a maximum value of 2355.7 kg [28].

From the Marshall test results, it can be seen that at the asphalt content of 9.6%, the flow value reaches a maximum value of 7.87 mm. It proves that the flow value continues to increase with the increase in the asphalt content used.

The test results show that the VMA value has decreased as the use of asphalt content increases. In soaked conditions, the minimum VMA value occurs at the asphalt content of 9.1 at 22.56% and asphalt content of 9.6%. The VMA value has increased again to 23.02%. Then, the higher the asphalt content used in the mixture, the higher the VFA value. At 9.6% asphalt content, the maximum VFA value was 90.68 for dry conditions and 86.06 for soaked conditions. The VIM value decreases as the amount of asphalt used in the mixture increases. Increasing the level of asphalt used, the MQ value will increase; at the asphalt content of 8.6%, the MQ value reaches a maximum of 330.10 kg/mm.

The utilization of rice husk ash in the HRS-WC mixture has advantages, such as increased stability, durability, and flexibility. However, it also has disadvantages, namely the use of asphalt, which tends to be higher due to the nature of rice husk ash, which easily absorbs asphalt [27]. Other advantages of the use of raw water as a filler material include the availability of raw milk as a raw material, which gives the prospect of the availability of filler material which is more economical compared to other materials that are relatively expensive and usually difficult to obtain [29].

4.6 Oil Palm Shells

Another alternative material for coarse aggregate substitution in the asphalt mixture is the waste of oil palm shells. Oil palm shell waste contains a high amount of grit (SiO_2), increasing the strength of asphalt concrete [30]. Oil palm shells have a very thick and hard skin structure and contain high levels of grit (SiO_2) [31]. Palm shell waste is available in large quantities at a more economical price [32].

The utilization of oil palm shell waste is generated from oil palm processing. One way to utilize palm oil waste is to use palm kernel shells to substitute some of the primary aggregates in an asphalt mixture [31].

The aims of their investigation was to determine the strength and durability of palm oil waste as a substitute for coarse aggregate in the HRS-WC Mixture. The parameters reviewed in this study are Marshall parameters and Wheel Tracking testing. Variations in the levels of waste oil palm shells used were 0, 30, and 50% [30].

In the variation of 0 and 30% oil palm shell waste, stability values are above the minimum required Marshall specifications, namely 988.48 and 869.67 kg. Meanwhile, for the variation of oil palm shell waste, 50% of the stability value does not meet the minimum requirements set, which is 800 kg. So, it can be concluded that the recommended use of oil palm shell is only at 30% oil palm shell content.

All the oil palm shell content variation in the HRS-WC Mixture still meets the minimum specifications of flow value of 3 mm. The HRS-WC Mixture with oil palm shell content of 0 and 30% meet the minimum MQ specifications required, while the Mixture with an oil palm shell content of 50% does not meet the minimum requirements for the MQ value of a minimum of 250 kg/mm.

At 0% oil palm shell waste, the VIM value obtained was 5.76%. At the 30% variation in oil palm shell content, the VIM value is still within the required minimum specification limit, which is 5.88%. The HRS-WC Mixture at 0 and 30% oil palm shell content has good durability properties. Whereas in the variation of oil palm shell content, 50%, the VIM value obtained was 6.13%. This value exceeds the required maximum value limit of 6%.

From the above research results, it can be concluded that for the use of oil palm shells with a content of 30% in the HRS-WC pavement mixture, the values of stability, Flow, Marshall Quotient and Void in Mix are within the required range of SNI 03-1737–1989. In comparison, the use of oil palm shells at a level of 50% as coarse aggregate is not recommended because the test results show that the Marshall parameter values are outside the predetermined specification range.

4.7 Latex

Latex or a natural rubber comes from the sap of the Hevea Brasiliensis tree, which comes from the Amazon region. It is formed by clumping and drying [33]. Natural rubber (polysoprene) is classified as an elastomer, a material that can be stretched and can return to its original form [34].

Natural rubber functions to increase bonds between aggregates, provide elasticity, and reduce vibrations that arise due to loads [35].

The use of natural rubber in road pavement, especially in the HRS-WC mixture, has been widely studied by researchers and experts.

The lumpy rubber (lumb), namely rubber or latex, which is directly formed into crepe (solid rubber) without being given any treatment. Using the same bitumen content at 7.3% for a standard mixture of HRS-WC specimens was made without the addition of natural rubber. The comparison between the standard HRS-WC Mixture with the HRS-WC Mixture plus raw rubber (natural rubber) was obtained with a variation of 5, 10, and 15% [33].

Moreover, the HRS-WC Mixture with the addition of raw natural rubber has higher Stability, Flow, VFA, and Marshall Quotient values than the HRS-WC mixture without the addition of raw natural rubber. However, the HRS-WC Mixture's VIM and VMA values with the addition of raw natural rubber are lower than the standard HRS-WC mixture without the addition of raw natural rubber.

The addition of natural rubber to oil asphalt in the HRS-WC mixture shows a better Marshall stability value, indicating that the interlocking between aggregates is getting better, the flow value is getting lower, and the Marshall Quotient is increasing [2].

The asphalt mixture with the addition of natural rubber has a higher strength to withstand loads than conventional asphalt mixtures and proved that using rubber asphalt in the mixture can increase the stability [36].

4.8 Coal Ash

In general, coal ash can be interpreted as the residual or contaminant material of coal which is not burnt and involve in the combustion process of carbon, hydrogen, sulfur, oxygen and evaporation of water contained in coal. There are two types of coal ash, fly ash and bottom ash [3].

Coal ash is very easy to find in Indonesia. It is available in large quantities, which is the smallest part of a mixture of solids produced from processing coal in mining, which is no longer used. For this reason, coal ash will be used in the making of a road pavement mixture [37]. The use of coal ash in the road pavement, especially in the HRS-WC mixture, has been widely studied. It was conducted to determine the effect of fly ash on the performance of the HRS-WC mixture. Samples have been made using a variation of fly ash 0, 6, 10 and 12% of the total aggregate mixture, with a total sample of 60 samples. It can be concluded that the higher the ash content of the coal used results in a decrease in the value of stability, flow and MQ. However, the stability, flow and MQ values obtained still meet the specifications. The use of fly ash as a filler in the HRS-WC mixture can still be used. The use of coal ash as a filler in the HRS-WC mixture has a positive effect on the stability and the residual strength index of the mixture using conventional fillers, namely rock ash alone. Hence, coal ash is suitable for use as an alternative filler material in the HRS-WC mixture. The higher the ash content of the coal used in the mixture, the values of VIM and VMA tend to increase, while the values of Stability, Flow, and MQ decrease [3].

5 Conclusions

Based on the analysis and discussion that has been described. In this literature study, conclusions can be drawn:

1. Several alternative materials were obtained that could be used to increase the HRS-WC Mixture in terms of the Marshall test results. Among them are used tire powder, concrete waste, Tras, bagasse ash, rice husk ash, palm oil shells, natural rubber (latex) and coal ash.
2. From several alternatives obtained, the most recommended alternative for use in the HRS-WC mixture is bagasse ash. This is based on the results of research that shows all Marshall parameters meet specifications, and almost all Marshall parameters have increased from the standard HRS-WC mixture. Sugarcane ash has a very high silt content, so it is suitable for pavement mixtures. Besides that, the material is also easy to obtain. If it is used, it will reduce the impact of environmental pollution because there is no large-scale utilization of bagasse ash waste.
3. In terms of cost, making mixtures using alternatives is more economical because it utilizes waste, thereby reducing/replacing new aggregates (fresh aggregate).

Acknowledgements This research was supported by Engineering Faculty of University of Andalas grant no. T/14/UN.16.17/PT.01.03/IS-RD/2021.

References

1. Darunifah N (2007) Pengaruh Bahan Tambahan Karet Padat Terhadap Karakteristik Campuran hot rolled sheet wearing course (HRS-WC) (the effect of solid rubber additives on the characteristics of the hot rolled sheet wearing course). Pascasarjana Universitas Diponegoro
2. Amiruddin AA, Sasmita SA, Ali N, Renta I (2012) Kajian Eksperimental Campuran HRS-WC Dengan Aspal Minyak Dan Penambahan Aditif Lateks Sebagai Bahan Pengikat (experimental study of HRS-WC mixture with oil asphalt and addition of latex additive as binder) proceeding Konferensi Nasional Teknik Sipil, 6
3. Ambarwati L, Arifin MZ, Bawono H (2012) Pengaruh Kadar Abu Batubara Sebagai Filler Terhadap Karakteristik dan Indeks Kekuatan Sisa (IKS) pada Campuran Hot Rolled Sheet (HRS) (Effect of Coal Ash Content as a Filler on Words and Residual Strength Index in Hot Rolled Sheet Mixtures). Rekayasa Sipil 3(2):131–140
4. Roberts FL, Kandhal PS, Brown ER, Lee D-Y, Kennedy TW (1991) Hot mix asphalt materials, mixture design and construction

5. Sukirman S (1999) Perkerasan Lentur Jalan Raya (Highway Flexible Pavement) Nova, Bandung, 2
6. Azizah N, Rahardjo B (2017) Kinerja campuran hot rolled sheet-wearing course (HRS-WC) dengan filler abu ampas tebu (performance of a hot rolled sheet-wearing courseix with bagasse ash filler). Bangunan: Teori, Praktek, Penelitian, dan Pengajaran Teknik Bangunan 22(2)
7. Sulaksono S (2001) Rekayasa Jalan (highway engineering), Departemen Teknik Sipil, Penerbit ITB, Bandung
8. Departemen Pekerjaan Umum.: Spesifikasi Umum Binamarga 2010 Revisi 3 (General Specifications Binamarga 2010 Revision 3). Dinas Pekerjaan Umum Bidang Binamarga, Jakarta (2010)
9. Suryaman F (2009) Pengaruh Penggunaan Limbah Botol Plastik Sebagai Bahan Tambah (Additive) Terhadap Karakteristik Beton Aspal (the effect of use of plastic bottle waste as an additive on the characteristics of asphalt concrete) In: Skripsi teknik sipil Universitas Atma Jaya. Yogyakarta
10. Dena T (2013) The literature review: a few tips on conducting it. Health Sciences Writing Centre, University of Toronto
11. Septiawan TD (2018) Pengaruh Penggunaan Bahan Tambah Serbuk Karet Ban Pada Campuran Lapis Aspal Beton (the influence of using additional material for tire rubber powder on concrete asphalt mixture), Jurnal Rekayasa Sipil, Univ Islam Malang
12. Indriyati EW, Susanto HA (2015) Kajian Sifat-Sifat Reologi Aspal Dengan Penambahan Limbah Ban Bekas (study of the rheological properties of asphalt with the addition of waste tires)
13. Oktaviastuti B, Wijaya HS, Indrawan P (2018) Pengaruh penambahan serbuk ban bekas untuk bahan tambah campuran ATB (asphalt treated base) (the effect of adding used tire powder to add material to the ATB mixture) Reka Buana. Jurnal Ilmiah Teknik Sipil dan Teknik Kimia 3(1):16–29
14. Hardwiyono S (2012) Pengaruh Penambahan Parutan Karet Ban Gradasi Tipe 2 Terhadap Parameter Marshall Pada Campuran hot rolled sheet wearing course (the effect of addition of type 2 grated rubber tire to marshall parameters in hot rolled sheet wearing course mixture) Jurnal Ilmiah Semesta Teknika
15. Kiswara CP, Saiful AA (2014) Studi Perilaku Campuran Lataston (HRS-WC) Dengan "Limbah Karet Ban Luar" Sebagai Bahan Pengganti Sebagian Aspal (behavior study of mixed lataston (HRS-WC) with "waste rubber tire" as a substitute for partial asphalt), Media Teknik Sipil
16. Gosali G, Jaya H, Wulandari PS, Padmadjaja H (2016) Pengaruh Penambahan Serbuk Ban Karetmesh# 80 Pada Campuran Laston Untuk Perkerasan Jalan Raya (the effect of addition of Rubbermesh # 80 tire powder to Laston mixture for highway pavement). Jurnal Dimensi Pratama Teknik Sipil 5(1)
17. Faisal S (2014) Karakteristik Marshall Campuran Aspal Beton AC-BC Menggunakan Material Agregat Basalt Dengan Aspal Pen. 60/70 Dan Tambahan Parutan Ban Dalam Kendaraan Roda 4 (characteristics of marshall asphalt concrete mixture AC-BC using basalt aggregate material with asphalt pen. 60/70 and additional grater tires in 4-wheeled vehicles) Jurnal Teknik Sipil 3(3):38–48
18. Jatmiko M (2019) Pengaruh Penambahan Limbah Serbuk Ban Luar Sebagai Bahan Pengganti Sebagian Agregat Halus Pada Campuran HRS-WC (hot rolled sheet-wearing course) Ditinjau Dari Karakteristik Marshall (the effect of addition of outer tire powder waste as a substitute for partial fine aggregate in the HRS-WC (hot rolled sheet-wearing course) mixture in terms of marshall characteristics), Skripsi Tesis, ITN Malang
19. Suwastika PMW, Wedyantadji B, Erfan M (2019) Pemanfaatan Limbah Beton Sebagai Pengganti Agregat Dalam Campuran Lapis Tipis Aspal Beton (hot rolled sheet-wearing course, HRS-WC) (utilization of concrete waste as a substitute for aggregates in an HRS-WC mixture), Skripsi Tesis, ITN Malang
20. Rahman ZA, Harnaeni SR (2016) Pemanfaatan Limbah Beton Pada Campuran hot rolled sheet base Ditinjau Dari Aspek Propertis Marshall (utilization of concrete waste in hot rolled sheet base mixtures in terms of marshall properties), Universitas Muhammadiyah Surakarta

21. Utomo B, Harnaeni SR (2017) Analisis Marshall properties asphalt concrete Dan hot rolled sheet Menggunakan Limbah Beton Sebagai Pengganti Agregat Kasar (analysis of marshall properties of asphalt concrete and hot rolled sheet using concrete waste as a substitute for coarse aggregate). Universitas Muhammadiyah Surakarta
22. Manoppo MRE (2011) Pemanfaatan Tras Sebagai Filler Dalam Campuran Aspal Panas HRS-WC (utilization of tras as filler in hot asphalt mixture HRS-WC). Jurnal Ilmiah Media Eng 1(2)
23. Achmad F, Maksud R (2014) Kajian Penggunaan Tras Lompotoo Sebagai Agregat Halus Pada Lapis Pondasi Bawah Ditinjau Dari Spesifikasi Umum 2007 dan 2010 (study of the use of Lompotoo tras as fine aggregate in the subbase layer)
24. Wijanarko FH, Mulia MB, Wicaksono Y, Purwanto D (2013) Pemanfaatan Limbah Ampas Tebu Sebagai Pengganti Filler Untuk Campuran Aspal Beton Jenis HRS-WC (utilization of sugarcane dregs waste as a filler substitute for HRS-WC for asphalt concrete mixtures) Jurnal Karya Teknik Sipil 2(2):363–369
25. Fauziah M, Kushari B, Ranski F (2014) Pengaruh Abu Ampas Tebu Sebagai Filler Pengganti Terhadap Karakteristik Marshall Campuran Superpave (effect of sugarcane bagasse ash as a substitute filler on Marshall characteristics of superpave)
26. Hadiastari IG, Fauziah M (2018) Kinerja Campuran Split Mastic Asphalt (SMA) 0/11 Dengan Filler Abu Sekam Padi Akibat Lama Rendaman Air Laut (performance of split mastic asphalt (SMA) 0/11 with rice husk ash filler due to long seawater soaking)
27. Susanto HA, Indriyanti EW, Edison B (2014) Permeability Campuran Hot Rolled Sheet Wearing Course (HRS-WC) Dengan Filler Abu Sekam Padi Untuk Jalan Perkotaan (permeability mixture HRS-WC with rice husk ash filler for urban roads). Jurnal APTEK 6(1):17–32
28. Ismadarni R, Kasan M (2013) Karakteristik Beton Aspal Lapis Pengikat (AC-BC) Yang Menggunakan Bahan Pengisi (Filler) Abu Sekam Padi (characteristics of AC-BC using rice husk ash filler). MEKTEK 15(2)
29. Senolinggi RD (2018) Pengaruh Penambahan Serbuk Crumb Rubber Menggunakan Filler Abu Sekam Padi pada AC-WC ditinjau dari Karakteristik Marshall (effect of crumb rubber powder using rice husk ash filler on AC-WC in terms of Marshall characteristics). ITN Malang
30. Putri EE, Adrian A, Hariadi M (2014) Penggunaan Limbah Cangkang Kelapa Sawit (CKS) sebagai agregat Kasar Pada Campuran Aspal HRS-WC (use of palm oil shell waste (CKS) as coarse aggregate in HRS-WC asphalt mixture). In: Andalas Civil Engineering (ACE) conference
31. Ali S (2009) Karakteristik Marshall Campuran Hot Rolled Sheet (HRS) Yang Mengandung Cangkang Kelapa Sawit Sebagai Agregat Kasar (characteristics of Marshall hot rolled sheet (HRS) mixtures containing palm oil shells as coarse aggregates). Rekayasa Sipil 5(1):9–19
32. Mukhlis L, Suardi E, Adibroto F (2018) Kinerja Marshall Immersion Pada Campuran asphalt concrete wearing course (AC-WC) Dengan Penambahan Cangkang Sawit Sebagai Substitusi Agregat Halus Marshall immersion performance in asphalt concrete wearing course (AC-WC) mixture with the addition of palm shells as a substitute of fine aggregate. Jurnal Ilmiah Rekayasa Sipil 15(2):99–105
33. Pataras M, Dewi R, Prasetya AD, Bazidno FD (2017) Pemanfaatan Karet Mentah Pada flexible pavement Laston AC-WC dan Lataston HRS-WC (utilization of raw rubber in flexible pavement Laston AC-WC and Lataston HRS-WC). Cantilever: Jurnal Penelitian dan Kajian Bidang Teknik Sipil 6(1)
34. Agusomal R, Hasan MW, Taufik (2019) Pengaruh Penggunaan Getah Karet alam Terhadap Karakteristik Campuran hot rolled sheet wearing course (HRS-WC) (the effect of the use of natural rubber latex on the characteristics of the hot rolled sheet wearing course (HRS-WC)). Abst Undergr Res Faculty Civ Plann Eng Bung Hatta Uni 1(1)
35. Susanto I (2019) Pemanfaatan Karet Alam Pada Campuran Beraspal AC-WC (utilization of natural rubber in AC-WC asphalt mixture). Prosiding SNST Fakultas Teknik 1(1)

36. Syaiful, Mulyaman S (2013) Studi Penambahan Abu Batubara sebagai Filler Pada Campuran Beraspal (study of addition of coal ash as filler to asphalt mixtures)
37. Shalahuddin M (2014) Kinerja Abu Batu Bara Bercampur Kayu Pada Stabilitas dan Rongga Pori Aspal Hot Mix HRS-WC (the performance of coal ash mixed with wood on stability and cavity of hot mix asphalt pores HRS-WC). Jurnal APTEK 2(1):96–106

Marshall Characteristics of Asphalt Mixture with Water Hyacinth Ash as Filler

Dony Rohmad Dony, Bagus Hario Setiadji, and Bambang Riyanto

Abstract Road pavement is a quite important to support the mobilization of goods or services. Asphalt pavement is the main type of road pavement constructed in Indonesia. This kind of pavement consists of several materials, such as coarse aggregate, fine aggregate, asphalt as binder, and filler. Cement as filler is generally used in Indonesia, however, because of high cost, ones are more preferable to seek the substitute of cement. One type of material that could be used as cement substitute is water hyacinth, due to chemical properties it possesses. Water hyacinth contains chemical cellulose from organic polymers, and in the form of ash, it has a potential to increase the quality of asphalt. The purpose of this study was to evaluate the possibility of the use of water hyacinth ash (WHA) as a part of material of asphalt mixture. The method used was an empirical design method by conducting laboratory works to obtain primary data. The coarse and fine aggregates used came from Gringsing District, Batang Regency, Central Java Province, while the asphalt was the one with penetration grade 60/70. In this study, the content of WHA used as filler was 0, 25, 50, 75 and 100% of the total weight of filler. The results showed that the increasing use of WHA will decrease the stability and durability. Therefore, it is recommended to limit the use of WHA to a maximum of 25% as filler of the asphalt mixture, as a part of the implementation of green technology in practice.

Keywords Water hyacinth ash · Asphalt mixture · Stability · Durability

1 Introduction

An asphalt mixture is generally designed to be composed of constituent materials such as aggregates, asphalt, and fillers and made using standardized methods. The asphalt mixtures also have to be tested to ensure that the properties could fulfill the required specifications. It indicates that to obtain a road pavement structure that meets minimum performance standards, the required specifications to the material

D. R. Dony · B. H. Setiadji (✉) · B. Riyanto
Universitas Diponegoro, Jl. Prof. Soedarto, S.H., Semarang 50275, Indonesia
e-mail: bhsetiadji@ft.undip.ac.id

S. A. Kristiawan et al. (eds.), *Proceedings of the 5th International Conference on Rehabilitation and Maintenance in Civil Engineering*, Lecture Notes in Civil Engineering 225,
https://doi.org/10.1007/978-981-16-9348-9_98

and supported by standard manufacturing procedures should be sufficient to pro-duce an asphalt mixture that is recognized for its quality.

However, with the reason to reduce materials that are not renewable and not environmentally friendly, there have been many studies conducted to add additives to an asphalt mixture with various objectives, such as increasing the mixture's stability and increasing the durability of the mixture. Likewise, the materials used are generally processed materials, such as rice husk ash, crumb rubber, and fly ash [1–3]. The performance of asphalt mixtures with used material which must be removed is generally compared to that of asphalt mixtures without additive material as a reference, with the main hypothesis is that the additive can improve the performance of one or more properties of the asphalt mixture.

In this study, water hyacinth ash, or called WHA in the rest of the paper, processed from water hyacinth plants, i.e., a plant that is often considered a nuisance at a pond or lake, is used as a partial replacement of cement fillers in asphalt mixtures. Compared to the use of WHA in asphalt mixtures, research on the use of WHA in concrete construction has been widely carried out [4–6]. The high potential for the use of WHA was found in work by Widyaningsih and Sutanto [7], in which a simulation was conducted to find the optimum asphalt content of the mixtures with Portland cement, WHA, or a combination of both as filler. However, the research did not provide detailed analysis results regarding the suitability of the WHA as a filler material in asphalt mixtures. Therefore, this study aims to explore in-depth the opportunities for using WHA as filler material in asphalt mixtures in terms of the physical aspects expressed in Marshall characteristic.

2 Water Hyacinth Ash

Water hyacinth (Eichhornia crassipes) is a floating tropical plant that grows in still or slow-moving fresh water bodies discovered by Carl Friedrich Philipp von Martius, a German botanist, in 1824 while on an expedition on the Brazilian Amazon River [8] Water hyacinth has a high growth speed, so that this plant is considered a weed that can damage the aquatic environment, as shown in Fig. 1.

Water hyacinth plants have the benefit of absorbing nutrients, metals, and trace organic substances from water [8], but due to uncontrolled growth of water, hyacinth causing adverse impacts, such as: (i) increasing evapotranspiration due to their broad leaves and rapid growth; (ii) decreasing the level of dissolved oxygen due to the reduced amount of oxygen entering the waters; (iii) accelerating the silting process, especially by dead water hyacinth plants. In order to overcome this negative impact, some actions are taken such as eradication using herbicides, predatory fish or lifting the water hyacinth and using it to support other useful things; one of them is by processing water hyacinth into water hyacinth ash (WHA) for added materials in construction in the field of Civil Engineering.

WHA was produced by collecting, washing, and drying water hyacinth in an oven for a certain period to convert the organic matter into an inorganic substance. The

(a) (b)

Fig. 1 Water hyacinth plant: **a** at Rawa Pening lake, Indonesia [9]; and **b** in the form of ash (WHA) [10]

dried water hyacinth then was ground and passed through a 150-micron sieve to enable the material used as the replacement for cement [4] (see Fig. 1b).

Based on the results of research by Sukarni et al. [11], the top five chemical proper-ties of water hyacinth are O (49.5 ± 6.71%), C (14.4 ± 6.81%), K (8.26 ± 2.62), Si (5.33 ± 4.52) and Fe (4.71 ± 4.32). After drying, one of the major constituents found in WHA is silicon dioxide or silica [10], which is one of the raw materials for Portland Cement. It is a strong reason why WHA has the potential to be investigated for its use as a partial replacement for Portland Cement in Civil Engineering structures. Although WHA has some similar constituent materials like cement, the specific gravity of WHA is much smaller than cement, which is less than 2.5, while the specific gravity of cement is equal to or more than 3.1 [5], and this could affect the calculation of voids in the mixture.

3 Research Methodology

The methodology of this research consists of several steps as follows:

a. Testing of material properties

In this research, an asphalt mixture called asphalt course wearing course (AC-WC) was used. The mixture was composed of dense-graded aggregate and asphalt Penetration 60/70. All materials have to be tested to ensure that their properties could conform to the specification used [12].

b. Mixture preparation

All materials were used to prepare the mixture briquettes in which the optimum asphalt content of the original asphalt mixture (i.e., with no WHA) was determined as the references for the subsequent works. Similar to Widyaningsih and Sutanto [7], four percentages of WHA (25, 50, 75, and 100 as ratio to ordinary cement) were selected in this study.

iii. Testing of the mixtures

The mixture were tested to determine the Marshall characteristics (Marshall stability, flow, VMA, VIM, and VFA) as indicators of the performance of the mixture with the addition of WHA. The mixtures also underwent a durability test by immersing the mixtures in the water at 60° for 1, 3, and 7 days, respectively.

iv. Analysis

The analyses conducted in this study were: (i) comparison of mechanistic and volumetric characteristics of each mixture with WHA against those of the original mixture, (ii) deep analysis of Marshall characteristics, especially volumetric ones, to see a possibility to answer the phenomenon occurred, and (iii) recommendation of the use of WHA in the actual project.

4 Research Methodology

Three fractions of local aggregate from Gringsing District, Central Java Province, Indonesia, i.e., ¾ and ½ in. maximum-size coarse aggregate fraction and fine aggregate fraction, and one type of asphalt, i.e., Pertamina Pen 60/70 asphalt, were used in this study. To ensure that aggregate and asphalt properties conform to the specifications, all materials were tested using ASTM/AASHTO or SNI (Indonesian national standards). The results of the tests are shown in Tables 1 and 2.

Table 1 Results of aggregate tests

No.	Properties	Result	Specification
	Fine aggregate		
1	Passing sieve no. 200 (%)	3.31	Max. 8
2	Water absorption (%)	2.48	Max. 3
3	Specific gravity	2.57	Min. 2.5
4	Sand equivalent value (%)	98.21	Min. 50
	Coarse aggregate		
1	Loss Angeles abrasion value (%)	12.52	Max. 40
2	Affinity for asphalt (%)	97.9	Min. 95
3	Flakiness and elongation indices (%)	8.12	Max.10
4	Passing sieve no. 200 (%)	0.39	Max. 1
5	Water absorption (%)	1.46	Max. 3
6	Specific gravity	2.63	Min. 2.5
7	Angularity with one/two fractured faces (%)	98.21/95.49	Min. 95/90

Table 2 Results of asphalt tests

No.	Properties	Result	Specification
1	Penetration 60/70 Asphalt (0.1 mm)	68.33	60–70
2	Softening point (°C)	50.5	Min. 48
3	Ductility (cm)	127.5	Min. 100
4	Solubility of Trichloroethylene (%)	99.67	Min. 99
5	Specific gravity	1.015	Min. 1
6	Loss on heating (%)	0.4	Max. 0.8
7	Penetration after loss on heating (%)	77.52	Min.54
8	Ductility after loss on heating (cm)	91.5	Min. 50

Tables 1 and 2 indicate that the aggregate and asphalt properties can conform to the specifications. An aggregate gradation has to be selected carefully based on the aggregate gradation specification used; that is, aggregate gradation for the mixture of Asphalt Concrete Wearing Course (AC-WC) [12]. The aggregate gradation selected in this study is shown in Fig. 2.

Using the selected aggregate gradation, the mixtures can be designed using the following variation of WHA—PC compositions (by weight): 0–100%, 25–75%, 50–50%, 75–25%, and 100–25%, which is the similar composition to Widyaningsih and Sutanto [7]. The obtained optimum asphalt content and the Mar-shall characteristics of the mixture with 0–100% composition was used as the reference for the rest of the mixtures.

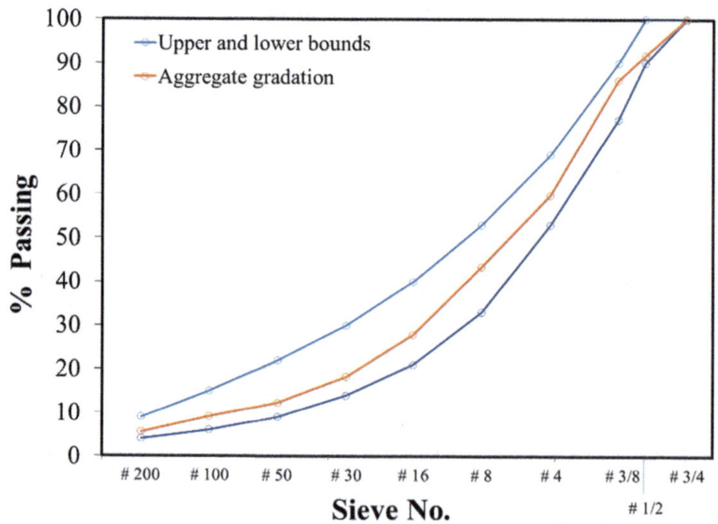

Fig. 2 Aggregate gradation used in this study

The optimum asphalt content of the mixture with 0–100% of WHA—PC composition is 5.5%, and the Marshall characteristics of that mixture at optimum asphalt content are as follows: Marshall stability 1157 kg, Flow 3.77 mm, VMA 17.49%, VIM 5%, and VFA 71.48%. All Marshall properties of the mixture can fulfill the Indonesian specification for the AC-WC mixture [12]. Next, using asphalt optimum content 5.5%, four sets asphalt mixture with different percentages by weight of WHA were prepared. Five Marshall properties (Marshall stability, flow, VMA, VIM, and VFA) were measured and calculated. The comparison among the four-set asphalt mixtures' property values against the reference mixture (i.e., mixture with no WHA) is shown in Fig. 3.

Figure 3 shows that only a mixture with WHA content equals 25% among mixtures with four different WHA content, which can satisfy all the Marshall properties. VMA is the only Marshall property that can be fulfilled by all mixtures containing WHA. Surprisingly, this result was supported by the similar findings by Widyaningsih and Sutanto [7], although the similarity of Marshall property trends between this study and the research by Widyaningsih and Sutanto [1] is only on the stability property, i.e., the highest the WHA content, the lowest the Marshall stability will be. This fact is also supported by the decrease of mixture density with the increase of WHA content.

From Fig. 3, it can be summarized that the mixture with a WHA content equal or more than 50% is not recommended due to the Marshall stability of the mixture is lower than the threshold specified by the specification. Marshall stability is one of the important parameters for AC-WC mixture because the advantage of this mixture type is the high Marshall stability value due to the ability to withstand high loads. Besides, high WHA levels also cause high VMA, but this finding is unfavorable, because the voids between aggregates are mostly filled with air voids. The high air void is a consequence of the low VFA in the asphalt mixture with high WHA at the same bitumen content, and this can cause the asphalt mixture to be less durable. The low VFA is also an indicator that the aggregates are not properly bonded to produce sufficient stiffness.

A durability test was also conducted on the mixtures with different WHA content. In this test, the mixtures were immersed for three different immersion periods (1, 3, and 7 days), and at the end of the periods, the mixtures were tested to measure the Marshall stability (see Fig. 4). Asphalt mixtures with a WHA content of 25% surprisingly experienced a significant decrease in stability value initially; however, the stability value did not change much after seven days of immersion and still met the mini-mum requirement of 800 kg. Meanwhile, asphalt mixtures with a WHA content of more than 50% experienced a slow decline in stability; however, due to the high air voids, it contributed to the lack of stiffness, therefore, it is not recommended to be used in practice.

As a closing remark, the use of WHA as a substitute for cement is a good option because it will reduce the cement material, which is an unrenewable material and a significant contributor to greenhouse gases. However, this study's results indicate that the use of WHA as a substitute for cement should be limited because excessive

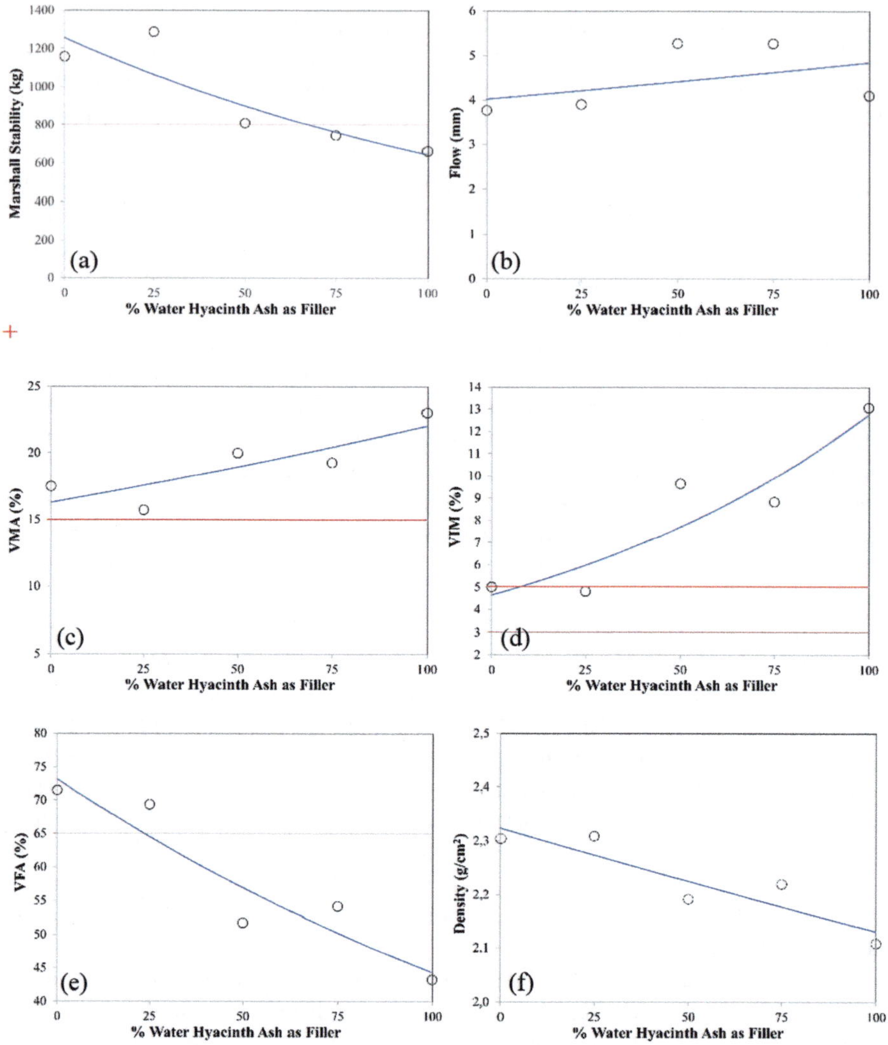

Fig. 3 Comparison of Marshall properties for asphalt mixtures with various WHA as filler: **a** Marshall stability; **b** Flow; **c** VMA; **d** VIM; **e** VFA; **f** density

use of WHA will reduce the stiffness of the mixture and absorb more asphalt causing increased air voids so that it reduces the durability of the mixtures.

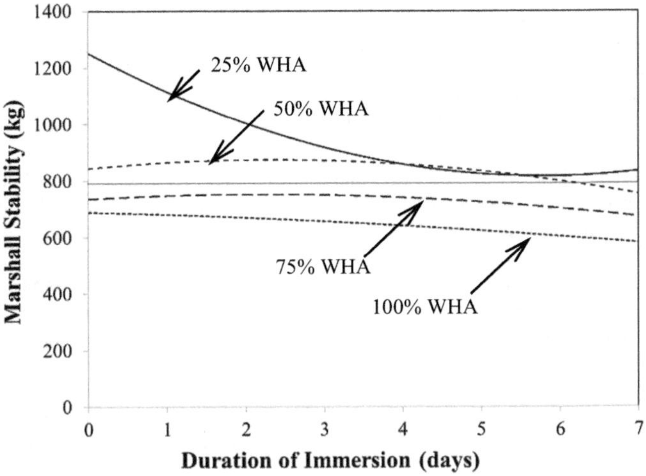

Fig. 4 Comparison of the results of durability test

5 Conclusions

This paper presented an evaluation of water Hyacinth Ash (WHA) used as a filler in asphalt mixture. To find out the influence of WHA on Marshall mixture's properties, five combination WHA—PC, i.e., 0, 25, 50, 75, and 100% WHA, were used. The results of this study indicated that the increasing use of WHA will decrease the stability. The use of WHA also could make the mixtures less durable due to high asphalt absorption and causing the increase of VIM. As a part of the implementation of green technology in practice, this study recommended a limit use of WHA to a maximum of 25% as filler of the asphalt mixture.

References

1. Hoe GB, Sunasee P, Hon CK, Gyoo KB, Ti KS (2014) Utilisation of rice husk ash in asphaltic concrete pavement. Adv Mater Res 1030–1032:961–964
2. Mashaan NS, Ali AH, Karim MR, Abdelaziz M (2014) A review on using crumb rubber in reinforcement of asphalt pavement. Sci World J
3. Mistry R, Roy TK (2016) Effect of using fly ash as alternative filler in hot mix asphalt. Perspect Sci 8:307–309
4. Murugesh V, Balasundaram N (2019) Experimental behavior of water hyacinth ash as partial replacement of cement on short column. Int J Recent Technol Eng 7(65S):648–651
5. Murugesh V, Balasundaram N (2017) Experimental investigation on water hyacinth ash as the partial replacement of cement in concrete. Int J Civ Eng Technol 8(9):1013–1018
6. Neelu D, Shashikant S (2016) Evaluation of water hyacinth stem ash as Pozzolanic material for use in blended cement. J Civ Eng Sci Technol 7:1–8

7. Widyaningsih N, Sutanto B (2018) Influence of hyacinth plant as filler on mixed AC-WC (Asphalt concrete—wearing course) with Marshall Test. In: International conference on design, engineering and computer sciences on IOP conference series: materials science and engineering, vol 453. IOP Publishing, Jakarta
8. National Research Council (1981) Food, fuel and fertilizer from organic waste. National Academy Press, Washington, D.C.
9. Media Indonesia Homepage. https://mediaindonesia.com/. Last accessed 29 Feb 2020
10. Larra-Serrano JS, Rutiaga-Quinones OM, Lopez-Miranda J, Fileto-Perez HA, Pedraza-Bucio FE, Rico-Cerda JL, Rutiaga-Quinones JG (2016) Physicochemical characterization of water hyacinth ((Eichhornia crassipes (Mart.) Solms). BioResources 11(3):7214–7223
11. Sukarni S, Zakaria Y, Sumarli S, Wulandari R, Permanasari AA, Suhermanto M (2019) Physical and chemical properties of water Hyacinth (*Eichhornia crassipes*) as Sustainable Biofuel Feedstock. In: International conference on condensed matters and advanced materials (IC2MAM 2018) on IOP conference series: materials science and engineering, vol 515. IOP Publishing, Malang
12. Directorate General of Highway (2020) 2018 General specification for road and bridge construction (Revision 2). Ministry of Public Works and Public Housing, Jakarta

The Setting Time of Portland Composite Cement Mixed with Calcium Stearate

A. Maryoto, P. Hardini, and R. Setijadi

Abstract This study aims to determine the effect of calcium stearate in setting time of Portland Composite Cement setting time. Tests carried out include the consistency of normal PCC, initial setting time, final setting time, mortar compressive strength and phase hydration. The calcium stearate content is used as additive for PCC is 0, 1, 2 and 3% of the weight of cement. Used cement is PCC type without added fly ash and with additional 10% fly ash by cement weight. Mortar compressive strength specimens are cubes with size of 50 × 50 × 50 mm. Mortar compressive strength test was performed at 3, 7, 14 and 28 days. The phase of hydration test object is a powder which is drilled from a harden cement paste. This powder is tested for its compound content using an X-Ray Diffraction (XRD) tool. The test results found that the use of calcium stearate accelerated the initial and final setting time of PCC around 78 min. Mortar compressive strength decreases when the use of calcium stearate in the mortar mixture increases. This tendency is in line with the results of the XRD test that the amount of Crystalline formed at 28 days becomes smaller when the amount of calcium stearate added to PCC increased.

Keywords Calcium stearate · Fly ash · Final setting time · Initial setting time · Portland composite cement

1 Introduction

Concrete damage caused by corrosion attacks endangers the existence of reinforced concrete structures and prestressed concrete [1]. Corrosion attacks can be caused by corrosive ion infiltrations such as chloride ions, sulfates and carbonation. These

A. Maryoto (✉) · P. Hardini
Department of Civil Engineering, Universitas Jenderal Soedirman, Purwokerto, Indonesia

P. Hardini
e-mail: probo.hardini@unsoed.ac.id

R. Setijadi
Department of Geology Engineering, Universitas Jenderal Soedirman, Purwokerto, Indonesia

S. A. Kristiawan et al. (eds.), *Proceedings of the 5th International Conference on Rehabilitation and Maintenance in Civil Engineering*, Lecture Notes in Civil Engineering 225,
https://doi.org/10.1007/978-981-16-9348-9_99

corrosive ions enter the concrete through concrete capillaries which form during the hydration process. There are two methods of reinforcement in concrete, i.e. treatment of reinforcement and treatment of the concrete. Treatment on the reinforcement can be done by coating on reinforcement or by cathodic protection [2]. While treatment on concrete can be done by making the concrete more waterproof. The increase in water resistance can be treated in several ways, namely by using the coating method on the concrete surface, by installing a layer of membrane on the concrete surface or by adding admixture or additive concrete.

Treatment for increasing permeability of concrete can be performed by adding waste material such as fly ash and waster rubber [3] to the concrete matrix. Particle size of fly ash is smoother when compared to particle size of cement. The use of fly ash in concrete can increase concrete compressive strength [4, 5], reducing the penetration of chloride ions contained in seawater [6, 7], decreasing the corrosion rate [8], decreasing the amount of pore in the concrete [9] and reducing water absorption [10]. Porosity in concrete can be reduced by reducing the amount of water used in concrete. Unfortunately, using a small amount of water in concrete causes a decrease in the workability of fresh concrete.

Meanwhile, poor workability of fresh concrete can be reduced by adding superplasticizers [11]. It can be produced using either lignosulfonate or naphtalene sulfonate. The addition of lignosulfonate compounds can accelerate the setting time of concrete [12]. Transportation and handling of the concrete, and also removal of form work are very dependent on the setting time of concrete. Some concrete additives and chemical compounds have an influence on the concrete time setting. The long duration of fresh concrete transportation requires added ingredients which can slow down the initial concrete time setting. Triethanolalamine [13], Carbamide [14], and styrene butadiene [15] have been studied to slow down the initial concrete setting time.

The use of lime in mortar [16], calcium stearate and superhydrophobic nanosilica [17] as additive in concrete can increase compressive strength and reduce water absorption [18], improve water repellents of concrete [19], chloride ion infiltration [20–22]. In this ragard, the accelerated corrosion attack on concrete reinforcement by chloride ions in concrete with calcium stearate addition, calcium stearate added to concrete is equal to 0, 1, 5 and 10 kg per 1 m^3 of concrete. It concluded then calcium stearate usage in concrete reduced significant corrosion rate ion attack. It can be observed in the Fig. 1a, b concrete without and with calcium stearate. The higher the calcium stearate, the higher the resistance of the concrete to the attack of corrosion by chloride ions. Unfortunately, the effect of calcium stearate in the initial and final setting time, phase of hydration and compressive strength in mortar have not been well known.

Therefore, this study aims to determine the effect of calcium stearate utilized with Portland Composite Cement (PCC) and fly ash in properties of mortar. The cement paste properties tested included the initial and final settings time of cement paste. In addition, the compressive strength of the mortar with variations in the addition of calcium stearate was also tested at 3, 7, 14 and 28 days. The type of solid constituents of PCC paste due to addition of fly ash and calcium stearate is investigated as well.

(a) (b)

Fig. 1. Concrete without calcium stearate (**a**), concrete with calcium stearate (**b**)

Table 1 Chemical content of cement and fly ash

Chemical composition	SiO_2	Al_2O_3	Fe_2O_3	CaO	MgO	SO_3	Na_2O	K_2O
Cement (%)	18.5	4.9	2.9	59.2	1.2	2.0	0.2	0.3
Fly ash (%)	45.2	17.9	10.9	10.0	2.8	0.3	1.6	0.7

2 Method

2.1 Material and Equipment

Materials for research include PCC, fly ash, calcium stearate, sand and water. The chemical content of PCC, fly ash and calcium stearate properties are shown in Table 1. The used calcium stearate contains 6.65 of calcium, 70.35 of carbon, 11.65 of hydrogen, and 10.6% of oxygen.

The equipment utilized consists of cement paste/mortar mixer, beaker glass, vicat apparatus, weighing balance 1000 gr with accuracy 1 gr, vicat mold, glass plate, the plunger of 10 mm diameter, hand trowel, stop watch, cube mold with the size 50 mm, compactor with the dimension. 13 × 25 × 120 mm, flow table, compression machine, drill and X-Ray Diffraction machine. The used Vicat apparatus is based on SNI 20,149:2015 [23] requirements.

2.2 Specimens

The amount of test specimens are 3 pieces for each type of code and testing. The code of specimens are COF0, C1F0, C2F0, C3F0, COF10, C1F10, C2F10, and C3F10. The type of testing are normal consistency of cement, initial and final setting time, phase of hydration and compressive strength. C represents calcium stearate, the number after C is the percentage of calcium stearate by weight of cement, F is a symbol for fly ash, and number after F is the content of fly ash by weight of cement. There are 5 types of testing namely normal consistency of cement, initial setting time, final setting time, compressive strength of mortar, phase of hydration.

2.3 *Testing Procedures*

Normal Consistency of Cement

This test is conducted based on the Indonesia National Standard, SNI 2049:2015.

Initial and Final Setting Time of Cement

The initial setting time is the time needed by cement and water to react to form cement paste and begin losing its plasticity. It can be measured by penetration of the needle. The initial setting time for penetration is the 1 mm diameter reaches 25 mm depth. Procedures of the those test is conducted and refer Indonesia National Standard, SNI 2049:2015.

Compressive Strength of Mortar

Specimen for compressive strength of mortar is cube with dimension 50 × 50 × 50 mm. The compressive strength test was carried out when the mortars were 3, 7, 14 and 28 days old. The proportion material for compressive strength of mortar is shown in the Table 3. The procedurs of making and testing the specimens of compressive strength of mortas is based on the Indonesia National Standard, SNI 2049:2015. Table 2 shows the proportion of material for setting time test.

The compressive strength test was conducted at 3, 7, 14 and 28 days old. Compression test is performed using UTM. The amount of compressive strength of mortar (MPa) is obtained by dividing the maximum load (N) and surface area of the specimen (mm^2). Figure 2 shows the process of compressive strength test.

Table 2 Proportion material for setting time test

Code	Weight (gr)			
	Cement	Water	Calcium stearate	Fly ash
C0F0	650.0	148.5	0	0
C1F0	643.5	148.5	6.5	0
C2F0	637.0	148.5	13.0	0
C3F0	630.5	148.5	19.5	0
C0F10	585.0	148.5	0	65
C1F10	578.5	148.5	6.5	65
C2F10	572.0	148.5	13.0	65
C3F10	565.5	148.5	19.5	65

Table 3 Proportion material for mortar

Code	Weight (gr)				
	Cement	Sand	Water	Calcium stearate	Fly ash
C0F0	500	1375	235	0	0
C1F0	495	1375	235	5	0
C2F0	490	1375	235	10	0
C3F0	485	1375	235	15	0
C0F10	450	1375	235	0	50
C1F10	445	1375	235	5	50
C2F10	440	1375	235	10	50
C3F10	435	1375	235	15	50

Phase of Hydration

Cement paste is poured in the molding 50 × 50 × 50 cm. After the specimen was 48 h old, then demolding the cement paste the specimen. Next, specimens are cured for 28 days. Drilling the hard specimen of cement paste (Fig. 3). The drilling results are in the form of cement paste powder (Fig. 4). Perform testing using XRD (Fig. 5) machine to obtain compound content and the type of solid in the cement paste powder.

Fig. 2 Compression test

Fig. 3 Drilling process

Fig. 4 Powder of cement paste

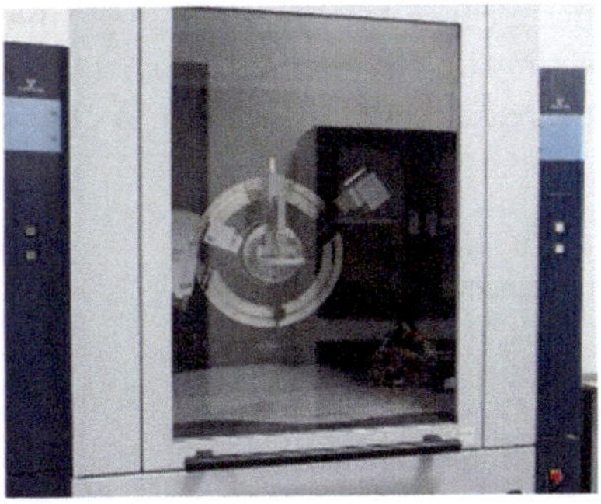

Fig. 5 X-ray diffraction

3 Result and Discussion

3.1 Normal Consistency of Cement

Needle penetration depth on cement paste with 20, 25 and 30% water amount from the cement weight shown in Fig. 6. By relating the required water and penetration depth in Fig. 6, the water content for consistency of normal cement is obtained when the penetration of needle reaches 10 mm by interpolating the data. The required water in the normal consistency of cement is 22.9%. This water content of 22.9% is then used to investigate the initial and final setting, phase of hydration as well.

3.2 Initial and Final Setting Time

Initial and final setting time cement without fly ash and with fly ash is shown in Fig. 7. It is found that additional of calcium stearate in cement reduces initial setting time of cement. The higher content of calcium stearate, the lower of initial setting time. Additional of calcium stearate 3% by weight of cement reduce the initial setting time from 129 to 111 min. This tendency is also occured in the final setting time of cement paste with calcium stearate. The usage of calcium stearate in the cement paste accelerate the final setting time. It can be observed in the Fig. 7 that the setting time of cement paste with 3% of calcium stearate is faster around 35 min than the final setting time of cement paste without calcium stearate.

The descent of setting time in the cement paste containing fly ash is obtained due to the added of calcium stearate. The initial and final setting time of cement paste with calcium stearate 3% is declined around 50 min and 78 min compared to the setting time of cement and fly ash paste without calcium stearate. This trends may be caused by cluster of stearate react with calcium hydroxide to form and mix with calcium silicate hydrate that the properties is like a wax [20]. Those reaction require some higher water and as the effect is the setting time is accelerated.

It is also can be confirmed from the Fig. 8 that the additional of fly ash in the cement paste cause the longer setting time. It is commonly understood that the prolonged setting time is affected by calcium rich gel. Appearance of calcium rich gel formation decrease the silica content. As the effect is the setting time to be retarded [24].

Fig. 6 Normal consistency of cement

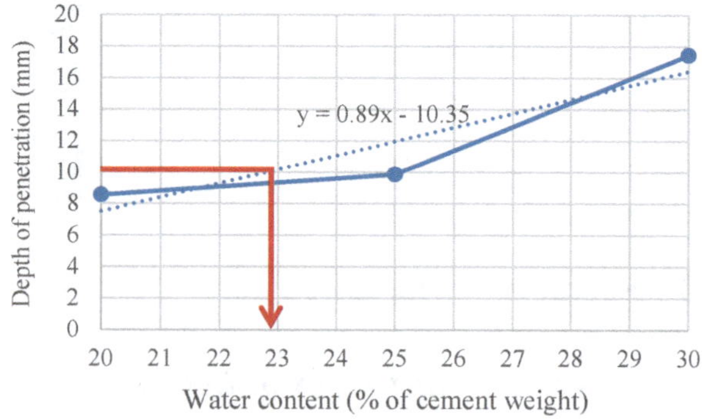

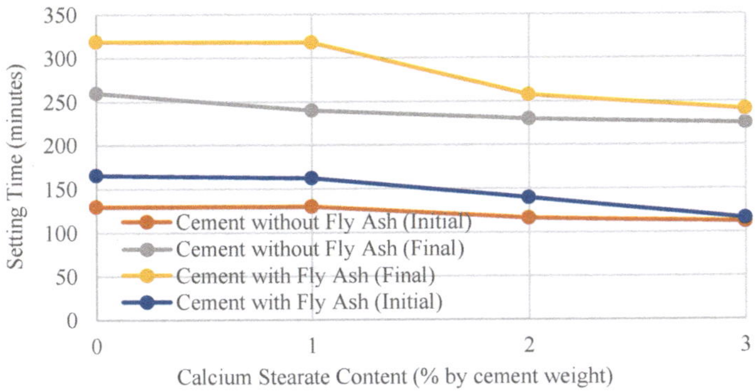

Fig. 7 Setting time of cement

3.3 Compressive Strength of Mortar and Phase of Hydration

Figure 8 shows the relationship between calcium stearate content as a horizontal axes and compressive strength of mortar without fly ash as a vertical axes. It can be found that the additional of calcium stearate in the mortar reduce the compressive strength not only at 3 days old but also at t 7, 14 and 28 days old. The usage of calcium stearate of 3% of cement weight decrease the compressive strength of mortar at 3 days old around 16.1% compare to mortar without calcium stearate. The reduction of compressive strength of mortar is also occurred at 7, 14 and 28 days old of 19.1, 49.0 and 54.5% respectively.

Figure 9 presents the correlation between calcium stearate and compressive strength of the mortar with fly ash 10%. Usage of calcium stearate in the mortar with fly ash 10% decrease the compressive strength. The higher calcium stearate content in the mortar increase a little of compressive strength of mortar. It is may be caused by reaction between silicon content in fly ash with calcium in the calcium stearate to form more calcium silicate hydrate (CSH) in the mortar.

Fig. 8 Compressive strength of mortar without fly ash

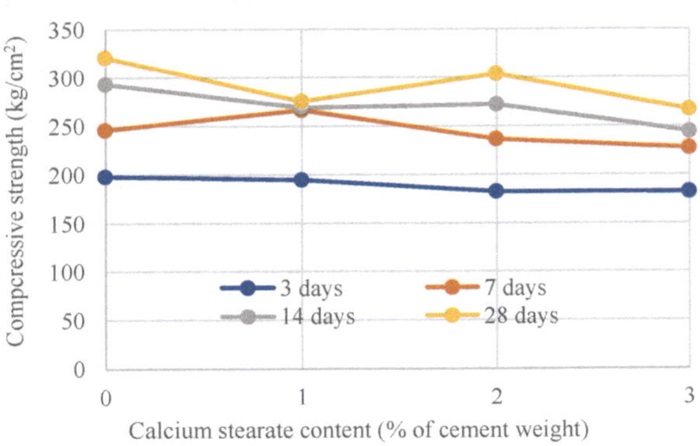

Fig. 9 Compressive strength of mortar with fly ash 10%

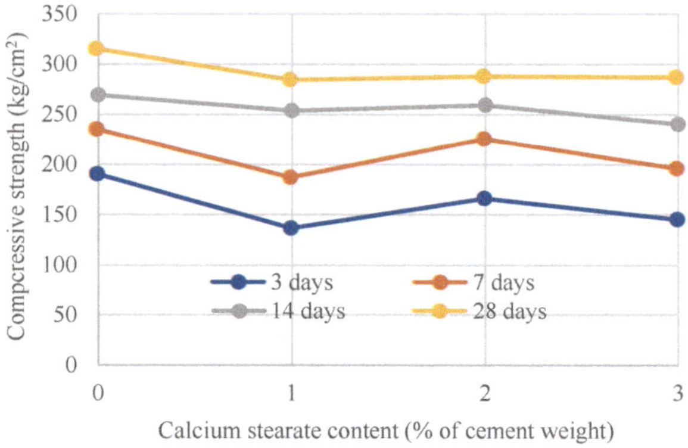

The result of X-ray diffraction of cement paste non fly ash with calcium stearate 0, 1, 2 and 3% at 28 days age as an example. The raw data in the Fig. 10 is then analyzed using Xpowder software and the result is shown in the Table 4.

According to Table 4 on cement paste with and without fly ash shows that the percentage of crystalline solid has an effect of compressive strength of mortar. Crystalline solid is a material in the solid form. Increasing of calcium stearate content tend to decrease the content of crystalline solid. Moreover, when the tobermorite (calcium silicate hydrate/C-S-H) in the cement paste increase, the compressive strength of mortar containing calcium stearate is reduced. This may be caused by some alike wax constituent united with C-S-H in the cement paste. The alike wax constituen have a better effect in the concrete to avoid the corrosion attack by chloride ion, but reduce the compressive strength of concrete.

On the other hand, specimens of cement paste containing fly ash with and without calcium stearate show that the percentage of crystalline solid have a similar tendency to the compressive strength. The crystalline solid is around 60%. It may be affected by Portland Composite Cement (PCC). This cement have been already mixed with fly ash during the production process. When PCC is then mixed again with fly ash, the some silika in the fly ash can not react with calcium hydroxide. As the effect, the compressive strength of mortar containing fly ash at this case little go down or same. This is also influenced by that the percentage of amorphous. A higher content of amorphous, the compressive strength of mortar tend to decrease, because amorphous solid usually is in the form of gel. The usage of calcium stearate both in concrete and mortar around 1% of cement weight have many advantages to reduce water absorbsion, corrosion rate, infiltration of chloride ion in the concrete. Even though the application calcium stearate in concrete with high dosage decrease the compressive strength, the dosage of 1% of calcium stearate in the concrete is still safe to be used.

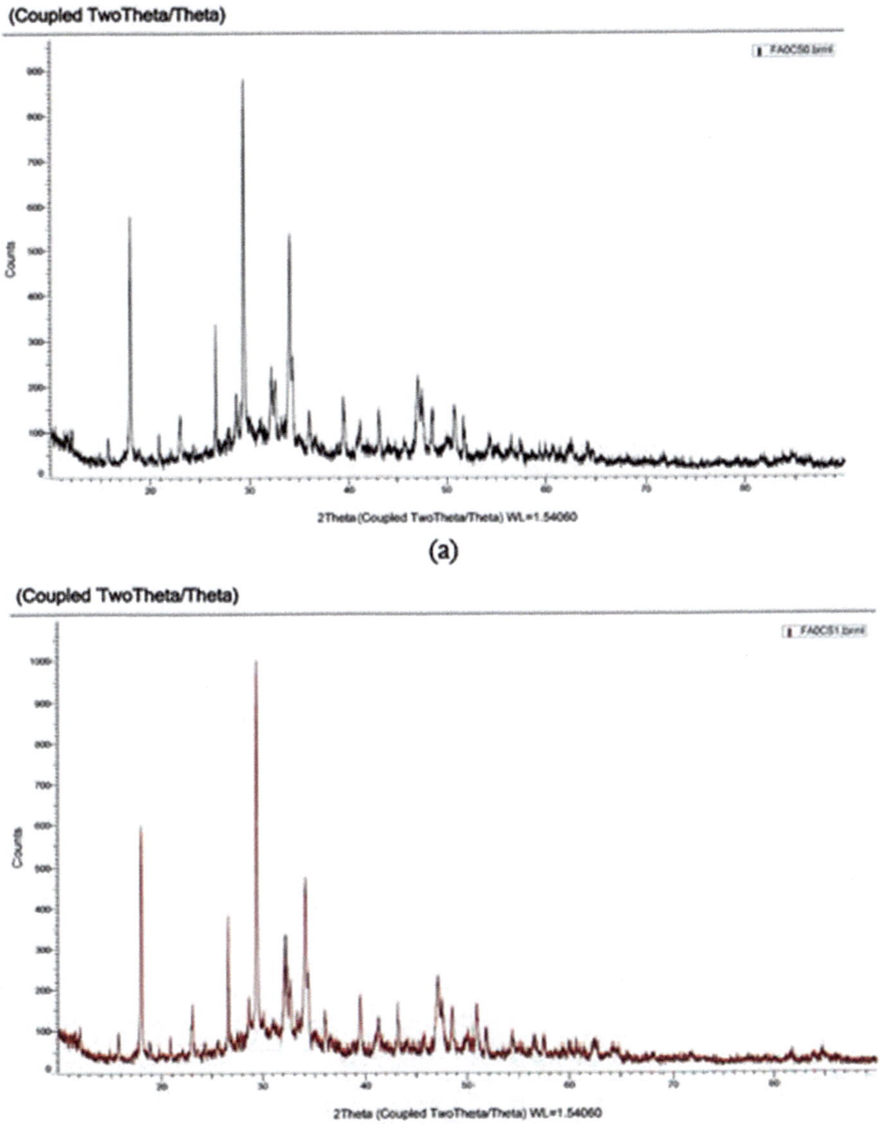

Fig. 10 X-ray diffraction result of cement paste powder: without calcium stearate (**a**), with calcium stearate 1, 2 and 3% (**b**, **c**, **d**)

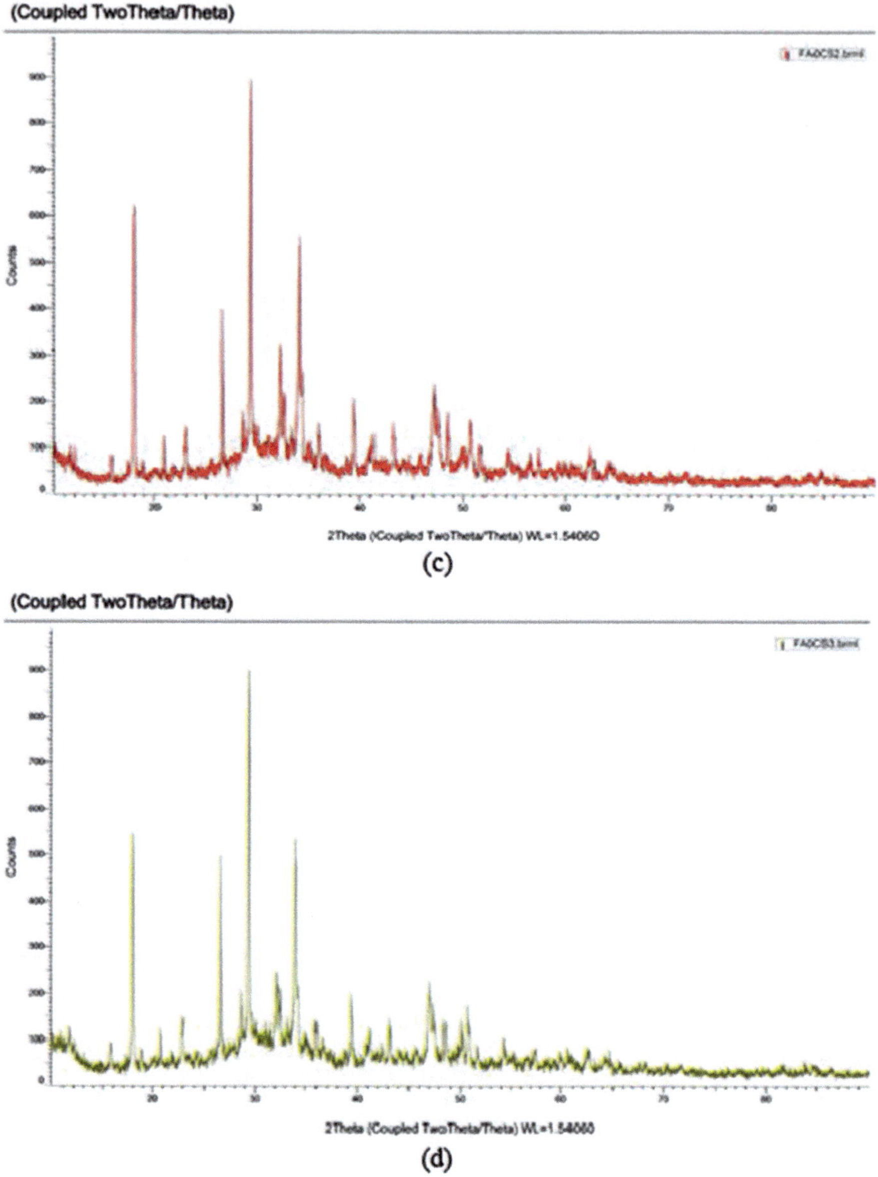

Fig. 10 (continued)

Table 4 Mineral content of cement and fly ash

Compound	Code of cement paste							
	C0F0	C1F0	C2F0	C3F0	C0F10	C1F10	C2F10	C3F10
Calcium sulphate hydrate (%)	17.4	17.4	24.9	22.8	19.8	30.2	26.0	23.2
Tobermorit (%)	3.6	4.6	7.1	4.0	4.1	5.8	4.8	3.7
Portlandite (%)	3.1	2.2	4.4	2.4	2.9	4.2	2.9	2.7
Ikaite (%)	75.8	75.8	63.5	70.8	73.2	59.8	66.3	70.4
Crystalline (%)	64.5	62.5	63.6	38.1	60.9	60.9	60.0	60.6
Amorphous (%)	35.5	37.5	36.4	41.9	39.1	39.1	39.1	39.4

4 Conclusions

According to the results and discussion above, the main conclusions are summarized as follows.

a. The consistency of normal paste of portland composite cement (PCC) is reached by adding of water around 22.9% by weight of PCC.
b. The addition of calcium stearate in cement paste accelerates the initial setting time of PCC. By adding 3% of calcium stearate by weight of cement, the initial setting time is faster around 19 min compared to non calcium stearate cement paste. The usage of calcium stearate 3% can reduce final setting time of PCC from 165 min to be 115 min.
c. In the cement paste of PCC with fly ash, addition of calcium stearate 3% decrease the inital setting time from 260 min to be 225 min. And the final setting time is reduced from 318 min to be 240 min.
d. In the mortar with PCC without fly ash, addition of calcium stearate 3% reduces the compressive strength around 17%. It also happens in the mortar with PCC and fly ash, addition of calcium stearate 3% in the mortar, the compressive strength is decreased 8.9%.
e. Reduction of compressive strength due to addition of calcium stearate is supported by forming of smaller crystalline solid in the cement paste. The higher calcium stearate in the mortar, the more reduction of compressive strength of mortar is produced.

Acknowledgements This study was conducted completely due the assistance of Jihan, Deni and PT. Holcim Indonesia—Cilacap plant. Also, thank you to Universitas Jenderal Soedirman, Purwokerto for funding this research through *Hibah Riset Unggulan Terapan*, 2018, Contract No. 3787/UN23.14/PN/2018.

References

1. Maryoto A, Shimomura T (2017) Effect of prestressed force and size of reinforcement on corrosion crack width in concrete member. J Eng Sci Technol 12(10):2664–2676
2. Laurens S, Francois F (2017) Cathodic protection in reinforced concrete structures affected by macrocell corrosion: a discussion about the significance of the protection criteria. RILEM Tech Lett 2:27–32
3. Maryoto A, Nastain N, Supriyanto H (2019) The bond response of concrete brick with recycled tire chip as partial replacement of aggregate applied in the non-structural masonry wall. Int Rev Civ Eng 10(1):33–44
4. Ashadi HW, Aprilando BA, Astutiningsih S (2015) Effects of steel slag substitution in geopolymer concrete on compressive strength and corrosion rate of steel reinforcement in seawater and an acid rain environment. Int J Technol 2:227–235
5. Hefni Y, Zaher YA, Wahab MA (2018) Influence of activation of fly ash on the mechanical properties of concrete. Constr Build Mater 172:728–734
6. Moffatt EG, Thomas MDA, Fahim A (2018) Performance of high-volume fly ash concrete in marine environment. Cem Concr Res, Article in Press
7. Yu Y, Yu J, Ge Y (2016) Water and chloride permeability research on ordinary cement mortar and concrete with compound admixture and fly ash. Constr Build Mater 127:556–564
8. Wang W, Lu C (2018) Time-varying law of rebar corrosion rate in fly ash concrete. J Hazard Mater, Article in Press
9. Jiang P, Jiang L, Zha J, Song Z (2017) Influence of temperature history on chloride diffusion in high volume fly ash concrete. Constr Build Mater 144:677–685
10. Nagrockiene D, Daugela A (2018) Investigation into the properties of concrete modified with biomass combustion fly ash. Constr Build Mater 175:369–375
11. Khudair MH, Youbi MS, Elharfi A (2018) Data on effect of a reducer of water and retarder of setting time admixtures of cement pastes and mortar in hardened stat. Accepted Article, Elsevier
12. Colombo A, Geiker MR, Justnes H, Lauten RA, Weerdt KD (2017) On the effect of calcium lignosulfonate on the rheology and setting time of cement paste. Cem Concr Res 100:435–444
13. Yaphary YL, Yu Z, Lam RHW, Lau D (2017) Effect of triethanolamine on cement hydration toward initial setting time. Constr Build Mater 141:94–103
14. Gong C, Zhou X, Dai W, Liu Y, Lu L, Cheng X (2018) Effects of carbamide on fluidity and setting time of sulphoaluminate cement and properties of planting concrete from sulphoaluminate cement. Constr Build Mater 182:290–297
15. Li L, Wang R, Lu Q (2018) Influence of polymer latex on the setting time, mechanical properties and durability of calcium sulfoaluminate cement mortar. Constr Build Mater 169:911–922
16. Mydin MAO (2017) Prelimenary studies on the development of lime-based mortar with added egg white. Int J Technol 5:800–810
17. Chen H, Feng P, Du Y, Jiang J, Sun W (2018) The effect of superhydrophobic nano-silica particles on the transport and mechanical properties of hardened cement pastes. Constr Build Mater 182:620–628
18. Ma C, Chen B (2016) Properties of foamed concrete containing water repellents. Constr Build Mater 123:106–114
19. Legazzo A, Vicini S, Cattaneo C, Botter R (2016) Effect of fatty acid soap on microstructure of lime-cement mortar. Constr Build Mater 116:384–390
20. Maryoto A, Gan BS, Hermanto NIS, Setijadi R (2018) The compressive strength and resistivity toward corrosion attacks by chloride ion of concrete containing type I cement and calcium stearate. Int J Corros 2018:1–9
21. Maryoto A, Gan BS, Aylie H (2017) Reduction of chloride ion ingress into reinforced concrete using a hydrophobic additive material. J Teknologi 79(2):65–72
22. Maryoto A, Gan BS, Hermanto NIS, Setijadi R (2017) Protection of corrosion attack in reinforced concrete due to chloride ion using calcium stearate. J Eng Appl Sci **12**(special issue 6):7965–7970

23. Indonesia National Standard, SNI 2049:2015. Cement Portland, National Standardization Bureau
24. Kaja AM, Lazaro A, Yu QL (2018) Effects of Portland cement on activation mechanism of class F fly ash geopolymer cured under ambient conditions. Construction and Building Material 189:1113–1123

Evaluation of Silt Content in Natural Sand Used as Building Materials: A Statistical Analysis Approach

Jauhar Fajrin, Mulyadi, Hariyadi, and Agung Prabowo

Abstract The excessive amount of silt within natural sand may reduce the bonding between sand and other constituent materials when used as building materials. The maximum amount allowed varies from 4 to 10%, and the allowable amount according to the Indonesian Standard (SNI S-04-1998-F) should be less than 5%. This study aims to evaluate the amount of silt in natural sand from several sand quarries in Lombok, Indonesia. The natural sand samples were collected from eight sand quarries in Lombok and the silt content was determined based on weight measurement. The study was designed to assess the silt content of each sand quarry against the SNI and also compared the silt content among the quarries. Two appropriate statistical analysis methods were applied; one-sample t-test and analysis of variance (ANOVA). The main conclusion drawn was that most of the natural sand in Lombok contains high silt content. The one-sample t-test results showed that the P-value of seven out of eight quarries was greater than the significance level (α), except for the sample taken from Sukadana. This value indicates that the silt content pretty close to the maximum value allowed by the SNI, which is already critical to be used as a building material. Meanwhile, the ANOVA result shows that the silt content from all sand quarries does not have a significant difference, which is indicated by the P-value greater than 0.05. Nevertheless, the result of a further analysis using Fisher's test shows that the sand collected from Sukadana quarry has fairly low silt content and differs significantly from the others. It is strongly recommended to give a proper pre-treatment such cleaning process before using sand in this area as a building material.

J. Fajrin (✉) · Hariyadi · A. Prabowo
Department of Civil Engineering, Mataram University, Mataram City, Indonesia
e-mail: Jauhar.fajrin@unram.ac.id

Hariyadi
e-mail: hariyadi@unram.ac.id

A. Prabowo
e-mail: agung.prabowo@unram.ac.id

Mulyadi
Ministry of Public Works, Mataram, West Nusa Tenggara Province, Indonesia

S. A. Kristiawan et al. (eds.), *Proceedings of the 5th International Conference on Rehabilitation and Maintenance in Civil Engineering*, Lecture Notes in Civil Engineering 225,
https://doi.org/10.1007/978-981-16-9348-9_100

Keywords Silt content · Inferential statistics · Material assessment

1 Introduction

One of the issues related to infrastructure development in Indonesia is the quality of building materials. Although the regulations in the form of standards have been issued, the implementation has not been well managed. Based on field observations, a lot of building materials especially those originating from nature or traditionally handmade, are directly delivered to users without sufficient quality checking process. As a result, many infrastructures built up using those materials were damaged earlier than their intended service life. Silt content is one of the important parameters to define the quality of sand. The maximum amount allowed varies between 4 and 10%, with the Indonesian Standard (SNI S-04-1998-F) stating that the maximum amount authorized should be less than 5%. Meanwhile, the maximum amount allowed by the British Standard (BS) is 4%, but the American Standard for Testing and Materials (ASTM) allows up to 10% [1].

Research related to the assessment of sand quality has not been widely reported, especially in Indonesia. Among a few reported studies, Qomaruddin et al. [2] who conducted a comparative study of river sand characteristics in Jepara reported that the silt content ranges from 10.55 to 14.37%, which means 2–3 times exceeds the maximum standard required by SNI. In addition to silt content, the study also evaluated several other parameters such as fine aggregate gradation, organic material content, density, and specific gravity. Endroyo [3] reported that the quality of sand depends on the time and the location of the sand quarry. Further, Koraila and Joshi [4] reported the results of their research on sand quality and its management in Nepal, which indicated that the quality of sand is also greatly influenced by how it is being managed. It was also observed in the study reported in Ref. [4] that the awareness of using sand that meets the technical requirements is very low among local contractors. Likewise, the attention of the local government to ensure the use of appropriate sand is also very low.

Some previous research reports the contribution of sand when it has been used as a certain building material, for example after being used as a mortar or concrete. Triadi et al. [5] conducted a compressive strength analysis of concrete made of fine aggregate from Rokan river sand, which showed that the compressive strength of concrete produced ranged from 12 to 14 MPa. Meanwhile, Ginting [6] conducted a comparative study of the utilization of Ciliwung, Cisadane, and Cikeruh river sand as normal fine concrete aggregates. The results reported in reference [6] show that concrete made with Cikeruh river sand has the highest compressive strength, which is 18.74 MPa. Furthermore, Ngugi et al. [7] reported that the high content of silt and organic material within sand reduces the bonding strength between concrete and reinforcement. It was also found that 86.2% of the observed sand samples in the study reported in Ref. [7] did not meet the requirements stated in the BS 882 standard. Gashahun [8] studied the impact of sand quality on concrete strength and found that using sand with silt levels over the permitted limit affected concrete strength significantly. In addition, Obafaye et al. [9] stated that the percentage of silt content that is unfit for engineering usage, notably concrete, is around 7%. Meanwhile, Kirthika et al. [10] proposed a fine aggregate content of less than 5% for alternative fine aggregates to make concrete with satisfactory performance.

The application of statistical methods in research in the field of civil engineering materials is still limited to descriptive statistics where the data are presented in such a way that they are easily read and understood by readers. The use of statistical methods for decision-making or inference is still limited. Fajrin et al. [11] reported the results of their research related to the development of composite materials by applying inferential statistical methods, namely the t-test method. The application of other inferential statistical tools, which is the F-test method, in the development of civil engineering materials was comprehensively reported in Refs. [12–14]. Moreover, the analysis of variance method was also applied by Fajrin et al. [15] to study the physical and mechanical properties of mortars with silica fume as additional material. Furthermore, Parmar et al. [16] applied the independent sample t-test method in their study of ready mix materials in the Gujarat region, India.

To further ensure the quality of the sand used as building material currently in Lombok, it is important to conduct a comprehensive investigation and laboratory testing. This article reports the results of an investigation into the silt content within the sand used as a building material in Lombok as a place for case studies. Furthermore, the results of laboratory tests in this study were statistically analyzed using Minitab software.

2 Materials and Method

The initial stage of this research is determining the sampling locations that represent the area in Lombok. Lombok is divided into 5 regencies/cities. Based on the mapping results it was decided that the sampling locations were in 4 districts namely: Gebong and Keling (West Lombok), Srojan and Pondok Omak (Central Lombok), Lenek Daya and Ijo Balit (East Lombok), and Luk and Sukadana (North Lombok). This study was conducted in the form of field surveys and laboratory testing which were further analyzed using inferential statistical methods. There are two research questions in this study that also become the focus of research. First, how much is the silt content at each sand quarry location, and how significant is the difference with the maximum content required by SNI. Second, how significant the difference of silt content among sand quarry locations. The measurement of silt content within the sand can be done in two ways, which are based on volume or weight. In this study, the second method was chosen which is based on weight measurement.

The suitable statistical method to elaborate the first research question is the one-sample t-test method. This method is commonly used to compare the average of a single variable or a set of data with a certain value [17]. In the case of this study, the average silt content at each sand quarry location was compared to the maximum value specified in SNI S-04-1998-F, which is 5%. There are two prerequisites for applying this statistical test method [17]; the data should be quantitative and satisfied a normal distribution assumption. The decision to accept or reject the null hypothesis (H_0) is based on the following rules; (1) The null hypothesis (H_0) is accepted if the P-value is greater than the confidence level (α). If the value is smaller, then the alternative hypothesis (H1) should be accepted. (2) The null hypothesis (H_0) is accepted if the calculated t value (t_{score}) is smaller than the t value obtained from the table of t distribution (t_{table}).

The second research question is related to several (more than two) sets or groups of data, so the appropriate statistical method used is the analysis of variance (Anova). This method is commonly used to test mean differences in more than two groups. According to Reference [17], three important assumptions must be satisfied to apply this method; (1) The groups must be independent, (2) There has an equal variance among the groups (homogeneous data), and (3) Data are normally distributed. The null hypothesis (H_0) in Anova states that there is no difference between the means of the n groups. In contrast, the alternative hypothesis (H_1) states that there is a difference between the means of the n groups. ANOVA results are expressed in the form of F values, which are further called the F-test or F-score. An inference or decision is drawing by comparing the F-score to the F value obtained from the F distribution table (F_{table}). The conditions are; if the F-score is greater than F-table, then H_0 is rejected, and as a consequence, the alternative hypothesis (H_1) is accepted. Another way to draw inference is to observe the P-value; if the value is less than the confidence level (α), then H_0 is accepted.

3 Results and Discussion

The average percentage of silt content for each sand quarry location is shown in Fig. 1. It is clearly shown that the percentage of silt content lies at around 5%, which is the maximum limit allowed according to SNI S-04-1998-F. There are four sand quarry locations with the percentage of the silt content exceeds this limit, while the other three are pretty close to this number. There has only a single location that has a fairly low level of silt content, which is safe to be used as a building material.

Overall, the average silt content from the eight sample locations in Lombok is 4.8%. The silt content at 4 locations has exceeded the maximum limit, with an average percentage of 5.58%. The other 3 locations are pretty close to the maximum limit determined by SNI, with an average value of 4.56%. Although the silt content at 3 sand quarry locations is currently still below 5%, the average value is very close to the limit and if the samples are checked one by one, there are samples whose values have exceeded 5%. To further confirm the results of the analysis, it is also necessary to do a more in-depth analysis using inferential statistics. As stated in the research methods section, the statistical method that can be used to assess whether the average silt content at a sand quarry location is the same or not with the maximum limit allowed in the SNI is the one-sample t-test method.

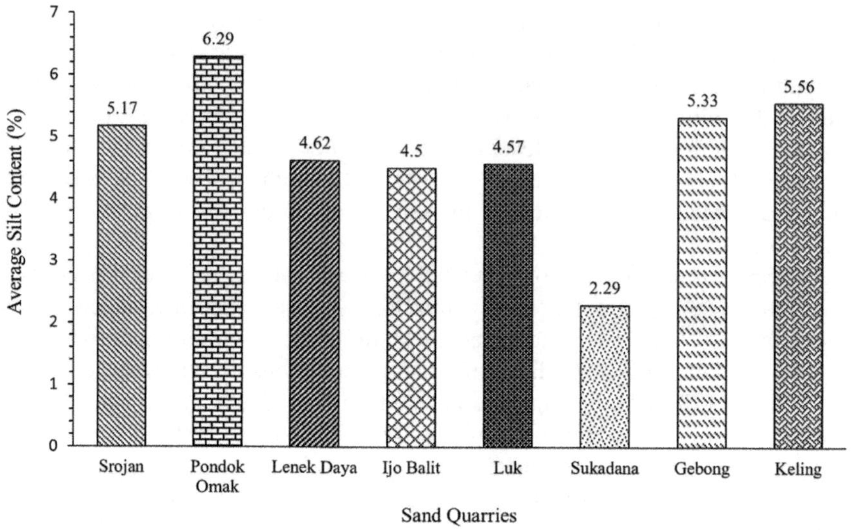

Fig. 1 The average percentage of silt content at each sand quarry location

Like many other statistical analysis methods, the first thing that needs to be done as a condition for applying the one-sample t-test method is that the data must be normally distributed. For this reason, a normality test had been performed using Minitab software based on three models; namely Anderson–Darling (AD), Ryan-Joiner (RJ), and Kolmogorov–Smirnov (KS). The results are presented in Table 1, which clearly shows that the data are normally distributed that indicated by higher P-values over the significance level (α).

Having been convinced that the data were normally distributed, a one-sample t-test was performed for all sample locations and the summary of results is shown in Table 2. The results of the one-sample t-test show that almost all locations have

Table 1 The summary of normality test results

Sample location	Anderson–Darling		Ryan-Joiner		Kolmogorov–Smirnov	
	AD	P-value	RJ	P-value	KS	P-value
Srojan	0.341	0.321	0.954	>0.1	0.239	>0.150
Pondok Omak	0.256	0.539	0.968	>0.1	0.218	>0.150
Lenek Daya	0.432	0.170	0.931	>0.1	0.255	>0.150
Ijo Balit	0.600	0.053	0.892	0.075	0.349	0.046
Luk	0.365	0.272	0.936	>0.1	0.286	>0.150
Sukadana	0.316	0.378	0.957	>0.1	0.227	>0.150
Gebong	0.275	0.497	0.968	>0.1	0.212	>0.150
Keling	0.227	0.640	0.974	>0.1	0.186	>0.150

Table 2 The summary of one sample t-test results

Sample location	Significance (α) 5%		Significance (α) 1%	
	t-score	P-value	t-score	P-value
Srojan	0.24	0.824	0.24	0.824
Pondok Omak	1.27	0.275	1.27	0.275
Lenek Daya	−0.39	0.715	−0.76	0.488
Ijo Balit	−0.39	0.716	−0.39	0.716
Luk	−0.67	0.538	−0.67	0.538
Sukadana	−4.59	0.010	−4.59	0.010
Gebong	0.27	0.800	0.27	0.800
Keling	0.47	0.662	0.47	0.662

a P-value greater than the significance level (α), except the sand quarry location in Sukadana, North Lombok. It has a meaning that the H_0 hypothesis which states that the silt content is equal to 5% must be accepted or the alternative hypothesis H_1 is rejected. Another way to conclude is by looking at the value of the t-score with the t-table. Based on the t distribution table, the value of the t-table is 2.31 and 3.75 for the value of 5 and 1%, respectively. It can be seen clearly in Table 2 that, except for the Sukadana location, the t-score value for all other locations is smaller than the value of the t-table, so that the H_0 is accepted. As the hypothesis null (H_0) is accepted, this means that the level of silt within the sand in Lombok is mostly pretty close to the maximum value tolerated by SNI, hence the sand must be cleaned up before be further used as a building material. The results of the homogeneity test which is based on two models; Bonett's test and Levene's test, are displayed in Table 3.

Ideally, a building material should have the same quality although it is provided or obtained in different ways. Hence, it is necessary to further analyze the consistency of the silt content observed at all sand quarries. A quick look at the data presented in Fig. 1 indicates that the silt content differs from each other. But how significant the differences have to be further analyzed using the appropriate method. The most relevant statistical method to apply for this case is the analysis of variance method. One of the prerequisites for data to be analyzed by the Anova method is that variants between groups must be homogeneous. To find out whether the data meets these requirements, it is necessary to do a data homogeneity test.

The homogeneity test result demonstrates that the variations (variances) of the silt content data are equal which is indicated by a P-value greater than the significance level ($\alpha = 0.05$) for all locations. Another requirement for applying the ANOVA method is that the sample must be independent of each other. This requirement has been satisfied because the samples were obtained from different locations. The third requirement is the data have a normal distribution and this has been confirmed as shown in Table 1. The results of statistical analysis using Minitab software are presented in Table 4.

Table 3 The summary of homogeneity test results

Sample location	Homogeneity test	
	P-value Bonett	P-value Levene
Srojan - P Omak	0.412	0.531
Srojan – L Daya	0.380	0.586
Srojan – I Balit	0.154	0.532
Srojan – Luk	0.795	0.636
Srojan – Sukadana	0.580	0.733
Srojan – Gebong	0.182	0.260
Srojan - Keling	0.307	0.447
P Omak - L Daya	0.898	0.980
P Omak – I Balit	0.550	0.786
P Omak - Luk	0.335	0.306
P Omak - Sukadana	0.246	0.318
P Omak - Gebong	0.645	0.607
P Omak - Keling	0,730	0.827
L Daya – I Balit	0.442	0.780
L Daya – Luk	0.300	0.362
L Daya – Sukadana	0.182	0.391
L Daya – Gebong	0.525	0.617
L Daya – Keling	0.636	0.820
I Balit - Luk	0.116	0.395
I Balit - Sukadana	0.077	0.429
I Balit - Gebong	0.876	0.967
I Balit - Keling	0.854	0.906
Luk – Sukadana	0.859	0.812
Luk – Gebong	0.139	0.141
Luk – Keling	0.250	0.273
Sukadana - Gebong	0.091	0.131
Sukadana - Keling	0.194	0.288
Gebong - Keling	0.962	0.809

Table 4 The summary of statistical analysis using Minitab software

One-Way ANOVA	Tukey Test	Fisher LSD Test
F-Value: 1.42 P-Value: 0.23	All sand quarries labeled as a single letter A	4 quarries labeled as A 3 quarries labeled as AB 1 quarry labeled as B

The most important information from the data shown in Table 4 is the P-value and the F-score, which are 0.23 and 1.42, respectively. As the P-value is greater than the significance level of 0.05, then the hypothesis H_0 is accepted. The conclusion to accept the null hypothesis is also confirmed by the value of the F_{score} (1.42), which is smaller than the value of the F_{table} (2.26). This indicates that the average sand content at all sand quarry locations is just basically the same. However, there is a little doubt about the Anova results because they equate the highest average of mud content at the Pondok Omak (6.29) with the lowest one (2.29) at Sukadana. For this reason, it is essential to carry out further tests (simultaneous tests) which usually employing Tukey, Fisher, and Dunnet tests. The simultaneous tests provide two information in the form of numbers and grouping information. The result of means grouping based on the Tukey and Fisher test is also shown in Table 4. The result of Tukey's test laid all the sand quarries within only one group labeled with the A letter, which means that there has no distinct difference in the silt content for all sample locations. The LSD Fisher test showed a more comprehensive result where the sand quarries were categorized into 3 levels that were labeled with a different letter. Meanwhile, the result of another simultaneous test using Dunnet's test provides a similar result with the Tukey test. It seems that the Fisher LSD method provides more accurate results.

4 Conclusions

The main conclusion drawn from this study is that most of the sand in Lombok has a silt content of approximately 5%, which is a maximum limit allowed by the National Indonesian Standard (SNI) S-04-1998-F. More specific findings are as follows:

(1) Most of the sand in Lombok has a fairly high silt content with an overall average of 4.8%. The result of inferential statistical analysis using the one-sample t-test method proves that the silt content in 7 out of 8 sand quarry samples already within the maximum limit allowed by SNI S-04-1998-F.

(2) The result of ANOVA shows that the silt content in the sand collected from 8 sand quarry location do not have any significant difference. However, the results of the simultaneous or pairwise test using the LSD Fisher method show that the sand sample taken from the Sukadana region in North Lombok has a fairly low silt content and differs significantly from the rest sand quarry locations.

(3) As the sand in Lombok contains fairly high silt, in which some of which have exceeded the maximum limit allowed by SNI, the sand must be cleaned up before being used as a building material.

References

1. Demissie T (2019) Evaluation of the quality of river sand on properties of concrete in the context of Addis Ababa, Ethiopia, A thesis of Master of Science in Addis Ababa Science and Technology University
2. Qomaruddin M, Ariyanto A, Umam K, Saputro YA (2018) Studi komparasi karateristik pasir sungai di Kabupaten Jepara. Jurnal Ilmiah Teknosains 4(1):6–10
3. Endroyo B (2015) Kualitas pasir Muntilan (Jawa Tengah) ditinjau dari tempat pengambilan dan musim. Wahana Teknik Sipil 12(1):1–8
4. Koirala MP, Joshi BR (2017) Construction sand, quality and supply management in an infrastructure project. Int J Adv Eng Sci Res 4(4):1–15
5. Triadi Y, Marliansyah J, Rani A (2017) Analisa kekuatan tekan beton dengan bahan agregat pasir dan kerikil sungai Rokan kanan Kabupaten Rokan Hulu. Jurnal Teknik Sipil UPP 2(1):1–8
6. Ginting ER, Erizal (2014) Pemanfaatan pasir sungai Ciliwung, Cisadane dan Cikeruh yang melalui wilayah Bogor sebagai agregat halus pembuatan beton normal. Skripsi, Departemen Teknik Sipil dan Lingkungan, Institut Pertanian Bogor
7. Ngugi HN, Mutuku RN, Gary ZA (2014) Effects of sand quality on bond strength of concrete: a case study in Nairobi and its environments Kenya. Int J Civ Struct Eng Res 2(1):119–129
8. Gashahun AD (2020) Investigating sand quality effect on concrete strength: a case of Debra Markos and its vicinities. Int J Constr Manag. https://doi.org/10.1080/15623599.2020.1774838
9. Obafaye BJ, Olukotun A, Audu MT, Ndububa EE (2020) Determination of tolerance limit of silt and clay impurities in the strength of concrete made with Abuja sand. Int J Eng Sci Invention 9(3):10–15
10. Kirthika SK, Surya M, Singh SK (2019) Effect of clay in alternative fine aggregates on the performance of concrete. Constr Build Mater 228
11. Fajrin J, Zhuge Y, Bullen F, Wang H (2011) The implementation of statistical inference to study the bending strength of sustainable hybrid sandwich panel composite. Adv Mater Res 250–253:956–961
12. Fajrin J, Zhuge Y, Bullen F, Wang H (2011) Flexural strength of sandwich panel with ligno-cellulosic composites intermediate layer-a statistic approach. Int J Prot Struct 2(4):453–464. ISSN 2041-4196
13. Fajrin J, Zhuge Y, Bullen F, Wang H (2013) significance analysis of flexural behavior hybrid sandwich panels. Open J Civ Eng 3:1–7
14. Fajrin J (2016) The application of statistical design of experiments to study the in-plane shear behavior of hybrid composite sandwich panel. J Civ Eng Dimension 18(1):25–30
15. Fajrin J, Pathurahman P, Pratama LG (2016) Aplikasi metode analysis of variance (Anova) untuk mengkaji pengaruh penambahan silica fume terhadap sifat fisik dan mekanik mortar. Jurnal Rekayasa Sipil 12(1):11–23
16. Parmar CR, Makwana AH, Pitroda J, Bhavsar JJ (2013) A study on ready mixed concrete using independent t-test through SPSS software in Charotar region of central Gujarat. J Int Acad Res Multidisc 1(11):292–300
17. Irianto A (2012) Statistik; konsep dasar, aplikasi dan pengembangannya. In: Handbook of statistics, Kencana Prenada Media Group, Indonesia

Compressive and Flexural Strength Behavior of Banana Tree Fiber Hybrid Concrete

Fadillawaty Saleh, Fanny Monika, Hakas Prayuda, Bella Lutfiani Al Zakina, Martyana Dwi Cahyati, Adira Aldi, and Feri Adri Wibowo

Abstract The purpose of this study is to discuss the utilization of banana tree fiber as additional material to improve hardened concrete properties. The existence of banana tree waste from the agriculture business has not been optimally utilized. Using banana tree waste as a constituent of concrete is expected to improve the quality of the concrete as well as to reduce waste. The fibers used in this study are placed in the tensile part of the specimen because concrete has low resistance to tensile forces, so it is hoped that the fibers can increase the tensile capacity of the concrete. The test consisted of checking the workability level of fresh concrete, while the hardened properties test consisted of compressive and flexural strength. Banana tree fibers are varied from 0 to 0.24% of the weight of cement. The results of the fresh properties show that the workability of concrete is reduced when the amount of fiber is increased. The hardened properties result show that the increased amount of

F. Saleh · F. Monika · H. Prayuda (✉) · M. D. Cahyati · A. Aldi · F. A. Wibowo
Department of Civil Engineering, Faculty of Engineering, Universitas Muhammadiyah Yogyakarta, Tamantirto, Kasihan, Bantul, Yogyakarta 55182, Indonesia
e-mail: hakasprayuda@umy.ac.id

F. Saleh
e-mail: fadillawaty@umy.ac.id

F. Monika
e-mail: fanny.monika.2007@ft.umy.ac.id

M. D. Cahyati
e-mail: martyana.dc@ft.umy.ac.id

A. Aldi
e-mail: adira.aldi.2016@ft.umy.ac.id

F. A. Wibowo
e-mail: feri.adri.2016@ft.umy.ac.id

B. L. Al Zakina
Department of Civil Engineering, Faculty of Science and Engineering, Universitas Bojonegoro, Glendeng, Kalirejo, Bojonegoro, East Java 62119, Indonesia
e-mail: bella@unigoro.ac.id

S. A. Kristiawan et al. (eds.), *Proceedings of the 5th International Conference on Rehabilitation and Maintenance in Civil Engineering*, Lecture Notes in Civil Engineering 225,
https://doi.org/10.1007/978-981-16-9348-9_101

fiber is not affect the compressive strength. However, the bending test results show that the flexural strength is improved when the amount of fiber increased.

Keywords Fiber concrete · Banana tree · Hybrid concrete · Compressive strength · Flexural strength

1 Introduction

Concrete is one of the most widely used construction materials in the world. Concrete has advantages including high ability to withstand compressive forces, durable, and is economical compared to other construction materials. However, one of the disadvantages of concrete is its low ability to withstand tensile forces. When the tensile strain exceeds the tensile strain capacity, the concrete will begin to crack, causing a decrease in the ability of the concrete to withstand loads, as such, it increases the corrosion of reinforcement. One method that can be used to increase the ability of concrete to withstand tensile forces is to add fiber, especially to areas of concrete that resist tensile forces [1].

High-strength concrete is known to have high brittle properties making it extremely dangerous if the load exceeds its capacity. Utilization of fiber in high-strength concrete can increase the tensile strength thereby reducing the brittle properties which is one of the weaknesses of high-strength concrete [2]. Normal fiber concrete can be defined as a composite material consisting of cement, water, coarse aggregate, fine aggregate, and fiber in the form of small pieces [3]. The fiber used in the manufacture of concrete comes from various resources, generally divided into two parts, namely artificial fiber, and natural fiber [4]. Examination of concrete properties using natural fiber has been carried out by many previous researchers whose study involved the use of bamboo fiber [5, 6], henequen fiber [7], Pineapple leaf fiber [8, 9], Abaca fiber [10, 11], and several classifications of other fibers [12].

In this research, banana tree fiber is used as natural fiber in the process of making concrete. Examination of the characteristics of fiber concrete using banana tree fiber has been done by many previous researchers, this is because banana tree fiber is easily available in tropical countries such as Indonesia, Malaysia, the Philippines, and Thailand [13]. Other researchers explained that the use of banana fiber as a reinforcement material significantly increased resistance to cracking and spalling and increased flexural strength [14, 15]. The utilization of banana fiber as additional materials of concrete can also reduce the use of cement in the manufacturing, thereby reducing the release of carbon dioxide into the atmosphere [16, 17].

Excessive use of natural fiber in concrete can also cause several problems, especially reducing the durability of concrete. This causes the service life of the concrete to be shorter. Thus, one possible method to increase the flexural and tensile capacity of composite concrete using banana fibers is by implementing a hybrid system. The hybrid system is achieved by placing fiber concrete only in the tensile part, while the compressive part is filled with normal concrete with the same compressive strength

planning. Additionally, it is believed that the fibers in the concrete do not significantly affect the compressive capacity of the concrete but have a significant effect in increasing its tensile strength. Although the properties of concrete using a hybrid system have been discovered by several previous researchers [18–22], none had utilized fibers from banana trees.

In this study, a fiber concrete properties test was carried out using a hybrid system. Tests carried out on concrete with a compressive strength and flexural strength were tested on concrete with ages 3, 7 and 28 days. The binders used in this study consisted of cement and silica fume. All specimens were cured using water curing.

2 Experimental Program

The concrete material in this study consisted of a binder in the form of cement with the type of Portland pozzolan cement based on ASTM C595 [23] and Silica fume from PT. Sika Indonesia with a classification from ASTM C1240 [24]. Coarse aggregate, fine aggregate, water, Sikament-NN type superplasticizer aims to increase workability in fresh concrete to facilitate the processing process [25], and banana tree fiber. The fiber used in this study comes from a banana tree with a length of 5 cm and a diameter of 1 mm as shown in Fig. 1. Before the fiber is used, the fiber must be dried to remove the water contained in the fiber, then the surface of the fiber is coated with resin so that the fiber becomes waterproof and durable. The composition of the fiber used in this study varied from 0, 0.08, 0.16, and 0.24% out of the weight of cement. The mix proportions used can be seen in Table 1. The water to binder ratio used in this study was 0.65. The type of curing used is water curing carried out 1 day after the casting process. The flexural strength and compressive strength tests were carried out on concrete of 3, 7, and 28 days.

Fig. 1 Banana tree fiber

Table 1 Mix proportions for 1 m^3

Materials	Fiber variations			
	S1 (0%)	S2 (0,08%)	S3 (0.16%)	S4 (0.24%)
Sand (kg/m^3)	836.29	836.29	836.29	836.29
Cement (kg/m^3)	287.98	287.98	287.98	287.98
Coarse aggregate (kg/m^3)	916.64	916.64	916.64	916.64
Silica fume (kg/m^3)	44.02	44.02	44.02	44.02
Fiber (kg/m^3)	0.00	0.265	0.531	0.797
Water (l/m^3)	217.05	217.05	217.05	217.05
Superplasticizer (l/m^3)	12.54	12.54	12.54	12.54

Table 2 Properties of fine and coarse aggregate

Properties	Unit	Fine aggregate	Coarse aggregate
Specific gravity	-	2.58	2.63
Water absorption	%	2.10	1.80
Mud content	%	4.70	1.50
Water content	%	2.20	2.00
Mass density	g/cm^3	1.34	1.35
Roughness content	%	-	20.90

Properties examination of fine aggregate and coarse aggregate is also carried out to determine the quality of the material used. The results of the mechanical properties of the aggregates can be seen in Table 2. The results show that both types of aggregate can be used as a mixture of concrete. The compressive strength test was carried out on cube-shaped specimens with specimen sizes of 100 × 100 × 100 mm referring to the ASTM C39 standard [26] as shown in Fig. 2. Meanwhile the flexural strength test was carried out referring to ASTM C239 [27] with specimen sizes of 150 × 150 × 600 mm. The flexural strength testing used the 4-point method without a reinforcement inside the concrete. Variation of 0% fiber was a test object as a control to determine the effectiveness of adding fiber to concrete. Fiber placed at the bottom of the specimen up to 50% of the total height of the specimen. The compressive strength test was taken from the average results of 3 specimens, while the flexural strength was taken from the average results of 2 specimens. Apart from the compressive strength and flexural strength tests, slump testing was also carried out to determine the workability level of fresh concrete.

Fig. 2 Setting up for compressive strength test for hybrid concrete (bottom: fiber concrete; top: normal concrete)

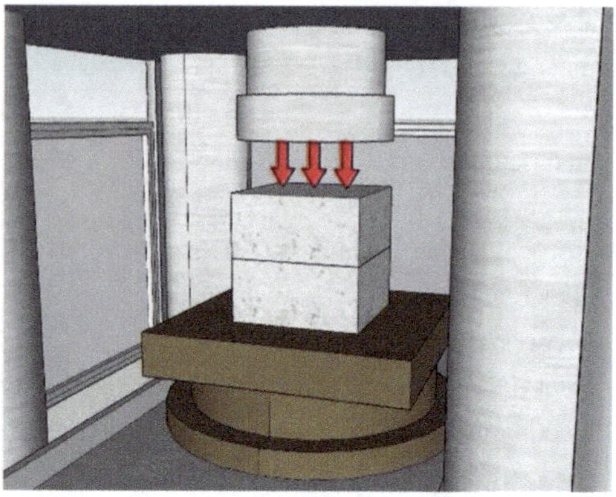

3 Results and Discussion

3.1 Slump Test

Slump testing is carried out on all variations of specimens to determine the level of workability in concrete using fiber as additional materials. The results of the slump test for each variation can be seen in Fig. 3a the test result data and Fig. 3b the process of the slump inspection. The test results showed that normal concrete without fiber obtained a slump value of 8.5 cm, while normal concrete used fiber 0.08%, 0.16%, and 0.24% of 7.0 cm, 5.5 cm, and 4.5 cm, respectively. The more increasing amount of fiber used, the less slump value will result, causing the level of workability in the concrete to consequently be lower. The decrease in slump value in concrete with a high amount of fiber is due to is due to the ongoing water absorption by the fiber during the casting process, causing the resulting workability to be reduced.

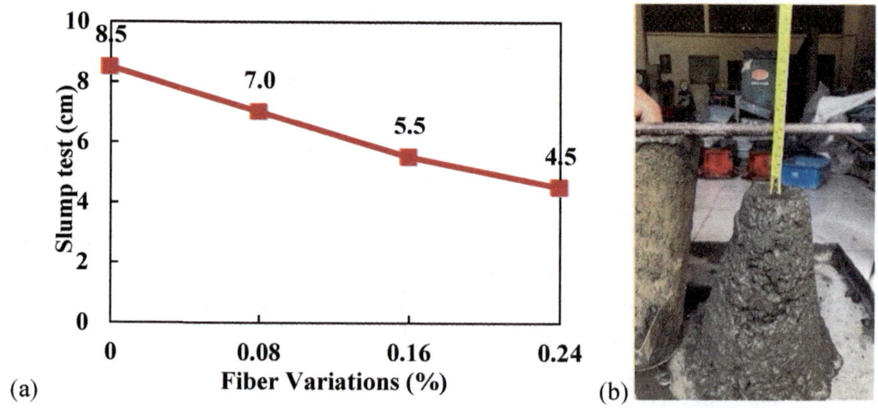

Fig. 3 **a** Relationship between fiber variations and slump test value; **b** View of slump test process

3.2 Compressive Strength

One of the hardened properties tests of concrete is the compressive strength test. Tests were carried out on test concrete aged 3, 7, and 28 days. Concrete testing was carried out at the age of 3 and 7 days to determine the characteristics of the concrete at the initial age. Such measure was done because the bond between the concrete and the fibers at the early-age concrete is feared to affect the compressive strength of the concrete. The type of curing used in this test is water curing during the waiting period of the concrete test. The results of the concrete compressive strength test can be seen in Fig. 4. The results show that the increasing age of the concrete results in higher compressive strength, this is because the bond between the aggregate or the fiber and the paste is getting stronger.

The test results at the initial age of normal concrete are 27.03 MPa, while in hybrid concrete it is 26.16 MPa, 27.26 MPa, and 27.88 MPa correspond to 0.08%, 0.16%, and 0.24%, respectively. The results of the early-age concrete testing done on day 3 show that the concrete has developed high resistance to withstand compressive forces due to the addition of Silica Fume to the concrete which can increase the compressive strength of the early-stage concrete. Silica fume particles which are much smaller than cement be able to cover small pores in the Allowing the concrete to become denser and hardens faster so that it can increase the compressive strength of the concrete at the early-age stage. The function of the high compressive strength in the initial age is to help speed up the work process in concrete construction, with high initial compressive strength. It does not require a long waiting time for the concrete construction process because the resulting compressive strength is considered to be able to withstand loads during the construction process.

In normal concrete, the one without fiber, a test on concrete aged 28 days produces a compressive strength of 38.40 MPa. Whilst when done with the concrete containing

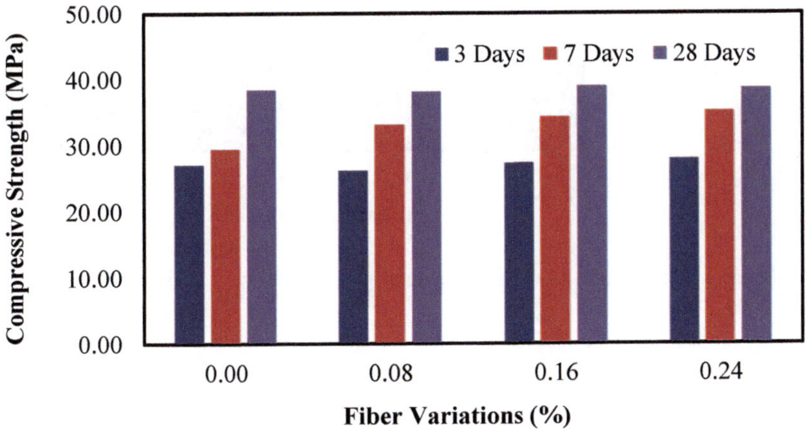

Fig. 4 Compressive strength of concrete with different amount of fiber

0.08, 0.16, and 0.24% fiber, it produces compressive strength of 38.11; 38.88; and 38.55 MPa. In the 28-day concrete test, it can be concluded that the hybrid system in concrete does not affect the compressive strength as seen from the results of the compressive test for each variation which does not have a striking difference. However, it is necessary to pay attention to the duration during the casting process of fiber concrete with normal concrete. It is feared that the use of too lengthy duration between the two types of concrete will result in cold joints in hybrid concrete. Cold joints in concrete can reduce the value of compressive strength and flexural strength in concrete. The results of cold joint testing on concrete can be seen in the following studies [28, 29].

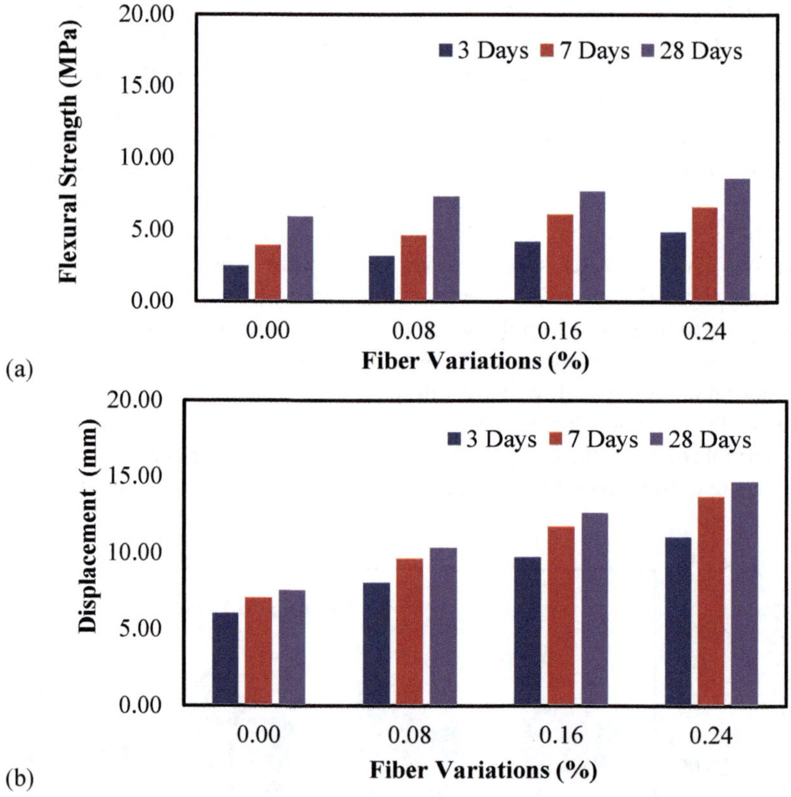

Fig. 5 Bending test results with different amount of fiber **a** Flexural strength; **b** displacement at the peak load

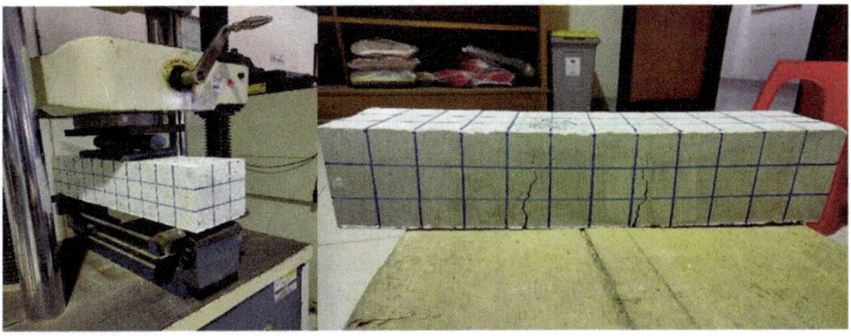

Fig. 6 Before and after bending test

3.3 Flexural Strength

The variation carried out in the flexural strength test is the same as the compressive strength test. The results of the flexural strength test can be seen in Fig. 5a, while the resulting maximum displacement can be seen in Fig. 5b. The results of the flexural strength test show that the increased fiber content results in higher flexural strength and the higher displacement value obtained. Figure 5a shows that the flexural strength at the initial age (3 days) of normal concrete is 2.44 MPa, while the concrete with the addition of fiber produces a successive flexural strength of 3.15 MPa (0.08% fiber), 4.15 MPa (0.16% fiber), and 4.81 MPa (0.24% fiber). However, the flexural strength results of concrete at 28 days are 5.89 MPa (0% fiber), 7.28 MPa (0.08% fiber), 7.65 MPa (0.16% fiber), and 8.55 MPa (0.24% fiber). It can be concluded that the hybrid system in flexural testing is very effective in increasing the flexural strength of concrete. This is because the concrete is weak in resisting the tensile force if the fibers are placed in the tensile area allowing the fibers to increase the capacity of the concrete. Figure 6 shows the result of the test object before and after testing.

4 Conclusion

Based on the aforementioned result and discussion, conclusion can be presented as follows:

a. The addition of fiber to concrete affects the workability of fresh concrete, a high amount of fiber reduces the level of workability in fresh concrete.
b. A hybrid system in fiber concrete combined with normal concrete does not affect the compressive strength value for the fiber is only placed in the bottom area of the specimen.
c. The high amount of fiber in the hybrid system increases the flexural strength of the concrete. This is because the fibers work effectively in resisting the tensile forces that occur during the bending test.

References

1. Afriandini B, Monika F, Saleh F, Prayuda H, Cahyati MD, Djaha SIK (2020) Fresh and hardened properties of self fiber compacting concrete (SFCC) incorporated with zeolite and nylon. IOP Conf Ser Mater Sci Eng 771:1–9
2. Saad M, Agwa IS, Abdelsalam A, Amin M (2020) Improving the brittle behavior of high strength concrete using banana and palm leaf sheath fibers. In: Mechanics of advanced materials and structures, pp 1–10
3. Navilesh J, Rahul BK, Shankar BK, Patil S, Gutteder VA (2017) A study on hybrid fiber reinforced concrete. Int Res J Eng Technol 4(06):1647–1656
4. Chandak A, Agrawal N, Thakur D, Titiksh A (2016) Analysis of self-compacting concrete using hybrid fibers. Int J Trend Res Dev 3(2):641–645
5. Manoharan S, Kumar GS, Balasubramainan K, Dhanasakkaravarthi B (2016) Investigation on properties of banana and bamboo fiber composites by experimentation. Middle East J Sci Res 24:16–18
6. Kavitha S, Kala TF (2016) Effectiveness of bamboo fiber as a strength enhancer in concrete. Int J Earth Sci Eng 9(3):192–196
7. Castillo-Lara JF, Flores-Johnson EA, Valadez-Gonzalez A, Herrera-Franco PJ, Carrillo JG, Gonzalez-Chi PI, Li QM (2020) Mechanical properties of natural fiber reinforce foamed concrete. Materials 12(3060):1–18
8. Abirami R, Vijayan DS, John SJ, Albert A, Alex AK (2020) Experimental study on concrete properties using pineapple leaf fiber. Int J Adv Res Eng Technol 11(6):913–920
9. Todkar SS, Patil SA (2019) Review on mechanical properties evaluation of Pineapple Leaf Fiber (PALF) reinforce polymer composites. Compos B 174:1–16
10. Tampi R, Parung H, Dhamaluddin R, Amiruddin AA (2020) Reinforced concrete mixture using abaca fiber. IOP Conf Ser Earth Environ Sci 419:1–7
11. Suhelmidawati E (2016) Tensile tests of abaca fiber as one of alternative materials for retrofitting of Unreinforced Masonry (URM) Houses. Rekayasa Sipil 13(2):22–29
12. Geremew A, Winne PD, Demissie TA, Backer HD (2021) Treatment of natural fiber for application in concrete pavement. Adv Civ Eng 2021:1–13
13. Shankar PS, Reddy KT, Sekhar VC (2013) Mechanical performance and analysis of banana fiber reinforced epoxy composites. Int J Recent Trends Mech Eng 1(4):1–10
14. Elbehiry A, Elnawawy O, Kassem M, Zaher A, Uddin N, Mostafa M (2020) Performance of concrete beams reinforced using banana fiber bars. Case Stud Constr Mater 13:1–13
15. Kumaat EJ, Mondoringin MRIAJ, Manalip H (2018) Basic behavior of natural banana stem fiber reinforced concrete under uniaxial and biaxial tensile stress. Int J Geomate 14(44):166–175
16. Tirkey N, Ramesh GB (2018) Experimental study on the banana fiber reinforced concrete. Int J Pure Appl Math 119(18):2053–2056
17. Musthafa S, Basha SM, Bhavani S, Anudeep P, Chakravarhy P, Vali SN (2019) Experimental study on partial replacement of cement by banana leaves ash and glass fiber. Int J Res Advent Technol (Special Issue):650–655
18. Libre NA, Shelarchi M, Mahoutian M, Soroushian P (2011) Mechanical properties of hybrid fiber reinforced lightweight aggregate concrete made with natural pumice. Constr Build Mater 25:2458–2464
19. Dawood ET, Ramli M (2011) Contribution of hybrid fibers on the hybrid fibers on the properties of high strength concrete having high workability. Procedia Eng 14:814–820
20. Oucief H, Habita MF, Redjel B (2006) Hybrid fiber reinforced self-compacting concrete: hardened properties. Int J Civ Eng 4(2):77–85
21. Yuan T, Lee J, Min K, Yoon Y (2019) Experimental investigation on mechanical properties of hybrid steel and polyethylene fiber reinforced no-slump high strength concrete. Int J Polym Sci 2019:1–11
22. Fard SG (2020) Hybrid fiber reinforced concrete with pozzolanic materials. J Cement Based Compos 1:16–24

23. ASTM International, ASTM C595/C595M-20 (2020) Standard specification for blended hydraulic cements, American Standards Testing Materials, West Conshohocken, United States
24. ASTM International, ASTM C1240-15 (2013) Standard specification for silica fume used in cementitious mixtures, American Standards Testing Materials, West Conshohocken, United States
25. ASTM International, C494/C4949M-13 (2015) Standard specification for chemical admixtures for concrete, American Standards Testing Materials, West Conshohocken, United States
26. ASTM International, C39/C39M-18 (2019) Standard test method for compressive strength of cylindrical concrete specimens, American Standards Testing Materials, West Conshohocken, United States
27. ASTM International, C293/C293M-16 (2016) Standard test method for flexural strength of concrete (using simple beam with centre point loading), American Standards Testing Materials, West Conshohocken, United States
28. Zega BC, Prayuda H, Monika F, Saleh F, Wibowo DE (2021) Effects of cold joint and its direction on the compressive and flexural strength of concrete. Int J Geomate 20(82):86–92
29. Illangakoon GB, Asamoto A, Nanayakkara A, Trong LN (2019) Concrete cold joint formation in hot weather conditions. Constr Build Mater 209:406–415

Investigation of Polypropylene Fiber Reinforced Concrete After Elevated Temperature Using Color Quantification and Alkalinity Method

Ni Nyoman Kencanawati, Suryawan Murtiadi, and Zul Aida Nur

Abstract Concrete properties change when it is exposed to high temperature. Fire can reduce the strength and the stiffness of the concrete. Another visible thing is the change in the color of the concrete. Furthermore, a decrease in the alkaline degree of concrete can also occur at high temperatures. This study combines quantifying the value of the color change, quantifying the alkaline level, and evaluating the residual strength of post-fire concrete. Two concrete grades were prepared: 25 and 45 MPa and exposed to a high temperature of 300 and 700 °C. The amount of 1.8, 2 and 2.5 kg/m^3 of polypropylene fiber were added to the concrete. According to the temperature rise analysis, the farther the position from the concrete surface causes a decrease in temperature. The 2.5 kg/m^3 fiber content shows more excellent post-fire mechanical performance because it provides the pores needed by concrete when water vapor pressure occurs due to high temperature; therefore, cracks in the cement paste can be avoided. Visually, the color change between grade 25 and 45 MPa are the same in each temperature; however, they have different coordinates as observed by the color test. Furthermore, the addition of 2.5 kg/m^3 fiber to the concrete contributes a superior effect to prevent the decrease of the alkaline level during the fire.

Keywords Concrete color quantification · Concrete pH · Elevated temperature · Polypropylene fiber

1 Introduction

Concrete is often used as material construction worldwide because of many advantages. Concrete is relatively easy in terms of producing and maintenance and is more

N. N. Kencanawati (✉) · S. Murtiadi · Z. A. Nur
Postgraduate Study Program of Civil Engineering, Mataram University, Jl. Majapahit 62, Mataram 83125, Indonesia
e-mail: nkencanawati@unram.ac.id

S. Murtiadi
e-mail: s.murtiadi@unram.ac.id

S. A. Kristiawan et al. (eds.), *Proceedings of the 5th International Conference on Rehabilitation and Maintenance in Civil Engineering*, Lecture Notes in Civil Engineering 225,
https://doi.org/10.1007/978-981-16-9348-9_102

resistant to climate change. However, the quality of the ingredient materials determines the strength of concrete. Fiber is also often added to concrete mixtures to improve its properties [1, 2]. Polypropylene fiber is one of the most widely used fibers in concrete mixtures. These fibers are in the form of filaments that will spread into the mixture when mixed in the concrete mix. When the temperature increases, polypropylene fibers can increase permeability by forming pore cavities to release concrete moisture. The existence of this water vapor release path can reduce spalling in the concrete when a fire occurs. European standards require the addition of a minimum of 2 kg/m^3 polypropylene fibers to reduce spalling in fire-induced concrete [3, 4].

After being exposed to high temperatures, the concrete will change its properties depending on the fire's temperature. Fires can cause cracking or spalling of the concrete, which can reduce its compressive strength. Another visible thing is the change in the color of the concrete. In general, the concrete's color becomes slightly reddish when the temperature reaches 300 °C, turns greenish grey when the temperature becomes 600 °C and even turns yellow when the temperature increases to 1200 °C [5–8]. A decrease in the alkaline degree of concrete can also occur at high temperatures. A decrease in the concrete's pH is a sign that there is carbonation of the concrete due to a fire [9].

Assessment of post-fire concrete condition, primarily to determine the residual strength, is mostly done using non-destructive tools on the concrete. The hammer test and ultrasonic pulse velocity methods are often used to estimate the residual strength after a fire. The rebound number indicating the surface hardness of the concrete is related to the compressive strength of the existing concrete based on the conversion value or the tool's calibration graph. Likewise, the velocity of the ultrasonic wave propagation that is transmitted into the concrete body is collected by the receiver sensor. It enables us to describe the quality of the material in the concrete body, whether cracks have occurred or are still sound after being exposed to high temperatures. Research on the application and analysis of non-destructive testing on concrete at elevated temperatures has been well established [10–12].

During assessment after a fire, we also need to investigate the temperature that has been reached by the concrete. Visual assessment by evaluating the color change can lead to the demand. Various methods have been used to describe the color of post-fire concrete. Spectrophotometer method, polarization microscopy, colorimetry, and digital applied concepts have been studied. Samples are prepared either from a concrete surface, heated concrete powder, or image capture directly on the concrete surface. These methods are continually evolving to quantify post-fire concrete color [13–16].

Examination of the alkaline content of the concrete also benefits in the assessment. Concrete that has experienced a decrease in pH certainly indicates carbonation in the concrete and requires proper follow-up repair. To date, checking the concrete's pH has been done by spraying phenolphthalein on the surface of the concrete. If the concrete surface does not change color to magenta, carbonation has occurred in the concrete [9].

Table 1 Concrete mixture proportion

Grade (MPa)	Weight (kg/m^3)			
	Water	Cement	Fine aggregate	Coarse aggregate
25	205	410	694	1041
45	205	586	624	936

To the end, research has only emphasized each method on the behavior of post-burn concrete mainly the investigation on mechanical properties. This study enriches the investigation by applying the series methods of quantifying the value of the color change, the method of quantifying the alkaline level in concrete, and evaluating the residual strength value of post-fire concrete. Combining these methods is useful to assess concrete after a fire and to produce more detailed conclusions.

2 Methodology

2.1 Material

Two concrete grades were prepared, which were 25 and 45 MPa. The water-cement ratios were 0.46 and 0.35, respectively, for grade 25 and 45 MPa. Natural aggregate was utilized as fine aggregate with a maximum size of 4.75 mm. Meanwhile, crushed stone was used as coarse aggregate with a maximum size of 20 mm. The mixture proportion of each concrete grade is shown in Table 1.

The mixture proportion shown in Table 1 is regarding as the control mixture proportion without polypropylene fiber. Then the series of 1.8, 2 and 2.5 kg/m^3 of polypropylene fiber were added to each control mixture. The fiber length was 12 mm and the diameter was 0.006 mm. Specimens were cylinder concretes with a size of 150 mm in diameter and 300 in height.

2.2 Temperature Growth

After curing time for 60 days, the concrete was heated to 300 and 700 °C. Concrete subjected to elevated temperature can be carried out if the water content in the concrete pores has decreased significantly. When there is still excess water in the concrete pores, a water vapor pressure explosion will inevitably occur during exposed high temperatures. This explosion can cause cracks in the cement paste due to the high tensile stress. Therefore, exposure should be carried out at a concrete age of more than 28 days.

After reaching the intended temperatures, they were kept until one hour. The growth of temperature was according to ISO 834 standard furnace test [17] as shown

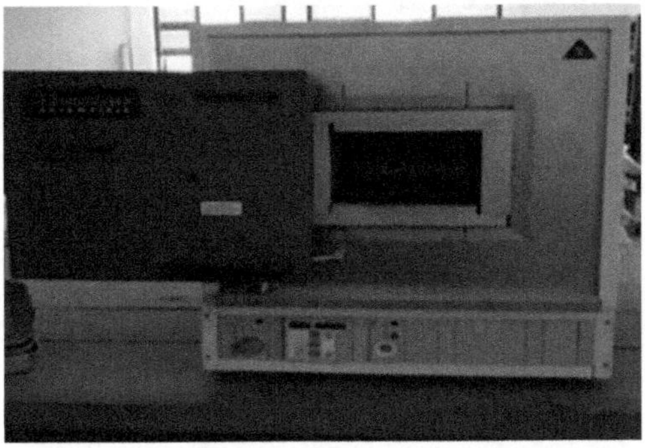

Fig. 1 Electric furnace

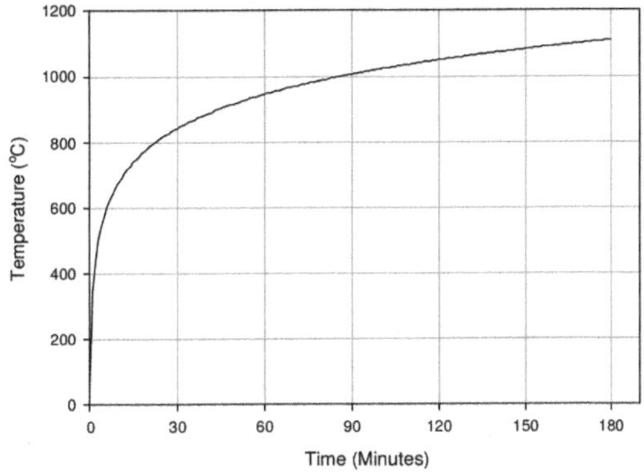

Fig. 2 Temperature growth [17]

in Eq. 10. The furnace was an electric furnace, as shown in Fig. 1 The maximum temperature capacity was 1200 °C. The relation between time and temperature from the standard furnace test is given in Eq. 1 and Fig. 2.

$$\theta_g = 20 + 345\log(8t + 1) \tag{1}$$

where *t* is time measured in minute.

2.3 Procedures of pH Level Measurement

The digital pH meter using the Ezdo brand was used to measure the alkaline content of concrete after the fire. This tool can measure the pH of the material from a range of 0–14. The pH testing process began with extracting concrete samples after the compression test with a thickness of approximately 2 cm from the surface. A sample of 30 g of concrete cement paste was crushed to a powder form and dissolved in 50 ml distilled water (Aquades).

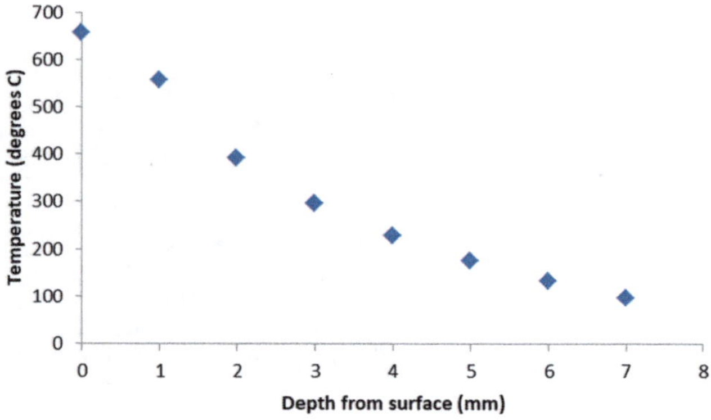

Fig. 3 Temperature rise in concrete for 700 °C exposed to fire

The solution is mixed with a stirrer to ensure homogeneity during measuring pH. At the same time, the pH measurement is carried out by inserting the sensor into the solution and the pH value was read on the display.

3 Result and Discussion

3.1 Temperature Rise

The temperature rise on the surface and some concrete depth using Wickstrom's method for a maximum temperature of 700 °C is given in Fig. 3. The temperature of the concrete surfaces approaches 657 °C; meanwhile, in the inner layer, the temperature is 82 °C. This temperature rise occurs in concrete, which has been exposed to the furnace test for one hour. The farther the position from the concrete surface causes a decrease in temperature.

3.2 Concrete Strength at Elevated Temperature

At ambient temperatures (20 °C), it appears that concrete with the addition of polypropylene fibers of 1.8 kg/m^3 reaches the highest compressive strength among other concrete with an additional proportion of other polypropylene fibers. This occurs either in the grade of 25 or 45 MPa. The addition of 1.8 kg/m^3 polypropylene fiber has a compressive strength greater as much as 1.4 times compared to that of control concrete without fiber, both on concrete grade 25 and 45 MPa.

However, when the temperature increased to both 300 and 700 °C, the concrete containing 2 kg/m^3 polypropylene fiber has the highest residual compressive strength compared to concrete with other fiber content either in grade 25 or 45 MPa. For grade 25 MPa, concrete with a content of 2.0 kg/m^3 only decreased by 30% from its initial strength when being exposed to a fire of 700 °C. For concrete with grade 45 MPa, it shows even better performance, which only decreases by 20%. Even at ambient temperatures, concrete with a fiber content of 1.8 kg/m^3 has the optimum strength, yet concrete with a fiber content of 2 kg/m^3 shows better performance when the temperature increases. When the temperature was high, the polypropylene fibers melted and provided more pores so that water subjected to vapor pressure due to high temperatures can pass and does not push the surrounding cement paste. This fiber content provides the pores needed by concrete when water vapour pressure occurs due to high temperature; therefore, cracks in the cement paste can be avoided. Thus, the concrete's performance with a fiber content of 2 kg/m^3 is better during the fire. Figure 4 and Fig. 5 present the concrete compressive strength at ambient and elevated temperature for concrete grade 25 MPa and 45 MPa respectively.

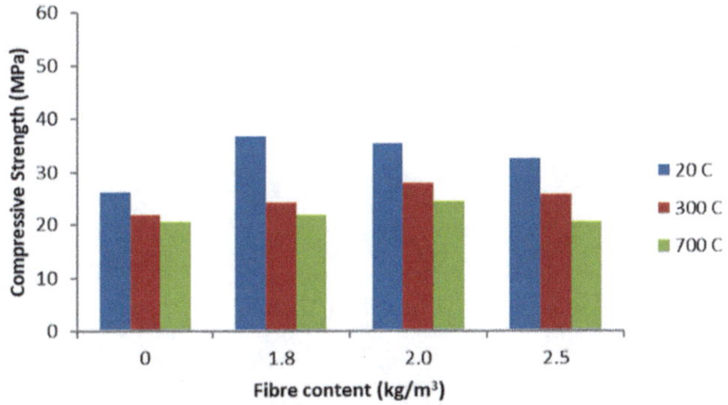

Fig. 4 Post-fire compressive strength of grade 25 MPa

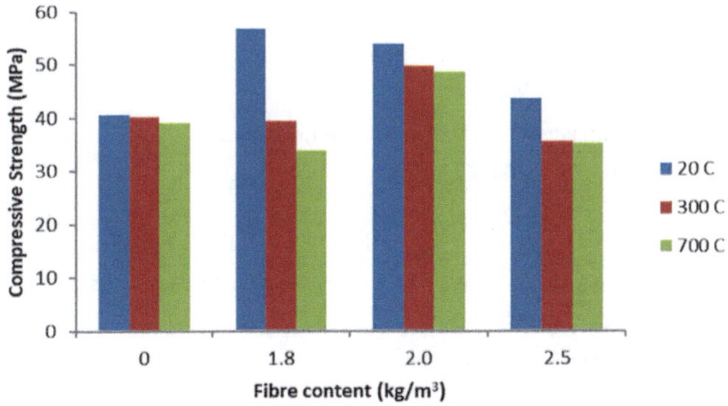

Fig. 5 Post-fire compressive strength of grade 45 MPa

3.3 Color Identification

The change of concrete surface color after exposed to 300 and 700 °C is presented in Fig. 6. The color test was carried out qualitatively and quantitatively. Qualitatively, the examination was conducted by visual inspection. Meanwhile, the quantitative test was performed using a spectrophotometer-based from the Color HunterLab tool. Color systems are represented by numbers in x, y, and z coordinates. L is a black to white color with a value range between black (0) and white (+100) and is located on the z-axis. The a value lies between the colors magenta (+100) and green (−100) and is located on the x-axis. The value of b is assigned between yellow (+100) and blue (−100) and lies on the y-axis. The color test results are presented in Table 2 and Table 3 for grade 25 MPa and 45 MPa respectively. Even though the color change between Grade 25 and 45 MPa are the same in each temperature visually, however they have different coordinates as observed by color test.

When the coordinates are plotted in a 3D curve (Figs. 7 and 8), the color pattern can be more understood clearly. The ambient temperature shows the color which lies mostly in the bottom, meanwhile due to exposed to the highest temperature, the concrete color is positioned more at top of the curve. This pattern is valid for

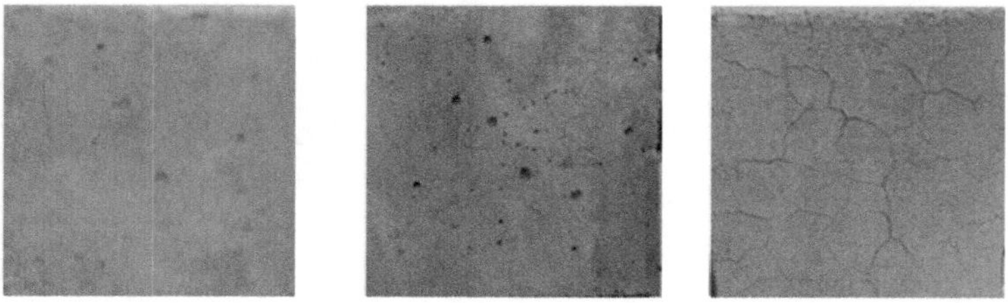

Ambient temperature 300°C 700°C

Fig. 6 Visualization of pre and post-fire concrete

Table 2 Color test for grade 25 MPa

Polypropylene (kg/m^3)	Visual examination			Color quantification								
	20 °C	300 °C	700 °C	20 °C			300 °C			700 °C		
				L	a	b	L	a	b	L	a	b
0	Grey	Brownish grey	Greyish white	34.83	0.17	3.98	62.71	0.70	5.22	68.50	0.07	3,29
1.8	Grey	Brownish grey	Greyish white	42.78	0.65	5.17	65.85	0.25	5.66	63.26	0.16	3,51
2	Grey	Brownish grey	Greyish white	43.45	0.28	6.41	64.33	0.17	3.24	66.61	0.24	3,43
2.5	Grey	Brownish grey	Greyish white	47.63	0.65	5.83	65.75	0.19	3.51	65.38	0.02	5,08

Table 3 Color test for grade 45 MPa

Polypropylene (kg/m^3)	Visual examination			Color quantification								
	20 °C	300 °C	700 °C	20 °C			300 °C			700 °C		
				L	a	b	L	a	b	L	a	b
0	Grey	Brownish grey	Greyish white	37.42	0.86	4.32	54.56	1.06	5.10	56.14	0.13	3.37
1.8	Grey	Brownish grey	Greyish white	48.65	0.88	5.52	57.97	0.33	3.13	57.95	0.26	5.55
2	Grey	Brownish grey	Greyish white	42.72	0.57	4.87	57.76	0.73	3.13	58.98	0.19	4.97
2.5	Grey	Brownish grey	Greyish white	51.03	0.22	6.00	56.28	0.97	5.63	58.14	0.30	6.02

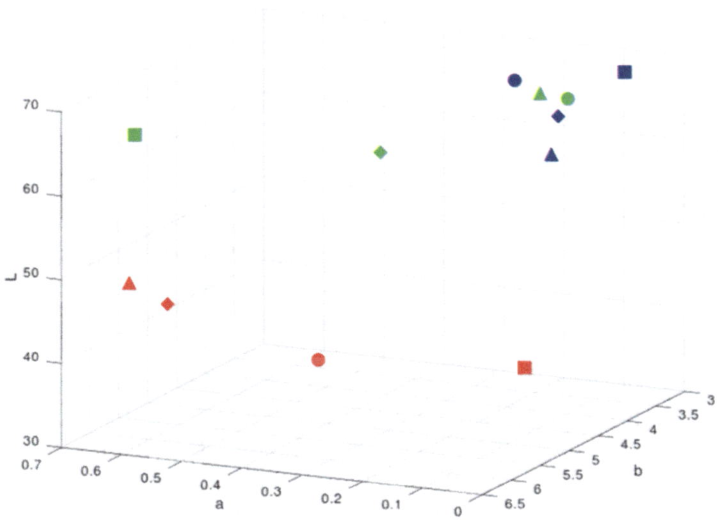

Fig. 7 Color visualization for grade 25 MPa

all concrete grades and each fiber contents. The color of red, green, and blue in in the figures refers to the temperature of 20 °C, 300 °C and 700 °C respectively. Meanwhile, the shape of square, diamond, circle, and triangle in the curves refers to the content of fiber of 0 kg/m^3, 1.8 kg/m^3, 2 kg/m^3 and 2.5 kg/m^3 respectively.

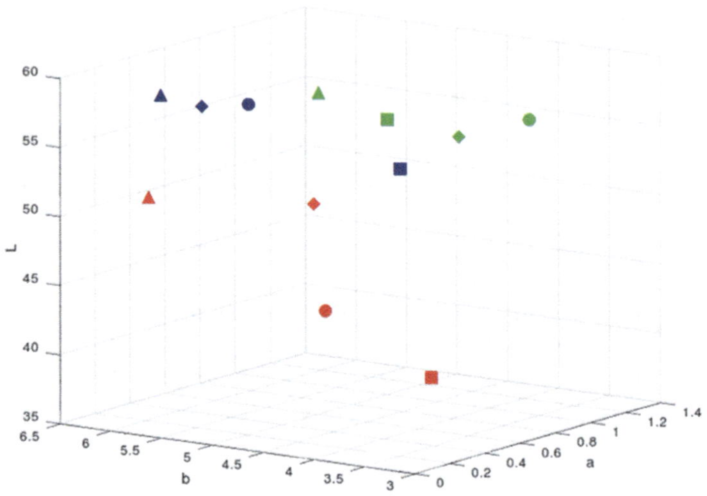

Fig. 8 Color visualization for grade 45 MPa

3.4 *Alkali Level Identification*

The alkaline level of concrete can be indicated by the pH value. The pH value is measured using a pH meter on concrete samples that have experienced high temperatures. The pH level of normal concrete was also measured as a control. The concrete alkalinity before and after fire is shown in Fig. 9 for grade 25 MPa and Fig. 10 for grade 45 MPa.

It appears that the properties of concrete are alkaline, with the top alkaline levels ranging from a pH of 12.3. However, with the increase in temperature, the concrete's alkaline content decreases due to the chemical process. Considering the polypropylene fiber content in concrete, the greatest the polypropylene content can be the best in maintaining the pH value at high temperatures compared to other polypropylene levels. It occurs either in grade 25 or 45 MPa. The addition of this fiber to the

Fig. 9 Alkaline level for concrete grade 25 MPa

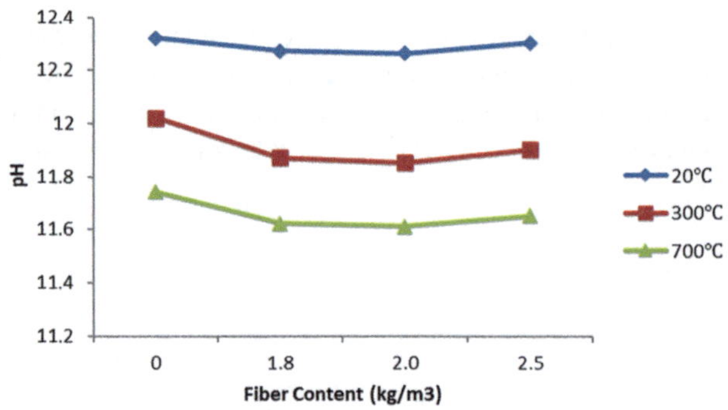

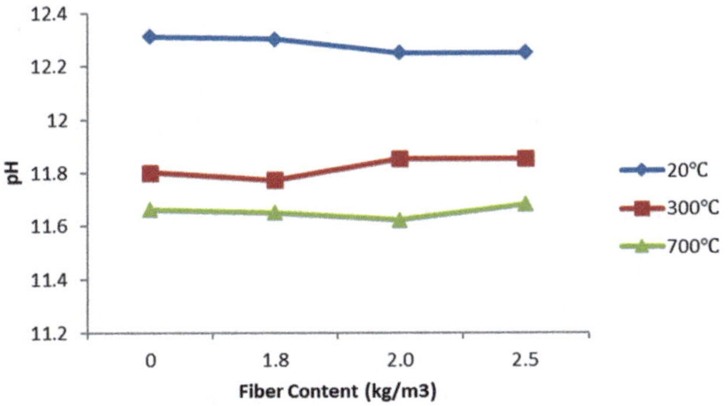

Fig. 10 Alkaline level for concrete grade 45 MPa

concrete is able to prevent the decrease of the alkaline level of the concrete because the polypropylene fiber is made of plastic; therefore, it does not affect the alkalinity level of the concrete during a fire. The alkalinity in concrete in concrete after a fire is very important property to provide passive protection to steel reinforcement continuously.

4 Conclusion

The 2 kg/m^3 fiber content shows greater performance because it provides the pores needed by concrete when water vapor pressure occurs due to high temperature; therefore, cracks in the cement paste can be avoided. Visually, the color change between grade 25 and 45 MPa are the same in each temperature; however, they have different coordinates as observed by color test. The addition of this fiber to the concrete is able to prevent the decrease of the alkaline level during the fire.

References

1. Tjokrodimuljo K (2007) Concrete technology (in Indonesian). Yogyakarta: Nafiri
2. Woodson RD (2009) Concrete structures: protection, repair and rehabilitation. Butterworth-Heinemann
3. Purkiss JA (2007) Fire safety engineering design of structures. Elsevier, Second
4. Serrano R, Cobo A, Prieto MI, de las Nieves González M (2016) Analysis of fire resistance of concrete with polypropylene or steel fibers. Constr Build Mater vol 122, pp 302–309. https://doi.org/10.1016/j.conbuildmat.2016.06.055
5. Yüzer N, Aköz F, Öztürk LD (2004) Compressive strength–color change relation in mortars at high temperature. Cem Concr Res 34(10):1803–1807. https://doi.org/10.1016/j.cemconres.2004.01.015
6. Hager I (2014) Colour change in heated concrete. Fire Technol 50(4):945–958

7. Lee J, Choi K, Hong K (2009) Color and material property changes in concrete exposed to high temperatures. J Asian Archit Build Eng 8(1):175–782. https://doi.org/10.3130/jaabe.8.175
8. Malik M, Bhattacharyya SK, Barai SV (2020) Thermal and mechanical properties of concrete and its constituents at elevated temperatures: a review. Constr Build Mater 121398. https://doi.org/10.1016/j.conbuildmat.2020.121398
9. Kencanawati N, Mahmud F, Merdana N, Ngudiyono N (2015) Alkaline onset detection of concrete after fire as a preliminary estimation of steel reinforcement corrosion. Spektrum Sipil 2:1858–4896
10. Amini K, Jalalpour M, Delatte N (2016) Advancing concrete strength prediction using non-destructive testing: development and verification of a generalizable model. Constr Build Mater 102:762–768. https://doi.org/10.1016/j.conbuildmat.2015.10.131
11. Uva G, Porco F, Fiore A, Mezzina M (2013) Proposal of a methodology for assessing the reliability of in situ concrete tests and improving the estimate of the compressive strength. Constr Build Mater 38:72–83. https://doi.org/10.1016/j.conbuildmat.2012.08.025
12. Lin Y, Hsiao C, Yang H, Lin Y-F (2011) The effect of post-fire-curing on strength–velocity relationship for nondestructive assessment of fire-damaged concrete strength. Fire Saf J 46(4):178–185. https://doi.org/10.1016/j.firesaf.2011.01.006
13. Annerel E, Taerwe L (2011) Methods to quantify the colour development of concrete exposed to fire. Constr Build Mater 25(10):3989–3997. https://doi.org/10.1016/j.conbuildmat.2011.04.033
14. Short NR, Purkiss JA, Guise SE (2001) Assessment of fire damaged concrete using colour image analysis. Constr Build Mater 15(1):9–15. https://doi.org/10.1016/S0950-0618(00)00065-9
15. Du S, Zhang Y, Sun Q, Gong W, Geng J, Zhang K (2018) Experimental study on color change and compression strength of concrete tunnel lining in a fire. Tunn Undergr Sp Technol 71:106–114. https://doi.org/10.1016/j.tust.2017.08.025
16. Annerel E, Taerwe L (2009) Revealing the temperature history in concrete after fire exposure by microscopic analysis. Cem Concr Res 39(12):1239–1249. https://doi.org/10.1016/j.cemconres.2009.08.017
17. Lim L, Wade C (2016) Experimental fire tests of two-way concrete slabs, 2020. J 2(5):99–110

The Effect of 12.5% Metakaolin and Variations of Silica Fume on Split Tensile Strength and Modulus of Rupture of High Strength Self-Compacting Concrete (HSSCC)

Endah Safitri, Wibowo, Halwan Alfisa Saifullah, and Farhan Gilang Septian

Abstract High-strength self-compacting concrete (HSSCC) is significantly used for the construction of roads and buildings due to its high compressive strength and ability to compact without a vibrator. This research aims to determine the effect of adding metakaolin and silica fume to increase HSSCC's tensile strength and modulus of rupture. The materials used were 12.5% metakaolin and 0, 9, 11, 13, and 15% variations of silica fume by weight of cement. For the split tensile strength test, a cylinder with a diameter of 15 cm and a height of 30 cm was used, while a beam with dimensions of 10 × 10 × 40 cm was used to test the modulus of rupture. The specimen was tested after 28 days. The result showed that the split tensile strength and modulus of rupture of HSSCC concrete with 12.5% metakaolin and variation of silica fume 0, 9, 11, 13, and 15% were 3.93, 4.76, 5.27, 4.94, 6.80, 7.66, 7.82, and 7.69 MPa.

Keywords HSSCC · Modulus of rupture · Split tensile strength

1 Introduction

High Strength Concrete is a concrete innovation with better workmanship and higher requirements than the conventional which is impossible to achieve only with the use of conventional materials and normal mixing [1]. Further research on admixtures and additives for concrete mixtures, especially water reducers or plasticizers and superplasticizers, has produced high-strength concrete capable of enhancing all concrete into a modern high-performance material [2]. Self-Compacting Concrete (SCC) is an innovation with a high slump value and is capable of compacting itself without using tools [3]. It has a high flowability value, able to flow under its weight, fill the formwork, and reach its own highest density [4].

E. Safitri (✉) · Wibowo · H. A. Saifullah · F. G. Septian
Department of Civil Engineering, Sebelas Maret University, Kota Surakarta, Indonesia

H. A. Saifullah
e-mail: halwan@ft.uns.ac.id

S. A. Kristiawan et al. (eds.), *Proceedings of the 5th International Conference on Rehabilitation and Maintenance in Civil Engineering*, Lecture Notes in Civil Engineering 225,
https://doi.org/10.1007/978-981-16-9348-9_103

Silica fume and metakaolin are pozzolanic admixtures often used in the manufacture of concrete to achieve certain characteristics. These materials are used as additives or substitutes for cement due to their ability to increase the resistance and strength of concrete compared to ordinary Portland cement [5]. Concrete added with metakaolin increases its compressive strength, split tensile strength, flexural strength, and modulus of elasticity [6].

From observation, the addition of metakaolin to the tensile strength of reinforced concrete prevents cracking due to shrinkage and temperature changes. Meanwhile, for unreinforced concrete, it is used in the planning and construction of roads, airports, and prestressed concrete. A direct tensile strength test needs to be carried out to determine its value on concrete. However, this test is difficult to conduct due to frequent failures during testing, thereby making it impossible to obtain the tensile strength value. Therefore, an indirect or split tensile strength commonly known as the Brazilian Test or Splitting Test is used [7]. Another method used to determine the value of the tensile strength of concrete is the Modulus of Rupture (MOR) test. Based on ASTM C293/C293M, the modulus of rupture is used to determine the tensile strength of concrete using the center point loading method due to its higher value [8].

2 Experimental Investigation

The mix design of HSSCC concrete with 12.5% metakaolin and variations of Silica Fume was made with 0.25 water-cement factors and 1.5% superplasticizer, as shown in Table 1.

The split tensile strength test of concrete is based on Indonesian National Standard (SNI) 03-2491-2002 [9]. The specimens were tested at the age of 28 days using the Brazilian or Splitting Test method as shown in Fig. 1. Meanwhile, Fig. 2 shows that the specimens used were concrete cylinders with a diameter of 15 cm and a height of 30 cm. A compression testing machine was used to test the split tensile strength of concrete.

Table 1 Recapitulation of mix design of high-strength self-compacting concrete (HSSCC)

Specimen code	Fine aggregate (kg/m^3)	Coarse aggregate (kg/m^3)	Cement (kg/m^3)	Metakaolin (kg/m^3)	Silica fume (kg/m^3)	Superplasticizer (l/m^3)	Water (l/m^3)
HSSCC-SF0%	915.30	777.90	525	75	0	9	150
HSSCC-SF9%	905.72	769.76	471	75	54	9	150
HSSCC-SF11%	903.59	767.95	459	75	66	9	150
HSSCC-SF13%	901.46	766.14	447	75	78	9	150
HSSCC-SF15%	899.33	764.33	435	75	90	9	150

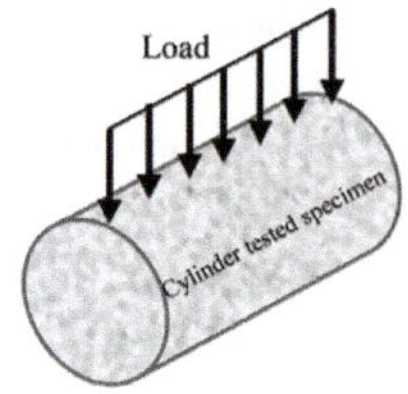

Fig. 1 The Brazilian splitting test

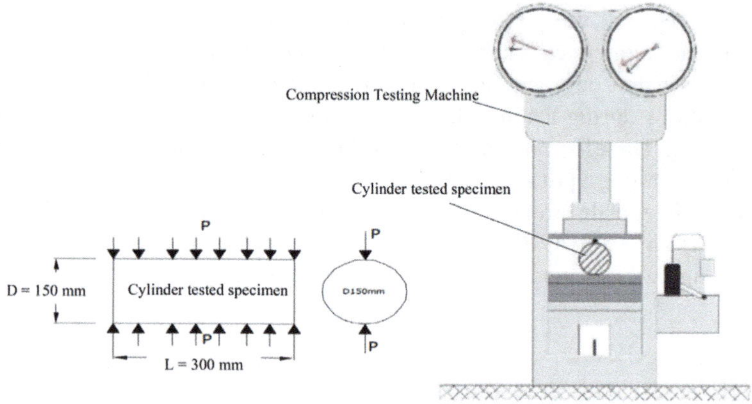

Fig. 2 Split tensile strength test setup

Fig. 3 Modulus of rupture test using Universal Testing Machine

The test object for the modulus of rupture was a concrete block with a size of 10 × 10 × 40 cm. The center point loading method was used to determine the effective span length of the beam at 30 cm. The test was carried out at the age of 28 days using the Universal Testing Machine as shown in Fig. 3. The modulus of rupture of concrete is determined by SNI 2847-2019 at 10–15% f'c [10].

3 Experimental Results and Discussion

Concrete is categorized as self-compacting concrete supposing it qualifies the requirements by EFNARC 2005 such as filling and passing abilities, and segregation resistance. These attributes are tested using the slump flow, L-Box, and V-Funnel methods [4]. The results of the recapitulation of the self-compacting concrete test are shown in Table 2.

Based on EFNARC 2005, the requirements for slump flow are T500 average between 2 and 5 s and d average of 650–800 mm. For the L-Box H2/H1 required is 0.8–1, with a V-Funnel of 6–12 s [4]. The test results of the three methods showed that the specimens with 12.5% metakaolin and variations of silica fume qualify the requirements as self-compacting concrete.

The split tensile strength of concrete is calculated using Eqs. 1 and 2.

$$f_{ct} = \frac{P}{\mathrm{A}} \text{ where } A = \frac{\pi L D}{2} \quad (1)$$

$$f_{ct} = \frac{2P}{\pi L D} \quad (2)$$

where f_{ct} is the split tensile strength (MPa), P is the maximum test load (N), A is the cross-sectional area (mm^2), L is the specimen length (mm) and D is the diameter (mm).

The results of the concrete split tensile strength test are shown in Table 3 and Fig. 4.

The addition of silica fume to HSSCC with 12.5% metakaolin is able to increase the split tensile strength of the concrete as shown in Table 3 and Fig. 4. Metakaolin and silica fume fill the voids between the aggregates, thereby making the concrete denser and able to support larger loads. The results showed that the maximum and optimum split tensile strengths of concrete were 5.78 MPa at 11% silica fume, and 6.02 MPa at 11.11% silica fume, respectively.

Table 2 Recapitulations of self-compacting concrete test result

Specimen code	Slump flow		L-box	V-funnel
	T_{500} average (s)	d average (mm)	H_2/H_1	Time (s)
HSSCC-SF 0%	4.7	690	0.81	12
HSSCC-SF 9%	4.6	710	0.9	10
HSSCC-SF 11%	4.1	725	0.91	9.5
HSSCC-SF 13%	3.9	730	1	7.5
HSSCC-SF 15%	3.5	740	1	6.5

Table 3 Split tensile strength test result

Specimen code	f_{ct} (MPa)	Percentage increase in split tensile strength of concrete (%)
HSSCC-SF0%	3.93	–
HSSCC-SF9%	4.76	21.25
HSSCC-SF11%	5.78	47.31
HSSCC-SF13%	5.27	34.16
HSSCC-SF15%	4.94	25.74

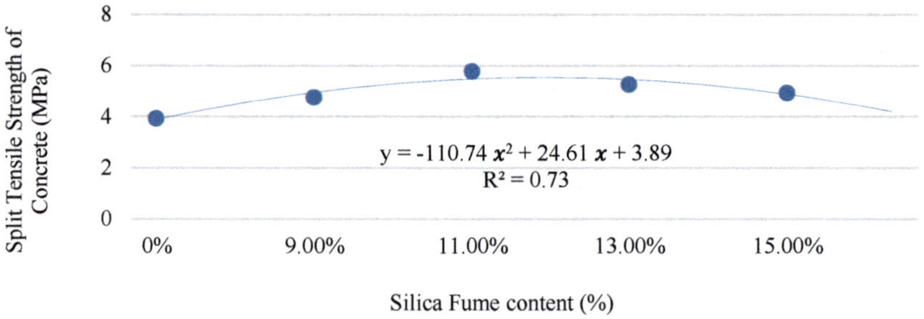

Fig. 4 Relationship between split tensile strength and silica fume content

The modulus of rupture is the maximum flexural strength of the concrete before it collapses due to loading. The tests showed the load limit that the concrete or structure is capable of supporting before the collapse. Furthermore, the center point loading method was used, where a 10 × 10 × 40 cm beam is loaded at one point in the middle of the span, as shown in Fig. 5.

Modulus of rupture can be calculated using Eq. 3.

$$R = \frac{3PL_{ef}}{2BH^2} \tag{3}$$

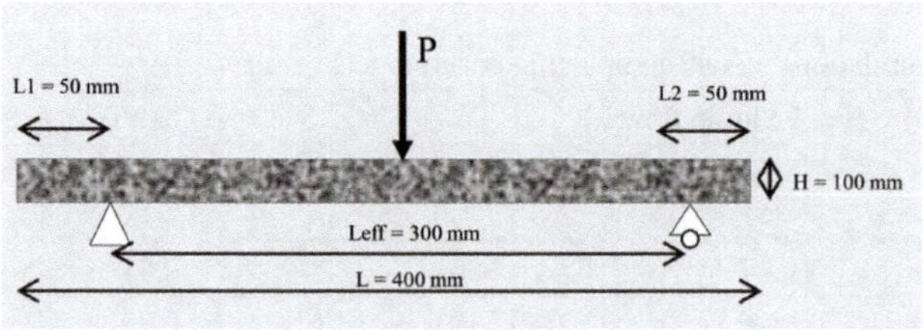

Fig. 5 Center point loading method

Table 4 Modulus of rupture and compressive strength test result

Specimen code	*R* (MPa)	Percentage increase in MOR (%)	*fc′* (MPa)	*R/fc′* × 100 (%)
HSSCC-SF0%	6.80	–	41.99	16.19
HSSCC-SF9%	7.66	12.64	55.39	13.83
HSSCC-SF11%	7.97	17.21	67.75	11.76
HSSCC-SF13%	7.82	15.08	61.80	12.65
HSSCC-SF15%	7.69	11.60	58.03	13.25

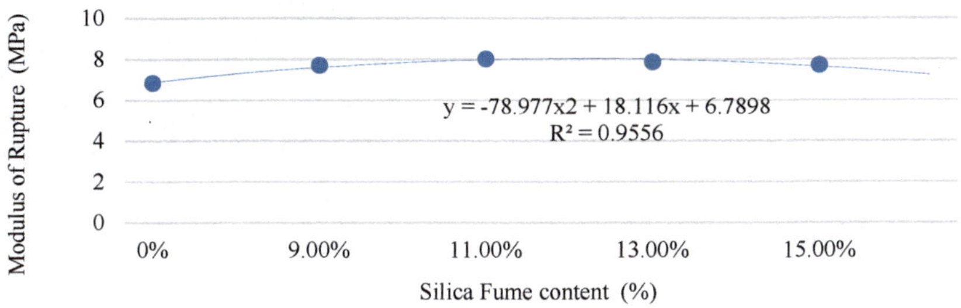

Fig. 6 Relationship between modulus of rupture and silica fume content

where *R* is the modulus of rupture (MPa), *P* is the maximum test load (N), L_{ef} is the effective length of test beam (mm), *B* is the width of test beam (mm) and *H* is the height of test beam (mm).

The result showed that the addition of silica fume to HSSCC with 12.5% metakaolin was able to increase the concretes modulus of rupture, as shown in Table 4 and Fig. 6. The greater the silica fumes content, the higher the modulus of rupture, which decreases in value after reaching a maximum point. Furthermore, the maximum modulus of rupture of HSSCC with 12.5% metakaolin occurred at 11% silica fume content of 7.97 MPa. The results of the regression analysis showed an optimum modulus of rupture of 7.83 MPa at 11.47% silica fume content. Table 4 shows the Modulus of Rupture value ranging from 11 to 16% fc′, which is close to the value of concrete that has been determined by SNI 2847-2019 of 10–15% of its compressive strength.

A type of collapse considered valid in the modulus of rupture test is a crack in the middle of the test object. It is found in both concrete with and without silica fume, which means that all test objects can be considered valid. The collapse of the modulus of the rupture test object is shown in Fig. 7.

Fig. 7 Concrete collapse when testing the modulus of rupture

4 Conclusion

In conclusion, the addition of 12.5% metakaolin and variations of silica fume up to 15% content still maintains the properties of the concrete as SCC. The optimum split tensile strength was achieved at 11.11% silica fume with a split tensile strength value of 6.02 MPa. Meanwhile, the optimum Modulus of Rupture at a value of 7.83 MPa was achieved at 11.47% silica fume with a value of compressive strength of 11–16%. The addition of metakaolin and silica fume increases the density of the concrete through the ball bearing effect, therefore, it has good workability and reduces the possibility of bleeding and segregation. This makes the concrete denser and increases its capacity to support loads.

Acknowledgements The authors are grateful to Sebelas Maret University for supporting this research through the Fundamental Research Scheme from the 2021 funds of non-APBN, under the contract number: 260/UN27.22/HK.07.00/2021.

References

1. Nawy EG (1996) Prestessed concrete a fundamental approach. PrenticeHill, New York
2. Joseph A, Mathew LA, John R (2017) Performance of metakaolin on high strength self compacting concrete. Int J Sci Technol Eng 3(12)
3. Wibowo, Mediyanto A, Dharmawan EA (2018) Study on effect of variations of metakaolin addition on selfcompacting parameter of high strength concrete. Int J Integr Eng 10(2):93–97
4. European Federation of National Association Representing Concrete Specification and Guidelines for Self-Compacting Concrete (EFNARC): The European Guidelines for Self-Compacting Concrete Specification—Production and Use, Europe (2005)
5. Siddique R, Khan MI (2011) Supplementary cementing materials. Springer Berlin Heidelberg, New York

6. Justice JM, Kennison LH, Mohr BJ, Beckwith SL, McCormick LE, Wiggins B, Zhang ZZ, Kurtis KE (2014) Comparison of two metakaolins and a silica fume used as supplementary cementitious materials. In: Proceedings seventh international symposium on utilization of high-strength/high performance concrete
7. Indriyantho BR (2014) Nuroji: finite element modeling of concrete fracture in tension with the Brazilian splitting test on the case of plane-stress and plane-strain. Procedia Eng 95(2014):252–259
8. American Standard Testing and Material: ASTM C293/C293M—16 standard test method for flexural strength of concrete (using simple beam with center-point loading), Amerika Serikat (2016)
9. Badan Standardisasi Nasional: SNI 03-2491-2002 Metode Pengujian Kuat Tarik Belah Beton, Jakarta, Indonesia (2002)
10. Badan Standardisasi Nasional: SNI 2847-2019 Persyaratan Beton Struktural untuk Bangunan Gedung dan Penjelasan, Jakarta, Indonesia (2019)

The Strength and Modulus of Elasticity of High Strength Self-compacting Concrete (HSSCC) with 12.5% Metakaolin and Variations of Silica Fume

Endah Safitri, Wibowo, Halwan Alfisa Saifullah, and Fernanda Sarwatatwadhika Perdana

Abstract The strength and modulus of elasticity of concrete are essential in determining the quality of concrete. In general, concrete with high strength receives a greater load that reduces the dimensions of the structural elements. The use of high-strength self-compacting concrete (HSSCC) with 12.5% metakaolin and variations of silica fume is perceived to be an innovation in construction. Metakaolin and silica fume work as pozzolans with a smaller size than cement to fill voids between aggregates and increase concrete density. This research used 12.5% metakaolin and variations of silica fume 0, 9, 11, 13 and 15% of weight of cement as an addition (admixture). A cylinder with a diameter of 15 cm and a height of 30 cm was used to test the strength and modulus of elasticity of concrete after 28 days. The test results showed that the concrete could be categorized as HSSCC using 12.5% metakaolin and variations of silica fume 0, 9, 11, 13, and 15%. The strength and modulus of elasticity of HSSCC concrete with 12.5% metakaolin and variation of silica fume 0%, 9%, 11%, 13%, 15% were 41.99 MPa; 55.39 MPa; 67.75 MPa; 61.80 MPa; 58.03 MPa and 25,495.7 MPa; 29,980.8 MPa; 34,952.5 MPa; 32,067.0 MPa and 31,147.5 MPa, respectively.

Keywords HSSCC · Modulus of elasticity · Strength of concrete

1 Introduction

The heavy loads supported by high-rise buildings require significant steel reinforcement in their structural elements, including columns, beams, and foundations. Most reinforcements lead to a small distance between reinforcement, meaning that fresh concrete can hardly flow to flow to all parts of the structural elements. Since concrete compaction is difficult, much air is trapped in the concrete, decreasing its strength

E. Safitri (✉) · Wibowo · H. A. Saifullah · F. S. Perdana
Department of Civil Engineering, Sebelas Maret University, Kota Surakarta, Indonesia

H. A. Saifullah
e-mail: halwan@ft.uns.ac.id

S. A. Kristiawan et al. (eds.), *Proceedings of the 5th International Conference on Rehabilitation and Maintenance in Civil Engineering*, Lecture Notes in Civil Engineering 225,
https://doi.org/10.1007/978-981-16-9348-9_104

to support the load. Therefore, high-strength self-compacting concrete (HSSCC) is used because of good workability [1]. Furthermore, high-strength concrete reduced the size of the structural element dimensions to lessen the structure's total weight. Self-compacting concrete can be compacted without tools [2] and depends on the proportion of aggregate size and superplasticizer admixture to attain the required viscosity. The use of superplasticizers allows water in the mixture to be reduced while maintaining workability and flowability. Metakaolin is a waste product of kaolin calcination at a temperature of 450–900 °C with a smaller size than silica fume with high SiO_2 and Al_2O_3 content [3]. Silica fume with 1/100 diameter of cement fills the space between aggregate and cement paste. Both elements are the main constituents of cement, hence metakaolin can be used as a substitute for cement [4, 5]. Metakaolin as a cement substitute with a concentration of 12.5% increases the compressive strength of the concrete [6].

The compressive strength and modulus of elasticity of the concrete are measured to define the performance of concrete. The compressive strength of concrete is the capability to receive compressive forces per unit area. The modulus of elasticity is the ratio of the pressure to the change in shape per unit length. The greater the compressive strength and modulus makes the performance of concrete better. The addition of metakaolin and silica fume is expected to increase the compressive strength and modulus of elasticity of HSSCC concrete.

2 Experimental Investigation

Mix design of HSSCC concrete with 12.5% metakaolin and variations of silica fume was made based on EFNARC [1]. Although the water-cement factor of 0.25 was used, the optimum superplasticizer content was 1.5% of the weight of the binder. Table 1 shows the proportions of each material.

The compressive strength and modulus of elasticity of concrete were tested using cylindrical specimens with a diameter of 15 cm and a height of 30 cm. Compression Testing machines were used to obtain the maximum load of concrete, as shown in Fig. 1.

Table 1 Recapitulation of mix design of high-strength self-compacting concrete (HSSCC)

Specimen code	Fine aggregate (kg/m^3)	Coarse aggregate (kg/m^3)	Cement (kg/m^3)	Meta-kaolin (kg/m^3)	Silica fume (kg/m^3)	Superplasticizer (l/m^3)	Water (l/m^3)
HSSCC-SF0%	915.30	777.90	525	75	0	9	150
HSSCC-SF9%	905.72	769.76	471	75	54	9	150
HSSCC-SF11%	903.59	767.95	459	75	66	9	150
HSSCC-SF13%	901.46	766.14	447	75	78	9	150
HSSCC-SF15%	899.33	764.33	435	75	90	9	150

Fig. 1 The test specimen and compression testing machine

To test the modulus of elasticity of concrete, a dial gauge was installed on the test object in addition to the Compression Testing Machine to obtain axial deformation data (Fig. 2).

Fig. 2 Dial gauge

3 Experimental Results and Discussion

Fresh concrete testing based on EFNARC [1] includes Slump Flow Test, L-Box, and V-funnel to define whether the test object qualifies the requirements as self-compacting concrete (SCC). Tables 2, 3 and 4, shows the test results.

Tables 2, 3 and 4 show that adding 12.5% metakaolin and variations of silica fume increased the workability. The test objects were still included in the self-compacting concrete category.

Table 2 Slump flow test results

Specimen code	Slump flow test					
	T_{500} average (second)	Requirements	Test results	d average (mm)	Requirements	Test results
HSSCC-SF 0%	4.7	2–5 s EFNARC 2005	Qualify	690	650–800 mm EFNARC 2005	Qualify
HSSCC-SF 9%	4.6		Qualify	710		Qualify
HSSCC-SF 11%	4.1		Qualify	725		Qualify
HSSCC-SF 13%	3.9		Qualify	730		Qualify
HSSCC-SF 15%	3.5		Qualify	740		Qualify

Table 3 L-box test results

Specimen code	h2 (mm)	h1 (mm)	h2/h1	Requirements	Test results
HSSCC-SF 0%	9	11	0.81	0.8–1 EFNARC 2005	Qualify
HSSCC-SF 9%	9	10	0.9		Qualify
HSSCC-SF 11%	10	11	0.91		Qualify
HSSCC-SF 13%	9	9	1		Qualify
HSSCC-SF 15%	9	9	1		Qualify

Table 4 V-funnel test results

Specimen code	V-funnel (s)	Requirements	Test results
HSSCC-SF 0%	12	6–12 s EFNARC 2005	Qualify
HSSCC-SF 9%	10		Qualify
HSSCC-SF 11%	9.5		Qualify
HSSCC-SF 13%	7.5		Qualify
HSSCC-SF 15%	6.5		Qualify

The compressive strength test was carried out at 28 days of concrete based on SNI 1974-2011 [7]. Equation 1 shows the formula used in calculating the compressive strength of concrete.

$$fc' = \frac{P}{\mathrm{A}} \tag{1}$$

where *fc'* is compressive stress (MPa), *P* is compression force (N), and *A* is cross-sectional area (mm^2).

The results of the concrete compressive strength test can be seen in Table 5.

Table 5 and Fig. 3 show that the addition of silica fume increases the compressive strength of high-strength self-compacting concrete with 12.5% metakaolin. Furthermore, the maximum compressive strength of 67.75 MPa occurred at 11% silica fume. The analysis shows that the optimum content of silica fume was at 11.54%, with a compressive strength of 61.62 MPa.

Load and axial deformation of the test object was obtained from CTM and dial gauge. From these data modulus of elasticity of each specimen can be calculated using Eq. 2 to Eq. 4.

Table 5 Compressive strength test result

Specimen code	fc' (MPa)	Percentage increase in compressive strength of concrete (%)
HSSCC-SF0%	41.99	–
HSSCC-SF9%	55.39	31.91
HSSCC-SF11%	67.75	61.35
HSSCC-SF13%	61.80	47.19
HSSCC-SF15%	58.03	38.20

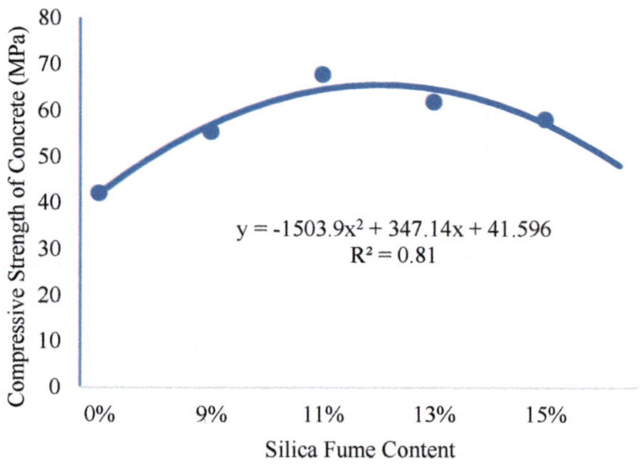

Fig. 3 Relationship between compressive strength and silica fume content

Table 6 The modulus of elasticity test results

Specimen code	Modulus of elasticity	
	Eurocode 2-1992	ASTM C469
HSSCC-SF0%	19,827.02	19,907.00
HSSCC-SF9%	24,430.58	25,443.33
HSSCC-SF11%	33,600.07	29,571.67
HSSCC-SF13%	28,242.92	26,616.00
HSSCC-SF15%	27,178.14	26,161.67

$$Stress\,\sigma = \frac{P}{\mathrm{A}} \tag{2}$$

$$Strain\,\varepsilon = \frac{\Delta L}{\mathrm{L}} \tag{3}$$

$$\text{Modulus of Elasticity } E = \frac{\sigma}{\varepsilon} \tag{4}$$

where σ is stress (MPa), ε is strain (mm/mm), ΔL is axial deformation (mm), L is specimen length (mm), and E is modulus of elasticity of concrete (MPa).

The value of the modulus of elasticity can be calculated using the standard Eurocode 2-1998 and ASTM C469 (Eqs. 5 and 6), as shown in Table 6 and Fig. 4 [8, 9].

$$\text{Eurocode 2-1992 } E_c = \frac{0{,}4fc'}{\varepsilon_{(0{,}4fc')}} \tag{5}$$

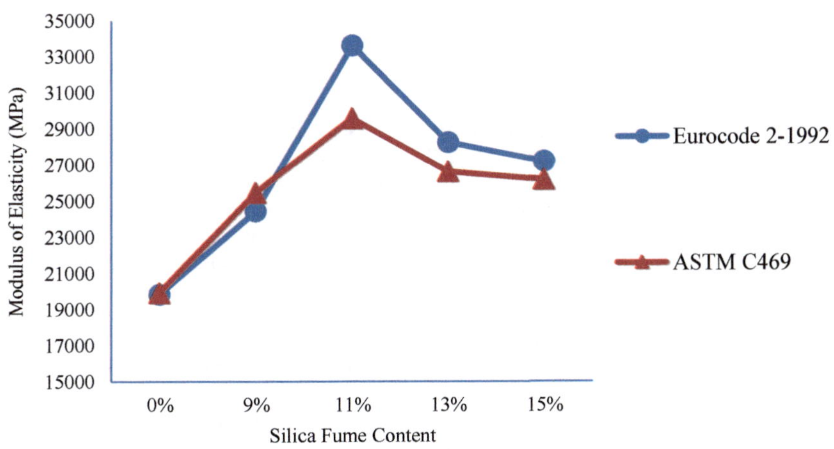

Fig. 4 Relationship between modulus of elasticity and silica fume content

$$\text{ASTM C469 } E_c = \frac{0{,}4fc' - \sigma_1}{\varepsilon_{(0{,}4fc')} - \varepsilon_1} \tag{6}$$

where *fc'* is maximum compressive stress (MPa), $\varepsilon_{(0{,}4\ fc')}$ is strain at 0,4*fc'* (mm/mm), σ_1 is stress at specific strain (MPa), and ε_1 is specific strain (mm/mm).

Table 6 and Fig. 4 show that the addition of silica fume increases the modulus of elasticity of concrete. The maximum modulus of elasticity occurred at 11% silica fume content. After the maximum replacement level, the value of modulus of elasticity is reduced. However, it was greater than the modulus of elasticity of HSSCC without the addition of Silica Fume.

4 Conclusions

High-strength concrete with 12.5% metakaolin and up to 15% silica fume is still included in the self-compacting concrete (SCC) category with good workability. Optimum compressive strength occurred with the addition of 12.5% metakaolin and 11.54% silica fume with a compressive strength of 61.62 MPa. The optimum modulus of elasticity occurred at 11.47% silica fume with a modulus of elasticity of 32,354.51 MPa. The physical properties of silica fume with a cement diameter of 1/100 fill the space between the aggregate and cement paste. Therefore, silica fume increases the density of the concrete for a greater load.

Acknowledgements The authors express gratitude to Sebelas Maret University for supporting this research through the Fundamental Research Scheme from the 2021 funds of non-APBN, under the contract number: 260/UN27.22/HK.07.00/2021.

References

1. European Federation of National Association Representing Concrete Specification and Guidelines for Self-Compacting Concrete (EFNARC): The European Guidelines for Self-Compacting Concrete Specification—Production and Use, Europe (2005)
2. Wibowo Mediyanto A, Dharmawan EA (2018) Study on effect of variations of metakaolin addition on self- compacting parameter of high strength concrete. Int J Integr Eng 10(2):93–97
3. Barbhuiya S, Chow P, Memon S (2015) Microstructure, hydration and nanomechanical properties of concrete containing metakaolin. Constr Build Mater 95:696–702
4. Sambowo KA (2001) Engineering properties and durability performance of metakaolin and metakaolin-PFA concrete.Thesis, Faculty of Engineering at University of Sheffield, Sheffield
5. Siddique R, Khan MI (2011) Supplementary cementing materials. Springer Berlin Heidelberg, New York
6. Joseph A, Mathew LA, John R (2017) Performance of metakaolin on high strength self compacting concrete. Int J Sci Technol Eng 3(12)
7. Badan Standardisasi Nasional: SNI 1974:2011 Cara Uji Kuat Tekan Beton dengan Benda Uji Silinder, Jakarta, Indonesia (2011)

8. European Committee for Standardization (2004) Eurocode 2-design of concrete structures—part 1-1 general rules and rules for buildings. Europe, Brussels
9. ASTM International: ASTM C469/C469M-14e1, Standard test method for static modulus of elasticity and Poisson's ratio of concrete in compression, West Conshohocken, PA, www.astm.org (2014)

GPSR Compliance

The European Union's (EU) General Product Safety Regulation (GPSR) is a set of rules that requires consumer products to be safe and our obligations to ensure this.

If you have any concerns about our products, you can contact us on ProductSafety@springernature.com

In case Publisher is established outside the EU, the EU authorized representative is:

Springer Nature Customer Service Center GmbH
Europaplatz 3
69115 Heidelberg, Germany

Batch number: 10165653

Printed by Printforce, the Netherlands